ENCYCLOPEDIA OF PHYSICS

EDITOR IN CHIEF

S. FLÜGGE

VOLUME LV

GENERAL INDEX

SPRINGER-VERLAG
BERLIN HEIDELBERG NEW YORK
LONDON PARIS TOKYO

HANDBUCH DER PHYSIK

HERAUSGEGEBEN VON

S. FLÜGGE

BAND LV

GENERALREGISTER

SPRINGER-VERLAG
BERLIN HEIDELBERG NEW YORK
LONDON PARIS TOKYO

ISBN-13:978-3-642-82504-0 e-ISBN-13:978-3-642-82502-6
DOI: 10.1007/978-3-642-82502-6

Softcover reprint of the hardcover 1st edition 1988
Library of Congress Catalog Card Number A 56-2942.

2156/3150-543210 – Printed on acid-free paper

Group 1

Mathematical Methods

Abelian group 2(72)
Abelian theorem(s) 1(55); 2(241)
Abelian theorems for the Laplace transform 2(242)
Abel's integral equations 1(257)
Abscissa, mean 2(409)
Access time 2(494)
Accumulator 2(482)
Acoustical problem 2(446)
Adams extrapolation 2(425)
Adder 2(492)
Adding glar 2(474)
Adding unit 2(475)
Addition theorems 1(181, 203)
Address 2(484)
Adjunction 2(27)
Affinity 2(140)
Affinors 2(128)
Algebra 2(23, 85)
Algebra, commutative 2(97)
Algebra, nilpotent 2(89)
Algebraic addition theorem 1(124)
Algebraic equation 2(375)
Algebraic number 1(5)
Algebraic relation 1(124)
Algebroid functions 1(211, 215)
Alignment nomogramm 2(354)
Amount of a matrix 2(389)
Analogue computer 2(472)
Analytic 1(48)
Angular momentum operator 1(206)
Anti-automorphism 2(33)
Approximation, mean 1(221)
Approximation, stepwise 2(441, 459)
Approximation, successive, of propervalues and properfunctions 1(260)
Arc length = Arcus 2(146)
Area 1(30, 37); 2(120, 149)
Area, element of 2(192)
Arithmetic geometric mean 1(145)
Arithmetic unit 2(481)
Asymptotic behavior 1(186)
Asymptotic behavior of special functions 1(174)
Asymptotic representations 1(174)
Atomic system 2(289)
Atomic units 2(288)
Ausstrahlungsbedingung 1(305, 306)
Automorphism 2(13)
Automorphism, inner 2(20)
Axis of rotation 2(134)

Backward difference 2(400)
Banach space 2(205, 206, 211, 331)
— real 2(208)
— separable 2(207)
Banach-adjoint transformation 2(212)
Banachiewicz's scheme 2(385)
Banach-Steinhaus theorem 2(209)
Base 1(5)
Basepoint 2(192)
Basic (or fundamental) points 2(141)
Basic region 1(220)
Basis, reciprocal (or inverse) 2(122)
Basis vector 2(121, 126, 182)
Beginning calculations 2(428)
Bernoulli's numbers 1(54)
Bessel differential equation 1(152)
Bessel formula 2(402)
Bessel function, modified 1(351)
Bessel functions 1(152, 321, 323, 332, 351)
Bessel inequality 1(222)
Beta-function 1(68)
Bianchi 2(176)
Bilinear form 2(60, 131)
Binary numbers 2(487)
Binormal 2(147)
Bisector, bisectrix 2(124)
Bivector 2(178)

Block relaxation 2(393, 448)
Bound, greatest lower 1(1, 42)
Bound, least upper 1(1, 42)
Bound, lower 1(1)
Bound, upper 1(1)
Boundaries, free 1(275)
Boundary collocation 2(439)
Boundary condition(s) 2(436)
— Cauchy's 1(283)
— Dirichlet's 1(283)
— homogeneous 1(282)
— inhomogeneous 1(283)
— mixed 1(283)
— natural 1(276, 329)
— Neumann's 1(283)
Boundary problems 2(478)
Boundary, treatment of the 2(447)
Boundary value problem 1(90)
— adjoint 1(298)
— first 2(440, 456)
— mixed 2(248)
— proper 1(299)
— third 2(440, 456)
— two-point 2(257)
Bounded 1(1, 16)
Boussinesq's first problem 2(304)
Boussinesq's second problem 2(306)
Brachistochrone 1(262)
Branch 1(62)
Branch-point 1(59)
Braun tube, cathode ray tube 2(473)
Brigg's logarithms 1(6)

Calculation for adjust for errors 2(361)
Calculus of observation, method of least squares 2(361)
Cam 2(474)
Canonical 1(106)
Cards, punched 2(480)
Carrying (over) 2(163, 173)
Cauchy condition 2(206)
Cauchy-Riemann differential equations 1(48)
Cauchy integral formula 1(50)
Cauchy main value 2(467)
Cauchy principal value 1(33)
Cauchy principle 2(204)
Cauchy theorem 1(50)
Center 2(11, 21, 192)
Center-of-gravity 2(142)
Central projection 2(139)
Change of variable 1(24)
Character of a representation 2(65)
Characteristic 1(92, 95, 117, 283, 293)
Characteristic equation 1(76, 78)
Characteristic exponent 1(167, 193, 200)
Characteristical number 2(395)
Characteristics, method of 2(435)
Charge-distribution, electric 1(237)
Check by back substitution 2(384)
Check column method 2(384)
Cholesky's method for symmetric systems of equations 2(386)
Christoffel symbols of the first kind 2(179)
Christoffel symbols of the second kind 2(153, 179)
Churchill, inversion formula of 2(273)
Churchill's convolution theorem 2(322)
Circle of convergence 1(53)
Circuital relations 1(167)
Circular functions 1(6, 52)
Circulation 2(160)
Class number 2(21)
Class of conjugate elements 2(21)
Class sums 2(107)
Classification 2(12)
Classification of the Euclidian surfaces of 2nd order 2(139)
Clifford algebra 2(86)
Closure 1(85)
Coherence, metric 2(178)
Collineations 2(139, 142)
Collocation 2(369, 438)
Collocation point 2(439)
Combination 2(3)
Comparable function 2(459)
Compatibility 2(435)
Compatibility conditions 1(90)
Complementary function 2(252)
Complementary modulus 1(143)
Complete elliptic Integral of the first (second) kind 1(140)
Completeness relation 1(222)
Complex conjugate 1(47)
Complex number 1(46)
Component 2(119)
Compton line 2(288)
Computer, digital 2(471)
Computing machines 2(471)
Conduction of heat 2(258, 260)

— in a finite slab 2(311)
— in a semi-infinite solid 2(280)
— in cylinders 2(317)
Cone 2(139)
Confluent hypergeometric functions 1(160)
— of second kind 1(161)
Confocal 1(60)
Conformal mapping 1(56)
Conic (section) 2(143)
Conjunction 2(490)
Co-normal 2(440)
Constant of integration 1(23, 69)
Contact transformation 1(101, 108); 2(192, 193)
Content of mechanics, algebraic 2(110)
Continuation, analytic 1(62)
Continued fraction representation 1(192)
Continuing computation 2(423)
Continuous 1(3)
Contour integral 1(49)
Contraction 2(168, 191)
Control unit 2(481)
Converge 1(2)
Convergence in mean 1(88, 221); 2(205)
Convergent 1(42, 43, 45)
Convergent, absolutely 1(32, 43, 46)
Convergent, conditionally 1(43)
Convergent, uniformly 1(34, 45, 46, 51)
Convolution 2(412)
Convolution, classical 2(221)
Convolution integral 2(221)
Convolution, Stieltjes 2(221)
Convolution theorem for Fourier transforms 2(267)
Convolution theorems for the Mellin transform 2(294)
Coordinate 2(119)
Coordinate system 2(119, 121)
— barycentric of Möbius 2(142)
— cartesian 2(122)
— geodesic 2(175)
— projective 2(141)
— right-handed 2(120)
Coordinates, contragredient 2(123)
Coordinates, contravariant 2(123)
Coordinates, covariant 2(123)
Coordinates, curvilinear 2(163)
Coordinates, cylindrical 2(184)
Coordinates, elliptical of Lamé 2(185)
Coordinates, homogeneous 2(142)
Coordinates, natural 2(148)
Coordinates, normed 2(142)
Coordinates, oblate spheroidal 1(200)
Coordinates of the elliptic cylinder 1(198)
Coordinates of the parabolic cylinder 1(197)
Coordinates, orthogonal 1(275)
Coordinates, orthogonal 2(182)
Coordinates, orthogonal curvilinear 1(193)
Coordinates, parabolic 2(185)
Coordinates, spherical 2(183)
Coordinates, spheroidal 1(198)
Correlation 2(142)
Correspondence, one-to-one 1(21)
Coset 2(19)
Cosine transforms, finite 2(309)
Cotes' quadrature formulae 2(406)
Covariant, bilinear 1(110)
Cramer's rule 1(252); 2(125)
Criterion of convergence of Cauchy-Riemann 1(32, 42)
Cross point 1(59)
Cross ratio 1(57)
Curl 2(159, 172, 182, 183)
Curvature 2(147, 164, 175)
Curvature, geodesic 2(154)
Curvature, geodesic circles of 2(154)
Curvature, lines of 2(152)
Curvature of a line 2(180)
Curvature of the direction 2(180)
Curvature, scalar 2(181)
Curvature tensor 2(176, 181)
Curve length 1(36); 2(146)
Curved boundary 2(447)
Curves, isotropic 2(150)
Cylinder coordinates 1(195, 275, 323)
Cylinder, elliptic 2(139)
Cylinder functions 1(152, 195, 305, 321, 323, 326, 332)
Cylinder, hyperbolic 2(139)
Cylinder, parabolic 2(139)

Darboux 2(186)
Darboux vector 2(147)
Debye potentials 1(207)
Decimal number, binary ordered 2(489)
Decomposition into partial fractions 1(24)

Defect 2(387)
Definite, positive 2(466)
Deflection of a beam 2(257)
Deformation tensor 2(159)
Degenerate kernel 2(465)
Degeneration 1(239)
Degeneration, degree of 1(315)
Delay line, acoustic 2(495)
Delta 2(182)
Delta function, derivative of the 2(232)
Density of the propervalues 1(334)
Density, scalar 2(171)
Dependent, linearly 1(75, 77)
Derivative 1(11, 48)
Derivative, left (right) hand 1(11)
Derivative of higher order 1(12, 49)
Derivative, partial 1(17, 48)
Determinant 2(49, 131)
Determining function 2(217)
Determining function, uniqueness 2(220)
Developable (surface) 2(155)
Diagonal form 2(130, 137, 138)
Difference, central 2(400)
Difference differential equations 1(176)
Difference equations 1(186)
Difference quotient 1(9, 48)
— backward 2(430)
— central 2(430)
— forward 2(430)
Difference schema method 2(422)
Differences, calculus of 2(399)
Differences, method of central 2(422)
Differentiable 1(11, 48)
Differential 1(11)
Differential, absolute 2(173)
Differential analysers 2(476)
Differential analysers, set up for 2(476)
Differential equation(s)
— elliptic 1(117, 286, 292, 328)
— homogeneous system 1(75, 77)
— hyperbolic 1(117, 285, 286, 328)
— inhomogeneous system 1(75, 77)
— isotropic 1(328)
— of Clairaut 1(72, 100)
— of d'Alembert-Lagrange 1(71)
— of Mathieu functions 1(171, 172, 192, 199, 208)
— of Riccati 1(72)
— of second order with rational coefficients 1(165)
— of spheroidal functions 1(171)
— ordinary 1(68); 2(477)
— parabolic 1(117, 285, 290, 328)
— partial 1(90); 2(435)
— undisturbed 2(444)
— with variable coefficients, ordinary linear 2(255)
Differential operator(s) 1(206, 281); 2(161)
Differential operator, hermitian 1(300, 314)
Differential, total 1(17); 2(158, 166, 195)
Differentiation 2(161)
Differentiation, covariant 2(173)
Differentiation homomorphism 2(34)
Differentiation ideal 2(34)
Differentiation in a ring 2(34)
Differentiation of a tensor 2(172)
Differentiation of distributions 2(343)
Diffusion equation 1(286, 291, 294, 295, 339)
Diffusion problems 2(273, 282)
Dimension 2(133)
Dimension, negative 2(133)
Dimension, positive 2(133)
Diminuation factor 2(414)
Dipole 1(308)
Dirac transformation theory 2(288)
Dirac's δ-function 1(231); 2(229, 231)
Dirac's δ-function, unsymmetrical 2(231)
Direct observation 2(361)
Direction of rotation (or revolution) 2(120)
Direction of screwing 2(120)
Directional differentiation 2(159)
Directions, invariant 2(130)
Dirichlet's conditions 2(202, 309)
Dirichlet's integral formula 2(235)
Dirichlet's problem for a sphere 2(323)
— Churchill's solution of the 2(323)
Discriminant of an algebra 2(105)
Disjunction 2(490)
Dissection, canonical 2(155)
Distance, hyperbolical 2(157)
Distance-circles, geodesic 2(154)
Distributive law 2(4, 7)
Disturbance calculation 2(443)

Disturbed differential equation 2(444)
Divergence 2(158, 172, 182)
Divergent 1(42, 43)
Divided differences 2(399)
Division algebra 2(86, 96, 97)
Domain 2(2)
Domain collocation 2(439)
Domain of definition 1(1)
Domain, simple 2(13)
Double integral 1(38)
Double layer 1(310ff.)
Double (or cross) ratio (of four collinear points) 2(143)
Dual integrations 2(294)
Dummies 2(165)
Dupin 2(186)

Edge of regression 1(98)
Eigenfunction 2(437)
Eigenfunctions, properfunctions 1(241, 313, 328)
Eigenvalue(s) 1(313, 328)
Eigenvalue problem 2(48, 437, 449, 458, 478)
Eigenvalue problem for matrices 2(395)
Eigenvector, proper vector 2(48, 130, 138)
Einstein, summation convention of 2(165)
Einstrahlungsbedingung 1(305, 306)
Elastic waves, generation of 2(284)
Electronic integrators 2(475)
Element, idempotent 2(89)
Element, nilpotent 2(87)
Element, proper nilpotent 2(87)
Elementary divisor, theory of 2(130)
Elements of constancy 2(34)
Elimination method 2(383)
Ellipsoïd 2(139)
Elliptic differential equation 2(439)
Elliptic function 1(120)
Elliptic integral 1(124)
End around carry 2(488)
Endomorphism 2(13)
Endomorphism ring of a vector space 2(55)
Enumerable 1(82)
Envelope 1(69)
Envelopping, condition of 2(192)
ε-scheme 2(431)
ε-tensors 2(171, 431)
ε-toheme 2(431)
Equally spaced points 2(400)
Equation, characteristic 2(130, 138)
Equation, trigonomial 2(350)
Equations, homogenous system of 2(50)
Equations, infinite system of 2(391)
Equations, inhomogenous system of 2(50)
Equations, monotonous system of linear 2(394)
Equations of motion 2(108)
Equations of motion in non-commutative rings 2(110)
Equations, system of linear 2(125, 383)
Equations, system of nonlinear 2(390)
Equianharmonic points 1(130)
Equipotential surfaces 2(158)
Equivalence relation 2(12)
Error 2(349)
Error, consumation of 2(438)
Error function 1(165); 2(275, 276)
Error integral 1(165)
Error, medium 2(361)
Error, orthogonality of 2(438)
Error, propagation of 2(350)
Error, seeming 2(361)
Euler angles 2(135)
Euler condition 1(111)
Euler constant 1(66)
Euler differential equation 2(453)
Euler equation 2(465)
Euler formula 1(52); 2(406)
Euler, theorem of 2(151)
Euler-Cauchy method, improved 2(418)
Eulerian integral of the first kind 1(68)
Eulerian integral of the second kind 1(66)
Euler-Lagrange equation 1(265, 269, 272, 281, 327)
Expansion, binomial 1(54)
Expansion theorems 1(190, 243, 248)
Exponential function 1(4, 51)

Factor group 2(20)
Factor ring 2(15)
Factorization method 1(176)

Falkner's extrapolation formula 2(426)
Families of curves, orthogonal 2(152)
F-equation 1(179)
Fermat's principle 1(263)
Field 2(5, 157)
Fields, irrotational 2(159)
Fields, solenoidal 2(159)
Filon's formulae 2(328, 329)
Finite difference method 2(444)
Finite equation 2(432, 445)
Finite expression 2(450)
Fitz-Gerald potential 1(208)
Flip-Flop 2(493)
Floquet's theorem 1(168, 209)
Flow 2(161)
Foot of a perpendicular 2(124)
Form, hermitian 2(61)
Form, positive-definite 2(61)
Form, quadratic 2(59, 131, 137)
Form, semi-definite 2(61)
Form, to principal axes transformed, quadratic 2(138)
Formal logic, functions of 2(489)
Forward difference 2(400)
Fourier coefficient 2(411)
Fourier cosine transforms 2(271, 275, 276, 277, 327)
Fourier inversion theorem 2(267, 268)
— Titchmarsh's complex form of 2(270)
Fourier integral 1(228)
Fourier kernel 2(214, 216)
Fourier series 1(61, 225); 2(241)
Fourier sine, double, of a symmetrical function 2(302)
Fourier sine, *n*-dimensional of a symmetrical function 2(303)
Fourier sine transform 2(216, 272, 273, 276, 277, 329)
Fourier transformation 1(230)
Fourier transforms 2(266)
— applications of 2(280)
— exponential 2(266,273, 274)
— finite 2(308, 310)
— formal properties of 2(273)
— in quantum mechanics 2(288)
— multiple 2(278)
— multiple finite 2(313)
— multiple, of derivatives 2(279)
— of the derivatives of a function 2(277)
Four-vectors 2(190)
Fredholm's integral equation of the second kind 2(286)
Fredholm's solution method 1(252)
Frenet formulae 2(147)
Frequency equation 2(395)
Friction disc 2(475)
Friction wheel 2(475)
Frontier point 1(16)
Fubini's, theorem 2(203, 214)
Function, even 1(2)
Function, generating 2(194, 196)
Function, homogeneous 1(19)
Function, implicit 1(20)
Function, integral 1(63)
Function, integral rational 1(2, 17)
Function, many-valued 1(62)
Function, meromorphic 1(63)
Function, monogenic 1(62)
Function, odd 1(2)
Function of several variables 1(15)
Function, one-valued 1(1, 62)
Function, periodic 1(2)
Function, rational 1(2, 17)
Function, rational integral 2(28)
Function space 1(219)
Function systems, orthogonal 1(218, 222)
Functional 1(262); 2(209)
Functional derivative 2(346)
Functional determinant 1(273)
Functional equation 2(468)
Functional equations 1(176)
Functional, linear 2(209)
Functional matrix 1(21)
Functional-differential equation 2(469)
Functions, normable 1(220)
Functions of the parabolic cylinder 1(164, 197)
Fundamental form, first 2(149, 163)
Fundamental form, metric 2(131)
Fundamental form, second 2(151)
Fundamental periodic parallelogramm 1(120)
Fundamental system 1(166)
Fundamental tensor 2(169, 182)
Fundamental tensor, covariant constant 2(181)
Funicular, construction of the, (or link) polygon 2(410)

Galerkin's equation 2(439, 453)
Gamma function 1(66)
Gamma function, incomplete 1(165)
Gas vibration 2(446)
Gauss curvature 2(155)
Gauss' elimination method 2(383)
Gauss' mechanical quadrature 2(406)
Gauss-Bonnet formula 2(155)
Gaussian curvature of a surface 2(151)
Gaussian theorem 1(296)
Gegenbauer's polynomials 1(304)
General 2(458)
Generating function 1(182); 2(217, 219)
Generating function of a Laplace-Stieltjes transform 2(229)
Genus 1(127)
Genus (of the surface) 2(155)
Geometry, elliptic 2(145, 157)
Geometry, hyperbolic 2(145, 157)
Geometry, non-euclidian 2(145, 157)
Geometry of coincidence 2(442)
Geometry, parabolic 2(145)
Geometry, projective 2(139)
Gibbs' phenomenon 1(226)
Gradient 2(158, 166, 182)
Graeffe's method 2(376)
Graphical integration 2(409)
Green function 1(237, 297, 300, 301, 339)
Green theorem (for a plane) 1(42, 296)
Grid point 2(429)
Group 1(57); 2(16, 134)
Group, Abelian 2(16)
Group, affine 2(167)
Group algebra 2(24, 107)
Group, alternating 2(72)
Group, centered affine 2(167)
Group, cyclic 2(18)
Group, full (or complete) linear 2(61)
Group method 2(364)
Group of translation 2(144)
Group, orthogonal 2(61, 75)
Group ring cf. group algebra 2(24, 107)
Group, special unitarian 2(80)
Group, symmetrical 2(18, 72)
Group, unitary 2(61)
Gudermann 1(144)

Hadamard determinant theorem 1(253)
Halfadder 2(492)
Hamilton, differential equation of, ordinary 2(196)
Hamilton mechanics 2(108)
Hamiltonian 2(196)
Hamiltonian system of differential equations 1(105)
Hamilton-Jacobi, partial differential equation of 2(196)
Hamilton-Jacobi theory 2(109)
Hamilton's partial differential equation 1(105)
Hankel functions 1(152, 174, 306, 351)
Hankel integral representation of the Γ-function 1(66)
Hankel inversion theorem 2(299)
Hankel transform(s) 2(299)
— application 2(304)
— finite 2(314)
— of the derivatives of a function 2(301)
— properties of the 2(300)
Harmonic points 1(130)
Heat equation 2(430, 433)
Heat pole 1(344)
Heaviside unit function $H(x)$ 2(225, 230, 233, 343)
Heaviside's expansion theorem 2(254)
Heaviside's partial fraction rule 2(254)
Heine-Borel theorem 1(83)
Heisenberg rings 2(32, 33)
Hermitian polynomials 1(162, 224, 320)
Hermitian scalar product 1(219)
Hessenberg's method 2(395)
Hilbert space 2(330)
— abstract 2(331)
— operators in 2(333)
— realizations of 2(332)
Hilbert transform 2(248)
Hill's differential equation 1(208)
Hölder's inequality 2(204)
Homomorphism 2(12)
Homomorphism to the complex number field 2(111)
Horner's scheme 2(372, 378)
Hurwitz's equation 2(380)
Hurwitz's reduction scheme 2(379)
Hyperbolic functions 1(9, 52)
Hyperboloid of one sheet 2(139)
Hyperboloid of two sheets 2(139)
Hypergeometric differential equation 1(148)
Hypergeometric function 1(148)

Hypergeometric function of second kind 1(151)
Hypergeometric series 1(148)
Hypersurface of 2nd order 2(138)

Ideal 2(13)
Ideal of an algebra 2(85)
Ideals as representation spaces 2(105)
Ideals, minimal 2(92)
Ideals, nilpotent 2(88)
Idempotents, primitive 2(94)
Identity operator 2(333)
Identity theorem 1(63)
Ikehara's theorem 2(243)
Image function 2(244)
Imaginary (pure) 1(47)
Inclusion theorem 2(397, 460)
Indefinite 1(17)
Index 2(433)
Indicial equation 1(81, 166)
Inequality, Schwarz's 1(221)
Inertia, index of 2(61)
Inertia, law of 2(133)
Inhomogeneity 1(282)
Initial part 2(422)
Initial value problem 1(90, 286, 288); 2(415)
Input 2(481)
Input table 2(472)
Integrable 1(29)
Integral 1(74)
Integral, (absolutely) convergent 1(32)
Integral, complete 1(97, 103)
Integral, definite 1(29)
Integral domain, domain of integrity 2(11)
Integral equation(s) 2(262, 330, 463)
— Abel's 2(262)
— dual 2(296)
— inhomogeneous of second kind 1(249)
— linear 1(235)
— non-linear 2(264)
— of first kind 1(248)
— of first kind, Fredholm's 1(236)
— of second kind, Fredholm's 1(236, 238)
— of Volterra 1(255); 2(263)
— singular 2(467)
— solution of 2(270)
Integral, general 1(98, 104)
Integral, improper 1(32)
Integral, indefinite 1(23)
Integral invariants 1(113)
Integral, multiple 1(41)
Integral of motion 2(109)
Integral relations 1(200)
Integral representations 1(151)
Integral, singular 1(99, 104)
Integral theorem(s) 2(160)
— of Gauss 2(160)
— of Green 2(160)
— of Stokes 2(160)
Integral transformation, linear 1(230)
Integral transformation (or: transform) 2(212, 216)
Integral transformation, reciprocal 1(249)
Integral transforms, approximate methods of evaluating 2(325)
Integral transforms, finite 2(307)
Integral transforms in Hilbert space 2(336)
Integrals, multiple 2(408)
Integrals, particular 1(92)
Integrals, singular 2(407)
Integrand 1(29)
Integration, numerical 2(404)
Integration per parts 1(24)
Integrations, successive 2(203)
Integrators 2(475)
Integro-differential equation 2(463, 465)
Interpolation 2(399)
Interpolation, trigonometric 2(411)
Intersection of two sets 2(199)
Interval, closed 1(1)
Interval, open 1(1)
Invariant 2(131, 133, 166)
Inverse 2(334)
Inverse circular function 1(8)
Inverse function 1(4)
Inversion 1(58)
Inversion theorem 2(214)
Involution system, maximal 2(109)
Irreducible constants 1(169)
Isolated 1(16, 63)
Isomer 2(168)
Isomorphism 2(12)
Isoperimetric problems 1(263, 269)
Iterated 2(469)
Iteration 2(415, 448)
Iteration method 1(250, 259); 2(370, 389, 396, 441, 464, 466)

Jacobi bracket 1(108)
Jacobi elliptic function 1(142)
Jacobi identity 1(107)
Jacobi polynomials 1(224, 320)
Jacobian, Funktionaldeterminante 1(20)
Jordan algebra 2(26)
Jordan curve 1(16)

Karamat's theorem 2(242)
Kernel 1(85, 230); 2(212, 463, 466)
Kernel, adjoint 1(234)
Kernel index method 2(167)
Kernel, normable 1(241)
Kernel of the integral equation 1(236)
Kernels, definite 1(244)
Kernels, Hermitian 1(240)
Kernels, iterated 1(244)
Kernels, non-bounded 1(256)
Klein, "Erlanger Programm" 2(146)
Klein-Gordon equation 1(306, 344)
Klein-Gordon equation, static 1(304, 321)
Kronecker product 2(167)
Kronecker product representation 2(71)
Kummer function 1(160)

Lagrange bracket 1(111)
Lagrange factor 1(270, 271)
Lagrange's bracket relation 2(194)
Lagrangian 1(262, 327)
Lagrangian density 2(345)
Lagrangian formalism for continuous systems 2(345)
Lagrangian function 2(345)
Laguerre polynomial 2(237)
Laguerre polynomials 1(224, 321)
Landen's transformation 1(144)
Laplace integrals 1(173)
Laplace operator 1(274, 293, 296)
Laplace, potential equation of 2(162)
Laplace transform 2(216, 219, 329)
— bilateral 2(243, 291)
— double 2(246)
— fundamental properties of the 2(222)
— Inversion formula for the 2(235)
— iterated 2(247)
— of definite integrals 2(227)
— of elementary functions 2(223)
— repeated 2(247)
— two-sided 2(243)
Laplace's differential equation 1(50)
Laplace-Stieltjes transform 2(217, 229)
Laplace-Stieltjes transform, inversion theorem for the 2(237)
Laplacian 2(159)
Lattice 1(120)
Lattice, singular 1(137)
Laurent series 1(167)
Laurent's series expansion theorem 1(60)
Law of the mean 1(13)
Least mean square deviations 1(218, 221)
Least squares method 2(368, 438)
Lebesgue concept of integral 1(223)
Lebesgue integral 1(86); 2(201)
Lebesgue space 2(203, 208, 211)
Lebesgue's theorem 2(202)
Legendre condition 1(267)
Legendre differential equation 1(156); 2(321)
Legendre functions, associated 1(159, 320, 325)
Legendre functions of second kind 1(158)
Legendre polynomial 2(240)
Legendre polynominals 1(156, 162, 224, 319)
Legendre transform(s) 2(320, 321)
— even 2(320)
— finite 2(319, 320)
— odd 2(320)
— of particular functions 2(323)
Length 2(144)
Length, integrable transfer of the 2(180)
Lerch's theorem 2(221)
Level line 1(57)
Lie ring 2(25)
Limit 1(2, 16, 42, 48)
Limit, improper 1(3)
Limit, iterated 1(17)
Limit, left (right) hand 1(2)
Limit, lower 1(1, 42, 82)
Limit, proper 1(3)
Limit, upper 1(1, 42, 82)
Limit, upper (lower) of an integral 1(29)
Limited to the left 1(1)

Limited to the right 1(1)
Limiting point 1(1, 16)
Limiting set 1(82)
Lindenbaum's postulates 2(206)
Line coordinates 2(142)
Line element 1(70); 2(166)
Line, geodesic 2(154, 175)
Line, ideal 2(139, 141)
Line index 2(127)
Line, parametric representation of a (straight) 2(119)
Line segment 2(144)
"Linear programming" 2(497)
Lines, geodesic parallel 2(154)
Lines, one-parameter family of 2(124)
Lines, skew 2(125)
Linkages 2(473)
Liouville's theorems 1(120)
Lipschitz condition 1(68); 2(370)
Lipschitz constant 2(390)
Logarithm 1(5)
Lorentz group, full 2(190)
Lorentz group, proper 2(189)
Loxodromie line 1(58)

Maclaurin's formula 2(406)
Magnetic core memory 2(495)
Magnetic drum 2(493)
Magnetism of the earth 1(197)
Magslip (engl.) 2(474)
Majorant 1(44)
Manifold, *r* dimensional 1(22)
Manybody problem in nuclear theory 2(290)
Mapping, conformal 2(150)
Mapping, congruent 2(163)
Mapping, equiangular, isogonal, angle-preserving 2(150)
Mapping, equiareal, area-preserving, equivalent 2(150)
Mapping, improper ortogonal 2(134)
Mapping, infinitesimal 2(136, 178)
Mapping, length-preserving 2(150, 163)
Mapping, optical 1(237)
Mapping, proper orthogonal 2(134)
Mapping, spherical 2(152)
Mapping, unimodular 2(187)
Mappings, linear 2(128)
Mappings, orthogonal linear 2(134)
Marking off, parallel 2(163)
Mash size 2(429, 445)
Mathieu differential equation, modified 1(199)
Mathieu functions 1(208)
Mathieu functions, modified 1(213)
Matrices, congruent 2(59)
Matrices, equivalent 2(44)
Matrices, hermitian congruent 2(61)
Matrices, similar 2(129)
Matrices, unitary similar 2(57)
Matrix 2(42, 127)
Matrix, adjoint 2(46)
Matrix algebra 2(98, 99)
Matrix algebra, full 2(43)
Matrix, alternating 2(137)
Matrix, characteristic polynomial of a 2(51)
Matrix, conjugated 2(132)
Matrix, diagonal 2(47)
Matrix, hermitian 2(47)
Matrix, idempotent 2(47)
Matrix, invertible 2(127)
Matrix, monotonous 2(394)
Matrix, nilpotent 2(47)
Matrix, normal 2(47)
Matrix of a transformation 1(230)
Matrix, orthogonal 2(47)
Matrix, self-adjoint 2(47)
Matrix, skew-hermitean 2(47)
Matrix, symmetrical 2(59, 137)
Matrix system, irreducible 2(52)
Matrix system, reducible 2(52)
Matrix, transposed 2(46)
Matrix types, canonical 2(45, 46)
Matrix, unitarian 2(47)
Maxwell equations, solutions in spherical coordinates 1(207)
Measurable 1(84, 85)
Measure 1(83)
Measure, coefficient of 2(177)
Measure, upper (lower) 1(83)
Measurement of angle 2(149)
Measurement of length 2(149)
Mehrstellen method 2(433, 450)
Mellin transform 2(215, 290)
— applications of the 2(249)
— inversion theorem for the 2(293)
Mellin's inversion integral 2(293)
Membrane 2(449)
Membrane, rectangular, vibrations of 2(313)
Memory 2(481)

Memory, electrostatic 2(495)
Mensuration 2(163, 169)
Mensuration, projective, of Cayley 2(144)
Method of least squares, calculus of observation 2(361)
Metric 2(131, 144, 177)
Metric determination 2(163, 169)
Metric determination, projective, of Cayley 2(144)
Metric, euclidian 2(133)
Metric, pseudo-euclidian 2(133)
Meusnier, theorem of 2(151)
Minimal curves 2(150)
Minimal polynomial 2(112)
— of a matrix 2(54)
— of a transformation 2(55)
Minkowski space 2(133, 188, 189)
Minkowski's inequalities 2(204)
Minor 2(49)
Minorant 1(44)
Mirror image 2(124)
Mirror point 2(124)
Mittag-Leffler identity, 2(229)
Modular equation 1(136)
Modular function 1(136)
Modulus 1(143); 2(21)
Moivre's theorem 1(47)
Molecular system 2(289)
Momentum density function 2(288)
Momentum space 2(289, 290)
Monge cone 1(94)
Monotonously decreasing 1(2)
Multiplication theorems 1(181)
Multiplier 1(23, 71, 92)
Multiplier, last 1(93)
Multiplier-quotients-register 2(482)
Multipliers, domain of 2(21)
Multipliers, Lagrange's 1(23, 71, 92)
Multipole potential 1(327)
Multipoles 1(308)
Multivector 2(170)
Murphy's formula 2(240)

Nabla 2(161)
Negation 2(490)
Neighborhood of a point 1(1, 16)
Neperian logarithms 1(5, 51)
Net chart 2(352)
Neumann functions 1(152, 324, 351)
Neumann problem for a sphere 2(324)
Neumann series 1(250)
Newton interpolation formula 2(399)
Newton method 2(371, 391)
Nodal line 2(135)
Nomography 2(352)
Norm 2(203, 206, 209)
Normal direction 2(120)
Normal form 1(287, 290, 292); 2(130, 132)
Normal of surface 1(39)
Normal (or right) section 2(151)
Normal (or standard) equation 2(138, 181, 362)
Normal oriented 2(120)
Normal plane 2(147)
Normal vector 2(123, 192)
Null element 2(205)
Null operator 2(333)
Null representations 2(104)
Nyström extrapolation 2(425)

Object, geometrical 2(165, 168)
Observation, conditional 2(365)
Observation, mediating 2(362)
Operation, invariant 2(168)
Operational calculus 2(244, 251, 254, 255)
Operator 2(289)
Operator, adjoint 2(334)
Operator domain 2(21)
Operator, Hermitian 2(335)
Operator, invertible 2(334)
Operator isomorphism 2(103)
Operator, normal 2(335)
Operator, unitary 2(335)
Order 2(417)
Order counter 2(481)
Orders 2(482)
Ordinary differential equation 2(415)
Ordinary finite difference method 2(429, 444)
Orientation 2(120)
Original function 2(244)
Orthogonal functions 1(218); 2(330)
Orthogonal system, associated to the kernel 1(247)
Orthogonal system, closed 1(222)
Orthogonal system, complete 1(222)
Orthogonalisation method 1(223)
Orthogonality method 2(369)
Orthogonality relations of the first kind 2(67)

Ring with Poisson brackets 2(24, 36)
Ritz's method 1(278, 330); 2(452)
Root element of a ring 2(87)
Root of unity 1(47)
Rotary reflection 2(80, 134)
Rotation 2(134, 177)
Rotation group 2(75)
Rotation, infinitesimal 2(178)
Roughening 2(424)
Routine 2(483)
Row-matrix 2(128)
Runge-Kutta method 2(418)
Runge's scheme 2(412)

Saddle point, elliptical 2(151)
Saddle point, hyperbolical 2(151)
Scalar 2(158, 166)
Scalar field 2(157)
Scalar product 1(220); 2(131)
Scale 1(56); 2(177, 180, 351)
Schmidt's orthogonalisation process 2(332)
Schrödinger equation 2(288, 443, 445)
Schrödinger wave equation 1(206)
Schwartz's theory of distributions 2(340)
Schwarz's constant 2(459)
Schwarz's inequality 2(204)
Schwarz's lemma 1(65)
Schwarz's quotient 2(459)
Schwarz's reflection principle 1(65)
Schwarz's theory of distributions 2(232)
Schwinger's variational method 1(337)
Secondary conditions 1(269)
Section 2(2)
Secular equation 2(130, 395)
Security digit 2(384)
Self-adjoint 2(458, 459)
Selsyn (amer.) 2(474)
Semidefinite 1(17, 247)
Semigroup 2(16)
Sensitivity measure 2(420)
Sentence, dual (or reciprocal) 2(142)
Sentences, affine geometric 2(117)
Sentences, Euclidean geometric 2(117)
Sequence 1(42)
Serial machine 2(494)
Series, infinite 2(442)
Series, alternating 1(43)
Series, geometric 1(54)
Series, infinite 1(43)
Servomechanism 2(473)
Servomotor 2(473)
Set 1(1)
Set, Borel 1(83, 85)
Set, complementary 1(82)
Set, derived 1(82)
Set, empty 1(82); 2(199)
Set of points 1(1)
— closed 1(16, 82)
— connected 1(16)
— dense 1(16, 82)
— perfect 1(16, 82)
Set, open 2(199)
Set, overlapped 1(83)
Share, alternating 2(158)
Share, symmetric 2(158)
Signature 2(133)
Similarity, geometry of (general) 2(146)
Simply connected 1(19)
Simpson's rule 2(404)
Sine transforms, finite 2(308)
Single step method 2(390)
Singular, essential 1(63)
Singular point 1(63, 80)
Singularity, irregular 1(167)
Singularity, regular 1(166)
Skerv field 2(8)
Slide rule 2(350)
Smoothing 2(369)
Solution, most general 1(282)
Solution surface 1(284)
Source density 1(238)
Source layer 1(310, 312)
Source point 1(238)
Space, abstract 2(205)
Space, adjoint 2(211)
Space, affine 2(181)
Space, affine connected 2(174, 181)
Space, complementary 2(204)
Space, complete 2(206)
Space curve 2(146)
Space element 2(170)
Space, enumerably infinite dimensional 2(207)
Space, euclidean 2(181)
Space, euclidean affine 2(176, 181)
Space, euclidean metric 2(181)
Space, finite dimensional 2(207)
Space, flat 2(165, 176)
Space, linear 2(205)
Space, metric 2(206)

Space reflection 2(190)
Space rotations 2(187)
Space, second adjoint 2(211)
Space, separable 2(206)
Space, unitary 2(331)
Spaces, in the large flat 2(181)
Spaces, locally-flat 2(165)
Special 2(458)
Special functions, expansion of 1(190)
Special slide rule 2(351)
Spectral function 1(228)
Spectrum of a linear transformation 2(49)
Spectrum of a matrix 2(49)
Spectrum of an operator 2(335)
Spherical Bessel functions 1(154, 196)
Spherical coordinates 1(195, 275, 324)
Spherical harmonics 1(155, 196, 310, 325f.); 2(77)
Spherical harmonics, zonal 1(325)
Spherical wave 1(294)
Spheroidal coordinates, prolate 1(199)
Spheroidal functions 1(200, 214)
Spiegelungsprinzip 1(307)
Spinor 2(190)
Spinor field 2(190)
Spurious periods 2(425)
Squarable 1(37)
Squares, method of the least 2(364)
Stability 2(379)
Stability of the finite difference method 2(431)
Step function 1(232f., 349)
Step size 2(423, 445)
Stieltjes integral 1(38); 2(201)
Stirling's formula 1(67, 151); 2(402)
Störmer's extrapolation formula 2(426)
Straight edge 1(16)
Straight line 2(139)
String, vibrating 1(288); 2(257)
Strip 1(94, 116, 264)
Strip, characteristic 1(116)
Sturm-Liouville's propervalue problem 1(317, 332)
Subalgebra 2(85)
Sub-group, invariant 2(19)
Subroutine 2(485)
Subscripts, free (and superscripts) 2(165)
Subsequence 1(42)
Subseries 1(43)
Subset 1(1)
Subspace, invariant 2(51)
Subspace, linear 2(207)
Substitution, unimodular 1(132)
Sum 1(43)
Sum, direct 2(23)
Sum of linear transformations 2(54)
Summation formula 2(406)
Superposition principle 1(239, 282)
Surface 2(148)
Surface area 1(39)
Surface charge 1(302)
Surface, closed 1(40)
Surface harmonics 1(196)
Surface integral 1(40)
Surface, mean curvature of a 2(151)
Surface, non-orientable 2(140)
Surface, normal of a 2(149)
Surface of revolution 1(40)
Surface, one-sided (or unilateral) 2(140)
Surface, orientable 2(155)
Surface oriented 2(120)
Surface, two-sided 2(155)
Surfaces, deformable 2(150)
Surfaces, developable 2(150, 156)
Surfaces of negative curvature 2(155)
Surfaces of positive curvature 2(155)
Svartholm's method 2(290)
Symmetrics 2(134)
System, hypercomplex 2(23)
System, physical 2(109)

Tangential plane 2(149)
Tangential vector 2(146, 147, 148)
Tape, punched 2(481)
Tauberian theorems 2(241)
Tauberian theorems for the Laplace transform 2(242)
Taylor's expansion 1(14, 19, 52, 55)
Tchebyshev polynomials 1(224, 320)
Tchebyshev's formula 2(406)
Telegraph equation 2(430)
Temple's quotient 2(462)
Tensor 2(136, 167, 190)
Tensor algebra 2(168)
Tensor, alternating 2(169)

Tensor, antisymmetrical 2(169)
Tensor density 2(171)
Tensor density, linear 2(171)
Tensor, dual 2(172)
Tensor field 2(172)
Tensor, inverse 2(170)
Tensor, linear 2(171)
Tensor, mixed 2(168)
Tensor, skew-symmetrical 2(169)
Tensor, symmetrical 2(169)
Theorem of the mean, first (second) 1(30)
Theorema egregium 2(152, 153, 164)
Theta function 1(140)
Torque amplifier 2(479)
Torsion 2(147, 439)
Total curvature 2(156)
Total step method 2(390)
Trace 2(131, 168)
Trace of a matrix 2(51)
Trace, regular 2(104)
Tractrix 2(156)
Trajectories, orthogonal 2(186)
Trajectory, isogonal 1(69)
Trajectory, orthogonal 1(69)
Transcendental number 1(5)
Transfer, affine 2(174)
Transfer, integrable 2(164)
Transformation, additive 2(208)
Transformation, adjoint 2(47, 212)
Transformation, canonical 2(193)
Transformation, closed linear 2(211)
Transformation, continuous 2(208)
Transformation group 2(17)
Transformation, hermitian 2(48)
Transformation, homogeneous 2(208)
Transformation, infinitesimal 1(112)
Transformation, infinitesimal canonical 2(196)
Transformation, inverse 2(211)
Transformation, Legendre 1(101)
Transformation, linear 2(41, 127)
Transformation, normal 2(48)
Transformation of order *n* 1(136)
Transformation, projective 2(356)
Transformation, self-adjoint 2(48)
Transformation, unitarian 2(48)
Transformation, unitary 1(234)
Translation group, non-Euclidean 2(189)
Transposed 2(127)
Transversality condition 1(277)
Trapezoidal rule 2(404)
Trefftz's method 2(457)
Tricomi's method 2(237)
Tricomi's method, accuracy of 2(238)
Trihedral, maving 2(147)
Trihedral, reciprocal 2(158)

Uniformly continuous 1(3)
Union of sets 1(82)
Union of two sets 2(199)
Unit circle 1(47)
Unit (or identity) tensor 2(170)
Unit point 2(141)
Unit sphere 1(47)
Unit vector 2(118)

Variable, dependent 1(1)
Variable, independent 1(1)
Variation 1(264, 267)
Variation derivative 1(264, 267)
Variation, limited 1(35)
Variation of constant 1(75, 78)
Variation, positive (negative) 1(35)
Variation, second 1(266)
Variational, calculus 1(262)
Variational method 1(261)
Variational methods in functional analysis 2(345)
Variational principles 2(345)
Variational problem 1(264)
Variational problem, higher derivatives in the 1(280)
Vector 2(22, 117, 126, 163)
Vector field 2(157, 163)
Vector gradient 2(158)
Vector model 2(85)
Vector potential 1(208); 2(162)
Vector space(s) 2(22, 206)
— complex 2(206)
— euclidian 2(40)
— irreducible 2(51)
— linear 2(118)
— reducible 2(51)
— unitarian 2(40)
Vectors, bound 2(157, 163)
Vectors, covariant 2(166)
Vectors, isotrope 2(133)
Vectors, linear independent 2(118)
Vectors, space-like 2(133)
Vectors, time-like 2(133)

Vertex 2(192)
Vertical-row-index 2(127)
Vertical-row-matrix 2(128)
Volterra's integral equation 2(463)

Watson transforms 2(339)
Wave equation 1(193, 286, 287, 288, 293, 294, 334, 344)
— optical 1(351)
— separation of the 1(194)
— static 1(304, 321)
Wave function, spatial 2(289)
Wave functions in spherical coordinates 1(205)
Wave functions, momentum space 2(288, 289)
Weddle's rule 2(406)
Weierstrassian elliptic function 1(120)
Weight factor 1(220)
Weyl space 2(178)
Whittaker differential equation 1(198)
Whittaker functions 1(163)
Wiener's form of Fubini's theorem 2(203)
Wooley's formula 2(408)
Wronskian 1(77, 153, 210)

Yukawa potential 1(307)

Zemanian's formulae 2(326)
Zero 1(2, 52)
Zero divisor 2(11)
Zero system 2(144)

Group 2

Principles of Theoretical Physics

A priori probabilities, equal, hypothesis 3/2(190, 211, 283, 286)
Abampere 4(6)
Abnormality of a field 3/1(400, 819, 825, 826)
Absolute activity 3/2(25, 64–65, 316)
Absolute dimensions 3/1(799–801)
Absolute position 3/1(6)
Absolute space 3/1(6, 43)
Absolute temperature 3/1(652)
Absolute temperature determination 3/2(23–25)
Absolute temperature scale of Kelvin 3/2(3, 6)
Absolute tensor 3/1(661)
Absolute thermometer 3/2(158)
Absolute time 3/1(6)
Absolute time scale 4(115)
Absolute zero, crystal behaviour 3/2(87–88)
Absolute zero, unattainability 3/2(22–23)
Absorption coefficient 4(88, 89, 91)
Absorption coefficient of a metal 4(38)
Absorption coefficient of sound 3/2(480)
Absorption isotherm of Gibbs 3/2(46)
Absorption of waves 3/3(276)
Absorption region 4(88, 92)
A.c. convergence, *see* Almost certain convergence
Accelerated charge, radiation 4(66–68)
Acceleration 3/1(26, 30, 201)
Acceleration, d'Alembert-Euler condition 3/1(388)
Acceleration, d'Alembert-Euler formula 3/1(374)
Acceleration energy 3/1(63)
Acceleration field 3/3(39)
Acceleration gradients 3/3(54)
Acceleration, Hankel-Appell condition 3/1(388)
Acceleration, intrinsic formulae 3/1(375–376)
Acceleration, Lagrange's formula 3/1(377)
Acceleration of higher order 3/1(383, 439–440)
Acceleration, physical component 3/1(27)
Acceleration potential 3/1(389)
Acceleration vector 3/1(140, 201f.)
Acceleration vector in four dimensions 3/1(461)
Acceleration wave(s) *see also* Waves 3/1(523–524); 3/3(267–272)
— growth and decay 3/3(384)
— in hypo-elastic materials 3/3(413–415)
Accelerationless homogeneous motion 3/1(434–436)
Accelerationless motion 3/1(432–433)
Accessibility 3/2(209, 254)
Accumulative material 3/1(733–734, 740)
Acoustic tensor, symmetry 3/3(384)
Acoustical axes 3/3(268, 414)
Acoustical tensor 3/3(268)
Action 3/1(698)
Action and reaction 3/1(37)
Action at a distance, relativistic 4(153–155)
Action line element 3/1(139)
Action principle 4(176, 179, 215)
Action principle of Toupin 3/3(392)
Action upon a system 3/2(122)
Action variables 3/1(175, 177)
Activity, absolute 3/2(25, 64–65, 316)
Addition of velocities 4(123–124)
Additive process 3/2(591)

Adhering of material to a surface 3/1(386, 394–396, 401)
Adiabatic, *see also* Isentropic
Adiabatic, *see* Isentropic
Adiabatic invariance of phase-space volume 3/2(244–245)
Adiabatic isolation 3/2(134–135)
Adiabatic parameter change 3/2(244, 304, 321)
Adiabatic perturbation 5(82–87)
Adiabatic process 3/1(644); 3/2(2)
Adiabatic stability 3/3(330)
Adiabatic switching on 5(233, 297, 336)
Adiabatic theorem 5(212, 233)
Adiabatically arrested variable 3/2(135)
Adkins' reciprocal theorem 3/3(207)
Admissible process 3/3(44)
Admissible thermodynamic process 3/3(365)
Admittance 3/2(511)
Advanced function 5(192)
Advanced potentials 4(47–48)
Advanced solution 4(176)
Aeolotropic solid 3/3(82)
Aeolotropic solid, stored energy 3/3(312)
Aeolotropy 3/1(700–701); 3/3(197)
Aeolotropy, curvilinear 3/3(91)
Aether relations of Maxwell and Lorentz 3/1(660, 677–683)
Affine connection 4(209, 211)
Affinity 3/1(643, 646); 3/2(16, 17, 429)
Affinity, generalised 3/2(486)
Affinity, temperature dependence 3/2(18)
Airy's stress function 3/1(583); 3/3(204)
Almost certain convergence 3/2(574)
Alpha particle 5(113, 119)
α-type forces 3/2(426)
α-variables 3/2(426)
Alternators 4(137, 186)
Ambigenous process 3/2(648)
Amount of shear 3/1(293)
Ampère-Lorentz principle 3/1(683)
Ampère's law 4(6)
Amperian equivalent current 3/1(686)
Amplitude of wave 3/3(268, 283)
Angle characteristics 3/1(131)
Angle of shear 3/1(293)
Angle variables 3/1(175, 177)
Angle variables, periodic property 3/1(177)
Angle-eikonal 3/1(131)
Angular distribution of radiation 4(67, 74)
Angular distribution of scattered radiation 4(94)
Angular momentum, *see also* Moment of momentum 3/1(34, 44, 56, 66, 213, 221, 482); 3/3(40)
— commutation relations 5(102, 103)
— conservation 4(144–145)
— conservation law in Dirac's theory 5(159)
— conservation law of field theory 5(180)
— matrix elements 5(103)
— of a spherical wave 4(52–53, 61)
— of the electromagnetic field 4(31)
— operator 4(42); 5(44, 45, 102, 103)
— relativistic 4(149)
Angular oscillations of second-grade fluids 3/3(510)
Angular variable 5(96)
Angular velocity 3/1(350, 355, 437)
Angular velocity of a rigid body 3/1(27, 28)
Anholonomic components 3/1(311, 801); 3/3(15)
Anisotropic fluids
— of Coleman 3/3(88)
— of Ericksen 3/3(523–537)
— of Green 3/3(521–523)
— of Wang 3/3(86–88, 151, 314–316, 473–475)
Anisotropic solids 3/1(314–315)
Anisotropy, *see also* Aeolotropy
Anisotropy 3/2(465, 468)
Annihilation operator 4(195); 5(187)
Anomalous absorption 4(89)
Anomalous first-order transformation 3/2(373, 389, 403)
Anomalous magnetic moment of the electron 5(299)
Anticommutator 5(217)
Anti-ferromagnetism 4(26)
Antisymmetric (eigen)functions 5(112, 114)
Antisymmetric states 5(119)
Antisymmetric tensor 3/1(660–662)
Aphelion 3/1(51)

Apparent circulation in relative motion 3/1(442)
Apparent force (or fictitious force) 3/1(4, 45, 535)
Apparent moduli and elasticities 3/3(239ff., 251ff.)
Apparent stress 3/1(550)
Apparent torque 3/1(535)
Appell's equations of motion 3/1(63)
Appell's theorem on vortex line stretching 3/1(429–430)
Approach to equilibrium 3/2(58–59)
Approximate theory 3/1(231)
Apsides of periodic orbits 3/1(50, 51)
Arc, deformation of an element of 3/1(247–249)
Arc, material derivative of an element of 3/1(341–342)
Arrest 3/2(254)
Assigned couple 3/1(538, 546)
Assigned force 3/1(537)
Astatic load 3/1(574); 3/3(127)
Astrophysical stochastic problems 3/2(603–611)
Atmospheric scattering 4(94)
Atomic weight 3/1(473)
Autocorrelation function 4(165)
Average field in a dielectric 4(17)
Average, quantum mechanical 5(25, 26)
Axes moving 3/1(29)
Axial *k*-vector density 3/1(661)
Axial oscillations of second-grade fluids 3/3(509ff.)
Axial relative tensor 3/1(661)
Axial tensor 3/1(795)
Axially symmetric, *see* Rotationally symmetric
Axiom of continuity 3/1(243, 325)
Axiomatics 3/1(5)
Axioms and experiments 3/1(228–230)
Axis of equilibrium 3/3(136, 258)
Axis of shear 3/1(292)
Azeotropic mixture 3/2(40–41)

Back-scattering in stochastic processes 3/2(646–648)
Backward differential equation 3/2(569, 571)
Backward propagation 3/2(647)
Baker-Erickson (B-E) inequalities 3/3(158–171, 175, 183, 281, 293)
Balance between deformation and rotation 3/1(397–399)
Balance equations in integral form 3/1(594)
Balance equations of irreversible thermodynamics 3/2(414, 419–422, 442)
Balance equations, relativistic 3/2(503)
Balance laws 3/1(232)
Balance of energy 3/1(607–615, 691–697)
Balance of entropy 3/2(422–424)
Balance of entropy, relativistic 3/2(498–499, 503)
Balance of internal energy, relativistic 3/2(498)
Balance of moment of momentum 3/1(531–532, 546, 554, 692–693, 697)
Balance of momentum 3/1(531, 545, 554, 691–698)
Balance of momentum in a mixture 3/1(567)
Ballistics, exterior 3/1(48)
Barotropic flow 3/1(388, 712)
Barycentric Lorentz frame 3/2(496)
Barycentric velocity 3/2(419, 495)
Barygyroscope 3/1(92)
Basic laws of electrodynamics 4(3–11)
Basic random process, *see also* Blended process 3/2(563, 565–566, 576, 591, 593)
Basic representation theorem of Cauchy 3/3(29)
Basic stability theorem 3/3(253)
Basis vectors 3/3(14ff., 23)
Bayes theorem 3/2(644)
Bending 3/1(300–302); 3/3(186, 194, 211)
Bending couple 3/1(566)
Bending of a block 3/3(186–188, 211, 342ff.)
Bending of light rays at the limb of the Sun 4(234)
Bernoullian theorems 3/1(402–409)
Bernoulli's binomial distribution 3/2(543)
Bernoulli's law of large numbers 3/2(527)
Bernoulli's theorem of classical hydrodynamics 3/1(402–403)
Bernstein's theorem on waves 3/3(332)
Berthelot formula of second virial coefficient 3/2(26)

Characteristic function 3/1(117, 120, 121, 137, 209); 3/2(540, 549, 552)
Characteristic temperature, rotational 3/2(75, 79, 81)
Characteristic temperature, translational 3/2(71)
Characteristic temperature, vibrational 3/2(77)
Charge and current as fundamental entities 3/1(683)
Charge conjugation 5(221)
Charge conservation 3/1(660, 667)
Charge, definition 3/1(666)
Charge density 3/1(672); 4(4, 11)
Charge, discovery of its two signs 4(1)
Charge equation 3/1(674)
Charge potential 3/1(674)
Charge renormalization 5(282, 344–348)
Charge renormalization of the one-electron state 5(301–305, 355–358)
Charge-current density 4(167)
Charge-current field 3/1(666–667)
Charge-current potential 3/1(668)
Chasles' theorem 3/1(14)
Chemical equilibrium 3/2(62–63)
— involving crystals 3/2(92–93)
— of gases 3/2(91–92)
Chemical interpretation of mass balance in the heterogeneous medium 3/1(473–474)
Chemical potential 3/1(621, 636, 638, 648–649); 3/2(14, 25, 316, 346, 614)
Chemical production 3/2(421, 427)
Chemical viscosity 3/2(428)
Christoffel condition for shock waves 3/1(711)
Christoffel symbols 3/1(26, 109, 804); 3/3(19ff.); 4(212, 216)
C-inequalities (classical) 3/3(153)
Circle diagram of Mohr for a tensor 3/1(253–254)
Circle diagram of Mohr for stress 3/1(552)
Circuit 3/1(103, 104, 156, 814); 4(6)
Circuit, irreducible 3/1(103, 104, 122)
Circular cylinder, typical deformations 3/1(299–300)
Circular frequencies 3/1(180, 182)
Circular motion of a charge, radiation 4(67)
Circular pipes, flow in them 3/3(448–452)
Circulating motion, Ertel's theorem 3/1(424)
Circulation 3/1(356–357, 361, 385–389)
Circulation, apparent, in relative motion 3/1(442)
Circulation of a tensor field 3/1(814)
Circulation of action 3/1(156, 163, 169)
Circulation preserving motion 3/1(388–389, 392–393, 405, 408, 432, 476, 711)
Circulation, rate of change 3/1(410)
Circulation-preserving motion 3/3(75)
Clapeyron equation 3/2(12, 41)
Clapeyron-Clausius relation 3/2(35, 372, 386)
Class division 3/2(122)
Class division, ordered 3/2(125)
Classical electron radius 4(93)
Classical mechanics as a limiting case 5(87)
Classical physics, its definition 3/1(234)
Classical statistics, definition 3/2(67)
Classification of spaces 3/1(100)
Clausius' entropy concept 3/2(4–5)
Clausius' form of the second law 3/2(5, 157)
Clausius-Duhem inequality 3/3(295, 361, 363ff.)
Clausius-Mossotti relation 4(18, 87)
Clebsch's transformation 3/1(424–427)
Climbing in steady Couette flow 3/3(452–455)
Clock, its rate 4(121, 123)
Clock paradox 4(124–125)
Closed loop 5(253, 254)
Closed phase 3/2(13)
Closed systems of equations 3/1(232)
Cluster approach 3/2(329)
Cluster integral 3/2(615)
Clustering of galaxies 3/2(610)
Coarse-grained density 3/2(225)
Coarse-grained density matrix 3/2(294)
Coarse-grained density operator 3/2(284)
Coefficient of thermal expansion 3/1(623)
Coexistence of phases 3/2(403–404)
Coherence properties of radiation 5(133)
Coleman and Noll's inequality 3/3(319–324)

Coleman and Noll's stability theorem 3/3(330)
Coleman and Noll's theory of fading memory 3/3(101–117)
Coleman-Noll (C-N) condition 3/3(157, 321, 333)
Coleman-Noll^{C+} condition 3/3(323)
Coleman's thermodynamics of simple materials 3/3(363–381)
Coleman's work theorem 3/3(306)
Collision 3/1(71, 94, 132, 217); 4(147, 148, 149)
Collision, elastic 3/1(95, 219)
Collision, inelastic 3/1(95, 219)
Collision line breadth 4(86)
Collision parameter 3/1(71)
Collision problem 5(73)
Collision radiation 4(67, 74–76, 78–80)
Combinatorial analysis 3/2(553–558)
Comets 4(33)
Common frame 3/1(241, 246–247, 325)
Commutation relations
— canonical 5(36, 101, 102, 103)
— canonical, of field theory 5(174)
— covariant formulation 5(190)
— of current components 5(225)
— of the electromagnetic field components 5(182, 183)
Commutative group 4(130)
Commutator 4(129, 130, 269)
Commutator of covariant differentiation 4(214)
Commuting observables 4(185)
Compatibility condition(s) 3/3(127)
— for a moving surface 3/1(500–501)
— for principal waves 3/3(286–288)
— for singular surfaces 3/1(493–494)
— for strain 3/1(271, 307)
— for stretching 3/1(351)
— geometrical iterated 3/1(496–498)
— in curved spaces 3/1(351–352)
— kinematical, for discontinuities 3/1(503–506, 515)
— of Toupin and Bernstein 3/3(277, 335)
Compatibility theorem of Signorini 3/3(225)
Complementary 5(1, 7)
Complementary strain-energy 3/3(327)
Complete transition relation 3/2(123)
Completeness relation 5(13, 48)
Complex sound velocity 3/2(481)
Complex-lamellar field 3/1(824–825)
Complex-lamellar motion 3/1(390, 400, 417–420, 431)
Complex-screw motion 3/1(417–418, 431)
Composite system 3/2(126–127)
Composite system, empirical entropy 3/2(136–139)
Composition of shears 3/1(296)
Composition of strains 3/1(282–283)
Compressibility 3/1(623)
Compressible elastic bodies
— all possible deformations 3/3(171–179, 336ff.)
— homogeneous strain 3/3(171–179)
Compression period 3/1(95)
Compression speed 3/1(95)
Compressive wave 3/1(525)
Compton effect 3/1(220); 5(263–266, 269)
Concentrated load 3/1(538–539)
Concentration 3/1(470)
Concepts 3/1(3, 6, 98)
Condensation at a shock 3/1(522)
Condensation theory of Mayer 3/2(373, 376)
Conditional probability 3/2(507, 530)
Conditions of compatibility
— for strain 3/1(271, 307)
— for stretching 3/1(351)
— in curved spaces 3/1(351–352)
Conducting electrons 3/2(97)
Conduction current 3/1(672)
Conductivity, electric 4(10, 11)
Conductors of heat 3/3(384)
Cone of shearing stress 3/1(552)
Cone-and-plate flow 3/3(461–464)
Configuration, definition 3/3(37)
Configuration integral 3/2(264)
Configuration integral in quantum statistics 3/2(326)
Configuration, local 3/3(52)
Configuration space 3/1(2, 13, 100, 134)
Connected circuit, simple and multiply 3/1(103)
Connection, material 3/3(89)
Conninuity equation 5(15, 25, 89)
Conninuity equation of the Dirac theory 5(139, 140)
Consael's lemma 3/2(544–546)

Conservation law(s) 3/1(232); 4(137)
— in irreversible thermodynamics 3/2(419–422)
— nonrelativistic 4(144–145)
— of angular momentum in field theory 5(180)
— of electromagnetism 4(172–173)
— of energy and momentum in field theory 5(175)
Conservation of angular momentum 3/1(692–693, 697)
Conservation of charge 3/1(660, 667)
Conservation of density in phase 3/2(181)
Conservation of energy, including electromagnetic fields 3/1(691–698)
Conservation of extension in phase 3/2(181)
Conservation of flux 3/1(346–347)
Conservation of magnetic flux 3/1(660, 667)
Conservation of mass 3/1(464, 466–474)
Conservation of mass in a heterogeneous medium 3/1(472)
Conservation of mechanical energy 3/1(572–573)
Conservative forces 3/1(571)
Conservative *k*-vector field 3/1(665–666)
Constant invariant tensor 3/1(662)
Constant stretch history 3/3(438–440)
Constitutive equation(s) 3/1(232, 233); 3/3(2, 36, 44)
— dimensionally reduced form 3/3(65ff., 120)
— for electromechanics 3/1(742–744)
— general mechanical 3/3(57–58)
— general properties 3/1(700–704)
— of a hyperelastic material 3/3(302)
— of an elastic fluid 3/3(149)
— of an elastic material 3/3(119)
— of an incompressible isotropic elastic material 3/3(148)
— of a rigid body 3/1(704–705)
— of Coleman 3/3(364ff.)
— of differential-type materials 3/3(93ff.)
— of electromagnetic materials 3/1(736–742)
— of rate-type materials 3/3(95)
Constitutive Maxwell equations 3/2(502)
Constraint hypersurface 4(183)
Constraints 3/1(183, 600–602, 605–606); 4(157, 159, 160, 182, 184, 261, 263, 264, 268, 269)
Constraints, internal 3/3(69–73)
Constraints, thermal 3/3(329)
Contact equilibrium 3/2(147)
Contact equilibrium, internal 3/2(147)
Contact force, *see also* Stress 3/3(40)
Contact loads 3/1(536)
Contact relation 3/2(133, 152)
Contact, thermal 3/2(171)
Contact transformation 3/1(124, 144)
Contact variable 3/2(148)
Continuity axiom 3/1(243, 325)
Continuity concept 3/2(120)
Continuity equation(s) 3/2(420); 3/3(39); 4(137, 161)
— in space-time 3/1(480)
— its solution 3/1(474–481)
— of electric charge 4(166, 167)
— of mass 3/1(467)
— of masses 4(203)
— spatial 3/1(342, 467)
Continuity of path assumption of Maxwell 3/2(192)
Continuity of state 3/2(33)
Continuous group 5(100)
Continuous metric variable 3/2(130, 131)
Continuous motion 3/1(325)
Continuous random variable 3/2(533, 540–542, 546–549)
Continuous spectrum of eigenvalues 5(42)
Continuous variable 3/2(125)
Continuum and molecular theories compared 3/3(5)
Continuum mechanics, fundamental laws 3/1(545–547)
Continuum mechanics in special relativity 4(159–166)
Continuum theories, physical basis 3/1(228)
Contraction of indices 4(133)
Contravariant components 3/3(14ff.)
Contravariant derivatives 3/3(16)
Contravariant vector 4(132)
Control surface 3/1(540)
Convected co-ordinates 3/1(457, 460)
Convected stress-rate 3/3(96, 97)
Convected time flux 3/1(448–452, 675)
Convection of a quantity 3/1(337)

Convection of stretching and spin 3/1(380–381)
Convection of vorticity 3/1(409–430)
Convection potential 4(65)
Convection theorems for vorticity 3/1(413–421)
Convention on concentrated loads 3/1(539)
Convention regarding description of motion 3/1(338)
Conventional entropy 3/2(90)
Convergence in probability 3/2(574)
Convergence modes in random functions 3/2(574–576)
Converse of the principle of virtual work 3/1(602–603)
Conversion factors of units 4(107)
Convolution 3/2(540)
Coolidge's extension of the Lexian scheme 3/2(543)
Coordinate conditions of de Donder 4(218)
Coordinate schemes of motion 3/1(327)
Coordinate system in space-time 4(109)
Coordinate systems 3/3(19ff.)
Coordinate systems, overlapping 3/1(103)
Coordinate transformation 3/1(164)
Coordinates, convected 3/1(457, 460)
Coordinates, curvilinear 3/1(25, 26, 43)
Coordinates, generalized 3/1(38, 59)
Coordinates, ignorable 3/1(58, 61, 102, 176, 179, 191)
Coordinates, material 3/1(326)
Coordinates, material and spatial 3/3(37, 39)
Coordinates, spatial 3/1(326)
Coriolis acceleration 3/1(30, 438, 462)
Coriolis force 3/1(45); 4(205)
Co-rotational frame 3/1(444–446)
Co-rotational time flux 3/1(445, 451–452)
Correlation function 3/2(629)
Correlation matrix 3/2(507)
Correlation of random variables 3/2(534)
Correspondence of Einstein's with Newton's theory 4(216, 219)
Correspondence principle 5(24, 96)
Correspondence principle of general relativity 4(208)
Cosmic ray cascades 3/2(600)
Cosserat material 3/3(397)
Cosserats' theory of polar elastic materials 3/3(389ff.)
Couette flow 3/3(449ff.)
Coulomb gauge 4(48)
Coulomb interaction and Hamiltonian 4(104)
Coulomb law 5(43)
Coulomb potential of a point charge 4(11–12)
Coulomb scattering 3/1(77)
Coulomb scattering radiation 4(77–78)
Coulomb-Hopkins' theorem on maximum shearing stress 3/1(551, 845–846)
Coulomb's experiments on torsion 3/3(236)
Coulomb's law, discovery 4(1–2)
Coulomb's law, formulation 4(3)
Counter problems 3/2(582)
Couple 3/1(531, 536)
Couple density 3/3(394)
Couple stress 3/1(538, 546, 547)
Couple-stress tensor 3/3(396, 525)
Covariance principle 4(206)
Covariant components 3/3(14ff.)
Covariant derivative, partial 3/1(794, 804)
Covariant derivative, total 3/1(246–247, 810–813)
Covariant derivatives 3/3(16)
Covariant differentiation 3/1(811, 813); 4(211–212)
— commutator 4(214)
— of a double tensor field 3/1(809)
Covariant metric tensor 4(209, 210)
Covariant radiation reaction formula 4(84)
Covariant vector 4(132)
Creation, *see* Production
Creation operator 4(195)
Critical point 3/2(32–33)
Critical temperature of complete miscibility 3/2(43)
Cross force 3/1(558)
Cross product 3/1(662)
Cross section for capture 3/1(75)
Cross section for scattering
— differential 3/1(75, 77)
— total 3/1(76)
Cross-viscosity 3/1(721)
Crystal classes 3/3(83)

Crystal, ideal mixed 3/2(86)
Crystal, simple 3/2(82–83)
Crystal-gas relations 3/2(87)
Crystals, exceptional 3/2(89)
Curie law 4(26)
Curie point in ferromagnetics 4(26)
Curie's law 3/2(113)
Current and charge as fundamental entities 3/1(683)
Current density 3/1(672); 4(7, 21)
— in the Dirac theory 5(140)
— statistical 5(16, 25)
Current, electric, discovery 4(2)
Current equation 3/1(674)
Current of dielectric convection 3/1(686)
Current of polarization 3/1(685, 686)
Current operator, radiative corrections 5(290–295)
Current potential 3/1(674)
Curvature, least 3/1(141–143)
Curvature of streamlines 3/1(376)
Curvature tensor, *see also* Riemann-Christoffel curvature tensor 4(211, 212, 213–214)
Curvilinear Bernoulli theorem 3/1(406–407)
Cyclic coordinates, *see* Coordinates, ignorable
Cyclic processes, general theorem 3/3(381)
Cyclic processes, inequalities 3/3(369ff.)
Cylinder, extension by twisting 3/3(193, 352)
Cylinder, radial oscillations 3/3(214–217)
Cylinder, rotation 3/1(550)
Cylinder, various deformations 3/3(189–195)
Cylinders, coaxial, flow between them 3/3(446–452)
Cylindrical coordinates 3/1(27); 3/3(19)
Cylindrical sector, straightening and stretching 3/3(188ff., 343)
Cylindrical tube, extension, inflation and shear 3/3(353)

D function 5(190)
— advanced 5(192)
— retarded 5(192)
Da Silva's theorem on torques 3/3(128)
D'Alembert-Euler condition for the acceleration 3/1(388)
D'Alembert-Euler motions 3/1(432)
D'Alembert-Euler spatial continuity equation 3/1(467)
D'Alembert-Euler vorticity equation 3/1(413)
D'Alembert's equation 4(225)
D'Alembert's principle 3/1(57, 59, 96, 532)
D'Alembert's stream function 3/1(474)
D'Alembert's vorticity theorem for plane motion 3/1(421)
Dalton's law of partial pressure 3/2(81)
Damped oscillator 4(85)
Darwin term 5(161)
Darwin-Fowler method
— for localised particles 3/2(309–314)
— for non-localised particles 3/2(314–318, 327)
de Broglie wavelength 5(6, 10)
de Broglie waves 3/1(215, 216); 5(4, 6)
de Donder conditions 4(218)
Debye characteristic temperature of a crystal 3/2(84)
Debye cube law 3/2(85)
Debye potentials 4(41, 55, 58–59)
Decomposition of motion 3/1(362)
Definite conductor of heat 3/3(384)
Deformation 3/1(243–325)
Deformation coordinates 3/2(168)
Deformation function, relative 3/3(49)
Deformation gradient 3/3(49)
Deformation gradients 3/1(245–249)
Deformation, local 3/3(52)
Deformation of an oriented body 3/1(311)
Deformation of areas 3/1(263)
Deformation tensor of Cauchy 3/1(257–274, 282, 370)
Deformation tensor of Green 3/1(257–274, 370)
Deformations possible in all
— compressible elastic bodies 3/3(171–179, 336ff.)
— incompressible elastic bodies 3/3(337–342)
Deformed material 3/1(244)
Degeneracy 3/2(55, 69, 318)
Degeneracy in mechanics 3/1(180, 183, 191)

Degeneracy in thermodynamics 3/1(633–634)
Degenerate eigenfunctions 5(47)
Degenerate eigenvalue, perturbation removal of degeneracy 5(78)
Degenerate ideal gas 3/2(93)
Degenerate systems 5(56)
Degree of degeneracy 3/2(69)
Degrees of freedom 3/1(38)
Degrees of freedom
— classical 3/2(70)
— electronic 3/2(76)
— of a linear molecule 3/2(74)
— of a non-linear molecule 3/2(79)
— of a phase 3/2(16)
— unexcited 3/2(69)
Delbrück scattering of photons 5(333)
δ, explanation of the symbol 4(135)
Delta function 3/2(531–533); 4(8)
Density 3/1(342)
Density in phase 3/2(179, 225)
— conservation law 3/2(181)
— invariance 3/2(183)
Density matrix 3/2(270, 272); 5(69)
Density matrix of canonical ensemble 3/2(319)
Density of magnetic flux 3/1(672)
Density of mass 3/1(465); 3/3(38)
Density of polarization 3/1(685)
Density of states 3/2(633, 636–638)
Density operator 3/2(273)
Density operator of canonical ensemble 3/2(326)
Detailed balancing 3/2(216, 297)
Determinant of a tensor 3/3(15, 23ff.)
— gradient 3/3(25)
Determinism 3/3(12, 36, 56)
Deuterium molecules, statistical behaviour 3/2(78–79, 90)
Deviation equation 3/1(141)
Deviator 3/1(834)
Deviator of stretching 3/1(349)
Diamagnetism 3/2(113); 4(2, 25)
Dielectric constant 4(5, 16)
Dielectric, constitutive equations 3/1(736–740)
Dielectric effects, discovery 4(2)
Dielectric media 4(14–19)
Dielectric polarization 4(6, 15)
Dielectric with elastic properties 3/1(742–744)
Differential equations involving random functions of time 3/2(615–621)
Differential equations of motion for superimposed infinitesimal displacement 3/3(247)
Differential form of general balance 3/1(468)
Differentiation, exterior 3/1(165)
Differentiation, stochastic 3/2(576–577)
Diffraction of corpuscular rays 5(4)
Diffusion coefficient after Einstein 3/2(598)
Diffusion equation 3/1(706)
Diffusion flow 3/2(419, 435)
Diffusion, framework of theories 3/3(537–541)
Diffusion in a mixture 3/1(705–708)
Diffusion in force fields 3/2(458)
Diffusion kinematics 3/1(469)
Diffusion of circulation 3/1(441)
Diffusion of stretching and spin 3/1(380–381)
Diffusion of vorticity 3/1(409–430, 441)
Diffusion potential 3/2(461)
Diffusion thermoeffect 3/2(427, 452, 455, 472)
Diffusion velocity 3/1(470)
Dilatation 3/1(350, 431)
Dilatation group 3/3(86)
Dilatation of hypo-elastic material 3/3(416)
Dilatation, uniform 3/3(173)
Dimension of action 3/1(698)
Dimension of electromagnetic quantities 3/1(671, 672, 680–681, 698–699)
Dimension of energy quantities 3/1(615)
Dimension of entropy 3/1(620)
Dimension of temperature 3/1(619)
Dimensional invariance 3/1(701–702); 3/3(64–65)
Dimensional transformation 3/1(798)
D-invariant 4(264–270)
Dipole, electric, oscillating 4(49)
Dipole, magnetic 4(22)
Dipole moment 4(12)
Dipole radiation 5(129)
Dirac canonical field variables 4(264, 270)
Dirac delta function 3/2(531–533); 4(8)
Dirac equation for free particles 5(140, 143, 214)

Dirac equation with external electromagnetic field 5(155–158, 226–230)
Dirac field, quantized 5(216)
Dirac matrices 5(140, 141, 143)
Dirac radiation theory 5(123ff.)
Dirac spin matrices 4(143–144, 192)
Dirac theory of the electron 5(137ff.)
Dirac theory of the spinning electron 4(189–193)
Director gradient 3/1(312)
Director quadric of a tensor 3/1(251)
Directors of an oriented body 3/1(311)
Directors of an oriented medium 3/3(389, 392)
Discontinuities, various definitions 3/1(529)
Discontinuity surface, *see* Singular surface
Discontinuity types 3/1(492)
Discovery of electromagnetic phenomena 4(1–3)
Discrete random variable 3/2(539–540, 543–546)
Dislocation 3/1(512); 3/3(126)
Dislocation density 3/3(92)
Dislocations, continous, in elastic bodies 3/3(125–126)
Dislocations, continuum theory of 3/3(88–92)
Dispersion 5(123, 129)
Dispersion, elementary theory 4(86–89)
Dispersion, justification of elementary theory 4(99)
Dispersion of waves 3/3(276)
Dispersion relation 4(91, 93)
Dispersion, rigorous theorie 4(96–100)
Displacement current 4(8–9)
Displacement derivative 3/1(501–503)
Displacement electric 3/1(674)
Displacement gradient 3/3(130)
Displacement gradients, small 3/1(306)
Displacement, infinitesimal 3/1(25, 29)
Displacement parallel to a plane 3/1(12, 199)
Displacement potential 3/1(277)
Displacement speed of a surface 3/1(498–499, 506)
Displacement vector 3/1(247); 4(5–6)
Dissipation function 3/2(428, 486)
Dissipation vinciple 3/3(295)
Dissipative absorption 4(88, 94)
Dissipative linear systems 3/2(484–486, 510)
Dissipative stress 3/1(571, 640)
Dissipative systems 3/1(185)
Dissipativity 3/2(485)
Distance between two particles 4(165)
Distant zone 4(50, 52)
Distortion tensor 3/1(350)
Distribution function
— and frequency function 3/2(531)
— convergence 3/2(574)
— kinetic 3/2(213, 491)
— of a random variable 3/2(528)
Distribution law for non-localised particles 3/2(316)
Distribution law of Maxwell 3/2(73)
Distribution laws of Fermi and Bose *see also* Bose-Einstein and Fermi-Dirac statistics 3/2(62, 317)
Distribution modulus 3/2(262)
Distribution of energy, Maxwell-Boltzmann law 3/2(65, 67, 256–261, 313, 318)
Divergence 4(134)
Divergence difficulties 5(232)
Doppler breadth of a spectral line 4(86)
Doppler effect of second order 4(123)
Double tensor 3/1(242, 805)
Double tensor equations 3/1(246)
Doublet 3/1(334)
Dual of a tensor 3/1(661); 4(137)
Dual space 4(143)
Dualism of waves and particles 5(4)
Duality principle 3/1(513)
— first 3/1(242, 246)
— second 3/1(263)
Duhem-Margules relation 3/2(43)
Duhem's acceleration formula 3/1(378, 425)
Duhem's identity for acoustical tensors 3/3(335)
Duhem's theorem on wave speeds 3/3(335)
Duhem's theorem on waves 3/3(272)
Dulong-Petit law 3/2(83)
Dyadic product of two vectors 3/3(15)
Dynamical conditions at a singular surface 3/1(545–547)
Dynamical equation of state 3/2(479–482)

Dynamical theory, general 3/1(98)
Dynamical vorticity number 3/1(379)
Dynamics 3/1(1, 2, 3, 8, 108, 198, 204)
Dynamics, geometrical representations 3/1(98)
Dynamics, isoenergetic 3/1(100, 136)
Dynamics of a particle 3/1(43)
Dynamics of particle systems 3/1(56ff.)
Dynamics of rigid bodies 3/1(56ff., 65)
Dyson operator 3/2(586)

Efflux of energy 3/1(609)
Ehrenfest relations for second-order transformation 3/2(373)
Ehrenfest urn model 3/2(218–222)
Ehrenfest's adiabatic law 5(82)
Eigenfrequencies 3/1(2, 81)
Eigenfunctions 5(46, 112)
Eigen-time, *see* Proper time
Eigenvalue problem 5(40, 41)
Eigenvalue spectrum 5(49)
Eigenvalues, discrete 5(41, 42)
Eikonal 3/1(117, 131)
Eikonal equation 4(113)
E-inequalities 3/3(171)
Einstein characteristic temperature of a crystal 3/2(83)
Einstein condensation 3/2(102–105, 373, 377, 393–404)
Einstein effects 4(234, 249)
Einstein field equations 4(216)
Einstein general theory of relativity 3/1(2)
Einstein-Bose statistics 5(119)
Einstein-Infeld-Hoffmann theory 4(244–245)
Elastic aether model 3/1(727–730)
Elastic body 3/3(125)
Elastic collision 3/1(95, 219)
Elastic dielectric 3/1(742–744); 3/3(385–388)
Elastic fluid 3/3(148, 480)
Elastic fluid, stored energy 3/3(313)
Elastic material 3/2(441); 3/3(106, 117)
— as approximation 3/3(111)
— definition 3/3(13, 117, 119ff.)
— isotropy group 3/3(122)
— polar 3/3(389–401)
— reinforced 3/3(122)
Elastic solid 3/3(149, 170)
Elastic stability 3/3(252–260)
Elastic stress 3/1(571)
Elastic subfluid 3/3(151)
Elastic theories of fluids 3/3(480ff.)
Elasticities, apparent 3/3(239ff., 251ff.)
Elasticities, number of independent ones 3/3(309ff.)
Elasticities of an elastic material 3/3(132)
Elasticity, classical infinitesimal theory 3/3(117)
Elasticity, linear 3/1(723–727)
Elasticity of *n*th order 3/3(228)
Elasticity, reduced constitutive equations 3/3(113, 120ff.)
Elastostatics, differential equation 3/3(127)
Elastostatics of simple materials 3/3(123)
Electric current, *see* Current
Electric dipole radiation 4(49–50, 52)
Electric displacement, *see* Displacement
Electric displacement 3/1(674)
Electric field intensity 4(4)
Electric field (strength) 3/1(672)
Electric potential 3/1(674)
Electric quadrupole radiation 4(53–54)
Electric susceptibility 4(16)
Electric susceptibility, dependence on temperature 4(18–19)
Electricity and heat, analogy 3/3(388)
Electrochemical cell 3/2(29–30, 462)
Electrochemical effects 3/2(458–464)
Electrochemical potential 3/2(459)
Electrodynamics in special relativity 4(166–181)
Electrodynamics in vacuo 5(181ff.)
Electromagnetic conservation laws 3/1(689–700)
Electromagnetic constitutive equations 3/1(736–742)
Electromagnetic energy 3/1(689)
Electromagnetic field 3/1(667); 3/2(501–505)
— analogy with quasi-elastic aether 3/1(729)
— axially symmetric 3/1(54)
— equations, four-dimensional form 3/1(669)
— equations, three-dimensional form 3/1(676–677)
— in vacuo 5(181)

Electromagnetic field, interaction with matter 3/1(692–693)
— quantization 4(199)
Electromagnetic integral laws
— four-dimensional form 3/1(667)
— three-dimensional form 3/1(673–674)
Electromagnetic mass 4(65, 80–82, 177)
Electromagnetic momentum 3/1(689)
Electromagnetic potential 3/1(212, 668)
Electromagnetic stress tensor 3/1(689)
Electromagnetic units 4(6, 29)
Electromagnetism, linear 3/1(233)
Electromechanical theories 3/3(385–389)
Electromechanics, constitutive equations 3/1(742–744)
Electromotive force 3/2(462); 4(10, 26)
Electromotive intensity 3/1(672)
Electron 5(31, 107, 109)
Electron, anomalous magnetic moment 5(299)
Electron conduction 3/2(97)
Electron, existence 4(85)
Electron gas, zero point energy 3/2(317)
Electron in motion 4(63–66)
Electron microscope 3/1(54)
Electron model of Lorentz 4(81–82)
Electron radius 4(93)
Electron scattering by a Coulomb field 5(259–261)
Electron scattering by protons 5(310)
Electron theory, historical remarks 4(2–3)
Electron-electron scattering 5(271–274, 331)
Electronic spin 4(189)
Electrostatic field 3/1(213)
Electrostatic field in vacuo 4(11–14)
Electrostatic potential 4(4, 11)
Electrostatic units 4(4, 5)
Electrostatics, early experiences 4(1)
Electrostriction 4(20)
Elektrostatic field 3/2(110–111)
Elementary charge, existence 4(84)
Elementary particle 5(31, 107, 109, 110, 113, 114, 150, 168)
Ellipsoid of inertia 3/1(33, 486)
Ellipsoid of polarization 3/3(332)
Elliptic polarization by metallic reflection 4(39)
Elliptical tube, secondary flow 3/3(503)
Elongation 3/1(256–268)
Elongation, principal axes 3/1(281)
Elongation quadric 3/1(268)
Elongation tensor 3/1(267–268)
Emergent spectrum 3/2(644)
Emission, induced 5(133)
Emission of radiation 5(127ff.)
Emission, spontaneous 5(133)
Empirical entropy 3/2(134–135, 168)
Empirical entropy of a composite system 3/2(136–139)
Empirical temperature 3/2(243, 313, 320–321)
Encounter, *see also* Collision 3/1(132, 217)
Energy balance 3/1(607–615, 691–697); 3/3(294ff.)
Energy balance in a heterogeneous medium 3/1(612–614)
Energy class 3/2(128, 130, 142)
Energy concept 3/2(1)
Energy conservation
— in radiation 4(82)
— including electromagnetic fields 3/1(691–698)
— law 4(114–145)
— law in irreversible thermodynamics 3/2(422)
— law of field theory 5(175)
Energy distribution law of Maxwell and Boltzmann 3/2(65, 67, 256–261, 313, 318)
Energy equation 3/1(58, 60, 109, 207, 212, 215)
Energy exchange between radiation and matter 5(123ff.)
Energy flow 3/2(422, 435)
Energy fluctuations 3/2(266)
Energy flux in an electromagnetic wave 4(32)
Energy flux in the electromagnetic field 4(29–30, 66)
Energy function 3/1(143)
Energy functional 3/3(298)
Energy, general theory 3/1(607–615)
Energy impulse 3/1(612)
Energy integral 3/1(58, 60, 63, 161, 213)
Energy isolation 3/2(142)

Energy, kinetic 3/1(35, 36, 39, 44, 365–369, 377, 402, 482–491)
Energy levels, degenerate 3/2(69)
Energy minimum, Kelvin's theorem 3/1(368)
Energy of the electromagnetic field 3/1(689)
Energy of the electrostatic field 4(19–21)
Energy of the magnetostatic field 4(26–29)
Energy of transfer 3/2(436)
Energy per radiation mode 3/2(106–107)
Energy, potential 3/1(42, 210, 212, 571)
Energy radiation 4(43)
Energy radiation of an oscillator 4(50, 52)
Energy, reduction to normal 3/1(180)
Energy, relativistic 4(146, 148)
Energy separation in Dirac's theory 4(191)
Energy surface 3/1(143, 620)
Energy tensor of the electromagnetic field 4(30)
Energy, total 3/1(44, 49, 57, 58, 63, 136)
Energy transmission across a boundary 4(36)
Energy zero 3/2(18)
Energy-momentum
— density 5(181)
— space 3/1(130)
— tensor 3/2(496–497, 504); 4(161)
— vector 3/1(110, 114, 202, 696)
Energy-stress complex 4(237)
Ensemble 3/2(48)
Ensemble average 3/2(179)
Ensemble average, physical significance 3/2(210)
Ensemble, canonical 3/2(52, 262, 265)
Ensemble, canonical, of quantum statistics 3/2(318ff.)
Ensemble, grand canonical 3/2(343–350)
Ensemble, microcanonical 3/2(239, 255, 265, 303)
Ensemble, mixed 3/2(269)
Ensemble, pure 3/2(269)
Ensemble, representative 3/2(190, 209–212)
Ensemble, stationary 3/2(186)
Ensemble, uniform 3/2(287)
Ensemble, virtual 3/2(179)
Enthalpy 3/1(627); 3/2(9, 615)
Entropy 3/2(4–5, 245, 248, 251); 3/3(133, 294ff.); 5(72)
Entropy, adiabatic invariance 3/2(305, 321)
Entropy analog, statistical 3/2(305, 322)
Entropy analogs of Gibbs 3/2(246, 250, 264)
Entropy and Massieu-Planck function 3/2(358)
Entropy balance 3/2(422–424)
Entropy balance, relativistic 3/2(498–499, 503)
Entropy concept 3/1(616)
— historical development 3/2(114–117)
— statistical 3/2(57)
Entropy conventional 3/2(90)
Entropy definition of quantum statistics 3/2(306)
Entropy, empirical 3/2(134–135, 136–139, 168)
Entropy flow 3/2(415, 423, 436)
Entropy for the grand canonical ensemble 3/2(348)
Entropy function 3/3(301)
Entropy, general theory 3/1(615–647)
Entropy inequality 3/1(643–647); 3/3(295, 357)
Entropy law of Boltzmann 3/2(256)
Entropy, metrical 3/2(139–141)
Entropy of a single substance 3/2(31–32)
Entropy of mixing 3/2(37)
Entropy of transfer 3/2(464)
Entropy production 3/1(638–647); 3/2(415, 423, 433, 443)
— bilinear form 3/2(443)
— minimal 3/2(458)
— quadratic form 3/2(444, 445)
Entropy relation 3/3(302, 368)
Entropy zero 3/2(18)
ε-substitution, invariance 3/2(437–438)
Equation of continuity, *see* Continuity equation
— in phase space 3/2(420); 3/3(39); 3/2(180)
Equation of heat conduction, general 3/2(488)
Equation of state 3/3(133)
— caloric 3/3(298, 361)

Equation of state, dynamical 3/2(479–482)
— for a mixture 3/1(634–636)
Equation of thermal conduction 3/1(710)
Equations of motion, *see also* Cauchy and Euler 3/1(6, 7, 43, 65, 203)
Equations of motion 3/2(178, 421)
Equilibrium approach 3/2(58–59)
Equilibrium between simple mixture and solid 3/2(41–42)
Equilibrium, chemical 3/2(62–63)
— involving crystals 3/2(92–93)
— of gases 3/2(91–92)
Equilibrium class 3/2(133)
Equilibrium, complete 3/2(426)
Equilibrium conditions 3/2(11)
Equilibrium configurations 3/1(180, 182)
Equilibrium constant 3/2(91)
Equilibrium, frozen 3/2(426)
Equilibrium, mechanical 3/2(450–451)
Equilibrium of liquid and vapour 3/2(35, 44)
Equilibrium of rods 3/1(564–567)
Equilibrium of shells 3/1(556–564)
Equilibrium of two phases 3/2(36)
Equilibrium properties, macroscopic 3/2(212)
Equilibrium, statistical 3/2(185–187, 274)
Equilibrium, thermal 3/2(3, 6)
Equilibrium, thermal, of simple systems 3/2(171)
Equilibrium, thermodynamic 3/1(647–660); 3/2(251–255, 264–267, 324–325)
Equimomental mass distributions 3/1(33)
Equipartition theorem, classical 3/2(242, 263)
Equipartition theorem, generalized 3/2(241)
Equipotential surfaces of a moving point charge 4(112)
Equipresence principle 3/1(703–704); 3/3(359ff., 364)
Equivalence of axioms 3/2(173)
Equivalence of motions and processes 3/3(42, 43–45)
Equivalence of operators and matrices 5(52)
Equivalence of strain measures 3/1(268–269)
Equivalence of surface and volume sources 3/1(469)
Equivalence principle of Einstein 4(204–206, 234)
Equivalence problem 4(219, 247)
Equivalence theorem for higher-order waves 3/3(273)
Equivalence theorem of Cosserat-Toupin 3/3(395)
Ergodic hypothesis of Boltzmann 3/2(188, 192, 193)
Ergodic problem 3/2(191–194)
Ergodic property of a Markovian stochastic process 3/2(625)
Ergodic system 3/2(584)
Ergodic theorem 3/2(194, 198, 207–208)
Ergodic theorem of quantum statistics 3/2(279–283)
Ergodic theory 3/2(188, 207, 623)
Ericksen's analysis of deformations possible
— in all compressible elastic bodies 3/3(336ff.)
— in all incompressible elastic bodies 3/3(337–342)
Ericksen's anisotropic fluids 3/3(523–537)
Ericksen's work theorem 3/3(307)
Ertel's commutation formula 3/1(413)
Ertel-Rossby convection theorem 3/1(414)
Ertel's potential theorem 3/1(423)
Ertel's theorem on circulating motions 3/1(424)
Escape velocity 4(217)
E-T inequalities (extension-tension) 3/3(157)
Eucken equation 3/2(471)
Euclid invariance 3/3(392, 395)
Euclidean action 3/3(389, 392)
Euclidean frame 3/1(455, 670)
Euclidean space 3/1(241)
Euclidean space-time and kinematics 3/1(453, 454–455)
Euler acceleration 3/1(438)
Euler-d'Alembert paradox 3/1(541–542)
Euler-d'Alembert principle 3/1(532)
Euler-Ertel vorticity theorems 3/1(413–414)

Eulerian angle(s) 3/1(18, 22, 24, 65); 5(46, 106)
Eulerian parameters 3/1(17, 20, 21, 22, 24, 28)
Euler-Lagrange equation 3/1(107, 204, 206)
Euler's criterion for isochoric motion 3/1(343)
Euler's criterion for rigid motion 3/1(350)
Euler's diagram in thermodynamics 3/1(623)
Euler's dynamical equation for fluids 3/1(710)
Euler's equations for a rigid body 3/1(67, 705)
Euler's expansion formula 3/1(342, 347, 373)
Euler's laws of classical mechanics 3/1(530, 531)
Euler's material and spatial continuity equations 3/1(467)
Euler's tensor of inertia 3/1(485, 490, 573)
Euler's theorem on irrotational motion 3/1(359)
Euler's theorem on spherical surface displacements 3/1(13)
Event 3/1(5, 453)
Event (= point in space-time) 3/3(41)
Events, mutually exclusive 3/2(525)
Evolution 3/2(572, 622)
Exactness of theory 3/1(231–232)
Excess functions in mixtures 3/2(38–39)
Exchange 5(116–118)
Exchange effect in scattering 5(274)
Exchange integral 5(116–118)
Exclusion principle 5(110, 114, 115, 117, 119, 218)
Existence and uniqueness theorems in elasticity 3/3(133–139)
Expanding universe 4(208)
Expansion coefficient, thermal 3/1(623)
Expansion, free 3/2(10)
Expansive wave 3/1(524, 525)
Expectation value 3/2(534); 5(37, 53, 69)
Expectation value of perturbation energy 5(76)
Experimental tests of general relativity 4(234–235)
Experiments and axioms 3/1(228–230)
Explosion of a particle 3/1(217ff.)
Extension 3/1(255–256, 257, 260–262, 266–267, 274)
Extension and torsion interrelated 3/3(264ff.)
Extension in configuration 3/2(185)
Extension in phase 3/2(181)
— conservation law 3/2(181)
— invariance 3/2(183)
Extension in velocity 3/2(185)
Extension of a cylinder by twisting 3/3(193, 352)
Extension of hypo-elastic material 3/3(417ff.)
Extension, steady 3/3(472ff.)
Extensive parameters, fluctuations 3/2(361–364)
Extensive property 3/2(3, 14)
Extensive variable 3/1(647)
External field 4(176)
External non-mechanical energy supply 3/1(638)
External power 3/1(638)
Extinction theorem 4(99)
Extra stress 3/3(71)
Extremal 3/1(123)
Extremal properties of tensors 3/1(845)
Extremal property of the steady state 3/2(455)
Extrinsic loads 3/1(536)

Factor group 4(261)
Factorisation of the partition function 3/2(68)
Fading memory 3/3(36, 101–106, 112ff., 365)
Faraday (unit) 3/2(29)
Faraday's law of induction 3/1(674)
Fast approximation 4(217, 244)
F.D.F., *see* Fluctuating density field
Ferel's law 3/1(55)
Fermi distribution 3/2(317)
Fermi level 3/2(95)
Fermi-Dirac gas 3/2(95–97)
Fermi-Dirac statistics 3/2(60–62, 93, 289, 307, 314, 327, 553, 556, 558, 639); 5(119)
Fermions 4(197)
Fermi's lemma on canonical normal systems 3/2(193, 203)

Ferrites 4(26)
Ferromagnetism 4(26)
Feynman diagrams (graphs) 5(251)
Feynman's operator calculus 3/2(587)
Feynman's solution of the Bloch equation 3/2(340–343)
Fick's law of diffusion 3/1(706)
Fictitious forces (or apparent forces) 3/1(4, 45, 535)
Field, continuous 3/1(227)
Field, electric 4(4)
Field energy, electrostatic 4(19–21)
Field energy, magnetostatic 4(26–29)
Field energy tensor 4(30)
Field equations 3/1(232, 233)
— of Einstein 4(216)
— of electromagnetism in four dimensions 3/1(669)
— of electromagnetism in three dimensions 3/1(676–677)
Field intensities in relativity 4(168)
Field, magnetic 4(8)
Field quantization 4(197)
Field strength, *see* Field
Fields in dielectrics 4(16–18)
Fields in magnetic media 4(23–26)
Fields in Minkowski universe 4(134)
Fields of a moving point charge 4(63–66)
Figuratrix 3/1(116)
Fine structure constant 4(72)
Fine structure of positronium 5(328)
Fine-grained density 3/2(225)
Finger's theorem on strain and rotation 3/1(278)
Finite elasticity, differential equation of motion 3/3(125)
Finite rotation 3/1(274–283)
Finitely elastic body 3/1(730)
Finsler space 3/1(106)
First entropy analog of Gibbs 3/2(264)
First law of thermodynamics 3/2(1, 245, 322)
— history of discovery 3/2(113–114)
First-class constraints 4(269)
First-order fluids 3/3(100)
First-order transformation 3/2(372)
Fisher's Z-distribution 3/2(548)
Flexibility, perfect 3/1(565)
Flow along a curve 3/1(357)
Flow between coaxial cylinders 3/3(446–452)
Flow between rotating parallel plates 3/3(458–464)
Flow of a tensor field 3/1(814)
Flow of anisotropic solids 3/3(520–523)
Flow of subfluids 3/3(473–475)
Fluctuating density field 3/2(592)
Fluctuation dissipation theorem 3/2(510, 513)
Fluctuation force 3/2(605)
Fluctuation theory, time dependent 3/2(516–518)
Fluctuations 3/2(506)
Fluctuations in energy 3/2(266)
Fluctuations in Galaxy brightness 3/2(608–610)
Fluctuations of extensive parameters 3/2(361–364)
Fluctuations of intensive parameters 3/2(364–369)
Fluctuations of internal parameters 3/2(369–371)
Fluctuations of occupation numbers 3/2(370)
Fluctuations of pressure 3/2(366)
Fluctuations of the index of refraction 3/2(369)
Fluidity 3/3(13)
Fluids of different grades 3/3(494)
Fluids of different orders 3/3(491)
Fluids of first order 3/3(100)
Fluids of rate type 3/3(482–484)
Fluids of second grade 3/3(504–513)
Fluids of second order 3/3(100)
Fluids, simple 3/3(427–513)
Fluids, special theories 3/3(475–513)
Flux of a tensor field 3/1(815)
Flux of electromagnetic energy 4(29–30, 66)
Fluxes 3/2(424, 432–435)
Fokker-Planck equation 3/2(491, 521)
Fokker-Planck equation, Chandrasekhar's treatment 3/2(594)
Force 3/1(1, 6, 531)
Force density 4(28)
Force exerted by a rigid obstacle 3/1(540)
Force field, external 5(24, 64)
Force function, *see* Potential function
Force function of Gibbs 3/2(8)

Force system 3/1(37)
Forces 3/2(424, 426, 432–435)
Forces and stresses 3/3(39–41)
Forces, effective 3/1(57)
Forces, external 3/1(37)
Forces, fictitious (or apparent) 3/1(4, 45, 535)
Forces, generalized 3/1(42)
Forces, internal 3/1(37)
Forces of inertia 3/1(57)
Forces, reversed effective 3/1(57)
Forward differential equation 3/2(561, 569)
Foucault rotation 3/1(53, 56)
Foucault's pendulum 3/1(55)
Four-component spinors 4(142–144, 190)
Four-current density 5(143, 148)
Four-current, Gordon's decomposition 5(162)
Four-force 3/1(7, 203)
Fourier analysis of oscillator damping 4(85–86)
Fourier expansion of radiation 4(68–71, 74)
Fourier integral for wave functions of free particles 5(13)
Fourier transform 3/2(541)
Fourier's heat conduction equation 3/2(489)
Fourier's law of heat conduction 3/1(709–710, 719)
Four-momentum 3/1(200, 202, 217)
Four-potential 3/1(212, 668)
Fracture 3/1(510)
Frame, change of 3/3(41)
Frame, co-rotational 3/1(444–446)
Frame, irrotational 3/1(446–447)
Frame of reference 4(109, 207)
Frame of reference, moving 3/1(30, 45, 68, 200)
Frame, preferred 3/1(443)
Frame-indifference, material 3/3(12, 36, 44 ff., 296)
Frame-indifferent quantity 3/3(42)
Frames of reference 3/1(25, 71, 78, 200, 218, 241, 246–247, 325)
Fréchet differential 3/3(104)
Free charge(s) 3/1(684); 4(5)
Free electromagnetic field 5(181)
Free energy 3/1(627); 3/2(7–8, 264, 324, 325); 3/3(301); 4(19, 27)
— analog 3/2(264, 324)
— for the grand canonical ensemble 3/2(349)
— in classical statistics 3/2(67–68)
— minimum 3/3(370)
— of mixing 3/2(81–82)
Free enthalpy 3/1(627)
Free expansion 3/2(10)
Free particles 5(335)
Free shape 3/3(130)
Frequency 3/1(179, 180, 182)
Frequency dependence of thermodynamical coefficients 3/2(482–487)
Frequency function 3/2(530, 531, 535, 537)
Fresnel formulas 4(36)
Fresnel quadric 3/3(268)
Fresnel-Hadamard theorem on wave propagation 3/3(168)
Fresnel's dragging coefficient 3/1(739)
Fresnel's ellipsoid 3/3(332)
Friction 3/1(85)
Frog's leg experiment of Galvani 4(2)
Fugacity 3/2(27–28)
Full Lorentz group 4(121, 126)
Function group 4(259)
Functional derivative 5(353)
Functions of matrices 3/1(838–840)
Functions of several world points 4(135)
Fundamental equations of thermodynamics 3/1(628)
Fundamental identities for strain invariants 3/1(265)
Fundamental identity for heterogeneous media 3/1(471)
Fundamental strain theorems of Cauchy 3/1(259–261, 274, 277–280)
Fundamental theorem
— of Hadamard on waves 3/3(271, 332 ff.)
— of motion decomposition 3/1(362)
— of Noll on simple fluids 3/3(81)
— on hyperelastic solids 3/3(311)
— on simple materials 3/3(66)
— on strain of shells 3/1(322–323)
— on the center of mass 3/1(483)
Furry process 3/2(581, 633)
Future 3/1(200)

Galilean frame of reference 3/1(200, 456)

Galilean invariance 3/1(491)
Galilean observer 3/1(200)
Galilean principle of relativity 3/1(491, 535)
Galilean space-time and kinematics 3/1(453, 455–456, 461)
Galilean transformation 4(110–112, 145)
Galilean transformation, generalization 3/1(681)
Galilean transformation, generator 4(149)
Galvanomagnetic effects 3/2(464ff.)
Gamma ray microscope 5(69)
Γ-space statistics 3/2(176)
Gas and liquid states 3/2(32ff.)
Gas constant 3/2(25)
Gas degeneracy 3/2(318)
Gas mixtures, kinetic theory 3/1(567–568, 612–614)
Gas-crystal relations 3/2(87)
Gauge condition 4(175)
Gauge covariance 4(170)
Gauge group 5(30, 156)
Gauge invariance 4(48); 5(30, 156)
Gauge invariance (vacuum polarization) 5(281)
Gauge transformation(s) 3/1(669, 675); 4(47, 56–57, 248); 5(181)
Gauss' flux theorem 4(4, 15)
Gauss' theorem, *see also* Green's transformation 4(160)
Gaussian probability distribution 3/2(506, 547, 593)
Gaussian process 3/2(593)
Gaussian system of units 4(3, 4, 5, 29, 107, 169)
Gauss-Weingarten equations 3/1(812)
GCN condition (generalized Coleman-Noll condition) 3/3(162–169, 321)
GCN_0 condition 3/3(160)
GCN_0^+ condition 3/3(162)
GCN^+ condition 3/3(166, 249, 257, 271, 281)
Géhéniau-Komar scalars 4(252, 269)
General balance 3/1(468)
General balance at a discontinuity surface 3/1(526–527)
General conservation law 3/1(468)
General co-ordinates in space-time 3/1(453)
General theory of relativity 4(203)
Generalised H-theorem of quantum statistics 3/2(297, 300–302)
Generalised H-theorem of Tolman 3/2(227)
Generalized convection vector 3/1(416–417)
Generalized Galilean transformation 3/1(681)
Generalized partition function 3/2(354, 356)
Generalized shear 3/1(297)
Generating function(s) 3/1(146, 169); 3/2(539)
Generator of an infinitesimal Galilean transformation 4(149)
Generator of an infinitesimal Lorentz transformation 4(151–153)
Generic phase 3/2(249)
Geometrical mechanics 3/1(98, 215)
Geometrical motion classes 3/1(328–329, 430–431)
Geometrical representation of stress 3/1(552)
Geometrical representation of tensors 3/1(250–255)
Gibbs' absorption isotherm 3/2(46)
Gibbs' chemical potential 3/2(14, 25, 316, 346, 614)
Gibbs, Coleman and Noll's stability theorem 3/3(331)
Gibbs' diagram 3/1(620)
Gibb's entropy analogs 3/2(246, 250, 264)
Gibb's equation 3/2(418)
— of thermodynamics 3/1(621, 635–638)
— relativistic 3/2(498, 503)
Gibb's free energy 3/2(7)
Gibbs function 3/1(628); 3/2(9, 10–11)
Gibbs paradox 3/2(249)
Gibbs phase rule 3/2(17–18, 374)
Gibbs relations 3/3(372, 381)
Gibbs-Duhem equation 3/1(648)
Gibbs-Duhem relation 3/2(15–16, 40–41, 43)
Gibbs-Helmholtz relation 3/2(13, 23, 30, 352)
Gibbs-Helmholtz relation, simple applications 3/2(28, 29, 30)
Good quantum numbers 4(201)
Grade of a fluid 3/3(494, 504)

Grade of a material of differential type 3/3(111)
Gradient of a tensor function 3/3(24–27)
Gradient of a vector field 3/3(15)
Gradient tensor 4(134)
Gradients of invariants, representations 3/3(34)
Gradients of principal invariants 3/3(25ff.)
Grand canonical ensemble 3/2(343–350)
Grand partition function 3/2(64–65, 348, 382)
— for photons 3/2(108)
— of quantum statistics 3/2(95, 97–98)
Graphs of Feynman 5(251)
Gravitation law of Newton 4(203)
Gravitation potential of Newton 4(215, 217)
Gravitational field 3/2(109)
Gravitational force, equivalence to inertial force 4(205)
Gravitational radiation 4(222)
Gravitational waves 4(224–226, 235)
Green and Rivlin's multipolar elasticity 3/3(400ff.)
Greenhill's acceleration formula 3/1(378)
Green's anisotropic fluids 3/3(521–523)
Green's deformation tensor 3/1(257–274, 370)
Green's formula 4(14)
Green's function of the Feynman method 3/2(341–343)
Green's functions in Minkowski universe 4(135, 155)
Green's theorem 4(13, 22, 98)
Green's transformation 3/1(815–816)
Green-St. Venant strain tensor 3/3(221)
Grioli's theorem for homogeneous strain 3/1(290)
Grioli-Toupin material 3/3(400)
Gromeka-Beltrami theorems on screw motion 3/1(392–393, 418)
Group character of Lorentz transformation 4(121, 126)
Group theory 5(98)
Group velocity 4(90); 5(5, 18, 152)
Group velocity in Dirac waves 5(152)
Gwyther-Finzi stress functions 3/1(585–587)
Gyration vector 3/3(386)
Gyrocompass 3/1(90, 91)
Gyroscopic couple 3/1(91)
Gyroscopic stability 3/1(185)
Gyroscopic stiffness 3/1(90)

$\hbar$: Planck constant 5(4)
Hadamard's fundamental theorem on waves 3/3(271, 332ff.)
Hadamard's inequality 3/3(253, 257)
Hadamard's lemma on singular surfaces 3/1(492–493)
Hagen-Poiseuille efflux formula 3/3(466)
Hamilton formalism, relativistic 4(155–158)
Hamilton-Cayley theorem 3/3(26)
Hamiltonian 3/2(178)
Hamiltonian action 3/1(110, 204, 205)
Hamiltonian density 5(171, 174)
Hamiltonian dynamics 3/1(112, 113, 204)
Hamiltonian for a particle in a magnetic field 5(27)
Hamiltonian for a system of point masses 5(24)
Hamiltonian formulation of radiation theory 4(100–102)
Hamiltonian four-vector (of momentum) 3/1(205, 215)
Hamiltonian function 3/1(62, 109, 110, 206, 213, 214)
Hamiltonian in special relativity 4(158, 160)
Hamiltonian method 3/1(2, 3)
Hamiltonian of a field 5(174)
Hamiltonian of charged particles 4(102–104)
Hamiltonian of the Dirac theory 5(156, 216)
Hamiltonian operator 3/2(271); 5(33)
Hamiltonian operator in spherical polar-coordinates 5(40)
Hamiltonian relativistic dynamics 3/1(204–206)
Hamiltonian surface 3/1(116)
Hamiltonian three-momentum 3/1(213)
Hamilton-Jacobi equation 3/1(117–120, 122, 125, 130, 138, 157, 170, 175, 209, 213, 215); 4(113)

Hamilton-Jacobi partial differential equation 5(88)
Hamilton's canonical equations 3/1(62, 63, 110, 207)
Hamilton's characteristic function, *see* Characteristic function
Hamilton's method in optics 3/1(107, 110, 117, 131)
Hamilton's partial differential equations 3/1(119)
Hamilton's principle 3/1(107, 110, 204, 205, 604–605); 4(149–151)
Hamilton's two-point characteristic or principal function 3/1(117, 120, 137, 209)
Hankel-Appell condition for the acceleration 3/1(388)
Harmonic analysis, generalised 3/2(629, 631)
Harmonic oscillation 3/1(529)
Harmonic oscillator 3/1(46, 210); 5(44)
Heat absorption 3/3(294)
Heat and electricity, analogy 3/3(388)
Heat and work, interconvertibility 3/1(608, 614)
Heat concept 3/2(2)
Heat conduction 3/1(233); 3/3(294ff.)
— Fourier's equation 3/2(489)
— Fourier's law 3/1(709–710, 719)
— general equation 3/2(488)
Heat conductivity 3/2(471ff.)
— of reacting mixtures 3/2(473)
— with delayed equilibrium restoration 3/2(475)
Heat flow, reduced 3/2(436)
Heat flux 3/3(294)
Heat function for constant pressure 3/2(9)
Heat of reaction 3/2(93)
Heat of transfer 3/2(437, 459)
Heaviside-Lorentz system of units 4(5)
Heisenberg representation 5(172)
Heisenberg representation, perturbation calculation 5(234–239)
Heisenberg's uncertainty principle 5(1, 20)
Helical flow 3/3(445–452)
Helium spectrum, Heisenberg's theory 5(117)
Helmholtz free energy 3/2(8, 264, 324, 325)
Helmholtz relation 3/2(30)
Helmholtz-Föppl formula 3/1(366)
Helmholtz' vorticity theorem
— first 3/1(386–387)
— second 3/1(389, 409)
— third 3/1(389, 410)
Helmholtz-Zorawski criterion 3/1(341)
Hemitropic tensor function 3/3(24)
Hemitropy 3/3(358)
Henry's law 3/2(40)
Hereditary materials 3/1(733–734, 740)
Hermitian matrix 5(51, 69)
Hermiticity 5(24, 33)
Herpolhode 3/1(83, 85)
Hertz vector 4(48)
Heterogeneous media 3/1(469–474, 567–568, 612–614, 634–638, 645–650)
— conservation of mass 3/1(472)
— fundamental identy 3/1(471)
Higher-order elasticities, experimental 3/3(230)
Higher-order waves 3/3(272ff., 283)
High-temperature approach 3/2(330, 333–337)
Hilbert space 4(158, 181, 188, 189, 256)
— of free photons 5(184–189)
— representations 5(170–173)
History of a function 3/3(52)
History of dynamics 3/1(1)
History of electromagnetic phenomena 4(1–3)
History of field description of motion 3/1(327)
History of finite hyperelasticity 3/3(304)
History of stretching and shearing theory 3/1(349)
History of the theory of spin 3/1(356)
Hodograph 3/1(26)
Hölder's principle 3/1(139)
Hole theory 5(168, 218)
Holohedry, *see* Isotropy
Holonomic constraints 3/1(600–602)
Holonomic systems, *see also* Non- and Semi-holonomic system 3/1(38, 60)
Holtsmark distribution 3/2(608)
Homentropic waves 3/3(272, 335, 383ff.)
Homogeneity and isotropy of space and space-time, axiom 3/1(9, 10)
Homogeneity of material lost by strain 3/3(251)

Homogeneous body 3/3(59, 90)
Homogeneous flow of Ericksen's anisotropic fluids 3/3(530–533)
Homogeneous Lorentz transformation 4(149)
Homogeneous motion 3/1(434–436)
Homogeneous motions, isochoric 3/3(74)
Homogeneous pure-stretch history 3/3(66)
Homogeneous state in thermodynamics 3/1(647–650)
Homogeneous strain 3/1(285–298); 3/3(171–183)
Homogeneous strain history 3/3(62)
Homogeneous wave equations of field vectors 4(31)
Homothermal waves 3/3(272, 335, 383ff.)
Hooke's law 3/1(724)
H-theorem 3/2(216)
H-theorem, generalised, of Tolman 3/2(227)
H-theorem of quantum mechanics 3/2(288ff.)
H-theorem of quantum statistics, generalised 3/2(297, 300–302)
Hugoniot-Duhem theorem on the speed of displacement 3/1(506)
Hugoniot's theorem on sound waves 3/1(712)
Huygens' construction 3/1(123)
Hydrogen atom 5(43)
Hydrogen molecules, statistical behaviour 3/2(77–79, 90)
Hydrostatic pressure 3/1(549, 574)
Hydrostatic pressure, effect on wave speed 3/3(284ff.)
Hydrostatic stress 3/1(544)
Hydrostatic tension states in elastic materials 3/3(170)
Hyperbolic motion in space-time 3/1(210)
Hyperelastic body 3/1(725, 730)
Hyperelastic material(s) 3/3(117, 294ff.)
— definition 3/3(302)
— of second grade 3/3(397–400)
— special properties of solutions 3/3(342–344)
Hyperelastic solids, fundamental theorem 3/3(311)
Hyperelastic stability 3/3(328–332)
Hyperelastic subfluids 3/3(314–316)
Hyperelasticity, definition 3/3(13)
Hyperfine structure of hydrogen 5(310–316)
— measurements 5(315, 316)
Hypergeometric distribution 3/2(544)
Hyperquantization 4(193–197)
Hyperstress tensor 3/3(395, 398)
Hypersurface of constraints 4(183)
Hypo-elastic material, definition 3/3(401ff.)
Hypo-elastic response functions 3/3(404)
Hypo-elastic waves 3/3(413–415)
Hypo-elastic yield 3/3(420–426)
Hypo-elasticity 3/1(731–733); 3/3(13, 117, 401–426)
Hypo-elasticity, relation to elasticity 3/3(406–411)
Hysteresis loop 4(27)

Ideal degenerate gas 3/2(93)
Ideal gas 3/1(633)
— in gravitational field 3/2(109)
— mixture 3/2(36–37, 38, 81–82)
— monatomic 3/2(70–73)
— thermodynamic functions 3/2(27)
Ideal material 3/1(700); 3/3(2, 44)
IFS condition (invertibility of force-stretch) 3/3(156)
Ignorable co-ordinates 3/1(58, 61, 102, 176, 179, 191)
Imbedding theory of strain 3/1(314)
Impact parameter 3/1(71)
Impedance matrix 3/2(485)
Impenetrability of matter 3/1(244, 325)
Impossible event 3/2(526)
Improper Lorentz transformation 4(140–141)
Impulse 3/1(536)
Impulsive balance at a singular surface 3/1(527–528)
Impulsive force 3/1(92)
Impulsive moment 3/1(92)
Impulsive motion 3/1(92)
Impulsive work 3/1(92)
Inaccessibility 3/2(254, 307)
Inaccessibility axiom of Carathéodory 3/2(167, 174–175)
Incoming fields 5(207)
Incoming particles 5(335)

Incompatibility of reference configurations 3/1(273)
Incomplete miscibility 3/2(42)
Incompressibility 3/3(71)
Incompressible, *see* Isochoric
Incompressible elastic bodies, all possible deformations 3/3(337–342)
Incompressible elastic material 3/3(121)
— homogeneous strain 3/3(179–183)
Incompressible isotropic elastic material 3/3(147ff.)
Incompressible isotropic material, second-order elasticity 3/3(241–246)
Incompressible material, wave propagation 3/3(291–294)
Indefinite metric 4(126); 5(200)
Independence of random variables 3/2(534)
Independent circuits 3/1(103)
Indeterminacy, *see* Uncertainty
Index of refraction 3/1(740); 4(87, 89, 99)
Indicatrix 3/1(116)
Individual motion of constituents 3/1(469)
Individuals, indistinguishable 5(31)
Induced current 4(10)
Induction, discovery 4(2, 10)
Induction law of Faraday 3/1(674); 4(10)
Induction lines 3/1(674)
Induction zone, *see* Near zone
Inelastic collision 3/1(95, 219); 4(148)
Inequalities between tensor invariants 3/1(835–836)
Inequalities for cyclical processes 3/3(369ff.)
Inequalities for elastic materials 3/3(153–171)
Inequalities, thermodynamic 3/1(652–659)
Inequality of Coleman and Noll 3/3(319–324)
Inert mixture 3/1(472)
Inertia, ellipsoid of 3/1(33, 486)
Inertia, moments of 3/1(32, 33, 485–486)
Inertia, principal axes 3/1(32, 33, 36, 485–486)
Inertia, principal moments of 3/1(32, 33, 485)
Inertia products of 3/1(31, 486)
Inertia, tensor of 3/1(485–486, 490)
Inertial force, equivalence to gravitational force 4(205)
Inertial frame 3/1(531, 533, 535, 555); 4(110, 118)
Inertial mass 4(203)
Inextensibility 3/3(72)
Infinitesimal canonical transformation 4(149, 158)
Infinitesimal Lorentz transformation 4(151–153)
Infinitesimal oscillation 3/1(529)
Infinitesimal plane waves 3/3(273–278)
Infinitesimal rotation tensor 3/3(115)
Infinitesimal stability 3/3(252–260)
Infinitesimal strain, historical remarks 3/1(270)
Infinitesimal strain superimposed upon a given strain 3/3(250, 325)
Infinitesimal strain tensor 3/3(115)
Infinitesimal theory of elasticity 3/3(117)
Infinitesimal transformation of a field 4(134)
Infinitesimal visco-elasticity 3/3(116)
Inflation of a circular cylinder 3/1(299–300)
Inflation of a sphere 3/1(302)
Inflation of a tube 3/3(189–195, 211ff., 343ff., 353)
Influx 3/1(468–469)
Infrared problem 5(309, 349)
Inhomogeneous wave equations 4(46–48)
Injected spectrum 3/2(644)
Instantaneous space 3/1(457)
Integrability conditions 4(137)
Integral, first, of motion 3/1(158, 170)
Integral form, four-dimensional, of electromagnetic laws 3/1(667)
Integral form, three-dimensional, of electromagnetic laws 3/1(673–674)
Integral invariant in phase space, relative and absolute 3/1(157, 172)
Integral of a random function 3/2(566–568)
Integral of a tensor field 3/1(808, 813–817)
Integral representation 4(175)
Integrals over Minkowski universe domains 4(135–137)

Integrating denominator 3/2(164)
Integration stochastic 3/2(578)
Intensity balance between deformation and rotation 3/1(396–397)
Intensity of a tensor 3/1(834)
Intensity of field, *see also* Field 4(4)
Intensive parameters, fluctuations 3/2(364–369)
Intensive property 3/2(3)
Intensive variable 3/1(647)
Interaction between systems 3/2(127)
Interaction Hamiltonian 3/1(196)
Interaction in Newtonian dynamics 3/1(9)
Interaction in relativistic dynamics 3/1(10)
Interaction of Dirac field and radiation field 5(230ff.)
Interaction of matter and electromagnetic field 3/1(692–693)
Interaction of several particles 5(30)
Interaction representation 5(172)
Interaction representation for natural line width 5(275)
Interactions in a relativistic gas 4(163–166)
Interface of liquid and vapour 3/2(44–46)
Interface of two dielectrics 4(14)
Interior part of stress 3/1(568)
Internal constraints 3/3(69–73)
Internal energy 3/1(609, 619–620, 627); 3/3(294)
Internal energy and strain energy 3/1(611–612)
Internal stability of a phase 3/2(18–21)
Internal variables 3/2(470)
Internal vibrations of a molecule 3/2(76–77)
Intrinsic coordinates 4(251, 252–255)
Intrinsic theory of strain 3/1(314)
Invariance, adiabatic, of phase-space volume 3/2(244–245)
Invariance, canonical 3/2(183–185)
Invariance in space-time 3/1(452)
Invariance of constitutive equations 3/1(701–703)
Invariance of Galileo 3/1(491)
Invariance of stretching and spin in relative motion 3/1(440–441)
Invariance of the density in phase 3/2(183)
Invariance of the extension in configuration 3/2(185)
Invariance of the extension in phase 3/2(183)
Invariance of the Hamiltonian 5(98)
Invariance, relativistic 5(142ff.)
Invariance with respect to ε-substitution 3/2(437–438)
Invariance with respect to moving observers 3/1(443)
Invariant, bilinear 3/1(144)
Invariant decomposition 3/1(840–842)
Invariant directions in a deformation 3/1(280–282)
Invariant directions in motion 3/1(344–345)
Invariant integral and differential equations 3/1(662–666)
Invariant part of the phase space 3/2(180)
Invariant subgroup 4(260)
Invariant transformation 4(150, 259)
Invariants, lists 3/3(30ff.)
Invariants, multilinear 3/3(30)
Invariants of a tensor 3/1(832–836); 3/3(22ff.)
Invariants of strain 3/1(264–266)
Invariants of stress 3/1(551)
Invariants of stretching 3/1(349)
Invariants of vector lines 3/1(818–819)
Invariants, polynomial 3/3(28–30)
Invariants, representation 3/3(27–32)
Inverse, *see* Convergence in probability
Inverse probability 3/2(643–646)
Inversion temperature of the Joule-Kelvin throttling experiment 3/2(26)
Irreconcilable circuits 3/1(103)
Irreducible circuit 3/1(103, 104, 122)
Irreducible representation 5(99)
Irreversibility 3/1(644)
Irreversible process 3/2(222, 413ff.)
Irreversible thermodynamics 3/1(233, 616, 706–707, 734–735); 3/3(373)
Irreversible transition 3/2(122)
Irrotational, *see* Lamellar field
Irrotational frame 3/1(446–447)
Irrotational homogeneous history 3/3(67)

Irrotational motion 3/1(357–360, 391, 404, 408, 431); 3/3(67, 75)
— theorem of Kelvin and Helmholtz 3/1(393)
— theorem of Kirchhoff 3/1(394)
— theorem of maximum speed 3/1(404)
— theorems of Supino 3/1(395)
Irrotational vortex 3/1(362, 432)
Isentropic change 3/1(622)
Isentropic elasticity 3/3(133)
Isentropic path 3/1(624)
Ising model 3/2(373, 377, 390, 393, 404)
Isobaric compressibility 3/1(623)
Isochoric, *see* Incompressible
Isochoric deformation 3/1(283)
Isochoric homogeneous motions 3/3(74)
Isochoric motion 3/1(343–344, 431, 525); 3/3(39, 55)
Isolation, axiomatic concept 3/2(134, 142)
Isostatic coordinates 3/1(848)
Isostatic surfaces 3/1(847–850)
Isotherm, metastable 3/2(20, 34)
Isotherm, stable 3/2(20, 34)
Isothermal compressibility 3/1(623)
Isothermal cyclic processes 3/3(372)
Isothermal elasticity 3/3(133)
Isothermal path 3/1(624)
Isothermal stability 3/3(330)
Isotherms of a one-dimensional system 3/2(410)
Isotherms showing gas and liquid state 3/2(32)
Isotope effects 3/2(88–89)
Isotropic elastic material 3/3(139)
— response coefficients 3/3(140ff.)
Isotropic elasticity 3/1(724, 730)
Isotropic functions of strain 3/1(265, 283)
Isotropic hyperelastic material 3/3(313 ff., 317–319)
Isotropic incompressible elastic material 3/3(147ff.)
Isotropic material(s) 3/3(78ff.)
— of differential type 3/3(94)
— wave propagation 3/3(278–291)
Isotropic tensor function 3/3(22–24, 34)
— representation 3/3(32, 34)
Isotropy 3/1(700)
Isotropy and homogeneity of space and spacetime, axiom 3/1(9, 10)
Isotropy, definition 3/3(358)
Isotropy group(s) 3/1(701); 3/3(76–79)
— of an elastic material 3/3(122)
— of hyperelastic subfluids 3/3(314–316)
— of hyperelastic materials 3/3(310–313)
Isotropy of fluids 3/1(715–716)
Isotropy, thermal 3/1(709)
Iterated geometrical conditions of compatibility 3/1(496–498)
Iterated kinematical compatibility conditions for discontinuities 3/1(505–506)

Jacobi relations 4(213)
Jacobi's complete integral of the Hamilton-Jacobi equation 3/1(125–128)
Jacobi's equation 3/1(79)
Jacobi's principle of least action 3/1(137, 139)
Jacobi's theorem 3/1(127, 128)
Janossy density 3/2(549, 552)
Janossy's cascade equations 3/2(601)
John's special theory of elasticity 3/3(349)
Joint distribution function 3/2(529)
Jordan-Wigner commutators 5(121)
Joule experiments 3/2(2, 10)
Joule-Thomson coefficient 3/2(26)
Jump 3/1(492)
Jump condition 3/1(233)
Jump continuation 3/3(374)
Jump of solenoidal field 3/1(495)

Kelvin effects 3/3(176, 236, 345)
Kelvin-Helmholtz's theorem on irrotational motion 3/1(393)
Kelvin's absolute temperature scale 3/2(3, 6)
Kelvin's extension of Helmholtz's first vorticity theorem 3/1(387)
Kelvin's form of the second law 3/2(5, 157)
Kelvin's proofs of Helmholtz's vorticity theorems 3/1(409–410)
Kelvin's theorem
— of hydrodynamics 3/1(711)
— of minimum energy 3/1(368)

— on impulsive motion 3/1(97)
— on irrotational motion 3/1(357–358)
Kelvin's transformation, *see* Stokes' theorem
Kelvin-Tait theorem 3/1(280)
Kepler problem 3/1(49, 129)
Kepler problem, relativistic 3/1(213)
Khinchin's proof of the ergodic theorem 3/2(208)
Khinchin's structure functions 3/2(627–629)
Khinchin's theory 3/2(538)
Killing equation 4(227)
Killing field 4(248, 253)
Kinematic effects of Lorentz transformation 4(123)
Kinematical compatibility condition for discontinuities 3/1(503–506, 515)
Kinematical line element 3/1(13, 139, 200)
Kinematical motion classes 3/1(431–432)
Kinematical vorticity number 3/1(363–364, 382, 403–404)
Kinematics 3/1(1, 12, 25)
Kinematics in space-time 3/1(200)
Kinematics of continuous media 3/1(240–530)
Kinematics of diffusion 3/1(469)
Kinematics of line integrals 3/1(345–346)
Kinematics of surface integrals 3/1(346–347)
Kinematics of volume integrals 3/1(347)
Kinestate 3/1(383)
Kinetic energy 3/1(35, 36, 39, 44, 365–369, 377, 402, 482–491)
— relativistic 4(148)
— theorems of König 3/1(36, 65, 484)
Kinetic equations 3/2(491–494)
Kinetic stability 3/3(259)
Kinetic theory of gas mixtures 3/1(567–568, 612–614)
Kinetic theory of gases 3/1(2); 3/2(430, 623)
Kinetics 3/1(1)
Kinosthenic coordinate, *see* Ignorable coordinate
Kirchhoff's special theory of elasticity 3/3(348)
Kirchhoff's theorem on irrotational motion 3/1(394)
Kirkwood's derivation of the Boltzmann equation 3/2(235–239)
Kirkwood's solution of the Bloch equation 3/2(333)
Klein-Gordon equation 4(187)
Klein-Nishina formula 5(158, 159, 160, 265)
Klein-Nishina formula, radiative corrections 5(330)
Klein's lemma 3/2(299, 300)
Klein's paradoxon 5(166)
Klein's principle of geometry 3/1(454)
König's theorems on the kinetic energy 3/1(36, 65, 484)
Korteweg's theory of capillarity 3/3(513–515)
Kotchine's theorem on singular surfaces 3/1(527)
Kowalewski's integral 3/1(86)
Kramers equation 3/2(491)
Kramers-de Kronig relations 4(92–93)
Kronecker delta 3/1(26)
Kronecker symbol as a tensor 4(133, 210)
k-vector 3/1(661)
k-vector density 3/1(661)
k-vector field 3/1(675)

Laboratory system 3/1(71)
Lagrange brackets 3/1(151, 170)
Lagrange function 3/1(44, 60, 105–106, 208, 212)
Lagrange matrix 3/1(152)
Lagrange-Cauchy theorem on irrotational flow 3/1(421, 712)
Lagrange-d'Alembert principle 3/1(595–600)
Lagrange-Hamiltonian correspondence 3/1(115)
Lagrange's criterion for a material surface 3/1(339, 509)
Lagrange's equations 3/1(58, 59, 93, 107, 109, 204ff., 208)
Lagrange's particles (three-body problem) 3/1(80)
Lagrangian 5(173ff.)
Lagrangian action 3/1(105, 107)
Lagrangian central equation 3/1(604)
Lagrangian derivative 3/1(107)
Lagrangian dynamics 3/1(108, 113, 204)

Lagrangian for interaction radiation-electron 5(231)
Lagrangian form of relativistic dynamics 3/1(204)
Lagrangian formalism 4(156)
Lagrangian, homogeneous 3/1(105, 137–139, 204, 212)
Lagrangian, infinitesimal generator 4(150)
Lagrangian of a quantized field 4(197)
Lagrangian of general relativity 4(215, 236)
Lagrangian of the Dirac theory 5(215)
Lagrangian of the electromagnetic field 4(100); 5(181)
Lagrangian, ordinary 3/1(44, 60, 105–106, 208, 212)
Lagrangian surface 3/1(116)
Lamb curve 3/1(405)
Lamb plane 3/1(404–405)
Lamb shift 5(316–322)
Lamb shift, measurements 5(321, 322)
Lamb surface 3/1(404–405)
Lamb vector 3/1(378, 402)
λ point of helium 3/2(371, 391, 394)
Lamb-Masotti form of Bernoulli's theorem 3/1(408)
Lamb's vorticity average theorem 3/1(399)
Lamé coefficients of elasticity 3/3(116–117)
Lamé coefficients of viscosity 3/3(113)
Lamellar acceleration field 3/1(388)
Lamellar field 3/1(824–825)
Lamé's cone of shearing stress 3/1(552)
Lamé's stress director quadric 3/1(552)
Lamé's stress ellipsoid 3/1(552)
Landé's splitting factor 3/2(112)
Langevin equation 3/2(514, 518, 594–595)
Langevin theory of paramagnetism 4(25)
Langevin's function 3/2(111)
Laplace equation 4(223)
Laplace operator 5(14)
Laplace transform 3/2(541, 542)
Laplace's frequency curves 3/2(549)
Laplacian expression for the speed of sound 3/1(657, 713)
Laplacian of a vector 4(21)
Laporte rule 5(107)
Large numbers, Bernoulli's law 3/2(527)
Latent heat 3/1(625)
Lattice velocity 3/2(441)
Lattice vibrations 3/1(81)
Laws of classical mechanics 3/1(531)
Laws of conservation (or of balance) 3/1(232)
le Chatelier-Braun principle 3/1(659)
Lebesgue integral 3/2(537, 578)
Legendre transform 3/2(30, 351)
Lenz' rule 4(10)
Lepton 5(110)
Level systems, non-combining 5(112)
Levi-Civita symbol 4(137, 211)
Lexian distribution 3/2(543)
Libration 3/1(103)
Lie algebra 4(159, 199)
Lie derivative 3/1(449, 458, 460, 501)
Lie group 4(259)
Liénard-Wiechert potentials 4(63, 65, 71, 176, 219)
Life time of positronium 5(329)
Light cone 4(126)
Light emission 5(123, 127, 135)
Light ray bending at the limb of the sun 4(234)
Light scattering 4(93–96)
Light velocity 4(32, 107)
Line element, kinematical 3/1(13, 139, 200)
Line element of action 3/1(139)
Line integral, Bernoulli theorem 3/1(405–406)
Line integrals, kinematics 3/1(345–346)
Line, material 3/1(339)
Line of magnetic flux 3/1(674)
Line width 4(86)
Line width, natural 5(274–279)
Lineal flow 3/3(429–432, 441–445)
Lineal flow of second-grade fluids 3/3(510–513)
Lineal motion 3/1(329, 430, 584)
Lineal oscillation flow 3/3(444ff.)
Linear balance between deformation and rotation 3/1(397–399)
Linear chain model 3/2(404–410)
Linear dissipative systems 3/2(484–486, 510)
Linear molecules, degrees of freedom 3/2(74)
Linear molecules, rotation 3/2(74–75)

Linear molecules, thermodynamic functions 3/2(77)
Linear momentum, *see* Momentum
— conservation 4(144–145)
— relativistic 4(146, 147)
Linear tensor function 3/3(22, 33)
Linearity of an operator 5(14, 23)
Linearization of gravitation theory 4(217, 219)
Linearly elastic body 3/1(723–727)
Linearly viscous fluid 3/1(715–720)
Linearly viscous material 3/3(114)
Liouville type, system of 3/1(129)
Liouville's equation 3/2(492, 520–521)
Liouville's theorem 3/1(172–174); 3/2(48–50, 180–182, 233, 624); 4(160, 164)
Liouville's theorem of quantum statistics 3/2(273–274)
Liquid and gas states 3/2(32ff.)
Liquid crystals (subfluids, anisotropic fluids) 3/3(86–88, 151, 314–316, 473–475, 523–537)
Liquid mixture 3/2(38)
Liquid-vapour equilibrium 3/2(35, 44)
Lissajous figures 3/1(52)
Local action, principle 3/3(12, 36, 56)
Local average 4(166)
Local configuration 3/3(52)
Local deformation 3/3(52)
Local rigidity 3/1(259)
Local speed of propagation 3/1(508, 513)
Local stress 4(162)
Local transformation of a vector field 4(134)
Local velocity 4(162)
Localisation of particles 3/2(308, 309)
Localization of a motion 3/3(51ff.)
Locally viscometric motion 3/3(432)
Longitudinal acceleration wave 3/1(524)
Longitudinal discontinuity 3/1(492)
Longitudinal parts of the Hamiltonian 4(104)
Longitudinal wave 3/3(268, 280)
Longitudinal wave in an elastic body 3/1(726–727)
Lorentz condition 4(46, 57, 175); 5(194)
Lorentz contraction 4(123)
Lorentz electron 4(81–82)
Lorentz force 3/1(212, 690); 3/2(502); 4(28)
Lorentz frame, barycentric 3/2(496)
Lorentz gauge 4(46, 57, 175)
Lorentz group 4(121, 126–129); 5(102)
Lorentz invariance, *see also* World invariance 5(137, 142–150, 177–180)
Lorentz invariance of electromagnetic theory 3/1(681)
Lorentz invariant constitutive equations of a moving dielectric 3/1(738–739)
Lorentz transformation 4(117–119, 122)
— generator 4(151–153)
— group character 4(121, 126)
— homogeneous 4(149)
— improper 4(140–141)
— infinitesimal 4(129–131, 151–153)
— proper and improper 3/1(10, 14, 198, 199, 200, 682)
— self-consistency 4(120–122)
— transitivity 4(121)
Lorentz-Lorenz relation 4(88)
Loschmidt's reversibility paradox 3/2(217)
Lowering of indices 4(132)
Low-temperature approach 3/2(330, 337–339)

Macromedium 3/3(391)
Macroquantities in polar elastic materials 3/3(392)
Macroscopic equilibrium properties 3/2(212)
Macroscopic measurable property 3/2(231)
Macroscopic system 3/2(54–57)
Macroscopic variable 3/2(275–278)
Magnetic dipole 4(22)
Magnetic dipole radiation 4(53–54)
Magnetic field 3/2(111–113)
— external 5(27)
— intensity 3/1(674); 4(8)
— irreversible processes 3/2(426)
Magnetic flux 3/1(667)
Magnetic flux, conservation 3/1(660, 667)
Magnetic flux density or magnetic induction 3/1(672)

Magnetic induction 4(6–7, 21)
Magnetic moment 4(22)
— anomaly of the electron 5(299)
— of the electron, measurements 5(300, 301)
Magnetic multipole 4(22)
Magnetic potential 3/1(674)
Magnetic scalar potential 4(24)
Magnetic susceptibility 4(24)
Magnetic susceptibility, dependence on temperature 4(25)
Magnetization 3/2(501)
Magnetization potential 3/1(683, 685–686); 4(8, 23, 55)
Magnetohydrodynamics 3/1(744)
Magnetostatic field 4(21–29)
Magnetostriction 4(28)
Magnitude of a tensor 3/3(15)
Mainardi = Codazzi-Gauss equations 3/1(812)
Many-body problem, nonrelativistic wave equation 5(34)
Many-time formalism 4(182)
Marginal distribution 3/2(530)
Markovian chain 3/2(559, 627)
Markovian process 3/2(222, 509, 515, 561, 569–571, 579, 591)
Mass 3/1(240)
Mass and proper mass 3/1(6, 202, 203, 219)
Mass as a thermodynamical coordinate 3/2(155)
Mass center, *see also* Center of mass 3/1(31, 221, 481, 532)
Mass center, frame of reference 3/1(71, 218)
Mass conservation in a heterogeneous medium 3/1(472)
Mass conservation law 3/1(464, 466–474); 4(144–145)
— in irreversible thermodynamics 3/2(420)
Mass, definition and dimension 3/1(464)
Mass density 3/1(465); 3/3(38)
Mass distribution 3/1(31); 3/3(38)
Mass, electromagnetic 4(65, 80–82, 177)
Mass flow 3/1(467)
Mass point mechanics in special relativity 4(144–159)
Mass, reduced 3/1(71, 72, 73)
Mass, relativistic 4(146, 148)
Mass renormalization 4(178); 5(296, 337–341)
Mass velocity 3/1(467)
Massieu-Planck function 3/2(351, 352, 356)
Massieu-Planck function and entropy 3/2(358)
Massieu's free energy definition 3/2(8)
Mass-point 3/1(465, 532)
Mass-point and center of mass 3/1(483)
Master equation 3/2(294, 297, 521)
Material 3/1(243)
Material connection 3/3(89)
Material constants 3/1(701)
Material coordinates 3/1(326, 337–339)
Material coordinates and description 3/3(37, 39)
Material derivative 3/1(337–339, 341–343)
Material derivative in four dimensions 3/1(463)
Material diffusion vector 3/1(381)
Material equation of continuity 3/1(467)
Material frame-indifference (principle) 3/3(36, 44, 296)
Material functionals 3/3(431)
Material indifference principle 3/1(702)
Material isomorphism 3/3(59)
Material line 3/1(339)
Material of grade n *see also* Grade 3/3(63)
Material representation of a surface of discontinuity 3/1(506–528)
Material singular surfaces 3/1(519)
Material surface 3/1(339–340, 510–512)
Material time derivative 3/3(39)
Material uniformity of a body 3/3(59, 90)
Material vector line 3/1(340–341)
Material volume 3/1(339)
Material vortex sheet 3/1(515–517)
Materially uniform elastic body 3/3(126)
Materials of differential type 3/3(93–95, 111)
Materials of integral type 3/3(98–101)
Materials of mixed integral-differential type 3/3(101)

Materials of rate type 3/3(95–98)
Mathematical concepts in physics 3/1(231)
Mathieu's canonical transformation 3/1(148)
Matrices 5(52)
Matrices, orthogonal 3/1(15, 199)
Matrix function 3/1(838–840)
Matrix mechanics 5(1, 56)
Matrix polynomial 3/1(838)
Matrix power series 3/1(838)
Matrix transformation 5(51)
Matrix, unitary 5(55)
Maupertuis action 3/1(137)
Maximum entropy, theorem on 3/3(380)
Maximum stress 3/1(577–580)
Maximum-minimum theorems for vorticity 3/1(403–404)
Maxwell equations 3/2(501); 4(9, 10, 29)
Maxwell field, covariant notation 4(236)
Maxwell relation(s) 3/2(13)
— generalised 3/2(352, 358)
— of thermodynamics 3/1(629)
Maxwell-Bateman laws of electromagnetism 3/1(667)
Maxwell-Boltzmann law of energy distribution 3/2(256–261, 313, 318)
Maxwell-Boltzmann statistics 3/2(290, 553, 555, 557)
Maxwell-Boltzmann statistics, quantised 3/2(308)
Maxwellian dielectric 3/1(736–740)
Maxwellian distribution 3/2(73, 595–597, 642)
Maxwellian fluid 3/1(735)
Maxwellian fluid, Truesdell's theory 3/3(515–520)
Maxwell-Lorentz aether relations 3/1(660, 677–683)
Maxwell-Lorentz theory under Galilean transformation 4(111)
Maxwell-Morera stress functions 3/1(587)
Maxwell's assumption of continuity of path 3/2(192)
Maxwell's construction of streamlines 3/1(475)
Maxwell's distribution law 3/2(73, 595–597, 642)
Maxwell's energy-stress tensor 4(173, 236)
Maxwell's stress tensor 4(20–21, 29)
Maxwell's theorem on discontinuities 3/1(494)
Mean pressure 3/1(545)
Mean rotation tensor 3/1(275–277)
Mean square convergence 3/2(574)
Mean square deviation 3/2(534, 535, 543–544)
Mean value 3/2(534)
Mean values of stress 3/1(568–569, 574–580)
Measure theoretical meaning of mass 3/1(465)
Measure theory 3/2(182, 525, 529, 572)
Measurement concept 5(64, 73)
Measurement of field strengths 5(204–206)
Measurements of the first kind 5(73)
Measurements of the second kind 5(73)
Measures of acceleration 3/1(378–379)
Mechanical equilibrium 3/2(450–451)
Mechanical equilibrium of a mixture 3/1(650)
Mechanical work, flux 4(163)
Mechanics, axioms of 3/1(230)
Mechanics, geometrical 3/1(98, 215)
Mechanics in non-Euclidean spaces 3/1(606–607)
Mechanics of mass points in special relativity 4(144–159)
Mechanics of the continuum in special relativity 4(159–166)
Mellin's transform 3/2(542, 603)
Membrane 3/1(556)
Membrane stress 3/1(558, 560)
Membrane stress functions 3/1(589–592)
Memory, *see* Fading memory
Mercury precession 4(234)
Metallic reflection 4(37, 39)
Metastable isotherm 3/2(20, 34)
Metastable phases 3/2(374)
Metric in a surface 3/1(811)
Metric operator 5(200)
Metric tensor 4(209, 210)
Metrical entropy 3/2(139–141)
Metrical entropy, analytical construction 3/2(141)
Metrical transitivity 3/2(198, 201–203)

Metrical variable 3/2(126, 132, 154)
Metrical variables of the same type 3/2(154)
Michelson experiment 4(81, 112, 114)
Microcanonical ensemble 3/2(239, 255, 265, 303)
Micromedium 3/3(391)
Microquantities in polar elastic materials 3/3(392, 395)
Microscopic field in the dielectric 4(16–17)
Microscopic reversibility 3/2(505, 507)
Mie's electromagnetic theory 3/1(741)
Milne's kinematical cosmology 3/1(2)
Minimal theorems in impulsive motion 3/1(96)
Minimum free energy, theorem on 3/3(370)
Minimum internal energy, theorem on 3/3(380)
Minimum stress intensity, theorems of Pratelli 3/1(603)
Minkowski geometry 4(125–127)
Minkowski universe 4(126, 131)
Minkowskian coordinates 3/1(208, 209)
Minkowskian space and time coordinates 3/1(198, 199, 208, 209)
Mixed crystal, ideal 3/2(86)
Mixed ensemble 3/2(269)
Mixing length 3/2(592)
Mixture of ideal gases 3/2(36–37, 38, 81–82)
Mixture of interdiffusing materials 3/3(537–541)
Mixture of liquids 3/2(38)
Mixtures 3/1(469–474, 567–568, 612–614, 634–638, 645–650)
Models, mathematical, of nature 3/1(3, 8, 231)
Modes of radiation 3/2(105–107)
Moduli, apparent 3/3(239ff.)
Molecular and continuum theories compared 3/3(5)
Molecular beam experiment of Stern and Gerlach 5(65, 67)
Molecular distribution functions of different order 3/2(232)
Molecular polarizability 4(87, 97)
Molecular system 3/2(59ff.)
Molecular weight 3/1(473)
Molecules in electrostatic field 3/2(110–111)
Molecules, linear, rotation 3/2(74–75)
Molecules, linear, thermodynamic functions 3/2(77)
Molecules, localised 3/2(66)
Møller scattering 5(273)
Moment condition 3/3(258)
Moment ellipsoid 3/2(590)
Moment, linear 3/1(31)
Moment of a random variable 3/2(534)
Moment of force 3/1(531)
Moment of momentum 3/1(482); 3/3(40)
Moment of momentum, balance 3/1(531–532, 546, 554, 692–693, 697)
Moment, quadratic 3/1(31)
Momental ellipsoid 3/1(33)
Moments of a tensor 3/3(23)
— field 3/1(815)
— their gradients 3/3(27)
Moments of inertia 3/1(31, 485–486)
Moments of stress 3/1(576–580)
Momentum 3/3(40)
Momentum balance 3/1(531, 545, 554, 691–698)
Momentum conservation law 4(144–145)
— in irreversible thermodynamics 3/2(421)
— in the electromagnetic field 4(30–31)
— of field theory 5(175)
Momentum density 3/2(496)
Momentum, electromagnetic 3/1(689)
Momentum, generalized 3/1(62, 607)
Momentum, linear 3/1(33, 56, 66, 110, 481–491)
Momentum measurement 5(8, 11, 73, 107)
Momentum production 3/2(421)
Momentum, relativistic 4(146, 147)
Momentum transport 3/2(421)
Momentum-energy space 3/1(130)
Momentum-energy vector 3/1(110, 114, 202, 696)
Momentum-space integral 5(13)
Monatomic ideal gas 3/2(70–73)
Monge's potentials 3/1(828)
Monotone transformation 3/3(161)
Mooney-Rivlin theory of rubber 3/3(349–354)
Mössbauer effect 4(123, 235)

Motion 3/1(240, 325–463)
Motion of a body, general description 3/3(38)
Motions, one-dimensional 3/1(46, 210)
Motions, three-dimensional 3/1(53)
Motions, two-dimensional 3/1(48)
Motor 3/1(37, 68)
Motor symbolism 3/1(68)
Moving dielectric, constitutive equations 3/1(738–739)
Moving electron, fields 4(63–66)
Moving electron, potentials 4(61–63)
m.s. convergence, *see* Mean square convergence
Multilinear invariants 3/3(30)
Multilinear tensor function 3/3(22, 34ff.)
Multiple production in a stochastic process 3/2(583)
Multiplication, exterior 3/1(165)
Multipolar elasticity 3/3(400ff.)
Multipole, electric 4(12–13)
Multipole expansion of the Hertz potential 4(50–52, 55, 72–74)
Multipole, magnetic 4(22)
Multipole radiation 4(59–61); 5(128)
Multipole radiation from point charge systems 4(72–74)
Murnaghan's coefficients of second-order elasticity 3/3(230)
Mutual forces 3/1(533–534)
Mutual loads 3/1(536)
μ-space statistics 3/2(176, 255–256)

Natural line width 4(86)
Natural process 3/2(3)
Natural state 3/3(306, 311, 322)
Natural time (relaxation time) of a viscous fluid 3/1(721)
Natural units of elasticity, time-lapse etc. 3/3(64)
Navier-Poisson law of viscosity 3/1(716)
Navier-Stokes equations 3/3(476)
Navier-Stokes theory of viscometry 3/3(113ff., 467)
n-body problem 3/1(79)
Near zone 4(52)
Nearest neighbour interaction 3/2(607, 640)
Needle radiation 5(136)
Negative absolute temperature 3/2(144, 145, 153, 157, 160)
Negative energy levels 5(166, 168)
Neo-Hookean theory of elasticity 3/3(350)
Nernst equation 3/2(471)
Networks, electrical 3/2(486, 511)
Neutrino 4(186–189); 5(150)
Newton potential 4(215, 217)
Newtonian absolute time 3/1(4)
Newtonian dynamics 3/1(1, 2, 3, 108)
— of a particle 3/1(5, 6)
— of a system 3/1(8, 108)
Newtonian relativity 3/1(45)
Newton's first and second law 3/1(43, 45)
Newton's law of gravitation 3/1(8)
Newton's third law 3/1(9, 10, 37, 38, 57, 533–534)
Noll's theorem on flows with constant stretch history 3/3(438)
Noll's theorem on simple fluids 3/3(81)
Noll's theorem on simple materials 3/3(66)
Non-branching transition relation 3/2(124)
Non-conductor of heat 3/3(384)
Non-holonomic constraints 3/1(606)
Non-holonomic system 3/1(39, 59, 64, 65)
Non-linear molecules, degrees of freedom 3/2(79)
Non-linearly viscous fluid 3/1(720–722)
Non-Markovian process 3/2(561, 570, 581)
Non-polar case of stress 3/1(538, 546)
Normal component 3/1(844)
Normal co-ordinates 3/1(182)
Normal dispersion 4(88, 94)
Normal distribution (Gaussian) 3/2(506, 547, 593)
Normal end effects 3/3(452–458)
Normal form transformation 3/1(191)
Normal frequencies 3/1(2, 81, 182)
Normal matrix 3/1(842)
Normal modes 3/1(2, 81, 180, 182)
Normal product 5(247)
Normal projection 3/1(796)
Normal strains 3/1(267)

Normal stress(es) 3/1(537, 544, 574)
— in cone-and-plate flow 3/3(463)
— in simple shear 3/3(175–178)
Normal system, canonical 3/2(193–194, 203–207)
Normal tensor 3/1(810)
Normal vectors in space-time 4(129)
Normalization of eigenfunctions 5(47)
Normal-stress difference
— in a second-order fluid 3/3(493)
— in Couette flow 3/3(449)
— in flow between coaxial cylinders 3/3(448)
— in flow between sliding cylinders 3/3(451)
Normal-stress function 3/3(436)
Normal-stress function, modified 3/3(446)
Notation for derivatives in thermodynamics 3/1(622)
Notation in field theory 3/1(235–239)
Notation of thermodynamic quantities, table 3/2(9)
Notations 3/3(13–16)
Nucleon 5(109, 110, 113, 114)
Null cone 3/1(199)
Null stress 3/1(547)
Null vector 4(129, 183)

Objective quantity, *see also* Frame-indifferent quantity 3/3(42)
Objectivity, *see* Frame-indifference
Obliviator (influence function) 3/3(102)
Observables 4(159, 181, 250–252, 255); 5(53)
Observer moving 3/1(443)
Occupation numbers 4(194, 201)
Occupation numbers, fluctuations 3/2(370)
Occupation state 3/2(559, 564, 583)
O.C.R.B., *see* Onsager Casimir reciprocal relations
Octahedral invariant 3/1(834)
Octupole 4(12)
O-F inequalities (ordered forces) 3/3(157)
Ohm's law 4(9, 11)
Ohm's law for moving conductors 3/1(741–742)
ω space 3/2(527, 529)
One-dimensional fluid 3/2(612–613)
One-dimensional problems 3/3(10)
One-dimensional random walk 3/2(579, 587)
One-dimensional system 3/2(404–410)
One-electron atom 5(43)
One-electron state, charge renormalization 5(301–305, 355–358)
One-event characteristic function 3/1(215)
One-point characteristic function 3/1(121–123)
Onsager coefficients 3/2(426)
Onsager reciprocity relations 3/1(707)
Onsager-Casimir reciprocity relations 3/2(426)
— statistical foundation 3/2(505, 512)
Onsager-Machlup function 3/2(516–518)
Onsager-Machlup principle 3/2(514–519)
Open phase 3/2(13–14)
Operational method of Bridgman 3/1(4)
Operator 5(14, 23, 24, 51, 55, 152–155)
Operator calculus 5(30, 51)
Operator, hermitian 5(33, 51)
Operator, linear 5(14, 23)
Operator, transformations 5(51, 54)
Operator, unitary 5(55)
Opposite polarization 4(45–46)
Orbital angular momentum 5(107, 160, 165)
Order of a fluid 3/3(491)
Order of a material of integral type 3/3(99)
Order of a simple material 3/3(109)
Order of elasticity 3/3(228)
Order-disorder transition 3/2(371)
Ordinary convergence 3/2(574)
Oriented bodies 3/1(309–325)
Oriented media 3/3(389)
Origins of classical field theories 3/1(234)
Ortho states of a molecule 3/2(76, 78–79)
Orthochronous Lorentz group 4(121)
Orthogonal group 4(121)
Orthogonal matrices, definitions 3/1(15, 199)
Orthogonal tensor 3/3(15)
Orthogonality of eigenfunctions 5(46)
Orthogonality relation 5(48)
Oscillation 3/1(529)

Oscillations about steady motion 3/1(191)
Oscillations, forced 3/1(47, 189)
Oscillations of second-grade fluids 3/3(509ff.)
Oscillations, small 3/1(180)
Oscillator, damped 4(85)
Oscillator, harmonic 3/1(46, 210); 5(44)
Oscillator strength 4(88)
Outgoing fields 5(208)

P symbol of time ordering 5(245)
Pair production by an external field 5(261)
Pair production by photons 5(271)
Palatini procedure 4(215, 238)
Para states of a molecule 3/2(76, 78–79)
Paradox of Euler and d'Alembert 3/1(541–542)
Parallel displacement of a vector 4(211)
Paramagnetic atoms in magnetic field 3/2(111–113)
Paramagnetic substance 3/2(30)
Paramagnetism 4(2, 25)
Parameter transformation 4(151)
Parity 5(106)
Parseval's formula 3/2(632)
Partial entropy 3/1(634)
Partial gradient of a function of tensors 3/3(25)
Partial internal energy 3/1(613)
Partial pressure 3/2(81)
Particle explosion 3/1(217ff.)
Particle numbers, definition 5(334)
Particle velocity 3/1(216)
Particles, charged, in an electromagnetic field 3/1(53, 212)
Particles, definition 3/3(37)
Particles, free 3/1(9, 207, 216, 217)
Particles in a potential field 3/1(210ff.)
Particles, indistinguishable 5(31, 110)
Particles of spin zero 4(185–187)
Particles, pair of, carrying electric charge 3/1(11)
Particles with spin 3/1(222)
Partition function 3/2(53, 264)
— factorization 3/2(68)
— generalised 3/2(354, 356)
— grand 3/2(348, 382)
— mathematical origin 3/2(539–540)
— of a single molecule 3/2(313)
— of hydrogen 3/2(78)
— of the potential energy 3/2(264)
— of the total system 3/2(325)
— transformation theory 3/2(352)
Past 3/1(200)
Path line 3/1(330–333,, 340)
Pauli approximation, relativistic 5(160–163)
Pauli exclusion principle 5(110, 114, 115, 117, 119)
Pauli principle 3/2(639, 640)
Pauli spin matrices 4(138, 139, 191); 5(109, 141, 214)
P-C inequality (pressure-compression) 3/3(155)
Pearson's distributions 3/2(546)
Peltier coefficient 3/2(468)
Peltier heat 3/2(468)
Pendulous body 3/1(576)
Pendulum, circular 3/1(47)
Pendulum, cycloidal 3/1(47)
Pendulum, spherical 3/1(47, 52, 56)
Penetration probability 5(167)
Perfect elasticity, *see* Hyperelasticity 3/3(304)
Perfect fluid 3/1(233, 710–715); 3/3(149)
Perfect gas, *see* Ideal gas
— equation 3/2(615)
— relativistic 4(159–161)
Perfect materials, thermodynamics 3/3(296–298)
Perfectly elastic body 3/1(233)
Perfectly plastic body 3/1(233, 722–723)
Perfectly soft body 3/1(723)
Perihelion 3/1(51)
Permanence of matter 3/1(243)
Permanence of vortex lines and tubes 3/1(387–389, 411)
Permittivity, *see also* Dielectric constant 4(5)
Permutation 5(111)
Perturbation, adiabatic 5(82–87)
Perturbation calculations in Heisenberg representation 5(234–239)
Perturbation energy, expectation value 5(76)
Perturbation series 3/1(309)
Perturbation theory 3/2(635); 5(73, 75ff.)

Perturbation, time dependent 5(79)
Perturbations 3/1(196)
Petrov-Pirani notation 4(219, 221)
Pfaffian form 3/1(163); 3/2(163–167)
p.f.f., *see* Probability frequency function
Phase average 3/1(227)
Phase equilibrium 3/2(12)
Phase integral 5(94)
Phase integral of Gibbs 3/2(262)
Phase rule of Gibbs 3/2(17–18, 374)
Phase space 3/1(2, 100, 143, 167, 191); 3/2(49, 126, 178, 623)
Phase space, invariant part 3/2(180)
Phase, thermodynamic, definition 3/2(3)
Phase transformation 3/2(371)
Phase velocity 4(89, 90)
Phenomenological coefficients 3/2(444, 465)
Phenomenological equations 3/2(414, 425, 426, 432ff., 443)
Photon 4(200); 5(1, 2, 3, 4, 137)
Photon, frequency four-vector 3/1(203, 217)
Photons as a consequence of wave quantization 5(186)
Photons scattered by photons 5(334)
Physical basis of continuous theories 3/1(228)
Physical components 3/3(15)
— of a tensor 3/1(799, 802–805)
— of stress 3/1(544)
Physical coordinate systems 3/2(145, 146)
Physical dimensions 3/1(799)
Physical particles 5(335)
Physical system, axiomatic concept 3/2(134)
Physicochemical equilibrium 3/2(16–17)
Piezo-caloric effect, its non-existence 3/3(356)
Piezotropic material 3/1(634)
Piola-Kirchhoff stress tensor(s) 3/1(553); 3/3(120, 124ff., 144, 169, 300, 320)
Piola's theorem on virtual work 3/1(596–597)
Pipe, *see also* Tube
Pipes, flow inside 3/3(448–452)
Planck constant $\hbar$ 5(4)
Planck's classification of thermodynamic processes 3/2(3)
Plane infinitesimal waves 3/3(273–278)
Plane motion 3/1(329, 430)
— stream function 3/1(474–477)
— vorticity theorem of d'Alembert 3/1(421)
Plane of shear 3/1(292)
Plane problems in elasticity 3/3(204–208)
Plane problems in second-order elasticity 3/3(246)
Plane strain 3/1(284–285)
Plane stress 3/1(582–584)
Plane wave in a conductor 4(37–40)
Plane wave in a dielectric 4(32, 35)
Plane waves as solutions of the Dirac equation 5(214, 215)
Plane-wave solutions of the Einstein field equations
— linearized 4(224–226)
— rigorous 4(231–234)
Plasma 3/2(494)
Plastic flow 3/1(722)
Plastic potential 3/1(723)
Plasticity 3/1(722–723); 3/3(8, 420)
Plate 3/1(556)
Plate condenser as a thermodynamic system 3/2(28–29)
Poincaré cycle period 3/2(217)
Poincaré recurrence theorem 3/2(217)
Poincaré vorticity average theorem 3/1(399)
Poinsot ellipsoid 3/1(83, 85)
Point process 3/2(581)
Point transformation 3/1(149)
Poiseuille flow 3/3(451ff.)
Poiseuille flow of Ericksen's anisotropic fluids 3/3(536)
Poisson binomial distribution 3/2(543)
Poisson bracket(s) 3/1(151, 170); 4(155, 158, 182, 198, 261, 267–271); 3/2(180)
Poisson distribution 3/2(544, 557, 580)
Poisson equation 4(11)
— of gravitational field 4(215, 217, 219)
— of vector potential 4(21)
Poisson matrix 3/1(152)
Poisson process 3/2(580)
Poisson theorem 3/1(153)
Poisson-Jacobi identity 3/1(151, 172)

Poisson-Kelvin equivalent charge distribution 3/1(685)
Polar decomposition theorem 3/3(52)
Polar elastic materials 3/3(389–401)
Polarisation, electrical 3/2(501)
Polarizability 4(18, 87, 97, 99)
Polarization charge 3/1(684); 4(15)
Polarization current 3/1(685, 686)
Polarization, dielectric 4(6, 15)
Polarization, electrical 3/3(385)
Polarization ellipse 4(45)
Polarization, longitudinal 5(184)
Polarization of a dielectric 3/1(683–685)
Polarization of a plane wave 4(32)
Polarization of electron waves 5(166)
Polarization of radiation 5(183–189)
Polarization of skylight 4(95)
Polarization parameters 4(44, 45)
Polarization potential, *see* Hertz vector
Polarization, scalar 5(184)
Polarization vector (for photons) 5(183)
Polarized light 4(36)
Polhode 3/1(83, 85)
Polya-Eggenberger distribution 3/2(545)
Polynomial invariants 3/3(28–30)
Polynomial of matrices 3/1(838)
Polynomials, interlocked 3/1(187)
Ponderomotive force 3/2(502)
Ponderomotive force density 4(28)
Ponderomotive law 4(170–171)
Ponderomotive laws in general relativity 4(242)
Population processes 3/2(582)
Position measurement 5(8)
Positron 5(168)
Positronium 5(323–330)
Positronium, fine structure 5(328)
Positronium, mean life 5(329)
Potential barrier 5(94)
Potential chemical 3/2(316, 346)
Potential deformation 3/1(277, 283)
Potential, electrochemical 3/2(459)
Potential electrostatic 4(4, 11)
Potential energy 3/1(42, 210)
— of stress 3/1(571)
— partition function 3/2(264)
Potential field 3/2(414)
Potential flow past a cylinder 3/1(332–336)
Potential, magnetostatic 4(24)
Potential of a conservative field 3/1(665–666)
Potential series, thermoelectric 3/2(469)
Potential, thermomechanical 3/3(299)
Potential transformations 3/1(666)
Potentials, electromagnetic 3/1(212, 674)
Potentials, electromagnetic, under Galilean transformation 4(111–112)
Potentials in relativity 4(169)
Potentials of a moving point charge 4(61–63)
Potentials, thermodynamic 3/1(627–629); 3/2(351)
Power 3/1(44, 569–570)
Power, non-mechanical 3/1(608)
Power series of matrices 3/1(838)
Power series of tensors 3/3(21)
Power spectrum 3/2(629)
Powers of a tensor 3/1(837)
Powers of a vector 3/1(796)
Poynting effect 3/3(177, 235, 345)
Poynting vector 3/1(690); 4(29–30, 32, 66, 74)
Poynting's experiments on torsion 3/3(237)
Poynting's problem 3/3(193, 352)
Precession, steady 3/1(88)
Predecessors of Einstein's theory 4(115–116)
Preferred directions 3/1(309)
Preferred frame 3/1(443)
Pressure 3/1(537, 545); 3/3(149)
Pressure co-efficient 3/1(623)
Pressure effect on wave speed 3/3(284 ff.)
Pressure ensemble 3/2(357)
Pressure fluctuations 3/2(366)
Pressure in the Lorentz electron 4(82)
Pressure viscosity 3/2(428)
Primary constraints 4(264)
Principal axes
— of a tensor quadric 3/1(251)
— of elongation 3/1(281)
— of inertia 3/1(32, 33, 36, 485–486)
— of strain 3/1(260, 274, 281)
— of stress 3/1(551)
Principal coordinate system of a tensor 3/1(832)
Principal curvatures 3/1(813)

Principal extension of a shear 3/1(294)
Principal extensions, small 3/1(303–308)
Principal forces 3/3(144, 317)
Principal function, *see* Two-point characteristic function
Principal invariants of a tensor 3/1(832–836); 3/3(23, 25)
Principal moments of inertia 3/1(32, 33, 485)
Principal stresses 3/1(551); 3/3(143)
Principal stretches 3/1(261)
Principal stretchings 3/1(349, 441); 3/3(472)
Principal torsion-flexure axes 3/1(318)
Principal trajectories of a tensor field 3/1(847)
Principal waves 3/3(279)
— compatibility conditions 3/3(286–288)
— fundamental theorems 3/3(280ff.)
Principle of detailed balancing 3/2(216, 297)
Principle of determinism for constrained materials 3/3(70)
Principle of determinism for stress 3/3(56)
Principle of duality, first 3/1(242, 246)
Principle of duality, second 3/1(263)
Principle of equipresence 3/1(703–704); 3/3(359ff., 364)
Principle of equivalence 4(204–206, 234)
Principle of fading memory 3/3(36, 101, 103ff.)
Principle of general covariance 4(206)
Principle of impenetrability 3/1(244, 325)
Principle of Lagrange and d'Alembert 3/1(595–600)
Principle of least action 3/1(107, 110, 204, 205, 604)
Principle of least constraint 3/1(605–606)
Principle of local action 3/3(12, 36, 56)
Principle of material frame-indifference 3/3(44, 296)
— its history 3/3(45–47)
Principle of microscopic reversibility 3/2(505, 507)
Principle of relativity, classical 4(110)
Principle of virtual work 3/1(595–600)
— its converse 3/1(602–603)
Principles for constitutive equations 3/1(700–704)
Principles of mechanical behavior 3/3(36)
Probability 5(12, 14, 15, 16)
Probability, conditional 3/2(507, 530)
Probability current 5(16, 25)
Probability density 3/1(175)
— in coordinate space 5(10, 15, 23)
— in momentum space 5(12, 15)
Probability frequency function 3/2(530, 531)
Probability index 3/2(252)
Process, dynamical 3/3(41)
Process fields 3/2(414)
Process function, linear 3/2(162)
Product density 3/2(549, 550–552)
Product of tensors 3/3(15)
Production 3/2(420, 421)
Production, chemical 3/2(421, 427)
Production of entropy 3/1(638–647); 3/2(415, 423, 433, 443–445, 458); 3/3(295)
Production of momentum 3/2(421)
Production of rest mass 3/2(497)
Production operator 5(187)
Products of inertia 3/1(31, 486)
Projection operator 5(49)
Projection, stereographic 3/1(21, 24)
Propagation condition 3/3(268)
Propagation kernel 4(135, 155)
Propagation speed 3/3(268)
Propagation speed, second-order effects 3/3(288–291)
Propagation vector 4(34, 113)
Propagation vector of gravitational waves 4(225)
Propagation velocity of waves 4(89–91)
Propagator 3/2(341–343)
Proper functions and proper values, *see* Eigenfunctions and eigenvalues
Proper Lorentz group 4(121, 126, 127, 139)
Proper mass, *see* Mass
Proper mass, flux 4(163)
Proper numbers of a tensor 3/1(829)
Proper orthogonal group 4(121)
Proper rate of a clock 4(121)
Proper time 3/1(6, 200); 4(119)
Proper vector 3/1(829)

Proton 5(31, 107)
Pseudo-lineal motion 3/1(329, 430)
Pseudo-lineal motion, stream function 3/1(477)
Pseudo-plane motion 3/1(329, 430)
Pseudoscalar 4(133)
Pseudo-stresses 3/1(553)
Pseudotensor 4(133)
Pseudo-thermostatics 3/2(415)
Ptolemean cosmogony 4(109)
Pure ensemble 3/2(269)
Pure normal stress 3/1(537–538)
Pure shearing stress 3/1(537, 545, 551)
Pure strain 3/1(275, 281, 283)
Pure strain, small 3/1(305)

q-process 3/2(592)
Quadric of a tensor 3/1(250–253, 259)
Quadric of polarization 3/3(268)
Quadric of stretching 3/1(349)
Quadrupole moment 4(12)
Quadrupole potential 4(13)
Quadrupole radiation 5(129)
Quantised Maxwell-Boltzmann statistics 3/2(308)
Quantum conditions 5(93)
Quantum field theory in special relativity 4(193–202)
Quantum mechanical *H*-theorem 3/2(288ff.)
Quantum mechanics 3/1(1–3)
Quantum mechanics in special relativity 4(181–193)
Quantum number 5(51, 94, 96, 97)
Quantum theory and stochastic processes 3/2(634)
Quantum theory, relativistic 5(137)
Quasi-equilibrated motions of incompressible bodies 3/3(208–214)
Quasi-ergodic hypothesis 3/2(193)
Quasi-Markovian process 3/2(568, 569, 570)
Quasistatic transition 3/2(140)
Quasistatic work 3/2(162)
Quaternions 3/1(21, 25, 28)

Radial oscillations of cylinders and spheres 3/3(214–219, 354)
Radiation 5(129, 133, 158)
Radiation damping 5(135, 137)
Radiation during Coulomb scattering 4(77–78)
Radiation from an accelerated charge 4(66–68)
Radiation modes 3/2(105–107)
Radiation pattern 4(67)
Radiation pressure 4(33)
Radiation processes, treatment using the correspondence principle 5(123)
Radiation reaction 4(82–84)
Radiation reaction, covariant treatment 4(84)
Radiation, spectral resolution 4(68–71, 78–80)
Radiation theory 5(123ff., 133ff.)
Radiation, thermodynamic functions 3/2(107–108)
Radiation width 4(86)
Radiation zone, *see* Distant zone
Radiativ damping 4(85)
Radiative corrections of the Klein-Nishina-formula 5(330)
Radius of gyration 3/1(31)
Raising of indices 4(132)
Random flight 3/2(589)
Random force 3/2(604)
Random function 3/2(506)
Random function, calculus 3/2(573)
Random parameters 3/2(642)
Random phase distribution 3/2(283)
Random variable 3/2(528)
— continuous 3/2(533, 540–542, 546–549)
— discrete 3/2(539–540, 543–546)
— moment 3/2(534)
Random walk, one-dimensional 3/2(579, 587)
Random walk with barriers 3/2(580)
Rank of a tensor 4(133)
Rankine-Hugoniot equation 3/1(711, 713)
Raoult's law 3/2(40)
Rate of a clock 4(121, 123)
Rate of deformation tensor 3/1(348)
Rate of strain 3/1(369–374)
Rationalized units 4(5)
Rayleigh scattering 4(94, 95)
Rayleigh's dissipation function 3/2(428)
Rays, *see* Trajectories
Reaction rate 3/1(473)
Realisation function 3/2(572)
Realised trajectory of a stochastic process 3/2(561, 575)

Rearranging operator 3/2(586)
Reciprocal directors 3/1(311)
Reciprocal quadric of a tensor 3/1(251)
Reciprocal relations of Onsager and Casimir 3/2(426, 505, 512)
Reciprocal relations of thermodynamics 3/1(629)
Reciprocal theorem of Adkins 3/3(207)
Reciprocal theorem of Betti 3/3(324ff.)
Reciprocity theorem of Hamiltonian and Lagrangian dynamics 3/1(116)
Recoil through light emission 5(135, 137)
Recollection of a history 3/3(102)
Reconcilable circuits 3/1(103)
Recoverable work 3/1(639, 641–642)
Rectilinear shearing 3/1(360–361, 432)
Rectilinear vortex 3/1(361–362, 389, 432)
Recurrence paradox of Zermelo 3/2(218)
Recurrence theorem of Poincaré 3/2(217)
Recurrence time 3/2(221, 564)
Red shift of spectral lines 4(234, 235, 249)
Reduced constitutive equations 3/3(66–69, 113, 120)
Reduced free enthalpy 3/1(638)
Reduced heat flow 3/2(436)
Reduced system 3/2(148)
Reducible circuit 3/1(103, 104)
Reducible representation 5(99)
Reduction of energies to normal form 3/1(180)
Reduction of order 3/1(158, 161, 162, 163, 170)
Reference configuration 3/3(48)
Reflection coefficient 4(36)
Reflection invariance 5(146)
Reflection of a plane wave 4(34–36)
Refraction index 4(87, 89, 99)
Refraction index, fluctuations 3/2(369)
Refraction of a plane wave 4(34–36)
Regenerative process 3/2(571, 581)
Regression of fluctuations 3/2(507)
Regularization 5(288)
Reiner, basic contributions of 3/3(9)
Reiner-Rivlin fluids 3/1(720–722); 3/3(477–480, 486)
Reissner's variational principle 3/3(328)
Relation 3/2(122)
Relative Cauchy-Green tensor 3/3(53)
Relative deformation function 3/3(49)
Relative deformation gradient 3/3(49)
Relative frame 3/1(71)
Relative motion 3/1(437–463)
Relative rotation tensor 3/3(53)
Relative spin 3/1(353, 441)
Relative stretch tensor 3/3(53)
Relative tensor 3/1(661)
Relative wryness of an oriented body 3/1(312–313)
Relativistic catastrophes 3/1(217)
Relativistic dynamics 3/1(1, 2, 108, 198)
— interaction 3/1(10)
— of a particle 3/1(5, 6, 198)
— of a system 3/1(8, 108, 204)
Relativistic electrodynamics 4(166ff.)
Relativistic mechanics 4(144ff.)
Relativistic quantum mechanics 5(137)
Relativistic thermodynamics of irreversible processes 3/2(494–505)
Relativity, general theory 4(203)
— of Einstein 3/1(2)
Relativity, Newtonian 3/1(45)
Relativity principle of Galileo 3/1(491, 535)
Relativity, special theory 4(109)
Relaxation phenomena 3/2(440, 487–491)
Relaxation time 3/2(475, 510)
Relaxation time of a conductor 4(31)
Removal of degeneracy by perturbation 5(78)
Renormalization of charge 5(282, 301ff., 347)
Renormalization of mass 4(178); 5(296, 337–341)
Renormalization of the Dirac field 5(341)
Representation in Hilbert space 5(170–173)
Representation of a transformation group 5(99)
Representation of a vector-valued function 3/3(35)
Representation of the gradient of an invariant 3/3(34)
Representation theorem
— basic, of Cauchy 3/3(29)
— for isotropic tensor functions 3/3(32)

— for linear isotropic tensor functions 3/3(33)
— for tensor invariants 3/3(28)
Representative ensemble 3/2(190)
Resistance 4(11)
Resolution of a displacement with small gradients 3/1(306)
Resonance 3/1(47, 189)
Resonance absorption 4(92)
Response function 3/3(296)
— hypo-elastic 3/3(404)
— of an elastic material 3/3(119)
Response functional 3/3(36, 58, 61)
Rest mass 4(147)
— density 3/2(495)
— of the photon 5(196)
Restitution coefficient 3/1(94, 95)
Restitution period 3/1(95)
Restricted theory of relativity, *see* Special theory
Retardation formulae of Coleman and Noll 3/3(110–111)
Retardation of a history 3/3(109)
Retarded function 5(192)
Retarded potentials 4(47, 62)
Retarded solution 4(175)
Reversibility paradox of Loschmidt 3/2(217)
Reversible motion 3/2(447)
Reversible process, definition 3/2(3–4)
Reversible transition 3/2(122)
Rheogoniometer in Weissenberg 3/3(464)
Rheology 3/3(8)
Rheonomic system 3/1(35, 39)
Riemann integral 3/2(530, 537, 578)
Riemann-Christoffel curvature tensor 4(211, 213–214)
— Bianchi identities 4(214)
— contractions 4(214)
— independent components 4(214)
Riemann-Christoffel tensor 3/1(271–273)
Riemannian space 3/1(106); 4(208–210)
Rigid body 3/1(8, 12, 82–92, 233)
— displacement 3/1(14, 65, 199)
— its constitutive equations 3/1(704–705)
— turning about a fixed point 3/1(68, 82–92)
— under no forces 3/1(82)
Rigid deformation 3/1(259, 265, 283)
Rigid deformation of an oriented body 3/1(312–313)
Rigid motion 3/1(350–352, 355, 431)
— body-fixed coordinates 3/1(486)
— world-invariant definition 3/1(457)
Rigid rotating frame 3/1(437)
Rigid rotation 3/1(274)
Rigidity 3/3(72)
Rigor, mathematical, in physics 3/1(231)
Rivlin and Topakoglu's interpretation of Signorini's expansion 3/3(227–229)
Rivlin, basic contributions of 3/3(9)
Rivlin-Ericksen fluids 3/3(481 ff.)
Rivlin-Ericksen tensor(s) 3/1(383), 3/3(54)
— generalization 3/3(516)
Rivlin's formula for the torsional modulus under extension 3/3(192 ff.)
Rivlin's solutions in elasticity 3/3(171–197)
Rod 3/1(309–310)
Rods, equilibrium state 3/1(564–567)
Rods, history of strain description 3/1(319–320)
Rods, strain of 3/1(315–320)
Rolling contacts 3/1(38)
Rotating cylinders, flow between them 3/3(449 ff.)
Rotating rigid frame 3/1(437)
Rotating system, irreversible processes 3/2(426)
Rotation 3/1(12, 15, 17, 20, 23, 25, 36, 104)
Rotation about a point 3/1(13)
Rotation, finite 3/1(274–283)
Rotation, infinitesimal 3/1(25, 27)
Rotation of a cylinder 3/1(550)
Rotation of a rigid body 3/1(82, 575)
Rotation of linear molecules 3/2(74–75)
Rotation, small 3/1(305–308)
Rotation tensor 3/3(52)
Rotational characteristic temperature 3/2(75, 79, 81)
Rotational group, three-dimensional 5(46, 101)
Rotationally symmetric motion 3/1(329, 431)
— stream function 3/1(478–479)
— vorticity theorem of Svanberg 3/1(421)

Rotationally symmetric stress 3/1(587–588)
Routhian function 3/1(61)
Rubber, experimental results 3/3(171, 181ff.)
Rubber theory of Mooney and Rivlin 3/3(349–354)
Russell-Saunders coupling 5(115)
Rutherford scattering cross section 5(261)
— exchange correction 5(274)
Rutherford scattering formula 3/1(77)
Rydberg constant 5(44)

S function 5(222–225)
Sample function 3/2(572)
Satellite experiments 4(235)
Scalar field in Minkowski universe 4(134)
Scalar potential 4(111)
Scalar superpotentials 4(57)
Scalar wave part of the Hamiltonian 4(104)
Scattering 3/1(71, 76)
Scattering angle 3/1(72)
Scattering cross section, differential 3/1(75, 77)
Scattering cross section, total 3/1(76)
Scattering in a Coulomb field, radiation 4(77–78)
Scattering of electrons by a Coulomb field 5(259–261)
Scattering of electrons by electrons 5(271–274, 331)
Scattering of electrons, radiative correction 5(306–309)
Scattering of light 3/2(369); 4(93–96)
Scattering of light by electrons 5(263–266)
Scattering of photons by photons 5(334)
Scattering of radiation 5(130, 131, 158)
Scattering, zero-scattering 3/1(76)
Schrödinger equation 3/1(3); 5(24)
— of the hydrogen atom 5(43)
— stochastic meaning 3/2(634, 648)
— time dependent 5(59)
Schrödinger representation 5(171)
Schwarz inequality 4(126)
Schwarzschild metric 4(253)
Schwarzschild radius 4(230, 233, 235)
Schwarzschild solution 4(222, 248)
Schwarzschild's line element 4(230)
Scleronomic system 3/1(39)
Screw field 3/1(826–828)
Screw motion 3/1(392–393, 404, 408, 431)
S-E condition (strong ellipticity) 3/3(129, 132, 147, 168ff., 248, 254, 271, 281, 333)
Second entropy analog of Gibbs 3/2(246)
Second fundamental form of surface theory 3/1(812)
Second law of thermodynamics 3/1(643–647); 3/2(4–5, 245, 322)
— Clausius' form 3/2(157)
— history of discovery 3/2(114–117)
— Thomson's form 3/2(157)
Second quantization 5(121)
Second quantization of relativistic electrons 5(216)
Second virial coefficient 3/2(25, 26, 27)
Second viscosity 3/2(428)
Secondary constraints 4(264, 268)
Secondary flow 3/3(463ff.)
Secondary flow in tubes 3/3(497–504)
Second-class constraints 4(269, 270)
Second-grade fluids 3/3(504–513)
Second-grade hyperelastic materials 3/3(397–400)
Second-order Doppler effect 4(123)
Second-order effects in elasticity 3/3(229–246)
Second-order effects in isotropic hyperelastic materials 3/3(344–346)
Second-order elasticities, determination by wave speeds 3/3(290ff.)
Second-order fluids 3/3(100)
Second-order transformation 3/2(373)
Second-order wave speed 3/3(288–291)
Secular determinant 5(78)
Selection rules for radiation transitions 5(129)
Self charge 5(289)
Self energy of the photon 5(289)
Self induction 4(10)
Self mass of the electron 5(296, 341)
Self stress of the electron 5(331)
Self-consistency of Lorentz transformation 4(120–122)
Self-energy 4(174)
Self-potential 4(176)

Semi-classical approximation 3/2(249, 264, 317, 336)
Semi-holonomic system 3/1(39)
Separating transition relation 3/2(123)
Separation column 3/2(455)
Separation of variables 3/1(128, 175)
Servo-mechanism 3/1(65)
Several world points, functions of them 4(135)
Shaking effect 5(83, 85)
Shear 3/1(256–257, 258–259, 267)
Shear components of a tensor 3/1(844)
Shear composition 3/1(296)
Shear cone of a tensor 3/1(251)
Shear, generalized 3/1(297)
Shear intensity 3/1(266)
Shear, maximum and minimum 3/1(263)
Shear modulus, generalized 3/3(174, 182)
Shear of a circular cylinder 3/1(299–300)
Shear of a sphere 3/1(302)
Shear strains 3/1(267)
Shear trajectories of a tensor field 3/1(847)
Shear viscosity 3/1(718); 3/2(428)
Shearing 3/1(348, 431); 3/3(436)
Shearing force 3/1(566)
Shearing function 3/3(436)
Shearing of hypo-elastic material 3/3(418–426)
Shearing plane 3/1(292)
Shearing, rectilinear 3/3(76)
Shearing stress 3/1(537, 544)
Shearing stress, maximum 3/1(551)
Shear-stress function 3/3(436)
Shell 3/1(310, 311)
Shell theory, relation to three-dimensional theory 3/1(560–564)
Shells, equilibrium state 3/1(556–564)
Shells, strain of 3/1(320–325)
Shifter 3/1(807); 3/3(49)
Shock surface 3/1(514)
Shock wave 3/1(512–513, 519–522, 546, 610, 711, 714–715); 3/3(267)
Signal velocity 4(90–91, 117)
Signorini's expansion 3/3(219–227)
Signorini's special theory of elasticity 3/3(348)
Signorini's stress inequality 3/1(578)
Simple extension 3/1(291–292); 3/3(173, 180)
Simple fluids 3/3(13, 78, 79–81, 427–513)
Simple material(s) 3/3(36, 60–63, 78)
— basic properties 3/3(66–69)
— of order n 3/3(109)
— thermal equilibrium 3/3(298–301)
Simple mixture, definition 3/2(39)
Simple shear 3/1(292–296)
Simple shear of elastic bodies 3/3(174)
Simple shearing 3/1(360, 432); 3/3(442ff.)
— of second-grade fluids 3/3(506)
— stress 3/1(551)
Simple solids 3/3(78, 81–86)
Simple system 3/2(168, 171)
Simple tension 3/1(551)
Simultaneity concept of Einstein 4(115–117)
Simultaneous invariant of several tensors, *see also* Invariant 3/3(22, 24, 25, 31)
Single gas, thermodynamic functions 3/2(26)
Singlet states 5(116, 117)
Singular charge distribution 4(174)
Singular functions of field theory 5(189–194)
Singular surfaces
— balance theorems 3/1(526–528)
— compatibility conditions 3/1(493–494, 496–498, 503–506, 515)
— definition 3/1(492)
— general classification 3/1(517–519)
— material 3/1(519)
— of higher order than second 3/1(524–525)
— of second order 3/1(523–524)
— transport theorem 3/1(525–526)
Singularities 5(42, 43)
Singularities of gravitation fields 4(230–231)
Singularity of mass density 3/1(466)
Six-dimensional bivector space 4(221)
Skew symmetric, *see* Antisymmetric
Skin depth 4(39)
Sky blue 4(94)
Skylight polarization 4(95)
Slater determinant 5(114)
Slater sum 3/2(326)
Slip surface 3/1(512)

Slippage tensor 3/1(372–373)
Slow approximation 4(217, 219, 244)
Slow motion 3/2(450–451)
Small deformation 3/1(303–309)
Small displacement 3/1(308)
Small displacement gradients 3/1 (306)
Small elastic deformations superimposed on large ones 3/3(116–117)
Small principal extensions 3/1(303–308)
Small pure strain 3/1(305)
Small rotation 3/1(305–308)
Small strain, historical remarks 3/1(270)
Smallness, the meaning of 3/1(303, 309)
S-Matrix 4(250); 5(240–242)
Snell's law of refraction 4(35)
— for a conductor surface 4(38)
Solenoidal field 3/1(819–824, 827)
Solenoidal vector field 4(7, 21)
Solutions in finite elasticity, approximate 3/3(219–267, 347–354)
Solutions in finite elasticity, exact 3/3(171–219)
Soret effect 3/2(461)
Sound absorption 3/2(479–482)
Sound, speed, *see* Speed of sound
Sound velocity, complex 3/2(481)
Sound waves, Hugoniot's theorem 3/1(712)
Source equivalence 3/1(469)
Sources and vortices as elements of motion 3/1(364–365)
Space, absolute 3/1(6, 43)
Space and space-time, homogeneity and isotropy 3/1(9, 10)
Space, configuration space 3/1(2, 13, 100, 134)
Space coordinates 5(115)
Space metric 3/1(455)
Space of events (space-time) 3/1(105)
Space of Finsler 3/1(106)
Space of momentum and energy 3/1(130)
Space of states 3/1(100, 163)
Space of states and energy 3/1(100, 143, 169)
Space, Riemannian 3/1(106)
Space tensor derived from world tensor 3/1(456)
Space-like 4(119)
Spaces, definitions and special kinds 3/1(100, 105, 130, 134, 143, 163, 169)
Spaces, representative (listed) 3/1(100, 104)
Space-time 3/1(105, 666); 3/3(41)
— curved 3/1(1)
— flat 3/1(1)
— interval 4(119–120)
Spatial Bernoulli theorem 3/1(408–409)
Spatial continuity equation 3/1(342, 467)
Spatial coordinates 3/1(326, 337–339)
Spatial coordinates and description 3/3(37, 39)
Spatial diffusion vector 3/1(381)
Spatial distribution of galaxies 3/2(610–611)
Spatial equation of continuity 3/1(342, 467)
Spatial gradient 3/3(39)
Spatial representation of a surface of discontinuity 3/1(506–528)
Spatial time derivative 3/3(39)
Special motion types 3/1(328–329, 430–437)
Special relativity 3/1(697–698)
Special theories of elasticity 3/3(347–354)
Special theory of relativity 4(109)
Specific energy, Poisson equations 3/1(402–404)
Specific entropy 3/1(619)
Specific entropy, total, of a mixture 3/1(634)
Specific heats 3/1(624–627, 630–632)
Specific heats, generalized theory 3/1(625–627)
Specific inductive capacity 4(5)
Specific inner energy, total, of a mixture 3/1(634)
Specific internal energy 3/1(609)
Specific kinetic energy 3/1(377, 402)
Specific phase 3/2(249)
Spectral function 3/2(631)
Spectral resolution of radiation 4(68–71, 78–80)
Speed 3/1(25, 201, 216, 328)
Speed of a surface 3/1(676)
Speed of displacement of a surface 3/1(498–499, 506)

Speed of light in a dielectric 3/1(739)
Speed of light in vacuum 3/1(679)
Speed of propagation 3/1(508–509, 514)
Speed of propagation of a shock 3/1(521)
Speed of sound 3/1(657, 713)
Speed of wave propagation 3/3(268)
— second-order effects 3/3(288–291)
Speeds of sound in an elastic body 3/1(727)
Spencer-Rivlin theorem on tensor invariants 3/3(31)
Spheres, radial oscillations 3/3(217–219)
Spherical coordinates 3/3(20)
Spherical harmonics 5(45)
Spherical pendulum 3/1(47, 52, 56)
Spherical polar coordinates 3/1(27); 5(40)
Spherical shell, inflation and eversion 3/3(195, 344)
Spherical tensor 3/1(834)
Spherical top 5(46)
Spherical wave 4(41–44, 51–54)
Spin 5(98, 107, 157, 158, 160, 165)
Spin coordinates 5(115, 170ff.)
Spin matrices of Dirac 4(143–144, 192)
Spin matrices of Pauli 3/1(23, 24); 4(138, 139, 191); 5(109, 141, 214)
Spin moment 5(107, 160, 165)
Spin of a photon 5(189)
Spin of a vector field 4(44–46)
Spin of higher order 3/1(384–385)
Spin partition function 3/2(78)
Spin space 4(138)
Spin tensor 3/3(54)
— of Cauchy 3/1(353–360, 441)
— of rigid motion 3/1(355)
Spin zero particles 4(185–187)
Spinning electron 4(189)
Spinor 5(148)
Spinor calculus 5(149)
Spinor notation of van der Waerden 4(140)
Spinors of four components 4(142–144, 190)
Spinors of two components 4(137–141, 187)
Spring extension as a thermodynamic problem 3/2(28)
St. Venant's compatibility conditions for small strain 3/1(306)
St. Venant's special theory of elasticity 3/3(348)
Stability 3/1(141); 3/3(153)
Stability criteria 3/2(375, 391–393)
Stability criterion of Gibbs 3/2(374)
Stability, elastic 3/3(252–260)
Stability, internal of a phase 3/2(18–21)
Stability, kinetic 3/3(259)
Stability of a sleeping top 3/1(89)
Stability of equilibrium 3/1(653–659); 3/2(11)
Stability of hyperelastic bodies 3/3(328–332)
Stability theorem of Hadamard 3/3(253)
Stability, thermodynamic 3/2(374–375, 391–393)
Stability with respect to continuous changes 3/2(374)
Stable isotherm 3/2(20, 34)
Staeckel type system 3/1(129)
Stagnation point 3/1(328)
Stark-effect 3/1(130)
State 3/3(329)
State, axiomatic description 3/2(121)
State continuity 3/2(33)
State density 3/2(633, 636–638)
State function 3/2(124)
State vector 4(181, 256)
States of negative energy 5(166, 168)
Static continuation 3/3(106, 369, 371)
Static solutions of the Einstein field equations
— linearized 4(222–224)
— rigorous 4(226–230)
Static stability 3/3(328)
Statically determinate problem 3/1(580–582)
Statics 3/1(1)
Statics of Ericksen's anisotropic fluids 3/3(528–530)
Stationary boundary of motion 3/1(330)
Stationary ensemble 3/2(186)
Stationary states as eigenvalue problem 5(40ff.)
Stationary stochastic process 3/2(622–632)

Statistical applications of quantum theory 5(118–120)
Statistical ensemble 3/2(48)
Statistical entropy analog 3/2(305, 322)
Statistical equilibrium 3/2(185–187, 274)
Statistical mechanics 3/1(175)
Statistical mechanics, historical development 3/2(116–118)
Statistical thermodynamics 3/2(70–93)
Steady current 4(6, 21)
Steady extension 3/3(76)
Steady motion 3/1(328, 339, 343, 431, 442–443, 467)
Steady motion with steady density 3/1(343, 431, 487, 489)
Steady quantity 3/1(328)
Steady streamlines 3/1(341, 431)
Steady vortex lines 3/1(391)
Steady vorticity motion 3/1(391, 431)
Steepest descent method 3/2(51, 61, 63)
Stefan-Boltzmann constant 3/2(109)
Stefan-Boltzmann law 3/2(108); 4(33)
Stefan's relations for diffusion 3/1(707)
Stellar dynamics 3/2(604ff.)
Stereographic projection 3/1(21, 24)
Stern-Gerlach experiment 5(65, 67)
Stieltjes integral 5(42)
Stirling's formula 3/2(56, 67)
Stochastic differentiation 3/2(576–577)
Stochastic integration 3/2(578)
Stochastic process
— definition 3/2(559)
— measure theoretical approach 3/2(572–573)
— physical approach 3/2(558–572)
Stoechiometric coefficients 3/1(473)
Stokes' law 3/2(594)
Stokes' parameters 4(45)
Stokes' potentials 3/1(828–829)
Stokes' principle of viscosity 3/3(475ff.)
Stokes' theorem, *see also* Kelvin's transformation
Stokes' theorem 3/1(356–357, 386, 816–817); 4(10)
Stokes-Christoffel condition for shocks 3/1(522)
Stokesian fluid 3/1(721)
Stokesian fluid, Truesdell's theory 3/3(485–488)
Stoppelli's theorem 3/3(138)
Stored energy 3/3(302, 312)
Stored-energy function 3/3(302, 312ff., 315)
Stored-energy function, convex 3/3(322)
Stoßzahlansatz 3/2(214, 224); 4(166)
Straightening of a cylindrical sector 3/3(188ff., 343)
Strain 3/1(240, 255); 3/2(53, 54)
Strain composition 3/1(282–283)
Strain ellipsoid 3/1(259, 261)
Strain energy 3/1(571, 598–599, 641–642)
Strain energy and internal energy 3/1(611–612)
Strain impulse 3/3(107–108)
Strain invariants 3/1(264–266)
Strain magnitude 3/1(266)
Strain measures, equivalence 3/1(268–269)
Strain measures, logarithmic 3/1(269–270, 271)
Strain measures, special 3/1(269, 271)
Strain of orientation 3/1(313, 314)
Strain of volume 3/1(266)
Strain tensor 3/1(266); 3/2(441)
Strain theory, historical remarks 3/1(270–271)
Strain-energy function 3/3(302, 307–316, 397ff.)
Strain-energy function, no experimental test of existence 3/3(344)
Streak line 3/1(330–333)
Stream function 3/1(474–481)
— for plane motion 3/1(474–477)
— for pseudo-lineal motion 3/1(477)
— for rotationally symmetric motion 3/1(478–479)
— for three-dimensional motion 3/1(479–480)
— for two-dimensional motion 3/1(477–479)
Stream line 3/1(330–333)
Stream potentials 4(49, 55)
Stream sheet 3/1(331)
Strength of a vortex 3/1(362, 387–389, 411)
Stress 3/2(53); 3/3(39ff.)
Stress deviator 3/1(545)
Stress, dissipative part 3/1(571, 640)
Stress ellipsoids of Cauchy and Lamè 3/1(552)

Stress function(s) 3/1(582–594)
— for flat spaces 3/1(585–586)
— for general motion 3/1(592–594)
— for membranes 3/1(589–592)
Stress, geometrical representations 3/1(552)
Stress impulse 3/1(549)
Stress invariants 3/1(551)
Stress means 3/1(568–569, 574–580)
Stress notations 3/1(548)
Stress power 3/1(570, 638)
Stress, principal axes 3/3(142)
Stress principle 3/1(537–538); 3/3(40)
Stress principle for rods 3/1(565)
Stress principle for shells 3/1(556–557)
Stress quadrics 3/1(552)
Stress relation 3/3(297, 300–302, 313, 368)
Stress relation, inversion 3/3(327)
Stress relaxation 3/3(47, 106ff., 369)
Stress resultant 3/1(565)
Stress tensor 3/1(542–544); 3/2(441, 496); 4(162); 5(25, 35)
— convected 3/3(67)
— electromagnetic 3/1(689)
— non symmetrical 3/3(389)
— of Maxwell 4(20–21, 29)
Stress, thermodynamic definition 3/1(651)
Stress trajectories 3/1(552)
Stress vector 3/1(537, 542–543, 546); 3/3(40)
Stress-energy-momentum tensor 3/1(696)
Stress-entropy relation 3/3(367ff.)
Stress-entropy tensor 3/3(365)
Stress-momentum-tensor 3/1(555)
Stress-relaxation function 3/3(115)
Stress-relaxation function, Coleman's theorem 3/3(378)
Stress-strain relation(s) 3/1(641); 3/3(119)
Stretch 3/1(255–256, 257, 260–262, 266–267, 274)
Stretch, finite 3/1(369)
Stretch history, constant 3/3(438–440)
Stretch tensors, right and left 3/3(53)
Stretching 3/1(348, 369)
Stretching and spin, basic identity 3/1(380)
Stretching of higher order 3/1(384–385)
Stretching of vortex lines 3/1(429–430)
Stretching tensor 3/3(54)
Stretching tensor of Euler 3/1(347–352)
Strict causality 4(93)
String 3/1(565)
String loaded 3/1(81, 82)
Structural and continuum theories compared 3/3(5)
Structure functions of Khinchin 3/2(627–629)
Structure, mathematical, of mechanics 3/1(98)
Structure of fluids, statistical theory 3/2(611–615)
Structures, periodic 3/1(81)
Student's *t*-distribution 3/2(547–548)
Subfluids (liquid crystals) 3/3(86–88, 151)
— flow problems 3/3(473–475)
— hyperelasticity 3/3(314–316)
— stored energy 3/3(315)
Submerged body 3/1(540)
Sub-space with a finite number of co-ordinates 5(56, 78)
Substantial, *see* Material
Substantial constant 3/1(339)
Substantial time derivative 3/2(420)
— Lorentz invariant 3/2(496)
Sudden change of a parameter 5(85, 86)
Sum of tensors 3/1(808)
Summation convention 3/1(100)
Summation convention of Einstein 4(110, 127)
Superficial Bernoulli theorem 3/1(407–408)
Superficial conditions of compatibility 3/1(493–494, 496–498, 503–506, 515)
Superimposed elastic strain (theorem) 3/3(250)
Superposable motions 3/1(396)
Superposition principle 3/2(485, 510); 5(1, 23)
Superposition principle of Kirkwood 3/2(390)
Superposition theorem for strain 3/1(308)
Superpotentials 4(41, 48, 72)
Superstability 3/3(255, 257)
Supino's theorems on irrotational motion 3/1(395)

Supply 3/1(468–469, 472)
Supply of energy 3/1(609)
Supply of momentum in a mixture 3/1(567)
Surface charge 3/1(683–684)
Surface charge density 4(5)
Surface current 3/1(683–684)
Surface element, deformation 3/1(247–249, 263)
Surface element, material derivative 3/1(341–342)
Surface integrals, kinematics 3/1(346–347)
Surface, material 3/1(339–340)
Surface metric 3/1(811)
Surface of constant curvature 3/1(590)
Surface of discontinuity 3/1(233, 610)
Surface of revolution 3/1(591)
Surface reaction 3/2(477)
Surface tension 3/2(46–47)
Surface theory 3/1(812–813)
Surfaces of maximum shear of a tensor field 3/1(848, 850)
Surroundings of a system 3/2(143)
Susceptibility 3/2(113)
Susceptibility, electric 4(16)
Susceptibility, magnetic 4(24)
Svanberg's vorticity theorem for rotationally symmetric motion 3/1(421)
Swelling at exit from a pipe 3/3(455–458)
Symbols frequently used 3/1(235–239)
Symbols, index of 3/3(17–19)
Symmetric (eigen)functions 5(112, 114)
Symmetric kernel function 4(176)
Symmetric states 5(119)
Symmetric transition relation 3/2(123)
Symmetrical top molecule 3/2(81)
Symmetry number of a molecule 3/2(75, 80)
Symmetry of the stress tensor 3/1(546, 569)
Symmetry of wave functions 3/2(640)
Symmetry properties of tensors 4(133)
Symplectic transformation 3/1(146)
System, closed 3/1(9)
System, conservative 3/1(100, 167)
System, degenerate 3/1(179, 183, 191)
System, holonomical 3/1(38, 60)
System, isolated 3/1(9)
System, non-conservative 3/1(170)
System, non-holonomic 3/1(39, 59, 64, 65)
System of Liouville type 3/1(129, 172)
System of localised molecules 3/2(66)
System of Staeckel type 3/1(129, 172)
System, rheonomic 3/1(35, 39)
System, scleronomic 3/1(39)
System, semi-holonomic 3/1(39)
Systems without constraints 3/1(69–82)
Szebehely's unsteadiness number 3/1(379, 432)

Tangent vector 3/1(810)
Tangent vector space 4(212)
Tangential projection 3/1(796)
Tautochrone 3/1(48)
T-E inequalities (tension-extension) 3/3(155)
Temperature 3/1(619, 621); 3/2(245, 262); 3/3(133, 294)
Temperature, absolute 3/2(3, 6, 23–25, 52)
Temperature, axiomatic definition 3/2(156)
Temperature concept 3/2(1)
Temperature dependence of affinity 3/2(18)
Temperature dependence of susceptibility 4(18–19, 25)
Temperature, empirical 3/2(243, 313, 320–321)
Temperature equation 3/2(444)
Temperature relation 3/3(297, 301, 361)
Tension 3/1(537, 551, 566, 574)
Tensor components 3/3(14)
Tensor densities of Levi-Civita 4(137, 211)
Tensor density 4(133)
Tensor density in Riemannian space 4(211)
Tensor function, definition 3/3(21)
Tensor function, linear 3/3(22, 23)
Tensor function of one variable 3/3(32–35)
Tensor invariants 3/1(832–836)
Tensor lines 3/1(847)
Tensor of deformation, *see* Deformation gradient
Tensor of inertia 3/1(31, 33, 485–486, 490)
Tensor polynomial 3/3(21)

Tensor power series 3/3(21)
Tensor sheet 3/1(847)
Tensor sum 3/1(808)
Tensors, geometrical representation 3/1(250–255)
Tensors in Minkowski universe 4(131, 132, 133)
Tensors in Riemannian space 4(210)
Theorem of König 3/1(36, 65, 484)
Theorem of parallel axes 3/1(32)
Thermal conductivity 3/1(709)
Thermal constraints 3/3(329)
Thermal contact 3/2(171)
Thermal diffusion 3/2(427, 452–455, 472)
— factor 3/2(453)
— potential 3/2(460–461)
Thermal equation of state 3/1(623)
Thermal equilibrium 3/2(3, 6)
— in a simple material 3/3(298–301)
— of simple systems 3/2(171)
Thermal isotropy 3/1(709)
Thermal potential 3/2(461)
Thermal system 3/2(156)
Thermodynamic coefficients, frequency dependence 3/2(482–487)
Thermodynamic coordinates 3/2(145 ff., 154, 155)
Thermodynamic coordinates, Gibbs' characterisation 3/2(151)
Thermodynamic degeneracy 3/1(633–634)
Thermodynamic diagrams 3/1(620, 623)
Thermodynamic equilibrium 3/1(647–660)
Thermodynamic flux 3/1(643, 646)
Thermodynamic force 3/1(643)
Thermodynamic functions for radiation 3/2(107–108)
Thermodynamic functions of linear molecules 3/2(77)
Thermodynamic functions of non-linear molecules 3/2(80)
Thermodynamic identities 3/1(629–632)
Thermodynamic inequalities 3/1(652–659)
Thermodynamic paths 3/1(624)
Thermodynamic potential(s) 3/1(627–629); 3/2(9, 351)
Thermodynamic pressure 3/1(621, 636, 638)
Thermodynamic process 3/3(295)
Thermodynamic stability 3/2(374–375, 391–393)
Thermodynamic state 3/1(620)
Thermodynamic substate 3/1(618–619)
Thermodynamic tensions 3/1(621)
Thermodynamics, classical, historical development 3/2(113–116)
Thermodynamics of diffusion 3/3(541)
Thermodynamics of irreversible processes 3/3(373)
Thermodynamics of simple materials 3/3(363–381)
Thermoelasticity 3/2(442); 3/3(355–363)
Thermoelasticity with linear viscosity 3/3(357)
Thermoelectric effects 3/2(464 ff.)
Thermoelectric potential series 3/2(469)
Thermoelectric power 3/2(469)
Thermomagnetic effects 3/2(464 ff.)
Thermomechanical potential 3/3(299)
Thermometer 3/2(3)
Thermometer, absolute 3/2(158)
Thermostat 3/2(3)
Thermostatic limit theorem 3/3(371)
Thermostatics 3/1(615–660)
Thermostatics of simple materials 3/3(300)
Thin body 3/1(309)
Third entropy analog of Gibbs 3/2(250)
Third law of thermodynamics 3/2(21–22)
— generalisation 3/2(144)
Thomas correction 5(161)
Thomson coefficient 3/2(468)
Thomson cross section 5(266)
Thomson heat 3/2(468)
Thomson relation 3/2(505)
Thomson scattering 4(93–94)
Thomson's form of the second law 3/2(5, 157)
Three-body problem 3/1(80)
Three-dimensional fluid 3/2(614)
Three-dimensional motion, stream function 3/1(479–480)
Throttling experiment of Joule and Kelvin 3/2(10, 26)
Time as a *c*-number 5(60)
Time average 3/2(188, 207, 209)

Time derivative(s)
— of an operator 5(59, 60)
— substantial 3/2(420, 496)
— material and spatial 3/3(39)
— of scalar integrals over moving regions 3/1(675)
Time flux, convected 3/1(448–452)
— co-rotational 3/1(445, 451–452)
— invariant, in four dimensions 3/1(459)
Time, introduction into field theory 3/1(325)
Time of first passage 3/2(564)
Time ordering operator 5(245)
Time reversal invariance 3/2(505, 507)
Time-dependent deformations of tubes 3/3(212ff.)
Time-lapse, natural 3/3(437)
Time-like 4(119)
Time-ordered product 4(212)
Tolman's generalised *H*-theorem 3/2(227)
Tolman's hypothesis of equal a priori probabilities 3/2(190, 211)
Top, spinning 3/1(85)
Top, symmetrical 3/1(85, 87, 88)
Top, unsymmetrical 3/1(85)
Topological remarks 3/1(102)
Torque 3/1(531)
Torque exerted by a rigid obstacle 3/1(540)
Torque in cone-and-plate flow 3/3(463)
Torque in Couette flow 3/3(449)
Torsion 3/1(298–300, 318)
Torsion balance 4(113)
Torsion of a circular cylinder 3/1(299–300)
Torsion of a cylinder 3/3(191–193, 343ff., 352ff.)
Torsion of a sphere 3/1(302)
Torsion of a tube 3/3(191–193, 232–237)
Torsion of streamlines 3/1(376)
Torsion of vector lines, *see* Abnormality
Torsion, second-order theory 3/3(232–237)
Torsion, wrong theories 3/3(236)
Torsional flow 3/3(458–464)
Torsional rigidity 3/3(266)
Torsional stress 3/1(588)
Torsionless stress 3/1(588)
Torsor 3/1(37, 68)
Total covariant, derivative 3/1(246–247, 810–813)
Total energy 3/1(609); 3/2(2)
Total entropy 3/1(642)
Total of a tensor field 3/1(815)
Total reflection 4(37)
Total squared speed 3/1(365–369)
Toupin's electromechanical theories 3/3(385–389)
Toupin's oriented hyperelastic materials 3/3(391–397)
Trace of a matrix 5(70)
Trace of a tensor 3/3(15, 23ff.)
Trace of a tensor, its gradient 3/3(25)
Trajectories 3/1(107, 111, 121, 134, 215)
Trajectories of a tensor field 3/1(847)
Trajectory of a stochastic process 3/2(561, 563, 575)
Trajectory, parabolic 3/1(48)
Transfer of energy, non-mechanical 3/1(608)
Transfer of momentum 3/1(549–550)
Transference number 3/2(459)
Transformation, canonical 3/1(144, 154, 157, 168, 169, 170)
Transformation, contact 3/1(124, 144)
Transformation, continuous (definitions) 3/1(103)
Transformation group 5(98)
Transformation in normal form 3/1(191)
Transformation, infinitesimal 5(100)
Transformation of a field, infinitesimal 4(134)
Transformation of first order 3/2(372)
— anomalous 3/2(373, 389, 403)
Transformation of fluxes 3/2(432–435)
Transformation of forces 3/2(432–435)
Transformation of matrices 5(51)
Transformation of *n*th order 3/2(372)
Transformation of operators 5(51, 54)
Transformation of random variables 3/2(536)
Transformation of second order 3/2(373)
Transformation operator 5(54)
Transformation, point (extended) 3/1(149)
Transformation properties of field theory 5(177)

Transformation, symplectic 3/1(146)
Transformation theory of partition functions 3/2(352)
Transforms associated with frequency functions 3/2(538–542)
Transient phenomena 3/2(485)
Transition, axiomatic description 3/2(121)
Transition probability 5(81, 133)
Transition probability, infinitesimal 3/2(632, 644)
Transition process 5(69, 82)
Transition, quasistatic 3/2(140)
Transition relations 3/2(122–124)
Transitivity, metrical 3/2(198, 201–203)
Transitivity of Lorentz transformation 4(121)
Transitivity of transition relations 3/2(122, 152)
Translation 3/1(12, 274, 350)
Translational characteristic temperature 3/2(71)
Translational group 5(101)
Transmission coefficient 4(36)
Transport equation, *see* Boltzmann equation
Transport of momentum 3/2(421)
Transport phenomena 3/2(297, 487–491)
Transport terms 4(113)
Transport theorem 3/1(347)
Transpose of a tensor 3/3(15)
Transposition 4(140)
Transversal 3/1(123)
Transversal discontinuity 3/1(492)
Transverse acceleration wave 3/1(524)
Transverse contraction ratio 3/1(291)
Transverse electric wave 4(44, 58)
Transverse isotropy 3/3(82)
Transverse magnetic wave 4(44, 58)
Transverse parts of the Hamiltonian 4(104)
Transverse wave 3/1(525); 3/3(268, 280); 4(32)
Transverse wave in an elastic body 3/1(726–727)
Transversely isotropic fluids of Ericksen 3/3(527–528)
Tri-axial stress 3/1(551)
Triclinic solids 3/3(85)
Triplet states 5(116, 117)
Trouton-Noble experiment 4(65, 112, 113)
True charge, *see* Free charge
True charges 4(5)
True current 4(8)
Truncation number 3/1(721)
Tscherenkov, *see* Čerenkov
Tube, *see* Pipe
Tube, inflation and bending 3/3(189–195, 212ff., 343)
Tube, non circular, flow through it 3/3(468–471)
Tube, radial oscillations 3/3(354)
Tube, secondary flow 3/3(497–504)
Tube, torsion and extension 3/3(232–237, 353)
Tube, various deformations 3/3(353)
Turbidity 3/2(369)
Turbulence 3/2(641)
Twin paradox 4(124–125)
Twist per unit length 3/1(298, 318)
Twisting moment 3/1(566)
Two-body problem 3/1(69)
Two-component spinors 4(107–141, 187)
Two-dimensional motion, stream function 3/1(477–479)
Two-event characteristic function 3/1(209)
Two-phase equilibrium 3/2(36)
Two-point characteristic function 3/1(117–121, 137, 138)

Ultrastability 3/1(658)
Uncertainty relations of Heisenberg 5(4, 12, 20)
Uncertainty relations, quantitative formulation 5(20)
Undeformed material 3/1(244)
Undistorted state of a solid 3/3(85)
Undistorted state of a subfluid 3/3(87)
Undistorted states of an isotropic material 3/3(78, 82, 139, 151)
Uniform dilatation 3/1(290–291); 3/3(173)
Uniform ensemble 3/2(287)
Uniform integral 3/2(179)
Uniform reference 3/3(89)
Unimodular condition 3/1(22, 150)
Unimodular matrix 4(138)
Unimodular matrix transformations in spin space 4(138)

Uniqueness, its failure 3/3(128ff.)
Uniqueness of potential 4(13–14)
Uniqueness theorem
— for free vibrations 3/3(256)
— for potential deformations 3/3(165)
— of Ericksen and Toupin 3/3(255)
— of Signorini 3/3(225)
— of Truesdell and Toupin 3/3(249)
Unit of charge 3/1(666)
Unit tensor 3/3(23)
Unit transformation 3/1(660)
Unitary 5(55, 105)
Unitary transformation of two complex variables 5(105)
Units 4(3, 4, 5, 6, 29, 107, 169)
Units, natural 5(170)
Unnatural process 3/2(3)
Unsteadiness number of Szebehely 3/1(379, 432)

v. Neumann-Birkhoff lemmas 3/2(194–201)
Vacuum polarization 5(281, 285, 352)
van der Waals formula of second virial coefficient 3/2(26)
van der Waals gas 3/1(632–633)
van't Hoff relation 3/2(42)
Vapour-liquid equilibrium 3/2(35, 44)
Variable, adiabatically arrested 3/2(135)
Variable, axiomatic description 3/2(125)
Variable, internal 3/2(470)
Variable, macroscopic 3/2(275–278)
Variance, *see* Mean square deviation
Variation of constants 3/1(197)
Variational condition of thermodynamic equilibrium 3/1(650–652)
Variational principle(s)
— for electromagnetic field equations 4(171–172)
— for hyperelastic materials 3/3(326–328)
— of mechanics 3/1(107, 111, 139, 594–607)
— of Schrödinger 5(50, 51)
Vector, absolute and relative rates of change 3/1(29)
Vector components 3/3(14)
Vector, contravariant 4(132)
Vector, covariant 4(132)
Vector line 3/1(817)
Vector line, material 3/1(340–341)
Vector model 5(104)
Vector potential, *see also* Magnetic potential 3/1(674); 4(7, 21, 46, 111)
— multipole expansion 4(73)
Vector sheet 3/1(817)
Vector spherical harmonics 4(42–43, 104–107)
Vector tube 3/1(817)
Vectors in Minkowski universe 4(131, 132)
Vectors in Riemannian space 4(210)
Velocities, addition 4(123–124)
Velocities, composition 3/1(202)
Velocity 3/1(6, 25, 200, 201, 328)
Velocity field 3/3(39); 4(162)
Velocity field in four dimensions 3/1(456–457)
Velocity gradient in four dimensions 3/1(461)
Velocity of light 4(32, 107)
— in a dielectric 3/1(739)
— in vacuum 3/1(679)
Velocity of sound, *see* Speed of sound
Velocity of the center-of-mass 3/2(419, 495)
Velocity of transport 3/1(30)
Velocity of wave propagation 4(32, 89–91)
Velocity, physical components 3/1(26)
Velocity potential 3/1(357)
Velocity potential theorem of Lagrange and Cauchy 3/1(421, 712)
Velocity profile of channel flow 3/3(444)
Velocity profile of Poiseuille flow 3/3(452)
Velocity vector 3/1(139, 200, 201)
Vessel, reaction of its content upon it 3/1(539)
Vibrational characteristic temperature 3/2(77)
Vibrational modes of a cavity 4(40–41, 44)
Vibrations, infinitesimal, uniqueness theorem 3/3(256)
Vibrations, internal, of a molecule 3/2(76–77)
Virial 3/2(242)
Virial coefficients 3/2(27)

Virial theorem 3/1(573); 3/2(242, 304); 5(26)
Virtual displacement 3/1(596–597)
Virtual ensemble 3/2(179)
Virtual work 3/1(595–600); 3/3(306)
Visco-elastic theories 3/1(733–734, 740)
Visco-elasticity, classical theory 3/3(116)
Visco-elasticity, linear 3/3(109, 378)
Viscometric flow 3/3(432, 435–440)
— asymptotic theory 3/3(488–490)
— of Ericksen's anisotropic fluids 3/3(533–537)
Viscometric functions 3/3(13, 426, 495, 504)
Viscometric functions for poly-isobutylene 3/3(467 ff.)
Viscometry, theories 3/3(465–468)
Viscosities of a fluid 3/1(717)
Viscosity 3/2(428); 3/3(65)
Viscosity, classical theory 3/3(113–117)
Viscosity function 3/3(436)
Viscosity, natural 3/3(437)
Viscosity of a fluid, linear 3/1(715–716)
Viscosity of a fluid, non-linear 3/1(720–722)
Viscosity tensor 3/2(442)
Viscous fluid 3/1(233)
Viscous stress 3/2(427)
Voltaic battery 4(2, 6, 10)
Volterra's electromagnetic constitutive equations 3/1(740)
Volume change at a shock 3/1(521)
Volume change in second-order theory 3/3(240, 346)
Volume discharge in channel flow 3/3(444)
Volume discharge in flow between co-axial pipes 3/3(451)
Volume discharge of Poiseuille flow 3/3(452)
Volume element, deformation 3/1(247–249, 266)
Volume element, material derivative 3/1(341–342)
Volume integrals, kinematics 3/1(347)
Volume, material 3/1(339)
Volume strain 3/1(266)
Volume viscosity 3/2(428, 475, 477–479)
von Mises' theorem on circulation preserving motion 3/1(430)
Vortex line 3/1(386)
Vortex sheet 3/1(512, 514, 519–522)
Vortex sheet, material 3/1(515–517)
Vortex tube 3/1(386)
Vortices and sources as elements of motion 3/1(364–365)
Vorticity 4(213)
Vorticity averages 3/1(396–402)
Vorticity formula of Cauchy 3/1(421)
Vorticity moments 3/1(400–401, 415)
Vorticity number, dynamical 3/1(379)
Vorticity number, kinematical 3/1(363–364, 382, 403–404)
Vorticity tensor in four dimensions 3/1(461)
Vorticity theorem of d'Alembert for plane motion 3/1(421)
Vorticity theorem of Svanberg for rotationally symmetric motion 3/1(421)
Vorticity theorems of Helmholtz 3/1(386–387, 389)
Vorticity vector 3/1(354, 385–386)

Wang's theories of fading memory 3/3(105, 112–113)
Ward identity 5(304)
Wave, definition 3/1(508)
Wave equation(s)
— for several particles, unrelativistic 5(34)
— in curvilinear coordinates 5(39)
— of electromagnetic potentials under Galilean transformation 4(112)
— of field vectors 4(31, 46)
— of Schrödinger, *see* Schrödinger equation
— of second order, relativistic 5(141)
— relativistic, of Dirac 5(137 ff.)
Wave function
— canonical conjugate 5(174, 177)
— of a free particle 5(12)
— of a particle dependent on external forces 5(23)
Wave packet 4(89); 5(2, 3, 17, 36, 37, 96)
— reduction 5(69)
— relativistic 5(150 ff.)
Wave propagation 3/3(267–294, 332–335)
— in a conductor 4(37–40)
— in materials with memory 3/3(382–385)

Wave, velocity 3/1(134, 201, 215, 216)
Wavelength of de Broglie waves 5(6, 10)
Waves, *see* Acceleration waves
Waves, homentropic and homothermal 3/3(272, 335, 383ff.)
Waves in a coherent system 3/1(134)
Waves in a viscous fluid 3/1(719–720)
Waves in an elastic body 3/1(726–727)
Waves in incompressible materials 3/3(291–294)
Waves in isotropic materials 3/3(278–291)
Waves of constant action 3/1(123)
Waves of finite amplitude 3/3(283ff.)
Waves of higher order 3/3(272ff., 283)
Waves, plane infinitesimal 3/3(273–278)
Waves, various definitions 3/1(529)
Weber's transformation 3/1(422–424)
Weight functions in the renormalization formalism 5(349)
Weight, statistical, of a state 5(47, 68)
Weingarten's first theorem on discontinuities 3/1(494)
Weingarten's second theorem on discontinuities 3/1(496)
Weissenberg effect 3/3(453ff.)
Weissenberg's assertion 3/3(484)
Welding, *see* Fracture
Weyl's tensor 4(220, 252, 253, 254)
Whitehead's theory of gravitation 3/1(2)
Whittaker potentials 4(57–59)
Wien displacement law 4(33)
WKB approximation 5(88, 92, 93)
Work 3/1(42)
Work done during a homogeneous deformation 3/3(299)
Work in hypo-elasticity 3/3(411–413)
Work, mechanical 3/2(2)
Work, quasistatic 3/2(162)
Work theorems for elastic materials 3/3(304–307)
Workless reaction of constraints 3/1(38, 43, 57)
World acceleration vector 3/1(461)
World invariant conservation laws 3/1(695–697)
World invariant form of Cauchy's laws of motion 3/1(555)
World invariant kinematics 3/1(452–463)
World line (or curve) 3/1(6); 4(126)
World material derivative 3/1(463)
World vectors, tensors etc., *see* Vectors, Tensors etc. in Minkowski universe
World velocity field 3/1(456–457)
World velocity gradient 3/1(461)
World vorticity tensor 3/1(461)
World tensor 3/1(456)
— form of energy balance 3/1(611)
— of stretching 3/1(458, 463)
— of the aether relations 3/1(679–680)
— representation of polarization and magnetization 3/1(686–689)

Yield 3/1(722)
Yield function 3/1(722)
Yield in hypo-elasticity 3/3(420–426)
Yield stress 3/1(722)

Zeeman effect 4(85)
Zener's formula for second-order change of volume 3/3(346)
Zermelo's recurrence paradox 3/2(218)
Zero point energy 5(185)
— of the electron gas 3/2(317)
Zeros of energy and entropy 3/2(18)
Zorawski's criterion of flux conservation 3/1(346)
Zorawski's theorem on relative motion 3/1(440)

Group 3

Mechanical and Thermal Behaviour of Matter

A priori bounds of existence theory 9(401, 415)
Abbreviated notion 6a/4(125)
Abe's rule 13(531–532)
Ablating boundaries 6a/3(3)
Abnormal structure of phosphate glasses 13(535–537)
Absolute discontinuity 6(342); 6a/3(427)
Absolute discontinuity of stress 6(343, 366)
Absolute discontinuity of tangential velocity 6(342, 366)
Absolute stress discontinuity 6a/3(434)
Absolute temperature 6a/3(174)
Absolute temperature scale 12(2, 3)
Absorbing cross section of a sphere 11/1(66)
Absorption coefficient
— classical 11/1(279–281)
— curve of systems of polar liquids 11/1(333–336, 338)
— dependent on temperature and pressure 11/1(314, 315)
— due to compressional relaxation 11/1(291)
— due to thermal relaxation 11/1(286)
— for shear waves 11/1(300)
— for sound waves 11/1(136)
— in a viscoelastic liquid 11/1(298)
— of a surface 11/1(16, 19)
— of heat radiation in glass 13(574)
— per wavelength 11/1(206, 211, 232, 286)
Absorption cross section 11/1(212, 232)
Absorption due to heat radiation 11/1(281–282)
Absorption due to paramagnetic resonance 11/2(242)
Absorption formula of Stokes and Kirchhoff 11/1(151)
Absorption in binary mixtures of liquids 11/1(328–339)
Absorption in highly dilute gases 11/1(153, 155–157)
Absorption in liquid helium 11/1(305)
Absorption in liquid with strong viscoelastic effects 11/1(323–327)
Absorption in liquids helium 11/1(306)
Absorption in liquids with temperature dependent structural effects 11/1(315–318)
Absorption in liquified mono- and diatomic gases 11/1(306)
Absorption in low viscosity liquids with pressure dependent structural effects 11/1(318–323)
Absorption in unassociated liquids 11/1(306–315)
Absorption in water 11/1(202, 320, 321)
Absorption in water after standing 11/2(273–274)
Absorption in electrolytes
— dependence on concentration 11/1(237–239)
— dependence on frequency 11/1(233–236)
— dependence on pressure 11/1(241)
— dependence on temperature 11/1(240)
— dependence on valency 11/1(233–237)
— mixtures 11/1(242–245)
Absorption loss 11/2(162)
Absorption measurement in electrolytes, methods 11/1(226)

Absorption measurement of ultrasound 11/2(146)
Absorption measurements in liquids 11/1(303–304)
Absorption of a phonon 11/2(216)
Absorption of high-frequency sound waves 8/2(612)
Absorption of sound, experimental methods of determination 11/1(165–171)
Absorption of sound in gases, theory 11/1(135ff.)
Absorption of sound in mercury 11/1(305)
Absorption of sound, molecular 11/1(157–165)
Absorption of sound waves in liquids 11/1(275ff.)
Absorption spectrum of a reaction sequence 11/1(212)
Absorption, total 11/1(217)
Absorption volume 11/1(212, 216–221, 232)
Abstract unilateral problems
— non-symmetric 6a/2(395)
— symmetric 6a/2(391)
Abundance 15(371)
Accelerated creep 7/2(197)
Accelerated flow with cavity of constant shape 9(358–360)
Accelerated fluid, stability problems 9(648, 651–653)
Acceleration of fluid 9(236, 238)
Acceleration of the piston 9(246–247)
Acceleration potential 9(284, 298)
Acceleration 8/1(130, 142, 151); 8/2(2, 3)
Acceleration, definition 6a/2(43)
Acceleration identities 8/1(168)
Acceleration in rod theories 6a/2(649, 652)
Acceleration in shell theories 6a/2(494)
Acceleration potential 8/1(78); 8/2(404)
Acceleration terms in the equations of motion 8/2(501–503)
Acceleration theory 15(287)
Acceleration waves 6a/2(253, 254)
— amplitude 6a/4(26, 59–62, 64ff., 68ff., 73, 98)
— compressive 6a/4(26, 61, 66–69)
— definition 6a/3(309, 313)
— expansive 6a/4(26, 61–66)
— growth and decay, mechanical theory 6a/4(97–99)
— growth and decay, thermodynamic theory 6a/4(97–99)
— in bodies of materials with memory 6a/3(334)
— in elastic bodies 6a/3(356)
— in general 6a/3(97, 309, 313, 428–432)
— in inhomogeneous elastic bodies 6a/3(378)
— in non-conductors 6a/3(362)
— in one dimension 6a/4(26, 29ff., 50, 66, 75, 94)
— in plasticity 6a/3(434, 435)
— in three dimensions 6a/4(72)
— local direction 6a/4(73)
— local speed 6a/4(73)
— local trajectory 6a/4(73)
— longitudinal 6a/4(73)
— longitudinal wave velocity 6a/4(73ff.)
— steady 6a/4(40ff.)
— transverse wave velocity 6a/4(73ff.)
— velocity (instantaneous sound velocity) 6a/4(41, 50, 60–66, 98)
Accolade (shape of obstacle) 9(374, 379, 408)
Accommodation coefficients 12(278)
— describing boundary substance 12(242)
— for reflections of molecules 8/2(597–600)
Accumulation region of borate network 13(529, 557, 569)
Accuracy of turbulence measurements 8/2(524)
Acetate solutions, absorption measurements 11/1(235)
Acetates 11/1(248)
Acetic acid, sound absorption and dispersion 11/1(316)
Acetone-chloroform mixture, entropy of dilution 13(334–335)
Acoustic dispersion data 6a/4(31ff., 44ff., 68, 87)
Acoustic efficiency 11/1(97)
Acoustic field quantities 11/1(6, 7)
Acoustic (frequency) branch 7/1(168, 176, 196, 316)
Acoustic impedance 11/2(4)

Acoustic impedance (impedance) 6a/4(15, 41, 56)
Acoustic lens, *see* Lenses
Acoustic loss 11/1(21)
Acoustic phase inverter 11/2(14)
Acoustic pressure 11/2(35–36, 115)
Acoustic pressure measurement for determination of sound absorption 11/1(303)
Acoustic radiation 11/1(8)
Acoustic sources 11/1(166)
Acoustic state equation 11/1(144)
Acoustic tensor 6a/2(243–248); 6a/3(98, 316, 324, 326)
Acoustic waves, *see* Infinitesimal sinusoidal progressive waves
Acoustical tensor, complex 6a/4(213, 256)
Acoustical tensor in a cubic crystal 6a/4(248)
Acoustical tensor, in thermoviscoelastic medium 6a/4(213)
Acoustoelectric effect 11/2(251–253)
Action maps 6a/3(131)
Action, remarked upon 6a/3(140)
Activation energies
— for diffusion of gas through fused silica 13(618)
— for self-diffusion of ions in glass 13(622)
— in mechanical relaxation 13(638, 640–641)
— of cations in glasses 13(586, 589–590)
Activation energy 7/2(221, 224); 11/1(215)
— for compressional viscosity 11/1(400)
— for cross-slip 7/2(156, 157)
— for intersection jog formation 7/2(117)
— for self-diffusion 7/2(120)
— for static viscosity measurements 11/1(381)
— for vacancy formation 7/2(119)
— for viscous flow 11/1(388)
— pure liquid fracture 6(552, 555, 556)
Activation, energy of 7/1(397, 399, 416, 474)
— dependence on temperature 7/1(397)
— experimental determination of 7/1(404, 416)
— self diffusion 7/1(474, 476)
— spectrum of 7/1(467)
Activation, enthalpy of 7/1(397)
Activation, entropy of 7/1(397, 405)
Activation, entropy of, dependence on temperature 7/1(397)
Activation heat, *see* Activation energy
Activation, quantum mechanical, *see* Tunnel effect
Activation stress 7/2(115, 120)
Activation volume 7/2(114, 121, 201)
Activity coefficient 11/1(207, 209, 213, 259); 12(18, 64)
Activity coefficient, average, of an electrolyte 13(370)
Activity coefficients of electrolytes 13(379, 380)
Activity differences 12(161–164)
Activity (fugacity) 10(194)
Activity, thermodynamical 10(30, 49, 69)
Actual mode 6a/3(416)
Actuator 11/2(64–66)
Added masses in ship theory 9(567)
Additions to electrolytes, effect on absorption 11/1(242–245)
Additivity of average force potential in pairs 13(20, 49)
Additivity of expansion coefficients 13(563–564)
Adiabatic 6a/2(304)
Adiabatic approximation 7/1(109, 410)
Adiabatic changes 7/1(237)
Adiabatic coefficient of bulk compressibility 12(9)
Adiabatic condition 6a/3(460)
Adiabatic conditions of inelastic continuum 6(231, 233, 236, 321)
Adiabatic deformation 6(14)
Adiabatic demagnetization, cooling effect 15(40, 53)
Adiabatic elastic potential 6(247, 250)
Adiabatic ellipse 12(570)
Adiabatic factor 11/1(197, 198)
Adiabatic flow of an ideal gas 9(6–8, 61–62)
Adiabatic Joule-Thomson coefficient 12(21, 36, 38, 40)
Adiabatic magnetization of a superconductor 15(46)

Adiabatic wall 8/1(421, 423)
Adjacent equilibrium, method of
— criticism 6a/3(202, 203–206)
— discussion 6a/3(224–226)
— follower forces 6a/3(282–284)
— history and application 6a/3(226–228)
— weakly conservative forces 6a/3(263)
Adjacent theory 6a/1(3)
Adjoint, *see also* Self-adjoint and Non-self-adjoint
Adjoint fields corresponding to Kelvin state 6a/2(175)
Adjoint function 11/1(31)
Adjoint operator 6a/2(355); 6a/3(170, 276, 280, 284, 285)
Adjoint traction field corresponding to Stokes process 6a/2(240)
Adjoint transformation 6a/2(369)
Admissible displacement field 6a/2(59, 312)
Admissible heat flux field, time-dependent 6a/2(326)
Admissible motion 6a/2(43, 326)
Admissible process 6a/2(215, 327, 484, 512, 531, 532)
Admissible process for a body: $\mathscr{B}$ [$\boldsymbol{u}$, $\boldsymbol{E}$, $\boldsymbol{S}$] 6a/3(42)
Admissible process, thermodynamic 6a/2(302, 304)
Admissible roughness 8/1(441)
Admissible state 6a/2(94, 313)
Admissible state for a body: $\mathscr{B}$ [$\hat{\boldsymbol{u}}$, $\hat{\boldsymbol{E}}$, $\hat{\boldsymbol{S}}$] 6a/3(42)
Admissible stress field 6a/2(59, 313)
Admissible stress field, time-dependent 6a/2(49, 326)
Admissible temperature difference field, time-dependent 6a/2(326)
Admittance 11/1(19, 35, 41, 111)
ADP, electric constants, temperature dependence 11/2(47)
Adsorption 10(168, 169, 231, 235, 276); 12(632); 15(450, 454)
Adsorption at plane interface 10(142)
Adsorption equation of Gibbs 10(140, 150, 166, 205, 268)
Adsorption formula of Gibbs 10(142, 170, 230, 267ff.)
Adsorption heat 10(169)
Adsorption in porous media 8/2(630–631)
Adsorption isotherm 10(205)
Adsorption isothermal 15(451)
Adsorption potential 10(226)
Adsorption pump 15(189)
Adsorption, relative 10(144)
Adsorption spectrometer of Hagstrum 12(628)
Aelotropy, *see* Anisotropy
Aerodynamic centre 8/2(392, 417, 430)
Affine transformation 9(30)
Affinity 11/1(142)
Affinity of adsorption 10(168, 169)
After-cooler 14(47)
After-effect, *see* Elastic after-effect, Thermal after-effect, Creep, recoverable
After-effect, elastic 6(464); 7/2(3); 11/2(173)
AgBr-crystal 7/1(643)
Age-hardening 6(602); 7/2(101, 220, 228, 235)
— at elevated temperatures 7/2(228, 239)
— reversion 7/2(235)
— time-rate 7/2(238)
Aggregate predicted from a knowledge of the free crystal 6a/1(666–676)
Aggregate ratios 6a/1(535, 560, 564, 667, 668, 671, 675, 704, 705, 707, 714)
Aggregate theory 6a/1(520, 560, 666, 668)
Aggregation of water molecules 11/1(203)
Aging 7/2(44, 185)
Air driven sound sources 11/2(24–32)
Air, humid, absorption 11/1(192–195)
Air jet generators of sound 11/2(27–32)
Air liquefaction 14(76)
Air liquefied, work done per mole of 14(49)
Air liquefiers 14(57, 58)
Air liquefiers at low pressure 14(75, 81)
Air liquefiers, two-stage 14(55)
Air, liquified 12(27)
Air rectifiers 14(111)
Air, wavelength 8/2(130)
Airfoil, induced drag 8/1(105–106)

Airfoil, lift and circulation 8/1(85–87, 89, 105–106)
Airfoil, luminar, *see* Luminar airfoil
Airfoil nose-shock 9(244, 271, 274)
Airfoil of delta shape 9(52–54)
Airfoil, skin friction 8/1(347)
Airfoil, supersonic flow 9(244, 253, 274)
Airfoil theory 8/1(101, 105–106)
Airfoil, thin, in linearized subsonic flow 9(32, 33–34)
Airfoil, thin, in linearized supersonic flow 9(43–45)
Airfoil, thin-wing integral equation 9(584, 590)
Air-water interface, parameters 9(645, 650)
Airy function 6a/2(157)
Airy functions 9(487, 511, 600)
Airy stress function 6(43–47, 50, 353, 424)
Airy stress function in polar coordinates 6(45–46)
Airy's solution 6a/2(54, 156)
Airy's stress function 6a/3(441, 496, 522); 7/1(533, 561f., 578)
Alcohol addition to $MgSO_4$ solutions 11/1(270)
Alfrey-Graffi theorem on elastic-viscoelastic correspondence 6a/3(72, 73)
Alkali borate glasses 13(531–533, 556, 559–560, 569)
Alkali borate glasses, viscosity 13(590, 592)
Alkali halide(s) 7/1(105, 106, 141, 197, 216, 223, 256, 284, 315, 400); 14(224)
Alkali halides, conductivity of 7/1(404, 417)
Alkali metal specimens, preparation of 14(165)
Alkali metals 7/1(105, 112, 127, 128, 143, 144, 224, 226, 227, 228, 249, 254, 322, 441, 474)
Alkali phosphate glasses 13(538–539)
Alkali silicate glasses 13(523, 524, 551, 553, 555, 558, 560)
Alkali silicate systems, limits of glass formation 13(612–614)
Alkaline earth borate glasses 13(530, 556, 557, 559–560, 567, 569)
— viscosity 13(591)
Alkaline earth salt 7/1(86, 140)
Allen dense air system 14(9)
Alloying contribution to critical shear stress 7/2(39)
Alloying, effect on work-hardening 7/2(50)
Alloys, *see* Binary alloys or systems and Solid solutions
Alloys, annealing of 14(167)
Alloys, binary 6a/1(152, 230–238, 391, 415, 416, 625, 647, 648, 691, 694)
Alloys, copper 6a/1(130)
Alloys, critical shear stress 7/2(37)
Alloys, first study of the deformation of 6a/1(230–238)
Alloys, melting and heat treatment of 14(167)
Alloys of transition with non-transition metals 14(324)
Alloys, ordered 7/2(4, 188–191)
Alloys, preparation of dilute 14(167)
Alloys, primary 7/2(167)
Alloys, superconductive 15(225, 268)
Alloys, tertiary 6a/1(230–238)
Alloys, variation with composition for annealed 14(265)
Alloys with long range order 7/1(456)
Alloys with precipitations 7/2(191–193)
Almansi measure of deformation 6(499)
α-brass
— electron microscopic investigation 7/2(99–100, 180)
— interpretation of yield point 7/2(179)
— recrystallization twins 7/2(180)
— single crystals, critical shear stress 7/2(37–39)
— single crystals, workhardening 7/2(62, 64)
α-creep 7/2(199)
α-curve, *see* Characteristic
Alphatron 12(517, 538–539)
Alter and Curtis' incremental wave experiment using dynamic prestress 6a/1(618)
Alternating load deformation of metals 7/2(131)
Alum, aluminium-ammonium 15(87)
Alum, aluminium-potassium 15(87)
Alum, ammonium-iron 15(95f., 123, 143f., 162)

Alum, chromium-potassium 15(87, 89f., 121, 137f., 142f., 158)
Alum structure 15(85)
Alum, titanium-cesium 15(100)
Alumina, sintered 14(225)
Aluminium alloys, glide-band formation 7/2(102, 181)
Aluminium alloys under the electron microscope 7/2(102)
Aluminium crystal, slip lines 7/2(82)
Aluminium phosphate glasses 13(535, 536)
Aluminium single crystals
— flow stress 7/2(117)
— work hardening 7/2(51, 54, 57)
Aluminium-copper alloy, microslip 7/2(102)
Aluminum diaphragm 11/2(18)
Aluminum, first measurement of the *E* modulus 6a/1(73)
Aluminum, plastic deformation 11/2(198–203)
Amagat units 12(28–29)
Amagat's law of additive volumes 12(58, 62)
Ambiguous boundary conditions 6a/2(405, 414)
Ammonia 14(16, 23, 24)
Ammonia, dissociation 11/1(254–257)
Ammonium dihydrogen phosphate (ADP) 11/2(47, 76)
Amorphous part in melting 13(296)
Amorphous regions in polymers 13(425, 450)
Amorphous structure 7/2(5)
Amplification of anemometer signals 8/2(585)
Amplification of disturbance 8/1(288ff., 322ff., 369–371, 373, 381, 396, 407, 434)
Amplification of ultrasonics in cadmium sulphide 11/2(250–251)
Amplitude distortion in direct loudspeakers 11/2(12)
Amplitude fluctuation 11/1(87, 89)
Amplitude of shock 6a/3(313)
Amplitude resolution of anemometer 8/2(584–585)
Amplitude (*see also* Strength) of a process or history 6a/3(7)
Amplitude vector 6a/3(309)
Amplitude, vibration 6a/2(261)
Amplitude, wave of order *n* 6a/2(254)
Analog of principle of minimum potential energy 6a/2(229)
Analytic continuation across a free streamline 9(348)
Analytic continuation of solutions 9(65, 73)
Analytic field 6a/2(11)
Analyticity, elastic displacement fields 6a/2(131)
Analyticity, thermoelastic displacement and temperature fields 6a/2(318)
Anderson bridge 15(74, 76)
Anelastic measurements 7/1(474, 657)
Anelastic media, response to periodic forces 6(268)
Anelastic medium, generalized 6(292)
Anelasticity 7/2(3)
Anelasticity (Poynting-Thomson body) 6(263–269, 471–472)
Anemometer, advantages 8/2(558–559)
Angle in a fluid lens 10(299)
Angle of attack 8/2(386, 408, 426, 428)
Angle of incidence 11/1(19)
Angle of refraction 11/1(19)
Angle of sweep 8/2(386)
Angle of twist per unit length 6a/3(496, 582, 583)
Angular momentum *see under* Momentum
— balance 6a/4(119)
— conservation law 8/1(136)
— functions, unitary transformation 12(473, 485)
— of the molecule 12(469, 489, 491)
— of the molecule, isotropic orientation 12(489, 491)
Angular velocity 8/1(23–25, 28)
Angular velocity, effect on hydrostatics 10(283)
Anharmonic effects in stress waves 11/2(160–162)
Anharmonic lattice forces 11/2(230, 231)
Anharmonicities 14(206)
Anharmonicities, cubic 14(206)
Anharmonicity 7/1(273, 278, 284, 287, 290, 297, 299, 315, 321)
Anharmonicity, constant of 7/1(274, 289)
Anionic polymers 13(492, 494)
Anisotropic 6a/3(18)

Anisotropic body, elastic constants 6(21)
Anisotropic material, thermoelastic effects 11/2(175)
Anisotropic Maxwellian expansion 12(268)
Anisotropic porous media, Darcy's law 8/2(636–638)
Anisotropic turbulence, *see* Non-isotropic turbulence
Anisotropies, chromium methylamine alum 15(152)
Anisotropies, chromium-potassium alum 15(159)
Anisotropy 6a/1(17, 230, 242, 248, 254, 257, 268, 355–357, 395, 396, 406–411, 503, 504, 676–677); 11/1(10–11); 15(9, 13, 30)
Anisotropy, effect on penetration depth 15(247)
Anisotropy, effect on transition temperature 15(229)
Anisotropy, elastic 7/1(128, 144, 253, 256, 292)
Anisotropy factor 7/1(364); 11/2(157)
Anisotropy in polymers 13(450)
Anisotropy in surface zone of polar liquids 10(80)
Anisotropy of attenuation in magnetized nickel 11/2(225)
Anisotropy of cleavage 7/1(100)
Anisotropy of the elastic properties 7/1(487)
Anisotropy of the electrical resistance 7/1(462)
Anisotropy of the penetration depth in tin 15(284)
Anisotropy, optical, of liquids 10(22)
Anistropic material 6a/2(7, 87–89, 530)
Annealed copper, behavior under irradiation 11/2(196)
Annealed materials, absence of Bordoni peaks 11/2(207)
Annealing, *see* Fully annealed solid
Annealing after plastic strain 7/2(194)
Annealing at elevaterd temperatures 7/2(129)
Annealing of photoelastic materials 6(163–164)
Annealing point of a glass 13(579, 580)
Annealing range 13(513)
Annealing range, viscosity of glasses 13(585–589)
Annihilation operator 14(204)
Annular throat structure of loudspeaker 11/2(19)
Annulus, plane 6(56–57)
Anomalies, cryomagnetic 15(2, 6)
Anomalous diffraction of light by ultrasound 11/2(132)
Anomalous mobility of protons 11/1(251)
Anorganic gases, dispersion and absorption 11/1(181–182)
Antenna potential 6a/2(420)
Anti plane strain 7/1(534)
Anti-cyclone 8/1(47)
Antiferromagnetism 15(28, 53)
Antifluorite lattice 7/1(88)
Antimony 14(262)
Antimony, modulus measured by vibration 6a/1(222)
Anti-phase boundaries 7/2(189)
Anti-phase domain 7/2(25)
Antiresonance frequency 6a/4(263)
Anti-resonant impedance 11/2(10)
Anti-thixotropy 6(540–541)
Aperture of a surface wave system 9(488–489)
Approach 11/1(212, 217, 222–224)
Approach to equilibrium 12(219, 225)
Approximate solution 6a/3(491)
Approximate solution in plane plasticity 6(362, 396–398)
Approximation, first 14(147)
Approximation theorem 6a/2(128)
Approximations for lift and moment 8/2(395)
Approximations in shell and plate theory 6a/2(476, 522, 566–568, 570, 587)
Aquo-complex 11/1(221, 251, 261)
AR, *see* Accumulation region
Aragonite 6a/1(408)
Archimedes' law 9(565)
Archimedes' principle for a fluid lens 10(298–299)
Area load 11/1(112)
Argon 14(228)
Argon, coordination numbers 13(233)
Argon, critical data 13(230, 232, 234)
Argon, critical opalescence 13(250)

Argon, deviations from the Clausius-Mossotti equation 13(252, 256)
Argon, liquid 10(65, 67, 117)
Argon, melting parameter 13(289, 300)
Argon, radial distribution function 13(275, 282)
Argon, surface tension 10(217)
Argon, vapour pressure 13(232, 234)
Argon, x-ray scattering 13(250)
Arrangement of dislocations 7/2(111)
Arrhenius' hypothesis of dissociation of electrolytes 13(370)
Artificial disturbances 8/1(386, 425)
As-lattice (structure) 7/1(83)
Aspect ratio 8/2(386)
Assembly of conduction electrons 14(193)
Associated and unassociated liquid, binary mixture 11/1(330–333)
Associated elastic body 6a/3(60)
Associated elastic material 6a/3(112)
Associated elastic problem 6a/3(60, 111)
Associated elastic solution 6a/3(60)
Associated liquids 10(4, 23, 25, 38, 80, 92)
Association, *see also* Aggregation 11/1(207, 221, 225)
Astatic equilibrium 6a/2(63, 64, 98, 193)
Asterism of Laue spots 7/2(4, 77–79, 195)
Asymmetric, *see* Non-symmetric turbulence
Asymmetric channel flow 9(413)
Asymptotic behavior
— of free streamlines 9(375–378)
— of incompressible potential flow 8/1(158)
— of subsonic potential flow 8/1(199)
— of the velocity potential 9(98–101)
— of viscous flow 8/1(261)
— theorem on 6a/3(32)
Asymptotic distribution of eigenvalues 6a/2(420)
Asymptotic expansion for late times 9(472, 510, 512)
Asymptotic expansion techniques for shells 6a/2(586, 587)
Asymptotic form of flow along a pervious wall 8/1(348)
Asymptotic form of the Green's function 11/1(32)
Asymptotic methods for rods 6a/2(664)
Asymptotic shape of axially symmetric cavity 9(377)
Asymptotic stability 6a/3(32)
— and boundedness 6a/3(169)
— and snap-through 6a/3(288)
— definition 6a/3(142)
— Hale's invariance principle 6a/3(169, 250, 254, 255)
— linear thermoelasticity 6a/3(249, 254, 255)
— necessary conditions for (footnote) 6a/3(200)
— of null solution with dissipative loads 6a/3(287)
— region of 6a/3(142, 158, 169)
— remarked upon 6a/3(129)
— sufficiency theorem 6a/3(152)
— theorem (corollary) 6a/3(158, 169)
— theorem of Hsu and Lee 6a/3(277)
— without eigenfunction decomposition 6a/3(275)
Asymptotic state after transition to turbulent flow 8/2(541)
Asymptotic suction profile 8/1(410, 411)
Asymptotic width zero of cavity 9(332, 376)
Athermal nucleation 13(595, 596)
Atmosphere formation after Cottrell 7/2(173, 183)
Atmospheric propagation of sound 11/1(81–82, 192–195)
Atmospheric region of free molecule flow 8/2(593)
Atom distribution and small-angle scattering 13(185)
Atomic, distances and valence angles, calculation of 7/1(71)
Atomic gases, energy levels 12(101)
Atomic heat, *see* Specific heat
Atomic heat, electronic 14(292)
Atomic radius 7/1(105)
Atomic weight, elasticity and 6a/1(218, 230, 394, 395)
Atoms, distance of 7/1(609)
Attenuated 6a/2(343)
Attenuation 6a/1(81, 176, 299, 387, 392, 393, 413–417); 6a/4(193)

Attenuation along a cylindrical tube 11/1(22)
Attenuation coefficient of viscoelastic waves 6(300)
Attenuation constant 11/1(14)
Attenuation factor 11/2(162)
Attenuation for polymers 11/1(399)
Attenuation matrix, reduced 6a/3(102, 103)
Attenuation of a plane wave 11/1(10, 14)
Attenuation of damped, propagating singular surfaces 6a/3(106, 107)
Attenuation of first sound 15(433, 434)
Attenuation of ultrasonic beam
— as a function of irradiation 11/2(197)
— by conduction electrons 11/2(208–223)
— by dislocations 11/2(179, 181)
— by phonon-phonon interaction 11/2(231–235)
— by scattering 11/2(164–166)
— definition 11/2(162)
— in a linear viscoelastic solid 11/2(173)
— in cadmium sulphide 11/2(249)
— in magnetic fields 11/2(225)
Attenuation of waves 6a/3(1)
Attenuation operator 6a/3(110)
Attenuation tensor 6a/3(101)
Auto-correlation function 11/1(41, 94)
Auxiliary functions *see* Liapounov functions 6a/3(153)
Auxiliary solution 6a/3(61)
Average displacement of a dislocation 11/2(179)
Average end-to-end distance 13(404, 408, 439)
Average forces, their potentials 13(9, 11, 16, 25)
Average molecular weight 13(401)
Average position of wave crest 9(508–509)
Average potential in liquids 13(222, 226)
Average potential model of liquid mixtures 13(360–365)
Average potential of forces 13(16)
Average speed 8/2(15)
Avogadro's hypothesis 12(1)
Axes, principal mean stress 6a/3(216, 240–243)
Axes, wave 6a/3(97, 98)
Axial collision, *see* Impact, Time of contact
Axial collision of a bar with a sphere 6a/1(316, 338)
Axial collision of rods 6a/1(313–352, 605–612, 623–649, 690–702)
Axial flow past bodies of revolution 8/1(311)
Axial flow turbines 14(80)
Axial force 6a/3(507)
Axial ratio 7/2(27)
Axial ratio, calculation of 7/1(64)
Axial ratio, graphic determination 7/1(66)
Axial symmetric steady flow along parallel straight lines 8/2(47–52)
Axial symmetry 6a/3(422, 525)
Axial test from tension through zero stress to compression 6a/1(138)
Axial vector 6a/2(6)
Axially dissimilar waves 6a/3(102)
Axially similar waves 6a/3(101)
Axially symmetric
— cavity, asymptotic shape 9(377)
— cavity flow, approximation 9(366–368)
— cavity flow, existence theory 9(394, 405)
— flow 8/1(155, 193)
— free surface flow 9(353)
— jet flow 9(434)
— potential flow 9(11, 13–14, 22–24)
— Riabouchinsky cavity 9(337)
— supersonic flow 9(274)
Axis of equilibrium 6a/3(216, 217, 219, 229, 239, 242)
Axis of principal stress 6(13, 137)
Axis of principal stress, secondary 6(138)
Axis of symmetry 6a/2(73, 248)
Axisymmetric flow, Oseen's problem 8/2(341, 345)
Axisymmetric steady flow, linear approximation 8/2(206)
Axisymmetric strain of grid turbulence 8/2(535)
Axisymmetric unsteady motion 8/2(149–159, 184)
Azeotropic point 13(323, 326)

Babinet compensator 6(136)
Babuška and Hlavaček's theorem 6a/3(61)
Bach-Schüle law 6(490); 6a/1(111–115, 118)
Backdiffusion of oil vapor 12(649)
Bäcklund transformation 7/1(568)
Backstreaming in pumps 12(630)
Backward 6a/3(486)
Bactericidal effects of cavitation 11/2(272)
Baffle 12(603–604)
Bainite 7/2(251)
Bakker equation 10(159)
Balance, angular momentum 6a/4(119)
Balance, cyclic 12(360)
Balance, detailed 12(360, 488, 490, 492)
Balance, energy 6a/4(120)
Balance, equations 6a/4(119)
Balance, mass 6a/4(119)
Balance of average turbulent kinetic energy 8/2(546)
Balance of energy 6a/3(174, 175); 6a/4(76, 90)
Balance of energy for steady motions 6a/4(82, 85)
Balance of linear and angular momentum 6a/3(42)
Balance of linear momentum in one-dimensional motion 6a/4(25ff., 90)
Balance of linear momentum in three dimensions 6a/4(70)
Balance of linear momentum steady one-dimensional 6a/4(39, 82)
Balance of mass in one-dimensional motion 6a/4(24)
Balance of mass in three dimensions 6a/4(70)
Balance principles for 3-dimensional theories
— forces and moments 6a/2(300)
— forces, for singular states 6a/2(181)
— linear and angular 6a/2(44, 52)
— moments 6a/2(306)
— momentum 6a/2(42, 51)
— momentum in space-time 6a/2(67)
— momentum of a wave 6a/2(254)
Balance principles for rods
— angular momentum 6a/2(666)
— director momentum 6a/2(666)
— energy 6a/2(667)
— linear momentum 6a/2(666)
Balance principles for shells 6a/2(483)
— (*see also* Conservation laws) for 3-dimensional theories, energy 6a/2(511)
Balance radiometer 11/2(122)
Balance, total energy 6a/4(161)
Ballantine's electrostatic actuator 11/2(64–66)
Ballistic senstitivity 15(74)
Balloon, scattering of sound 11/1(101)
Banach space 9(119–121, 400)
Band theory 14(290)
Band-band scattering 14(194)
Bands of secondary slip 7/2(74, 79–80)
Bar, *see also* Beam and Rod
Bar of circular cross section 6(405)
Bar of oval cross section 6(406)
Barium silicate glasses, structure 13(524)
Barium titanate 11/2(83–84)
Barium titanate cylinders in microphones 11/2(48)
Barkhausen effect 15(133)
Barnett effect 12(302, 489)
Barometric equation 12(363)
Barotropic flow 8/1(150)
Barrier in a long canal 9(527, 529)
Bars 6a/3(279)
Baryta 6a/1(407)
Base vectors, in general 6a/2(439, 618)
Base vectors on a surface 6a/2(621)
Basic equations
— linear thermoelasticity 6a/2(310, 311)
— of elasticity 6a/3(173–176)
— of perturbed motion *see* Perturbed motion
— of plasticity 6a/3(403)
— shell theory 6a/2(565)
Basic principles for shells 6a/2(479)
Basic singular solutions, elastodynamics 6a/2(239)
Basic singular solutions, elastostatics 6a/2(173)
Basic structure of unstrained crystal 7/2(111–112, 155, 220)
Basic work-hardening 7/2(141, 144, 179)
Basin in circular planform, shallow-water approximation 9(672)
Basin of rectangular form, shallow-water 9(671)

Basin with elastic restoring forces 9(628)
Basin with sloping side-walls, waves 9(624–627)
Basins, infinitesimal waves 9(620–631)
Basins of different shape 9(623)
Bass-reflex enclosure 11/2(14)
Bateman-Dirichlet principle 8/1(204)
Bateman-Kelvin principle 8/1(204)
Bateman's unsteady flow 8/2(145, 148)
— derived from Taylor flow 8/2(183)
Bausch arrangement 7/2(29, 30, 70)
Bauschinger effect 6(283); 6a/1(87, 460–472); 7/2(130–134, 148)
Bauschinger's experiments 6a/1(83–87, 90, 462–474)
Bauschinger's mirror extensometer 6a/1(83, 462)
Bayard and Alpert ionization gauge 12(612–615)
B-body, *see* Bingham body
B-coefficients of osmotic pressure, definition 13(451)
Beach, waves on it 9(537–542, 545, 547–551)
Beam, *see also* Bar and Rod
Beam 6a/3(278, 503)
Beam bending, elastic 6(38–39)
Beam, elastic-plastic behaviour 6(314)
Beam loading treated by Airy's method 6(44–45)
Beam of visco-elastic material 6(311, 312)
Beam, pre-stressed 6(223)
Beattie-Bridgeman equation 12(54, 55, 65)
Beatty's criterion 6a/3(201, 205)
Becker gas 11/1(151)
Becker-Döring theory of droplet formation 10(303)
Beer's law 13(574)
Bell's diffraction grating experiment 6a/1(621–640)
Bell's incremental wave experiment 6a/1(612–621, 659, 696–699)
Bell's optical displacement experiment 6a/1(628, 629)
Beltrami flow 8/1(43, 188)
Beltrami-Donati-Michell equations, generalized 6a/3(45, 46)
Beltrami's diffusion equation 8/1(151)
Beltrami's solution 6a/2(54)
Beltrami-Schaefer solution 6a/2(58)
Bending of a beam 6(38–39, 311–314)
Bending of glide lamellae 7/2(28)
Bending stiffness 11/1(395, 402, 406)
Bending strength of glasses 13(624, 629)
Bending theory
— of shells, classical 6a/2(582)
— of thin shells 6a/2(425)
— plates, classical theory 6a/2(577)
— plates, v. Kármán's nonlinear theory 6a/2(588)
Benedict-Webb-Rubin equation 12(54, 55, 56, 65)
Benzene-cyclohexane mixture, entropy 13(339, 355)
Benzene-methanol mixture, orientation effects 13(343–344)
Berg-Barrett method 7/2(80)
Bergman's series development of hodograph equation 9(67, 73–77)
Bernoulli equation 6a/3(339); 6a/4(60, 69)
Bernoullian theorems 8/1(153, 168, 260, 261)
Bernoulli's equation 8/1(82–83); 8/2(40, 54, 91, 388); 9(5, 6, 56, 93, 164–165, 179, 449, 454)
— for irrotational flow 8/1(154)
— for isentropic flow 8/1(184)
— for non-isentropic steady flow 8/1(186)
— for steady barotropic flow 8/1(153)
— in the gravity field 9(320)
— linearization 9(28)
— validity across shock fronts 8/1(221)
Bernoulli's law for non-steady flow 9(285)
Bernoulli's parabolic law 6a/1(13, 52–56, 72, 112, 115, 118–122, 126)
Berthelot's equation 12(54)
Berthelot's state equation 11/1(153–154)
Beryllium 14(250, 263)
Beryllium copper 6a/1(132)
Beryllium phosphate glasses 13(537)
Bessel transform 11/1(69)
B.E.T. (Brunauer-Emmet-Teller) areas 10(236)
BET equation of adsorption 8/2(630)
β-brass 7/1(450)

β-creep 7/2(199, 202)
β-curve, *see* Characteristic
Bethe method 15(133)
Bethe parameter 13(89, 90, 335, 337)
Bethe's approximation for superlattice transformations 13(88–94, 104, 285, 335)
Betti's integration method 6(95)
Betti's reciprocal theorem 6a/2(98, 99, 170, 211, 419)
Betti's reciprocity-theorem 7/1(517, 518, 520)
Betti's theorem 8/2(265)
Betti's theorem in viscoelasticity 6a/3(55)
Betz inversion formula 9(36)
Betz-Keune profiles 8/2(390)
Bi-characteristic, definition 9(25)
Bifurcation 6a/3(125)
Biharmonic equation 6(43, 50); 6a/3(75)
Biharmonic field 6a/2(20)
Biharmonic functions solving the Stokes equation 8/2(203–204)
Bilateral constraints 6a/2(403)
Bilinear integro-differential form 6a/2(361)
Bimorph micorphone 11/2(47–48)
Binary alkali silicate glasses 13(555, 559, 566)
Binary alkali silicate systems, limits of glass formation 13(612–614)
Binary alkaline earth borate glasses 13(530)
Binary alloys, small-angle scattering of x-rays 13(180–189)
Binary collisions 12(233–237)
Binary glasses 13(515)
Binary liquids, experimental data 13(326–328)
Binary liquids, results of theory of free volume and experimental data 13(355)
Binary mixture, *see* Solutions of polymers
Binary mixtures 12(60)
— critical region 12(65–70)
— diffusion and heat conduction 12(412–415, 423)
— of unassociated liquids, sound absorption 11/1(328–330)
— viscosity 12(447–451)
— with structural effects, sound absorption 11/1(330–339)
Binary polymer systems 13(480–481)
Binary system, *see also* Solid solutions
— experimental curves 13(167–175)
— Fuchs theory 13(143–153, 159–167)
— generalization of the Ornstein-Zernike integral equation 13(191)
Binary thermodiffusion factor 12(412, 425)
Binding 7/1(107, 154, 172, 201, 202, 218, 278, 284)
Binding agent 15(171)
Binding, energy, cohesive energy 7/1(105, 129, 143, 144, 319)
Binding, heteropolar 7/1(105)
Binding, homopolar 7/1(75, 105, 179)
Binding, van der Waals, bond 7/1(105)
Bingham body 6(455, 457–463)
Biot's entropy equation 6(123)
Bipotential equation 7/1(518)
Bipotential equation, principal solution of the 7/1(523)
Birefringence of deformed glass as a means to study delayed elasticity 13(639)
Birnbaum series 8/2(394, 421)
Bismuth 14(250, 262)
Bismuth, modulus measured by vibration 6a/1(222)
Bismuth single crystals, critical shear stress 7/2(35)
Bismuth, superconducting 15(239)
Bismuth wire technique 15(225, 226, 250)
Bitumen (sol or gel) 6(522–523)
Bjerknes' circulation theorem 8/1(190)
Bjerrum's theory of conductivity 7/1(401)
Blasius' theorems 8/2(389)
Bloch equation 14(233)
Bloch wall motion 11/2(223, 227)
Bloch-Grüneisen equation 14(177)
Bloch-Grüneisen formula for resistance 14(170)
Bloch-Nordsieck transformation 15(363, 365)
Blockage effect in water tunnels 9(344)
Blocking effect to calcium ions 13(622)
Blood viscosity 11/1(375)

Blowing of laminar boundary layer 8/1(409, 411)
Blunt body, transition from laminar to turbulent flow 8/1(358–360)
Bodies of revolution, axial flow past them 8/1(311)
Body $\mathscr{B}$, definition 6a/3(16)
Body, definitions 6a/2(12, 14, 299)
Body force(s) 6(10, 15, 72, 98); 6a/3(42, 174, 178, 263, 269–271, 273, 278, 283); 6a/4(157)
— analogy in thermoelasticity 6a/2(316)
— definition 6a/2(44, 299, 309, 485)
— generating elastic waves 6(116–118)
— moments for rods 6a/2(650)
— two-dimensional 6a/2(513)
Body of revolution, submerged 9(577)
Body with closure $\bar{\mathscr{B}}$ and boundary $\partial\mathscr{B}$ 6a/3(42)
Body-centred cubic lattice 7/2(25, 26)
Body-force equations, *see* Equations of motion
Boggio's theorem 6a/2(237)
Boggio's theorem, viscoelastic counterpart 6a/3(94)
Bogoliubov-Green technique 10(199)
Bogolubov's derivation of the Boltzmann equation 12(231)
Bohnenblust equation 6a/1(641)
Bohr magneton 15(4, 42)
Boiling point 13(231, 232); 14(20, 36)
Boiling process in liquids 10(18)
Bolts, photoelastic fringe patterns 6(211–215)
Boltzmann electron-transport equation 14(195)
Boltzmann equation 7/1(295, 296, 305, 307, 309, 313); 10(104–107); 14(184, 205, 230)
— approximate 12(385, 402–403)
— as integral equation 12(244, 245)
— asymptotic solutions 12(253, 262–266)
— derivation 12(216, 217, 229–233)
— for a gas of rotating molecules 12(481, 484)
— for monatomic gas mixtures 12(348)
— for monatomic gases 12(347)
— linear approximation 12(364)
— linearized 12(244, 281)
— linearized, eigentensor expansion 12(378–381)
— modifications 12(284–286)
— normal solutions 12(252, 257, 265)
— relativistic 12(345)
— solubility conditions 12(386, 404–405)
— solutions in the large 12(249–251)
— space dependent solutions 12(247–249)
— time-independent solutions 12(246)
— validity 12(205, 214)
Boltzmann gas 12(213, 214–216)
Boltzmann H-function, *see* H-function
Boltzmann impact experiment 6a/1(313–331)
Boltzmann laws 6a/3(12, 18)
— additional properties 6a/3(31)
— characterization of 6a/3(14, 15)
— creep 6a/3(43, 89)
— reduction of differential operator law 6a/3(36)
— relaxation 6a/3(43, 89, 91)
— with (strong) response function 6a/3(13)
Boltzmann limit 12(79–85)
Boltzmann operator, kernel 6a/3(10)
Boltzmann statistics 12(454, 457)
Boltzmann's equation 11/2(211, 212)
Bond, binding, van der Waals 7/1(105)
Bond, metallic 7/1(105)
Bond, valency 7/1(127)
Bonds between chain elements 13(402)
Bone, deformation of 6a/1(60, 61, 183)
Booster 12(515, 547, 559, 560)
Borate glass, specific heat 13(572)
Borate glass under high pressure 13(558)
Borate glasses 13(529–534)
— absorption spectra 13(532)
— thermal expansion 13(568–570)
Borate systems, limits of glass formation 13(615)
Borda mouthpiece 9(340, 342, 431)
Border 6a/2(402)
Bordoni peaks in stress wave damping 11/2(203–208)
Born and Cheng microscopic theory 15(343)
Born and Green's derivation of the Boltzmann equation 12(229)

Born approximation
— for inhomogeneous media 11/1(80)
— for scattering of slow neutrons 13(197)
— for scattering of sound 11/1(29–30, 32, 69, 89–93)
— rectangular duct 11/1(123)
Born-Green computation method of thermodynamic functions 13(11)
Born-Green equation 13(20–21)
Born-Green equation, as a specialization of Mayer's equations 13(39–40)
Born-Green-Yvon equation 10(180, 185, 197, 210, 215ff.)
Born-Haber cycle 7/1(142)
Born's approximation in scattering theory of x-rays or neutrons by liquids 10(43)
Born-von Kármán function 7/1(334)
Born-von Kármán one-dimensional crystal model 13(58)
Boron anomaly 13(590)
Borosilicate glasses 13(534)
Borosilicate glasses, hardness 13(634–635)
Borosilicate glasses, viscosity 13(584)
Bose-Einstein condensation 12(178, 179, 201); 15(386, 454)
Bose-Einstein perfect gases 12(174ff.)
Bose-Einstein statistics 10(122–123, 133); 12(81, 454, 458)
Bose-Stoner hypothesis 15(2, 8)
Bottom covered with dipole sources 9(571)
Bottom, irregular 9(569–571)
Boundaries, external 14(211, 220)
Boundaries of basins, special selection 9(622)
Boundary, ablating 6a/3(3)
Boundary condition at confluence of three fluid phases 10(286, 288, 294)
Boundary condition at intersection of two fluid phases with a solid phase 10(289, 295)
Boundary conditions
— asymptotic 12(265)
— at a beach 9(537, 548)
— at a free surface 9(320, 454)
— at a rigid surface 9(454–455)
— at an interface 9(451–455)
— at infinity 9(471–472)
— clamped 6a/2(263, 314)
— constraints 6a/2(403)
— derived from boundary layer theory 8/1(269)
— diffuse 12(242)
— dynamical 6(439, 448)
— elasticity 6a/2(102, 129, 314, 403, 405, 414, 420)
— Euler equations 12(258)
— for a Knudsen gas 12(241, 245)
— for capillary-gravity waves 9(631–632)
— for sound waves 11/1(15–22, 24)
— general viewpoints 8/1(61)
— in non-local theories 15(315)
— in viscous flow 8/1(240)
— Navier-Stokes equations 12(277)
— normal solutions 12(253, 258, 264)
— of Debye and Hückel 13(376, 377)
— on wing surface 9(288–291)
— or data, general 6a/3(46, 47)
— or data, homogeneous 6a/3(234)
— or data, mixed 6a/3(180, 184)
— or data, separable and synchronous body forces 6a/3(73)
— or data, stress 6a/3(175, 215, 258, 268)
— or data, temperature 6a/3(184, 188, 250)
— polynomial expansions 12(277)
— regularization 6a/2(401)
— rod theories 6a/2(648, 659, 667, 674, 678)
— shell theory, linear 6a/2(583)
— shell theory, restricted 6a/2(553)
— specular 12(242, 278)
— stochastic 12(242)
— thermoelasticity 6a/2(314, 321)
— thirteen moment equations 12(276)
Boundary effects 15(321)
Boundary energies, theory of 15(324)
Boundary in wave propagation 6a/3(3)
Boundary layer 8/2(1, 42–43, 122); 12(253, 263, 265, 266, 288)
— and shock waves, interaction 8/1(341–343)
— approximation 8/2(525, 544–548)
— around elliptical cylinder 8/1(400, 403)
— concept 8/1(62, 264)
— control 8/1(347–349)

— growth 8/1(288ff., 322ff., 369ff., 381, 396, 407, 434)
— in compressible flow 8/1(419–421, 422)
— in visco-plastic flow 6(305)
— limitation of this concept 8/1(328–334)
— Pohlhausen's method 8/2(107)
— separation 8/2(555)
— separation in three-dimensional flow 8/1(318–322)
— separation in two-dimensional flow 8/1(266, 271, 276, 277, 306)
— thickness 8/1(265, 268, 279, 281, 347); 11/1(12)
— transition laminar-turbulent 8/1(358–366)
— turbulent, *see* Turbulent boundary layer
— vorticity 8/1(269–270)
Boundary layer equations
— in three-dimensional flow derivation 8/1(63, 307–310)
— in three-dimensional flow, exact solutions 8/1(314–317)
— in two-dimensional compressible flow 8/1(335–336)
— in two-dimensional steady flow, derivation 8/1(266–268)
— in two-dimensional steady flow, solutions 8/1(271–277)
Boundary of body 6a/2(102, 440)
Boundary point lemma of elliptic equations 9(381)
Boundary points 6a/3(47)
Boundary resistance 14(222); 15(430, 431)
Boundary scattering 14(220)
Boundary stress 6(170, 182)
Boundary value problem 6a/3(3, 47, 69, 95, 495)
— first, of plane stress 6(62–64)
— homogeneous 6(80)
— of Frankl 9(137)
— of Tricomi 9(137)
— on characteristics 9(233, 255, 257, 268)
— second, of plane strain 6(64)
Boundary value problems 6a/1(5)
— displacement and temperature 6a/2(335)
— displacement, traction 6a/2(321)
— for the heat conduction equation 8/2(113–119)
— harmonic functions 6a/2(419)
— in elastic-plastic material 6(399–425)
— in visco-elasticity 6(294ff.)
— in visco-plasticity 6(305–308)
— of dynamic elasticity 6(188–122)
— of elasticity, ambiguous 6a/2(405, 414)
— of elasticity, displacement ("first") 6a/2(103, 220, 380)
— of elasticity, general 6a/2(129, 185)
Boundary-initial-value problems of elasticity 6a/2(335)
Boundary-value operator 6a/2(402)
— of elasticity, mixed ("third") 6a/2(102, 381, 420)
— of elasticity, traction ("second") 6a/2(103, 220, 381)
— rods 6a/2(658)
— shells, general theory 6a/2(550)
— shells, linear bending 6a/2(584)
— shells, restricted linear theory 6a/2(607, 609)
— thermoelasticity 6a/2(321)
Bounded basin, surface waves 9(620–631)
Bounded histories 6a/3(5)
Bounded regular region 6a/2(13)
Boundedness and asymptotic stability 6a/3(169)
Boundedness and Hale's invariance principle 6a/3(169)
Boundedness of map, definition 6a/3(148)
Boundedness of null solution, equiboundedness 6a/3(149)
Boundedness of solution 6a/3(148)
Boundedness, ultimate 6a/3(149)
Boundedness, uniform 6a/3(149, 231)
Bounds for strain energy 6a/2(113)
Boussinesq-Papkovich solution 6(89–91, 267)
— completeness 6(94)
— simple examples 6(91)
Boussinesq-Papkovich-Neuber solution, generalized 6a/3(75–78)
Boussinesq-Papkovich-Neuber stress processes 6a/3(78)

Boussinesq-Papkovitch-Neuber solution
— elimination of scalar potential 6a/2(143)
— general form 6a/2(139–142, 235)
— in elasticity, elimination of component of vector potential 6a/2(146)
Boussinesq's problem 8/2(121)
Boussinesq's second type solution 6(91)
Boussinesq's solution in elasticity 6a/2(147)
Boussinesq-Somigliana-Galerkin solution
— generalized 6a/3(75–77)
— in elasticity 6a/2(141, 142, 235, 236)
Box turbulence 8/2(533)
Boyle points 12(4, 20, 22, 23)
Boyle temperature 12(22)
Boyle-Charles law 12(1)
Boyle's law 14(40)
Bragg reflection 7/2(109)
Bragg's equation 7/1(69)
Bragg's soap bubble model 7/1(569, 629)
Bragg-Williams approximation 15(132)
— for binary solutions 13(346, 347)
— for rotation transformations 13(211–212)
— for superlattice transformations 13(85–88, 104, 285)
Branch line 6(389); 9(61, 87, 235, 239, 261, 267, 269)
Branch lines 6a/3(483)
Branch type singularities 6a/3(477)
Branched laminar model 15(337)
Branched macromolecule 13(399, 406)
Branches of the phonon spectrum 11/2(232)
Branching into two jets 9(342)
Branching model of the intermediate state 15(250)
Branch-type singularity 9(236)
Brass, *see also* α-brass
Brass 6a/1(56, 74, 103–105, 120, 152, 172–176, 183, 189, 233–236, 302, 304, 319, 398, 628, 648, 695)
Brass, isotropy of 6a/1(254, 411)
Brass, Poisson's ratio for 6a/1(261)
Brass single crystals (α-brass)
— critical shear stress 7/2(37–39)
— work-hardening 7/2(62)
Brass superlattice structure 13(81–82)
Bravais lattice 7/1(19, 385, 387, 389, 513, 661); 7/2(24)
Bravais lattices 7/1(16)
Break-away of dislocation 11/2(178, 179, 187, 189)
Breaking of oxygen bridges 13(587)
Breaking of wave crest 9(681)
Break-through of glide 7/2(156)
Breakwaters 9(544, 545)
Brewster effect of double refraction 6(128)
Brewster (unit) 6(149)
Bridge method 12(313)
Bridges, ballistic, photographic registration 15(73)
Bridging oxygen 13(519, 587, 591)
Bridgman's experiments
— on hydrostaticity of solids 6a/1(490–499)
— on the plastic deformation of steels under high pressure 6a/1(513–514)
— using high pressure apparatus 6a/1(490–494)
Bright-field illumination of glide bands 7/2(88, 97)
Brillouin conditions of cavity flow 9(322, 373)
Brillouin function 15(4, 56, 78)
Brillouin zone 7/1(74, 355); 14(203, 229)
Brillouin's planes 7/1(638)
Brinell hardness 7/1(102)
Brittle fracture 7/2(6, 7, 36, 67, 184)
Brittleness 6(574, 586, 589); 7/2(3)
Broadening of the energy levels 15(4)
Bronze 6a/1(96, 97, 275, 410, 411)
Brouwer's fixed point theorem 9(121)
Brownian motion 8/2(515, 516, 519, 520, 645, 646); 10(85, 112, 115); 13(439)
Brownian motion of polymer chains 11/1(384)
Brun's theorem 6a/2(217, 223)
Bubble and drop formation 10(154, 157)
Bubble formation 11/2(262)
Bubble in water, scattering of sound 11/1(101)
Bubble lifetime 11/2(273)
Bubble point 12(66, 67)
Bubbles 10(12)

Buckingham potential 13(275)
Buckingham-Corner potential 12(151)
Buckingham-Reiner equation 6(459)
Buckled states for rods 6a/2(694)
Buckling load 6a/3(225)
Buckling mode 6a/3(226, 287, 288)
Bulk crystallization of glass 13(593, 594)
Bulk liquid formation 15(446, 447, 448)
Bulk modulus 6(25, 143, 266); 6a/4(135, 230); 11/2(165); 13(433)
— effective 11/1(297)
— *K*, *see* Modulus, bulk
— measuring device 11/1(379)
— of glass 13(635)
— of inclusions in a cubic crystal 7/2(215)
— variability 6(489)
Bulk phase 10(282, 283)
Bulk phase, elimination 10(290)
Bulk scattering 11/1(29–32)
Bulk viscosity 8/1(49–50); 8/2(601); 11/1(3, 362, 379–380, 400)
Bulk viscosity in gases 11/1(131, 133, 134, 135, 150)
Bulk viscosity, kinetical explanation 11/1(157–158)
Buna rubber 11/1(410, 411)
Burgers body (=visco-elastic material), *see also* Visco-elasticity 6(456, 472–474)
Burgers circuit 7/1(647)
Burgers equation 8/2(295–298)
Burgers vector 7/1(387, 388, 477, 479, 480, 519, 521); 7/2(16, 24); 11/2(205, 207)
Burnett equations 8/2(595, 610–612, 614)
Burnett's method for compressibility measurements 12(33, 36, 371, 422)
Burnett's theory 11/1(157, 172)

Cabannes factor 13(262, 267)
Cadmium crystals, work-hardening 7/2(48–49)
Cadmium single crystals
— critical shear stress 7/2(116)
— surface effects 7/2(75)
Cadmium sulphide, acoustical properties 11/2(243, 249–251)
Caesium 14(239)
Caesium-chloride-structure 7/1(114, 116, 135, 137, 139, 217, 232, 233, 247)
Cailletet-Mathias law 12(43, 44)
Calcium phosphate glasses 13(534)
Calcium silicate glasses, structure 13(524)
Calculus of variations
— and stability 6a/3(158, 163–167, 195, 205)
— choice of positive-definite function (footnote) 6a/3(211)
— Legendre-Hadamard condition 6a/3(166)
— minimum problem 6a/3(164)
— proper minimum 6a/3(205)
— van Hove's theorem 6a/3(167)
Caldonazzo flow 8/2(101, 172)
Calibration curves for double quartz-thread pendulums 12(529)
Calibration curves for radiometer-vacuummeters 12(532)
Calibration curves for the alphatron 12(539)
Calibration of an ionization gauge 12(536, 602, 619–620)
Calibration of microphones 11/2(62–73)
Calibration of photoelectric materials 6(164–166, 180)
Caloric equation of state 13(11, 17, 27)
Caloric measurements 15(111, 113, 116, 141)
Calorimeter 14(294)
Calorimeter materials 14(334)
Camber, mean 8/2(387, 392, 429)
Campbell's impact experiment 6a/1(608–612)
Canal theory of tides 9(669)
Canal, wave resistance of a ship in it 9(581)
Canal waves of long wavelength 9(669)
Canal waves of long wavelength, bounded canal 9(670, 681)
Canonical distribution function 13(64)
Canonical ensemble 10(177, 178, 179, 237); 12(74, 79); 13(7–8)
Canonical ensemble, grand 10(191, 192, 194)
Canonical form of a hyperbolic equation 9(691)
Canonical transformation 15(352, 356)

Cantilevered rod 6a/3(282)
Capacitance displacement technique of Davies 6a/1(331–337, 346–349, 592)
Capacitance gauge 6a/4(9)
Capacitive leaks in bridges 15(76)
Capillaric models 8/2(641–643)
Capillarity 6a/1(248)
Capillarity, boundary conditions 10(286, 288, 289)
Capillarity, differential equation of Laplace 10(286, 288, 294)
Capillarity, meniscus formation 10(301)
Capillarity theory, historical remarks 10(281)
Capillarity, thermodynamical formulation of Gibbs 10(289–293)
Capillary 8/2(152–156); 15(169, 174, 175, 178, 179)
Capillary condensation 8/2(630, 631)
Capillary constant 10(300)
Capillary pressure 8/2(631–633)
Capillary waves, existence theorem 9(757)
Capillary waves of large amplitudes 9(746–749)
Capillary waves, orbits on second-order infinitesimal wave approach 9(661)
Capillary-gravity waves 9(511, 614, 631–638)
— generated by a point sources 9(636–637)
— initial-value problem 9(614, 631, 636)
— velocity 9(514, 614)
Capsule 15(168, 173, 174)
Capture of electrons 7/1(434)
Carbon content of iron 7/2(183)
Carbon dioxide 14(24)
Carbon dioxide, deviations from the Clausius-Mossotti equation 13(251–252)
Carbon disulphide, sound absorption 11/1(308, 309)
Carbon in α-Fe, diffusion of 7/1(425)
Carbon resistors 14(297)
Carman equation 8/2(644)
Carnot cycle 14(24); 15(54, 201)
Carriage containing water, periodic motion 9(628)
Cartesian coordinate system 6a/2(438)
Cascade compressed vapor system 14(30)
Cascade demagnetization 15(198, 203)
Cascade system of Keesom and Huguenin 14(33)
Cast steel, longitudinal and transverse vibrations in 6a/1(222, 251)
Catalogues for shell theory
— linear kinematic measures for Cosserat surface 6a/2(458)
— notations 6a/2(432–438)
— resultants and equations of motion 6a/2(526)
Catalyzed nucleation in glasses 13(593, 604)
Catenoidal horn 11/2(21)
Cationic polymers 13(494)
Cation-oxygen bond, glass forming ability 13(526)
Cauchy data 6a/3(417, 425, 450)
Cauchy elastic bodies 6a/3(173)
Cauchy equations of motion 6a/2(510)
Cauchy initial data 6(333, 336ff., 340, 360ff., 365, 396)
Cauchy initial value problem 6a/3(105); 9(232, 254–257, 268)
Cauchy integral 6(58, 59–62)
Cauchy measure of deformation 6(438, 486, 498, 544)
Cauchy principal value of a surface integral 6a/2(181)
Cauchy principal value of an integral 9(476)
Cauchy problem 6a/3(104, 421, 491)
Cauchy reciprocal theorem 6a/2(45)
Cauchy representation theorem for isotropic functions 6a/2(535, 551)
Cauchy singular integrals 6a/2(419)
Cauchy, solution of initial value problem 6a/3(105, 106)
Cauchy stress 6a/3(175, 180, 183, 185, 326, 536)
Cauchy stress tensor 6a/2(51, 53, 305)
Cauchy-Green tensor 6a/4(111)
Cauchy-Green tensor, representation in rod theories 6a/2(649)
Cauchy-Kovalevski-Somigliana solution in elasticity 6a/2(235–237)
Cauchy-Poisson initial value problem 9(603, 607–614)
— for superposed fluids 9(614)
— with surface tension 9(614, 631)

Cauchy-Poisson law of viscosity 8/1(232, 236–237)
Cauchy-Poisson theorem 6a/2(44); 6a/3(42)
Cauchy-relation 7/1(192, 195, 198, 201, 202, 206, 208, 216, 222, 229, 232, 233, 253, 276, 281, 284, 287)
Cauchy-Riemann equations 9(12, 470)
Cauchy's basic vorticity formulae 8/1(37)
Cauchy's integral 9(534)
Causality 6a/3(12)
Cavitating hydrofoil 9(586–587)
Cavitation 9(312); 11/2(260–263)
Cavitation, bubble formation 11/2(262, 264)
Cavitation experiments in water tunnel 9(312, 315, 344)
Cavitation, noises 11/2(264)
Cavitation number or cavitation parameter 9(312, 323)
Cavitation, shock wave 11/2(266)
Cavitation, signs 11/2(262)
Cavitation, theory 11/2(267–271)
Cavity 9(312)
Cavity dimensions 9(314)
Cavity, finite 9(324)
Cavity flow 9(312)
Cavity flow, plane, linearization 9(364–366)
Cavity growth in liquids 6(551, 555)
Cavity, infinite 9(322, 328–331)
Cavity of asymptotic width zero 9(332, 376)
Cavity of constant shape in accelerated flow 9(358–360)
Cavity pressure 9(312, 316)
Cavity resonance 11/2(100–101)
Cavity shape 9(323)
Cayley-Hamilton theorem 6(258, 261, 444)
$CdCl_2$ lattice 7/1(90)
CdJ_2 lattice 7/1(90)
Cell cluster, free volume 13(244)
Cell formation 7/2(153, 194, 206)
Cell model of Eyring and Hirschfelder 11/1(311)
Cell model of large molecules 13(477, 478)
Cell model of liquids 10(7–10, 47–51, 61–65, 242)
Cellular distribution function 13(225)
Cellular vortex 8/1(431, 433)
Cement stone, rheological behaviour 6(473)
Cementation factor 8/2(629)
Center of compression or rotation 6a/2(179)
Centered simple wave 6a/4(280)
Centered wave 6a/3(486)
Central axis 6a/2(417)
Central limit theorem 8/2(645)
Central line of a beam 6(38)
Central-forces 7/1(120, 127, 202, 216, 217, 228, 231, 232, 233, 240, 247, 276, 280, 281, 286, 287, 299)
Centre of mass, motion of the 7/1(166, 199, 304)
Centre of symmetry 7/1(13, 216, 287)
Centre of symmetry, influence on tensors 7/1(49)
Centred face 7/1(17)
Centred wave 6(391)
Centrifugal compressor 12(559, 563)
Centrifugal field 10(283)
Centrifugal force, effect on the transition to turbulence 8/1(415–417, 430)
Ceramic cylinder microphones 11/2(48)
Ceramics 13(606–607)
Ceramics as transducer materials 11/2(83)
Ceramics, fatigue effects 6(593)
Čerenkov radiation of electrons 7/1(623)
Cerium magnesium nitrate 15(117, 173)
Cerruti's solution for the half-space 6(97–98)
Cesium-chloride structure 7/2(25, 26)
Chain element 13(400, 402)
Chain ends, average separation 11/1(381)
Chain molecule 13(399)
Chain molecules 6(18)
Chain of spin particles 13(109)
Chain, partition function 13(443–444)
Chain segment, equivalent sphere 11/1(396)
Chain segment, length 11/1(395)
Chains of PO_4 tetrahedra 13(538–539)
Chains of SiO_4 tetrahedra 13(521, 613–614)
Change of orientation 7/2(28)

Change of temperature, *see* Temperature
Change of volume 7/1(230)
Channels of different cross sections 8/2(549–550)
Chaplygin-Molenbroek transformation 9(58–61)
Chaplygin's correspondence principle 9(346)
Chaplygin's equation 9(60)
Chaplygin's equation, solution by separation 9(62–65)
Chaplygin's equation, solution depending on an arbitrary analytic function 9(65–67)
Chapman-Enskog expansion 12(253, 259–262, 384)
Chapman-Enskog expansion, compatibility condition 12(260)
Chapman's Ω-integrals 12(393, 422)
Character of the physical properties 7/1(56)
Characteristic, *see also* Slip line 6(355, 357ff., 399)
Characteristic boundary value problem 6a/3(530); 9(233, 255, 257, 268)
Characteristic condition 6(341); 6a/3(424, 426)
Characteristic coordinates 6(377); 6a/3(447, 468, 469, 490)
Characteristic curves 6a/4(278)
Characteristic determinant 6(372)
Characteristic diagram of Prandtl and Busemann 9(90)
Characteristic direction 6a/3(451)
Characteristic displacement for vibration problem 6a/2(262)
Characteristic functional of Hopf 8/2(444)
Characteristic impedance 6a/4(293)
— complex 6a/4(195)
— for a single relaxation frequency 11/1(373)
— in Maxwell's model 11/1(366, 394)
Characteristic initial-value problem 6a/3(450, 491)
Characteristic invariants 9(234)
Characteristic line 6(334); 6a/3(418, 419, 443, 485)
Characteristic line diagram after Prandtl and Busemann 12(569–571)
Characteristic motions 7/1(587)
Characteristic plane 6(337)
Characteristic quadrangle 6a/3(491)
Characteristic segment 6a/3(430)
Characteristic solution for free vibration problem 6a/2(262)
Characteristic space 6a/2(8)
Characteristic strip 9(15)
Characteristic surface 6(333, 340–341); 6a/3(416, 424)
Characteristic surface elements 6a/3(423)
Characteristic temperature 7/1(247); 14(168)
Characteristic times of an anemometer 8/2(556, 567–568, 572, 578)
Characteristic values, elasticity tensor 6a/2(76)
Characteristic values, vibration problem 6a/2(263)
Characteristic variable 9(17, 246)
Characteristic velocities 8/2(546)
Characteristic vibration 7/1(106, 160, 166, 290, 587)
Characteristics 8/1(68–69, 74, 106, 207); 9(16, 22, 229, 254)
Characteristics and characteristic conditions
— for general flows 8/1(214)
— for steady axi-symmetric flow 8/1(210)
— for steady plane flow 8/1(207–210)
— for steady potential flow 8/1(211)
— in the hodograph plane 8/1(209)
— relation to singular surfaces 8/1(214)
Characteristics, fixed 6a/3(471)
Characteristics for the plastic tube 6(414)
Characteristics in plane torsion 6(402)
Characteristics in shallow-water theory 9(676, 677ff.)
Characteristics of refrigerants 14(24)
Characteristics of the complete plane problem 6a/3(464)
Characteristics of the v. Mises plasticity equations 6a/3(422)
Characteristics, real 6(339)
Characteristics, their role in linearized theory 9(226, 252, 279, 281)
Charcoal catalyst 14(67)
Charge of ionic clouds in electrolytes 13(379)

Charged dislocation line 7/2(125)
Chemical bond 12(338)
Chemical bond in crystals 7/1(72)
Chemical composition of a liquid, effect on sound velocity 11/1(340–346)
Chemical effects of cavitation 11/2(272, 277, 282–283)
Chemical equilibrium in real gases 12(64, 65)
Chemical interaction between foreign atoms and dislocations 7/2(169, 171, 176, 179)
Chemical linkage 10(23)
Chemical potential(s) 10(82); 12(73)
— and molecular distribution function 13(17)
— complete 10(169)
— definition 13(16)
— difference 7/2(217)
— lateral 10(170, 171, 272)
— of a gas mixture 12(61)
— of a liquid 13(274)
— of vapour 13(274)
Chemical reactions in liquids 10(40, 82, 92)
Chemical resistance of ceramics 13(606)
Chemical resistance of technical glasses 13(543, 544)
Chézy formula of hydraulics 9(697)
Chilling 7/1(455, 464)
Chladni process, *see* Vibration
Chloroform-acetone mixture, entropy of dilution 13(334–335)
Chord of a supersonic profil 9(48)
Christoffel symbol(s) 8/1(142)
— of the first kind 6(7, 85)
— of the second kind 6(4, 11, 85)
— of the first and second kinds 6a/2(619)
— physical 6(442)
Chromatography of alkali phosphate glasses 13(538)
Chrome alumina catalyst 14(67)
Chromic alums 15(21, 23, 85)
Chromium alums, diluted 15(93)
Chromium methylamine alum 15(86f., 121, 133f., 151)
Chromium nitrate 15(94f.)
Chromium potassium alum, diluted 15(142f.)
Cinematography of cavitation bubbles 11/2(263–264)
Circuit analog for polymer chains in a solvent 11/1(390–391)
Circuit analog of Maxwell's model 11/1(364, 367)
Circuit analog of Voigt's model 11/1(369)
Circuit, equivalent, *see* Equivalent circuit
Circular arc profile 8/2(390)
Circular cross section of a bar 6(405)
Circular cylinder
— horizontal reflection of surface waves 9(528)
— in molecular flow 8/2(606)
— moving and fixed planes 8/2(80–82)
— submerged 9(574–577)
— torsion 6(27–28, 36–38)
— translation in a canal 8/2(253)
— translation perpendicular to its axis 8/2(257)
— uniform translation, Faxén's solution 8/2(310–311)
— uniform translation, Lamb's solution 8/2(305–310)
— vertical, diffraction of surface waves 9(544)
Circular disk, resistance 8/2(383)
Circular disk, scattering of a sound wave 11/1(68–73)
Circular disk, uniform translation in Oseen approximation 8/2(314)
Circular hole 6a/3(519)
— in elastic-plastic plane stress 6(422)
— in plastic material 6(365)
— photoelastic fringe patterns 6(205–206)
Circular orbits in ocean waves 9(499, 501)
Circular orbits, modifications by second-order infinitesimal wave theory 9(660, 665)
Circular orbits of first-order infinitesimal wave theory 9(499, 501)
Circular piston, directivity 11/2(104–105)
Circular piston, its radiation 11/1(38)
Circular piston, radiation impedance 11/2(2)
Circular polarisation 6(134, 151–153)
Circular polariscope 6(153)

Circular ring 6a/3(499)
Circular ring, directivity 11/2(104)
Circular rods 6a/2(690)
Circular tube 6a/3(192)
Circular wing 9(297)
Circular-cylinder basin
— shallow-water approximation 9(672)
— vertical motion 9(651, 653)
— waves 9(623)
Circularly polarized sound waves 11/2(229)
Circulation 8/1(34–35, 85); 8/2(132–134); 9(4, 71, 101)
Circulation and lift 8/1(85–87, 161)
Circulation around an airfoil 8/1(105–106)
Circulation distribution over span 8/2(407)
Circulation process 15(451)
Circulation theorem of Helmholtz and Kelvin 9(4)
Circulation theorem of Thomson 8/1(35)
Cisotti's differential-difference equation 9(611)
Clam shell markings of fatigue crack 6(591)
Clamped boundary 6a/2(314)
Clamped plate condenser microphone 11/2(45)
Clamped plate, partially supported on a subdomain 6a/2(406)
Clapeyron's equation 12(37)
Class $C^{M,N}$ 6a/2(24)
Class C^N 6a/2(11)
Classes C^N and C^∞ 6a/3(5)
Classes H^{ac} 6a/3(6)
Classes, Heaviside 6a/3(6)
Classes H^N and H^∞ 6a/3(6)
Classical boundary condition 11/1(15, 16)
Classical elastic law for isotropic materials 6a/3(29)
Classical solutions 6a/3(138, 171, 173, 178–180, 211, 224, 231, 250, 258)
Classical theories of plates and shells 6a/2(575–577, 582, 585)
Classical thermostatic relations 6a/4(124)
Classification of discontinuities 6a/3(434)
Classification of known solutions 8/2(12)
Claude air liquefier 14(77)
Claude cycle 12(26, 27); 14(71)
Claude-Heylandt air liquefier 14(75)
Claude-Heylandt system 14(58)
Clausius virial 12(160)
Clausius-Clapeyron equation, generalization 13(296)
Clausius-Clapeyron's relation 7/1(319)
Clausius-Duhem inequality 6a/2(301, 484, 512, 532); 6a/3(537, 539, 540)
Clausius-Mossotti equation 13(251, 255)
Clay, rheological behaviour 6(461–462)
Cleaning of a glass surface by ultrasonics 11/2(277)
Cleavage 7/1(97)
Cleavage of diamond 7/1(102)
Cleavage surface 6(574)
Clebsch's transformation 8/1(170–171)
Climbing of dislocations 7/2(195, 205)
Closed convex set 6a/2(391)
Closed regular surface 6a/2(13)
Closed strain path 6a/2(82)
Closed-end condition 6a/3(506, 508)
Closed-end conditions of a tube 6(409, 410, 414, 418)
Cluster expansion 13(33, 34, 145)
Cluster expansion for electrolytes 13(392, 394)
Cluster expansion, limitation below critical point 13(217)
Cluster functions 12(125)
Cluster functions for two molecules 12(154)
Cluster integral(s) 12(126, 148)
— for an electrolyte 13(372)
— generalized 10(202)
— of the second kind 13(35)
— superficial 10(202, 204ff.)
Cluster model of liquids 10(10–11)
Cluster sums 13(146, 160)
Cluster theory of non-electrolyte solutions 13(302)
Cluster theory of the free volume 13(241, 244)
Clusters in liquids 10(6, 30, 50)
Clusters of molecules 12(125, 131–134)
Cnoidal wave 9(462, 654, 667, 702, 707)
— existence theorems 9(751–753)
— limiting height 9(712)

— second approximation to its profile 9(709)
CO_2 lattice 7/1(91)
Coacervation 13(502, 504)
Coagulation by ultrasonics 11/2(278)
Coarse graining, *see* Smoothing
Coaxial cylinders in unsteady rotation 8/2(222)
Cobalt 7/1(660, 662)
Cobalt 60 gamma rays 11/2(191)
Cobalt ammonium tuttonsalt 15(10)
Cobalt lattice, transformation at high temperature 7/2(239–240)
Cobalt precipitation 7/2(102)
Coefficient of expansion 6a/1(372)
Coefficient of friction 6a/3(452)
Coefficient of linear expansion 6a/4(145)
Coefficient of restitution 6a/1(313, 320, 642–645)
Coefficient of thermal expansion 6a/2(311, 316)
Coefficient of viscosity 8/1(49, 50)
Coefficient of work-hardening, *see* Work-hardening
Coefficients b_{ijkl} 6a/3(182, 183)
Coërcitive field 15(127)
Coercive force 7/2(230)
Coercive inequality 6a/3(225)
Coerciveness 6a/2(422)
Coexistence of two phases 10(252)
Coexisting phases in binary polymer systems 13(480)
Coexisting phases in ternary polymer systems 13(482)
Cofactors 6a/2(618)
Coherence of a network 13(520, 531, 569)
Coherent nuclei 7/2(212)
Cohesion 6(372); 6a/3(463)
Cohesion of solids under pressure 6a/1(474–477)
Cohesive energy, binding energy 7/1(105, 129, 143, 144, 319)
Cohesive properties of lattice 7/1(97ff.)
Coil model of polyelectrolytes 13(494–495)
Cold-work 7/1(460); 7/2(4, 101)
Cole diagram 11/1(146–147, 178)
Coleman's canonical free energy 6a/3(190)
Collapse load 6(315–318)
Collective description of electron-ion interaction 15(355)
Collective electron treatment 14(293)
Collimation of light 6(156)
Collision cross section, differential 12(336–338)
Collision factor 11/1(341, 342)
Collision integrals 12(270, 271, 275)
Collision invariant 12(351, 354, 370)
Collision, inverse 12(346)
Collision matrix 7/1(309, 313)
Collision operator, eigenfunctions 12(367, 369)
Collision operator, linearized 12(366)
Collision parameter, classical mechanics 12(342)
Collision processes in gas mixtures 11/1(185–187)
Collision time, mean, in gases 11/1(133, 139)
Collisions, inelastic 7/1(432)
Colloidal systems 13(492)
Colloids 15(242, 260)
Colonnetti's theorem 7/1(527, 545)
Colour centres 11/2(194)
Colour of a glass 13(540)
Coloured image of an ultrasonic field 11/2(141)
Columbium 14(275)
Combinatorial factor 13(286, 321, 412–418, 456)
Combinatorial factor for orientation effects 13(341)
Communal entropy 13(220, 223, 233)
Commutation theorem 6a/2(9)
Compaction factor 8/2(629)
Comparison theorem of elasticity 6a/2(269)
Comparison theorems 9(380–383, 407)
Compatibility 6a/2(39); 7/2(216)
Compatibility condition(s) 6a/3(426, 432, 467, 471); 7/1(518, 520, 524f., 572)
— Chapman-Enskog expansion 12(260)
— extended 6a/3(430)
— geometrical 6a/3(307, 429, 430)
— geometrical, iterated 6a/3(307)
— Hilbert expansion 12(255)
— kinematical 6a/3(308, 429, 430)
— kinematical, iterated 6a/3(308)

Compatibility condition(s) (cont.)
— kinematical, one-dimensional 6a/3(309)
Compatibility equation(s) 6(7–8, 238, 239, 298, 341, 345ff., 356ff., 366, 373ff., 398); 6a/3(45, 456); 8/2(3, 4, 47, 66, 201)
— 3-dimensional bodies 6a/2(39–41, 62, 63, 211, 318)
— for axial symmetric motion 8/2(11)
— for plane elastic states 6a/2(155)
— for plane motion 8/2(9)
— shells 6a/2(463, 479)
— shells, restricted theory 6a/2(466)
— for traceless symmetric tensor process 6a/3(71)
Compatibility relations 6a/3(411, 466, 494)
Compatibility theorem 6a/3(45)
Compatibility theorem for plane displacements 6a/2(41)
Compatible values 6a/3(426)
Compensation of anemometer signals 8/2(578–580, 585)
Competition principle of Dietzel 13(541)
Complementary energy 6(253); 6a/3(583, 584, 586, 587)
Complementary energy function 6a/2(570)
Complementary regular subsurfaces 6a/2(14)
Complete problem for plane 6a/3(438)
Complete problem for plane strain 6a/3(511)
Complete randomness in glass structure 13(524, 528)
Complete solution(s)
— elasticity 6a/2(138)
— elastodynamics 6a/2(232)
— of plastic wedge problem 6a/3(527)
— thermoelasticity 6a/2(329)
Complete stabilization in supersonic flow 8/1(425)
Complete wave 6(393); 6a/3(488)
Completely plastic torsion 6a/3(560, 564)
Completeness, minimum principle for the vibration problem 6a/2(267)
Completeness, normal modes 6a/2(271)
Completeness of dynamic displacement generating functions 6a/3(92)
Completeness of stress function solutions 6a/3(76, 93)
Completeness, solutions
— of Beltrami 6a/2(56)
— of Beltrami-Schaefer 6a/2(58)
— of Boussinesq-Papkovitch-Neuber and Boussinesq-Somigliana-Galerkin 6a/2(141)
— of Cauchy-Kovalevski-Somigliana 6a/2(235)
— of Green-Lamé 6a/2(233)
— of Love-Boussinesq 6a/2(147, 149)
Complex acoustic tensor 6a/3(107)
Complex acoustical tensor 6a/4(213, 256)
Complex characteristic impedance 6a/4(195)
Complex elastic modulus 6a/4(195)
Complex formation in the lattice 7/2(229)
Complex modulus 6a/3(106)
Complex of approach 11/1(212, 217, 222)
Complex phase factor 6a/4(195)
Complex phase velocity 6a/4(193)
Complex potentials for plane stress 6(48–58)
Complex rigid displacement 6a/2(39)
Complex variables for infinitesimal surface waves 9(470–471)
Complex variables in dynamic elasticity 6(118–121)
Complex velocity 9(35, 471)
Complex velocity potential 8/2(388); 9(470, 480, 489, 505)
— for finite depth 9(482, 490, 494–495)
Complex wave number 6a/3(107)
Complexes (= nuclei) 7/2(235–237)
Compliance tensor 6a/2(69, 71, 77, 310)
Compliances 6a/4(126)
Compliances, generalized 6(248, 258)
Components, linearly independent 7/1(54)
Components of stress 8/2(3, 6)
— principal 6(13, 137)
Composite wave, second-order infinitesimal wave theory 9(662)
Composition of technical glasses 13(543–544)
Composition plane 7/1(95)

Compounds 14(333)
Compounds, superconductive 15(268)
Compressed gas gun 6a/4(4, 6, 55)
Compressibility 6a/1(83–86, 245, 274–278, 303, 486–501, 504, 634–637, 707, 734); 6a/4(135, 230); 7/1(131, 132, 141, 143, 144, 183, 218, 221, 224, 232, 247, 277, 278, 312); 11/1(9, 13); 13(29, 262, 298, 359); 15(239)
Compressibility, absolute 6a/1(276, 493)
Compressibility, additional, due to equilibrium shift 11/1(205)
Compressibility, cubic 6a/1(275, 303)
Compressibility, dependence on pressure 7/1(278)
Compressibility factor 12(4, 5, 50, 51)
Compressibility, hydrostatic, atomic dependence of 6a/1(394, 491)
Compressibility integral 13(31, 67)
Compressibility integral, determination of virial coefficients 13(19)
Compressibility, isothermic 11/1(143)
Compressibility law 9(173)
Compressibility modulus 7/1(529, 530)
Compressibility of gases
— measurements at high pressures 12(30–34)
— measurements at low pressures 12(34–37)
Compressibility of glasses 13(558–560)
Compressibility of liquids 10(19, 20, 66, 69, 119, 126)
Compressibility of solids as a function of temperature 6a/1(373, 374, 409)
Compressibility, relaxation part 11/1(206, 209–212, 217)
Compressibility, total 11/1(206)
Compressible atoms 7/2(9)
Compressible flow 6(369)
— effect of wall roughness 8/1(446–447)
— laminar boundary layer 8/1(334–343)
— method of small disturbances 8/1(422)
— observation of density fluctuations 8/1(364)
— turbulent boundary layer 8/1(419–421, 422)
Compressible fluid flow 6a/3(417, 459)
Compressible fluid jet 9(311)
Compressible porous media 8/2(636)
Compression, axial test in tension through zero stress to compression 6a/1(138)
Compression buckling 6a/1(162, 201)
Compression, centre of 7/1(528, 530, 556)
Compression, first experiment in 6a/1(83, 201, 457)
Compression gauge 12(517, 518)
Compression gauge after McLeod 12(520–522)
Compression gauge, turnable 12(522–523)
Compression, heat of 14(57)
Compression modulus 6a/3(30)
Compression, multi-stage 14(28)
Compression of lead cylinders 6a/1(421)
Compression ratio 6a/4(71–73)
Compression shock 6a/3(312)
Compression shocks 6a/4(295)
Compression test 7/2(30, 197)
Compression, two-dimensional 6a/1(514, 706)
Compression, work of 14(49, 74)
Compressional relaxation 11/1(205, 288–291, 297)
Compressional viscosity 11/1(3, 131–135, 150, 362, 379–380, 400)
Compressional waves 11/2(74, 155, 156)
Compressional waves, pure 11/2(157)
Compressional waves, scattering coefficients 11/2(169–170)
Compressive acceleration wave 6a/3(311, 314)
Compressive clearing 7/1(99)
Compressive twin-formation 7/1(96)
Computation of axially symmetric flows 9(421–438)
Computation of plane flow past obstacles 9(416–420)
Concave corners, local plastic behaviour 6(400)
Concave reflectors for ultrasound waves 11/2(112–113)
Concave wall, boundary layer stability 8/1(433–436)
Concentrated load 6a/2(174, 179); 6a/3(77, 78)
Concentrated stresses 6a/2(416)

Concentration dependence
— of G-values 11/1(219)
— of measured absorption in electrolytes 11/1(237–239)
— of relaxation absorption in ammonia 11/1(255)
— of relaxation absorption in $MgSO_4$ solutions 11/1(264–265)
— of relaxation absorption in sulphuric acid 11/1(253)
Concentric circles, steady flow along them 8/2(16–19, 33, 79)
Concentric circles, unsteady flow along them 8/2(131–138)
Concentric spinning spheres, flow between them 8/2(216)
Concrete 6a/1(143)
Concrete, rheological behaviour 6(435, 466, 536)
Condensation 7/2(218); 10(6, 33); 12(37–42, 126, 133, 141–146)
Condensation, retrograde 12(66, 67, 68)
Condenser microphone 11/2(38–46)
Condition for pure longitudinal waves 11/2(157)
Condition for pure transverse waves 11/2(158)
Condition of Kutta and Joukowski 8/2(391, 420)
Conditional stability, examples 6a/3(147, 217, 219, 220)
Conditional stability in general 6a/3(147, 284)
Conditions of validity for Rayleigh approximation 11/2(165)
Conduction coefficient of heat transfer through glasses 13(574)
Conduction, "dipole" theory of 14(140)
Conduction electrons, interaction of, with transverse and longitudinal waves 14(241)
Conduction electrons, interaction with stress waves 11/2(208–223)
Conduction losses near a boundary 11/1(21)
Conductivity, electric 10(81)
Conductivity, electrical 14(137, 230)
Conductivity, electrical, quasi-equilibrium effect 14(256)
Conductivity electrons 7/1(427, 473, 639)
Conductivity, isotropic material 6a/2(311)
Conductivity, perfect 15(214)
Conductivity, simultaneous measurement of thermal and electrical 14(201)
Conductivity tensor 6a/2(307, 309); 6a/3(250, 251, 254, 255)
Cone hypothesis 6a/2(354)
Cone, impact on a water surface 9(361–362)
Cone of axial symmetry, flow inside it 8/2(57, 224–227)
Cone of normals 6a/3(424)
Configuration integral 10(179, 182, 188, 241ff., 249ff., 253ff., 263ff.); 13(7, 245)
— for ionic gas model of an electrolyte 13(371, 373)
— for the free-volume theory of solutions 13(344)
— in the theory of the free volume 13(219–220, 222–223)
Configurational free energy of polyelectrolytes 13(494)
Configurational relaxation in polymers 11/1(385)
Configurations of macromolecules 13(402–409)
Confluence of two jets 9(342)
Confluent hypergeometric functions 9(613)
Confluent zone of three fluid phases 10(282, 283, 285)
Conformal mapping 6a/3(521); 8/2(76)
Conformal mapping for the construction of airfoil profiles 8/2(389–390)
Conformal solutions 13(356–360, 368)
Conical flow 8/1(75)
Conical flow, shock waves 9(188, 192)
Conical horn 11/2(21)
Conical supersonic flow 9(22–24, 49–52)
Conjugate slip system 7/2(62)
Conjugate tensor 6a/2(618)
Connectivity of a network 13(437)
Conservation equations (mass, momentum, energy) 12(238–239)
Conservation law(s) 7/1(157, 304); 11/1(3)
— for gas mixtures 12(353)
— for pure gases 12(350)
— in subsonic flow 9(121–122)

— in terms of a reference state 6a/2(490)
— of energies 6a/3(179, 207, 211, 236)
— of mass 6a/2(440, 483); 6a/3(174)
— (*see also* Balance principles) for a Cosserat surface 6a/2(479, 482, 487)
Conservation of energy 6(243, 293); 10(35, 120, 128)
Conservation of mass 6(234)
Conservation of momentum 6(235); 10(33, 120, 128)
Conservative motions 7/1(480, 657)
Conservative system of forces 10(26, 107)
Consistency condition 6a/2(129)
Consistency curve 6(461, 494, 497)
Consistency factor in the free-volume theory 13(220)
Consistency of a material 6(461)
Consistency variables 6(461)
Consistent initial past history 6a/3(51, 89)
Consolidation 6(263)
Constant process 6a/3(5)
Constantan 14(296, 298)
Constants of classical elasticity 6(20–26)
Constitutive coefficients in linear theory
— of Cosserat plates 6a/2(600, 603)
— of Cosserat surfaces 6a/2(598, 606)
— of shells, restricted 6a/2(607)
Constitutive equations for elastic materials 6a/2(67)
Constitutive equations for rods 6a/2(654, 657, 659, 669, 673)
Constitutive equations for shells and plates
— Cosserat surface, elastic 6a/2(528)
— Cosserat surface, in-extensible 6a/2(549)
— elastic shells 6a/2(528, 533, 537)
— elastic shells, from 3-dimensional theory 6a/2(561, 564)
— in terms of material coordinates 6a/2(529)
— in terms of reference state 6a/2(544)
— in terms of relative measures 6a/2(540)
— linear, for Cosserat surface 6a/2(553)
— linear, for shells of uniform thickness 6a/2(556)
— linear, in terms of Gibbs function 6a/2(571)
— mechanical theory 6a/2(544)
— restricted linear theory 6a/2(560)
Constitutive equations
— of a viscous fluid 8/1(230–243)
— of inelastic continuum 6(236, 294)
— or relations for elasticity 6a/3(19, 175)
— or relations for thermoelasticity 6a/3(175)
— or relations for viscoelasticity 6a/3(16, 18)
Constitutive relations, *see* Constitutive equations
Constrained materials 6a/3(2, 71)
Constraint response 6a/2(548)
Constraints, unilateral 6a/2(403)
Constriction energy 7/1(611, 612)
Constriction model of porous media 8/2(643)
Constriction of a extended dislocation 7/1(610ff.)
Contact angle in a capillary 8/2(632)
Contact couple 6a/2(504)
Contact director couple 6a/2(480, 491)
Contact discontinuities 6a/3(434)
Contact discontinuity, *see* Material discontinuity
Contact force 6a/2(480, 491, 510)
Contact gettering 12(637)
Contact point 13(521)
Contact points of macromolecules 13(411)
Contact problem 6a/2(129)
Contact problem, viscoelastic solution 6(296)
Contact problems 6a/3(3)
Contact resistances 14(154)
Contained plastic domain 6a/3(494, 519, 523, 524)
Continuity, *see* Equation of continuity
Continuity, definition 6a/3(12)
Continuity equation 6(234, 323); 6a/3(404); 8/2(444, 447, 545); 9(2, 228, 249, 283, 448, 450); 11/1(137, 276); 12(298, 351, 354)
— Eulerian form 8/1(17, 23, 33)
— integral form 8/1(4–5)
— invariant forms 8/1(141)

Continuity equation (cont.)
— Lagrangian form 8/1(31, 133)
— material form 8/1(31, 133)
— of plane steady flow 8/1(66, 78)
Continuity, equation of 6a/2(448)
Continuity method in uniqueness theory 9(409, 413)
Continuity of map 6a/3(133, 140)
Continuous at infinity 6a/3(5)
Continuous dependence of the solution on its data
— definition 6a/3(145)
— linear thermoelasticity 6a/3(523)
— relation with Liapounov stability 6a/3(134, 145, 146)
— with time-dependent, position-dependent data 6a/3(273–275)
Continuous dependence of the solutions on the initial data 6a/3(134, 253)
Continuous in time 6a/2(24)
Continuous processes 6a/3(35)
Continuous transition 6a/3(414)
Continuous transition from elastic to plastic state 6(331ff.)
Contour lines 6(400ff.); 6a/3(496, 497)
Contraction by mixing 13(351, 364)
Contraction coefficient 9(317–318, 340, 342, 383, 423)
Contraction, effect on fluctuations 8/2(554–555)
Contraction, effect on isotropy of turbulence 8/2(533, 535, 554)
Contraction, effect on steady flow irregularities 8/2(552–554)
Contraction in wind tunnel 8/2(551)
Contraction mapping 6a/2(396, 422, 618)
Contraction ratio 8/2(536, 553)
Contraction, relativistic 7/1(567, 620)
Contraction, theoretical design 8/2(555)
Contravariant tensors 6a/2(617, 622)
Control contour 9(129)
Control of boundary layer 8/1(347–349)
Convected coordinates 6a/2(431, 439, 449)
Convected time derivative 6a/4(151)
Convection 7/1(295); 8/2(503)
Convection of vorticity 8/1(151, 163, 188)
Convective flow 8/1(396)
Convective heat loss 8/2(559, 567)
Convective transport of kinetic energy 8/2(546)
Conventional response equation of anemometer 8/2(571)
Convergence of approximate solutions 6a/2(125)
Convergent channel, boundary layer 8/1(272)
Convergent flow 8/2(22, 23, 26–29)
Converging channels or tubes 8/2(550)
Converse principle of minimum complementary energy 6a/2(117)
— potential energy 6a/2(116)
Convex 6a/3(410)
Convex hull 6a/2(417)
Convex region with respect to $\mathscr{S}$ 6a/2(16)
Convex set 6a/2(392, 417)
Convolution 6a/2(25, 332); 7/1(14)
Convolution and Heaviside processes 6a/3(28)
Convolution, definition 6a/3(8)
Convolution for vectors and tensor processes 6a/3(9)
Convolution integral 7/1(15)
Convolution, properties 6a/3(9)
Convolution theorem 9(492)
Cooling coefficient, isentropic 14(70)
Cooling, magnetic 15(40, 53)
Cooling methods 12(319)
Cooling process in glass manufacture 13(544–546, 548)
Cooling rate and damage time, relation 13(630)
Cooperative effects at lowest temperatures 15(126)
Cooperative orientation 10(80)
Cooperative orientation effects in solutions 13(335–344)
Cooperative phenomena in crystals 13(1, 77ff.)
Cooperative phenomena, statement of problem 13(77)
Coordinate choice for surface waves 9(447, 456, 466, 475, 485)
Coordinate system, orthogonal (Cartesian) 7/1(45)
Coordinate systems 6a/2(438)
Coordinate transformation 6a/2(531, 615); 7/1(42)

Coordinates 6a/3(536)
Coordinates, convected 6(447)
Coordinates, optimal, in boundary layer flow 8/1(329–331, 333)
Coordinates, orthogonal curvilinear 6(84–86)
Coordination lattice 7/1(116)
Coordination number 7/1(73, 116)
Coordination numbers for argon 13(233)
Coordination shells 10(17, 89, 124, 125)
Copolymer 13(399)
Copper 6a/1(32, 44, 72, 103, 152); 14(239)
Copper chloride 15(30)
Copper crystals, kink bands 7/2(75)
Copper crystals, slip lines 7/2(91)
Copper foil trap 12(656)
Copper, irradiation 11/2(195–197)
Copper phtalo-cyanide in the electron microscope 7/2(81)
Copper single crystals
— electrical resistivity under deformation 7/2(165)
— flow stress discontinuity under temperature change 7/2(163)
— work hardening 7/2(53, 55)
Copper-gold alloy 13(80–81)
Copper-gold single crystals 7/2(175, 188–190)
— with long-range order 7/2(189)
— work-hardening 7/2(65–66, 191)
Copper-nickel single crystals
— critical shear stress 7/2(38)
— work-hardening 7/2(61)
Copper-nickel-alloys 14(325)
Coriolis force(s) 6a/3(286); 8/2(109, 110, 175)
Cork 6a/1(111)
Corner, supersonic flow around it 9(20–22)
Cornu's optical interference experiment 6a/1(264–266, 281–283, 286)
— Mallock's non-optical version 6a/1(267)
Correlation 7/1(266, 267, 269)
Correlation, binary 10(65)
Correlation coefficient 7/1(268)
Correlation distribution functions 8/2(442)
Correlation function(s) 10(26, 47, 62, 79, 181, 216); 11/1(86, 91); 13(28)
— and x-ray scattering 13(249, 250)
— depending on time 13(200)
— general shape 8/2(448, 449)
— in binary alloys 13(179, 186, 187, 194, 195)
Correlation length 15(319, 320)
Correlation radius 13(74)
Correlation radius in the van der Waals gas 13(76)
Correlation tensor of two velocities 8/2(444, 452)
Correlation tensor of two velocities, defining scalars 8/2(447)
Correlations in the hole model of liquids 13(242)
Correspondence, elastic-viscoelastic 6a/3(68, 69)
Correspondence principle 6a/3(69)
— differential operator laws 6a/3(69)
— for simple free vibrations 6a/3(111, 112)
— of Chaplygin 9(346)
Corresponding states 13(361, 365–369, 476–479)
— for the transport coefficients 12(397)
— principle of 10(172, 174, 175, 177, 186, 187), 12(48–51, 62, 98)
— quantum mechanical principle of 10(176)
Corresspondence principle of Alfrey-Graffi 6a/3(72)
Corrosion by intense ultrasonics 11/2(276)
Corrosive effects upon endurance limit 6(593)
Cortrell-Lomer's dislocation reaction 7/1(512)
Cosine law for the anemometer 8/2(566)
Cosserat rod 6a/2(665)
Cosserat surface, in general 6a/2(430, 446, 449)
Cosserat surface, relation to 3-dimensional theory 6a/2(594)
Cosserat-Ericksen-Toupin theorem 6a/2(106)
Cottrell effect 7/2(173, 178)
Cottrell interaction 11/2(188–190)

Couette flow 8/1(47–49, 55–56, 119, 255, 257); 8/2(16, 39, 509, 525, 550); 12(313)
— cylindrical 12(292)
— plane 12(290)
— stability 8/1(374, 416)
Coulomb forces between ions in electrolytes 13(371, 373, 374)
Coulomb friction analog for discontinuous deformation 6a/1(655)
Coulomb's condition 6(372, 380); 6a/3(463, 464, 473)
Coulomb's experiments on plasticity and viscosity 6a/1(174)
Counter-current interchangers 14(101)
Counter-examples of Hellinger, Hamel, Shield and Green 6a/3(139, 160, 204, 210)
Counterflow 15(390)
Counterions surrounding a polyelectrolyte molecule 13(496–497, 498)
Couple assigned 6a/2(504)
Couple director, assigned 6a/2(480)
Couple resultant vector 6a/2(651)
Coupled acoustic systems 11/1(99–126)
Coupled heat equation 6a/2(328, 335)
Coupling 6a/4(262)
Coupling between conduction electrons and phonons 14(241, 253)
Coupling between longitudinal and transverse waves 14(214, 253)
Coupling of lattice vibrations 7/1(291, 293)
Coupling parameters in the theory of electrolytes 13(373)
Coupling stiffness 11/1(395)
Covariant derivative 6(4)
— with respect to A 6a/2(456)
— with respect to $a_{\alpha\beta}$ 6a/2(450)
— with respect to Euclidean metric 6a/2(619)
— with respect to first fundamental form (metric tensor) 6a/2(431, 624)
Covariant tensors 6a/2(617, 622)
Covolume, *see also* Excluded volume
Crack division 6(578, 579, 580)
Crack formation 7/2(7, 45)
Crack, photoelastic investigation 6(207–208)
Crack problem in plane elasticity 6(76–78)
Crack problem in three-dimensional elasticity 6(102, 106)
Crack products in diffusion pumps 12(601–604)
Crack propagation 7/2(7)
Crack-extension-force 6(559–561, 573, 574, 575)
Cracks, co-linear two-dimensional 6(565, 566)
Cracks, disc-shaped 6(557, 558)
Cracks, limiting velocity 6(579)
Cracks, running 6(479, 575, 576)
Crank driven pistonphone 11/2(62, 63)
Creation operator 14(204)
Creep 6(271, 292, 310, 463, 466, 472); 11/1(361, 370)
Creep compliance and relaxation function 6a/3(28)
Creep compliance for relaxation function $\boldsymbol{G}$ 6a/3(26)
Creep compliance $\boldsymbol{J}$ 6a/3(24)
Creep deformation 6a/1(32, 421, 422, 571)
Creep effects on photoelastic measurements 6(162, 177)
Creep function 6(273); 6a/4(197)
Creep law 6a/3(23)
Creep problems, variational principle 6a/3(66, 67, 114)
Creep rate 7/2(206)
Creep, recoverable 6a/1(145)
Creep, steady state 7/2(71, 197, 204–208)
Creep, survey 7/2(197)
Creep test 7/2(31, 107)
Creep type order of pair of differential operators 6a/3(34)
Creep-buckling 6(311)
Creep-damage 6(594)
Cristobalite 13(608, 610)
Cristobalite lattice 7/1(90)
Cristobalite-like structure 13(523)
Critical acceleration 6a/4(49–54, 66, 94, 96–98)
Critical amplitude (of an acceleration wave) 6a/4(50, 61, 64–66, 94, 98)
Critical constants, measurement 12(42–45)
Critical crack-extension-force 6(559–561, 573, 574, 575)
Critical current density 15(336)

Critical density 12(42, 43, 141–146)
Critical depth for capillary-gravity waves 9(633)
Critical distance of ions 11/1(251, 256)
Critical enthalpy 8/1(223)
Critical field curve 15(130, 157, 165)
Critical fluctuations 13(71, 75)
Critical free stream Mach number 9(121, 125)
Critical frequencies 6a/4(264)
Critical frequencies of canal waves 9(502)
Critical hyperbola 15(33)
Critical initial amplitude 6a/3(342, 367, 368, 381)
Critical jump in strain gradient 6a/3(353, 361, 374, 384); 6a/4(49, 93–95)
Critical layer of main flow 8/1(374, 376, 377)
Critical load 6a/3(146, 277, 284, 285, 287)
Critical maximum shear stress 6(575)
Critical normal tensile stress 6(575)
Critical opalescence 13(71, 75, 76, 189–210, 250)
— of binary liquid systems 13(316)
— of ferromagnetics 13(196–210)
— of solid solutions 13(189–196)
Critical point 6a/3(288, 289); 10(19, 66, 172, 248, 297); 12(37–42, 141–146)
— definition in solutions 13(309)
— of a miscibility gap 13(307)
— of phase separation in polymer solutions 13(480)
— of precipitation 13(149–153, 161–166)
Critical points in ternary polymer systems 13(482)
Critical pressure 14(20, 36, 38)
Critical rate of mass transport 15(451)
Critical ratio (*pc Vc/RTc*) for several substances 12(49, 50)
Critical region for binary mixtures 12(65–70)
Critical Reynolds number 8/1(118, 353, 355, 360, 366, 376, 380, 383–385, 401, 417)
— effect of heat transfer 8/1(421)
— effect of suction or blowing 8/1(411)
— effect of wall temperature 8/1(427)
— experimental 8/1(371)
— for a flat plate at zero incidence 8/1(424)
— for concave walls 8/1(435)
— in supersonic flow 8/1(426)
— theoretical 8/1(371)
Critical roughness 8/1(441)
Critical scattering of slow neutrons 13(196–210)
Critical scattering of x-rays 13(189–196)
Critical shear strain 7/2(33)
Critical shear stress 7/2(7, 14, 31, 32–45, 110, 112)
— concentration dependence in alloys 7/2(39, 178)
— dependence on impurities 7/2(40)
— dependence on strain-rate 7/2(34)
— dependence upon orientation 7/2(34)
— dependence upon temperatures 7/2(34)
— in metals of large stacking-fault energy 7/2(115–119)
— in metals of small stacking-fault energy 7/2(120–124)
— law of Schmid, *see also* Schmid's law 7/2(33)
— of alloys, short-range order theory 7/2(174–179)
Critical solidity of grid 8/2(532)
Critical speed 8/1(184, 223); 9(7)
Critical state in a valve 12(567, 568)
Critical state of flow 8/1(94)
Critical temperature(s) 7/2(115); 12(42, 141–146); 14(36, 38)
— of binary polymer systems 13(480, 481)
— of precipitation 13(324)
— of gases 13(230, 234)
Critical tensile stress 7/2(33)
Critical time for acceleration wave growth 6a/4(61, 66–69, 98)
Critical value 6a/3(225, 226, 278, 283, 284)
Critical velocity 7/1(586); 15(411, 412, 413, 414, 415, 416, 427, 428, 429)
Critical velocity for long waves 9(669, 676, 713)
Critical volume 10(66); 12(44)
Crocco's equation 8/1(41–43, 186); 9(4, 250)

Crocco-Vazsonyi equation 8/1(185–186, 191)
Cross bonding of polymer chains 11/1(361, 409)
Cross section(s)
— multiply connected 8/2(69)
— of a spherical obstacle 11/2(160, 164)
— of critical magnetic scattering 13(205)
— of different shapes 8/2(75)
Cross slip 7/1(483, 713); 7/2(73, 92–94, 97–100, 158, 161, 180, 188, 194, 202, 204)
— hindering in alloys 7/2(180–181)
— line 7/2(92)
— mechanism 7/2(150)
— of screw dislocations 7/2(150)
Cross-elasticity 6(502ff., 507)
Cross-section 7/1(432, 441)
Cross-stream inhomogeneity 8/2(532)
Crowdion 7/1(423)
Crudeli's method 8/2(187, 266)
Cryostat 15(61)
Cryostat and magnet, separation of 15(63)
Crystal class 6a/2(72)
Crystal classes 7/1(25ff., 28)
Crystal diameter, effect on work-hardening 7/2(56)
Crystal domains, bad 7/1(386)
Crystal domains, good 7/1(386)
Crystal formation in glasses 13(592–616)
Crystal, glass 6a/1(240)
Crystal growth 7/1(70); 7/2(212)
Crystal growth in glass 13(594–604)
— calculation of its rate 13(600–601)
— dependence on viscosity 13(602)
— effect of chemical composition 13(607–612)
Crystal holders for piezoelectric transducers 11/2(84–85)
Crystal lattice 6(18)
Crystal morphology 7/1(66)
Crystal, one-dimensional model 13(58–64)
Crystal optics 6(132–136)
Crystal potential 7/1(565)
Crystal powder 15(49, 56)
Crystal powder, filling factor 15(50)
Crystal structure 7/2(4)
Crystal structure(s)
— of Al, Ga, In 7/1(87)
— of alkalihalides 7/1(86)
— of Cu-, Ag-halides 7/1(87)
— of elements 7/1(75)
— of Zn-, Cd-, Hg-salts 7/1(87)
Crystal surface 7/1(95)
Crystal system(s) 6a/4(178); 7/1(19ff.)
— cubic 7/1(25)
— hexagonal 7/1(23)
— monocline 7/1(20)
— rhombic 7/1(20)
— rhombohedral 7/1(24)
— tetragonal 7/1(23)
— triclinic 7/1(19)
— trigonal 7/1(24)
Crystal tuning fork 11/1(371, 377, 379)
Crystal zones 7/1(62)
Crystalline electric field 15(2, 5, 6, 14, 27)
Crystalline state 10(6, 7, 17, 22)
Crystalline state, nucleation 7/2(219–221)
Crystallinity of a polymer 13(402, 425, 450)
Crystallinity of rubber 11/1(413)
Crystallite 13(425)
Crystallite theory of glasses 13(527)
Crystallization limits in glasses 13(612–616)
Crystallization rate 13(513, 595)
Crystallization tendency of glasses 13(543)
Crystallographic axes 6a/4(236)
Crystallography of glide 7/2(20)
CsCl lattice 7/1(101)
C^∞-smooth 6a/2(370)
CuAu-alloys 7/1(446)
Cubic crystals 6a/4(246, 247)
— acoustical tensor 6a/4(248)
— pure modes 6a/4(248–250)
— thermoelastic effects 11/2(175, 177)
Cubic lattice, theory of melting 13(293)
Cubic space-centered lattice, results of Fuchs theory 13(162, 165)
Cubic structure, Bordoni peaks 11/2(203)
Cubic symmetry 11/2(156)
Cubical dilatation 6a/4(113)
Curie constant 15(107)

Curie point 6a/1(387); 13(205, 207–209); 15(1, 128, 136, 137, 144, 146, 149, 159)
Curie's law 15(1, 42, 78, 108, 110, 115)
Curie-Weiss law 15(1, 2, 59, 80, 82, 105, 107, 109, 129)
Curl 6a/2(11)
Curl components 8/2(7)
Current, electric 14(230)
Current flow from a normal conductor into a superconductor 15(294)
Curvature conditions (assumptions) 6a/4(28, 39, 49, 61, 80, 84, 91ff.)
Curvature equation 9(400, 401)
Curvature, macroscopic 7/1(559)
Curvature of a free streamline at detachment 9(372)
Curvature of a stream line 9(259)
Curvature of a streamline at detachment 9(372)
Curvature of a surface 10(284)
Curvature of boundary 6a/2(378)
Curvature of free streamline 9(322)
Curvature of slip lines 6(358, 364, 378)
Curvature of surface, Gaussian 6a/2(471)
Curvature of surface, mean 6a/2(471, 624)
Curvature of velocity profile 8/1(374, 376, 397, 422)
Curvature of wall 8/1(267, 415, 430)
Curvature, principal radii of 6a/1(264–268, 281–283, 286)
Curvature tensor 6a/2(620, 625)
Curvature, tensor of 7/1(559f.)
Curvature term of Gibbs 10(290)
Curve of neutral stability 8/1(371, 373, 376–379, 382, 387, 401, 412, 423)
Curve on the surface 6a/2(622)
Curved boundary, method of continuity 9(413)
Curved channel 8/1(433)
Curved shock 9(274)
Curved wall, boundary layer 8/1(267–268)
Curvilinear coordinates 8/1(141–144, 308–311)
Cusp 6(384, 385, 387, 389, 393)
Cusped cavity 9(338, 397)
Cusps 6a/3(477, 479, 483)
Cut-off frequency in a duct 11/1(45, 50)
Cut-off frequency of anemometer 8/2(578)
Cut-off of short waves in fluctuation spectrum 13(70–71)
Cut-off-radius 7/1(582)
Cut-off-radius, internal 7/1(548)
C^v-smooth 6a/2(369)
Cybotaxis 10(7, 10)
Cyclic balance 12(360)
Cyclical loading 6a/1(28, 30, 31, 68, 97–99, 123, 147, 150–152, 221, 223, 297, 366, 369, 567, 682–686, 688)
Cyclohexane-benzene mixture, entropy 13(339, 355)
Cycloids 6a/3(453, 456, 529, 532)
Cycloids as slip lines 6(363, 366, 381, 384, 429)
Cyclone 8/1(47)
Cyclotron frequency 11/2(218, 221)
Cylinder, *see* Circular cylinder
Cylinder 6a/3(411)
Cylinder, boundary layer 8/1(272–275, 313)
Cylinder, elliptical, boundary layer 8/1(400, 403)
Cylinder, impulsive motion and boundary layer 8/1(291–293, 324–328)
Cylinder of finite length 8/2(51)
— in oscillating rotation 8/2(223)
Cylinder of Mises yield condition 6(328)
Cylinder, rotating 6(423)
Cylinder spinning in presence of a wall 8/2(218–220)
Cylinder under axial moment 6(399)
Cylinder waves 6(112–116)
Cylinders, co-axial, steady flow between them 8/2(17–19, 49–52, 79)
Cylinders of any cross section, submerged 9(576)
Cylinders of different cross sections, torsion 6(27–38)
Cylinders, rotating coaxial, transition to turbulence 8/1(415, 416, 430)
Cylindrical coordinates 8/2(1, 58)
Cylindrical coordinates, equations of motion 8/2(5)
Cylindrical coordinates for surface waves 9(485)
Cylindrical distortion 6(504–507)

Cylindrical model of polyelectrolytes 13(496)
Cylindrical polar coordinates 8/1(143, 238)
— in elasticity 6(85–86, 105)
Cylindrical resonators for sound waves 11/1(166)
Cylindrical tube
— elastic-plastic expansion 6(408–418)
— fully plastic expansion 6(365)
— initial value problem 8/2(149–152)
— meniscus of a liquid surface 10(301–302)
Cylindrical wave produced by linear discontinuity 11/1(111)

Dafermos' theorem 6a/3(106)
d'Alembert solution of the wave equation 9(16, 669)
d'Alembert-Lagrange variational principle 8/1(145–146)
d'Alembert's paradox 8/1(53, 91, 105, 160, 201); 9(31)
Daltons's law of partial pressures 12(58, 62)
Dam breaking 9(681)
Damage, fracture strength 6(570, 571, 572)
Damage rate of fatigue 6(609, 611)
Damped 6a/2(343)
Damped oscillation, method for viscosity measurements 12(314)
Damping 7/2(3)
Damping, air 6a/1(177, 299, 314, 417)
Damping by thin fluid layer covering surface 9(646)
Damping coefficients 6a/1(174, 176, 313–317, 386, 387, 392)
Damping coefficients in ship theory 9(567)
Damping of disturbance 8/1(369–371)
Damping of sound in metals 11/2(178)
Damping of waves in viscous fluids 9(639, 646, 651)
Damping theories of fatigue 6(603)
Darcy's law 6(265); 8/2(634–641, 645, 646)
Darcy's law, limitations 8/2(647–653)
Dark-field illumination of glide bands 7/2(95)
Dashpots 6a/3(2)
Data given along two intersecting lines 6a/3(492)
Data of thermoelasticity 6a/2(335)
Davies' capacitance displacement technique 6a/1(331–337, 346–351)
Davis' experiments on combined stress in polycrystals 6a/1(509–513, 671, 703)
Dead space 12(547, 550)
Dead weight loading 6a/1(125, 181, 400–403, 550, 652)
Dead-weight gauge, *see* Piston gauge
Debye function 14(286)
Debye parameter 14(287)
Debye spectrum 7/1(249, 250, 257, 262, 279, 281, 319)
Debye temperature 7/1(248, 250, 254, 257, 263, 286, 321); 13(571); 14(240, 254)
Debye term 7/1(258)
Debye theory 14(286)
Debye-(cut-off-) frequency 7/1(251, 258, 263, 277, 321)
Debye-Falkenhagen effect 11/1(250)
Debye-Hückel approximation for polyelectrolytes 13(494–495, 503)
Debye-Hückel parameter 13(376, 388)
Debye-Hückel theory 10(223)
— extension to colloidal systems 13(492)
— of electrolytes 13(375–381)
— of electrolytes, justification 13(381, 383–389, 392)
Debye's charging process 13(374, 377, 393)
Debye's formula for the correlation radius 13(76)
Debye's theory 7/1(248, 258, 266, 269, 290, 311, 327)
Debye-Scherrer pattern of vitreous silica 13(522)
Debye-Sears effect 11/1(229); 11/2(124–125)
Debye-Waller theory 7/1(377)
Decay constant of pure water molecules 11/1(256)
Decay law in final period of turbulence 8/2(461, 482)
Decay method 11/2(151)
Decay of energy-containing eddies 8/2(466)

Decay of turbulence 8/2(459–460, 469–470, 482, 528–530, 552)
Decay of turbulence energy balance 8/2(461–462)
Decay or growth of the strength of waves
— elastic 6a/3(320–323, 328, 330–333)
— in elastic non-conductors 6a/3(362–376)
— in general 6a/3(341–344)
— in inhomogeneous elastic bodies 6a/3(378–385)
— in materials with memory 6a/3(346–348, 352, 354)
— thermo-elastic 6a/3(333, 334)
— visco-elastic 6a/3(110)
Decibel (unit) 11/2(162–163)
Decomposition theorem
— pure strains 6a/2(35)
— pure strains, isochoric 6a/2(36)
— simple shears 6a/2(36)
— singular states 6a/2(180)
— stress fields 6a/2(59)
Decrement decrease by irradiation 11/2(196)
Deep tone of vibrating rods 6a/1(254–256)
Deexcitation 11/1(196)
Deexcitation of the vibrational mode 11/1(284)
Defect generation by moving dislocations 7/2(113)
Defects, chemical, in metals 7/1(430)
Defects, electric 7/1(433)
Defects, Frenkel 7/1(390, 393, 436, 437, 439, 441, 442, 451)
Defects in crystals, statistical treatment 13(78)
Defects, Shottky 7/1(390, 392, 395, 399, 400, 402, 405, 451, 452)
Deflection due to shear in transverse vibration 6a/1(384)
Deflection multiplier, optical 14(156)
Deflection of flap, pressure distribution 8/2(394, 412)
Deformable region 6a/3(416)
Deformation, *see also* Discontinuous deformation, Finite deformation, Resolved shear deformation 6a/2(29); 6a/4(111); 11/2(154)
Deformation bands 7/2(74–77, 108)
Deformation energy 6a/1(426, 565, 586, 588, 591, 600, 603, 645, 667, 703, 735)
Deformation gradient(s) 6a/2(29, 299); 6a/3(305, 536); 6a/4(111)
— in one-dimensional motion 6a/4(24, 111)
— in three dimensions 6a/4(70, 111)
Deformation history 6a/3(537)
Deformation, homogeneous 7/1(183, 199, 234, 242)
Deformation, infinitesimal 6a/1(37, 116–124, 147, 154, 357); 6a/2(28, 456)
Deformation, inhomogeneous 6a/1(257, 410, 650, 668, 716)
Deformation, isotropic 11/1(134)
Deformation, mechanism of 7/1(462)
Deformation, microplastic 6a/1(22, 25, 27, 140, 148, 223, 458)
Deformation modes 6a/1(535, 551, 553, 564, 694, 703, 705–716)
Deformation modes, quantized distribution of 6a/1(553–562, 639, 703)
Deformation of liquids by fields of forces 10(67–69)
Deformation of molecular structure in liquids 10(20, 67)
Deformation of polymers, thermodynamical treatment 13(432–451)
Deformation of single crystals 7/2(28–32)
Deformation, plastic 7/1(464)
Deformation point of a glass 13(580)
Deformation rate 6a/4(113)
Deformation, shearing 11/1(133)
Deformation tensor 8/1(138, 139)
Deformation theory 6a/3(549)
Deformation type 6a/3(415)
Deformation under hydrostatic pressure 6a/1(486, 490, 499, 514)
Deformations, permanent 7/2(3, 130)
Deformations, plastic 7/2(3)
Deformation-type law 6(332)
De-gassing 11/2(261)
Degeneracies 6a/3(37, 447)
Degeneracy of energy levels of paramagnetic salts 15(42)
Degeneracy temperature 14(143)
Degenerate solution 6(359, 391)
Degree of crystallinity 13(402, 425)
Degree of crystallization 13(598–599)
Degree of depolarization 13(262)
Degree of dissociation 11/1(209, 210)

Degree of dissociation of a polyelectrolyte 13(492)
Degree of dissociation, total 11/1(215)
Degree of order, effect on plastic properties 7/2(190)
Degree of polymerisation 13(399)
Degrees of freedom of macromolecules 13(410)
Degrees of freedom, separation 12(85–91)
DeHaas-van Alphen effect 14(182, 251)
Delay leg, *see* Laser interferometry
Delay time, *see* Laser interferometry
Delay line 11/1(230)
Delayed elasticity in glass 13(637–640)
δ function 9(289, 599, 602, 637); 11/1(23, 26, 58)
Delta wing 9(52–54)
Demagnetization, adiabatic 15(38f., 40)
Demagnetization coefficient 15(47, 221)
Demagnetization, isentropic 15(43)
Demagnetization process, suitability of salts 15(43)
Demagnetizing field 15(142, 143, 147)
Demixing of polymers in solution 13(484)
Density 6a/2(43, 299, 309); 6a/3(42); 6a/4(18); 15(372, 373, 404)
Density, acoustic 11/1(6, 9)
Density, critical 12(42, 43, 141–146)
Density fluctuation(s)
— in compressible turbulent flow 8/1(364)
— in liquids 10(9, 11, 12, 15, 19, 30, 32, 43, 47)
— near the critical point 12(46–48)
— total 13(30)
Density formulae for glasses 13(549, 552, 554)
Density in the reference configuration 6a/4(24, 70)
Density matrix 10(126, 237, 240); 12(476); 13(64)
— for perfect gases 12(192–196)
— general formulation 12(182–192)
— with a potential 12(196–201)
— with a potential, symmetrization 12(201–204)
Density, molecular, reduced 10(133)
Density of glasses
— and their thermal history 13(555)
— at high temperatures 13(555–557)
— dependence on composition 13(548–554)
— dependence on pressure 13(557–560)
— dependence on temperature 13(561–570)
Density of pairs 13(5)
Density of particles 13(4)
Density, orthobaric 12(43, 45, 53)
Density, present 6a/4(24, 70)
Density transition curves 10(222)
Density variation, effect on the transition to turbulence 8/1(415, 417–419, 436)
Dependence domain 9(40)
Dependence of penetration depth on mean free path 15(301)
Depolarization degree 13(262)
Depolymerization by intense ultrasonics 11/2(284, 287)
Depth growth of glide steps 7/2(94, 107)
Depth of field, diffuse interferometry 6a/4(14)
Depth of field, displacement interferometer 6a/4(11)
Depth of water, *see* Finite depth
Depth of water, determination 9(673)
Deresiewicz-Zorski solution 6a/2(330)
Derivative, material, in curvilinear coordinates 8/1(141)
Derivatives, two types of 8/1(20, 129, 130)
Design problems for rods 6a/2(698)
Designations, *see* Nomenclature
Desorption at very low pressure 12(638, 642)
Destruction of order 7/2(4)
Destruction of short-range order in alloys 7/2(172, 174)
Destruction of superconductivity by currents 15(338)
Destruction region of borate network 13(530, 569)
Deswelling of a gel in different solvents 13(489, 507)
Detachment of a free streamline 9(322, 323, 371–375)
Detailed balance 12(360, 488, 490, 492)
Detection of leaks 12(541–542, 661)
Detection of ultrasound 11/2(114–142)
Determinant 6a/2(6)

Determination of elastic constants 6a/4(251)
Determinism 6a/3(12)
Detonations as sources of sound 11/2(32–35)
Developable boundary surface 8/1(309, 314)
Deviator 6(438)
Deviator of permanent deformation tensor 6a/3(550)
Deviator of strain 6a/3(29, 71)
Deviator of stress 6a/3(29, 71, 546)
Deviator of stretching tensor 6a/3(548)
Devitrification 13(543, 592–616)
Devitrification, effect on specific heat 13(573)
Devitrite, formation in glass 13(601, 602, 608)
Dew point 12(66, 67)
Diagonal flow 6a/3(455, 456)
Diagram of state 15(372)
Diamagnetic properties, derivation from energy gap model 15(303)
Diamagnetic rotation 15(15)
Diamagnetism of an electron gas 15(285)
Diamagnetism of superconductors 15(212)
Diamond 14(223, 224)
Diamond pyramid hardness 13(633–635)
Diamond, stress birefringence 7/2(45)
Diamond structure 7/2(26)
Diamond-structure 7/1(80, 114, 116, 117, 127, 603, 631, 639)
Diaphragm in circular duct 11/1(116–119)
Diaphragm of a loudspeaker 11/2(9, 18)
Diatomic liquids 10(66)
Diatomic molecules, rotation and vibration levels 12(101–108)
Die extrusion 6a/1(431, 477, 478)
Dielectric behaviour of liquids,
— non polar liquid 10(71–74)
— polar liquids 10(74–77)
Dielectric constant(s) 7/1(217, 220ff.); 11/1(362); 13(251–256)
— high-frequency 7/1(223)
— of liquids 10(22, 74, 81)
— of piezoelectric materials 11/2(47)
— of solvent 11/1(245–246, 269)
— static 7/1(221)
— very high 15(342)
Dielectric relaxation 11/1(325)
Dieterici's equation 12(54)
Difference of partial volumes 11/1(209, 210)
Differentiable $(r-1)$-cell of Class C^n 6a/2(402)
Differential collision cross section 12(336–338)
Differential equation
— of the elliptical type 8/1(66); 9(16, 29, 93, 94, 99, 209, 667, 683)
— of the hyperbolic type 8/1(67); 9(16ff., 22, 29, 42, 93, 135, 209, 667, 673, 683, 691)
— of the parabolic type 8/1(72)
— of elliptic-hyperbolic type 9(93, 101, 135)
— of mixed type 9(93, 101)
— of the velocity potential 9(6)
— of the velocity potential, its three-dimensional explicit solutions 9(8–11)
— of the velocity potential, plane flow 9(11)
— of Tricomi 9(81, 131, 133, 150)
Differential inequality for velocity fields 9(123)
Differential models for viscoelastic behavior 6a/3(2)
Differential operator
— of creep type 6a/3(34)
— of relaxation type 6a/3(34)
— relation to relaxation function 6a/3(34)
Differential operator laws
— and initial conditions 6a/3(44)
— degenerate cases 6a/3(37)
— general 6a/3(34, 35)
— special 6a/3(39)
— standard 6a/3(39)
— tensor 6a/3(37)
Differential viscosimeter 8/2(152–156)
Differential-difference equation 9(611)
— Levi-Civita 9(726–727)
Diffraction grating experiment of Bell 6a/1(597, 621–627, 634, 648, 690, 695, 698)
Diffraction in liquids 10(15, 19, 43)

Diffraction of light by an ultrasonic wave 11/2(124–125)
Diffraction of sound by a Rayleigh disc 11/2(59)
Diffraction of water waves
— by a freely floating body 9(566–567)
— by a horizontal cylinder 9(545)
— by a vertical cylinder 9(544)
— by a vertical half-plane 9(544, 560)
— general obstruction 9(543)
Diffraction of x-rays 7/1(353)
Diffraction pattern from a periodically supported membrane 11/1(107)
Diffraction spectrum 11/1(229)
Diffuse first-order transition 13(430)
Diffuse reflection of molecules from surfaces 8/2(600, 610)
Diffuse surface interferometry 6a/4(14)
Diffusion 6(229, 231, 233); 11/1(212, 215)
Diffusion by continuous movements 8/2(510–512)
Diffusion coefficient(s) 7/1(408, 420); 8/1(60); 10(84)
— dependence on mixing ratio 12(321)
— dependence on the molar fraction 12(431)
— determination from the formation period of thermodiffusion 12(325)
— first approximation 12(422)
— measurements 12(320–326)
— of the Lorentz gas 12(417)
— of turbulent diffusion 8/2(509)
— second approximation 12(432)
Diffusion constant 7/2(207, 219, 238)
Diffusion, effect on sound absorption 11/1(132, 138–139, 174)
Diffusion equation 6a/4(255); 12(303, 408)
— for pressure relaxation 6(266)
— for the entropy 6(264)
— of Beltrami 8/1(151)
Diffusion in a binary mixture 12(412–415, 423)
Diffusion in a ternary mixture 12(423)
Diffusion method for the determination of molecular weights 13(401, 496)
Diffusion, negative 7/2(219, 233–234)
Diffusion of gases through glass 13(616–621)
Diffusion of helium through glass 12(634)
Diffusion of interstitials 7/2(6)
Diffusion of ions through glass 13(621–623)
Diffusion of kinks in dislocations 11/2(208)
Diffusion of vancancies 7/2(6)
Diffusion of vorticity 8/1(188–190, 248); 11/1(12)
Diffusion path, mean 7/1(469)
Diffusion problems 6a/2(373, 377, 386)
Diffusion pump, *see also* Oil diffusion pump 12(515, 547, 559)
Diffusion pump, theory 12(576–590)
Diffusion, quasi-stationary 12(320, 321)
Diffusion rate 12(357, 403, 407, 411)
Diffusion, representation of the cross section by scattering phases 12(458–461)
Diffusion, stationary 12(322)
Diffusion, thermal effect 12(304, 323–325, 330–334)
Diffusion, thermal, in liquids (Soret effect) 10(2, 24, 82)
Diffusional jogs 7/1(495, 498, 612)
Diffusional jogs, charged 7/1(498)
Diffusionless transformations 7/2(239–252)
Diffusion-relaxation 6(263)
Diffusivity equation and seepage flow 8/2(638, 647)
Dihedral 8/2(386)
Dilatancy (=volume change by shear) 6(261, 503)
Dilatation 6(24, 143, 438, 481); 6a/1(86, 245, 275, 292, 303, 394); 10(89, 118)
Dilatation, centre of 7/1(528, 556)
Dilatation, cubical 6a/4(113)
Dilatation, definition 6a/2(31)
Dilatation, determination by Betti's method 6(95–97)
Dilatation, uniform 6a/2(34, 77)
Dilatational wave equation 6a/2(213, 328)
Dillon's experiment on thermoplasticity 6a/1(566–570)
Dilute gases, sound absorption and dispersion 11/1(153, 155–157)
Dilute sol 6(524)

Dilute solutions of polymers 13(463, 465)
Diluted crystals 15(26)
Diluted salts 15(84)
Dilution entropy 13(172–174, 334, 339)
Dilution, infinite 13(304, 305)
Dilution of a crystal 15(44)
Dimensional analysis 8/1(57–60, 183, 243–246)
— in initial value problems 9(610, 611)
— in seepage flow 8/2(649)
Dimensional perturbation 9(421)
Dimensionality 6a/1(3)
Dioxane addition to $MgSO_4$ solutions 11/1(269)
Dipolar forces in liquids 10(23)
Dipole approximating bow and stern of a ship 9(582)
Dipole formation 7/2(131)
Dipole gases 12(500)
Dipole interaction 10(75); 15(43)
Dipole moment
— and sound velocity 11/1(346)
— electric 10(72)
— induced in scattering 11/1(30)
— of a source 11/1(27)
— of superconductive specimen 15(221)
Dipole nature of a jet-edge source 11/2(28)
Dipole singularity 9(484, 487, 489, 577)
Dipole source 11/2(1, 3, 13)
Dipole source covering bottom 9(571)
Dipole wave 11/1(5, 27)
Dipole-dipole interaction, induced 12(339)
Dirac δ-function 8/2(118); 11/1(23, 26, 58)
Direct approach to shell theory 6a/2(448)
Direct correlation function 13(73)
Direct notation 6a/2(297)
Direct-inspection-method 7/1(51)
Direction 6a/3(484)
Direction condition, *see* Characteristic condition
Direction condition 6a/3(426)
Direction of displacement 6a/2(343)
Direction of motion of a progressive wave 6a/2(245)
Direction of propagation 6a/2(245, 248, 343); 6a/3(97, 306)
Directional derivative 6a/3(444)
Directional distribution of gamma radiation 15(297)
Directional permeability in anisotropic porous media 8/2(637)
Directional sensitivity of anemometer 8/2(565–566, 585)
Directivity index 11/2(2)
Directivity of an ultrasound generator 11/2(103–105)
Directivity of condenser microphone 11/2(43, 44)
Directivity of ribbon microphones 11/2(57)
Directivity of sound radiation from a gun 11/2(33)
Director couple, assigned 6a/2(480)
Director, definition 6a/2(446, 467, 666)
Director inertia coefficient 6a/2(594–596)
Director momentum 6a/2(482, 483)
Director velocity 6a/2(449, 467)
Dirichlet boundary conditions 6a/2(370)
Dirichlet growth condition 6a/3(565)
Dirichlet integral 9(97)
Dirichlet problem 6a/2(375, 449)
Dirichlet problems in elliptic equations 6a/3(211)
Dirichlet's boundary problem 9(12, 108)
— of potential theory 6(32)
Dirichlet's boundary value problem 8/2(78)
Dirichlet's principle 8/1(161–162)
Discharge coefficient 9(317–318)
Discharge gettering 12(638–640)
Discharge rate 9(569)
Discontinuities on characteristics 9(15–16, 22)
Discontinuities, second order, *see* Transitions, second order
Discontinuity, compatibility 6a/3(494)
Discontinuity conditions, kinematical 6(344)
Discontinuity distance 6a/4(279)
Discontinuity, general 6a/3(429, 456)
Discontinuity interaction 6(367)
Discontinuity line as a transition zone 6(348)
Discontinuity line (in plane strain) 6(348, 428)

Discontinuity, material or stationary 6(343)
Discontinuity of order *n* 6a/3(427)
Discontinuity of order one 6a/3(429)
Discontinuity of order two 6a/3(430)
Discontinuity of stress 6a/3(457)
Discontinuity order 6(342)
Discontinuity (*see also* Waves), classification 6a/3(434)
Discontinuity, stationary 6a/3(429)
Discontinuity surface, *see also* Singular surface 6(333, 343); 6a/3(416, 433)
Discontinuity surface vanishing 8/2(124)
Discontinuous deformation, *see also* Portevin-Le Chatelier effect, Savart-Masson effect
— McReynolds' experiment 6a/1(41, 653–666)
— McReynolds' slow waves 6a/1(655, 661)
Discontinuous glide 7/2(156)
Discontinuous processes 6a/3(35)
Discontinuous solution 6(341–343)
Discontinuous solutions 6a/3(416, 426)
Discrete curvature equations 9(416–420)
Discrete systems, *see also* Lagrange-Dirichlet stability concept 6a/3(131)
— conservative, discussion of Lagrange's work 6a/3(153, 154, 163)
— energy criterion 6a/3(203, 204)
— Lagrange-Dirichlet criterion 6a/3(153)
— Lagrange-Dirichlet criterion and energy criterion (footnote) 6a/3(164, 195, 196, 203)
— Liapounov theorems 6a/3(155, 156)
— measures in 6a/3(141)
— stability, comparison with continuous systems 6a/3(140, 159)
— stability theorem 6a/3(155)
— stability theorem, proof 6a/3(159)
Disk, circular, scattering of a sound wave 11/1(68–73)
Disk, rotating 6(423)
Disk spinning in a steady flow 8/2(217–218)
Dislocation 6(53, 230, 263, 603, 604); 7/1(388, 485); 7/2(2, 14)
Dislocation, accelerated 7/1(621)
Dislocation arrangement 7/2(111)
Dislocation arrangement in a kinked crystal 7/2(75)
Dislocation, character of the 7/1(481, 482)
Dislocation, Cortrell's 7/1(512)
Dislocation damping of sound waves 11/2(177–208)
Dislocation density 7/2(111, 216); 11/2(184)
Dislocation density and impurity concentration 14(266)
Dislocation, density of 7/1(540, 558, 560, 627)
Dislocation energy 7/1(533, 545)
Dislocation, extended 7/1(510, 511)
Dislocation forest 7/2(113, 121)
Dislocation jog 7/1(493)
Dislocation kernel 7/1(546, 563, 571)
Dislocation kernel, structural of the 7/1(652)
Dislocation line(s) 7/2(17, 110)
— carrying positive charge 7/2(125)
— density of 7/1(559)
— flexibility of a 7/1(553)
— hollow 7/1(558)
— jogs of 7/1(453)
— quasi-momentum of a 7/1(620)
— scattering by a 7/1(622)
— self- and back-inductivities 7/1(552)
— spontaneous generation of 7/1(555)
— straight 7/1(531, 535, 592)
— string approximation of a 7/1(621)
Dislocation, mean free path 7/2(128, 141, 144, 146)
Dislocation mechanism 7/2(15)
Dislocation migration 7/2(15)
Dislocation model of the grain boundary 7/1(652)
Dislocation models 7/1(514)
Dislocation models, non-linear 7/1(563)
Dislocation motion, damping of the 7/1(591)
Dislocation, moving 7/2(113)
Dislocation multiplication 11/2(198)
Dislocation network 7/2(111)
Dislocation, partial 7/2(24)
Dislocation, partial = half dislocation 7/1(510)
Dislocation potential 7/1(560)

Dislocation reaction 7/1(483, 484, 511); 7/2(141, 148)
Dislocation, rest mass of a 7/1(620)
Dislocation ring, circular 7/1(548)
Dislocation rings 7/1(538)
Dislocation, Somigliana's 7/1(520f., 524, 620)
Dislocation source, spiral 7/1(492)
Dislocation sources 7/1(489, 538, 551); 7/2(112)
Dislocation spiral 7/1(538)
Dislocation, stable 7/1(485)
Dislocation strain 11/2(178)
Dislocation strength of 7/1(97)
Dislocation velocity 7/2(186)
Dislocation, zigzag 7/1(549)
Dislocations 6a/1(521, 525, 528); 14(210, 221)
Dislocations, damping of 7/1(622)
Dislocations, density and array of the 7/1(642ff.)
Dislocations, dynamics of 7/1(564, 619)
Dislocations, electric interaction of, with foreign atoms or vacant lattice sites 7/1(627)
Dislocations, electric resistance of 7/1(625)
Dislocations, generation, of 7/1(645)
Dislocations, geometric theory of 7/1(477)
Dislocations in a plate 7/1(563)
Dislocations in alloys 7/1(513)
Dislocations in an elastical semi-space 7/1(562)
Dislocations in close packed slip planes 7/1(604)
Dislocations, interaction of, and imperfections 7/1(555)
Dislocations, interaction with dissolved interstitials 7/2(35)
Dislocations, mathematical theory of 7/1(477, 513)
Dislocations, maximum velocity of 7/1(624)
Dislocations nearly straight-lined 7/1(549)
Dislocations, number of 14(265)
Dislocations of higher order 7/1(600)
Dislocations, optic effects of 7/1(629)
Dislocations, partial 7/1(477, 503)
Dislocations, pole 7/1(661)
Dislocations, proofs for the existence of 7/1(629)
Dislocations, radiographic effects of 7/1(628)
Dislocations, sequences of 7/1(583)
Dislocations, sessile 7/1(509, 599)
Dislocations, thermal production of 7/1(554)
Dislocations, thermodynamic stability of 7/1(553)
Dislocations, uniformly moved 7/1(563)
Dislocations, velocity of 7/1(567)
Dislocations, Volterra's 7/1(620)
Disorder, one-dimensional 7/1(97)
Disorder-order transitions 7/2(188)
Dispersed system 6(435)
Dispersed waves 6a/2(343)
Dispersion 7/1(216)
Dispersion by continuous movements 8/2(512–515)
Dispersion coefficient, dependence upon scale 8/2(510, 514)
Dispersion, curve of 7/1(220, 316)
Dispersion, degree of 7/2(169–171, 192)
Dispersion due to heat radiation 11/1(281–282)
Dispersion factor of a porous medium 8/2(646, 647)
Dispersion forces 11/1(351); 13(349)
Dispersion formula 11/1(145, 291)
Dispersion frequency 7/1(220ff.)
Dispersion frequency region 11/1(226)
Dispersion in highly dilute gases 11/1(153, 155–157)
Dispersion law of capillary-gravity waves 9(614, 633)
Dispersion measurement in electrolytes 11/1(226–227)
Dispersion of high concentration 6(529)
Dispersion of high-frequency sound waves 8/2(612)
Dispersion of sound in gases, theory 11/1(135ff.)
Dispersion of sound, molecular 11/1(157–165)
Dispersion of sound waves in liquids 11/1(275ff.)
Dispersion of the one-dimensional crystal 13(60)

Dispersion of the speed of sound 7/1(623)
Dispersion relations of surface waves 9(502, 507, 513–515, 518, 550)
Dispersion tensor 8/2(519)
Dispersion wave 6a/1(331, 338, 341, 343, 347, 621)
Dispersions in liquids 6(526–527)
Dispersions in solids 6(528–529)
Displacement 6a/4(24, 38ff., 82, 111); 7/1(145, 260, 261, 287, 299, 322)
Displacement, acoustic 11/1(9)
Displacement, boundary data 6a/3(47)
— in terms of stresses 6a/3(48)
Displacement, boundary value problem 6a/3(47, 83)
Displacement condition 6a/2(102, 220, 321, 335)
Displacement, definition 6a/2(29, 43, 299)
Displacement derivative 6a/3(308, 430)
Displacement difference 7/1(266, 267, 720)
Displacement distribution function 10(47–50)
Displacement equation(s) of equilibrium 6a/2(90); 6a/3(44)
— for isotropic and homogeneous elastic body 6a/3(45)
— for isotropic viscoelastic bodies 6a/3(44)
Displacement equations of motion 6a/2(212); 6a/3(91, 92)
Displacement field 6a/2(31); 7/1(178, 186, 193, 194, 213, 234)
— processes for a body 6a/3(8, 16, 42)
— solving mixed problem 6a/2(103)
Displacement, general 6a/3(8)
Displacement gradient(s) 6a/2(29); 11/2(154)
— processes 6a/3(8)
— tensor 6(437)
Displacement inequality 6a/3(541)
Displacement intrinsic 7/1(195, 199, 202, 206, 216ff.)
Displacement (Michelson) interferometer 6a/4(10–16, 18–20)
Displacement of the centre of mass 7/1(164, 167, 195, 197, 214, 261, 264)
Displacement problem 6a/2(102, 220, 321)
— characterization in terms of displacements 6a/3(51)
— of incompressible viscoelasticity 6a/3(72)
Displacement relation 6a/3(91, 173, 249, 257, 263, 275, 285–389)
Displacement spike 7/1(439, 440, 442)
Displacement, theorem on single-valued 6a/2(161)
Displacement thickness of boundary layer 8/1(279, 281)
Displacement transducer
— capacitance gauge 6a/4(9)
— displacement interferometer 6a/4(9–12)
— streak camera 6a/4(9)
Displacement vector 6(2, 437); 11/2(154)
Displacement vector, physical components 6(5)
Displacement-free states 6a/2(322)
Displacement-Temperature equation
— of equilibrium 6a/2(317, 321)
— of motion 6a/2(328, 335)
Displacement-time measurements 6a/1(332, 343, 346–351, 629)
Displacer piston 14(12, 15)
Displacer type 14(12)
Dissipation 8/1(19, 51, 52–53, 65, 250); 11/1(4, 13–14, 20)
Dissipation function 6(451); 8/1(178, 236, 335, 343)
Dissipation function of a viscous fluid 6(250, 251, 252)
Dissipation length 8/2(451)
Dissipation mechanism 6(230, 231)
Dissipation of energy in boundary layer 8/1(270, 280)
Dissipation of energy in homogeneous turbulence 8/2(450–451, 461)
Dissipation of internal energy 6(244)
Dissipation of turbulent energy 8/2(529, 534, 547)
Dissipation of ultrasonic energy 11/2(162)
Dissipation of waves in viscous fluids 9(639)
Dissipation potential 6(250, 252, 256, 259, 264, 285, 291, 296, 305–306)
Dissipation potentials, quasi-linear relations 6(260)

Dissipation theorem of Helmholtz and Rayleigh 8/1(258–259)
Dissipative eddies 8/2(467)
Dissipative force 6a/3(258, 285)
Dissipative material 6a/3(22)
Dissipative processes 10(112)
Dissociated gases, sound propagation 11/1(188)
Dissociation 10(92)
Dissociation equilibrium 11/1(248, 249)
Dissociation equilibrium, simple 11/1(206)
Dissociation field effect 11/1(257)
Dissociation of ammonia 11/1(254–257)
Dissociation of non-hydrolyzing salts 11/1(258–260)
Dissociation of sulphuric acid 11/1(252)
Dissociation, stepwise, of 2-2-valent electrolytes 11/1(260–270)
Dissociation through intermediary steps 11/1(216, 221)
Distorsion tensor 7/2(21, 216)
Distorting duct 8/2(535)
Distortion 6(438, 481)
Distortion by second harmonic 11/2(8, 40)
Distortion, finite amplitude 6a/4(276, 286)
Distortion of a plane wave 11/1(124)
Distortion of amplitudes in direct loudspeakers 11/2(12)
Distortion waves 6a/4(275–302)
Distortionless viscoelastic medium 6a/4(199, 200)
Distributed roughness 8/1(448–449)
Distributed section of boundary layer 8/1(347, 348)
Distributed system of Maxwell elements in parallel 6a/3(41)
Distribution 7/1(266)
Distribution function(s) *see also* Singlet, pair, Triplet distribution function, Boundary conditions, Velocity, Displacement, Molecular, Radial distribution function 8/2(442–443); 10(25ff.); 12(152–161, 345); 14(286, 288)
— eigentensor expansion 12(390)
— F_1 12(208, 212)
— F_n 12(207, 212)
— F_r 12(208, 210)
— F_1^σ 12(209, 211)
— F_2^σ 12(209)
— locally Maxwellian 12(240, 241)
— number density 12(219)
— of the Ising model 13(79, 83)
— of the Maxwellian gas 12(391)
— probability density 12(218)
— reduced 12(208)
— series expansion 12(419)
— truncated 12(208)
Distribution matrix 12(476, 486)
Distribution of angle of attack 8/2(386)
Distribution of dislocations in a grain boundary 7/1(648)
Distribution of energy transfer 8/2(494–497)
Distribution of radiated intensity for special sources 11/2(104–105)
Distribution of relaxation times 6(274ff.)
Disturbance, artificial 8/1(386, 425)
Disturbance equation, *see* Perturbation equation
Disturbance, natural 8/1(386)
Disturbance of main flow 8/1(366, 367)
Disturbance, three-dimensional 8/1(430–440)
Divergence 8/1(23)
Divergence, definition 6a/2(11)
Divergence difficulties 14(218)
Divergence difficulties at low frequencies 14(253)
Divergence, instability 6a/3(282–284)
Divergence theorem 6a/2(16, 28); 6a/3(585); 8/1(17, 128)
Divergence theorem in two dimensions 10(284, 286)
Divergent flow 8/2(23, 24–26)
Diverging channels or tubes 8/2(550)
Dividing surface 10(135, 136, 139, 142–144, 146, 147, 152, 158, 160–162, 166, 167, 171, 210, 213, 214, 216, 225, 232, 244, 248, 259, 264, 267, 274)
— equimolecular 10(137, 139, 142, 152, 186, 213, 216, 217, 218, 244, 257)
— of Gibbs 10(282, 286, 287, 289)
Dock problem 9(527, 536–537, 545)
Dodecahedral slip 7/2(69)
Domain boundaries 7/2(188)
Domain, definition 6a/2(348)

Domain of a Boltzmann law 6a/3(14)
Domain of dependence 8/1(67); 9(40)
Domain of influence 6a/2(257); 9(19, 40, 233, 256)
Domain walls, interaction with stress waves 11/2(223, 227)
Domains 15(132, 133, 155, 162)
Dominating function 9(74)
Donati, *see* Beltrami
Donnan equilibrium 13(495, 499, 507, 508)
Donnell's bilinear hypothesis for plastic waves 6a/1(594, 598)
Doppler effect, displacement interferometer 6a/4(10)
Doppler effect for diffracted light 11/2(133–136)
Doppler effect, velocity interferometer 6a/4(12)
Doppler shift 11/1(90)
Double cross-slip 7/2(108, 151)
Double deflections (of a galvanometer) 15(73, 120, 135, 139)
Double demagnetization technique 15(190)
Double limit point 6(387); 6a/3(480)
Double molecule formation 12(311)
Double quartz-thread pendulum after Coolidge and Brüche 12(528)
Double refraction under pressure 6(128)
Double slip, *see also* Duplex slip 6a/1(515–520)
Double velocity correlation tensor 8/2(444)
Double-crystal interferometer for sound waves 11/1(167)
Double-layer potential of Odqvist 8/2(268)
Doublet process 6a/2(241)
Doublet source, *see* Dipole source
Doublet state 6a/2(177, 178)
Double-voids 7/1(401)
Dough, *see* Flour dough
Downstream homogeneity 8/2(538)
Downwash 8/2(408)
D.P.H., *see also* Diamond pyramid hardness 13(633–635)
D.R., *see* Destruction region
Drag, *see also* Wave resistance, Viscous resistance 9(34, 48)
Drag 8/1(89, 90–93, 106); 8/2(385)
Drag coefficient, dependence on cavitation parameter 9(313)
Drag coefficient of an airfoil 8/2(387)
Drag coefficient of the disk, computation 9(425–427)
Drag in infinite cavity flow 9(329–331, 376)
Drag measurements, cavitation influence 9(312)
Drag minimum from cavity formation 9(386)
Drag of a flate plate 9(329–330, 333)
Drag of a hydrofoil 9(586)
Drag of an airfoil 8/1(105–106)
Drag of Riabouchinsky cavity 9(336, 389)
Drag reduction in the laminar airfoil 8/1(365, 408, 413)
Drag theory of permeability 8/2(644–645)
Drag-lift ratio 8/2(385)
Drilled crack, *see* Crack
Driving impedance 11/1(108)
Drop and bubble formation 10(154, 157)
Droplet formation 10(302–303)
Droplet formation from a jet 9(326)
Droplet, spherical 10(293, 302–303)
Dry-compression type machines 14(18)
Dryden's method of boundary layer theory 8/1(282–283)
Dual lattice 13(118, 119)
Dual problem 6a/3(264, 266–269, 271, 274)
Duality between rods and shells 6a/2(664)
Duality of material and spatial descriptions 6a/3(428)
Duct 11/1(44–61, 116–126)
Duct of fixed cross section 8/2(549)
Duct wall periodically supported 11/1(125–126)
Duct with diaphragm 11/1(116–119)
Duct with flexible walls 11/1(119–121)
Ductile fracture 7/2(6, 7)
Ductile material 7/2(3)
Ductility of marble and sandstone 6a/1(486–490)
Duhamel-Neumann law of thermoelasticity 6(123)
Duhamel's paradox 6a/1(16)
Duhem's theorem 10(295)

Dulong Petit law 7/1(107, 245)
Dunn experiment 6a/1(583–586)
Dunn hypotheses for the "quasi-static" impact test 6a/1(583–597)
Dupin's hyperbola 6a/1(21)
Dupin's nonlinear law for wood 6a/1(16, 20)
Duplex slip 7/2(31, 47, 60, 63, 66)
Dupuit and Forchheimer's assumptions 8/2(639)
Duraluminium alloys 7/2(191, 193, 211)
Duwez's impact experiment 6a/1(600–604)
Dye experiment of Reynolds 8/1(352)
Dynamic boundary value problem 6a/3(89)
Dynamic conditions of compatibility 6a/3(431, 563)
Dynamic linear viscoelasticity 6a/3(89)
Dynamic microphone 11/2(37)
Dynamic modulus 6(268–269, 277–278)
Dynamic modulus **G** 6a/3(33)
Dynamic prestress 6a/1(618, 619, 698, 699)
Dynamic problems, other formulations 6a/3(90)
Dynamic reduced variables of polymers 11/1(388)
Dynamic similarity 8/2(542)
Dynamic stability 6a/3(32)
Dynamic stress, photoelastic investigation 6(227)
Dynamic test 7/2(30)
Dynamic variational principles 6a/3(112)
Dynamic viscoelastic process for $\mathscr{B}$ corresponding
— to **G**, **b**, and ρ: DVP (**G**, **b**, ρ) 6a/3(89)
— to **J**, **b**, and ρ: DVP (**J**, **b**, ρ) 6a/3(89)
Dynamic viscosity 6(277)
Dynamic viscosity η 6a/3(34)
Dynamical equation, *see* Equation of motion
Dynamical equations in isotropic turbulence 8/2(491–492)
Dynamical equations of rheology 6(439)
Dynamical matrix 7/1(150)
Dynamical problem for rods 6a/2(699)
Dynamical problems of elasticity 6(107–122)
Dynamical process 6a/2(44, 49)
Dynamical similarity 8/1(57–60, 183, 243–246)
— for perfect fluids 8/1(181)
— for viscous fluids 8/1(243–246)
— relation to dimensional analysis 8/1(183)
Dynamical systems
— definition of 6a/3(128)
— differential equations 6a/3(127, 129, 157, 172)
— equilibrium solution, uniqueness of 6a/3(130)
— examples 6a/3(127, 128, 160)
— general features 6a/3(127, 131)
— of Hale 6a/3(130, 168, 169, 255)
— of Zubov 6a/3(130)
— stability of, definition 6a/3(134)
— stability of, influences on 6a/3(135, 288)
— stability of, influences on, elasticity 6a/3(171, 207, 211, 250, 254, 255)
— stability of, influences on, examples 6a/3(136–139)
Dynamically admissible stress field 6a/2(230)
Dynamics of homogeneous turbulence 8/2(445–446)
Dynamics of isotropic turbulence 8/2(448–450)

E moduli for 31 elements as a function of ambient temperature 6a/1(389–393)
E moduli, Riccati's ratio of 6a/1(161, 190)
E modulus, definition of 6a/1(74)
E modulus, height of the 6a/1(48, 60, 97, 186)
E modulus, weight of the 6a/1(186)
Ear, detection of pressure amplitudes 11/2(35)
Easy glide region 7/2(52)
Echo-pulse transmission technique 6a/4(75)
Eddy currents 15(258)
Eddy currents in magnets 15(73, 75, 76, 203)
Eddy viscosity transfer 8/2(471)

Edge 6a/3(483)
Edge dislocation(s) 7/1(96, 481, 499, 536, 572, 596, 611); 7/2(16)
Edge dislocation as line-dipole 7/1(625)
Edge dislocation, stress-field of an 7/1(531)
Edge effects on photoelastic measurements 6(162, 164)
Edge, flow around it 8/1(103)
Edge hit by a jet 11/2(97)
Edge wave of Stokes type 9(550)
Edge wave of Ursell type 9(551)
Effective elastic constants 6a/4(239)
Effective incidence 8/2(408)
Effective mass 14(292)
Effective pore space, *see also* Porosity 8/2(626)
Effective stress 8/2(629)
Efficiencies 14(53)
Efficiency, adiabatic 14(78)
Efficiency changes of electrostatic loudspeakers 11/2(17)
Efficiency of a horn loudspeaker 11/2(20)
Efficiency of a siren 11/2(25)
Efficiency of an electrodynamic loudspeaker 11/2(10)
Efficiency of Saint Clair generator 11/2(24)
Efficiency of sound generation by cannons 11/2(34)
Efficiency of the Hartmann generator 11/2(31)
Efficiency, practical 14(56, 74)
Efficiency, relative 14(9, 18)
Efflux of a plane jet, *see also* Discharge 9(341)
Ehrenfest-Oppenheimer law 12(458)
Ehrling lemma 6a/2(354)
Eigenfunction expansions 6a/3(170, 258, 279, 280, 284, 285)
Eigenfunctions, complete set of 6a/2(369)
Eigenmode 6a/3(225, 226)
Eigen-temperature 8/1(420)
Eigentensor expansion 12(378–381, 390)
Eigentensors of the Maxwellian gas 12(390)
— mixture 12(419)
Eigenvalue, definition 6a/2(369)
Eigenvalue method for the one-dimensional Ising model 13(111)
Eigenvalue method for the two-dimensional Ising model 13(113–115, 121–133)
Eigenvalue problem of the perturbation equation 8/1(370–372, 378)
Eigenvalue problems 6a/2(386)
Eigenvalues and wave speeds 6a/3(98, 101–104)
Eigenvalues of basin waves 9(623–627, 630, 652)
Eigenvalues of modes in a bounded canal 9(670, 671)
Eigenvalues, spectrum of differential operator law 6a/3(38, 41, 42)
Eigenvector, *see also* Wave axis 6a/3(98)
Eikonal equation 11/1(80)
Eikonal function 11/1(80)
Einstein formula 7/1(326)
Einstein function 13(571); 14(285, 286)
Einstein relation of diffusion theory 13(623)
Einstein term 7/1(248, 258)
Einstein-Bose gas of charged particles 15(314)
Einstein's convention 7/1(515)
Einstein's model 7/1(396, 414)
Einstein's relation 7/1(408, 418)
Einstein's theory
— of dilute sols (dispersions in liquids) 6(524, 526)
— of light scattering for binary systems 13(312, 315)
— of light scattering, justification 13(262, 267)
Ekman flow on the rotating earth 8/2(109–111, 127)
— generalisation 8/2(175–176)
Elastic after-effect 6a/1(38, 44, 48, 49, 67, 70, 77, 81, 82, 97, 99, 102, 103, 118, 145, 228, 279, 286, 299, 305, 732); 7/2(3); 11/2(173)
Elastic anisotropy 7/1(572, 615)
Elastic body, material 6a/3(19, 175)
Elastic body, mathematical definition 6(14)
Elastic coefficients 6a/4(116, 185, 226, 230, 231, 239); 7/1(180, 609); 11/2(154)

— effective 6a/4(226, 228, 230, 232, 234, 235)
— fourth-order 6a/4(222)
— mixed 6a/4(225)
— third-order 6a/4(221)
Elastic collision, equations 12(334–336)
Elastic compliance 6a/3(42)
Elastic constants 7/1(179, 180, 182, 185, 187, 192, 193, 195, 196, 197, 200, 207, 226, 229, 245, 246, 273, 285, 286); 14(287)
— adiabatic 7/1(237, 238)
— adiabatic vs isothermal, *see* Moduli
— and atomic volume 6a/1(218, 230, 394)
— as function of temperature 6a/4(240)
— comparative study of 6a/1(218, 291, 292, 304, 375, 385, 394, 397–406, 410)
— dependence on temperature 7/1(286)
— determination of 6a/4(251)
— effective 6a/4(239)
— first study of the temperature dependence of 6a/1(218, 227)
— Fuchs' calculation of 6a/1(385)
— isothermal 7/1(236, 238, 280, 284)
— master 6a/4(243)
— measurement of 6a/1(179)
— of anisotropic solids 6a/1(230)
— of the classical theory 6(20–26)
— pressure, coefficients of 6a/1(356, 409)
— quantized distribution of 6a/1(149–152, 397–406)
— temperature dependence of 6a/1(360–396)
— third-order 6a/1(155, 409, 410)
— ultrasonic determination of 6a/1(345, 352–357, 377–380)
Elastic continuation 6a/3(539, 540)
Elastic core 6a/3(568, 575)
Elastic curve of beams 6a/1(20, 25)
Elastic defect 6a/1(26, 58, 72, 94, 123, 140, 148, 457)
Elastic deformation 6a/3(545, 550)
— gradients 6a/3(538, 550)
— history 6a/3(539)
— of fluids 10(91)
— of polymers 13(432–451, 487)
Elastic differential equation 7/1(516, 522, 533)
Elastic differential operator law 6a/3(39)
Elastic displacement fields 6a/2(131)
Elastic distortional potential 6(291)
Elastic energy 11/2(158–159)
Elastic energy flux 11/2(158, 159)
Elastic flow 7/1(478, 522)
Elastic force 13(442, 443)
Elastic fundamental integral 7/1(516)
Elastic interaction between foreign atoms and dislocations 7/2(168)
Elastic layer hypothesis in impact of bars 6a/1(314, 319, 322)
Elastic limit 6a/1(87, 174, 223, 229, 425, 457, 458, 465, 467, 471, 484, 489, 501–508, 630, 642, 646–649, 682); 7/2(32)
— Bauschinger's definition of 6a/1(458, 682)
— dynamic 6a/1(641, 646–649)
— elevation of 6a/1(453–457, 463, 577, 578, 631, 647, 648, 733)
— natural 6a/1(467, 472)
Elastic material 6a/2(301)
Elastic medium 6a/4(199, 214, 215)
Elastic medium and fluid, interface 9(455)
Elastic mixtures 6a/3(286)
Elastic moduli 7/1(180, 237)
Elastic modulus, *see also* E modulus, Shear modulus, Bulk modulus
— and cohesion, effect of electric current and magnetic fields upon 6a/1(89, 90, 92, 238, 329, 367–371, 685)
— complex 6a/4(195)
— first measurement of μ by Coulomb 6a/1(173)
Elastic motion 6a/2(233)
Elastic nonlinearity 6a/4(275)
Elastic potential 6(13–14, 244, 256); 6a/2(380, 386, 544); 11/2(159)
— adiabatic 6(247, 250)
— combined of solid and liquid phases 6(265)
— existence 11/2(154)
— isothermal 6(247, 250)
— variation 6(252)
Elastic process 6a/2(215)
Elastic progressive wave 6a/2(246)

Elastic properties of wood 6a/1(15–21, 24, 157, 183, 186, 243)
Elastic recovery 6a/1(30, 70, 81, 97, 102, 118, 228, 279, 299)
Elastic region 6a/3(501, 563, 578, 580)
Elastic response, equilibrium 6a/3(18)
Elastic response, initial, instantaneous 6a/3(18)
Elastic restoring forces of a movable basin 9(628)
Elastic shear moduli
— first measurement of 6a/1(173)
— quantized distribution of 6a/1(397–400, 406)
— variation with stress 6a/1(90, 94, 129–131, 137)
Elastic singularities 7/1(517, 525f., 528)
Elastic sol (Lethersich body) 6(467–468)
Elastic solid 6a/4(197)
Elastic sphere in an elastic medium 11/2(166)
Elastic state 6a/2(95, 102); 6a/3(49)
— properties 6a/2(133)
— with displacement vanishing at infinity 6a/2(167)
— with stress vanishing at infinity 6a/2(168)
Elastic stiffness 6a/4(126)
Elastic stress surface 6a/3(497)
Elastic wave(s) 6(107–118)
— generation by body forces 6(116–118)
— of order *n* 6a/2(255)
— radiation of 7/1(623)
— transverse and longitudinal 10(104)
Elastica 6a/2(697, 698)
Elasticities A_{iAkB} 6a/3(178, 182)
Elasticities c_{ijkl} 6a/3(180, 215, 218, 244)
Elasticities d_{ijkl} 6a/3(180, 206, 209, 211, 213, 216, 217, 239, 255, 264)
Elasticities, elasticity, *see* Tangent modulus
Elasticities, elasticity field 6a/2(68, 75)
Elasticities, modulus 6a/3(412)
Elasticities, positive longitudinal 6a/3(317)
Elasticities, tensor 6a/3(316)
Elasticity 7/1(178)
Elasticity and atomic weight, *see also* Elastic constants and atomic volume 6a/1(218, 394)
Elasticity and elastic constants of liquids 10(20, 69–71)
Elasticity, classical theory 7/2(2, 18)
Elasticity, defect of, *see* Elastic defect
Elasticity, density dependence of 6a/1(186, 229)
Elasticity, finite, of rubber 6a/1(115, 245, 309, 716, 720, 726–741)
Elasticity, first nonlinear law of 6a/1(13)
Elasticity, hyperbolic law of 6a/1(71, 115)
Elasticity, imperfect 6a/1(274)
Elasticity law, non-linearity of the 7/1(569)
Elasticity, maximum 6a/1(99)
Elasticity, non-linear theory 7/2(2, 17)
Elasticity of rubberlike materials 13(435)
Elasticity tensor, general 6a/2(67–69, 309, 313)
Elasticity tensor, isotropic materials 6a/2(75, 317)
Elasticity, theory of linear 7/1(563)
Elasticity, theory of non-linear 7/1(563)
Elasticity, theory of, of anisotrope mediums 7/1(515)
Elastico-viscosity (Maxwell body) 6(465–467)
Elastic-plastic 6a/3(495)
— behaviour 6(313–322)
— boundary 6(400, 403); 6a/3(580)
— boundary experiments 6a/1(635, 692, 695)
— boundary-value problems 6(399–425)
— torsion 6a/3(499, 562–564)
— torsion problem 6a/2(405, 422)
Elastic-viscoelastic correspondence 6a/3(68)
Elastodynamic Lamé and Somigliana stress functions 6a/3(93, 94)
Electric conductivity 15(189)
Electric field 7/1(208ff., 217)
— longitudinal 7/1(219)
— mean 7/1(213)
Electric resistance strain gage 6a/1(329, 333, 342, 347, 610)
Electrical conductivity of glass 13(584, 591)

Electrical detection interferometer for ultrasound 11/2(142–143)
Electrical discharge in the cavitation void 11/2(281)
Electrical free energy of polyelectrolytes 13(494, 495)
Electrical impedance 11/2(5)
Electrical impedance of a vibrating piezo-electric plate 6a/4(262)
Electrical interaction between foreign atoms and dislocations 7/2(168, 177)
Electrical properties of ceramics 13(606)
Electrical quadrupole tensor 11/2(236)
Electrically driven flying foil 6a/4(4)
Electrocapillary phenomena 10(303–304)
Electrodynamic loudspeaker 11/2(9–12)
Electrodynamic pistonphone 11/2(63–64)
Electrodynamic potentials 6a/4(155)
Electrodynamic pressure microphone 11/2(36–38)
Electrodynamics of superconductivity 15(320)
Electro-elastic behavior 6a/1(89, 90, 238, 367–371)
Electrolytes of high concentration 13(392–395)
Electrolytes, potential of average force 13(306)
Electrolytes, solutions 13(369–395)
Electrolytes, surface tension calculation 10(288, 298)
Electromagnetic effects 6a/3(286)
Electromagnetic field of a lattice vibration 7/1(210)
Electromagnetic field, transverse 7/1(219)
Electromagnetic forces 10(67)
Electromagnetic induction meter 8/2(586)
Electromagnetic transducers 11/2(92–94)
Electromechanical coupling coefficient 6a/4(259)
Electromechanical interactions 6a/4(158, 165)
Electromechanical transformer 6a/4(266)
Electron diffraction 7/2(107–109)
Electron mean free path comparable with ultrasonic wavelength 11/2(208, 210)
Electron microscopy of surface effects 7/2(80–103, 180)
Electronegativity and glass formation 13(526, 541)
Electron-electron collisions 14(194)
Electron-electron interaction 14(193)
Electro-neutrality 11/1(208)
Electron-fluid, collective oscillations of the 14(193)
Electronic excitation of monatomic gases 12(101)
Electronic microscope 7/1(69)
Electronic modes 14(283, 288)
Electronic specific heat 15(230, 235, 238, 269, 362)
Electronic states, density of 14(291)
Electron-phonon interaction energies 15(343, 345, 347, 362, 363)
Electron-phonon interaction parameter 14(233, 240, 252, 254)
Electrons and crystal structures of elements, configurations of (table) 7/1(79)
Electrons and lattice waves, interaction between 14(232)
Electrons and phonons, interaction between 14(251)
Electrons and plasma, interaction between 15(358)
Electrons, interaction with superconductors 15(245)
Electrons, normal 15(245)
Electrons per atom, number of free 14(234, 241, 254)
Electrons, scattering of 7/1(426)
Electrons, superconducting 15(245)
Electrons with small effective mass 15(312)
Electro-optic coefficients 6a/4(166, 181)
Electrostatic actuator of Ballantine 11/2(64–66)
Electrostatic bond strength 13(565)
Electrostatic loudspeaker 11/2(15–17)
Electrostatic probe 11/2(116)
Elementary curvilinear triangle 6a/2(493)
Elementary events 7/2(221)

Elementary (lattice) cell 7/1(111, 260, 305, 310, 385)
Elements, integral 7/1(49)
Elements of area and volume 6a/2(441)
Elements of translation 7/2(24–28)
Elements of translation, determining factors 7/2(24)
Elements of translation, survey (table) 7/2(26)
Elements, superconductive 15(229)
Ellipsoid spinning in a steady flow 8/2(216–217)
Ellipsoid, submerged 9(577)
Ellipsoid, superconducting 15(220)
Ellipsoid, translation in a fluid 8/2(233–235)
Ellipsoid, uniform translation in Oseen approximation 8/2(313–315)
Elliptic cross section of a tube 8/2(69, 73)
Elliptic cylinder, uniform translation in Oseen approximation 8/2(311–313)
Elliptic cylinder with moving plane inside 8/2(79)
Elliptic differential equation 9(16, 29, 93, 94, 99, 209, 667, 683)
Elliptic, ellipticity 6a/3(211, 213, 460, 462, 478)
Elliptic functions 8/2(20, 93)
Elliptic functions of Jacobi 9(706ff.)
Elliptic hole in an infinite plate 6(57–58, 64)
Elliptic integrals 8/2(25, 36)
Elliptic integrals in capillarity theory 10(301)
Elliptic loading 8/2(409)
Elliptic operator 6a/2(355, 374)
Elliptic problem of differential equations 6(339, 369, 371, 384)
Elliptic region 6(384)
Elliptic strong 6a/3(64, 99)
Elliptic vortex of Kirchhoff 8/2(141)
Elliptical cylinder, boundary layer 8/1(400, 403)
Elliptical cylinder, torsion 6(30, 34)
Elliptical differential equation 8/1(66)
Elliptical lift distribution 8/1(106)
Elliptical orbits, modifications by second-order infinitesimal wave theory 9(661)
Elliptical orbits of first-order infinitesimal wave theory 9(500, 501)
Elongation of a crystal, measurement 7/2(28)
Elongation of a glass fibre 13(579)
Elongation resolution 6a/1(37, 124)
Embedded (internal) mirror 6a/4(15–17, 56, 63)
Embedding 6a/2(35, 352)
Embouchure 6a/1(253)
Embryo 7/2(218)
Emission of a phonon 11/2(216)
Empirical equation of state 12(48–57)
Empirical equations of state of mixtures 12(65)
Empty interstices 13(553)
Emulsification by ultrasonics 11/2(277, 278)
Emulsion 6(435, 523)
Enamels 6a/1(126)
Enclosures of electrodynamic loudspeakers 11/2(13–15)
End loss in anemometer wire 8/2(563, 564, 571)
Endosity (=negative viscosity) 6(478)
End-to-end distance 13(404, 408, 439)
End-to-end vector 13(436, 437)
Endurance limit 6(592–593, 606)
Energies, higher order 6a/3(211, 212, 236–238)
Energy, *see also* Free energy, Internal energy, Power and Work
Energy attenuation 11/1(14)
Energy balance 6a/3(174–176); 6a/4(120)
Energy balance in decaying homogeneous turbulence 8/2(461–462)
Energy band 14(229)
Energy barrier for a dislocation 11/2(189)
Energy conservation law 8/1(177, 239)
Energy, conservation of 7/1(303, 307); 14(207)
Energy content 7/1(257)
Energy criterion for stability, *see* Lagrange-Dirichlet stability concept, Discrete systems, conservative
— Beatty's 6a/3(201, 205)
— criticisms 6a/3(203–206)
— Hadamard stability 6a/3(197)
— history 6a/3(196–199)

— Lagrange-Dirichlet criterion 6a/3(154, 159, 195, 203)
— of Truesdell and Noll 6a/3(202)
— Pearson's condition 6a/3(202)
— relation to Liapounov stability (footnote) 6a/3(164)
— relation to Liapounov stability, necessity conditions 6a/3(200–202)
— relation to Liapounov stability, sufficiency conditions 6a/3(203, 206)
— statement of 6a/3(196)
Energy decay 8/2(529)
Energy density 10(35)
— acoustic 11/1(7, 9)
— elastic 7/1(178, 186, 187, 189, 192, 198, 200, 224)
— of fluid motion 11/1(3)
— of free 7/1(234)
— of internal 7/1(234)
Energy deposition 6a/4(3, 85, 99)
Energy dissipation 6(249, 260); 8/2(527, 529)
— in boundary layer 8/1(270, 280)
— in viscous flow 8/1(250)
— mechanism 11/2(162)
Energy, elastic 7/1(273, 275, 276, 279, 281, 287); 11/2(158–159)
Energy, electrostatic 7/1(128, 136, 139, 232)
Energy equation 8/2(548); 12(351, 355)
— Eulerian form 8/1(18–19)
— general 6a/2(326)
— integral form 8/1(8)
— Lagrangian form 8/1(32)
— of Carl Neumann 8/1(177)
— of compressible flow 8/1(335, 343)
— of disturbance velocity 8/1(115)
— rods 6a/2(653, 658)
— shells, derived from 3-dimensional theory 6a/2(515)
— shells, direct approach 6a/2(481, 482)
Energy exchange between rotational and translational degrees of freedom 11/1(196)
Energy exchange between vibrational and translational degrees of freedom 11/1(197–199)
Energy field 6a/4(161)
Energy flow 7/1(293,295); 8/1(18)
— acoustic 11/1(7, 9)
— vector (or flux) 11/1(3, 7, 152)
Energy flux 6a/4(118); 10(35, 82, 96)
— density 7/1(293)
— elastic 11/2(158, 159)
— in surface wave 9(519)
— partial 10(41)
Energy, free 6a/2(497, 511, 555, 570); 7/1(234, 243, 271, 272, 273, 288, 289, 553)
Energy functional 6a/2(385, 391, 403)
Energy gap in superconductors 11/2(221–223)
Energy integral 9(135)
Energy integral for free surface motion 9(718–721)
Energy, internal 6a/2(481, 510); 6a/3(362); 6a/4(120, 121); 7/1(233, 236, 243, 244, 245)
Energy levels of paramagnetic salts 15(41)
Energy losses 11/1(13)
Energy, mean thermal 7/1(273, 284)
Energy method 6a/3(153); 8/1(367)
Energy momentum theorems 15(288)
Energy of kink formation in dislocations 11/2(205–206)
Energy of surface waves 9(458–460)
Energy of the boundary between normal and superconducting phases 15(327)
Energy parameter in binary systems 13(145)
Energy per atom 7/1(121)
Energy per elementary cell 7/1(119, 121, 129, 226, 239, 280ff.)
Energy, potential 7/1(106, 109, 145, 269, 270, 272, 275, 281, 282, 285)
Energy propagation in surface waves 9(515–522)
— second-order infinitesimal wave theory 9(661)
Energy spectrum in Heisenberg's transfer theory 8/2(475, 478–481)
Energy spectrum in Kolmogoroff's theory 8/2(464, 475)
Energy spectrum of turbulence 8/2(449, 453, 464)
Energy, superficial internal 10(78, 79, 80)
Energy test 6a/3(154)
Energy thickness of boundary layer 8/1(280)

Energy, total 6a/3(137, 138, 156, 160, 162, 163, 172, 207, 211, 222, 224, 231–236, 250, 265, 267, 269, 287, 288)
Energy, total stored 6a/4(162)
Energy transfer
— at small wave numbers 8/2(497–498)
— by neutron scattering 13(197, 198)
— distribution 8/2(494–497)
— equation 8/1(138, 247)
Energy transport 10(87)
Energy variation 7/1(596)
Energy, zero-point 7/1(106)
Energy-containing eddies 8/2(464, 466)
Energy-flux vector 6a/2(214)
Energy-gap model 15(276, 283)
Energy-strain relaxation function 6a/4(78, 95ff.)
Energy-temperature relaxation function 6a/4(78, 95ff.)
Enskog's approximation 12(384)
Entanglements 13(429, 448)
Enthalpy 6a/4(127); 8/1(12, 18–19, 32, 94); 9(3, 250); 12(300); 15(51)
Enthalpy, critical 8/1(223)
Enthalpy, free 7/1(271)
Enthalpy of adiabatic ideal gas flow 9(6)
Entire boundary, clamped 6a/2(263)
Entropy 6a/4(76ff., 80, 83, 90, 92, 94, 123); 7/1(233, 235, 237, 243); 8/1(12, 32, 33, 173, 175); 9(3, 167, 193); 10(67, 78, 81, 82, 120, 121); 12(359); 15(111, 112, 118, 133, 375, 389, 390, 391, 392, 401, 405, 409, 410, 424, 452, 455, 457, 458)
Entropy, acoustic 11/1(6)
Entropy balance 6(245, 246, 294)
Entropy, communal 10(242ff., 248, 250, 263, 273)
Entropy density 7/1(234); 12(487)
Entropy, equilibrium response function 6a/4(78)
Entropy filter 15(379)
Entropy fluctuations 11/1(5)
Entropy flux 6(245, 264); 10(83)
Entropy inequality, *see* Clausius-Duhem inequality, Stress-entropy 6a/3(174)
— (Clausius-Duhem), general 6a/2(301, 510, 511)
— rods 6a/2(667)
— shells, derived from 3-dimensional theory 6a/2(515)
— shells, direct 6a/2(483)
Entropy, instantaneous response function 6a/4(79)
Entropy jump 6a/4(290)
Entropy of an inviscid gas 9(240, 242, 272, 283)
Entropy of dilution 13(172–174, 334, 339)
Entropy of evaporation 13(231, 232)
Entropy of melting 13(289, 299)
Entropy of mixing 13(172–174, 325, 333, 338, 350, 351)
Entropy of network deformation 13(436, 440, 442)
Entropy of surface, *see* Surface entropy, Superficial entropy
Entropy of variation in gas flow 9(240–249, 272–274)
Entropy principle for constitutive relations,
— general 6a/2(302)
— rods 6a/2(669)
— shells 6a/2(512)
Entropy production 6(245, 246, 248, 251, 264); 6a/2(483, 492, 511); 6a/3(539, 540); 10(83); 11/1(4); 12(496)
— local 12(414)
— minimum rate 6(253)
Entropy relation(s) 6a/2(303, 306); 6a/4(79)
Entropy, residual inequality 6a/3(183, 188)
Entropy temperature diagram 12(26)
Entropy terms in polymer solutions 13(457)
Envelope of characteristics, *see* Limit line
Envelope of characteristics 6a/3(477, 479, 483)
Envelope of Mohr circles 6(352, 376); 6a/3(439, 458)
Eötvös formula 10(174)
Epicycloids 6a/3(453)
Epicycloids as principal curves 9(89)
Epicycloids as slip lines 6(363)
Epicycloids in hodograph plane 9(686, 692–694, 701)
ε-system 6a/2(618)

Equal modulo a rigid displacement 6a/2(104)
Equation of continuity, *see also* Continuity equation 8/2(3, 8, 10, 47, 66, 200, 444, 447, 545); 9(2, 228, 249, 283, 448, 450); 11/1(137, 276); 12(298, 351, 354)
Equation of hydrostatics 10(283, 287)
Equation of hydrostatics, one- and two-dimensional analogues 10(285, 298)
Equation of motion 6a/4(120, 163); 11/1(4, 137, 276); 12(298, 343, 351, 355)
— general Cauchy form 8/1(135)
— in Euler representation 8/1(18, 33)
— in Lagrange representation 8/1(31)
— intrinsic 8/1(155–157, 190, 250)
— invariant forms 8/1(142)
— linearization 6a/4(217)
— material form 8/1(135–136)
Equation of Navier and Stokes, *see also* Navier 8/1(51–52)
Equation of state 6(236); 6a/1(688); 9(229, 240, 249, 251, 273, 283, 449); 11/1(4, 137, 276); 12(48–57, 65)
— caloric 13(11, 17, 27)
— derived by Tonks 13(55)
— for a gas of rigid spheres 13(247, 276)
— for liquids 10(13, 19, 52, 55)
— in α-creep region 7/2(199)
— reduced 13(229)
— thermal 13(13, 27, 65, 229, 240, 247, 276)
Equation of variations 9(414)
Equations of balance 6a/4(119)
Equations of boundary layer
— flow in three dimensions, derivation 8/1(307–310)
— flow in three dimensions, exact solutions 8/1(314–317)
— two-dimensional steady flow, derivation 8/1(266–268)
Equations of compressible boundary layer flow in two dimensions 8/1(335–336)
Equations of motion 6(10–12, 139, 235, 323); 9(228, 249, 283, 448, 450)
— elasticity 6a/2(45, 64, 212)
— for an elastic medium 11/2(155–156)
— rods 6a/2(653, 658, 672)
— shells, catalogue of 6a/2(526)
— shells, derived from 3-dimensional theory 6a/2(520)
— shells, direct approach 6a/2(494)
— shells, restricted theory 6a/2(552)
— thermoelasticity 6a/2(326)
Equations of Stokes, *see* Stokes
Equilateral triangle beam 6a/1(199)
Equilibrated surface tractions 6a/3(63)
Equilibrium, astatic 6a/2(63, 64, 98, 193)
Equilibrium, axis of 6a/3(216, 217, 219, 229, 239, 242)
Equilibrium between droplet and vapor 10(303)
Equilibrium between electrons and phonons 14(255)
Equilibrium bulk modulus 6a/4(72)
Equilibrium, chemical, in real gases 12(64, 65)
Equilibrium coefficient of linear thermal expansion 6a/4(86)
Equilibrium condition(s) 7/1(516, 525, 572); 11/1(208)
— for instability 6a/3(199, 200)
— for stability of 6a/3(192, 194)
— in statement of energy criterion 6a/3(196)
— necessary 6a/3(200–202)
— sufficient 6a/3(203)
Equilibrium conductivity matrix 6a/4(208)
Equilibrium configuration in Cottrell potential 11/2(189)
Equilibrium constant 11/1(208); 12(64)
— dependence on pressure 11/1(209, 218)
— of ammonia dissociation 11/1(255)
— of ion cloud 11/1(213, 224)
— of sulphuric acid dissociation 11/1(252)
Equilibrium criterion of Gibbs 10(291)
Equilibrium, deviation of a distribution from 14(231)
Equilibrium, energy equation 6a/2(312)
Equilibrium equation(s) 6(235, 293); 6a/3(42, 43)
— covariant form 6(11, 15, 28)
— general 6a/2(49, 59, 90, 312)

Equilibrium equation(s) (cont.)
— in lines of curvature coordinates 6a/2(633)
— in rectangular, cylindrical, and polar coordinates 6a/2(93)
— Kelvin's solution 6(87–88)
— of rods, integrals 6a/2(696)
— solution 6a/2(53)
— vector form 6(87)
Equilibrium heat equation 6a/2(317)
Equilibrium history
— (constant history), hydrostatic compression 6a/4(71)
— of the strain and entropy (one dimension) 6a/4(81)
— of the strain and temperature (one dimension) 6a/4(79)
— of the strain (one dimension) 6a/4(27, 29ff., 44, 63ff.)
Equilibrium Hugoniot stress-strain curve 6a/4(84ff.)
Equilibrium Hugoniot temperature-strain curve 6a/4(85ff.)
Equilibrium in each load region 6a/2(193)
Equilibrium isentropic second-order modulus 6a/4(81)
Equilibrium isentropic tangent modulus 6a/4(81)
Equilibrium isothermal bulk modulus 6a/4(86)
Equilibrium isothermal second-order modulus 6a/4(79)
Equilibrium isothermal tangent modulus 6a/4(79, 83)
Equilibrium, Lagrange instability and existence 6a/3(151)
Equilibrium, Liapounov instability of 6a/3(200)
Equilibrium modulus 6a/3(18)
Equilibrium of diffusion 11/1(212, 213, 215)
Equilibrium of forces 6a/2(63)
Equilibrium of hydratization 11/1(225)
Equilibrium of longitudinal modes 14(214)
Equilibrium of phonons 14(211)
Equilibrium points 6a/3(131)
Equilibrium position 7/1(145, 269, 270, 273, 276, 281, 283, 287)
Equilibrium pressure-density curve 6a/4(70–75)
Equilibrium problem, elasticity 6a/2(380)
Equilibrium problem, heterogeneous elastic medium 6a/2(386, 421)
Equilibrium problem, thin plate 6a/2(377)
Equilibrium response 6a/3(13)
Equilibrium second-order modulus 6a/4(28)
Equilibrium solution 6a/3(132)
— boundary-value problem 6a/3(151)
— definition 6a/3(129)
— non-uniqueness 6a/3(144)
— uniqueness and dynamic problem 6a/3(130)
Equilibrium sound velocity
— (equilibrium longitudinal sound velocity) 6a/4(40, 44)
— isentropic 6a/4(82, 85)
Equilibrium specific heat 6a/4(79, 83, 86ff.)
Equilibrium spectrum of Heisenberg 8/2(473)
Equilibrium, stability
— of buckling mode 6a/3(287, 288)
— of cantilevered rod 6a/3(282)
— of convexity 6a/3(234, 273)
— of curved panel in windstream 6a/3(281)
— of divergence, non-uniqueness in follower force problems 6a/3(282, 283)
— of non-uniqueness and instability 6a/3(224)
— of non-uniqueness and neutral equilibrium 6a/3(226)
— of under dead load in class of small incremental displacements, fixed surface 6a/3(206)
— of under dead load in class of small incremental displacements, movable surface 6a/3(215)
— of under dead load in class of small incremental displacements, theorem 6a/3(219)
— of under time- and position-dependent data 6a/3(268, 269, 274)
Equilibrium state 7/1(240); 11/1(6)
— natural 6a/4(28, 39, 80, 82, 84, 90)
— static 6a/4(122)
Equilibrium stress-energy curve 6a/4(98ff.)

Equilibrium stress-energy modulus 6a/4(79)
Equilibrium stress-strain response
— (equilibrium stress-strain law, equilibrium stress-strain curve), mechanical theory, dilatational motion 6a/4(70)
— mechanical theory, one-dimensional motion 6a/4(27–29, 39ff., 44ff., 48, 51, 53ff., 64ff., 75)
— thermodynamic theory, isentropic 6a/4(81)
— thermodynamic theory, isothermal 6a/4(79, 80, 84, 86)
Equilibrium stress-temperature modulus 6a/4(79, 83, 86)
Equilibrium structure and glass structure 13(544, 555)
Equilibrium tangent modulus 6a/4(28)
Equilibrium, thermal 7/1(293, 294, 309); 14(193)
Equilibrium, thermodynamic 7/1(390)
Equilibrium viscosity 6(293)
Equipresence 6a/2(530)
Equivalence principle for attenuation, infinite frequency damped progressive waves and propagating singular surfaces 6a/3(110)
Equivalence principle for wave speeds and wave axes, infinite frequency damped progressive waves and propagating singular surfaces 6a/3(110)
Equivalence relations 6a/3(139, 157)
Equivalence, semiequivalence 6a/3(139)
Equivalent chain 13(405)
Equivalent circuit(s)
— for acoustic systems 11/2(3–4)
— for bass-reflex enclosures 11/2(14–15)
— for electrodynamic loudspeakers 11/2(10–12)
— for horn loudspeakers 11/2(18)
— of a piezoelectric transducer 6a/4(264)
— of a siren 6a/4(267, 269); 11/2(24)
— of actuator 11/2(65)
— of condensor microphone 11/2(39, 41)
— of magnetostrictive ultrasound generator 11/2(88–90)
— of moving coil microphone 11/2(37)
— of quartz transducer 11/2(79–80)
— of the probe microphone 11/2(51)
Equivalent processes 6a/2(487)
Equivalent sphere 13(453, 454, 473, 474)
Equivalent systems of equations, classical linear theory of shells 6a/2(583)
Equivalent turbulence 8/1(446)
Ergodicity 13(281)
Erosion by intense ultrasonics 11/2(276)
Error estimate in the linear theories of plates and shells 6a/2(584)
Error function 8/2(120, 123)
Ertel-Rossby theorem 8/1(38)
Ertel's vortex theorem 8/1(43–45)
Escape velocity (into vacuum) 9(6)
Eshelby's solution 6(121)
Estimates in subsonic flow 9(125–128)
Etch figures 7/1(68)
Etching 7/1(630); 7/2(111)
Etching of a glass surface 13(626)
Ethylene 14(26, 30)
Ethylene-di-bromide, molecular heat 13(210)
Ethyl-sulfates of the rare earths 15(8, 17, 18)
Eucken relation 12(499–501)
Euclidean point space 6a/3(4)
Euler critical load 6a/3(277)
Euler equations 12(216, 252, 258)
Euler limit 8/1(328)
Euler modulus 6a/1(161, 187)
Eulerian coordinates 6a/3(428, 536)
Eulerian derivative 9(2)
Eulerian description 9(447)
Eulerian differential equation 8/2(43)
Eulerian equations 7/1(157)
Eulerian picture of fluid motion 11/1(2–3)
Eulerian representation 8/1(3, 17–19, 20)
Euler's buckling formula 6a/1(162, 196, 198, 201)
Euler's equation 8/1(33, 135)
Euler's expansion formula 6a/4(115)
Evaporation heat 13(231)
Evaporation rate 12(541)
Evaporation, retrograde 12(66, 67)
Evaporator 14(16)

Evapor-ion pump 12(646)
Eventually monotone 6a/3(20)
Even-valent groups 14(180)
Eversion experiments 6a/1(740)
Evolution of steady shock waves 6a/4(58ff.)
Evvard type equation 9(305, 306)
Ewald's method 7/1(134, 136, 226)
Exact solution for the two-dimensional Ising model 13(121–133)
Exact solutions in surface wave theory 9(714–757)
Exceptional direction 6(385, 389–390); 6a/3(479, 483, 484)
Exceptional line 6a/3(418)
Exceptional line, plane etc., *see* Characteristic
Excess entropy at a surface 10(296)
Excess free energy at an interface 10(286, 296)
Excess (non ideal) free-energy of mixing 12(61)
Excess thermodynamic functions at an interface 10(290)
Excess volume 10(232, 233)
Exchange energy 14(293)
Exchange gas 15(198)
Exchange interaction 15(43, 82f., 86, 101, 108, 128)
Exchange narrowing 15(28)
Exchange of dependent and independent variables 6a/3(443)
Exchange of stabilities 6a/3(226)
Exchange reactions in the thermodiffusion column 12(329)
Exchange terms 15(353)
Excitation in solid 14(226)
Excitation methods for ultrasonic vibrations 11/2(147–149)
Excitation of sound by friction 11/2(100)
Excitation of vibrations 11/1(249)
Excitation reaction 11/1(209, 210, 211)
Excluded volume, *see also* Covolume 13(407, 408)
— effect on second virial coefficient 13(471)
— model of polymer solutions 13(451, 452–456, 461–463, 487, 491)
— of polyelectrolytes 13(494)
Exhausting power 12(545)
Exhaustion hypothesis of work-hardening 7/2(127)
Exhaustion rate, definition 12(545)
Exhaustion rate, measurement 12(546)
Exhaustion rate of diffusion pumps 12(584–590)
Exhaustion theory of creep 7/2(199–200)
Existence and nonuniqueness for elastostatic boundary-value problems 6a/2(544)
Existence of an elastic core 6a/3(568)
Existence of approximate solutions 6a/2(126)
Existence of boundary value problem solution 8/2(186–199)
Existence of jet flow 9(392)
Existence of quasi-static viscoelastic processes 6a/3(59)
Existence of solutions
— approached by infinitesimal waves 9(654)
— of the Oseen equation 8/2(322–323)
— of the Stokes equation for steady flow 8/2(262–276)
— of the Stokes equation for unsteady flow 8/2(276–287)
Existence of viscoelastic waves 6a/3(97)
Existence proof for the cnoidal wave 9(462, 668)
Existence proof for the initial value problem 9(19)
Existence proof for the solitary wave 9(462, 668)
Existence theorem(s)
— for capillary waves 9(757)
— for rods 6a/2(680)
— for subsonic flow 9(108, 111, 115, 119)
— for surface waves 9(749–757)
— for transonic flow 9(138–145)
— general 6a/2(368–387)
— generalized Signorini problem 6a/2(407)
Existence theory in free boundary flow 9(391–406)
Existence, uniqueness, and asymptotic stability theorem for dynamic problems in visco-elasticity 6a/3(104–106)

Expansion, coefficient(s) 6a/4(139, 145); 7/1(236, 380)
— isobaric 11/1(143)
— of a gas 12(5, 8–9)
— of ceramics 13(606)
— of glasses 13(544, 547, 557, 561–570)
— of glasses, additivity 13(563–564)
— of glasses, dependence on composition 13(566–568)
— of glasses, dependence on fictive temperature 13(562)
— of glasses, dependence on temperature 13(562)
Expansion, double 14(55)
Expansion engine(s) 14(2, 71, 78, 125)
Expansion engine, reciprocating 14(81 f.)
Expansion engine, two-stage 14(77)
Expansion, Euler's formula 6a/4(115)
Expansion, expansive wave 6a/3(51)
— acceleration 6a/3(311, 314)
— shock 6a/3(312)
Expansion, isenthalphic 14(16, 33, 70)
Expansion, isentropic 14(69)
Expansion, method 14(87)
Expansion, multiple 14(28)
Expansion of a Green's function in spherical harmonics 11/1(23)
Expansion parameter of shallow-water theory 9(462, 667, 704)
Expansion rate of a Laval valve 12(568)
Expansion shocks 6a/4(295)
Expansion, single isentropic 14(87)
Expansion, thermal 7/1(273, 276, 289)
Expansivity, linear 7/1(236)
Expansivity, volume 7/1(238, 277)
Experimental investigation of capillary dispersion 9(633–634)
Experimental results for (6a/1)
aluminum 41–43, 128, 129, 140, 147, 150, 259, 291, 292, 306, 321, 326, 343–345, 348, 350, 351, 359, 373–375, 379, 380, 385, 387–389, 391, 392, 394–396, 405, 406, 411, 496, 497, 500, 501, 505–507, 518–521, 524, 531–533, 538–540, 542, 545, 546, 552–554, 556, 564, 567–570, 578, 606–609, 615, 624–627, 629, 631, 632, 635–638, 642–644, 647, 652, 654–657, 659–665, 673–675, 681–686, 688–702, 705, 708–711, 713–715, 719
antimony 223, 259, 388, 389, 495
barium 388, 389, 391, 495
beryllium 388, 390, 391, 398
beryllium copper 133, 134, 136, 138
binary alloys 167, 232, 233–235, 520, 658
bismuth 223, 259, 292, 373, 387–389, 406, 411, 480
brass 80, 81, 96, 103–105, 120, 131, 152, 167, 172, 176, 183, 184, 186, 210, 211, 216, 223, 227, 236, 248, 251, 268, 277, 291, 292, 302, 304, 317, 319, 320, 326, 345, 363, 405, 406, 411, 424–426, 520, 628, 648, 692, 695, 724
bronze 405, 406, 411, 426
butter 420
cadmium 224, 231, 259, 292, 387–389, 398, 399, 405, 406, 411, 480, 506
calcium 388, 389, 391, 398, 399
cardboard 268
cement 112, 113
cement-plaster 112
cesium 388, 394, 395, 495
chromium 259, 388, 394, 395
clay 439, 441
cobalt 259, 370, 388, 390, 392, 398
concrete 112, 113, 144, 145
constantan 291, 292
copper 46, 47, 93, 96, 104, 105, 112, 113, 120, 123, 141, 151, 183, 184, 186, 216, 222, 225–227, 231, 236, 239, 259, 268, 277, 280, 291, 292, 302, 304, 306, 317, 321, 363, 366, 371, 374, 375, 385, 387–389, 391, 393–396, 402, 405, 406, 411, 440, 441, 445, 447, 452, 496, 502, 505, 506, 508, 511, 512, 520, 522, 530–533, 544, 545, 549, 560, 561, 563, 575, 578, 585, 590, 593, 595, 596, 601, 602, 604, 611, 612, 617, 627, 650, 651, 672, 677, 678, 679, 689, 693, 712, 724
cork 110, 268
crystal 241, 248, 277
delta metal 277
gallium 388, 394, 395
germanium 388, 394, 395
glass 75, 76, 78, 79, 183, 186, 241, 242, 249, 265, 271, 277, 284, 285, 318, 498, 506
gold 95, 224, 227, 231, 236, 239, 259, 291, 292, 302, 385, 388, 389, 391, 396, 405, 406, 411, 520, 522, 527, 536
gut 14, 54, 55

Experimental results for (6a/1; cont.)
hafnium 388, 394, 395
hair 63, 70, 170
human body tissues, arteries 59, 61–63
human body tissues, bones 59
human body tissues, muscles 59–62, 64, 66–69, 143
human body tissues, nerves 59–62, 64
human body tissues, tendons 59, 60, 62
human body tissues, veins 59–62
ice 186, 482, 483
indium 388, 389, 395
iridium 373, 388, 390, 398, 496
iron 26, 28, 29, 33, 35, 36, 38–40, 46, 47, 71, 86, 90, 91, 93, 95, 96, 106, 107, 112, 113, 119–122, 138, 150, 167, 172, 175, 176, 183, 184, 186, 193, 199, 200, 203, 204, 210–212, 216, 221, 225, 227, 231, 237, 239, 251, 259, 280, 291, 292, 302, 306, 363, 374, 375, 382, 387, 388, 390, 392, 401–403, 405, 406, 411, 424–426, 440, 441, 445, 447, 450, 452, 455, 466, 468–471, 473, 496, 497, 502, 522, 540, 545, 575–578, 582
ivory 268
lanthanum 388, 389, 391, 398, 399
lead 146, 211, 216, 224, 227, 231, 259, 268, 277, 292, 306, 317, 373–375, 385, 387–389, 391, 394–396, 405, 422, 423, 425, 426, 430, 431, 433–445, 480, 506, 593, 620, 633
leather 109, 113
lithium 388, 394, 395, 495
lucite 345
magnesium 259, 340–342, 387–389, 392, 396, 402, 406, 411
manganese 388, 390, 392
manganin 291, 292
mercury 186, 388
molybdenum 387, 388, 390, 391, 398
nickel 259, 280, 291, 292, 373, 387, 388, 390, 392, 396, 402, 406, 411, 496, 502
niobium 388, 394, 395
paints and varnishes 127
palladium 225, 227, 231, 259, 291, 292, 373, 388, 390, 391
perspex 593
phosphor bronze 97, 98, 108, 128
plaster of Paris 268
platinum 46, 47, 225, 227, 231, 237, 291, 292, 306, 374, 388, 390, 391, 496
polythene 593
potassium 388, 395, 481, 495
quartz 353–355
rhenium 388, 394, 395
rhodium 259
rocksalt 355, 497
rubber 110, 127, 246, 268, 310–313, 497, 593, 717, 720, 721, 727, 730, 731, 733, 736–739
rubidium 388, 390
ruthenium 388, 394, 395
selenium 495
silicon 388, 394
silk 48, 50, 51, 63, 98–100
silver 46, 47, 95, 105, 120, 183, 184, 186, 221, 224, 227, 231, 236, 259, 280, 291, 292, 302, 306, 317, 374, 375, 388, 389, 391, 393–396, 405, 406, 411, 496, 520, 528, 556, 557
sodium 388, 394, 395, 481
steel 31, 95, 104, 105, 120, 128, 129, 178, 186, 200, 201, 210–212, 214, 216, 217, 223, 225, 227, 231, 232, 237, 239, 251, 263, 268, 273, 277, 279, 291, 292, 294, 295, 297, 302, 304, 317, 320, 321, 326, 336, 338, 339, 345, 368, 382, 383, 403, 406, 411, 426, 452, 455, 459, 461, 464, 485, 505–507, 512, 513, 545, 577, 588, 603, 672, 677, 723, 724
stone, granite 112
stone, marble 488
stone, sandstone 489
strontium 388, 395
tantalum 259, 388, 390, 391, 398
tellurium 259
tertiary alloys 235
thallium 388, 389, 394, 395, 398, 399, 481
thorium 388, 391, 398, 496
tin 183, 184, 186, 216, 224, 231, 259, 292, 306, 317, 374, 375, 387–389, 391, 406, 411, 426, 440, 441, 445, 481, 482, 571–573
titanium 388, 390, 398
topaz 354
tungsten 388, 390, 398
uranium 558
vanadium 388, 394, 395
wax 439, 441

whalebone 183
wood 19–21, 183, 186, 243, 268
zinc 216, 222, 224, 231, 259, 268, 302, 317, 387–389, 391, 406, 411, 426, 445, 481, 525, 559, 628, 637, 677
zirconium 388, 390, 398
Experimentist 6a/1(2)
Experiments on polycrystals at finite strain: uniaxial tests 6a/1(547–550)
Exploding foil 6a/4(4, 56)
Explosion 9(211)
Explosion under water 9(619)
Explosions as sources of sound 11/2(32–35)
Explosive loading 6a/1(332, 346, 358, 499, 591)
Explosive plane wave generator 6a/4(4)
Exponential horn 11/1(48); 11/2(20, 22)
Exponential law, *see* Bernoulli's parabolic law
Exponential potential model 12(341, 397)
Exponentially damped flow 8/2(127)
Exponentially growing flow 8/2(128)
Extended uniqueness theorem for the mixed problem 6a/2(223)
Extensibility in the direction of the unit normal to a reference surface 6a/2(460)
Extension 7/1(187, 270)
Extension, thermal 7/1(236, 238, 282)
Extensional deformation of a plate 6a/2(600)
Extensional motion 6a/2(460)
Extensional theory of plates, classical 6a/2(577)
Extensional vibration, methods of excitation 11/2(147)
Extensional vibration of a glass bar 11/2(94)
Extensional vibrations of plates 6a/2(575)
Extensional wave along a rod 11/2(75, 147)
Extensions of the fundamental lemma 6a/2(115, 116, 223)
Extensometer, mechanical 6a/1(28, 34–37, 124, 128, 139, 157, 214, 248, 450)
Extensometer, optical 6a/1(84, 102, 117, 130, 133, 145, 288–291, 364, 457, 629, 723)
Exterior boundary value problem 8/2(187, 190, 195)
Exterior derivative(s) 6(341); 6a/3(420, 426)
Exterior domain 6a/2(14, 165, 261)
Exterior region 6a/3(81)
External body force 6a/2(509)
External force 8/2(2, 5, 200)
External force system 6a/2(95, 215, 327)
External force-temperature system 6a/2(313)
External thermal system 6a/2(327)
Extinction law 13(260)
Extreme-value statistics of strength 6(580, 581, 582)
Extrusion 6a/3(525)

Face in general position 7/1(67)
Face-centred cubic alloys
— surface phenomena 7/2(99–103)
— work-hardening 7/2(57–66)
Face-centred cubic lattice 7/2(24, 26)
Face-centred cubic metals, theory of work-hardening 7/2(139–167)
Face-centred cubic metals, work-hardening 7/2(51–57)
Fading memory 6a/3(2, 334)
Fairbairn's lever 6a/1(26)
Fall of a drop of water through air 8/2(232)
Fall of a sphere in viscous fluid 8/2(232–233)
Family of load regions contracting to a boundary point 6a/2(191)
Family of singular elastic states 6a/2(192)
Fan, *see* Simple wave
Fanning equation 14(100)
Fanning of slip-lines 7/2(104–105)
Faraday effect 15(15)
Fatigue crack 6(591)
Fatigue cracks, propagation 6(596, 601, 603, 606)
Fatigue damage 6(571, 752)
Fatigue damage, initiation 6(596, 606)
Fatigue effects in ceramics and glasses 6(593)
Fatigue effects in high polymers 6(593)
Fatigue fracture 7/2(6, 7)

Fatigue limit 6(592)
Fatigue of steel 6a/1(457, 461, 472)
Fatou's lemma 6a/2(384)
Faxén's resistance formulae 8/2(235–238)
Faxén's theorems concerning two spheres 8/2(247–248)
Federov-Stippes theorem 6a/2(247)
Fermat's principle 11/1(85)
Fermi distribution 11/2(217, 218)
Fermi energy 7/1(403, 406, 473, 475); 14(229, 289, 290, 291)
Fermi sphere 7/1(74)
Fermi statistics 10(122)
Fermi surface 11/2(209, 210, 219–221)
Fermi surface, deviations from spherical shape 14(241)
Fermi-Dirac distribution in electrolytes 13(393)
Fermi-Dirac perfect gases 12(174ff., 179–182)
Fermi-Dirac statistics 12(81, 454, 458); 14(139)
Ferrites as magnetostrictive materials 11/2(92)
Ferromagnetic crystals, spin waves 11/2(228–229)
Ferromagnetic model of Ising 13(79, 109–112)
Ferromagnetic resonance 11/2(229)
Ferromagnetics, critical opalescence 13(196–210)
Ferroxcube 11/2(92)
FeS_2 lattice 7/1(91)
Fiber model of porous media 8/2(644–645)
Fictive temperature 13(544, 546–548, 561)
Fictive temperature, effect on the expansion coefficient of glasses 13(562)
($\mathscr{X}$)-field 6a/3(7)
($\mathscr{X}$)-field process for $\mathscr{D}$ 6a/3(7)
Field, demagnetizing 15(14)
Field, effective 7/1(209ff.)
Field emission 15(271)
Field emission microscope 12(610)
Field energy 6a/4(161)
Field equations 6a/4(152)
— elasticity 6a/2(89–94, 212–215)
— nonlinear thermoelasticity 6a/2(307)
— of viscoelasticity, dynamic 6a/3(91)
— of viscoelasticity, quasi-static 6a/3(42)
— shells, derived from 3-dimensional theory 6a/2(508, 519)
— shells, direct approach 6a/2(492, 498)
— shells, in terms of reference state 6a/2(502)
— shells, linearized 6a/2(500, 523)
— shells, restricted theory 6a/2(503)
Field quantities 15(47, 51)
Fields, force-like and flux-like 6a/4(150)
Fields of forces in liquids, deformation 10(67–69)
Field-strength of cation
— effect on glass structure 13(541, 554, 561)
— effect on thermal expansion of glasses 13(564–566, 568)
Fifth-order infinitesimal wave approach for progressive waves 9(663, 665)
Film 15(381, 382, 383, 384, 385)
Film creep 15(178)
Film formation 15(450, 451)
Film formation between two fluid phases 10(299)
Film of Helium II 15(380)
Film, superconducting thin 15(240, 243, 260)
Film thickness 15(437, 438, 439, 440, 441, 446)
Film, unsaturated 15(450)
Filon theory of generalised plane stress 6(129, 143–145)
Filter cake 8/2(640)
Filtration theory and Darcy's law 8/2(640–641)
Final period of decay 8/2(460, 470, 482)
Finite amplitude wave propagation
— directly measured strain profiles 6a/1(621–640, 690–702)
— experiments before 1956 6a/1(597–621)
— in single crystals 6a/1(694)
Finite amplitude wave theory 6a/1(580, 594, 605)
Finite cavity 9(324)
Finite cavity for the disk 9(427)

Finite deformation(s) 6a/1(419–472); 6a/2(28); 6a/3(3); 11/2(161)
— discontinuous 6a/1(41–44, 649–666)
— first graphical presentation of 6a/1(424)
— of strings 6a/1(718–721)
— or strain 6(414, 489–501)
Finite depth
— complex velocity potential 9(482, 490, 494–495)
— effect on wave profile 9(659, 660)
— for progressive ocean waves 9(499)
— for standing ocean waves 9(496, 498)
Finite difference approximation 9(435)
Finite elasticity 6a/2(80)
Finite extension 6(237)
Finite histories 6a/3(5)
Finite horns 11/2(21–22)
Finite linear viscoelasticity
— dilatational motion 6a/4(69–75)
— one-dimensional motion 6a/4(37, 44ff., 48, 51, 63ff., 66)
Finite part of a divergent integral (Hadamard definition) 9(42, 303)
Finite rigid deformation 6a/2(30)
Finite strain tensor 6a/2(29, 80, 306)
Finite thickness of a plate, *see* Generalized plane stress
Finite-amplitude waves, *see also* Solitary waves, Cnoidal waves
Finite-amplitude waves, limiting height 9(703, 712ff.)
Finn-Gilbarg extension of the maximum-minimum principle 9(95–97, 107)
Finn's method 8/2(191–193)
Firmo-viscosity of the Kelvin body 6(464)
First and second fundamental forms of a surface 6a/2(623)
First law of thermodynamics 6a/2(300); 8/1(175)
First order quantities in sound waves 11/2(35)
First Piola-Kirchhoff stress tensor 6a/4(117, 123)
First sound in liquid Helium 15(431, 432, 433)
First-order shallow water approximation 9(467, 469)
First-order transitions in polymers 13(423, 424, 430)
Fishline problem 9(637)
Fission 7/1(432)
Fission products 7/1(431)
Fitzgerald's particle wave hypothesis 6a/1(655)
Fixed point 9(117, 119, 121)
— of the inversion axis 7/1(13)
— method of Leray and Schauder 9(392, 400–402)
— theorem 6a/3(558)
Flash-filament method 12(627)
Flat along a pervious wall 8/1(348, 409)
Flat around a parabola 8/1(101–104)
Flat around an edge 8/1(103)
Flat, axial, past bodies of revolution 8/1(311)
Flat jet 8/2(63)
Flat past yawed infinite cylinder 8/1(313, 314)
Flat plate at zero incidence
— critical Reynolds number 8/1(424)
— laminarization by suction 8/1(410–412, 414)
— transition from laminar to turbulent flow 8/1(360–365, 379–382, 394)
Flat plate
— critical Reynolds number 8/1(383)
— formation of cusped cavity behind it 9(338)
— impulse motion and boundary layer 8/1(293–294)
— in molecular flow 8/2(605)
— infinite cavity formation behind it 9(328–330)
— steady boundary layer flow 8/1(272, 330–331, 337)
— unsteady boundary layer flow 8/1(289–290)
Flat profile, pressure, distribution 8/2(394)
Flat ring in elastic-plastic plane stress 6(418–421)
Flaws in glass 13(625, 626)
Flaws in liquids 6(555–556, 580–584)
Flexural motion 6a/2(460)
Flexural vibrations of plates 6a/2(575)
Flexural wave along a rod 11/2(75, 149, 150)
Flexural waves in cylinders 6(115–116)

Flexure of beams, quasi-static 6a/1(18–30, 177, 187–189, 199–201, 213–215, 264, 282)
Flexure of beams, Saint-Venant's theory of 6a/1(264)
Floating body 9(563ff.)
Floating boundary 9(271, 273)
Floating radiometer 11/2(122)
Flory temperature 13(460, 468, 469, 471, 481)
Flour dough, rheological behaviour 6(475–478)
Flow, *see* Crep
Flow about submerged obstacles 9(571–573)
Flow along logarithmic spirals 8/2(30, 33–35, 93)
Flow around
— a body, Osseen's limiting case 8/2(329–333)
— a circle, method of Pérès 8/2(361–362)
— a circle, method of Zeilon 8/2(359–360)
— a circle, velocity field 8/2(362–367)
— a plane plate, method of Zeilon 8/2(373)
— a plane plate, velocity field 8/2(375)
— an edge 9(85)
Flow birefringeance 12(302, 489)
Flow condition, *see also* Yield condition
Flow condition of Hencky 6(520)
Flow condition, of Mises 6(521)
Flow, definition 6(434)
Flow in multiply connected bodies 15(294)
Flow of constant vorticity 8/2(39, 140–141)
Flow of mixed type, *see also* Transonic flow 9(226, 256)
Flow of solids 6a/1(427–446, 478–483)
Flow over an uneven bottom 9(569–571)
Flow (plastic)
— equations of Prandtl and Reuss 6a/3(412)
— rule 6a/3(547)
— theories 6a/3(549)
— type 6a/3(415)
— velocities 6a/3(405)
Flow point of a glass 13(579)
Flow rate, increase with temperature 6a/1(479)
Flow stress, *see* Yield
Flow type pumps 12(559–605)
Flow velocity tensor, plastic 6(325, 326)
Flow with circulation 9(71, 83)
Flow-tensor 6(438)
Flow-type law 6(332)
Fluctuating static pressure 8/2(546)
Fluctuation 7/1(177)
Fluctuation, absolute 7/1(264, 265)
Fluctuation in turbulent flow behind a grid 8/2(443)
Fluctuation of density in compressible turbulent flow 8/1(364)
Fluctuation of pressure 8/2(498–501)
Fluctuation of temperature 8/2(534)
Fluctuation, of velocity in turbulent flow 8/1(352, 384)
Fluctuation, relative 7/1(264, 265)
Fluctuation, thermal 7/1(261, 262, 264, 601)
Fluctuational quantity 13(28)
Fluctuations 11/1(5, 8, 87, 89, 98)
Fluctuations in density near the critical point 12(46–48)
Fluctuations in nucleation 7/2(218, 219)
Fluctuations in reverse direction 7/2(126)
Fluctuations of local density 13(68)
Fluctuations of mass density 10(282)
Fluctuations of surplus protons 11/1(251)
Fluctuations of the distance between two atoms 7/1(266)
Fluctuations of the index of refraction 13(310)
Flügge-Lur'e-Byrne-Truesdell equations in the linear theory of shells 6a/2(592)
Fluid blast pump 12(559, 563–565)
Fluid character of one-dimensional crystal 13(61, 63)
Fluid gauge 12(517, 519–520)
Fluid lens embedded in two fluid phases 10(282–283, 285, 298, 299)
Fluid motion, Eulerian and Lagrangian pictures 11/1(2–3)
Fluid, normal 10(120)

Fluid velocity potential 10(41)
Fluid wave 11/1(109, 113)
Fluidity at rest 6(496)
Fluidity coefficient 6(450)
Fluids, *see* Liquids
Fluids 6a/3(2)
Fluids used in compressed vapor refrigerators 14(20)
Fluid-solid analogy 6a/1(250, 253)
Fluorescence 7/1(94)
Fluorite 7/1(101)
Fluorite lattice 7/1(89)
Fluorspar 6a/1(407)
Flush pins 6a/4(6–8, 22)
Flutter 6a/3(283, 284)
Flux trapping 15(270)
Flux-like 6a/4(150)
Fluxoid 15(295)
Flyer plate (noseplate, projectile plate) 6a/4(3–7, 15ff., 20, 22, 41, 55ff., 63, 65)
Focusing equations 9(259, 275)
Focusing of waves at origin 6a/3(138)
Fog formation by ultrasonics 11/2(279, 286)
Fokker Planck equation 8/2(514, 519); 12(233)
Folded horns 11/2(22–23)
Follower force 6a/3(257, 258, 261, 278–285)
Follower load problem 6a/2(693)
"Forbidden regions" 14(291)
Force at infinity 6a/2(170)
Force constant 7/1(160, 170, 273, 274, 299)
Force constant (atomic) 7/1(145, 146, 147, 148, 150, 152, 153, 155, 156, 157, 186, 188, 191, 200, 202, 203, 208, 222, 245, 274, 278)
Force multipoles 6a/3(213)
Force on a profile 9(128, 130)
Force on body entering into water 9(362)
Force on obstacle 9(370)
Force potential 8/2(3, 9, 201)
Force potential in porous media 8/2(635)
Force vector, assigned 6a/2(480)
Forced motion of coupled boundary 11/1(108–116)
Forced standing waves in viscous fluids 9(643)
Force-like 6a/4(150)
Forces exerted
— by sound waves 11/1(65, 68)
— on a body by the fluid 8/2(8)
— on a body by the fluid in molecular flow 8/2(604–606)
— on a body by the fluid, Oseen's limiting case 8/2(343)
Forces on a hydrofoil 9(586)
Forces on a ship 9(567, 580)
Forces on a submerged body 9(573–574)
Forces on a submerged cylinder 9(576)
Forces, parallel and non-tangential to the boundary 6a/2(193)
Forces, system 6a/2(43)
Forchheimer equation 8/2(648)
Fore-effect, elastic, *see* Creep
Foreign atom 7/1(528, 529, 557)
Foreign atoms in lattice sites 7/2(168)
Foreign nuclei 7/2(218)
Formation of shock waves 9(253, 265–268, 279, 280)
Formic acid, sound absorption 11/1(318)
Förster's experiment for support free vibrations 6a/1(385, 386)
Foundations of linear theory of viscoelasticity 6a/3(11)
Fountain effect of ultrasound 11/2(120)
Fountain-effect 15(178, 378)
Fourier analysis of initial surface profile 9(507, 613–614)
Fourier and Laplace transforms 6a/3(28)
Fourier coefficients in short-range order theory 7/2(174)
Fourier expansion method of determining relaxation spectra 11/1(370)
Fourier integral 8/2(116, 118); 9(507, 523)
Fourier integral theorem 6a/3(33)
Fourier series for surface waves 9(522ff.)
Fourier transform(s) 9(475); 11/1(41, 57, 70, 110)
— in elasticity 6(72–78)
— of a distribution function 8/2(443)
— of correlation tensors 8/2(445, 452)

Fourier transform(s) (cont.)
— of the velocity correlation function 11/1(91)
— of three-point correlations 8/2(489)
— physical meaning 13(69)
Fourier's law of heat conduction 6a/4(205, 214, 255)
Four-tensor 6a/2(27)
Four-terminal networks 11/2(66)
Fourterminal networks 11/2(67)
Fourth-order elastic coefficients 6a/4(167, 222)
Fourth-order tensor 6a/2(9)
Four-vector 6a/2(27)
Four-velocity correlations 8/2(441, 443, 488)
Fractional melting point temperature 6a/1(378, 393–397, 409, 535, 574, 638)
— definition of 6a/1(397)
Fractionation of polymers 13(485–487)
Fracture 7/2(3, 6)
Fracture condition of Huber 6(520)
Fracture origins 6(567, 569, 685)
Fracture, roughness and texture 6(577, 578)
Fracture size-effect, measurements 6(584, 585, 586)
Fracture strength, *see also* Strength
— definition 6(563)
— frequency dependence 6(593ff., 600)
— measurements (table) 6(566)
Fracture stress 6(567, 573, 574, 575)
Fracture stress, in fatigue experiments 6(594)
Fragmentation of glide bands (slip lines) 7/2(95–96, 151–152, 158, 161)
Frame-indifference 6a/3(544, 545)
— material, rods 6a/2(655, 669)
— material, *see also* Objective) 6a/2(80, 305, 306)
Franck-Condon principle 11/1(200)
Frank partial dislocation 7/1(507f.)
Frankl's boundary value problem 9(137, 139)
Frank-Read dislocation source 7/1(490, 491, 492)
Frank-Read source 7/1(538, 543, 661); 7/2(112, 128, 170, 192)
Fraunhofer's region 11/2(105, 106)
Fréchet derivative 6a/3(165, 194, 195, 203, 205)
Fréchet differential 9(413)
Fredholm equation for surface waves 9(535, 543)
Fredholm equation, homogeneous 11/1(123)
Fredholm integral equation(s) 6a/2(419)
— of the first kind 11/1(25, 65, 117)
— of the second kind 11/1(29, 32, 107)
Fredholm's functions 7/1(517)
Free body 6a/2(315)
Free boundaries 6a/3(501, 580, 581, 583, 585, 586)
Free boundary 9(320)
Free detonations 11/2(32)
Free draining coil model 13(473)
Free electron gas, one-dimensional model of Fröhlich 15(366)
Free energy 6(16–17, 247); 6a/3(3, 179, 180, 183, 191, 192); 6a/4(76ff., 127); 10(62, 67, 78, 285, 286); 13(13, 221, 225, 229, 237, 287, 296, 319); 15(41, 132, 233, 455, 458)
— 3-dimensional 6a/2(301)
— and long-range ordering 13(86)
— equilibrium response function 6a/4(79, 85)
— functional 6a/3(539)
— functional 6a/4(79, 83)
— instantaneous response function 6a/4(78)
— intrinsic 6(440)
— minimal property 6a/4(79ff., 83)
— of a gel 13(487–489)
— of a rigorously regular solution 13(322)
— of dilution 13(315, 319)
— of electrolytes 13(374)
— of electrolytes in the Debye-Hückel approximation 13(377, 378)
— of iron modifications 7/2(249)
— of network deformation 13(435, 442)
— of nucleation 7/2(217, 218, 234–237)
— rods 6a/2(649, 657)
— shells, elastic Cosserat surface 6a/2(557)
— shells, linear theory 6a/2(556)
— superficial 10(137, 149, 285, 288, 296)
Free enthalpy 6a/4(127)
Free flowing 6a/3(494)
Free molecule flow 8/2(592, 601–609)
Free molecule probe 8/2(613)

Free path 12(295, 307, 395, 417)
Free path mean 7/1(291, 292, 309, 312, 441)
Free shear layer 8/2(549)
Free stream Mach number, *see* Mach number
Free streamline(s) 9(63, 320, 322)
— analytic continuation across it 9(348)
— asymptotic behavior 9(375–378)
— behavior at detachment 9(371–375)
Free surface 6a/4(294); 9(320, 446 ff.)
— boundary conditions 9(454)
— flow in the gravity field 9(350)
— flow, numerical methods, *see also* Computation 9(415–438)
— under tangential force 8/2(125)
— velocity 6a/4(8)
— velocity, measurement 6a/4(9–14)
— with surface tension 9(352)
Free vibration problem 6a/2(261)
Free vibrations 6a/3(106, 111)
Free volume 10(240, 242 ff., 247, 249 ff., 262, 265, 273)
— calculated from Lennard-Jones' and Devonshire's theory 13(228)
— cluster theory 13(241, 244)
— definition 13(225)
— generalized 10(250); 13(235, 241)
— in lattice model of liquid solutions 13(316)
— of a cell cluster 13(244)
— principles of its theory 13(219–226)
— smeared 10(243, 247 ff.)
— sphericalized 10(243, 250)
— theory, application to solutions 13(344–356)
Free vortex layer 8/1(271, 302–307)
Free-boundary problem 6(404)
Free-electron model, comparison with 14(179)
Free-field pressure response 11/2(38, 44)
Free-field reciprocity calibration of microphones 11/2(70–73)
Freely floating body 9(563 ff.)
— diffraction of water waves 9(566–567)
— initial displacement 9(619)
Freely floating sphere 9(568)
Freezing point 14(20)
Freezing procedure 6(179)
Freezing-in temperature 7/2(175)
Frenkel defects 7/2(222); 13(77–78)
Frenkel's equation 7/1(585)
Frenkel's model 7/1(514, 564 f., 568, 583, 623 f.); 7/2(9)
Frenkel's probability density 13(55)
Freon 12 14(24, 26, 27, 30, 53)
Frequency 7/1(279)
Frequency dependence of measured absorption in electrolytes 11/1(233–236)
Frequency distribution 14(286)
Frequency distribution function 7/1(159, 163, 164, 168, 201, 202, 244, 245, 247, 262)
Frequency moments 7/1(349)
Frequency number 11/1(278)
Frequency of cyclic stress, effect on strength 6(593 ff., 600)
Frequency of vibration 6a/2(261)
Frequency response of a ribbon microphone 11/2(57)
Frequency response of a thermoelectric microphone 11/2(62)
Frequency response of condensor microphones 11/2(43)
Frequency spectrum of gun fire 11/2(34–35)
Frequency tensor 7/1(213, 214, 243)
Fresnel integrals 9(545, 613)
Fresnel-Hadamard propagation condition 6a/2(246, 256); 6a/3(98, 105, 316, 324, 326)
Fresnel's ellipsoid 6(133, 146, 186)
Fresnel's region 11/2(106)
Friction analog, Coulomb's 6a/1(655)
Friction coefficient, Ω-integral representation 12(392–394)
Friction constant 10(85, 112, 118)
Friction drag 8/1(90)
Friction factor 14(100)
Friction factor in porous media 8/2(647)
Friction gauge 12(517, 527–530)
Friction generator of ultrasound 11/2(100)
Friction heating 6(600)
Friction, internal 7/2(3, 111)
Friction, mutual 15(418, 427, 428)
Friction tensor 10(114)
Friction velocity 8/1(442)
Friction vortex 8/1(55)

Frictional force 13(473, 474)
Frictional force on a cylinder 8/1(325)
Frictional forces 7/2(132)
Friedel's condition 7/1(428)
Friedrichs' shallow-water expansion *see* Shallow-water approximation
Fringe count, interferometer 6a/4(11–13, 19)
Fringe frequency, *see* Doppler effect
Fringe value (of a material) 6(149, 164)
Fringes, isochromatic 6(149)
Fringes isoclinic 6(148, 166)
Frocht's method of stress determination 6(184–186)
Frog muscle 6a/1(64–69, 142, 731)
Fröhlich's shell distribution 15(359)
Fröhlich's theory 15(345)
Froude number 8/1(58); 9(581, 675, 676, 695, 699, 701)
Froude number for a cnoidal wave 9(713)
Froude number for a solitary wave 9(713)
Froude-Krylov hypothesis 9(566)
Frozen stress, *see also* Stress freezing 6(130, 177–180)
Frustrated reflection effect 6a/4(23)
Fuchs' calculation of elastic constants 6a/1(385)
Fuchs theory of solid solutions, *see also* Solid solutions 13(143–153, 159–167)
— definition of critical point 13(149)
— experimental test 13(167–175)
— special lattices 13(161–166)
Fugacity 12(17–18, 21, 64); 13(32)
Fugacity (activity) 10(194)
Fulcher-Tammann equation 13(581)
Fully annealed solid
— definition of 6a/1(551)
— general constitutive equations for 6a/1(703)
Fully developed turbulence 8/2(541)
Fully elastic or plastic 6a/3(495)
Functional equations 9(395–400)
Functions of position and time 6a/3(7)
Fundamental equation of potential flow 9(6)
— linearization 9(26–29)
— three-dimensional explicit solutions 9(8–11)
Fundamental equation of potential plane flow 9(11)
Fundamental equations for first-order shallow-water approximation 9(667–668)
Fundamental form of a surface
— first and second 6a/2(623, 624)
— third 6a/2(626)
Fundamental forms of surface theory 10(284)
Fundamental lemma 6a/2(20)
Fundamental solution matrix 6a/2(419)
Fundamental solution of the heat conduction equation 8/2(115)
Fundamental solution of the Stokes equation 8/2(264)
Fundamental system of field equations 6a/2(89)
Fundamental tensor of the crystal space 7/1(44)
Fundamental tensor of the reciprocal lattice 7/1(44)
Funicular region 8/2(632)
Fused silica, *see also* Quartz glass
Fused silica, hardness 13(633)
Fused silica, structure 13(528–529)

G belongs to the pair *P*, *Q* 6a/3(35)
Gadolinium anthraquinone sulphonate 15(115)
Gadolinium nitrobenzene sulphonate 15(114)
Gadolinium phospho-molybdate 15(115, 126)
Gadolinium sulfate (8 aq) 15(14, 20, 21, 23)
Gaede's mol-vacuummeter 12(533)
Gaede's vacuscope 12(523)
Galerkin, *see* Boussinesq
Galerkin procedure 6a/3(285)
Galileo's beam experiment 6a/1(162, 166)
Galileo's rule 6a/1(17, 22)
Galin's solution for the moving punch 6(121)
Galton's whistle 11/2(95–96)
Galvanometer amplifier 14(159)
Galvanometer amplifier, oscillatory instability 14(161)
Galvanometer, superconducting 14(162); 15(220, 267)

Gamma ray heat supply 15(89, 103)
γ-space (gas-space) definition 12(80, 81)
Garabedian's iteration method 9(427–431)
Gardner's reduction method 9(304–305)
Garett parameter 15(59, 105, 109, 123, 150)
Garett's method 15(59)
Gas adsorption 10(205, 231 ff.)
Gas adsorption on a solid surface 10(231)
Gas analysis 12(319)
Gas constant 12(26, 363)
Gas crystal transitions 12(170–172)
Gas degeneration 12(345)
Gas dynamic shock loading 6a/4(3)
Gas dynamics, connection with shallow-water theory 9(677, 681)
Gas dynamics, linearized 9(669)
Gas dynamics, variational principles 8/1(203–206)
Gas evolution in high vacuum 12(631–636)
Gas, ideal 7/1(317)
Gas liquefaction 14(40, 76)
Gas liquefaction, using expansion engines 14(76)
Gas mixtures 12(57 ff.)
— critical region 12(65–70)
— dispersion and absorption 11/1(173, 183–187)
— empirical equations of states 12(65)
— perfect 12(57–59)
— real 12(59–64)
Gas purification 14(63)
Gas refrigerating machines
— closed-cycle 14(9)
— of Philips 14(10, 14)
— open-cycle 14(2)
Gas thermometers 14(200)
Gasdynamic equation 8/1(68)
Gaseous cavitation, *see* De-gassing
Gases of excitable molecules 12(490–497)
Gases of rough spheres 12(497–499)
Gas-gas immiscibility 12(69, 70)
Gateau derivative 6a/3(165, 194, 198, 203, 205)
Gauge invariance in non-local theories 15(315)
Gauss, equations of 6a/2(46, 627)
Gauss' formulae 6a/2(625, 626)
Gauss' quadrature formulae 8/2(433)
Gaussian approximation for highly elongated chains 13(405, 442)
Gaussian chain 13(443)
Gaussian curvature of a surface 6a/2(624); 10(284)
Gaussian distribution 8/2(442, 443, 498)
Gaussian distribution in permeability models 8/2(645)
Gel 6(435, 522)
Gel, free energy 13(487–489)
Gellerstedt's differential equation 9(133, 135)
General functions 13(5)
General position 7/1(39)
Generalized cluster integrals 13(35)
Generalized flux 6(248, 249)
Generalized force 6(248, 249, 267)
Generalized free volume 13(235)
Generalized plane stress 6(129, 143–145, 352–353); 6a/3(440)
Generalized potential 6a/3(439)
Generalized solutions 6a/3(128, 138, 254, 255)
Generalized Somigliana solution 6a/3(94)
Generalized strain-rates 6(255)
Generalized strains 6a/3(409)
Generalized stress(es) 6(255, 327); 6a/3(409)
Generating element of a crystal class 7/1(47 ff.)
Generating equation for a viscoelastic relaxation function 6a/3(36)
Generation of shock waves 6a/3(97)
Generation of sound 11/1(6)
Generation of sound in turbulent flow 11/1(28, 93–99)
Generation of turbulence 8/2(525–526)
Generation of waves by oscillating bodies 9(553 ff.)
Generator of Saint Clair 11/2(23–24)
Geometric approximation for inhomogeneous media 11/1(79–82)
Geometric approximation for small-scale inhomogeneities 11/1(86–89)
Geometric dispersion 6a/3(107)
Geometric multiplicity 6a/2(369)
Geometrical and kinematical discontinuity conditions 6a/3(429)
Geometrical boundary 9(459)

Geometrical condition of compatibility 6a/3(307, 430)
Geometrical correlation of orientations 13(340)
Geometrical incidence 8/2(408)
Geometrical non-linearity 6(487, 488, 498–507)
Geometrical symmetries 6a/2(297)
Geometry of a surface in a Euclidean space covered by normal coordinates 6a/2(628)
Geometry of free boundary 9(378–380)
Geometry of glide 7/2(20–24)
German silver 6a/1(93)
Germanium 7/2(44); 14(263, 298)
Germanium, acoustoelectric effect 11/2(252)
Germanium, temperature dependence of ultrasonics 11/2(230, 231, 233)
Gerstner's law 6a/1(30, 31, 108, 423)
Gerstner's wave(s) 8/1(167–168); 9(742–744)
Gerstner's wave, uniqueness theorem 9(756)
Gettering 12(636–640)
Giant-cell model 10(11)
Gibbs adsorption
— equation 10(140, 150, 166, 205, 255, 268, 288, 292, 303)
— equation, generalization 10(150, 168, 170)
— formula 10(142, 170, 230, 267ff., 269)
— formula for a regular solution 10(268)
Gibbs' dividing surfaces 10(282, 286, 287, 289)
Gibbs' free energy 7/2(221, 223); 10(293); 12(73); 13(296, 319)
— effect on crystallization of glasses 13(593, 595, 600)
Gibbs fundamental equation 10(139, 170)
— for the interface layer 10(140)
Gibbs' thermodynamic potential 10(26, 107–108)
Gibbs' thermodynamical formulation of capillarity 10(289–293)
Gibbs-Duhem equation(s) 10(143, 150, 165, 166, 167, 219, 292, 297); 13(29)
Gibbs-Helmholtz equation 6(440)
— for a plane interface 10(142)
— surface analogue 10(296)
Gibbs-Kelvin formula 10(288, 292)
Gibbs-Thomson formula, generalized 10(155)
Gibbs-Tolman-Koenig-Buff equation 10(153)
Ginsburg-Landau theory 15(324)
Gladstone-Dale model, *see* Index-of-refraction effect
Glass 6a/1(74, 84, 89, 240, 254, 264, 269, 282, 318); 14(216)
Glass ceramics 13(606–607)
Glass families 13(514–515)
Glass formation 13(515, 525)
Glass, isotropy of 6a/1(254, 269)
Glass, network concept 11/1(403, 413–414)
Glass, Poisson's ratio for 6a/1(248, 261, 264, 276, 284)
Glass rod, vibrations 11/2(94)
Glass structure and equilibrium structure 13(544, 555)
Glass transition 13(431, 432, 447)
Glass transition temperature 6a/4(32ff.)
Glasses 10(5, 23, 69)
Glasses, fatigue effects 6(593)
Glass-forming oxides 13(514, 518)
Glassiness 13(610)
Glassy material, shift factor 13(447)
Glauert rule for airfoil theory 9(29, 34)
Glide, *see* Slip
Glide band formation 7/2(93–95, 151–152, 180)
Glide bands 6(232, 597); 7/2(83, 91, 102, 150, 180)
Glide bands containing large numbers of lamellae 7/2(84, 97)
Glide, discontinuous 7/2(156)
Glide geometry 7/2(20–24)
Glide lamellae 7/2(83)
Glide lamellae, bending 7/2(28)
Glide on pyramidal planes 7/2(48)
Glide plane, *see also* Slip 7/1(13, 96, 599)
Glide propagation 7/2(15, 107)
Glide relaxation 7/2(153, 155)
Glide systems, conjugate 7/2(62)
Glide systems in aluminum 11/2(198, 199)
Glide systems, latent 7/2(60)
Glide (=translation) 7/2(5, 6, 19)

Glide vector 7/2(16)
Gliding angle 9(48)
Gliding on a water surface 9(343–344, 587–592)
Global reaction exerted by the constraint of support 6a/2(416)
Globule of liquid, floating 10(299)
Glow discharge anemometer 8/2(586)
Glycerine, sound absorption 11/1(324–325)
Gold 14(239)
Gold-copper alloy 13(80–81)
Gold-silver single crystals
— critical shear stress 7/2(39, 177)
— work-hardening 7/2(58)
Goldstein's solution for the uniform translation of a sphere 8/2(303–304)
Goodier's thermoelastic potential 6(125)
Gorsky's treatment of superlattice transformations 13(84)
Gorter-Casimir theory 15(280, 284)
Görtler's principal function 8/1(277)
Görtler-Taylor vortex 8/1(119–121, 434, 436, 438)
Gough effect in rubber 6a/1(44, 308, 411, 726, 728, 734)
Gough effect Schmulewitsch's conjecture 6a/1(309–311)
Goursat's theorem 6a/2(160)
G.P., *see* Guinier-Preston zones
Grad 6a/3(408, 465)
Gradient, deformation 6a/2(29, 299)
Gradient, director 6a/2(452)
Gradient, director velocity 6a/2(452)
Gradient, displacement 6a/2(29)
Gradient field 6a/2(10)
Gradient, two-parametric, in a surface 10(284)
Grad's method of moments 12(266, 268–276)
Graffi's reciprocal theorem 6a/2(218, 243)
Graffi, *see* Alfrey
Grain boundaries 7/1(646); 14(210, 225)
— in face-centered cubic lattice 7/1(650)
— in the soap bubble model 7/1(649)
— mobility of the 7/1(656)
— relaxation of 7/1(657)
— specific energy of 7/1(651)
— structure of 7/1(659)
Grain boundary 7/2(227)
— diffusion 7/1(659)
— diffusion or migration 6(267, 598)
— energy of 7/1(652ff.)
— flow 7/2(197, 206)
— migration 7/2(195)
— model, Kê's 7/1(657)
Grain size 8/2(628)
Grain size effect on slip processes 6(602, 606)
Grain size of a surface 11/1(42)
Grains, twisted 7/1(649, 651)
Grand canonical ensemble 12(74); 13(23)
Grand partition function 10(293)
Grand potential 10(140, 192, 193)
Graphical method 6a/3(493)
Graphite 14(225)
Graphite lattice 7/1(80)
Graphite, specific heat and thermal conductivity 14(226)
Grashof number 8/2(560, 562)
Gravitational field 10(283, 287, 295)
Gravitational field, influence on liquids 10(15, 67)
Gravity effect upon a jet 9(326)
Gravity field, free surface flow 9(350)
Gravity flow through porous media 8/2(638)
Gravity wave 9(508, 510)
Gravity waves, velocity 9(513–514)
Gravity-capillary wave 9(511, 614, 631–638)
Gravity-capillary wave, velocity 9(514, 614)
Green measure of deformation 6(499)
Green-Lamé solution 6a/2(233, 236, 237)
Green's deformation tensor 6a/3(180)
Green's formula for the Laplace operator 6a/2(419)
Green's function 6a/3(8, 564); 11/1(22–23, 32, 36, 57, 83)
— for a duct with diaphragm 11/1(116)
— for a half-space 11/1(39)
— for a periodic structure 11/1(104, 114, 125)
— for an enclosure 11/1(57, 74)

Green's function (cont.)
— for rectangular membrane duct 11/1(121–122)
— for surface waves 9(524, 543, 547, 559, 579, 604ff., 614, 616, 621)
— for the interior of a duct 11/1(46)
— for the interior of a rectangular duct 11/1(53)
— of second kind 9(291ff.)
— of the Dirichlet problem 6a/2(421)
— subsonic and supersonic cases 11/1(79, 97)
Green's identity 9(40–41)
Green's integration method 8/2(119)
Green's method of solving
— of the Oseen equation 8/2(320–322)
— the Stokes equation 8/2(263, 276)
Green's process(es) 6a/3(81)
— integral solutions 6a/3(81)
— of the first kind 6a/3(82)
— of the second kind 6a/3(83)
Green's states 6a/2(186)
Green's tensor 7/1(516); 8/2(271, 275)
Green's theorem 6(95)
Green-Saint-Venant tensor 6a/3(550)
Grid Reynolds number 8/2(528)
Grid solidity 8/2(531, 533)
Grid-generated shear flow 8/2(539–540)
Griffith theory 6(557)
Grimm-Sommerfeld compound(s) 7/1(639); 7/2(44)
Grimm-Sommerfeld's rule 7/1(87)
Group velocity 7/1(161, 170, 190, 293, 312, 316)
Group velocity and energy propagation 9(519–520)
Group velocity of surface waves 9(506, 509, 513, 514, 518)
Grouping of prominent slip-lines 7/2(92)
Grouping of slip-lines 7/2(82)
Group-velocity 14(203, 226)
Growth, *see* Decay
Growth conditions 7/1(645)
Growth in depth of glide steps 7/2(94, 107)
Growth of boundary layer 8/1(288ff., 322ff., 369ff., 381, 396, 407, 434)
Growth of energy transfer 8/2(495)
Growth of nuclei 7/2(225, 226, 240, 250)
Growth of the second phase 7/1(661)
Growth of vorticity 8/2(492–494)
Growth spiral 7/1(97, 661)
GRS rubber 11/1(411, 412)
Grüneisen numbers 6a/4(132, 139)
Grüneisen's constant 7/1(276, 277, 278, 285, 312, 315, 321, 370, 380); 11/2(232, 234)
Grüneisen's experiments using an interferometer 6a/1(116–124, 289–292)
Guest experiment for determining the yield surface 6a/1(483, 501–509, 676–684)
Guest-Tresca hypothesis 6a/1(433–445, 484–486, 508)
Guggenheim's formula 10(155)
Guggenheim's mixing law 10(268)
Guinier-Preston zone 7/1(389, 430)
Guinier-Preston zones 7/2(191–193, 229, 235, 239)
Guitar strings 6a/1(56)
Gun sound 11/2(33)
Güntelberg's charging process 13(374, 382, 393)
Gut, deformation of 6a/1(13, 52–56, 419)
G-values, dependence on concentration 11/1(219)
G-values in simple reactions 11/1(209–212)
G-values in stepwise dissociation 11/1(218–221)
Gyromagnetic effect 15(274)
Gyromagnetic ratio 15(228)
Gyroscopic force 6a/3(257, 260)

Haar and von Kármán's hypothesis 6a/3(526)
Haar-Kármán hypothesis 6(427)
Haber Kerschbaum equation 12(528)
Habit plane 7/2(245)
Hadamard stability 6a/3(198, 199, 204, 224, 246)
Hadamard superstability 6a/3(224)
Hadamard's criterion of stability 6a/3(197)
Hadamard's criterion, sufficient conditions 6a/3(246, 248)
Hadamard's finite part of an integral 9(42, 303)
Hadamard's method 9(305)

Hadamard's theory of singular surfaces 6a/3(428)
Hagen-Poiseuille equation 6(459, 497); 8/2(49, 642)
Hagen-Poiseuille law 12(606)
Hagstrum adsorption spectrometer 12(628)
Hair 6a/1(64, 67, 70, 169)
Hale's dynamical system 6a/3(168, 169)
Hale's instability theorem 6a/3(169, 199)
Hale's invariance principle 6a/3(158, 169, 191, 255, 288, 289)
— and asymptotic stability 6a/3(254)
Hale's necessary conditions for stability 6a/3(199)
Half dislocation = partial dislocation 7/1(510, 654)
Half-dislocation 7/2(139, 188)
Half-fine quantities of Massignon 13(68, 69)
Half-plane moving between two parallel planes 8/2(83)
Half-plane moving parallel to a fixed plane 8/2(82)
Half-range expansion of the distribution function 12(267, 284)
Half-range Fourier transforms 6a/3(21, 33)
Halide glasses 13(515)
Hall effect 15(249, 290)
Hall field 14(178)
Hamel flow 8/2(30–34)
Hamel spirals 8/2(33, 144)
Hamiltonian equations of motion 7/1(157)
Hamiltonian (function) 7/1(156, 166, 296); 14(195)
Hamilton-Kirchhoff principle 6a/2(226, 255)
Hamilton's principle 8/1(146, 147)
Hampson air liquefier 14(57)
Hampson-type interchanger 14(63)
Hankel functions 9(296)
Hankel transform 6(100, 101); 11/1(83)
Hard sphere molecules 10(221)
Hard testing machines 6a/1(125, 578, 651, 653, 676, 678, 679)
Hardness 7/1(102)
Hardness of ceramics 13(606)
Hardness of glasses 13(632–635)
Hardness of sodium borate glasses 13(532)
Harkins-Jura equations of adsorption 8/2(631)
Harmonic approximation 7/1(290, 296)
Harmonic distribution 7/1(272, 288)
Harmonic equation 6a/3(75)
Harmonic field 6a/2(20)
Harmonic growth 6a/4(284)
Harmonic oscillator model 13(349, 571)
Harmonic plane progressive wave 6a/2(342)
Harmonic vibrations, *see* Oscillations
Harmonic vibrations in subsonic compressible flow 9(297–299)
Harmonic vibrations in supersonic compressible flow 9(306)
Harnack inequalities 6a/3(587)
Harnack's convergence theorem 6a/2(21)
Hartfree-Fock potential 7/1(403)
Hartig's law 6a/1(91, 105, 107, 118, 120, 127, 131, 136, 137, 148)
Hartmann's whistle 11/2(31, 96)
Hartshorn bridge, ballistic 15(74)
Hartshorn bridge for alternating current 15(75)
Havelock's method for the investigation of wave crests 9(735–736)
H-body, *see* Hooke solid
He^3 and He^4, mixtures of 15(181, 183, 184)
He^3, nuclear susceptibility 15(182)
Healing effect of surface etching 13(626)
Heaslet-Lomax formula 9(304)
Heat absorption 6a/2(510)
Heat absorption coefficient 15(56, 73)
Heat addition to subsonic flow 9(240, 247)
Heat capacity 7/1(325)
Heat capacity at constant
— pressure 12(7, 28)
— volume 12(7, 28, 47)
Heat conducting wall 8/1(423)
Heat conduction 7/1(290, 291); 8/1(8, 65–66, 239); 11/1(11, 13); 15(166, 375, 376, 423–430)
— at different gas pressures in vacuo 12(523–525)
— by convection 12(317)

Heat conduction (cont.)
— effect on gas dynamics 9(264)
— effect on sound absorption 11/1(132, 138–139, 150)
— equation 6a/2(312, 326); 6a/3(184, 185); 8/2(113)
— for a gas of rough spheres 12(498)
— in a binary mixture 12(412–415, 451)
— inequality 6a/2(303, 307)
— losses near a boundary 11/1(21)
— neglected 9(225, 240, 263)
Heat conductivity 8/2(113)
— at low temperatures 12(467)
— for ortho- and paramodifications 12(467)
— measurements 12(316–320)
— of solid helium 15(180)
Heat conductivity coefficient
— definition 12(301)
— for polyatomic gases 12(495)
— higher approximations 12(401)
— measured values 12(395)
— of the Lorentz gas mixture 12(418)
— of the Maxwellian gas 12(383)
— of the monatomic gas 12(388)
— of the monatomic gas mixture 12(410, 423, 451)
— Ω-integral representation 12(392–394)
— pressure dependence 12(310)
— temperature dependence 12(400)
Heat current 14(230)
Heat dissipated 14(298)
Heat exchangers 14(47, 49, 55, 58, 62, 89, 120)
Heat flow 7/1(293, 294, 296, 309); 12(352, 355)
— density 7/1(293)
— due to two-fluid circulation 14(268)
— one-dimensional 12(290)
— vector 11/1(3); 12(270)
Heat flux 6a/3(174, 175)
— condition 6a/2(335)
— definition 6a/2(327)
— moments for rods 6a/2(649)
— relaxation functions 6a/4(205)
— vector 6a/2(300, 510)
Heat fusion per monomer 13(428)
Heat leak 15(65–68, 142, 200, 202)
Heat of activation, *see* Activation energy
Heat of dilution 13(319)
Heat of melting for polymers 13(427)
Heat of sublimation 7/1(319)
Heat of transport 15(390, 391, 409, 410)
Heat operators 6a/2(329)
Heat radiation, effect on sound absorption 11/1(140, 150)
Heat source moments for rods 6a/2(651)
Heat supply 6a/2(300, 510, 513); 15(57)
Heat transfer 15(166, 167, 170, 203)
— coefficient 8/2(560); 14(92, 99)
— effect on the transition to turbulence 8/1(419–422, 426)
— in compressible boundary layer 8/1(422–428)
— in incompressible boundary layers 8/1(343–347)
— on a body in molecular flow 8/2(604–606)
— through glasses 13(574–577)
Heat transmitted by radiation and by conduction 6a/2(581)
Heat transport 7/1(292)
Heat-conduction gauge 12(517, 523–527)
Heat-flashes causing striation 6(600, 604ff.)
Heating by alternating current 15(137, 140, 146, 150)
Heating by gamma rays 15(137, 141, 146, 148)
Heating by hysteresis 15(141)
Heat-loss equation 8/2(561)
Heaving circular cylinder generating waves 9(563)
Heaving hemisphere generating waves 9(562)
Heaving motion of a floating body 9(562, 619)
Heaviside classes 6a/3(6)
Heaviside processes, approximation of 6a/3(16)
Heaviside type 6a/3(5)
Heaviside unit step process 6a/3(6)
Heavy water, absorption of sound 11/1(246–247)
Heavy water, effect on relaxation in $MgSO_4$ solutions 11/1(270)
Heisenberg's microscopic theory of superconduction 15(343)
Heisenberg's transfer theory 8/2(471ff.)

Hele Shaw flow 8/2(209–211)
Helical trajectories 8/2(99, 101, 166, 172)
Helium atoms, interaction between two 13(2)
Helium film 15(168, 169)
Helium I 15(374)
Helium II 15(374)
Helium II, liquid, thermal conduction by the phonons 14(228)
Helium II, solid 14(222)
Helium isotopes, viscosity 12(463)
Helium liquefiers 14(127)
Helium liquefiers, Collins-type 14(97)
Helium, liquid 10(119ff.)
Helium, liquid, sound absorption 11/1(305, 306)
Helium, liquified 12(27)
Helium permeation through glass 12(634)
Helium reemission from molybdenum 12(642)
Helium valve 15(169)
Hellinger-Prange-Reissner principle 6a/2(124, 125, 228)
Helmholtz equation 8/2(4); 11/1(22)
Helmholtz equation for the rate of velocity change 8/2(452)
Helmholtz' free energy, *see also;* Free energy 6a/3(175); 10(137, 149, 293, 296); 12(75); 13(13, 221, 225, 229, 237, 287, 297, 319)
Helmholtz free energy for isothermal deformations 6a/3(191)
Helmholtz theory, its limitations 8/2(124)
Helmholtz-Kelvin circulation theorem 9(4)
Helmholtz-Rayleigh dissipation theorem 8/1(258–259)
Helmholtz's equation 9(284, 291, 298)
Helmholtz's theorem 6a/2(19)
Helmholtz's vorticity theorems 8/1(34–36, 163)
Hemisphere, resistance 8/2(383–384)
Hemp rope 6a/1(56)
Hencky measure of deformation 6(486, 499–500, 544)
Hencky theory of plasticity 6a/3(549)
Hencky's flow condition 6(520)
Hereditary elasticity 6a/2(423)
Hereditary integral equation 6a/3(61, 64)
Hereditary problems 6a/2(377, 386)
Hereditary response 6a/3(1)
Hereditary wave operator ($\Box_K$) 6a/3(92)
Hermite coefficients in terms of moment 12(269)
Hermite expansion 12(267)
Hermite expansion, absolute Maxwellian 12(267, 280–284)
Hermite expansion, locally Maxwellian 12(267, 268–276)
Hermite polynomials, eigenfunctions 12(272, 281, 283)
Hermite polynomials, inversion formula 12(269)
Hermite polynomials, tensor form 12(269)
Hertz relation for the evaporation rate 12(541)
Hertzian vector 13(257, 263)
Heterogeneous fluid 9(506)
Heterogeneous fluids, theorems on wave motion 9(722–725)
Heterogeneous media 6a/2(421, 422)
Heterogeneous nucleation in glasses 13(593, 605)
Hevea rubber 11/1(410, 412)
Hexagon of Tresca 6(371, 379–380)
Hexagonal axis 7/1(116)
Hexagonal close-packed structure 7/2(26)
Hexagonal condition 6a/3(472)
Hexagonal coordinates 7/1(11)
Hexagonal crystals, thermoelastic effects 11/2(176)
Hexagonal metals 7/2(6, 30)
— kinking 7/2(74)
— surface phenomena 7/2(103)
— theory of work-hardening 7/2(137–139)
— work-hardening 7/2(45–50)
Hexagonal prism 6a/3(411)
Hexagonal prism of Tresca yield condition 6(328, 371)
Hexagonal symmetry, elastic constants 6(22–23)
H-function $\bar{H}$ 12(239)
H-function $H^{(0)}$ 12(241)
H-function $H_r^{(0)}$ 12(241)
H-function $H^{(13)}$ 12(275)

H-function $H_r^{(1)}$ 12(241)
H-function H_Γ 12(208, 213, 226)
H-function H_γ 12(213, 225, 226, 228)
H-function H_i 12(239)
H-function H_x 12(225, 226)
H-function of Kochin 9(556–558, 573–574)
H-function of Kochin, time dependent motion 9(617)
Hidden variables 6a/3(3)
High frequency
— behavior of a viscoelastic material 6a/3(22)
— effects 15(340)
— limit of the attenuation 6a/3(110)
— limits of oscillatory viscoelastic processes 6a/3(106)
— resistance of superconductors 15(247)
— stress waves 11/2(153ff.)
High hydrostatic stress 6a/1(490–499)
High pressure, Savart-Masson effect at 6a/1(475, 477)
High pressures, measurement 12(29–30)
High temperature range of glass viscosity 13(582)
High temperature tests 6a/1(281, 293–296, 385–393, 636)
High temperatures, glass density 13(555–557)
High viscosity liquids, sound absorption 11/1(323)
Higher order energies 6a/3(211, 212, 236–238)
Higher order Kelvin processes 6a/3(81)
Higher order waves (mild discontinuities) 6a/4(60ff.)
High-frequency horn loudspeaker 11/2(18–20)
High-frequency sirens 11/2(26)
Highly dilute gases, solution of the Boltzmann equation for 12(364)
High-temperature creep 7/2(194)
High-voltage electrostatic loudspeaker 11/2(16)
Hilbert class of solutions of the Boltzmann equation 12(257)
Hilbert expansion 12(252, 254–258)
— boundary condition 12(261)
— compatibility condition 12(255)
— continuation 12(258–259)
Hilbert space 6a/3(551)
Hilbert uniqueness theorem 12(252, 257)
Hilbert's linear integral equation 12(366)
Hildebrand's equation 10(268, 269)
Hilsch's vortex tube 14(6)
($\mathscr{X}$)-history 6a/3(4)
History, bounded 6a/3(5)
History, definition 6a/3(4)
History dependence of the solutions to the visco-elastic problem 6a/3(61)
History, finite 6a/3(5)
History gradient 6a/3(540)
History, infinite 6a/3(5)
History, integrable 6a/3(5)
History of shell theory 6a/2(527, 589)
History of the deformation gradient 6a/3(305)
History, restricted 6a/3(5)
Hodgkinson's nonlinear law for tension and compression 6a/1(26, 71, 115, 138, 140, 188)
Hodgkinson's nonlinear response in flexure 6a/1(25)
Hodograph characteristics 8/1(209)
Hodograph differential equation, *see also* Chaplygin's equation 9(60, 123)
Hodograph equation
— alternate forms 8/1(195)
— Chaplygin's solutions 8/1(198)
— Crocco's solutions 8/1(198)
— derivation 8/1(194)
— Manwell's solution 8/1(197)
— Ringleb's solution 8/1(197)
Hodograph method 9(27, 56ff., 70, 226, 326ff., 345)
Hodograph plane 8/1(195, 209–210); 9(59, 123, 683–684)
Hodograph plane, logarithmic 9(71)
Hodographic graphical construction 9(692)
Hoeppler viscosimeter 12(315)
Hölder continuity 6a/3(146, 274); 9(99, 104, 400)
Hölder stability
— continuous dependence 6a/3(253)
— definition 6a/3(146)
— elastic bodies 6a/3(171)
— logarithmic convexity 6a/3(231)

— null solution in linear thermo-elasticity 6a/3(252)
— sufficient condition for 6a/3(153)
— theorem 6a/3(253)
Hole in an infinite plate 6(58)
Hole model of liquids 10(11–13, 242, 249ff., 258, 261); 13(234, 242)
Holes 14(292)
Holes in *d*-band 14(265)
Holes of different shapes, photoelastic fringe patterns 6(205–211)
Holomorphic 6(60)
Homentropic motion 9(4, 9, 228–229, 249–272)
Homogeneity in turbulence 8/2(525)
Homogeneous body 6a/3(17)
— elastic 6a/2(68)
— isotropic elastic 6a/2(380)
— isotropic thermoelastic 6a/2(317, 327)
— thermoelastic 6a/2(327)
Homogeneous boundary conditions and initial data 6a/3(270)
Homogeneous displacement field 6a/2(33)
Homogeneous glide 7/2(101)
Homogeneous isotropic body 6(17–19, 21, 23–26)
Homogeneous isotropic incompressible viscoelastic body 6a/3(71)
Homogeneous isotropic viscoelastic body for relaxation type 6a/3(46)
Homogeneous nucleation in glasses 13(593, 595)
Homogeneous shear 7/2(14, 15)
Homogeneous shear flow 8/2(508)
Homogeneous strain of turbulence 8/2(535)
Homogeneous turbulence 8/2(438, 441)
Homogeneous turbulence, dynamics 8/2(445–446)
Homogeneous turbulence, kinematics 8/2(444–445)
Homogeneous turbulent shear flow 8/2(540)
Homomorphism 6a/3(130)
Homopolar crystals, critical shear stress 7/2(41)
Homopolar crystals, work-hardening 7/2(68)
Homovalent solute atom 14(151)
Honeycomb 8/2(551–552)
Hooke solid (=elastic material) 6(441, 448–450)
Hooke's experiments 6a/1(156–159)
Hooke's law 6(15, 17, 24, 142, 329, 400, 401, 409); 6a/1(12, 70, 93, 156); 6a/3(496); 7/1(178, 187); 11/2(154)
— contested 6a/1(32, 102, 111, 114, 118, 123)
— defect of 6a/1(94, 123)
— for glasses 13(623, 635, 637)
— for seepage flow 8/2(636)
— "repeal" of 6a/1(70)
Hooke's tensor 7/1(515f., 523)
Hopf's theory of turbulence 8/2(443–444)
Hopkinson pressure bar 6a/1(324, 325, 332, 591, 605–609)
Hopkinson's experiment 6a/1(448, 579–583)
Hoppmann's impact experiments 6a/1(599, 600, 603, 604, 621)
"Horizontal" movement 14(231, 232)
Horizontal pendulum 11/1(378)
Horn loudspeaker 11/2(17–22)
Horn shapes 11/2(20–23)
Horns 11/1(48)
Hot wire microphone 11/2(58–59)
Hot-wire anemometer
— conventional response equation 8/2(571)
— dynamic operation 8/2(567–585)
— effect of extraneous temperature fluctuations 8/2(534)
— non-linear case 8/2(576–578)
— operation 8/2(555–559)
— resolving power 8/2(558, 578–585)
— steady state operation 8/2(559–567)
Hot-wire in liquids 8/2(586)
Houston ionization gauge 12(625–626, 646)
H-S Mollier diagram 14(20, 49)
Hsu's criterion 6a/3(288, 289)
H-theorem 12(239–241, 358, 484–489)
H-theorem of liquids 10(107)
Huber's fracture condition 6(520)
Huggins' density formula 13(549–550, 554)
Hugoniot relation 6a/3(372); 6a/4(90, 95, 289); 8/1(11, 221, 222)
Hugoniot stress-strain curve 6a/4(82, 94)

Hugonoit curve, *see* Rankine-Hugonoit curve
Hull shape, effect on wave resistance 9(580–581)
Human tissue, elastic properties of 6a/1(56–62, 244)
Hume-Rothery phases 7/1(87)
Hund's stability rules 15(79)
Huntsman steel 6a/1(215, 262)
Hu-Washizu principle 6a/2(122, 125, 226)
Huygens' principle 11/2(103)
Hybrid binding 7/1(75)
Hybridisation 7/1(75)
Hydratation of ions 13(393)
Hydrate 11/1(221, 225)
Hydrate envelopes, mutual effect 11/1(250, 252, 260, 266)
Hydratization 11/1(222, 225, 248–249, 261)
Hydratization equilibrium 11/1(225)
Hydraulic analogy of shallow-water theory 9(668, 673)
Hydraulic diameter 14(99)
Hydraulic friction 9(696–697)
Hydraulic jump 9(685, 689, 692, 696ff.)
Hydraulic jump, oblique 9(700–701)
Hydraulic radius 8/2(643)
Hydraulics 9(446)
Hydraulics in open channels, steady one-dimensional case 9(695–701)
Hydrodynamic interaction between polymer segments 13(473, 476)
Hydrodynamic interaction of chain segments 11/1(384)
Hydrodynamic paradox of d'Alembert 8/1(53, 91, 105, 160, 201)
Hydrodynamic potentials of Faxén 8/2(322)
Hydrodynamic potentials of Odqvist 8/2(267–271)
Hydrodynamic potentials of Wolska-Bochenek 8/2(285–286)
Hydrodynamic stability 8/1(253–257)
Hydrodynamical approximation 12(296, 297, 384, 402–415, 493–497)
Hydrodynamical derivative 9(2)
Hydrodynamical equations 11/1(275–277)
Hydrofluoric acid, etching of glass 13(626)
Hydrofoil close to surface 9(586)
Hydrofoil of finite span 9(586, 590)
Hydrofoil, thin body approximation 9(583–587)
Hydrogen bonds 11/1(251, 254)
Hydrogen bonds in nylon 11/1(415, 417)
Hydrogen, dispersion and absorption 11/1(176)
Hydrogen liquefaction 14(88)
Hydrogen liquefiers 14(59, 61, 67, 68, 85)
Hydrogen liquefiers, two-circuit 14(66)
Hydrolysation 11/1(248, 252, 257–258)
Hydrostatic approximation 9(468)
Hydrostatic behaviour of porous media 8/2(630–633)
Hydrostatic fundamental equation 10(283, 287)
— one- and two-dimensional analogues 10(285, 298)
Hydrostatic interaction between foreign atoms and dislocations 7/2(178)
Hydrostatic load 6a/3(244–249, 262, 263)
Hydrostatic model of capillarity 10(281)
Hydrostatic pressure 6a/2(678, 690); 6a/3(261); 8/2(5)
Hydrostatic tension 6(143)
Hydroxides of metals 11/1(258)
Hydroxyle ions 11/1(248, 249, 255)
Hydroxyle ions in ion reactions 11/1(254–258)
Hydroxyle ions, mobility 11/1(216)
Hyperbolic differential equation 8/1(67); 9(16ff., 22, 29, 42, 93, 135, 209, 667, 673, 683)
— canonical form 9(691)
Hyperbolic law 6a/1(20, 71, 115)
Hyperbolic point(s) 6(369, 372, 377); 6a/3(465)
Hyperbolic problem 6a/3(439, 460, 462, 526)
Hyperbolic problem of differential equations 6(339, 341, 352, 360, 362, 369, 371, 384)
Hyperbolic region 6(384–385); 6a/3(478)
Hyper-conical horn 11/1(48)
Hyperelastic 6a/3(173)
Hyperelastic rods 6a/2(673)

Hyperfine structure 15(25, 43f., 81f., 85, 110f., 149)
Hypergeometric differential equation 9(63, 69, 150)
Hypergeometric functions 9(240, 612, 672)
Hypersonic flow 8/1(71–72); 9(242)
Hypersound generation 11/2(100–102)
Hypocycloids 6a/3(453)
Hypocycloids as slip lines 6(363)
Hypo-elasticity 6(262)
Hypotheses of Lennard-Jones and Devonshire 13(223–226)
Hysteresis effects 15(57, 127, 132, 135, 429)
Hysteresis effects in dislocation damping 11/2(187)
Hysteresis, elastic 7/1(563)
Hysteresis heat 15(137)
Hysteresis in the transition 15(335)
Hysteresis loop 15(72, 73, 127, 139, 145)
Hysteresis loop, sheared 15(139, 145)
Hysteresis loops 6a/1(97, 99, 123)
Hysteresis of bending stiffness 11/1(402)
Hysteresis of capillary pressure 8/2(632)
Hysteresis of rubber 11/1(361)
Hysteresis type loss 11/1(399)

Iberall equation 8/2(645)
Ice 6a/1(187, 428, 474, 482)
Ice sheet on water 9(455)
Ice, structure 11/1(202)
Ideal crystal 7/1(383, 384, 385)
Ideal fluid 9(448, 450)
Ideal gas 8/1(174–175, 179)
Ideal gas, adiabatic flow 9(6–8, 61–62, 93)
Ideal gas, sound absorption and dispersion 11/1(148–149)
Ideal jet 9(271)
Ideal liquid 6(441)
Ideal material 6(440)
Ideal plastic body 6a/3(403)
Ideal solution, definition 13(320)
Idealized fluid of Kármán and Tsien 9(77, 80, 81–84, 93)
Idealized fluid of Tomotika and Tamada 9(77, 145–147, 150, 172–173)
Idealized thermoelastic medium 6a/4(124, 215)
Identities 6a/4(114, 116)
Identity of Euler, Piola, and Jacobi 6a/4(114)
Impact, *see also* Axial collision, Time of contact
Impact cleaving 7/1(99)
Impact conditions 6a/4(41ff., 47, 51ff., 56, 63, 65, 88)
Impact, Dunn's experiment 6a/1(583–597, 599, 604)
Impact, duration of 6a/1(316–327, 644)
Impact, energy balance 6a/1(426, 565–570, 591, 603, 604, 645)
Impact experiments of J. Hopkinson 6a/1(579)
Impact force 9(362)
Impact, Hertz's theory of 6a/1(314, 317, 319–323, 325, 330)
Impact, Kolsky's version of Dunn's experiment 6a/1(591–595)
Impact, nonsymmetrical 6a/1(607, 691)
Impact of a body on a water surface 9(360–363)
Impact of bars as a function of the radius 6a/1(319, 326)
Impact of spheres 6a/1(316, 319)
Impact of spheres on rods 6a/1(316, 332, 338, 358)
Impact parameter 12(234, 236)
Impact strength 13(624)
Impact, symmetrical free flight 6a/1(317–330, 613, 624, 625, 634)
Impact tests, Dunn's hypotheses for 6a/1(584, 589, 591–595)
Impact tests, extended quasi-static 6a/1(579, 586–596)
Impact, transverse of strings 6a/1(718–721)
Impact velocity, *see* Projectile velocity
Impact (zero) time 6a/4(21–24)
Impact (zero) time, optical detection 6a/4(21–24)
Impedance, *see also* Radiation impedance, Specific impedance etc. 11/2(4)
Impedance, acoustic 11/1(9, 16, 19, 35, 99, 132)
Impedance, complex mechanical 6(276ff.)
Impedance disk 11/1(69, 76)
Impedance due to lattice motion 11/2(211)

Impedance, electrical 6a/4(262)
Impedance factor for a flexible sphere 11/1(103)
Impedance factor for membrane and plate 11/1(101, 107)
Impedance factors for membrane and plate 11/1(100)
Impedance, normal 11/1(16)
Impedance of a radiating sphere 11/1(63)
Impedance transformation ratio 6a/4(268)
Impenetrability 6a/3(536, 538)
Imperfect gases 12(91–100)
Imperfections, dimension of 7/1(386ff.)
Imperfections of crystal 7/1(383)
Imperfections, one-dimensional 7/1(388)
Imperfections, systematics of 7/1(386ff.)
Imperfections, transient 7/1(381)
Imperfections, two-dimensional 7/1(389)
Imperfections, zero-dimensional 7/1(387)
Impermeable equivalent sphere model 13(473, 474)
Impermittivity 6a/4(166)
Impulse methods 11/1(230)
Impurities 7/2(124, 125); 15(224, 246, 269)
Impurities in crystals 7/1(646)
Impurities in glasses, effect on devitrification 13(609–610)
Impurity additions 7/2(40)
Incident amplitude 11/1(16, 29)
Incipient 6a/3(525)
Incipient plastic flow 6(427)
Inclusions in the crystal 7/1(551)
Incoherent nuclei 7/2(212)
Incompatibilities 7/1(525)
Incompatibility of polymers in solution 13(484)
Incompatibility tensor 7/1(523)
Incompatible rotations 7/1(510)
Incomplete gamma function 8/2(55)
Incompressibility condition 6a/2(404)
Incompressible 6a/3(2, 71, 176, 183, 248, 264, 512)
Incompressible body, material 6a/2(210)
Incompressible flow 9(294)
Incompressible matter, strain invariants 6(6)
Incompressible matter, stress-strain relations 6(18)
Incompressible plane visco-plastic flow 6(302)
Incremental body force 6a/3(259, 286)
Incremental displacement 6a/3(176, 183)
Incremental quantities 6a/3(176)
Incremental rate of work equation 6a/3(178, 180, 182, 183)
Incremental state 6a/3(188, 189)
Incremental stress vector 6a/3(259–261, 286)
Incremental temperature variable 6a/3(183)
Incremental wave experiment, *see* Bell's incremental wave experiment
Incremental wave in prestressed solids 6a/1(308, 356, 409, 613, 614, 617, 618, 698, 699, 716)
Incubation period 7/2(43, 126, 187, 228)
Incubation period of fatigue 6(609, 610)
Indentation 6(101, 432)
Indentation hardness of glass 13(632)
Independent measurement of E, μ and ν 6a/1(287, 288, 289–292)
Independent scatterers 11/2(165)
Independent segments of a chain, their number 11/1(385, 395)
Index of refraction 11/1(29, 86); 13(256–262)
— effect 6a/4(15–21)
— effect, Gladstone-Dale model 6a/4(18)
— fluctuations 13(310)
— of glasses 13(544)
Indexes of frequently used symbols 6a/2(2–5, 298, 299, 432–438)
Indexing of crystal faces and zones 7/1(61)
India rubber, Poisson's ratio for 6a/1(245, 274, 732)
Indicator for a test gas 12(542–544)
Indices, four-member- 7/1(67)
Indices, hexagonal 7/1(67)
Indium 14(271f.)
Indium-thallium alloys 14(275)
Individual particle model 15(313)
Induced angle of attack 8/2(408)

Induced dipole moment 11/1(30)
Induced discontinuity 6a/3(311)
Induced drag 8/1(89, 105); 8/2(406, 409, 416)
Induced jump of a plane wave 6a/3(100)
Induced map 6a/3(131)
Induced mass of a fluid 9(362)
Induction forces 11/1(351)
Induction heater 15(116)
Inductive leaks in bridges 15(76)
Inelastic collision, equations 12(469)
Inelastic scattering of a conduction electron 11/2(216)
Inelasticity, physical basis 6(229–231)
Inequality for velocity fields 9(123)
Inertia forces 6a/3(525)
Inextensional theory of shells 6a/2(546, 548)
Infinite cavity 9(322, 328–331)
Infinite dilution, limiting laws 13(304, 305)
Infinite growth of the characteristic values 6a/2(270)
Infinite histories 6a/3(5)
Infinite horn 11/2(20)
Infinite slab 6a/3(529)
Infinite strip 6(84, 303, 306–307)
Infinite wake 9(321)
Infinitely deep fluid 9(456)
Infinitesimal deformation 6a/2(28, 456); 6a/3(3)
Infinitesimal deformation (or strain) 6(6–7, 12, 438)
Infinitesimal director displacement 6a/2(456)
Infinitesimal linearized theory of viscoelasticity 6a/3(1)
Infinitesimal motion of 3-dimensional continuum 6a/2(473)
Infinitesimal rigid motion 6a/3(17)
Infinitesimal rigid-body displacements 6a/2(30, 462, 463)
Infinitesimal rotation 6a/2(461)
Infinitesimal rotation process 6a/3(17)
Infinitesimal sinusoidal progressive waves
— (acoustic waves), amplitude 6a/4(30)
— attenuation 6a/4(30–34)
— echo-pulse transmission technique 6a/4(75)
— frequency 6a/4(30–34)
— longitudinal velocity 6a/4(75)
— phase velocity 6a/4(30–34)
— pulse transmission technique 6a/4(31–33)
— relation to acceleration waves 6a/4(61 ff., 75, 98)
— transverse velocity 6a/4(75)
Infinitesimal strain process 6a/3(17)
Infinitesimal strain tensor 6a/2(461)
Infinitesimal uniqueness 9(410)
Infinitesimal volume change 6a/2(30)
Infinitesimal wave approximation 9(456, 462, 463–466, 469 ff.)
— for viscous fluids 9(638–646)
— limits of application 9(522)
Infinitesimal waves, higher-order approximation 9(653–667)
Infinitesimal waves in irrotational flow 9(464)
Inflection 6a/3(484)
Inflection point in velocity profile 8/1(372, 375, 379, 396, 398, 422, 439)
Inflection point on free boundary 9(379)
Influence domain (or region) 9(19, 40, 233, 256)
Influx (from leaks, etc.) 12(630, 631 ff.)
Information rate of anemometer 8/2(585)
Infrared absorption spectra of borate glasses 13(532)
Infrared spectra of polymers, change by crystallization 13(599)
Inhomogeneities in glass structure 13(527)
Inhomogeneities of small extension 11/1(86–89)
Inhomogeneity 6a/2(530)
Inhomogeneity induced by decay of turbulence 8/2(528–530)
Inhomogeneous body 6a/3(17)
Inhomogeneous deformation 6a/1(410, 650, 661, 668, 670)
Inhomogeneous glide 7/2(101)
Inhomogeneous medium 11/1(6–7, 10, 76–99)
Inhomogeneous terms in acoustic wave equations 11/1(6, 24)
Inhomogeneous waves 6a/4(194)
Initial boundary value problem 6a/3(151)

Initial conditions 6a/2(219, 335)
Initial conditions or data 6a/3(35, 90, 95, 112, 130, 178, 179, 186, 211, 232, 234, 235, 237, 264, 271, 274, 275, 417, 420)
Initial curve 6(333)
Initial data 6a/2(218)
Initial elastic compliance 6a/3(24)
Initial elasticity 6a/3(18)
Initial isotropy of the relaxation function, relation to wave propagation 6a/3(99)
Initial mechanical response 6a/3(49)
Initial past-history
— consistent 6a/3(51, 89)
— definition 6a/3(50)
— problem 6a/3(61, 63, 64)
— vanishing 6a/3(52, 63)
Initial period of decay 8/2(485–487)
Initial perturbation 6a/3(207)
Initial quasi-static viscoelastic state 6a/3(49)
Initial response 6a/3(13, 49)
Initial state 6a/4(216)
Initial stress 6a/3(197, 227–240)
Initial value 6a/3(246)
Initial value problem(s) 6a/3(104)
— for capillary-gravity waves 9(614, 631, 636)
— for surface waves 9(603ff.)
— for the cylindrical tube 8/2(149–152)
— for the supersonic wing 9(43–45)
— for vortices 8/2(132, 134–138)
— method of images 9(610)
— of Cauchy 9(232, 254–257, 268)
— of Cauchy and Poisson 9(603, 607–614, 631)
— of hyperbolic equations 9(16–20)
— of the wave equation 9(40–43)
— two-dimensional waves 9(610–614)
Initial-boundary value problems 6a/2(371)
Initially flat Cosserat surface 6a/2(460)
Initial-value method of Havelock 9(472, 513, 515)
Initial-value problem
— characteristic 6(361, 396, 397)
— for rods 6a/2(700)
— mixed 6(362, 397)
— of Cauchy type 6(333, 336ff., 340, 360ff., 365, 396)
Initiation of fatigue damage 6(596, 606)
Initiation of stage III of work-hardening curve 7/2(155–157)
In-material gauge 6a/4(9, 14–21)
— electromagnetic 6a/4(9)
— interferometry, *see* Laser interferometry
— manganin wire 6a/4(8)
Inner product 6a/3(4, 539)
— of four-vectors 6a/2(27)
— of second-order tensors 6a/2(7)
— of tensors 6a/2(618)
— of vectors 6a/2(5)
Instability, definitions
— asymptotic 6a/3(150)
— exponentially asymptotic 6a/3(150)
— Lagrange 6a/3(150)
— latent (adjacent equilibrium technique) 6a/3(144)
— Liapounov 6a/3(135, 147)
Instability, general 6a/3(135)
Instability of flow 8/1(116, 118)
Instability theorems 6a/3(169, 199)
— discrete conservative systems 6a/3(167)
— discussion 6a/3(168)
— from non-uniqueness 6a/3(151, 167, 224)
— Liapounov 6a/3(168)
— Liapounov conditions sufficient for equilibrium solution 6a/3(199, 200)
Instability, time dependent 6(310, 311)
Instantaneous bulk modulus 6a/4(72)
Instantaneous elastic response function 6a/3(350)
Instantaneous elasticity 6a/3(335)
Instantaneous entropy function 6a/4(78)
Instantaneous free energy function 6a/4(78)
Instantaneous Hugoniot entropy-strain 6a/4(84ff., 91ff., 95)
Instantaneous Hugoniot temperature-strain curve 6a/4(85ff., 95)
Instantaneous internal energy function 6a/4(78)
Instantaneous isentropic second-order modulus 6a/4(81, 94)

Instantaneous isentropic tangent modulus 6a/4(81, 94, 96)
Instantaneous isothermal second-order modulus 6a/4(77ff.)
Instantaneous isothermal tangent modulus 6a/4(77ff., 83)
Instantaneous pressure-density curve 6a/4(70, 72–75)
Instantaneous second-order modulus 6a/4(27, 50, 64)
Instantaneous second-order tangent modulus 6a/3(336, 350)
Instantaneous sound velocity
— instantaneous longitudinal sound velocity 6a/4(39–41, 44, 54ff., 60)
— isentropic 6a/4(82, 85, 92)
Instantaneous specific heat 6a/4(77ff., 83)
Instantaneous stress-energy curve 6a/4(98ff.)
Instantaneous stress-energy modulus 6a/4(78, 85, 96, 99)
Instantaneous stress-entropy modulus 6a/4(81, 91)
Instantaneous stress-particle velocity curve 6a/4(54)
Instantaneous stress-strain response
— (instantaneous stress-strain law, instantaneous stress-strain curve), mechanical theory, dilatational motion
— mechanical theory, one-dimensional motion 6a/4(27–29, 39–41, 44ff., 48, 51, 53ff., 58, 61, 64–66, 75)
— thermodynamic theory isentropic 6a/4(81, 90–92)
— thermodynamic theory, isothermal 6a/4(78, 80)
Instantaneous stress-temperature modulus 6a/4(77ff., 83)
Instantaneous tangent modulus 6a/3(336, 350); 6a/4(27, 50ff.)
Insulated wall, *see* Adiabatic wall
Insulator 7/1(290, 292)
Integrability conditions 6a/3(444)
Integrable histories 6a/3(5)
Integral equation(s)
— method 9(271, 299–306)
— method for three-dimensional surface waves 9(545)
— method for two-dimensional surface waves 9(533, 535, 537, 543)
— of Abel 8/2(242)
— of Born and Green 13(20–21, 39–40)
— of Fredholm 8/2(269, 270, 281, 284)
— of Kirkwood and Monroe 13(42)
— of lifting surface 8/2(421)
— of linear acoustics 11/1(22–34)
— of Mayer 13(37, 39)
— of Nekrasov 9(729–730)
— of Ornstein and Zernike 13(72–73, 203)
— of Ornstein and Zernike, generalization for binary systems 13(191)
— of Oseen 8/2(278–279)
— of the first kind of Kirkwood 13(21–23, 41, 267)
— of the second kind of Kirkwood 13(43, 45)
— of Trefftz 9(431–434)
— of Villat 9(730, 742)
— solution 14(235)
Integral form of energy balance equation 6a/3(176)
Integral forms 6a/4(154)
Integral formulae 6a/3(81)
Integral identity for the displacement gradient 6a/2(184)
Integral laws 6a/3(34)
Integral operators fo Bergman 9(67)
Integral representation for the solution
— of displacement boundary value problem 6a/3(82)
— of mixed boundary value problem 6a/3(84)
— of traction boundary value problem 6a/3(84)
Integral representation theorem for solutions of mixed problem 6a/2(187)
Integral theorems 6a/3(53)
Integral transforms in dynamic elasticity 6(121–122)
Integral transforms in three-dimensional elasticity 6(98–102)
Integral transforms in two-dimensional elasticity 6(72–78)
Integrals across transition zone 10(285)
Integrals of the equilibrium equations for rods 6a/2(696)
Integrating factor 6(382)
Integration of two-dimensional stress equations numerical methods 6(173–176)

Integrative method of Scott Blair 6(480)
Integro-differential equation(s) 6a/2(373); 9(271)
Intense ultrasonics 11/2(259)
Intensities of strain and stress 6(240, 241, 284, 292)
Intensity, absolute, of ultrasonics 11/2(259–260)
Intensity of cavitation 11/2(271)
Intensity of sound, *see also* Power 11/1(9)
Intensity spectrum of sound generated in turbulent flow 11/1(94)
Interaction between dissolved interstitials and dislocations 7/2(35)
Interaction between foreign atoms and dislocations 7/2(168–169, 173–179)
Interaction between sound waves and dislocations 11/2(177–208)
Interaction between two helium atoms 13(2)
Interaction energies, variational method of calculation 15(362)
Interaction energy 7/1(527, 548, 552)
Interaction energy in liquids 10(3, 39)
Interaction energy pro unit of surface 7/1(576)
Interaction, nearest-neighbour 7/1(121, 277, 282, 286, 314)
Interaction, non-linear 7/1(571)
Interaction of discontinuities 6(367); 6a/3(458)
Interaction of ions in electrolytes 11/1(203)
Interaction of lattice waves 7/1(110, 120, 121, 131, 147, 154, 177, 230, 233, 236, 241, 273, 275, 280, 282, 283, 284)
Interaction of pressure wave and shock wave 9(242, 245, 247)
Interaction of waves 7/1(295, 304)
Interaction parameter 13(456)
Interaction potential in the ionic gas model of an electrolyte 13(371, 372)
Interaction potential of lattice atoms 7/2(175, 177)
Interaction types between stress waves and the lattice 11/2(163)
Inter-chain connection 11/1(401)
Interchange of sites 7/2(4)
Interchanger efficiency 14(97)
Interchanger, Hampson type 14(91, 101)
Interchanger, ideal 14(98)
Interchanger, Linde 14(90)
Interchangers 14(93, 101)
Interchangers, pressure drop in simple 14(100)
Intercoolers 14(47)
Interface 10(134, 136)
Interface, adsorption at a plane 10(142)
Interface normal, sign convention 10(283–284)
Interface of two immiscible fluids 9(451–455)
Interface, plane 10(142, 157, 165, 181, 184, 194, 210ff., 264ff.)
Interface, plane, Gibbs-Helmholtz equation 10(142)
Interface, plane, quasithermodynamic relations 10(165)
Interface, spherical 10(145, 146, 148, 149, 150, 159, 166, 213ff., 214, 215)
— fundamental thermodynamic equations 10(148)
— quasithermodynamic relations 10(166ff.)
— surface tension 10(146, 161, 187, 214)
Interface, stress continuity 9(451)
Interfacial waves 9(502–506, 634–635)
Interfacial waves, existence theorems 9(755)
Interfacial waves, viscous fluids 9(644–646)
Interference methods for sound waves 11/1(166–167)
Interference optics, experiments employing 6a/1(58, 116, 145, 264, 281, 283, 288, 289, 352, 364, 622)
Interference ring pattern, first use of photography for 6a/1(265)
Interferometer 6a/1(116, 118, 145, 282, 288, 289, 364)
Interferometer for sound velocity and absorption measurements in liquids 11/1(302, 303)
Interferometric detection of ultrasound 11/2(142–146)
Interior boundary value problem 8/2(186, 188, 194)
Interior derivatives 6(341, 345); 6a/3(420, 426)

Interior equations 6a/2(586)
Interior regularity 6a/2(355, 401)
Interior variations 9(405)
Intermediary state 11/1(216, 221)
Intermediary step of dissociation 11/1(249)
Intermediate ions 13(520, 540, 542, 543, 614)
Intermediate plastic region 6a/1(440, 445, 484, 580)
Intermediate state 7/2(252); 14(273)
Intermediate state of superconductors 15(197, 214, 222, 250, 254, 265, 321, 322, 336)
Intermediate state of superconductors, hollow specimens 15(226)
Intermediate step 7/2(217)
Intermediate stress 6a/3(514)
Intermediate temperatures 14(243)
Intermetallic compounds 7/2(24)
Intermittency factor 8/1(355, 357, 390, 393)
Intermittent flow 8/1(355)
Intermolecular forces 10(2, 3)
— (6; Exp) potential 12(341, 397)
— and sound velocity 11/1(351–352)
— Lennard-Jones potential 12(92, 151, 341, 396, 400, 428, 438–440)
— power law 12(236, 396, 427, 438)
— rigid spheres 12(340, 343, 394–395, 427, 437)
— virial 13(12)
Internal coherence of a network 13(520, 531, 569)
Internal concentrated loads 6a/2(179)
Internal conical refraction 6a/4(251)
Internal constraints 6a/2(403, 548)
Internal degrees of freedom in molecules 8/2(601)
Internal dissipation, mechanical theory 6a/4(28, 39, 54, 61)
Internal dissipation, thermodynamic theory (second law of thermodynamics) 6a/4(76, 79ff., 90)
Internal energy 6(243, 244); 6a/2(300, 481); 6a/3(174); 6a/4(76ff., 80, 82ff., 90, 92, 95, 120, 127); 8/1(172, 176); 10(20, 35–36, 68, 78, 82, 107, 119); 15(50)
— as a function of other variables 6a/4(131, 140)
— as an independent variable 6a/4(131, 140)
— equilibrium response function 6a/4(79, 81, 84ff.)
— instantaneous response function 6a/4(78, 81, 90)
— of fluid motion 11/1(3)
— superficial 10(137, 141, 166, 167, 189, 190, 216, 244ff., 257ff.)
Internal friction 6(372); 6a/3(463); 7/2(3)
Internal friction of glass 13(641, 643)
Internal friction of rubber 11/1(361)
Internal modes 10(95)
Internal parameters 13(421, 423, 431, 446)
Internal partition function 12(100ff., 121–123)
Internal state variables 6a/3(3)
Internal stress 7/2(213, 216)
Internal surface 8/2(627)
Internal temperature 12(497)
Interplanar spacing 7/1(61, 71)
Interreflection between surface elements 8/2(608)
Intersection jogs 7/1(495, 498, 612)
Intersection of dislocation lines 7/2(113, 117, 121)
Intersection process 7/2(115)
Intersection property of the plastic part 6a/3(580)
Interstices 7/1(94)
Interstitial 7/1(87)
Interstitial atoms in glasses 11/1(405)
Interstitial carbon atoms in iron 7/2(183)
Interstitial diffusion 7/2(6)
Interstitial formation 7/2(222)
Interstitial impurities 7/2(168)
Interstitial migration 7/2(222)
Interstitialcy 7/1(413, 429, 447)
Interstitials 7/1(391, 409, 412, 444, 502, 530, 557, 599, 657); 7/2(165)
Intra-chain connection 11/1(401)
Intrinsic director couple 6a/2(482, 491)
Intrinsic equations of motion 8/1(155–157, 190, 250)
Intrinsic velocity 6a/4(25)
Intrinsic viscosity, *see* Limiting viscosity number
Intrinsic viscosity 11/1(380)
Invariable line, *see* Neutral line

Invariance conditions 6a/2(484, 485, 487)
Invariance of wave equation 9(286)
Invariance principles 6a/3(129, 169, 191, 255, 288, 289)
Invariance with respect to rotation 8/1(28–31)
Invariant from of theory 6(335, 337, 338)
Invariant interpretations 6a/3(419)
Invariant plane of an allotropic transformation 7/2(243)
Invariant set 6a/3(129, 153, 169, 288)
Invariants I_1, I_2, I_3 6a/3(180)
Invariants of a tensor 6(239–243,, 258, 443ff.)
Invariants of strain tensor 6(6)
Inverse collisions 12(235)
Inverse method of boundary condition construction 9(736)
Inverse methods 9(67)
Inverse solution 9(350)
Inversion 7/1(298)
Inversion center 6a/4(175, 179)
Inversion, centre of 7/1(118)
Inversion curve 14(37, 38, 40, 55)
Inversion, method of 7/1(589)
Inversion of Boltzmann laws 6a/3(23)
Inversion operations 6a/4(176)
Inversion point 12(22)
Inversion symmetry 12(359)
Inversion temperature 12(22, 27); 14(36, 37, 48)
Inversion theorem for response functions 6a/3(27, 64)
Inversion theorem for strong response functions 6a/3(25, 30)
Inviscid equations 9(225)
Inviscid gas
— entropy 9(240, 242, 272, 283)
— steady flow in two dimensions 9(249ff.)
— unsteady flow in one dimension 9(228ff.)
— unsteady flow in three dimensions 9(283ff., 297)
— unsteady incompressible flow in two dimensions 9(294)
Ion cloud 11/1(203, 224)
— effect 11/1(250, 261, 262)
— equilibrium 11/1(213, 224)
— relaxation 11/1(263–264)
Ion complex 11/1(251)
Ion pump 12(515, 621, 640–646)
Ion reactions in electrolytes
— general survey 11/1(250–251)
— special reactions 11/1(251–270)
Ionic bond 7/1(73)
Ionic crystal(s) 7/1(105, 132, 208ff., 285, 400, 404)
— conductivity of 7/1(398)
— critical shear stress 7/2(41)
— plastic properties 7/2(18)
— work-hardening 7/2(69)
Ionic deformation 7/1(74)
Ionic gas model of an electrolyte 13(370, 371)
Ionic lattice 7/1(139, 168, 247, 287)
Ionic liquids 10(23)
Ionic radius 7/1(73, 105, 139, 144)
Ionic strength in an electrolyte 13(379)
Ionic strength of a polyelectrolyte 13(493, 499, 500, 502)
Ionization, *see* Specific ionization
Ionization constant 7/1(434)
Ionization gauge 12(517, 534–539, 601)
— calibration 12(619–620)
— conventional 12(611–612)
— linearity 12(617–618)
— of Bayard and Alpert 12(612–615)
— of Houston 12(625–626, 646)
— pumping action 12(620–623, 638, 640–646)
— pumping speed measurement 12(644–645)
— sensitivity 12(616–617, 623)
Ionized gas, shock propagation 9(220–221)
Iron 6a/1(22, 25–29, 32, 34, 37, 70, 72, 82, 83, 89, 93, 104, 106, 115, 118, 127, 176, 183, 191, 196, 202, 210–215, 272, 363–375, 493, 540, 545, 576, 580, 603, 612)
Iron, critical scattering of neutrons 13(208–209)
Iron, crticial shear stress 7/2(186)
Iron crystals, work-hardening curve 7/2(188)
Iron, effect of dissolved carbon 7/2(183)
Iron, isotropy of 6a/1(254, 291, 410)

Iron, magnetostrictive properties 11/2(87)
Iron polycrystals, yield stress 7/2(36)
Iron single crystal
— work-hardening 7/2(67)
— yield stress 7/2(36)
Iron, yield phenomena 7/2(183–184, 187)
Irradiation of solids 11/2(191)
Irradiation with gamma rays 15(57)
Irreducible and reducible cluster-integrals 13(302–303)
Irreducible cluster sums 13(146, 160)
Irreducible integrals 12(135, 148)
Irregularities in steady flow, effect of contraction 8/2(552–554)
Irreversibility 12(360)
Irreversible change of flow stress 7/2(161, 162)
Irreversible processes in liquids 10(24, 25, 38, 81–89, 92)
Irreversible processes, velocity of 14(189)
Irreversible temperature dependence 7/2(55)
Irreversible thermodynamics 7/2(3)
Irrotational fields 6a/2(17)
Irrotational flow 9(4, 449, 462)
— Bernoulli's equation 8/1(154, 184)
— impossibility in viscous fluid 8/1(248)
— of a perfect gas 8/1(199–211, 215, 218)
— of incompressible fluid 8/1(157–161)
— relation to isentropic and isoenergetic flow 8/1(186)
Irrotational vector field 6a/3(73)
Irrotational velocity 6a/2(231)
Irrotational waves, existence theorems 9(749–755)
Irrotationality 8/1(52, 66, 152, 188)
Isentropic 6a/2(304)
Isentropic compressibility 11/1(9)
Isentropic elasticity tensor 6a/3(324)
Isentropic flow 8/1(12, 94, 150, 184, 186); 9(3, 164, 174, 197, 686)
Isentropic irrotational flow 8/1(193)
Isentropic second-order elasticity tensor 6a/3(324)
Isentropic states 6a/3(187–189)
Isentropic tangent modulus 6a/3(362, 369)
Ising model
— application to polyelectrolytes 13(497)
— application to superlattice structure 13(83–84)
— matrix theory 13(108ff., 122ff.)
— of solid solutions 13(153–159)
— one-dimensional 13(108–113)
— principles 13(79)
— rotation transformation 13(211, 335)
— three-dimensional 13(139)
— two-dimensional 13(113–139)
Isobaric expansion coefficient 11/1(143)
Isobars 6(384, 387, 390); 6a/3(476, 477, 484)
Isochores of the inert gases 12(17)
Isochoric 6a/3(548)
Isochoric dynamic deformation 6a/1(634, 636)
Isochoric motion 6a/2(210)
Isochoric stress coefficient 11/1(143)
Isochoric velocity 6a/2(213)
Isochromatic fringes 6(149)
Isoclines 6(384, 387, 390); 6a/3(477, 484)
Isoclinic fringes 6(148, 166)
Isoelectric temperature 7/2(125)
Isoenergetic flow 8/1(186–187)
Isolated, general 6a/3(478, 482)
Isolated point 6a/3(480)
Isometric coordinates 8/2(30)
Isometric curves 8/2(30, 40)
Iso-mixture, definition 12(462)
Isoperimetric condition 6a/3(564)
Isothermal 15(453)
Isothermal changes 7/1(234)
Isothermal coefficient of bulk compressibility 12(6)
Isothermal conditions of inelastic continuum 6(233, 236)
Isothermal deformation 6(14, 16–17)
Isothermal elastic potential 6(247, 250)
Isothermal gas 9(62)
Isothermal Joule-Thomson coefficient 12(21, 37, 38, 50)
Isothermal problems 6a/3(178)
Isothermals of the osmotic pressure 13(308)
Isothermic compressibility 11/1(143); 13(29, 262, 359)

Isothermic propagation of sound in solutions 11/1(203)
Isothermics calculated by Monte Carlo procedures 13(282, 283)
Isothermics of melting theory 13(288)
Isotherms in liquids, superposition approximation 10(53)
Isotherms of adsorption in porous media 8/2(630)
Isotope effect 15(237, 275)
Isotope effects on the partition function 12(118–121)
Isotopic orientation of angular momenta of molecules 12(491)
Isotopic thermodiffusion factor 12(436, 447)
Isotropic body, elastic constants 6(23–26)
Isotropic deformation 11/1(134)
Isotropic elastic solid 11/2(155)
Isotropic function 6a/3(545)
Isotropic homogeneous body 6(17–19, 21, 23–26)
Isotropic materials 6a/3(2, 18, 29, 180, 185, 244–249, 544, 545, 550)
Isotropic materials and bodies
— elastic 6a/2(71, 74, 85, 86)
— elastic plates 6a/2(573)
— thermoelastic 6a/2(311, 317)
Isotropic orientation of angular momenta of molecules 12(489)
Isotropic point 6(166, 168)
Isotropic tensor function(s) 6a/3(546); 6(444)
Isotropic turbulence 8/1(384); 8/2(441, 524, 532); 11/1(91–92)
— dynamical equations 8/2(491–492)
— dynamics 8/2(448–450)
— kinematics 8/2(446–448)
Isotropy 7/1(180, 182, 184, 192, 265, 267, 293, 320)
Isotropy criteria 8/2(532)
Isotropy, elastic 7/1(179, 185, 266, 292)
Isotropy groups 6a/4(174, 177)
Isotropy, measurement of 6a/1(254, 270, 291, 410)
Isotropy of glass 6a/1(254, 269)
Isotropy of stress tensor 10(283)
Isotropy, transverse 6a/2(72)
Isovel 9(245)
Iterated conditions of compatibility 6a/3(307)
Iterated Laplace operator 6a/2(377)
Iteration method(s)
— of determining relaxation spectra 11/1(370)
— of Garabedian 9(427–431)
— of the discrete curvature equations 9(416–420)
— to compute the pressure distribution 8/2(398–399)

J_2 lattice 7/1(83)
Jacobian 6a/3(445, 477); 6a/4(113)
Jacobian elliptic functions 9(706ff.)
Jacobian of transformation 6a/2(619)
Janovsky-Pohlmann whistle 11/2(98, 260, 278)
Janzen-Rayleigh iteration 9(115–117)
Jeffery flow, generalisation 8/2(102, 104, 174)
Jeffery flow, steady 8/2(38, 40–41, 55, 57, 93, 182)
Jeffery flow, unsteady 8/2(140, 142–143)
Jeffreys body (=elastic sol), *see also* Lethersich body 6(455, 468–471)
Jelly, *see also* Gel 6(435, 522)
Jenne's equations 6a/3(459)
Jet 8/2(549); 9(64)
Jet boundary 9(256, 268)
Jet, emerging from a tube 8/2(60)
Jet entering fluid at rest 8/2(548, 549)
Jet flow 9(317, 325, 339–342)
Jet formation in valves, thermodynamics 12(566)
Jet into a fluid at rest, boundary layer 8/1(304)
Jet of compressible fluid 9(311)
Jet thrown out by a glider 9(588)
Jet, turbulent, generating sound 11/1(97–99)
Jet width 9(334)
Jet-edge system 11/2(27)
Jog formation in dislocations 7/2(112, 113, 206)
Jogs 7/1(610)
Jogs, generation of 7/1(611, 612)
Jogs in special lattices, production of 7/1(494)
Jogs, non conservative motion of 7/1(499, 510)
John's inverse method 9(740–742)
Joints for ultrahigh vacuum 12(654)

Jones' theory for low-aspect-ratio wings 8/2(415–417)
Jongenburger-Abéles potential 7/1(403, 428, 475)
Jordan domain 6a/3(561)
Joukowski, *see* Youkowski
Joule coefficient 12(22, 23, 36)
Joule effect of magnetostriction 11/2(86)
Joule expansion 12(8)
Joule's experiments 6a/1(308, 411, 728–731)
Joule's law 12(7); 14(40)
Joule-Thomson coefficient(s) 12(21–23, 36, 37, 38, 40, 50); 14(34, 36, 38)
Joule-Thomson effect 14(33)
Joule-Thomson expansion 12(9); 14(48, 57, 59)
Jump 6a/4(287)
Jump across a singular surface 6a/2(249, 421)
Jump discontinuity 6a/3(14, 96, 97)
Jump discontinuity (discontinuity), *see* Wave front
Jump discontinuity for plane waves 6a/3(97)
Jump history (instantaneous jump history)
— hydrostatic compression 6a/4(72)
— of the strain and temperature, one-dimension 6a/4(78, 83, 95)
Jump history of the strain
— and entropy, one-dimension 6a/4(81, 90)
— one-dimension 6a/4(27, 49, 52, 65)
Jump of a function 6a/3(97)
Jump of the stress 6a/3(435)
Jump of the velocity 6a/3(494)

Kapitza H_2 14(63)
Kapitza thermal resistance 15(178, 181)
Kármán flow 8/2(99, 104)
Kármán-Howarth equation 8/2(449–450, 452, 489)
Kármán-Pohlhausen method of boundary, layer theory 8/1(280–282, 297, 318)
Kármán's method of singularity superposition 9(55)
Kármán's notation of stress components 6(26)
Kármán's vortex path 8/1(122–123)
Kármán-Trefftz profiles 8/2(390)
Kármán-Tsien fluid 9(77, 80, 81–84, 93)
Kármán-Tsien method 8/1(196)
Katayama-Guggenheim formula 10(174, 175, 176, 297)
— modified 10(177)
K-body, *see* Kelvin body
Keesom forces 11/1(351)
Kelvin, *see* Thomson
Kelvin body (= delayed elasticity) 6(312–313,, 454ff., 463–464)
Kelvin doublet process
— normalized 6a/3(79)
— properties of 6a/3(79, 80)
— with moment or without moment 6a/3(80)
Kelvin effect (dilatancy) 6(261)
Kelvin elements 6a/3(40)
Kelvin equation for the capillary pressure 8/2(631)
Kelvin operators 6a/3(39)
Kelvin problem 6a/2(173, 176)
Kelvin problem in elastostatics, viscoelastic counterpart 6a/3(77)
Kelvin process 6a/3(78, 79)
Kelvin relation 10(134, 147, 154, 160, 161, 162)
Kelvin state 6a/2(174, 176, 178–180)
Kelvin theorem 6a/2(208)
Kelvin-Helmholtz circulation theorem 9(4)
Kelvin's elastic sphere 6(102–104)
Kelvin's minimum energy theorem 8/1(161)
Kelvin's solution of equilibrium equations 6(87–88)
Kelvin's two wire experiment 6a/1(92, 102, 123, 127–130, 167)
Kelvin-Voigt
— elements 6a/3(40)
— elements in series 6a/3(41)
— model 6a/4(199, 200)
— operators 6a/3(39)
Kernel of the function *F* 6a/2(392)
Kernel of the nonnegative quadratic form 6a/2(392)
Kernel of the operator *T* 6a/2(391)
Kerr cell 11/2(137)
Keyes and Beattle apparatus for compressibility measurements 12(31)

Kihara approximation 12(392, 401, 436)
Kikuchi bands and lines 7/2(107–108)
Kinematic approximations in plates and shells 6a/2(478)
Kinematic boundary condition 9(451, 452, 455)
Kinematic measures for Cosserat surface 6a/2(452)
Kinematic measures for shells
— catalogue 6a/2(458)
— derived from 3-dimensional theory 6a/2(468)
— linearized, derived from 3-dimensional theory 6a/2(474)
— restricted theory 6a/2(460)
Kinematic results for shells 6a/2(471)
Kinematic viscosity 9(447)
Kinematical condition of compatibility 6a/3(308, 309, 429, 430)
Kinematical discontinuity conditions 6(344)
Kinematical theorems for free surface motion 9(715–716)
Kinematical viscosity 8/1(49); 12(300)
Kinematically admissible
— displacement field 6a/2(112)
— process 6a/2(225)
— state 6a/2(111, 212, 324)
— strain field 6a/2(112)
Kinematically and thermally admissible process 6a/2(341)
Kinematically determined 6a/3(526)
Kinematics of Cosserat surface 6a/2(449)
Kinematics of homogeneous turbulence 8/2(444–445)
Kinematics of isotropic turbulence 8/2(446–448)
Kinematics of oriented media 6a/2(445)
Kinematics of shells, derived from 3-dimensional theory 6a/2(466)
Kinematics of shells, restricted theory 6a/2(455)
Kinemtical condition of compatibility 6a/4(25)
Kinetic curves 7/2(228, 238)
Kinetic energy 6a/2(65, 216, 481)
— decay 8/1(255, 259)
— definition 6a/3(95)
— Lamb's formula 8/1(250)
— *n*th order 6a/3(236)
— of a vortex-system 8/1(165)
— of an irrotational motion 8/1(159)
— of elastic deformation 11/2(159)
— of fluid motion 11/1(3)
— of molecules 10(27)
— of turbulence, balance 8/2(546)
— rate of change 8/1(138, 247)
— spectrum of turbulence 11/1(92)
Kinetic model of liquids 10(13–15, 52–61, 65–67)
Kinetic temperature 12(297)
Kinetic theory of liquids, superposition approximation 10(52, 53, 59, 66, 74ff., 111)
Kinetics of dislocation break-away 11/2(189–190)
Kinetics of phase transitions 15(258, 340)
Kink band density 7/2(154)
Kink band formation, orientation dependence 7/2(76)
Kink bands 7/2(19, 74–79, 108, 150, 195)
Kink bands, asymmetric 7/2(77)
Kink bands, half 7/2(77)
Kink diffusion 11/2(208)
Kink pair formation in dislocations 11/2(204–205)
Kinking 7/2(5, 6, 19, 47)
Kinking of hexagonal metals 7/2(74)
Kirchhoff boundary conditions 6a/2(577, 586)
Kirchhoff's approach to beach waves 9(537–542, 624–627)
Kirchhoff's experiment for determining Poisson's ratio 6a/1(259–263, 278, 292–296)
Kirchhoff's theorem 6a/2(32); 9(300, 305)
Kirchhoff's theories for stretching and bending of plates 6a/2(586)
Kirkendall-effect 7/1(419, 423)
Kirkwood and Monroe's theory of melting 13(290–300)
Kirkwood and Yvon's theory of dielectrics 13(252–255)
Kirkwood closure 12(157–159)
Kirkwood-Müller equation 10(234)
Kirkwood's computation method of thermodynamic functions 13(14)

Kirkwood's derivation of the Boltzmann equation 12(230)
Kirkwood's equations
— of the first kind 13(21–23, 41, 267)
— of the first kind derived from Mayer's equations 13(41)
— of the second kind 13(43, 45)
— of the second kind, application to an one-dimensional system of rigid spheres 13(56)
Kirkwood's method for electrolytes 13(373)
Kirkwood's method of semi-invariants 13(98–103, 354, 366)
Kirkwood's theory of polyampholyte interaction 13(506)
Klein and Tisza's phenomenological theory 13(201)
Knife-edge vacuum seal 12(654–655)
Knowles' theorem on Saint-Venant's principle 6a/2(200–206)
Knudsen formula for pressure force 12(531)
Knudsen gas 12(212, 214–216, 245, 288)
Knudsen gas equation 12(241)
Knudsen layer 8/2(609)
Knudsen number 8/2(592)
Knudsen radiometer-vacuummeter 12(533)
Knudsen's laws of gas flow 12(617)
Kochin's formulas for derivatives of Bessel functions 9(609)
Kochin's *H*-function 9(556–558, 573–574)
Kochin's *H*-function, time dependent motion 9(617)
Koiter's generalization 6a/3(439)
Kolmogoroff microscale 8/2(527, 528, 568, 572–573)
Kolmogoroff's theory of turbulence 8/2(461, 462–464)
Kolmogorov's differential equation 7/2(222)
Kolosov's theorem 6a/2(164)
Kolsky's experiment 6a/1(591–595)
Koppe's theory 15(280)
Korn's constant 6a/3(218, 221, 240, 241)
Korn's inequality 6a/3(218, 221, 241)
— alternative 6a/2(96)
— first 6a/2(38, 381, 384)
— second 6a/2(381, 384)
Kozeny constant 8/2(642)
Kozeny equation 8/2(642, 643)
Kozeny-Carman theory 8/2(643–644, 648)
Kramers degeneracy 15(2)
Kramers doublet 15(108, 117, 119)
Kramers theorem 15(44, 79)
Kronecker delta 6a/2(618, 623)
Kuhn-Treloar theory of rubber-like substances 6(18)
Kupffer's experiments 6a/1(296–303, 363)
Kutta-Joukowski condition 8/2(391, 420); 9(34, 106–107, 287, 560, 584, 588)
Kutta-Joukowski formula 9(34, 130)
Kutta-Joukowski lift formula 8/1(84, 161, 201)
Kutta-Joukowski theorem 8/2(389)

*

$\mathscr{L}$ is a creep law for $\mathscr{L}$ 6a/3(23)
L wave, *see* Longitudinal wave
L_2 inner product 6a/3(552)
Labil 6a/3(147)
Lagerström und Cole's problem 8/1(326–328)
Lagrange instability
— asymptotic 6a/3(150)
— choice of definition in elasticity 6a/3(171)
— definition 6a/3(150)
— discussion 6a/3(150, 151)
— elastic bodies 6a/3(171)
— example in elasticity 6a/3(202, 223)
— logarithmic convexity 6a/3(239)
— relation to boundedness 6a/3(150)
— relation to Liapounov 6a/3(150)
— sufficient conditions 6a/3(170)
Lagrange multiplier 6a/3(564)
Lagrange stability, definition 6a/3(149)
Lagrange stability, Liapounov stability 6a/3(156)
Lagrange stability, relation to boundedness 6a/3(149)
Lagrange-Cauchy theorem on irrotationality 8/1(152)
Lagrange-Dirichlet stability concept, *see also* Discrete conservative systems
— definition 6a/3(131, 134)
— definition, ingredients 6a/3(132)
— definition, observations 6a/3(134)

— discussion 6a/3(153, 154, 163)
Lagrange-Dirichlet stability concept (cont.)
— energy criterion (footnote) 6a/3(203, 204)
— criterion or energy test 6a/3(154)
— theorem 6a/3(155)
— theorem, proof 6a/3(159)
Lagrangian coordinates 6a/3(428, 524, 536)
Lagrangian density 7/1(179, 186, 187, 189, 192, 198, 304)
Lagrangian description 9(447)
Lagrangian (function) 7/1(156)
Lagrangian integral scale 8/2(543)
Lagrangian of elastic stress 6(15)
Lagrangian picture of fluid motion 11/1(2)
Lagrangian representation 8/1(3, 19–20, 31–32)
Lagrangian statistical functions in turbulent flow 8/2(586–587)
λ point 13(107)
λ transitions 14(284)
Lambda-line 15(374, 375)
Lambda-phenomenon 15(373)
λ-phenomenon 10(94, 119, 122)
Lambda-point 15(374)
λ-point of helium 10(119, 122, 133)
Lamé constants 6a/3(45, 181)
Lamé constants, ratio of 6a/1(272, 275)
Lamé functions 9(297, 299)
Lamé moduli 6a/2(76, 311); 6a/3(29)
Lamé stress functions 6a/3(93)
Lamé-Maxwell equations 6(140, 174, 182)
Lame's elastic coefficients 11/2(155)
Lamé's elastic constants 6(24, 449)
Laminar airfoil 8/1(348, 365, 407–409)
Laminar boundary layer, *see* Boundary layer
Laminar flow 8/1(64)
Laminar flow, transition to turbulence 8/1(351, 357, 358ff., 383)
Laminar instability 8/2(541)
Laminar jet, stability 8/1(388)
Laminar structure of intermediate state 15(251, 337)
Laminarization by suction 8/1(348, 409–414)
Landau symbol 6a/2(356)
Landau theory 15(321, 336, 391, 392, 457)
Langevin approximation for higly elongated chains 13(405)
Langevin function 13(405)
Langmuir-Dushman molecular gauge 12(529–530)
Laplace equation 6a/3(136); 8/1(52, 66, 77, 97); 9(294, 319, 462, 667)
— determining shape of fluid lens 10(298)
— for plane stress 6(144, 173)
— of capillarity 10(286, 288, 294)
— of capillarity, dimensionless form 10(300)
— of capillarity, planar case 10(295)
— of capillarity, simple solutions 10(299–302)
— two-dimensional, in the theory of torsion 6(29)
Laplace formula 11/1(130)
Laplace operator Δ_2 6a/2(375)
Laplace transform 6a/2(25, 231); 6a/3(28, 68); 9(296, 483, 491)
Laplacian 6a/2(11)
Large deformation solutions 6a/2(546)
Large-angle-grainboundary 7/1(633, 646, 657)
Large-scale structure of turbulence 8/2(456–460)
Largest invariant set 6a/3(169)
Laser beam 6a/1(352, 358)
Laser interferometry 6a/4(9–24)
— diffuse 6a/4(14)
— displacement (Michelson displacement) 6a/4(10–20)
— in-material gauge 6a/4(9, 14ff., 41)
— velocity 6a/4(12–14, 41)
Latent glide system 7/2(60)
Latent heat 15(231, 234, 373, 374)
Latent heats of fusion 6a/1(412)
Latent work-hardening 7/2(60, 130, 134–137, 151, 182)
Lateral extensometer of Coker 6(173)
Lateral quadrupole source 11/2(3)
Lattice 7/1(3)
Lattice, bravais, simple lattice 7/1(110, 119, 280)
Lattice complex 7/1(39)
Lattice component of alloys 14(264)

Lattice component of pure metals 14(262)
Lattice conductivity enhanced in the superconducting state 14(270)
Lattice constants 7/1(61)
Lattice constants, cubic 7/1(112, 117)
Lattice, cubic 7/1(180, 183, 231, 238, 254, 276, 294, 314)
Lattice, cubic, body-centered 7/1(78, 111, 114, 116, 117, 152, 154, 193, 201, 232, 485)
Lattice, cubic, face-centered 7/1(77, 112, 114, 115, 117, 152, 171, 193, 200, 232, 246, 265, 282, 299, 312, 485, 576)
Lattice defect (cf. imperfection)
Lattice defects 7/2(13); 14(148)
Lattice defects in cubic metals 7/1(557)
Lattice defects, thermally excited 14(173)
Lattice distance 7/1(132)
Lattice energy 7/1(129, 132, 141)
Lattice energy, calculation of 7/1(74)
Lattice energy, electrostatic 7/1(135)
Lattice frequency 13(410)
Lattice gas 13(218)
Lattice, hexagonal 7/1(180, 183)
Lattice, homogeneously strained 7/1(228)
Lattice imperfection(s) 7/1(108, 292, 383, 387, 389)
Lattice influence 15(4)
Lattice, isotropic face centered 7/1(263)
Lattice, linear 7/1(133)
Lattice model
— of binary liquids 13(316–328)
— of liquids 10(7–10, 24, 240, 242)
— of macromolecular materials 13(413, 419)
— of polymer solutions 13(451, 456–462, 488)
Lattice motion, impedance 11/2(211)
Lattice, normal modes 11/2(215)
Lattice parameter 13(411)
Lattice, plane 7/1(133)
Lattice planes, equipotential 10(252)
Lattice potential, electrostatic 7/1(132, 209)
Lattice, reciprocal 7/1(59, 112, 171, 260, 305, 310, 628)
Lattice relaxation 15(19)
Lattice, simple cubic 7/1(111, 116, 118, 149, 152, 154, 266, 267, 570)
Lattice, single primitive 7/1(11)
Lattice specific heats 15(87, 93, 98, 114)
Lattice spectrum 7/1(248, 250)
Lattice structure 10(123)
Lattice structure of a glass 13(517)
Lattice structures, ordered 14(147)
Lattice, symmetry of 7/1(385)
Lattice theory 7/1(257, 291, 320, 339); 14(287)
Lattice thermal conductivity 14(253)
— variation with electron concentration 14(254)
Lattice, twofold primitive 7/1(11)
Lattice vacancies 14(150)
Lattice vectors, inverse 14(202)
Lattice vibration(s) 7/1(159); 15(22, 42, 45)
Lattice vibrations, anharmonicity 14(173)
Lattice waves 7/1(290); 14(201)
Lattice waves, scattered 14(265)
Lattice-point method 9(691)
Lattices for which the Fuchs theory has been evaluated 13(161–163)
Laue groups 6a/4(174, 177)
Laue symmetry 7/1(49)
Laval nozzle 8/1(95); 9(8, 695)
Laval valve 12(567, 568, 573, 575–576, 586)
Laves phases 7/1(92)
Law of corresponding states 12(48–51, 62, 98)
Law of rational indices 7/1(61)
Laws of balance, energy 6a/3(174, 175)
Laws of balance, linear and angular momentum 6a/3(42)
L-beam 6a/3(495)
Lead 6a/1(32, 145, 191, 224, 231, 388, 399, 405, 429–445, 632); 14(154, 269, 275, 296)
Lead crystals, work hardening 7/2(54, 57)
Lead silicate glasses, structure 13(524)
Lead-bismuth alloys 14(273)
Leaded-copper 14(296)
Leak detection 12(541–542, 661)
Leak measurement 12(542–544)
Lebesgue decomposition 6a/2(415)
Lebesgue derivative 6a/2(415)

Lee's theory of unloading waves 6a/1(642, 645)
Left Cauchy-Green tensor 6a/3(326); 6a/4(70)
Left inverse 6a/3(23)
Legendre polynomials 9(479, 671)
Legendre transformation 6(370); 6a/3(461)
Legendre transformation for the velocity potential 9(56)
Legendre-Hadamard condition 6a/3(166)
Length correction of an anemometer 8/2(556, 559, 580)
Length of anemometer wire 8/2(556, 559)
Length parameter in capillarity theory 10(285, 286, 291–293)
Length parameters 9(259)
Length-preserving transformations 6a/2(538)
Length-to-diameter ratio of anemometer wire 8/2(557, 559, 560)
Lennard-Jones' and Devonshire's model 10(242)
— of liquids 13(222–232)
— of liquids, basic hypotheses 13(223–226)
— of liquids, modifications 13(233–241)
Lennard-Jones' and Devonshire's theory of melting 13(284–290)
Lennard-Jones model 7/2(13)
Lennard-Jones potential 10(173, 217, 244, 246); 12(92, 151, 341, 396, 400, 428, 438–440); 13(227, 243, 361, 367)
— in polymer solutions 13(469)
— law 11/1(197, 200)
— modified 10(181)
— Rushbrooke's modification 10(217)
Lens of liquid embedded in two fluid phases 10(282–283, 285, 298–299)
Lens system in photoelastic equipment 6(155–157)
Lenses for ultrasound waves 11/2(112, 113–114)
Leray's method 8/2(189–191, 196–198, 285)
Leray-Schauder fixed-point method 9(392, 400–402)
Lethersich body (=elastic sol) 6(456, 467–468)
Level broadening in paramagnetic salts 15(42, 44)
Level oscillations 15(439)
Leverett function 8/2(633)
Levi-Civita's differential-difference equation 9(726–727)
Levi-Civita's representation 9(369)
Levi-Civita's tensor 7/1(518)
Lévy's theorem 6a/2(156)
Liapounov boundary 6a/2(419)
Liapounov functions (auxiliary functions)
— Coleman's canonical free energy V_3 6a/3(189, 190)
— elasticity, generation or choice (footnote) 6a/3(171, 172, 177, 193, 207, 212, 223, 250, 255, 259)
— entropy production inequality 6a/3(190, 222)
— entropy production inequality V_3 6a/3(189)
— entropy production inequality V_4 6a/3(190)
— finite thermoelasticity, from energy balance V_1 6a/3(188)
— finite thermoelasticity, from energy balance V_2 6a/3(189)
— follower forces 6a/3(280–282)
— $F_{\tau,t}$ 6a/3(154, 157, 163, 164, 168, 169)
— in sense of Lasalle 6a/3(168, 169)
— isothermal primary state, thermal perturbations 6a/3(188)
— Movchan's function 6a/3(195)
— Movchan's theorem 6a/3(156)
— of Hsu and Lee for asymptotic stability 6a/3(275, 276)
— potential energy V 6a/3(205)
— total energy 6a/3(172, 272)
Liapounov instability
— and non-uniqueness 6a/3(151, 152)
— by divergence 6a/3(282)
— choice of definition in elasticity 6a/3(171)
— definition 6a/3(147)
— discussion of example 6a/3(139)
— relation to Lagrange instability 6a/3(150)
— sufficient conditions 6a/3(199, 200)
— theorems 6a/3(168, 169)

Liapounov instability in elasticity (6a/3)
and nonuniqueness 224
dead loads for linear thermo-elastic displacements, null solution 251, 252
dissipative forces 286
equilibrium solution under dead loads for linear incremental displacements 222
follower forces 282–285
incompressible elastic media 248, 249
negative-definite total energy V_4 222
special traction boundary conditions, null solution 246–248
time- and position-dependent data 266
traction boundary conditions, sufficient conditions on mean stresses 242
weakly conservative forces 259
Liapounov stability (6a/3)
and boundedness 148–150
and calculus of variations 158, 163–167
and Lagrange stability 148, 149, 156
and maximum principles 152
definition 133
dynamical system 134
equivalence with uniqueness 151
equivalent defintions 140
examples 160–163
geometrical, representations 140
measures 135
non-uniqueness of equilibrium solution 144
relation to continuous dependence 146
relation to Lagrange-Dirichlet 153, 154
theorems 149, 150, 152, 155, 157
theorems, discrete systems 155, 159
theorems, discussion 156–160
theorems, uniform 157
theorems, variations 155
theorems, version of Zubov and Movchan 156
weakly asymptotic 158
Liapounov stability in elasticity (6a/3)
analysis for traction boundary problem 239–242
analysis for traction boundary problem for incompressible media 244
analysis for traction boundary problem, sufficient conditions 239–241
analysis for traction boundary problem, sufficient conditions with axis 242
and calculus of variations 195, 205
and energy criterion (criticism) 203–206
discussion and history 195, 196, 199
dynamical system 171
in class of non-linear perturbations 191
Movchan's conditions 195
necessity theorems
necessity theorems for equilibrium solution in non-linear thermoelasticity 200–202
necessity theorems, relation to sufficiency conditions 203
of natural state 195
of null solution
of null solution, choice of measures 210–213
of null solution, in incompressible media 213, 214
of null solution, theorem 219
of null solution under dead loads in class of small incremental displacements (fixed surface) 206
of null solution, use of Liapounov theorem 212
of null solution with movable surface 215
of null solution with time-dependent elasticities d_{ijkl} 209
of primary state 177
related to linear problems 194
special traction boundary value problems 244
special traction boundary value problems for incompressible elastic materials 248

Liapounov stability in elasticity (6a/3); (cont.)
special traction boundary value problems, isotropic compressible material under hydrostatic stress 244–248
special traction boundary value problems, necessary conditions, null solution 246
sufficiency theorems in non-linear thermoelasticity 192
theorem on uniqueness of equilibrium solution 224
under dead loads in class of linear thermoelastic displacements 249–255
under dead loads in class of linear thermoelastic displacements, Liapounov function 250
under dead loads in class of linear thermoelastic displacements, null solution 251
under follower forces 278–285
under follower forces, equilibrium solutions 281, 282
under follower forces, examples, null solution 280, 281
under follower forces with dissipative forces 285–289
uniform stability 207
with non-dead loads 256–257
with non-dead loads, persistent stability 256
with non-dead loads with motion-dependent data 257
with time- and position-dependent data 264–278
with time- and position-dependent data, asymptotic stability of null solution 275
with time- and position-dependent data, asymptotic stability of null solution, Liapounov function of Hsu and Lee 276
with time- and position-dependent data, asymptotic stability of null solution, theorem 257
with time- and position-dependent data, non-zero initial data 266
with time- and position-dependent data, null solution 266
with time- and position-dependent data under prescribed body force and dead loads
with time- and position-dependent data under prescribed body force and dead loads, changes in initial data 272, 273
with time- and position-dependent data under prescribed body force and dead loads, convexity arguments 273
with time- and position-dependent data under prescribed body force and dead loads, equilibrium solution 269–271
with time- and position-dependent data, under prescribed body force and dead loads, further arguments 275
with time- and position-dependent data, under prescribed body force and dead loads, null solution of mixed problem with linear incremental diplacements 273
with time- and position-dependent data, under prescribed body force and dead loads, variation in elasticities, with linear incremental displacements 271, 272
with time- and position-dependent data, under prescribed surface displacements and linear incremental displacements, equilibrium solution 268, 269
with time- and position-dependent data, under prescribed surface displacements and linear incremental displacements, non-zero initial data 266
with time- and position-dependent data, under prescribed surface displacements and linear incremental displacements, null solution 266
with time- and position-dependent data, under prescribed surface tractions and linear incremental displacements and zero initial data 266
with weakly conservative loads 258–264
with weakly conservative loads, linear 263

with weakly conservative loads, stability analyses 263, 264
Lichtenstein's method 8/2(187)
Lifetime of a bubble 11/2(273)
Lift 8/1(84, 106, 161, 201); 9(34, 36, 47, 130)
Lift and circulation 8/1(85–87, 161)
Lift coefficient of an airfoil 8/2(387, 395, 426, 428)
Lift distribution 8/2(430, 431)
Lift of a glider 9(592)
Lift of a hydrofoil 9(586)
Lift of an airfoil 8/1(106); 8/2(385, 406, 408)
Lifting line 8/2(407)
Lifting line, generalized theory 8/2(413)
Light diffraction method for sound absorption 11/1(169)
Light microscopy of surface effects 7/2(72–80, 180)
Light scattering
— as a means to determine molecular weights 13(401, 475–476, 497)
— by liquids 13(262)
— by multi-component systems 13(310–316)
— in polyelectrolytes 13(500–501)
Lighthill-Eppler method 8/2(400–402)
Lighthill's approximation in boundary layer theory 8/1(287–288)
Lighthill's series development solution of hodograph equation 9(66, 68–71)
Lighthill's solution for the flow up a step 8/1(305–307)
Limit analysis 6a/3(408, 431)
Limit definition of solution to Kelvin's problem 6a/2(174)
Limit design 6(315)
Limit line 6a/3(477, 478, 480); 9(85, 92, 147, 235, 261, 262, 267, 279, 282)
— between plastic and rigid material 6(383)
— envelope of characteristics 6(384–385, 408)
— singularities 6(384–390)
Limit of stability 6a/3(202, 225, 244)
Limit of stability of laminar flow 8/1(371)
Limit point 6(385); 6a/3(478); 9(243)
Limit sets 6a/3(126)
Limit surface 9(9)
Limit type singularities 6a/3(477)
Limiting height of finite-amplitude waves 9(703, 712ff.)
Limiting law of Debye and Hückel 13(378, 380, 381)
Limiting laws for infinite dilution 13(304, 305)
Limiting line 8/1(96)
Limiting streamline in boundary layer separation 8/1(278, 319)
Limiting, velocity 7/1(567, 586)
Limiting viscosity number 13(473, 474)
Limits of crystallization in glasses 13(612–616)
Limit-type singularity 9(236)
Linde air liquefier 14(47)
Linde cycle 12(26, 27); 14(47)
Linde system, two-stage 14(58)
Lindeck potentiometer 14(158)
Lindemann electron-lattice 14(140)
Lindemann relation 7/1(376/77)
Lindemann's formula for the melting point 7/1(320)
Line energy 7/1(547f., 612)
Line force driving a membrane 11/1(108–112)
Line integral for the displacement process 6a/3(46)
Line of passage from tension to compression, *see* Neutral line
Line of rupture, *see* Stress discontinuity line
Line source 8/2(59, 63)
Line stress 7/1(546f., 550)
Line width 15(25, 28)
Lineage structure 7/1(96, 641)
Linear acoustics 8/1(109)
Linear approach to the Navier-Stokes equation 8/2(200)
Linear approximation 6a/1(156)
Linear approximation to acoustic waves 11/1(5)
Linear chain, *see also* One-dimensional fluid or crystal 13(245)
Linear chain 7/1(121, 159, 186, 264, 330)
Linear chain, monatomic 7/1(269ff.)
Linear constitutive equations for shells 6a/2(553, 555, 556, 571)
Linear elasticity, rod problems 6a/2(661)
Linear hereditary laws 6a/3(11)
Linear integral equations 6a/3(77)

Linear ionic chain 7/1(356)
Linear, linearized approximation 6a/3(2)
Linear, linearized, theory of elasticity 6a/3(1, 176–183)
Linear linearized theory viscoelasticity 6a/3(1)
Linear macromolecule, *see also* Chain molecule 13(399)
Linear medium 6a/4(286)
Linear medium, piezoelectric 6a/4(301)
Linear momentum 6a/2(43, 51, 482, 483)
Linear operation limit
— of condenser microphone 11/2(39)
— of moving coil microphones 11/2(38)
Linear relation between elasticity and composition of alloys 6a/1(233)
Linear superposition 6a/3(2, 12)
Linear systems 6a/3(178)
Linear theory
— based upon finite deformations 6a/3(3)
— elastic plates 6a/2(585, 596)
— elastic shells 6a/2(595, 597)
— elastic shells, derived from 3-dimensional theory 6a/2(578)
— elastic shells, remarks on 6a/2(613)
Linear thermoelasticity
— asymptotic stability 6a/3(254, 255)
— basic equation 6a/3(250)
— basic equations 6a/3(183–185, 249)
— Hale's invariance principle 6a/3(255)
— Hölder stability 6a/3(252–254)
— instability 6a/3(251, 252)
— Liapounov stability 6a/3(250, 251)
— weak (generalized) solutions 6a/3(254, 255)
Linear transformations 6a/3(3, 4)
Linear viscoelasticity (linear viscoelastic materials), mechanical theory 6a/4(28, 30ff., 37, 50, 60ff., 196)
Linear viscoelasticity, thermodynamic theory 6a/4(77)
Linearity 6a/3(12)
Linearity of an ionization gauge 12(617–618)
Linearization based upon infinitesimal deformations 6a/3(3)
Linearization for infinitesimal surface waves 9(463, 469, 554)
Linearization of plane plasticity 6(367–370)
Linearization of reaction equations 11/1(213)
Linearization of shallow-water approximation 9(668ff.)
Linearization of the equation of motion 6a/4(217)
Linearization of the fundamental equation 9(26–29)
Linearization of the Poisson-Boltzmann equation 13(376, 388, 392, 394)
Linearization parameter 9(463, 466, 526, 554)
Linearized Bernoulli equation 9(28)
Linearized circuit for condensor microphone 11/2(41)
Linearized equations 9(285–286)
Linearized field equations for shells 6a/2(500, 523)
Linearized gas dynamics 9(669)
Linearized kinematic measures for Cosserat surfaces and shells 6a/2(456, 460, 474)
Linearized supersonic flow 9(38–40)
— two-dimensional 9(45–49)
Linearized theory of fluid dynamics, definition 9(28)
Linearized theory of plane cavity flow 9(364–366)
Linearized theory, role of characteristics 9(226, 252, 279, 281)
Linearly elastic element (spring) 6a/3(40)
Linearly elastic incompressible body 6a/2(210)
Linearly independent constants, number of the 7/1(57)
Linearly polarized plane waves 6a/4(245)
Linearly polarized waves 6a/3(108)
Linearly viscoelastic body
— of creep type 6a/3(23)
— of relaxation type 6a/3(16, 17)
Linearly viscoelastic material 6a/3(1, 16)
Linearly viscous element (dashpot) 6a/3(40)
Liouville theorem 12(84, 85, 470)

Liouville theorem for collision 12(335, 336)
Liouville's equation 10(97); 12(208, 252)
Liouville's theorem 9(480, 610)
Lipowitz's alloy 6a/1(281)
Lippmann equation 10(304)
Lipschitz constant 6a/3(587)
Liquefaction 15(371)
Liquefaction coefficient 14(50, 56, 72)
Liquefaction coefficient for hydrogen 14(54)
Liquefaction coefficient, maximal 14(51)
Liquefaction, minimum work required for 14(47)
Liquefaction of gases 12(26–28)
Liquefiers, combination Linde-type H_2 and He 14(69)
Liquefiers, combined H_2 and He 14(69)
Liquid at flat vertical plates 10(300)
Liquid crystals 10(22, 81)
Liquid He^3 15(396)
Liquid He^3, specific heat 15(181, 198)
Liquid He^4, heat conduction 15(173)
Liquid He^4, second sound velocity 15(175)
Liquid He^4 sound velocity 15(175)
Liquid He^4, specific heat 15(174)
— attenuation of sound 15(175)
Liquid helium 15(168, 190)
Liquid helium pump 12(657)
Liquid helium trap 12(657)
Liquid lenses 11/2(114)
Liquid mercury, compressibility of 6a/1(276)
Liquid mixtures, sound velocity 11/1(355–359)
Liquid model(s)
— cell-cluster model 10(10–11, 64–65)
— hole model 10(11–13)
— kinetic model 10(13–15)
— lattice-cell model 10(7–10, 24, 53)
— of the jet-edge system 11/2(28)
Liquid, Newtonian 6(441, 450–452)
Liquid, non-Newtonian 6(488, 497, 544)
Liquid phases, molecular density 13(6)
Liquid state as disordered state 10(5)
Liquid whistle 11/2(278)
Liquid-gas interface, its stability 9(321)
Liquids, associated 10(4, 23, 25, 38, 80)
Liquids, classification 11/1(305)
Liquids, ionic 10(23)
Liquids, nucleation 7/2(217–218)
Liquids, one-dimensional model 13(49–58)
Liquids, polar 10(21, 38, 80)
Liquids, solutions of electrolytes 13(369–395)
Liquids, solutions of non-electrolytes 13(300–369)
Liquids, systems of partially soluble, sound absorption 11/1(336–338)
Liquids, theory of pure liquids 13(217–300)
Liquids with classical absorption 11/1(305–306)
Liquids with pressure dependent structural effects, sound absorption 11/1(318–323)
Liquids with similar molecular constitution 10(2)
Liquids with strong visco-elastic effects, sound absorption 11/1(323–327)
Liquids with temperature dependent structural effects, sound absorption 11/1(315–318)
Liquidus temperature 13(511, 516, 543, 594, 595)
Liquified gases, sound absorption 11/1(306)
Lithium 14(243)
Lithium phosphate glasses 13(539)
Lithium sulphate 11/2(76)
Lithium-aluminium silicate ceramics 13(607)
Load bar experiment 6a/1(324, 607, 691–694)
Load parameter 6a/3(225, 226, 284, 289)
Loading, cyclical 6a/1(28, 30, 31, 68, 97–99, 123, 147, 150–152, 221, 223, 297, 366, 369, 567, 682–686, 688)
Loading, dead weight 6a/1(125, 149, 181, 400–403, 539, 550, 652)
Loading, explosive 6a/1(332, 346, 358, 499, 591)
Loading, explosive pulse 6a/1(358)
Loading function 6(280, 283)
Loading histories, short time 6a/1(357)

Loading histories with more than one non-zero stress component 6a/1(483–490, 509–513, 671–675, 677, 681, 702–716, 734–739)
Loading in steps 7/2(32)
Loading, non-radial paths of 6a/1(673, 679–684, 703–711, 715)
Loading paths, non-radial 6a/1(671, 673, 679–684, 703–711, 715)
Loading paths, radial 6a/1(508–513, 671–675, 702–716)
Loading, shock tube 6a/1(341, 342)
Loads, conservative 6a/3(201, 202, 204, 226, 284)
Loads, dead 6a/3(202, 215, 222, 226, 244, 249, 260, 272, 273, 286)
Loads, follower forces 6a/3(257, 258, 261, 278, 279, 285)
Loads, hydrostatic 6a/3(262, 263)
Loads, non-dead 6a/3(256–258)
Loads, path-dependent 6a/3(193)
Loads, velocity and time-varying 6a/3(257, 278)
Loads, weakly conservative 6a/3(193, 194, 257–264, 286)
— definitions 6a/3(259)
Loam as a Kelvin body 6(453–455)
Local density fluctuations 13(68)
Local differentiation 6a/3(524)
Local distribution function 13(89)
— of solution 13(336)
— of solvent 13(335)
Local entropy production 12(414)
Local second-order tangent modulus 6a/3(378, 382)
Local speed of propagation 6a/3(307, 308)
Local tangent modulus 6a/3(378, 382)
Local transfer theory of Kovásznay 8/2(471)
Local units 6a/1(55)
Localized suction of boundary layer 8/1(347)
Locally integrable 6a/3(5)
Locus of incompressible yield 6(279)
Loewner estimates 9(121–122)
Logarithmic continuity 6a/3(146)
Logarithmic convexity 6a/3(228–239, 253, 273)
Logarithmic convexity and Hölder stability 6a/3(231)
Logarithmic convexity of function F $(t;\ \alpha,\ t_0)$ 6a/3(229)
Logarithmic creep 7/2(199–201)
Logarithmic decrement and attenuation 6a/1(176, 367, 386, 392, 415, 416)
Logarithmic decrement of attenuation 11/2(163, 181)
Logarithmic hodograph plane 9(71)
Logarithmic homologues of consistency curves 6(497)
Logarithmic spiral(s) 6a/3(452); 8/2(30, 33–35, 41, 93)
— generalised 8/2(35, 38)
— as slip lines 6(363–365, 380)
— as streamlines 6(364–365)
Logarithmic strain 6(237)
Loitsiansky's integral 8/2(455)
Loitsiansky's integral, generalization 8/2(457)
Lomer-Cottrell dislocations, stabilizing 7/2(132, 141, 144, 146, 148)
London approach to superconductivity 15(295f.)
London constant 10(234)
London equations 15(241, 275, 286)
London equations, solutions for special cases 15(290, 291, 293)
London forces 10(233)
London tensor 15(290)
London theory 15(284)
London's r^{-6} law 12(339)
Long range order 7/1(282, 322, 447, 456); 15(246)
Long wavelenghts in canal 9(669)
Long wire in non-uniform velocity field 8/2(573–574)
Long wire in uniform velocity field 8/2(569–570)
Longitudinal discontinuity 6a/3(430)
Longitudinal elasticity 6a/3(317)
Longitudinal modes 6a/4(245)
Longitudinal motion of a piezoelectric 6a/4(295–302)
Longitudinal one-dimensional energy spectrum 8/2(454, 461)
Longitudinal piezoelectric effect 11/2(78)
Longitudinal quadrupole source 11/2(3)
Longitudinal velocity correlations in Heisenberg's theory 8/2(477)

Longitudinal wave technique 11/1(362, 399)
Longitudinal waves, *see also* Compressional waves 11/2(74)
— definition 6a/3(99, 310)
— in cylinders 6(113–114)
— in elastic bodies, anisotropic 6a/3(315–321)
— in elastic bodies, isotropic 6a/3(327, 328)
— in elastic bodies, principal 6a/3(327)
Long-range effect of pressure forces 8/2(455–456)
Long-range interaction in polymer solutions 13(464–465)
Long-range order 7/2(65, 102)
— in copper gold alloys 7/2(189–190)
— in one dimension 13(61)
— in the Ising model 13(112)
— in the three-dimensional Ising model 13(141–142)
Long-range ordering degree
— calculated by the two-dimensional Ising model 13(139)
— of Bethe 13(91)
— of Bragg and Williams 13(83, 87)
Lorentz contraction 7/1(586)
Lorentz force 14(178)
Lorentz gas mixture 12(416)
Lorentz theory for dielectrics 15(48, 52, 56, 128)
Lorentz transformation 9(286)
Lorentz-Lorenz formula 13(256)
Lorentz-Lorenz formula, generalization 13(257, 261)
Lorentz's formula for the forces exerted on a sphere in presence of a wall 8/2(245)
Lorenz number 14(138, 231)
Lorenz ratio, effect of magnetic field 14(249)
Loschmidt method 12(320)
Loss angle 6(268, 277–278)
Loss modulus 6a/3(33)
Loss rate, total, of anemometer in operation 8/2(556, 559)
Losses by hysteresis 11/1(399)
Lost solutions 6a/3(485)
Loudspeaker 11/1(28, 48)
Loudspeaker, electrodynamic 11/2(9–12)
Loudspeaker, electrostatic 11/2(15–17)
Love waves in stratified media 6(111–112)
Love's first approximation 6a/2(478, 582, 589, 590)
Love's integral identity in elasticity 6a/2(243)
Love's solution in elasticity 6a/2(147)
Low pressure system 14(58, 79)
Low temperature data 6a/1(374, 384, 396–399, 524–533, 556–560)
Low temperature range of glass viscosity 13(582, 584–589)
Low viscosity liquids with pressure-dependent structural effects, sound absorption 11/1(318–323)
Lower relaxation maximum 11/1(263)
Lower semi-continuous with respect to weak convergence 6a/2(393)
Lowest helium temperatures 15(40)
Low-frequency cut-off of piezoelectric microphones 11/2(47)
Low-frequency horn loudspeaker 11/2(18)
Low-frequency sensitivity of condensor microphone 11/2(41)
Low-frequency sirens 11/2(27)
Low-voltage electrostatic loudspeaker 11/2(16–17)
L-shaped beam 6(400, 406)
Luders' lines 6a/1(449, 450)
Lueders band(s) 6(232); 7/2(62, 158, 174)
Lueders band, propagation 7/2(183–184)
Lueders slip 7/2(62)
Lueders strain 7/2(32, 62)
Lueders streaks or bonds 7/2(183)
Luminescence due to cavitation 11/2(272, 279–282, 287)

Mach angle 8/1(13, 14, 109, 208); 9(39, 179, 250, 692); 12(570)
Mach cone 8/1(211); 9(38); 11/1(77)
Mach directions 9(88)
Mach line 8/1(13, 14, 69, 107, 209)
Mach lines 9(16, 20, 22, 45, 61, 130, 138, 229, 241, 245, 250, 273–275, 687)
Mach net of a flow 9(88, 89)

Mach number 6a/3(281); 8/1(13, 60, 65, 71, 94, 182, 184, 245–246, 335, 339, 419, 423); 8/2(592, 602, 611); 9(6, 7, 167ff., 184, 198, 251); 12(567, 569)
— critical 9(121)
— effect on critical Reynolds number 8/1(426)
— estimates 9(128)
Mach quadrangle 9(232–233, 255)
Machining of photoelastic materials 6(164)
Mackenzie's computation of shear strength 7/2(9–13)
Macrocanonical ensemble 12(74, 79)
Macromolecular solutions, orientation effects 13(329)
Macromolecule, definition 13(399)
Macro-rheology 6(435)
Madelung constant 7/1(74, 135, 137, 138, 139)
Madelung's constants of CsCl, NaCl, ZnS 7/1(86)
Madelung's method 7/1(137, 225)
Magnesium 14(247)
Magnesium crystals, twinning 7/2(48)
Magnesium phosphate glasses 13(537)
Magnesium single crystals
— critical shear stress 7/2(35)
— work-hardening 7/2(49)
Magnesium sulphate, quantitative theory 11/1(262)
Magnet of Bitter 15(71)
Magnet of Weiss 15(69)
Magnetic coupling between ions 15(43)
Magnetic crystal classes 6a/4(185)
Magnetic dipole interaction 15(100, 110, 117, 120, 128, 132)
Magnetic domains 11/2(223, 227)
Magnetic field, effect on ultrasonic attenuation by conduction electrons 11/2(218–219, 220)
Magnetic fields, influence on magnetic moments 15(120, 127, 129, 151)
Magnetic induction technique for particle velocity 6a/1(630–633)
Magnetic interaction 15(44, 82f., 86, 96, 100, 111, 112)
Magnetic interaction between electrons 15(344)
Magnetic measurement of pressure 12(533)
Magnetic model of defect treatment 13(79)
Magnetic moment(s) 14(226); 15(42, 155, 165)
Magnetic moment along an isentropic 15(53)
Magnetic moment, frozen-in 15(217, 224, 270)
Magnetic susceptibility and spin correlation function 13(207)
Magnetically driven flying foil 6a/4(4)
Magnetization 6a/1(89)
Magnetization calculated by the Ising model 13(112, 137, 139, 141)
Magnetization curve 15(72, 120, 155, 159, 163, 222)
Magnetization curve under plastic deformation 7/2(102)
Magnetization, isothermal, decrease of entropy 15(54)
Magnetomechanical losses 11/2(223)
Magneto-resistance 14(178, 180, 247, 298)
Magneto-resistance, longitudinal and transverse 14(182)
Magneto-resistive effects, similarity relations of the 14(248)
Magnetostriction 11/2(85–86)
Magnetostrictive cores 11/2(89, 90)
Magnetostrictive probe 11/2(116)
Magnetostrictive ultrasound generator 11/2(87)
Magnets of iron 15(69)
Magnets of iron, cooling 15(70)
Magnets of iron-free coils 15(70)
Magnets of iron-free coils, cooling 15(71)
Magnets, types 15(68f.)
Magnitude of a tensor 6a/2(7)
Main flow, disturbances 8/1(366, 367)
Mainardi-Codazzi equations 6a/2(626)
Mallock's method for determining the bulk modulus 6a/1(303–306)
Manganese resonance 11/2(243)
Manganin 14(296, 298)
Manganin wire gauge, *see* Stress transducer
Mangler's transformation 8/1(311–312, 340–341)
Mann impact experiment using a rotating flywheel 6a/1(588–590)

Manometer, *see also* Vacuummeter 12(655)
Manometer 8/2(558, 587)
Many-body-forces 7/1(127, 193, 201, 222, 281, 287)
Map T_τ 6a/3(130, 133, 140, 148, 149)
Mapping of regions onto a circle 6(62–64)
Mapping of regions onto a half-plane 6(66–71)
Margules equation 6(460, 497)
Mariotte's experiments 6a/1(160)
Markoff process 8/2(514)
Markovian chains 13(279–280)
Martensite 7/2(243–246)
Martensite curve 7/2(251)
Mason's theory 11/1(326)
Mass balance 6a/4(119)
Mass centre 6a/3(217, 218, 247)
Mass conservation law 8/1(132, 149)
Mass density 6a/3(536); 10(33); 12(208, 238)
— Cosserat surface 6a/2(447)
— deformed configuration 6a/2(440)
— reference configuration 6a/2(440, 472)
Mass diffusion equation 12(303)
Mass flow 8/1(5, 94); 11/1(3)
Mass flow, acoustic 11/1(7)
Mass fraction 10(41, 82)
Mass spectrometer for measuring ultrahigh vacuum 12(626–627, 635, 643)
Mass spectrometer method to determine diffusion permeability 13(617)
Mass transport in free surface motion 9(716–718)
Mass transport in second-order infinitesimal wave theory 9(661)
Mass transport in viscous fluid 9(666)
Mass, virtual 8/1(93)
Massau semi-graphical procedure 9(20)
Massey-Mohr formulae 12(459)
Master elastic constants 6a/4(243)
Master equation 12(232)
Master state 6a/4(216, 243)
Matched impedance 6a/4(294)
Material coefficients 6a/4(165, 179)
Material coordinates 8/1(141)
Material derivative 6a/3(404, 440, 540); 6a/4(113, 114); 8/1(20, 129, 130); 9(451)
Material derivative in curvilinear coordinates 8/1(141)
Material description 6a/4(113)
Material discontinuities 6a/3(429, 434)
Material discontinuity 6(343)
Material, elastic 6a/3(19, 175, 315)
Material, elastic, related to viscoelastic 6a/3(19)
Material, elastic-plastic 6a/3(537)
Material electric field 6a/4(166)
Material frame-indifference 6a/2(80, 305, 487)
Material instability 6a/1(714, 716)
Material memory phenomena 6a/1(472, 550–511, 649, 688, 703)
Material moduli, initial experimental study of 6a/1(161)
Material objectivity 6a/4(44)
Material, plastic 6a/3(403, 412–415)
Material point 6a/2(229, 439, 446, 530); 6a/3(16)
Material representations 6a/4(150, 152)
Material resonances 6a/1(417)
Material sample, *see* Target plate
Material standpoint 6a/3(428, 524)
Material surface 6a/2(439)
Material symmetry 6a/3(2, 18, 29); 6a/4(174)
Material symmetry in 3-dimensional theory 6a/2(69–74, 87–89)
Material symmetry in shells 6a/2(537, 539, 541)
Material, thermoelastic 6a/3(183)
Material time derivative 6a/2(431, 449, 453)
Materials with memory, *see also* Simple materials 6a/2(373); 6a/3(172)
Mathematical models 6a/3(131, 150, 213)
Mathematical notation 6a/3(125)
Mathieu functions 9(298, 652–653)
Mathieu-Hill equation 6a/3(257)
Matrix notation and tensor notation of elastic coefficients 11/2(155)
Matrix, reciprocal 7/1(43)
Matrix theory of the Ising model 13(108ff., 122ff.)
Matter constants 7/1(41)

Matthiessen's rule 7/1(427); 14(145, 147, 238)
Maximal absorption volume 11/1(216–221)
Maximum amplitude waves 9(731–736, 749)
Maximum and minimum shearing stress 6a/3(423, 465)
Maximum complementary energy 6a/2(125)
Maximum crystal growth 13(602)
Maximum elastic modulus 6a/2(85)
Maximum energy of distortion hypothesis 6a/1(440–445, 486, 506)
Maximum hardness of glasses 13(634)
Maximum of dislocation damping 11/2(197)
Maximum potential energy 6a/2(125)
Maximum principle(s) 6a/2(120, 375); 6a/3(152); 9(380)
Maximum shear 6a/3(442)
Maximum shear criterion, Tresca 6a/1(440–445)
Maximum shear stress and strain 6a/1(512)
Maximum shearing stress hypothesis 6a/1(484, 513)
Maximum speed of propagation 6a/2(245, 256, 257)
Maximum strain hypothesis 6a/1(484)
Maximum stress hypothesis 6a/1(484)
Maximum-minimum principle for elliptic differential equations 9(94ff.)
Maxwell body (=elastico-viscous material) 6(290, 301, 312–313, 455, 465–467, 516–518)
Maxwell body, propagation of a shear wave 6(301)
Maxwell coefficients 6a/4(128, 129, 138)
Maxwell element 6a/3(40)
Maxwell elements in parallel 6a/3(41)
Maxwell field 7/1(208ff.)
Maxwell field equations 6a/4(147)
Maxwell model 6a/4(199, 200); 11/1(363–368)
Maxwell operators 6a/3(39)
Maxwell relation(s) 6a/2(304, 310); 6a/4(127, 130)
Maxwell relaxation effect 12(272)
Maxwell, theorem 6a/3(98, 101)
Maxwell theory of relaxation, its limitations for glass 13(638–639)
Maxwell-Boltzmann distribution function 12(363)
Maxwell-Boltzmann equation 8/2(593, 600)
Maxwellian equilibrium distribution 8/2(594)
Maxwellian gas 12(370–378, 390–391)
— distribution function 12(391)
— eigentensors 12(390)
— eigenvalues 12(377)
— mixture 12(418, 419)
Maxwellian materials 6a/4(37, 45, 48, 51, 53, 58, 60, 65)
Maxwellian materials, similarity solutions 6a/4(37)
Maxwellian molecules 12(237, 270, 284, 344)
Maxwellian relaxation time 11/1(363)
Maxwellian transport equation 12(351)
— for a gas of rotating molecules 12(480, 483)
Maxwellian velocity distribution 12(361)
— experimental verification 12(363)
Maxwell's compatibility theorem 6a/2(250)
Maxwell's equations 7/1(209)
Maxwell's problem 8/2(84)
Maxwell's relaxation model 8/1(50)
Maxwell's solution in elasticity 6a/2(54)
Maxwell-von Mises criterion 6a/1(486, 506, 508, 563, 667)
Mayer's integral equations 13(37, 39)
M-body, *see* Maxwell body
McLeod gauge 12(520–523)
McLeod's equation 10(174, 175)
McReynolds' slow waves 6a/1(655, 661)
Mean bubble radius in cavitation 11/2(264, 267–268, 274)
Mean camber 8/2(387, 392, 429)
Mean camber line 8/2(387)
Mean collision time in gases 11/1(133, 139)
Mean curvature 6a/2(471, 624)
Mean curvature of a surface 10(284, 288, 300)
Mean flow 8/2(439)
Mean free path 7/1(291, 292, 309, 312, 441); 12(214); 14(138)

— of a dislocation 7/2(128, 141, 144, 146)
— of molecules 8/2(592)
— of phonons 11/2(233)
Mean mass velocity 12(303, 355)
Mean molecular velocity 11/1(139)
Mean momentum equations 8/2(544–546)
Mean normal stress and strain 6a/3(71)
Mean particle velocity 12(303, 357)
Mean pressure 8/1(234, 241)
Mean specific heat of a glass 13(571)
Mean strain 6a/2(37, 38, 97, 98)
Mean stress 6a/2(96, 97); 6a/3(216, 240–243)
Mean stress and strain in terms of external force-temperature 6a/2(315)
Mean thickness of the ionic atmosphere 10(224)
Mean value of a function 6a/2(314)
Mean value theorem
— for elastic states 6a/2(133–138)
— for harmonic and biharmonic fields 6a/2(21)
Mean-square vorticity 8/2(452, 492, 497)
Measurable quantities in sound waves 11/2(35)
Measure 6a/2(414, 415)
Measurement of hardness by corrosion 7/1(103)
Measurement of the crystal faces 7/1(68)
Measurements below 1° K 14(172, 201, 247)
Measures and boundedness and existence 6a/3(135)
Measures as positive-definite functions on $X \times X$ 6a/3(135)
Measures, choice 6a/3(134–139, 163)
Measures, choice, in elasticity 6a/3(171, 172, 210–213)
Measures, definition of positive-definite function 6a/3(132)
Measures, discussion 6a/3(135, 158)
Measures, equivalence relations (semi-equivalence) 6a/3(139)
Measures, examples 6a/3(133)
Measures, examples in elasticity 6a/3(192–195, 207, 209, 210, 212, 219, 220, 222, 241, 245, 251, 254, 266, 268, 270, 271, 273, 280)
Measures in discrete systems 6a/3(141)
Measures, influence on stability, examples 6a/3(135, 136, 138)
Measures of finite deformation 6(498–501)
Measuring procedures 14(298)
Mechanic, statistical 7/1(410)
Mechanical effects of ultrasound 11/2(115–123)
Mechanical forcing 6a/3(31)
Mechanical impedance 11/2(4)
Mechanical losses in glass 13(642, 644)
Mechanical models of materials 6(452ff.)
Mechanical power 6a/2(496)
Mechanical properties of glasses 13(544)
Mechanical pumps 12(547–559)
Mechanical relaxation 13(446)
— function 6a/4(83ff., 87)
— in glasses 13(635–644)
— in glasses, atomic theory 13(640–644)
Mechanical strength of glasses, *see also* Strength 13(623–635)
Mechanics, molecular 10(33–47)
Mechanism of plastic deformation 7/2(18)
Mechano-caloric effect 15(379, 388, 390)
Mechano-caloric effect in liquid Helium II 15(46)
Mechano-electric impedance 11/2(5)
Meeting of two solutions 6a/3(480)
Meissner effect 15(211, 220, 228, 274, 278, 322)
Melting 7/2(218)
Melting curve 15(372, 399, 400, 403)
Melting heat 7/1(546); 15(375)
Melting heat per atom 7/1(503)
Melting phenomenon = order-disorder transition 10(5, 7)
Melting point 7/1(282); 10(66)
Melting point depression of a polymer 13(426, 428)
Melting temperature 7/1(263, 312, 315, 320, 321)
Melting theory of Kirkwood and Monroe 13(290–300)

Melting theory of Lennard-Jones and Devonshire 13(284–290)
Membrane 6a/3(208, 436, 497); 8/2(68)
Membrane behavior of dividing surface 10(286)
Membrane driven by line force 11/1(108–112)
Membrane, fixed along boundary and stretched over obstacle 6a/2(405)
Membrane gauge 12(517, 518–519)
Membrane generating waves 9(558–559, 616)
Membrane model of Prandtl 6(401, 424)
Membrane supported periodically 11/1(104–108)
— forced motion 11/1(114–116)
Membrane theory of shells 6a/2(547)
Membrane, transmission of sound 11/1(99)
Membrane, unilateral constraint 6a/2(422)
Membrane wall of a rectangular duct 11/1(121)
Membrane wave 11/1(109, 110)
Memory functions of stress and strain 6(273)
Memory, materials with 6a/3(172, 334–339, 344–356)
Meniscus 10(301)
Meniscus, radius of curvature 8/2(632)
Mercury 14(269, 275)
Mercury crystals, critical shear stress 7/2(40)
Mercury manometer 12(519)
Mercury, sound absorption 11/1(305)
Mercury vapour pump 12(547, 559, 591, 646)
Mesh length 7/2(112)
Mesh width 7/2(112)
Mesh width, effect on isotropy of turbulence 8/2(533, 536)
Metacenter 9(565)
Metal hydroxides 11/1(258)
Metallurgy, powdered 6a/1(476)
Metals 7/1(143, 247, 254, 285, 292, 312)
Metals, conduction properties
— of monovalent 14(242)
— of polyvalent 14(245)
Metals, interaction of stress waves with conduction electrons 11/2(208–223)
Metals, monovalent 14(238f.)
Metals, treatment by intense ultrasonics 11/2(286)
Metal-to-salt contact 15(171)
Metamagnetism 15(19)
Metarheology 6(436)
Metastable currents
Metastable lattice 7/2(217)
Meteorological problems 8/1(47)
Methane 14(26, 30)
Methanol-benzene mixture, orientation effects 13(343–344)
Method of characteristics 9(254–255)
Method of small disturbances 8/1(367, 422)
Methyl chloride 14(24, 27)
Metric tensor of space 6a/2(618)
Metric tensor of surface 6a/2(624)
Meyer's type of transonic nozzle flow 9(147–148)
m-fold Dirichlet norm 6a/3(551)
Micelle 13(425, 492)
Michell, *see* Beltrami
Michell's integral 9(580–581)
Michell's method for the investigation of wave crests 9(732–734)
Michell's theorem 6a/2(163)
Michel's apparatus for compressibility measurements 12(32, 33)
Michelson displacement interferometer, *see* Displacement interferometer
Micro states, definitions 12(80)
Micro-beam technqiue 7/2(153)
Microcanonical distribution 12(221)
Microcanonical ensemble 12(76, 79)
Micro-cinematography 7/2(106–107, 119)
Microcracks 6(567)
Microcrystals 7/1(463)
Micro-mechanism of fatigue 6(596–597)
Microphone, ideal properties 11/2(36)
Microplasticity 6a/1(22, 25, 27, 140, 148, 223, 458, 682)
Micro-rheology 6(436, 522–535)
Microscopy, *see* Light microscopy and electron microscopy
Microseismology 6a/1(343–351, 356)
Micro-slip 7/2(84)
Micro-slip in stage I 7/2(85, 87)
Microsporous system 8/2(626)
Micro-stirring effect 11/2(282)

Micro-strain 7/2(183)
Microstrain, first resolution of 6a/1(34)
Microwave technique 15(44)
Mie's lattice model of liquids 13(221)
Migration of dislocations 7/2(222, 240)
Migration of interstitials 7/2(222)
Migration of vacancies 7/2(222)
Mild discontinuities 6a/2(253)
Miller indices 6a/4(249)
Miller, indices of 7/1(46, 62)
Milne-Thomson's circle theorem 9(574)
Minerals, plastic properties 7/2(18)
Minimal property, *see* Free energy
Minimal surface 9(81, 93)
Minimax principle 6a/2(268)
Minimizing sequence 6a/2(393); 6a/3(564, 565)
Minimum cavity drag 9(386)
Minimum complementary energy 6a/2(125, 324, 325)
Minimum elastic modulus 6a/2(85)
Minimum energy 7/1(437)
Minimum life in fatigue experiments 6(593, 607)
Minimum of wave velocity for capillary-gravity waves 9(633)
Minimum potential energy 6a/2(111, 125)
— converse 6a/2(116)
Minimum pressure determining point of transition 8/1(365, 373, 404, 415)
Minimum principle
— displacement problem 6a/2(209)
— elastodynamics 6a/2(230)
— lowest characteristic value 6a/2(264)
— stress field 6a/2(232)
— vibration problem 6a/2(262, 266, 267)
Minimum size of nucleating agent 13(605)
Minimum suction quantity 8/1(413)
Minimum transformed energy 6a/2(231)
Minimum-maximum principle of free streamline 9(390)
Minus wave 6(391 ff.)
Mirror plane 7/1(8, 580)
Miscibility gap 13(307)
Miscibility gap in ternary polymer systems 13(482, 483)
Mises flow equations 6(339)
Mises plasticity equations 6(323, 404)
— characteristics 6(337–339)
Mises yield condition 6(282, 289, 320, 322, 324, 328, 370, 521)
Mises-Lévy equations 6(284, 286, 323)
Mist formation by ultrasonics 11/2(279, 286)
Mixed boundary value problem 6(58, 70, 81–84, 101–102)
— of elasticity, dynamic 6a/2(219)
— of elasticity, static 6a/2(102, 377, 379, 385)
— of thermoelasticity, dynamic 6a/2(335)
— of thermoelasticity, static 6a/2(321)
Mixed cavity problem 9(392–393)
Mixed chemical bonding in glass formation 13(525, 542)
Mixed coefficients 6a/4(225)
Mixed crystals 7/1(94)
Mixed crystals, compl. series of 7/1(94)
Mixed phosphate glasses 13(539)
Mixed problem 6(362, 381, 397, 430); 9(226, 256–257, 270–272)
Mixed problem, nonlinear 6a/3(474)
Mixed region 6a/3(518)
Mixed tensor 6a/2(617, 622)
Mixed type of differential equations 9(93, 131)
Mixed-mixed boundary data 6a/3(47, 57)
Mixing entropy 13(172–174, 325, 333, 338, 350, 351)
Mixing length 8/2(508)
Mixture of noble gases, sound propagation 11/1(173–174)
Mixture rule for absorption in electrolytes 11/1(242, 243, 268–269)
Mixtures 6a/3(172)
Mixtures, deformation behavior of 6a/1(230, 393)
Mixtures of liquids 10(2, 23–25, 40–43, 67)
Mobility 11/1(216)
Mobility of protons 11/1(248, 251)
Möbius crystal 7/1(386)
Mode deformation, *see* Deformation modes

Mode of deformation 6a/3(416)
Model equation to the Maxwell-Boltzmann equation 8/2(600, 610)
Model laws 8/1(56–60)
Model of excluded volume, *see* Excluded volume model
Models of dislocation damping 11/2(178)
Models of liquids, *see* Liquid models
Models of materials 6(452ff.)
Models of polymers 11/1(363–370)
Modes 6a/4(231, 245)
Modes, lattice 14(283, 284)
Modes, longitudinal 6a/4(245)
Modes, molecular 14(283)
Modes of waves in a basin 9(623–627)
Modes, pure 6a/4(246)
Modes, transverse 6a/4(245)
Modified Maxwell model 11/1(369)
Modulator, superconducting 14(164)
Moduli, adiabatic vs isothermal 6a/1(121, 122, 228, 229, 241, 289, 320, 338, 339, 355, 378)
Moduli, decrease of, with permanent deformation 6a/1(90, 92, 131)
Moduli, generalized 6(248, 258)
Moduli, linear temperature dependence of 6a/1(227, 294, 297, 366–371, 375, 379–380, 391, 396)
Moduli of glass, *see* Experimental results for glass
Modulus, *see* *E* modulus, Shear modulus, Tangent modulus
Modulus, bulk 6a/1(303–305, 373)
Modulus, comparison of longitudinal and transverse, measured by vibrations 6a/1(222)
Modulus, height of the 6a/1(48, 60, 97, 186)
Modulus of compression, *see also* Elasticities 6a/2(74, 97)
Modulus of cross-elasticity 6(507)
Modulus of rigidity, *seer* Shear modulus
Modulus of torsional rigidity of quartz fibers 6a/1(366)
Modulus, secant, definition of 6a/1(105, 137)
Modulus, tangent, definition of 6a/1(105)
Modulus, weight of the 6a/1(186)
Modulus, Young's 6a/1(185)
Mohr circle(s) 6(241, 292, 325ff., 352, 375–376, 446); 6a/3(407, 468, 477)
Mohr circles of stresses on both sides of a discontinuity line 6(367)
Mohs' hardness scale 7/1(102)
Mohs' scale of hardness 13(632)
Moiré patterns in metals 7/2(81)
Molar fraction 12(303)
Molar volume of gases 12(5)
Molecular absorption and dispersion of sound 11/1(135, 157–165)
Molecular air pump 12(559, 560–563)
Molecular association 11/1(350)
Molecular beam methods of vacuum investigation 12(540–541, 603–604)
Molecular chaos 12(216, 221–224, 228, 345, 347)
Molecular crystal 7/1(127)
Molecular density in liquid phases 13(6)
Molecular distribution 13(401)
Molecular distribution function 8/2(593); 10(25, 27, 39, 47, 179, 182, 194, 195); 13(3)
— and chemical potential 13(17)
— definition 13(5)
— for binary alloys 13(175–180)
— in quantum statistics 13(64)
— nonlinear integral equation 10(79)
— normalization 13(5, 7)
Molecular field 15(1, 30)
Molecular flow 12(515, 530–531)
Molecular force
— (6; Exp) potential 12(341, 397)
— elastic sphere 12(236)
— finite cut-off 12(211)
— Lennard-Jones potential 12(92, 151, 341, 396, 400, 428, 438–440)
— Maxwellian 12(237)
— power law 12(236, 396, 427, 438)
— pseudo-Maxwellian 12(244)
— rigid spheres 12(340, 343, 394–395, 427, 437)
Molecular gauge after Langmuir and Dushman 12(529–530)
Molecular mean free path 8/2(592)
Molecular mechanics of liquids 10(33–47)
Molecular model, rough spheres 12(484, 497–499)
Molecular polarizability 10(72)
Molecular radius 11/1(339)

Molecular relaxation times, distribution 11/1(361)
Molecular rotation 7/1(95)
Molecular sound absorption 11/1(135, 157–165)
Molecular speed ratio 8/2(602)
— infinite 8/2(604)
Molecular spin 12(301)
Molecular statistics of liquids 10(28–33)
Molecular thermal pressure 12(530–531)
Molecular velocity 8/1(64)
Molecular velocity mean 11/1(139)
Molecular volume and sound velocity 11/1(346–350)
Molecular weight
— determination 13(305, 315, 401, 497)
— effect on second virial coefficient 13(465–466)
— measurements 12(25–26)
— of macromolecular materials 13(400)
— of polymer, average 11/1(380–381)
Molecule flow 8/2(592, 601–609)
Molecules, rigid spherical 10(48, 59)
Mollier diagram 14(17, 22)
Mollifiers 6a/2(420)
Mol-vacuummeter of Gaede 12(533)
Molybdenum single crystals, critical tensile stress 7/2(35)
— work-hardening 7/2(66–67)
Moment at infinity 6a/2(170)
Moment coefficients 8/2(387, 395, 428)
Moment equations for the 13 state variables 12(271, 273, 276)
Moment of momentum 6a/2(482)
Moment on a profile 9(128)
Moments 13(99, 367, 384)
Moments, balance of 6a/2(306); 6a/3(42, 216, 242, 243)
Moments of constitutive variables in rod theories 6a/2(649)
Momentum, angular 6a/2(44, 52); 6a/4(119)
Momentum density 10(33)
Momentum equation, Eulerian form 8/1(17–18)
Momentum equation, integral form 8/1(5–8)
Momentum flow tensor (=momentum flux)
— acoustic 11/1(7–8, 9)
— total 11/1(3)
Momentum flux 10(33)
Momentum integral equation of boundary layer 8/1(278–280, 318)
Momentum integrals for free surface motion 9(718–721)
Momentum, linear 6a/2(43, 44, 51, 254, 482, 483)
Momentum of a surface wave 9(461)
Momentum, rods (angular, director, linear) 6a/2(666)
Momentum, shells 6a/2(483)
Momentum, space-time 6a/2(67)
Momentum tensor 8/1(7)
Momentum thickness of boundary layer 8/1(279, 281)
Momentum transfer equation 8/1(138–139)
Monatomic liquids, surface tension calculation 10(288)
Monge-Ampère equation 6a/3(441)
Monge-Ampère equation for Airy function 6(353)
Monolayer model 10(269, 270, 275)
Monomer 13(399)
Monopole source 11/2(1, 102, 103)
Monopole wave 11/1(5, 27, 83)
Monotonicity theorems 9(383–386)
Monovalent metals, experimental comparison with 14(173)
Monte Carlo method 8/2(608); 13(217, 277–284)
Mooney strain energy function 6a/1(740)
Mooney's equation for elastic rubber potential 6(19)
Morera's solution in elasticity 6a/2(55)
More-stage demagnetization 15(198, 199)
Morphologic extination 7/1(70)
Morse potential 13(243)
Mortar 6a/1(143)
Mosaic structure 7/1(96, 641)
Motion 6a/4(110)
Motion, definition 6a/3(304)
Motion, elastic equation of 7/1(179, 191)
Motion equation 11/1(137, 276)

Motion, equation of 6a/3(42, 89, 91, 176–181, 404); 6a/4(120); 7/1(145, 156)
Motion in dynamical system 6a/3(127, 128, 131)
Motion, in three dimensions 6a/4(70)
Motion, longitudinal (one-dimensional motion, one-dimensional stretch) 6a/4(24, 68, 71)
Motion of body 6a/2(43, 299)
Motion of Cosserat surface 6a/2(449)
Motion, relative, of fluid and boundary 11/1(18–19)
Motion reversibility 12(475)
Motion, steady compressive 6a/4(37 ff.)
Motional impedance 11/2(4, 11)
Mott model 7/2(9, 12, 13)
Mott's island model 7/1(658, 660)
Movable basin, surface waves 9(627–631)
Moving coil microphone 11/2(36–38, 54)
Moving coordinates 6a/3(448)
Moving coordinates, Stokes equation 8/2(201–202)
Moving discontinuity 6(342, 346)
Moving medium 11/1(11, 18, 76–99)
Moving plane inside an elliptic cylinder 8/2(79)
Moving surfaces 6a/3(428)
Moving wall 8/2(119, 549)
Multhopp's method 8/2(410)
Multidimensional singular integral equation 6a/2(421)
Multiindex 6a/2(348)
Multiple elasticities 6a/1(150–152, 174, 195, 201, 308, 400–405, 567, 640, 685, 686, 716)
Multiple elasticities during unloading cycles 6a/1(685)
Multiple elasticities from the 19th century 6a/1(404)
Multiple jet instability behind grid 8/2(531)
Multiple Maxwell model 11/1(364, 367, 369)
Multiple scattering 11/2(165, 166–169)
Multiple scattering of polarized electrons 12(481)
Multiple screens 8/2(551)
Multiple slip 7/2(22, 31, 60)
Multiple Voigt model 11/1(369)
Multiplicity 6a/2(369)
Multipolar elasticity 6a/3(172, 213)
Multistage system 14(28)
Multi-valued displacement 6(53)
Munk's theorem 8/2(405)
Murnaghan's second order elastic constants 6a/3(181)
Mushkhelishvili potentials 6(48–58)
Musschenbroek, experimental apparatus of 6a/1(162–164)
Mutative process 6a/4(122)
Mutually inverse 6a/3(26)
μ-space (molecule-space) definition 12(80)

Nabarro-Herring theory of creep 7/2(208)
NaCl lattice 7/1(84, 101, 116, 135, 137 ff., 217, 223, 247)
Narrow non-uniform ducts 11/1(47–50)
Natta-Baccaredda's formula 11/1(355)
Natural configuration 6a/2(380)
Natural direction 6a/4(219, 220)
Natural disturbances 8/1(386)
Natural equilibrium state 6a/4(28, 39, 80, 82, 84, 90)
Natural state 6a/3(195); 6a/4(216)
Natural velocity 6a/4(25, 219, 220)
Naumann's indices 7/1(62)
Navier's equation of elasticity
— dynamic 6a/2(213)
— static 6a/2(90)
Navier-Stokes equation 8/1(51–52, 237, 247); 8/2(1, 3, 200, 439, 445, 452, 601, 611)
— existence and uniqueness of the solutions of boundary value problems 8/2(186–199)
— existence of solutions 8/1(252)
— for axisymmetric motion 8/2(10)
— for plane motion 8/2(8)
— linear approach 8/2(200)
— pseudo-plane steady solutions 8/2(91)
— rigorous solutions 8/2(11–13)
— solutions obtained by transformation of other solutions 8/2(176–186)
— solutions with constant vorticity 8/2(39)
— uniqueness theorems 8/1(251)

Navier-Stokes equations 6(451); 9(225, 448); 12(216, 259, 272, 273, 277, 292)
Navier-Stokes formula, *see* Stokes Kirchoff formula
N-body, *see* Newtonian liquid
Near free molecule flow 8/2(608–609)
Nearest neighbour coupling 7/1(154)
Nearest neighbour interaction 7/1(121, 277, 282, 286, 314)
Néel point 15(29, 32)
Negative attenuation 11/2(250)
Negative cnoidal wave 9(714)
Neighborhood of infinity 6a/2(165)
Neighbourhood 6a/3(132, 133)
Neighbours, effective number of 7/1(130, 135)
Nekrasov's transformation and integral equation 9(728–730)
Nekrasov-Yamada method for the investigation of wave crests 9(734–735)
Nelson's law for paints and varnish 6a/1(126, 127)
Neon liquefier 14(66)
Neon, liquid 14(66)
Neper (unit) 11/2(163)
Nernst-Eucken method 14(294)
Nernst-Lindemann formula 7/1(327)
Nernst's law 15(39)
Net force on a sphere 11/1(65)
Network 13(399, 448, 487, 489, 490)
Network concept of glass 11/1(403, 413–414)
Network, connectivity 13(437)
Network equivalence 6a/4(269)
Network, free energy of deformation 13(435, 442)
Network, internal coherence 13(520, 531, 569)
Network of dislocations 7/2(111)
Network parameters 11/1(363)
Network structure of a glass 13(517, 524)
Network, three-dimensional 11/1(407, 409)
Network-forming ions 13(518, 524, 537, 540, 542, 572)
Network-modifying ions 13(519, 524, 540, 542)
Networks of linearly elastic and viscous elements (springs and dashpots) 6a/3(34)
Neumann problem 6a/2(375, 419)
Neumann triangle 10(286, 288, 292, 294, 298–299)
Neumann's boundary problem 9(12, 622)
— of potential theory 6(28)
Neumann's function of a profil 9(116)
Neutral disturbance 8/1(371, 373, 379)
Neutral equilibrium 6a/3(144, 170, 202, 204, 205, 226, 227)
Neutral line 6a/1(16, 23, 24, 196)
Neutral line, definition of 6a/1(196)
Neutral loading 6(290)
Neutral point, thermal, in rubber 6a/1(309–311)
Neutral stress change 6(280, 288)
Neutral surface 7/1(559)
Neutron damage in silicon 11/2(169)
Neutron diffraction 15(29, 271)
Neutron diffraction in liquids 10(43–47, 124)
Neutron irradiation 11/2(196)
Neutron irradiation of quartz 13(577)
Neutron polarization 15(206)
Neutron scattering in alloys 7/2(178)
Neutrons 7/1(431)
Neutrons, critical scattering 13(196–210)
Newtonian flow 8/2(604)
Newtonian liquid (=ideal viscous liquid) 6(441, 450–451, 452)
Newtonian liquids 10(15)
Newtonian potential 6a/2(19); 6a/3(76)
Newtonian viscosity 6a/4(214, 256)
Newton's viscosity concept 8/1(50)
n-fold symmetry of the axis 7/1(7)
NiAs lattice 7/1(84)
Nickel 7/1(476)
Nickel, attenuation in magnetized state 11/2(224–227)
Nickel crystals, work hardening 7/2(53, 162)
Nickel, magnetostrictive properties 11/2(86–87)
Nickel single crystals, change in flow stress under temperature change 7/2(163)
Nickel-cobalt single crystals
— stacking faults 7/2(182)
— work-hardening 7/2(64, 65)

Nickel-iron alloys, surface phenomena 7/2(99–100)
Nicol's prism 6(135)
Nikolsky-Taganov theorem 8/1(215–217)
Nitrogen 14(228)
Nitrogen, dispersion and absorption 11/1(177–178)
Noble gases, liquid 10(1, 2, 48, 53, 55, 65, 67, 80, 93, 117)
Noble gases, propagation of sound 11/1(172–173)
Noble metal 7/1(105, 112, 127, 143, 144, 228, 232)
Noise of cavitation 11/2(264)
Noise threshold for condensor microphone 11/2(41–42)
Noise voltage of an anemometer 8/2(558)
Nomenclature for macromolecules 13(399)
Non-bridging oxygen ion 13(519, 529, 591)
Non-coercive non-symmetric bilinear forms 6a/2(422)
Non-conservative motion of jogs 7/2(132)
Non-conservative motions 7/1(481, 493, 499)
Non-conservative problems 6a/3(258, 279)
Non-continuum theories 8/2(592–593)
Non-cooperative orientation effects in solutions 13(329–335)
Non-dimensional form of boundary layer equations 8/1(268–269)
Non-electrolytes, solutions 13(300–369)
Non-equilibrium for molecular flow 8/2(607–608)
Nonexistence theorem of elasticity 6a/2(109)
Non-homogeneous condition 6a/3(403)
Non-isotropic material 6(370)
Non-isotropic turbulence 8/2(533–540)
Non-isotropy 6a/3(461)
Nonlinear acoustics 8/1(109)
Nonlinear constitutive equations
— rods 6a/2(654)
— shells 6a/2(528)
— shells, membrane theory 6a/2(566, 588)
— shells, restricted theory 6a/2(549)
— thermoelastic materials 6a/2(297)
Nonlinear dissipative media 6(255)
Nonlinear effects 7/1(618)
— in airfoil theory 8/2(424–425)
— in seepage flow 8/2(649)
— in stress waves 11/2(160–162)
Nonlinear elasticity 6(508–509)
Nonlinear elasticity, theory 7/2(2, 17)
Nonlinear Liapounov functions 6a/3(187–191)
Nonlinear Liapounov stability, conditions 6a/3(192)
Nonlinear Liapounov stability, related to linear problems 6a/3(194)
Nonlinear natural state, conditions for stability 6a/3(195)
Nonlinear response functions (6a/1)
Bach-Schüle law 111
Bell's parabolic response functions for crystals 543, 551, 563, 639, 704
Bernoulli's parabolic law for gut 13, 72, 112, 126
Cox's hyperbolic law for iron 71, 115
Dupin's hyperbola for elastic line in wood 21
Dupin's nonlinear law for flexure of wood 16, 20
Gerstner's law 30, 31, 108, 423
Hartig's law 91, 105, 107, 118, 120, 127, 131, 136, 137
Hodgkinson's nonlinear response in flexure 25
Hodgkinson's parabolic response in tension and compression 26, 151
Hooke's law, introduced 156
Nelson's law for paints and varnish 126, 127
Rivlin's response function for rubber 735
Sayre's nonlinear law 128
Wertheim's law for organic tissue 60, 115
Nonlinear shear 6(19)
Nonlinear theories of mechanics 6a/3(1, 11)
Nonlinear thermoelasticity, basic equations 6a/3(173–176)
Nonlinear thermoelasticity, equations of perturbed motion 6a/3(176, 177)
Nonlinear viscosity 8/1(231, 241)
Nonlinear viscous flow 6(289)

Nonlinearity in electrostatic transduction 11/2(8)
Nonlinearity in rheology 6(487)
Nonlinearity of damage rate in fatigue 6(609–611)
Nonlinearity parameters 6a/4(277, 278)
Non-local theories 15(314, 315)
Non-local theory, version of Schafroth and Blatt 15(314)
Nonmechanical energy flux 6a/4(118)
Nonmutative processes 6a/4(122)
Non-Newtonian liquid 6(488, 497, 544)
Nonoverlapping 6a/2(402)
Non-self adjoint equations 6a/3(258, 279)
Non-self adjoint problems 6a/3(170, 264, 280)
Nonsimple motions 6a/3(112)
Non-singular initial and equilibrium response 6a/3(26)
Non-steady flow in plane plastic strain 6(427, 432)
Non-superconducting metals 15(195)
Nonsymmetric form 6a/2(395)
Nonsymmetric turbulence 8/2(539–540)
Non-symmetrical cylinder, boundary layer 8/1(273, 275)
Non-trivial solution of the mixed problem with null data 6a/2(109)
Non-uniform liquids 10(81ff.)
— steady states 10(99)
— unsteady states 10(101)
Non-uniform translation of a sphere in a fluid 8/2(238–242)
Non-uniform velocity field along anemometer wire 8/2(566–567, 573)
Non-uniqueness 6a/3(151, 157, 224)
Non-viscous instability of laminar flow 8/1(373, 402)
Norm 6a/2(348)
Normal band 14(292)
Normal coordinate system 6a/2(442, 628)
Normal coordinates 7/1(158, 165, 174, 260)
Normal curvature of a curve 6a/2(628)
Normal deuterium, scattering cross section 12(463)
Normal fluid 15(387)
Normal hydrogen, scattering cross section 12(462)
Normal impedance of a surface 11/1(16)
Normal modes 6a/2(269, 271); 14(202)
Normal modes of lattice 11/2(215)
Normal of slip plane 7/2(22)
Normal, outward 6a/3(47)
Normal solutions of the Boltzmann equation 12(252, 257, 265)
Normal state 13(32, 144, 318)
Normal state in an electrolyte 13(371)
Normal strain 6(140)
Normal strain rate 6a/3(442)
Normal stress 6(137); 6a/3(441); 8/1(6)
Normal stress, critical 7/2(33)
Normal stress, maximum direction 6(354)
Normal system 6a/3(421)
Normal system of differential equations 6(336ff., 340)
Normal to interface, sign convention 10(283–284)
Normal unit of curve 6a/2(623, 627)
Normal unit of surface 6a/2(440, 622)
Normal variables in reaction equations 11/1(214, 217, 218)
Normalized Kelvin process 6a/3(78)
Normalized vector field 6a/2(186)
Normals 6a/3(497)
Nose-shock of an airfoil 9(244, 271, 274)
No-slip condition 8/1(265, 369)
Not hyperbolic 6a/3(526)
Notation 6a/3(3); 6a/4(125)
Notation, tables of 6a/2(2–5, 298, 299, 432–438)
Notations 8/2(2)
Notched bar impact test 6a/1(579, 597)
Novozhilov's derivation of classical linear theory of shells 6a/2(593)
Nozzle design 9(685, 693, 695)
Nozzle design for supersonic flow 9(264)
Nozzle flow 9(147–149)
Nozzle, simple waves in it 9(253)
Nuclear alinement, *see also* Orientation 15(205)
Nuclear demagnetization 15(203)
Nuclear ferromagnetism 15(204)
Nuclear magnetic moments 15(25, 203, 209)

Nuclear magnetization changed by ultrasonics 11/2(238–239)
Nuclear polarization 15(205)
Nuclear quadrupole resonance 7/1(430)
Nuclear reactors 7/1(431)
Nuclear spin 15(85, 209)
— contribution to the partition function 12(108–111)
— resonance 7/1(430, 474)
— systems, interaction with stress waves 11/2(235–241)
Nucleation 10(157, 294, 303)
Nucleation as a first-order reaction 7/2(223–225)
Nucleation, definition 7/2(212)
Nucleation, fluctuations 7/2(218, 219)
Nucleation, free energy 7/2(217, 218)
Nucleation in allotropic transformation 7/2(220)
Nucleation in glass, definitions 13(593, 595)
Nucleation in liquids 7/2(217–218)
Nucleation in recrystallization 7/2(220, 224, 226)
Nucleation in vapours 7/2(217–218)
Nucleation near martensite point 7/2(252)
Nucleation process 15(256)
Nucleation rate, calculation 13(599–600)
Nucleation, Volmer theory 7/2(217–218)
Nuclei, preformed 7/2(221, 240, 253)
Nucleus formation by cavitation 11/2(272–276)
Null boundary data 6a/3(56)
Null solution 6a/3(129, 134, 149, 150, 207, 209, 211, 224, 231, 239, 246–248, 251, 266, 268, 273, 277, 280–284)
Number average molecular weight 13(401, 420)
Number density 10(26, 67); 12(219)
Number density, local variation in transition zone 10(282)
Number density, superficial 10(141, 142, 145, 166, 167, 171, 184, 185, 202, 227)
Numerical computation of free surface flow, *see also* Computation 9(415–438)
Numerical solution(s) 6a/4(58ff.); 14(237)
Numerical techniques 6a/3(3)
Nusselt equation 14(101)
Nusselt number 8/2(560); 14(99)
Nutting-Scott Blair equation 6(480)
Nylon structure 11/1(415)

Objective, *see also* Frame-indifference 6a/2(487)
Oblique derivative problem 6a/2(376)
Oblique entry of a body into water 9(363)
Oblique plane, steady flow on it 8/2(16)
Obstacle between two parallel plates 8/2(211)
Obstacle, forces on it, in Oseen approximation 8/2(292)
Obstacle in plane flow, linear approximation 8/2(204–205)
Obstructions in long canal 9(526ff.)
Obstructions originating diffraction of surface waves 9(543–545)
Occupancy probability 11/2(217)
Occupation (phonon) number, mean 7/1(294, 295ff.)
Ocean floor, waves from its sudden motion 9(617–619)
Oceanography 9(446)
Octahedral shear strain 6(240)
Octahedral shear stress 6(240, 241, 326)
Octahedral shearing stress 6a/3(407)
Oddvalentgroups 14(180)
Odqvist's method 8/2(188–189, 267, 282)
Oil air pump, rotary 12(516, 547, 549–553, 559, 593)
Oil diffusion pump 12(594–605, 647–649)
Oil paint, *see* Paint
Oil vapor, back diffusion 12(649)
Oil vapor, thermal dissociation 12(649)
Oil vapour pump 12(547, 559, 560, 592, 593)
Oldroyd number 6(462)
Ω-integrals 12(393, 422)
One electronmodel 14(193)
One- or two-circle goniometer 7/1(68)

One-dimensional counterpart of kinematical condition of compatibility 6a/3(309)
One-dimensional crystal 13(58)
One-dimensional energy spectra in turbulent flow 8/2(453–454)
One-dimensional flow 8/1(93)
One-dimensional fluid 13(49)
One-dimensional model of Ising 13(108–113)
One-dimensional wave equation 6a/3(160)
One-phase reaction 11/1(207)
One-step dissociation 11/1(261)
Onsager reciprocity law 12(283, 407, 410)
Onsager relation 10(110)
Onsager symmetry relations 6(249)
Onsager theory 15(48, 52, 56, 128)
Onsager theory of the two-dimensional Ising model 13(121–136)
Onsager-Casimir postulate 12(227–229)
Onsager's principle 6a/3(23)
Opalescence, critical 13(71, 75, 76, 189–210, 250)
Opalescence in liquids 10(47)
Opalescence of binary liquid systems 13(316)
Open ball 6a/2(12)
Open jet wind tunnel 8/2(530)
Open system 10(139)
Open-channel flow 9(466)
Open-end conditions 6a/3(506, 508)
Open-end conditions of a tube 6(409, 410, 418)
Operational calculus 6a/3(68)
Optical anisotropy of liquids 10(22)
Optical branches 14(203)
Optical detection interferometer for ultrasound 11/2(143–146)
Optical displacement technique 6a/1(628)
Optical effects of ultrasound 11/2(124–136)
Optical (frequency) branch 7/1(168, 176, 248, 316)
Optical interference experiments, *see* Interference optics experiments
Optical methods for sound absorption measurement 11/1(229)
Optical methods for sound velocity measurements in liquids 11/1(302, 304)
Optical microscopy, *see* Light microscopy
Optical properties 15(271)
Optimal coordinates in boundary layer flow 8/1(329–331, 333)
Optimal design of a rod 6a/2(699)
Optimum suction 8/1(414)
Orbit 6a/3(128)
Orbital stability 6a/3(144, 172)
Orbits, *see* Circular orbits, Elliptical orbits, Particle orbits, Trajectories
Order destruction 7/2(4)
Order of a pair of operators 6a/3(34)
Order of discontinuity 6(342)
Order parameter 15(236, 244, 250)
Order to transformations 13(421)
Order-disorder phenomena in polymers 13(423)
Order-disorder transformation 7/2(25, 188)
Order-disorder transition(= phenomenon of melting) 10(5, 7)
Ordered alloys 7/2(4, 24, 188–191)
Ordering domains 7/2(188)
Ordering transition, effect on work-hardening curve 7/2(65)
Ordinary (normal) processes 7/1(305)
Ordinary sonic point 6a/3(481)
Ordinary spirals 6a/3(472)
Organ pipes 11/2(29)
Organic binary solutions, experimental data 13(326–328)
Organic crystals, critical shear stress 7/2(41)
Organic gases, dispersion and absorption 11/1(182)
Organic tissue, deformation of 6a/1(57–70, 115, 142, 170)
Organic tissue, Wertheim's law for 6a/1(60–65)
Orientation change 7/2(28)
Orientation change of specimen axis in a tensile test 7/2(136)
Orientation, cooperative 10(80)
Orientation, crystallographic 6a/1(516–520)
Orientation dependence
— of critical shear stress 7/2(34)
— of kink band formation 7/2(76, 154)

Orientation dependence (cont.)
— of work-hardening curve 7/2(45–48, 54–57, 59)
— of work-hardening parameters 7/2(55, 59)
Orientation effects in solutions 13(329–344)
Orientation factor 7/2(23, 24, 29)
Orientation fusion 10(81)
Orientation of nuclear moments 15(205)
Orientation of nuclear moments, Gorter-Rose method of 15(205)
Orientation of nuclei 7/2(217)
Orientation of specimen axis with respect to crystal axes 7/2(22, 59, 61)
Orientation ordering 13(215)
Orientation relations of allotropic transformations 7/2(243–246)
Orientation triangles 7/2(24)
Oriented media 6a/2(445)
Orifices for static pressure readings 8/2(587)
Ornstein and Zernike's compressibility integral 13(31, 67)
Ornstein and Zernike's integral equation 13(72–73, 203)
Ornstein and Zernike's integral equation, generalization for binary systems 13(191)
Ornstein and Zernike's method, generalization for ferromagnetics 13(201)
Ornstein-Zernike's scattering formula 13(76)
Orr-Sommerfeld perturbation equation 8/1(116, 370–382)
Orthobaric density 12(43, 45, 53)
Orthobaric volume 12(52, 53)
Orthogonal 6a/2(7)
Orthogonal complement 6a/2(385, 395)
Orthogonal coordinates 8/1(142, 237)
Orthogonal curvilinear coordinates 6(84–86)
Orthogonal cycloids 6a/3(456)
Orthogonal group 6a/2(7)
Orthogonal net of slip lines 6(355, 357–360, 397)
Orthogonal projector 6a/2(392)
Orthogonal transformations 7/1(45)
Orthogonality of characteristic displacements 6a/2(264)
Orthogonality of elementary solutions 9(522, 523)
Orthohydrogen, partition function 12(110)
Orthohydrogen, scattering cross section 12(462)
Ortho-para-H_2 conversion 14(67)
Orthotropic material 6a/2(72)
Orthotropy, elastic constants 6(21–22)
Oscillating basins, resonance waves 9(627, 629)
Oscillating basins, stability 9(651–653)
Oscillating boundaries, wave formation 9(553–563)
Oscillating rotations of a finite cylinder in a fluid 8/2(223)
Oscillating rotations of a sphere filled with liquid 8/2(220)
Oscillating rotations of a sphere in a fluid 8/2(220)
Oscillating submerged body generating waves 9(554–563)
Oscillation of the atoms in the dislocation kernel, possibilities of 7/1(553)
Oscillations of a freely floating body
— oncoming waves 9(566–567)
— steady case 9(563ff.)
Oscillator in a wall generating waves 9(558–559, 616)
Oscillator lattice model of liquids 13(221)
Oscillator, linear 7/1(244)
Oscillator model of relaxation 6(275ff.)
Oscillator model of specific heat 13(571)
Oscillator without wall generating waves 9(560–562)
Oscillatory behaviour of attenuation 11/2(220, 221)
Oscillatory displacement processes 6a/3(106)
Oscillatory flat plate, boundary layer 8/1(294–295)
Oscillatory mainstream around a body of general shape 8/1(296–301)
Oscillatory motion 9(297, 306, 309)
Oseen approximation 8/2(560–561)
Oseen equation 8/2(1, 248, 287–329)
— axisymmetric flow 8/2(291, 294)
— existence and uniqueness of solutions 8/2(320–323)

— obtained from Navier-Stokes equation 8/2(287–288)
— plane flow 8/2(291, 293, 294)
Oseen flow (generalized spirals) 8/2(35–37, 94)
Oseen flows in three dimensions 8/2(102)
Oseen's integral equation applied to the flow along concentric circles 8/2(136)
Oseen's integral equations 8/2(278–279)
Oseen's limit of zero velocity 8/2(1, 329)
Oseen's problem 8/2(331, 333, 340)
— axisymmetric case 8/2(341, 345)
— of the viscosity limit zero 8/2(1, 329)
— plane case 8/2(351)
Oseen's unsteady flow 8/2(148)
Osmotic coefficient for dilute electrolytes 13(378)
Osmotic conditions 10(196ff., 200)
Osmotic pressure 10(196ff., 201, 228); 13(304–306, 315)
— and second virial coefficient 13(315, 316)
— isothermics 13(308)
— normal 10(228)
— of polyelectrolyte solutions 13(498–499)
— of polymer solutions 13(451, 468, 471)
— tangential 10(228)
Osmotic solution 10(198, 201)
Ostwald's rule 7/2(217)
Outer product of tensors 6a/2(617)
Oval cross section bar 6(406)
Oval cross-section 6a/3(503)
Overall yielding 6a/3(516)
Overburden stress 8/2(629)
Overhanging ends 6a/3(531)
Overheat of anemometer wire 8/2(560, 563–564)
Overheat small, linear approximation 8/2(570)
Overshooting 7/2(60, 62, 66, 182)
Oxide films in fatigue striations 6(600)
Oxide glasses 13(514, 518)
Oxide layer 7/2(109)
Oxygen 14(228)
Oxygen bridges 13(519, 587, 591)
Oxygen, dispersion and absorption 11/1(177, 178)
Oxygenic acids 11/1(254, 260, 267)

Packing of grains 8/2(628–629)
Paint as a Bingham body 6(457, 462)
Paints 6a/1(126)
Pair density 13(5)
Pair distribution function 10(179ff., 184, 186ff., 194, 196, 200, 204, 208, 215, 238); 13(3, 13, 198)
— and thermodynamic quantities 13(27)
— for liquids 13(218–219)
— in an one-dimensional crystal 13(61–63)
— in an one-dimensional fluid 13(51)
Pair distribution matrix 12(483)
Pair of voids, diffusion of a 7/1(416)
Pair ⟨*P*, *Q*⟩ belongs to the relaxation function *G* 6a/3(34)
Pairs of kinks in dislocations 11/2(204–205)
Palladium 14(245)
Papkovich solution, *see* Boussinesq-Papkovich solution
Parabola coefficients 6a/1(523, 529, 534–543, 550–565, 634, 639, 666)
— discrete distribution of 6a/1(534–543, 550–562, 563, 639, 672, 695, 702–716)
— linear temperature dependence of 6a/1(535, 556–560, 626, 638)
— quantized distribution of 6a/1(534–543, 550–562, 563, 639, 672, 695, 702–716)
Parabola condition 6a/3(473)
Parabola, flow around it 8/1(101–104)
Parabola limit 6(371, 379–380, 394–395); 6a/3(462, 488)
Parabolic creep 7/2(202)
Parabolic differential equation 8/1(72)
Parabolic point 6(388); 6a/3(481)
Parabolic problem 6a/3(460)
Parabolic problem of differential equations 6(369, 384, 422)
Parabolic response function 6a/1(529, 535, 543, 551, 563, 602, 632, 634, 639, 646, 662, 664, 670, 689, 691, 694, 704, 706)
Parabolic velocity distribution 8/2(14, 15, 49)

Parabolic velocity profile, curve of neutral stability 8/1(382)
Paraboloïd, flow around it 8/2(56)
Paraboloïd, flow inside it 8/2(58)
Parachor 10(174)
Paradox of d'Alembert 8/1(53, 91, 105, 160, 201)
Paradox of Stokes 8/2(257–262)
Para-hydrogen, partition function 12(110)
Parahydrogen, scattering cross section 12(462)
Parallel flow 8/1(368)
Parallel layers of flow 8/2(126–130)
Parallel planes, steady flow between them 8/2(14, 15)
Parallel straight lines
— steady flow along them 8/2(13, 30, 33)
— unsteady axisymmetric flow along them 8/2(149–156)
— unsteady flow along them 8/2(112, 119–127, 149–156)
Parallel type capillaric models 8/2(641)
Paramagnetic absorption 15(3, 19)
Paramagnetic alums 15(44)
Paramagnetic dispersion 15(3, 19)
Paramagnetic effect 15(255, 339, 340)
Paramagnetic molecules 15(10)
Paramagnetic relaxation 15(3, 19, 44, 92, 95, 104, 105, 113, 117, 166)
Paramagnetic resonance 11/2(241–243); 15(88, 92)
Paramagnetic resonance experiments 15(80, 104, 107, 108, 110)
Paramagnetic rotation 15(16)
Paramagnetic samples, data on the thermal insulation 15(65)
Paramagnetic saturation 15(15, 17)
Paramagnetic solids 14(226)
Paramagnetic susceptibility 15(8, 11)
Paramagnetism, normal 15(77)
Parameter of linearization 9(463, 466, 526, 554)
Parameters characterizing glass structure 13(520–521)
Parameters of Heisenberg's theory 8/2(474, 478, 482)
Parametric excitation 6a/3(256, 278)
Parametric stability 6a/3(187)
Parseval theorem 6a/2(366, 381)
Part 6a/2(14, 300)
Parthasarathy's rule 11/1(340)
Partial differentiation with respect to surface coordinates θ^{α} 6a/2(431)
Partial equilibria of diffusion and reaction 11/1(212, 213, 215, 217)
Partial equilibrium 11/1(271)
Partial fluctuation 8/1(369, 371)
Partial pressure of a gas 12(58, 64)
Partial volume per molecule 13(29)
Partial volumes 11/1(209, 210)
Partial volumes of free ions 11/1(263)
Partially permeable coil 13(474)
Particle acceleration in one-dimensional motion 6a/4(26, 49, 53, 55, 58, 66, 96ff.)
Particle acceleration in three dimensions 6a/4(70)
Particle diffusion equation 12(303)
Particle orbits in surface waves 9(496–497, 499, 501, 660, 661, 665)
Particle orbits, second-order infinitesimal wave approximation 9(660–661)
Particle path 8/1(4, 79)
Particle velocity 6a/4(148, 288)
— in one-dimensional motion 6a/4(15, 21, 24, 26, 38, 40–43, 46, 49, 52, 54–57, 66, 82, 84, 88, 96)
— in three dimensions 6a/4(70)
— wave profiles 6a/1(629–633)
Particular solution of the equation of equilibrium 6a/2(59)
Partition 6a/2(27)
Partition function 7/1(235, 243, 269, 270, 287, 289, 316, 317); 10(12, 79, 131, 177, 178, 237, 238, 240, 241, 242); 12(73, 78, 79, 100); 15(41, 42, 80, 84)
— cell method for calculating 10(240)
— grand 10(190, 191ff., 194ff., 206)
— of a chain 13(443–444)
Partly plastic tube 6(411)
Pascalian liquid, *see* Ideal liquid
Past history of a process 6a/3(4, 5)
Past history problem 6a/3(50–53)
($\mathscr{X}$)-past-history 6a/3(4)
Pauli exclusion principle 14(139)
Payne's potential functions 6(78–80)
Pearson's formulation 6a/3(202)
Péclet number 8/1(60)
Peierls' equation 7/1(577, 580, 585)
Peierls' equation, approximation methods 7/1(588)

Peierls' integral equation 7/1(571)
Peierls' model 7/1(514, 546, 563f., 569, 583, 586, 623f., 654)
Peierls', modified integral equation of 7/1(593)
Peierls potential 7/2(222)
Peierls stress 7/1(589, 590, 591, 592, 615, 617); 7/2(124, 132); 11/2(204, 205)
Peierls stress in element- and compound-semiconductors with the coordination number 4 7/1(603)
Peierls stress, temperature dependence of 7/1(602)
Pendular region 8/2(632)
Pendulum bob, containing fluid 9(631)
Pendulum for measuring radiation pressure 11/2(121, 146)
Penetrating twins 7/1(95)
Penetration depth 6a/4(255); 15(241, 243, 244, 245, 246, 282, 290, 330)
— dependence on field 15(245, 246, 330)
— dependence on mean free path 15(301)
— effect of impurities 15(246)
— temperature dependence 15(244)
Penetration law 15(243)
Penetration of butter 6a/1(420)
Penetration of sound into the shadow zone 11/1(82–85)
Penning vacuummeter 12(518, 533–534)
Pérès method to find the plane flow for viscosity limit zero 8/2(356–359)
Perfect fluid, definition 8/1(135)
Perfect fluid, variational principles 8/1(146–149)
Perfect gas 8/1(180)
Perfect gas behaviour, deviation from it, *see* Residual quantities
Perfect gas mixtures 12(57–59)
Perfect gases 12(1–11)
— density matrix 12(192–196)
— measurement of properties 12(28)
— of Bose-Einstein type 12(174ff.)
— of Fermi-Dirac type 12(174ff., 179–182)
— separation of degrees of freedom 12(85–91)
— thermodynamic functions 12(5–8)
Perfect plasticity 6(278, 323, 327, 425)
Perfect-gas scale 12(2)
Performance, coefficient of 14(3, 7, 11, 17, 21, 24, 27, 30)
Performance index of shock test 13(577)
Period 6a/3(129)
Periodic boundary condition(s) 7/1(162, 243, 270, 275); 13(278)
Periodic boundary layer solutions 8/1(295–301)
Periodic duct 11/1(50)
Periodic grid 8/2(525)
Periodic motions 6a/3(129)
Periodic potential, electron in a, 14(229)
Periodic sinusoïdal flow 8/2(129)
Periodic stress and strain processes 6a/3(21, 31, 32)
Periodic structure of a scattering surface 11/1(42)
Periodic system 7/1(117)
Periodic Table and glass formation 13(514, 518, 526)
Periodic volume 7/1(162, 171, 173, 182, 243, 245, 287)
Periodic wave of maximum height 9(732)
Periodic waves in a basin 9(621–622)
Peripheral tension 10(286)
Permalloy, surface phenomena 7/2(99–100)
Permanent deformation(s) 6a/3(538, 542, 550); 7/2(3, 130)
— gradient history 6a/3(539)
— gradients 6a/3(538)
Permanent finite-amplitude wave forms 9(702, 703)
Permanent set, early measurement of 6a/1(175, 223)
Permanent stretching tensor 6a/3(541, 545)
Permeability coefficient in a porous medium 6(266)
Permeability of fused silica to gas diffusion 13(617, 619)
Permeability of porous media 8/2(634, 636)
Permeability of porous media, theories 8/2(641–647)
Permeation of air through envelope of vacuum system 12(633ff.)
Permeation rate of helium through glass 12(634)

Permitted wave number vector, propagation vector 7/1(170, 181, 293, 298, 305)
Permutation symbols 6a/2(453, 455, 618)
Peroxide formation in water 11/2(282)
Perpendicular projection 6a/2(9, 129)
Persistence length 13(406)
Persistent current 15(214, 218, 294)
Persistent stability 6a/3(187, 256, 257)
Perturbation character of acoustical motion 11/1(1)
Perturbation equation(s) 8/1(113–114)
— frictionless 8/1(372, 376)
— of Orr and Sommerfeld 8/1(116, 370, 382)
— of Orr, eigenvalue problem 8/1(370–372, 378)
Perturbation Hamiltonian 14(206, 209)
Perturbation method 9(27)
Perturbation method for surface waves 9(462)
Perturbation of an eigenvalue 11/1(74)
Perturbation of potential flow 8/2(45–47)
Perturbation of pressure 9(28)
Perturbation of visco-plastic flow 6(306–308)
Perturbation sensitive properties 7/2(1)
Perturbations 6a/3(171)
Perturbations, causes of, initial data 6a/3(186)
Perturbations, causes of, loading 6a/3(186)
Perturbations, causes of, material properties 6a/3(187)
Perturbations, method 6a/3(526)
Perturbed motion, basic equations, elasticity
— linear isothermal 6a/3(177–183)
— linear isothermal, isotropic homogeneous 6a/3(180, 181)
— linear isothermal, isotropic homogeneous incompressible 6a/3(181–183)
— linear thermal 6a/3(183–185)
— linear thermal, isotropic homogeneous 6a/3(185)
— linear thermal, isotropic homogeneous incompressible 6a/3(185)
Perturbed motion
— basic equations, elasticity, non-linear isothermal 6a/3(176, 177)
— incremental displacement 6a/3(176, 183)
— incremental rate of work equation 6a/3(178, 180)
— incremental state 6a/3(187–189)
— incremental temperature variable 6a/3(183)
— primary motion 6a/3(179, 186, 256)
— primary state 6a/3(176, 178–180, 182, 184, 188, 189)
— secondary motion 6a/3(178, 179, 186, 187)
— secondary state 6a/3(176, 177)
Pervious wall, suction of boundary layer 8/1(348)
p-H Mollier diagram 14(22)
Phase comparison in dispersion measurement 11/1(227)
Phase contrast, method of 7/1(69)
Phase density 13(7)
Phase diagram 15(34)
Phase dispersion 6a/3(107, 108)
Phase equilibrium in polymers 13(479–491)
Phase factor 6a/4(195)
Phase fluctuation 11/1(87, 89)
Phase function, definition 12(207)
Phase interfaces 7/1(646)
Phase interfaces of different cristalline phases 7/1(660)
Phase jump in critical layer 8/1(377, 387)
Phase lag 6a/3(22)
Phase map 6a/3(131)
Phase separation in polyelectrolyte solutions 13(502–507)
Phase shift between density and pressure 11/1(131, 135)
Phase shift between pressure and displacement in viscous fluids 9(643)
Phase space, definition 12(206)
Phase transformation(s) 7/2(211); 13(46)
Phase transitions 12(96)
Phase velocity 6a/3(107); 6a/4(193); 7/1(160, 161, 170, 567); 11/1(45, 136, 206); 14(203)
— in flexible duct 11/1(120)
— of an edge wave 9(550)
— of disturbance 8/1(370, 371, 373)
— of progressive ocean waves 9(499)
— of viscoelastic waves 6(300, 301)

— of waves at an interface 9(503)
— of waves in a canal 9(502)
Phenakite 13(537)
Phenomenological capillarity equations 10(284–286)
Phenomenological models of polymers 11/1(363–370)
Phenomenological theory of relaxation 11/1(142)
Philips vacuummeter, *see* Penning vacuummeter
Phonon 7/1(304); 14(141, 204); 15(175, 436, 454, 457, 458)
Phonon entropy 15(407, 408, 426)
Phonon gas 14(204, 214)
Phonon-phonon interaction 11/2(230–235)
Phonons 10(123); 11/2(216)
Phonons in glass 13(576)
Phonons in glasses, mean free path of 14(216)
Phonons, polarisation of 14(252)
Phosphate glasses 13(534–540)
Phosphate systems, limits of glass formation 13(615–616)
Phosphor bronze 6a/1(96, 129); 14(296)
Photochemical properties 7/2(43)
Photoconductance of cadmium sulphide 11/2(249)
Photoelastic bench 6(159–161)
Photoelastic materials for stress freezing 6(178, 180)
Photoelastic materials, properties 6(163)
Photoelasticity 6a/1(58, 89, 244, 339); 6a/2(156)
Photoelasticity applied to shock waves 6(227)
Photoelasticity detection and history 6(128–130)
Photoelasticity in a rotating body 6(227)
Photoelasticity, theory 6(145–155)
Photoelectric cell in photoelastic observations 6(161)
Photographs of ultrasonic waves 11/2(137)
Photo-nucleation 13(607)
Physical basis of turbulent diffusion 8/2(515–518)
Physical boundary 9(459)
Physical components of displacement vector 6(5–6)
Physical components of stress tensor 6(10, 12)
Physical components of tensors 6a/2(631)
Physical interpretation of Airy function 6a/2(158)
Physical non-linearity 6(487–488)
Physical plane 6a/3(445, 471, 477)
Physical transfer theories 8/2(471–487)
Piecewise continuous function 6a/2(14)
Piecewise regular 6a/2(14, 25)
Piecewise smooth 6a/2(14)
Piezo-crystal measurement of strain 6a/1(338, 340)
Piezoelectric axes 11/2(78)
Piezoelectric ceramics 6a/4(268)
Piezoelectric coefficients 6a/4(181)
Piezoelectric constants 11/2(78–79)
Piezoelectric crystals, stress waves 11/2(243–251)
Piezo-electric effect 6a/1(352–357, 379, 690)
Piezoelectric effect in liquids 10(70)
Piezoelectric effects 7/1(198, 216)
Piezoelectric equations 6a/4(173)
Piezoelectric *h*-coefficient 6a/4(166)
Piezoelectric material, longitudinal motion of 6a/4(295–302)
Piezoelectric material, shock in 6a/4(297–302)
Piezoelectric microphone 11/2(46–50)
Piezoelectric probe 11/2(115–116)
Piezoelectric transducers 11/2(76–85)
Piezoelectrically excited vibrations 6a/4(257–275)
Piezometer, cylindrical glass 6a/1(247–250, 303–306, 373)
Piezometer, "Mallock method" for measuring the bulk modulus using 6a/1(303, 304, 373)
Piezometer, spherical 6a/1(250, 275–278)
Piezo-optic coefficients 6a/4(166, 185)
Piezotropic fluid 8/1(150)
Piola-Kirchhoff stress 6a/2(51, 53, 229, 305, 491); 6a/3(174, 181, 183, 185, 316, 542)
Piola-Kirchhoff stress, first 6a/4(117, 123)

Piola-Kirchhoff stress, second 6a/4(118)
Piola's theorem 6a/2(60)
Pipe flow 8/1(351, 383, 393)
Pipe flow, resistance coefficient 8/1(354)
Pippard generalization of the London equations 15(285, 299, 301)
Pippard's concept of coherence 15(285)
Pirani resistance gauge 12(526)
Pistol report 11/2(35)
Piston acceleration 9(246–247)
Piston, circular, as a source of sound 11/1(38)
Piston gauge 12(29–30)
Piston generating a wave 9(230, 236, 243, 245)
Piston generating shock wave 9(196)
Piston pump 12(547, 548)
Piston radiator 11/2(2, 104)
Piston, rectangular, inserted in a plane 11/1(79)
Piston, spherical, inserted in a plane 11/1(67)
Pistonphone 11/2(62–64)
Pitching moment 8/2(386, 416)
Pitot tube 8/2(587); 15(437)
π-transformation 7/1(566)
Plait point of a binary mixture 12(65, 66, 67)
Planar coupling coefficient 6a/4(273)
Planar impact 6a/4(3–8)
Planar interface 10(142, 157, 165, 181, 184, 194, 210ff., 264ff., 295–298)
Planck's differential equation 7/1(408)
Plane boundary layer 8/2(549)
Plane boundary surface 11/1(15, 35–44)
Plane cavity flow theory, linearized 9(364–366)
Plane convex set 6a/2(417)
Plane displacement 6a/2(39)
Plane elastic state 6a/2(154)
Plane flow 8/1(66ff., 154)
Plane flow, Oseen's limiting case 8/2(350–359)
Plane jet 8/2(549); 9(341)
Plane of polarisation 6(131)
Plane of vibrations 6(131)
Plane plate, flow around it 8/2(372–383)
Plane plate in presence of a wall 8/2(384)
Plane plate, uniform translation in Oseen approximation 8/2(312)
Plane polarised light, methods of production 6(135, 158)
Plane potential flow 9(11–12)
Plane problem 6a/2(150, 151)
Plane problem of plasticity, complete 6(352)
Plane problems for rods 6a/2(670)
Plane rigid displacement 6a/2(39)
Plane steady flow 8/2(8–47)
— along straight lines 8/2(13)
— linear approximation 8/2(204–206)
Plane strain 6(40–41, 339, 347, 349–350, 353–367, 409–410); 6a/2(151); 7/1(534)
— and stress in visco-elasticity 6(272)
— in plasticity, complete set of equations 6(349)
— of a tube 6(409, 410, 418)
— solution 6a/2(152)
— under general yield condition 6(350ff.)
— under general yield conditions 6a/3(437)
Plane stress 6(40–42); 6a/3(424, 427, 436)
— generalised 6(129, 143–145)
— generalized 6a/2(152, 577)
— with quadratic yield conditions 6a/3(438)
— with qudratic yield condition 6(351)
Plane symmetry 7/1(68)
Plane visco-plastic flow 6(302)
Plane wake 8/2(549)
Plane wall, boundary layer 8/1(266–267)
Plane wave(s) 6a/3(100); 11/2(156)
— longitudinal 6a/3(103)
— of order *N* propagating in a homogeneous, isotropic, viscoelastic material 6a/3(103)
— of sound 11/1(15)
— propagation 6a/3(100)
— representation 12(185)
— transverse 6a/3(103)
— travelling 6a/3(107, 109)
— linearly polarized 6a/4(245)

Planing on a water surface 9(343–344, 587–592)
Plasma oscillations 14(193); 15(355)
Plastic body 6a/3(403)
Plastic compliance 6(290)
Plastic cubic dilatation 6(286)
Plastic deformation 7/2(3); 15(240)
— gradient 6a/3(550)
— mechanism 7/2(18)
— of aluminum 11/2(198–203)
Plastic equilibrium 6(418ff., 425); 6a/3(523)
Plastic flow 6(279, 425, 434)
Plastic flow velocity tensor 6(325, 326)
Plastic foils in electrostatic loudspeakers 11/2(17)
Plastic gel (Schwedoff body) 6(474–475)
Plastic lenses 11/2(114)
Plastic mass between two curved plates 6(383)
Plastic mass between two rigid rough plates 6(366, 429–432)
Plastic mass pressed between two rough rigid plates 6a/3(526)
Plastic potential 6(254, 284, 285, 326, 350, 373); 6a/3(408, 465)
Plastic potential, piece-wise continuous 6(254)
Plastic power 6a/3(546)
Plastic properties 7/1(610)
Plastic shear 7/2(21)
Plastic strength 6(481)
Plastic torsion 6(339, 399–408); 6a/3(424)
Plastic viscosity 6(457, 462)
Plastic work 6(325); 6a/3(407)
Plastic yield in photoelastic materials 6(176, 225–226)
Plastic-elastic boundary 6(400, 403); 6a/3(500, 517)
Plasticity, Coulomb's experiments in 6a/1(174)
Plasticity, dynamic 6a/1(573, 579, 586, 597–649)
Plasticity, general theory of 6a/1(702–716)
Plasticity, incremental theories of 6a/1(440, 679–682, 707)
Plasticity, mathematical theory 7/2(3, 18)
Plastic-rigid boundary 6(366, 383, 399, 422); 6a/3(494, 520, 528)
Plastic-rigid problems 6a/3(523)
Plastic-rigid region 6a/3(501)
Plate, *see* Flat plate
Plate analogy 6(424); 6a/3(521)
Plate, circular, directivity 11/2(104)
Plate driven by point force 11/1(112–114)
Plate equation, visco-elastic 6(310)
Plate, general 6a/3(136, 137, 172, 278, 279, 281)
Plate impact 6a/4(3)
Plate impact, experiments
— characteristic time scale 6a/4(36)
— for acceleration wave studies 6a/4(67)
— steady shock waves 6a/4(41–44)
Plate pressed between two pistons (plastic stress) 6(362)
Plate, rectangular, directivity 11/2(104)
Plate, transmission of sound 11/1(99, 101)
Plate under hydrostatic pressure 6a/2(602)
Plate wave 11/1(113)
Plates, elastic-plastic behaviour 6(320)
Plates of visco-elastic material 6(310, 311)
Plating 7/2(109)
Platinum-silver 6a/1(93)
Playing card model of glide 7/2(71)
Plemelj-Sokhotskii theorem 9(535, 583, 613)
Plexiglass lens 11/2(114)
Pohlmann whistle 11/2(98, 260, 278)
Poincaré inequality 6a/2(350)
Poincaré recurrence theorem 12(219, 226)
Point defects 11/2(188, 189, 193)
Point field 6a/3(7)
Point force driving a plate 11/1(112–114)
Point groups 6a/4(174, 177); 7/1(25)
Point imperfection 14(209, 221)
Point of neutral stability 8/1(371, 380, 394)
— airfoil section 8/1(404)
— calculation for given bodies 8/1(402–408)
Point of transition 8/1(360, 362, 365, 371, 373, 380, 388, 394, 405)

Point of transition, dependence on wall roughness 8/1(441, 443)
Point symmetry 7/1(39)
Point-force 6(76, 88–89)
Points 6a/3(4)
Points, equivalent 7/1(4, 37)
Points, identical 7/1(4)
Poiseuille equation, *see* Hagen-Poiseuille equation
Poiseuille flow 8/1(166, 242, 255, 258)
Poiseuille flow in a cylindrical tube 8/2(48, 152)
Poiseuille flow through porous media 8/2(642)
Poisseuille formula 12(312)
Poisson equation 8/2(66, 67)
Poisson equation for an electrolyte 13(375)
Poisson equation for the pressure 8/2(440)
Poisson ratio 6(25, 142, 177, 212, 329, 449); 13(433)
Poisson ratio, modified 6(50)
Poisson ratio of glass 13(635)
Poisson-Boltzmann equation 10(223); 13(376, 389, 494, 496)
Poisson-Boltzmann equation, linearization 13(376, 388, 392, 394)
Poisson-Kirchhoff theory for bending of plates 6a/2(524)
Poisson's decomposition theorem 6a/2(234)
Poisson's formulae 8/2(431–432)
Poisson's probability law applied on fatigue fractures 6(610–611)
Poisson's ratio 6a/1(85, 91, 135, 180, 248, 258, 259, 261, 262, 266, 267, 272, 274, 281, 283, 287, 293, 378, 393, 732); 6a/2(78, 318, 570); 6a/3(412)
— Cauchy relations 6a/1(245, 259, 274)
— Cornu's direct measurement of 6a/1(264–266, 281–283, 286)
— for generalized plane stress 6a/2(153)
— Kirchhoff's experiment 6a/1(259, 278, 293)
— Poisson-Cauchy theory for 6a/1(247, 249, 254, 256, 264, 268, 269, 272)
— Stokes-Bock hypothesis 6a/1(278–282, 286)
— temperature dependence of 6a/1(278, 293–297)
Poisson's relation between pressure and density 9(6, 61, 93, 284)
Polanyi machine 7/2(32, 118)
Polar coordinates for surface waves 9(475, 476)
Polar curve of an airfoil 8/2(427)
Polar decomposition theorem 6a/4(111)
Polar liquids 10(4, 21, 38, 80)
Polar liquids, sound absorption 11/1(333–336, 338)
Polarisation plane of light 6(131)
Polariscope, circular 6(153)
Polarised light, passage through a crystal 6(133–135)
Polarised light, passage through a stressed plate 6(147–150)
Polarising angle 6(135)
Polarizability 7/1(73); 7/2(177)
Polarizability, molecular 10(72)
Polarizability (of ion-cores) 7/1(222, 287)
Polarization 14(202)
Polarization direction 6a/4(245)
Polarization, electric 7/1(198, 215ff.)
Polarization, linear and circular, of γ-radiation 15(207, 208, 209)
Polarization, microscope 7/1(69)
Polarization of a dielectric 13(251, 257)
Polarization (of a wave) 7/1(171, 173, 181, 266, 293)
Polarization vector 7/1(174, 311)
— of a plane travelling wave 6a/3(107)
— of a singular surface 6a/3(97)
Polarizing power 7/1(73)
Polaroid material 6(135, 158)
Pole dislocation 7/1(661)
Polyampholytes 13(492, 494, 502–503, 506)
Polyatomic gases, propagation of sound 11/1(174–175, 179–180)
Polyatomic liquids 10(65)
Polyatomic molecules, rotation and vibration levels 12(101–108)
Polycrystal, work-hardening curve 7/2(45)
Polycrystalline material 7/1(648)
Polycrystalline samples, molecular distribution functions 13(175)

Polycrystals 10(10)
Polyelectrolyte 13(400, 491)
Polyelectrolyte solutions 13(491–509)
Polyethylene 11/1(414–415)
Polygonal boundary of a jet 9(409)
Polygonization, *see also* Annealing 7/2(193, 194, 195)
Polygonization 6(598, 602)
Polygonization, atomistic interpretation 7/2(195)
Polyisobutylene, measurements 11/1(380–381, 396–399)
Polymer chain 11/1(361)
Polymer molecules in dilute solutions 11/1(362)
Polymer solutions, *see* Solutions of polymers
Polymerization by intense ultrasonics 11/2(285)
Polymers 6a/1(416)
Polymers, fatigue effects 6(593)
Polymers, treatment by intense ultrasonics 11/2(283–285)
Polymethyl metacrylate 11/1(415–417)
Polytropic relation 9(345)
Pore 8/2(626)
Pore size distribution 8/2(627, 633)
Pore volume 6(230)
Pore-fluid 6(231, 265)
Pore-fluid anelasticity 6(263, 265–267)
Porosity 8/2(627, 629)
Porous disk 11/1(68, 75)
Porous media, description 8/2(626)
Porous surface 11/1(15)
Port basin waves 9(623)
Portevin-le Chatelier effect, *see also* Savart-Masson effect 6a/1(41, 230, 649–666, 707, 712)
Position 6a/3(7)
Position, representation in rod theories 6a/2(647)
Position-dependent data 6a/3(264–278)
Position-vector field 6a/2(5, 439)
Positive definite elastic constants 6a/2(153)
Positive definite elasticity field 6a/2(104)
Positive definite elasticity tensor 6a/2(69, 85–87)
Positive definite strain energy 6a/4(247)
Positive longitudinal elasticity 6a/3(317)
Positive operator 6a/2(369)
Positive semi-definite elasticity tensor 6a/2(69)
Positive semi-definite tensor 6a/2(84)
Positive work, theorem of 6a/2(95)
Positive-definite elasticity 6a/3(62)
Positive-definite functions 6a/3(134, 135, 168)
— and continuous dependence 6a/3(145)
— and sets, definition 6a/3(133)
— as measures 6a/3(135)
— discussion 6a/3(186)
— examples 6a/3(133)
— $F_{\tau,t}$ 6a/3(154, 155, 157, 163, 164, 168, 169, 207, 212)
— in calculus of variations 6a/3(164)
Possio's equation 9(302)
Post-buckling behavior 6a/3(125, 195, 202)
Potassium 14(239, 242)
Potassium borate glasses 13(533)
Potassium chrome alum 14(223)
Potassium phosphate glasses 13(539)
Potassium silicate glasses, structure 13(524)
Potential barriers and time dependence of rubber elasticity 13(448)
Potential, chemical 12(73)
Potential, chemical, of a gas mixture 12(61)
Potential density 6a/3(259, 261)
Potential energy of elastic strain 11/2(159)
Potential energy of elastic stress 6(13–14)
Potential flow 9(448, 450); 15(396)
Potential flow, *se also* Irrotational flow 8/1(77)
Potential function 6a/3(259, 260, 287)
— complex, for plane stress 6(48–58)
— of Airy 6(43–47, 50)
— of Boussinesq-Papkovich 6(91–93)
— of Kelvin 6(87–88)
— of Mushkhelishvili 6(48–58)
— of Payne 6(78–80)
— of torsion 6(29–33)
— real, for plane stress 6(78–84)
Potential, grand 10(140, 192, 193)

Potential law of Lennard-Jones 11/1(197, 200)
Potential lines 6a/3(460)
Potential motion 15(392)
Potential of average force 12(156, 157)
— in electrolytes 13(306)
Potential of average forces 10(196, 200ff., 223, 226); 13(9, 11, 16, 25)
Potential of double and simple layers 6a/2(419)
Potential of interaction in the ionic gas model of an electrolyte 13(371, 372)
Potential, plastic 6a/3(408)
Potential, thermodynamic 10(26, 107–108)
Potential vortex 8/1(39, 97–98)
Potential well 13(242, 352)
Potentials, limits to their application to real fluids 8/2(124)
Power and energy 6a/3(95)
Power and energy, theorem of 6a/2(216, 331)
Power, expended per cycle 6a/3(34)
Power expended, theorem of 6a/2(65); 6a/3(95)
Power identity 6a/3(54)
Power input, specific 6(439)
Power law model 12(236, 396, 427, 438)
Power laws to be expected in homogeneous turbulence 8/2(456)
Power radiated by an electrodynamic loudspeaker 11/2(10)
Power radiated from piezoelectric quartz 11/2(81)
Power radiated from turbulent flow 11/1(95, 97)
Power reflected and transmitted by a membrane 11/1(101)
Power scattered by a flexible sphere 11/1(103)
Power scattered by a sphere 11/1(66)
Power, stress 6a/3(95, 174)
Poynting effect 6(261, 509–512); 6a/1(2, 88, 89, 721–726, 739)
Poynting's correction 12(70)
Poynting's theorem 6a/4(156)
Poynting-Thomson body (=anelastic material) 6(456, 471–472)
Prandtl angle 9(250)
Prandtl body (=elastic-plastic body) 6(285, 290, 451)
Prandtl limit in boundary layer theory 8/1(268, 328)
Prandtl number 8/1(60, 239, 245, 246, 336); 8/2(560, 562); 12(301, 394); 14(99)
Prandtl relation for shock waves 9(698)
Prandtl-Busemann characteristic line diagram 12(569–571)
Prandtl-Busemann graphical procedure 9(90–91)
Prandtl-Meyer expansion 9(20, 254, 263, 693)
Prandtl-Meyer flow 8/1(110)
Prandtl-Meyer function 9(688, 690)
Prandtl-Reuss body 6(285, 290, 451)
Prandtl-Reuss equations 6(330, 331, 340, 404, 412, 416); 6a/3(424, 548)
Prandtl-Reuss theory 6(329ff.)
Prandtl's boundary layer approximation 8/2(525, 544–548)
Prandtl's boundary layer equation 8/1(63)
Prandtl's correspondence rules 9(29, 31)
Prandtl's lifting line theory 8/2(407, 418, 427)
Precipitation 7/2(167, 191–193, 198, 211, 226)
Precipitation by room-temperature aging 7/2(220, 228)
Precipitation, cellular 7/2(227)
Precipitation, critical point 13(149–153, 161–166)
Precipitation curve of a binary liquid 13(309)
Precipitation curve of a rigorously regular solution 13(324)
Precipitation fractionation of polymers 13(485)
Precipitation hardening 7/2(192)
Precipitation, inhomogeneous 7/2(227)
Precipitation, kinetics 7/2(227–239)
Precipitation, microscopically homogeneous 7/2(229)
Precooling 14(52)
Preferred direction 15(30)
Preferred reference state 6a/2(530)
Preformed nuclei 7/2(221, 240, 253)
Present state 6a/4(216)
Present state equilibrium values 6a/4(203)

Present state of deformation and stress 6a/3(1)
Present value 6a/3(5)
Pressure, *see also* High pressure
Pressure 6(323, 439, 448); 6a/3(404, 480); 8/1(135, 234, 241); 10(19, 30, 35, 49, 62, 67, 107, 119); 13(12)
Pressure, acoustic 11/1(6, 8)
Pressure at a free surface 9(454, 460, 472)
Pressure at an interface 9(453)
Pressure box 8/2(551)
Pressure calibration by reciprocity method 11/2(68–73)
Pressure coefficient 8/1(103)
Pressure coefficients of elastic constants 6a/1(356, 409)
Pressure, critical 10(66, 133)
Pressure dependence
— of concentration 11/1(211)
— of glass density 13(557–560)
— of reaction equilibrium 11/1(209, 218)
— of relaxation absorption in ammonia 11/1(256–257)
— of relaxation frequency 11/1(270)
— of sound absorption 11/1(318–323)
— of sound absorption in electrolytes 11/1(241)
— of sound velocity 11/1(352–354)
— of sound velocity in gases 11/1(188–191)
— of thermodiffusion 12(333)
Pressure derivatives 6a/4(144)
Pressure determination in vaccum, accuracy 12(518–519)
Pressure determination in vaccum, magnetic method 12(533)
Pressure diffusion 10(84); 12(409)
Pressure distribution
— along a glider 9(592)
— any dependence on time 9(614–615)
— moving with constant velocity 9(597–602, 637)
— on an airfoil 8/2(392–399)
— second-order infinitesimal wave theory 9(665)
— suddenly applied 9(615, 619)
Pressure drag 8/1(90)
Pressure drop 8/1(373, 397–402, 407, 428)
Pressure effect on polymer relaxation 11/1(387)
Pressure effect on superconductivity transitions 15(238)
Pressure equation 8/1(169)
Pressure estimates 9(125, 128)
Pressure fluctuations 8/2(498–501, 546)
Pressure fluctuations on a boundary 11/1(98)
Pressure force 6(323)
Pressure forces 8/2(455)
Pressure gradient 9(236, 242, 259, 263, 268, 279, 281, 282)
Pressure gradient, influence on transition 8/1(365, 372, 396–408, 425, 449)
Pressure, homogeneous 7/1(183)
Pressure in a sound wave 11/2(115)
Pressure, internal, expanding
— a spherical shell 6(422)
— a tube 6(365, 408–418)
Pressure, kinetic 10(132)
Pressure, mean 8/1(234, 241)
Pressure on a wall 8/2(7)
Pressure originating free plastic flow 6(425)
Pressure periodic in time, generating waves
— three-dimensional case 9(593–595)
— two-dimensional case 9(595–597)
Pressure perturbation 9(28)
Pressure point (δ function) 9(599, 602, 637)
Pressure, Poisson equation 8/2(440)
Pressure ratio 11/2(52)
Pressure, reduced 10(133)
Pressure relaxation 11/1(205, 288–291, 297)
Pressure rise 8/1(373, 397–402, 407)
Pressure sensitivity of barium titanate 11/2(48)
Pressure solidification 6a/1(475)
Pressure, Spring's experiments on cohesion of solids under 6a/1(474–477, 482)
Pressure, static 8/1(5)
Pressure tensor 10(33, 35, 82, 157, 159, 161, 164, 167, 208, 209, 210, 211, 212, 213); 12(298, 302, 352)
— normal component 10(158, 159, 164, 210, 218)
— partial 10(41)

Pressure tensor (cont.)
— statistical-mechanical expression 10(208ff., 212, 215)
— tangential component 10(158, 159, 164, 167, 210, 212, 214, 220, 222, 260, 261, 262)
Pressure, thermodynamic 10(132)
Pressure transitions 6a/1(476)
Pressure, uniform 6a/2(50, 77)
Pressure units (table) 12(516)
Pressure values in mechanical pumps 12(556–558)
Pressure variations 15(279)
Pressure-displacement phase shift in viscous fluids 9(643)
Pressure-flow velocity, data on 6a/1(479–483)
Pressure-relaxation 6(265, 266)
Pressure-velocity correlation 8/2(440, 455)
Pressure-volume measurements in several solids 6a/1(490–497)
Pre-stress 6a/3(219)
Pre-stressed beam 6(223)
Prigogine's theory of corresponding states 13(476–479)
Primary chain 13(399)
Primary motion 6a/3(179, 186, 256)
Primary state 6a/3(176–182, 184, 185, 188, 206, 215, 217, 219, 240, 256)
Primary stress 6a/3(198, 215, 239, 241)
Primary wave (*P*-wave) 6(107–109)
Primitive plane of motion 8/2(90–94, 164)
Principal axes of mean stress 6a/3(216, 240–243)
Principal curvatures of a surface 6a/2(628); 10(284, 289–290)
Principal curves in the hodograph 9(89)
Principal direction(s) 6a/2(8, 628); 6a/3(406, 441)
Principal direction of strain 6a/2(35)
Principal direction of stress 6a/2(50)
Principal equations 7/1(564)
Principal function of Görtler 8/1(277)
Principal net of a flow (Mach net) 9(89)
Principal strains 6a/2(35)
Principal stress axis 6(13, 137, 327, 354)
— secondary 6(138)
Principal stress-difference 6(149, 166)
Principal stresses 6a/2(50); 6a/3(406)
Principal tangential stresses 6a/3(407)
Principal trajectories 6a/3(484, 486)
Principal value 6a/2(8)
Principal value of an integral 9(476)
Principal waves 6a/3(326, 328)
Principle of corresponding states 12(48–51, 62, 98)
Principle of decomposition of thermoelastic states 6a/2(313)
Principle of equipresence 6a/2(530)
Principle of frame-indifference 6a/3(544, 545)
Principle of impenetrability 6a/3(536, 538)
Principle of linear momentum 6a/2(483)
Principle of linear superposition 9(55, 57, 63)
Principle of material frame-indifference 6a/2(80, 305, 487)
Principle of maximum
— complementary energy 6a/2(121)
— potential energy 6a/2(121)
— complementary energy 6a/2(112, 130, 212)
— potential energy 6a/2(111, 112, 130, 212, 231)
— virtual mass 9(387, 403)
Principle of production of entropy 6a/3(539)
Principle of superposition for elastic states 6a/2(95)
Principle of superposition for thermoelastic states 6a/2(313)
Prior thermal and mechanical history 6a/1(96, 121, 149, 177–179, 181, 201, 213, 224, 225, 240, 262, 279, 294, 362, 514, 547, 685, 703)
Probability, conservation of 12(207)
Probability density 12(153, 160, 218)
Probability density function 10(178, 191, 237)
Probability distributions in turbulent flow 8/2(442–443)
Probability viewpoints in fatigue fracture 6(595, 607ff.)
Probe microphone 11/2(50–56)
Problem of fixed detachment 9(323, 373)
Problem of smooth detachment 9(323, 372–373, 408)

($\mathscr{X}$)-process 6a/3(4)
Process, class C^{∞} 6a/3(5)
Process, class C^{N} 6a/3(5)
Process, constant 6a/3(5)
Process, continuous 6a/3(5)
Process, continuous, at infinity 6a/3(5)
Process, definition 6a/3(4)
Process, discontinuous 6a/3(6)
Process, discontinuous approximation by continuous 6a/3(16, 35)
Process, dynamic viscoelastic 6a/3(89)
Process, dynamical 6a/2(44, 49)
Process, Heaviside type 6a/3(5, 16)
Process, locally integrable 6a/3(5)
Process, quasistatic viscoelastic 6a/3(42–44)
Process, restricted 6a/3(5)
Process, strong 6a/3(6)
Process, thermodynamic 6a/2(484, 497, 510)
Process, thermoelastic 6a/2(327)
Product of two tensors 6a/2(6)
Production of entropy 6a/2(483, 511)
Production of kinetic energy 8/2(546)
Production of vorticity 8/2(493)
Profile geometry 8/2(387)
Profile of given velocity distribution, calculation 8/2(399–402)
Profile thickness of an airfoil 8/2(387, 392, 429)
Progressive surface wave 9(456)
Progressive wave 6a/2(245)
— second-order infinitesimal wave approach 9(654–662)
— third-order infinitesimal wave approach 9(657–658)
Progressive waves in a canal 9(502)
Progressive waves in an infinite ocean 9(498–501)
Projectile 6a/4(4–7, 56)
Projectile plate, *see* Flyer plate
Projectile, velocity (impact velocity) 6a/4(4, 6, 15ff., 23, 41ff., 47, 56, 89)
Projection, gnomonic 7/1(5)
Projection methods for rod theories 6a/2(663)
Projection, perpendicular 6a/2(9, 129)
Projection property 6a/2(31, 32)
Propagation condition, *see also* Wave 6a/3(98, 105, 316, 324, 326)
Propagation condition 6a/4(219, 221)
— for isotropic bodies 6a/2(247)
— for progressive waves 6a/2(246)
— thermoviscoelastic medium 6a/4(213)
Propagation constant, complex 11/1(135)
Propagation constants of ultrasound, determination 11/2(142–152)
Propagation direction 6a/4(245); 11/2(156)
Propagation direction, misorientation 11/2(160)
Propagation factor 11/2(162)
Propagation method 11/1(229–231)
Propagation methods for sound absorption measurements 11/1(168–169, 172)
Propagation of an initial elevation on a free surface 9(507–513)
Propagation of elastic waves 6a/2(386)
Propagation of energy in surface waves 9(515–522)
— second-order infinitesimal wave theory 9(661)
Propagation of fatigue cracks 6(596, 601, 603, 606)
Propagation of glide 7/2(15)
Propagation of light in a crystal 6(132–135, 151–153)
Propagation of motion from the surface into the interior of a fluid 8/2(126–130)
Propagation of slipe-lines 7/2(15, 107)
Propagation of sound 11/1(6)
— in a visco-elastic fluid 11/1(298–299)
— in anorganic gases 11/1(181–182)
— in gas mixtures 11/1(173, 183–187)
— in highly dilute gases 11/1(155–157)
— in hydrogen 11/1(176)
— in mixtures of noble gases 11/1(173–174)
— in nitrogen 11/1(177–178)
— in organic gases 11/1(182)
— in oxygen 11/1(177, 178)
— in polyatomic gases 11/1(174–175)
— in pure noble gases 11/1(172–173)
— in real gases 11/1(153)
— in the atmosphere 11/1(192–195)
— in three-atomic gases 11/1(179–180)

Propagation of stress waves 11/2(153, 156–159)
— in piezoelectric crystals 11/2(244–246)
Propagation problems 6a/2(371, 377)
Propagation theorems 6a/2(256)
Propagation vector, permitted wave number vector 7/1(170, 181, 293, 298, 305)
Propagation velocity 6(343); 11/2(156)
Propagation velocity of a solitary wave 9(710, 711)
Proper frequency, characteristic frequency 7/1(106, 157, 158, 160, 172, 176, 243)
Proper minimum of *V* 6a/3(205)
Proper orthogonal group 6a/2(7)
Proper vector 7/1(158, 159, 160, 163, 165, 174)
Properly regular 6a/2(14, 38, 351)
Properties, central 7/1(49/51, 56)
Properties, perturbation insensitive physical 7/1(93)
Properties, structure sensitive 7/1(93)
Propionic acid, sound absorption and dispersion 11/1(317)
Proportional limit 6a/1(458, 682)
Proportional loading 6(294)
Propyl-alcohol, sound absorption 11/1(326)
Proton capture in electrolytes 11/1(248)
Proton resonance 15(26, 33)
Protons in ion reactions 11/1(251–254)
Protons, mobility 11/1(216, 248, 251)
Prow problem 9(323, 373, 392)
Pseudo body force field 6a/2(66, 218, 332)
Pseudo heat supply field 6a/2(332)
Pseudoanalytic function 9(104)
Pseudo-axisymmetric steady flows
— of the first kind 8/2(95)
— of the second kind 8/2(96, 98)
Pseudo-axisymmetric unsteady flows
— of the first kind 8/2(170–171)
— of the second kind 8/2(171–174)
Pseudo-critical volume 12(49)
Pseudo-differential operators 6a/2(421)
Pseudo-exact solutions of Davies and Packham 9(736, 744–746)
Pseudo-Maxwellian molecule 12(244)
Pseudo-plane steady flows
— linear approximation 8/2(207–209)
— of the first kind 8/2(84)
— of the second kind 8/2(84, 89)
Pseudo-plane unsteady flows
— of the first kind 8/2(161–163)
— of the second kind 8/2(163–169, 185)
Pseudosteady 6a/3(525)
Pseudo-symmetry 7/1(39)
Pseudo-tension operator 6a/2(420)
Pseudo-turbulence 8/2(37)
Pulsating source in three dimensions 9(475)
Pulsating source in two dimensions 9(479)
Pulsating sphere, radiation impedance 11/2(1)
Pulsating vortex in two dimensions 9(479, 483)
Pulse front 6a/3(96)
Pulse method 11/2(150)
Pulse methods for sound absorption measurements 11/1(303, 304)
Pulse profiles 6a/1(331, 358)
Pulse propagation in the visco-elastic medium 6(300, 313)
Pulse transmission technique 6a/4(31–33)
Pumping action of ionization gauge 12(620–623, 638, 640–646)
Pumping speed 12(629)
Pumping speed of ionization gauge, measurement 12(644–645)
Pumps 12(544–605)
Pumps, general choice of parameters 12(515)
Punch 6a/3(455)
Punch, circular, in plastic material 6(365)
Punch problem in plane elasticity 6(70, 71–72, 76–78, 81)
Pure bending of a plate 6a/2(599)
Pure cavitation, *see* Cavitation
Pure dilatation 6a/3(30)
Pure modes 6a/4(246)
Pure modes in cubic crystals 6a/4(248–250)
Pure shear 6a/2(50, 77)
Pure shear stress 6a/3(30)
Pure shearing motion 6a/3(30)
Pure strain 6a/2(33)

Pure tension 6a/2(50, 78)
Purity of electrolytic solutions 11/1(231)
Push-pull electrostatic transducer 11/2(8, 9, 16)
P-waves (primary waves) 6(107–109)
Pyrex glass 13(453)
Pyrites 6a/1(408)
Pyroelectric coefficient 6a/4(170)
Pyroelectric effects 6a/4(165)
Pyroelectricity 6a/4(181)

Q function 13(181)
Quadratic electro-optic coefficients 6a/4(166)
Quadratic yield condition, *see* Mises yield condition
Quadratic yield condition 6a/3(403, 405, 461, 471, 487, 549)
Quadrupole interaction of an atomic nucleus 11/2(236)
Quadrupole moment of a source 11/1(27)
Quadrupole moment of the nucleus 15(81)
Quadrupole radiation 11/1(5, 27–28)
Quadrupole source 11/2(1, 3)
Quadrupole-quadrupole coupling 15(120)
Qualitative behavior for rods 6a/2(697)
Quantization of lattice waves 14(204)
Quantized distribution of deformation modes 6a/1(553)
Quantized distribution of elastic
— constants 6a/1(151, 308, 397–406, 639)
— shear moduli 6a/1(397–400, 406)
Quantized parabola coefficients and second order transitions 6a/1(534–543, 550–562, 639, 666, 672, 675, 689, 695, 703)
Quantum effects in liquids 10(20, 28, 93, 119–133)
Quantum mechanical effects in transport phenomena 12(452ff.)
Quantum mechanical parameter 10(176)
Quantum statistics 10(122)
Quantum statistics, molecular distribution function 13(64)
Quantummechanical parameter 12(453)
Quarter-wave plate 6(135, 151, 153, 157–158)
Quartz 13(523, 525, 608, 617); 14(222, 224)
Quartz as piezoelectric transducer 11/2(76, 77)
Quartz bar, excitation of hypersound 11/2(101)
Quartz crystal 11/1(371)
Quartz detector 11/1(229, 230)
Quartz, electric constants, temperature dependence 11/2(47)
Quartz gange, *see* Stress transducer
Quartz glass, *see* Fused silica 14(217)
— produced by neutron irradiation of quartz crystals 13(577)
— specific heat 13(572)
— under high pressure 13(558)
Quartz, temperature dependence of ultrasonics 11/2(230, 231)
Quartz-thread pendulum 12(527–528)
Quasi-chemical approximation 10(8)
Quasi-chemical approximation for superlattice transformations 13(94–98, 104)
Quasi-chemical equation 13(93, 96)
Quasi-chemical equation, generalization for oriented molecules 13(337, 339, 341)
Quasi-chemical equation of a rigorously regular solution 13(322, 337)
Quasi-chemical method, application to binary liquids 13(320–326)
Quasi-chemical method for liquids 13(236)
Quasi-chemical method in the free-volume theory of solutions 13(345)
Quasi-chemical theory of ordering phenomena 7/2(178)
Quasi-equilibrium 14(212, 255)
Quasi-equilibrium-distribution 7/1(559)
Quasi-Gaussian hypothesis 8/2(498, 507)
Quasi-homogeneous material 6(435)
Quasi-isotropic material 6(436)
Quasi-linear 6a/3(440)
Quasi-linear partial differential equation 9(14)
Quasi-Lorentz-Gas 12(416)
Quasi-static assumption 6a/3(42)

Quasi-static boundary value problem 6a/3(47, 69)
Quasi-static linear elasticity, first experimental study of 6a/1(197)
Quasi-static linear viscoelasticity 6a/3(42)
Quasi-static test, extended to impact 6a/1(585, 589)
Quasi-static theory 6a/2(311); 6a/3(2)
Quasi-static variational principles 6a/3(64)
Quasi-static viscoelastic process
— corresponding to creep function ***J*** and body force *b* 6a/3(43)
— corresponding to relaxation function ***G*** and body force *b* 6a/3(43, 77)
Quasi-stationary diffusion 12(320, 321)
Quasisteady flow 8/1(62)
Quasithermodynamic theory 10(163, 172, 218, 219ff., 260ff.)
— expression for equilibrium 10(165, 166, 260, 261)
— fundamental postulate 10(163, 165, 166)
— relations for plane interface 10(165)
— relations for spherical interface 10(166, 167)
Quasiviscous flow 7/2(3, 6)
Quenching method 13(598)
Quenching of the orbital moments by crystalline fields 15(8, 79)
Q-value of dissipation 11/2(163, 181)

Radial distribution function 10(15–17, 19, 26, 47–61, 65, 125); 13(6)
— calculation 13(268–277)
— for a gas of rigid spheres 13(10)
— for electrolytes 13(389)
— of liquid argon 13(275, 282)
Radial flow 9(9, 58)
Radial loading 6(281, 287)
Radial motion of thin circular piezoelectric ceramic disks 6a/4(271–275)
Radially symmetric potential flow 9(9, 10)
Radiated wave 11/1(26)
Radiation absorption in glass 13(574)
Radiation annealing 7/1(467)
Radiation condition 9(298, 471, 475, 493, 595, 616)
Radiation damage 7/1(431)
Radiation damage in solids 14(224)
Radiation damping 11/1(228)
Radiation field of ultrasound source 11/2(102–114)
Radiation from a sphere 11/1(61–64)
Radiation from the open end of a duct 11/1(119)
Radiation impedance 11/1(107)
Radiation impedance of a circular piston 11/2(2)
Radiation impedance of a pulsating sphere 11/2(1)
Radiation induced defects 11/2(191)
Radiation mass reactance 11/1(63)
Radiation pressure 11/2(115, 116–123)
Radiation pressure measurement for determination of sound absorption 11/1(303)
Radiation pressure of sound waves 11/1(9, 229, 230)
Radiation resistance 11/1(62)
Radicals, free 15(10, 28)
Radii of curvature 6a/3(446)
Radio resistors 15(186)
Radioactive tracer studies of diffusion through glass 13(621)
Radiometer for ultrasound waves 11/2(121–122)
Radiometer-vacuummeter 12(517, 530–533)
Radius of cavitation bubbles 11/2(264, 267–268, 274)
Radius of gyration of a chain molecule 13(404, 409, 476)
Radius of ionic clouds in electrolytes 13(379)
Radok's solution 6(120)
Raman and Nath's formula 11/2(128–131)
Random chain model 13(403, 419, 453)
Random coil model of polyelectrolytes 13(494–495)
Random loading in fatigue 6(609)
Random network theory 13(527)
Random structure of glass at high temperature 13(524)
Random walk 7/1(408)
Range of correlation of local fluctuations 13(71–76)
Rankine-Hugoniot conditions 8/1(219, 225, 226, 228)

Rankine-Hugoniot curve 8/1(11, 222–223, 225)
Rankine-Hugoniot equation 6a/4(90)
Rao's formula 11/1(346–350, 358)
Raoult's law 13(305, 319, 412, 429)
Rare earth ethylsulfates 15(8, 17, 18)
Rare gas 7/1(105, 112, 247, 285)
Rare gas crystals 7/1(402)
Rarefied gas dynamics 8/2(591 ff.)
Rate at which work is done 6a/3(95)
Rate change, effect on work-hardening 7/2(161)
Rate independence 6a/3(543)
Rate independence of the generalized stress 6a/3(544)
Rate independent materials 6a/3(542)
Rate method of colling 13(548)
Rate of crystal growth in glass 13(513, 595, 597, 600)
— calculation 13(600–601)
— experimental determination 13(597–598, 604)
Rate of dilatation 6a/3(474)
Rate of extension 6a/3(420)
Rate of growth of nuclei 7/2(227–228)
Rate of homogeneous nucleation 13(595)
Rate of reaction 7/2(224)
Rate of work by contact
— and assigned forces and couples 6a/2(481)
— force and director couple 6a/2(544)
Rate of work-hardening, *see* Work-hardening coefficient
Rate of working 6a/2(300)
Rate of working of surface tractions and body force 6a/3(54)
Rate-of-deformation tensor 6(239)
Ratio of the two principal curvatures 6a/1(264–267, 281–283, 287)
Ray 6a/2(249, 257)
Ray of sound 11/1(81)
Rayleigh disc 11/2(59–61); 15(437)
Rayleigh line 6a/4(40ff., 53, 91)
Rayleigh scattering of stress waves 11/2(165, 166, 170)
Rayleigh wave 11/2(76)
Rayleigh waves 6(109–111)
Rayleigh-Ritz procedure 6a/2(125–129)
Rayleigh's higher-order infinitesimal wave approximations 9(663)
Rayleigh's scattering ratio 13(262)
Reaction equation 7/1(484)
Reaction equations, linearized 11/1(213)
Reaction equilibrium 11/1(205)
Reaction equilibrium, dependence on pressure 11/1(209, 218)
Reaction field 15(49)
Reaction force on a piston 11/1(68)
Reaction kinetics of precipitations 7/2(232)
Reaction of first order 7/2(223, 225)
Reaction order 7/2(212, 219, 221)
Reaction product 11/1(208)
Reaction sequences 11/1(212–222)
Reaction time of turbulence 8/2(451, 487)
Reaction velocity 11/1(207, 213, 222–226)
Read's model 7/1(655)
Real crystal 7/1(92ff., 383, 390)
Real gas mixtures 12(59–64)
Real gas, sound absorption and dispersion 11/1(153)
Real gases, critical state 11/1(188–191)
Real media 6a/4(286)
Real potentials for plane stress 6(78–84)
Rearrangement of alloy atoms 7/2(170)
Reasonableness 6a/3(146)
Reciprocal base vectors 6a/2(618)
Reciprocal lattice 13(292)
Reciprocal space 7/1(339, 353)
Reciprocal theorem 6a/3(55)
— for elastodynamics 6a/2(66)
— for elastostatics 6a/2(101, 207)
— for singular states 6a/2(182)
— for synchronous external force 6a/2(219)
— for thermoelasticity 6a/2(320, 332)
Reciprocity condition for Green's functions 11/1(23)
Reciprocity parameter 11/2(71)
Reciprocity relations 9(307)
Reciprocity relations for electromechanical transducers 11/2(66–68)
Reciprocity theorem for the solutions of the Stokes equation 8/2(265)
Recoil vector 11/1(30)
Recombination 7/1(469); 11/1(207)
Recoverable work 6a/3(3)

Recovery, *see also* Annealing 7/2(193)
Recovery accompanying fatigue 6(601)
Recovery, dynamical 7/2(202, 204)
Recovery factor 8/1(421, 426)
Recovery from creep 6(464)
Recovery (mechanical, static, dynamic) 7/2(129–130)
Recovery of electrical resistivity 7/2(194)
Recovery of flow stress 7/2(194)
Recovery rate 7/2(139)
Recovery time of delayed elasticity in glass 13(637)
Recrystallization, *see also* Annealing 6(601); 7/1(470); 7/2(4, 193, 195–197)
Recrystallization in situ 7/2(195)
Recrystallization, nucleation 7/2(220, 224, 226)
Recrystallization temperature 7/2(129)
Recrystallization twins in α-brass 7/2(180)
Rectangular basin, waves 9(623)
Rectangular canal, infinitesimal wave approximation 9(502)
Rectangular canal, shallow-water approximation 9(670–671, 696)
Rectangular coordinates, equations of motion 8/2(5)
Rectangular cross section of a tube 8/2(70, 71)
Rectangular duct 11/1(51–56)
Rectangular duct, coupling with transmitted wave 11/1(121–125)
Rectangular hole, photoelastic fringe pattern 6(209)
Rectangular plate, directivity 11/2(104)
Rectification of air 14(58)
Rectified diffusion 11/2(261, 286)
Rectilinear characteristics 6(377, 392)
Rectilinear diameter, *see* Cailletet-Mathias law
Reduced body force 6a/2(316)
Reduced distribution function 12(208)
Reduced dynamic variables of polymers 11/1(388)
Reduced equation of state 8/1(59)
Reduced equations 8/1(57–60); 12(209–212)
Reduced forms 6a/2(306)
Reduced frequency 8/1(57)
Reduced pressures and temperatures 14(37)
Reduced stress 6a/2(316)
Reduced stress of Hencky 6(489)
Reduced variables 10(172, 173, 182, 186, 187, 188)
Reducible and irreducible cluster-integrals 13(302–303)
Reducible equations 6a/3(459)
Reducible system 6a/3(440)
Reduction method 9(525, 545, 546)
Reduction of constitutive equations under superposed rigid body motions 6a/2(534)
Reemission of helium from molybdenum 12(642)
Re-entrant jet 9(314–315, 323, 397)
Re-entrant jet cavity 9(324, 332–335)
Reference configuration 6a/2(299); 6a/3(173); 6a/4(111)
Reference director 6a/2(447)
Reference pressure 6a/4(137)
Reference spectrum 13(72, 75)
Reference state 6a/3(182)
Reference states, definition 12(11)
Reference surface 6a/2(425)
Reference temperature 6a/2(308, 312)
Reference temperatures of glass viscosity 13(579)
Reflected amplitude 11/1(16)
Reflection coefficient 6a/4(293); 11/1(19, 65)
— of a membrane 11/1(100)
— of a plate 11/1(101)
— of surface wave at obstruction in canal 9(527, 528)
Reflection factor of circular diaphragm within duct 11/1(117)
Reflection from horizontal circular cylinder 9(528)
Reflection method of Shiffman 9(347–349)
Reflection of a continuous disturbance 6a/4(292)
Reflection of a shock wave 6a/4(294, 295)
Reflection of shock waves 9(182, 204–211)
Reflection of simple waves 9(693–694)
Reflection of sound from a plane boundary 11/1(35–44)

Reflection of thermal molecules from surfaces 8/2(596–600)
Reflection of ultrasound waves 11/2(106–108, 110)
Reflection of wave fronts 9(276)
Reflection planes 6a/4(175)
Reflection producing standing waves 11/2(111)
Refraction index, *see* Index of refraction
Refraction index of light in an ultrasonic wave 11/2(125)
Refraction of ultrasound waves 11/2(106–108, 110)
Refraction term in the wave equation 11/1(5)
Refreezing 6a/1(474)
Refrigerant, choice of 14(113)
Refrigerating machines, compressed vapor 14(16, 28)
Refrigerator, magnetic 15(171, 200f.)
Regenerator 14(12, 15)
Regenerator, temperature variations in a 14(106)
Region of determinateness 8/1(68)
Region of influence 8/1(67); 9(19, 40, 233, 256)
Regular domain 6a/2(349)
Regular hypersurface 6a/2(252)
Regular oblique derivative problem 6a/2(376)
Regular point of boundary 6a/2(14)
Regular solution 13(325)
Regular subsurface 6a/2(14)
Regular surface 6a/2(13)
Regularity, preservation of 6a/3(14)
Regularity, theorems 6a/3(74)
Regularization, regularity
— at or near the boundary 6a/2(357, 401, 411)
— in general 6a/2(370, 422)
— interior 6a/2(408)
— rods 6a/2(680)
Rehbinder effect 7/2(109)
Reiner-Rivlin equation 6(460)
Reiner-Rivlin fluid 6(512)
Reissner's plate theory 6a/2(575, 578, 584)
Relations between elastic constants 6(26)
Relative displacement 7/1(214)
Relative motion of source and medium 11/1(76–79)
Relative square fluctuation 7/1(266)
Relative tensors 6a/2(618)
Relative viscosity of solution and solvent 11/1(380)
Relaxation, *see* Glide relaxation
Relaxation 7/1(457); 11/1(14); 15(433)
Relaxation absorption in ammonia 11/1(255)
Relaxation absorption in sulphuric acid 11/1(253)
Relaxation and creep laws 6a/3(23)
Relaxation and transport phenomena, relation 12(381–384)
Relaxation body, simple 11/1(147)
Relaxation, compressional 11/1(288–291)
Relaxation constant 15(20)
Relaxation, dielectric 11/1(325)
Relaxation distance 8/2(542–543)
Relaxation effects 15(145, 147, 162)
Relaxation equation 6(291, 292); 11/1(144, 284, 289)
Relaxation formula for sound propagation 11/1(285–287)
Relaxation frequency 6(275); 10(103, 114); 11/1(206)
Relaxation frequency in electrolytes
— dependence on concentration 11/1(215, 216, 220)
— dependence on kind of ions 11/1(239)
Relaxation function 6(273, 274ff.); 6a/4(196); 11/1(145–146)
— and creep compliance 6a/3(28)
— for creep function (compliance) J 6a/3(26)
— $\boldsymbol{G}$ 6a/3(18)
Relaxation functional 6a/3(17)
Relaxation in an ideal gas 11/1(148–149)
Relaxation in electrolytes 11/1(202)
Relaxation in glasses, *see also* Mechanical relaxation 13(635–644)
Relaxation in strained metals 7/2(223)
Relaxation in the standard solid 6(268)
Relaxation law at x: $\mathscr{L}^x$ 6a/3(17)
Relaxation law for body $\mathscr{B}$ 6a/3(17)
Relaxation leading to Bordoni peaks 11/2(203)

Relaxation lengths 14(215)
Relaxation method of approximation 9(255, 434, 436)
Relaxation model of Maxwell 8/1(50)
Relaxation modes 6(266)
Relaxation of phonon distribution 11/2(234)
Relaxation part of compressibility 11/1(206, 209–212, 217)
Relaxation, phenomenological theory 11/1(142)
Relaxation problems, variational principles 6a/3(65, 66, 113)
Relaxation processes 10(82, 95); 11/1(282–283)
Relaxation processes, complicated 11/1(161–165)
Relaxation spectra 6(273–278, 312)
Relaxation spectrum 13(445–448)
Relaxation spectrum of polymers 11/1(361, 370)
Relaxation, thermal 11/1(283–285)
Relaxation, thermal, due to slow chemical processes 11/1(287–288)
Relaxation, third 15(23)
Relaxation time(s) 6(465, 479, 481); 7/1(309); 8/2(542–543); 10(71); 11/2(174); 14(184, 205, 230); 15(19, 21, 43, 57, 72, 126, 135, 148, 203)
— and differential operator 6a/3(37, 38)
— (characteristic relaxation time) 6a/4(36, 45ff., 51ff., 64)
— combination of 14(220)
— direct determination 11/1(170–171)
— distribution 11/1(291, 299, 326)
— effective 14(213, 252)
— for a Maxwell element 6a/3(41)
— for establishing equilibrium in a network 11/1(392, 406)
— in gases 11/1(133, 143, 159, 160)
— in molecular flow 8/2(600, 601)
— mechanical theory (mechanical relaxation time) 6a/4(87, 96)
— of a gas 12(368)
— of a single mode 14(212)
— of inner molecular motion 12(497, 501)
— of ion cloud 11/1(224)
— theory 11/1(195–201)
— thermal 14(295)
— thermodynamic theory (thermal relaxation time) 6a/4(87ff., 96)
— variable 6(291)
Relaxational equations 11/1(403)
Relaxing gel (Jeffreys body) 6(468–471)
Relaxing media 6(269)
Rellich selection principle 6a/2(352); 6a/3(576)
Rellich's lemma 6a/2(20)
Remaining equations for shear response 6a/3(71)
Remaining equations of plasticity 6a/3(426, 456, 494)
Remanent magnetic moments 15(72, 73, 127, 132, 135, 144, 148, 150, 152, 162, 164)
Remanent polarization 6a/4(270)
Removal 11/1(223)
Reorientation of crystallites 7/2(195)
Representation of biharmonic fields at infinity 6a/2(22)
Representation of dilatation and rotation 6a/2(188)
Representation of displacement field 6a/2(160)
Representation of displacement gradient 6a/2(187)
Representation of elastic displacement fields at infinity 6a/2(165)
Representation of harmonic fields at infinity 6a/2(22)
Repulsive potential 7/1(130)
Rescaling function 6a/3(542, 543)
Residual displacements 6a/3(219)
Residual energy 12(53, 54, 55)
Residual entropy production inequality 6a/3(175, 183, 188)
Residual equation of energy balance 6a/3(174, 175, 183)
Residual free-energy 12(60)
Residual gases in high vacuum 12(634–635)
Residual heat capacity at constant volume 12(47, 56)
Residual pressure 12(12)
Residual rays 7/1(375)
Residual resistance 14(145, 232, 246)
Residual resistance in the superconducting state 14(270)
Residual stress 6a/2(81, 308)
Residual stress pattern 6(130)

Residual volume 12(13–16)
Resistance anomalies and thermal conductivity 14(247)
Resistance coefficient as function of Reynolds number 8/2(303–304, 308, 310)
Resistance coefficient for pipe flow 8/1(354)
Resistance due to impurities, defects and structural changes 14(145)
Resistance due to physical and chemical defects 14(147)
Resistance due to scattering by electrons 14(253)
Resistance, electrical 7/1(379)
Resistance from zero-point vibrations 14(144)
Resistance gauge after Pirani 12(526)
Resistance, ideal 14(232)
Resistance, ideal, in the superconducting state 14(269)
Resistance, lattice model for electrical 14(168)
Resistance measurement by bridge methods 14(154)
Resistance measurement of single crystals 14(165)
Resistance measurement with potentiometers 14(155)
Resistance model of porous media 8/2(643)
Resistance of a homogeneous (random) solid solution 14(151)
Resistance of a plane plate 8/2(378)
Resistance of gold 14(152)
Resistance of interstitials, electric 7/1(429)
Resistance of metals 7/1(444)
Resistance of silver 14(152)
Resistance, Stokes' law 8/2(229)
Resistance, survey of formulas 8/2(323–329, 383)
Resistance thermometer(s) 8/2(586); 14(201)
— of carbon 15(185)
— of phorphorbronze 15(185, 189, 190)
Resistance unter pressure 14(142)
Resistance variation by irradiation 14(150)
Resistance-minimum at low temperatures 14(152, 188)
Resistive transition of superconductors 15(288)
Resistively terminated probe microphone 11/2(53–54)
Resistivity, electrical, of dislocation 7/2(166, 167)
Resistivity, electrical, recovery 7/2(194)
Resistivity, electrical, under deformation 7/2(165)
Resistivity in a magnetic field 14(243)
Resistivity minimum, electrical 7/2(1)
Resolution of an anemometer 8/2(558, 578–585)
Resolved shear deformation 6a/1(515, 521)
— defined 6a/1(516)
Resolved shear response function 6a/1(529, 543)
Resonance, antiferromagnetic 15(32)
Resonance frequency 6a/4(263)
Resonance in enclosures 11/1(56–61)
Resonance methods 11/1(227–229)
Resonance methods of ultrasound measurement 11/2(146–150)
Resonance, paramagnetic 15(3, 24)
Resonance waves for oscillating basins 9(627, 629)
Resonant bar transducer 11/2(23)
Resonant cavity 11/2(100–101)
Resonant frequency of a quartz crystal 11/1(373)
Resonant vibration in fatigue experiments 6(591)
Resonator containing absorbing electrolyte 11/1(227–228)
Response function
— in elasticity 6a/2(80)
— in shell theory 6a/2(529, 531)
— in thermoelasticity 6a/2(301, 302, 304)
— or functional 6a/3(11–13)
— prediction of aggregate from single crystal 6a/1(666–676)
Response of fiber and whole muscle 6a/1(142)
Response of isotropic materials 6a/2(74)
Rest history (zero strain history) 6a/4(31, 66)
Restrained deformation 6a/3(523)
Restricted cone-hypothesis 6a/2(382)

Restricted form of the constitutive equations for an isotropic material 6a/2(557)
Restricted histories 6a/3(5)
Restricted theory (kinematic) of shells 6a/2(455)
Restriction of response functions under reflection of director 6a/2(540, 542)
Restriction on stress due to uniform dilatation 6a/2(73)
Restrictions placed on elastic materials by the second law 6a/2(302)
Resultant assigned couple 6a/2(504)
Resultant assigned director couple 6a/2(481)
Resultant assigned force 6a/2(481)
Resultant body force 6a/2(509)
Resultant contact couple 6a/2(504)
Resultant contact director couple 6a/2(480)
Resultant contact force 6a/2(480, 510)
Resultant couple vectors 6a/2(514)
Resultant force 6a/2(514)
Resultants, catalogue of definitions 6a/2(526)
Resultants, specific internal energy and entropy 6a/2(515)
Retardation 7/1(208, 216, 219)
Retardation of strain, *see* Creep 6(463)
Retardation (of strain after load), *see* Incubation period
Retardation, relative, of optical paths 6(149, 153–155, 171)
Retardation spectrum 13(445)
Retardation time 6(269, 463, 481)
Retardation times 6a/3(38, 42)
Retarded elastic media 6(269ff.)
Retarded potenial 11/1(26, 77)
Retrograde condensation 12(66, 67, 68)
Retrograde evaporation 12(66, 67)
Return to isotropy 8/2(538)
Reuss equations, theory etc., *see* Prandtl-Reuss equations theory etc.
Reverberation 11/2(152)
Reverberation in electrolytes 11/1(228–229)
Reverberation in enclosures 11/1(56–61)
Reverberation method 11/1(168)
Reverberation method for sound absorption measurements 11/1(303, 304)
Reverse creep 7/2(133)
Reverse flow theorem 8/2(423)
Reverse fluctuations 7/2(126)
Reverse strain 7/2(133)
Reversibility, microscopic 10(83)
Reversible change of an interface 10(289–290)
Reversible change of flow stress 7/2(161)
Reversible elastic phenomena 6(14, 15)
Reversible expansion of a gas 12(8, 27, 28)
Reversible (quasi-static) process 15(43)
Reversing switch, superconducting 14(163)
Reversing switch with mechanical contacts 14(163)
Reversion of age-hardening 7/2(235)
Revolution solid, symmetrical strain 6(105)
Reynolds' experiment 8/1(352)
Reynolds' hypothesis 8/1(366, 388)
Reynolds number(s) 6(462); 8/1(58, 64, 113ff., 181, 244, 246, 265, 268, 288); 8/2(526, 560, 592, 611); 9(316); 11/1(155); 14(99)
— behavior along flow 8/2(548–550)
— critical 8/1(118, 353, 355, 360, 366, 376, 380, 383–385, 401, 417)
— critical, effect of heat transfer 8/1(421)
— critical, effect of suction or blowing 8/1(411)
— critical, effect of wall temperature 8/1(427)
— critical, for a flat plate at zero incidence 8/1(424)
— critical, for concave walls 8/1(435)
— critical, in supersonic flow 8/1(426)
— critical value for Darcy's law 8/2(647)
— experimental critical 8/1(371)
— for a tube 8/2(48)
— for boundary layer thickness 8/2(544)
— for plane flow 8/2(14, 16, 21, 25)
— in fully turbulent regions 8/2(546)
— large 8/2(1, 28, 29)

— limits for validity of Lamb's solution 8/2(307, 308)
— limits for validity of Oseen equation 8/2(288, 299)
— limits for validity of Stokes equation 8/2(200)
— of turbulence 8/2(461)
— problem of the rotating disk 8/2(107, 109)
— series expansion in its powers 8/2(304)
— small 8/2(1, 27, 29)
— small, approximations 8/2(200ff., 209, 214, 227)
— spectrum 8/2(527)
— theoretical critical 8/1(371)
— variation along flow 8/2(542)
Reynolds stress(es) 8/1(112); 8/2(439–440, 466, 525); 11/1(5)
Rheological bodies, complex 6(478ff.)
Rheological bodies, definitions 6(454–456)
Rheological bodies, superposition 6(478ff.)
Rheological equation of a material 6(440)
Rheological equation, quasi-linear 6(488, 502)
Rheological formula of a material 6(455)
Rheological fundamental properties 6(447)
Rheological tree 6(456)
Rheology of porous media 8/2(629)
Rheology, phenomenological 6(435)
Rheometry 6(535–542)
Riabouchinsky cavity 9(323, 324, 335–337, 398, 403)
Riabouchinsky cavity, general case of axial symmetry 9(337)
Riabouchinsky flow past a circular disk 9(427)
Riabouchinsky's solution 8/2(41, 44)
Ribbon loudspeaker 11/2(12)
Ribbon microphone 11/2(56–58)
Riccati equation 9(174, 204)
Riccati's equation 8/2(59, 62)
Riccati's ratio of E moduli 6a/1(161, 190)
Richardson number 8/1(418)
Richardson's law of dispersion 8/2(509–510, 514–515)
Richardson's law of dispersion, generalization 8/2(519)
Riegels factor 8/2(397)
Riegels-Truckenbrodt method 8/2(399)
Riemann domains 9(65)
Riemann function 6(360, 361); 6a/3(449, 450); 9(239, 240, 269, 272, 277, 354)
Riemann integration method 6(360ff.)
Riemann invariant(s) 6a/4(278); 9(677–679, 687ff.)
Riemann method 6a/3(449)
Riemann representation 6a/3(104)
Riemann-Christoffel tensor 6(7); 6a/2(620)
— in space 6a/2(620)
— of a surface 6a/2(462)
Riemannian space 8/1(143–144)
Riemann-Lebesgue lemma 6a/3(21)
Riesz-Fredholm theory of compact operators 6a/2(369)
Right Cauchy-Green tensor 6a/4(70, 111)
Right inverse 6a/3(23)
Right material derivative 6a/3(541)
Rigid atoms 7/2(9)
Rigid body 6(441)
Rigid boundary 6a/4(294)
Rigid boundary surface 11/1(15)
Rigid displacement 6a/2(31–33, 61, 68, 104)
Rigid field on ∂B with force and moment 6a/2(185)
Rigid kernel of slab between rough plates 6(431–432)
Rigid motion 8/1(139)
Rigid motions, rotation 6a/4(111)
Rigid, rigid-body motions 6a/3(17, 18, 23, 216, 246, 262, 266, 274)
Rigid rotations 6a/3(215, 216, 219, 239, 240, 242, 244, 246)
Rigid sphere model 12(340, 343, 394–395, 427, 437)
Rigid sphere, scattering of stress waves 11/2(164)
Rigid surface, boundary conditions 9(454–455)
Rigid translational velocities 6a/3(218, 220)
Rigid translations 6a/3(215, 217, 219, 220, 239)

Rigid-body displacement, infinitesimal 6(7)
Rigidity modulus 6(25–26, 142, 448)
Rigidity modulus, anomalous decrease 7/2(131)
Rigidity of liquids 10(5, 69–71, 81)
Rigid-plastic analysis 6a/3(524)
Rigid-plastic approximation 6(281, 316, 319ff.)
Rigid-plastic problem 6a/3(495)
Rigid-sphere gas, interaction and average force potentials 13(11)
Rigid-sphere gas, radial distribution function 13(10)
Rigid-sphere model, an approximate solution of the Kirkwood equation 13(267)
Rigid-sphere model of a liquid, equation of state 13(247, 276)
Rigid-sphere model of one-dimensional fluid 13(52)
Rigorous solutions of the Navier-Stokes equation 8/2(11–13)
Rigorously regular solution 13(317, 320, 322–324, 345)
— quasi-chemical equation 13(322, 337)
Ring 6a/2(415); 6a/3(424, 515)
Ring, circular, directivity 11/2(104)
Ring (rod) 6a/2(659, 697)
Ring, superconducting 15(216)
Ring vortex distribution 9(366)
Ringleb flow 8/1(197)
Ring-mechanism 7/1(423)
Ring-shaped basin, shallow-water approximation 9(672)
Ring-shaped basin, waves 9(624)
Ripples (capillary waves) 9(633)
Rise time of a pressure wave 11/2(32)
Ritz's representation theorem 6a/3(558)
Rivets, *see* Bolts
Rivlin and Saunders' experiments 6a/1(734–739)
Rivlin's theory for the finite elasticity of rubber 6a/1(734–739)
Rochelle salt 11/2(76)
Rock salt, irradiation 11/2(191–195)
Rocking beam oscillator 11/1(379)
Rocksalt 6a/1(355, 407)
Rock-salt crystals, critical shear stress 7/2(41)
Rock-salt crystals, work-hardening 7/2(69–71)
Rock-salt, effect of cadmium impurities 7/2(125)
Rock-salt structure 7/2(25, 26)
Rod, *see* Bar and Beam
Rod 6a/3(172, 279, 282, 283, 285)
Rod, definition 6a/2(646)
Rod, definition of theory 6a/2(641)
Rods of visco-elastic material 6(313)
Rolling moment of an airfoil 8/2(386, 408, 416)
Rolling of a ship 9(555)
Roof 6a/3(500)
Roots pump 12(515, 553–555, 559)
Rotary alternating axis 7/1(7)
Rotary inertia in flexure of beams 6a/1(384)
Rotary inversion axis 7/1(7)
Rotary piston pump 12(552, 553)
Rotating body, photoelastic behaviour 6(227)
Rotating bucket 15(319)
Rotating coaxial cylinders, transition to turbulence 8/1(415, 416, 430)
Rotating cylinder 6(423); 8/2(18)
Rotating cylinder in presence of a wall 8/2(218–220)
Rotating cylinders and disks 6a/3(521)
Rotating disc, boundary layer 8/1(316–317)
Rotating disc, transition to turbulence 8/1(437–439)
Rotating disk 6(423); 8/2(104)
Rotating disk in a steady flow 8/2(217–218)
Rotating ellipsoïd in a steady flow 8/2(216–217)
Rotating of the earth 8/2(109, 110)
Rotating sphere in a steady flow 8/2(214)
Rotation 8/1(23–25)
Rotation and stretch of material line and surface elements 6a/4(112)
Rotation axes 6a/4(175)
Rotation axis 7/1(7)
Rotation field 6a/2(31)
Rotation, infinitesimal 6(438)
Rotation levels of diatomic and polyatomic molecules 12(101–108)
Rotation of chain elements 13(405)

Rotation of the plane of polarization 15(15)
Rotation, rigid 6a/4(111)
Rotation tensor 6(238); 6a/4(111)
Rotation transformation 13(210–216, 335)
Rotation vector 6a/2(31, 39)
Rotational balance 6a/3(42, 216, 242, 243)
Rotational invariance 8/1(28–31)
Rotational isomers, sound absorption 11/1(313–314)
Rotational partition functions for diatomic molecules 12(111–113, 122)
Rotational partition functions for polyatomic molecules 12(113–115, 122)
Rotational wave equation 6a/2(328)
Rotational waves, existence theorems 9(755–756)
Rotation-inversion axes 6a/4(175)
Rotatory inertia 6a/2(578, 596)
Rotons 10(123); 15(175, 391, 392, 423, 433, 454, 457, 458)
Rottery's rule 7/1(430)
Rough sphere model 12(484, 497–499)
Rough surface, scattering 11/1(42–44)
Roughness element 8/1(442–445)
Roughness of walls 8/1(440–450)
Round jet 8/2(549)
Round wake 8/2(549, 551)
Rouse-Bueche-Zimm theory of polymer solutions 11/1(369, 383–386, 389, 395, 401)
Royal Iron Commission 6a/1(70)
Rubber 6a/1(44, 53, 109, 115, 245, 274, 308, 497, 720, 726–741, 746); 11/1(361, 409–413)
Rubber, compressibility of 6a/1(246, 274, 497, 732)
Rubber elasticity 6(16, 18, 492–494); 13(435, 442, 488)
Rubber, finite elasticity of 6a/1(726–741)
Rubber, Rivlin and Saunders' experiments 6a/1(734–739)
Rubber structure 6(18, 450, 493)
Rubber, tensile deformation of 6a/1(53, 109, 115, 733)
Rubber, wave propagation in 6a/1(308–313, 716, 720, 733)
Rubidium 14(239)
Rudzki's inverse method 9(737–738)
Rudzki's transformation 9(727)
Rupture 6a/1(5, 11, 17, 59, 167, 186, 211, 234, 239, 273, 442)
Rupture, Coulomb's comparisons in 1773 of rupture in tension with rupture in shear 6a/1(167)
Rupture strength, theoretical 7/2(14)
Rutgers' relation 15(279)

S matrix 12(471)
S wave, *see* Transverse wave
Saddle point, configuration of 7/1(472)
Sage and Lacey apparatus for compressibility measurements 12(32)
Saint Clair generator 11/2(23–24)
Saint Clair's elektromagnetic transducer 11/2(92–93)
Saint-Gobain glass 6a/1(264)
Saint-Senant-Tresca yield condition 6(286ff.)
Saint-Venant body (=rigid-plastic body) 6(441, 451–452)
Saint-Venant problem 6a/2(661)
Saint-Venant theory of rod impact 6a/1(313, 314, 319, 320, 327)
Saint-Venant-Lévy-Mises relations 6(323)
Saint-Venant's conjecture 6a/3(562)
Saint-Venant's principle 6(203); 6a/1(264, 270, 351); 6a/2(190–207); 6a/3(85)
— and potential theory 6a/3(87)
— bodies of general shape 6a/3(87)
— decay of stresses 6a/3(87)
— for a filament 6a/3(88)
— for viscoelastic bodies 6a/3(86)
— of elastostatics 6a/3(88)
— pointwise estimate 6a/3(88)
— preferred direction 6a/3(87)
Saint-Venant's solution 6a/3(87)
Saint-Venant's theory of torsion 6a/1(88, 202)
Saint-Venant-Tresca yield condition 6(283, 320, 328)
Saint-Venant-Wantzel formula 9(7)
Salts, anhydrous 15(28)
Samoilov's method 15(192)
Samples, paramagnetic, *see also* Specimens 15(63, 65)
Sand hill analog of Nadai 6(402, 424)
Sand roughness 8/1(449)
Sandstone 6a/1(83, 86, 486–490)

Sapphire 14(222)
Sapphire, synthetic 14(223)
Saturation, *see* Paramagnetic 15(15, 17, 26)
Saturation region 8/2(632)
Sauders' viscosity constant 11/1(348)
Sautreaux' inverse method 9(736–737, 742)
Savart-Masson effect 6a/1(41, 230, 459, 475, 477, 578, 649–666, 699, 707, 710, 712)
Savart-Masson effect at high pressure 6a/1(475, 477)
Sayre's nonlinear law 6a/1(128)
Scalar field 6a/2(10); 6a/3(7)
Scalar product 6a/2(399)
Scalar solid spherical harmonic 6a/2(21)
Scalars defining correlation tensors 8/2(447, 449, 491)
Scale of turbulence 8/1(360, 384–386, 426, 445)
Scatter in fatigue fracture 6(595)
Scattered light method of stress determination 6(193–202)
Scattering amplitude matrix 12(472)
Scattering by static imperfections 14(208)
Scattering by the internal boundaries 14(275)
Scattering cross section 11/1(30)
— effective 12(490)
— of a flexible sphere 11/1(103)
— of a sphere 11/1(66)
— of particles in gases 12(457, 461)
— for electrons 14(151)
Scattering, "diffuse" 14(185)
Scattering, "elastic" 14(185)
Scattering, energy-dependent 14(192)
Scattering from a disk 11/1(68–73)
Scattering from a flexible sphere 11/1(101–104)
Scattering from a rough surface 11/1(42–44)
Scattering from a sphere 11/1(64–68)
Scattering from an inhomogeneity in surface impedance 11/1(40–42)
Scattering in enclosures 11/1(73–76)
Scattering, inelastic 14(144)
Scattering matrix, symmetry relation 12(476)
Scattering of a compressional wave by an elastic sphere 11/2(169–170)
Scattering of a transverse wave by an elastic sphere 11/2(170–172)
Scattering of elastic waves 11/2(159–160, 163–172)
Scattering of electrons in superconductors 11/2(222)
Scattering of light
— as a means to determine molecular weights 13(401, 475–476, 497)
— by liquids 13(262)
— by multi-component systems 13(310–316)
— in liquids 10(30, 32, 47)
— in polyelectrolytes 13(500–501)
— near critical point, *see also* Opalescence 13(76–77)
— near the critical point 12(47, 48)
Scattering of particles in gases, classical limit 12(460)
— quantummechanical treatment 12(455–485)
Scattering of phonons by conduction electrons 11/2(216)
Scattering of sound in a turbulent region 11/1(86, 90)
Scattering of sound waves 11/1(28–34)
Scattering of spin waves by lattice vibrations 11/2(229)
Scattering of x-rays at binary alloys 13(180–196)
Scattering of x-rays by liquids 13(248–251)
Scattering phases 12(456)
Scattering, representation of the cross section for viscosity and diffusion 12(458–461)
Scattering term in the wave equation 11/1(5)
Scattering, thermal 7/1(354)
Schaaffs' theory of sound velocity 11/1(340–346)
Schauder's fixed-point theorem 6a/3(577)
Schleiermacher's method of a coaxial wire 12(316)
Schlieren method 9(38); 11/2(139–142)
Schlieren photograph 11/2(29)
Schlieren photograph of a cavitation shock wave 11/2(266)

Schmid's law of critical shear stress 7/2(33)
Schmid's law of critical shear stress, deviations 7/2(34)
Schmulewitsch's conjecture on the Gough effect 6a/1(309–311)
Schneidentöne 11/2(27)
Schofield-Scott Blair body 6(475–478)
Schottky transitions 14(283, 331)
Schottky vacancies 13(77)
Schrödinger wave equation for quantum liquids 10(127)
Schwarz alternating method 6a/2(421)
Schwarz-Christoffel transformation 9(326)
Schwarz's inequality 6a/3(559)
Schwedoff body (=plastic gel) 6(456, 474–475)
Scratch hardness of glass 13(632)
Scratch resistance of ceramics 13(606)
Screen of fine mesh 8/2(533, 536)
Screened Coulomb potential 14(176, 193)
Screening 7/1(473, 628)
Screening in metals 15(346)
Screening of ionic charge 14(170, 175)
Screening of the foreign ion 14(151)
Screening temperature 14(176)
Screens producing uniform flow 8/2(551–552)
Screw acis 7/1(10)
Screw dislocation 7/1(96, 481f., 499, 535, 572, 596, 611)
— cross-slip of 7/1(613)
— extended 7/1(613)
— in a plate 7/1(584)
— in stage II 7/2(88)
— stress-field of the 7/1(534, 556)
s-d band scattering 14(173)
s-derivative 6a/2(350)
Sea water, absorption of sound 11/1(202, 204)
Seals for ultrahigh vacuum 12(653–655)
Secant modulus, defined 6a/1(105)
Second harmonic distortion of condensor microphone 11/2(40)
Second law of thermodynamics, *see also* Internal dissipation 6a/3(537, 539, 540); 8/1(175); 15(54, 58)
— (Clausius-Duhem inequality) 6a/2(301, 484, 512, 532)
Second or characteristic problem 6a/3(492)
Second order effects in viscous flow 6(512–514)
Second order elasticity 6(502ff., 507–509)
Second order infinitesimal wave approach for progressive waves 9(654–662)
Second order quantities in sound waves 11/2(35)
Second order transformation 13(107)
Second order transitions, *see* Transitions, second order
Second Piola-Kirchhoff stress 6a/3(542); 6a/4(118)
Second shear viscosity 11/1(399)
Second sound in He II 10(94)
Second sound in liquid Helium II 15(388, 393, 394, 430, 434, 435, 437, 458)
Second virial coefficient 13(2, 227, 315, 451, 457)
— dependence on molecular weight 13(465–466)
— dependence on temperature 13(466–469)
Second virial coefficients 12(19, 21, 29, 35, 50, 54)
— of a binary mixture 12(60, 63)
Secondary flow 8/1(439, 440)
Secondary interferences 11/2(136–139)
Secondary motion 6a/3(178, 179, 186, 187, 279)
Secondary principal stress axis 6(138)
Secondary state 6a/3(176, 177, 256)
Secondary wave (*S*-wave) 6(107–109)
Second-order elastic constants 6a/3(181)
Second-order elasticity tensor 6a/3(316)
Second-order isentropic tangent modulus 6a/3(362)
Second-order shallow-water approximation 9(468, 469)
Second-order tangent modulus 6a/3(356)
Second-order transformations in polymers 13(423, 424, 431–432)
Sectionally holomorphic function 6(60)
Sector-shaped basin, waves 9(624)
Secular equation 13(111)

Sedimentation as a means to determine molecular weights 13(401, 473)
Seepage velocity 8/2(635)
Segment interaction parameter 13(407)
Segment length 11/1(395)
Seismic waves (reference) 6(107)
Seismological measurement, early consideration of 6a/1(255)
Se-lattice (structure) 7/1(83)
Selenium 6a/1(281)
Self consistent field 13(224)
Self diffusion 7/1(420, 474, 657); 7/2(205); 12(427, 430, 431)
— coefficient 12(310)
— coefficient, measured values 12(427)
— for a gas of rough spheres 12(498)
— in tin 7/1(426)
— of ions in glass 13(622, 642)
— temperature dependence 12(429)
Self-adjoint 6a/2(385); 6a/3(264)
Self-annealing 6(598)
Self-dual lattice 13(118)
Self-energy 7/1(525, 548)
Self-equilibrated Kelvin process 6a/3(81)
Self-equilibrated stress field 6a/2(55, 59)
Self-equilibrated surface tractions 6a/3(83)
Self-excited Saint Clair generator 11/2(93)
Self-noise of microphone 11/2(45)
Self-potential 7/1(135, 136, 224, 225)
Self-preserving solutions of turbulent flow 8/2(467–470)
Self-similar unsteady flow 9(360)
Sell transformer 11/1(168)
Semicircular groove 6(423)
Semi-circular groove 6a/3(521)
Semi-coercive bilinear form 6a/2(422)
Semi-coerciveness hypothesis 6a/2(392)
Semiconducting piezoelectric crystals 11/2(243, 246–249)
Semiconductors, use in resistance vacuummeters 12(526)
Semi-infinite channel 9(559)
Semi-invariant 10(202)
Semi-invariants, method of Kirkwood 13(98–103, 354, 366)
Semi-invariants of Thiele 13(99, 367, 384)
Semi-inverse method 6a/3(560)
Semi-inverse solution 9(350)
Sénarmont method of measuring retardations 6(154–155, 160–161, 172)
Senftleben effect 12(484, 490)
Sensitivity of an ionization gauge 12(616–617, 623)
Sensitivity of thermoelectric microphone 11/2(62)
Separable 6a/3(170, 557)
Separate surface elements 6a/3(46)
Separation by alloying 14(260, 264ff.)
Separation by magnetic field 14(261, 263)
Separation of boundary layer
— in three-dimensional flow 8/1(318–322)
— in two-dimensional flow 8/1(266, 271, 276, 277, 306)
Separation of degree of freedom 12(85–91)
Separation of electronic and lattice components 14(259)
Separation of variables 6a/3(255, 256, 284)
Separation of variables for infinitesimal surface waves 9(472–475)
Separation of variables in Chaplygin's equation 9(62–65)
Separation surface 6(341)
Separation, surface of 6a/3(426)
Sequence of nucleation 7/2(229)
Series approaches to boundary layer solutions 8/1(272–277)
Series developments for imperfect gases 12(123ff., 131ff., 134ff., 146–152)
Set of all linear transformations 6a/3(4)
Set of initial data 6a/3(130)
Set $\mathscr{B}(T, X)$ 6a/3(127, 128)
Set X 6a/3(127, 128, 130)
s'Gravesande experiment 6a/1(101, 163–167)
Shade theorem 12(473, 480, 483, 487, 490, 491)
Shadow zone 11/1(81, 85)
Shadow zone, penetration into it 11/1(82–85)
Shaft 6a/3(505)
Shake down 6(317–318)
Shallow curved panel 6a/3(281)

Shallow-water approximation 9(456, 462, 466–469, 667ff.)
— first-order fundamental equations 9(667–668)
— non-linear theory 9(676–701)
— of higher order 9(701ff.)
Shallow-water results, connection with infinitesimal wave approximation 9(673)
Shallow-water second-order approximation 9(668, 703, 704)
Shallow-water third order approximation 9(703, 705, 707)
Shape factor of velocity profile 8/1(399)
Shattering 6(578, 579)
Shear 6(437)
Shear, deflection due to, in flexure of beams 6a/1(384)
Shear deformation, transverse 6a/2(460, 575, 578, 584)
Shear, finite 7/1(322)
Shear flow 8/2(508); 9(273, 274)
Shear flow, generation by a grid 8/2(539–540)
Shear flow, unsteady 12(291)
Shear flows, special cases 8/2(548–551)
Shear, homogeneous 7/1(184, 185, 224, 230, 277)
Shear layer 8/1(271, 302–307)
Shear modulus 6(25–26, 142, 448); 6a/2(74, 76, 210, 311, 574); 6a/3(30); 11/2(166); 13(433, 441); 15(240)
Shear modulus, Coulomb's discovery of 6a/1(161)
— definition of 6a/1(83)
— μ (0) vs atomic number 6a/1(394)
— of elasticity 10(92)
— of glass 13(635)
— of liquids 10(20, 71)
— quantized distribution of 6a/1(149, 378, 397–406, 567)
Shear relaxation time 11/1(296)
Shear, simple 6a/2(34, 36, 77)
Shear stiffness 11/1(382, 384, 397)
Shear strain 6(141); 6a/3(442); 7/2(29)
Shear strain, critical 7/2(33)
Shear strain, octahedral 6(240)
Shear strength, theoretical 7/1(266, 320, 322, 323); 7/2(7–17)
— of ideal crystals 7/2(7–9)
Shear stress 6(137); 6a/2(50, 51); 6a/3(442); 7/2(28); 8/1(6, 48)
Shear stress, critical, *see* Critical shear stress
Shear stress, maximum direction 6(354, 373)
Shear stress, octahedral 6(240, 241, 326)
Shear transfer theory of Obukhoff 8/2(464, 471)
Shear viscosity 6(450); 11/1(296)
— effect on sound waves 11/1(140–142, 150)
— for a gas of rough spheres 12(498)
— for polyatomic gases 12(496)
— kinetical explanation 11/1(157–158)
— of ideal gases 11/1(141)
Shear wave 11/1(362)
Shear wave reflectance method 11/1(375–376)
Shear waves, *see* Transverse waves
Shear waves in liquids 11/1(299–300)
Shear-free turbulence 8/2(525–526)
Shearing deformation in gases 11/1(133)
Shearing modulus, isotropic 7/1(263)
Shearing modulus, rigidity 7/1(185, 201, 224, 226, 227, 231, 232, 233, 321, 323, 467, 609)
Shearing stress 10(91)
— critical 7/1(552)
— lines 6(32)
— mean 7/1(269, 323)
— thermal 7/1(265, 320, 322)
Shearing viscosity 10(83, 89, 91, 92)
Shearing viscosity, coefficient 10(91)
Shear-strain tensor 11/1(3)
Sheet 6a/3(515)
Shell 6a/3(172, 287)
Shell-like body 6a/2(411, 425, 438, 440, 442)
Shells reinforced with cords 6a/2(547)
Shell-theory 6a/2(444, 445)
Shell-theory thickness 6a/2(443, 447)
Shield's theorem 6a/3(213)
Shiffman's theorem for subsonic flow 9(108, 111)
Shift factor 13(447)
Shinbrot's fixed-point theorem 6a/3(558)
Ship, dipole approximation 9(582)
Ship floating on water 9(563ff.)

Ship, forces on its body 9(567, 580)
Ship heading into waves 9(567)
Ship rolling 9(555)
Ship, transient problems 9(616–617)
Ship waves 9(446)
Shock 8/1(11, 13–16); 12(570–576)
Shock amplitude equation 6a/4(48ff., 51, 53–55, 92–94)
Shock angle 8/1(15)
Shock, compression 6a/4(295)
Shock conditions 6(346); 6a/3(433)
Shock curvature 8/1(16)
Shock equations 9(166–168, 169)
Shock, expansion 6a/4(295)
Shock expansion 9(245–247, 274)
Shock formation 9(253, 265–268, 279, 280)
Shock in a piezoelectric material 6a/4(297–302)
Shock in solids 6a/4(287)
Shock layer 8/1(15, 226); 12(266, 288)
Shock of compressible fluid flow 6(342, 346)
Shock polar 9(701)
Shock polaric diagram after Busemann 12(572–574)
Shock propagation 6(333)
Shock propagation in ionized gases 9(220–221)
Shock pulse attenuation (thin pulse problem) 6a/4(53–59, 66)
Shock, reflection of 6a/4(294, 295)
Shock relations 8/1(218ff.)
Shock transition, properties of 8/1(224)
Shock velocity 6a/4(288)
Shock wave 6a/2(253, 254); 6a/4(25ff., 31, 36, 45, 60ff., 63, 298); 8/1(13–16, 218ff.); 9(130, 162, 227, 242); 15(435)
— amplitude (shock strength) 6a/4(26, 41ff., 48–50, 92, 94)
— compressive 6a/4(26, 48–53, 91)
— connection with shallow-water theory 9(681, 693, 698)
— detachment 9(174ff., 177)
— evolution, *see* Evolution of steady shock waves
— expansive 6a/4(26, 91)
— experimental observation 6a/4(41–44)
— formation 9(171ff., 203)
— formation of 6a/4(61, 68ff.)
— from cavitation 11/2(266)
— growth and decay, mechanical theory 6a/4(48–59)
— growth and decay, thermodynamic theory 6a/4(89–97)
— in plane flow 9(180ff., 180–187)
— in three-dimensional flow 9(188–195)
— reflection 9(182, 204–211)
— spherical 9(211)
— steady, *see* Steady waves
— subsonic case 9(171)
— supersonic case 9(171)
— velocity 6a/4(41–44, 48ff., 50ff., 91ff., 96)
— weak 6a/4(49ff., 54, 59, 61, 92, 94)
Shock waves 6a/1(499–501, 716–718); 8/2(612); 10(35); 11/1(140, 171); 12(347, 364)
— and boundary layer, interaction 8/1(341–343)
— compression or expansion 6a/3(312)
— definition 6a/3(97, 311)
— in bodies of material with memory 6a/3(349–356)
— in connection with weak solutions 6a/3(171)
— in elastic bodies 6a/3(357–361)
— in inhomogeneous elastic bodies 6a/3(382–385)
— in non-conductors 6a/3(369–377)
— in plasticity 6a/3(416)
— photoelastic investigation 6(227)
Shock-free entry 8/2(393)
Shockley's partial dislocation 7/1(507f., 510, 610)
Short circuits 7/1(426)
Short range order 7/1(322, 456f.); 15(32)
Short time relaxation 11/1(390)
Short wire in uniform velocity field 8/2(571–573)
Short-crested wave 9(545, 547, 549)
Short-range disorder 7/2(189)
Short-range interaction in polymer solutions 13(464–465)
Short-range order 7/2(101, 252)
— decrease 7/2(188)
— destruction in alloys 7/2(172, 174)
— in alloys 7/2(170–174)
— in glasses 13(517)

— theory of critical shear stress of alloys 7/2(174–179)
Short-range ordering parameter 13(186)
— of Bethe 13(91)
Short-range two-body forces 13(8–9)
Shrinkage of material in glass manufacture 13(546)
Side groups of polymer chains 11/1(417)
Side surface 7/2(73)
Sigma-monogenic functions of Bers-Gelbart 9(67)
σ-ring 6a/2(415)
Sign conventions in photoelasticity 6(169–170)
Sign conventions of curvature 6(358, 378)
Sign of cavitation number 9(323)
Sign of volume of mixing 13(351, 364)
Signal-to-noise ratio of anemometer 8/2(578, 584)
Signorini, generalized 6a/2(404, 408)
Signorini in viscoelasticity 6a/2(423)
Signorini problem, definition 6a/2(404, 413)
Signorini's theorem 6a/2(61)
Silica, *see* Quartz and Fused silica
Silica, internal friction 11/1(404)
Silicate systems, limits of glass formation 13(612–614)
Silicon 7/2(44)
Silicon, neutron damage 11/2(169)
Silicone rubber 11/1(410)
Silk 6a/1(30, 47–49, 56, 64, 97–101)
Silsbee effect 15(214, 253)
Silsbee's hypothesis 15(338)
Silver 6a/1(46, 103, 105, 120, 186, 221, 231, 302, 374, 389, 398, 496, 528, 555–557); 14(239)
Silver alloys 14(265)
Silver chloride crystals, work-hardening 7/2(69)
Silver halides, latent photograph in 7/1(634)
Silver single crystals, work-hardening 7/2(54)
Silver-gold single crystals, short-range order parameters 7/2(176)
Similarity, dynamical 8/1(57–60, 183, 243–246)
Similarity of boundary layer solutions 8/1(271, 290)
Similarity principle of Bers 9(104)
Similarity spectra 8/2(473–478, 485)
Simon expansion liquefier 14(119)
Simple backward-facing wave 9(679–680)
Simple compression wave 9(689, 694)
Simple expansion wave 9(689, 694)
Simple extension 6a/2(34)
Simple forward-facing wave 9(679)
Simple glide 7/2(21, 22)
Simple material 6a/3(17)
Simple materials with fading memory (materials with memory)
— mechanical theory 6a/4(24, 26–30, 39, 48, 51, 59ff.)
— thermodynamic theory 6a/4(75–81, 92, 97ff.)
Simple (reversible) expansion of a gas 12(8, 27, 28)
Simple shear 6(437); 6a/2(34, 77)
Simple vector field 6a/3(73, 112)
Simple wave 6(359, 390–396); 6a/3(447, 484, 485); 6a/4(54ff., 58, 278, 280); 8/1(106–109); 9(227, 230–231, 236, 243, 247, 252–254, 263, 678)
— definition 9(22)
— in shock waves 9(196–198)
— region 6a/4(279)
Simple-harmonic wave 11/1(22)
Simply connected 6a/3(45)
Simply supported elastic plate 6a/3(136)
Simply-connected 6a/2(13)
Sing-around method 11/1(169)
Single crystal, plastic deformation of a 7/1(634)
Single crystals
— deformation 7/2(28–32)
— experiments on 6a/1(355, 406–411, 495, 514–534, 663, 694)
— growth of 6a/1(515)
— hexagonal 6a/1(523)
— lead 6a/1(145, 543)
— of copper, irradiation 11/2(196)
— zinc 6a/1(525)
Single dislocations of an infinite crystal 7/1(577)
Single intersection theorem 9(379)
Single roughness 8/1(447–448)
Single slip 6a/1(515–521)

Single slip, macroscopic 6a/1(518, 541)
Single wave 8/1(110)
Single-liquid model of liquid mixtures 13(360, 368, 483)
Singlet density 10(282)
Singlet distribution function 10(179ff., 184ff., 194, 197, 200, 208, 210, 215ff., 238, 240)
Singular elastic state 6a/2(179)
Singular integral equations on a curve 6a/2(420)
Singular line 9(261, 267)
Singular set of the measure γ 6a/2(415)
Singular solutions 6a/3(77)
Singular solutions in infinitesimal surface wave theory 9(475, 478, 479)
Singular solutions including surface tension 9(635–638)
Singular surface(s) *see also* Acceleration waves, Shock waves 6a/3(96, 97, 305); 6a/4(281); 8/1(212)
— of order 6a/2(250)
— of order zero 6a/2(249)
— related to high frequency waves 6a/3(109)
Singular yield conditions 6a/3(411)
Singularities in second-order, infinitesimal wave approach 9(656–657)
Singularities of constant strength in uniform motion 9(489)
Singularities of higher order 9(481, 484, 495)
Singularity, branch lines 6a/3(483)
Singularity, branch type 6a/3(477)
Sinking point of a glass 13(580)
Siren 11/2(24–27)
Sirens for ultrasound 11/2(98–99)
Size effect 14(332)
Size effect on easy-glide 7/2(141)
Size, high frequency 14(183)
Size, magneto-resistance 14(186)
Size of complex 7/2(237)
Size-effects, fracture 6(580–586, 589)
"Size-factor" of the solute atom 14(152)
Skalak theory of impacting rods 6a/1(336, 340–342)
Skew 6a/2(6, 617)
Skew (infinitesimal) rotation field process W 6a/3(17)
Skew tensors, subspace of 6a/3(4)
Skew-symmetric $r \times r$ matrix 6a/2(384)
Skin depth 11/2(213)
Skin effect, anomalous 15(248)
Skin effect, anomalous Chambers' derivation of current density 15(300)
Skin friction 8/1(331, 339, 347)
Skin round a bubble 11/2(273)
Skin-effect 14(184)
Skin-effect, anomalous 14(183, 188)
Slater sum 13(64)
Slender body 9(28, 573, 577, 578–579)
Slender body of revolution 9(54–56)
Slender profile in shallow water 9(675)
Slip, *see* Glide
Slip 6(230, 321)
Slip avalanche 6(597–598)
Slip bands 7/1(632)
Slip conditions 8/1(240)
Slip direction 7/2(20, 22, 24, 25)
Slip ellipses 7/2(71, 97)
Slip field 6a/3(445, 450)
Slip flow 8/2(592, 593, 609–614)
Slip flow, fundamental experiments 8/2(612–614)
Slip in fatigue 6(597)
Slip line(s) 6(355, 357ff., 399); 7/2(5, 71, 72, 82ff.)
— definition 6a/3(442, 443)
— field 6(357–360, 361, 366, 431)
— field of the telegraph equation 6(363)
— pattern 7/2(21)
— plastic boundary 6a/3(494)
— properties of 6a/3(447)
Slip motions 7/1(479)
Slip plane 7/2(20)
Slip plane blocking 6(602)
Slip plane in the body-centered cubic lattice 7/1(615)
Slip planes, secondary 7/2(180)
Slip surface(s) 6a/3(424); 7/1(480)
Slip velocity 8/1(265); 8/2(600, 612)
Slip zones 7/1(537f., 544)
Slip-band distribution 6(597)
Slip-band spacing 6(597, 606)
Slip-flow 12(265)
Slip-line grouping 7/2(82, 92)
Slip-line length, active 7/2(90, 105, 146)
Slippage along the wall 6(461)
Slow neutrons, critical scattering 13(196–210)

Small deformation from a stress free isotropic reference state 6a/3(191)
Small fluctuations 8/2(556, 557)
Small-angle grain boundary 7/2(153–154)
Small-angle scattering 7/2(101)
Small-angle scattering of x-rays by binary alloys 13(180–189)
Small-angle-grainboundary 7/1(630, 646, 656)
Smekal's theory of glasses 13(525–527)
Smoluchowski formula for pressure force 12(531)
Smoluchowski's solution 8/2(249–251)
Smooth field 6a/2(11)
Smooth in time 6a/2(24)
Smooth structured wave 6a/4(26, 40)
Smoothed (half-fine) correlation function 13(69, 70)
Smoothed potential in liquids 13(222, 226)
Smoothed potential model of Prigogine and Mathot 13(352–353)
Smoothing 12(224–227)
Smoothly propagating surface 6a/2(248)
Snail form of tube cross section 8/2(72)
Snap-through 6a/3(148, 287–289)
Sneddon's solution 6(120)
Snoek effect 7/2(186)
Soap bubble model 7/1(649, 650)
Soap film analog of Prandtl 6(401, 424)
Sobolev embedding theorems 6a/3(140, 212)
Sobolev lemma 6a/2(354); 6a/3(578)
Sobolev norms 6a/3(224)
Sobolev space 6a/3(212)
Sodium 14(239)
Sodium borate glasses 13(532–533)
Sodium borate glasses, hardness 13(634)
Sodium chloride, irradiation 11/2(191–195)
Sodium disilicate 13(608, 610)
Sodium, liquid 10(56)
Sodium metasilicate 13(610)
Sodium orthosilicate 13(610)
Sodium phosphate glasses 13(539)
Sodium silicate glasses
— composition dependence of density 13(551–553)
— internal friction 13(643)
— structure 13(523)
— under high pressure 13(558)
— viscosity 13(581, 589)
Sodium sulphate 11/1(215, 259–260)
Sodium-calcium silicate glasses, crystal growth 13(598, 601)
Softening 7/2(179)
Softening, geometrical 7/2(48)
Softening point of a glass 13(579, 594)
Softening temperature of sodium borate glasses 13(532)
Sokolovsky condition 6a/3(473)
Sokolovsky yield condition 6(381)
Sol 6(435, 522)
Sol, dilute 6(524)
Soleil-Babinet compensator 6(136)
Solenoïdal conditions for the correlation tensor 8/2(444, 447)
Solenoidal fields 6a/2(18)
Solenoidal vector field 6a/3(73, 557)
Solid boundary 10(69)
Solid dielectric condenser microphone 11/2(43–44)
Solid He^3, melting pressure 15(182)
Solid solution, nucleation 7/2(215)
Solid solution, supersaturated 7/2(192)
Solid solutions, *see also* Binary systems and Alloys 7/2(183–188)
— comparison of theory and experiment 13(167–175)
— critical opalescence 13(189–196)
— theory of Fuchs, *see also* Fuchs theory 13(143–153, 159–167)
Solid spherical harmonic 6a/2(22)
Solidification 15(371, 372)
Solidity of grid 8/2(531, 533)
Solids under high explosive shock 6a/1(500)
Solitary wave 9(462, 654, 667, 702)
— existence theorems 9(751–753)
— limiting height 9(712)
— propagation velocity 9(710, 711)
— second approximation to its profile 9(710, 711)
Solubility in compressed gases 12(70–72)
Solubility of helium in glass 12(634)

Solute atom diffusion-vacancy hypothesis 6a/1(655)
Solution, auxiliary problem for E^* 6a/3(61)
Solution, Cauchy initial value problem for dynamic viscoelasticity 6a/3(105)
Solution, dynamic viscoelastic boundary value problem 6a/3(89)
Solution in dynamical system 6a/3(128, 131)
Solution of associated elastic problem 6a/3(60)
Solution of equation of elasticity 6a/2(102)
Solution of general problem of elasticity 6a/2(129)
Solution of generalized mixed problem 6a/2(185)
Solution of initial past history problem in terms of displacements 6a/3(52)
Solution of mixed problem 6a/2(102, 220, 321, 335)
Solution, quasi-static, in terms of
— displacements, relaxation function, or creep compliance 6a/3(47)
— stresses 6a/3(48)
Solution, viscoelastic, initial past history problem 6a/3(51)
Solution, viscoelastic problem 6a/3(60)
Solutions, athermal polymer 10(272ff.)
Solutions of boundary layer equations
— exact 8/1(271–272, 314–318)
— in series form 8/1(272–277)
Solutions of non-electrolytes 10(262ff.)
Solutions of polymers
— critical point of phase separation 13(480)
— excluded volume model 13(451, 452–456, 461–463)
— lattice model 13(451, 456–462)
— models 13(449, 451ff.)
— osmotic pressure 13(451, 468, 471)
— second virial coefficient 13(451, 457)
— visco-elasticity 13(449–450)
Solutions, polymer 10(272ff.)
Solutions, regular 10(262, 264ff.)
Solutions, regular, with plane interface 10(264)
Solutions, surface tension 10(298)
Solvent 11/1(222)
Solvent aggregate 11/1(226)
Solvent, definition 13(159)
Solvents of small dielectric constant 11/1(245–246)
Somigliana, *see* Boussinesq
Somigliana fundamental matrix 6a/2(419)
Somigliana stress functions, viscoelastic 6a/3(94)
Somigliana's theorem 6a/2(183)
Sommerfeld theory 14(289)
Sommerfeld's radiation condition 9(298)
Sonic limit line 6a/3(482)
Sonic limit point 6a/3(482)
Sonic line 6(387–388); 6a/3(480)
Sonine expansion 12(267)
Sonine polynomials 12(283, 373, 419)
Soret effect in liquids 10(2, 24, 82)
Sound absorption 12(496, 501)
Sound absorption in monatomic gases 12(384)
Sound absorption, sound impedance, sound propagation, etc., *see* Absorption, Impedance, Propagation, etc.
Sound dispersion 12(287, 288, 497)
Sound intensity 11/1(9)
Sound, propagation and absorption in liquids 10(91, 92, 95)
Sound, speed of 8/1(13, 60, 180, 213, 222, 228)
Sound velocity
— and dipole moment 11/1(346)
— and intermolecular forces 11/1(351–352)
— and molecular volume 11/1(346–350)
— and van der Waals' equation 11/1(339–340)
— experimental methods of determination 11/1(165–171)
— in a visco-elastic fluid 11/1(298)
— in gases, dependence on temperature and pressure 11/1(188–191)
— in liquid mixtures 11/1(355–359)
— in liquids without absorption 10(95)
— longitudinal 7/1(181, 251, 256, 263)
— mean 7/1(291, 311)
— measurements 11/1(301–303)
— of, *see* Speed of sound

— Schaaffs' theory 11/1(340–346)
— (sound velocity) 7/1(161, 181, 187, 189, 190, 267, 290, 293, 620)
— table of values 11/1(345)
— transverse 7/1(181, 251, 256)
— variation with temperature and pressure 11/1(352–354)
Sound wave 8/1(12–13, 214)
— in uniformly moving fluid 11/1(18–19)
— momentum of a 7/1(304)
— plane, in an ideal gas 11/1(136–138)
Source and vortex, combined 8/1(99)
Source distribution replacing ship's hull 9(581)
Source function, *see* Green's function
Source in a parallel flow 8/1(98–100)
Source method 8/2(117)
Source of constant strength in uniform motion 9(483)
Source of position and strength varying with time 9(490–495)
Source of pulsating strength
— in three dimensions 9(475)
— in two dimensions 9(479)
Source speed 11/1(77)
Source strength 8/1(96); 9(101)
Source term in the wave equation 11/1(4)
Source terms in the integral equation 11/1(24–25)
Source type solutions including surface tension 9(635–638)
Sources 6a/1(9)
Sources of sound
— air driven 11/2(24–32)
— classification 11/2(1–3)
— electrical 11/2(4–23)
— explosive 11/2(32–35)
Source-sink representation of cavity flow 9(314, 430)
Space $\mathscr{C}_0$ ($\mathscr{R}^+$; $\mathscr{X}$) of continuous ($\mathscr{X}$)-histories which vanish at infinity 6a/3(7)
Space $\mathscr{L}^{11}$ ($\mathscr{R}^+$; $\mathscr{X}$) of integrable ($\mathscr{X}$)-histories 6a/3(7)
Space, Euclidean point 6a/3(4)
Space filling 11/1(340)
Space group(s) 7/1(4)
— deduction of 7/1(5, 27ff.)
— number of coordinates 7/1(28ff.)
— properties of the 7/1(37ff.)
— symbols 7/1(13, 32ff.)
— table of 7/1(32ff.)
Space, inner product 6a/3(4)
Space of linear maps in $\mathscr{T}$ (the space of fourth-order tensors): $[\mathscr{T}]$ 6a/3(4)
Space of (second-order) tensors $[\mathscr{V}] \equiv \mathscr{T}$ 6a/3(4)
Space resolution of anemometer 8/2(580–584)
Space translation 6a/3(4)
Space vector 6a/3(4)
Spaces $H_m(A)$ and $\mathring{H}_m(A)$ 6a/2(349)
Space-time 6a/2(27)
Space-time resolution of anemometer 8/2(582–584)
Space-time transformation 8/2(529)
Spalerite lattice 7/1(101)
Spannungskoeffizient 12(6)
Spark generator of ultrasound 11/2(94–95)
Spatial description 6a/3(428, 536); 6a/4(113, 114)
Special functions 13(5–6)
Special heat 7/1(107, 244, 245, 247, 258, 259, 272, 273, 274, 276, 287, 291, 310)
— at constant pressure 7/1(236, 274)
— at constant volume 7/1(235, 236, 274, 289)
— at high temperatures 7/1(369)
— curves, inversion of 7/1(350)
Special position 7/1(39)
Special theories of shells 6a/2(546)
Specific admittance 11/1(19, 35)
Specific entropy 6a/2(483, 510); 6a/3(536)
Specific free energy 6a/3(536)
Specific Gibbs free energy 6a/2(555, 570)
Specific heat 6a/2(304); 6a/3(193, 201); 8/1(173); 10(123); 15(21, 22, 28, 31, 32, 51, 80, 84, 85, 90, 94, 105, 107, 109, 111, 112, 113, 117, 128, 142, 193, 374, 392, 401, 406, 407, 408, 454, 455, 457, 458)
— and long-range ordering 13(88)
— anomalies in the 14(227)
— at constant magnetic moment 15(53)
— in the three-dimensional Ising model 13(141)

Specific heat (cont.)
— in the two-dimensional Ising model 13(117, 135)
— influence of Stark field 15(97, 98, 104)
— linear variation 14(193)
— of $AuCu_3$ 13(81)
— of brass 13(82, 106)
— of cobalt sulphate 15(111)
— of copper sulphate 15(107, 108)
— of ethylene-di-bromide 13(210)
— of gadolinium sulphate 15(112ff.)
— of glasses 13(570–573)
— of iodine hydride 13(216)
— of metals 15(187f.)
— of polymers 13(421–432)
— of polymers, temperature dependence 13(422)
— of solid helium 15(179)
— of superconductors 15(230, 234)
— of the single vibrational mode 11/1(287)
— ratio 2 9(677, 681)
— relaxation 11/1(401)
— supply 6a/2(481)
Specific heats 6a/1(38, 47, 82, 94, 122, 228–229, 320, 372, 381); 6a/4(127, 136–139, 170)
Specific Helmholtz free energy 6a/2(497, 511)
Specific impedance 11/1(16, 35); 11/2(4)
Specific internal energy 6a/2(481, 510); 6a/3(536)
Specific ionization of several gases 12(535)
Specific production of entropy 6a/2(483, 511)
Specific radiation impedance 11/2(102)
Specimens, hollow 15(226)
Specimens, non-ellipsoidal 15(226)
Specimens, small 15(251)
Spectra in crystalline electric fields 15(14)
Spectral average 7/1(245, 279)
Spectral decomposition and theorem 6a/2(8)
Spectral distribution of local fluctuations 13(70)
Spectral function of total energy 8/2(449, 453, 464, 470)
Spectral function of vorticity 8/2(452, 468)
Spectral relaxation time 8/2(543)
Spectroscopic stability 15(6)
Spectroscopy at centimeter wavelenghts 15(3, 25)
Spectrum 6a/2(269); 9(507)
Spectrum, elastic 7/1(182, 248, 256, 258)
Spectrum of activation energies 7/2(201)
Spectrum of differential operator law 6a/3(38)
Spectrum of Reynolds numbers 8/2(527)
Spectrum of sound generated by a subsonic jet 11/1(98)
Spectrum, statistical 7/1(260)
Spectrum tensor of double velocity correlation 8/2(445)
Specular reflection of molecules from surfaces 8/2(599, 600, 610)
Speed of displacement 6a/3(307, 308, 429)
Speed of propagation, *see also* Wave 6a/3(96, 97, 306, 308, 429)
Speed of propagation 6a/2(248, 343); 8/2(129)
Speed of sound, *see* Sound velocity
— complex 11/1(206)
— in magnetized nickel 11/2(226)
— in the air 11/1(194)
— local 9(5, 7, 209, 228, 231, 235, 250, 285)
Sphere floating freely 9(568)
Sphere heaving on water surface 9(562)
Sphere in molecular flow 8/2(605)
Sphere in oscillating rotation 8/2(220)
Sphere model of polyelectrolytes 13(495)
Sphere, non-uniform translation in a fluid 8/2(238–242)
Sphere, resistance 8/2(383)
Sphere spinning in a steady flow 8/2(214)
Sphere, submerged 9(577)
Sphere, transition from laminar to turbulent flow 8/1(358, 384)
Sphere, uniform translation
— along the tube 8/2(246, 318)
— between two parallel walls 8/2(245–247)

— in a fluid 8/2(227)
— in presence of a wall 8/2(242–245)
— Oseen approximation 8/2(298–305)
Spheres, stable packing 8/2(629)
Spherical Bessel functions 11/1(23, 61, 65)
Spherical cavity, scattering of stress waves 11/2(164)
Spherical cavity, viscoelastic solution 6(295)
Spherical coordinates, equations of motion 8/2(6)
Spherical Hankel functions 11/1(61)
Spherical harmonic 6a/2(22)
Spherical harmonics 11/1(23, 61)
Spherical interface, *see also* Droplet 10(145, 148, 150, 159, 166, 213ff., 292)
Spherical obstacle, scattered energy 11/2(159, 164)
Spherical polar coordinates 6(86, 102–104); 8/2(59)
Spherical resonators for sound waves 11/1(167)
Spherical shell 6a/3(192, 521)
Spherical shell in elastic-plastic expansion by internal pressure 6(422)
Spherical source of sound 11/1(61–64)
Spherical vortex of Hill 8/2(54, 141)
Spheroid, submerged 9(577, 578)
Spheroidal coordinates, oblate, in elasticity 6(106–107)
Spheroidal functions 11/1(68)
Spin alignment (or alinement), *see also* Orientation 15(400, 401)
Spin chain 13(109)
Spin correlation function 13(201, 207)
Spin lattice relaxation 15(166)
Spin orbit coupling 15(79)
Spin relaxation 15(22)
Spin system 15(166)
Spin walls 11/2(227)
Spin wave method 15(133)
Spin waves 14(227)
Spin waves, interaction with ultrasonics 11/2(228–229)
Spinel 7/2(217)
Spin-lattice relaxation 11/2(239)
Spinodal curve 7/2(219, 233–235)
Spiral flow 8/1(196)
Spiral of Ekman 8/2(112)
Spiral source 7/1(538, 661)
Spirals 6a/3(456)
Spirals of Hamel 8/2(33, 144)
Split 14(104)
Split-Hopkinson bar test, *see* Kolsky's experiment
Splitting of dislocations 7/2(180)
Splitting parameter 15(87, 89, 91, 94, 95, 100, 104, 113)
Spontaneous magnetization calculated by the Ising model 13(112, 137, 139, 141)
Spontaneous nucleation 13(595)
Spread of hump on a free surface 9(508)
Spreading of a liquid film 10(299)
Spring's experiments, Bridgman's comments on 6a/1(476)
Spring's experiments on the compaction of solids 6a/1(474–477, 482)
Square lattice, orientation effects 13(340–342)
Square lattice, results of Fuchs theory 13(161, 164)
Square mesh grid 8/2(526, 532)
S-shaped profile 8/2(390, 394)
Stability 6a/3(3); 7/1(155, 156, 158)
Stability and buckling 6a/3(130)
Stability, atomic 7/1(233)
Stability, choice in elasticity 6a/3(171)
Stability, definitions 6a/3(146)
— asymptotic 6a/3(142)
— asymptotically orbital 6a/3(144)
— conditional 6a/3(147)
— contractive 6a/3(144)
— early (footnote) 6a/3(132)
— exponentially asymptotic 6a/3(142)
— finite-time, (practical or Lasalle and Lefschetz) 6a/3(143)
— globally asymptotic (completely) 6a/3(142)
— Hadamard 6a/3(198)
— infinitesimal superstability 6a/3(202)
— Lagrange 6a/3(149)
— Lagrange-Dirichlet 6a/3(131)
— Lagrange-Dirichlet, essential ingredients 6a/3(132)
— Liapounov 6a/3(133)
— Liapounov, equivalent definitions 6a/3(132, 140)
— Liapounov, observations 6a/3(134)
— local (Krasovskii) 6a/3(143)
— neutral 6a/3(144)

Stability, definitions (cont.)
— of a given motion 6a/3(134)
— of dynamical system 6a/3(134)
— of equilibrium 6a/3(131)
— orbital 6a/3(144)
— parametric 6a/3(187)
— persistent 6a/3(187)
— perturbative (footnote) 6a/3(146)
— quasi-contractive 6a/3(143)
— region of stability 6a/3(142)
— superstability 6a/3(202)
— trigonometric (footnote) 6a/3(146)
— uniform 6a/3(141)
— uniformly asymptotic 6a/3(142)
— weakly asymptotic 6a/3(142)
Stability, elastic 7/1(178, 233)
Stability limit of laminar flow 8/1(371)
Stability of boundary layer along a concave wall 8/1(433–436)
Stability of condenser microphones 11/2(44)
Stability of deformed state 7/2(130–131)
Stability of laminar jets 8/1(388)
Stability of liquid-gas, interface 9(321)
Stability of motion
— interface between moving fluids 9(648–650)
— interface of stationary system 9(646–648)
— interfacial waves 9(634, 635)
— moving basin 9(630–631)
— second-order infinitesimal wave theory 9(663)
— surface tension effects 9(647)
— vertical oscillation of a basin 9(651–653)
— viscosity effects 9(647–648)
Stability of similarity spectra 8/2(485–487)
Stability of the permanent deformation 6a/1(32, 38, 43, 86, 453, 458, 463, 475, 477, 578, 649–666, 714, 716)
Stability of visco-plastic flow 6(307)
Stability of viscous flow 8/1(253–257)
Stability theorems
— and calculus of variations 6a/3(163–167, 195)
— discussion of 6a/3(156–160)
— discussion of, examples 6a/3(160–163)
— energy criterion (footnote) 6a/3(164, 195)
— further Liapounov, null solution 6a/3(209, 212, 219)
— necessity, for equilibrium solution 6a/3(200–202)
— Shield's theorem 6a/3(213)
— sufficiency in non-linear thermoelasticity 6a/3(192–194)
— sufficiency, natural state 6a/3(195)
Stability theory of laminar flow 8/1(366–374)
Stability with respect to three-dimensional disturbances 8/1(430–440)
Stabilization, complete, in supersonic flow 8/1(425)
Stabilization process of a glass 13(512, 544–548, 584)
Stabilized glasses, delayed elasticity 13(639)
Stable disturbance 8/1(371)
Stable packing of spheres 8/2(629)
Stable solids 6a/1(152, 181, 399)
Stacking 7/1(636ff.)
Stacking fault(s) 7/1(507f., 627, 636); 7/2(25, 109, 169, 221, 240)
— energy of 7/1(636); 7/2(150, 155, 180–181, 188)
— energy of, indirect determination 7/1(639)
— formation tendency 7/2(137)
— in the face-centered cubic lattice 7/1(508)
— of type I 7/1(508, 637)
— of type II 7/1(508, 637)
— width 7/2(166)
Stacking operators 7/1(506)
Stage I of work-hardening curve (easy glide) 7/2(53)
— surface phenomena 7/2(85–86, 87, 91)
— theory 7/2(140–144)
Stage II of work-hardening curve 7/2(53–54)
— surface phenomena 7/2(86–92, 93)
— theory 7/2(145–149)
Stage III of work-hardening curve 7/2(54)
— β-creep 7/2(202)
— surface, phenomena 7/2(91, 92–99)
— theory 7/2(149–155)

Stage II-stage III transition stress 6a/1(528)
Stages of finite deformation in single crystals 6a/1(526, 529, 535)
Stagnation point 8/1(98); 9(107)
Stagnation point, boundary layer around it 8/1(272, 301–302, 315, 330, 339)
Stagnation pressure 11/2(33)
Staircase phenomenon, *see* Savart-Masson effect
Standard configuration for orientation effects 13(329–332)
Standard differential operator law 6a/3(39)
Standard heat treatment of a glass 13(598)
Standard linear solid 6a/4(197, 199, 200)
Standard material 6(440)
Standard models 6a/3(40)
Standard solid 6(268, 271)
Standard states, definition 12(11)
Standing capillary-gravity waves 9(632–634)
Standing surface wave 9(458)
Standing time of water, effect on absorption 11/2(273–274)
Standing ultrasound waves 11/2(110–111, 134, 137)
Standing wave(s) 8/1(109)
— decomposition of thermal motion 13(69)
— in a canal 9(502)
— in a liquid column 6a/1(251)
— in an infinite ocean 9(495–498)
— in basins 9(622–627)
— in viscous fluids 9(641–646)
Stark effect (in crystals) 15(43, 44, 79, 85, 94, 96, 111)
Stark splitting 15(111, 115)
Star-shaped region 6a/2(14)
Starting vortex 8/1(85–86)
State, density of 14(230)
State equation 11/1(137, 276)
— caloric 7/1(234, 287)
— empirical 12(48–57)
— empirical, of mixtures 12(65)
— of 14(37)
— of Born 7/1(282)
— of Mie and Grüneisen 7/1(277)
— thermal 7/1(235, 244, 274, 279, 281, 285)
State functions of the inelastic continuum 6(244, 245, 248–251)
State, homogeneously strained 7/1(275)
State of equilibrium 7/1(240)
Static electromechanical coupling coefficient 6a/4(260, 301)
Static equilibrium state 6a/4(122)
Static fatigue of glass 13(627–629)
Static pressure 8/1(5)
Static siren 11/2(99)
Static strength 13(624)
Static test 7/2(30)
Static work-hardening curve 7/2(32)
Statically admissible state 6a/2(114)
Statically admissible stress field 6a/2(112, 324)
Statically determinate 6a/3(526)
Static-geometric ananlogy 6a/2(613)
Stationary, diffusion 12(322)
Stationary discontinuity 6(343)
Stationary flow, orders of magnitude 8/2(547)
Stationary isotropic turbulence 8/2(505)
Stationary source in moving medium 11/1(11)
Stationary state 6a/3(524)
Stationary value problem 6a/3(66)
Statistical averages in turbulent flow 8/2(442–443)
Statistical description of turbulence 8/2(438)
— conventional approaches 8/2(440–442)
Statistical theories of permeability 8/2(645–647)
Statistics, molecular 10(28–33)
Steady flow
— along concentric cycles 8/2(16–19, 33, 182)
— around a spinning obstacle 8/2(214–220)
— as balance of work-hardening and recovery 7/2(204)
— between two co-axial cylinders 8/2(17–19, 49–52, 79)
— between two fixed parallel planes 8/2(14)

Steady flow (cont.)
— between two gliding parallel planes 8/2(15)
— between two walls 8/2(22, 35)
— between two walls around an obstacle linear approximation 8/2(209)
— in plane plastic strain 6(427)
— in three dimensions along parallel straight lines 8/2(65)
— on a oblique plane with free surface 8/2(16)
— reciprocity theorem for the solutions 8/2(265)
— through porous media 8/2(638)
Steady Jeffery flow 8/2(38, 40, 41, 55, 57, 93, 182)
Steady motion 8/2(4)
Steady plastic flow 6a/3(524)
Steady shock waves 6a/3(354)
Steady state as asymptotic limit state 9(472, 506ff.)
Steady state creep 7/2(71, 197, 204–208)
Steady states of a non-uniform liquid 10(99–101)
Steady supercritical flow, shallow-water first-order approximation 9(682–695)
Steady waves 6a/4(36–48)
— acceleration 6a/4(40ff.)
— existence 6a/4(36–41, 81)
— experimental observation 6a/4(41–44)
— in thermoviscoelastic solids 6a/4(82–88)
— shock 6a/4(41, 45–52, 59, 63, 66, 94, 96)
— velocity 6a/4(38–44, 46, 82, 87ff.)
Steady-state response 6a/4(207)
Steel 6a/1(30, 83, 86, 89, 95, 103–105, 120, 129, 177, 186, 210–217, 262, 272, 293–296, 304, 320, 326, 336, 403, 426, 463–474, 505–507, 512, 577, 603, 612, 672)
Steel piano wire 6a/1(30)
Steel, Poisson's ratio for 6a/1(261, 273, 293–296)
Step height 7/2(97)
Step reactions, *see* Reaction sequences
Stepwise dissociation of 2-2-valent electrolytes 11/1(260–270)
Stepwise loading 7/2(32)
Stereographic projection 7/2(23, 59)
Stereographic projection of stress axes 6(189)
Sternberg's theorem on Saint-Venant's principle 6a/2(193)
Stevel's density formula 13(552–554)
Stiff chains 13(406)
Stiffness modulus 6(491)
Stiffness of polymers, different types 11/1(382, 384, 395, 397, 402, 406)
Stochastic methods 13(217, 277–284)
Stochastic process 12(220)
Stockmayer potential 12(151)
Stokes' edge waves 9(550)
Stokes equation 8/2(1, 200)
Stokes equation, existence and uniqueness of solutions 8/2(262–276)
— for unsteady flow 8/2(276–287)
Stokes equation, second form 8/2(201)
Stokes' formula 12(292)
Stokes formula for force of friction 13(474)
Stokes' law 11/1(391)
Stokes' law of resistance 8/2(229)
— generalised for a sphere between two walls 8/2(245)
— generalised for a sphere in presence of a wall 8/2(245, 246)
— generalised for an ellipsoïd 8/2(234–235)
— generalised for two spheres 8/2(251–252)
— limits of its validity 8/2(231–232)
Stokes number 11/1(278)
Stokes' paradox 8/2(257–262)
Stokes' principle 8/1(231–234)
Stokes process 6a/2(239, 240)
Stokes relation 6(300, 464, 482); 8/1(238)
Stokes' second method 9(654, 726)
Stokes' theorem 6a/2(17)
Stokes' theorem on wave crests 9(731)
Stokes-Bock hypothesis in the temperature dependence of Poisson's ratio 6a/1(286)
Stokes-Bock hypothesis on the temperature dependence of Poisson's ratio 6a/1(278–282)
Stokesian fluid 8/1(231)
Stokes-Kirchhoff absorption formula 11/1(151)

Stokes-Navier equations 12(300)
Stone 6a/1(18, 22, 28, 86, 143, 167, 486)
Storage and transport vessels, Helium 14(135)
Storage modulus 6a/3(33)
Stored energy 6a/3(173)
Stored energy (strain energy)
— function in elasticity 6a/2(82, 94)
— function in rods 6a/2(673, 677, 680)
— function in shells 6a/2(545)
Straight rods 6a/2(690)
Strain 6a/4(24, 49, 53, 63, 79, 88, 90, 95–97); 8/1(21–23, 25)
Strain aging 7/2(172)
Strain displacement equation 6a/3(43, 71, 89)
Strain, elastic 7/1(195)
Strain energy 6a/3(87, 173, 225)
Strain energy function 11/2(161)
— third order terms 11/2(161–162)
Strain energy of martensite formation 7/2(247–249)
Strain energy of nucleation state 7/2(213–217)
Strain energy (total stored energy)
— elasticity 6a/2(94, 110, 208, 216)
— elasticity, bounds for 6a/2(113)
— shells 6a/2(545)
— thermoelasticity 6a/2(324)
Strain field 6a/2(31, 39, 43, 461)
Strain field process 6a/3(16)
Strain, finite 7/1(238ff.)
Strain, finite, wave profiles 6a/1(629)
Strain, generalized 6(327)
Strain history 6(280, 287); 6a/3(305)
Strain, history (entire history) 6a/4(24, 26–28, 76)
Strain, homogeneous 7/1(186, 188, 233ff., 235, 238, 243, 280)
Strain in rheology 6(434)
Strain invariants 6(6); 13(433)
Strain, logarithmic 6(237)
Strain measure 6a/1(543)
Strain, natural 6a/1(543)
Strain, nominal 6a/1(509, 543, 551, 666, 688, 703)
Strain, octahedral shearing 6a/1(511)
Strain, past history 6a/4(24, 76ff.)
Strain path 6a/2(82)
Strain point of a glass 13(579, 580)
Strain, present value 6a/4(24, 27, 76, 78)
Strain quadric 6(141)
Strain rate 6(328, 426); 6a/1(570–578, 586–596, 629, 634, 664); 6a/2(43)
Strain rate, averaged through plate thickness 6(353)
Strain rate variation 6(233)
Strain resolution 6a/1(37, 124, 128, 148, 154)
Strain, resolved shear, *see* Resolved shear deformation
Strain, small relative 6a/4(28ff., 51, 77, 95)
Strain, steady 6a/4(38–41, 46, 82, 87)
Strain tensor 6(3–5); 6a/4(111); 7/1(178, 236, 239, 283, 515); 7/2(21); 10(83, 90); 11/2(154)
Strain tensor, non-symmetric 7/1(559)
Strain, true 6a/1(509, 543, 545, 546, 556, 561, 668, 671)
Strain velocity 8/1(26)
Strain-amplitude effects in dislocation damping 11/2(187)
Strain-displacement relation 6a/2(31, 89, 212, 312, 326)
Straining frame experiment 6(158–159)
Strain-rate 7/2(99, 115)
— before reloading 7/2(204)
— dependence of critical shear stress 7/2(34)
— tensor 6a/3(499)
Strains 6(140–141)
Strain-spin interaction 11/2(242)
Strain-work 6(440, 448, 520)
Strakhovitch solution 8/2(55, 99, 101, 166)
Stratford's method of boundary layer theory 8/1(286–287)
Stratification parameter 8/1(418)
Stratified viscous fluids, interfacial waves 9(644–646)
Streak camera 6a/4(9, 56)
Stream direction 9(250)
Stream function 8/1(78–82); 8/2(8, 10); 9(12, 319)
— for axially symmetric flow 8/1(155, 193)
— for compressible flow 8/1(191)
— for partial fluctuation 8/1(369)
— for plane flow 8/1(154)
— for plane viscoplastic flow 6(303, 306)

Stream function (cont.)
— for viscous flow 8/1(250)
— linear 8/2(41–45)
Stream line 6(386, 400, 426); 8/1(3, 78, 151)
Stream tube 8/1(3); 9(8)
Stream tubes in contraction 8/2(553)
Streamline curvature 9(259, 322)
Streamline, free 9(63, 320, 322, 375–378)
Streamline of discontinuity 9(312, 321)
Streamlines 6a/3(455, 460, 479, 531)
Streamlines, Busemann image 9(24)
Streamlines for standing ocean waves 9(496)
Streams superposed, boundary layer 8/1(305)
Strength 6(448, 519–522)
Strength effect of mean stress upon it 6(593)
Strength of a monopole source 11/1(27)
Strength of a plane wave, *see also* Amplitude 6a/3(100)
Strength of glasses 13(623–635)
— effect of chemical composition 13(629)
— experimental determination 13(631–632)
— surface etching 13(626, 627)
— temperature dependence 13(627–628)
Strength scale for glasses 13(630)
Stress, *see* High hydrostatic stress
Stress 6a/4(117); 7/1(122, 178, 191, 217, 233ff., 242, 265)
Stress, alternating 6a/1(28, 31, 96, 97, 223, 472, 686, 688)
Stress around disc-shaped complexes 7/2(216)
Stress at boundary 8/1(270)
Stress axis, principal 6(13, 137, 327)
Stress birefringence 7/2(42)
Stress birefringence of diamond 7/2(45)
Stress, Cauchy 6a/1(130, 543); 6a/3(175, 180, 183, 185, 325, 536); 6a/4(24, 70–72, 76, 80, 90, 92, 95)
Stress characteristics 6(354, 373)
Stress coefficient, isochoric 11/1(143)
Stress, combined 6a/1(483, 501, 509, 670, 674, 676–684, 702–716)
Stress components at a wall 8/2(7)
Stress concentration 6(202)
Stress conic 6(138, 166)
Stress continuity at an interface 9(451)
Stress definition 6a/1(543–546)
Stress determination in three-dimensional models 6(180–186)
Stress deviation tensor 6(323)
Stress deviator force 6a/3(404)
Stress deviator tensor 6a/3(404)
Stress discontinuities 6a/3(433, 457)
Stress discontinuity 6(366, 399)
Stress discontinuity line 6a/3(498)
Stress energy 6a/2(111, 324)
Stress entropy modulus 6a/3(362, 369)
Stress entropy vector 6a/3(537)
Stress equation of compatibility 6a/2(92); 6a/3(45, 46)
Stress equation of equilibrium 6a/3(46)
Stress equation of motion 6a/2(214)
Stress field 6a/2(45–64)
Stress field for solution of mixed problem 6a/2(103)
Stress field process 6a/3(16)
Stress freezing 6(130, 177–180)
Stress freezing materials 6(178, 180)
Stress fringe pattern(s) 6(149, 166ff.)
— three-dimensional examples 6(215–224)
— two-dimensional examples 6(203–215)
Stress function(s) 6(31–33, 400); 6a/2(53–59); 6a/3(496, 522, 561, 584); 7/1(518, 524, 572)
— for the viscoelastic medium 6(298)
— for the viscoplastic medium 6(303, 304, 306)
— of Kröner 7/2(217)
— viscoelastic 6a/3(73, 93)
Stress graph, *see also* Stress plane 6(357, 368–369, 388–389)
Stress graph 6a/3(445, 477)
Stress in porous media 8/2(629)
Stress, independent of elastic constants 6a/2(163)
Stress, initial high peak 6a/1(690)
Stress lines 6(400); 6a/3(496)
Stress moments for rods 6a/2(649)
Stress multipoles 6a/3(213)
Stress, nominal 6a/1(135, 509, 543, 557–561, 562, 667, 703)
Stress, octahedral shear 6a/1(511, 512)

Stress on a free boundary, *see* Boundary stress
Stress or traction boundary data 6a/3(47)
Stress or traction boundary value problem 6a/3(47)
Stress, Piola-Kichhoff 6a/1(130, 545, 667, 703)
Stress, Piola-Kirchhoff 6a/3(174, 181, 183, 185, 316, 542)
Stress plane, *see also* Stress graph 6(357, 368–369, 388–389)
Stress plane 6a/3(445, 471)
Stress power 6a/2(65, 216, 300, 496); 6a/3(95, 174)
Stress principle of Cauchy 8/1(133)
Stress process 6a/3(42)
Stress profile (stress pulse, stress-time profile), *see* Wave profile
Stress quadric 6(137, 146)
Stress relation 6a/2(89, 212, 302, 305, 306); 6a/4(79)
Stress relaxation 6a/1(74); 6a/3(1, 18, 20, 32); 11/1(361, 370)
Stress relaxation function 6a/3(335, 350, 351)
Stress relaxation property 6a/3(20)
Stress, resolved shear, *see* Resolved shear deformation
Stress response functional
— curvature conditions 6a/4(28)
— for finite linear viscoelastic materials 6a/4(44, 51)
— for simple materials with fading memory 6a/4(26–28)
— thermodynamic theory 6a/4(77)
Stress resultant 6a/2(512, 513, 651)
Stress space 6a/3(406)
Stress state, uniaxial 6(272)
Stress, steady 6a/4(38, 82ff.)
Stress tensor 6(8–10, 234, 323); 6a/2(45–64, 513); 6a/3(404); 7/1(178, 515, 523f., 527, 560); 8/1(135, 136, 230); 8/2(3, 6); 9(452); 10(283, 286); 11/1(3, 7); 11/2(153); 12(238, 239, 270)
Stress tensor, physical components 6(10, 12)
Stress theory, maximum 6a/1(484)
Stress threshold for a dislocation 11/2(189)
Stress trajectory 6(139, 168–170)
Stress transducer 6a/4(8ff.)
Stress transducer, manganin wire gauge 6a/4(8ff.)
Stress transducer, quartz gauge 6a/4(8ff., 37)
Stress, transverse normal 6a/2(584)
Stress, true 6a/1(135, 509, 543, 546, 557–561, 671, 688)
Stress, ultimate 6a/1(5, 11, 60, 167, 186, 442, 456, 571, 574)
Stress vector 6a/2(43, 485, 510); 6a/3(258–260, 261); 8/1(133–135, 230, 241)
Stress wave 6a/2(254)
Stress waves of high frequency 11/2(153ff.)
Stress-change, neutral 6(280)
Stress-couples 6a/2(512, 513, 595)
Stress-difference 6(149, 166)
Stress-difference determination from fringe pattern 6(170–171)
Stress-energy response 6a/4(98ff.)
Stresses in a plane of symmetry 6(180)
Stresses, inner 7/1(455)
Stress-free state 6a/2(323)
Stressless deformation 6(434)
Stressless volume change 6(434)
Stress-momentum field 6a/2(67, 253)
Stress-optical coefficient 6(149)
Stress-relaxation 6(270, 271)
Stress-relaxation along slip bands 6(604)
Stress-strain behavior, first quantitative diagrams of 6a/1(210, 424)
Stress-strain curve 7/2(31, 45)
Stress-strain curve for rubber, temperature dependence 13(442–444)
Stress-strain function, a nonlinear late 19th century tabulation of 6a/1(114, 115)
Stress-strain relation, *see also* Stress relation 6a/2(89, 212)
Stress-strain relations 13(433)
Stress-strain relations in an isotropic elastic material 6(15–26, 142–143)
Stress-strain relaxation function 6a/4(28–30, 45, 64)
— determination from acoustic measurements 6a/4(31–36)
— in three dimensions (bulk stress relaxation) 6a/4(71–72, 74)

Stress-strain relaxation function (cont.)
— initial slope 6a/4(49ff., 64, 66)
— thermodynamic theory 6a/4(78, 94ff.)
Stress-strain temperature relation 6a/2(312, 326)
Stress-temperature equation of compatibility 6a/2(219, 318, 321)
Stress-temperature modulus 6a/2(311)
Stress-temperature relaxation function 6a/4(78, 95ff.)
Stress-temperature tensor 6a/2(309, 313)
Stretch 6a/4(111, 112)
Stretching power 6a/4(118)
Stretching tensor 6a/3(548); 6a/4(113)
Striae (= bands of secondary slip) 7/2(74, 79–80)
Striations in fatigue 6(598–601, 606, 610)
Strings 6a/3(278)
Strip condition 9(14)
Strip, infinite, elastic solution 6(84)
Strip, infinite, visco-plastic solution 6(303, 306–307)
Strip load 11/1(112)
Strong condition for existence of solution of Signorini problem 6a/2(417)
Strong derivative 6a/2(349, 357)
Strong discontinuities, *see also* Shock 6(342, 347); 6a/3(427, 434)
Strong discontinuities in conical flow 9(23)
Strong electrolytes 11/1(248, 250); 13(369–395)
Strong electrolytes, absorption measurements 11/1(234–235)
Strong electrolytes, definition 13(370)
Strong ellipticity 6a/3(64, 99, 211, 213, 321)
Strong quasi-static viscoelastic process 6a/3(44)
Strong response function 6a/3(13)
Strong sense 6a/2(416)
Strongly dissipative material 6a/3(22)
Strongly elliptic elasticity
— field 6a/2(105)
— tensor 6a/2(86, 87, 247, 256)
Strongly elliptic mapping 9(124)
Strongly elliptic operator 6a/2(365, 422)
Structons 13(550)
Structural breakdown, *see also* Yielding 6(230, 231)
Structural changes 14(149)
Structural inhomogeneities in glass 13(527)
Structural relaxation 11/1(202, 401)
Structural shrinkage 13(546)
Structural viscosity 6(488, 494–498, 531–535)
Structure, close-packed 7/1(115, 505f.)
Structure, cubic closed-packed 7/1(115, 116, 203, 207)
Structure, hexagonal close-packed 7/1(77, 114, 116, 198, 202, 205, 207, 576)
Structure insensitive properties 7/2(1)
Structure of liquids, quantitative description 10(25–47, 47–81, 81–113)
Structure of turbulent field 8/2(441)
Structure of water 11/1(202)
Structure sensitive properties 7/2(1)
Structurised fine-slip 7/2(86, 88–89, 92)
Strutt's diagram 6a/3(257)
Struve function 9(533); 11/1(72)
Sturm-Liouville problem 9(523)
StV-body, *see* Saint-Venant body
Subcritical flow in an open channel 9(673, 675, 676)
Subcritical state of flow 8/1(359)
Subgrain formation 7/2(194)
Subgrain growth 7/2(194)
Sublattice 15(29)
Sublattice, spontaneous magnetization 15(129)
Submarine explosion 9(617–619)
Submarine, wave resistance 9(577)
Submerged moving dipole 9(487, 489)
Submicroscopic catalyst crystals 13(605)
Subsonic compressible flow
— harmonic vibrations 9(297–299)
— solution by Green's function 9(291ff.)
Subsonic edge 8/1(76)
Subsonic flow 8/1(13, 69, 94)
Subsonic flow, definition 9(7)
Subsonic infinite cavity flow, existence theory 9(394)
Subsonic jet, acoustic spectrum 11/1(98)
Subsonic jets generating sound 11/2(27–29)

Subsonic potential flow 8/1(198–206)
— existence and uniqueness 8/1(201)
— properties 8/1(199–201)
— variational principles 8/1(203)
Subsonic source speed 11/1(77)
Subsonic whistle 11/2(29)
Subspaces of symmetric and skewsymmetric tensor $\mathcal{T}_{sym}$ and $\mathcal{T}_{skw}$ 6a/3(4)
Substitution, continuous, of electrolyte by solvent 11/1(227, 231)
Substitutional solid solutions, work-hardening curve 7/2(179–182)
— yield point 7/2(171–179)
Subsurfaces $\mathcal{S}_1$ and $\mathcal{S}_2$ 6a/2(12)
Successive approximation of solution 9(19)
Sucking away of boundary layer 8/1(347, 409–414)
Suction along wall 8/2(549)
Sudden transition 6a/3(414)
Sulphate, Cerium ethyl- 15(119)
Sulphate, cobalt 15(110f., 125)
Sulphate, cobalt ammonium 15(110)
Sulphate, copper 15(107f., 124)
Sulphate, copper potassium 15(104f., 149f., 188)
Sulphate, gadolinium 15(111f.)
Sulphate, Manganese ammonium 15(102f., 146f.)
Sulphate solutions, absorption measurements 11/1(234–245)
Sulphates of bivalent metals 11/1(248, 258)
Sulphide glasses 13(514)
Sulphur dioxide 14(16, 24)
Sulphuric acid, activation energy 11/1(253)
Sulphuric acid, dissociation 11/1(252)
Sulphuric acid, relaxation absorption 11/1(253)
Summational invariant 12(235)
Superaerodynamics, *see* Rarefied gas dynamics
Superconducting circuits 15(219)
Superconducting nucleus 15(341)
Superconducting sphere oscillating in a magnetic field 15(275)
Superconductive cylinder 15(221)
Superconductive sphere 15(221)
Superconductivity 14(140, 171, 188, 272, 325); 15(172, 394)
Superconductor 15(191, 196, 197, 429)
Superconductors, attenuation of ultrasonics 11/2(221–223)
Superconductors, temperature dependence of attenuation 11/2(209, 221)
Supercooled liquid 13(291, 511, 595)
Supercooled state in liquids 10(9)
Supercooling 15(256, 340)
Supercritical flow in an open channel 9(673, 675–676, 696)
Supercritical state of flow 8/1(359)
Supercritical steady flow, shallow-water first-order approximation 9(682–695)
Super-exchange interaction 15(83, 84)
Superficial activity coefficient 10(205, 208)
Superficial density
— of internal energy 10(245)
— of molecules 10(267, 269)
— of potential energy 10(245, 246)
— of the Helmholtz free energy 10(139)
Superficial entropy 10(78, 137, 166, 167, 177)
Superficial free energy 10(137, 149, 285, 288, 296)
Superficial internal energy 10(137, 166, 167, 185)
Superficial number density 10(141, 142, 145, 166, 167, 171, 184, 185, 202, 227)
Superficial pair density 10(184)
Superficial quantities 10(137, 139, 149, 166)
Superfluidity 15(379, 380, 387, 394, 397, 410)
Superfluidity of helium 10(119, 122)
Superfluidity onset 15(419, 451, 452)
Superheating 15(256, 340)
Superlattice structure in brass 13(81–82)
Superlattice structure of $AuCu_3$ 13(80–81)
Superlattice transformations 13(80, 81)
Superlattice transformations, comparison of different approximations 13(104, 105)
Superlattice x-rays lines 13(81)
Superposed rigid body motions 6a/2(452, 532)
Superposition approximation of kinetic theory 10(52, 53, 59, 66, 74, 111, 181, 197, 199, 201)

Superposition, method 6a/3(448)
Superposition of ionic clouds 13(383, 389)
Superposition of motions 8/2(90)
Superposition of rheological bodies 6(478ff.)
Superposition of singularities on the axis 9(55)
Superposition principle 9(55, 57, 63)
— of Kirkwood 13(19, 27, 39, 41, 369)
— of Kirkwood for an electrolyte 13(372)
— validity for liquids 13(219)
Superposition, property 6a/3(12)
Supersaturated solutions of copper in aluminium, slip lines 7/2(102)
Supersaturation 7/2(217)
Supersonic air-stream 6a/3(281)
Supersonic compressible flow, harmonic vibrations 9(306)
Supersonic conical flow 9(22–24, 49–52)
Supersonic edge 8/1(76)
Supersonic flow 8/1(13, 69, 94, 206)
— around a corner 9(20–22)
— complete stabilization 8/1(425)
— critical Reynolds numbers 8/1(426)
— definition 9(7)
— linearized approximation 9(38–40)
— past an airfoil 9(244, 253, 274)
— standing waves 8/1(109–111)
— steady and axially symmetrical 9(274)
— tables 9(23)
Supersonic jet 11/2(29–32)
Supersonic linearized two-dimensional flow 9(45–49)
Supersonic pockets in the flow around an edge 9(86)
Supersonic profile 9(48)
Supersonic source speed 11/1(77)
Supply of moment of momentum 6a/2(482)
Support impedance 11/1(105)
Support of *u* 6a/2(349)
Support resonance 11/1(107)
Suppression of crystallization 13(611)
Surface angle, diffraction grating measurement of 6a/1(622–625, 635–637)
Surface angle, measurement of 6a/1(622–625, 635–637)
Surface angle vs time histories 6a/1(634)
Surface base vectors and their reciprocals 6a/2(623)
Surface charge 10(226)
Surface condition(s) 7/2(109); 8/1(137, 240)
Surface contribution 14(332)
Surface crystallization of glass 13(593)
Surface damage of glass 13(630)
Surface displacement 6a/4(11–13)
Surface effects on glass fracture 13(626–627)
Surface energy 7/2(212); 10(141, 185, 190, 216, 244, 245, 246, 247, 248, 257ff., 259); 15(254, 259, 446, 447)
Surface energy of argon 10(217)
Surface energy parameter 15(249, 260)
Surface entropy 10(141, 171, 248, 249, 259, 260)
Surface force 6a/3(174, 278); 8/1(5); 10(286)
Surface force field 6a/2(59, 215)
Surface forces 7/1(234)
Surface fraction 13(458)
Surface, free 9(320)
Surface geometry 6a/2(621)
Surface gloss of glasses 13(544)
Surface heat flux moments for rods 6a/2(651)
Surface impact 9(360–363)
Surface impedance at microwave frequencies 15(341)
Surface impedance, scattering at an inhomogeneity 11/1(40–42)
Surface integrals representing sources of sound 11/1(24, 26)
Surface interaction of rarefield gases 8/2(595)
Surface interreflection 8/2(608)
Surface loads 6a/3(85, 278)
Surface normal 9(451)
Surface of constant slope 6(401, 402); 6a/3(497)
Surface of discontinuity 6a/2(248); 10(216ff.)
Surface of revolution, boundary layer 8/1(311, 314)
Surface of seepage 8/2(639)
Surface phenomena in plastic deformation 7/2(19, 71–109)

Surface, potential density 6a/3(259, 260)
Surface problems 7/1(534, 560)
Surface rate of deformation 6a/2(451)
Surface scattering 11/1(29, 32–34, 40, 42)
Surface scratches 6(598)
Surface spherical harmonic 6a/2(21)
Surface spin tensor 6a/2(451)
Surface stretch 6a/4(113)
Surface tension, *see* Surface energy
Surface tension 6a/2(547); 7/2(220); 9(352, 451, 452, 460, 614, 631–638); 10(23, 77, 81, 138, 139, 148, 149, 158, 161ff., 166, 167, 171, 172ff., 177, 179, 182, 186, 190, 192, 193, 198, 200, 201, 205, 211, 213, 215, 216, 218, 219, 237–239, 240, 246, 248, 257ff., 259, 260, 266, 268, 270, 271, 275, 276, 277); 11/1(343)
— curvature dependence 10(152, 153, 215)
— definition 10(288, 290)
— difference 10(197, 202, 227ff., 230ff.)
— dilute solution 10(298)
— dynamic 10(271ff.)
— effect on cnoidal waves 9(714)
— effect on second-order infinitesimal wave approach 9(656ff.)
— effect on stability 9(647)
— empirical formula 10(172, 174)
— γ 7/1(604)
— high temperature corrections 10(288)
— in viscous fluids 9(638, 641ff., 646)
— increase 10(229)
— mechanical definition 10(158, 161, 166, 167, 261)
— of a perfect solution 10(268)
— of a plane interface 10(181, 184)
— of a plane interface mechanical definition 10(157)
— of argon 10(217)
— of athermal polymer solutions 10(273)
— of polymer solutions 10(272)
— of solutions 10(201)
— of strong electrolytes 10(223)
— of the spherical interface 10(146, 159, 161, 187, 214)
— of the spherical interface, mechanical definition 10(159, 160)
— quantum effects 10(237)
— reduced 10(172, 173, 175, 176, 187)
— static 10(271ff.)
— temperature dependence 10(151, 259, 297)
Surface theory, basic theorems 10(284)
Surface traction 6a/2(44, 93, 95, 211, 313, 319, 327); 6a/3(42, 85, 178, 259)
Surface vectors 6a/2(450)
Surface wave(s) 11/1(109–110, 113, 124); 11/2(76)
— dispersion relations 9(502, 507, 513–515, 518, 550)
— initial value problem 9(603ff.)
— of the Rayleigh type 6(109–111)
Surface zone in polar and associated liquids 10(80)
Surface-active material 7/2(109)
Surfaces of constant phase 11/1(78)
Surfaces of constant pressure amplitude 11/1(78)
Surplus protons, fluctuations 11/1(251)
Susceptibilities 15(47, 80, 85, 104, 112, 117, 126, 129, 133, 137, 142, 143, 147, 149, 150, 151, 158, 162, 165, 191)
Susceptibility and spin correlation function 13(207)
Susceptibility, anisotropy 15(152, 159)
Susceptibility, anomaly 15(130)
Susceptibility, ballistic 15(126)
Susceptibility changes during precipitation 7/2(230)
Susceptibility, complex 15(20)
Susceptibility, dynamic 15(72, 126)
Susceptibility maximum 15(76, 126, 134, 137, 147, 149, 159)
Susceptibility, polar diagrams 15(152, 159)
Susceptibility, static 15(72)
Suspension 6(435)
Sutherland model 12(340)
Sutherland temperature 12(395, 396, 428)
Sutherland's formula 8/1(336)
Suzuki effect 7/2(176, 178)
Swainger measure of deformation 6(499)
S-waves (secondary waves) 6(107–109)
Swelling equilibrium 13(487–491, 507)
Swelling of a gel in different solvents 13(489, 507)

Swelling parameter of a polymer solution 13(471–472)
Sylvite 6a/1(408)
Symbols, short, Hermann-Mauguin 7/1(27)
Symmetric bounded bilinear form 6a/2(391)
Symmetric elasticity tensor 6a/2(69, 84, 85, 247, 256)
Symmetric gradient 6a/2(11)
Symmetric impact 6a/4(16, 41, 58, 63, 65)
Symmetric infinitesmial strain field process 6a/3(17)
Symmetric initial elasticity 6a/3(62)
Symmetric operator 6a/2(369)
Symmetric tensor 6a/2(6, 617)
Symmetric tensors, subspace of ($\mathscr{T}_{\text{sym}}$) 6a/3(4)
Symmetric turbulence 8/2(535)
Symmetrical cylinder, boundary layer 8/1(272–274)
Symmetrical Joukowski profile 8/2(390)
Symmetrical of stress tensor 8/1(136)
Symmetrization 9(404)
Symmetrization of the wave function 12(458)
Symmetry condition for rods 6a/2(654)
Symmetry conditions of elastic properties 11/2(154)
Symmetry, geometrical 6a/2(297)
Symmetry group 6a/2(70, 210); 6a/3(18)
Symmetry groups, enantiomorphic 7/1(11)
Symmetry line 7/2(60)
Symmetry number 12(104–108, 122)
Symmetry of a velocity field 8/2(4)
Symmetry of acoustic tensor 6a/3(99)
Symmetry of Cauchy stress 6a/3(542)
Symmetry of Green's processes
— of the first kind 6a/3(82)
— of the second kind 6a/3(84)
Symmetry of Green's states 6a/2(187)
Symmetry operation 7/1(3ff., 118)
— generating 7/1(4)
— graphic symbols of 7/1(7)
— of the lattice 7/1(118, 146, 147, 148, 298)
— two-dimensional 7/1(16)
Symmetry principle 7/1(3, 76)
Symmetry, properties of 7/1(515)
Symmetry relations 12(237–238)
Symmetry symbols of, Hermann and Mauguin 7/1(8)
Symmetry symbols of Schönflies 7/1(8)
Symmetry transformation 6a/2(70); 6a/3(18)
Synchronous and separable boundary data and body forces 6a/3(48, 55)
Synchronous external force systems 6a/2(219)
System of concentrated loads 6a/2(179)
System of forces 6a/2(43)
System of *n* fluids 9(506)
System of two fluids, *see* Interface
Systems of differential equations 6a/3(420)

T^3 region 14(288, 307)
Tangent modulus 6a/1(90, 91, 103, 105, 106, 118, 129, 136, 732, 734); 6a/3(356, 358)
Tangent modulus, defined 6a/1(105)
Tangent modulus, second-order transitions in 6a/1(400–406)
Tangent plane 6a/2(622)
Tangent unit of a curve 6a/2(623)
Tangential and normal components of director and of director gradient 6a/2(535)
Tangential force on a free surface 8/2(125)
Tantalum 14(275)
Taper 8/2(386, 429)
Tapered network 11/1(407)
Tardy method of measuring retardations 6(153–154, 160–161, 172)
Target alignment 6a/4(7)
Target plate (material sample, specimen) 6a/4(3–10, 14–22, 41, 55ff., 65)
Taylor and Elam's single crystal parabola coefficients for aluminum and gold 6a/1(536)
Taylor and Quinney's experiments 6a/1(503–509, 521–522, 559, 561–563, 677)
Taylor dissipative scale 8/2(528, 547, 568)
Taylor flow, transform leading to Bateman flow 8/2(183)

Taylor instability 11/2(271)
Taylor type of transonic nozzle flow 9(147–148, 151)
Taylor vortex 8/1(119–121, 431–432)
Taylor work-hardening 7/2(128)
Taylor-Batchelor theory of turbulence 8/2(536–537)
Taylor-Görtler vortex 8/1(119–121, 434, 436, 438)
Taylor-Maccoll flow pattern 9(23)
Taylor's hypothesis on turbulent flow 8/2(453)
Taylor's vortex system 8/2(138–139, 146, 168)
Taylor-von Kármán, nonlinear theory of waves 6a/1(598, 609, 610, 629, 645)
Technical glasses, properties and composition 13(543–544)
Technological properties of materials 6(540–542)
Technology of metals 7/2(17)
Telegraph equation 6(363, 369ff.)
Telegraph equations of viscoelastic waves 6(301, 313)
Telegraph wires, wave propagation in 6a/1(306)
Telegraphic equation 8/2(647)
Temperature, *see also* High temperature, Low temperature, Temperature dependence 6a/3(174, 536); 6a/4(123)
Temperature, absolute 6a/2(301, 483, 510); 6a/4(32ff., 76–80, 82, 88, 90, 92, 95ff.)
Temperature, absolute, definition 15(54)
Temperature, acoustic 11/1(6, 9)
Temperature alternation, tests 7/2(118)
Temperature change, effect on work-hardening 7/2(161–166)
Temperature change under load 6(130, 177–180)
Temperature coefficients from ultrasonic measurement 6a/1(409)
Temperature, Compressibility of solids as a function of 6a/1(373, 374)
Temperature condition 6a/2(335)
Temperature, critical 10(66, 133, 297); 12(42, 141–146)
Temperature dependence
— of critical field 15(277)
— of critical shear stress 7/2(34–38, 116)
— of dielectric constants in piezoelectric materials 11/2(47)
— of elastic constants 6a/1(217, 218, 278, 293–296, 305, 311, 360–396)
— of glass viscosity 13(581–583)
— of long-range ordering 13(87, 88)
— of magnetostriction 11/2(86)
— of measured absorption in electrolytes 11/1(240)
— of orientation ordering 13(215)
— of overshooting 7/2(182)
— of parabola coefficients for finite strain 6a/1(511, 535, 536, 625, 638, 703)
— of relaxation frequency 11/1(270)
— of second virial coefficient 13(466–469)
— of sound velocity 11/1(352–354)
— of sound velocity in gases 11/1(188–191)
— stress-strain relations in rubber 13(442–444)
— of superlattice structure 13(81, 87)
— of surface tension 10(151, 259, 288, 297)
— of the heat conductivity 12(400)
— of the specific heat of polymers 13(422)
— of the viscosity 12(399)
— of the volume of polymers 13(422)
— of ultrasonic absorption 11/1(315)
— of ultrasonics in quartz and germanium 11/2(230, 231, 233)
— of work-hardening curve 7/2(48–49, 51–54, 66–71)
— of yield point 7/2(173, 174)
Temperature determination 15(137, 140, 143, 146, 148, 150)
— bridge methods 15(71)
— calorimetric 15(89, 90, 92, 99, 111)
— theoretical method 15(58)
Temperature difference field 6a/2(312)
Temperature effect(s)
— in inelastic media 6(292, 293)
— on dislocation damping 11/2(187–191)
— on surface healing of glass 13(627)
— on the propagation of sound 11/1(81)
— upon endurance limit 6(593)

Temperature equation 6a/2(328)
Temperature, equilibrium response function 6a/4(81)
Temperature field, non-homogeneous 6(293)
Temperature, first study of effect upon elastic constants 6a/1(218, 227)
Temperature fluctuations 8/2(534)
Temperature, functional 6a/4(80)
Temperature gradient 6a/2(301); 7/1(294ff., 314)
Temperature, history 6a/4(76)
Temperature increase, local 6(602, 604)
Temperature, instantaneous response function 6a/4(81, 90)
Temperature, internal 12(497)
Temperature irregularities in steady flow 8/2(553)
Temperature jump at a surface 8/2(600, 611)
Temperature, Kelvin scale of 15(55, 56)
Temperature, magnetic 15(55, 118)
Temperature of anemometer wire 8/2(563)
Temperature of Bordoni peak 11/2(203)
Temperature of glass transition 13(431, 447)
Temperature of phase transformation in the two-dimensional Ising model 13(118)
Temperature of precipitation 13(149–153, 161–166)
Temperature of wall, effect on critical Reynolds number 8/1(427)
Temperature, past history 6a/4(76ff.)
Temperature, present value 6a/4(76, 78)
Temperature, reduced 10(133)
Temperature relations 6a/2(305)
Temperature, representation in rod theories 6a/2(649)
Temperature scale, absolute 12(2, 3)
Temperature sensitive detectors of sound 11/2(61–62)
Temperature sensitivity of viscosity 6(296)
Temperature, small relative 6a/4(77, 95)
Temperature, steady 6a/4(82, 87)
Temperature, suitable, of anemometer wire 8/2(557)
Temperature thickness of boundary layer 8/1(347)
Temperature, translational 12(496)
Temperature variation with magnetic field 15(121)
Temperature-entropy diagram for helium 14(116)
Temperature-entropy moment for rods 6a/2(649)
Temperature-gradient 14(196)
Tempering, effect of, Coulomb's conjecture 6a/1(177–178, 213, 214, 217)
Tensile deformation of aluminium 11/2(198)
Tensile pulse 6a/1(717)
Tensile strength 7/1(98); 7/2(14); 10(22)
— liquids 6(551, 555)
— measurements, pure liquids 6(555, 556)
— of ceramics 13(606)
— of glass 13(577–578, 625)
— of glass, theoretical relations 13(624)
Tensile stress 7/2(28)
Tensile stress, critical 7/2(33)
Tensile test 7/2(21, 22, 28, 32)
Tensile waves 6a/1(695–697, 716)
Tension cleaving 7/1(99)
Tension, homogeneous 7/1(193)
Tension of glasses, effect on gas diffusion 13(620)
Tension, surface of 10(147, 148, 155, 160, 161, 163, 212, 213, 216, 217, 218, 287, 290)
Tension test through zero stress to compression 6a/1(138)
Tension versus compression 6a/1(467)
Tensions, thermodynamic 6a/4(124)
Tension-torsion experiments 6a/1(485, 501–508, 513, 670–675, 678–684, 702–716, 739)
Tensor 6a/3(3, 4)
Tensor components 6a/2(498)
Tensor components, physical 6(441ff.)
Tensor condition, for cylindrical symmetry 7/1(53)
Tensor condition for isotropy 7/1(53)
Tensor, covariant, contravariant, mixed 6a/2(617, 622)

Tensor, definition 6a/2(6)
Tensor field 6a/3(7)
Tensor fields 6a/2(10)
Tensor invariants 6(6, 239–243, 258, 443ff.)
Tensor laws of differential type 6a/3(37)
Tensor notation and matrix notation of elastic coefficients 11/2(155)
Tensor of angular velocity 6a/3(415)
Tensor of momentum 8/1(7)
Tensor of permeability 8/2(636)
Tensor of stress 8/1(135, 136, 230)
Tensor of surface forces 8/1(5)
Tensor of valence *s* 7/1(46, 56)
Tensor product 6a/2(7)
Tensor, relative 6a/2(618)
Tensor, surface 6a/2(631)
Tensor transformation laws 6a/2(616)
Tensor type 6a/2(617)
Tensorial non-linearity 6(487, 507–509)
Tensors, axial 7/1(47)
Tensors, polar 7/1(47)
Tensors, proper symmetries 7/1(56)
Terezawa's solution for the half-space 6(99–101)
Ternary alkali silicate systems, limits of glass formation 13(613)
Ternary borate glasses 13(570)
Ternary mixture 12(60)
Ternary mixture, diffusion 12(423)
Ternary polymer systems 13(481–487)
Ternary silicate glasses 13(609)
Ternary systems with one polymer component 13(482–484)
Ternary systems with two polymer components 13(484–485)
Test gas method of leak investigation 12(542–543)
Tetrahedra of BO_4 13(531)
Tetrahedra of SiO_4 13(521, 613–614)
Thallium 14(275)
Theodorsen's method to compute the pressure distribution 8/2(398)
Theorem of conformal states 13(229, 356, 358)
— for mixtures 13(361, 362)
— for multicomponent systems 13(365–369)
Theoretical design for contractions 8/2(555)
Thermal after-effect 6a/1(38, 102, 118, 120, 152, 411–413, 565–570)
Thermal anomaly of stretched rubber; the Gough effect 6a/1(44, 308, 309, 312, 411, 728–733)
Thermal boundary layer 8/1(420); 11/1(12)
Thermal conduction of liquids 10(14, 20, 38, 82, 86–89, 111, 115)
Thermal conductivity 7/1(291, 292, 294, 296, 307, 310, 311, 312, 313, 314, 315, 316); 14(137, 204, 214, 221, 230, 239); 15(166, 167, 195, 261)
— anomalies 14(227)
— at high temperatures 14(219)
— at intermediate temperatures 14(238)
— at low temperatures 14(221)
— effect of neutron irradiation on the 14(224)
— effect of three-phonon interaction on the 14(220)
— (electronic), quasi-equilibrium effect 14(258)
— in the intermediate state 14(273)
— lattice component of the 14(251)
— lattice component of the 14(272)
— methods of measurement 14(200)
— minimum of 14(196)
— of alloys 14(263)
— of amorphous solids 14(216)
— of crystals 14(218, 221)
— of ferromagnetic metals 14(227)
— of glass 13(577–578)
— of liquids 14(227)
— of rubber 14(217)
— of superconductors 14(266, 269)
— stationary 14(237)
Thermal contact 15(165)
Thermal diffusion in liquids (Soret effect) 10(2, 24, 82)
Thermal dissipation potential 6(264)
Thermal dissociation of oil 12(649)
Thermal distribution 7/1(294, 307, 313)
Thermal distribution of the amplitudes of an oscillator 7/1(261)
Thermal effects 6a/4(201)
Thermal effects in dislocation damping 11/2(187–191)
Thermal effects of ultrasound 11/2(123–124)

Thermal equation of state 13(13, 27, 65, 229, 240, 247, 276)
Thermal equilibrium 12(361, 489); 15(166)
— in a demagnetized sample 15(67)
— jogs in 7/1(553)
— of a glass 13(544, 555)
Thermal excitation, modes of 14(283)
Thermal expansion 6a/4(146, 240); 14(208)
— coefficient 15(239)
— coefficient, cubic 12(5)
— coefficient of a liquid 13(298, 359)
— coefficients 6a/4(125)
— coefficients of 6a/1(372)
— of ceramics 13(606)
— of glasses, *see also* Expansion coefficient 13(561–570)
— tensor 6a/2(310)
— wave 6a/4(255, 256)
Thermal fluctuations 6(553)
Thermal flux vector 10(36, 42, 82)
Thermal generation of vacancies 7/2(115, 121, 123)
Thermal gradient-temperature relation 6a/2(326)
Thermal histories, *see also* Fully annealed solid, Prior thermal and mechanical histories
Thermal histories during deformation 6a/1(44–47, 86, 103, 412, 565–570, 687, 688, 728–731)
Thermal history of a glass
— and crystallization 13(610–611)
— effect on its density 13(555)
— effect on its specific heat 13(572)
Thermal inertia of anemometer 8/2(557, 578)
Thermal molecules, reflection from surfaces 8/2(596–600)
Thermal motion 11/1(2)
Thermal nucleation 13(595)
Thermal phonon distribution changed by sound 11/2(233–234)
Thermal pressure coefficient 12(6)
Thermal pretreatment in glass manufacture 13(545)
Thermal properties 15(275)
Thermal pulse 6a/1(358)
Thermal relaxation 6a/4(85); 11/1(14, 283–285)
Thermal relaxation due to slow chemical processes 11/1(287–288)
Thermal relaxation function 6a/4(83ff.)
Thermal resistance 7/1(291)
Thermal resistance, additivity of 14(220, 238, 244)
Thermal resistance, anisotropy of 14(245)
Thermal shearing stresses 7/1(622)
Thermal shock resistance of glass 13(577)
Thermal softening 6(600)
Thermal spikes 7/1(438)
Thermal stresses, photoelastic investigation 6(228)
Thermal switch 15(197, 198)
Thermal time constant of a sphere 11/2(270)
Thermal valve 15(197, 203)
Thermal vibrations 15(166)
Thermalcreep 8/2(610)
Thermistor vacuummeter 12(526)
Thermo-anelasticity 6(263–264)
Thermocouple microphone 11/2(61–62)
Thermodiffusion 11/1(139, 174); 12(304, 325)
Thermodiffusion at low temperatures 12(467–469)
Thermodiffusion coefficient 12(407)
Thermodiffusion coefficient of the Lorentz gas 12(417, 418)
Thermodiffusion column 12(328)
Thermodiffusion column, exchange reactions 12(329)
Thermodiffusion factor
— change of sign 12(444)
— correction for the first approximation 12(438, 439)
— dependence on the molar fraction 12(441–442)
— for a binary mixture 12(412, 425, 435–437)
— for polynary isotope mixtures 12(436)
— for special intermolecular forces 12(437–441)
— Kihara approximation 12(436)
— measurements 12(326–334)
— pressure dependence 12(310, 333)
— temperature dependence 12(443–447)

Thermodiffusion ratio 12(304, 408, 423, 424)
Thermodynamic basic parameters of capillarity 10(287)
Thermodynamic behaviour of liquids 10(61–67)
Thermodynamic coefficients 6a/4(232)
Thermodynamic coupling stiffness 11/1(395)
Thermodynamic data 14(19, 46)
Thermodynamic elastic coefficients 6a/4(234, 235, 239)
Thermodynamic entropy 12(241, 253)
Thermodynamic equations of state 12(6)
Thermodynamic equilibrium 11/1(271)
Thermodynamic equilibrium of a network 11/1(391–392, 406)
Thermodynamic functions
— Born and Green's computation method 13(11)
— Kirkwood's computation method 13(14)
— near the critical point 12(39)
— of two-dimensional oriented solutions 13(342)
Thermodynamic potential(s) 6a/4(127, 133, 168); 10(26, 107–108)
Thermodynamic potentials including electrical variables 6a/4(169)
Thermodynamic principal curvature terms 10(290)
Thermodynamic process 6a/2(484, 497, 510)
Thermodynamic properties of the two-dimensional Ising model 13(133)
Thermodynamic quantities and pair distribution functions 13(27)
Thermodynamic relations 15(277)
Thermodynamic relations across a shock 6a/4(289)
Thermodynamic stability 12(10)
Thermodynamic state 12(212)
Thermodynamic stiffness 11/1(406)
Thermodynamic substate variables 6a/4(124)
Thermodynamic temperature for a liquid at rest 10(28)
Thermodynamic tension 6a/4(124, 166)
Thermodynamic variables 6a/4(169); 11/1(3)
Thermodynamics, first law 8/1(175)
Thermodynamics, local formulation 10(163)
Thermodynamics of adiabatic demagnetization 15(52)
Thermodynamics of irreversible processes 7/2(3); 10(24, 25, 81, 168, 169, 271)
Thermodynamics of irreversible processes, validity 12(414)
Thermodynamics of jet formation in valves 12(566)
Thermodynamics of magnetization 15(50)
Thermodynamics of precipitations 7/2(232–238)
Thermodynamics of the superconductive phase transition 15(233)
Thermodynamics of viscoelasticity 6a/3(3)
Thermodynamics, second law 6a/3(537); 8/1(175)
Thermoelastic anisotropy 13(450)
Thermoelastic attenuation 11/2(177)
Thermoelastic coupling 6a/4(255)
Thermoelastic effects 11/2(172–177)
Thermoelastic effects in cubic crystals 11/2(175, 177)
Thermoelastic effects in hexagonal crystals 11/2(176)
Thermoelastic inversion 13(443)
Thermoelastic medium, idealized 6a/4(124, 215)
Thermoelastic potential of Goodier 6(125)
Thermoelastic process 6a/2(327)
Thermoelastic rod 6a/2(656, 669)
Thermoelastic shell 6a/2(561)
Thermoelastic state 6a/2(313)
Thermoelastic wave velocity 13(576)
Thermoelasticity 6(123–126); 6a/1(44, 103, 411–413, 446, 728–731); 6a/3(173–176, 183–185, 187–195, 249–255)
Thermoelasticity, Joule's experiments 6a/1(411, 728–731)
Thermoelectric effect, lattice component of the 14(257)
Thermoelectric e.m.f. 15(267)
Thermoelectric power 14(137, 146)
Thermoelectric power, absolute 14(191)
Thermoelectric probe 11/2(123–124)

Thermoelectric vacuummeter 12(526–527)
Thermo-mechanical behavior (response)
— of bodies 6a/2(512)
— of elastic Cosserat surface 6a/2(529)
— of shells 6a/2(497)
Thermo-mechanical cycle 15(389)
Thermo-mechanical effect 15(377, 378, 388, 390, 408, 424, 444, 449)
Thermometer problem 8/1(61)
Thermometer, secondary 15(55)
Thermometers 14(295)
Thermometric parameter 15(55, 136, 140, 141, 143, 146, 148, 150, 162, 187)
Thermometry 12(23–25)
Thermoplasticity 6a/1(86, 565–570, 588)
Thermoplasticity, Tresca on 6a/1(446)
Thermorheologically simple materials 6a/3(3); 13(447)
Thermostatic approximation 6a/4(163)
Thermostatic assumptions 6a/4(123)
Thermostatic relations, classical 6a/4(124)
Thermostatics 6a/4(162)
Thermostatics under hydrostatic pressure 6a/4(162)
Thermoviscoelastic medium 6a/4(202, 209)
Thermoviscoelastic medium, acoustical tensor in 6a/4(213)
Thermoviscoelastic medium, propagation condition in 6a/4(213)
Thermoviscoelastic waves 6a/4(251)
Thermoviscoelasticity 6a/4(204)
Thermoviscous number 11/1(278)
Theta temperature of Flory 13(460, 468, 469, 471, 481)
Θ_D value 7/1(364)
$\Theta_D - T$ curves, types of 7/1(372)
$\Theta_D - T$ plot 7/1(361)
ϑ-function-transformation 7/1(134, 213)
Θ-value 14(254)
Thickening 6a/3(519)
Thickness distribution 8/2(387)
Thickness of an airfoil 8/2(387, 392, 429)
Thickness of laminar boundary layer 8/1(63, 265, 268, 279, 281, 347)
Thickness of shell 6a/2(443)
Thickness of shock layer 8/1(15)
Thickness of turbulent boundary layer 8/1(361)
Thickness shear 6a/4(258)
Thickness shrink 6a/4(113)
Thickness swell 6a/4(113)
Thickness vibrations 6a/4(257, 271)
Thickness-longitudinal vibrations 6a/4(268)
Thickness-shear vibration 6a/2(575)
Thickness-shear vibrations 6a/4(257)
Thick-walled tube 6a/3(506)
Thiele's semi-invariants 13(99, 367, 384)
Thin airfoil in linearized subsonic flow 9(32, 33–34)
Thin body 9(28, 46)
Thin film 6a/2(547)
Thin films 14(183)
Thin floating body, heaving motion 9(620)
Thin hydrofoil 9(583–587)
Thin shells and plates 6a/2(438, 524)
Thin ship approximation 9(572, 579–583)
Thin ship, transient problems 9(616–617)
Thin supersonic wing 9(43–45)
Thin wing approximation 9(32–34, 43–45, 572, 584, 590)
Thin wires 14(183)
Third virial coefficient 13(451)
Third-order elastic coefficients 6a/4(166, 221)
— mixed 6a/4(225)
Third-order elastic constants, *see also* Elastic constants, third order 6a/4(185)
Third-order infinitesimal wave approach for progressive waves 9(657–658)
Thirteen moment equations 8/2(595, 610–612, 614); 12(271, 273, 276)
Thixotropy 6(540–541)
Thomas-Fermi model 14(177)
Thomson body, *see* Poynting-Thomson body
Thomson coefficient 15(267)
Thomson's circulation theorem 8/1(35, 162)
Three-atomic gases, dispersion and absorption 11/1(179–180)
Three-dimensional body *B* 6a/2(438)

Three-dimensional disturbances 8/1(430–440)
Three-dimensional Euclidean point-space 6a/3(4)
Three-dimensional flow 9(24–26)
Three-dimensional inner product space 6a/3(4)
Three-dimensional Ising model 13(139)
Three-dimensional laminar boundary layer 8/1(307–328)
Threedimensional lattice 7/1(133)
Three-dimensional problem of plasticity 6a/3(403)
Three-dimensional stress, photoelastic investigation 6(130, 177–202)
Three-dimensional subspace of skew-symmetric tensors 6a/3(4)
Three-dimensional surface waves
— diffraction 9(543–547)
— higher-order infinitesimal wave theory 9(662)
— in a canal 9(551–553)
— on the beach 9(545, 547–551)
Three-dimensional turbulent boundary layer 8/1(437–439)
Three-dimensional wave equation 6a/3(137, 162)
Three-phase confluent zone, *see* Confluent zone
Three-phonon processes 14(206, 212, 213)
Three-point correlations 8/2(488)
Three-step dissociation 11/1(262, 270)
Threshold energy 7/2(221)
Threshold field 15(213, 222)
Threshold field curve 15(213, 234, 235)
Threshold field of colloids 15(242)
Threshold of cavitation 11/2(271–272, 286)
Thrust 8/1(87)
Thurston's autographic testing machine 6a/1(96, 106, 450–455)
Thwaites' method of boundary layer theory 8/1(283–286, 318)
Tidal waves 9(466, 668, 669)
Tilt 6a/4(4, 9, 12ff., 56)
Tilt, measurement 6a/4(4, 7)
Tilting stage 6(187ff.)
Time constant 12(272, 275)
— for complexes of approach 11/1(223–224)
— of ion cloud 11/1(224)
— of relaxation 11/1(205, 207)
— thermal, of a sphere 11/2(270)
Time dependent correlation function 13(200)
Time effect on elasticity 13(445–450)
Time lag of anemometer 8/2(578)
Time law of nucleation 7/2(226–227)
Time of contact, *see also* Impact duration of 6a/1(34, 321, 325–327, 642–645)
Time of contact, Hertz's theory of 6a/1(116, 314, 319)
Time of contact, Pouillet's technique 6a/1(315, 317, 320, 325, 583)
Time of life of a bubble 11/2(273)
Time of travel of a sound signal 11/1(82, 85)
Time reversal 6a/3(23); 12(227–229, 475)
Time reversal, symmetry relation 12(488, 492)
Time, role played by present 6a/3(13)
Time scale of cavitation 11/2(263–264, 273)
Time *t* 6a/3(4, 7)
Time translation invariance 6a/3(12)
Time-creep, *see* Creep
Time-dependent data 6a/3(264)
Time-edge effect in photoelastic measurements 6(162, 164)
Time-interval 6a/2(23)
Time-temperature superposition, shift factor 6a/4(32ff.)
Time-temperature superposition (thermorheologically simple materials) 6a/4(31ff., 35)
Tin 14(271f.)
Tin, allotropic transformation 7/2(214, 220, 225)
Tin-canning 6a/3(287)
TiO_2 lattice 7/1(90)
Titania as a means of nucleation in ceramics 13(607)
Titchmarsh's theorem 6a/3(74)
Titration curves of polyelectrolytes 13(493)
Tobolsky-Leaderman-Ferry reduction formula 11/1(386–387)
Toepler pump 12(548, 549)
Toepler's schlieren method 9(38)
Tollmien-Schlichting wave 8/1(436, 438)

Tolman-Buff equation 10(162, 163)
Toluene, sound absorption 11/1(312, 313)
Tomonaga theory, application to superconductivity 15(363)
Tomotika-Tamada fluid 9(77, 145–147, 150, 152, 172–173)
Ton (unit) 14(25)
Tonks' equation 10(221)
Tonks equation of state 13(55)
Top surface 7/2(73, 80)
Topaz 6a/1(407)
Topological nature of stability 6a/3(139, 141)
Topology of $H_m(A)$ 6a/2(401)
Torque 6a/3(496, 564, 582)
Torque exerted
— on a Rayleigh disc 11/2(59)
— on a sphere by a fluid 8/2(245, 246)
— on an ellipsoïd by a fluid 8/2(237)
— on two spheres by a fluid 8/2(248)
Torsion 6a/1(74, 87–91, 95, 168–177, 202, 251, 291, 298, 562–564, 652, 661, 670, 680, 705, 721–726, 738–740); 6a/3(495)
Torsion, elastic-plastic 6(403–408)
Torsion, finite 6(36–38)
Torsion, first nonlinear relation between torque and twist 6a/1(204)
Torsion, fully elastic 6(399)
Torsion, fully plastic 6(399)
Torsion function 6(28)
Torsion function, complex 6(31)
Torsion function for an elliptic cylinder 6(30)
Torsion of a circular cylinder 6(27–28, 36–38)
Torsion of a circular shaft with circular groove 6(35–36)
Torsion of a cylinder of rectangular cross section 6(35)
Torsion of a cylinder of triangular cross section 6(34)
Torsion of a hollow cylinder 6(33)
Torsion of a non-circular cylinder (general) 6(28–31)
Torsion of a prism 8/2(67)
Torsion of an elliptic cylinder 6(30, 34)
Torsion of glass 13(641, 643)
Torsion of rectangular plate 6a/2(604)
Torsion of rectangular tubes 6a/1(89, 202)
Torsion of silk 6a/1(171)
Torsion pendulum as radiometer 11/2(121)
Torsion, simple 6(506, 509)
Torsion studies of Coulomb 6a/1(168–177)
Torsion with extension 6(507, 508)
Torsional flow 6(513)
Torsional pendulum 11/1(376)
Torsional quartz crystal 11/1(371, 372)
Torsional rigidity 6(28); 6a/3(585, 586)
Torsional vibration gauge 12(530)
Torsional vibration of a glass cylinder 11/2(94)
Torsional vibration of a quartz cylinder 11/2(82–83)
Torsional wave 11/2(75, 149)
Torsional wave pulse method 11/1(374)
Torsional waves in cylinders 6(112–113)
Tortuosity 8/2(627, 643)
Total absorption 11/1(217)
Total compressibility 11/1(206)
Total degree of dissociation 11/1(215)
Total differential 6(382)
Total energy 6a/2(216, 332)
Total force 6a/2(44)
Total moment 6a/2(44)
Total scattering cross section 11/2(160, 164)
Total stored energy 6a/4(162)
Toupin's theorem on Saint-Venant's principle 6a/2(196)
Trace 6a/2(6, 353)
Trace of a tensor 8/1(30, 234)
Traceless symmetric tensor process 6a/3(71)
Traction, *see also* Stress boundary value problem 6a/3(83)
Traction condition 6a/2(102, 220, 321, 335)
Traction problem 6a/2(102, 220, 321)
Traction problem for incompressible media 6a/3(72)
Trailing edge of a profile 9(106)
Trajectories of particles in surface waves 9(496–497, 499, 501, 660, 661, 665)
Trajectory 6a/3(128, 151)

Trajectory (material trajectory), *see* Waves trajectory
Trajectory/orbit 6a/3(129)
Transducer, electrodynamic 11/2(4–5)
Transducer, electrostatic 11/2(6–9)
Transfer effect of Helium II 15(382)
Transfer function 8/2(451, 461, 470)
Transfer function in Heisenberg's theory 8/2(476, 479–481)
Transfer medium 15(166, 168)
Transfer rate 15(441, 442, 443, 444, 445, 448, 452, 453)
Transfer theory of Heisenberg 8/2(471 ff.)
Transference experiments 13(496)
Transformation 7/2(211)
Transformation of the second order 15(375)
Transformation of v. Mises 8/1(270)
Transformation of wave type 11/2(109–110)
Transformation range of a glass 13(511, 562, 580, 593)
— diffusion 13(621)
Transformation temperature in the two-dimensional Ising model 13(118)
Transformation temperature of vitreous state 13(511)
Transformations of atom arrangements (order-disorder) 7/2(252–253)
Transformations of Nekrasov and Rudzki 9(728, 729)
Transformations, orientation relations 7/2(243–246)
Transformations without diffusion 7/2(239–252)
Transforms of Navier-Stokes solutions 8/2(176–186)
Transforms of plane flows 8/2(181)
Transforms of Taylor flows 8/2(183)
Transient creep 7/2(197, 198–204)
Transit time 6a/4(21)
Transition between stages I and II of work-hardening curve 7/2(144–145)
Transition, crystalline 15(88, 91, 134)
Transition curve 15(191)
Transition field 15(130)
Transition field for a cylindrical wire 15(336)
Transition field for a sphere 15(336)
Transition flow 8/2(592, 593, 600)
Transition from laminar to turbulent flow 8/1(64, 351, 357–366, 383); 8/2(540–541)
Transition from turbulent to laminar flow 8/1(427–430)
Transition induced by experiment 8/2(541)
Transition map 6a/3(131)
Transition metals 14(323)
Transition point 7/2(182)
Transition point between laminar and turbulent flow 8/1(118)
Transition probability 7/1(301, 302, 303, 427)
— and relaxation time 11/1(160)
— between electron spin levels 11/2(241)
— between nuclear spin levels 11/2(236–238)
— due to binary collisions 11/1(310)
Transition range 8/1(390)
Transition strains for second order transitions 6a/1(539, 560, 564, 674, 675, 705, 712)
Transition temperature 15(210, 230)
Transition zone 6a/3(433, 435); 10(282, 284, 286)
Transition zone, form of stress tensor 10(283, 286)
Transition zone, integrals across it 10(285)
Transitions 6a/1(479)
Transitions, first-order 6a/1(494, 495)
Transitions in thin films 15(333, 336)
Transitions, polymorphic 6a/1(387, 499)
Transitions, second-order 6a/1(152, 406, 495, 497, 534, 537–539, 553–561, 640, 663, 666, 672, 674, 708, 710, 714–716)
Transitions, third-order 6a/1(498)
Translate of a function 6a/3(128)
Translate process 6a/3(12)
Translation, *see also* Glide 7/1(2ff., 118, 298); 7/2(5, 6, 19); 9(286)
Translation continuity 6a/3(15)
Translation groups 7/1(15, 19 ff.)
Translation lattice, hexagonal 7/1(485)
Translation of a circular cylinder
— in a canal 8/2(253)
— perpendicular to its axis 8/2(257)

Translation of a solution of the Navier-Stokes equation 8/2(179)
Translation relaxation 11/1(155)
Translation, secondary 7/2(47)
Translation space 6a/3(4)
Translation, tertiary 7/2(47)
Translational partition function 12(90)
Translational temperature 12(496)
Transmission coefficient(s) 6a/4(293); 11/1(19); 11/2(108)
— for surface wave, passing obstruction in canal 9(527, 528)
— of a membrane 11/1(100)
— of a plate 11/1(101)
Transmission factor of circular diaphragm within duct 11/1(117, 118)
Transmission line without losses 11/1(112)
Transmission of sound through ducts 11/1(44–61, 116–126)
Transmission of ultrasound through a plate 11/2(108–109)
Transmission through a periodically supported membrane 11/1(104–108)
Transonic flow 8/1(71, 215–217)
— definition 9(130)
— discovery of branch lines 9(87)
— in pumps 12(566, 568–576)
— through a nozzle 9(147–149)
Transonic region, extension of a solution into it 9(73)
Transport and relaxation phenomena, relation 12(381–384)
Transport coefficients 12(305)
— corresponding states 12(397)
— field dependence 12(484, 490)
— mass dependence 12(308)
— signs 12(414)
Transport equation 11/2(211, 212); 12(351)
Transport equation for a gas of rotating molecules 12(480, 483)
Transport equation for a mixture 12(354)
Transport phenomena 11/1(11)
Transport theorem 6a/4(115, 116, 154); 8/1(131–132)
Transport theorem, two-dimensional 6a/2(456)
Transpose of a tensor
— fourth-order 6a/2(10)
— second-order 6a/2(6)
Transverse impedance 11/1(114)
Transverse isotropy 6a/2(72)
Transverse modes 6a/4(245)
Transverse normal stress 6a/2(584)
Transverse piezoelectric effect 11/2(78)
Transverse principal waves 6a/3(328)
Transverse shear deformation 6a/2(460, 575, 578, 584)
Transverse, transversal 6a/3(430)
Transverse velocity correlation 8/2(448)
Transverse wave(s) 6a/2(246–248, 254, 343); 6a/3(99, 310); 11/1(99, 101); 11/2(75, 155, 156)
— in elastic bodies, anisotropic 6a/3(321–323)
— in elastic bodies, isotropic 6a/3(328)
— pure 11/2(158)
— scattering coefficients 11/2(170–172)
— system 9(488)
Trapping modes 9(551)
Traps for ultrahigh vacuum production 12(642, 656)
Traube's rule 10(276ff.)
Travelling resonance 11/1(100, 101, 107, 108)
Tredgold's apparatus 6a/1(213)
Trefftz' integral equation 9(431–434)
Trefftz' plane 8/2(405)
Trefftz-Glauert method 8/2(410)
Trennschaukel 12(330)
Tresca's intermediate region 6a/1(440, 484, 580)
Tresca's maximum shear criterion 6a/1(440–445, 484)
Tresca's yield condition 6(328, 339, 350, 371, 411ff., 417)
Tresca's yield criterion 6a/3(411, 423, 462)
Triangle inequalities 6a/3(561, 565)
Triangle lattice 13(118, 121)
Triangles of BO_3 13(531)
Triangular cross section of a tube 8/2(71)
Triangular elastic-plastic torsion solution 6(407)
Triclinic symmetry 11/2(154)
Tricomi equation 9(81, 131, 133, 150)
Tricomi's boundary value problem 9(137)
Tridymite 13(608)

Trigonal component in the electric field 15(85, 97)
Trigonometric interpolation and quadrature 8/2(432–433)
Trinitrotoluol 11/2(32)
Triode as ionization gauge 12(534–538)
Triple point 14(36)
Triple pressure 14(36)
Triple velocity correlations 8/2(440, 456, 477, 488)
Triplet density 13(5)
Triplet distribution function 10(181, 197)
Tripping wire 8/1(359)
Trkal flow 8/2(161, 172)
Trouton's rule 13(231, 232); 14(26); 15(373)
Truesdell and Noll's criterion 6a/3(202)
Truesdell number 6(517)
Truesdell's vorticity measure 8/1(166)
Truncated distribution function 12(208, 209)
TS-diagram of a paramagnetic salt 15(38, 45)
Tsunami 9(617–619)
Tube, *see* Cylindrical tube
Tube, horizontal cylindrical 8/2(48)
Tube of elliptical cross section 8/2(69, 73)
Tube with multiple connected cross section 8/2(69)
Tubes of cross section other than circular 8/2(67)
Tubes of diverse cross section shapes 8/2(70–75)
Tuckermann optical extensometer 6a/1(130, 133)
Tungsten 14(250)
Tuning fork as quadrupole source 11/2(3)
Tuning fork crystal 11/1(371, 377, 379)
Tunnel effect 7/2(118, 155)
Tunnel model 13(247–248)
Turbine 14(71, 76)
Turbine expansion engines 14(78, 79, 80)
Turbo-expansion engines 14(59)
Turbulence 8/1(64, 111ff.); 11/1(1, 28, 86)
Turbulence, equivalent 8/1(446)
Turbulence in porous media 8/2(649)
Turbulence, isotropic 8/1(384)
Turbulence level 8/2(554)
Turbulence, non-symmetrical 8/2(539–540)
Turbulence of viscous fluid surface waves 9(639)
Turbulence, scale of 8/1(360, 384–386, 426, 445)
Turbulence theory of Hopf 8/2(443–444)
Turbulence theory of Kolmogoroff 8/2(461, 462–464)
Turbulence-produced radiation of sound 11/1(28, 93–99)
Turbulent boundary layer, thickness 8/1(361)
Turbulent diffusion 8/2(505, 509–520)
Turbulent diffusivity 8/2(543)
Turbulent energy equation 8/2(529)
Turbulent flow 8/1(351)
— decay 8/2(528–530)
— generation 8/2(121, 525–526, 535)
— isotropy criteria 8/2(532)
— transition to laminar flow 8/1(427–430)
Turbulent jet, sound generation 11/1(97–99)
Turbulent region, scattering of sound 11/1(86, 90)
Turbulent shear flow 8/2(540–551)
Turbulent spot 8/1(389–392)
Turbulent wedge 8/1(447)
Turmalin 11/2(76, 100)
Tutton salts, crystalline structure 15(101)
Twin 7/1(505, 506)
Twin boundary, coherent 7/1(637)
Twin crystal 7/1(389)
Twin crystal, composition surface of a 7/1(505)
Twin energy 7/1(640)
Twin formation 7/1(95)
Twinning 7/2(5, 6, 19, 45, 47, 67, 184, 240)
Twinning and basal glide alternating 7/2(47, 48)
Twinning energy 7/2(180)
Twist 6a/3(496, 582, 583)
Twisting, *see* Torsion
Twisting moment 6a/3(496)
Two layer model 10(269)

Two-dimensional
— body forces 6a/2(513)
— compressible steady flow 9(249)
— ε-system 6a/2(455)
— heat supply 6a/2(513)
— incompressible non-steady flow 9(294)
— Ising model 13(113–139)
— linearized supersonic flow 9(45–49)
— model of a glass 13(519, 524)
— model of orientation effects in solutions 13(340–342)
— permutation symbols 6a/2(455)
— roughness 8/1(441–447)
— steady flow, shallow-water first-order approximation 9(682–695)
Two-dimensional surface waves
— dock problem 9(527, 536–537)
— integral equation methods 9(533, 535, 537)
— obstruction 9(526ff.)
— on the beach 9(537–542)
— reflection coefficient 9(527, 528)
— transmission coefficient 9(527, 528)
Two-dimensional transport theorem 6a/2(456)
Two-fluid model 14(266); 15(236, 244, 250, 263, 276, 280, 387, 388, 389, 392, 426)
Two-liquid model of liquid mixtures 13(364)
Two-network model 13(448)
Two-phase transition zone, *see* Transition zone
Two-relaxation network theory of polymers 11/1(401)
Typical surface particle 6a/2(530)

Ultimate pressures attainable by ionization gauges 12(645)
Ultracentrifuge 13(401, 473)
Ultrahigh vacuum 12(609–663)
Ultrasonic absorption in liquids 10(35, 91)
Ultrasonic attenuation 6a/4(30, 62); 15(240)
Ultrasonic determination of elastic constants 6a/1(345, 352, 377, 409)
Ultrasonic diffraction, determination of Poisson's ratio 6a/1(266)
Ultrasonic quartz generator 11/2(82)
Ultrasonic relaxation time 11/2(239)
Ultrasonic velocity 6a/4(30, 62)
Ultrasonic wind method for sound absorption measurements 11/1(169)
Ultraviolet catastrophe, turbulence analog 8/2(463)
Ultra-Visconson 11/1(375)
Umbilics 6a/2(628)
Umklapp processes 7/1(305, 307, 308); 11/2(231, 233–234); 14(176, 199, 208, 219, 221, 233, 252, 262)
Umklapp processes, low-frequency longitudinal waves 14(219)
Umklapp-resistance (or intrinsic resist) 14(222)
Unassociated liquids 11/1(284)
Unassociated liquids, sound absorption 11/1(306–315)
— binary mixtures 11/1(328–330)
Unbounded bodies 6a/3(77)
Unbounded regular region 6a/2(13, 257)
Uncertainty effects 14(206)
Uncoupled 6a/2(311, 312)
Uncoupled quasi-static theory 6a/2(325)
Undeformed state 6a/3(538)
Undercooled liquid 13(291, 511, 595)
Undercooling 7/2(217)
Underpressure wake 9(343)
Underwater acoustics 11/2(95)
Undistorted states 6a/4(174, 179)
Uneven bottom 9(569–571)
Uniaxial stress state 6(272)
Uniaxial tension 6a/3(249)
Uniform acceleration in boundary layer 8/1(290)
Uniform dilatation 6a/2(34, 77)
Uniform first-order approximation 9(251, 280–282)
Uniform flow, generation by screens 8/2(551–552)
Uniform hydrostatic loads 6a/3(262, 263)
Uniform liquids 10(47–80)
Uniform pressure 6a/2(50, 77); 6a/3(30)
Uniform shear 8/2(508)
Uniform stability
— definition 6a/3(141)
— sufficiency theorem 6a/3(152)
— theorem (corollary) 6a/3(157)
Uniform states 6a/2(546)
Uniform states for rods 6a/2(698)
Uniform translation

— of a body, case of viscosity limit zero 8/2(329–333)
— of a circular cylinder, Lamb's solution 8/2(305)
— of a plane plate in Oseen approximation 8/2(312)
— of a sphere along a tube 8/2(246, 318)
— of a sphere between two parallel walls 8/2(245–247)
— of a sphere in a fluid 8/2(227)
— of a sphere in Oseen approximation 8/2(298–305)
— of a sphere in presence of a wall 8/2(242–245)
— of an ellipsoïd in a fluid 8/2(233–235)
— of an ellipsoïd in Oseen approximation 8/2(313–315)
— of an elliptical cylinder in Oseen approximation 8/2(311–313)
— of two spheres in a fluid 8/2(247–252, 318)
Uniformity in the loudspeaker frequency band 11/2(11)
Uniformly elliptic equation 9(118)
Uniformly subsonic flow 9(98, 104)
Unilateral constraints 6a/2(403)
— differential operator 6a/2(422)
— elastic body 6a/2(402)
— elastostatics 6a/2(423)
— elliptic operators 6a/2(399)
— linear operators of applied mathematics 6a/2(422)
— membrane 6a/2(422)
— problem, abstract 6a/2(39, 395)
— symmetric coercive bilinear form 6a/2(422)
Unilaterally loaded wedge 6a/3(526)
Uniqueness 6(253, 254, 255, 284, 316)
Uniqueness (and existence) of viscoelastic processes 6a/3(58)
Uniqueness and stability for elastic-plastic bodies 6a/3(523)
Uniqueness, Cauchy initial value problem 6a/3(105)
Uniqueness, equivalence with Liapounov stability 6a/3(151)
Uniqueness for rigid plastic solids 6a/3(416, 523)
Uniqueness in general 6a/3(131)
Uniqueness in linear elasticity 6a/3(152, 231)
Uniqueness, incompressible bodies 6a/3(72)
Uniqueness, initial past-history problem 6a/3(57)
Uniqueness, isotropic bodies 6a/3(57)
Uniqueness, non-uniqueness and Liapounov instability 6a/3(151, 152, 167, 224)
Uniqueness, non-uniqueness in follower force problems, divergence 6a/3(282)
Uniqueness of boundary value problem solution 8/2(186–199)
Uniqueness of quasi-static viscoelastic processes 6a/3(55)
Uniqueness of solution(s) 9(14, 40, 102–104, 106, 113, 131, 132)
— of the Oseen equation 8/2(322–323)
— of the Stokes equation for steady flow 8/2(262–276)
— of the Stokes equation for unsteady flow 8/2(276–287)
Uniqueness theorem(s) 9(227, 231–234, 255, 266, 275)
— displacement boundary data 6a/3(96)
— elasticity, displacement ("first") problem 6a/2(105)
— elasticity, elastodynamics 6a/2(222, 261)
— elasticity, exterior domain 6a/2(171)
— elasticity, general problem 6a/2(130)
— elasticity, incompressible body 6a/2(212)
— elasticity, internal concentrated loads 6a/2(190)
— elasticity, mixed ("third") problem 6a/2(104, 112)
— elasticity, plane problem 6a/2(154)
— for elastic states 6a/3(49)
— for linear elasticity, thermoelasticity, initial boundary value problem 6a/3(231, 254)
— for processes with infinite histories 6a/3(59)
— for quasi-static viscoelastic processes 6a/3(56)
— in dynamic viscoelasticity 6a/3(95, 96)
— Laplace transform formulation 6a/3(96)

Uniqueness theorem(s) (cont.)
— mixed problem 6a/3(96)
— rods 6a/2(692)
— shells 6a/2(610)
— thermoelasticity, dynamic 6a/2(338)
— thermoelasticity, static 6a/2(322)
Uniqueness theory for mixed differential equations 9(131, 135ff.)
Uniqueness theory in free boundary flow 9(406–415)
Unit cell 7/1(14)
— *m*-fold primitive 7/1(14)
— vector(s) 7/1(3, 43, 102, 111, 275, 305)
— vectors of the reciprocal lattice 7/1(44, 112)
— volume of 7/1(14)
Unit doublet state 6a/2(178)
Unit Kelvin state 6a/2(178)
Unit normal to a surface 6a/2(440, 622)
Unit principal normal to a curve 6a/2(627)
Unit tangent and unit normal to a curve lying in the surface 6a/2(623)
Universal 3 axis stage 7/1(69)
Universal displacement field 6a/2(91)
Universal equilibrium of Kolmogoroff 8/2(464, 477)
Unloading 6(280, 285)
Unloading wave, absorption of 6a/1(641)
Unloading wave (expansive acceleration wave) 6a/4(15, 18, 41, 53–58)
Unloading wave, velocity 6a/4(42, 62ff., 65)
Unloading waves, Lee's theory of 6a/1(642, 645)
Unstable, *see* Stability of motion
Unstable disturbance 8/1(371)
Unsteady boundary layer flow 8/1(288–302, 322–324)
Unsteady flow
— along concentric circles 8/2(131–138, 211–214)
— always in the neighbourhood of steady flows 8/2(139–140, 156–157)
— around a spinning obstacle 8/2(220–224)
— in three dimensions 8/2(160–176)
— of an inviscid gas 9(228ff., 283ff.)
— shallow-water first-order approximation 9(677)
— through porous media 8/2(638)
— with free boundary 9(356)
Unsteady Jeffery flow 8/2(140, 142, 143)
Unsteady plane flow 8/2(112ff.)
Unsteady states of a non-uniform liquid 10(101–104)
Unsteady two-dimensional incompressible flow 9(294)
Unstrained crystal, basic structure 7/2(111–112)
Unsymmetric, *see* Non-symmetric turbulence
Unsymmetrical extrusion 6a/3(525)
Upper and lower bounds for strain energy 6a/2(113)
Upper relaxation maximum 11/1(263, 268)
Ursell edge wave 9(551)
U-tube manometer 12(519)

Vacancies 7/2(165)
Vacancies, aggregates of 14(150)
Vacancies in β brass, mobility of the 7/1(450)
Vacancies, separation of 7/1(454)
Vacancies, thermal generation 7/2(115, 121, 123)
Vacancy 7/1(391, 502)
Vacancy concentration 6(603)
Vacancy diffusion 7/2(6)
Vacancy diffusion as a creep process 7/2(206)
Vacancy formation 7/2(222)
Vacancy migration 7/2(222)
Vacancy probability 11/2(217)
Vacuscope after Gaede 12(523)
Vacuum, classification 12(515)
Vacuum connections, flow processes 12(605–607)
Vacuum determination, accuracy 12(518–519)
Vacuum equations 12(628–631)
Vacuum equipment, general device 12(516)
Vacuummeter after Penning 12(518, 533–534)
Vacuummeter, thermoelectric 12(526–527)
Valence angle 13(402)
Valency dependence of measured absorption in electrolytes 11/1(233–237)

Validity of Rayleigh's approximation 11/2(165)
Value of a process a time t 6a/3(4)
Values of Θ 14(240)
van der Waals binding 7/1(75)
van der Waals crystal 7/1(129)
van der Waals equation 10(172, 219, 220, 221); 12(40, 52–55, 65)
van der Waals equation and sound velocity 11/1(339–340)
van der Waals forces 10(23, 119, 223)
van der Waals loop 10(172, 244, 252)
van der Waals potential 7/1(126, 127)
van der Waals theory of liquids 13(217)
van Laar and Guggenheim's formula 13(309)
Vane technique 15(171, 187, 189, 190, 192)
Vanishing near a surface 6a/2(20, 223)
van't Hoff's equation for the osmotic pressure 13(304)
Van Vleck theory 15(49, 52, 56, 82, 128)
Vaporous cavitation, *see* Cavitation
Vapour, chemical potential 13(274)
Vapour exhaustion by mechanical pumps 12(555–559)
Vapour pressure 7/1(316, 319, 320); 10(23, 133); 12(52, 53); 13(221, 232, 240); 15(371, 372, 404)
— curvature dependence 10(154)
— equilibrium 13(230, 273)
— equilibrium for a rigorously regular solution 13(323)
— measurement as vacuum gauge 12(540–541)
— of a drop 10(302, 303)
Vapour pump, *see also* Oil vapour pumps, Mercury vapour pumps 12(515, 559, 591)
Vapour pump, theory 12(576–590)
Vapours, nucleation 7/2(217–218)
Variation, definition 6a/3(64, 65)
Variation in amplitude with time 6a/3(96)
Variation of a functional 6a/3(112)
Variation of constants 9(65)
Variational characterization of solutions to the quasi-static boundary value problem with prescribed initial past history 6a/3(64)
Variational equation 9(27, 28)
Variational method(s) 11/1(117, 124)
— elastic rod theory 6a/2(668)
— for the two-dimensional Ising model 13(115)
— in existence theory 9(394, 403)
— in scattering 11/1(30–31, 71)
— 3-dimensional elasticity 6a/2(125–129)
Variational principle(s) 7/1(551, 577, 588); 14(235)
— and theorems, elasticity 6a/2(122–125, 223–232)
— and theorems, hyperelastic rods 6a/2(676)
— and theorems, shell theory 6a/2(476)
— dynamic creep problems 6a/3(114)
— dynamic relaxation problems 6a/3(113)
— first, for creep problems 6a/3(66)
— first, for relaxation problems 6a/3(65, 68)
— for free surface flow 9(387–391)
— for general fluids 8/1(144–146)
— for perfect fluids 8/1(146–149)
— for subsonic flow 8/1(203–206)
— in elasticity 6(15)
— of Bateman 8/1(204)
— of capillarity 10(293–295)
— of Clebsch and Bateman 8/1(149)
— of d'Alembert-Lagrange 8/1(145, 146)
— of Dirichlet 8/1(161)
— of gas dynamics 8/1(203–206)
— of Hamilton 8/1(146, 147)
— of Herivel 8/1(147)
— of Herivel-Lin 8/1(148)
— of hydrodynamics 8/1(144–150)
— of Kelvin 8/1(161)
— of the inelastic continuum 6(251–256)
— second, for creep problems 6a/3(67)
— second, for relaxation problems 6a/3(66)
Variational problem for two-dimensional subsonic flow with circulation 9(114–115)
— without circulation 9(92, 107–111)
Varnish 6a/1(126)
Vector field 6a/2(10); 6a/3(7)
Vector in a surface 6a/2(622)

Vector product 6a/2(5)
Vector solid spherical harmonic 6a/2(22)
Vectors 6a/3(3)
Vectors, contragredient 7/1(43)
Vectors, contravariant 7/1(43)
Vectors, covariant 7/1(43)
Velocities, phase and group 6a/1(333)
Velocity, acceleration wave 6a/4(41)
Velocity, acoustic 11/1(6, 9)
Velocity changes caused by dislocations 11/2(181)
Velocity characteristics 6(354)
Velocity, characteristics 6a/3(445)
Velocity compatibility relations 6(356–357)
Velocity, compatibility relations 6a/3(444)
Velocity constant of approach 11/1(222)
Velocity constant of removal 11/1(223)
Velocity, constant on characteristics 6a/3(489)
Velocity correction of Kármán and Tsien 8/1(196)
Velocity correlation function 11/1(91)
Velocity, critical 6a/1(588, 599, 603)
Velocity, definition 6a/2(43, 467); 6a/3(8, 173, 305)
Velocity, director 6a/2(449)
Velocity distribution, calculation of corresponding profile 8/2(399–402)
Velocity distribution function 10(27–28, 39, 42, 127)
Velocity distribution of laminar and turbulent flow 8/1(352, 360)
Velocity distributions 6a/3(438)
Velocity (electromagnetic) gauge 6a/4(9)
Velocity ellipse in the hodograph plane 9(692)
Velocity equations 6a/3(449)
Velocity, equilibrium sound, *see* Equilibrium sound velocity
Velocity field 8/1(130)
Velocity field in plane plastic flow 6(364)
Velocity field process 6a/3(8)
Velocity fluctuations 11/1(8)
Velocity fluctuations in turbulent flow 8/1(352, 384)
Velocity gradient 6a/3(541); 9(259)
Velocity, instantaneous sound, *see* Instantaneous sound velocity
Velocity, interferometer 6a/4(12–17, 41)
Velocity, intrinsic 6a/4(25)
Velocity, longitudinal wave, in three dimensions 6a/4(73)
Velocity microphone 11/2(56)
Velocity minimum of capillary-gravity waves 9(633)
Velocity, notation of components 8/2(2)
Velocity of a wave, *see* Phase velocity, Group velocity
Velocity of displacement of a wave surface 6(343)
Velocity of disturbance propagation 8/1(370, 371, 373)
Velocity of equilibrium approach 11/1(208, 209, 214)
Velocity of flow 10(33, 41)
Velocity of gravity waves 9(513–514)
Velocity of gravity-capillary waves 9(514, 614)
Velocity of molecules, mean 8/1(64)
Velocity of propagation of a wave 6a/2(246)
Velocity of propagation of the normal-superconducting boundary 15(341)
Velocity of reaction 11/1(207, 213, 222–226)
Velocity of slip 8/1(265)
Velocity of sound, *see also* Sound velocity, Speed of sound 8/1(13, 60, 180, 213, 222, 228); 12(9–10, 21)
— first direct measurement in a solid 6a/1(191)
— in a column of water 6a/1(252)
— in water, Canton's experiments 6a/1(187)
— near the critical point 12(48)
Velocity of the mediums 6a/3(429)
Velocity, particle *see* Particle velocity
Velocity, phase 6a/4(30–34)
Velocity plane, *see* Hodograph method
Velocity plane 6(357, 375); 6a/3(445, 467)
Velocity plane, image of streamlines 9(24)
Velocity potential 8/1(77, 155, 158); 8/2(388, 403); 9(5, 172, 319, 449, 470)
— asymptotic behavior 9(98–101, 104)
— differential equation 9(6)

— differential equation, three-dimensional explicit solutions 9(8–11)
— in porous media 8/2(635)
— Legendre transformation 9(56)
— series expansion at infinity 9(104–106)
Velocity profile(s), *see also* Wave profile 8/2(14, 15, 49)
— in suction 8/1(409)
— shape factor 8/1(399)
— with point of inflection 8/1(372, 375, 379, 396, 398, 422, 439)
— of three-dimensional boundary layer 8/1(439)
Velocity, projectile *see* Projectile velocity
Velocity sensitive detectors of sound 11/2(56–61)
Velocity spectrum in isotropic turbulence 8/2(468)
Velocity (speed) of displacement 6a/3(307, 308, 429)
Velocity, steady wave, *see* Steady waves
Velocity transducer, electromagnetic gauge 6a/4(9)
Velocity transducer, velocity interferometer 6a/4(12–14, 56)
Velocity, transverse wave, in three dimensions 6a/4(73)
Velocity vector 11/1(3)
Velocity, wave front 6a/4(21–24)
Vena cava of an ox 6a/1(67, 70)
Vena contracta 9(423–425, 431–434)
Verdet's constant 15(17)
Vertical barrier in a canal 9(529)
Vertical cylinder cutting water surface, diffraction 9(544)
Vertical entry of a cone into water 9(361–362)
"Vertical" movement 14(232)
Vertical plates, capillarity effect of liquid 10(300)
Very high frequencies, *see* Hypersound
Very slow motion 8/1(64)
Vibrating plate, first and second mode frequencies of 6a/1(254, 272)
Vibrating rods, deep tone in 6a/1(254)
Vibration, Chladni's experiment 6a/1(182, 192–195, 221, 255, 404, 405)
Vibration, elastic 7/1(168, 245)
Vibration excitation 11/1(249)
Vibration, free energy of 7/1(273, 275, 279, 283, 319)
Vibration levels of diatomic and polyatomic molecules 12(101–108)
Vibration, longitudinal 7/1(171, 172, 219, 251, 292, 313)
Vibration of a dislocation line 11/2(180)
Vibration of plates 6a/1(254)
Vibration, optical 7/1(168)
Vibration spectrum 14(286, 288)
Vibration, transverse 7/1(171, 172, 219, 313, 2451)
Vibrational partition functions 12(116–118, 123)
Vibrational spectrum 7/1(339, 345)
Vibrational spectrum of a lattice, analytic discontinuities in the 7/1(359)
Vibrational spectrum, the exact form of the 7/1(356)
Vibrations, longitudinal vs transverse 6a/1(222)
Vibrations, modulus of bismuth measured by 6a/1(222)
Vibrations of glass rods 13(641, 643)
Vibrations, Pochhammer-Chree theory 6a/1(333, 336, 340)
Villari effect of magnetostriction 11/2(86)
Villat's integral equations 9(730, 742)
Violoncello string 6a/1(55)
Virial coefficient(s) 10(52, 66); 12(18–21, 29, 35, 50, 54, 150)
— determination by superposition principle 13(19)
— in polyelectrolytes 13(499, 501)
— in polymer solutions 13(451, 457, 465)
— of a binary mixture 12(60, 61)
— of higher order 13(2)
— second 10(66); 13(2, 227, 315, 451, 457)
— second reduced 13(239)
Virial equation of state 12(71)
Virial law 13(12)
Virial of a system, definition 13(11)
Virial of Clausius 12(160)
Virtual displacement 6a/3(197, 198, 202, 203, 261)
Virtual mass, *see also* Induced mass 9(362, 387, 403)
Virtual mass 8/1(93)

Virtual modes 6a/3(416)
Virtual polymer molecules 13(461)
Virtual work 6(15)
Viscoelastic analogs of work-energy identities of elastostatics 6a/3(53)
Viscoelastic analogue to Boggio's theorem 6a/3(94)
Viscoelastic body 6a/3(17)
Viscoelastic boundary value problem 6(294ff.)
Viscoelastic center of compression 6a/3(80)
Viscoelastic center of rotation 6a/3(81)
Viscoelastic counterpart of Beltrami-Donati-Michell equations 6a/3(46)
Viscoelastic displacement boundary value problem 6a/3(81)
Viscoelastic field equations (dynamic) 6a/3(42, 91)
Viscoelastic foundations of linear theory 6a/3(11)
Viscoelastic generalization of Boussinesq-Somigliana-Galerkin and Boussinesq-Papkovitch-Neuber solutions 6a/3(74)
Viscoelastic Kelvin process 6a/3(78)
Viscoelastic linear solid 11/2(173)
Viscoelastic mixed boundary-value problem 6a/3(59)
Viscoelastic moduli 6(309–310)
Viscoelastic potential function 6a/3(3)
Viscoelastic process corresponding to differential operator law 6a/3(57)
Viscoelastic solution 6a/3(60)
Viscoelastic stress analysis 6a/3(68)
Viscoelastic stress functions 6a/3(93, 94)
Viscoelastic stress-strain relations in one or two dimensions 6(272)
Viscoelastic theory 11/1(294ff.)
Viscoelastic wave 6a/4(256)
Viscoelastic wave equations 6(299–301)
Viscoelasticity 6(269–278, 293–301, 308–313, 448); 6a/1(48, 74, 173–177, 413–416); 6a/2(386)
Viscoelasticity (Burgers body) 6(472–474, 491)
Viscoelasticity, nonlinear 6a/1(74)
Viscoelasticity of polymer solutions 13(449–450)
Visco-plastic boundary layer 6(305)
Visco-plastic boundary value problems 6(305–308)
Visco-plasticity 6(289, 301–308)
Viscoplasticity 6a/1(454, 570–578, 589, 597)
Viscosity 7/2(3); 8/1(47ff., 52ff.); 9(447); 12(299, 383); 15(419, 420, 421, 422, 427, 428, 430)
Viscosity at low temperatures 12(463–467)
Viscosity average molecular weight of a polymer 11/1(380–381)
Viscosity, Cauchy-Poisson law 8/1(232, 236)
Viscosity coefficient(s) 6(450, 459); 6a/4(214, 256); 8/1(49, 50, 236); 11/1(3)
Viscosity coefficient, second 11/1(292–294)
Viscosity constant of Sauders 11/1(348)
Viscosity difference of ortho- and paramodifications 12(464–466)
Viscosity, effect on gas dynamics 9(264, 265, 287)
Viscosity, effect on sound absorption 11/1(132, 150)
Viscosity, effect on stability 9(647–648)
Viscosity, fictitious 9(479)
Viscosity, higher approximations 12(401)
Viscosity integral, evaluation for special models, *see also* Intermolecular forces 12(394–399)
Viscosity, kinematic 12(300)
Viscosity, measured values 12(395, 399)
Viscosity measurement as a means to determine molecular weights 13(401, 472, 475–476, 496, 497)
Viscosity measurements 12(312–316)
Viscosity, neglected 9(225, 263, 448, 462)
Viscosity, neglected for cavitation 9(319)
Viscosity, non-linear 8/1(231, 241)
Viscosity number 11/1(278)
Viscosity of a monatomic gas mixture 12(414, 426)
Viscosity of binary mixtures 12(447–451)
Viscosity of electron gas 11/2(209)

Viscosity of glasses 7/2(4); 13(544, 578–592)
Viscosity of glasses as a function of stabilization 13(583–584)
Viscosity of glasses, effect of chemical substitutions 13(587–589)
Viscosity of glasses, relation to electrical conductivity 13(584, 591)
Viscosity of glasses, temperature dependence 13(581–583)
Viscosity of liquids 10(14, 20, 81, 82, 89–91, 110, 111, 115, 119)
Viscosity of polyelectrolytes 13(501–502)
Viscosity of solids 6a/1(476)
Viscosity of the monatomic gas 12(389)
Viscosity, pressure dependence 12(308, 310)
Viscosity pump 8/2(71)
Viscosity, representation of the cross section by scattering phases 12(458–461)
Viscosity, temperature dependence 12(399)
Viscosity, temperature sensitivity 6(296)
Viscosity transport 11/1(11)
Viscosity, zero limit (Oseen's limit) 8/2(329)
Viscothermal equations for small amplitude sound waves 11/1(277–279)
Viscothermal theory 11/1(278, 279)
Viscous boundary layer 11/1(12)
Viscous correction to curve of neutral stability 8/1(377)
Viscous damping resistance 11/2(42)
Viscous dissipation 11/1(4, 13–14, 20)
— function 11/1(14)
— of energy 8/2(450–451, 461, 471)
— rate 8/2(534, 547)
Viscous effects in slip flow 8/2(611–612)
Viscous flow 6(434); 6a/3(405); 8/1(47–56, 230–262)
Viscous flow of glass 13(582, 584)
Viscous fluid, boundary conditions 9(451)
Viscous fluid, surface waves
— in linear approximation 9(638–646)
— in second-order infinitesimal wave approximation 9(666)
Viscous instability of laminar flow 8/1(373, 402)
Viscous liquids, sound absorption 11/1(323–327)
Viscous losses near a boundary 11/1(20)
Viscous resistance, separation from wave resistance 9(581)
Viscous stress 11/1(3)
Viscous traction coefficient 6(451)
Viscous transport of kinetic energy 8/2(546)
Visualization of ultrasonic waves 11/2(136–142)
Vitreous state, basic properties 13(511)
Vitreous state, characteristic parameters 13(520–521)
Vitreous state, definition 13(514)
Voice coil 11/2(9, 19, 20)
Voids 7/1(94, 391, 444)
Voids, concentration of, at the melting-point 7/1(405)
Voids, diffusion of 7/1(423, 657)
Voids, energy of generation for 7/1(473)
Voids, jump frequency of the 7/1(459)
Voigt model 11/1(368–370)
Voigt's elastic constants 7/1(516)
Volmer theory of nucleation 7/2(217–218)
Volterra dislocation 6a/2(42)
Volterra integral equations in Banach spaces 6a/2(374)
Volterra's distortion 7/1(520, 524)
Volterra's equation 6a/3(62)
Volterra's lemma 6a/3(56)
Volterra's uniqueness theorem 6a/3(56)
Volume change 6a/2(30, 31, 37, 97); 11/1(217–218, 220)
— by shear (=dilatancy) 6(261, 503)
— in plastic deformation 6a/1(83, 86, 504, 710–712)
— of hollow tubes in tension 6a/1(247)
Volume contraction by mixing 13(351, 364)
Volume, critical 10(66); 12(44)
Volume dilatation 6(24, 143, 438, 481)
Volume expansivity, *see also* Expansion coefficient 13(557)
Volume flow 6(482)

Volume forces 7/1(231)
Volume fraction 13(456)
Volume integrals representing sources of sound 11/1(24, 26)
Volume of polymers, temperature dependence 13(422)
Volume potential density 6a/3(259)
Volume preserving 6a/3(550)
Volume variations 15(279)
Volume viscosity 10(83, 89, 91, 92, 118); 11/1(400); 12(299, 353, 501–502)
— coefficient 10(91)
— for a gas of rough spheres 12(498)
— for polyatomic gases 12(496)
Volume-anelasticity 6(267)
Volumetric expansion coefficient 6a/4(135)
Volumetric flow 6(438)
Volumetric strain 6(438)
Volume-viscosity 6(231, 249, 267, 269)
von Kármán equations for nonlinear bending of thin plates 6a/2(588)
von Kármán experiment for unaxial tests at high pressure 6a/1(486–490)
von Kármán's critical velocity 6a/1(588, 599, 603)
von Mises extremum principle 6a/3(409)
von Mises general equations 6a/3(410)
von Mises linear equations 6a/3(459)
von Mises Mises-Reuss theory 6a/3(413)
von Mises theory of plastic potential 6a/3(408)
von Mises transformation 8/1(270)
von Mises yield condition 6a/3(403, 405, 461, 471, 487, 549)
von Mises-Sternberg version of Saint-Venant's principle 6a/2(190)
Vortex and source, combined 8/1(99)
Vortex distribution 8/1(100)
Vortex filament 8/1(34)
Vortex flow in two dimensions 9(10)
Vortex formation at an edge 11/2(97)
Vortex generation 8/2(121)
Vortex in a parallel flow 8/1(98–100)
Vortex line 8/1(34, 151, 163); 9(284)
Vortex lines as material lines 8/1(152, 163)
Vortex lines in turbulent flow 8/2(453)
Vortex motion 8/1(163–166); 15(392)
Vortex of pulsating strength in two dimensions 9(479, 483)
Vortex path of Kármán 8/1(122–123)
Vortex sheet 8/1(87); 9(366)
Vortex stretching mechanisms 8/2(453, 461–462)
Vortex theorem of Crocco 8/1(41–43, 186)
Vortex theorem of Ertel 8/1(43–45)
Vortex tube 8/1(34, 163); 14(4, 7)
Vortex varying with time 9(495)
Vortex whistle 11/2(97)
Vortex with curved axis 8/2(87)
Vorticity 8/2(2, 441, 451–453); 9(3, 4, 284); 11/1(12)
Vorticity at a wall 8/2(7)
Vorticity behind a shock 8/1(221)
Vorticity components for circulating flow 8/2(131)
Vorticity, constant 8/2(39, 140–141)
Vorticity convection 8/1(151, 163, 188)
Vorticity diffusion 8/1(188–190, 248)
Vorticity effect on shock formation 9(265)
Vorticity for axial symmetric motion 8/2(11)
Vorticity formula
— for viscous fluids 8/1(248)
— of Beltrami 8/1(151)
— of Cauchy 8/1(37, 152)
— of Ertel 8/1(189)
— of Helmholtz 8/1(152)
— of Lamb 8/1(165)
— of Nanson 8/1(151)
— of Truesdell 8/1(166)
Vorticity growth 8/2(492–494)
Vorticity measures 8/1(46, 166)
Vorticity measures of Truesdell 8/1(166)
Vorticity of boundary layer 8/1(269–270)
Vorticity of Gerstner's waves 9(743)
Vorticity spectrum 8/2(452, 468)
Vorticity tensor 6(239); 8/1(140)
Vorticity theorems of Helmholtz 8/1(34–36, 163)
Vorticity, variations in the wake 8/2(368)
Vorticity vector 8/1(140)

Wada's formula 11/1(347, 358)
Wake 8/1(303–304, 347, 388); 8/2(549); 9(316); 11/1(46)

Wake behind a plane plate 8/2(379)
Wake in real fluid 9(311)
Wake, infinite 9(321, 325)
Wake parameter 9(116, 343)
Wake pressure 9(316)
Wake structure, effect of rod shape in grid 8/2(531)
Wake variations of vorticity 8/2(368)
Wake with negligible mean velocity gradients 8/2(539)
Wall corrections in the windtunnel 9(297)
Wall curvature 8/1(267–268, 415, 430)
Wall effects 8/2(549)
Wall effects in rheology 6(461)
Wall effects in water tunnel cavitation experiments 9(344)
Wall influence on one-dimensional fluid 13(49, 50)
Wall interference 8/2(530)
Wall roughness 8/1(440–450)
Wall temperature, effect on critical Reynolds number 8/1(427)
Warping, elastic 6a/3(496)
Warping (in torsion) 6(401, 405–408)
Warping, plastic 6a/3(501, 505)
Watch spring 6a/1(156, 217)
Water as a rheological object 6(435)
Water blast pump 12(565–566)
Water content of glasses 13(521, 533)
Water entry 9(312, 360ff.)
Water hammer 11/2(277)
Water, kinematical viscosity 8/2(124)
Water pipes, wave propagation in 6a/1(191)
Water, pressure dependence of sound absorption 11/1(320, 321)
Water, structure 11/1(202)
Water tunnel experiments on cavitation 9(312, 314, 344)
Water, wavelength 8/2(130)
Water waves, diffraction 9(543–545, 560, 566–567)
Water-air interface, parameters 9(645, 650)
Wave, acceleration 6a/2(253, 254); 6a/4(26, 40)
Wave, acouistic, *see* Infinitesimal sinusoidal progressive waves
Wave amplitude, *see also* Variation of amplitude 6a/3(309–311, 315)
Wave arrival time 6a/4(21ff., 24)
Wave axes 6a/3(97, 98)
Wave, axially dissimilar 6a/3(102)
Wave, axially similar 6a/3(101)
Wave, balance of momentum for 6a/2(254)
Wave, centered 6a/3(486)
Wave, classification 6a/4(25ff.)
Wave, compressive 6a/4(26)
Wave crest 9(487, 494)
— average position 9(508–509)
— breaking 9(681)
— form, Stokes' theorem 9(731)
Wave, definition 6a/3(96, 306)
Wave, dispersed 6a/2(343)
Wave drag 8/1(92)
Wave equation(s) 6a/2(328); 6a/3(137, 160, 162); 8/1(67); 9(16, 285); 11/1(4, 8)
Wave equations, visco-elastic 6(299–301)
Wave, expansive 6a/4(26)
Wave, flexural 6a/1(336)
Wave forms from guns 11/2(33–34)
Wave front 6(132); 9(234, 235, 243, 260, 262, 267, 272, 276, 278, 281)
— amplitude (wave amplitude) 6a/4(8–21)
— in three dimensions 6a/4(72)
— (jump discontinuity, discontinuity) 6a/4(21, 25, 41)
— velocity (wave velocity, propagation velocity) 6a/4(21–24)
Wave generation by oscillating bodies 9(553ff.)
Wave generation by pressure distribution 9(592–603)
Wave groups 7/1(568f.)
Wave, high frequency, related to decay 6a/3(109)
Wave in bodies
— elastic 6a/3(315–334, 356–385)
— elastic inhomogeneous 6a/3(378–385)
— elastic isotropic 6a/3(325–334)
— elastic non-conductors 6a/3(362–377)
— of material with memory 6a/3(334–339, 344–356)
— viscoelastic 6a/3(1, 92, 96–104, 106–112)
Wave length 7/1(160, 170, 181)
Wave length of surface wave 9(487)
Wave, longitudinal 6a/3(99, 310, 315–321, 327, 328)

Wave of finite amplitude 8/1(106–109)
Wave of order *n* 6a/2(253)
Wave operators 6a/2(213, 328)
Wave packet 9(509, 518)
Wave, principal 6a/3(326, 328)
Wave profile(s)
— during shock formation 6a/4(68ff.)
— from the Hopkinson bar experiment 6a/1(324)
— in the Boltzmann experiment 6a/1(329)
— of acceleration waves 6a/4(66–69)
— of steady shock waves 6a/4(42, 45–48, 52, 82, 87–89, 94, 96)
— of thin shock pulses 6a/4(53–58)
— unloading 6a/4(65)
— (wave shape) 6a/4(6, 8–21, 23ff.)
Wave propagation, *see* Incremental wave, finite amplitude wave
— 3-dimensional bodies 6a/2(243–261)
— in rubber 6a/1(308–312, 716)
— nonlinear 6a/1(621–664, 690–699)
— one-dimensional 6a/1(306–352, 634, 716)
— straight rods 6a/2(663)
Wave, reflected 6a/1(331, 608, 640–645)
Wave resistance 9(576ff.)
Wave resistance of a ship 9(580–581)
Wave (*see also* Discontinuity), acceleration 6a/3(97, 309, 313, 334, 356, 362, 428–432, 434, 435)
Wave, shear, in water 6a/1(253)
Wave, shock 6a/3(97, 171, 311, 313, 349–361, 369–377, 382–385, 416); 6a/4(25, 31)
Wave, simple 6a/4(54–58)
Wave, singular surface 6a/4(25)
Wave, smooth structured 6a/4(26, 40)
Wave speed(s) 6a/1(700); 6a/3(92, 96, 97, 197, 306–339, 345–347, 351–372, 378–380, 382–385)
Wave speed, dilatational 6a/1(228, 253, 345, 352)
Wave speed, incremental in the prestressed field 6a/1(611)
Wave, steady, *see* Steady waves
Wave strength 6a/3(100)
Wave surface 6(343)
Wave theory, nonlinear 6a/1(597, 634, 699)
Wave, trajectory (material trajectory) 6a/4(25)
Wave transit time 6a/4(21, 41, 68)
Wave transit time for ultrasonic experiments 6a/4(32ff.)
Wave, transverse 6a/2(246–248, 254, 343); 6a/3(99, 310, 321–323, 328, 332, 334)
Wave, two, structure following impact 6a/1(699)
Wave, variation of amplitude with time (attenuation, decay, growth) 6a/3(1, 96, 98, 106, 107, 110, 320–323, 328, 330–333, 341–348, 353, 364–376, 378–385)
Wave velocity, definition 9(457)
Wavelength in a fluid 8/2(129)
Wavelength measurement of ultrasound 11/2(142–146)
Wavelength of sound waves 11/1(136)
Wave-profile, second-order infinitesimal wave approximation 9(656, 659)
Wave-profile, third-order infinitesimal wave approximation 9(658, 660)
Waves, *see* Shock waves, Standing waves, Tensile waves
Waves at an interface of two fluids 9(502–506, 634)
Waves, centered simple 6a/4(280)
Waves, elastic 6(107–118); 7/1(181)
Waves, finite amplitude 6a/4(275–302)
Waves from a submerged oscillating body 9(554–563)
Waves, long (method of) 7/1(186, 189)
Waves of finite amplitude, *see* Finite amplitude waves
Waves of infinitesimal amplitude, *see* Infinitesimal waves
Waves on the beach 9(537–542, 545, 547–551)
Waves, progressive 7/1(161, 166, 179)
Waves, shock 6a/4(287, 288, 294, 295, 297–302)
Waves, simple 6a/4(278–280)
Waves, standing 7/1(161, 166, 175, 260, 261, 568)
Waves, thermoviscoelastic 6a/4(251)
Waves, traveling 7/1(568, 587)
Waves, viscoelastic 6a/4(256)
Wave-vector 14(202, 229)
Wave-vector, conservation of 14(206, 211)
Wavy slip-lines 7/2(158)

Weak compactness 6a/3(559)
Weak convergence 12(226)
Weak differential 6a/3(552)
Weak discontinuities on characteristics 9(22)
Weak discontinuity 6(342, 347)
Weak electrolyte 11/1(248, 250, 258–259)
Weak electrolytes, absorption measurements 11/1(235)
Weak quasi-static viscoelastic process 6a/3(44)
Weak scattering approximation 11/2(168)
Weak shock discontinuity, *see* Shock wave
Weak solution 6a/2(355)
Weak solutions 6a/3(162, 171, 173, 178, 179, 211, 215, 224, 229, 258)
Weakly asymptotic stability
— definition 6a/3(142)
— theorem 6a/3(158)
Weakly closed 6a/2(393)
Weakly compact 6a/2(394, 398)
Weakly conservative forces 6a/3(193, 258–264, 286)
Weakly conservative stress vector 6a/3(263)
Weber's transformation 8/1(32–33, 169–170)
Weber's transformation, generalization 8/1(171)
Wedge 6a/3(521, 526)
Wedge, cavity formation behind it 9(330–331)
Wedge, impact on a water surface 9(360, 362)
Wedge shaped surface wave system 9(488)
Wedge unilaterally loaded, incipient plastic flow 6(427–429)
Weeping tin 6a/1(217)
Weight average molecular weight 13(401, 420)
Weingarten's formulae 6a/2(625, 626)
Weingarten's theorem on singular surfaces 6a/2(252)
Weiss constant 15(82, 84)
Weiss molecular field 15(128, 129f.)
Weissenberg effect 6(261, 514–517)
Well-posed 6a/3(136, 146)
Wertheim controversy 6a/1(257)
Wertheim's anomaly 6a/1(228, 360, 366–369)
Wertheim's atomic law 6a/1(230, 231, 393)
Wertheim's law in photoelasticity 6a/1(244)
Wet-compression machine 14(16)
Wheeler-Sternberg lemma 6a/2(257)
Whiskers 7/2(16)
Whistle 11/2(29, 95–97)
White spectrum 13(72, 191)
Widmannstedt figures 6a/1(449, 450)
Wiedemann-Franz-Lorenz law 14(231)
Wiener approach 8/2(530)
Wiener-Hopf integral equation 9(537, 545)
Wiener-Hopf method 11/1(55, 70, 119)
Wigner-Seitz' atomic sphere 7/1(429, 529)
Wind effect on the propagation of sound 11/1(81)
Wind generating water waves 9(650)
Wind tunnel 8/2(530, 551, 554)
Window plate (window material) 6a/4(15ff., 18, 21)
Window type of magnetostrictive transducer 11/2(90)
Wind-tunnel wall corrections 9(297)
Wing, *see* Airfoil, Thin wing
Wing chord 8/2(386)
Wing profile 8/2(386)
Wing span 8/2(386, 402)
Wing surface, boundary condition 9(288–291)
Wire, *see* Long wire, Short wire
Wire diameter in anemometer 8/2(556, 559)
Wire temperature fluctuations 8/2(574)
Wires, superconducting thin 15(242, 252)
Witoszyński-Szymański flow 8/2(57, 100, 173)
Wittich's method to compute the pressure distribution 8/2(398–399)
WKB approximation for inhomogeneous media 11/1(80)
w-lines 6a/3(485)
Wood 6a/1(15, 17, 22, 24, 25, 183, 240)
Wooden beams 6a/1(16, 21–25)
Wood's alloy 6a/1(281)
Work, *see also* Energy, Power 6a/3(3, 23)

Work, definition 6a/2(60)
Work done along a strain path 6a/2(82)
Work done by an isotropic body 6a/2(207)
Work done by surface displacement 10(286, 294)
Work done starting from rest 6a/2(217)
Work expended 6a/2(60)
Work of formation of a droplet from vapor 10(302–303)
Work theorem in thermoelasticity 6a/2(320)
Work theorems in elasticity 6a/2(95, 208, 211)
Workhardening 6(280, 285, 598, 603); 7/2(31, 45–71, 110)
Work-hardening coefficient 7/2(31, 110, 128, 137–138, 143–144, 147, 188)
Work-hardening curve *see also* Stage I, Stage II, Stage III of work-hardening curve 7/2(30, 31, 45–71)
Work-hardening curve of substitutional solid solutions 7/2(179–182)
Work-hardening curve, static 7/2(32)
Work-hardening general theory 7/2(126–139)
Work-hardening, latent 7/2(60, 130, 134–137, 151, 182)
Work-hardening mechanism in the easy-glide region 7/2(140)
Work-hardening parameters 7/2(55)
Work-hardening parameters, reproducibility 7/2(56)
Work-hardening, theory for face-centred cubic metals 7/2(139–167)
Work-hardening, theory for hexagonal metals 7/2(137–139)
Working point of a glass 13(580)
Working range, viscosity of glasses 13(579, 589–592)
Work-softening, *see also* Softening 7/2(157–160, 162)
Wulff, construction by 7/1(70)
Wulff, net of 7/1(65)
Wurtzite lattice 7/1(85, 603, 639)
Wurtzite structure 7/1(116, 135, 139)

X-cut 11/2(77, 230)
X-cut quartz 6a/4(268)
x_i-convex region 6a/2(15)
X-ray analysis 15(386)
X-ray diffraction 15(270)
— in glass 13(522, 524)
— in liquids 10(7, 16, 19, 43–47, 55, 56, 81, 124)
— method 7/1(68)
X-ray effect in ionization gauge 12(612)
X-ray investigation of liquids 13(218)
X-ray investigations 7/2(4, 229–231)
X-ray scattering by liquids 13(248–251)
X-ray small angle scattering by binary alloys 13(180–189)
X-rays for structural transformations 14(149)

Yawed infinite cylinders, flow past them 8/1(313–314)
Yawed wing, transition to turbulence 8/1(440)
Yawing moment of an airfoil 8/2(387, 409)
Yield condition 6(278, 281–283, 370–372)
— for an isotropic, incompressible, perfectly plastic body 6(327)
— of Sokolovsky 6(381)
— of Tresca 6(328, 339, 350, 371, 411 ff., 417)
— or criterion 6a/3(403, 405, 411, 461, 463, 464, 471, 473, 487, 548, 549, 561)
— parabolic form 6(282)
— parametric representation 6(368)
Yield function 6(327, 350, 352); 6a/3(410)
Yield limit 6a/1(425, 457, 458, 465, 467, 471, 646)
Yield locus 6(279, 283)
Yield mechanism 6a/3(439)
Yield, overall 6a/3(516)
Yield phenomenon 7/2(44, 134, 172, 185)
Yield point 6(434); 7/2(32, 126, 167)
— dependence on domain size 7/2(189)
— of substitutional solid solutions 7/2(171–179)
— phenomenon 7/2(134)
— temperature dependence 7/2(173, 177)
Yield strength 6a/3(584, 586)

Yield stress 6(230, 451, 458, 462); 7/2(31, 110)
— change under temperature change 7/2(163)
— in metals of large stacking-fault energy 7/2(115–119)
— in metals of small stacking-fault energy 7/2(120–124)
— in pre-strained material 7/2(124)
— in reverse direction 7/2(133)
— in simple shear 6(324)
— in simple tension 6(324); 6a/3(405)
— ratio, corrected 7/2(164)
— recovery 7/2(194)
Yield surface(s) 6(278, 283); 6a/1(483, 490, 510–508, 548, 562, 676–690)
— von Mises hypothesis 6a/1(486, 506, 508, 563, 667)
— Lode variables 6a/1(502)
— Lode's experiments 6a/1(501–503, 506)
Yielding boundary surface 11/1(15)
Yoffé effect 7/2(14, 42, 69, 109)
Yoffé's solution for the moving Griffith crack 6(121)
Youkowski profile 8/2(389)
Youkowski transformation 8/2(389)
Young equation 10(289, 295)
Young's modulus 6(24–25, 142, 329, 449, 451); 6a/2(78, 570); 7/1(184); 13(433); 15(240)
— after irradiation 11/2(196)
— for adiabatic deformation 11/2(175)
— for isothermal deformation 11/2(174)
— of a glass 13(624, 635)
— of a glass, calculation from chemical composistion 13(636)
— reduced, for several polymers 11/1(398)
— variability 6(490)
Yvon's integral equation 13(17)

Zachariasen's theory of glasses 13(516–520)
— confirmation by experience 13(522)
— criticism 13(524)
Zanaboni-Robinson theorem on Saint-Venant's principle 6a/2(206)
Zeilon's method to find the plane flow for viscosity limit zero 8/2(350–356)
Zener's theory of entropy of activation 7/1(406, 415, 459)
Zernike-Prins equation of scattered x-ray intensity 13(249)
Zernike's formula for the scattering of light by multi-component systems 13(311)
Zero extension 6a/3(506)
Zero moment 6a/3(217)
Zero-point energy 7/1(177, 244, 273, 275, 281, 287, 319); 14(169); 15(373)
Zero-point motion 7/1(177)
Zinc 6a/1(44, 93, 222)
Zinc blende lattice 7/1(84)
Zinc blende structure 7/1(116, 135, 139)
Zinc crystal, surface phenomena 7/2(104–106)
Zinc crystals 6a/1(523)
Zinc phosphate glasses 13(537)
Zinc single crystals
— critical shear stress 7/2(116)
— work-hardening 7/2(46, 49)
Zinc, ultrasonic attenuation 11/2(177)
Zone indices 7/1(62)
Zone of silence, *see also* Shadow zone 11/1(81–85)
Zubov, general system of 6a/3(130)

Group 4

Electric and Magnetic Behaviour of Matter

Absolute zero 23 (373)
Absorbed gases 21 (641)
Absorber, perfect 16 (478)
Absorptance 23 (441)
Absorption 18/1 (6, 12, 476); 22 (38, 39); 23 (89)
Absorption by free charge carriers 20 (210)
Absorption by imperfections 20 (213)
Absorption cell 23 (132, 142, 168)
Absorption coefficient 16 (642); 20 (210); 22 (356, 358)
Absorption coefficients and quantum yields 21 (359)
Absorption continuum 21 (330)
Absorption currents 16 (669)
Absorption, deviative and non-deviative 16 (559)
Absorption edge 20 (208)
Absorption factor 16 (633, 635)
Absorption index 16 (634, 642)
Absorption limit, optical 19 (321)
Absorption line 18/1 (217)
Absorption, magnetic 18/1 (482)
Absorption method using two waveguides 16 (693)
Absorption, molecular 19 (389)
Absorption spectrum of Eu^{2+} ions 18/1 (474)
Absorption, total 22 (360)
Accelerating fields, influence of 21 (28)
Acceptor 20 (18)
Acceptor level 19 (346)
Accumulation region 20 (382)
Acetone 17 (62, 95)
Acoustic branch 18/1 (37)
Acoustic gyrator 18/2 (213)
Acoustic paramagnetic resonance 18/1 (91)
Acoustical spin wave branch 18/2 (161)
Action at a distance 17 (2)
Activation 21 (380)
Activation energies 19 (203, 330)
Activation energy 17 (129); 20 (277, 355); 21 (53, 87, 116)
— dependence on the nature of the cation 20 (359)
— dependence on the stiffness of the network 20 (357)
— specific 20 (359)
Activation, heat of 21 (206, 214)
Activation, thermal 19 (323)
Active filter 23 (357)
Actual value 23 (458)
Actuating variable 23 (458)
Actuator 23 (458)
Adiabatic approximation 18/1 (43, 51)
Adiabatic demagnetization 18/1 (1)
Adiabatic susceptibility 18/1 (5, 10, 17)
Adiabatic transformation theorem of Boltzmann and Ehrenfest 16 (412)
Adler tube 23 (66)
Admittance, characteristic 16 (303)
Admittance matrix 16 (397)
Admittance of driving-point 16 (476)
Admittance per unit length 16 (210)
Admittance properties 16 (216, 260)
ADP = ammonium dihydrogen phosphate 17 (337, 339, 373)
ADP = ammonium, Ferroelectricity 17 (267)
ADP = ammonium, phase transitions 17 (328)
Adsorbed gas 22 (85)
Adsorbed layers 21 (647)
Adsorption 21 (195, 208, 214, 364, 377)
After-glows 22 (228)
Agents, poisoning 21 (153)
Agents, reducing 21 (153)
Aging effect 18/1 (10)

Air blast methods 21 (479)
Air discharges 22 (577, 596)
Airy's rainbow integral 16 (449, 450)
Alkali halides 17 (249); 19 (352, 353)
Alkali metal ions 21 (649, 653)
— of low energy 21 (644)
Alkyl-substituted benzene 18/1 (176)
Alloys 19 (206, 231)
Alloys, liquid 19 (221)
Alloys, resistivity of binary 19 (269)
Allyl radical 18/1 (189)
Alnico alloy 18/2 (457)
α, determination in rare gases 21 (593)
α-Fe_2O_3 18/1 (268)
Alternating current conductivity 20 (313)
Alternating electric current 16 (165)
Alti-electrograph 22 (598)
Aluminium and gallium antimonides 19 (386)
Ambient temperature measurement 23 (396)
Ambiguity (inherent one-count) 23 (331, 333, 336, 340, 363, 364)
Ambipolar diffusion 21 (397, 479, 482)
Ammonia 17 (51, 150)
Ammonia maser 23 (327)
Ammonium sulphate, ferroelectricity 17 (273)
Amorphous substances 17 (232)
Amplification factor 22 (446)
Amplifier, circulator type 23 (41, 46, 72)
Amplifier, common-base 23 (19)
Amplifier, common-collector 23 (19)
Amplifier, common-emitter 23 (19, 26)
Amplifier, cyclotron wave 23 (65, 66)
Amplifier, grounded-anode 23 (19)
Amplifier, grounded-cathode 23 (19)
Amplifier, grounded-grid 23 (19)
Amplifier, hybrid-type 23 (69)
Amplifier, linear two-port 23 (4, 5)
Amplifier, negative-resistance-type 23 (41)
Amplifier, one-port 23 (41)
Amplifier, parametric 23 (39)
Amplifier, tunnel diode 23 (67, 74)
Amplifiers in cascade 23 (2, 4)
Analogue computer 23 (219)
Analogue recording 23 (285)
Analogue transistor 20 (175)
Anderson mechanism, Kramers- 18/1 (379)
Angular frequency 16 (286)
Angular spectra 16 (465)
Angular variation 18/1 (135, 140)
Anhysteretic remanence 18/2 (509, 538)
Anhysteretic remanent initial susceptibility 18/2 (534)
Anhysteretic susceptibility 18/2 (533)
Anion 18/1 (185)
Anion, anthracene 18/1 (156)
Anion, benzene 18/1 (186)
Anion, benzosemiquinone 18/1 (180)
Anion, butadiene 18/1 (160)
Anion, cyclo-octatetrene 18/1 (176)
Anion radical 18/1 (155)
Anisotropic conductivity, influence of pressure on the 19 (195)
Anisotropic coupling between electron spin and magnetic field 18/1 (133)
Anisotropic effect shown by polarisation experiments 19 (369)
Anisotropic exchange 18/2 (36, 110)
Anisotropic magnetoelastic effect 18/2 (207)
Anisotropic magnetostriction 18/2 (308)
Anisotropy 17 (38, 73)
Anisotropy coefficient 18/2 (66, 280)
Anisotropy constant 18/2 (66, 70, 279, 281, 284, 323, 325)
Anisotropy constant, pressure dependence of 18/2 (335)
Anisotropy, crystalline 18/2 (279, 291, 441)
Anisotropy, dipolar 18/2 (62)
Anisotropy, electrical 16 (287)
Anisotropy, exchange 18/2 (182, 288, 439, 440, 443, 444)
Anisotropy, ferromagnetic 18/2 (62, 65)
Anisotropy field 18/2 (67)
— magnetocrystalline 18/2 (324)
— surface 18/2 (92)
Anisotropy, hyperfine 18/2 (260)
Anisotropy in relaxation times 18/1 (63)
Anisotropy in spin-orbit coupling 18/1 (64)
Anisotropy in state functions 18/1 (63)
Anisotropy in the strain 18/1 (85)
Anisotropy, macroscopic and microscopic 21 (487)
Anisotropy, magnetical 16 (21, 287)

Anisotropy of the magnetization 18/2 (77, 283)
Anisotropy on microscale 21 (472, 497)
Anisotropy, quadrupolar 18/2 (69)
Anisotropy, shape 18/2 (283, 451)
Anisotropy, stress 18/2 (369, 442)
Anisotropy, surface 18/2 (182, 288, 367, 373, 442)
Anisotropy, uniaxial 18/2 (286, 394, 491)
Anisotropy, unidirectional 18/2 (283, 287, 451)
Anisotropy, vector 18/2 (446)
Annealing 21 (381)
Annealing temperatures for total recovery 19 (197)
Annihilation operator 18/1 (38)
Annular ring 16 (345)
Anode and anode-plasma temperature 22 (424)
Anode, contraction at the 22 (425)
Anode, current density at the 22 (425)
Anode dark space 22 (57)
Anode evaporation 22 (431)
Anode fall 22 (61, 83, 145, 146, 148, 165, 279, 422)
— negative 22 (146)
— normal 22 (436)
— positive 22 (146)
— theories 22 (434)
Anode glow 22 (57, 146, 149, 247)
Anode glow, oscillations of the 22 (144)
Anode, heat at the 22 (150)
Anode, movable 22 (115)
Anode rays 22 (147)
Anode region 22 (145, 152)
Anode, size and shape of the 22 (146)
Anode, spherical 22 (146)
Anode spots 22 (146, 151, 165)
Anomalies in higher order coefficients 17 (359)
Anomalous skin effect 18/2 (179)
Antenna arrays 16 (455, 485, 486)
Antenna, asymmetrical 16 (240)
Antenna, biconical 16 (468)
Antenna, collinear 16 (248, 252)
Antenna current, integral equation 16 (237)
Antenna, definition 16 (234)
Antenna, forming a square loop 16 (253)
Antenna gain 23 (182)
Antenna, horizontal 16 (484, 527)
Antenna, linear 16 (467, 482, 488)
Antenna, reflecting 16 (487, 495)
Antenna, rhombic 16 (485, 486)
Antenna, symmetrical, center-driven 16 (235, 238)
Antenna, symmetrical, elements of the terminal network 16 (236)
Antenna temperature 16 (454)
Antenna terminating a two-wire line 16 (235)
Antenna, vertical 16 (484, 516)
Antennas as probes 16 (267 ff.)
Antennas, folded and closely spaced 16 (246)
Antennas, parallel coupled 16 (242, 247)
Antennas uniformly spaced around the circumference of a circle 16 (245)
Anthracene 18/1 (185)
Anthracene anion 18/1 (156)
Antiferroelectricity 17 (273, 335, 368, 374, 384, 385)
Antiferromagnet 18/2 (95, 100)
Antiferromagnet, magnetoelastic effect in 18/2 (338)
Antiferromagnet, molecular field in 18/2 (100)
Antiferromagnet, orthorhombic 18/2 (108)
Antiferromagnet, phase diagram of 18/2 (137)
Antiferromagnet, phase transition in 18/2 (134)
Antiferromagnet, sinusoidal 18/2 (141)
Antiferromagnet, spin-wave spectrum of 18/2 (136)
Antiferromagnet, thermodynamic properties of 18/2 (108)
Antiferromagnetic axis 18/2 (445)
Antiferromagnetic coupling 18/1 (267)
Antiferromagnetic, linear chain 18/2 (116)
Antiferromagnetic ordering 18/2 (155)
Antiferromagnetic phase 18/2 (138)
Antiferromagnetic resonance 18/2 (187)
Antiferromagnetic spin wave 18/2 (106)
— dispersion 18/2 (127)
— spectrum 18/2 (106)

Antiferromagnetic spiral 18/2 (141, 144, 147)
Antiferromagnetism 19 (123)
Antiferromagnetism, band model of 18/1 (417)
Antiferromagnetism, compensated 18/1 (267)
Antiferromagnetism, magnetostriction in 18/2 (338)
Antiferromagnetism, Néel theory of 18/2 (98)
Antiferromagnetism, not-compensated 18/1 (267)
Antiholes 16 (627)
Antimony 19 (340)
Antimony trisulphide 19 (388)
Antipode 16 (606, 611)
Antiresonance in crystal vibrations 17 (298)
Antiresonance point 18/2 (178)
Aperture antennas 16 (453, 463–465, 473–475, 489)
Appearance potential 21 (462, 464)
Appleton-Hartree formula 16 (547)
Approximation, adiabatic 18/1 (43, 51)
Approximation, non-adiabatic 18/1 (43, 52)
Arc 22 (1, 2, 9)
Arc, carbon 22 (291)
Arc cathode, temperature of the 22 (407)
Arc characteristic, dynamical 22 (286)
Arc, characteristic of an 22 (276)
Arc, chemical reactions in 22 (284)
Arc column 22 (366)
Arc diameter 22 (402)
Arc discharge 21 (634); 22 (55)
Arc furnaces 22 (303)
Arc, Gerdien- 22 (300, 371)
Arc, ignition of 22 (262)
Arc initiation 21 (198, 200, 217)
Arc lines 22 (112)
Arc, matter transport 22 (283)
Arc mechanism 22 (257)
Arc on metals of low boiling point 21 (636)
Arc physics 22 (255)
Arc, radiation properties of 22 (281)
Arc, rotating 22 (299)
Arc, self-sustained 22 (255, 304)
Arc, stabilization of an 22 (265)
Arc, temperatures 22 (282)
Arc, thermal 22 (257)
Arcs, alternating current 22 (286)
Arcs, electrode stabilized 22 (276, 391)
Arcs in rare gases 22 (296)
Arcs, metallic vapour 22 (294)
Arcs, non-stationary 22 (290)
Arcs, reignition of, after extinguishing 22 (291)
Arcs, switch 22 (302)
Arcs, switched off 22 (382)
Arcs, types of 22 (291)
Area, effective 16 (453, 477, 487, 488, 503, 506)
Argon 17 (75)
Arsenic 19 (78, 340)
Arsenic trisulphide 19 (388)
Ashkin's tube 23 (67)
Association 20 (298–300)
Aston or primary dark space 22 (56, 96, 97)
Asymmetrical fields 22 (449)
Atmospherics 16 (565, 619 ff.)
Atom, α-carbon 18/1 (159)
Atom, β-carbon 18/1 (159)
Atom energy level 18/1 (401)
Atom in solids 18/1 (245)
Atom, ligand 18/1 (132)
Atomic beam techniques 23 (322)
Atomic conductance 19 (162)
Atomic defects in the lattice 19 (321)
Atomic frequency standard 23 (321)
Atomic resistivity increase 19 (209)
— pressure coefficient of the 19 (213)
— temperature coefficient for the 19 (211)
Atomic resistivity values 19 (210)
Atomic resonator 23 (321)
Atomic vibrations, equal 19 (162)
Attachment 22 (178)
Attachment coefficient 21 (550)
Attachment, decline of, with increasing energy 21 (462)
Attachment, dissociative 21 (454, 459, 462, 551, 553)
Attachment, free path for 21 (447, 448)
Attachment frequency 21 (412); 22 (566)
Attachment in fields causing ionization by electron impact 21 (453)
Attachment in high fields 21 (446)
Attachment, mono-energetic, techniques 21 (457)

Attachment of electrons 21 (445f., 453, 458, 475)
Attachment of electrons, near thermal energies 21 (457)
Attachment, probability 21 (446, 447, 451)
Attachment to molecules 21 (482)
Attenuation 17 (145, 148)
Attenuation constant 16 (295, 688)
Attenuation constants in waveguides of various cross-sections 16 (320, 325, 327, 334, 376)
Attenuation function 16 (351)
Attenuation region 16 (351, 356, 358)
Auger transition 21 (651)
Aureole 22 (57, 110, 112, 372)
Auto-correlation function 16 (577, 579, 592)
Automatic reading of data 23 (224, 226)
Auxiliary actuating variable 23 (517)
Auxiliary anode 22 (152)
Auxiliary controlled variable 23 (515)
Available gain 23 (7)
Avalanche 22 (4, 28, 29, 47, 446)
Avalanche fluctuation, statistical 22 (447)
Avelanche breakdown 17 (228)
Avelanche formation 17 (245)
Averaging time 23 (307, 311, 315)
Azimuthal modes 16 (605, 610ff.)

B change 22 (610)
Babinet's principle 16 (490, 507)
Back diffusion 22 (154, 512)
— losses 22 (447)
— of electrons 21 (262, 523)
Back scattering 16 (537, 541, 586, 631)
— of electrons 22 (4, 15)
— of electrons to the cathode 21 (578)
Back-scattered electrons 21 (518)
Backward traveling wave 16 (390)
Backward-wave amplifier 23 (31)
Balance method 16 (696)
Ball lightning 22 (577)
Ball point 23 (223)
Ballistic measurements 23 (206)
Band, filled 19 (321)
Band, forbidden 19 (321)
Band model 20 (9ff.)
Band model of antiferromagnetism 18/1 (417)
Band spectra 22 (98, 100, 184)
Band structure 18/1 (392)
— for $Eu^{++}S$ 18/1 (418)
— model of Cs_3Sb 21 (358)
— of Eu chalcogenides 18/1 (477, 511)
— RE metals 18/1 (472)
Band theory 18/1 (397)
Band with of a cavity resonator 16 (409)
Band-model of ferromagnetism 18/2 (85, 86)
Band-theory of the Bloch wall 18/2 (87)
Bandwidth 18/1 (109, 113)
Bandwidth of c.r.o. amplifiers 23 (242)
Bandwidth of recorders 23 (221)
Barium borate glasses 20 (359)
Barium oxide 19 (374)
Barium titanate 17 (342, 361, 382, 386)
— ferroelectricity 17 (268)
— hexagonal 17 (309, 381)
— infra-red vibration bands 17 (295)
— spontaneous polarization 17 (330)
— transducer 23 (207)
— type, ceramics of 17 (282)
Barkhausen jumps 18/2 (344)
Barretter 23 (97)
Barrier 21 (342, 350, 363, 367, 369)
Barrier at surface, lowering by an electric field 21 (364)
Barrier layer 20 (178)
— magnetic 20 (137ff.)
— theory 19 (330)
Bateman vector 16 (434, 435, 442, 451, 452)
Bead (or chain) lightning 22 (577)
Beat-frequency method 18/1 (12)
Beck-arc 22 (293, 428, 429)
Beer's law 21 (312)
Bend of a waveguide in the *E*-plane 16 (367)
Bend of a waveguide in the *H*-plane 16 (366)
Bennett distribution 21 (391)
Bent rectangular waveguide; longitudinal section waves 16 (365)
Bent waveguides 16 (364–369)
Benzene 17 (93, 95, 151, 152); 18/1 (219)
Benzene, alkyl-substituted 18/1 (176)
Benzene anion 18/1 (186)
Benzo-semiquinone 18/1 (104)
Benzosemiquinone anion 18/1 (180)

Benzosemiquinone, dibutyl- 18/1 (106)
Beryllium 19 (44, 46)
Bethe-Slater curve 18/1 (307)
Bias 23 (460, 493)
Binary system 23 (219)
Binary-coded-decimal (BCD) output 23 (330)
Binding energy 19 (83)
Binding force 17 (381)
Biochemical 18/1 (95)
Biot-Savart law 16 (20)
Biradical 18/1 (205, 226)
Birefringence 17 (323, 328, 329, 336)
Bismuth 19 (46, 263)
Bismuth sulphide, selenide and telluride 19 (388)
Bit 23 (219)
Bizette-Erickson-Yosida array 18/2 (147)
Black body 23 (442)
Bloch equation 18/1 (8)
Bloch functions 17 (183)
Bloch line 18/2 (401)
Bloch or tight-binding approximation 19 (25)
Bloch sum 19 (26)
Bloch wave 18/1 (321)
Bloch-Bloembergen equation 18/2 (174)
Bloch's integral equation 19 (281, 286)
Bloch's theorem 19 (12)
Bloch's theory of spin waves 18/2 (13)
Bloch's wall 18/2 (79, 81, 381, 384, 394, 395, 453)
— band-theory of the 18/2 (87)
— parameter 18/2 (81)
— spin wave in the 18/2 (88)
Bloch's wave function 18/2 (350)
Block method of Shortley and Weller 16 (151, 152)
Bloembergen-Rowland coupling 18/1 (411)
Bode-diagram 23 (464)
Body-centered structure 19 (15)
Bogolyubov's perturbation theory 18/1 (322, 365)
Bohr magneton 18/1 (95, 404); 18/2 (2)
Boiling point, effect of pressure on 23 (374)
Bolometer 22 (69); 23 (97, 98, 99, 143, 447)
Bolton relation, Colpa- 18/1 (158)
Boltzmann distribution 17 (221); 18/1 (72); 20 (36); 21 (331); 22 (70, 106, 107, 126, 306)
Boltzmann energy 17 (40, 54)
Boltzmann equation 19 (234, 271); 20 (50); 21 (391, 404); 22 (121)
Boltzmann equilibrium 20 (148)
Boltzmann factor 17 (40, 41, 55); 18/1 (96)
Boltzmann interferometer 23 (128)
Boltzmann law 23 (90)
Boltzmann relation 21 (19, 28)
Boltzmann relaxation, Maxwell- 18/1 (30)
Boltzmann statistics 20 (32)
Bombardment of the cathode 22 (13)
Bond moments 17 (151, 152)
Bonding, covalent 18/1 (67, 69)
Bonding, crystal 18/1 (67)
Bootstrap circuit 23 (244)
Born approximation 16 (580)
Born-Haber cycle 18/1 (548)
Boron 19 (340)
Boron, photoconductivity in 19 (340)
Boron, semiconducting 20 (235)
Böttcher formula 17 (59, 61, 67)
Böttcher theory 17 (57, 83)
Boundary between two dielectrics, numerical treatment 16 (153, 154)
Boundary conditions 16 (169, 170)
Boundary conditions in propagation problems 16 (424, 518, 520, 527, 531, 532, 542)
Boundary conditions in wave guides 16 (288, 292)
Boundary, intercrystalline 19 (330)
Boundary value problems of potential theory 16 (26)
Boys camera 22 (581)
Boys drum camera 22 (583)
Bradbury's theory 21 (498)
Bragg reflection 17 (183)
Bragg's law 19 (22)
Branch, acoustic 18/1 (37)
Branch and *M* components 22 (595)
Branch, optical 18/1 (37)
Branching and stroke-order 22 (580)
Branching of lightning 22 (579)
Breakdown 22 (1, 8, 17, 20, 27, 28, 35, 38, 39, 40, 49)
Breakdown at low pressure 21 (637)
Breakdown by field emission 22 (450)

Breakdown, collective 17 (233, 243)
Breakdown condition 22 (6)
Breakdown condition, static 21 (605)
Breakdown criterion 17 (160, 228); 22 (5, 9, 16, 32, 33, 35, 37, 45)
Breakdown criterion, physical significance of the 22 (9)
Breakdown, definition 21 (607)
Breakdown, directional 17 (163)
Breakdown, electron attachment controlled 22 (566)
Breakdown, electronic 17 (239)
Breakdown events, time sequence of 22 (2)
Breakdown field 17 (238, 240, 241)
Breakdown field-strength 20 (370, 371, 372)
Breakdown field-strength, intrinsic 20 (369)
Breakdown in air 22 (524, 566)
Breakdown in amorphous media 17 (161, 242)
Breakdown in asymmetrical fields 22 (506)
Breakdown in non-uniform fields 22 (480)
Breakdown in oxygen 22 (567)
Breakdown in pure argon 22 (509, 513)
Breakdown in spark suppressing gases 22 (522)
Breakdown, influence of photons 21 (606)
Breakdown, intrinsic 17 (160)
Breakdown mechanism 22 (39)
Breakdown, mobility-controlled 22 (568)
Breakdown potential 22 (1, 5, 10, 12, 15, 26, 451)
Breakdown potential of hydrogen 21 (598)
Breakdown probability 22 (469)
Breakdown, secondary electron resonance 22 (570)
Breakdown, thermal 17 (157, 166, 253)
Breakdown threshold 22 (450 f.)
— accurate formalation 22 (456)
— with electron attachment 22 (468)
Breakdown time 22 (28, 454)
Breakdown time, true 22 (472)
Breit and Tuve's law 16 (550)
Brewster angle, *see also* Pseudo Brewster angle 16 (525)
Bridge 23 (117, 420–425)
Bridge, balanced 23 (142)
Bridge circuit 18/1 (11)
Bridge, circularly polarized 23 (155)
Bridge compensator 23 (414)
Bridge for measuring spectral absorption 23 (144, 145)
Bridge, impedance 17 (139)
Bridge methods 16 (642–658)
Bridge, RC 23 (345, 358)
Bridge, six-arm 23 (118)
Bridge, transmission line 23 (135)
Bridge, twin-T 23 (134)
Bridgman relation 20 (94)
Brigde method 17 (139, 300)
Brigde of Cole and Gross 17 (141)
Brillouin function 18/1 (405); 18/2 (2)
Brillouin zone(s) 17 (183); 19 (10, 17, 22, 230, 246, 303)
— for antiferromagnets 18/2 (135)
— for bismuth 19 (306)
— for ferromagnets 18/2 (135)
— for lattice waves 19 (247)
— for polyvalent metals 19 (261)
— in *q*-space 19 (287)
Broadened line, homogeneously 18/1 (125)
Broadened line, inhomogeneously 18/1 (125)
Broadening, dipole 18/1 (122)
Broadening, modulation 18/1 (117)
Broadening, statistical 22 (213)
Broadside array 16 (487)
Bromobenzene 17 (62)
Brons-van Vleck form 18/1 (74, 81)
Brown differential equation 18/2 (455)
Brown paradox 18/2 (413, 457, 474)
Brownian displacement law 21 (525)
Brownian motion 17 (126, 130)
Brownian movement 19 (150)
Buckling 18/2 (417, 418, 455, 456)
Bulk diffusion 21 (206, 214)
Burst pulse 22 (515, 518)
Burst, threshold for a localized 22 (484)
Busch's theorem 21 (387)
Butadiene anion 18/1 (160)

^{13}C splitting constant 18/1 (165)
Cable of elliptical cross section 16 (80)
Cadmium and zinc sulphides 19 (317, 368)

Cadmium arsenide 19 (388)
Cadmium ions 21 (645)
Cadmium niobate 17 (320)
Cadmium selenide and telluride cf. zinc
Cadmium sulfide 17 (252)
Cage antenna 16 (248)
Calcium 19 (39)
Calcium difluoride 20 (328)
Calibration of photographic plates 21 (319)
Calorimeter 23 (100, 101)
Calorimeter, low-temperature 23 (140)
Calorimeter method 18/1 (12)
Camera for technical photography 23 (231, 263)
Camera (high speed) 23 (231)
Camera observations, correlations with 22 (608)
Camera, types of 22 (580)
Canal, *see* Streamer or Leader
Canal ray discharges 22 (162)
Canal rays 22 (57, 59, 96)
Canting, hidden 18/2 (153, 168)
Canting of sublattices 18/2 (163)
Canting, overt 18/2 (153)
Capacitance 16 (210)
Capacitance modulation factor 23 (41)
Capacitance of a coil 16 (200)
Capacitor 16 (9)
Capacity 16 (9)
Capacity, mutual 16 (10)
Capacity of an ellipsoid 16 (136)
Capacity, proper 16 (10)
Capillar tubes 22 (91, 118, 163, 174, 199)
Capillary electrometer 22 (598)
Capture cross section of the ion 21 (486)
Carbon 19 (340)
Carbon arc, hissing 22 (425, 427)
Carbon atom, α- 18/1 (159)
Carbon atom, β- 18/1 (159)
Carbon dioxide 17 (51, 61, 74)
Carbon disulfide 17 (74)
Carbon paper tape 23 (223)
Card punch 23 (276)
Card reader 23 (276)
Carnot cycle 23 (373)
Carrier diffusion length 19 (393)
Carrier life-time 19 (349)
Carrier, minority 19 (322)
Carrier mobility 19 (318)
Carrier replenishment 19 (325)
Carrier support 17 (246, 248)
Carriers, diffusion, mobility and life-time of 19 (393)
Carriers, diffusion of liberated 19 (348)
Carriers in dynamic equilibrium 21 (473)
Cascade compensator 23 (410)
Cascade control 23 (515)
Cascading constants 22 (234)
Casimir and Du Pré formalism 18/1 (77)
Casimir and Du Pré relaxation curve 18/1 (7)
Casimir and Du Pré, theory of 18/1 (4)
Cataphoresis 22 (195)
Cathode, area of the 22 (94)
Cathode dark space 22 (56, 62, 81, 165)
— length 22 (92, 165)
— length in gas mixtures 22 (183)
— thickness of 22 (106)
Cathode, deformation of the 21 (638)
Cathode effects 22 (10)
Cathode emission 22 (33)
Cathode, energy balance at the 22 (99)
Cathode evaluation 21 (48, 55, 94, 134)
Cathode, evaporating 22 (420)
Cathode fall 22 (81, 82, 83, 87, 89, 101, 103, 106, 108, 165, 193, 279, 405, 412, 420)
— abnormal 22 (84)
— defects of the theories 22 (107)
— in gas mixtures 22 (183)
— normal 22 (84)
— striations 22 (100)
Cathode, field at the 22 (105)
Cathode geometry 22 (97)
Cathode glow 22 (96, 246)
Cathode glow, ion current at the 22 (103)
Cathode, heating of the 22 (99, 100)
Cathode, highly stressed 22 (526)
Cathode instability 22 (506)
Cathode irradiated 22 (1)
Cathode layer, first 22 (56, 60, 96, 146)
Cathode light 22 (59, 96)
Cathode material 22 (12, 59, 88, 156)
Cathode material, influence of the 21 (600)
Cathode mechanism of arcs 22 (403)

Cathode metal, sputtering of the 22 (86, 87)
Cathode, movable 22 (120)
Cathode, oxidation of the 22 (33)
Cathode, photo-sensitive 22 (202)
Cathode, positive ions on the 22 (6)
Cathode processes 22 (14)
Cathode property determination 21 (45)
Cathode ray(s) 22 (59, 147)
— oscilloscope 23 (238)
— tube 23 (239)
— tube screen 23 (245)
Cathode region, magnetic field influences 22 (165)
Cathode sputtering 21 (202); 22 (154)
Cathode surface 22 (15, 108, 408)
Cathode temperature 21 (26); 22 (90, 157)
Cathode, thermionic 22 (243)
Cathode, virtual 22 (59)
Cathodes, complex 21 (354, 358, 377)
Cathodes, concave and convex 22 (97, 99)
Cathodes, multilayered 21 (343)
Cathodes, oxide coated 21 (363, 378)
Cation radical 18/1 (155)
Cauchy-Riemann equations 16 (40)
Cauchy's integral formula 16 (54)
Cauchy's theorem 16 (54)
Caustic surface 16 (447–450, 510, 631)
Cavity field of Onsager 17 (108)
Cavity, resonance 18/1 (109)
Cavity resonator 16 (406–422, 578); 18/1 (97, 111, 112, 113)
— cylindrical 16 (410)
— small perturbations 16 (411–415)
CdS 18/1 (255)
Cellular method 19 (28)
Centigrade 23 (374, 378)
Central limit theorem 16 (591)
Centre, electron donor 18/1 (80)
Centre, F 18/1 (78, 144, 228)
Centre, H 18/1 (241, 244)
Centre interaction F- 18/1 (231)
Centre, interstitial 18/1 (245)
Centre, local electron 18/1 (78)
Centre, M- 18/1 (235)
Centre, modified, F- 18/1 (233)
Centre, odd-electron 18/1 (259)
Centre, V_1 18/1 (239)
Centre, V_F 18/1 (241)
Centre, V_K 18/1 (240, 241, 243)
Centre, V_{OH} 18/1 (245)
Centre, V_t 18/1 (241)
Centres, F-aggregate 18/1 (235)
Centrosymmetric crystals 17 (305)
Ceramic capillary 21 (317)
Ceramics 17 (268, 283, 301, 308, 318)
Cesium films on tungsten, properties 21 (109)
CH_2 18/1 (101)
CH group 18/1 (101, 102)
CH_2 group 18/1 (102)
$CH9_3$ group 18/1 (101)
CH_3 group 18/1 (102, 103)
Chalcogenide, Eu- 18/1 (476, 481, 506, 512, 554)
Chalcogenide, Gd- 18/1 (518)
Chalcogenides, band structure of Eu 18/1 (477, 511)
Chalcogenides, magnetic redshift of Eu 18/1 (483)
Characteristic 22 (10, 15, 16, 147)
Characteristic impedance 23 (91)
Characteristic, negative 17 (167, 253); 22 (506, 513)
Characteristic temperature of Debye 19 (169, 171, 243, 250)
Characteristic temperature of Einstein 19 (242, 250)
Characteristic temperature, table of 19 (265, 290)
Characteristics of the second kind 22 (176)
Charge carriers, generation 20 (110ff.)
Charge carriers, life time 20 (111)
Charge conservation 16 (2)
Charge density 16 (2, 167)
Charge density, effective 19 (85)
Charge distortion 22 (503)
Charge distortion, space 22 (30)
Charge, effective 17 (162)
Charge, excess 18/1 (158)
Charge (in ligthning), quantities of 22 (600)
Charge multiplication 21 (505, 511, 526)
— in non-uniform fields 21 (558)
— in rare gases 21 (529)
— in space-charge distored fields 21 (564)
— in the presence of negative ions 21 (550)

Charge multiplication (cont.)
— in the presence of positive ions 21 (553)
— in two concentric cylinders 21 (558)
Charge neutralization 21 (484)
Charge transfer 21 (624, 659); 22 (77, 79, 80, 99, 106, 109, 163)
Charge transport, theory of 21 (512)
Charge, true 16 (8)
Chemical bond in semiconductors 20 (6ff.)
Chemical potential 22 (317)
Chemical statistics 20 (42)
Chemical transfer 18/1 (202)
Child-Langmuir law 16 (17)
Child-Langmuir space charge 21 (9, 34, 40)
Chlorine superhyperfine structure 18/1 (252)
Chlorobenzene 17 (94, 95, 152)
Chlorobenzosemiquinone 18/1 (105)
Chloroform 17 (93, 94, 95)
Chopper 23 (413)
Chopper amplifier 23 (55)
Chopper bar controller 23 (496)
Chromium halide 18/1 (481)
Circle diagram of impedance 23 (91, 120)
Circle diagram (Smith chart) 16 (216)
Circuit, electric, conventional 16 (195)
Circuit, electric, quasiconventional 16 (195)
Circuit, equivalent, of a waveguide junction 16 (396–401)
Circuit problems 16 (233)
Circuit properties of receiving antennas 16 (264, 268)
Circular current 16 (109)
Circular cylinder of finite length, potential 16 (111)
Circular disc, potential 16 (124, 134, 140–142)
Circular disc with a hole 16 (134, 140–142)
Circular polarization 21 (394)
Circular ring, potential 16 (102, 124)
Circular sector waveguide 16 (343)
Circulator 18/2 (169); 23 (71)
Cl_2 spectrum 18/1 (241)
Clamped crystal 17 (294, 324, 328, 332, 335, 339, 341, 343, 348, 363, 364, 375, 381, 387)
Clausius-Clapeyron relation 17 (357)
Clausius-Mosotti expression 17 (53, 59, 107)
Clausius-Mosotti formula 17 (1, 50, 67, 75, 83, 105, 117)
Clausius-Mossotti relation 16 (712)
Clearing time 22 (455, 519, 520)
Closed shell 19 (49)
Cloud and air discharges 22 (617)
Cloud discharge 22 (577, 597)
Cloud luminosity 22 (584)
Coastal refraction 16 (543, 607)
Coating activation 21 (152)
Coating conductivity 21 (120, 152)
Coaxial cylinders 22 (517)
— breakdown in air and pure oxygen 22 (516)
— breakdown in pure gases 22 (514)
Coaxial line 16 (326, 685); 17 (146)
Coaxial line sector 16 (344)
Coaxial regions 16 (293)
Code for information interchange 23 (273)
Code, progressive 23 (219)
Coercitive field 16 (21)
Coercive force 18/2 (342, 453)
— of the remanence 18/2 (511)
— particle size dependence of the 18/2 (467)
Coercivity 17 (277)
Coherence distance 16 (597, 623, 638)
Coherent scattering 16 (538, 557, 558, 628, 636)
Coil 16 (24)
Coil, flat 16 (110)
Coil, long 16 (110)
Coil of spherical shape 16 (126)
Cole and Cole plot of complex permittivity 16 (719)
Collapse of voltage 22 (41, 45, 50)
Collective electron model 19 (320)
Collector 20 (131)
Colli-Facchini effect 22 (463)
Collision cross section 22 (344, 348)
— for capture 21 (447)
Collision damping 22 (361)
Collision frequency 21 (392, 423); 22 (80, 538)
Collision frequency for momentum transfer 21 (410)
Collision frequency, total 21 (410)
Collision integral 21 (408, 409)

Collision number 21 (396)
Collision, propability of 21 (410)
Collision theory of Lorentz 17 (135)
Collision theory of van Vleck and Weisskopf 17 (136)
Collisional multiplication of carriers 19 (337)
Collisions between molecules and positive ions 21 (510)
Collisions, elastic 22 (125)
Collisions, inelastic 22 (125, 133, 194)
Collisions, lattice 19 (318)
Collisions of the first kind 22 (218)
Collisions of the second kind 22 (75, 85, 112, 136, 183, 184, 236)
Collodion films 21 (315)
Colour centre(s) 18/1 (72, 78); 19 (352, 353); 21 (153)
Colpa-Bolton relation 18/1 (158)
Column, constricted 22 (178)
Column, contraction of the 22 (57, 175, 193)
Column gradient 22 (279)
Column, length of the 22 (114)
Column, linear 22 (128, 129)
Column, low current (subnormal) 22 (130)
Column, magnetic field on the positive 22 (168)
Column of rectangular section 22 (128)
Column, plane positive 22 (169)
Column, positive 22 (57, 114, 120, 165, 174, 247)
Column, striated 22 (57, 137, 142, 252)
Column, uniform, in atomic gases 22 (114, 115, 116)
Column, uniform, in molecular gases 22 (118)
Column, unstable 22 (194)
Columnar recombination 21 (480)
Common-mode-rejection-ratio 23 (243)
Comparator (VLF) 23 (303, 306)
Compensation method 23 (207, 408–416, 424)
Compensation point, magnetic 18/2 (96)
Compensator 19 (147, 148)
Complex ions 21 (647)
Complex potential 16 (41)
Compliance 17 (341, 347)
Compliance, adiabatic elastic 17 (349)
Compliance, electrostrictive 17 (360)
Compliance, isopolar 17 (294)
Compliance matrix 17 (353)
Composite crystal 17 (293)
3–5-compounds 23 (208)
Compressibility 19 (92)
Compression, self-magnetic 22 (396)
Compressional vibrations, longitudinal 17 (294)
Concentration 18/1 (70); 22 (209)
Concentration effect 18/1 (78)
Concentric rings 22 (151)
Condon effect, Franck- 18/1 (80)
Conductance 19 (137)
Conductance matrix 16 (397)
Conductance per unit length 16 (210)
Conduction band 17 (227); 19 (321); 21 (359, 360)
Conduction current density 17 (144)
Conduction electron polarization 18/1 (433)
Conduction electrons 17 (183)
Conduction in glasses of more complicated composition 20 (363)
Conduction in porous semiconductors 21 (85)
Conduction losses 20 (373 f.)
Conductivities, ratio for axial 19 (158)
Conductivity, *see* Alternating current c., Surface c., Grain boundary c.
Conductivity 16 (15, 287); 19 (137, 156, 158, 160, 162, 180); 22 (159)
Conductivity at low temperatures 20 (83)
Conductivity, bombardment induced 19 (340)
Conductivity changes at the melting point 19 (177)
Conductivity characteristic of an oxide cathode 21 (115)
Conductivity counters 19 (395)
Conductivity decay 19 (326)
Conductivity, effective 16 (540)
Conductivity, electric 22 (326, 352, 378)
Conductivity, electrical 19 (61); 20 (68)
Conductivity, heating curve of 19 (329)
Conductivity, intrinsic 20 (255, 258, 279, 285)
Conductivity, mean 19 (157)
Conductivity of elements 19 (155)
Conductivity of gases 21 (504)
Conductivity optimum of the earth 16 (606)

Conductivity pulses 19 (353)
Conductivity, specific 20 (56)
Conductivity, structure-sensitive 20 (255, 279)
Conductivity tensor 18/2 (170); 19 (139)
Conductivity, theory 19 (160, 161)
Conductivity, thermal 22 (134)
Conductivity, Valence exchange 18/2 (222)
Conductor, good 16 (172)
Conductor, perfect 16 (1)
Configuration interaction 18/1 (337); 19 (49, 86, 98)
Configurations, equivalent 17 (285, 311, 315, 329, 352, 354)
Conformal mapping
— by a power 16 (59–61)
— by elliptical functions and integrals 16 (65–68)
— by exponential functions 16 (61–65)
— by linear functions 16 (57–59)
— general remarks 16 (39, 56)
— of a polygon 16 (71–79)
— of waveguide cross-sections 16 (358–364)
— special cases 16 (57–71)
Congruence, assumption of 21 (398)
Congruence principle 18/2 (480)
Conical functions 16 (117, 129, 132)
Conservation of mass 22 (316)
Constitutive parameters 16 (286)
Constrictions, narrow 22 (147, 163)
Contact potential 19 (348); 20 (180); 21 (48, 52, 344, 368); 22 (65); 23 (40)
— as a function of the temperature 21 (48, 52, 106, 129)
— at a CdS layer 19 (369)
— Fermi 18/1 (72)
Contact semiconductor-metal 20 (188 ff.)
Contact semiconductor-semiconductor 20 (205 ff.)
Contact thermometer 23 (378, 380)
Continuity equation 16 (169, 170, 423, 518); 21 (404); 22 (36, 42)
Continuity equation, harmonic form 16 (286)
Continuum 22 (229)
Contour vibrations 17 (302)
Control system 23 (458)
Controlled system 23 (458)
Controlled variable 23 (457)
Controller 23 (458)
Convergence factor 16 (497)
Convergence of numerical integration 16 (150)
Converter, analogue to analogue (a-a) 23 (219)
Converter, analogue to digital (a-d) or digital to analogue (d-a) 23 (221, 342)
Converter, digital-to-digital (d-d) 23 (220)
Converter, frequency-to-current 23 (346)
Converter, frequency-to-voltage 23 (347)
Conwell-Weisskopf-formula 20 (65 ff.)
Coordinates, internal 17 (10, 15, 36, 40, 72)
Coordinates, oblate spheroidal 16 (134–138)
Coordinates of a flat ring 16 (144)
Coordinates of the asymmetric ellipsoid 16 (145–148)
Coordinates of the paraboloid of revolution 16 (143, 144)
Coordinates, prolate spheroidal 16 (134–138)
Coordinates, sphero-conical 16 (145)
Coordinates systems of rotational symmetry 16 (114)
Coordination, eight-fold 18/1 (69)
Coordination, nine-fold 18/1 (47)
Coordination, six-fold 18/1 (36)
Copper 19 (38, 80)
Copper nucleus 18/1 (100)
Corona 22 (506)
Corona, burst pulse 22 (518)
Corona discharge 22 (252)
Corona, positive point-to-plane 22 (452, 521)
Corona, self-sustaining 22 (512)
Corona threshold, burst pulse 22 (482)
Correlation 19 (86, 116)
Correlation conductance 23 (5)
Correlation effects (Frenkel disorder) 20 (329)
Correlation effects (Schottky disorder) 20 (327)
Correlation, electron 18/1 (284, 398, 400, 402, 532)
Correlation energy 19 (86)

Correlation factor 20 (328, 330, 333, 334, 337)
Correlation function 16 (577, 597); 21 (443)
Correlation, π-σ 18/1 (164)
Correlation time 18/1 (90)
Correlations, long-range 19 (88)
Cotton scales 23 (215)
Coulomb interaction 17 (215, 219); 18/1 (35)
Coulomb law 16 (2)
Coulomb (unit) 16 (2)
Counting diamonds 19 (341)
Coupling, anisotropy in spin-orbit 18/1 (64)
Coupling, antiferromagnetic 18/1 (267)
Coupling, Bardeen's calculation of 19 (251)
Coupling between electron spin and magnetic field, anisotropic 18/1 (133)
Coupling between electrons and lattice 19 (249)
Coupling, Bloembergen-Rowland 18/1 (411)
Coupling coefficient 16 (195)
Coupling coefficient, magnetoelastic 18/2 (204)
Coupling constant 19 (249, 250)
Coupling constant, magnetoelastic 18/2 (204)
Coupling, hyperfine 18/1 (149)
Coupling of short range 17 (372)
Coupling parameter, hyperfine 18/1 (149, 160, 168)
Coupling, Russel-Saunders 18/1 (402)
Coupling, spin orbit 18/1 (36, 47, 79, 126, 129, 130, 132, 259, 460)
Coupling, spin-phonon 18/1 (91)
Covalency 18/1 (65, 82, 383)
Covalency of the 4*f* electrons 18/1 (510)
Covalent binding 19 (3, 98, 109)
Covalent bonding 17 (309, 313, 318); 18/1 (67, 69)
Covalent bonding of 4*f* electrons 18/1 (470)
Coverage, degree of 21 (215, 219, 222)
$CrBr_3$ 18/1 (267)
Cr-chalcogen spinel 18/1 (522)
Creation operator 18/1 (84)
Creeping, magnetic 18/2 (466)
Creeping wave 16 (604, 609)
Critical distance 18/1 (399)
Critical frequencies 16 (372)
Critical point 17 (352, 361)
Critical temperature 21 (48)
Crookes (or cathode) dark space 22 (246)
Crookes (or Hittorf) dark space 22 (56)
Cross drift 21 (385)
Cross modulation 16 (573)
Cross section 22 (5, 219)
— discontinuities of a line 16 (224)
— electrical 22 (219)
— for absorption 22 (37)
— for ionization 22 (221)
— in collisions 22 (219)
— optical 22 (220)
Cross-sectional data 22 (37)
Cross-tie-wall 18/2 (401, 402)
Cr-spinels, magnetic red shift 18/1 (487)
Crystal bonding 18/1 (67)
Crystal, choice of the wave function for the 18/1 (290, 300, 304)
Crystal, concentrated 18/1 (65)
Crystal detector 18/1 (98, 112)
Crystal field 18/1 (22, 34, 35, 47, 107, 131, 408, 463, 466)
— in rare-earth salts 18/1 (47)
— parameter 18/1 (47, 60, 62, 63)
— potential 18/1 (48)
— splitting 18/1 (407, 534)
— theory, strong crystal field 18/1 (376)
— theory, weak crystal field 18/1 (374)
Crystal, molecular 18/1 (68)
Crystal noise, excess 18/1 (110)
Crystal, non-polar 19 (336)
Crystal, oblique cut 17 (305)
Crystal periodicity 19 (5)
Crystal planes 21 (195, 207)
Crystal, polar 19 (336)
Crystal spectrum in RE 18/1 (463)
Crystal structures 19 (2)
Crystal symmetry 19 (15)
Crystal triode or transistor 19 (340)
Crystalline anisotropy 18/2 (279, 291, 441)
Crystalline energy 18/2 (353)
Crystallographic transitions 19 (174)
Crystals, additively coloured 19 (371)

Crystals of preferred orientation, method for growing 19 (158)
Crystals, organic 19 (389)
Crystals with excess halogen 19 (355)
Cubic lattice, body-centered 19 (10)
Cubic lattice, face-centered 19 (10)
Cubic point group 19 (15)
Cubic structure, body-centered 19 (6, 34, 37)
Cubic structure, face-centered 19 (6)
Cuccia coupler 23 (65)
Cuprous oxide 19 (377)
Cuprous oxide on copper 19 (316)
Curie constant 17 (388); 18/1 (2, 3, 405); 18/2 (3)
Curie law 18/1 (2); 18/2 (480)
Curie point(s) 17 (359, 366)
— anomalies at 17 (352)
— change with pressure 18/2 (330, 334)
— ferroelectric 17 (51)
Curie temperature 16 (724); 18/1 (265); 18/2 (1, 4, 343); 19 (218)
Curie temperature, paramagnetic 18/2 (3)
Curie-Weiss law 16 (725); 18/2 (3)
Curie-Weiss law for ferromagnetics 17 (266)
Curie-Weiss temperature 18/1 (409)
Curling 18/2 (416, 417, 455, 456)
Curling, anti- 18/2 (423)
Curling -mode, modified 18/2 (494)
Current, a.c. component of 19 (374)
Current change, undirectional irreversible 22 (453)
Current density 16 (14, 288); 22 (2, 3, 16, 17, 18, 20, 21, 49, 326, 330)
— at cathode 22 (407)
— electrical 20 (53)
Current distribution in an antenna 16 (238, 260)
Current, electric, definitions 16 (14, 165)
Current, electric, through vacuum 16 (16)
Current, field intensified 22 (446, 450, 452)
Current gain 19 (337)
Current gain as for carrier replenishment 19 (325)
Current glow curve 19 (361)
Current growth 22 (4, 7, 33, 499)
Current in unloaded loop 16 (273)
Current matrix 16 (397)
Current measurements 16 (668)
Current multiplication 19 (388)
Current noise 23 (31)
Current sources within waveguides 16 (301)
Current, space charge limited 22 (93)
Current, stability of the 22 (10)
Current stabilizing 19 (151)
Current, surface density 16 (167)
Current voltage characteristics, influence of magnetic field 22 (175)
Current, volume density 16 (167, 172, 185)
Currie's hollow beam gun 23 (31, 33)
Curvature of the boundaries 22 (57)
Curve follower, magnetic 23 (258)
Curve follower, optical 23 (256)
Curvilinear orthogonal coordinates 16 (90, 148)
Cut-off attenuator 23 (106, 107, 172)
Cut-off frequency or wavelength 16 (293, 295, 310, 311, 317, 339, 341, 342, 671)
Cut-off wavelength 23 (92)
Cycloid, conical 18/2 (141)
Cyclo-octatetrene anion 18/1 (176)
Cyclotron resonance 20 (215); 23 (149, 150)
Cyclotron resonance experiments 19 (67, 343)
Cyclotron vector 21 (384)
Cylinder carrying space charge 16 (52)
Cylinder condenser 17 (178)
Cylinder with slots 16 (71)
Cylinders, coaxial 22 (510, 557)
— in air 22 (511)
— in pure nitrogen 22 (514)
— of different diameters 16 (106)
— of equal diameters 16 (105)
Cylinders of arbitrary length 22 (552)
Cylindrical waveguide 16 (315)

"*D*" cathode 21 (12)
d orbital 18/1 (127, 128, 129)
3 *d* orbital 18/1 (127)
Damage, irradiation 18/1 (95, 98)
Damping 17 (133); 22 (360)
Damping ratio of galvanometer systems 23 (232)
Dark current 22 (189)

Dark discharge 21 (574, 588, 634); 22 (202)
Dark space, energy distribution of ions and electrons 22 (95)
Dark space, length of the 22 (105)
Dark space, optical properties 22 (96)
Dart leader, mechanism of the 22 (620)
Dart leaders 22 (592)
Dart-stepped leaders 22 (590, 591, 593)
Data reduction 23 (218)
Davidson's growth equation 22 (42, 44)
Davydov splitting 18/1 (483)
d-band 19 (266)
d-band for nickel 19 (268)
Dead band 23 (253, 460, 500)
Dead capacitance 16 (646)
Dead time 22 (455, 511, 522)
Dead zone, *see also* Skip distance 16 (631)
Deadtime in a control system 23 (462, 489)
Debye cut-off frequency 18/1 (82)
Debye distribution 18/1 (38, 49, 80)
Debye equation 16 (714)
Debye length 20 (109, 117); 21 (399, 434)
Debye potentials 16 (433, 435, 436)
Debye relation between relaxation and viscosity 16 (716)
Debye relaxation 18/1 (21, 42)
Debye temperature 18/1 (42)
Debye theory of dielectric
— polarization 17 (20, 38, 87)
— relaxation 17 (125)
Debye units 17 (46)
Debye's relaxation 23 (88)
Decade counter 23 (329)
Decade counting assembly (DCA's) 23 (329)
Decade divider assembly 23 (329)
Decay, exponential 18/1 (21, 22)
Decay of current, time constant 19 (323)
Decibel (unit of attenuation) 16 (321)
Deep donor impurity 18/1 (254)
Defect in insulating solids 18/1 (228)
Defect in semiconductors 18/1 (253)
Defect mobilities 20 (309)
Deflecting system, electron optical 16 (97)
Deflection blanking 23 (241)
Deflection sensitivity of cathode ray tube 23 (239)
Deflection tensor 21 (435)
Deformation losses 20 (377f.)
— as a function of temperature 20 (378)
— dependence on the chemical composition 20 (382f.)
Deformation, plastic 17 (166)
Deformation potential 19 (253); 20 (64)
Degeneracy 16 (292); 19 (321); 22 (38)
Degeneracy, Kramers 18/1 (36, 40, 69, 72)
Degeneracy, time-reversal 18/1 (43)
Degenerate state, doubly 18/1 (36)
Degree of association 20 (299, 300, 308)
de Haas-van Alphen oscillatory magnetostriction 18/2 (339)
Delay generator 23 (369, 370)
Delay line 23 (244, 371)
Delay time in a control system 23 (462)
3 *d*-element oxide 18/1 (550)
ΔE effect 18/2 (319)
Demagnetization, adiabatic 18/1 (1)
Demagnetizing factor 17 (28, 29)
Demagnetizing field 18/2 (343)
Dember effect 20 (119)
Dense gases and liquids 17 (47)
Densities, high 17 (73, 75)
Density matrix operator 18/1 (2, 15)
Density, measurement of 22 (376)
Density of in the *d*-band 19 (266)
Density of states 19 (60)
Density of states for noble metals 19 (277)
Density of states, phonon 18/1 (30)
Depolarizing factor 16 (707)
Deschamps graphical procedure 16 (224)
Desensitising dyes for the photographic process 19 (384)
Detachment of electrons 21 (456)
Detector noise 18/1 (109, 110)
Detector, phase sensitive 18/1 (113, 114, 117)
Deuterium, *see also* Isotope effect 17 (376)
Deuterium atom 18/1 (246)
Deuterium exchange 18/1 (188)
Diamagnetic behaviour 16 (21)

Diamagnetism, magnetostriction in 18/2 (339)
Diamond 17 (251); 19 (41, 317)
Diamond, semiconducting 20 (230)
Diamond structure 19 (7, 10, 11, 15, 40)
Dibutyl-benzosemiquinone 18/1 (106)
Dielectric 16 (1)
Dielectric, anisotropic 16 (8)
Dielectric anisotropy and free energy 17 (332)
Dielectric anomalies 17 (273, 319)
Dielectric behaviour of materials 16 (640)
Dielectric, boundary of two 16 (7)
Dielectric constant(s) 17 (7, 294, 324, 333, 337, 363, 365, 385); 20 (210, 211)
— bridge methods 16 (651)
— complex 16 (171, 641)
— direct current methods 16 (666)
— equivalent 18/2 (170)
— high frequency measurements: free wave methods 16 (698–705)
— high frequency measurements: resonance methods 16 (671–686)
— high frequency measurements: waveguide methods 16 (686–698)
— low frequency measurements 16 (642)
— measuring methods of high sensitivity 16 (663)
— of glass 20 (360, 367, 388 f.)
— of vacuum 16 (2)
— or permittivity 16 (2, 18, 170, 286, 306, 640)
— ratio of 19 (372)
— relative 16 (2, 8)
— resonance methods 16 (658)
— static 16 (713); 17 (34)
Dielectric, definition of 17 (1)
Dielectric dispersion at acoustic resonances 17 (297)
Dielectric, effect of immersion in a 17 (28)
Dielectric, good 16 (172)
Dielectric image line 16 (375)
Dielectric layer 21 (350, 377, 378)
Dielectric, linear 17 (7)
Dielectric losses 16 (306, 310); 20 (311 f., 372 f.)
Dielectric materials 16 (710)
Dielectric measurements, resonance method 23 (136)
Dielectric, nonpolar 16 (711)
Dielectric properties of mixtures 16 (706)
Dielectric relaxation time 20 (108, 126)
Dielectric rod, thin, inside a cavity resonator 16 (414)
Dielectric, small charged or polarized body in 17 (23)
Dielectric sphere, small, inside a cavity resonator 16 (413)
Dielectric strength in AC fields 20 (389)
Dielectric tensor 16 (287)
Dielectric transformer in a line 16 (226)
Dielectric waveguide 16 (369–375); 23 (93)
— asymmetric hybrid waves 16 (374)
— waves circularly symmetrical 16 (370, 374)
Dielectronic capture 21 (486)
Difference equation 16 (148)
Differential amplifier 23 (242)
Differential bridge 19 (149)
Differential bridge of Bowen and Rodenbach 19 (150)
Diffraction zone 16 (605, 627)
Diffusion 20 (115 ff.); 22 (16, 77, 80, 169, 323)
Diffusion, ambipolar 22 (122, 129, 170, 174, 332)
Diffusion coefficient 19 (335, 339)
— effective 21 (398)
— for radioactive tracers 20 (325)
Diffusion constant 21 (412, 418, 611)
Diffusion constant for metastable atoms 21 (614)
Diffusion equation(s) 20 (121); 21 (414)
Diffusion frequency 21 (416)
Diffusion in ionic conductors 20 (248, 324)
Diffusion in velocity tensor 21 (430, 438)
Diffusion, lateral 22 (169)
Diffusion length 21 (396, 399); 22 (533)
Diffusion method of Bailey for measuring the attachment probability 21 (450)
Diffusion of divalent ions in NaCl-type crystals 20 (337)
Diffusion of electrons 22 (143)

Diffusion of resonance radiation 21 (305, 616)
Diffusion potential 20 (178)
Diffusion process 18/1 (75)
Diffusion, radial 22 (169)
Diffusion tensor 21 (396, 397)
Diffusion theory 22 (128, 537)
Diffusion, transition from free to ambipolar 21 (478)
Digital code disk 23 (256)
Digital recording 23 (273, 285)
Dilute system 18/1 (72)
Diode, construction of 21 (139)
Diode, planar 21 (145)
Diode, surface barrier 23 (39)
Diphenylpicrylhydrazyl 18/1 (182)
Dipolar anisotropy 18/2 (62)
Dipolar ferromagnetism 18/2 (93, 140)
Dipolar interaction 18/1 (73)
Dipolar materials 16 (712)
Dipolar orientation 16 (711)
Dipole axis 16 (4)
Dipole broadening 18/1 (122)
Dipole density, polarization 16 (5, 7)
Dipole, electric 16 (4, 29, 86, 121, 458, 461, 475, 479, 507, 530, 552, 553, 562, 601)
— horizontal 16 (529)
— vertical 16 (519–522)
Dipole, interaction of two 16 (14)
Dipole layer 21 (108)
— constant 16 (30)
— electrical 16 (5, 30)
— magnetic 16 (19)
Dipole liquids and solids 16 (714)
Dipole, magnetic 16 (18, 460, 461, 476, 485, 508, 535, 552, 553, 562, 605, 620)
Dipole, magnetic, vertical 16 (527, 529, 530)
Dipole modes, diagonal, in loop antennas 16 (258)
Dipole moment 16 (4, 713); 17 (3, 8)
Dipole narrowing 18/2 (236)
Dipole relaxation losses 20 (375f.)
Dipole scattering 18/2 (202)
Dipole surface density 16 (5)
Dipole transition, electric 18/1 (37, 459)
Dipole transition, magnetic 18/1 (37)
Dipole-dipole effect 18/1 (122)
Dipole-dipole interaction 18/1 (125)
Dipole-dipole interaction, magnetic 18/1 (14, 34, 35, 75)
Dipole-dipole operator 18/1 (207)
Dirac delta function 16 (173, 426, 451, 539)
Dirac spin Hamiltonian 18/1 (266)
Dirac-van Vleck Hamiltonian 18/1 (267)
Direct current 16 (14)
Direct photoionization measurement 21 (310, 314)
Direct process 18/1 (39, 121)
Direct process in non-Kramers salts 18/1 (48)
Direct process of relaxation 18/1 (39)
Direct relaxation process in rare-earth salts 18/1 (48)
Direction of arrival 16 (578, 599, 601)
Directional coupler 23 (112, 122)
Directive gain 16 (477, 487, 503)
Directivity 16 (477, 480)
Dirichlet's boundary value problem 16 (26, 33)
Dirichlet's boundary value problem for a circle 16 (54, 55)
Disc, elliptical 16 (147)
Discharge, abnormal 22 (55)
Discharge by thermal electrons 22 (247)
Discharge, characteristic colour 22 (58, 114)
Discharge, characteristic curve of the 22 (201)
Discharge, cold cathode 22 (246)
Discharge, constricted 22 (163, 175)
Discharge, effective resistance of the 22 (189)
Discharge, electrodeless 22 (144, 195, 251)
Discharge, flashing 22 (191, 577)
Discharge, hot cathode 22 (90)
Discharge, intermittent 22 (187, 625)
Discharge, magnetic fields 22 (73)
Discharge, mechanism of the continuing 22 (627)
Discharge, normal 22 (55, 82)
Discharge, obstructed 22 (56, 57, 161, 168)
Discharge, photographic studies of luminosity in the 22 (580)
Discharge, self capacity of the 22 (188)
Discharge sequence, interruption of it 22 (455)

Discharge, special types of 22 (161, 598)
Discharge, spontaneous 17 (248)
Discharge spread uniformly over the electrodes 22 (505)
Discharge, stability of 22 (201)
Discharge, subnormal 22 (55, 82)
Discharge, throbbing 22 (190)
Discharge, time of development of the 22 (188)
Discharge, time of formation of the 22 (190)
Discharge to ground, total duration of a 22 (579)
Discharges in electronegative gases 22 (177)
Discharges in magnetic fields 22 (164)
Discharges (lightning) to very high buildings 22 (596, 607)
Discharges, methods of measurement 22 (61)
Discharges, monoenergetic electron 22 (242)
Discharges, unstable 22 (506)
Disc-loaded rod 16 (383)
Disc-loaded waveguide, *see* Waveguide circular
Discontinuities of a line 16 (224, 315)
Discriminator characteristic 23 (167)
Dislocation 17 (278, 300)
Dislocations 19 (382)
Disorder 17 (312, 377, 380, 390)
Dispenser cathodes 21 (12, 154)
Dispersion 16 (228); 18/1 (12)
Dispersion, elastic-wave 18/2 (210)
Dispersion formula of Cauchy 17 (115)
Dispersion formula of Sellmeier 17 (116)
Dispersion of electromagnetic waves 16 (640)
Dispersion radio-interferometer 16 (524)
Dispersion relations 22 (193)
Dispersion, spin-wave 18/2 (13, 128, 158, 210)
Dispersion theory of the plasma 22 (356)
Displacement, atomic 17 (92, 93)
Displacement, dielectrical 16 (4, 18)
Displacement, electric 17 (7)
Displacement, electronic 17 (92)
Dissociation 21 (321)
Dissociation continuum 21 (321, 326)
Dissociation into negative ions 21 (464)
Dissociation of molecules in the polar phase by electron impact 21 (463)
Dissociation of polar molecules 21 (464)
— without attachment 21 (466)
Dissociation products 21 (621)
Dissociative capture 21 (486)
Dissociative ionization 21 (321)
Dissociative recombination 22 (232)
Dissymetry and pyroelectricity 17 (312)
Distant field 16 (436, 454, 455, 460, 463, 464, 477, 481, 482, 485, 510, 516)
Distortion, tetragonal 18/1 (130)
Distribution 22 (4)
Distribution coefficient 20 (24)
Distribution curve of voltage or current along a line 16 (221)
Distribution function 20 (50); 21 (404)
— displacement of, in electric field 19 (238, 292)
— general 19 (235)
— in galvomagnetic effects 19 (298)
— in the conduction band 17 (220)
— in thermal conduction 19 (280, 292)
— molecular 17 (64, 81)
— of Fermi 19 (228, 293)
— radial 17 (66, 74)
Distribution of positive ions in space 21 (513)
Disturbance, *see also* Sudden ionospheric disturbance = SID 16 (630, 636)
Disturbance in control system 23 (458)
Disturbance response 23 (480)
Disturbance variable feed-forward control 23 (515)
Divergence factor 16 (497, 510, 560, 609, 610)
Divergent field 22 (21)
Diversity-reception 16 (597)
DKT = dipotassium tartrate 17 (282)
Doehring's four-gauze-shutter method for measuring the attachment probability 21 (452)
Domain 18/2 (79, 343)
Domain behaviour, sub- 18/2 (476)
Domain, ferromagnetic 18/2 (79, 343)
Domain theory 18/2 (343, 345)
Domain wall, effective mass 17 (277)

Domain wall mass 18/2 (89)
Domain wall motion 18/2 (82, 344)
Domain, Weiss 18/2 (1, 453)
Domains and hysteresis loop 17 (273)
Domains, antiparallel 17 (330)
Domains in crystals and ferroelectricity 17 (269)
Dominant mode of a waveguide field 16 (292, 318, 332, 336)
Donor 20 (18)
Donor level(s) 19 (331); 21 (116, 118)
Donor migration 21 (154)
Doppler broadening 21 (493); 22 (215, 361)
Doppler broadening yielding no continuum 21 (493)
Doppler contribution 22 (372)
Doppler effect 22 (70, 98, 136)
Doppler effect frequency shift 23 (292)
Doppler frequency shift 16 (593, 636)
Double exchange 18/1 (351, 364, 367, 441, 446, 450, 520, 535); 18/2 (101)
Double layer(s), thin 22 (143, 147)
Double refraction 16 (567)
Double resonance 18/1 (142, 144)
Double-periodic field in the plane 16 (50)
Doublet, Kramers 18/1 (131)
DPPH, diphenylpicrylhydrazyl 18/1 (182)
Drawing wires at controlled temperatures, apparatus for 19 (198)
Drift energy 21 (393)
Drift velocity 21 (389, 426, 509); 22 (1, 169)
— of electrons 21 (542); 22 (18, 19)
— of positive ions 22 (18, 19)
Druyvesteyn distribution 21 (417, 547)
Druyvesteyn's theory 21 (546)
Dual-trace presentation 23 (368)
Ducts 16 (625, 626)
Dumbbell model 17 (42)
Du Pré formalism, Casimir and 18/1 (77)
Du Pré relaxation curve, Casimir and 18/1 (7)
Du Pré, theory of Casimir and 18/1 (4)
d-wave function 18/1 (132)
Dye films 19 (390)
Dyes, adsorbed 19 (383)
Dyes, effect of adsorbed 19 (373)
Dye-sensitisation 19 (383)
Dynamical friction 21 (430, 436, 441)

E_{010} resonator 16 (673)
Earnshaw's lemma 16 (34)
Earth flattening approximation 16 (613)
Earth potential 16 (9)
Earth radius, effective 16 (613 ff., 627)
Earth, rotation of 23 (184)
Eccle's relation 22 (357)
Eckhart theorem 18/1 (48)
Eddy-current damping 18/2 (220)
Edge capacitance 16 (644, 652)
Edison effect 21 (7)
EDT = ethylene diamine tartrate 17 (282)
Effective area 16 (453, 477, 487, 488, 503, 506)
Effective electron mass, low 19 (386)
Effective field 21 (395, 402)
Effective length, complex, of an antenna 16 (265, 267)
Effective mass of elctrons 19 (67, 321)
Effective number of free electrons 19 (264)
— from skin effect 19 (315)
e_g orbital 18/1 (128)
Eggert-Saha equation 22 (308)
Ehrenfest relations 17 (357, 389)
Eiconal equation 16 (440, 443, 514, 548)
Eigenfunctions of wave guide fields 16 (292, 297, 300, 341–344)
Eigenvalue problems in wave guides 16 (292, 341, 342, 343, 345, 347, 406)
Eigenvalues of a cavity resonator 16 (406)
Einstein coefficients 23 (90)
Einstein constant of spontaneous emission 22 (127)
Einstein relation 20 (116); 21 (343, 344, 397, 414)
Elapsed phase difference method 23 (314)
Elastic constants 19 (91)
Elastic vibrations 19 (94)
Elasticity constant 18/2 (354)
Elasticity module 18/2 (355)
Elastic-wave dispersion 18/2 (210)
Elatation 22 (209, 218)
Elatation cross-section, total 22 (245)
Electrets 17 (113)

Electric dipole transition 18/1 (37, 459)
Electric energy 16 (179)
Electric field, internal 18/1 (108)
Electric field near half-wave antenna 16 (276)
Electric field, radial 22 (171)
Electric quadrupole interaction 18/1 (135, 136)
Electric quadrupole moment 18/1 (135, 141)
Electric quadrupole, nuclear 18/1 (140)
Electric quadrupole transition 18/1 (37)
Electric solution 16 (431, 433, 434, 469, 518, 550, 551)
Electric wall 16 (288)
Electrical breakdown 19 (344)
Electrical conductivity 19 (237, 238, 249)
— at low temperatures 19 (293, 296)
— copper-nickel and silver-palladium alloys 19 (270)
— in magnetic fields 19 (299)
— in thio-spinels 18/1 (531)
— isothermal 19 (272)
— ratio at high and low temperature 19 (295)
Electrocaloric coefficient 17 (347, 354, 360)
Electrode influence 17 (249)
Electrode material 17 (163)
Electrode materials 22 (11)
Electrode, mechanical force on it 22 (62)
Electrodes 16 (644); 17 (365)
Electrodes, cleanliness of 22 (507)
Electrodes in piezoelectric experiments 17 (294)
Electrodes, polarisation of the 22 (190)
Electrodynamic force 22 (169)
Electroluminescence 19 (337)
Electrolyte solutions 17 (119)
Electrolytic solution 16 (723)
Electrolytic trough 16 (159–161)
Electromagnet 18/1 (98, 119)
Electromagnetic energy 16 (179)
Electromagnetic force 23 (87, 103)
Electromagnetic torque 23 (87, 105)
Electromechanical equations, non-linear 17 (343)
Electrometer measurements 16 (667)
Electron affinity 21 (17, 31, 97)
Electron atmosphere in an enclosed space 21 (18, 24)
Electron attachment 22 (77, 80, 85, 129, 130)
Electron attachment to neutral molecules 22 (19)
Electron avalanche 20 (370); 21 (514); 22 (9, 33, 35, 48)
Electron beam, deflection of an 22 (91)
Electron beam, monoenergetic 22 (243)
Electron bombardment 21 (370)
Electron centre, local 18/1 (78)
Electron centre, odd- 18/1 (259)
Electron concentration 21 (405)
Electron concentration, microwave measurement 22 (68)
Electron correlation 18/1 (284, 398, 400, 402, 532)
Electron density 22 (374)
— alternating 21 (407)
— unpaired 18/1 (155, 163)
Electron diffusion 21 (592)
Electron distribution 21 (189, 196, 226, 549)
Electron donor centre 18/1 (80)
Electron emission 22 (9)
Electron emission from the cathode 21 (580)
Electron emitter, metallic 21 (97)
Electron energy, anisotropic distribution 21 (509)
Electron filter 21 (446, 451)
Electron flow between concentric cylinders 21 (67)
Electron gas statistics 21 (181)
Electron gun 23 (31)
Electron, hydrated 18/1 (238)
Electron impact 21 (464)
Electron interaction 18/1 (541)
Electron interaction, short range 19 (227)
Electron mobility 21 (85, 86, 89, 119); 22 (121); 23 (208)
Electron model, localized 18/1 (400)
Electron nuclear double resonance (ENDOR) 18/1 (142, 143, 144, 229, 236, 241, 247, 252, 253, 262)
Electron optical system 16 (97)
Electron paramagnetic resonance (E.P.R.) 18/1 (36)
Electron phase diagram 18/1 (532)

Electron polarization, conduction 18/1 (433, 490)
Electron pressure 22 (192)
Electron reflection 21 (369)
Electron relaxation 18/2 (257)
Electron shells, radius of 18/1 (389)
Electron spin
— and magnetic field, anisotropic coupling between 18/1 (133)
— resonance 18/1 (9, 94); 23 (148, 149, 200)
— resonance of gases 18/1 (256)
— resonance spectrometer 18/1 (112)
Electron statistics 17 (236); 20 (31)
Electron temperature 22 (63, 117, 120, 121, 124, 130, 132, 171, 174)
Electron temperature in gas mixtures, theoretical 22 (184)
Electron transfer, intramolecular 18/1 (203)
Electron transitions 21 (380)
Electron, unpaired 18/1 (94, 98, 100, 109, 137)
Electro-negative gases 21 (524, 535, 647)
Electron-electron collisions 19 (228)
Electron-electron collisions, effective cross-section for 19 (227)
Electron-electron interaction 17 (226, 232, 248); 18/1 (289, 306, 341)
Electron-electron repulsion 18/1 (537)
Electron-electron scattering 21 (360, 369)
Electronic controller 23 (478)
Electronic current 17 (220)
Electronic excitation 21 (401)
Electronic oscillations 22 (191)
Electronic quadrupole 18/1 (136)
Electronic splitting 18/1 (107, 108, 133, 136, 139)
Electronic splitting, angular variation of 18/1 (107, 140)
Electron-ion recombination 21 (485, 486, 490, 492 f.)
Electron-ion recombination, preferential 21 (495)
Electron-ion wall recombination 21 (495)
Electron-magnon interaction 18/1 (398, 544)
Electron-phonon interaction 18/1 (398, 543)
Electrons and ions in gases, mean energies 22 (2, 14)
Electrons, covalent bonding of 4*f* 18/1 (470)
Electrons, energy distribution of 22 (96, 170)
Electrons, fast 22 (143, 173)
Electrons, "hot" 20 (68)
Electrons, mobility of 3*d*, 4*f* 18/1 (398)
Electrons, triggering 22 (473)
Electrophoresis 22 (195)
Electrostriction 17 (265, 329, 343, 351)
Electrostriction and piezoelectric coefficients 17 (326)
Electrostriction and thermodynamic potentials 17 (334)
Electrostriction as tensor 17 (304)
Electrostriction in liquids 17 (359)
Electrostrictive force 17 (32)
Electrostrictive stress 17 (32)
Elenbaas-Heller equation 22 (337)
Elinvar material 18/2 (321)
Ellipsoid, ideal 18/2 (454)
Ellipsoid, polarized 17 (28)
Ellipsoidal model 17 (60, 90, 108, 127)
Elliptic polarization 16 (543)
Elliptical coordinates, *see* Coordinates
Elliptical coordinates in the plane 16 (79)
Elliptical functions 16 (50, 144)
E.m.f. method 16 (239)
Emission 22 (527)
Emission by ion impact 21 (445)
Emission by neutral atoms 21 (655)
Emission capability 21 (134)
Emission coefficient 22 (353, 354)
Emission continuum 21 (318)
Emission current 21 (369)
Emission current, measurement methods 21 (121)
Emission due to incidence of photons 22 (6)
Emission due to ions at the cathode 21 (626)
Emission due to velocity of the charge 21 (653)
Emission, enforced 22 (210, 359)
Emission, measurement of the total 21 (639)
Emission of a phonon, spontaneously 18/1 (46)

Emission of electrons by positive ions 22 (447)
Emission of positive ions 21 (577)
— from the anode 21 (582)
Emission, spontaneous 18/1 (46)
Emission, stimulated 18/1 (121)
Emissive properties, stable 22 (34)
Emittance 23 (429, 442)
Emitter 20 (131)
Emitter, thoriated tungsten 21 (111)
E-mode analogy 16 (307ff., 311)
E-mode functions of waveguides with various cross-sections 16 (293, 316, 318, 322ff., 327ff., 330ff., 339ff., 350, 353ff.)
Empire State Building 22 (591, 607)
Emulsions, photographic 19 (380, 383)
End capacitance 16 (484)
End effects 22 (172, 174)
End-fire array 16 (487)
End-fire array, linear 16 (456)
End-fire couplet 16 (487)
ENDOR (Electron Nuclear DOuble Resonance) 18/1 (142, 143, 144, 229, 236, 241, 247, 252, 253, 262)
Energy balance 22 (122, 126, 132, 133, 381, 392, 401)
Energy band broadening with increase of temperature 19 (337)
Energy band, forbidden 17 (184)
Energy band structure 21 (381)
Energy conservation 17 (187, 191); 22 (317)
Energy current density 20 (53)
Energy density 16 (12, 289); 21 (405)
Energy distribution 17 (225)
— of electrons 21 (13, 18, 23, 105)
— of ions 21 (623)
— of secondary electrons 21 (649)
Energy, Eigenfeld- 18/2 (363)
Energy, electromagnetic 16 (179)
Energy equations 16 (177)
Energy, exchange 18/2 (347)
Energy, flux of 22 (322)
Energy gap 18/1 (413)
Energy level for lanthanide atoms 18/1 (458)
Energy level, orbital 18/1 (129)
Energy levels of barium oxide 21 (119)
Energy losses of slow electrons 21 (280)
Energy of ion formation 21 (457)
Energy reservoir, exchange- 18/1 (22)
Energy spectrum for a ferromagnetic insulator 18/2 (84)
Energy spectrum for a ferromagnetic metal 18/2 (86)
Energy states in crystals, electronic 19 (320)
Energy states in perfect crystals 19 (320)
Energy storage 16 (311, 408)
Energy theorem 16 (289, 312)
Energy transport 16 (311)
Enthalpy 17 (285, 346); 22 (342)
Enthrakometer 23 (98)
Entropy 17 (311, 344, 345, 374, 381, 388); 22 (342)
Entropy balance 22 (317)
Entropy, configurational 17 (312)
Entropy current (flux) 22 (318)
Entropy production 22 (383)
Entropy source 22 (318)
Enzyme 18/1 (95)
Ephemeris time 23 (162, 290)
E-plane bend 16 (366)
Epoch, determination of the 23 (292)
Equilibrium energy 21 (401, 403)
Equilibrium field 17 (223)
Equilibrium, labile 21 (494)
Equilibrium method 21 (472, 479)
Equilibrium ratio, dynamic 21 (475)
Equilibrium relations 17 (67)
Equilibrium, thermal 22 (305)
Equipotential surface 16 (4)
Equipotential surfaces 22 (121)
Error in control loop 23 (458)
Esaki diode 23 (68)
ESR of gases 18/1 (256)
ESR spectrometer 18/1 (112)
Etch patterns on crystals 17 (281)
Ethane 17 (109)
Ethyl ether 17 (46, 47, 93, 95)
Ethylene chloride 17 (109)
Ettingshausen coefficient 20 (60, 94)
Ettingshausen effect 20 (60, 94)
Eu chalcogenide 18/1 (476, 481, 506, 512, 554)
Eu chalcogenides, bandstructure of 18/1 (477, 511)
Eu chalcogenides, magnetic red shift 18/1 (483)
EuI_2 18/1 (267)
EuO 18/1 (267)

EuS 18/1 (267)
EuSe 18/1 (267)
Eu_2SiO_3 18/1 (267)
Evanescent and non-evanescent waves 16 (295, 464, 465, 481, 522, 538, 559)
Evaporation 22 (154, 160)
Evaporation energy 22 (158)
Evolute 16 (449)
E-wave = wave of electric type 16 (291)
Excess charge 18/1 (158)
Excess crystal noise 18/1 (110)
Exchange, anisotropic 18/2 (36, 110)
Exchange anisotropy 18/2 (182, 288, 439, 440, 443, 444)
Exchange anisotropy torque 18/2 (450)
Exchange charge 19 (51)
Exchange, deuterium 18/1 (188)
Exchange, direct 18/1 (432)
Exchange, double 18/1 (351, 364, 367, 411, 446, 450, 520, 535); 18/2 (101)
Exchange energy 18/2 (347)
Exchange field 18/2 (6)
Exchange force 18/2 (350)
Exchange Hamiltonian 18/2 (1)
Exchange, indirect 18/1 (410, 430, 440, 487, 515, 521)
Exchange integral 18/2 (5)
— between orthogonal functions 18/1 (276, 350)
— definition for 2 electrons 18/1 (266, 272, 275)
— definition for 3 electrons 18/1 (314)
— direct, sign 18/1 (306)
— H_2 problem 18/1 (276, 334)
— Heisenberg 18/1 (277, 330)
— indirect, for 3 centres and 4 electrons 18/1 (343, 347, 349, 358)
— indirect, in the crystal 18/1 (366)
— indirect, numerical calculations 18/1 (381)
— indirect, semiempirical rules for the estimation 18/1 (379)
— indirect, sign 18/1 (373)
— influence of electron nuclear interaction 18/1 (308, 310)
Exchange interaction 18/1 (14, 20, 34, 35, 71, 73, 75, 77); 18/2 (5, 64, 342, 349)
— direct 18/1 (268, 286)
— indirect 18/1 (268, 335)
— indirect, Anderson's new approach 18/1 (269, 368)
— indirect, in the crystal 18/1 (365)
— paramagnetic relaxation by modulated 18/1 (71)
Exchange inversion 18/2 (139)
Exchange, kinetic 18/1 (270, 330, 370, 432)
Exchange magnetoelastic effect 18/2 (208)
Exchange magnetostriction 18/2 (338)
Exchange mechanism, non-Kramers 18/1 (269)
Exchange, mechanism of indirect 18/1 (268, 337)
Exchange mode 18/2 (129)
Exchange narrowing 18/1 (123, 125); 18/2 (236)
Exchange of kinetic energy 21 (652)
Exchange, potential 18/1 (270, 330, 370)
Exchange pumping 18/2 (252)
Exchange resonance 18/2 (129, 160, 194)
Exchange scattering 18/2 (200)
Exchange, spin 18/1 (199, 203)
Exchange stiffness parameter 18/2 (81)
Exchange, super 18/1 (336)
Exchange torque 18/2 (450)
Exchange-energy reservoir 18/1 (22)
Excitation 21 (321); 22 (17, 209, 218)
Excitation by particle impact 22 (219)
Excitation energy, average 21 (403)
Excitation frequency 21 (418)
Excitation function 22 (60, 135)
Excitation number 21 (403)
Excitation probability 21 (506)
Excitation, probability of 22 (127)
Excitation, stepwise 22 (225)
Excitation temperature 22 (309)
Excitation, triplet 18/1 (219)
Excited atoms 22 (6)
Excited configuration 18/1 (321)
Excited molecular states of argon 22 (505)
Excited triplet system 18/1 (210)
Exciton 19 (69)
Exciton-magnon absorption 18/1 (483)
Exclusion principle 19 (228)
Experimental measuring of fields 16 (159–163)
Experimental techniques 19 (392)
Exponential decay 18/1 (21, 22)
Extinction coefficient 20 (210)
Extinction potential 22 (188)

Extraction of minority carriers 20 (133)
Extraordinary wave 16 (547)

^{19}F fluorine splitting constant 18/1 (167)
Fabry-Pérot interferometer 16 (704); 23 (128)
Fabry-Pérot resonator 23 (95)
Face-centered structure 19 (15)
Fading 16 (597, 628, 636)
Fading rate 16 (592, 595, 636)
Fading, selective 16 (580)
Fahrenheit scale 23 (378)
Farad (unit) 16 (2)
Faraday cage 16 (9)
Faraday effect 18/2 (169, 343)
Faraday rotation 18/2 (177); 23 (151–156, 214)
Faraday rotation in garnets 18/1 (481)
Faraday's law 20 (248, 250, 251)
Fast fading 16 (588, 589, 593)
Fast neutral atoms 21 (577, 581, 585)
Fast-relaxing mechanism 18/2 (224)
Fatigue, photoelectric 21 (378, 379)
Faultrecorders 23 (271)
F-centre 18/1 (78, 144, 228)
F-centre interaction 18/1 (231)
F-centre, modified 18/1 (233)
Feedback 23 (475, 498)
Feedback control 23 (457, 459)
Feedforward control 23 (459)
Feraday dark space 22 (56, 57, 60, 110, 224)
Fermat's principle 16 (572)
Fermi contact potential 18/1 (72)
Fermi degeneration 22 (314)
Fermi distribution 19 (328); 20 (36); 21 (369)
Fermi gas 18/1 (266)
Fermi integral 20 (36)
Fermi interaction 18/1 (72)
Fermi level 19 (349); 20 (33)
Fermi limit 21 (657)
Fermi potential 20 (47)
Fermi statistics 19 (55, 56, 228, 234, 322); 20 (31)
Fermi-Dirac statistics 21 (342, 366)
Ferrimagnet 18/2 (95, 97)
Ferrimagnet, molecular field in 18/2 (100)
Ferrimagnet, spin wave in 18/2 (15)
Ferrimagnetic array 18/2 (158)
Ferrimagnetic spin-wave theory 18/2 (104)
Ferrimagnetic spiral 18/2 (141, 153)
Ferrimagnetism 18/1 (265, 267); 18/2 (96)
Ferrimagnetism, Néel theory of 18/2 (96)
Ferrite 18/1 (335)
Ferrite junction 16 (402)
Ferrite rod in a waveguide 16 (404)
Ferrite slab in a waveguide 16 (405)
Ferrite sphere inside a cavity resonator 16 (415)
Ferro-antiferromagnetic system 18/2 (444)
Ferro-electric materials, dielectric properties 16 (724)
Ferro-electric properties 19 (375)
Ferroelectricity of small crystals 17 (277)
Ferromagnet, thermodynamic properties of 18/2 (1, 274)
Ferromagnet, two-dimensional 18/2 (120)
Ferromagnet, weak 18/2 (164)
Ferromagnetic alloys 19 (81, 121)
Ferromagnetic amplifier 18/2 (246)
Ferromagnetic anisotropy 18/2 (62, 65)
Ferromagnetic behaviour 16 (21)
Ferromagnetic linear chain 18/2 (119)
Ferromagnetic planar layer 18/2 (119)
Ferromagnetic resonance 18/2 (173, 177, 220, 243, 373)
Ferromagnetic resonance, high-power 18/2 (243)
Ferromagnetic resonance relaxation 18/2 (220)
Ferromagnetic spiral 18/2 (141)
Ferromagnetism 18/1 (265, 266); 19 (111)
Ferromagnetism, band model of 18/2 (85, 86)
Ferromagnetism, dipolar 18/2 (93, 140)
Ferromagnetism of band electrons 18/1 (538)
Ferromagnetism, statistical mechanics of 18/2 (58)
Ferromagnetism to the resistivity, contribution of the 19 (219)
Ferromagnetism, Weiss model of 18/2 (1)

Feynman's theorem 19 (96)
Field, asymmetrical 22 (455)
Field, demagnetizing 18/2 (343)
Field desorption 21 (218, 230)
Field, directing 17 (109)
Field discontinuity 16 (440, 450, 452)
Field distortion 22 (8, 29, 33)
Field effect 20 (185)
Field effect transistor 20 (176)
Field, effective 22 (532)
Field emission 21 (101, 577, 582); 22 (79, 80, 163, 449, 450)
— current formula 21 (184, 191)
— dependence on field strengths 21 (178, 192)
— dependence on temperature 21 (177, 188, 196)
— dependence on work function 21 (179, 194, 208, 209)
— electrodic 17 (207)
— external 17 (201, 205)
— from semiconductors 21 (201)
— in *p-n* junctions 20 (156)
— internal 17 (162, 201, 240)
— microscope 21 (180, 201, 215)
— microscope, resolution of the 21 (203)
— theory of 21 (181)
— thermal 22 (414)
— tubes, lifetime of 21 (199, 200)
Field energy 16 (11); 18/2 (360, 396)
Field equations 16 (285)
Field, exchange 18/2 (6)
Field in waveguides 16 (290)
Field, inhomogeneous 17 (235)
Field ion emission 21 (218)
Field ion emission with gas supply 21 (223)
Field ion mass spectroscopy 21 (227)
Field ion microscope 21 (227)
— resolution of the 21 (228)
Field limit, uniform 22 (537)
Field line 16 (4)
Field, longitudinal 22 (124, 125)
Field, molecular 18/2 (2, 3, 4)
Field, nonuniform 22 (552)
Field of v. Hippel 17 (223)
Field plate 23 (212)
Field problem 16 (233)
Field strength
— electrical 16 (2, 14, 18)
— electrical, at a boundary 16 (7, 21)
— magnetic 16 (18); 18/2 (341)
— magnetic, at a boundary 16 (25)
Field, uniform 22 (455)
Filament 17 (159)
Filament of Wagner 17 (169)
Filament, oxide coated 22 (90)
Filamentary transistor 20 (133)
Filling factor 18/1 (109)
Film, metallic 22 (13, 14)
Film, thin 22 (13)
Fine structure 22 (143)
Fitzgerald's vector 16 (431, 460, 461, 527, 530, 531)
Flaming arcs 22 (292)
Flash 22 (1, 51, 506)
Flash illumination 19 (319)
Flashover 17 (156, 165)
Flexowriter 23 (275, 279)
Flexure modes 17 (302)
Flicker noise 18/1 (109); 23 (143)
Floating *P-N* junction 19 (348)
Floating potential 20 (169)
Flow vector in velocity space 21 (338, 430, 434)
Fluctuation time 21 (443)
Fluctuations, local 19 (332)
Fluctuations, statistical 22 (454, 474)
Fluorescence 22 (37f.)
Fluorescence, quantum efficiency of 21 (319)
Fluorine superhyperfine structure 18/1 (252)
Flutter 23 (260)
Flux, electrical 16 (4, 156)
Fluxmeter 16 (162); 23 (206)
Fluxmeter, electrostatic 22 (602, 605)
Flying spot technique 19 (393)
Flying-clock method 23 (300)
Fock method, restricted Hartree- 18/1 (352)
Fock method, unrestricted Hartree- 18/1 (352, 369)
Fock-operator 17 (190)
Focus 16 (447, 449, 450)
Focussing 16 (582, 611)
Foils elastically, apparatus for deforming 19 (198)
Fokker-Planck equation 21 (430, 433)
Folded dipole 16 (248, 483)
Forbidden transition 18/1 (136)
Force density 16 (12)
Force equation 22 (317)

Force, exchange 18/2 (350)
Force, magnetic, upon an electric current 16 (25)
Force on a dipole 16 (14)
Force oscillations of a cavity resonator 16 (415–422)
Formative lag, fluctuation in the 22 (474)
Formative lag, observed and true 22 (472)
Formative time 22 (41, 50)
Formative time lag 21 (305, 526); 22 (45, 49, 471, 494)
— increased rapidly as overvoltages decreased 22 (492)
— long 22 (526)
Förster probes 23 (206)
Forward scattering 16 (586, 621, 635)
Foster's reactance theorem 16 (312, 401)
Fourier integral as solutions of the potential equation 16 (94, 96)
Fourier series as solution of the potential equation 16 (38, 93, 94)
Fourier spectrum 18/1 (90)
Fourier transform 16 (439, 463–465, 503, 510, 538–541, 578, 579, 583 ff.)
Fowler thermionic constant 21 (144)
Fowler-Nordheim theory 21 (184)
Fowler's emission equation 21 (77, 83, 128, 130, 133, 144)
Franck-Condon effect 18/1 (80)
Franck-Condon shift 19 (336)
Franck-Condon transitions 21 (379)
Free electron gas 22 (512)
Free energy 17 (340, 346); 18/2 (347, 352, 353, 354, 360)
Free radical 18/1 (77, 94, 95, 98, 116, 119, 126, 151)
Free radical *g*-value 18/1 (126)
Free radical, stable 18/1 (182)
Free-electron approximation 19 (20, 55)
Free-free electron transitions 21 (481, 492)
Free-free-radiation 22 (354)
Free-radical transient 18/1 (142)
Freezing, normal 20 (24)
Frenkel defects 20 (255, 258, 274, 319, 329, 341)
Freon 22 (523)
Frequency addition 23 (353)
Frequency constant 19 (367)
Frequency control, automatic 18/1 (108, 116, 117)
Frequency converter 23 (335)
Frequency counter 23 (329, 361)
Frequency division 23 (164, 353, 355)
Frequency limit, upper 22 (192)
Frequency measurement techniques 23 (328)
Frequency modulation 23 (259, 347)
Frequency modulation, high- 18/1 (113)
Frequency multiplication 23 (164, 353, 355)
Frequency response of a control system 23 (463, 480)
Frequency spectrum 19 (338)
Frequency stability 17 (282, 283)
Frequency stabilizer a.c. 23 (167)
Frequency standard
— beam-type 23 (170)
— caesium 23 (170, 171, 322)
— crystal-controlled 23 (163)
— figure of merit of 23 (315)
— primary 23 (301)
— spectral-line 23 (168, 169)
Frequency substraction 23 (353)
Frequency synthesis 23 (350)
Frequency translation 23 (354)
Fresnel reflection coefficient 16 (521, 528, 535, 553, 555, 557, 610, 625)
Fresnel zones 16 (440, 434–437, 597)
Frictional forces 22 (318)
Frictional resistance 18/2 (368)
Frozen-in defects 20 (284)
Fry-Langmuir profile 23 (33)
Functions of the cylindrical wedge 16 (99)
Functions of the parabolic cylinder 16 (113)
Fundamental wave 16 (380)

g component 18/1 (147)
g factor 18/1 (115)
g parameter 18/1 (95)
g tensor 18/1 (146)
g value 18/1 (99, 115, 120, 126, 131, 132, 133, 136, 139)
g value, free radical 18/1 (126)
Gain equation 21 (416, 417)
Gain factor 19 (325)
Gain of energy 21 (405, 414, 415, 439)
Gallium 19 (79)

Gallium arsenide 19 (386)
Galvani potential 20 (179)
Galvanomagnetic effects 20 (59)
Galvanometer 23 (236, 407)
Galvanometer amplifier 23 (416)
Galvanometer theory 23 (231)
Galvomagnetic effects 19 (298)
— condition for isotropic 19 (301)
— in cubic crystals 19 (301)
— of bismuth 19 (306)
γ at low X/p 21 (593, 597)
γ, dependence on irradiation 21 (588)
γ, dependence on the nature of the anode 21 (587)
γ, dependence on the nature of the cathode 21 (587, 601)
γ for complex ions on metal 21 (621)
γ for Hg ions 21 (596)
γ, Güntherschulze's method 21 (646)
γ in clean air 21 (599)
γ in impure air 21 (598)
γ, measurement of 21 (586)
γ, values for cylindrical nickel cathode in argon 21 (619)
γ, values for nickel in air 21 (599)
γ, values for nickel in nitrogen 21 (597)
γ, variation(s)
— over the surface 21 (592)
— with angle of incidence 21 (647)
— with field and pressure 21 (587, 593, 597)
— with ion mass 21 (652)
— with ionisation potential 21 (645)
— with length of gap 21 (633)
— with time 21 (587)
Gap between crystal surface and plane electrodes 17 (295)
Gap, energy 18/1 (413)
Gap, second 18/1 (413)
Garnet 18/1 (497)
Garnet structure 18/1 (253)
Garnets, Faraday rotation in 18/1 (481)
Gas absorption 22 (198)
Gas density 22 (70, 72)
Gas discharge at low pressure 22 (216)
Gas dynamics 22 (402)
Gas flow 22 (200)
Gas interferometer 22 (196)
Gas layer(s) 21 (370, 641); 22 (190)
Gas layer, adsorbed 22 (155, 157)
Gas mixtures 21 (629)
— discharges in 22 (181)
— normal current density 22 (183)
Gas pressure 22 (80)
Gas pressure, low 22 (2)
Gas process, permitted 22 (75)
Gas stream, longitudinal 22 (202)
Gas temperature 22 (69, 100, 136)
Gas treatment 21 (375)
Gas viscosity 22 (200)
Gaseous purity 22 (507)
Gases at low pressure 17 (34)
Gases, electronegative (negative ions) 22 (129, 146)
GASH = $(CN_3H_6)Al(SO_4)_2 \cdot 6H_2O$ (guanidine aluminium sulphate hexahydrate) antiferroelectric 17 (273)
Gastemperature measurement 23 (434)
Gasthermometer 23 (373)
Gauss' theorem 16 (3, 27)
Gaussian curvature 16 (441, 446, 447, 496)
Gaussian line 18/1 (124)
Gaussian line shape 18/1 (24, 123)
Gaussian units 17 (2, 148)
Gaussian-Lorentz line shape 18/1 (26)
Gauze cathode 22 (110)
Gd chalcogenide 18/1 (518)
Geiger counter 21 (621); 22 (180, 240, 483, 515, 522)
Geiger pulse spreading, threshold for a 22 (484)
Geissler or glow discharge 22 (452)
Geometrical-optics approximation 16 (439, 442, 497, 501, 506, 559, 608)
Geometrical-optics expansion 16 (444, 451)
Germanium 17 (252); 18/1 (254); 19 (43, 78, 317, 340)
Germanium blocking layers 17 (240)
Germanium doped with various impurities 19 (343)
Germanium, photo-electro-magnetic effect in 19 (343)
Germanium photoelement 23 (449)
Germanium, photo-voltaic and photomagnetic effects in 19 (348)
Germanium, semiconducting 20 (230)
Giacometti-Nordio-Pavan relation 18/1 (159)
Giant magneto-resistance effect 18/1 (554)

Giant moment 18/1 (516)
Gibbs equation 22 (317)
Gibbs function 17 (285, 346)
Gibbs potential 22 (342)
Giebe-Scheibe method for measurement of piezoelectric coefficients 17 (300)
Giorgi units 17 (8, 148)
Glass 17 (253)
Glow, boundary of the 22 (102)
Glow, diffuse 22 (28)
Glow discharge 21 (303, 316, 319, 575, 634); 22 (242, 506)
— alternating 22 (194)
— formation of a 22 (53)
— high pressure 22 (194)
— in metal tubes 22 (55)
— in moving gas 22 (202)
— obstructed 22 (190)
Glow, length of the negative 22 (113)
Glow, negative 22 (56, 110, 143)
Glow, negative, theory 22 (113)
Glow, spark flash in form of a 22 (1, 2, 9, 10, 33)
Gold doping 19 (345)
Goldstein's law 22 (138, 139)
Goodenough mechanism 18/1 (379)
Goodenough-Kanamori rules 18/1 (451)
Goubau line 16 (379)
Gradient probe 23 (206)
Gradient reflection in a stratified medium 16 (556, 573, 628, 636)
Grain boundary conductivity 20 (281)
Graphical construction of fields 16 (156–158)
Graphite 19 (47)
Gray-code 23 (220)
Green's formulae 16 (28)
Green's function 16 (35, 38, 57, 173)
— dyadic 16 (301, 459)
— for two spheres 16 (131)
— in toroidal coordinates 16 (134)
— of a circular disk 16 (141)
— of a cone 16 (121)
— of a hollow ring 16 (103)
— of a wedge 16 (104)
Green's reciprocity theorem 16 (10)
Green's symmetrical theorem 16 (175)
Green's tensor 16 (301, 459)
Ground and air discharges, general features of 22 (577)
Ground state triplet 18/1 (221)
Ground wave 16 (544, 552)
Group velocity 16 (296, 312, 383, 393, 548, 549, 620, 631); 19 (61)
Groups III and V elements 19 (344)
Growth curve 22 (364)
Growth curve, shape 22 (8)
Grüneisen-Borelius relation 19 (170)
Grüneisen's constant 19 (267)
Guard-ring 16 (648); 17 (140, 143)
Guide wavelength 16 (295); 23 (92)
Gyrator 18/2 (169)
Gyrator, acoustic 18/2 (213)
Gyrofrequency 16 (546); 22 (177)
Gyro-interaction 16 (577)
Gyrotropy 18/2 (169)

H 18/1 (241)
H centre 18/1 (241, 244)
Hafnates, ferroelectricity in 17 (270)
Hale or Mohler processes 22 (446)
Half-wave antenna 16 (276)
Half-wave dipole 16 (280, 461, 468, 493)
Hall angle 20 (75ff.)
Hall angle Θ 19 (339, 392)
Hall coefficient 19 (336); 20 (59, 75ff.)
Hall coefficient measurement 19 (392)
Hall effect 18/2 (170); 19 (302); 20 (59, 135); 23 (102, 207)
— adiabatic 20 (59)
— at high magnetic fields 20 (81)
— at low magnetic fields 20 (76)
— at low temperatures 20 (83)
— isothermal 20 (59, 73)
— of bismuth 19 (306)
— planar 20 (76)
Hall generator 23 (208, 209)
Hall measurement technique 19 (354)
Hall mobility 20 (78)
Halogen acceptor 19 (384)
Halogens 22 (522, 523)
Hamiltonian 18/1 (3, 14, 20, 235); 19 (48, 84)
Hamiltonian, Dirac spin 18/1 (266)
Hamiltonian, Dirac-van Vleck- 18/1 (267)
Hamiltonian, exchange 18/2 (1)
Hamiltonian, Heisenberg 18/1 (267)
Hamiltonian, Holstein-Primakoff 18/2 (55)
Hamiltonian, lattice 18/1 (38)
Hamiltonian matrix 18/1 (208)

Hamiltonian, spin 18/1 (4, 14, 39, 56, 74, 132, 133, 135, 136, 137, 141, 170, 227, 232, 242); 18/2 (5, 35)
Hamiltonian, Weiss 18/2 (6)
Harmonic generator 23 (337)
Harmonic oscillator 19 (287)
Harmonic selector 23 (336)
Harper-Langevin process 21 (498)
Hartree operator 18/1 (67)
Hartree-Fock equations 19 (1, 47, 48, 50)
Hartree-Fock method 19 (53)
— restricted 18/1 (352)
— unrestricted 18/1 (352, 369)
Hartshorn and Ward's resonance method 16 (658, 659)
Hartshorn bridge 19 (154)
He_2 spectrum 22 (166)
Heat at constant field, specific 18/1 (5, 17)
Heat at constant magnetization, specific 18/1 (5, 18)
Heat breakdown 20 (370)
Heat capacity 18/2 (131)
Heat conduction 22 (325, 338, 346, 351, 381)
Heat conduction error in temperature measurement 23 (425, 428–434)
Heat conductivity 20 (56)
Heat flow 22 (159)
Heat, magnetic specific 18/1 (10)
Heat of formation of defects 20 (263)
Heat of transfer 22 (322)
Heat of transport 20 (342)
Heat of transport of a defect 20 (340)
Heat, Peltier 18/1 (429)
Heat transfer coefficient 18/1 (6, 30)
Heat treatment 21 (372, 373, 380)
Heating by ion migration 20 (370)
Heating effects 21 (198, 203)
Heaviside units 17 (148)
Heaviside's unit function 16 (451, 527)
Heg gas 22 (535)
Hehl's law 22 (82)
Height determination (in lightning) 22 (617)
Height, virtual, of ionospheric reflection 16 (550)
Height-gain function 16 (551, 553, 604, 610, 612 ff.)
Heisenberg exchange integral 18/1 (277, 330)
Heisenberg Hamiltonian 18/1 (267)
Heisenberg magnet 18/1 (507)
Heisenberg model 18/2 (342)
Heisenberg theory 18/2 (4)
Heitler-London method 18/1 (271); 19 (98)
Heitler-London model 18/1 (409); 18/2 (7)
Heitler-London theory, asymptotically exact solution 18/1 (317)
Heitler-London theory for the crystal 18/1 (287)
Helical coil 16 (198)
Helical spin array 18/2 (122, 139)
Helical waveguide 16 (390–395)
Helimagnet 18/2 (147)
Helimagnetic effect 18/2 (147)
Helimagnetism 18/2 (139, 146)
Helium 17 (72, 75)
Helium, impure 22 (549)
Helium ions 21 (640, 648, 649)
Helium magnetometer 23 (203)
Helix, cycloidal 18/2 (141)
Helix, improper 18/2 (141)
Helix of tape-wire 16 (391)
Helix of thin wire 16 (391, 393)
Helmholtz coils 16 (127)
Helmholtz equations 16 (173, 291)
Hermite functions 16 (113)
Herreng, method of 21 (454)
Hershberger's method 16 (686)
Hertzian vector 16 (290, 429, 433, 437, 455, 458, 461, 516, 519–523, 529–531, 534, 551, 563, 564, 566, 601 ff., 610)
Heterocharge 17 (114)
Heterodyne beat method 17 (142)
Heterodyne converter 23 (292, 335, 341, 348, 354)
Heterodyne frequency meter 23 (348, 351)
Heterodyne method 16 (664)
Heterodyne systems 23 (3, 120, 142, 163)
Hexagonal structure 19 (7, 10, 15, 44)
Hg-arc 22 (390)
High current carbon arc 22 (372, 427, 432)
High frequency discharge 22 (177, 247, 251)
High pressure 22 (27, 527)
High pressure arc
— water stabilized 22 (379)
— Xenon 22 (390)

High pressure glow discharge 22 (193)
High vacuum processes 22 (79)
High-loss materials 23 (131, 133, 134)
High-pressure data 17 (51)
High-speed, 10-Volt effect 21 (127, 136, 143)
Hissing mechanism 22 (425)
HMO approach 18/1 (147, 151, 159)
HMO method 18/1 (175)
H-mode 16 (307, 311)
H-mode, dominant 16 (332, 336, 340, 341, 342, 343, 344, 356, 375)
H-mode functions of wave guides with various cross sections 16 (293, 317, 318, 322ff., 327ff., 330ff., 340ff., 350, 353ff.)
H_{01n} resonator 16 (679)
Hole(s) 16 (627); 20 (5, 11)
Holes, positive 19 (326)
Hollow cathode discharge 22 (91, 97, 108)
Hollow cathodes 21 (18)
Holstein-Primakoff diagonalization 18/2 (42)
Holstein-Primakoff Hamiltonian 18/2 (55)
Holstein-Primakoff spin wave theory 18/2 (37)
Holstein-Primakoff transformation 18/2 (38)
Holst-Oosterhuis discharge 22 (241)
Holtsmark probability function 22 (213)
Homocharge 17 (114)
Homogeneous electric field 16 (42, 121)
Homologous regions 22 (18)
Homologous system 22 (17)
Homologous volumes 22 (19)
Homopolar 19 (362)
Hopping 18/1 (410, 427)
Horn, electromagnetic 16 (493)
Horn, sectorial, *see* Sectorial horn
Horn waveguide, conical 16 (356)
— field components 16 (357ff.)
Horn waveguide, quasipyramidal 16 (352)
Hornbeck-Molnar effect 22 (446)
Houston functions 17 (209)
H-plane bend 16 (366)
Hückel Molecular Orbital approximation (HMO) 18/1 (147, 151)
Hund's rule 18/1 (377, 404)
H-wave = wave of magnetic type 16 (291)
Hybrid junction 16 (204)
Hybrid tee 23 (134)
Hybrid waves 16 (374, 390)
Hybrid-tee junction 16 (696)
Hydrocarbon, "alternant" 18/1 (154)
Hydrogen 17 (72); 22 (545)
Hydrogen bonding 17 (309, 373, 377)
Hydrogen bromide 17 (45, 104)
Hydrogen chloride 17 (45, 103)
Hydrogen iodide 17 (45)
Hydrogen maser 23 (301, 326)
Hydrogen molecule 19 (98, 117)
Hydrogen, pure, breakdown between coaxial cylinders 22 (508)
Hydrogene atom 18/1 (246, 259)
Hydrogene molecule, asymptotically exact solution 18/1 (283)
— Heitler-London theory 18/1 (271)
— orthogonalization of one-electron functions 18/1 (280)
— treatment by Bogolyubov's method 18/1 (332)
Hydrostatic pressure chamber 19 (190)
Hyperconjugation 18/1 (172, 176)
Hyperfine anisotropy 18/2 (260)
Hyperfine constant, one-electron 18/1 (149)
Hyperfine coupling 18/1 (149)
Hyperfine coupling parameter 18/1 (149, 160, 168)
Hyperfine effect, second-order 18/1 (149)
Hyperfine interaction 18/1 (100, 102, 132, 143)
— anisotropic 18/1 (187)
— modulation of 18/1 (194)
— second-order 18/1 (147, 148)
Hyperfine line 18/1 (103, 116)
Hyperfine line, super- 18/1 (103)
Hyperfine pattern, super- 18/1 (107)
Hyperfine splitting 18/1 (14, 99, 104, 108, 133, 135, 139, 140, 148)
— constant, sign of the 18/1 (168)
— super- 18/1 (132, 184, 255)
— temperature-dependent 18/1 (198)
— tensor 18/1 (147, 148, 187)
Hyperfine structure 18/1 (101, 132)
— chlorine super- 18/1 (252)
— fluorine super- 18/1 (252)
— mechanism of 18/1 (162)
— splitting 18/1 (14)

Hysteresis 17 (328, 339, 364, 372); 18/2 (345); 22 (189, 194)
Hysteresis, ferromagnetical 16 (21)
Hysteresis in ferroelectrics 17 (272)
Hysteresis integral 18/2 (506)
Hysteresis loop 18/2 (342, 345)
Hysteresis loop, minor 18/2 (447)
Hysteresis loss 18/2 (453, 504)
Hysteresis of an actuator 23 (461, 488, 503)
Hysteresis, physical origin of 18/2 (406)
Hysteresis, rotational 18/2 (447, 449, 522)

Ice 17 (373)
Idler frequency 23 (4, 39)
Image converter camera 23 (262)
Image force 17 (203); 21 (8, 28, 33, 363)
Image formation, latent 19 (384)
Impact broadening 22 (212)
Impact ionization 17 (158, 190, 195, 198)
Impact ionization in *p-n* junctions 20 (156)
Impact ionization rate 17 (225, 230, 241)
Impact, massive particle 22 (223)
Impact rate 17 (194)
Impact with molecules 22 (448)
Impedance 16 (210, 286)
Impedance, apparent terminal 16 (212)
Impedance, characteristic 16 (303, 376)
Impedance, internal 16 (185, 189, 194, 209)
Impedance matching 23 (12)
Impedance matrix 16 (222, 312, 397)
Impedance, measurement of 23 (113, 119)
Impedance, mutual 16 (194, 198, 244)
Impedance, normalized 16 (215, 314)
Impedance of a center driven antenna 16 (238)
Impedance of driving-point 16 (476, 492)
Impedance properties 16 (216, 261)
Impedance, reduced 16 (314)
Imperfections 17 (277, 282, 300, 332, 364)
Imperfections and domain wall displacements 17 (275)
Impuls function response of a control system 23 (462)
Impulse breakdown 22 (526)
Impulse fields 22 (45)
Impulse, sharply peaked 22 (27)
Impurities 20 (16); 22 (85, 197)
Impurities, aliovalent 20 (285)
Impurities, effect of, at low temperatures 19 (295)
Impurities, monovalent 19 (256)
Impurities, polyvalent 19 (259)
Impurities, solubility 20 (290, 292)
Impurities, valency 20 (295)
Impurity atoms 19 (65, 67, 255)
Impurity band conduction 20 (83)
Impurity diffusion 20 (335)
Impurity ion 18/1 (78)
Impurity, longitudinal relaxation of 18/2 (222)
Impurity pair 18/1 (254)
Impurity problems for semi-conductors 19 (78, 81)
Impurity semiconductor 20 (5, 39)
Impurity spin-deviation state 18/2 (23)
Impurity state, magnetic 18/1 (516)
Impurity-vacancy pairs 20 (311, 313, 332)
Incoherent scattering 16 (540, 584, 621, 628, 636)
Indirect interstitial mechanism, *see also* Interstitialcy mechanism 20 (247)
Indium antimonide 19 (385); 23 (211)
Indium phosfide filter 23 (453)
Indium selenide, telluride and oxide 19 (387)
Induced emission 23 (89, 90)
Inductance 22 (201)
Inductance, effective 22 (201)
Inductance (external) 16 (210)
Induction 18/2 (341)
Induction, internal, of moments 17 (152)
Induction, magnetic 16 (18)
Induction methods 19 (153)
Induction terms 16 (458)
Inductivity 16 (23)
Inductivity, mutual 16 (23, 24)
Inelastic frequencies 21 (412)
Influence of a discharge on a neighbouring system 21 (588)
Influence of space charge on the current 21 (588)
Infrared 17 (377, 388)

Infrared detectors 19 (317, 356)
Infrared emission spectra 22 (250)
Infrared radiation 19 (319)
Infrared removal of space charges 19 (319)
Infrared spectrometer 23 (128)
Initial breakdown 22 (618)
Initial current interrupted 21 (610)
Injection of minority carriers 20 (130)
Inkjet 23 (223)
Inkpen 23 (223, 249)
Input admittance 16 (214, 217, 260, 453)
Input admittance of a cavity resonator 16 (420)
Input impedance 16 (214, 217, 453, 456, 463, 465, 468, 470)
Input range of a controller 23 (472)
Insertion loss 16 (220); 23 (106)
Instability 22 (543)
Instability, classical 18/2 (244)
Instability, critical field for magnon 18/2 (251)
Instability, first-order Suhl 18/2 (244)
Instability, higher-order 18/2 (254)
Instability of a control system 23 (481)
Instability, phonon-pumped 18/2 (253)
Instability, second-order Suhl 18/2 (246, 247)
Instability, suppression of 18/2 (248)
Insulator 16 (15)
Insulator, perfect 16 (1)
Insulators, thin 17 (244)
Integral action controller 23 (474)
Integral equations
— for current distributions in antennas 16 (465, 467, 473, 499)
— for the propagation of electric waves 16 (532, 533, 542, 580, 607)
— for the propagation over the earth's surface 16 (532, 533, 607)
— in potential theory 16 (36)
Integral of error criterion 23 (485)
Integral, overlap 18/1 (276, 278, 311, 323, 354, 359, 362)
Integrating systems 23 (254)
Integrator, Kelvin ball and disk 23 (255)
Intensities, absolute measurement of 22 (368, 369)
Interaction 17 (328, 341, 352, 355, 357)
Interaction among defects 20 (298)
Interaction, anisotropic hyperfine 18/1 (187)
Interaction between magnon and conduction electron 18/2 (254)
Interaction between magnon and nucleus 18/2 (257)
Interaction, configuration 18/1 (337)
Interaction, Coulomb 18/1 (35)
Interaction, dipolar 18/1 (73)
Interaction, dipole-dipole 18/1 (125); 18/2 (347)
Interaction, direct exchange 18/1 (268, 286)
Interaction, dynamical 18/2 (20)
Interaction, electric quadrupole 18/1 (135, 136)
Interaction, electron 18/1 (541)
Interaction, electron-electron 18/1 (289, 306, 341)
Interaction, electron-magnon 18/1 (398, 544)
Interaction, electron-phonon 18/1 (398, 543)
Interaction, electrostatic 17 (20, 47)
Interaction, exchange 18/1 (14, 20, 34, 35, 71, 73, 75, 77); 18/2 (64, 342, 349)
Interaction, *F*-centre 18/1 (231)
Interaction, Fermi 18/1 (72)
Interaction, hyperfine 18/1 (100, 102, 132, 143)
Interaction in RE-elements, magnetic 18/1 (491)
Interaction in thio-spinels, superexchange 18/1 (531)
Interaction, indirect exchange 18/1 (268, 335)
Interaction, interionic 18/1 (72)
Interaction, magnetic dipole-dipole 18/1 (14, 34, 35, 75)
Interaction, magnetostatic 18/2 (481, 523)
Interaction, magnon-magnon 18/2 (196)
Interaction, magnon-phonon 18/2 (196, 203)
Interaction, modulation of hyperfine 18/1 (194)
Interaction, multi-nuclear 18/1 (104)
Interaction, nuclear quadrupole 18/1 (34, 35, 72)
Interaction of a magnon with beam of particles 18/2 (261)

Interaction, orbit-lattice 18/1 (48)
Interaction, paramagnetic relaxation by modulated exchange 18/1 (71)
Interaction, phonon-phonon 18/1 (87)
Interaction, piezoelectric 17 (324)
Interaction, π-σ 18/1 (163)
Interaction potential 17 (186)
Interaction, relativistic 18/1 (286)
Interaction, second-order hyperfine 18/1 (147, 148)
Interaction, spin-orbit 18/1 (35, 462); 18/2 (351)
Interaction, spin-lattice 18/1 (4, 48, 64, 99, 120, 124)
Interaction, spin-phonon 18/1 (84, 85)
Interaction, spin-spin 18/1 (99, 120, 122); 18/2 (351)
Interaction, spin-wave 18/2 (20, 50, 132)
Interaction time 21 (443)
Interaction via semiconducting carriers, magnetic 18/1 (512)
Interaction with the lattice 21 (652)
Interatomic distance 19 (4)
Interfacial polarization 16 (709)
Interference fading 16 (588)
Interference photograph 22 (396)
Interference spectrogram 22 (373)
Interferometer 23 (161, 182)
Interferometer, four-horn 23 (129)
Interferometer, waveguide 23 (125, 126)
Interionic force fields, enhancement of α 21 (481)
Intermetallic compounds 19 (317, 384); 21 (380)
Internal electric field 18/1 (108)
Internal energy 22 (341)
Internal reflection in a stratified medium 16 (556, 568, 603, 605, 612)
Interpolation equipment 23 (163)
Interpolation oscillator 23 (350)
Interpolation techniques in frequency measurement 23 (361, 364)
Interstitial atoms 19 (321)
Interstitial centre 18/1 (245)
Interstitialcy mechanism 20 (247, 258)
Interstitialcy migration 20 (330)
Interstitials 20 (15)
Inter-valley scattering 20 (64)
Inter-zone transitions 19 (263)
Intra-valley scattering 20 (64)
Intrinsic semiconductor 20 (5, 39)
Inversion 17 (309)
Inversion at a circle 16 (46, 58)
Inversion at a sphere 16 (40, 87–90, 114, 119)
Inversion layer 20 (48, 178, 200)
Inverted Orbach relaxation process 18/1 (46, 65)
Iodine 19 (340, 352)
Ion bombardment 21 (370); 22 (81)
Ion cross-section 22 (349, 380)
Ion current to the walls 22 (126)
Ion density 22 (28, 44)
Ion drift velocity 22 (136)
Ion formation, critique of data from different methods of 21 (456)
Ion formation, measurement of energy 21 (455)
Ion production 22 (121)
Ion production by secondary processes 21 (473)
Ionic binding 19 (4)
Ionic concentration, effective 16 (723)
Ionic conductivity, effects of radiation 20 (345)
Ionic conductors
— alkali halides 20 (251 ff., 260, 268, 273, 283, 288, 291)
— alkaline halides 20 (257)
— lead halides 20 (251)
— silver halides 20 (312, 317, 322, 323, 329–331, 338, 339)
— thallous halides 20 (251)
Ionic crystals 17 (99)
Ionic energy level 18/1 (35)
Ionic force, strong 21 (492)
Ionic interaction, inter- 18/1 (72)
Ionic migration 21 (29)
Ionic state 18/1 (282, 293)
Ion-ion recombination 21 (485, 487, 496)
— in oxygen and air 21 (483)
— measurement 21 (479)
— preferential 21 (496)
Ionization 22 (1)
Ionization and excitation probabilities 22 (18, 89, 196)
Ionization by collision 21 (457, 515); 22 (446)
Ionization by positive ions 22 (6, 448)
Ionization by single collisions 22 (75, 80)
Ionization chamber 21 (316, 317, 318, 505); 22 (28, 29, 113)

Ionization coefficient(s) 21 (510, 519, 530, 536, 539, 540, 543); 22 (533)
— in diatomic gases 21 (532)
— in gases with complex molecules 21 (535)
— in mixtures of diatomic gases 21 (535)
— in mixtures of monatomic gases 21 (532)
— in monatomic gases 21 (530)
Ionization continuum 21 (320, 323, 328, 329, 331, 335)
Ionization cross sections 21 (508)
Ionization current in air 21 (517)
Ionization current in irradiated gas 21 (505)
Ionization, direct measurement of 21 (638)
Ionization, dissociative 21 (329, 331)
Ionization efficiency 21 (508, 539)
Ionization energy 18/1 (294, 319, 382); 21 (511)
Ionization energy for RE 18/1 (458)
Ionization equilibrium 22 (308)
Ionization, external source of 22 (54)
Ionization frequency 21 (412, 418)
Ionization function 22 (135, 162)
Ionization growth 22 (1, 2, 14, 30, 43, 47, 49)
— of, in nonuniform fields 22 (21)
— of, spatial 22 (32)
Ionization in combined electric and magnetic fields 21 (570)
Ionization in gases by electrons in electric fields 21 (504)
Ionization in non-uniform fields 21 (561)
Ionization in two stages 22 (194)
Ionization in uniform electric fields 21 (510)
Ionization limit 21 (320, 321, 323, 325, 328)
Ionization on positive and negative points 21 (564)
Ionization per unit light intensity 21 (316)
Ionization, photoelectric, *see* Photoionisation
Ionization potential 22 (89, 196, 526)
Ionization probability 21 (506, 507)
Ionization, secondary 22 (33, 34)
Ionization, spatial growth
— of, in static fields at high pressure 22 (27)
— of, in static uniform fields 22 (2, 5)
Ionization, stepwise 22 (7, 19, 75, 80, 128, 146)
Ionization, temporal growth of 22 (40)
Ionizing wave 22 (526)
Ionizing zone 22 (513)
Ionosphere 21 (482, 492)
Ionospheric conductivity 16 (546, 630)
Ions 22 (9, 33)
Ions, arrangement of 18/1 (267)
Ions, destruction of negative 22 (179)
Ions, energy distribution of the 22 (64)
Ions, motion of 22 (135)
Ions, multiply charged 21 (641)
Ions, negative 22 (93, 130, 143, 154, 178)
Ions, number balance for 22 (121)
Ions, positive ion bombardment 22 (31)
Ions, positive: space charges 22 (523)
Ions, radial distribution of 22 (171)
Ions, slow 22 (14)
Iris-loaded waveguide, *see* Waveguide, circular corrugated
Iron 19 (36)
Irradiation damage 18/1 (95, 98)
Irradiation, external 22 (202)
Irradiation, internal 22 (110)
Irreversible processes (thermodynamics) 17 (25); 22 (315)
Isoelectronic row 20 (7)
Isolator 18/1 (265)
Isotope effect(s) 17 (328, 369, 376); 21 (646)
Isotropy 17 (308)
Isotropy, partial 16 (567)

Jaffé, analysis of 21 (487)
Jahn-Teller effect 18/1 (69, 70)
Jahn-Teller theorem 18/1 (36)
Jet oscillograph 23 (250)
Jitter 23 (354)
Joule heat 16 (17); 19 (273); 22 (339)
Jump distance of ions in glasses 20 (354, 355, 356, 360, 366)
Jump mechanism of ions in glasses 20 (360)
Junction, nonreciprocal 16 (401–406)
Junction of waveguides 16 (395–406)

Kanamori rules, Goodenough- 18/1 (451)
Kapitza's experiments 19 (183)
KCl 18/1 (144)
KDP, antiferroelectricity in 17 (267, 268, 273)
KDP, disorder in 17 (372)
KDP, isotope effect in 17 (268)
KDP, phase transitions 17 (328)
KDP = potassium dihydrogen phosphate 17 (355)
Kelvin, degree 23 (374, 378)
Kelvin transform 16 (88)
Kerr effect 17 (323, 328); 18/2 (343)
Kerrcell shutter 23 (262)
Kinetic coefficients 19 (272, 274, 282, 283, 285)
Kirchhoff's formula for edge capacitance 16 (653)
Kirchhoff's formula of diffraction theory 16 (444)
Kirchhoff's law 22 (358, 359); 23 (442)
Kirkwood theory 17 (75, 84)
Kirkwood-Yvon method 17 (63)
Kittel function, Ruderman- 18/1 (434)
Klystron 18/1 (97, 108, 110, 115, 116)
Knife-edge effect 16 (608)
Kohler relation 19 (184)
Kohler's principle 19 (284)
Kohn effect 18/2 (149)
Koopmans' theorem 19 (49, 57)
Kramers degeneracy 18/1 (36, 40, 69, 72)
Kramers doublet 18/1 (131)
Kramers exchange mechanism, non- 18/1 (269)
Kramers salt 18/1 (39, 40)
Kramers salt, non- 18/1 (39, 40, 41)
Kramers salt, Raman process in non- 18/1 (41, 50)
Kramers theorem 18/1 (36)
Kramers-Anderson mechanism 18/1 (379)
Kramers-Kronig relation 18/1 (19)
Kronig relation, Kramers- 18/1 (19)
k-space 19 (230)
Kunzmann source 22 (145)

"*L*" cathode 21 (12, 155)
Lambert's law 21 (312)
Lamé equation and functions 16 (146)
Landau trapping 18/1 (411, 543)
Landau-Lifshitz equation 18/2 (169)
Landé splitting factor 18/1 (126, 404)
Langevin coefficient 21 (496)
Langevin equation 21 (392, 421, 443)
Langevin function 17 (38); 18/2 (2)
Langevin model for ion-ion recombination 21 (484f., 487, 498f.)
Langevin-Curie theory of paramagnetism 18/2 (2)
Langevin-Debye formula 17 (39, 44, 45, 49)
Langevin's mobility equation 22 (122, 131)
Langmuir dark space 22 (224)
Langmuir probes 22 (61, 62, 110, 115, 138)
Langmuir's space-charge theory 21 (34)
Lanthanide atoms, energy level for 18/1 (458)
Laplace equation 16 (4, 15); 18/1 (47)
— for a *T*-wave 16 (294)
— in the plane 16 (40)
— solution by Fourier series 16 (38, 93, 94, 133)
— solution by separation of variables, *see also* Separation 16 (37, 83, 91)
Laplace transform 16 (229); 23 (468)
Larenz and Bartels, method of 22 (371)
Larmor frequency 18/1 (8, 22, 24); 18/2 (16)
Laser 23 (90, 161)
Laser, ring-type 23 (184)
Lattice binding 19 (362)
Lattice disorder 20 (15)
Lattice Hamiltonian 18/1 (38)
Lattice imperfections 20 (15); 21 (381)
Lattice interaction, spin- 18/1 (4, 48, 64, 99, 120, 124)
Lattice of wires finite, its electric field 16 (47, 48)
Lattice of wires, infinite, its electric field 16 (49)
Lattice operator, spin- 18/1 (30)
Lattice parameters 17 (332, 336)
Lattice perturbation, spin- 18/1 (34)
Lattice potential, orbit- 18/1 (48)
Lattice relaxation operator, effective spin- 18/1 (56)
Lattice relaxation, spin- 18/1 (1, 30, 56, 59, 78, 120)

Lattice relaxation time, spin- 18/1 (6, 9, 120, 121, 132)
Lattice state 18/1 (38)
Lattice temperature 18/1 (31); 18/2 (207)
Lattice vibrations 17 (183); 20 (262)
— acoustical and optical branch 20 (12)
— broadening due to 19 (337)
Lattice waves, longitudinal 19 (246, 286, 290)
Lattice waves, transverse 19 (246)
Laval jet 22 (403)
Laves phase 18/1 (492)
Layers, chemical deposited 19 (359)
Layers, inhomogeneous 22 (365)
LCAO 18/1 (67, 151, 378, 409)
LCAO, MO- 18/1 (321)
Lead borate glasses 20 (363)
Lead chromate 17 (385)
Lead selenide 19 (317)
Lead silicate glasses 20 (361, 362)
Lead sulfide type semiconductors 20 (239)
Lead sulphide 19 (40, 317, 355)
Lead sulphide, vacuum evaporation of pure 19 (356)
Lead telluride 19 (355)
Lead titanate 17 (318, 322, 334, 336, 382)
Lead titanate, antiferroelectric displacements 17 (316)
Lead zirconate 17 (336, 339)
Lead zirconate, antiferroelectricity in 17 (270, 316)
Leader field change 22 (608, 610)
Leader, stepped 22 (586, 597)
Leader, tip-field and velocity while advancing 22 (618)
Leaders, α and β type, first 22 (586, 587, 588, 589)
Leaders, distribution of effective velocities of stepped 22 (591)
Leadsulfide photoresistor 23 (449)
Leakage currents 16 (668)
Le Chatelier-Guertler rules 19 (207)
Lecher line, *see* Transmission line
Lecher wave = (TEM-wave or *T*-wave) 16 (294)
Lecher wires 22 (68, 137)
Lennard-Jones potential 17 (75)
Lens, artificial dielectric 16 (504, 507)
Lens configuration 16 (505, 513)
Lens, dielectric 16 (504, 505)
Lens, electron optical
— electrical 16 (97, 101)
— magnetic 16 (133)
Lens for microwaves 16 (504)
Lens, inhomogeneous 16 (508 ff.)
Lens, metal plate 16 (504, 505)
Lens of Luneberg 16 (509, 510)
Lens, path length 16 (505, 514)
Lens, zoned for microwaves 16 (504, 512)
Leontovich condition 16 (532)
Leplace equation, separation in different coordinate systems, *see* Separation
LiF 19 (39)
Life-time 19 (325)
Life-time of the phonon 18/1 (86, 87)
Ligand atom 18/1 (132)
Ligand field theory 18/1 (127, 270, 379, 406)
Ligand nucleus 18/1 (132)
Light absorption, cf. Photoabsorption 21 (367); 22 (113)
Light absorption, mechanisms 21 (321)
Light amplification 19 (337)
Light beam oscillograph 23 (231)
Light, modulated 19 (374)
Light, partially polarized 22 (81)
Light, polarised 19 (351)
Light probe 19 (358)
Light probe techniques 19 (319)
Light source 21 (350)
Light, ultraviolet 22 (198)
Lightning, daylight photography of 22 (584)
Lightning discharge, historical 22 (574)
Lightning flash 16 (566)
LiH 19 (103)
Limiting sensitivity 19 (338)
Line intensities 22 (353)
Line printer 23 (279)
Line shape 18/2 (173)
Line shape for non-oriented systems 18/1 (214)
Line shift 22 (375)
Line spectrum 21 (318); 22 (135)
Line width 18/1 (98); 18/2 (173, 220)
— and shape 17 (135)
— anomaly 18/1 (192)
— effect 18/1 (192)

Linear chain 18/2 (13, 44)
— antiferromagnetic 18/2 (116)
— spin-wave spectrum of 18/2 (118)
Linear charge 16 (28, 30, 42)
Linear combination of atomic orbitals LCAO 18/1 (67, 151, 378, 409)
Linearizing mechanisms 23 (250)
Liouville operator 18/1 (16)
Liouville theorem 21 (409)
Liquid surfaces 22 (158)
Liquids, polar, under high pressure 17 (87)
Lissajous figure 23 (292, 349)
Lithium borate glasses 20 (359)
Lithium ions 21 (645)
Lithium silicate glasses 20 (358)
Lithium tantalate 17 (381)
Li_xMn_2 18/1 (520)
Local field, *see* Lorentz
Local free energy 17 (76, 81)
Local heating 21 (652)
Localized defekt 18/1 (82)
Localized electron model 18/1 (400)
Logarithmic recorder 23 (254)
Log-normal distribution 16 (588)
Long times lags, interpretation of 22 (493)
Long wavelength response in crystals 19 (337)
Longitudinal phonon 18/1 (82)
Longitudinal propagation 16 (568)
Longitudinal relaxation time 18/1 (8)
Longitudinal section waves 16 (317)
Loop antenna 16 (253 ff.)
— as a probe 16 (270)
— in a plane-wave 16 (272)
— in an elliptically polarized field 16 (275)
— shielded 16 (274)
— unloaded 16 (271, 273)
Loops as probes, simple and shielded 16 (278)
Lorentz catastrophe 18/2 (93)
Lorentz condition 16 (172, 173)
Lorentz factor 17 (371, 381)
Lorentz internal field 20 (324)
Lorentz line shape, Gaussian- 18/1 (26)
Lorentz local field 17 (8, 47, 87, 95, 100, 107, 112, 117, 127, 370, 377)
Lorentz model 17 (10)
Lorentz reciprocity theorem 16 (180, 398)
Lorentz reciprocity theorem, generalization 16 (401)
Lorentz-forces 22 (326, 396)
Lorentz-Heaviside units 17 (2, 148)
Lorentzian line shape 18/1 (26, 123, 124)
Lorentz-Lorenz formula 17 (177); 19 (338)
Loss, dielectric 17 (119)
Loss, dielectric, arc plot of 17 (124)
Loss factor 16 (641); 20 (373)
Loss tangent 16 (641); 23 (88)
Lossy guide 16 (310)
Low pressure column 22 (174)
Low voltage arc 22 (242, 257, 304)
Löwdin's perturbation theory 18/1 (344, 349, 361)
Lower sideband up-converter 23 (41, 45)
Luminescence 17 (253); 19 (324)
Luminescence emission 19 (352)
Luminescence emission of Cu_2O 19 (378)
Luminescence in the infra-red region 19 (360)
Luminosity 22 (44)
Luminosity, fluctuating 22 (191)
Luminosity, propagation of 22 (51)
Lumped constants 16 (200)
Luneberg's lens 16 (509, 510)
Luxemburg effect 16 (573)
Lyman α-line of hydrogen 21 (327)
Lyman continuum 21 (318)

M band 19 (355)
M centre 18/1 (235)
Mach-number 22 (403)
Mach-Zehnder interferometer 23 (131)
Madelung energy 20 (264)
Madelung potential 18/1 (548)
Magic-*T* 18/1 (119)
Magnesium 19 (46)
Magnesium antimonide 19 (388)
Magnesium oxide 19 (370)
Magnet, superconducting 18/1 (120)
Magnetic absorption 18/1 (482)
Magnetic alloys 19 (217)
Magnetic amplifier 23 (419)
Magnetic card 23 (283)
Magnetic charge 16 (18, 431)
Magnetic circuit 16 (25)
Magnetic core 23 (229)

Magnetic current 16(431)
Magnetic current density 16(288)
Magnetic deflection 22(95)
Magnetic digital recording 23(281)
Magnetic dipole transition 18/1(37)
Magnetic dipole-dipole interaction 18/1(14)
Magnetic disk 23(283)
Magnetic drum 23(229, 283)
Magnetic energy 16(179)
Magnetic field
— in the plane 16(41)
— longitudinal 22(166)
— measurement 16(162)
— modulation 18/1(110, 112, 113, 115)
— near half-wave dipole 16(280)
— strength 18/2(341)
— superimposed 22(559)
— transverse 22(166, 170)
Magnetic flux 16(19)
Magnetic gradient 21(387)
Magnetic high-frequency spectrum 18/1(136)
Magnetic impurity state 18/1(516)
Magnetic interaction in RE elements 18/1(491)
Magnetic interaction via semiconducting carriers 18/1(512)
Magnetic lenses 22(173)
Magnetic links 22(605)
Magnetic moment 18/1(94, 99)
Magnetic moment, nuclear 18/1(94)
Magnetic moment of RE 18/1(490)
Magnetic polaron 18/1(399, 419, 421)
Magnetic pole 16(18)
Magnetic quasiparticle 18/1(368)
Magnetic rays 22(177)
Magnetic recording 23(258)
Magnetic semiconductor 18/1(391, 396)
Magnetic solution 16(431, 433, 435, 513, 518, 550, 551)
Magnetic specific heat 18/1(10)
Magnetic susceptibility 20(213)
Magnetic tape 23(229, 259)
Magnetic tape equipment 23(283)
Magnetization 16(18, 21); 18/2(77, 341)
Magnetization charge 16(22)
Magnetization curve 18/2(281, 342, 439)
Magnetization, equation of motion of 18/2(357)
Magnetization process 18/2(441, 453)
Magnetization process, thermaly activated 18/2(459)
Magnetization, reversal of 18/2(96, 427)
Magnetization saturation 18/2(53, 346)
Magnetization sublattice 18/2(109, 116)
Magnetization, temperature dependence of 18/2(34, 370)
Magnetization, timedependent 18/2(466)
Magnetization, volume density 16(167)
Magnetoacoustic resonance 18/2(215)
Magnetocaloric effect 18/2(335, 336, 337)
Magnetocrystalline anisotropy field 18/2(324)
Magnetoelastic constant 18/2(289, 326)
Magnetoelastic constant, temperature dependence of 18/2(327)
Magnetoelastic coupling 18/2(288)
— coefficient 18/2(204)
— constant 18/2(204)
Magnetoelastic dispersion curve 18/2(214)
Magnetoelastic effect, anisotropic 18/2(207)
Magnetoelastic effect, exchange 18/2(208)
Magnetoelastic effect in antiferromagnet 18/2(338)
Magnetoelastic energy 18/2(288)
Magnetoelastic wave 18/2(209, 210, 212)
Magnetomechanical ratio 18/1(8)
Magneto-meter, low temperature 19(151)
Magneton, Bohr 18/2(2)
Magneton, nuclear 18/1(99)
Magneton number, effective 18/1(405)
Magneto-optic 18/1(480)
Magnetoresistance 19(303); 20(73)
Magnetoresistance, anisotropy in the 19(186)
Magnetoresistance at high magnetic fields 20(81)
Magnetoresistance at low magnetic fields 20(76)
Magnetoresistance at low temperatures 20(83)
Magnetoresistance effect 23(207, 210)
Magnetoresistance effect, two-zone theory of 19(304)

Magnetoresistance, longitudinal 19 (185, 304)
Magnetoresistance of binary magnetic alloys 19 (220)
Magnetoresistance of bismuth at low temperatures 19 (308)
Magnetoresistance of ferromagnetic metals 19 (188)
Magnetoresistance, transverse 19 (183)
Magnetostatic interaction 18/2 (481, 523)
Magnetostatic mode 18/2 (171, 174, 176, 228)
Magnetostatic mode, parallel pumping of 18/2 (251)
Magnetostatics 16 (18)
Magnetostriction 18/2 (204, 291, 294, 295, 327)
Magnetostriction, anisotropic 18/2 (308)
Magnetostriction constant 18/2 (299, 356)
Magnetostriction, de Haas-van Alphen oscillatory 18/2 (339)
Magnetostriction, exchange 18/2 (338)
Magnetostriction, external 18/2 (329)
Magnetostriction, forced 18/2 (295, 306–311)
Magnetostriction, forced volume 18/2 (306)
Magnetostriction in antiferromagnetism 18/2 (338)
Magnetostriction in dia- and paramagnetism 18/2 (339)
Magnetostriction, internal 18/2 (329)
Magnetostriction, linear 18/2 (338)
Magnetostriction, measurement of 18/2 (321)
Magnetostriction, shape effect of 18/2 (356)
Magnetostriction, spontaneous 18/2 (295)
Magnetostriction, temperature dependence of 18/2 (293, 322, 326–328)
— the forced 18/2 (330)
Magnetostriction, volume 18/2 (303, 306, 307, 310)
Magnification factor 16 (674)
Magnon 18/2 (196, 376, 377)
Magnon absorption 18/2 (254)
Magnon absorption, exciton- 18/1 (483)
Magnon and conduction electron, interaction between 18/2 (254)
Magnon and nucleus, interaction between 18/2 (257)
Magnon drag 18/1 (430)
Magnon instability, critical field for 18/2 (251)
Magnon interaction, electron- 18/1 (398)
Magnon magnon-interaction 18/2 (196)
Magnon process 18/2 (196)
2 Magnon process 18/2 (229)
4 Magnon process 18/2 (199)
Magnon radiation 18/2 (254)
Magnon, relaxation of 18/2 (198)
Magnon splitting process 18/2 (197)
Magnon, Suhl 18/2 (228, 229)
Magnon, thermal 18/2 (197)
Magnon, thermal, conduction by 18/2 (216)
Magnon with beam of particles, interaction of 18/2 (261)
Magnon-electron process 18/2 (254, 256)
Magnon-phonon interaction 18/2 (196, 203)
Magnon-phonon interaction process 18/2 (206)
Maintenance condition 22 (22)
Maintenance, equation of 22 (103)
Maintenance process 22 (23)
Malan camera 22 (584)
Malman alloys 19 (221)
Malonic acid 18/1 (259)
Malter effect 21 (270); 22 (164)
Manganate 18/1 (141)
Manganate hyperfine splitting 18/1 (140)
Manganese fluosilicate 18/1 (138)
Manganese nucleus 18/1 (138, 141)
Manley-Rowe power relations 23 (37)
Margenau distribution 21 (417)
Martyn's absorption theorem 16 (558)
MASD, antiferroelectric (methylammonium aluminium sulphate dodecahydrate) 17 (273)
Maser 18/1 (108); 23 (1, 37, 90)
Maser, ammonia 23 (169, 170, 183)
Maser, solid-state 18/1 (2); 23 (177)
Mass action law 20 (43)
Mass concentration 22 (316)

Mass, effective 18/1 (398); 20 (11, 14, 34)
Master clocks, transport of 23 (300)
Master curves 21 (60, 62)
Master equation, generalised 18/1 (18, 23)
Matched line 16 (213, 225)
Matched receiver 16 (454, 463, 477, 479)
Mathieu equation 19 (24)
Mathieu functions 16 (112, 331)
Matthiessen law 19 (207)
Matthiessen rule 19 (206, 255)
Matthiessen rule for alloys with polyvalent solvents 19 (263)
Mattuck and Strandberg's effective perturbation operator 18/1 (56)
Maue's integral equations for aperture antennas 16 (473, 474, 499)
Maxwell distribution 21 (411)
Maxwell equations 16 (168, 169, 285, 423, 442, 451, 517, 571 ff.)
— harmonic form 16 (186, 287)
— modified complex form 16 (288)
— symmetry 16 (287)
Maxwell relations 17 (346, 360)
Maxwell-Boltzmann relaxation 18/1 (30)
Maxwellian distribution 22 (63, 65, 111, 121, 135)
Maxwell equations 23 (86, 91)
Maxwell fish-eye 16 (509)
McConnel equation 18/1 (126, 165)
M-components 22 (579)
Mean energy 21 (418, 426)
Mean free path 19 (229, 318); 20 (14, 52); 22 (80)
— constant 21 (425)
— electron-electron 19 (228)
— electron-lattice 19 (230)
— for photoionizing radiation 22 (485)
— for semi-metals 19 (262)
— in center of gravity system 21 (424)
— laboratory 21 (424)
— Maxwellian 21 (426)
— scalar 21 (395)
— table of 19 (238)
Mean free time 20 (52)
Mean free time, constant 21 (423)
Mean life-time 19 (318)
Mean square free path, tensorial 21 (395)
Measuring cell 16 (644)
Mechanism of indirect exchange 18/1 (268, 337)
— double transition, d_1-p, $p'-d_2$ 18/1 (351, 359, 364)
— polarization mechanism 18/1 (352)
— transition to the occupied orbital 18/1 (348, 359, 369, 373)
— transition to the vacant orbital 18/1 (337, 371, 373)
— twofold excitation, $p-d_1$, $p-d_2$ 18/1 (349)
Melting point 22 (87)
Membrane method for the numerical integration of the Poisson equation 16 (161)
Membrane, vibrations 22 (128)
Memory effect 19 (391)
Memory, magnetic 18/2 (441)
Mercury arc 22 (297)
Mercury atoms 21 (655)
Mercury contact thermometer 23 (494)
Mercury electrode 16 (648)
Mercury ions 21 (645)
Mercury vapor rectifier 22 (304)
Merk's criterion 22 (29)
Meta-dichlorobenzene 17 (152)
Metal vapour 22 (185)
Metallic bond 19 (109)
Metals, anisotropic 19 (157, 158)
Metals, isotropic 19 (156)
Metals, liquid 19 (180)
Metals, monovalent 19 (160)
Metamagnet 18/2 (138)
Metamagnetic transition 18/2 (138)
Metamagnetism 18/2 (121, 137)
Metastability 21 (382)
Metastable A_2 molecules 22 (509)
Metastable atoms 21 (577, 580, 606, 610, 614); 22 (6, 15, 20f., 76, 79f., 80, 112, 448)
— decay of 22 (202, 448)
— destruction of 21 (556)
— diffusion and collision 21 (614, 661)
— emission by 21 (655)
— ionisation by 21 (660)
— ionization in the gas by 22 (448)
— irradiation of 21 (556)
— surface emission due to 21 (632)
Metastable gas molecules 22 (65, 509)
Metastable Neon 22 (87)
Metastable states, production of 21 (630)

Meteoric ionization 16 (593, 636, 637)
Meteoric light production 22 (224)
Methane 19 (104)
Methyl chloride 17 (62, 95)
Methyl ether 17 (46, 47)
Methyl green 19 (390)
Methyl group 18/1 (106)
Methylene 18/1 (221)
Meyer-Neldel rule 20 (21)
Mica 17 (253)
Michelson interferometer 16 (703); 23 (129, 130)
Michelson-Morley experiment 23 (183)
Microcrystalline layers 19 (330)
Micromagnetism 18/2 (346, 454)
Microstrip 16 (375)
Microwave bridge 18/1 (112, 113, 116)
Microwave discharges 22 (251)
Microwave gyrator 16 (404)
Microwave lenses, *see also* Lenses 16 (504)
Microwave method 17 (143)
Microwave printed circuit 23 (63)
Microwave techniques 21 (454)
Migration losses 20 (376, 379)
— dependence on the chemical composition 20 (379 f.)
— gradual chemical substitution 20 (369)
— immobile ions 20 (380)
— in borate and boro-silicate glasses 20 (382)
— influence of the coherence of the network 20 (380)
— influence of the network-modifying ions 20 (379)
Miller-integrator 23 (244)
Millimetre wavelength interferometry 16 (703)
Minimum principle 22 (104, 108, 151)
Mirror image, electrical 16 (39, 45, 84–90)
Mirror image force 21 (8, 28, 33, 363)
Mirror image function 21 (30, 31)
Mixing, turbulent 16 (587)
Mks units 17 (148)
Mn^{2+} ion 18/1 (137, 252)
Mn^{55} nucleus 18/1 (137)
Mn perovskite 18/1 (498)
MnO 18/1 (267, 268, 335, 380, 382)
Mobile ions in glass 20 (351, 360)
Mobility 18/1 (397); 20 (14); 22 (103, 178)
Mobility in high electric fields 20 (67)
Mobility of $3d$, $4f$ electrons 18/1 (398)
Mobility of lattice defects 20 (275)
Mobility of liberated carriers 19 (392)
Mobility tensor 21 (394, 413)
Mobility, theory 20 (62 ff.)
Mode 16 (466, 472, 489, 490, 493, 494, 511, 514, 515, 552, 561–563, 605, 613–619, 628, 633)
Mode admittance 16 (303)
Mode conversion 16 (325)
Mode impedance 16 (303)
Mode, normal vibrational 18/1 (48)
Mode, resonance 18/2 (192–194)
Model of solid, Debye 19 (145)
Model of solid, Einstein 19 (242)
Modulation broadening 18/1 (117)
Modulation, high-frequency 18/1 (113)
Modulation, magnetic field 18/1 (110, 112, 113, 115)
Molar polarization 17 (107)
MO-LCAO 18/1 (321)
Molecular absorption 21 (320, 325)
Molecular crystal(s) 17 (98); 18/1 (68)
Molecular crystal, paramagnetic relaxation in a 18/1 (2)
Molecular field 18/2 (2–4)
— in antiferromagnet 18/2 (100)
— in ferrimagnet 18/2 (100)
— parameter 18/2 (2)
— theory, Néel 18/2 (95)
Molecular gas ions 21 (642)
Molecular ion-electron recombination, dissociative 21 (482)
Molecular ions, formation 21 (494)
Molecular ions in inert gases 21 (482)
Molecular orbit, one-electron 18/1 (378)
Molecular orbit, orthogonal 18/1 (321)
Molecular orbital 18/1 (409, 470)
Molecular orbital, delocalised 18/1 (95, 101, 126)
Molecular orbitals, method of 19 (98)
Molecular species forming negative ions 21 (462)
Molecular structure, relation to dipole moments 17 (148)
Molecule, diatomic 17 (42)
Molecule formation, pseudo- 21 (486)
Molecule, nonpolar 17 (9, 58, 62)

Molecule, polar 17 (9, 57, 87)
Molybdenite group 19 (389)
Moment 17 (3)
Moment, induced 17 (35, 38, 80)
Moment, magnetic 18/1 (94, 99)
Moment, mean 17 (19)
Moment, nuclear magnetic 18/1 (94)
Moment of a dipole antenna 16 (458, 461, 475, 476, 508, 516, 517, 521, 524, 527, 566, 620)
Moment, permanent 17 (9, 36, 38, 80)
Moments of atom groups 17 (151)
Momentum and energy exchange of electrons with plasma waves 22 (192)
Momentum transfer collision 21 (426)
Monatomic solids 19 (317, 339)
Monochromator 21 (316, 317, 318, 319, 350)
Monolayer 21 (358)
Moon 16 (569, 593, 609, 633)
Moriya coupling 18/2 (156)
Morphic effect 18/2 (317, 318)
Moss relation 20 (20)
Mossotti's model of a molecule 17 (9, 26)
Motional narrowing 18/1 (122, 125)
Motive 21 (19, 28f.)
Mott localization 18/1 (411)
Mott problem 18/1 (536)
Mott transition 18/1 (397, 413, 414, 543)
Mott transition, critical distance 18/1 (397)
Mott trapping 18/1 (399, 543)
Muf = maximum usable frequency 16 (559, 632ff.)
Multibeam oscilloscopes 23 (243)
Multi-level system 18/1 (65)
Multi-level systems, relaxation in 18/1 (65)
Multimode propagation 16 (315)
Multi-nuclear interaction 18/1 (104)
Multiple period method 23 (307, 310)
Multiplet, 4*f* 18/1 (464)
Multiplication 22 (81, 102, 103, 197)
Multipole 16 (5)
Multipole field 16 (86, 87, 121)
Multipole, linear 16 (87)
Multi-position action controller 23 (493)
Multi-spin process 18/1 (74, 75)
Multitrace oscilloscope 23 (243)
Multi-variable control 23 (518)
Mutual inductance 16 (196)

^{14}N splitting constant 18/1 (166)
NaCl-type compound, metallic 18/1 (505)
Naphthalen triplet 18/1 (207, 224)
Narrow tubes 22 (62)
Narrowing, exchange 18/1 (123, 125)
Narrowing, motional 18/1 (122, 125)
National standard of time and frequency 23 (315, 325)
Natural line breadth 22 (211)
Néel molecular field theory 18/2 (95)
Néel temperature 18/2 (98, 104)
Néel theory of antiferromagnetism 18/2 (98)
Néel theory of ferrimagnetism 18/2 (96)
Néel two-sublattice model 18/2 (104, 140)
Néel wall 18/2 (394, 395, 401)
Negative feedback 23 (475)
Negative ions 21 (462, 468)
— by processes other than attachment 21 (463)
— created in gaseous phase 21 (460)
— creation by impact of positive ions on metal surfaces 21 (463)
— formation 21 (653)
— formation by positive ion impact on clean metal surfaces 21 (466)
— formation energy 21 (458)
— identified from oxide coated cathodes 21 (469)
— thermal liberation from surfaces 21 (468)
Negative thermions, liberated 21 (463)
Negative zones 22 (57)
Neon 22 (550)
Neper (unit of attenuation) 16 (321)
Nephelauxetic effect 18/1 (471, 504)
Nernst coefficient 20 (60)
Nernst effect 20 (60, 92)
Nernst-Einstein relation 20 (248, 324)
Network 16 (202, 204)
Network-forming ions in glass 20 (350)
Network-modifying ions 20 (350, 360, 379, 386)
Networks 20 (350, 359)
Neumann formula of mutual inductance 16 (196)

Neumann's boundary value problem 16 (26)
Neutrality condition 20 (38, 121)
Neutralization, structural 22 (233)
Neutron diffraction 17 (373, 384); 18/2 (140)
Neutron irradiation 19 (370)
Nickel 19 (38)
NiO 18/1 (268, 552)
Nitrobenzene 17 (62, 94, 95)
Nitrogen 22 (507)
Nitrogen ions 21 (643)
Nitromethane 17 (103)
Nodal shift technique 23 (114)
Noise 18/1 (109, 110); 22 (193)
Noise, detector 18/1 (109, 110)
Noise, excess crystal 18/1 (110)
Noise factor 23 (2, 76, 173)
Noise figure 23 (2)
Noise, flicker- 18/1 (109); 23 (1, 13, 28, 29, 57)
Noise, galactic 23 (18)
Noise measure 23 (2, 6, 26)
Noise measurements 22 (69)
Noise of electron-tubes 23 (178)
Noise of transistors 23 (26)
Noise resistance 23 (24)
Noise, shot- 23 (1, 13, 47)
Noise source 23 (8, 11–15, 174–177)
Noise temperature 22 (69); 23 (1–16, 172–177)
Noise voltage 22 (373)
Noise-influence on accuracy of frequency counting 23 (333)
Noisy amplifiers, equivalent circuit 23 (21)
Non-adiabatic approximation 18/1 (43, 52)
Non-Kramers exchange mechanism 18/1 (269)
Non-Kramers salt 18/1 (39, 40, 41)
— direct process in 18/1 (48)
— Raman process in 18/1 (41, 50)
Nonlinear effect 18/2 (242)
Nonlinear reactance elements 23 (39, 355)
Nonlinearity 17 (111)
Nonlinearity in control systems 23 (460, 488)
Nonlinear mixer 23 (348)
Non-orthogonality catastrophe 18/1 (290)
Non-orthogonality integral 18/1 (276, 278, 311, 323, 354, 359, 362)
Non-reciprocal device 18/2 (169)
Nonstationary states of current 16 (165)
Non-uniform fields 22 (40, 526)
Non-uniform form effect 18/2 (316)
Non-uniform symmetrical fields 22 (525)
Norbury's rule 19 (209)
Nordheim relation 19 (209)
Nordio-Pavan relation, Giacometti- 18/1 (159)
Normal coordinates 17 (11, 13, 17)
Normal distribution 16 (588, 589)
Normal spinel structure 18/2 (96)
Normal vibrations of the lattice 17 (184)
Normalization of eigenfunctions 16 (300)
Norton equivalent circuit 23 (76)
Notation for tensor components 17 (305)
n-p-n transistor 20 (169)
N-processes 19 (245)
N-processes, limiting angle for 19 (248)
N-type 19 (322)
n-type semiconductor 20 (5)
Nuclear electric quadrupole 18/1 (140)
Nuclear interaction, multi- 18/1 (104)
Nuclear magnetic moment 18/1 (94)
Nuclear magnetic resonance 18/1 (2, 119, 142); 23 (199, 214)
Nuclear magneton 18/1 (99)
Nuclear particles 19 (340)
Nuclear quadrupole relaxation 18/1 (82)
Nuclear resonance 17 (273, 300, 318, 377)
Nuclear spin 18/1 (99, 100)
— resonance 18/1 (9)
— wave 18/2 (258, 259, 260)
Nucleation 18/2 (345, 428)
Nucleation, field strength for 18/2 (429, 430, 497)
Nucleation, spontaneous 18/2 (426)
Nucleus 18/1 (99, 101)
Nucleus, copper 18/1 (100)
Nucleus, disappearence of 18/2 (432)
Nucleus, for reversal of magnetization 18/2 (427, 453)
Nucleus, manganese 18/1 (138)

Number of electrons released from a surface, *see* γ 21 (576)
Number of excitations per ionization 21 (415)
Numerical computation
— of fields 16 (150)
— of fields, convergence 16 (148–155)
— of fields in dielectrics 16 (153)
— of three-dimensional fields 16 (154)
— of two-dimensional fields 16 (149–154)
Numerical distance 16 (523)

O_2 bands 22 (119)
O_2 effect of small amounts 22 (510)
O_2 ion 18/1 (248)
Objects for FEM, atomic and molecular 21 (214, 228)
Oblong, current around it 16 (53, 95)
Obstacle gain 16 (537, 608)
Octahedral field 18/1 (129)
Octahedral site 18/2 (96)
Octahedral symmetry 18/1 (129)
Octahedron structure 17 (320, 382)
Octupole 16 (6, 87)
Odd-electron centre 18/1 (259)
Ohm (unit) 16 (14)
Ohm's law 16 (14); 19 (64, 137, 236, 312); 22 (326)
Ohm's law regime 22 (513, 520)
One-electron hyperfine constant 18/1 (149)
One-electron molecular orbit 18/1 (378)
One-electron transition integral 18/1 (330)
On-off-controller 23 (488, 508)
Onsager approximation 17 (83, 89, 95, 97)
Onsager correction 17 (371)
Onsager spherical model 17 (53, 59, 89, 95, 108, 112)
Onsager theory 17 (53, 83)
Onsager's reciprocity laws 22 (319)
Onsager's relation 19 (272, 274, 283)
Onset of ionization 21 (658)
Open band semiconductors 20 (241)
Open circuit voltage 16 (267)
Open loop 23 (459, 479)
Operator, annihilation 18/1 (38)
Operator equivalent 18/1 (466)
Optical branch 18/1 (37)
Optical constants 21 (350, 360, 361, 368)
Optical depth 22 (358)
Optical modes 17 (295, 372)
Optical phonon 18/1 (80, 81)
Optical pumping 23 (325)
Optical spectrum 18/1 (404)
Optical thick layer 22 (373)
Optical thickness of layer 22 (367)
Optical transitions 19 (321)
Orbach process, inverted 18/1 (46, 65)
Orbach process of relaxation 18/1 (43, 51)
Orbach process, transition probability for 18/1 (43)
Orbach relaxation process in rare-earth salts 18/1 (51)
Orbit interaction, spin- 18/1 (35, 462)
Orbit, one-electron molecular 18/1 (378)
Orbit, orthogonal molecular 18/1 (321)
Orbital angular momentum, quenching of the 18/1 (36)
Orbital, antibonding 18/1 (132)
Orbital, bonding 18/1 (132)
Orbital, *d* 18/1 (127, 128, 129)
Orbital, 3*d* 18/1 (127)
Orbital, delocalised 18/1 (99)
Orbital, delocalised molecular 18/1 (95, 101, 126)
Orbital, e_g 18/1 (128)
Orbital energy level 18/1 (129)
Orbital, 4*f* 18/1 (454)
Orbital level splitting 18/1 (129)
Orbital, nonbonding 18/1 (177)
Orbital, π 18/1 (127)
Orbital quenching 18/1 (408)
Orbital radius of RE 18/1 (452)
Orbital, *s* 18/1 (128)
Orbital, σ 18/1 (127)
Orbital splitting 18/1 (131)
Orbital, t_{2g} 18/1 (128)
Orbitals, equivalent 19 (105)
Orbitals, spatial distribution of 3*d* 18/1 (127)
Orbit-lattice interaction 18/1 (48)
Orbit-lattice potential 18/1 (48)
Order-disorder transition 16 (715)
Ordinary wave 16 (547)
Organic semiconductors 20 (242)
Ortho-dichlorobenzene 17 (152)
Orthoferrite 18/1 (498)
Orthogonality properties 16 (297, 299)

Orthorhombic antiferromagnet 18/2 (108)
Oscillation amplitude limit 22 (539)
Oscillations, forced 16 (466)
Oscillations, free 16 (466, 467)
Oscillator, anharmonic 17 (70)
Oscillator, harmonic 17 (12, 41, 80, 133)
Oscillator, marginal 23 (359)
Oscillator, phase-locked 23 (357, 340)
Oscillator strength 17 (137); 21 (313); 22 (211, 360)
Oscillator, virtual 17 (137)
Oscillators, coupled 17 (17, 42)
Oscillographic techniques 22 (140)
Oscilloscope 23 (239)
Oscilloscope comparisons for frequency measurement 23 (349)
Oscilloscope displays for time interval measurement 23 (368)
Oscilloscope, sampling- 23 (368)
Outgassing 21 (370, 371, 655)
Output range of a controller 23 (472)
Overhauser effect 18/1 (142, 143); 18/2 (261)
Overhauser technique 18/1 (142)
Overlap 18/1 (362)
Overlap integral 18/1 (276, 278, 311, 323, 354, 359, 362)
Overshoot in potential 22 (508, 511, 513)
Overvoltage 17 (182); 22 (42, 44f., 47, 50, 51, 470)
Oxide cathodes 21 (11, 114, 147)
Oxide coated thermionic cathode 19 (374)
Oxide film 22 (13)
Oxide layer 22 (25, 156, 163, 164)
Oxides 19 (370)
Oxygen, absorbed 19 (348, 373)
Oxygen displacement 17 (384)
Oxygen-octahedron ferroelectrics 17 (270)

Paraboloidal reflector 16 (502)
Para-dichlorobenzene 17 (152)
Parallel displacement 16 (444)
Parallel resonance in crystal vibrations 17 (298)
Parallel-plate line 23 (91)
Parallel-pumping in antiferromagnet 18/2 (250)
Parallel-pumping in ferrimagnet 18/2 (250)
Parallel-pumping in ferromagnet 18/2 (250)
Parallel-pumping of magnetoelastic mode 18/2 (251)
Parallel-pumping of magnetoelastic wave 18/2 (252)
Parallel-pumping threshold 18/2 (253)
Paramagnetic behaviour 16 (21)
Paramagnetic centre in silicon 18/1 (253)
Paramagnetic Curie temperature 18/2 (3)
Paramagnetic ion 18/1 (131)
Paramagnetic longitudinal relaxation 18/2 (223)
Paramagnetic relaxation 18/1 (2)
— by modulated exchange interaction 18/1 (71)
— in local electron centres 18/1 (78)
— in molecular crystals 18/1 (2)
— stereochemistry and 18/1 (68)
— Waller's mechanism of 18/1 (65, 70, 74, 77)
Paramagnetic resonance 18/1 (2, 8, 145)
— acoustic 18/1 (91)
— electron (E.P.R.) 18/1 (36)
— saturation experiment 18/1 (28)
Paramagnetic saturation 18/1 (47)
Paramagnetic susceptibility 18/1 (404)
Paramagnetic transition 18/2 (134)
Paramagnetism 18/1 (2); 19 (127)
Paramagnetism, collective 18/2 (476)
Paramagnetism, Langevin-Curie theory of 18/2 (2)
Paramagnetism, magnetostriction in 18/2 (339)
Paramagnetism, quasi- 18/2 (476)
Parameter, Bloch wall 18/2 (81)
Parameter, bottleneck 18/1 (86)
Parameter, *D*, splitting 18/1 (137)
Parameter, exchange stiffness 18/2 (81)
Parameter, *g* 18/1 (95)
Parameter, molecular field 18/2 (2)
Parameter, phonon 18/1 (63)
Parameter, relaxation 18/1 (21)
Parameter, saturation 18/1 (32, 86)
Parametric amplifier 23 (1, 37–65)
Parity 18/1 (30, 37, 47)
Partial potential 19 (228, 271)

Particle flow in configuration space 21 (405)
Particles, critical dimension of 18/2 (459)
Particles, group of 18/2 (457)
Particles, system of 18/2 (438)
Partition function 17 (378, 380, 382, 387); 22 (321, 334, 350)
Partition noise 23 (35)
Paschen curves 22 (23)
Paschen law 22 (10, 17, 462)
Paschen law, failure of carrier density 22 (463)
Patch areas 21 (29)
Patch effect 21 (24, 101, 123)
Patches on the surface 21 (350, 358, 363, 365)
Pauli exclusion principle 18/2 (5)
Pauli function 19 (47)
Pauli principle 19 (14, 47)
Pause times 22 (590)
Pavan relation, Giacometti-Nordio- 18/1 (159)
Peak current values 22 (605)
Pedersen rays 16 (631)
Peltier coefficient 20 (60)
Peltier effect 20 (60, 89)
Penning effect 21 (555); 22 (75, 76, 80, 85, 446, 465)
Penning gas 22 (46)
Penning mixture 21 (419)
Pentane 17 (73)
Perinaphthenyl radical 18/1 (162, 171)
Period measurement 23 (332)
Periodic potential 19 (20)
Periodic system (based on the electronic configuration of the elements) 19 (163)
Permanent magnet 16 (22)
Permeability 16 (18, 170, 286); 18/2 (342); 23 (88)
Permeability of vacuum 16 (18)
Permeability tensor 16 (287, 403); 18/2 (168, 170); 23 (151, 153)
Permeability tensor, antisymmetrical 16 (402, 403)
Permitted and forbidden processes 22 (80)
Permittivity 23 (88)
Permittivity, complex 16 (641)
Permittivity (dielectric constant) 16 (640)
Permittivity, instantaneous 16 (668)
Permittivity, static 16 (668)
Permittivity tensor 21 (386)
Permutation, factorable 18/1 (291)
Permutation group 18/1 (295, 303, 305, 313)
Permutation, intraatomic 18/1 (301)
Permutation, non-factorable 18/1 (291)
Perovskite 18/1 (253, 497)
Perovskite, adjacent layers 17 (317)
Perovskite, Mn 18/1 (498)
Perovskite, phase diagrams 17 (319)
Perovskite = calcium titanate 17 (268, 271, 313, 336)
Persistence of c.r.t. screen 23 (227, 245)
Perturbation operator, effective 18/1 (39)
Perturbation operator, Mattuck and Strandberg's effective 18/1 (56)
Perturbation theory 17 (189); 19 (240)
— Bogolyubov's 18/1 (322, 365)
— first-order 18/1 (66, 79)
— for transitions 19 (288)
— higher order 18/1 (74)
— Löwdin's 18/1 (344, 349, 361)
— second-order 18/1 (50, 66, 68, 80)
Phantastron 23 (245)
Phase change 17 (309)
Phase changer 16 (696)
Phase constant 16 (295, 688)
Phase detector 23 (302, 313)
Phase diagrams, antiferroelectrics 17 (321)
Phase function 16 (351)
Phase integral relation 16 (564, 617)
Phase measurement 23 (363)
Phase noise 23 (351, 356)
Phase shift 16 (450)
Phase shift determination 23 (349)
Phase space 21 (404)
Phase velocity 16 (214, 296, 382, 386, 393, 511, 524, 549, 607)
Phase-locked signals 23 (303)
Phases in tungsten trioxide 17 (317)
Phenol 17 (93)
Phonon(s) 18/1 (37); 18/2 (376); 20 (12)
Phonon bottleneck 18/1 (7, 46, 74, 86, 90)
Phonon coupling, spin- 18/1 (91)
Phonon, Debye distribution 18/1 (38)
Phonon density of states 18/1 (30)
Phonon drag 18/1 (430)

Phonon interaction, electron- 18/1 (398)
Phonon interaction, spin- 18/1 (84, 85)
Phonon, life-time of the 18/1 (86, 87)
Phonon, longitudinal 18/1 (82)
Phonon matrix element 18/1 (30, 38)
Phonon mode, local 18/1 (82)
Phonon, optical 18/1 (80, 81)
Phonon parameter 18/1 (63)
Phonon relaxation 18/1 (88)
Phonon relaxation process in rare-earth salts, two- 18/1 (50)
Phonon spectrum, real 18/1 (85)
Phonon, spontaneous emission of a 18/1 (46)
Phonon system 18/1 (4)
— relaxation in the 18/1 (80)
— spin- 18/1 (34)
— state functions of the spin- 18/1 (34)
Phonon, transverse 18/1 (82)
Phonon-phonon interaction- 18/1 (87)
Phonon-pumped instability 18/2 (253)
Phosphorescence, slow 19 (371)
Phosphors for c.r.t. 23 (246)
Phosphorus 19 (340)
Phosphorus, semiconducting 20 (235)
Photoabsorption bands 21 (320, 377)
Photoabsorption coefficient 21 (307, 312, 320, 358, 361)
Photoabsorption coefficient in molecular gases and vapors 21 (309, 337, 339)
Photoabsorption cross section 21 (313, 320, 507)
— in molecular gases and vapors 21 (308, 309)
Photoabsorption, results 21 (337)
Photocells 22 (140)
Photochemical reactions 21 (316)
Photochemical recording 23 (226)
Photoconduction 19 (321)
Photoconduction, enhanced 19 (370)
Photoconduction, *F*-band 19 (354)
Photoconduction in anthracene 19 (389)
Photoconduction in uncoloured crystals 19 (353)
Photoconduction, intrinsic 19 (322, 370)
Photoconduction, sensitivity 19 (332)
Photoconductivity 20 (123); 21 (380, 381)
Photoconductivity in binary compounds 19 (352)
Photoconductivity, negative 19 (383)
Photoconductivity, spectral dependence of 19 (335)
Photoconductivity, theories of 19 (317)
Photoconductor 20 (5)
Photocurrent, bimolecular form for the decay of 19 (319)
Photocurrent decay 19 (325)
Photocurrent, quenching of the 19 (341)
Photocurrent stimulation 19 (341)
Photocurrents from activated and unactivated BaO 21 (381)
Photodetachment 21 (456)
Photodielectric effect in solids 19 (355, 391)
Photodielectric techniques 19 (384)
Photodiffusion 20 (119)
Photodissociation 21 (306, 333)
Photoeffect in *p-n* junctions 20 (161)
Photoelectric action in the gas 22 (517)
Photoelectric amplifier 19 (150)
Photoelectric current 22 (2, 202)
Photoelectric effect 22 (14, 26, 44, 80, 81, 225)
Photoelectric efficiency 22 (33)
Photoelectric emission 21 (305, 306, 380, 620); 22 (3, 20, 25, 98, 447)
Photoelectric emission, complete 21 (345)
Photoelectric phenomena, survey of 21 (342)
Photoelectric sensitivity 22 (30)
Photoelectric yield 21 (343, 345, 350–355)
Photoelectric yield, spectral distribution 21 (345, 350, 368, 369, 370, 372)
Photo-electro-magnetic effect 19 (339)
Photoelectron energy distribution 21 (343, 345, 356–360, 369, 372)
Photoelectrons, mean energy of 21 (594)
Photoelements 20 (161)
Photoemission, *see* Photoelectric emission
Photoemission from metals 21 (654)
Photoemission, temperature dependence 21 (369)
Photoemission threshold 21 (342–345, 350, 355, 369, 370, 371, 379)

Photo-excitation cross section 21 (557)
Photogalvanomagnetic effect 20 (142)
Photographic measurements 22 (198, 585)
Photographic recording 23 (226, 230, 248, 261)
Photoionization 21 (305–308, 311, 316, 321, 324, 325, 329–333, 336, 659); 22 (7, 14, 20, 28, 35, 36, 37, 39, 77, 80)
Photoionization cross sections 21 (310, 312, 320, 321, 331, 332, 335)
Photoionization efficiency 21 (314, 318, 331, 337)
Photoionization in air 22 (481)
Photoionization in gases 21 (306); 22 (449, 480, 481, 511, 515)
Photoionization in monatomic gases and vapors 21 (323, 337)
Photoionization of alkali vapors 21 (307, 317)
Photoionization probability 21 (507)
Photoionization, results 21 (337)
Photomultiplier 21 (315, 318, 319, 481); 22 (140)
Photon absorption 22 (225)
Photons 21 (577, 580, 583); 22 (9, 14, 33, 35, 37)
Photons of low energy 21 (602)
Phototransistor 19 (343, 346); 20 (171)
Phototransistor, *P-N* junction 19 (346)
Phototransistor, point contact 19 (346)
Photovoltaic effect in CdS 19 (368)
Photovoltaic effect (photovoltage) 19 (316, 332)
Photovoltaic effect, theory of the 19 (332)
Photoyield changes in SrO induced by ultraviolet irradiation 21 (382)
Phthalocyanine, copper 18/1 (106)
π orbital 18/1 (127)
Pierce oscillator 23 (320)
Piezocaloric coefficient 17 (287, 347, 354)
Piezoelectric coefficient
— adiabatic 17 (349)
— and piezoelectric crystal classes 17 (290)
— measurement 17 (299)
— transverse, longitudinal and shear effect 17 (288)
— values 17 (327, 363)
Piezoelectric effect, converse 17 (265)
Piezoelectric effect, quadratic 17 (359)
Piezoelectric equations linear 17 (297, 347)
Piezoelectricity 23 (317)
Piezoelectricity as tensor 17 (304)
Piezoelectricity in different crystal classes 17 (289)
Piezoelectricity, phenomenological theory 17 (340, 363)
Piezoelectricity, technical applications 17 (281)
Piezoelectricity, units and notations 17 (284)
Piezomagnetic effect 18/2 (338)
Piezomagnetism 18/2 (168)
Pilot streamer 22 (590, 621)
p-i-n rectifier 20 (172)
Pinch effect 22 (175)
Pipe discharge 22 (181)
π-radical 18/1 (151)
π-radical in irradiated organic crystals 18/1 (187)
π-radical, inorganic 18/1 (190)
π-σ correlation 18/1 (164)
π-σ interaction 18/1 (163)
Planck's radiation law 23 (90, 375, 441, 443)
Plane wave method, augmented 19 (30)
Plane waves, orthogonalized 19 (30, 36)
Plasma 22 (256)
Plasma balance 21 (402, 414, 440)
Plasma beam 22 (400)
Plasma beam, cathodic 22 (409)
Plasma contraction, theory 22 (416)
Plasma diagnostics 23 (161)
Plasma dynamics 22 (402)
Plasma flow 22 (396, 401)
Plasma flow with supersonic speed 22 (403)
Plasma frequency 16 (546); 21 (386)
Plasma interaction 22 (314)
Plasma, magnetic properties of a 22 (175)
Plasma, neutral 22 (114)
Plasma oscillations 19 (227); 22 (66, 132, 142, 144, 191, 356)
Plasma, quasi-neutral 22 (120)
Plasma sphere 22 (390)
Plasma, thermal 22 (305)
Plasma wavelength 16 (586)
Plastic flow 19 (382)
Plate capacitor 16 (42)
Plate capacitor, edge stray field 16 (70)

Plate condenser 17 (171)
p-n junction(s) 20 (145)
— field emission 20 (156)
— impactionisation 20 (156)
— photoeffect 20 (161)
p-n rectifier 20 (151)
Pneumatic controller 23 (477)
p-n-p-n transistor 20 (171)
Point cathode 21 (179, 192, 201, 202)
Point charge 16 (2)
— two equal 16 (85)
— two opposite 16 (84)
Point contact 20 (202)
Point dipole 17 (3)
Point lattice 16 (149–155)
Point-discharge methods 22 (598)
Pointlike discharge, negative 22 (526)
Point-to-plane discharge
— in N_2, H_2 and Ar with 1% O_2 22 (508, 510)
— negative, in air 22 (518)
— positive 22 (524)
Point-to-point gap 22 (526)
Poisson equation 16 (4); 22 (327)
Poisson integral 16 (55)
Polar aurora 22 (224, 242)
Polar blackouts 16 (636)
Polar dielectrics 16 (712)
Polar liquid of low-viscosity 16 (720)
Polar molecules 17 (75, 80)
Polar molecules in solids 17 (103)
Polar state 18/1 (282, 293)
Polarisation effects in ionic conductivity 20 (318)
Polarisation energy 20 (264, 267)
Polarizabilities, principal 17 (9, 35)
Polarizability 17 (9, 45)
Polarizability, electronic 17 (315, 370, 381, 386)
Polarizability, ionic 17 (380, 385)
Polarizability, mean 17 (35)
Polarizability, quantummechanical formula 17 (20)
Polarizability tensor 17 (9)
Polarizability variation with density 17 (70)
Polarizability variation with frequency 17 (10)
Polarizable adsorbed layers 21 (107)
Polarization 16 (439, 526, 669); 22 (223)
Polarization, atomic 16 (711); 17 (114)
Polarization charge 16 (8)
Polarization, conduction, electron 18/1 (433, 490)
Polarization, dielectric 17 (5)
Polarization, differential 16 (707)
Polarization, electronic 16 (710, 712); 17 (114)
Polarization, elliptic 16 (567)
Polarization force 21 (427)
Polarization, ideal molar 17 (107)
Polarization index 16 (567)
Polarization, interfacial 17 (130)
Polarization, molecular 16 (712)
Polarization of a dielectric 16 (7, 18)
Polarization, orientational 17 (114)
Polarization selectivity of photoemission 21 (350, 362, 368)
Polarization, spin 18/1 (174, 311, 352, 380)
Polarization, spontaneous 17 (51, 269, 282, 286, 325, 327, 328, 331, 335)
Polarization, volume density 16 (167)
Polarized light, effect on photoemission 21 (360f.)
Polaron 18/1 (398)
Polaron, large 18/1 (543)
Polaron, magnetic 18/1 (399, 419, 421)
Polaron, small 18/1 (543)
Polder tensor 18/2 (169)
Poley's method for liquids 16 (692)
Polycrystalline samples 19 (140)
Polygon mapping 16 (71–79)
Polygon, regular 16 (78)
Polygon with truncated corners 16 (77)
Polymorphism 19 (93)
Polynomial, harmonic 16 (97)
Polyvalent metals 19 (261)
Poole relation 20 (355, 366)
Population 22 (209)
Population density unbalance 23 (321)
Pore conductivity 21 (91)
Position of the *d*-levels 18/1 (546)
Positive hole 19 (321)
Positive ions 22 (13)
Post deflection acceleration 23 (240)
Potassium borate glasses 20 (359)
Potassium chloride 20 (311, 332)
Potassium niobate 17 (318, 331, 332, 336)
Potassium silicate glasses 20 (358, 361, 362)
Potassium tantalate 17 (318, 381)
Potential barrier 19 (330)

Potential barrier at surface 21 (183, 217, 222)
Potential, chemical 20 (32)
Potential distribution, radial 22 (172)
Potential, electrical 16 (2, 3, 14, 18, 172, 428)
Potential, electrochemical 20 (46, 110)
Potential exchange 18/1 (270, 330, 370)
Potential functions 16 (172)
Potential liberation 22 (447)
Potential, logarithmic 16 (32, 42)
Potential, magnetic 16 (18, 19, 20)
Potential minimum 21 (42, 43)
Potential of a circular
— cylinder of finite length 16 (111)
— diaphragm 16 (140, 141–143)
— disc 16 (124, 134, 140–142)
— ring 16 (102, 124, 132)
Potential of a cylinder in a homogeneous field 16 (43)
Potential of a dipole ring 16 (132)
Potential of a sphere in a homogeneous field 16 (87, 121)
Potential of a spherical segment 16 (133, 134, 141–143)
Potential of a spindle shaped body 16 (132)
Potential of a toroid 16 (133)
Potential of a wedge shaped region 16 (103)
Potential of a wire bent in a circle 16 (102, 124, 132)
Potential of an asymmetric ellipsoid 16 (147, 148)
Potential of an ellipsoid of revolution 16 (135–137, 140)
Potential of an elliptical disc 16 (147)
Potential of two metallic cylinders 16 (44, 45)
Potential of two spheres 16 (129–131)
Potential, reduced 16 (114, 118, 129, 135)
Potential, transverse 16 (470, 471)
Potential well 17 (217)
Potentials, thermodynamic 17 (285, 344, 346, 367)
Potentiometer according to Brown and Clothier 19 (153)
Potentiometer method 23 (409, 412)
Potentiometers 19 (146)
Potentiometers, automatic 19 (150)
Pound stabilizing circuit 23 (148, 168, 172)
Power breakdown 23 (92)
Power, complex 16 (290)
Power divider 23 (111–113)
Power equations 16 (177)
Power flow along a waveguide 16 (311)
Power gain 23 (6)
Power maximum 16 (266)
Power series expansion of the potential 16 (97, 107, 108)
Power spectrum 16 (590)
Poynting vector 16 (179, 289, 427, 437, 438, 459, 462, 480); 23 (87)
Pre-breakdown ionization, growth of 22 (30)
Precision measurements 16 (649)
Preisach model 18/2 (523, 530)
Prescaler 23 (335)
Pressure broadening 22 (362)
Pressure coefficient, average 19 (193)
Pressure, electrostatic 16 (13)
Pressure gradient 21 (390)
Pressure, hydrostatic 17 (288, 289, 325, 327, 356, 361, 366, 369)
Pressure of the electron gas 22 (174)
Pressure, two-dimensional 17 (343, 366)
Pressure-sensitive paper 23 (224)
Primary action 22 (446)
Primary and secondary processes 19 (320)
Primary electrons 21 (232); 22 (111)
Primary electrons, fast 22 (147, 161)
Primary ionization 22 (4)
Primary pulse 22 (522)
Principal mode of a waveguide field 16 (292, 318, 332, 336)
Principle of detailed balance 18/1 (31)
Printer 23 (223, 280, 342)
Prism 23 (113, 126–131)
Probe antenna 16 (267–283)
Probe, area of the 22 (63, 64)
Probe curves 22 (65)
Probe, hot 22 (62, 67)
Probe in bridge circuit 22 (66)
Probe in characteristics in a magnetic field 22 (173)
Probes 22 (65, 82, 150, 381)
Probes, cylindrical 22 (66)
Probes, screened 22 (67, 132)
Probes, special types of 22 (67)

Problem of 3 centres and 4 electrons
— non-orthogonal orbitals 18/1 (359)
— orthogonal orbitals 18/1 (336)
Production function 22 (217)
Production of ions 21 (622)
Production of photons 21 (626)
Projection operator 18/1 (325)
Propagating wave 16 (295, 302, 315)
Propagation constant, complex 16 (688)
Propagation delay 23 (299)
Propagation parameter 16 (295, 297, 367, 369)
Propagation speed 17 (7, 145)
Proper variables 22 (536)
Proportional action controller 23 (472)
Proportional plus reset controller 23 (474)
Proportionality, assumption of 21 (398)
Propyl ether 17 (46, 47)
Proton, α- 18/1 (187)
Proton, β- 18/1 (188)
Proton magnetic resonance 23 (198)
Proton transfer 18/1 (202)
Pseudo Brewster angle 16 (525, 527, 618)
Pseudo *H*- and *E*-waves 16 (350)
p-s-n rectifier 20 (172)
P-type 19 (322)
p-type semiconductor 20 (5)
Pulse duration modulation 23 (259)
Pulse methods 21 (198, 200, 208)
Pulse saturation 18/1 (10, 12)
Pulse saturation technique 18/1 (10, 12, 32)
Pulse, secondary 22 (522)
Pulse traveling along a line 16 (230)
Pump source 23 (38)
Pump technique, parallel 18/2 (241)
Pumping, longitudinal 18/2 (376)
Pumping, parallel 18/2 (240, 249)
Pumping, perpendicular 18/2 (249)
Punched card 23 (273, 276)
Punched tape 23 (273)
Punched tape reader 23 (275)
Purcell's reflection method 23 (181)
Pyroelectric coefficient 17 (347)
Pyroelectric crystals 17 (349)
Pyroelectric effect 17 (331)
Pyrometer 23 (379, 441, 446–455)

Q-factor, *see* Quality constant
Q-factor, definition of 23 (91)
Q-factor, measurement of 23 (120–122)
Q-factor of spectral line 23 (146)
Q-meter 16 (662)
Quadripole radiation 22 (210)
Quadrupolar selection rule 18/1 (57)
Quadrupole 16 (5, 29); 17 (5, 8)
Quadrupole, electronic 18/1 (136)
Quadrupole interaction, electric 18/1 (135, 136)
Quadrupole interaction, nuclear 18/1 (34, 35, 72)
Quadrupole, linear 16 (6)
Quadrupole moment 16 (6)
Quadrupole moment, electric 18/1 (135, 141)
Quadrupole, nuclear electric 18/1 (140)
Quadrupole relaxation, nuclear 18/1 (81, 82)
Quadrupole transition, electric 18/1 (37)
Quality-constant of a cavity resonator 16 (407, 408, 672)
Quantization of electronic orbits 19 (307)
Quantization second 18/1 (294, 322, 369)
Quartz 17 (253, 279, 310)
Quartz crystal standard 18/1 (115)
Quartz resonators 23 (301–320)
Quartzcrystal thermometer 23 (401)
Quasi Fermi level 20 (110)
Quasi Fermi potential 20 (110)
Quasi maximum field 16 (587, 628)
Quasi minimum field 16 (587)
Quasi-neutrality 22 (312)
Quasiparamagnetism 18/2 (476)
Quasiparticle, magnetic 18/1 (368)
Quasi Raman process 18/1 (1)
Quasistationary states of current 16 (165)
Quasi-transverse propagation 16 (568)
Quasi-*T*-wave 16 (326)
Quenching of the orbital angular momentum 18/1 (36)
Quenching spectra 19 (345)

Radar antenna 16 (480)
Radar cross-section 16 (480, 501, 502)
Radar equation 16 (480)
Radial flow in velocity space 21 (405)

Radial integral 18/1 (460)
Radial modes 16 (605); 17 (301)
Radiating efficiency of an antenna 16 (247)
Radiating system, definition 16 (234)
Radiation, blockaded 22 (209, 217, 236)
Radiation condition of Sommerfeld 16 (176, 466, 485)
Radiation, continuous 22 (354)
Radiation density 23 (442)
Radiation diffusion 22 (365)
Radiation effects 22 (609)
Radiation efficiency 22 (388)
Radiation, fluctuation in 19 (338)
Radiation from transmission lines 16 (231)
Radiation, high energy 19 (353)
Radiation law 23 (90, 375, 441, 443)
Radiation pattern, antenna 23 (180)
Radiation pressure 23 (104)
Radiation pyrometer 23 (379, 441, 455)
Radiation resistance 16 (194, 231, 232, 454, 456–458, 462)
Radiation screen 23 (433)
Radiation standard 22 (368)
Radiation terms 16 (458, 517)
Radiation theory 22 (9, 202, 210)
Radiation-induced centre 18/1 (254)
Radiation-induced defects in solids 18/1 (145)
Radiative capture 21 (486)
Radiative recombination 21 (480, 484)
Radiative recombination of electrons 22 (229)
Radiator, isotropic 16 (478)
Radical, allyl 18/1 (189)
Radical, anion 18/1 (155)
Radical, cation 18/1 (155)
Radical, free 18/1 (77, 94, 95, 98, 116, 119, 126, 151)
Radical g-value, free 18/1 (126)
Radical in non-oriented irradiated solids 18/1 (192)
Radical, malonic acid 18/1 (165, 187)
Radical, monocyclic 18/1 (177)
Radical, perinaphthenyl 18/1 (162, 171)
Radical, π 18/1 (151)
Radical, π, in irradiated organic crystals 18/1 (187)
Radical, π, inorganic 18/1 (190)
Radical, semiquinone 18/1 (104)
Radical, σ 18/1 (204)
Radical, stable free 18/1 (182)
Radical transient, free- 18/1 (142)
Radical, vinyl 18/1 (205)
Radio echoes from the ionosphere 21 (478, 492)
Radio interferometer 16 (543)
Radio waves 22 (243)
Radius of electron shells 18/1 (389)
Raether's criterion 22 (29)
Rainbow expansion 16 (603, 604)
Rainbow integral of Airy 16 (449, 450)
Raman process(es)
— in non-Kramers salts 18/1 (50)
— of relaxation 18/1 (41, 53, 121)
— quasi 18/1 (1)
— transition probability for 18/1 (41)
Raman relaxation process in rare-earth salts 18/1 (50)
Ramp function response of a control system 23 (462)
Ramsauer cross-section 22 (348, 378)
Ramsauer effect 22 (72, 73, 132)
Ramsauer free path 21 (447)
Ramsauer scattering cross section 21 (447)
Ramsauer-Townsend effect 21 (544)
Random energy 21 (393, 400, 424 f.)
Random flight 21 (395)
Random impurities 19 (82)
Random velocity 21 (509)
Random walks 20 (326)
Rankin scale 23 (378)
Rare gas ions 21 (648, 649, 652)
— of low energy 21 (643)
— on molybdenum 21 (644)
Rare gas molecular ions 21 (622)
Rare-earth (RE)-chalcogenide 18/1 (553)
Rasch-Hinrichsen relation 20 (352, 356)
Rate action controller 23 (473)
Ratemeter 23 (345)
Ratio measurement of frequencies 23 (334)
Ray trajectory 16 (441, 444, 447, 509, 515, 547, 557, 561, 568)
Rayleigh and Jeans, radiation law 23 (443)
Rayleigh approximation 16 (508)
Rayleigh criterion 16 (536, 540)
Rayleigh distance 23 (181, 182)
Rayleigh distribution 16 (588, 593)
Rayleigh effect 22 (224, 252)

Rayleigh formula 16 (185)
Rayleigh-Carson reciprocal theorem 16 (181)
R-centre 18/1 (146, 236)
RE chemistry 18/1 (393)
RE, crystal spectrum in 18/1 (463)
RE, ground state of 4*f*-electrons in 18/1 (455)
RE, ionization energy for 18/1 (458)
RE, magnetic moment of 18/1 (490)
RE, melting point of 18/1 (456)
RE metals, band structure 18/1 (472)
RE, orbital radius of 18/1 (452)
RE salts
— crystal field in 18/1 (47)
— direct relaxation process in 18/1 (48)
— Orbach relaxation process in 18/1 (51)
— Raman relaxation process in 18/1 (50)
— two-phonon relaxation process in 18/1 (50)
RE^{3+} with pnictides, III-V-compounds of 18/1 (493)
Reactance matrix 16 (313, 397)
Reaction 16 (483)
Reaction field of Onsager 17 (108)
Reaction kinetics 20 (43 ff.)
Reactive waves 16 (464)
Reactor, temperature control of a 23 (469, 506)
Receiving antenna 16 (262–270)
— center loaded cylindrical 16 (262, 264)
— equivalent circuit 16 (264, 268)
Receiving circuit 16 (261)
Reciprocal and symmetrical network 16 (204)
Reciprocal lattice 19 (10, 230)
Reciprocity theorem(s) 16 (180, 398, 401, 476, 477, 479, 482, 546); 17 (22, 23, 77, 92, 97)
Recoil 21 (408, 410)
Recombination 19 (317); 21 (412); 22 (19, 36, 78, 80, 107, 112, 136, 139, 245)
Recombination at higher pressures 21 (487)
Recombination at very low pressure 21 (501)
Recombination center 20 (21, 110 ff.)
Recombination coefficient 19 (328); 21 (471); 22 (253)
— measurement of 21 (477)
— tables and curves 21 (483)
— theories of the 21 (483)
— variation of, with pressure and temperature 21 (483)
Recombination, columnar 21 (487, 497)
Recombination continuum 21 (480)
Recombination, criterion for 22 (227)
Recombination, critique of its ideal theory 21 (472)
Recombination, definition of 21 (471)
Recombination, dielectronic 21 (494); 22 (231)
Recombination, dissociative 21 (485, 492, 493, 494)
Recombination, generalized relations in its theory 21 (472)
Recombination, generalized statistical theory of 21 (487)
Recombination, generalized theory of 21 (485)
Recombination, initial 21 (487, 497)
Recombination of charge carriers 20 (110 ff.)
Recombination, preferential 21 (487, 495)
Recombination process, summary 21 (488)
Recombination rate 17 (225)
Recombination, small ion, large ion 21 (495)
Recombination surface 22 (227)
Recombination theories 19 (330)
Recombination theory, bimolecular 19 (389)
Recombination, three body electron 22 (232)
Recombination, volume 22 (227)
Recorder, analogue 23 (341)
Recorder, chopped bar 23 (224, 264)
Recorder, dielectric 23 (228)
Recorder, digital 23 (341)
Recorder, force balance 23 (256)
Recorder, helix balance 23 (265)
Recorder, level 23 (257)
Recorder, multichannel 23 (252, 265)
Recorder, potentiometric 23 (251, 257)
Recorder, segmental 23 (266)
Recorder, sweep balance 23 (264, 265)
Recording material 18/2 (441)

Recording, optical 23 (230)
Recording tape 18/2 (441)
Recovery effect over inhomogeneous soil 16 (541, 543)
Recovery processes 19 (203)
Recovery time 23 (156, 157, 160, 161)
Rectifier, diffusion theory 20 (191)
Rectifier, diode theory 20 (196)
Red shift, magnetic 18/1 (524)
— Cr spinels 18/1 (487)
— Eu chalcogenides 18/1 (483)
RE-elements, magnetic interaction in 18/1 (491)
Re-entrant cavity 16 (665)
Reference input 23 (457)
Reference load 23 (12)
Reference phase 17 (325, 340, 352)
Reflectance 23 (441)
Reflection at a dielectric boundary surface 16 (46, 47, 86, 688–690)
Reflection at a dielectric circular cylinder 16 (47)
Reflection at a metallic circular cylinder 16 (46)
Reflection coefficient 17 (147); 21 (640)
— for electrons 19 (311)
— of a line terminal 16 (212, 314)
Reflection, diffuse 16 (537)
Reflection effect 21 (18, 24)
Reflection factor 23 (91)
Reflection in a converging magnetic field 21 (387)
Reflection, incomplete 16 (46, 47, 86)
Reflection, internal, in a stratified medium 16 (568, 603, 605, 612)
Reflection, metallic 21 (368)
Reflection, multiple 16 (46, 85, 86)
Reflection of positive ions 21 (648)
Reflection of slow electrons 21 (107)
Reflection, partial 16 (553, 555, 557)
Reflectometer 23 (118)
Reflector of electromagnetic waves 16 (495, 496, 499)
Refraction index 16 (444, 445, 507–511, 515, 578, 612, 615, 618, 620); 17 (7, 92, 115, 370, 382); 19 (338); 20 (210); 22 (376)
Refractive index, real and complex 16 (642)
Refraction, substandard 16 (627)
Regenerative 22 (450)
Region, simply connected 16 (293)
RE-hydride 18/1 (512)
Relative error 19 (141)
Relativistic theory 22 (130)
Relaxation by modulated exchange interaction, paramagnetic 18/1 (71)
Relaxation curve, Casimir and DuPré 18/1 (7)
Relaxation, Debye 18/1 (21, 42)
Relaxation, dielectric, Debye theory 17 (125)
Relaxation, direct process of 18/1 (39)
Relaxation distance 21 (520)
Relaxation due to spin-orbit coupling 18/1 (47)
Relaxation, electron 18/2 (257)
Relaxation, formal theory of 18/2 (237)
Relaxation function 18/1 (14, 15)
Relaxation in local electron centres, paramagnetic 18/1 (78)
Relaxation in molecular crystals, paramagnetic 18/1 (2)
Relaxation in multilevel systems 18/1 (65)
Relaxation in S-state ions 18/1 (64)
Relaxation in the phonon system 18/1 (80)
Relaxation, Maxwell-Boltzmann 18/1 (30)
Relaxation mechanism 18/2 (220)
Relaxation method 16 (148)
Relaxation, nuclear quadrupole 18/1 (81, 82)
Relaxation of a magnon 18/2 (198)
Relaxation of ferromagnetic resonance 18/2 (173, 177, 220, 243, 373)
Relaxation of impurities, transverse 18/2 (224)
Relaxation operator, effective spin-lattice 18/1 (56)
Relaxation, Orbach process of 18/1 (43, 51)
Relaxation, paramagnetic 18/1 (2)
Relaxation, paramagnetic longitudinal 18/2 (223)
Relaxation parameter 18/1 (21)
Relaxation phenomenon 18/1 (120)
Relaxation, phonon 18/1 (88)
Relaxation polarization 16 (668)
Relaxation process in rare-earth salts
— direct 18/1 (48)
— Orbach 18/1 (48)

— Raman 18/1 (50)
— two-phonon 18/1 (50)
Relaxation process, spin-wave 18/2 (174)
Relaxation processes 17 (119)
Relaxation, Raman process of 18/1 (41, 53, 121)
Relaxation scheme 18/2 (238)
Relaxation, spin lattice 18/1 (1, 30, 56, 59, 78, 120); 18/2 (174, 206)
Relaxation, spin-spin 18/1 (1, 9, 13, 72)
Relaxation, stereochemistry and paramagnetic 18/1 (68)
Relaxation time(s) 16 (666); 17 (120, 221, 226); 20 (52, 375, 376); 21 (393); 22 (209)
— anisotropy in 18/1 (63)
— dielectric 20 (108, 126)
— distribution 16 (670); 17 (121)
— logarithmico-normal distribution 17 (133)
— longitudinal 18/1 (8)
— measurement of 18/2 (239); 23 (157)
— single 17 (119, 140)
— spin-lattice 18/1 (6, 9, 120, 121, 132); 23 (156, 158)
— transversal 18/1 (8)
Relaxation, valence-exchange 18/2 (222)
Relaxation, Waller's mechanism of paramagnetic 18/1 (65, 70, 74, 77)
Relay II satellite 23 (301)
Remanence 18/2 (342, 465, 467, 507)
Remanence, a.c. 18/2 (509)
Remanence, analysis of 18/2 (513)
Remanence, anhysteretic 18/2 (509, 538)
Remanence, coercive force of the 18/2 (511)
Remanence curve 18/2 (507, 521, 533, 537, 541)
Remanence, d.c. 18/2 (509)
Remanence, particle size dependence of 18/2 (467)
Remanence saturation 18/2 (507)
Remanence, time-dependent 18/2 (466)
Remanent anhysteretic initial susceptibility 18/2 (534)
Remanent polarization 16 (21)
RE-pnictide 18/1 (495)
Representation, irreducible 19 (16, 17, 18)
Repulsive potential energy between ions 20 (265)
RE-sesquioxide 18/1 (501)
Reset time of a controller 23 (474)
Residual resistance 19 (173)
Residue series 16 (182ff., 186, 604–608)
Resistance 16 (14); 19 (137, 156)
Resistance anomalies 19 (174)
Resistance, effective 22 (201)
Resistance, external 16 (194, 231 ff.)
Resistance lattice for the numerical integration of the Poisson equation 16 (149, 161, 162)
Resistance line 22 (189, 201)
Resistance matrix 16 (397)
Resistance measurement 23 (419–425)
Resistance, negative 22 (452, 453)
Resistance, pressure coefficient of 19 (191, 192)
Resistance thermometer 23 (393, 399)
Resistive termination 23 (11)
Resistive vane 23 (108)
Resistivities, temperature dependence of the ratio of axial 19 (174)
Resistivity 16 (15); 19 (137, 140, 156)
Resistivity after coldwork, recovery of electric 19 (202)
Resistivity at high temperatures 19 (266, 267)
Resistivity, effect of irradiation on electic 19 (201)
Resistivity, effect of plastic deformation on 19 (199)
Resistivity, effect of quenching on electric 19 (201)
Resistivity for a liquid metal 19 (181)
Resistivity increase caused by the defects 19 (205)
Resistivity increment due to impurities 19 (259, 270)
Resistivity, mean 19 (157)
Resistivity, mean temperature coefficient of 19 (157)
Resistivity of binary alloys, influence of coldwork on the 19 (216)
Resistivity of binary solution 19 (207)
Resistivity of ferromagnetic crystals 19 (189)
Resistivity of ferromagnetic metals 19 (186)

Resistivity of glass 20 (351, 366)
— dependence on composition 20 (361)
— dependence on temperature 20 (352)
Resistivity of intermediate phases of binary solid solutions 19 (217)
Resistivity of tungsten 19 (269)
Resistivity, temperature coefficient of 19 (138, 156)
Resistivity, variation of temperature coefficient of 19 (260)
Resistor with a positive temperature coefficient 23 (400)
Resonance absorption 17 (133)
Resonance, acoustic paramagnetic 18/1 (91)
Resonance, antiferromagnetic 18/2 (187)
Resonance bands 21 (319, 320, 321, 327, 328)
Resonance cavity 18/1 (109)
Resonance criteria 16 (672)
Resonance curve for a line 16 (220)
Resonance curve of a cavity resonator 16 (409)
Resonance, double 18/1 (142, 144)
Resonance, electron paramagnetic (E.P.R.) 18/1 (36)
Resonance, electron spin 18/1 (9, 94)
Resonance, exchange 18/2 (129, 160, 194)
Resonance, ferromagnetic 18/2 (173, 177, 220, 243, 373)
Resonance, high-power ferromagnetic 18/2 (243)
Resonance lines 22 (133)
Resonance, magnetoacoustic 18/2 (215)
Resonance method 17 (141, 148)
Resonance method by Wyman 17 (142)
Resonance mode 18/2 (192, 193, 194)
Resonance, nuclear magnetic 18/1 (2, 119, 142)
Resonance, nuclear spin 18/1 (9)
Resonance, paramagnetic 18/1 (2, 8, 145)
Resonance photons 22 (79)
Resonance radiation 21 (306); 22 (76, 112, 209, 225)
Resonance relaxation 18/2 (173)
Resonance relaxation, ferromagnetic 18/2 (173, 177, 220, 243, 373)
Resonance saturation 18/2 (243)
— effect 18/1 (124)
— experiment, paramagnetic 18/1 (28)
— technique 18/1 (10)
Resonance sharpness 16 (658)
Resonance, spin-wave 18/2 (178, 442)
Resonant and non-resonant technique 18/1 (76)
Resonant cavity,
— equivalent circuit 23 (120)
— measurements with 23 (135, 153, 159, 166, 182)
— r.f. fields in 23 (123)
— ring type 23 (155)
Resonant frequency of a cavity resonator 16 (407)
Resonant frequency variation of rochelle salt and KDP 17 (295)
Resonant length for a line 16 (220)
Resonator, cylindrical 16 (672)
RE-spectrum 18/1 (459)
Response time 19 (323)
Response time of thermometers 23 (434)
Restriction of a control system 23 (460)
Retarding fields, influence on thermionic emission 21 (22)
Retarding potential 21 (54, 196, 226, 344)
Reversal of magnetization 18/2 (96, 427)
Ribbon carrying an electric current 16 (53)
Ribbon lightning 22 (577)
Riccati equation 21 (415)
Richardson constant 21 (52, 145)
Richardson equations 21 (8, 16, 17, 128, 133)
Richardson work-function 21 (48, 52, 154)
Righi-Leduc coefficient 20 (60, 94)
Rigidity matrix 17 (335)
Ring discharge 22 (195)
Ring electrodes 22 (94)
Ring, hollow, circular cylindrical 16 (99–101, 103)
Roberts-von Hippel method 16 (690)
Rochelle salt 17 (342, 355, 377, 387, 389)

Rochelle salt (cont.)
— Curie points 17 (266)
— dielectric hysteresis loop 17 (265, 273)
— piezoelectric isotope effect 17 (266)
— twinning 17 (322)
Rock magnetism 18/2 (339)
Rocket lightning 22 (577, 597)
Root-branching 22 (580)
Rotating dipoles 17 (368, 380, 385)
Rotating mirror systems 23 (262)
Rotation, homogeneous 18/2 (455, 491)
Rotation, inhomogeneous 18/2 (493)
Rotation, intramolecular 17 (109, 152)
Rotation of the plane of polarization 16 (569, 594)
Rotation process 18/2 (344)
Rotation, quasihomogeneous 18/2 (494)
Rotational band 22 (370)
Rotational band spectra 22 (70)
Rotational freedom 17 (99, 103, 109, 152)
Rotational hysteresis 18/2 (447, 449, 522)
Rotational symmetric field, numerical computation 16 (155)
Roughness spectrum 16 (539)
Rowland coupling, Bloembergen- 18/1 (411)
Rubidium magnetometer 23 (201)
Rubidium vapor cell standard 23 (303, 327)
Ruderman-Kittel function 18/1 (434); 18/2 (11)
Running wave 16 (214)
Russel-Saunders coupling 18/1 (402)
Rutherford scattering 21 (434)
Rutile 18/1 (253)

s orbital 18/1 (128)
Saddlepoint approximation 16 (440, 444–447, 464, 501, 508–511, 523, 533, 535, 540, 548, 560, 581)
Sampling oscilloscope 23 (270)
Sampling theorem 23 (218, 220)
Satellite relay 23 (300)
Saturation 17 (51, 111); 18/1 (109, 121, 125, 143)
Saturation current 16 (12); 21 (505)
Saturation effect(s) 18/1 (124); 19 (317, 348)
Saturation effect, resonance 18/1 (124)
Saturation experiment 18/1 (9)
Saturation experiment, paramagnetic resonance 18/1 (28)
Saturation magnetization 18/2 (53, 346); 19 (120)
Saturation magnetization, temperature dependence of 18/2 (370)
Saturation moment of ferromagnetic alloys 19 (81)
Saturation, paramagnetic 18/1 (47)
Saturation parameter 18/1 (32, 86)
Saturation phenomena 21 (363, 364)
Saturation, pulse 18/1 (10, 12)
Saturation remanence 18/2 (507)
Saturation, steady state 18/1 (10)
Saturation technique 18/1 (31)
Saturation technique, pulse 18/1 (10, 12, 32)
Saturation technique, resonance 18/1 (10)
Saunders coupling, Russel- 18/1 (402)
Scalar potential, *see also* Potential 16 (172, 428)
Scale of turbulence 16 (578, 579)
Scatter fading 16 (588)
Scattering by dislocations 20 (67)
Scattering by imperfections 20 (65)
Scattering by ionized impurities 20 (65)
Scattering by neutral impurities 20 (67)
Scattering by spin fluctuations 18/1 (424)
Scattering coefficient 16 (583, 585)
Scattering, coherent 16 (538, 557, 558, 628, 636)
Scattering cross section 23 (180)
Scattering, dipole 18/2 (202)
Scattering, exchange 18/2 (200)
Scattering frequency 21 (410)
Scattering, incoherent 16 (540, 584, 621, 628, 636)
Scattering ionic 20 (55)
Scattering matrix 16 (222, 397); 19 (241)
— as function of scattering angle 19 (245)
— Bardeen's calculation of 19 (251)
— for polyvalent metals 19 (261)
Scattering of electrons 19 (238)
— by impurities 19 (258)
— effective potential for 19 (257)
Scattering of light, classical (Rayleigh) 21 (507)
Scattering of x-rays, thermal diffuse 19 (95)

Scattering thermal 20 (55, 63–67)
Schering bridge 16 (642)
Schmitt trigger 23 (311, 332, 346)
Schottky defects 20 (255, 259, 273, 319, 327, 341)
Schottky effect 21 (8, 9, 31)
Schottky effect, periodic deviations from the 21 (11, 33f., 100)
Schottky equation 21 (31, 32)
Schottky theory 21 (33, 100); 22 (120, 122, 124)
Schubweg 19 (318)
Schumann-Runge bands 21 (325)
Schumann-Runge dissociation continuum 21 (313)
Schwarz-Christoffel polygon mapping 16 (71–79)
Scintillation 16 (578)
Scintillation fading 16 (595)
Seasonal factor 16 (634)
Secondary action 22 (446)
Secondary burst 22 (522)
Secondary coefficient 22 (9)
Secondary currents 19 (381)
Secondary effect 21 (550, 576)
Secondary electron(s) 22 (23, 24, 78, 80)
— angular distribution 21 (275)
— effect of temperature on the yield 21 (262)
— emission from metals by positive ions 21 (638)
— emission of glasses 20 (390, 391)
— energy distribution of 21 (234, 235)
— range of primary absorption of the 21 (276)
— released by quanta 21 (627)
— theory of 21 (282, 650)
— time of liberation of the 21 (276)
— yield 21 (249, 250, 254)
— yield and periodic system 21 (265)
— yield, dependence on angle of incidence 21 (259)
— yield, dependence on work function 21 (257)
— yield from oxidized metals and alloys 21 (269)
— yield from photocathodes and oxide cathodes 21 (267)
— yield measurements with pure metals 21 (266)
— yield measurements with semiconductors and insulators 21 (270)
Secondary emission 21 (232)
Secondary emission coefficient, apparent 21 (580)
Secondary ionization 22 (5, 11)
Secondary potential 16 (86, 141)
Secondary processes 22 (6, 7, 14, 23, 28, 39, 41, 49, 78, 447, 503)
Sectorial horn 16 (348, 493)
Sectorial horn waveguide, pseudo *H*- and *E*-wave 16 (350)
Seebeck effect 19 (273)
Seeliger's rule 22 (60)
Selection rules 18/1 (459); 19 (321)
Selectivity of a resonator 19 (309)
Selenides and tellurides 19 (317, 360)
Selenium 19 (316, 340)
Selenium barrier layer photovoltaic cell 19 (350)
Selenium on iron 19 (316)
Selenium photoelement 23 (449)
Self absorption 22 (225, 358)
Self impedance 16 (194, 196, 197, 198, 244)
Self inductance 16 (23, 24, 42, 196)
Self inversion 22 (373)
Self modulation 16 (576)
Self repulsion 22 (77, 80)
Self stabilization 22 (276)
Self sustaining current 22 (450)
Self sustaining threshold condition of a discharge 22 (450)
Self sustaining Townsend discharge 22 (54, 449, 452)
Self-conjugate fields 16 (427)
Self-consistent field 19 (1, 47, 53, 78, 81)
Self-diffusion 20 (324)
Self-gyro-interaction 16 (577)
Semiconducting compounds 20 (7, 16, 19, 236–242)
Semiconducting elements 20 (6, 16, 18, 230–236)
Semiconductor(s) 16 (15); 18/1 (95); 19 (65, 78, 324); 21 (343, 378, 380, 381)
Semiconductor, amphoteric 20 (19)
Semiconductor, anisotropic 20 (95ff.)
Semiconductors, defects in 18/1 (253)
Semiconductor, definition 20 (2)
Semiconductor, degenerate 20 (6)
Semiconductor diode for temperature measurement 23 (400)

Semiconductor, isotropic theory 20 (12)
Semiconductor, magnetic 18/1 (389, 396)
Semiconductor, mixed 20 (5, 40)
Semiconductor, nondegenerate 20 (32, 36)
Semiconductor, *n*-type 20 (5)
Semiconductor of intrinsic type 19 (321)
Semiconductor, *p*-type 20 (5)
Semiconductor surface, free 20 (177)
Semiconductor, valency controlled 20 (19)
Semiconductor-metal contact 19 (332)
Semiquinone 18/1 (180, 202)
Semiquinone, benzo- 18/1 (104)
Semiquinone radical 18/1 (104)
Sensitivity 19 (141, 144, 148)
Sensitivity, photographic 19 (382)
Sensitization of alkali metals 21 (373, 377)
Separation of the Laplace equation in
— Cartesian coordinates 16 (38, 93, 94)
— coordinates of the circular cylinder 16 (98)
— coordinates of the elliptical and hyperbolical cylinder 16 (111, 112)
— coordinates of the parabolic cylinder 16 (112)
— spherical polar coordinates 16 (119)
Septate coaxial line 16 (345)
Series limit 21 (311)
Series resonance in crystal vibrations 17 (298)
Series slot 16 (490)
Setpoint 23 (458)
Setpoint change response 23 (480)
Shadow effects 22 (59)
Shallow acceptor impurity 18/1 (254)
Shallow donor impurity 18/1 (253)
Shape anisotropy 18/2 (283, 451)
Shear modes 17 (301)
Shear modulus 18/2 (355)
Shear, spontaneous 17 (323)
Sheath model 16 (390)
Shielded loop antenna 16 (274–278)
Shielded loop with two gaps 16 (278)
Shielding by a dielectric cylinder 16 (81)
Shielding, electromagnetic 16 (186)
Shielding factor 16 (123)
Shift, chemical 18/1 (169)
Shock 22 (377)
Shock excitation 23 (359, 370)
Shock wave 22 (253)
Short antenna 16 (267, 268)
Short circuit, ideal 16 (226)
Shot noise 23 (315, 325)
Shunt slot 16 (490)
Siemens (unit) 16 (14)
σ orbital 18/1 (127)
σ-radical 18/1 (204)
Signal-to-noise ratio 23 (5)
Silicon 17 (252); 18/1 (253); 19 (43, 317, 340)
Silicon carbide, semiconducting 20 (234)
Silicon crystal rectifier 18/1 (98)
Silicon, paramagnetic centre in 18/1 (253)
Silicon photocell 23 (449)
Silicon, photovoltaic effect in 19 (343)
Silicon, semiconducting 20 (230)
Silver bromide 20 (312, 317, 322, 323, 338, 339)
Silver chloride 20 (329, 330, 331)
Silver halides 19 (317, 319, 380)
Silver sulphide 19 (382)
Similarity 22 (17, 20, 82, 138, 153, 193)
Similarity laws 22 (70, 97, 99, 117, 166)
Similarity principle, physical significance of the 22 (18)
Similarity rules 21 (510)
Simple medium 16 (170, 286)
Single crystal specimens 19 (317)
Single crystal surfaces 22 (87)
Single domain particle 18/2 (459, 460)
Single sweep 23 (245)
Singlet-triplet splitting 18/1 (266, 272)
Si-O network 20 (350)
Skin depth 16 (184, 186, 288)
Skin effect
— anomalous 19 (308)
— classical theory of 19 (308)
— electron theory of 19 (310)
— in a circular plate capacitor 16 (186 ff.)
— in circular cylinders 16 (182 ff.)
— losses 16 (306)
— skin depth 19 (309)
Skip-distance, *see also* Dead zone 16 (447, 611, 631)
Sky wave 16 (544, 552, 559, 624 ff., 632)

Slater model 18/2 (342)
Sleeve dipole 16 (241)
Slit in an electric field 16 (69)
Slotted antenna 16 (489)
Slotted section 23 (343)
Slow fading 16 (589, 590)
Slow motion arrangement 22 (383, 398)
Slow-relaxing mechanism 18/2 (222)
Small particles, magnetization curve for 18/2 (438, 464)
Smith chart 16 (216, 314)
Snell's law 16 (506, 509, 522, 547, 568, 511 ff., 630)
Sodium 19 (35)
Sodium borate glasses 20 (359, 363)
Sodium bromide 20 (325, 332)
Sodium chloride 20 (310, 311, 313, 317, 318, 332, 334, 339)
Sodium chloride structure 19 (37)
Sodium niobate 17 (316, 336, 339)
Sodium niobate, antiferroelectric displacements 17 (316)
Sodium silicate glasses 20 (358, 361, 362)
Solar converter 20 (164)
Solar energy converters 19 (343, 386)
Solar time 23 (162)
Solenoid 16 (110)
Solid mixtures 17 (104)
Solidification curve 18/2 (512)
Solids, atom in 18/1 (245)
Solid-state maser 18/1 (2)
Solubility of impurities 20 (290, 292)
Solubility product 20 (259, 260, 286)
Solutions, dilute 17 (87, 96)
Solutions, polar 17 (87, 96)
Solutions, solid 17 (318)
Solvent as a continuum 17 (97)
Solvent effect 17 (88, 95)
Solvent, nonpolar 17 (87, 96)
Sommerfeld free electron theory 21 (9)
Sommerfeld line 16 (378)
Sommerfeld radiation condition 16 (176, 466, 485)
Sound waves 22 (192)
Space charge 20 (46); 21 (43, 194); 22 (2, 8, 15, 16, 33, 34, 35, 43, 45, 47, 50, 471, 499, 505, 526)
— accumulation 22 (481)
— by ion migration 20 (370)
— density 16 (2, 29)
— detector 21 (658)
— distortion 21 (633)
— effect 22 (454, 460)
— field 21 (397); 22 (486)
— layers 17 (277)
— limitation 16 (17)
— limited currents 17 (253); 22 (513)
— limited photocurrent 19 (320)
— minimum 21 (71)
— minimum, distance to the 21 (138)
— negative 21 (379); 22 (523)
— positive 22 (29, 523)
Space factor of an antenna array 16 (456)
Space waves 16 (525)
Spark 17 (155); 22 (1, 51, 506)
Spark breakdown 22 (451, 453)
Spark current 22 (2)
Spark diameter 22 (47)
Spark discharge 22 (252); 23 (225, 229)
Spark flash 22 (1, 51)
Spark in very short gaps 22 (527)
Spark lines 22 (112)
Sparking potential 22 (1, 5, 10, 11, 12, 15, 20, 45, 82)
— curve of Paschen 22 (161)
— minimum 22 (1, 5, 10, 12, 15, 464)
Sparks, time fluctuation of appearance of 22 (471)
Spatial distribution of $3d$ orbitals 18/1 (127)
Spatial harmonics 16 (380, 390)
Specific heat 17 (99, 104); 18/2 (27, 54, 111); 19 (58); 22 (159, 193)
— anomalies 17 (310, 349, 355, 388)
— anomalous 20 (261)
— at constant field 18/1 (5, 17)
— at constant magnetization 18/1 (5, 18)
Spectra, continuous 22 (251)
Spectral lines of lowest energy 22 (60)
Spectral measurements 22 (112)
Spectrograph 21 (318)
Spectrometer 23 (131, 141, 144)
Spectrometer, echelette grating 23 (127)
Spectrometer, electron spin resonance 18/1 (112)
Spectrometer, ESR 18/1 (112)
Spectrum analyser 23 (164–166)
Spectrum, magnetic high-frequency 18/1 (136)
Spectrum of excited ions 22 (136)
Spectrum, real phonon 18/1 (85)

Speed of sound 22 (344, 377)
Sphere, conducting 17 (26)
Sphere of active attraction 21 (485)
Sphere with space charge 16 (125)
Spherical coil 16 (126)
Spherical *E*- and *H*-waves 16 (353)
Spherical harmonics, definitions 16 (115–118, 354)
Spherical polar coordinates 16 (119–128)
Spheroid, prolate conducting, free oscillations 16 (466, 467)
Spin and magnetic field, anisotropic coupling between electron 18/1 (133)
Spin array, helical 18/2 (122, 139)
Spin configuration, triangular 18/2 (140, 142)
Spin density 18/1 (161, 162, 164, 173, 180, 225)
Spin density, negative 18/1 (189)
Spin disorder resistivity 18/1 (538)
Spin echo 18/1 (263)
Spin exchange 18/1 (199, 203)
Spin fluctuations, scattering by 18/1 (424)
Spin Hamiltonian 18/1 (4, 14, 39, 56, 74, 132, 133, 135, 136, 137, 141, 170, 227, 232, 242); 18/2 (5, 35)
Spin Hamiltonian, Dirac 18/1 (266)
Spin lattice relaxation 18/2 (174, 206)
Spin, nuclear 18/1 (99, 100)
Spin operator 18/1 (266, 279, 299)
Spin orbit interaction 18/2 (351)
Spin pattern 18/2 (96, 122, 139, 141)
Spin polarization 18/1 (174, 311, 352, 380)
Spin process, multi- 18/1 (74, 75)
Spin resonance, electron 18/1 (9, 94)
Spin resonance, nuclear 18/1 (9)
Spin system 18/1 (4, 5)
Spin temperature 18/1 (6, 72, 82, 86)
Spin temperature, negative 18/1 (87)
Spin wave 18/2 (79, 84)
— antiferromagnetic 18/2 (106)
— branch, acoustical 18/2 (161)
— degeneracy 18/2 (227)
— dispersion 18/2 (13, 128, 158, 210)
— dispersion, antiferromagnetic 18/2 (127)
— energy 18/2 (161)
— in ferrimagnet 18/2 (15)
— in spiral 18/2 (142, 144)
— in the Bloch wall 18/2 (88)
— instability 18/2 (137)
— interaction 18/2 (20, 50, 132)
— nuclear 18/2 (258–260)
— relaxation process 18/2 (174)
— resonance 18/2 (178, 442)
— resonance absorption 18/2 (185)
— spectrum 18/2 (48, 73, 186, 228)
— spectrum, antiferromagnetic 18/2 (106)
— spectrum of an antiferromagnet 18/2 (136)
— spectrum of linear chain 18/2 (118)
— standing 18/2 (372)
— state 18/2 (18)
— theory 18/2 (104, 122)
— theory, antiferromagnetic 18/2 (104, 122)
— theory, ferrimagnetic 18/2 (104)
— theory, Holstein-Primakoff 18/2 (37)
Spin-density wave 18/2 (154)
Spin-deviation state 18/2 (23)
Spinel 18/1 (253)
Spinel $CdCr_2SE_4$ 18/1 (481)
Spinel, Cr-chalcogen 18/1 (522)
Spinel structure, inverse 18/2 (96)
Spinel structure, normal 18/2 (96)
Spinels, Cr, magnetic red shift 18/1 (487)
Spinels, electrical conductivity in thio- 18/1 (531)
Spinels, superexchange interaction in thio- 18/1 (531)
Spin-flop 18/2 (134)
Spin-flop phase 18/2 (134, 137)
Spin-lattice interaction 18/1 (4, 48, 64, 99, 120, 124)
Spin-lattice operator 18/1 (30)
Spin-lattice perturbation 18/1 (34)
Spin-lattice relaxation 18/1 (1, 30, 56, 59, 78, 120)
Spin-lattice relaxation operator, effective 18/1 (56)
Spin-lattice relaxation time 18/1 (6, 9, 120, 121, 132)
Spin-orbit coupling 18/1 (36, 47, 79, 126, 129, 130, 132, 259, 460)
— anisotropy in 18/1 (64)
— coefficient 18/1 (121, 131)
— relaxation due to 18/1 (47)
Spin-orbit interaction 18/1 (35, 462)

Spin-phonon coupling 18/1 (91)
Spin-phonon interaction 18/1 (84, 85)
Spin-phonon system 18/1 (34)
Spin-phonon system, state functions of the 18/1 (34)
Spin-spin interaction 18/1 (99, 120, 122); 18/2 (351)
Spin-spin relaxation 18/1 (1, 9, 13, 72)
Spin-spin relaxation time 18/1 (120)
Spin-spin-time 18/2 (174)
Spin-wave linewidth 23 (158)
Spiral, antiferromagnetic 18/2 (141, 144, 147)
Spiral conducting current 16 (110)
Spiral configuration 18/2 (141)
Spiral, ferrimagnetic 18/2 (144, 153)
Spiral, ferromagnetic 18/2 (141)
Spiral, spin wave in 18/2 (142, 144)
Split beam c.r.t. 23 (243)
Splitting, angular variation of electronic 18/1 (107, 140)
Splitting constant
— ^{13}C 18/1 (165)
— ^{19}F fluorine 18/1 (167)
— ^{14}N 18/1 (166)
— sign of the hyperfine 18/1 (168)
Splitting, crystal field 18/1 (407, 534)
Splitting, electronic 18/1 (107, 108, 133, 136, 139)
Splitting factor, Landé 18/1 (126)
Splitting, hyperfine 18/1 (14, 99, 104, 108, 133, 135, 139, 140, 148)
Splitting, manganate hyperfine 18/1 (140)
Splitting, orbital 18/1 (131)
Splitting, orbital level 18/1 (129)
Splitting parameter, *D* 18/1 (137)
Splitting, Stark effect 18/1 (108, 127)
Splitting, superhyperfine 18/1 (132, 184, 255)
Splitting, temperature-dependent hyperfine 18/1 (198)
Splitting tensor, hyperfine 18/1 (147, 148, 187)
Spontaneous alignment 16 (724)
Spontaneous emission 18/1 (46); 23 (89, 90)
Spots 22 (507, 511, 514, 521)
Spray discharge 21 (634); 22 (163)
Sputtered atoms 22 (96)
Sputtered material 22 (157)
Sputtering 21 (370)
Sputtering, chemical 22 (509)
Sputtering of cathode 22 (161)
Sputtering of negative ions 21 (463, 467, 468)
Sputtering phenomena 22 (509)
Sputtering, radial variation of 22 (160)
Sputtering rate 22 (165)
Sputtering, surface changes 22 (158)
Sputtering, theories of 22 (158)
Sputters 16 (637)
Square loop antenna
S-state ion 18/1 (70, 71)
S-State ions, relaxation in 18/1 (64)
Stability condition of discharge 22 (187, 189)
Stability criterion, Nyquist 23 (482)
Stability criterion, Routh-Hurwitz 23 (481)
Stability, forms of 22 (201)
Stability of control systems 23 (481, 501)
Stability of oscillators 23 (301)
Stable free radical 18/1 (182)
Standard atmosphere 16 (614, 624)
Standard frequency and time stations 23 (293, 296)
Standing wave 16 (214, 222, 225); 23 (343)
— in a waveguide 16 (314, 688, 689)
— radial 16 (185)
— meter 23 (115, 133, 343)
Stannates, ferroelectricity 17 (270)
Star 19 (17)
Stark effect 21 (478, 481, 492); 22 (62, 101, 363); 23 (144–146, 169)
Stark effect broadening 22 (374)
Stark effect splitting 18/1 (108, 127)
Starting potential 22 (55, 85, 165, 176, 202)
State, doubly degenerate 18/1 (36)
State functions 18/1 (30)
— anisotropy in 18/1 (63)
— of the electrons 18/1 (59)
— of the spin-phonon system 18/1 (34)
Static terms 16 (458, 459, 517)
Stationary representation 16 (345)
Stationary states of current 16 (165)
Steady corona, onset of 22 (522)
Steady discharge 22 (513)
Steady state conditions 19 (328, 330)
Steady state response of a control system 23 (459, 480)

Steady state saturation 18/1 (10)
Steenbeck's minimum principle 22 (383)
Stefan-Boltzmann law 23 (443)
Step function response 23 (234, 253, 461)
Step process 22 (622)
Step lengths 22 (590)
Stereogram description of the crystal classes 17 (292)
Stiffness constant 18/2 (349, 368, 372)
Stiffness matrix 17 (353)
Stimulated emission 18/1 (46, 96)
Stoichiometry 19 (372)
Stokes phenomenon 16 (554)
Stoletow 22 (463)
Stoletow constant 21 (537)
Stoletow's measurements 21 (517)
Stoner band model 18/2 (85)
Stoner model 18/2 (342)
Stoner-Wohlfahrt model 18/2 (506)
Stopping potential method 21 (356)
Storage oscilloscope 23 (271)
Storage tube 23 (227, 272)
Strain 18/1 (48, 83)
Strain anisotropy 18/1 (85)
Strain, anomalous linear 18/2 (304)
Strain experiment, static 18/1 (92)
Strain, forced linear 18/2 (308)
Strain operator 18/1 (30, 49)
Strain, spontaneous 17 (269, 325, 327, 328, 332, 335, 337, 340)
Strain tensor 18/2 (277)
Strandberg's effective perturbation operator, Mattuck and 18/1 (56)
Straubel contour 17 (301)
Stray field 16 (9)
Stray field at the boundary of a plate capacitor 16 (70)
Strayfield, magnetic 18/2 (345, 403)
Streak or forked lightning 22 (577)
Streamer 22 (1, 28, 30, 477, 480, 481, 489, 517, 523)
Streamer, anode directed negative 22 (481)
Streamer, breakdown 22 (518, 524, 526)
Streamer, cathode 22 (526)
Streamer channel, diameter of 22 (597)
Streamer corona, anode 22 (515)
Streamer, current in a 22 (619)
Streamer, discharge, threshold 22 (489)
Streamer formation 22 (449, 456)
Streamer in uniform field breakdown 22 (490)
Streamer, main or return 22 (585, 593, 596)
Streamer, negative 22 (526)
Streamer, positive corona 22 (481)
Streamer, pre-onset 22 (484, 516, 524)
Streamer, pre-onset corona 22 (518)
Streamer propagation, mechanism of breakdown and of 22 (618)
Streamer, radar studies of interstroke 22 (615)
Streamer, recoil 22 (596)
Streamer, return 22 (597, 621)
Streamer spark 22 (454, 496, 513, 517)
Streamer, transition from, to Townsend spark 22 (491)
Strength, dielectric 17 (156)
Stress anisotropy energy 18/2 (369, 442)
Stress energy 18/2 (290)
Stress phase diagram 17 (366)
Stress tensor 18/2 (277, 355)
Stresses in medium 17 (2)
Striations 22 (92, 141, 142, 143, 169, 192, 252)
Striations, blue (CH_2) 22 (139)
Striations, broadened, in hydrogen 22 (138)
Striations, curvature of the 22 (137)
Striations, distance between 22 (138)
Striations, double 22 (139)
Striations in capillary tubes 22 (145)
Striations, moving 22 (115, 140)
Striations, stationary 22 (115, 137)
Strip-line 16 (375)
Stroboscopic sampling 23 (218, 242, 267)
Strokes, height determination of initiation of successive 22 (616)
Strokes of a flash 22 (577, 580)
Strokes, time-intervals between 22 (578, 579)
Strong collisions 17 (138)
Sturm-Liouville theorem 18/1 (281)
Stylus 23 (224, 225)
Subdomain behaviour 18/2 (476)
Sublattice 17 (335)
Sublattice magnetization 18/2 (109, 116)
2-Sublattice-model 18/2 (95, 104, 140)
Sudden ionospheric disturbance (SID) 16 (630, 636)
Suhl effect 20 (229)

Suhl instability, first-order 18/2 (244)
Suhl instability, second-order 18/2 (246, 247)
Suhl magnon 18/2 (228, 229)
Sulphides, selenides and tellurides of zinc 19 (360)
Sulphur 19 (340, 350)
Sulphur dioxide 17 (95)
Sunspot number 16 (633)
Superconducting magnet 18/1 (120)
Supercooling effect 18/2 (138, 151)
Superexchange 18/1 (336, 410, 448, 535); 18/2 (100, 101)
Superexchange interaction in thio-spinels 18/1 (531)
Superheating effect 18/2 (138, 151)
Superhyperfine line 18/1 (103)
Superhyperfine pattern 18/1 (107)
Superhyperfine splitting 18/1 (132, 184, 255)
Superhyperfine structure, chlorine 18/1 (252)
Superhyperfine structure, fluorine 18/1 (252)
Supermagnetism 18/2 (439, 440, 443, 466, 475, 478, 484)
Superparamagnetic behaviour, critical volume for 18/2 (466)
Superposition principle 16 (2, 19)
Superposition principle for control systems 23 (460)
Superrefraction 16 (626)
Superzone boundary 18/2 (149)
Surface admittance 16 (532)
Surface anisotropy 18/2 (182, 288, 367, 373, 442)
— field 18/2 (92)
— pinning 18/2 (92, 182)
Surface barriers 19 (317)
Surface charge 16 (3, 167, 288, 425, 432)
Surface cleaning 21 (370)
Surface, cleaning of the 22 (520)
Surface, composite or complex 21 (350, 370, 376, 377, 380)
Surface conduction of glasses 20 (367f., 374)
— dependence on the chemical resistance 20 (368)
— dependence on the relative humidity 20 (368)
— metal layers 20 (369)
— semi-conducting layers 20 (369)
Surface conductivity 19 (310); 20 (185, 281, 283)
Surface contaminations or impurities 21 (320, 350, 352, 354, 368, 370, 372)
Surface current 16 (424, 432, 470, 478)
Surface effect 18/2 (439)
Surface films 21 (364)
Surface, gas-covered 21 (648)
Surface, gas-free 21 (648)
Surface impedance 16 (186, 289, 377, 379, 380, 381, 383, 385)
Surface impedance, average or uniform 16 (381)
Surface layers 21 (612)
Surface migration 21 (202, 206, 210)
Surface photoelectric effect 21 (354, 368, 373, 629)
Surface pit 18/2 (232)
Surface, plane corrugated 16 (380, 381)
Surface reactance 16 (289)
Surface recombination 20 (182)
Surface recombination velocity 19 (339)
Surface resistance 16 (289)
Surface roughness 21 (351, 368)
Surface space-charge 19 (318)
Surface states 19 (330); 20 (177)
Surface temperature 23 (380, 426)
Surface treatment 21 (376)
Surface wave 16 (524)
Surface wave guiding structures 16 (378–384)
Surfaces, clean, gas covered 21 (643)
Surfaces, clean, gas-free 21 (639)
Surfaces of low workfunction 21 (594)
Susceptance matrix 16 (313, 397)
Susceptibility 17 (7)
Susceptibility, adiabatic 18/1 (5, 10, 17)
Susceptibility, complex 18/1 (6, 19)
Susceptibility, diamagnetic 22 (175)
Susceptibility, electrical 16 (8)
Susceptibility, high-frequency 18/2 (168)
Susceptibility, magnetic 16 (21)
Susceptibility, parallel 18/2 (110)
Susceptibility, paramagnetic 18/1 (404)
Susceptibility, perpendicular 18/2 (124)
Susceptibility, remanent anhysteretic initial 18/2 (534)
Susceptibility, static 18/1 (2, 5, 108, 109, 125)
Susceptibility tensor 18/2 (169)

Sutherland-factor 22 (348)
Sweep delay 23 (245)
Symmetrical components, method to treat square loop antennas 16 (256)
Symmetrical fields 22 (525)
Symmetry group, cubic 18/1 (375)
Symmetry, octahedral 18/1 (129)
Symmetry, relation to polarity 17 (149)
Synthesizers 23 (350–353)

T, magic- 18/1 (119)
Tantalates, ferroelectricity 17 (270)
Tape 18/2 (441)
Tape punch 23 (274)
Tape transport mechanism 23 (259, 283)
Tarnish layer 22 (15, 30)
Tarnished electrode surface 22 (21)
Teévenin's theorem 16 (264)
Teledeltospaper 23 (227)
Telegraph equations 16 (210ff., 302)
Teleprinter 23 (275, 279)
Telestar I satellite 23 (300)
Teller effect, Jahn- 18/1 (69, 70)
Teller theorem, Jahn- 18/1 (36)
Tellurium 19 (340)
Tellurium, semiconducting 20 (235)
Temperature coefficient 19 (181)
— of the Fermi level 21 (77, 98)
— of the true work-function 21 (18, 76, 82, 114)
Temperature control 23 (506)
Temperature, Curie 18/2 (1, 4, 343)
Temperature, Curie-Weiss 18/1 (409)
Temperature, Debye 18/1 (42)
Temperature dependence
— of breakdown 17 (249)
— of spectral sensitivity 19 (337)
— of the forced magnetostriction 18/2 (330)
— of the magnetization 18/2 (34, 370)
— of the magnetoelastic constant 18/2 (327)
— of the magnetostriction 18/2 (293, 322, 326–328)
— of volume 17 (327)
— upon radius in a cylindrical arc 22 (387)
Temperature difference measurement with thermo elements 23 (388)
Temperature distribution 22 (100, 134)
Temperature, lattice 18/1 (31)
Temperature, measurement of 22 (369)
Temperature, Néel 18/2 (98, 104)
Temperature, negative spin 18/1 (87)
Temperature, paramagnetic Curie 18/2 (3)
Temperature radiation 23 (379, 441)
Temperature ratio control 23 (513)
Temperature scale 23 (374–378)
Temperature, spin 18/1 (6, 72, 82, 86); 18/2 (207)
Temperature variation of dielectric constant 17 (93)
Temperature-dependent hyperfine splitting 18/1 (198)
Temperature-dependent moment 18/1 (406)
Temporal growth 22 (41f.)
Tension coefficient of resistivity
— longitudinal 19 (196)
— transverse 19 (196)
Tension, electrical 16 (9, 14)
Terminal surface of a waveguide 16 (396)
Terminal-zone network 16 (210, 211, 236)
Termination, coaxial-line 23 (109)
Termination, waveguide 23 (109)
Tetragonal distortion 18/1 (130)
Tetrahedal site 18/2 (96)
Tetrahedral field 18/1 (129, 130)
Tetrahedral group 19 (16)
TE-wave = transverse electric wave-type 16 (292)
t_{2g} orbital 18/1 (128)
Thallium beam standard 23 (327)
Thallous halides 19 (380)
Thallous sulphide 19 (388)
Thermal agitation 17 (20)
Thermal conduction 18/2 (219)
Thermal conduction by magnons 18/2 (216)
Thermal conductivity 19 (271, 274); 20 (87)
— at low temperatures 19 (294)
— effect of impurities at low temperatures 19 (295)
— in the magnetic field 19 (306)
Thermal e.m.f. 23 (402)
Thermal expansion, anomalous 18/2 (303, 304, 306); 20 (260)
Thermal ionisation 21 (577, 583)
Thermal magnon 18/2 (197)

Thermal vibration 18/1 (121)
Thermionic arc 21 (635)
Thermionic constant 21 (10, 17, 48, 52)
Thermionic emission 21 (1 ff., 373, 469, 577, 583); 22 (449, 450)
— properties of polycrystalline wire 21 (99)
— properties of single crystals of tungsten 21 (103)
— velocity analysis 21 (104)
Thermistor 23 (99, 100, 320, 399)
Thermium 20 (26)
Thermocouple 23 (143)
Thermocouple power monitor 23 (96)
Thermo-diffusion 22 (322)
Thermodynamic potential 18/2 (274, 277, 278)
Thermodynamic property 18/2 (21, 95)
— of antiferromagnet 18/2 (108)
— of ferromagnet 18/2 (1, 274)
Thermoelectric data 19 (386)
Thermoelectric effect(s) 19 (271); 20 (56, 87 ff.)
Thermoelectric effect in an oxide cathode 21 (148)
Thermoelectric force 18/1 (429)
Thermoelectric power 19 (273); 20 (60, 89, 339, 344)
— at low temperatures 19 (295, 297)
— effect of impurities on 19 (278)
— homogeneous 20 (341, 343)
— inhomogeneous 20 (343)
— measurements of 19 (358)
— of bismuth 19 (277)
— of noble metals 19 (277)
— of transition metals and alloys 19 (280)
— on the two-zone model 19 (276)
Thermoelement 23 (375, 380–393, 402)
Thermomagnetic effects 20 (60, 92)
Thermometer 23 (378)
Thermometer with digital display 23 (401)
Thermopile 23 (447)
Thermostat 23 (390)
Theta functions 16 (50)
Thin film 18/2 (90)
Thin films 21 (360)
Thomas-Fermi method 19 (54)
Thompson bridge 16 (655)
Thomson bridge 19 (145)
Thomson coefficient 19 (273); 20 (58, 89)
Thomson heat current 20 (56)
Thomson lemma 16 (34)
Thomson model 17 (9)
Thomson potential gradient 20 (56)
Thomson relation 20 (60, 90)
Thomson theory for ion-ion recombination 21 (484)
Thoriated filament 21 (11)
Thorium surfaces 22 (158)
Th_3P_4 compound 18/1 (502)
Three body collisions 21 (620)
Three carrier model 20 (103)
Three term controller 23 (475)
Three-electrode cell 16 (651)
Three-point method 23 (113)
Three-position action controller 23 (500, 511)
Threshold condition 22 (459)
Threshold lowering by positive space charge 22 (461)
Threshold potential, variation of 22 (517)
Threshold wavelength for photoconduction 19 (321, 336)
Thundercloud charges, distribution of 22 (598)
Thunderstorm 16 (517, 524, 565, 570)
Tight-binding calculation 19 (37)
Time, atomic 23 (290)
Time base 23 (329)
Time base generator 23 (244)
Time, characteristic 22 (471)
Time constant 21 (514)
Time, determination of the unit 23 (292)
Time interval 23 (289)
Time interval counter 23 (366)
Time lag(s) 17 (248); 22 (469)
— experimental study 22 (473)
— formative 21 (602)
— long, near threshold 22 (492)
— observational 22 (471, 474)
— statistical 21 (602); 22 (471).
Time, mean solar 23 (291)
Time of growth 22 (41)
Time of relaxation 19 (236, 294)
— condition for existence of 19 (237)
— due to impurity scattering 19 (296)

— for electrical conduction at low temperatures 19 (293)
— for thermal conduction at low temperature 19 (294)
Time reference for the local clock 23 (299)
Time required to establish space charges 21 (608)
Time, sidereal 23 (291)
Time, unit, constant 22 (471)
Time-reversal 18/1 (66)
Time-reversal degeneracy 18/1 (43)
Time-reversal symmetry 18/1 (30, 36)
Tin 19 (340)
Tin, semiconducting 20 (234)
Titanates, ferroelectricity 17 (270)
Titanium dioxide-rutile 19 (375)
$T^{3/2}$-law 18/2 (346)
TM-wave = transverse magnetic wave-type 16 (292)
Tonks and Langmuir theory 22 (125)
Tonks theorem 21 (393)
Toque curve 18/2 (442, 500, 522)
Toque exchange 18/2 (444)
Toque exchange anisotropy 18/2 (182, 439, 440, 443)
Toroidal coordinates 16 (131)
Toroidal functions 16 (102, 117, 118, 119, 132)
Torsion and flexure of crystals 17 (292)
Torsional modes 17 (304)
Total flux, electrical 16 (4, 156)
Total reflection 16 (552, 629)
Tourmaline, pyroelectricity 17 (264)
Townsend β effect 22 (162)
Townsend breakdown in highly over-volted gaps 22 (497)
Townsend breakdown, statistical fluctuations of 22 (469)
Townsend coefficient 21 (419, 515); 22 (446)
Townsend criterion 22 (5, 10, 29, 30)
Townsend (dark) discharge 22 (54, 240, 477)
Townsend discharge 21 (419, 574, 575)
— antecedent 22 (492)
— growth of the 22 (457)
— spectrum 21 (627)
— transition into dark discharge 21 (566)
Townsend first ionization coefficient 21 (515); 22 (81)
Townsend integral 22 (446)
Townsend law 22 (513)
Townsend relation for α 22 (4, 103, 104)
Townsend second ionization coefficient 21 (554); 22 (5)
Townsend theory of the ionization coefficient 21 (540)
Townsend type of predischarge 22 (505)
Tracer method, radioactive 20 (352, 358)
Traces of impurity 22 (518)
Transconductance 16 (48); 21 (134, 137)
Transfer, chemical 18/1 (202)
Transfer, intramolecular electron 18/1 (203)
Transfer oscillator 23 (336–341)
Transfer, proton 18/1 (202)
Transformer bridge 16 (654)
Transformer theorem of Weissfloch 16 (222)
Transient function of a control system 23 (461, 468, 494)
Transient intermediate 18/1 (98)
Transient response of control system 23 (459, 480)
Transient states 22 (195)
Transient time 22 (27, 448)
Transistor 20 (169–176)
Transistor for temperature measurement 23 (400)
Transistor, *N-P-N* photomultiplier 19 (347)
Transistor of type *A* 20 (204)
Transit time 19 (337)
Transition, electric dipole 18/1 (37, 459)
Transition, electric quadrupole 18/1 (37)
Transition elements 19 (344)
Transition, 4*f*-5*d* 18/1 (472)
Transition, forbidden 18/1 (136)
Transition from glow to arc 22 (16)
Transition integral, one-electron 18/1 (330)
Transition layer 16 (553, 555, 557, 624 ff.)
Transition, magnetic dipole 18/1 (37)
Transition, metal-insulator 18/1 (400)
Transition metals 19 (161, 260, 265)

Transition probability 18/1 (30, 33, 76, 81, 90); 19 (235, 240, 286, 295); 22 (210, 309, 353)
— and relaxation time 16 (721)
— for Orbach processes 18/1 (43)
— for Raman processes 18/1 (41)
— for the direct process 18/1 (40)
Transition temperature 18/2 (283)
Transitions, radiationless 21 (323)
Transmission coefficient 16 (481); 21 (23)
Transmission cross-section 16 (453, 480)
Transmission efficiency 16 (220)
Transmission factor, cavity 23 (94)
Transmission line 16 (295, 468, 470, 471, 483, 490)
— analogy 16 (302)
— equations 16 (210, 302)
— with nonsinoidal e.m.f. 16 (227)
Transmission region 16 (351, 356, 358)
Transmission-line relations 23 (90)
Transmittance 23 (441)
Transport coefficients 22 (328, 333)
Transport equation 21 (420, 421, 430)
Transport equation of geometrical optics 16 (444, 452, 510)
Transport integral 18/1 (42, 66)
Transport numbers 20 (251, 252, 253, 254, 287)
Transport property 18/1 (412)
Transposition 18/1 (290, 301, 313)
Transversal relaxation time 18/1 (8)
Transverse gradient 16 (301)
Transverse phonon 18/1 (82)
Transverse relaxation of impurities 18/2 (224)
Trap 19 (323, 324, 326, 347); 20 (21, 127)
Trapped modes 16 (616ff., 626)
Traveling wave 16 (214)
Traveling-wave tube 23 (1, 31, 36)
Travelling-wave-pump 18/2 (253)
Triangular spin configuration 18/2 (140, 142)
Tributary functions 22 (234)
Trichel pulse corona 22 (518)
Trichel pulses 22 (511)
Triggering agency 22 (449)
Triggering electrons 22 (506)
Triggering single discharge sequences by α-rays 22 (512)
Trilinear coordinates 16 (336)
Triple split of ionospheric soundings 16 (573)
Triplet excitation 18/1 (219)
Triplet, ground state 18/1 (221)
Triplet, naphthalene 18/1 (207, 224)
Triplet splitting, singlet- 18/1 (266, 272)
Triplet state system 18/1 (205)
Triplet system, excited 18/1 (210)
Tropical year 23 (290)
Truncated corner 16 (368)
Tube radius 22 (117f.)
Tubes, long 22 (140)
Tubes, narrow, *see also* Capillar tubes 22 (118)
Tubes, very wide 22 (120)
Tungsten trioxide 17 (317)
Tunnel diode 23 (2, 67–71, 79)
Turbulence, homogeneous 16 (586)
T-wave = TEM wave = Lecher wave 16 (294, 326)
Twilight zone 16 (584, 622, 627)
Twinning 17 (278, 295, 311, 335)
Twinning in antiferroelectrics 17 (312)
Twin-*T*-bridges 16 (656)
Twisted rectangular waveguide 16 (369)
Two-stage processes 21 (629)
Two-step photoelectric processes 21 (321)
Two-terminal-pair network 16 (202, 222)
Two-wire-line, differential equation 16 (208)
Two-zone theory 19 (261)
Typewriter, electric 23 (279)

U centre 18/1 (247)
Ultra violet light 22 (53)
Ultra-high-frequency breakdown 22 (535)
Umbrella array 18/2 (142)
Umklapp processes 17 (188)
Unambiguity of the potential 16 (33)
Uniform static field 22 (32)
Uniform-mode resonance 18/2 (178)
Uniform-mode strain 18/2 (276)
Unipolar transistor 20 (175)
Unit cell 19 (6)
Unit source 16 (174)
United States frequency standard 23 (303, 316)

Units 16 (1, 2, 167)
Universal time 23 (291)
Unorthodox propagation 16 (626)
Unstable region 22 (452)
Upper sideband up-converter 23 (39, 41, 44)
U-processes 19 (245)

V_1 18/1 (239)
Vacancies 19 (382)
Vacancy 20 (15)
Vacancy pairs 20 (302, 313, 332)
Vacuum spark 22 (527)
Valence band 21 (342)
Valence electron 21 (342, 360)
Valence-bond method 19 (108)
Valence-exchange conductivity 18/2 (222)
Valence-exchange-relaxation 18/2 (222)
Valency change 18/1 (142)
Valency, controlled 20 (17, 242)
van der Waals energy 17 (69, 70, 111); 20 (265)
Van Vleck method 17 (44, 79, 86)
Van Vleck cancellation 18/1 (37, 52, 57, 66)
Van Vleck form, Brons- 18/1 (74, 81)
Van Vleck Hamiltonian, Dirac- 18/1 (267)
Van Vleck theory 18/1 (56)
Vanadates, ferroelectricity 17 (270)
Vapours, organic 22 (181)
Varactor 23 (37–41)
Varactor, self-resonant frequency of 23 (53)
Variable attenuator 23 (106, 110, 111)
Variance 16 (577)
Variational method 19 (281, 283)
Variational principle 16 (474)
Vector model for more than one electron
— with orbital degeneracy 18/1 (304)
— without orbital degeneracy 18/1 (300)
Vector model for one *s*-electron per atom 18/1 (296)
Vector model with orthogonalized atomic wave functions 18/1 (321)
Vector potential 16 (19, 92, 172, 428)
— in cylindrical coordinates 16 (108–110)
— its components in spherical polar coordinates 16 (125)
— logarithmic 16 (32)
— of a current-carrying conductor 16 (190ff.)
— of a dipole 16 (20)
Vectorial effect of photoemission 21 (350, 360)
Vectorial selectivity 21 (361, 362)
Velocity feedback 23 (254)
Velocity noise 23 (31)
Velocity of a pulse 16 (228)
Velocity selector, magnetic 21 (356)
V_F centre 18/1 (241)
Vibration losses 20 (378f.)
— dependence on the chemical composition 20 (386f.)
— gradual chemical substitution 20 (387)
— in borate and borosilicate glasses 20 (387)
— influence of the coherence of the network 20 (386)
— influence of the network-modifying ions 20 (386)
— minimum 20 (387)
Vibration, thermal 18/1 (121)
Vibration wattmeter 23 (105)
Vibrational mode, normal 18/1 (48)
Vibrations, acoustical 17 (184)
Vibrations, optical 17 (185)
Video recorder 23 (259)
Vinyl radical 18/1 (205)
Viscosity 17 (127, 129)
V_K centre 18/1 (240, 241, 243)
V_{OH} centre 18/1 (245)
Volt (unit) 16 (2)
Voltage 16 (9, 14)
Voltage collapse 22 (27)
Voltage, fluctuations 22 (141)
Voltage matrix 16 (397)
Voltage reflection coefficient 16 (314)
Volume changes and phase transitions in antiferroelectrics 17 (314)
Volume changes in crystals 17 (332)
Volume conduction 20 (351)
Volume density of charge 16 (2, 167)
Volume effect in photoelectric yield 21 (352, 353, 354, 355, 360, 367, 369, 370, 373, 378)

Volume ionization by positive ions 21 (625)
Volume magnetostriction 18/2 (303, 306, 307, 310)
Volume recombination 21 (486, 498); 22 (128, 133, 194)
V_t centre 18/1 (241)

Wagner earth 16 (649)
Wagner ground 17 (141)
Walker mode 18/2 (171, 174, 176, 228, 251)
Walkinshaw's frequency equation 16 (387)
Wall 18/2 (343)
Wall, band theory of the Bloch 18/2 (87)
Wall, Bloch 18/2 (79, 81, 381, 384, 394, 395, 453)
Wall charge density 22 (80)
Wall charges 22 (54, 59, 62, 123, 190)
Wall, cross-tie- 18/2 (401, 402)
Wall current 22 (64, 124, 174, 202)
Wall, deformation in 18/2 (392)
Wall deformation of a cavity resonator 16 (413)
Wall displacement 18/2 (82, 344)
Wall effects 22 (92, 199)
Wall energy 18/2 (368, 370, 383)
Wall energy in thin films 18/2 (396)
Wall energy, specific 18/2 (386, 400)
Wall, equation of motion of 18/2 (89)
Wall friction 18/2 (453)
Wall in cubic crystal 18/2 (385)
Wall in thin film 18/2 (393)
Wall inertia 18/2 (89)
Wall loss 22 (133)
Wall losses 16 (310, 352)
Wall, Néel 18/2 (394, 395, 401)
Wall recombination 21 (487)
Wall, spin wave in the Bloch 18/2 (88)
Wall, strayfield of 18/2 (403)
Wall temperature 22 (90, 134, 185)
Wall thickness, effective 18/2 (387)
Wall, wedge-type reversed 18/2 (36, 433)
Waller's mechanism of paramagnetic relaxation 18/1 (65, 70, 74, 77)
Wannier function 19 (70)
Water 17 (86)
Water vapour 22 (190)
Wave antenna 16 (529)
Wave equation 16 (172, 440, 443, 514, 525, 527, 613)
Wave front 16 (230, 440, 441, 447–451, 496, 504, 600)
Wave function, *d*- 18/1 (132)
Wave function for the crystal, choice of the 18/1 (290, 300, 304)
Wave functions 19 (231, 257, 261)
Wave guide 17 (146)
Wave impedance, *see also* Impedance 16 (292, 314); 23 (86)
— characteristic 16 (689)
— normalized 16 (689)
Wave of magnetic or electric type 16 (291)
Wave packet 16 (296, 547); 19 (61, 231)
Wave packets in external fields 19 (232)
Wave, standing 17 (147)
Wave, transverse electric 17 (146)
Wave, transverse electromagnetic 17 (146)
Wave, transverse magnetic 17 (146)
Wave vector 19 (231)
Wave zone 16 (437, 455, 459–464, 475, 478)
Waveguide, *see also* Horn waveguide 16 (290)
Waveguide, circular 16 (321)
Waveguide, circular, corrugated 16 (384–390)
Waveguide, corrugated 16 (381)
Waveguide, cylindrical, of various cross-sections 16 (315, 321, 330, 334, 336, 340, 341, 343, 344ff., 348, 352)
Waveguide, elliptical 16 (330)
Waveguide, equilateral-triangular 16 (336)
Waveguide, helical 16 (390–395)
Waveguide, isosceles right-triangular 16 (340)
Waveguide junction 16 (395–406)
Waveguide methods for measuring dielectric properties 16 (686)
Waveguide, non-uniform 16 (358)
Waveguide, parabolic 16 (334, 335)
Waveguide, parabolically curved 16 (361, 364)

Waveguide, rectangular 16 (315)
— bends, corners and twists 16 (364–369)
— filled with an inhomogeneous medium 16 (360)
Waveguide, ridged 16 (347)
Waveguide, triangular, with angles 30°–60°–90° 16 (341)
Wavelength limit, energies below 21 (493)
Wavelength scale 21 (318)
Wavemeters, cavity 23 (124, 343, 359)
Wavemeters, coaxial 23 (124, 343)
Wavemeters, LC 23 (344, 345)
Waves, electromagnetic 17 (7, 143)
Waves filters, electric, formed by EDT and DKT 17 (282)
Wax film 23 (224)
Weights and Measures, International Committee on 23 (290)
Weiss domain 18/2 (1, 453)
Weiss Hamiltonian 18/2 (6)
Weiss model of ferromagnetism 18/2 (1)
Weissfloch's transformer theorem 16 (222)
Welding arc 22 (294)
Weston cell 23 (409)
Wheatstone bridge 19 (142, 144, 152); 23 (420)
Whirl-stabilization 22 (300)
Whisker 18/2 (390, 414, 426)
Whistlers 16 (570)
Wichert-Liénard potentials 16 (428)
Wiedemann-Franz law 20 (88); 22 (329)
Wiedemann-Franz ratio 19 (281)
Wiedemann-Franz ratio at low temperatures 19 (307)
Wien bridge 16 (642); 23 (345)
Wien, radiation law 23 (443)
Wiener's theorem 16 (579, 598)
Wien's displacement law 22 (69)
Wigner-Eckhart theorem 18/1 (48)
Wigner's rule 22 (224)
Wigner-Seitz cell 19 (7, 10, 15)
Williams' tube 23 (228)
Windows 21 (315, 327, 330)
Wire, bent in a circle, carrying current 16 (109)
Wire, bent in a circle, charged 16 (102, 124, 132)
Wire lattice, *see also* Lattice 16 (48–51)
W.K.B. approximation 16 (361, 552, 555, 558, 572, 611)
Wood's bridge 16 (65, 67)
Work function 20 (177); 22 (14, 15)
Work-factor 21 (9, 48, 51)
Workfunction 21 (8, 350, 364, 373)
Workfunction, changes 21 (27)
Workfunction, empirically determined 21 (99)
Workfunction, photoelectric 21 (346, 371)
Workfunction, reduction by Schottky effect 21 (104)
Workfunction true 21 (16, 17, 28, 83)
Workfunctions of elements, thermionic 21 (346)
Wow 23 (260)
Wurtzite structure 20 (236)

Xenon ions 21 (649)
Xerography 19 (350); 23 (228)
X-rays 19 (353); 21 (577, 582); 22 (202)
X-rays, absorption 22 (200)
X-rays produced at the anode 21 (585)
X-rays soft 21 (633); 22 (247)
XY_6 complex 18/1 (56, 68, 69, 81)

Yafet-Kittel instability 18/2 (140)
Yafet-Kittel-state, triangular 18/2 (140)
Yagi antenna 16 (456, 487)
Y-factor 23 (9)
YIG 23 (201)
Young's modulus 18/2 (320, 355)

Zeeman effect 23 (146, 171, 321)
— electronic 18/1 (35, 79, 126)
— nuclear 18/1 (35, 79)
Zeeman-energy reservoir 18/1 (22)
Zener effect 20 (157)
Zener emission 17 (162, 201, 208)
Zener emission, polar diagramm 17 (213)
Zener formula 17 (209)
Zener model 18/2 (342)
Zenerdiode 23 (412)
Zener-Ruderman-Kittel-Kasuya-Yosida theory 18/2 (11)

Zenneck waves 16 (524)
Zero offset 23 (312)
Zero point entropy 17 (373)
Zero-beat condition 23 (337, 338, 348)
Zero-crossing error 23 (333)
Ziman limit 18/2 (219)
Zinc and cadmium selenide, cadmium telluride 19 (369)
Zinc blende structure 19 (5, 10, 11, 15, 40, 43, 108)
Zinc blende type semiconductors 20 (236)
Zinc blende (ZnS) 19 (317)
Zinc oxide 19 (322, 372)
Zinc phosphide 19 (387)
Zinc sulphide 17 (367); 19 (361)
Zinc sulphide, calculations of piezoelectric coefficients 17 (367)
Zintl border 20 (8)
Zirconates, ferroelectricity 17 (270)
ZnS 18/1 (255)
Zone melting 20 (24)
Zone melting, floating 20 (25)
Zone refining 20 (24)

Group 5

Optics

Abbe-König prism 29(215, 217)
Abbe's theory of image formation in microscope 29(349)
Abelian group, translations as 25/2b(40)
Aberration 24(19)
Aberration coefficients 29(27, 146ff., 149ff., 152ff., 162, 163ff.)
Aberration experiment with water 24(21)
Aberration function 29(169ff., 181, 184ff.)
Aberration in concave gratings 29(478)
Aberration ordering of Hamilton 29(6)
Aberration tolerances 29(31ff.)
Aberrations in image formation
— different definitions 29(5)
— higher orders 29(18)
— lowest order 29(6, 15ff.)
Absolute intensity of absorption and emission 27(447)
Absolute term values, definition 27(123)
Absorbing crystals
— biaxial 25/1(89ff., 102ff.)
— dielectric tensor 25/1(86ff.)
— idiophanic rings 25/1(91)
— index tensor 25/1(86ff.)
— non-optically active 25/1(85ff.)
— optically active 25/1(97ff.)
— Poincaré sphere 25/1(91ff.)
— propagation along optic axes 25/1(90, 103ff.)
— singular axes 25/1(90ff., 95ff., 106ff.)
— unaxial 25/1(89, 101)
Absorption 25/1(13)
Absorption and emission continuum 27(387)
Absorption coefficient 25/1(85, 93ff.); 25/2a(8, 21, 157); 27(72, 332, 333, 446–449); 28(82, 97)
— effective 28(85)
— in the Lorentz theory 27(4, 6, 17)
— *k* 24(463)
Absorption cross section 25/2a(200); 28(102)
Absorption, cross-field magneto 25/2a(312)
Absorption edge 25/2a(187)
Absorption edge, intrinsic 25/2a(187)
Absorption ellipsoid 25/1(86)
Absorption, exciton 25/2a(191)
Absorption factor 26(131, 133, 134)
Absorption, free carrier 25/2a(167, 200)
Absorption, free carrier interband 25/2a(201)
Absorption index 25/2a(242)
Absorption, indirect magneto 25/2a(296)
Absorption, infrared, *see also* Infrared absorption 25/2a(72, 81, 82, 123)
Absorption, integrated 27(447, 448)
Absorption, interband magneto 25/2a(286)
Absorption lines, intensity 27(331, 447)
Absorption, magneto 25/2a(242)
Absorption of light in lenses 29(209)
Absorption probability 27(331)
Absorption rate 28(99)
Absorption spectra, continuous 28(167, 168, 169)
Absorption spectra of crystals 28(207)
Absorption spectra of stable polyatomic molecules 27(421)
Absorption spectrum of a crystalline phosphor 26(7, 8)

Absorption, strength of the lattice 25/2a(33)
Acceptable representation of $\mathfrak{G}(\boldsymbol{k})$ 25/2b(56)
Acceptor-donor pair 25/2a(221)
Acceptors 25/2a(181, 212, 213)
Accidental ruling errors 29(564)
Acetylene, Raman spectrum 26(231)
Achromatic lens 29(194, 256, 282)
Achromatic microscope objectives 29(357)
Achromatism 29(44, 255)
Acoustic modes 25/2a(67)
Acoustic modes, symmetry and total representation 25/2b(207)
Acoustic sum rule 25/2d(75)
Acoustically rigid body 25/1(227)
Acoustically soft body 25/1(227)
Actinide series, magnetic resonance 28(63)
Activation of phosphors 26(19–23, 34, 44, 48, 49, 50)
Activity in infrared absorption 26(206, 208)
Activity in Raman spectrum 26(197, 206, 208)
Adiabatic approximation 25/2a(71, 94, 140, 183)
— or Born-Oppenheimer approximation 25/2b(253); 25/2d(14–16, 112, 349, 363, 367, 378)
— general statement 25/2b(257)
— in the interruption theory 27(16–17)
Advance of ruling diamond 29(669)
Advanced Green's function 25/2a(151)
Afterglow 28(137, 195)
AgBr 25/2d(239)
AgBr: Li^+ 25/2d(422, 436)
AgCl 25/2d(32, 189, 199, 239)
AgCl: OH^- 25/2d(283)
Airy integral 25/1(505)
Airy integral, zeros 25/1(506, 509)
Airy pattern 29(31ff.)
Airy's determination of the velocity of light 24(21)
Al, aluminum 25/2a(110)
Algebraic equations for reduction coefficients 25/2b(95)
Alignment by interference 29(814)
Alignment by telescope 29(809–814, 825)
Alignment sphere 29(814ff.)
Alignment telescope of Huet 29(811)
Alkali chlorides 25/2d(230)
Alkali fluorides 25/2d(238)
Alkali halide crystals, spectra 28(284)
Alkali halides, *see also* Substanes 25/2d(22, 25, 27, 31, 51, 68, 71, 162, 199, 234, 280, 342, 343, 379)
Alkali halides 25/2a(73, 82, 90, 145, 147)
— Born's model of 25/2a(90)
— continuous spectra 28(178)
— defects 25/2d(269, 272, 286, 331)
— F center 25/2d(347)
— impurity activated 26(19, 23)
— infrared spectra 25/2d(158, 159, 163, 178, 180)
— lattice vibrations 25/2d(30ff., 69)
— luminescence 26(16–23)
— mass defect 25/2d(340)
— Raman spectra 25/2d(230, 237)
— U center 25/2d(264, 267, 269, 289, 295, 347, 368, 371, 405, 415, 418, 422, 436, 443, 461)
Alkali hydrides 25/2d(240)
Alkali metal 25/2a(105)
Alkali spectra, continuous 28(182)
Alkaline earth
— halides 25/2d(204, 269, 389)
— oxides 25/2d(231, 240)
— oxides, luminescence 26(29, 30)
— phosphates, halide free 26(45, 46)
— silicates, luminescence 26(43–45)
— spectra, continuous 28(183)
— sulphides, luminescence 26(38–40)
Allowable representation of $\mathfrak{G}(\boldsymbol{k})$ 25/2b(56)
Allowed transitions 25/2a(288); 27(330, 337–339)
α-bands of ammonia 27(432)
Altazimuth mounting 29(322)
Alternating intensity in rotation spectrum 26(212)
Aluminum coating of grating surfaces 29(601)
Aluminum coating of mirrors 29(298)
Amici prism 29(215ff.)
Ammonia, Raman spectrum 26(232)
Amorphous systems 25/2d(227, 384)
Amplification of noise 25/2c(120–124)

Amplification of signals 25/2c(124–125)
Amplitude and phase grating 24(596)
Amplitude fluctuations 25/2c(120, 132, 134, 143, 164)
Amplitude modulation 29(736)
Amplitude reflectance coefficient 25/2a(9)
Amplitude reflectance of the lamella 25/2a(19)
Amplitude, slowly varying 25/2c(179, 237)
Amplitude, stability 25/2c(134–138, 144, 238)
Amplitude, surfaces of constant amplitude 25/2a(4)
Amplitude transmittance coefficient 25/2a(9)
Amplitude transmittance of the lamella 25/2a(20)
Amplitudes, anticorrelation of amplitudes 25/2c(143)
Analyser, definition 25/1(2)
Analyser, transmitted intensity 25/1(3ff.)
Analysis of a, example BO_2 27(426–428)
Analysis of a spectrum, complete 27(420)
Analytical signal, complex 25/2c(74)
Anamorphotic image formation 29(30)
Anderson's determination of the velocity of light 24(13)
Anderson's line broadening theory 27(67–70)
Angle, Brewster 25/2a(18, 41)
Angle, characteristic 25/2a(18, 41)
Angle, complex 25/2a(11)
Angle, critical 25/2a(17, 41)
Angle determination in microscopy 29(405)
Angle measurement 29(818)
Angle of refraction 29(90, 130, 139)
Angle, principal angle of incidence 25/2a(18)
Angle, principal azimuth 25/2a(18)
Angular defects in prisms 29(225)
Angular dispersion of a grating 29(458, 460)
Angular momentum
— associated to a degenerate vibration, *see* Renner effect
— of the electrons 27(225, 251, 252, 302)
— of the nuclei 27(251)
— total 27(396)
Angular spectrum of plane waves 29(447)
Anharmonic crystals 25/2a(117)
Anharmonic effects 25/2a(146)
Anharmonic forces 25/2a(131)
Anharmonic interaction 25/2a(113, 131, 135, 141)
Anharmonic potential
— cubic terms 25/2b(234, 239)
— cubic terms and reduction coefficients 25/2b(242)
— cubic terms and transformations 25/2b(234, 242)
— terms of general degree 25/2b(243)
— terms of general degree and reduction coefficients 25/2b(243)
— terms of general degree and transformation 25/2b(243)
Anharmonic vibrations 26(195–196)
Anharmonicity 25/2d(6, 77, 377, 381, 406)
Anharmonicity constants 26(195)
Anharmonicity, Coulomb 25/2d(180)
Anharmonicity, cubic 25/2d(171, 183)
Anharmonicity, defects 25/2d(284, 286, 376, 378, 406, 446, 463)
Anharmonicity, force constant 27(385, 432)
Anharmonicity, Green function 25/2d(138, 376, 416, 480ff., 527)
Anharmonicity, potential 25/2d(142, 166, 187, 446, 478)
Anharmonicity, quartic etc. 25/2d(172)
Anharmonicity, quartic, overlap 25/2d(223)
Anion vacancy as emission centre 26(23, 33)
Aniticorrelation of amplitudes 25/2c(143)
Annihilation and creation operators for the harmonic oscillator 25/2c(294–297)
Anomalous waves 25/2a(194)
Anomalous Zeeman effect 28(302)
Anticathode luminescence 28(143)
Anti-ferroelectric 25/2a(137)

Antiferromagnetic resonance 25/2a(316); 28(68, 69)
Antiresonance 25/2d(320, 324, 371)
Anti-Stokes line 26(171, 202–203)
Antisymmetric stretching vibration 26(177)
Antisymmetrized product (AP) 27(267)
Antiunitary operator, *K* 25/2b(160)
Antiunitary operator, *K* commutes with unitary operators 25/2b(160)
Aperture 29(344, 368)
Aperture limitation 24(600)
Aperture stop 29(11, 231)
Aperture width of a grating 29(460, 462)
Aplanate 29(249)
Aplanatic two-mirror system 29(265)
Aplanatism 29(28)
Apochromatic lens 29(283)
Apochromatic microscope objectives 29(358ff.)
Apodization 29(241, 368, 370, 537)
Apodization test of a grating 29(649ff.)
Approximation of Born and Oppenheimer 27(222, 303, 311, 334, 342, 352, 383, 384)
Arc discharge 28(119)
Arc discharge continuum 28(189)
Arc in rare gases 28(130)
Arc, water stabilized, *see also* High pressure, High current arc 28(124, 127)
Arched-yoke mounting 29(336)
Area determination in microscopy 29(405)
Argon 25/2a(82, 89)
Argon continuum in the ultraviolet 29(646)
Argon high current arc 28(131)
Aromatic hydrocarbons, luminescence 26(54, 55)
Artificial doublet 29(612, 614)
Aspheric plate 29(39, 273)
Aspherical surfaces 29(17, 72, 232, 272)
Aspherical surfaces, test 29(307)
Astigmatism 29(16, 22, 234, 251ff., 260, 263, 266)
Astigmatism in concave gratings 29(480–485)
Astrograph lenses 29(291)
Asymmetrical rotator 27(355, 431)
Asymmetry parameter 28(15, 17)
Asymmetry parameter of asymmetric top molecules 26(215)
Asymptotic expansion, first term 25/1(287ff., 297)
Atmosphere, influence on the velocity of light 24(3)
Atmosphere, local 24(9)
Atomic molecules 27(267)
Atomic orbital 27(225)
Atomic screening 27(43)
Atomic spectra, historical remarks 27(82–83)
Atomic units 27(230)
Attenuated total reflection 25/2a(57)
Attenuation factor 25/2a(19)
Auger effect 28(114)
Autocollimation 29(435, 452, 458, 807)
Autocollimation telescope 29(757, 815ff., 822)
Autocorrelation function 29(563)
Autoionisation 27(146); 28(109, 114)
Automatic guider 29(339)
Automorphisms, inner, outer, of group 25/2b(52)
Average emissivity 25/2a(54)
Average reflectance 25/2a(50)
Average transmittance 25/2a(50)
Axial resolution 29(348)
Axial symmetry of the molecular field 27(225, 252)
Axially symmetric force constants 25/2a(104)
Axially symmetric interaction 25/2a(131, 135)
Axicon 29(814)
Azbel-Kaner cyclotron resonance 25/2a(295)
Azbel-Kaner resonance 25/2a(269)
Azimuth of polarisation state 25/1(2)
— methods of determination 25/1(37ff.)
Azimuth, principal angle of azimuth 25/2a(18)
Azimuthal quantum number 27(86)

Babinet compensator 25/1(46ff.)
Babinet's principle 25/1(262ff., 278, 316, 387, 401)
Back-Goudsmit effect 28(41, 43, 73, 320, 322f.)
— selection rules 28(323)

BaF_2 25/2d(204, 389)
BaF_2:H^- 25/2d(422, 434, 435)
Baird-Atomic infrared grating engines 29(711 ff.)
Baker-Schmidt System 29(279)
Balmer limit 28(123, 151, 153)
Balmer transition 28(124, 151, 153)
Band head 27(387, 389, 392, 394)
Band models 25/2a(259); 28(287)
Band modes in perturbed crystal 25/2b(440)
Band systems 27(384, 385)
Band systems, forbidden 27(339)
Band-pass filter 24(507)
Barometric changes, interferometric control 29(700)
Basis, crystallographic 25/2b(1)
Basis functions 25/2b(28)
Bateman's integral equation 25/1(435)
$BaTiO_3$ 25/2d(219)
$BaTiO_3$, barium titanate 25/2a(136)
Bauernfeind prisms 29(220, 221)
Bausch and Lomb ruling engines 29(706–712)
Beam partition 29(380)
Bearings for telescope mountings 29(323)
Becke method 25/1(175 ff.)
Becquerel phosphoroscope 26(126)
BeI-like spectra, isoelectronic relations 27(151, 152)
Bending control 29(817)
Bending force constant 26(187)
Bending frequencies, data 26(176, 179)
Bending vibration 26(176)
Benzene, calculation of electron levels 27(290)
Benzene, intensity of transitions 27(359)
Benzene, Raman spectrum and structure 26(239–242)
Bergstrand's determination of the velocity of light 24(16)
Berremann technique 25/2d(197, 391)
Berremann, method of 25/2a(33)
Beryllia, luminescence 26(27, 28)
Bessel functions 25/1(501 ff.)
— Debye representation 25/1(502)
— representation for index and variable nearly equal 25/1(505)
— Sommerfeld representation 25/1(501)
— zeros 25/1(502, 504, 509)
Betatron radiation 28(145)
Bethe splitting 28(210)
Bi, bismuth 25/2a(269, 311)
Biaxial crystals 25/1(69)
— absorbing 25/1(89 ff., 102 ff.)
— wave surface 25/1(75)
BI-like spectra, isoelectronic relations 27(151)
Billings filter 25/1(161 ff.)
Binding types of crystals 28(205)
Binormals 25/1(70)
Bi-radials, optic 25/1(71)
Birefringence 25/1(5–9, 12)
Birefringence and optical activity, superposition 25/1(81 ff.)
Birefringence circular 25/1(5 ff.)
Birefringence, elliptic 25/1(5–9)
Birefringence, linear 25/1(5 ff.)
Birefringent crystals, dispersion 25/1(166 ff.)
Birefringent filters 25/1(159 ff.)
Birefringent interference microscopes 29(653)
Birge's rule 27(387)
Bi_2Te_3, bismuth telluride 25/2a(279)
Black body
— artificial 26(137, 138)
— deviation from the ideal 26(137, 138)
— ideal, definition 26(134)
— radiation 26(129, 134, 141, 142, 143)
— radiation of gases 28(98, 133, 142)
— temperature 28(84, 142)
Black screen 25/1(219, 273)
Blank imperfections 29(534 ff., 541)
Blazing 29(429, 580 ff., 599, 653, 669, 714)
Blazing, triangular groove profile 29(597, 599)
Bleaching of colour centres 26(18)
Bloch equations 25/2c(235–236)
Bloch function, definition of $\psi^{(k)}$ 25/2b(46)
Bloch functions 25/2a(251)
Bloch vector, definition of $\psi^{(k)}$ 25/2b(46)
Bloch vector, equivalent 25/2b(51)
Bloch vector, transformed by rotation 25/2b(50)
Blue light mirror 24(511)

BO_2, example for the analysis of a spectrum 27(426–428)
Bolometer detection 28(7)
Boltzmann equation 27(72)
Boltzmann equation, phonon 25/2d(105)
Boltzmann factor 27(399)
Boltzmann law 28(85)
Boltzmann theory 25/2a(261, 271)
Bond polarizabilities 25/2d(60, 70, 157, 257, 262); 26(200)
Bond-charge model 25/2d(46, 47, 50, 59, 72, 76, 187, 213, 216, 367)
Bond-dipole model 25/2d(47)
Bond-orbital model 25/2d(72, 73)
Born approximation 25/2a(152)
Born approximation, first 25/2d(80)
Born method for adiabatic approximation 25/2b(256)
Born model 25/2d(22)
Born model of an alkali halide 25/2a(90)
Born-Huang conditions 25/2d(38, 39)
Born-Karman, (cyclic) boundary conditions 25/2b(12)
Born-Mayer potential 25/2d(22, 37, 52, 132, 379, 425)
Born-Oppenheimer, adiabatic approximation 25/2b(253)
Born-Oppenheimer approximation, *see* Adiabatic approximation
Born-Oppenheimer approximation 27(222, 303, 311, 334, 342, 352, 383, 384)
Born-Oppenheimer method, generalization by Born 27(300, 306, 308)
Born-Oppenheimer, parameter $\kappa \equiv (m/M)^{1/4}$ 25/2b(253)
Born-Oppenheimer, perturbation theory equations 25/2b(253)
Born-Oppenheimer, potential function for nucleii 25/2b(257)
Boro-silicate glasses 29(296)
Bose operators, commutation relations 25/2c(25, 113)
Bose operators, disentangling system 25/2c(301–304)
Bose operators, operator relations 25/2c(297–298)
Bose statistics 26(211)
Bosons, commutation relations 25/2c(25)
Bound, charges, dielectric constant of 25/2a(23)
Bound states, two-phonon 25/2d(190, 191, 254, 475)
Boundaries of anisotropic media, law of refraction 25/1(110ff.)
Boundaries of anisotropic media, total reflection 25/1(112ff.)
Boundary conditions 24(467)
— cyclic 25/2a(59, 66, 117)
— electromagnetic 25/2a(13)
— periodic 25/2a(59, 66, 117); 25/2d(497)
Boundary of shadow 25/1(238, 409ff., 417, 517ff., 527)
Boundary value problem(s)
— first term of asymptotic expansion 25/1(287ff., 297)
— Fourier method 25/1(217ff., 221ff.)
— integral equation theory 25/1(298ff.)
— method of surface currents 25/1(220ff.)
— of diffraction theory 25/1(219, 220)
— of Maxwell theory 29(583ff.)
— of Neumann 25/1(235)
— reduction to dual integral equations 25/1(316ff.)
— relation to discontinuity value problem 25/1(237ff.)
— separation of variables 25/1(222ff.)
— solution by theory of complex functions 25/1(222, 336)
— Wiener-Hopf method 25/1(222ff.)
Brace compensator 25/1(48)
Bradley determination of the velocity of light 24(19)
Branched solution
— Fourier amplitudes as solution of integral equation 25/1(413ff.)
— multiply branched 25/1(266)
— reflection 25/1(264ff., 408)
— representation by Fresnel integrals 25/1(407ff.)
— Sommerfeld's method 25/1(264ff., 404ff.)
Branches of lattice frequencies 25/2b(155, 221)
Branching rule 27(96)
Braukbek's method 25/1(285ff., 482ff.)
Braunbek's boundary values 25/1(285ff.)

Breathing shell model 25/2d(29, 30, 42, 46, 235, 245, 291, 292, 318, 345)
Breit-Rabi formula 28(41)
Bremsstrahlung 28(88, 141)
Brewster angle 25/1(34); 25/2a(18, 41)
Brewster case 24(474)
Brewster law 25/1(202)
Bright field image 24(603)
— for aperture limitation 24(615)
Bright field observation 29(376)
Bright spot of Poisson 25/1(294, 537, 540)
Brightening 24(609, 616)
Brightness 24(630)
Brightness (luminance), definition 29(492)
Brightness temperature, *see* Luminance temperature
Brillouin scattering 25/2a(74); 25/2d(147, 246)
Brillouin zone(s) 25/2a(60, 67, 110, 132);
— first, and Wigner-Seitz polyhedron 25/2b(44)
— first, defined 25/2b(43)
— first, Schönfliess cell 25/2b(44)
— table 25/2d(550)
Broadening 25/2a(164)
Broadening by dipole molecules 27(61)
Broadening, Doppler 25/2c(34, 195)
Broadening, homogeneous 25/2c(33, 34)
— in gases 25/2c(34)
— in semiconductors 25/2c(34)
— in solids 25/2c(34)
Broadening in molecular spectra 27(57ff.)
Broadening, inhomogeneous 25/2c(33)
— in gases 25/2c(34)
— in semiconductors 25/2c(34)
— in solids 25/2c(33)
Broken symmetry, imperfection in crystals 25/2b(436)
Brout's sum rule 25/2d(215)
Build-up of fluorescence 26(75–77)
Bulk emissivity 25/2a(54)

C (diamond) 25/2d(209, 252)
Cadmium iodide, luminescence 26(27)
Cadmium lines 29(770)
Cadmium spectra, continuous 28(183)
Cadmium sulphide phosphors 26(31, 32, 36–38)
— with divalent impurities (Mn^{2+}, Fe^{2+}, Co^{2+}) 26(36, 37)
CaF_2 25/2d(203, 204)
CaF_2, calcium fluoride 25/2a(126)
CaF_2:Eu^{3+} 25/2d(361)
CaF_2:H^- 25/2d(422)
Cage effect in solids 28(75)
Ca I-like spectra, isoelectronic relations 27(155, 156)
Calcite interferometer 24(640)
Calcite, twin plane reflection 25/1(119ff.)
CaO 25/2d(216, 241)
CaO:Cu^{2+} 25/2d(341)
Capacity coefficients of radiation fields 25/1(361)
Capture cross section of an empty centre 26(4)
Carbon arc as radiation source 26(163)
Carbon dioxide
— as an example for classical Raman theory 26(198)
— Fermi resonance 26(195–196)
— normal vibrations 26(184)
— Raman spectrum 26(229)
Carbon disulfide, Raman spectrum 26(229)
Carbon, emissivity 26(162)
Carbondisulfide 24(32)
Carotene (β-) 25/2d(350)
Carrier injection 26(13, 14, 107–110)
Cartesian displacement representation
— and mechanical representation 25/2b(204)
— as a direct product 25/2b(207)
— trace of 25/2b(206)
Cascades, laser 25/2c(259–264)
— discussion of an example 25/2c(262–264)
— field equations 25/2c(260)
— homogeneously broadened line, standing waves 25/2c(261)
— inhomogeneously braoded line, standing waves 25/2c(261–262)
— matter equations 25/2c(260)
Cassegrain system 29(39, 260)
Cathodoluminescence 26(2, 111–115)
Cathodoluminescence spectra 26(25, 29, 30, 40, 48)

$CaTiO_3$, calcium titanate 25/2a(137)
Cation vacancy 26(34)
Cauchy relation 25/2d(30, 38, 40–43, 52)
Cauchy's product sum 29(152, 157, 160, 161)
Caustic of diffracted rays 25/1(292ff.)
Cavity losses, anisotropic 25/2c(206)
Cavity oscillations, effect on solution outside cavity 25/1(351, 360ff.)
Cavity resonator 24(23)
$Ca_{1-x}Sr_xF_2$ 25/2d(396, 397, 404)
CdF_2:In^{3+} 25/2d(347)
CdO 25/2d(224)
CdS 25/2d(50, 224, 225)
CdS, cadmium sulphide 25/2a(278)
CdS:Be 25/2d(437)
CdSe:Be^{2+} 25/2d(437)
CdS:Mn 25/2d(448)
$CdS_{1-x}Se_x$ 25/2d(398, 399, 401, 405)
CdTe, cadmium telluride 25/2a(278)
CdTe:Be^{2+} 25/2d(422, 437)
Centering of lenses, test methods 29(824–827)
Centering of light source 29(762)
Centers, colour 25/2a(217)
Centers, F- 25/2a(217, 315)
Centers, imperfection centers 25/2a(181)
Central obstruction, *see also* Apodization 29(241, 368, 370)
Centrifugal distortion of a molecule 26(211, 213)
Centrifugal terms in the expression for rotational energy 27(304, 320, 396)
Čerenkov radiation 28(148)
CER-VIT 29(297)
Cesium arc 28(100)
Cesium halides 25/2d(199, 238)
Cesium selenate, crossed axial dispersion 25/1(189ff.)
CH_2, analysis of the spectrum 27(435–438)
CH_3, analysis of the spectrum 27(439, 440)
CH_2, application of the method of Walsh 27(437)
C–H bending vibration 26(176)
CH_2 frequencies 26(179, 180)
CH_3 frequencies 26(179–180, 235)
CH_3, geometrical configuration of excited states 27(439, 440)
CH_3, isotope effect 27(440)
CH_2, nature of electronic states 27(436, 437)
CH_3, nature of the electronic states 27(439)
CH_3, Rydberg series 27(439)
C–H stretching vibration 26(176)
Chalcogenides 25/2d(30, 200, 244, 245)
Channeled spectra 25/2a(46)
Channeled spectrum phase 25/2a(53)
Character of a matrix 28(214)
Character of a symmetry operation, definition 26(205)
Character of waves 25/2a(7)
Character system, complete subgroup 25/2b(105)
Characteristic angles 25/2a(18, 41)
Characteristic function of Hamilton 29(25)
Characteristic functions 25/2c(63, 65)
Characteristic parameter χ 25/2a(14)
Characteristic vibration 26(175)
Characters, dotted, of 𝔊 25/2b(61)
Characters, induced, of 𝔊, 25/2b(60)
Characters of representations 28(214)
Characters of several groups (tables) 26(206, 208–210)
Charge, bound 25/2d(29, 44, 75, 76)
Charge carrier injection 26(13, 14, 107–110)
Charge neutrality 25/2d(10, 12, 15, 75)
Charge of electron, sign determination from Zeeman effect 28(301)
Charge transfer, intermolecular and intramolecular 27(357, 358)
Charge transfer transitions 27(343, 358)
Charges, apparent 25/2d(332)
Charges, atomic (effective) 25/2d(73)
Charges, Born (transverse effective) 25/2d(52, 182)
Charges, bound, dielectric constant of 25/2a(23)
Charges, Callen (longitudinal effective) 25/2d(187)
Charges, defects 25/2d(265, 288, 333, 334, 338, 365)
Charges, effective 25/2d(61, 65–67, 69, 358, 359)
Charges, effective, defects 25/2d(334, 335, 373)

Charges, electronic (effective) 25/2d(70, 154)
Charges, free, conductivity of 25/2a(22)
Charges, ionic effective 25/2d(61, 102)
Charges, longitudinal 25/2d(66)
Charges, Lundqvist 25/2d(52)
Charges, overlap 25/2d(28, 68)
Charges, polarization 25/2d(61, 67, 74, 188)
Charges, renormalization 25/2d(76)
Charges, shell model 25/2d(373)
Charges, static 25/2d(25, 29, 30, 42, 61, 67, 69)
Charges, Szigeti (effective) 25/2d(27, 67–69, 74, 188, 215)
Charges, transfer 25/2d(63, 332, 338)
Charges, transverse (effective) 25/2d(52, 66, 69, 72, 75, 94, 187, 215)
Charts, reflectance 25/2a(35, 40)
Charts, Šimon 25/2a(40)
Charts, Smith 25/2a(38)
Chemical analysis using the Raman effect 26(219)
Chemical bond properties 28(34)
Chemoluminescence continuum 28(193)
Chopper disk measurement of luminescence decay times 26(126)
Christiansen phenomenon in birefringent powders 25/1(124ff.)
Chromatic aberrations 29(22ff., 38, 180, 253ff.)
Chromatic correction of microscope objectives 29(359)
Chromatic difference of magnification 29(287)
Chromosphere 28(157)
CI-like spectra, isoelectronic relations 27(154, 155)
Circle, constant phase 25/2a(37)
Circle diagram, resonance circle diagram 25/2a(27, 32)
Circle relations, constant reflectance 25/2a(37)
Circular aperture, diffraction problem complementary to circular disk 25/1(465)
Circular birefringence 25/1(5ff.)
Circular cylinder
— asymptotic representation of surface currents 25/1(516ff.)
— damped partial waves 25/1(496, 508, 511ff.)
— geometrical optics field 25/1(513ff.)
— intensity of back scattering 25/1(500, 525)
— Rayleigh limit 25/1(499ff.)
— Rayleigh series 25/1(498)
— residual waves 25/1(507ff.)
— scattering cross-section 25/1(526ff.)
— surface waves 25/1(495, 511ff.)
— Watson transformation 25/1(507ff.)
Circular dichroism, directional variation 25/1(98ff.)
Circular dichroism tensor 25/1(97)
Circular disk
— acoustic scattering 25/1(456ff.)
— Braunbek's method 25/1(482ff.)
— electromagnetic scattering 25/1(457ff.)
— Kirchhoff's solution 25/1(480ff.)
— Poisson diffraction 25/1(481)
— Rayleigh solution 25/1(466ff.)
— Ritz method 25/1(491)
— scattering 25/1(453)
— scattering coefficient 25/1(479, 490ff., 494)
— separation of variables in scattering problem 25/1(454ff.)
— variation methods 25/1(491ff.)
Circular waveguide 25/1(340ff.)
Circular wavefront 29(558)
Circularly curved slits 29(470–472)
Class I costar and corepresentations of $\mathcal{G}$ 25/2b(195)
Class I, II, III wave vectors, defined 25/2b(177)
Class I wave vectors and reality of representations 25/2b(177)
Class II costar
— and cases A, B, C 25/2b(193, 194)
— and corepresentations of $\mathcal{G}$ 25/2b(188)
— and dotted corepresentations matrices 25/2b(195)
— and $\mathcal{G}(\boldsymbol{k})$ 25/2b(189)
Class II wave vectors and reality of representation 25/2b(179)
Class III costar and corepresentations 25/2b(187)
Class III wave vectors and (lack of) reality of representations 25/2b(179)

Class symbols 28(216)
Classes of a group 28(212)
Classical model parameters 25/2a(32)
Classical oscillator model 27(3, 49, 77)
Classical path approximation 27(21, 37, 38, 70)
Clausius-Mosotti relation 25/2a(138); 25/2d(63, 67, 374)
Clebsch-Gordan coefficients 28(243)
— and Raman scattering tensor 25/2b(301)
— calculation of 25/2b(37)
— definitions 25/2b(34, 101)
— for space groups 25/2b(101)
— for space groups, (1, 1, 1) block 25/2b(103)
— for space groups, (σ, σ, σ'') block 25/2b(103)
— matrix of 25/2b(36)
— unitarity 25/2b(37)
Clebsch-Gordan series, analogy for space groups 25/2b(94)
Clebsch-Gordan series, coefficients, *see also* Reduction coefficients 25/2b(33)
Cl I-like spectra, isoelectronic relations 27(165)
Cluster expansion 25/2d(491, 492, 498, 500, 511, 531, 533)
CO_2, application of the method of molecular orbitals 27(283, 284, 285)
Coated sample 25/2a(55)
Coating of mirrors 29(295)
Coexistence of modes, *see also* Modes 25/2c(143, 187, 255–259)
— homogeneous line 25/2c(143, 187)
— homogeneously broadened line 25/2c(258)
— inhomogeneously broadened line 25/2c(259)
— space-dependent modes 25/2c(255–258)
— spatially inhomogeneous pumping 25/2c(258)
Coherence, classical 25/2c(73–76)
Coherence function, mutual (classical) 25/2c(75)
Coherence function, quantum theoretical 25/2c(78)
Coherence length, difficulties in length measurements 29(778 ff.)
Coherence, mutual (classical), complex degree 25/2c(75)
Coherence, mutual (quantum theoretical), complex degree 25/2c(78)
Coherence properties of spontaneous emission 25/2c(97–99)
Coherence properties of stimulated emission 25/2c(97–99)
Coherence, quantum theoretical 25/2c(76–83)
Coherence, spatial 25/2c(5)
Coherence, temporal 25/2c(6)
Coherent fields by classical sources 25/2c(80–83)
Coherent illuminating of grating 29(489)
Coherent potential approximation 25/2d(382, 388, 391, 393)
Coherent radiation 29(345)
Coherent scattering 25/2a(77)
Coherent states 25/2c(78)
Cohesion of ionic solids 25/2a(90)
Cohesive energy 25/2a(90); 25/2d(22, 25, 59)
Co I-like spectra, isoelectronic relations 27(160)
Coincidence 29(764, 772)
Coincidence focometer 29(823)
Coincidence method 24(559)
Collective electron model for the energy states of a perfect crystal lattice 26(5)
Collective lens 29(361)
Collimator attached to grating 29(454 ff.)
Collision cross section 27(10)
Collision time 27(9–11)
Collision time τ 25/2a(263)
Collision time, very short 27(68, 70)
Colour centres, *see also* *F* centres 25/2a(217); 26(16)
Colour centres, bleaching 26(18)
Colour coordinate 29(74 ff., 161)
Colour cycle system 24(561)
Colour grating 24(561)
Colour phase contrast method 24(630, 633)
Colour saturation 24(630)
Colour schlieren method 24(560, 633)
— continuous 24(562)
— twodimensional imaging 24(567)

Colour temperature 26(131, 160, 161); 28(84, 131, 152)
Coma 29(16, 22, 234, 248 ff., 259, 263, 276)
Combination differences, examples 27(402, 403, 405)
Combination differences, polyatomic molecules 27(420, 425)
Combination of refracting surfaces 29(9)
Combination of wave functions 27(248)
Combination principle 27(85)
Combination tones 25/2c(192–193)
Commutation relations
— for Bose operators 25/2c(25, 113)
— for Bosons 25/2c(25)
— for Fermi operators 25/2c(27, 114)
— for Fermions 25/2c(27)
— for phonons 25/2a(63)
— for spin $\frac{1}{2}$ operators 25/2c(32)
Comparator principle of Abbe 29(761, 791)
— circumvention 29(806)
Compatibility and subduced representations 25/2b(231)
Compatibility of representations 25/2b(230)
Compatibility of representations for adjacent stars 25/2b(86)
Compensation of tilt errors 29(793, 808)
Compensator for small ellipticity 25/1(47 ff.)
Compensator of Babinet 25/1(46 ff.)
Compensator of Brace 25/1(48)
Compensator of Soleil 25/1(46 ff.)
Compensator of Szivessy 25/1(49)
Compensators 29(400)
Complementary filter 26(173)
Complementary plane screens, transmission and scattering cross-section 25/1(387, 390)
Complete radiation 28(83, 142)
Complex analytical signal 25/2c(73)
Complex angle 25/2a(11)
Complex conductivity 25/2a(157)
Complex conjugation operator K
— and Leibfried convention 25/2b(163)
— and time reversal 25/2b(164)
Complex degree of mutual coherence 25/2c(75)
Complex dielectric constant 25/2a(157)
Complex filter 29(747)
Complex normal coordinates
— and corepresentations 25/2b(201)
— and physical displacements 25/2b(142)
— bases for irreducible representation of $\mathscr{G}$ 25/2b(165)
— defined 25/2b(141)
— Hamiltonian as function of 25/2b(141)
— symmetry transformation of 25/2b(158)
Complex perpendicular 29(91)
Complex propagation constant 25/2a(4)
Complex ray data 29(86 ff.)
Complex refractive index 25/2a(4, 241)
Complex representation of $\mathfrak{G}$
— test (Frei), modified 25/2b(176)
— test (Herring) 25/2b(176)
Complex wave vector 25/2a(4)
Complexes, exciton impurity complexes 25/2a(225)
Compounds, III–V 25/2d(25, 69, 258)
Compounds, II–VI 25/2d(25, 69, 216, 258)
Compounds, II–VI compounds 25/2a(278)
Compounds, mixed-valence 25/2d(43)
Compressibility 25/2d(24, 51, 194, 464)
Compton scattering 25/2a(76)
Compton scattering continuum 28(116)
Computer programmes for diffraction patterns 29(564 ff.)
Computer programmes for optical design 29(40 ff., 66)
Concave gratings 29(434)
— aberrations 29(478)
— astigmatism 29(480–485)
— basic equations 29(475 ff.)
— focussing properties 29(473 ff.)
— parameters 29(475)
— resolution 29(479 ff.)
— survey of different mountings 29(484–486)
Condensor 29(383 ff.)
Condensor diaphragm 24(557)

Condon approximation 25/2d(349)
Condon parabola 27(345)
Conduction electrons 25/2a(106)
Conductivity 24(462); 25/2a(157)
Conductivity, complex 25/2a(157)
Conductivity constant 25/2a(3)
Conductivity of free charges 25/2a(22)
Conductivity tensor 25/2a(240)
Conductors, microwave resonance 28(64)
Cone of diffracted rays 25/1(267, 290)
Configuration coordinates 25/2a(186)
Configuration, Faraday 25/2a(240, 242)
Configuration, Voigt 25/2a(241, 243)
Configurational interaction 27(144, 268, 289)
Configurational-energy diagram for luminescence centres 26(9, 10, 20, 22, 48, 50, 60–62)
Configurations, *see* Electron configurations
Configurations, relative position of different 27(218)
Conical refraction 25/1(69ff., 163)
— external 25/1(71)
— in naphthalene 25/1(165ff.)
— internal 25/1(71)
— Voigt's theory 25/1(164)
Conjugate coordinate and momentum, mass weighted displacements 25/2b(137)
Conjugate distances 29(8, 46)
Conjugate representations of $\mathfrak{T}$ 25/2b(51)
Connected diagrams 25/2a(146)
Connectivity of representations 25/2b(230, 374)
Conoscopic arrangement 25/1(126)
Conoscopic arrangement of a polarising microscope 25/1(170)
Conoscopic figure in biaxial inactive crystals 25/1(130)
Conoscopic figure in uniaxial inactive crystals 25/1(130)
Conoscopic illumination 29(401ff.)
Conservation of flux and brightness 29(494)
Consistency, quantum mechanical
— atoms 25/2c(45)
— field 25/2c(44)
Construction of Vincent-Geisse and Lecomte 25/2a(56)
Constructive interference 29(450)
Continuous spectra of Hg_2; Cd_2; Zn_2 28(181)
Continuous spectra of two-atomic molecules 28(173ff.)
Contrast 24(605, 606)
Convention, positive direction 25/2a(13)
Convention, positive sign 25/2a(15)
Convergence in multiple-limit series 27(130ff.)
Convergence limits of the Rydberg series 27(415–418)
Convergence limits of vibrational levels 27(413)
Convolution of ghosts 29(556)
Convolution of two Fourier transforms 29(522)
Cook lens 29(196)
Cooling of crystals 28(207)
Coordinates, configuration 25/2a(186)
Coordinates, normal 25/2a(61, 66)
Coordination number theory of Linwood and Weyl 26(50)
Copper activated zinc sulphide 26(34)
Copper activated zinc sulphide, magnetic measurements 26(35)
Copper, emissivity 26(146, 147)
Corepresentations of $\mathscr{G}$
— and complex normal coordinates 25/2b(199)
— and necessary degeneracy 25/2b(179)
— and Schur Lemma 25/2b(183)
— basis space $S^{(j)}$ 25/2b(180)
— multiplication for matrices in 25/2b(181)
Corepresentations of $\mathscr{T}$ induced from $\mathfrak{T}$ 25/2b(184)
Corepresentations of $\mathscr{T}$ induced from $\mathfrak{T}$, irreducible space for 25/2b(184)
Coriolis forces, rotation spectrum 26(217–218, 235)
Coronograph 24(621)
Correction of eyepieces 29(361)
Correction of objectives 29(357ff.)
Corrector plate 29(238, 267, 270, 272)

Correlation between
— atomic and molecular states 27(258, 260, 414)
— electronic states of different geometrical configuration 27(428)
— electronic states of different symmetry 27(328)
— energy levels, potential curves, and spectra 27(285, 286)
— molecular electronic states and products of dissociation 27(426, 428)
— molecular orbital energy and internuclear distances 27(228–230, 246–251)
— molecular orbitals 27(247, 249)
— molecular orbitals and the geometrical configurations (Method of Walsh) 27(292–299)
— molecular quantum numbers 27(227)
Correlation function(s) 25/2d(14, 479, 480); 27(13–16, 25, 41, 67)
— current-current 25/2d(82)
— density-density 25/2d(82)
— for Langevin forces 25/2c(46–49)
— intensity 25/2c(76)
— multi-time, evaluation 25/2c(59–60, 71–73)
— of the reduced Fokker-Planck equation 25/2c(163)
— static 25/2d(101)
— time autocorrelation 25/2d(81)
Correlation rule of Wigner and Witmer 27(358–360)
Cosine variables 29(86ff., 90, 110)
Costar co$\star\boldsymbol{k}$, class I, II, III, defined 25/2b(187)
Costar co$\star\boldsymbol{k}$, defined 25/2b(187)
Coudé focus 29(264)
Coudé systems 29(324ff.)
Coulomb anharmonicity 25/2d(166, 180, 182, 407)
Coulomb forces 25/2d(21, 23, 28, 37, 52, 180, 363, 379, 425)
Coulomb gauge 25/2c(25)
Coulomb integral 27(238, 246, 269, 275, 285, 354)
Coulomb interaction 25/2a(84); 27(261, 263, 275)
Coupled oscillator theory 27(44ff.)
Coupling between rotation and vibration 26(218)
Coupling coefficient between atom and mode 25/2c(114)
Coupling constants 28(34)
Coupling, intermediate 27(111–117)
Coupling of angular momenta 27(251)
Coupling of jj type 27(95)
Coupling of ΛS type 27(252, 253)
Coupling of LS type 27(94ff., 100ff.)
Coupling of $(\omega\omega)$ and (Ω_0, ω) type 27(252)
Coupling of spin and the molecular axis, polyatomic molecules 27(319, 320)
Coupling of the field components 24(465)
Coupling of the field strengths 24(464)
Coupling parameter 27(113)
Coupling schemes 27(93ff.)
Coupling, variation as a function of J 27(315, 338)
Coupling, variation through excitation 27(338)
Coupling with the molecular axis, *see also* Hund's cases 27(313)
Covalency 25/2d(44, 48, 59, 72)
Covalent crystals 25/2a(96)
Covalent states 27(267)
Covalent structure 27(264)
Covering group $\mathfrak{P}\star(\boldsymbol{k})$, also representation group 25/2b(71)
Covering group $\mathfrak{P}\star(\boldsymbol{k})$, defined 25/2b(68)
Creation and annihilation operators for the harmonic oscillator 25/2c(294–297)
Creation operator 25/2a(62, 113)
Creep current 25/1(221, 495, 517ff.)
Critical angle 25/2a(17, 41)
Critical angles for total reflection at boundaries of anisotropic media 25/1(113)
Critical illumination 29(378)
Critical point index, defined, $P_j(\mu)$ 25/2b(388)
Critical points 25/2a(70, 81, 107, 209); 25/2d(18, 141, 389)
— analysis 25/2d(167, 209)
— defect-activated 25/2d(289–294, 331, 389)

Critical points in density of states
— and "Morse theory" 25/2b(221)
— defined 25/2b(220)
— "dynamical" 25/2b(220)
— index of 25/2b(227)
— $P_j(\mu)$, defined 25/2b(388)
— possible point symmetry for 25/2b(230)
— sufficient condition for, and reduction coefficient 25/2b(227)
Critical points in density of states, "symmetry set" 25/2b(220)
Critical value of *J* for predissociation 27(377, 378)
Cross section, absorption cross section 25/2a(200)
Crossed axial dispersion 25/1(189ff.)
Crossed elliptic analysers 25/1(45)
Crossed rulings 29(641–643)
Cross-field magneto absorption 25/2a(312)
Crystal classes
— cyclic 28(235)
— cyclic-inverse 28(236)
— with inversion axes 28(227)
— with inversion center 28(239)
— without inversion axes 28(225)
Crystal covariants
— electric moment 25/2b(245)
— electric moment and reduction coefficient 25/2b(248)
— electric moment, transformation 25/2b(246)
— polarizability 25/2b(249)
Crystal double group 28(222, 228)
Crystal group 28(222)
Crystal hyperfine structure 28(265, 272, 280)
Crystal invariants, potential energy 25/2b(231)
Crystal levels, analytic treatment 28(232)
Crystal levels, number 28(229)
Crystal quantum numbers 28(232)
Crystal spectra
— application 28(293)
— experimental generation 28(207)
— group theoretical treatment 28(211)
— of alkali halides 28(284)
— of molecules, bichromate ions 28(283)
— of molecules, permanganate ions 28(282)
— of molecules, plutonyl ions 28(282)
— of molecules, uranyl radicals 28(281)
— of molecules, vanadate ions 28(283)
— of organic molecules 28(284)
Crystal spectrometer, triple-axis 25/2a(78)
Crystal structure 28(293)
Crystal systems and optical behaviour of crystals 25/1(67)
Crystalline field interaction
— intermediate field case 28(50)
— strong field case 28(50, 73)
— weak field case 28(49)
Crystalline field
— matrix elements 28(240, 246)
— splitting, amount 28(242)
— splitting of actinide group 28(276)
— splitting of cerium 28(271)
— splitting of europium 28(275)
— splitting of gadolinium 28(276)
— splitting of iron group 28(277)
— splitting of lanthanide group 28(268)
— splitting of mixed crystals 28(279)
— splitting of neodymium 28(273)
— splitting of praseodymium 28(271)
Crystals, anharmonic 25/2a(117)
Crystals, covalent 25/2a(96); 25/2d(48, 73, 208, 214, 215, 252)
Crystals, covalent, ionic etc., *see* Substances
Crystals, cubic, Raman oscillator 25/2d(145, 203)
Crystals, ferroelectric 25/2a(135); 25/2d(207, 359)
Crystals, group-IV 25/2d(267, 288, 331, 338, 342, 398)
Crystals, ionic 25/2a(90, 137); 25/2d(230, 247, 331)
Crystals, low-symmetry, infrared spectra 25/2d(224)
Crystals, mixed 25/2d(263, 394, 460)
Crystals, mixed ionic and covalent 25/2d(214)
Crystals, non-cubic 25/2a(191, 192, 193, 201)
Crystals, organic, luminescence 26(35)
Crystals, oxides 25/2d(199)
CsBr 25/2d(161, 162, 200, 292, 368)

CsCl 25/2d(284, 307, 332, 347)
CsI 25/2d(292)
CsI:K^+ 25/2d(293)
CSIRO Australian ruling, engine 29(721)
CsI:Tl^+ 25/2d(422)
CuI-like spectra, isoelectronic relations 27(163)
Cummulants of a random variable 25/2c(95)
Cumulants 25/2d(475, 492)
Cumulative ruling errors 29(541, 671)
Cuny's optical rule 29(812)
Cuprous halides, luminescence 26(24)
Cuprous oxide, luminescence 26(30, 31)
Curie temperature 25/2a(136)
Curie-Weiss law 25/2a(139)
Current glow curve 26(93)
Curvature of Petzval, *see* Petzval curvature
Curvature of spectral lines 29(451, 467–472)
Curvature radii of lenses, determination 29(820–822)
Curved slit 29(469ff., 628)
Curves of growth 27(447)
Curves of minimum intensity, construction 25/1(129)
Cyclic boundary conditions 25/2a(59, 66, 117)
Cyclic crystal classes, *see* Crystal classes
Cyclotron resonance 25/2a(234, 237, 239, 242, 251, 264); 28(66, 67, 68)
Cyclotron resonance, Azbel-Kaner cyclotron 25/2a(295)
Cyclotron resonance, infra-red 25/2a(267)
Cylindrical objects 24(612)
Cylindrical optical systems 29(31)
Cylindrical problems, diffraction case 25/1(305ff.)
Cylindrical problems, scattering case 25/1(304ff.)

Dall-Kirkham Systems 29(264)
Damping, *see* Relaxation
Damping constant 25/2a(23, 34)
Damping function 25/2d(140, 159, 166, 168, 169, 179, 181–183, 189, 196, 201, 418, 431)
Damping function, density-of-states approximation 25/2d(169)
Damping function, quartic anharmonicity 25/2d(175, 184)
Dark contrast method 24(555, 618)
Dark field method
— one-sided 24(610, 618)
— real symmetric 24(610)
— rigorous 24(593, 609)
Dark field observation 29(376, 383)
Data of construction 29(70ff., 102, 114ff., 158)
Davydov-splitting 28(266, 267, 284)
Debye distribution 25/2a(108)
Debye potential 25/1(228ff., 545)
Debye relaxation 25/2d(70)
Debye temperature 25/2a(144)
Debye-Hückel potential 27(42)
Debye-Waller factor, *see* Huang-Rhys factor
Debye-Waller factor 25/2a(75, 125)
Decay times in luminescence 26(122, 127)
Decentering 29(263)
Decentering of lenses, measurement 29(824–827)
Deep-level impurities 25/2a(215)
Defects, *see also* Critical points, Gap, Local, resonance modes 25/2d(262)
Defects, bond-charge model 25/2d(367)
Defects, charge changes 25/2d(265, 288, 333, 365, 388)
Defects, concentration effects, *see also* Coherent potential approximation, Virtual crystal approximation 25/2d(288, 289, 336, 351, 382, 394, 411, 451)
— self-consistent approximation 25/2d(388)
— single site scattering approximation 25/2d(386, 391, 393, 451)
Defects, effective charge 25/2d(334, 335, 373)
Defects, effective mass 25/2d(326, 327, 330, 340, 421, 429)
Defects, elastic constants, compliance tensor 25/2d(458)
Defects, elastic constants, local 25/2d(450, 459, 462, 464)
Defects, electric-field effects 25/2d(451)
Defects, force-constant changes 25/2d(264, 293, 300, 323, 334, 336, 344, 347, 362, 370–372, 440, 467)

Defects, Green function 25/2d(296, 315–318, 324, 366, 384)
Defects, hindered rotations 25/2d(351)
Defects, image strain 25/2d(454)
Defects, impurity space 25/2d(310, 311, 323, 324, 326, 330, 336, 382, 412, 452)
Defects in prism manufacture 29(225)
Defects, infrared absorption 25/2d(209, 266, 273, 289–295, 331–340, 371)
Defects, interstitials 25/2d(286, 351, 352)
Defects, isoelectronic defects 25/2d(264)
Defects, isotope effects 25/2d(264, 268, 289–291, 295, 383, 428, 434)
Defects, lattice distortion, *see* Lattice distortion
Defects, local modes, *see* Local modes
Defects, many-defect perturbation matrix 25/2d(382)
Defects, molecular 25/2d(263, 280, 283, 347, 351, 355)
Defects, molecular model 25/2d(288, 295, 369)
Defects, off-center 25/2d(280, 282, 406)
Defects, pairs 25/2d(288, 347, 383, 384, 405)
Defects, pressure induced effects, *see* Defects, stress
Defects, quasi-localized modes 25/2d(272)
Defects, Raman scattering, ionic 25/2d(341, 381)
Defects, Raman spectra, defect-induced 25/2d(266, 292, 341, 377, 384, 389, 393)
Defects, resonance modes, *see* Resonance modes
Defects, resonance Raman scattering 25/2d(347)
Defects, rigid-ion models 25/2d(332)
Defects, shell model 25/2d(332, 334, 358)
Defects, strain local, *see* Lattice distortion
Defects, stress 25/2d(449, 454, 466)
Defects, substitutional 25/2d(305)
Defects, symmetry considerations 25/2d(302, 334, 362)
Defects, thermal conductivity 25/2d(391)
Defects, U-center, *see* Substances
Definition of meter 29(766)
Deflection, registering of, automatic 24(571)
Deflection, registering of the 24(569)
Defocussing 29(245)
Deformability 25/2d(29, 31, 33, 34, 38, 40, 42, 60)
Deformability, breathing 25/2d(46, 245)
Deformability, shell modes 25/2d(33, 171, 365)
Deformation, *see* Lattice distortion
Deformation dipole 25/2a(96)
Deformation dipole model 25/2d(28, 29, 49, 74, 292, 345, 346)
Deformation frequencies, data 26(179)
Deformation, homogeneous 25/2a(114); 25/2d(129, 132, 452, 453, 459, 497)
Deformation of optical surfaces 29(232)
Deformation, potential 25/2d(125)
Deformation vibration types 26(180)
Degeneracy, suppression by vibrational excitation 27(307, 308, 325, 353)
Degenerate bands 25/2a(254)
Degenerate electronic states 27(225, 321, 353, 354)
Degenerate electronic states and geometrical configuration 27(310, 349)
Degenerate levels, splitting in solids 26(221)
Degenerate vibration 26(176, 184)
Degradation of energy, internal 27(381)
Degree 29(102ff., 123, 138, 179, 188)
Degree of coherence 25/1(22); 29(345)
Degrees of freedom of vibrations 26(174)
de Haas-van Alphen 25/2a(294)
d-electron configurations 27(213–218)
Density 24(576)
Density matrix 25/2a(150); 25/2d(97, 479); 27(17–20, 67)
Density matrix equation 25/2c(38–39)
— direct solution 25/2c(146–153, 290–294)
— exact elimination of atomic operators 25/2c(288)
— for atoms 25/2c(57)
— for atoms and fields 25/2c(58)

— for light modes 25/2c(56)
— general derivation 25/2c(51–56)
— general form 25/2c(272–273, 289)
Density of states 25/2a(80, 82, 160, 167); 25/2d(17)
— critical points 25/2d(289)
— defective crystals 25/2d(266, 289, 293, 328)
— joint 25/2a(209)
— phonon (squared frequency) 25/2b(232)
Deperturbation 27(369)
Deperturbation and Λ doubling 27(372)
Deperturbation, heterogeneous 27(372)
Deperturbation, homogeneous 27(371, 372)
Deperturbation, rotational 27(405, 406)
Depolariser 25/1(53ff.)
Depolarization effect in luminescence 26(85)
Depolarization factor 26(173, 174, 198–200)
Depolarization ratio and Raman scattering 25/2b(304)
Dereflection of *TH*-waves 24(479)
Design of optical instruments 29(34ff.)
Deslandres' tables 27(345, 389, 392, 393, 419)
Destruction operator 25/2a(62, 113)
Detailed balancing 27(11–13)
Detector, phase sensitive 28(4, 7)
Deterioration of phosphors under particle bombardment 26(115, 120)
Detuning dip, *see* Lamb dip
Diabaticity 27(36)
Diagram, connected diagrams 25/2a(146)
Diagram, one-phonon 25/2a(118)
Diagram, perturbation-theory 25/2a(153)
Diagram, self-energy 25/2a(118)
Diagrammatic techniques 25/2a(152)
Diagrams, defect scattering 25/2d(386–388)
Diagrams, free energy 25/2d(532)
Diagrams, non-linear dipole moments 25/2d(186, 377)
Diagrams, Raman scattering 25/2d(153, 380–381)
Diagrams, self-energy 25/2d(413, 502, 518)
Diagrams, susceptibility 25/2d(374)
Diagrams, vertex part 25/2d(512)
Diamagnetic resonance, *see also* Cyclotron resonance 28(67)
Diamond 25/2a(73, 74, 96, 130, 266)
Diamond (C) 25/2d(44, 48, 51, 73, 208, 209, 252, 257, 375, 554)
Diamond carriage control 29(702ff.)
Diamond, Clebsch-Gordan coefficients 25/2b(368)
Diamond, irreducible representations 25/2b(362)
— at Γ 25/2b(362)
— at $^{*}L$ 25/2b(364)
— at $^{*}X$ 25/2b(363)
— connectivity along Δ 25/2b(377)
Diamond, luminescence 26(11–13)
Diamond, phonons, critical points 25/2b(386, 398)
Diamond, phonons, density of states: 2 phonon 25/2b(400)
Diamond, phonons, symmetry 25/2b(382)
Diamond, space group $Fd3m$, O_h^7 25/2b(21, 343)
Diamond, subgroup $F\bar{4}3m$, T_d^2 25/2b(22)
Diamond types used for ruling 29(667ff.)
Diamond, vibration spectrum 26(222)
Diamond-ruled traces on a measure rod 29(785)
Diaphragms, complementary pairs of 24(564)
Diatomic molecules, spectra 27(383–404, 455–460)
Dichroism 25/1(13, 14)
Dielectric changes in phosphors 26(94, 95)
Dielectric constant 25/2a(3, 24, 72, 119, 140, 157, 160, 241); 25/2d(4, 24, 65, 68, 93, 158, 165, 217, 535)
— complex 25/2a(157)
— electrons 25/2d(86)
— ε 24(462)
— Hartree approximation 25/2d(257)
— hydrostatic pressure 25/2d(248)
— longitudinal 25/2a(162)
— of the bound charges 25/2a(23)
Dielectric films in x-ray gratings 29(435ff.)
Dielectric function 25/2a(106)

Dielectric properties 25/2a(137)
Dielectric susceptibility 25/2d(331, 374, 393)
Dielectric tensor 25/1(59); 25/2d(247)
Dielectric tensor of absorbing crystals 25/1(86ff.)
Difference, bands, three-phonon 25/2d(162)
Difference bands, two-phonon 25/2d(162)
Difference operators 29(171, 172, 177, 184, 187)
Differential approximation 29(177, 188, 190)
Differential operators 29(171, 177, 179, 181, 187)
Diffracted ray 25/1(219, 267, 290ff., 489, 495, 528)
— associated field 25/1(291ff.)
— caustic 25/1(292ff.)
— contribution to field 25/1(295ff.)
— multiple diffraction 25/1(269, 291, 297, 489)
Diffraction and scattering 29(433)
Diffraction at object 24(589)
Diffraction at the schlieren diaphragm 24(581)
Diffraction by a half-plane 25/1(404ff.)
Diffraction by a narrow slit 25/1(426ff.)
— far field 25/1(431ff.)
— transmission coefficient 25/1(436ff.)
Diffraction by a slit 25/1(419ff.)
— integral equations 25/1(419ff.)
— intensity of the far field 25/1(446, 452)
— polarisation properties 25/1(446, 448ff.)
— test of Kirchhoff's solution 25/1(442ff.)
— test of stationary representation 25/1(442ff.)
— transmission coefficient 25/1(446ff., 448ff.)
— transmission cross-section for grazing incidence 25/1(453)
Diffraction by a strip 25/1(419ff., 447ff.)
— problem complementary to slit 25/1(447)
Diffraction by a wedge 25/1(411ff.)
Diffraction by a wide slit 25/1(437ff.)
— asymptotic solution 25/1(436ff.)
Diffraction by smooth objects 25/1(351ff.)
Diffraction case, definition 25/1(252)
Diffraction coefficient 25/1(529)
Diffraction functions 24(589, 602, 612)
Diffraction grating(s)
— dispersion 29(458ff.)
— for metrological purposes 29(439ff.)
— for microwaves 29(436ff.)
— for the infrared 29(436)
— for the ultraviolett 29(435)
— for x-rays 29(429, 435)
— formulation of rigorous theory 25/1(370ff.)
— history 29(426ff.)
— interaction of elements 25/1(371ff.)
— parameters 29(446, 449)
— types 29(434ff.)
Diffraction image 24(590, 599)
Diffraction losses of the Fabry-Perot resonator 25/2c(17)
Diffraction pattern 29(512)
— for ring aperture 29(369)
— from a rectangular aperture 29(452)
— in microscope 29(343)
— in spectrographs 29(487)
— of circular pupil 29(238ff.)
Diffraction phenomena, intuitive interpretation 25/1(299)
Diffraction plane 24(599)
Diffraction problem
— and complementary scattering problem 25/1(387, 390, 401ff.)
— cylindrical 25/1(226ff., 304ff.)
— heuristic methods 25/1(266)
— method of stationary representation 25/1(391ff.)
— plane 25/1(226ff.)
— scalar 25/1(227)
— uniqueness of solution 25/1(245–250ff.)
Diffraction unit 29(565)
Dihedral classes 28(238)
Dimension of a representation 28(213)
Diodes, emission from semiconductor diodes in a magnetic field 25/2a(312)
Diodes, GaAs 25/2a(313)
Diodes, InSb 25/2a(313)

Dipole approximation 25/2a(137); 25/2c(28); 25/2d(26, 61, 84, 85, 92, 96, 115)
Dipole, deformation 25/2a(96)
Dipole forces 25/2d(28, 45, 57)
Dipole gauge 25/2d(86, 88, 92)
Dipole matrix elements 25/2d(87, 111)
Dipole model 25/2d(26, 44, 49, 53, 56, 57)
Dipole model, deformation 25/2d(28, 29, 49, 74, 292, 345)
Dipole moment(s)
— auto-correlation function 25/2d(100)
— equation 25/2c(175)
— induced 26(108, 201, 203–204)
— magnetic 25/2a(251)
— non-linear 25/2d(8, 59, 98, 135, 156, 186, 188, 208, 210, 263, 360, 376, 438)
— renormalized 25/2d(141, 377)
— second-order 25/2d(135, 200, 212, 360, 377, 439)
— static 25/2d(97)
Dipole operator 25/2d(122)
Dipole radiation, electric 28(241)
Dipole radiation, magnetic 28(241)
Dipole sum rule 25/2d(71, 87, 88)
Dipole sums, two-dimensional 25/2d(472)
Dipole transitions in luminescence 26(84)
Dipole-dipole broadening 28(48)
Dipole-dipole interaction 28(76)
Dirac comb 29(521)
Dirac's identity 25/2d(100)
Direct product 27(325, 326, 328, 352, 354)
Direct product of space group representations 25/2b(87)
— character 25/2b(87)
— powers 25/2b(88)
Direct transitions 25/2a(163)
Direction convention, positive 25/2a(13)
Direction uncertainty 24(578)
Directional effect in molecular broadening 27(57–60, 63)
Discontinuity value 25/1(224ff.)
Discontinuity value problem
— electromagnetic 25/1(243ff.)
— of diffraction theory 25/1(219)
— relation to boundary value problem 25/1(236ff., 269)
— scalar 25/1(236–299)
Discontinuous two-dimensional method 24(567)
Disentangling theorem 25/2c(299–301)
Disentangling theorem for Bose operators 25/2c(301–304)
Dislocations 25/2d(289)
Dispersion 25/2c(184); 29(73ff.)
Dispersion coefficients 29(77ff.)
Dispersion curves, phonon 25/2a(59, 88)
Dispersion, dependence on grating angle 29(615)
Dispersion formula of the plasma 28(144)
Dispersion in birefringent crystals 25/1(166ff.)
Dispersion increase by use of two gratings 29(616ff.)
Dispersion number of Abbe 29(254, 292)
Dispersion of diffraction gratings 29(458ff.)
Dispersion oscillators, *see also* Reststrahlen oscillator 25/2d(7, 21, 69, 95, 134, 136, 138–140, 159, 163)
— coupling between 25/2d(142)
— damping function 25/2d(172)
— line widths 25/2d(191, 195)
— volume-dependence 25/2d(195)
Dispersion relation, phonon 25/2a(60)
Dispersion relations 25/2a(28, 42, 245, 246); 25/2d(51, 86, 220)
Dispersion, spatial 25/2a(194)
Dispersive properties of glasses 29(24)
Displacements, Fourier transformation 25/2b(136)
Displacements, particle 25/2b(115)
— and complex normal coordinates 25/2b(142)
— and general symmetry 25/2b(119, 232)
— and translation symmetry 25/2b(117)
— matrix of 25/2b(116)
Displacive transition 25/2a(137, 143)
Dissipation of quantum systems 25/2c(33, 271–294)
Dissociated systems 27(228, 229, 249)

Dissociation 28(165)
Dissociation, adiabatic 27(267, 414)
Dissociation energy 27(243, 272, 273, 276, 412, 417)
— determination by extrapolation 27(395, 412, 413)
— of H_2 27(264)
— of H_2^+ 27(231, 232, 241, 264)
Dissociation of Rydberg states 27(417)
Dissociation process 27(412)
Distant collisions 27(9)
Distortion 29(16, 22, 61, 234, 793)
Distortion correction 29(205)
Distortions, *see* Lattice distortions
Distribution, Debye 25/2a(108)
Distribution, frequency 25/2a(80, 96, 144)
Distribution functions
— classical 25/2c(37)
— definitions 25/2c(60–63, 274)
— expectation values, calculation by distribution functions 25/2c(64)
— of electrons and fields 25/2c(65)
— of laser 25/2c(65)
— of single electron 25/2c(64)
— Wigner and related 25/2c(61)
Distribution, moments of frequency 25/2a(146)
Distribution of light among orders, *see also* Blazing 29(570)
Distribution of molecular population as function of *B*, *J* and *T* 27(347, 348)
Distribution, Poisson 25/2c(78, 95)
D-lines of sodium
— Paschen-Back effect 28(311)
— Zeeman effect 28(297)
— Zeeman effect of hyperfine structure 28(325)
Donor levels 26(105)
Donor-acceptor pair 25/2a(221)
Donors 25/2a(181, 212, 213)
Doppler broadening 25/2c(34, 195); 27(2, 3); 28(24)
Doppler line, inhomogeneously broadened laser transitions, semiclassical 25/2c(194–201, 203)
Doppler width 29(611)
Dotted matrix, defined 25/2b(58)
Double bond, restricted rotation 26(193–194)
Double field analyser 25/1(37)
Double Fizeau interferometer 29(777)
Double grating method 24(565)
Double images 24(640)
Double-anastigmatic objective 29(194, 196)
Double-layer 24(490)
Doublets, irregular 27(156)
Doublets, regular 27(165)
Doubly refracting half-shade 25/1(39)
Dove prism 29(215)
Dowell's comparator with single mirror 29(776, 781)
Dowell's comparator with two mirrors set at an angle 29(773ff.)
Dragging coefficient 24(34)
Driving mechanisms for telescopes 29(337ff.)
Drude part, infrared absorption 25/2d(218)
Drude theory of optically active crystals 25/1(76)
Drude's relation 24(463)
Dual integral equations
— for diffraction by a slit 25/1(421ff., 433ff.)
— for diffraction problems 25/1(316ff., 330ff.)
— reduction to algebraic problems 25/1(321ff., 332ff.)
— reduction to singular integral equations of Cauchy type 25/1(337)
Duane-Hunt law 28(138)
Duran 29(297)
Dye molecules, luminescence 26(58–60)
Dynamic form factor 25/2d(81, 82)
Dynamical matrix 25/2a(65, 117); 25/2d(17)
— $[\boldsymbol{D}(\boldsymbol{k})]$, crystal symmetry and 25/2b(145, 146)
— $[\boldsymbol{D}(\boldsymbol{k})]$, defined 25/2b(137)
— $[\boldsymbol{D}(\boldsymbol{k})]$, hermitian property 25/2b(137)
— perturbed crystal 25/2b(443)
Dynamical properties of ferroelectrics 25/2a(143)
Dynamics of phonons 25/2a(64)
Dyson equation 25/2d(138, 316, 386, 408, 474, 500, 503, 504, 507, 527)

Eccentricity 29(13)
Echelette in polarized light 29(575ff.)
Echelle, echellette, echelon: definitions 29(465)
Echelon and echelle in reflection 29(466ff.)
Echelon in transmission 29(466)
Edge 24(563)
Edge, absorption 25/2a(187)
Edge condition 25/1(220, 247ff., 252ff., 265, 301ff., 309ff., 313ff., 319, 338, 378, 394)
— analytical formulation 25/1(253)
— and separation of variables 25/1(257ff.)
— for Franz iteration method 25/1(284ff.)
— for Hertz vector 25/1(258ff.)
— Meixner's formulation 25/1(154, 230)
— using potential theory 25/1(254ff., 372)
Edge emission 25/2a(187, 197)
Edge emission spectra 26(28, 33)
Edge modes 25/2d(471)
Edge-diffraction law 25/1(291)
Effective charge (H_2^+) 27(241)
Effective combination of wave functions 27(249)
Effective field 25/2a(87, 138); 27(244)
Effective *g*-factor 25/2a(249, 260, 304)
Effective indices of refraction 25/2a(15)
Effective local field 25/2a(159)
Effective mass
— *m* 25/2a(253, 264, 267, 271, 273)
— temperature dependence of 25/2a(275)
— tensor 25/2a(161)
— theory 25/2a(174, 181, 251)
Effective nuclear charge 27(88–90)
Effective quantum number 27(87)
Effective thickness 25/2a(47)
Efficiency increase of gratings by blazing 29(580–583)
Efficiency of gratings, *see also* Blazing
Efficiency of gratings in the infrared 29(606–609)
Efficiency of the cathodoluminescence process 26(114)
Efficiency of x-ray gratings 29(647)
Efficiency of x-ray production 28(140)
Efficiency test of a grating 29(644)
Eigen current, electric 25/1(351, 356ff.)
Eigen current, magnetic 25/1(359ff.)
Eigenfrequencies of the hermitian dynamical matrix $[\boldsymbol{D}(\boldsymbol{k})]$ 25/2b(138)
Eigenfrequencies of the real dynamical matrix 25/2b(126)
Eigenvalue equations 25/2d(477)
Eigenvalue treatment 25/2d(323, 340, 411)
Eigenvalues 25/2a(66)
Eigenvectors 25/2a(66, 119)
Eigenvectors $\boldsymbol{e}_{j_\varrho}$
— matrix of 25/2b(128)
— necessary degeneracy 25/2b(132)
— of the dynamical matric, completeness 25/2b(129)
— of the dynamical matrix 25/2b(126)
— of the dynamical matrix and symmetry 25/2b(131)
— of the dynamical matrix, orthonormality 25/2b(127)
— representation based on 25/2b(134)
Eigenvectors of $[\boldsymbol{D}(\boldsymbol{k})]$ 25/2b(138)
— and corepresentations of $\mathcal{G}$ 25/2b(202)
— as basis for representation of $\mathcal{G}$ 25/2b(165)
— as basis for representation of $\mathfrak{G}(\boldsymbol{k})$ 25/2b(148, 150)
— as first rank tensor field 25/2b(145)
— effect of symmetry transformation 25/2b(144, 147)
— matrix of 25/2b(140)
— related to real eigenvectors $\boldsymbol{e}_j$ 25/2b(140)
— scalar products, defined 25/2b(139)
Eikonal 29(139ff., 143ff.)
Eikonal coefficients 29(141ff., 149ff., 163ff.)
Eikonal equation 29(4, 25)
Einstein coefficients 25/2c(91, 92); 27(331, 446–448)
Einstein oscillator 25/2d(2, 296, 312, 318, 320, 326, 414, 416, 426, 463, 465)
Elastic constants 25/2a(71, 79, 123, 134); 25/2d(42, 44, 449, 535)
— Born relation 25/2d(44)
— defects 25/2d(450, 459, 462, 464)
— relation to symmetry 25/1(204ff.)

Electric conductors, ideal 25/1(220, 223ff.)
Electric dipole moment operator
— phased, diagonal, definition 25/2b(277)
— phased, off diagonal 25/2b(287)
— symmetry 25/2b(278)
Electric dipole radiation 27(332)
Electric dipole transitions in luminescence 26(3)
Electric field
— effective 25/2d(62, 66)
— influence on optical behaviour of crystals 25/1(194ff.)
— internal 25/2d(61, 66, 361, 365, 367)
— local 25/2d(64, 70, 75)
— local-field correction 25/2d(249, 258)
— Lorentz field 25/2d(60, 62, 66)
— Lorentz-Lorenz formula 25/2d(75, 248)
— macroscopic 25/2d(4, 23, 25, 62, 63, 70)
— static external 25/2d(127, 132, 147, 451)
— strength, operator 25/2c(26)
Electric quadrupole transitions 27(340)
— in luminescence 26(3)
Electric surface polarisation 25/1(226)
Electric waves, transverse 25/2a(7)
Electrodynamic potential 25/1(315ff.)
Electroluminescence 26(2, 98–107)
Electroluminescence in a cathode ray tube screen 26(106)
Electroluminescence in single crystals 26(100)
Electroluminescence, modern theory 26(101)
Electromagnetic boundary conditions 24(464); 25/2a(13)
Electromagnetic boundary value problems in grating theory 29(512, 517)
Electromagnetic groove shape 29(524)
Electromagnetic units, ratio of 24(21)
Electromagnets for Zeeman measurements 28(299)
Electron affinity 27(199–201)
Electron affinity spectra 28(118, 121)
Electron beam at grazing incidence over grating 29(440)
Electron, bonding, anti-bonding or non-bonding 27(231, 243, 415, 417)
Electron configurations 27(91, 92)
Electron emission from phosphors 26(110)
Electron interaction 27(242, 244, 246, 265, 269)
Electron micrographs of gratings 29(654ff.)
Electron multiplet, energy 27(253)
Electron spin resonance 28(65)
— in ferromagnetic substances 28(68, 77)
Electron Stark broadening 27(39)
Electron traps 26(6, 8, 9, 36, 65–67, 75, 77)
Electronegativity 25/2d(70)
Electronegativity, relative of atoms in molecules 27(279)
Electron-electron interaction 25/2d(26, 51)
Electronic and ionic Stark broadening, combination 27(42)
Electronic configuration 27(242, 243, 255, 269, 414, 418)
— notations and symbols in the case of polyatomic molecules 27(414)
— of excited states 27(244)
— of H_2 27(242)
— of He_2 27(243)
Electronic core 27(274, 275)
Electronic density 27(232, 233, 234, 241, 265)
Electronic density (charge) distribution 27(232, 233, 234, 241, 265)
Electronic energy of H_2^+ 27(230, 239)
Electronic entropy 26(227)
Electronic heat capacity 26(225)
Electronic partition function 26(223)
Electronic polarizability 25/2d(28, 46, 74, 378)
Electronic propagator 25/2d(378, 381)
Electronic shells 27(243)
Electronic spectra 26(172); 28(287)
Electronic spin 27(90, 251, 252)
Electronic states, classification according
— to symmetry 27(254, 320–330)
— to the configuration 27(255, 256)
— to the nature of the coupling 27(251–253)

Electronic states, correlation between states
— of different geometrical configuration 27(428)
— of different symmetry 27(328)
Electronic states
— even and odd 27(255, 323)
— multiplicity 27(253, 329)
— of H_2^+ 27(227–242)
— positive and negative 27(322, 323)
— Σ^+ and Σ^- 27(254, 259)
— Σ, Π, Δ, Φ ... 27(252)
— symbols of symmetry 27(253, 323, 329)
Electronic susceptibility 25/2d(87, 88, 133, 143, 154, 341)
Electron-phonon interaction 25/2a(81, 107)
— adiabatic 25/2d(7, 15, 26, 32, 34, 51, 58, 67, 75, 110, 155, 208, 223, 249, 349, 378)
— covalent crystals 25/2d(213)
— microscopic 25/2d(151)
— model treatment 25/2d(152, 378)
— multiple expansion 25/2d(29, 86, 215)
— non-adiabitic 25/2d(113)
— unitary transformation 25/2d(114)
Electron-photon interaction 25/2d(79)
Electrons, conduction 25/2a(106)
Electrons, tunnelling 25/2a(79, 81, 130)
Electro-optic coefficients 25/1(194)
Electro-optic constants 25/1(193ff.)
Electro-optics 25/1(191ff.)
Electro-optics, experimental methods 25/1(197ff.)
Electrophotoluminescence 26(106, 107)
Electrostatic images of an electron beam 29(441)
Electrostatic integrals in isoelectronic systems 27(170–178)
Elementary wavelet 29(518)
Elimination of atomic operators, exact 25/2c(286, 288–289)
Elimination of atomic variables, adiabatic 25/2c(156–158)
Ellipsoid, electromagnetic scattering 25/1(465, 559)
Ellipsoidal energy surfaces 25/2a(253)
Elliptic birefringence 25/1(5–9)
Elliptically polarised light, measurement 25/1(36ff.)
Elliptically polarised light, production 25/1(35ff.)
Ellipticity 24(514, 515)
Ellipticity analysis 24(628, 638)
Ellipticity of polarisation state 25/1(2)
Ellipticity of polarisation state, determination 25/1(39)
e/m determination from Zeeman effect 28(302)
Emission and absorption continuum 27(387)
Emission below threshold 25/2c(120–125)
Emission coefficient 28(82, 97)
Emission, coherence properties
— of spontaneous emission 25/2c(97–99)
— of stimulated emission 25/2c(97–99)
Emission, edge 25/2a(187, 197)
Emission, exciton 25/2a(197)
Emission, forced 28(85, 156)
Emission from semiconductor diodes in a magnetic field 25/2a(312)
Emission independent of frequency 28(95, 120, 123, 128, 132)
Emission, intrinsic 25/2a(197)
Emission lines, intensity 27(331, 447)
Emission, recombination 25/2a(197)
Emission spectra of crystals 28(207, 209)
— excitation by electron collision 28(209)
Emission spectra of polyatomic molecules 27(421)
Emission spectrum
— continuous 28(182ff.)
— continuous, of the hydrogen molecule 28(179)
— of alkali halides 26(17, 18)
— of diamond 26(12, 14)
— of germanium 26(14)
— of halogens 26(15, 16)
Emission, spontaneous 25/2c(88–90)
Emission, stimulated 25/2c(4, 90–91, 120–125)
Emission, super-radiant emission, classical description 25/2c(231)
Emissivity 25/2a(53)
Emissivity as a function of wavelength and temperature 26(145–152)
Emissivity, average 25/2a(54)

Emissivity, bulk 25/2a(54)
Emissivity, hemispherical 26(154, 155, 156)
Emissivity, normal 26(154, 155, 156)
Emissivity of allays 26(158)
Emissivity of carbon 26(162)
Emissivity of glass 26(169)
Emissivity of globar 26(164)
Emissivity of oxides 26(164–167)
Emissivity of plastic materials 26(170)
Emissivity of several metals 26(146–152)
Emissivity, spectral 26(131)
Emissivity, total 26(131)
Emissivity, total, of metals 26(154, 156)
Empirical method of assignment of Raman spectra 26(176)
Energy band model for phosphors 26(6, 17, 21, 22, 67)
Energy, cohesive 25/2a(90); 25/2d(22)
Energy distribution among diffraction orders 29(517, 573, 575)
Energy flow density, time average 25/1(246)
Energy, free 25/2a(141, 144)
Energy gap 25/2a(169)
Energy level diagram for Mn^{2+} 28(56)
Energy levels in intermediate coupling 27(111–117)
Energy loss function 25/2a(209)
Energy loss, specific, per unit path in the scintillator 26(118)
Energy of an electronic multiplet 27(253)
Energy parameter
— F_2 (pp) 27(202, 203)
— G_2 (sd) 27(220)
— G_1 (sp) 27(202)
— ζ_{np} 27(204)
Energy states of luminescence centres 26(5, 13)
Energy surfaces, non-spherical 25/2a(277)
Energy transfer in non-photo-conducting phosphors 26(81–85)
Energy transfer in organic solid and liquid solutions 26(55–58)
Energy transfer in resonance broadening 27(49ff.)
Energy transport, velocity of 24(3)
Energy, zero-point 25/2a(90)
Energy-configurational diagram for luminescence centres 26(9, 10, 20, 22, 48, 50, 60–62)
English (cross-axis) mounting 29(332)
Entrance pupil 29(11, 48, 49, 95ff., 231)
Entrance slit of spectrograph, finite width effect 29(488ff.)
Entropy 25/2a(144, 147)
Entropy derived from molecular spectra 26(227–228)
Envelope construction 25/1(218)
Eppenstein optical circumvention 29(806)
ε_0 25/2a(24, 33)
ε_∞ 25/2a(24, 32)
Equation of motion 25/2d(16)
— normal coordinates 25/2d(495)
— perturbed lattice 25/2d(300)
— unitary transformation 25/2d(303)
— without losses 25/2c(24–31)
Equatorial mounting 29(322)
Equidensites 29(831ff.)
Equilibrium condition 25/2d(24, 38, 40)
Equilibrium configuration 25/2d(15)
Equilibrium distance 27(231, 418, 419)
Equilibrium Green function 25/2d(492)
Equilibrium positions 25/2d(298, 497, 498, 531, 534, 535)
Equivalent electrons 27(96)
Equivalent line width 27(447)
Ergodicity 25/2d(106)
Error limits 24(577, 581)
Error of period 29(552)
Error table 29(786)
Errors in measurement by imperfect focussing 29(765)
Essen, determination of the velocity of light 24(24)
Étendue 29(493)
Ethane derivatives, Raman spectra 26(237–239)
Ethane, Raman spectrum 26(235–236)
Ether, dragging of the 24(34)
Ether drift 24(37)
EuO 25/2d(244)
Europium activated alkaline earth silicates 26(44)
EuS 25/2d(245)
Evanescent wave 29(446, 570, 588, 594)

Even and odd levels 27(323, 324)
Ewald and Lindquist method 25/2a(39)
Ewald technique 25/2d(23)
Ewald Theta-function transformation 25/2d(130)
Ewald transformation 25/2a(85)
Exactly solvable problems 25/2c(201–203)
Exchange degeneracy 27(263)
Exchange forces 27(240, 263)
Exchange integral 27(238, 240, 246, 263, 269, 285); 28(269)
Exchange interaction 27(265); 28(49)
Exchange (resonance) effect 28(103)
Exchange-charge model 25/2d(53)
Excitation energy 27(270, 276)
Excitation potential 27(272)
Excitation, selective 27(407, 408)
Excited complex 27(382)
Exciton 25/2a(170); 28(288, 292)
Exciton absorption 25/2a(191)
Exciton bands 25/2a(176, 177)
Exciton emission 25/2a(197)
Exciton, Frenkel 25/2a(173)
Exciton impurity complexes 25/2a(225)
Exciton, indirect magneto 25/2a(296)
Exciton line shape 25/2a(179)
Exciton, longitudinal 25/2a(172, 193)
Exciton transitions, indirect 25/2a(178, 192)
Exciton, transverse 25/2a(172, 193)
Exciton, Zeemann effect of 25/2a(297)
Excitons 26(7, 56)
Exclusion principle 27(90, 91)
Exit pupil 29(11, 49, 61, 117, 120, 231, 285)
Exit slit of spectrograph, finite width effect 29(490)
Expansion, thermal 25/2a(113, 118, 123, 146)
Expectation values, calculation by distribution functions 25/2c(64)
Experimental techniques 25/2a(317)
Experimental techniques in Raman effect 26(172)
Exploded wire 28(133)
Exploded wire continuum 28(192)
External fields 25/2c(203–212)
External signals applied to laser 25/2c(124–125)
Extinction coefficient 25/1(85); 25/2a(5, 157)
Extra degeneracy and class
— I wave vectors 25/2b(177)
— II wave vectors 25/2b(179)
— III wave vectors 25/2b(180)
Eyepiece field of vision 29(363)
Eyepiece for microscope 29(361)
Eyepiece for spectacled persons 29(367)
Eyepiece micrometer 29(764)
Eyepiece of a telescope 29(285)
Eyepiece of measuring microscope 29(799)

F centre luminescence of fluorite 26(26)
F matrix in molecular vibration theory 26(189, 191)
Fabry-Pérot étalon 29(464, 532)
Fabry-Perot resonator 25/2c(3, 11–18)
Fabry-Perot resonator, diffraction losses 25/2c(17)
Factor set of extension of 𝔓 by 𝔗 25/2b(19)
Factor set of ray representation, definition 25/2b(68)
Factorial group 28(266)
Family tree of quantum theory of laser 25/2c(100)
Far field amplitude and static capacity of screen 25/1(380, 400)
Far field amplitude, stationary representation 25/1(399ff.)
Faraday configuration 25/2a(240, 242)
Faraday effect 25/2a(243); 27(411, 412); 28(296)
— and perturbations 27(412)
— free carrier 25/2a(269)
Faraday ellipticity 25/2a(243, 279)
Faraday rotation 25/2a(268)
— in anisotropic media 25/1(201ff.)
— interband 25/2a(298)
Fascicular gypsum, twin plane reflection 25/1(114ff.)
F-bands 28(289)
F-center 25/2d(347, 361, 448)
F-center resonance 28(77)
F-centers 25/2a(217, 315)
F-centers (color centers) 26(16); 28(290)

F-centers, luminescence emission 26(16–18)
FeO 25/2d(244)
Fermat's principle 29(2ff.)
Fermat's principle for diffracted rays 25/1(267, 528)
Fermi, golden rule 25/2d(80)
Fermi level 25/2a(262)
Fermi limit 26(68)
Fermi operators, commutation relations 25/2c(27, 114)
Fermi resonance 25/2d(190); 27(422, 424, 425, 427, 434)
— and intensity problems 27(425, 453)
— effect 26(195–196)
Fermi sphere 25/2a(102)
Fermi statistics 26(211, 212)
Fermi surfaces 25/2a(79)
Fermions, commutation relations 25/2c(27)
Ferroelectric, anti-ferroelectric 25/2a(137)
Ferroelectric crystals 25/2a(135)
Ferroelectricity 25/2d(219, 282, 359)
Ferroelectricity soft modes 25/2d(60)
Ferroelectricity, thermodynamic theory 25/2a(143)
Ferroelectrics 25/2d(207, 359)
Ferroelectrics, dynamical properties 25/2a(143)
Ferroelectrics, perovskite 25/2a(143)
Ferromagnetic materials 25/2a(316)
Ferromagnetic resonance 28(68)
Fiber optics 29(383)
Field absorption method 24(608)
Field curvature 29(16, 22, 234, 251ff.)
Field, effective 25/2a(87, 138)
Field, effective local 25/2a(159)
Field emission 26(104)
Field energy density, time average 25/1(246)
Field, external 25/2c(203–212)
Field flattener 29(275)
Field lens 29(361)
Field lense 24(566)
Field, local 25/2a(93)
Field, macroscopic 25/2a(87)
Field, magnetic 25/2c(205–212)
Field models for molecular vibrations 26(186)
Field of the bare nucleus 27(269)
Field of vision 29(363)
Field shielding 27(42)
Field strength probability function after Holtsmark 27(31, 33)
Fields, selection rules 25/2c(26)
Films, thin 25/2a(49)
Filter 26(173)
Filter combination 24(503)
Filter combination without coupling 24(504)
Filter operator 25/1(322ff.)
Filter with metal layers 24(500)
Filter without absorption 24(508)
Filtered diffraction pattern 29(639)
Fine structure constant 27(107)
Fine structure doublet 28(37)
Fine structure in a multiple-limit series 27(130ff.)
Finite crystals, continuum theory 25/2d(467)
Finite crystals, semi-infinite 25/2d(467)
Finite crystals, slabs 25/2d(467)
Finite slit width 29(528)
Finite strains 25/2d(448)
First-order transition 25/2a(142)
Fitzgerald's vector, representation of diffraction field 25/1(309)
Fizeau grating test 29(637)
Fizeau's formula 24(35)
Flame emission continuum 28(193)
Flash photolysis 27(422, 426, 432, 435, 436, 438, 448)
Flexure compensation of telescope tube 29(318)
Fluctuating forces, *see* Langevin forces
Fluctuation-dissipation theorem 25/2d(78, 101, 486)
Fluctuations, intensity
— below threshold 25/2c(105, 163–164)
— close vicinity of threshold 25/2c(105, 163–164)
— high above threshold 25/2c(106–109)
— of the gas laser 25/2c(146)
— well above threshold 25/2c(105, 146, 163–164)
Fluctuations of phase 25/2c(130–132, 142, 146)
Fluctuations of quantum systems 25/2c(33–75)
Fluctuations of the amplitude 25/2c(120, 132, 134, 143, 164)

Fluorescence 27(408, 409, 445)
Fluorescence build-up 26(75–77)
Fluorescence, definition 26(2)
Fluorescence microscope 29(397)
Fluorescence, quenching by a magnetic field 27(412)
Fluorine 24(101)
Fluorite objectives 29(358)
Fluorite phosphors 26(24–26)
— change in magnetic susceptibility 26(26)
— emission from cerium and manganese centres 26(25)
— emission from uranium centres 26(24)
Fluorometer 26(124)
Flux density 29(386)
Flux of light
— definition 29(492, 494)
— in gratings 29(462)
— through photographic objective 29(207)
Flying spot techniques 26(127)
f-number 29(232, 234)
Focal length 29(13, 230)
Focal length, measurement 29(822–824)
Fock functions 25/1(519, 543)
Focogram of a mirror 29(306)
Focus, diffraction pattern around 25/1(296)
Focussing properties of concave gratings 29(473 ff.)
Fokker-Planck equation
— classical 25/2c(37)
— generalized 25/2c(39, 70–71, 275–276)
— generalized, derivation 25/2c(276–284)
— generalized, projection onto macroscopic variables 25/2c(284–285)
— generalized, quasi-linear solution 25/2c(172–173)
— generalized, reduction 25/2c(153–158)
— multimode action 25/2c(168–169)
— multimode action, exact stationary solution 25/2c(169)
— multimode action, explicit form 25/2c(168)
— reduced 25/2c(159–168)
— reduced, correlation functions 25/2c(163)
— reduced, in the van der Pol approximation 25/2c(158)
— reduced, noise spectrum 25/2c(164)
— reduced, steady state solution 25/2c(159)
— reduced, transient solution 25/2c(166–168)
Foldy-Wouthuysen representation 25/2d(82)
Forbidden transitions 25/2a(288); 27(330, 333, 339–341, 349–354, 358–360)
— and perturbations 27(358–360)
— in crystals 28(206)
— in the spectrum of N_2 27(451, 452)
Force constant(s) 25/2a(64, 68, 94, 99, 110); 25/2d(16, 17, 364); 26(185 ff.)
— axially symmetric 25/2a(104)
— coupling parameters 25/2b(238)
— coupling parameters, defined 25/2b(238)
— defect-induced changes, *see* Defects
— effective 25/2d(361, 364)
— formal 25/2d(39, 41, 44, 364)
— interplanar 25/2a(102)
Force matrix for dynamical equations 25/2b(125)
Force matrix, symmetry and reality 25/2b(137)
Fork mounting 29(327)
Form factor 25/2a(74, 124)
Foster prism 25/1(172)
Foucault test 29(304)
Four-element lens 29(290)
Fourier analysis 25/2a(102)
Fourier integral 29(446)
Fourier lattice, defined 25/2b(44)
Fourier method 25/1(217 ff., 221 ff.)
— edge condition 25/1(319)
— integral equations for Fourier amplitudes 25/1(317, 328 ff.)
— integral equations of convolution type 25/1(322 ff., 333 ff.)
— intuitive interpretation 25/1(316)
— iteration method for the far field 25/1(325 ff., 335 ff.)
— relation between scattering and diffraction problem 25/1(318 ff., 331)
— Ritz method to evaluate Fourier amplitudes 25/1(321 ff.)

Fourier method (cont.)
— transform of Green's tensor 25/1(329ff.)
— transform of magnetic surface current density 25/1(329)
— transform of surface current 25/1(317, 328ff.)
Fourier series 25/2a(149)
Fourier transform 25/2a(45, 83, 106); 27(3, 4, 7, 67)
— hologram 29(743)
— of an interferogram 25/2a(51)
— of grating diffraction patterns 29(518)
— of Huygens' principle 29(506, 510)
Franck-Condon factors 27(344, 345)
Franck-Condon principle 27(20, 25, 344); 28(161, 163)
— and non intersection of potential curves 27(246–248, 257, 265, 267, 362, 369, 371, 376)
— and perturbations 27(362, 364, 368)
— and predissociation 27(376–378, 382)
— application to polyatomic molecules 27(349, 420, 424, 427)
— application to the study of the geometrical configurations 27(433, 452, 453)
Franz iteration method 25/1(282ff., 326)
— convergence difficulties 25/1(284ff., 326)
— edge condition 25/1(284ff., 326)
Fraunhofer diffraction field 25/1(221)
— angular dependence 25/1(316, 320, 328)
Fredholm's integral equation of second kind 25/1(235)
Free carrier
— absorption 25/2a(167, 200)
— effect 25/2a(162, 237)
— Faraday effect 25/2a(269)
— interband absorption 25/2a(201)
Free charges, conductivity of 25/2a(22)
Free energy 25/2a(141, 144); 25/2d(13, 14, 21, 103, 491, 494, 504, 530)
Free energy, perturbation theory 25/2d(532)
Free enthalpy 25/2d(491)
Free radicals
— mean life time 27(422, 432, 436, 438)
— production by flash photolysis 27(422, 426, 432, 435)
— recombination through a potential wall 27(380)
— spectroscopic investigation 27(455–464)
Free spectral range 29(463)
Free-bound transition spectrum 28(86, 88)
Free-free-radiation 28(87, 140, 143, 158)
Frenkel exciton 25/2a(173)
Frequency data for atom groups (tables) 26(178–179)
Frequency distribution 25/2a(80, 96, 144)
— moments of 25/2a(146)
— of the normal modes 25/2a(69, 107)
Frequency, lattice 25/2a(33)
Frequency, lattice, natural undamped 25/2a(24)
Frequency locking in a magnetic field 25/2c(141, 144, 193, 211)
Frequency, longitudinal optical 25/2a(34)
Frequency, longitudinal optical lattice 25/2a(26)
Frequency measurements of microwaves 28(5)
Frequency part, negative, of vector potential 25/2c(25)
Frequency part, positive, of vector potential 25/2c(25)
Frequency, plasma 25/2a(208, 282)
Frequency, pulling 25/2c(129, 142, 186, 188, 190, 199)
Frequency, pushing 25/2c(188, 190, 199, 201)
Frequency, quasi-harmonic 25/2a(113, 121, 130)
Fresnel ellipsoid 25/1(61, 64)
Fresnel equation 25/2d(95)
Fresnel formula of magneto-optics 25/1(199)
Fresnel formulae 24(468)
Fresnel integral 25/1(407ff., 439)
Fresnel number 25/2c(10)
Fresnel relations 25/2a(16)
Fresnel theory of diffraction 25/1(218)
Fresnel theory of optically active crystals 25/1(76)

Fresnel-zone gratings 29(437 ff.)
Fringe spacing 25/2a(47)
Fringe spacing defect 25/2a(48)
Fringe visibility 25/2c(76)
Fröhlich interaction 25/2d(79)
Fröhlich mode 25/2d(468)
f-sum rule 25/2d(88); 28(106)
Full radiator, *see* Black body
Full-group reduction coefficients, defined 25/2b(91)
Function, advanced Green's 25/2a(151)
Function, Bloch 25/2a(251)
Function, dielectric 25/2a(106)
Function, energy loss 25/2a(209)
Function, Green 25/2a(115, 148)
Function, partition 25/2a(144)
Function, retarded Green 25/2a(151)
Function, spectral 25/2a(125, 127)
Function, thermodynamic Green 25/2a(115, 148)
Functional derivatives 25/2d(475, 487, 489, 492)
— lineare responce 25/2d(487)
— method 25/2d(140, 487)
f-values 27(332, 343, 446–452)
f-values, numerical data 27(464, 465)

G matrix in molecular vibration theory 26(188, 191)
GaAs 25/2d(49, 58, 214, 215, 217, 258, 288)
GaAs diodes 25/2a(313)
GaAs, gallium arsenide 25/2a(98, 269, 274, 275, 303)
$GaAs_{0.94}P_{0.06}$ 25/2d(404)
Gabor's method of wavefront reconstruction 29(741 ff.)
Gain, linear 25/2c(124, 125, 192)
Gain, nonlinear 25/2c(140, 184, 192)
Galileo experiment 24(5)
GaP 25/2d(49, 216, 470)
Gap modes 25/2d(267, 288, 314, 318, 320, 322, 324, 400)
GaP:B 25/2d(437)
Gas, ionized 28(93, 98, 140, 142)
Gas laser 25/2c(194–201, 203)
— intensity fluctuations 25/2c(146)
— linewidth 25/2c(146)
— polarization of lasing modes 25/2c(201)
— quantitative treatment of single-mode operation above threshold (semiclassical) 25/2c(197–199)
— semiclassical equations 25/2c(194–197)
Gas nebula 28(157)
Gas spectra, continuous 28(182)
GaSb, gallium antimonide 25/2a(278, 294, 303)
Gauge, Coulomb 25/2c(25)
Gauge, dipole 25/2d(86, 88, 92)
Gauge invariance 25/2d(78, 82, 86)
Gauge, multipole 25/2d(86)
Gauge, radiation 25/2d(83, 91)
Gauge, transverse 25/2d(83, 86)
Gaunt factors 28(121)
Gauss type objective 29(197, 199)
Gaussian approximation 29(6, 7–14, 92 ff., 99 ff., 137, 228)
Gaussian coefficients 29(97 ff., 102, 124)
Gaussian colour aberrations 29(23)
Gaussian image formation 29(95, 98, 137)
Gaussian image plane 29(98)
Gaussian line, inhomogeneously broadened laser transitions, semiclassical 25/2c(187–190, 203)
Gaussian processes 25/2c(36)
$Ga_{1-x}In_xSb$ 25/2d(402, 403)
Ge, α- (amorph) 25/2d(228)
Ge, germanium 25/2a(80, 96, 99, 130, 132, 133, 145, 266); 25/2d(44, 208, 209, 212, 217, 228, 252, 257, 288)
Generalized series formula 27(137, 141)
Geometrical configuration as a function of the number of valence electrons 27(292–299)
Geometrical configuration, variation during excitation 27(295, 296, 321, 356, 420, 429, 433, 434)
Geometrical optics limit 25/1(238, 390 ff.)
Geometrical optics, limits of validity 29(3)
Geometrical theory of diffraction 25/1(219, 267 ff., 289 ff., 489)
— convex objects 25/1(297 ff., 527 ff.)
— for plane screens 25/1(290 ff.)
Geometrical wavefront 29(3)
German mounting 29(325)
Germanate phosphors 26(45)

Germanium, carrier injection emission 26(108)
Germanium, luminescence 26(13–15)
$Ge_{1-x}Si_x$ 25/2d(398)
g-factor 25/2a(259); 28(29, 42, 43, 298, 304, 315, 326)
g-factor, anisotropy 28(73)
g-factor, effective 25/2a(249, 260, 304); 28(44)
g-factor for Mn^{2+} 28(57)
g-factor, spectroscopic splitting 28(45, 68)
g-factors for palladium and platinum groups 28(60)
Ghost intensities 29(552)
Ghost wavelengths 29(555, 557)
Ghosts 24(621); 29(453, 535, 540, 547, 549ff., 555)
Ghosts of different order 29(550ff.)
Giant pulse laser 25/2c(267–271)
— quantitative treatment 25/2c(269–271)
— semiquantitative treatment 25/2c(268)
Glass 29(24)
Glass, emissivity 26(169)
Glass measure rod 29(785)
Glass, radiation-proof 29(375)
Glass, survey of optical glasses 29(292)
Glasses 25/2d(227)
Glow discharge continuum 28(188)
Gold, emissivity 26(147)
Goniometric eyepiece 29(800)
Gouy's theorem 29(511, 558)
g-permanence rule 28(315)
Granularity of plate 29(531)
Graphite arc 28(81)
Graphite, emissivity 26(162)
Graphs, reflectance graphs 25/2a(35)
Grating, *see* Diffraction grating, Concave grating
Grating diaphragm 24(562)
Grating equation 29(445, 448)
Grating imperfections 29(610ff.)
Grating in immersion 29(457)
Grating mosaic 29(432)
Grating of cylinders, boundary value problem 29(572, 584)
Grating of δ-function grooves 29(520ff.)
Grating replication methods 29(731, 734ff.)
Grating testing interferometers 29(635ff.)
Grazing incidence, optical diffraction gratings 29(458)
Grazing incidence, x-ray gratings 29(435, 603ff.)
Green function(s) 25/2a(115, 148)
— advanced 25/2a(151)
— eigenvalue treatment 25/2d(323)
— equilibrium 25/2d(492)
— for the half-space 25/1(278ff.)
— four-time 25/2d(510, 513)
— generalized retarded 25/2d(491)
— harmonic 25/2d(481)
— Hartree 25/2d(118, 525)
— Hartree-Fock 25/2d(510)
— integrated 25/2d(510)
— many-time 25/2d(511)
— mode mixing 25/2b(326)
— non-equilibrium 25/2d(487)
— of first and second kind 25/1(237, 279)
— of free space, scalar 25/1(232, 239)
— one-phonon 25/2d(137, 483)
— perturbed, *see* Defects
— phonon, anharmonic 25/2d(376, 416)
— Raman scattering 25/2d(116, 342)
— real 25/2d(315, 317)
— resonance mode 25/2d(326, 327)
— retarded 25/2a(151); 25/2d(98, 482, 484, 502, 506, 524)
— symmetry of 25/2b(326)
— thermal 25/2b(319)
— thermodynamic 25/2a(115, 148)
— time-ordered 25/2d(475, 480)
— treatment of grating diffraction 29(571)
— two-phonon 25/2d(135, 138, 518, 525)
Green's tensor, differential equation 25/1(241ff.)
Green's tensor of free space 25/1(241ff.)
— Fourier representation 25/1(329ff.)
"Grille" multislit spectrometer of Girard 29(499)
Groove angle suitable for blazing 29(581)

Groove profile 29(428, 436, 451)
— determination by electron micrographs 29(661–663)
— of rectangular shape 29(575, 581)
— piecewise linear, boundary value problem 29(594–597)
— sinusoidal, boundary value problem 29(586–594)
— triangular, boundary value problem 29(597–599)
Groove shape, effect upon efficiency 29(645)
Groove shape function 29(517, 523)
Groove shaping, prism effect 29(458)
Grooves infinitely narrow slits 29(520ff.)
Group algebra, projection operators 25/2b(31, 33)
Group algebra, two-sided ideals or idempotents 25/2b(31)
Group frequencies 26(176, 177, 178, 179)
Group of $\boldsymbol{k}$, $\mathfrak{G}(\boldsymbol{k})$, representations based on eigenvectors 25/2b(148)
Group theoretical treatment of crystal spectra 28(211)
Group theoretical viewpoints 26(180ff.)
Group theory 25/2d(18, 303, 311)
Group theory, application to molecular spectroscopy 27(322–329)
Group velocity 24(1)
Grüneisen approximation 25/2a(132)
Grüneisen parameter 25/2d(194)
g-sum rule 28(316)
Gudden and Pohl primary current 26(90)
Guided waves 25/1(338)
g-values 24(466)

H_2, application of the method of Heitler and London 27(261–266)
H_2^+, correlation between quantum numbers 27(227)
H_2^+, correlation with the levels of the H atom 27(222)
H_2^+, effective charges 27(241)
H_2, electronic configuration 27(242)
H_2, energy and dissociation 27(264)
H_2, heterogeneous perturbation 27(372)
H_2^+ limiting cases $r_{ab}\to 0$ and $r_{ab}\to\infty$ 27(225, 226)
H_2^+, potential curves 27(223, 230, 231)
H^- (U-center)
Haidinger's rings in birefringent crystals 25/1(161ff.)
Half-plane 25/1(222)
Half-plane, test of Kirchhoff's solution 25/1(416ff.)
Half-shade 25/1(38)
Half-shade analyser of Ramaseshan 25/1(43ff.)
Half-shade analyser of Richartz 25/1(42ff.)
Half-shade analyser of Tool 25/1(41ff.)
Half-width 27(3, 6, 23, 24, 52, 54, 65, 79)
Hall-constant 25/2a(277)
Halo phenomena 24(609, 616)
Halogen molecules, continuous spectra 28(173, 187)
Halogens, solid, luminescence 26(16)
Halophosphate phosphors 26(46)
Hamiltonian 25/2d(12)
Hamiltonian and complex normal coordinates 25/2b(237)
— demonstration of invariance 25/2b(237)
Hamiltonian and molecular model 27(362, 365)
Hamiltonian and real normal coordinates 25/2b(130)
Hamiltonian, approximate 27(269)
Hamiltonian as function of complex normal coordinates 25/2b(141)
Hamiltonian, electronic 27(268, 271)
Hamiltonian for the self-consistent field 27(270)
Hamiltonian, harmonic 25/2d(134)
Hamiltonian, interaction 25/2d(84, 88, 96, 117)
Hamiltonian, ionic (LCAO MO) 27(280)
Hamiltonian, lattice 25/2d(126, 127, 131, 476)
Hamiltonian, lattice dynamical 25/2a(61, 62, 71)
Hamiltonian, many electron – many ion 25/2b(252)
Hamiltonian, molecular 27(223, 244, 383)

Hamiltonian, monoelectronic 27(269)
Hamiltonian, non-relativistic 25/2d(91)
Hamiltonian of electrons 25/2c(27)
Hamiltonian of field 25/2c(26)
Hamiltonian of field interacting with electrons 25/2c(28)
Hamiltonian, radiation field 25/2d(117)
Hamiltonian, total Hamiltonian of the laser equations 25/2c(37–38)
Hankel approximation 25/1(506)
Hansen and Bol, determination of the velocity of light 24(25)
Hariharan-Sen interferometer 29(779ff.)
Harmonic adiabatic approximation
— and vibrational lattice energy 25/2b(262)
— defined 25/2b(258)
— lattice eigenfunctions as a product 25/2b(262)
— symmetrized eigenfunction 25/2b(264)
Harmonic approximation 25/2a(69, 70, 113, 143)
Harmonic oscillator(s) 25/2a(62)
— creation and annihilation operators 25/2c(294–297)
— forced quantum mechanical 25/2c(80–83)
— renormalized 25/2d(479)
— thermodynamic expectation value 25/2d(482)
Hartmann constant 29(302)
Hartmann, generalized 29(303)
Hartmann test 29(301)
Hartree method 28(103, 269)
Hartree-Fock approximation
— calculations 25/2d(21, 232)
— Green function 25/2d(510)
— polarizabilities 25/2d(74)
Havelock's formula 25/2a(33)
HCN, analysis of the spectrum 27(440, 441)
HCO, analysis of the spectrum 27(441, 442)
Head of a band sequence 27(385)
Heat capacity of gases derived from molecular spectra 26(225–226)
Heatbath, definition 25/2c(35)
Heatbath, interaction with field and electrons 25/2c(37–38)
Heavy-hole ladders 25/2a(257)
Heavy-hole levels 25/2a(290)
He I, singlet-triplet splitting 27(130)
Heights 29(90)
Heisenberg picture 25/2c(30–31)
Heisenberg representation 25/2a(115, 148)
Heitler-London approximation (HLA) 25/2d(52, 54)
Heitler-London method 27(260–265)
— equivalence, with the method of molecular orbitals 27(265, 266)
— higher order approximations 27(264, 265)
Helium, continuous spectrum of principal series 28(135)
Hellmann-Feynman theory 25/2d(58, 113, 530)
Helmholtz formula 25/1(233, 241)
Helmholtz wave equation 29(446, 450)
Helmholtz-Huygens equation 29(507)
Herapathite crystal 25/1(34)
Hermitian matrix 25/2a(66)
Hertz vector 25/1(229ff.)
— edge condition 25/1(258ff.)
— representation of scattering field 25/1(309)
— stationary representation of far field amplitude 25/1(402ff.)
Herzberg's rule concerning bonding and non-bonding electrons 27(256, 257)
Heterogeneous media 25/1(121ff.)
Hexagonal SiC 25/2a(279)
HgCdTe alloy 25/2a(269)
High current carbon arc 28(131)
High pressure arc, water stabilized 28(101, 123)
High pressure energy transfer 27(55–57)
High series members, shift 27(72–76)
Higher-order aberrations 29(70, 103, 138, 139, 176ff., 187ff., 190ff.)
Hi_i^- (U_1 center, KBr) 25/2d(286, 287)
Hindered internal rotation 26(193–194)
Historical survey 25/2d(2)
Hitachi ruling engine with interferometric control 29(724ff.)
HNO, analysis of the spectrum 27(442, 443)
Hole spectra 28(287)

Holeburning 25/2c(184, 192, 199)
Holograms 29(739–741, 743, 745)
Holographic filter 29(748)
Holography 29(738ff.)
Holtsmark field strength probability function 27(31, 33)
Homogeneity of optical glasses 29(293)
Homogeneous deformation 25/2a(114)
Homogeneous wave 25/2a(4, 6)
Homogeneously broadened laser transitions
— above threshold (quantum mechanical) 25/2c(125–138)
— semiclassical 25/2c(185–187, 202)
— below threshold (quantum mechanical) 25/2c(120–124)
Horseshoe mounting 29(334)
Houston, determination of the velocity of light 24(19)
Huang-Rhys factor 25/2d(420, 423)
Huet alignment telescope 29(811)
Huet prism 29(215ff., 218)
Hund's case(s)
— *a* 27(313, 396, 397)
— (*a*) and (*b*) 28(18, 20, 52)
— *b* 27(313, 396, 397)
— *c* 27(315)
— *d* 27(315)
— transition between *a* and *b* 27(315, 316)
Hunzinger's interferometer 29(779ff.)
Hüttel, determination of the velocity of light 24(15)
Huygens construction of total reflection at boundaries of anisotropic media 25/1(112ff.)
Huygens eyepiece 29(288, 367)
Huygens principle 25/1(218); 25/2c(19)
— and grating diffraction 29(445, 504ff.)
— approximate form for electromagnetic waves 29(508–510)
— difficulties in energy calculations 29(513)
— functional connection of boundary values 25/1(233ff., 242ff.)
— interaction of several objects 25/1(366ff.)
— limitations of its applicability 29(513ff.)
— rigorous form for electromagnetic waves 29(506–508)
— scalar formulation 25/1(231ff.)
— vector formulation 25/1(238ff.)
Huygens tensor 25/1(238, 241)
$H_{1-x}D_x$ 25/2d(290)
Hydrides 25/2d(234)
Hydrogen arc 28(122)
Hydrogen bond 26(177, 219)
Hydrogen, deformation frequencies 26(179)
Hydrogen halides, continuous spectra 28(173)
Hydrogen ion, negative 28(101, 111, 119, 122, 154)
Hydrogen molecule and hydrogen molecular ion, *see* H_2 and H_2^+
Hydrogen molecule, continuous emission spectrum 28(179)
Hydrogen, Raman effect in the liquid and solid state 26(220)
Hydrogen, rotation spectrum 26(212)
Hydrogen series in semi-conducting crystals 28(292)
Hydrogen spectra, continuous 28(182)
Hydrogen spectrum, terms and lines 27(85)
Hydrogen, stretching frequencies 26(178–179)
Hydrostatic pressure, effect on optical behaviour 25/1(209)
Hyperfine structure 28(320)
— as a resolution test 29(621)
— for the $[IrCl_6]^{2-}$ complex 28(59)
— in microwave spectra 28(30)
— multiplets 28(39)
— of V^{51} 28(55)
Hypergon 29(198)
Hypersurfaces of potential energy 27(350)

Ideal conductor, screening effect 25/1(358ff.)
Identity for *TH*-waves, special complex 24(536)
Identity, general complex 24(535)
Identity operator 25/2b(10)
Idiophanic rings 25/1(91)
II–VI compounds 25/2a(278)
Illumination (flux density) 29(35, 494)
Illumination of microscope objects 29(376ff., 398ff.)

Illumination of spectrometers 29(491)
Illumination, vertical 24(623)
Image correlation filtering 29(746ff.)
Image formation by gratings 29(428, 437, 452ff.)
Image formation, Fourier transform formulation 29(510)
Image formation on Rowland circle 29(477)
Image function 24(590, 601, 602)
Image function of cylindrical objects 24(612)
Image height aberration of skew rays 29(119, 123, 139, 143, 146ff., 148, 151)
Image height, reduced 29(119, 124, 151)
Image perpendicular aberration 29(92, 99ff., 123, 136, 166ff.)
Image perpendicular, reduced 29(92, 100, 112, 166ff.)
Image plane 24(557)
Imaging gratings 29(437ff.)
Immersion mirror objectives 29(371)
Immersion objects lens 29(346, 358)
Imperfect crystal, diamond and rock-salt
— local modes 25/2b(441)
— local modes, symmetry 25/2b(440)
Imperfect crystal, infra-red absorption 25/2b(449)
Imperfect crystal, Raman scattering 25/2b(453)
Imperfection centers 25/2a(181)
Imperfection variation law for diffraction patterns 29(543, 545, 547)
Imperfections of gratings 29(610)
Imperfections of wavefront 29(452, 454)
Imperfections of wavefront, effect on spectral image 29(538ff.)
Impurities, *see* Defects substances
Impurities, deep-level impurities 25/2a(215)
Impurity activation of phosphors 26(19–23, 34, 44, 48, 49, 50)
Impurity, exciton impurity complexes 25/2a(225)
Impurity magneto-optic effects 25/2a(314)
InAs 25/2d(216)
InAs, indiumarsenide 25/2a(267, 269, 275, 276, 303, 308)
Incident wave 24(465)
Incident-light illumination 29(380)
Incoherence of spectral lines 29(527)
Incoherent addition of light beams 25/1(15ff.)
Incoherent illumination of grating 29(459, 489)
Incoherent radiation 29(344)
Incoherent scattering 25/2a(78)
Index ellipsoid 25/1(61ff.)
— and crystal symmetry 25/1(66ff.)
— Poinsot's construction 25/1(62)
Index tensor 25/1(59)
Index tensor for absorbing crystals 25/1(86ff.)
Indirect exciton transitions 25/2a(178, 192)
Indirect interband transitions 25/2a(165)
Indirect magneto-absorption 25/2a(296)
Indirect magneto-excitons 25/2a(296)
Indirect transitions 25/2a(163, 190)
Indirect transitions, intraband 25/2a(168)
Induced dipole moment 26(198, 201, 203–204)
Induced representation of $\mathfrak{G}$ from $\mathfrak{G}(\boldsymbol{k})$ 25/2b(57, 60)
Induced representation of space group from subgroups 25/2b(82)
Inelastic neutron scattering 25/2a(76, 145)
Inert-gas solids 25/2a(88)
Infrared absorption 25/2a(72, 81, 82, 123); 25/2d(4, 7, 77, 95, 114, 149, 164, 186, 213, 331, 358, 373, 409)
— absorbed energy 25/2d(99)
— band mode 25/2d(290, 293)
— bibliography 25/2d(203)
— by phonons 25/2b(272)
— by phonons, Hamiltonian 25/2b(272)
— by phonons, Lax-Burstein approximation 25/2b(274)
— by phonons, rate 25/2b(277)
— by phonons, selection rule 25/2b(278)
— concentration effects 25/2d(382, 394)
— crystals covalent, ionic, etc., *see* Substances

— defect-induced, *see also* Critical points, Gap modes, Local modes, Resonance modes 25/2d(209, 266, 273, 289–295, 331–340, 371, 439)
— diagrams 25/2d(186)
— dispersion oscillators, *see* Dispersion oscillator
— Drude part 25/2d(218)
— electric-field-induced 25/2d(451)
— high-frequency 25/2d(178)
— integrated intensity 25/2d(333, 341, 375, 419)
— isotopes 25/2d(289)
— low-frequency 25/2d(183)
— low-symmetry 25/2d(224)
— many body theory 25/2b(322)
— many body theory, susceptibility 25/2b(324)
— mixed crystals 25/2d(349)
— model theory 25/2d(148, 358)
— multiphonon 25/2b(282)
— multi-phonon 25/2d(20, 162, 178, 180, 418, 420, 434)
— non-linear dipole moments 25/2d(8, 59, 98, 135, 156, 186, 188, 208, 210, 263, 360, 376, 438)
— one phonon 25/2b(279)
— one phonon, selection rule 25/2b(280)
— pressure-dependence 25/2d(195, 405, 439, 450, 463)
— Reststrahlen band, *see* Reststrahlen oscillator, Dispersion oscillator
— small particles 25/2d(469)
— spectra of crystals 25/2d(157ff.)
— sum rule 25/2d(321)
— temperature dependence, *see also* Self-energy 25/2d(195)
— two phonon 25/2b(282)
— two phonon in diamond, Ge, Si 25/2b(414)
— two phonon, selection rule 25/2b(282)
Infrared block 24(507)
Infrared cyclotron resonance 25/2a(267)
Infrared emission of phosphors 26(17, 29, 51, 79–81)
Infrared grating 29(436, 711)
Infrared grating efficiency 29(606)
Infrared luminescence of cuprous oxide 26(30)
Infrared microscopy 29(395)
Infrared reflectivity 25/2a(134)
Inhomogeneous wave 25/2a(4, 6)
Inhomogeneously broadened laser transitions above threshold
— Doppler line 25/2c(194–201, 203)
— Gaussian line 25/2c(187–190, 203)
— Lorentzian line 25/2c(190)
— quantum mechanical 25/2c(144–146)
— semiclassical 25/2c(187–194)
InP indium phosphide 25/2a(269, 274, 303)
$InP_{1-x}As_x$ 25/2d(398)
InSb 25/2d(216, 218)
InSb diodes 25/2a(313)
InSb, indium antimonide 25/2a(25, 267, 275, 276, 299, 303, 308, 311)
Instrument function, theoretical 29(528–530)
Integral equation for surface currents 25/1(299ff.)
Integral equation of convolution type for diffraction by a slit 25/1(422ff., 435ff.)
Integral equation, singular, of Cauchy type 25/1(337, 345ff.)
Integral intensities of band systems 27(330, 341, 342, 357, 359, 360, 446–452)
Integrated absorption 27(447, 448)
Integro-differential equation for surface current, vector problem 25/1(311)
Intensities in microwave spectra 28(22)
Intensities, ratio of 24(473)
Intensities, relative, of a hyperfine-structure 28(73)
Intensity alternation in rotation spectrum 26(212)
Intensity alternation in rotational branches 27(338, 348, 357, 437, 439)
Intensity correlation function 25/2c(76)
Intensity distribution in the rotational structure 27(330, 342, 347, 348, 357, 399)
Intensity distribution in the vibrational structure 27(330, 342, 344–347)

Intensity fluctuations
— below threshold 25/2c(105, 163–164)
— high above threshold 25/2c(106–109)
— in close vicinity of threshold 25/2c(105, 163–164)
— well above threshold 25/2c(105, 146, 163–164)
Intensity formulae, rigorous 24(518)
Intensity, influence of, on the velocity of light 24(4)
Intensity method 24(518, 540)
Intensity of a plane wave 25/2a(7)
Intensity of absorption and emission lines 27(331)
Intensity of forbidden transitions 27(352, 451, 452)
Intensity of light, definition 29(492)
Intensity of magnetic dipole transitions and electric quadrupole transitions 27(340)
Intensity of transitions and Jahn-Teller effect 27(360)
Intensity ratio of Stokes and anti-Stokes lines 26(203)
Intensity sum rule in Zeeman effect 28(318)
Interaction and intensity problems 27(306)
Interaction and isotope effect 27(306)
Interaction and transition probability 27(353)
Interaction, anharmonic 25/2a(113, 131, 135, 141)
Interaction, axially symmetric 25/2a(131, 135)
Interaction between
— electronic states 27(239, 248, 306, 376)
— electrons and the rotation 27(312, 318–320)
— neighbouring states 27(368)
— particles and radiation 27(330)
— radiation field and electron wave field 25/2c(28)
— three neighbouring states 27(372, 373)
Interaction, Coulomb 25/2a(84)
Interaction, electron-phonon 25/2a(81, 107)
Interaction in the case of degenerate states 27(354)
Interaction of diffracting objects
— Huygens' principle 25/1(366ff.)
— iteration method 25/1(366ff., 369)
— mutual irradiation of two cylinders 25/1(369ff.)
— Schwarzschild's iteration method 25/1(363ff.)
Interaction of higher order 27(305)
Interaction of potential curves 27(248)
Interaction, $\pi \cdot p$ 25/2a(259)
Interaction representation 25/2c(29)
Interaction with molecular axis 27(252)
Interband absorption, free carrier 25/2a(201)
Interband effect 25/2a(206, 238)
Interband Faraday rotation 25/2a(298)
Interband magneto-absorption 25/2a(286)
Interband magneto-reflection 25/2a(311)
Interband processes 25/2a(257)
Interband transitions 25/2a(159)
Interband transitions, indirect 25/2a(165)
Intercollision time 27(8, 14)
Interference comparator of Kösters 29(759, 768–773)
Interference contrast 29(407)
Interference filter 24(500)
Interference filter, thin layer as 24(508)
Interference fringes, position measurement 29(766)
Interference gratings 29(735–754)
Interference measurement of parallelity 29(777)
Interference method(s) 24(528)
— of the first kind 24(556, 638)
— of the second kind 24(556, 638)
— of the third kind 24(556, 640)
Interference microscope testing of groove shape 29(652ff.)
Interference microscopy 29(407ff.)
Interference phenomena
— in convergent light 25/1(128ff., 132ff., 138ff., 150ff.)
— in inactive crystals 25/1(129)

— in optically active crystals 25/1(132ff.)
— in parallel light 25/1(126ff.)
— on transparent crystals 25/1(126ff.)
Interference surface control, Saunders-Gross method 29(778)
Interference test of surfaces 29(827–832)
Interferential alignment 29(814)
Interferogram 25/2a(43); 29(559, 566ff., 632ff., 693ff.)
Interferogram, Fourier transform (=transformation) of 25/2a(51)
Interferometer 24(37)
Interferometer, Michelson 25/2a(43, 51)
Interferometer of Hariharan and Sen 29(779ff.)
Interferometer of Hunzinger 29(779ff.)
Interferometer of Michelson 29(408, 631, 635, 783)
Interferometer with mercury surfaces 29(777)
Interferometers 29(436ff.)
Interferometric control of diamond carriage 29(728–730)
Interferometric control of ruling engines 29(429ff., 674, 678ff.)
Interferometric modulation spectrometers 29(502)
Interferometric tests of gratings 29(621ff.)
Interferometric tests of telescopes 29(309ff.)
Intermediate coupling 27(111–117); 28(305)
Intermediate coupling, approximations 27(117–119)
Intermediate image in microscope 29(354)
Intermediate magnetic fields 28(314)
Internal conversion 27(381, 382)
Internal conversion and the principle of Franck-Condon 27(381)
Internal coordinates 26(187)
Interplanar force constants 25/2a(102)
Interruption distribution of lines 27(36)
Interruption theory for molecular broadening 27(62–65, 67)
Intersection of potential curves 27(248, 371)
Interval rule of Landé 27(107)
Intraband effect 25/2a(162, 206)
Intraband indirect transitions 25/2a(168)
Intramolecular tension 26(187)
Intrinsic absorption edge 25/2a(187)
Intrinsic efficiency of a luminescence centre 26(10)
Intrinsic emission 25/2a(197)
Invariance of the parameter β in the LCAO MO method 27(279)
Inversion at the origin 27(229, 245)
Inversion doubling 26(233)
Inversion, saturated, time-dependent 25/2c(140, 175, 180, 185)
Inversion, saturated, time-independent 25/2c(131, 138, 140, 185)
Inversion spectrum of ammonia 28(7, 8, 9)
Inversion, unsaturated 25/2c(121, 177)
Inverted terms 27(109)
Iodine absorption test of grating resolution 29(613)
Ionic and electronic Stark broadening, combination 27(42)
Ionic crystals 25/2a(90, 137); 25/2d(15, 21, 61, 157, 161)
Ionic crystals, lattice energy 25/2d(21)
Ionic molecules 27(266, 267)
Ionic plasma frequency 25/2d(23)
Ionic polarizabilities 25/2d(60, 190)
Ionic Stark broadening 27(35, 41)
Ionic structure 27(264)
Ionic terms in the wave function 27(265–267)
Ionicity 25/2d(59, 71, 72)
Ionicity of an electronic state 27(267)
Ionisation balance 28(97)
Ionisation energy, effective 28(96, 129)
Ionisation stages from Zeeman observations 28(330f.)
Ionization energies of neutral atoms 27(201)
Ionization energy 27(211, 270, 276)
Ionization potential 27(191ff., 212, 272, 418, 453–455)
Ionization potentials of systems of three to ten electrons 27(197)
Ions, negative 28(101, 104, 111)
Ions, positive 28(109)
Iridescence 25/1(116ff.)
Iron, emissivity 26(148, 149)

Iron group, magnetic resonance 28(50)
Irreducible representation(s) 25/2a(117, 128, 141); 28(213)
— block matrix structure 25/2b(53)
— completeness of all 25/2b(29)
— $D^{(*k)(m)}$ of space group $\mathfrak{G}$
— of symmetry groups 27(245, 321, 326)
— of $\mathfrak{T}$, subduced by $D^{(*k)(m)}$ of $\mathfrak{G}$ 25/2b(49)
— space group, elements 25/2b(60)
Irregular doublets 27(156)
Isoelectronic extrapolation of electron affinities 27(199–201)
Isoelectronic molecules 27(250, 251, 299, 418, 428)
Isoelectronic sequences 27(88–90, 147ff.)
Isoelectronic series 27(418)
Isoelectronic term systems 27(150–156)
Isomerism 26(193, 219, 237–239)
Isoplanatism 29(28ff., 360)
Isotope effect(s), *see* Defects 27(409, 410)
— and perturbations 27(410)
— and predissociation 27(410)
— examples 27(426, 427, 432)
— in Raman spectra 26(191)
Isotopic substitution 27(409–410, 419, 421, 436)
Iteration fromulae 29(108ff., 113ff., 121ff., 135ff.)
Ivey relation 25/2d(405)

Jablonski line shape theory 27(25–29)
Jahn-Teller effect 27(308–312, 354); 28(53)
Jahn-Teller effect and intensity of electronic transitions 27(360)
Jarrell-Ash ruling engine with interferometric control 29(703–706)
J-dependence in molecular broadening 27(57, 60, 65)
Jellium 25/2d(13)
Jig-transit square 29(824)
jj coupling 27(95); 28(305)
jl coupling 27(95)
Jobin et Yvon ruling engine with interferometric control 29(719ff.)
Johns Hopkins ruling engine 29(680–682)
Joint density of states 25/2a(209)
Jones matrix 25/1(24)
Jones matrix, experimental determination 25/1(26ff.)

K doupling 27(356)
Kane, $k \cdot p$ perturbation calculation 25/2a(259, 273)
Karolus and Mittelstaedt determination, velocity of light 24(12)
KBr 25/2d(161, 173, 175, 178, 184, 191, 195, 234, 237, 285, 292, 294, 345, 372, 443)
KBr, potassium bromide 25/2a(122, 124, 127, 134, 145)
$KBr:D^-$ 25/2d(434)
$KBr:H^-$ (U-center) 25/2d(434, 439, 440, 441, 444)
$KBr:H_i^-$ (U_1) 25/2d(286, 287)
$KBr:Li^+$ 25/2d(390, 422, 430)
$KBr:{}^6Li^+$ and ${}^7Li^+$ 25/2d(429)
$KBr:Tl^+$ 25/2d(344, 345)
KBr_xI_{1-x} 25/2d(406)
KCl 25/2d(161, 179, 184, 230, 267, 281, 292, 293, 294, 345, 347)
KCl, F centers 25/2d(348)
$KCl:CN^-$ 25/2d(351)
$KCl:D^-$ 25/2d(428, 433)
$KCl:H^-$ 25/2d(370, 418, 422, 423, 428, 433, 465, 466)
$KCl:{}^6Li$ 25/2d(281)
$KCl:Na^+$ 25/2d(288)
$KCl:Na^+$ (pair) 25/2d(422)
$KCl:Tl^+$ 25/2d(291)
$KCl_{1-x}Br_x$ 25/2d(395, 405)
$(KCl)_{1-x}(NH_4Cl)_x$ 25/2d(390)
Kellermann model 25/2d(25, 27, 30, 50, 165)
Kellner eyepiece 29(289)
Kerr cell 24(11)
KH_2PO_4, potassium dihydrogen phosphate or KDP 25/2a(143)
KI 25/2d(160, 292, 300, 301, 350, 443)
$KI:Ag^+$ 25/2d(422, 428, 431, 437)
$KI:Cl^-$ 25/2d(288)
$KI:H_i^-$ 25/2d(286)
$KI:Na^+$ 25/2d(288)
Kinetic energy of lattice, classical 25/2b(116)

Kinetics of luminescence decay 26(4, 5, 24)
KI:Rb^+ 25/2d(288)
Kirchhoff's integral 25/1(219, 236)
Kirchhoff's law 26(133, 134); 28(83)
Kirchhoff's solution 25/1(238, 269ff.)
— analytical continuation 25/1(273)
— electromagnetic analog 25/1(274)
— for circular disk 25/1(480ff.)
Kirchhoff's theory 25/1(218, 267, 268ff.)
— Braunbek's improvement 25/1(219)
— modification for plane screens 25/1(279ff.)
KI:Tl^+ 25/2d(347)
Knife-edge test 29(304ff.)
Köhler illumination 29(379)
Kohn anomaly 25/2a(100)
Kösters' interference comparator 29(759, 768–773)
Kösters' prism 29(812ff., 823)
$k \cdot p$ perturbation calculation of Kane 25/2a(259, 273)
Kramer effect 26(110)
Kramers degeneration 28(229)
Kramers doublet 28(53)
Kramers-Kronig relations 25/2a(29, 157, 206); 25/2d(5, 89, 101, 184, 193, 319, 320, 324, 325, 417, 486)
Kramers-Unsöld theory 28(98, 101, 141)
Kronecker powers, *see* Direct product
— cube 25/2b(90)
— square 25/2b(89)
— symmetrized 25/2b(89)
Kronecker product, *see* Direct product
Krypton 25/2a(89)
Krypton continuum in the ultraviolet 29(646)
Krypton lamp 29(767, 769)
$KTaO_3$ 25/2d(219, 245)
Kubo formula 25/2d(96, 101)
Kubo, linear-response theory 25/2d(490)
Kurosawa formula 25/2d(206, 221)
$K_{1-x}Rb_xI$ 25/2d(405)

l doubling 27(356, 423, 425)
Lagrangian, classical lattice 25/2b(116)
Lagrangian, function of complex normal coordinates 25/2b(137)
Lagrangian invariant 29(12, 15, 46, 49)
Lamb dip 25/2c(199)
Lamb shift 28(36, 38)
Λ doubling 27(306, 316, 317, 322, 356, 362, 397, 398, 423, 428)
Λ doubling and perturbations 27(372)
Lambert's cosine law 26(153)
Lamella 25/2a(19, 45)
Lamella, amplitude reflectance 25/2a(19)
Lamella, amplitude transmittance 25/2a(20)
Lamella, thick 25/2a(46, 50)
Lamella, thin 25/2a(49)
Landau levels 25/2a(237, 247, 256, 267)
Landé formula 27(129)
Lande *g*-factor, *see* *g*-factor
Landé interval rule 27(107)
Lange data of construction 29(114ff., 177, 187, 191)
Langevin equations 25/2c(156)
— classical 25/2c(36)
— electrons, two-level atom 25/2c(42–43)
— exact elimination of atomic operators within 25/2c(286)
— field equations 25/2c(50, 113)
— field (one mode) 25/2c(40–42)
— matter equations 25/2c(50, 115–116)
— quantum mechanical 25/2c(38, 39–43)
— quantum mechanical, qualitative discussion of single-mode operation 25/2c(116–120)
Langevin forces, correlation functions
— arbitrary quantum systems 25/2c(46)
— field 25/2c(46)
— *N*-level atom 25/2c(46–49)
Langevin forces, quantum mechanical, origin 25/2c(39–43)
LaO, analysis of the spectrum 27(389–394)
Laser cascades 25/2c(259–264)
— discussion of an example 25/2c(262–264)
— field equations 25/2c(260)
— homogeneously broadened line, standing waves 25/2c(261)
— inhomogeneously broadened line, standing waves 25/2c(261–262)
— matter equations 25/2c(260)

Laser condition 25/2c(4, 186)
Laser, distribution functions 25/2c(65)
Laser equations
— family tree 25/2c(100)
— field equations (fully quantum mechanical) 25/2c(50)
— field equations (semiclassical) 25/2c(51, 177–179)
— matter equations (fully quantum mechanical) 25/2c(50)
— matter equations (semiclassical) 25/2c(51, 177–181)
— quantum mechanical 25/2c(33–73)
— survey of methods of solution 25/2c(100)
— theoretical results (fully quantum mechanical treatment), comparison with experiments 25/2c(101)
— total Hamiltonian 25/2c(37–38)
Laser, external signals applied to 25/2c(124–125)
Laser, gas laser 25/2c(194–201, 203)
— intensity fluctuations 25/2c(146)
— linewidth 25/2c(146)
— polarization of lasing modes in a gas laser 25/2c(201)
— quantitative treatment of single-mode operation above threshold (semiclassical) 25/2c(197–199)
— semiclassical equations 25/2c(194–197)
Laser, giant pulse laser 25/2c(267–271)
Laser, 3-level, quantum mechanical treatment 25/2c(125–137)
Laser output, qualitative discussion 25/2c(102)
Laser, solid-state
— semicalssical equations 25/2c(51, 173–197)
— semiclassical treatment 25/2c(173–194, 201)
Laser transitions, homogeneously broadened
— above threshold (quantum mechanical) 25/2c(125–138)
— above threshold (semiclassical) 25/2c(185–187, 202)
— below threshold 25/2c(120–124)
Laser transitions, inhomogeneously broadened
— above threshold (Doppler line) 25/2c(194–201, 203)
— above threshold (Gaussian line) 25/2c(187–190, 203)
— above threshold (Lorentzian line) 25/2c(190)
— above threshold (quantum mechanical) 25/2c(144–146)
— above threshold (semiclassical) 25/2c(187–194)
Lateral resolution 29(343)
Lattice absorption edge 26(28, 33)
Lattice absorption, strength of 25/2a(33)
Lattice and nuclear vibration superposition in the uranyl spectrum 26(51)
Lattice defects in alkali halides 28(289, 291)
Lattice, dipole moment 25/2d(132)
Lattice, distortions, defect-induced 25/2d(297–300, 310, 337, 365, 405, 440, 445, 450, 451, 460)
Lattice dynamical, Hamiltonian 25/2a(61, 62, 71)
Lattice dynamics 25/2a(59); 25/2d(2)
Lattice dynamics, adiabatic condition 25/2d(34)
Lattice eigenfunction, transformation of 25/2b(269)
Lattice, energy 25/2d(22)
Lattice frequency 25/2a(33)
— longitudinal optical 25/2a(26)
— natural undamped 25/2a(24)
Lattice, Hamiltonian 25/2d(126, 127, 131, 476)
Lattice hypothesis 25/2b(9)
Lattice, polarizability 25/2d(143)
Lattice, potential 25/2d(16, 114, 126)
Lattice, reciprocal 25/2a(67)
Lattice statics 25/2d(297)
Lattice susceptibility 25/2d(133, 141)
Lattice vibration frequencies 28(260)
Lattice vibrations 25/2a(64); 26(221)
Layer systems 24(468, 501)
Layers, inhomogeneous 24(491)
Layzer's first-order theory of complexes 27(174, 175, 176, 177)
Lead activated alkali halides 26(23)
Lecomte and Vincent-Geisse construction 25/2a(56)
Ledermann theorem 25/2d(315)
Left circularly waves 25/2a(234)
Leman-Sprenger prism 29(216, 218)

Lemma of necessary degeneracy and eigenvectors 25/2b(132)
Length measurement by wavelength 29(755, 766)
Length measurement in microscopy 29(403ff.)
Length measurement with Pérot-Fabry plates 29(782)
Lens curvature, determination 29(820–822)
Lenz-Weisskopf interruption theory 27(8, 9)
Leontovich condition 25/2c(13)
Level diplacement by a crystalline field 28(210)
Level numbers 28(221)
Level structure in intermediate coupling 27(111–117)
Level structure in pair coupling 27(120, 121, 123)
Level structures determined by Zeeman measurements 28(326)
Levelling telescope 29(808)
Levels, definition 27(100)
Levels, electronic, rovibronic..., *see* Electronic levels, Rovibronic levels...
Levels of rare earth ions 28(269)
Levels of the chromium ion 28(279)
LiBr 25/2d(69)
Librations 25/2d(284)
LiCl 25/2d(390)
LiD 25/2d(232, 240, 241)
LiF 25/2d(59, 69, 161, 168ff., 179ff., 189, 196, 248, 290, 438, 469, 470)
Life-time of phosphors 26(122, 127)
$LiF:H^-$ 25/2d(269, 438)
Lifshitz formalism 25/2d(8, 296, 302, 309, 311, 363)
Light beam pointer 29(757, 760)
Light beams, non-orthogonally polarised 25/1(22)
Light beams, partially coherent 25/1(20ff.)
Light beams, partially polarised 25/1(15ff.)
Light duct 29(383)
Light scattering 25/2d(79)
— electrons 25/2d(90)
— inelastic, see Raman scattering
— polaritons 25/2d(123)
— transition rate 25/2d(118)
Light source(s)
— for Zeeman measurements 28(299)
— in microscopy 29(385)
— modulation in luminescence excitation 26(123, 125–126)
— moving 24(4, 40)
— slit 24(619)
Light-hole ladders 25/2a(257)
Light-hole levels 25/2a(290)
Light-intersection method 24(556, 633)
LiH 25/2d(232, 240)
$LiH_{1-x}D_x$ 25/2d(404)
Li I sequence ionization energy 27(193)
Li I sequence, polarization formula 27(187)
Li I, term system and line series 27(86)
Li I-like spectra, isoelectronic relations 27(89)
Limit of Rydberg series 27(415–418)
Lindholm theory for molecular broadening 27(9, 62–65)
Lindquist and Ewald, method of 25/2a(39)
Line asymmetry 27(9)
Line broadening due to predissociation 27(274, 279)
Line broadening due to tunnel effect 27(304, 379, 380)
Line narrowing 25/2c(122–123)
Line shape(s)
— according to the Stark theory 27(33)
— according to the statistical theory 27(24)
— natural 27(77–79)
— of lasing modes 25/2c(131–132)
— of microwave spectra 28(22, 24)
— theory of Jablonski 27(25–29)
Line shift 27(9)
Line strength (S_J) 27(347, 348, 357, 400)
Line width as function of rotational quantum number 27(66)
Line width in solids 28(47)
Line width, natural 28(24)
Linear birefringence 25/1(5ff.)
Linear chain 25/2a(59, 67, 69, 102, 108)
Linear combination of atomic orbitals (LCAO) 27(234, 244, 245, 250)

Linear dispersion of a grating 29(459, 461ff.)
Linear molecules, spectra 27(421–428)
Linear phase aberration in gratings 29(541)
Linear Stark effect 27(30, 34)
Linear wavefront aberrations 29(535ff., 541)
Linewidth at threshold 25/2c(166)
Linewidth below threshold 25/2c(122, 166)
Linewidth, natural 25/2c(33, 97–98)
Linewidth somewhat below threshold 25/2c(103, 166)
Linewidth, spontaneous 25/2c(97–99)
Linewidth well above threshold 25/2c(102–103, 131, 146, 166)
Liouville operator 25/2c(288)
Liquid state, Raman effect 26(219)
Lissajous figure 27(307, 379, 388, 441)
Lithium fluoride, absorption lines 28(293)
Little group of ***k***, *Π*(***k***)
— definition 25/2b(63)
— irreducible representation, allowable 25/2b(64)
— irreducible representation, non-allowable 25/2b(65)
$^6Li_x{}^7Li_{1-x}F$ 25/2d(290)
Local field 25/2a(93)
Local field, effective 25/2a(159)
Local modes 25/2b(440); 25/2d(267, 287, 288, 295, 314, 318, 320, 322, 324, 330, 391, 398, 400, 412, 421, 435, 436, 466)
— crystals with impurities, *see* Substances
— intensity 25/2d(327, 422)
— isotope effects 25/2d(267, 268, 295, 296)
— line width, *see* Self-energy
— shift under pressure 25/2d(463)
— sideband 25/2d(283, 286, 371, 423, 441, 444)
— U center 25/2d(296, 368, 389, 415, 428, 432, 437, 461)
Localised modes = local modes 25/2a(82, 186)
Long-distance objective 29(196, 199)
Longitudinal component of a wave 25/2a(5)
Longitudinal dielectric constant 25/2a(162)
Longitudinal excitons 25/2a(172, 193)
Longitudinal optical frequency 25/2a(34)
Longitudinal optical lattice frequency 25/2a(26)
Longitudinal sharpness of objective 29(204)
Lorentz field 25/2d(60, 62, 66)
Lorentz theory of line shape 27(4–6)
Lorentz theory of Zeeman effect 28(297)
Lorentz unit 28(297, 301)
Lorentzian line, inhomogeneously broadened laser transitions, semiclassical 25/2c(190)
Lorentzian line shape 28(24)
Lorentz-Lorenz formula 25/2d(75, 248)
Loss modulation 25/2c(215–216, 217–224)
Loupe 29(351, 787)
Loupe magnification 29(351, 373)
Low pressure discharge 28(134)
Low pressure energy transfer 27(50ff., 52ff.)
Low terms, relations between them 27(218)
LS coupling 27(94, 100ff.)
LS multiplets 27(98)
LS-coupling 28(303, 310)
l-type doublet 28(11)
Luminance temperature 26(130, 158, 159)
Luminescence activation 26(19–23, 34, 44, 48, 49, 50)
— copper 26(34, 35)
— divalent manganese 26(23, 25–27, 36, 37, 49, 50)
— thallium 26(20, 21)
— uranium 26(23, 24, 51)
Luminescence and photo-conductivity, relation 26(92)
Luminescence, bimolecular reactions 26(3, 4)
Luminescence centres 26(3, 9–11)
— energetic conditions 26(9, 10)
— metastable states 26(5, 62–65)
Luminescence decay, kinetics 26(4, 5, 24)
Luminescence decay, measurement of long decay times 26(127)

Luminescence decay, measurement of very short life times 26(122)
Luminescence, definition 26(1)
Luminescence, distinction from Raman and Compton effect 26(1)
Luminescence due to minority carrier injection 26(13, 14, 107–110)
Luminescence efficiency 26(10, 11)
Luminescence emission of *F* centers 26(16–18)
Luminescence, energy transfer 26(81–85)
Luminescence excitation of crystal powder 28(209)
Luminescence in monoatomic solids 26(11–16)
Luminescence in organic solids and liquids 26(53–60)
Luminescence in oxide phosphors 26(27–31)
Luminescence in silver halides 26(23)
Luminescence measurements, experimental techniques 26(121–127)
Luminescence, monomolecular decay 26(3, 66)
Luminescence of alkali halides 26(16–23)
Luminescence of alkaline earth sulphides 26(39–40)
Luminescence of halide phosphors 26(16–27)
Luminescence of phosphates 26(45–47)
Luminescence of silicates 26(40–45)
Luminescence of zinc and cadmium sulphide 26(31–38)
Luminescence, polarization 26(18, 26, 85–89)
Luminescence, polymolecular reactions 26(5)
Luminescence, quenching 26(8, 18, 42, 58, 60, 78, 80)
Luminescence, sensitisation 26(25, 56, 81–85)
Luminescence, stimulation 26(78–81)
Luminescence under periodic excitation conditions 26(122–127)
Luminescent chamber 26(120)
Luminosity of gratings 29(453, 461, 486–504)
Luminosity of spectrographs and spectrometers 29(495 ff.)
Luneberg-Kline method 25/1(298, 529)
Lyddane-Sachs-Teller relation 25/2a(27, 139); 25/2d(24, 66, 70, 76, 94, 95, 102, 204, 206, 395)
Lyman α line, broadening by electron impacts 27(39–41)
Lyman ghosts 29(555)
Lyot filter 25/1(160)

Mach-Zehnder interferometer 29(409, 747 ff.)
Macroscopic electric field
— and Lyddane-Sachs-Teller 25/2b(306)
— and symmetry lowering 25/2b(306)
Macroscopic field 25/2a(87)
Madelung constant 25/2a(91)
Madelung sums 25/2d(13, 22)
Magnesia, luminescence 26(27, 28)
Magnetic conductor, ideal 25/1(224, 372)
Magnetic dipole
— interactions 25/2d(85)
— moment 25/2a(251)
— radiation 27(339, 340)
— transitions 27(340)
— transitions in luminescence 26(3)
Magnetic field
— effect of longitudinal 25/2c(205–212)
— emission from semiconductor diodes 25/2a(312)
— influence on spectral line intensities 28(330)
— polarization 25/2c(205, 211)
— strength, operator 25/2c(26)
Magnetic interaction 28(39)
Magnetic materials 25/2a(315)
Magnetic moment, rotational 28(28)
Magnetic quantum number 27(90)
Magnetic resonance in solids 28(43)
Magnetic resonance of free atoms 28(40, 41)
Magnetic rotation spectra 27(410, 411)
Magnetic waves, transverse 25/2a(7)
Magneto-absorption 25/2a(242)
— cross-field 25/2a(312)
— indirect 25/2a(296)
— interband 25/2a(286)
Magneto-excitons, indirect 25/2a(296)
Magneto-optic effects, impurity 25/2a(314)

Magneto-optic rotation and birefringence 25/1(199ff.)
Magneto-optics 25/1(198ff.)
Magneto-plasma reflection 25/2a(281)
Magneto-reflection 25/2a(244)
Magneto-reflection, interband 25/2a(311)
Magnification law 29(8)
Magnification of a telescope 29(285)
Magnifying lens 29(350ff., 787)
Manganese activated
— alkali halides 26(23, 49)
— cadmium iodide 26(27)
— fluorides 26(26)
— fluorides, change in paramagnetic resonance spectrum 26(27)
Many-body forces 25/2d(5, 22, 38, 40, 41)
Many-body techniques 25/2a(148)
Many-electron configuration 27(96, 99ff.)
Many-electron jumps 27(144)
Markoffian process 25/2c(36)
Maschke's theorem, statement of 25/2b(30)
Maser principle 25/2c(1)
Masks put on slits 29(500ff.)
Mass defects 25/2d(312, 318, 321, 322, 340, 461)
Mass tensor 25/2a(260)
Mass tensor, effective 25/2a(161)
Master equation, *see* Density matrix equation
Matrix, density matrix 25/2a(150)
Matrix description of Gaussian optics 29(14)
Matrix, dynamical 25/2a(65, 117)
Matrix element of a crystal field 28(240, 246)
Matrix, Hermitian 25/2a(66)
Matrix method of solving Maxwell's equations 25/1(107ff.)
Matrix method, relationship to method of superposition 25/1(109ff.)
Matrix operator, differential 25/1(27ff.)
Matrix representation of a group 26(182–183)
Maximum f number 29(496)
Maxwell distribution of dipole moments after collision 27(70)
Maxwell's equation(s) 24(462); 25/1(54ff., 223); 25/2a(2, 3); 25/2c(24)
— boundary conditions 25/1(223ff.)
— separation 25/1(227ff.)
— separation for cylinder problems 25/1(226ff.)
— solution by matrix method 25/1(107ff.)
McKinley, determination of the velocity of light 24(19)
Mean life time of excited states 27(331, 446–452)
Mean life time of free radicals 27(422, 452)
Mean life time of metastable states 27(451, 452)
Mean ray data 29(109ff., 112, 125, 166ff.)
Measurability, limits 25/2c(83–88)
Measurement under microscope 29(403)
Measuring force 29(761)
Measuring loupe 29(787)
Measuring microscope 29(792, 799)
Mechanical design of ruling engines 29(669)
Mechanical model of vibration of a molecule 26(175)
Mechanical representation as a direct product 25/2b(206)
Mechanical representation or "total" representation 25/2b(204)
Mechanical representation, trace of 25/2b(206)
Medial telescope 29(284)
Meniscus systems 29(279ff.)
Mercury lamp 29(389)
Mercury lines test of grating resolution 29(611, 617ff., 621, 694)
Mercury spectrum, continuous 28(183)
Mercury ultraviolet standard 28(81, 129, 141)
Mercury vapor high pressure lamp 28(101, 120, 128, 140)
Mercury vapor lamp 26(172)
Meréchal integrator 29(564)
Meridional image height aberration 29(129, 135ff., 138ff., 146ff., 174ff., 176, 187, 190ff.)
Meridional rays 29(89ff., 169)

Merton's NPL ruling engine for metrology gratings 29(725–727)
Metal measure rod 29(785)
Metal mirrors 29(298)
Metal oxides and sulphides, continuous spectra 28(179)
Metal surface, interference test 29(830)
Metallicity 25/2d(59, 70, 72, 73)
Metastable state of luminescence centres 26(5, 9, 62–65)
Metastable states 27(451, 452)
Meter comparator 29(767)
Meter, definition 29(766)
Methane, Raman spectrum 26(234)
Method of Berremann 25/2a(33)
Method of Lindquist and Ewald 25/2a(39)
Method, sampling 25/2a(110)
Methods of solution, laser equations 25/2c(100)
Mètre prototype 29(755, 766)
MgO 25/2d(60, 69, 74, 216, 232, 239, 244, 283, 472)
MgO:OH^- 25/2d(283)
Michelson interferometer 25/2a(43, 51); 29(408, 631, 635, 783)
Michelson-Lorentz interruption theory 27(2, 3, 7–20)
Michelson-Twyman-Green interferometer 29(533ff., 652)
Microbeam 29(396)
Micrometric eyepiece 29(367)
Microphotography 29(390)
Microphotometry 29(411ff.)
Microprojection 29(390)
Microscope for hot cells 29(374)
Microscope stages 29(407)
Microscopic interference test of surfaces 29(827–832)
Microscopic theory 25/2d(51, 57, 473)
Microscopic theory, phonons 25/2d(74)
Microscopy and length determination 29(788)
Microspectrophotometry 29(411ff.)
Microwave absorption, absorption coefficient 28(23)
Microwave detectors 28(4, 7)
Microwave experiments with standing grating waves 29(579)
Microwave gratings 29(436ff.)
Microwave harmonic generator 28(3)
Microwave interferometry 24(26)
Microwave propagation, atmospheric 28(17, 18)
Microwave region 28(1, 36)
Microwave source 28(2)
Microwave spectrometers 28(2)
Microwave technique 28(138)
Microwave transitions 28(36, 37)
Microwave Zeeman effect 28(29)
Migration of excitation energy, intermolecular and intramolecular 27(382)
Millimeter wave spectroscopy 28(15)
Millimeter waves 28(140)
Mills' cross 29(437)
Minimization of the energy 27(235)
Minimum beam, definition of 24(582)
— marking of 24(544, 578, 582)
— marking of, as for light intersection 24(634)
Minimum schlieren method 24(610, 618)
Minimum usable slit width 29(496)
Minority carriers 26(13, 14, 107–110)
Mirror materials 29(295)
Mirror objective 29(200)
Mirror objectives for microscopes 29(367)
Mirror support in telescopes 29(314)
Mirror-like grating
— experimental verification 29(579)
— theory 29(572ff., 575–578)
Mirrors, rotating 24(7)
MIT ruling engine with interferometric control 29(687–703)
Mitscherlich phenomenon 25/1(189)
Mixed crystals 25/2d(263, 394)
— mode behaviour 25/2d(394, 399–404)
— strain 25/2d(405, 460)
Mixture of wave functions 27(243, 248, 361, 369, 371, 425)
Mode locking
— by a saturable absorber 25/2c(216)
— by gain modulation 25/2c(216)
— by loss modulation 25/2c(215, 217–224)
— principle 25/2c(213–214)
Model, mechanical, of vibrations 26(175)
Model theory, dynamics of perturbed lattice 25/2d(358)

Model theory, infrared absorption 25/2d(148, 358)
Model theory, Raman scattering 25/2d(148, 358)
Mode-locked pulses
— pulse width 25/2c(214, 222)
— shape 25/2c(222)
Modes, acoustic 25/2a(67)
Modes, collective 25/2c(125)
Modes, coupling, *see also* Mode locking 25/2c(127, 141, 183–185, 206–212)
Modes, density matrix equations for light modes 25/2c(56)
Modes, frequency distribution of the normal 25/2a(69, 107)
Modes, lasing, polarization in a gas laser 25/2c(201)
Modes, light (lasing and nonlasing) 25/2c(127)
Modes, localised (=local) 25/2a(82, 186)
Modes, multimode operation, coexistence 25/2c(143, 187, 255–259)
Modes, normal 25/2a(142)
Modes, optic 25/2a(67)
Modes, resonance condition 25/2c(13)
Modes, spatial distribution 25/2c(11–16)
Modes, two-mode operation
— above threshold (quantum mechanical, semiclassical) 25/2c(141–144, 184–185, 189–190, 199–201)
— below threshold (quantum mechanical) 25/2c(139)
— near threshold (quantum mechanical) 25/2c(140–141)
Modulation 28(5, 6)
Modulation interference gratings 29(735–754)
Moiré gratings 29(439ff.)
Moiré principle 29(764)
Moiré-fringe testing 29(641–643)
Molecular bands 27(385)
Molecular complex 27(358)
Molecular crystals 25/2d(43)
Molecular defects 25/2d(263, 283, 347, 351, 355)
Molecular dissociation 28(166)
Molecular emission spectra, continuous 28(170, 171)
Molecular ionisation spectra, continuous 28(112)
Molecular model(s) 25/2d(288, 295, 369); 27(361, 362, 365)
— of anharmonic oscillator and symmetrical rotator 27(362)
— of united atom 27(225–227, 436)
— of valence forces 27(260–265)
— with fixed nuclei 27(222, 383)
Molecular orbital(s) 27(267–299)
— bonding, antibonding 27(231, 246, 256, 273, 299, 426)
— definition 27(225, 227, 242)
— delocalized 27(292)
— energy 27(246)
— hybrid 27(249)
— in a self-consistent field 27(270)
— inner 27(243)
— method, application to π electrons 27(287–289)
— method, comparison with the method of Heitler and London 27(265, 266)
— method, LCAO 27(234, 244, 245, 250)
— method, LCAO MO 27(270–280)
— method, LCAO SCF MO of Roothaan, application to CO_2 27(283–285)
— method, LCAO SCF MO of Roothaan, application to N_2 27(285, 286)
— method, LCAO SFC MO of Roothaan 27(282–287)
— method, Moffit's method of atoms in molecules 27(290)
— method, Mulliken's population analysis 27(290)
— method, recent development 27(281–292)
— Mulliken's symbols 27(245, 250)
— normalized 27(246, 268)
— occupied, and non-occupied 27(251, 268)
— one-center orbitals 27(290, 291)
— orthogonal 27(268)
— polycentric 27(242)
— SCF method 27(282)
— Slater's orbitals 27(271, 276, 283)
— symmetrical and antisymmetrical 27(229)

Molecular recombination spectra 28(172)
Molecular spectra, discrete 28(159)
Molecular spectra, line broadening 27(57ff.)
Molecular stability 27(231, 239, 256, 257, 458–462)
Molecules two-atomic, continuous spectra 28(173)
Molecules with electronic moments (spin and orbital momentum) 28(18, 20)
Molybdate phosphors 26(47–49)
Molybdenum, emissivity 26(151, 156)
Moment as function of *v* and of the internuclear distance 27(345, 448)
Moment of inertia of a molecule 26(211, 213)
Moment of transition 27(332, 334)
Moments of frequency distribution 25/2a(146)
Monocentric eyepiece 29(289)
Monochromatic filter 24(511)
Monochromatic microscope objectives 29(358)
Monoelectronic energy as a function of the geometrical configuration 27(292–299)
Monoelectronic energy in a self-consistent field 27(269, 274)
Morphic effects, definition of 25/2b(456)
Morphic effects, definition of, symmetry effect 25/2b(458)
Morse relations 25/2d(19)
Moseley's law 27(88, 89, 164)
Mount Wilson Observatory ruling engines 29(682–687, 716–719)
Mounting of telescopes 29(322ff.)
Mueller matrix 25/1(24)
Mulliken's population analysis 27(290)
Multi-layer, "pseudo-*g*-value" of the 24(487)
Multi-layer-filters with coupling 24(505)
Multi-layer-polarizer 24(489)
Multi-layer-systems 24(485)
Multi-layers, non-absorbing 24(509)
Multi-layers, recurrence formulae for non-absorbing 24(510)
Multi-level system, quantum mechanical equations 25/2c(50)
Multi-level system, semiclassical equations 25/2c(51, 180)
Multimode operation, *see also* Modes, Coexistence 25/2c(143, 187, 255–259)
— above threshold (semiclassical) 25/2c(186–187, 191–192)
— below threshold (quantum mechanical) 25/2c(123–124)
Multi-phonon processes 25/2d(20, 162, 178, 180, 418, 420, 434)
Multiple periodic errors 29(555ff.)
Multiple slit spectrometers 29(498–503)
Multiple slits 29(432)
Multiple-limit series 27(130)
Multiplet splitting in rare earths 28(270)
Multiplet splitting of iron group ions 28(277)
Multiplets 27(97–99)
Multiplets, equidistance of levels 27(253)
Multiplets, regular and inverted 27(397)
Multiplication (composition) of operators 25/2b(10, 15)
Multiplicity 27(94)
Multiplicity of the electronic states 27(251, 253)
Multiplicity of the electronic states in polyatomic molecules 27(329)
Multipole radiation, mixed 28(275)
Mutual exclusion rule 26(208)
— broken at resonance 25/2b(340)
— falsity of converse 25/2b(298)
— in crystals with inversion 25/2b(298)
Mylar 25/2a(51)

N_2, application of the LCAO SCF MO method 27(285, 286)
N_2, forbidden transitions 27(451, 452)
Na, sodium 25/2a(100, 107, 111)
NaCl 25/2d(160, 161, 174, 184, 195, 230, 237, 273, 290, 292, 293, 300, 305, 336ff., 347)
NaCl, sodium chloride 25/2a(93, 133)
NaCl:Ag^+ 25/2d(288, 422)
NaCl:Cu^+ 25/2d(273, 422, 431, 437, 438, 448)
NaCl:F^- 25/2d(288, 337)

NaCl:H^- 25/2d(269)
NaCl:KF 25/2d(288)
$Na^{35}Cl_x{}^{37}Cl_{1-x}$ 25/2d(290)
NaI 25/2d(28, 312, 318, 327)
NaI sequence, polarization formula 27(188)
NaI:Cl^- 25/2d(422, 431)
NaI:^{35}Cl and ^{37}Cl 25/2d(429)
NaI:H^- 25/2d(269, 422)
NaI-like spectra, isoelectronic relations 27(89, 147, 148, 149, 161, 169)
NaI-like spectra, screening doublet 27(157)
NaJ, sodium iodide 25/2a(127, 134)
$NaNO_2$, sodium nitrite 25/2a(143)
Nasmyth focus 29(261)
Natural line shape 27(77–79)
Natural undamped lattice frequency 25/2a(24)
Nb, niobium 25/2a(104)
NCO, analysis of the spectrum 27(421–426)
Near-axis rays 29(85)
Necessary degeneracy and space-time group $\mathcal{G}$ 25/2b(166)
Necessary degeneracy of eigenvectors 25/2b(132, 152)
Needle effect of electrostatics 25/1(252, 314)
Negative rotational level 26(213, 216)
Nernst glower as radiation source 26(168)
Neumann's boundary value problem 25/1(235)
Neutral fibre of measure bar 29(786)
Neutron scattering 25/2a(76, 124, 130); 25/2d(78, 390)
Neutron scattering, inelastic 25/2a(76, 145)
Neutrons, thermal 25/2a(77)
Newton conjugate equation 29(8, 202)
Newton focus 29(259)
Newton integral 25/2d(13)
NH_2, analysis of the spectrum 27(432–435)
NH_3, analysis of the spectrum, NH_3 27(443)
NH_2, transitions with variation of molecular symmetry 27(432–435)
$NH_4Cl_{1-x}Br_x$ 25/2d(404)
Ni, nickel 25/2a(81, 102, 112, 145)
Nickel, emissivity 26(149)
Nitrogen, rotation spectrum 26(212)
NO, example of a deperturbation 27(371)
NO_2^- molecule 25/2d(351)
NO, Rydberg series 27(416–418)
Nobel Institute ruling engine 29(727ff.)
Nodal surfaces of wave functions and electronic energy 27(233, 239)
Noise, amplification 25/2c(120–124)
Noise, effect of noise on laser output, *see* Linewith, Amplitude fluctuations,
Noise, phase noise, *see also* Linewidth 25/2c(130, 142)
Noise sources, *see* Langevin forces
Noise spectrum of the reduced Fokker-Planck equation 25/2c(164)
Non self-luminous object 29(349)
Non-adiabaticity 27(36)
Non-cubic crystals 25/2a(191, 192, 193, 201)
Non-photoconducting phosphors 26(81–85)
Non-radiative transitions 27(330, 361–382)
Non-reflection at two boundary surfaces 24(475)
Non-Rydberg transitions 27(343, 357, 358, 414–418, 453–455)
Non-spherical energy surfaces 25/2a(277)
Non-spherical surfaces 29(17)
Normal colour system 24(631)
Normal construction 29(80ff.)
Normal coordinate(s) 25/2a(61, 66); 25/2d(17, 476, 495); 26(175, 205)
— real 25/2b(129)
— real, and mass-weighted displacements 25/2b(130)
— real, and quantization 25/2b(259)
Normal direction 24(556)
Normal modes 25/2a(142)
— frequency distribution 25/2a(69, 107)
— symmetry in a crystal 25/2b(202)
Normal schliere 24(563)
Normal vibration 26(174, 175, 176)
Normal vibrations of the benzene molecule 26(240–241)
Normal Zeeman effect 28(297)

Normal Zeeman triplet, quantum theory 28(308f.)
Notation of vectors 25/1(226)
NPL ruling engine 29(725–727)
n-point functions 25/2d(492, 498, 507, 514)
Nuclear and lattice vibration superposition in the uranyl spectrum 26(51)
Nuclear charge, effective 27(88–90)
Nuclear coupling constant for Mn^{2+} 28(57)
Nuclear coupling constant for palladium and platinum groups 28(60)
Nuclear events, stereo-photography 29(415)
Nuclear magnetic coupling in molecules 28(35, 39)
Nuclear quadrupole coupling 28(30, 31, 32, 34, 40)
Nuclear spin, *see also* Spin 27(335)
Nuclear spin effect on molecular rotation 26(211)
Null test 29(306)

O branch 26(217)
Object function 24(590)
Object height, reduced 29(96, 119)
Object perpendicular, reduced 29(96)
Object plane 24(557); 29(96)
Object preserving 24(596, 609, 617, 640)
Objectives for microscopes 29(357ff.)
Objectives for photography 29(192ff.)
Objects with finite volume, perturbation calculation 25/1(380ff.)
Oblate spheroidal coordinates 25/1(454)
Occupation number 25/2a(144)
OH, analysis of the spectrum 27(395–404)
OH^- molecule 25/2d(281–284)
Oil of immersion 29(346)
One-layer-filter 24(500)
One-layer-polarizer 24(480)
Opacity 28(155)
Operation function 24(591, 599, 600)
Operator, creation operator 25/2a(62, 113)
Operator, destruction 25/2a(62, 113)
Operator equivalence 28(248)
Operator, Hamiltonian
— lattice dynamical 25/2a(61, 62, 71)
— of electrons 25/2c(27)
— of field 25/2c(26)
— of field interacting with electrons 25/2c(28)
Operator, Liouville operator 25/2c(288)
Operator of the electric field strength 25/2c(26)
Operator of the magnetic field strength 25/2c(26)
Operator, phase operator 25/2c(85)
Operator relations, Bose operators 25/2c(297–298)
Operator techniques 25/2c(294–304)
Operators, Bose, disentangling system 25/2c(301–304)
Operators, commutation relations
— for Bose operators 25/2c(25, 113)
— for Fermi operators 25/2c(27, 114)
— for spin $\frac{1}{2}$ operators 25/2c(32)
Operators, exact elimination
— of atomic operators 25/2c(286, 288–289)
— of atomic within Langevin equations 25/2c(285)
Operators, harmonic oscillator, creation and annihilation operators 25/2c(294–297)
Optic axes 25/1(70)
Optic axes, absorbing crystals 25/1(90)
Optic modes 25/2a(67)
Optical activity 25/1(2, 12)
Optical activity and birefringence, superposition 25/1(81ff.)
Optical activity and symmetry of crystals 25/1(83–85)
Optical collision diameter 27(8, 37, 49, 68)
Optical constants 25/2d(4, 5, 27, 75, 89, 102, 171, 536)
— measurement of the 24(513)
— relation to symmetry 25/1(204)
Optical cosine law 29(28)
Optical direction 29(79, 153)
Optical elements, absorbing 25/1(14ff.)
Optical elements, non-absorbing 25/1(11–13)
Optical glasses, *see* Glass
Optical grating 29(433ff.)
Optical materials in the ultraviolet 29(24)
Optical palpating 29(756)

Optical path 28(83, 98); 29(52ff., 140, 509)
Optical rotation, measurement 25/1(184ff.)
Optical rotation of quartz normal to optic axes, measurement 25/1(187)
Optical rotation, Szivessy's method of measurement 25/1(186ff.)
Optical rule of Cuny 29(812)
Optical sign of a crystal 25/1(130ff., 182)
Optical stimulation 26(2, 6, 19, 21)
Optical stimulation of photo-conductivity 26(93)
Optical transfer function 29(257)
Optically active crystals
— absorbing 25/1(97ff.)
— dielectric tensor 25/1(76)
— Drude theory 25/1(76)
— Fresnel theory 25/1(76)
— index tensor 25/1(76)
— non-absorbing 25/1(76)
— refractive index 25/1(78ff.)
Optically active transparent crystals, wave surface 25/1(84ff.)
Orbital, *see* Molecular orbital
Order 29(85, 93, 102ff., 123, 138, 188)
Order of a ghost 29(551)
Orders of diffraction, energy distribution 29(517, 570)
Organic crystals and liquids, luminescence 26(53)
Organic solutions, energy transfer 26(55–58)
Origin of a band 27(387)
— missing lines 27(387, 401, 402)
— system 27(393)
Orthogonal transformation 25/2b(8)
Orthohydrogen 26(212, 226)
Orthoscopic arrangement of polarising microscope 25/1(170)
Orthoscopic eyepiece 29(289)
Orthoscopic illumination 29(398)
Oscillations, damped ("relaxation") 25/2c(134–138, 266–267)
Oscillations, small, approximate solution 25/2c(266–267)
Oscillations, undamped, *see* Spiking
Oscillator, creation and annihilation operators 25/2c(294–297)
Oscillator, quantum mechanical, harmonic 25/2c(294–297)
Oscillator strength 25/2a(161); 27(332); 28(105)
Oscillator strength, infrared 25/2d(66, 162, 399)
Oscillators, two coupled 27(47)
Overlap 27(243)
Overlap effects 25/2d(25, 30, 43)
Overlap integral 27(235, 240, 278, 334, 345, 346, 364, 377)
Overlap of orders in a grating 29(462)
Overlap polarization 25/2d(36, 54)
Overlap potential 25/2a(93)
Overlap shell model 25/2d(36, 37, 50, 260)
Overlap theory 25/2d(29, 42, 52, 74)
Overton absorption 25/2d(347, 418, 434, 436)
Oxide phosphors 26(27–31)
Oxides 25/2d(199, 231, 234)
Oxides, emissivity 26(164–167)
Oxygen molecule, photo-dissociation spectra 28(176)
Oxygen O^{2-}, polarizability 25/2d(60, 150, 233)
Oxygen, rotation spectrum 26(212)
Oxygen spectra, continuous 28(185)
Ozone spectra, continuous 28(186)

P branch 26(217)
p^2 configuration 28(315)
Pair coupling 27(119–123)
Palladiumgroup, magnetic resonance 28(58)
Pancratic systems 29(375)
Parabolic wavefront 29(558)
Parahydrogen 26(212, 226)
Parallactic error 29(762ff.)
Parallel laminated system 25/2a(12)
Parallel perspective 29(760)
Parallel plate waveguide 25/1(339ff.)
Parallelity control of axes 29(819)
Paramagnetism of the transition elements 28(49)
Parameters α, β in the method of molecular orbitals 27(272, 275, 276, 278–280)
Paraxial rays 29(92)
Paraxial raytracing 29(9ff.)
Paraxial region 29(228)
Parent molecules 27(421)
Parentage in *LS* coupling 27(104–106)
Parity 27(97)

Partial dispersion 29(292)
Partial waves 25/2a(12)
Partition function 25/2a(144); 25/2d(13, 487)
Partition functions of internal molecular motions 26(223ff.)
Paschen limit 28(151)
Paschen-Back effect 28(310f.)
— complete 28(314)
— partial 28(311)
Pauli principle 25/2d(10, 15); 27(90, 91, 242)
Pb, lead 25/2a(81, 100)
PbS, lead sulphide 25/2a(294, 303)
PbSe 25/2d(202)
PbSe, lead selenide 25/2a(294)
PbTe 25/2d(359)
PbTe, lead telluride 25/2a(99, 269, 294)
Peak power 25/2c(270)
Peak, reststrahlen 25/2a(33)
Peak, width of the reststrahlen 25/2a(33)
Pechan prism 29(217)
Peierls-Bogoljubov inequality 25/2d(13)
Pencil 29(5, 50)
Pendulum mirror 29(808, 825)
Penetration, *see* Screening defect
Penetration coefficient 24(468)
Penetration depth 24(464)
Penetration parameter 27(171, 173)
Penn's formula 25/2d(70)
Penn's gap 25/2d(250)
Pentagon prism 29(214, 218, 818ff.)
Perfect grating 29(428)
Perfect stigmatism in gratings 29(472)
Perflectometer 29(789ff.)
Perforated plates as microwave gratings 29(437)
Periodic boundary condition 25/2a(59, 66, 117)
Periodic errors 29(671)
Periodic structure 24(591)
Periodic system, electronic structure 27(92)
Perkin-Elmer ruling engines 29(716)
Permeability 24(462)
Permeability constant 25/2a(3)
Permittivity 25/2a(3, 241)
Permittivity constant, *see also* Dielectric constant 25/2a(3, 241)
Permittivity tensor 25/2a(240)
Pérot-Fabry étalon combined with Michelson interferometer 29(781)
Perovskite ferroelectrics 25/2a(143)
Perovskites 25/2d(37, 219, 245, 307)
Perpendicular (⊥) and parallel (||) transitions 27(326, 335, 336, 348)
Perpendicular(s) 29(90)
Perspective of photographic image 29(201ff.)
Perturbation calculation of Kane, $k \cdot p$ 25/2a(259, 273)
Perturbation matrix 25/2d(302, 365); 27(362, 369)
— as a function of J 27(347, 377)
— many defects 25/2d(382)
Perturbation method 27(223, 226, 239)
Perturbation, mutual, of two series 27(140–144, 153)
Perturbation theory 25/2a(146, 147); 27(239, 361)
— free energy 25/2d(532)
— thermodynamic 25/2d(488)
Perturbations 27(229, 261, 300, 301, 361–382, 405, 406, 419)
Perturbations, accidental 27(373)
Perturbations and non-radiative transitions 27(361–382)
Perturbations and potential curves 27(369, 371)
Perturbations and predissociation 27(375)
Perturbations and vibronic interaction 27(364, 368)
Perturbations, detailed calculations 27(419)
Perturbations, determination of molecular constants of the perturbed and perturbing states 27(419)
Perturbations, fictive character 27(363)
Perturbations for degenerate and non-degenerate states 27(365–367)
Perturbations, forbidden 27(364)
Perturbations, heterogeneous 27(372)
Perturbations, homogeneous 27(371)
Perturbations in the Rydberg series 27(417)
Perturbations of higher order 27(365, 366)
Perturbations of position and intensity 27(361)
Perturbations, rotational 27(368–372, 404, 406)

Perturbations, vibrational 27(371)
Perturbation-theory diagrams 25/2a(153)
Perturbed crystal, site group 25/2b(437)
Perturbed crystal, symmetry group 25/2b(437)
Perturbed crystals, *see also* Defects 25/2d(262)
Perturbed series 27(136ff.)
Perturbed spectra 27(405–407)
Petzval curvature 29(16, 51, 195, 360, 365)
Petzval objective 29(194)
Petzval sum 29(237, 251)
Phase, channeled spectrum phase 25/2a(53)
Phase, circles constant 25/2a(37)
Phase contrast
— combination 24(628)
— images 24(599)
— instruments 24(622)
— method 24(555)
— method of Zernike, rigorous 24(594, 605, 615)
— method, real 24(608, 616)
— microscope, universal 24(623)
— negative 24(594)
— positive 24(594)
— variable 24(624)
Phase correction 24(530)
Phase diffusion 25/2c(120)
Phase, fluctuations 25/2c(130–132, 142, 146)
Phase formulae 24(538)
Phase identity 24(538)
Phase locking, *see* Mode locking
— of many modes 25/2c(170–171)
— of three modes 25/2c(171–172)
— of two modes 25/2c(144)
Phase measurement, absolute 24(528)
Phase measurement in wavefront from grating 29(525)
Phase measurement of transmitted light 24(541)
Phase modulation 29(736)
Phase noise, *see also* Linewidth 25/2c(130, 142)
Phase operator 25/2c(85)
Phase shifts in confocal resonator 25/2c(19–21)
Phase spectrum 25/2a(53)
Phase, stability 25/2c(130)
Phase, surfaces of constant 25/2a(4)
Phase velocity 24(1)
Phase-contrast fluorescence microscope 29(397)
Phase-contrast grating test 29(637ff.)
Phase-contrast method 29(393, 407, 744, 746)
Phase-contrast microscope 25/1(11)
Phased arrays 29(437)
Phase-reversion by half-zone plates 29(438)
Phillips' model 25/2d(70, 72)
Philpot Svensson register curves 24(572, 585)
Phonon 25/2a(59, 63, 68); 25/2d(9)
Phonon, Boltzmann equation 25/2d(105)
Phonon, commutation relations 25/2a(63)
Phonon, creation operators 25/2d(126, 478)
Phonon, destruction operators 25/2d(126, 478)
Phonon dispersion curves 25/2a(59, 88)
Phonon dispersion relation 25/2a(60)
Phonon, dynamics 25/2a(64); 25/2d(21, 494)
Phonon, eigenvectors 25/2d(58, 301, 325, 476)
Phonon, field operators 25/2d(126)
Phonon, final-states interaction 25/2d(190)
Phonon, frequency, anharmonic 25/2d(199)
Phonon, Green function, *see also* Green function 25/2d(5)
Phonon, lifetime 25/2d(505)
Phonon, one-phonon diagram 25/2a(118)
Phonon, one-phonon scattering cross-section 25/2a(129)
— Phonon-phonon interaction 25/2d(113, 531)
— Phonon-photon interaction 25/2d(6, 8, 25, 117)
Phonon, polaritons 25/2d(93)
Phonon, pseudo-harmonic 25/2d(7, 20, 21, 40, 163, 165)
Phonon, quasi-harmonic 25/2d(21)

Phonon, renormalized harmonic 25/2d(500, 501, 505, 527, 533)
Phonon, self-consistent 25/2d(532)
Phonon, self-energy, *see* Self-energy
Phonon, transport 25/2d(510, 523)
Phonon, zero-phonon line 25/2a(187, 218)
Phonon, zero-phonon transitions 25/2a(226)
Phosphate phosphors 26(45–47)
Phosphor activation 26(19–23, 34, 44, 48, 49, 50)
Phosphor activation, electrostatic compensation 26(32, 35)
Phosphor screens 26(111, 113)
Phosphorescence 27(360)
Phosphorescence, definition 26(2)
Phosphorescence due to metastable states of luminescence centres 26(62–65)
Phosphorescence in photoconducting phosphors 26(65–72)
Phosphorescence of alkaline earth sulphides 26(39–40)
Phosphorescence of zinc silicate 26(42)
Phosphors, decay times 26(122, 127)
Phosphors, deterioration under particle bombardment 26(115, 120)
Phosphors, electron emission 26(110)
Phosphors for scintillation counters 26(117)
Phosphors, infrared emitting 26(17, 29, 51, 79–81)
Phosphors, microcrystalline layers 26(89, 90)
Phosphors, non-photoconducting 26(81–85)
Phosphors, photoconducting 26(6, 65–72, 89)
Phosphors, photodielectric effect 26(94–97)
Phosphors, semiconduction 26(97, 98)
Phosphors, synthesis 26(2, 3)
Phosphors, ultraviolet emitting 26(50)
Photoconducting phosphors 26(6, 65–72, 89–94)
Photoconduction, dependence on excitation intensities 26(91)
Photoconduction, spectral response 26(90)
Photoconductivity and luminescence, relation 26(92)
Photoconductivity in ZnS and CdS single crystals 26(90–93)
Photoconductivity, optical stimulation 26(93)
Photoconductivity, quenching 26(93)
Photodielectric effect in phosphors 26(94–97)
Photodissociation 27(375, 412); 28(112)
Photodissociation, continuous spectra 28(165)
Photo-echo 25/2c(232–234)
Photoelastic behaviour of cubic crystals 25/1(213)
Photoelastic constants defining classes 25/1(210, 212)
Photoelastic constants, relation to crystal symmetry 25/1(204ff.)
Photoelastic (elasto-optic) coefficients 25/2d(147, 246–249, 251, 255, 332, 342)
Photoelastic measurements 25/1(213ff.)
Photoelastic measurements on optically active media 25/1(215ff.)
Photoelasticity 25/1(202ff.)
Photoelectric analyser 25/1(49ff.)
Photoelectric analyser of Kent and Lawson 25/1(50ff.)
Photoelectric detection 29(482, 489ff.)
Photoelectric eyepiece 29(801ff.)
Photoelectric recording of spectral lines 29(621ff.)
Photographic measurement of Raman effect 26(173)
Photographic objectives 29(192ff.)
Photoionisation 28(86, 104)
Photoionization 27(412)
Photometer microscope 29(411ff.)
Photometric concepts and definitions 29(491ff.)
Photon counting, classical treatment 25/2c(95–97)
Photon counting, quantum mechanical treatment 25/2c(93–94)
Photon self energy 25/2d(118)
Photon spectral density 25/2d(119)
Photon statistics 25/2c(7, 108)
Photon-electron interactions 25/2d(8)
Photon-exciton interaction 25/2d(6)
Photon-phonon interactions 25/2d(6, 8, 25, 117)
Photosphere 28(152)

Physically irreducible representation
— and space-time group $\mathcal{G}$ 25/2b(167)
— defined 25/2b(132)
Piezo-electric constants 25/2a(124)
Piezo-electric crystals 25/2d(108, 448)
Piezo-optical coefficients 25/2d(247)
Pin hole camera 29(193)
Pioncaré representation of partial polarisation 25/1(16ff.)
Pioncaré sphere
— definition 25/1(2)
— for absorbing non-optically active crystals 25/1(91ff.)
— for absorbing optically active crystals 25/1(99ff.)
— properties 25/1(2ff.)
— representation of birefringence 25/1(12)
— representation of dichroism 25/1(14)
— representation of optical activity 25/1(12)
— representation of polarisation states 25/1(2, 3)
Pipe, evacuated 24(9)
$\pi \cdot p$-interaction 25/2a(259)
π-polarisation 28(305)
π-Pulse 25/2c(245–246)
π-Pulse, 2π-pulse 25/2c(246–247)
Placzek approximation 25/2d(109, 116)
Placzek's polarizability theory 26(204)
Planar achromat 29(361)
Planar symmetric tetratomic molecules 26(233)
Planck function 28(83)
Planck radiation law 26(135, 136)
Plane grating 29(434)
Plane gratings, basic equations 29(456)
Plane screen
— diffraction case 25/1(259, 300ff., 404ff.)
— diffraction case, vector problem 25/1(308ff.)
— Hertz vector 25/1(230ff., 310ff.)
— integro-differential equation for surface current, scalar problem 25/1(302ff.)
— modified Kirchhoff's theory 25/1(279ff.)
— rigorous solution for scalar problem 25/1(298ff.)
— rigorous solution for vector problem 25/1(307ff.)
— scattering case 25/1(259, 299ff.)
— scattering case, vector problem 25/1(307ff.)
— scattering problem 25/1(386ff.)
— symmetry properties of scattering 25/1(259ff.)
Plane transmission grating, basic equations 29(457)
Plane wave, intensity 25/2a(7)
Plane waves and Fourier transformation 25/2b(136)
Plane waves, orthonormality and completeness for 25/2b(136)
Plasma 28(96, 99, 120)
Plasma frequency 25/2a(208, 282)
Plasma, non-thermal 28(134)
Plasma oscillations 28(143, 157, 159)
Plasmon 25/2a(100)
Platinum coating of grating surface 29(602ff.)
Platinum, emissivity 26(150, 155)
Platinum group, magnetic resonance 28(58)
Pnictides 25/2d(245, 258)
Pockels coefficients, *see* Photo-elastic coefficients
Poinsot's index ellipsoid construction 25/1(62)
Point group(s) 26(181)
— examples in molecular structure 26(182)
— matrix equations into different irreducible representations 25/2d(303)
— $\mathfrak{P}$, definition 25/2b(13)
— $\mathfrak{P}$, subgroup of $\mathfrak{G}$ 25/2b(23)
Point spread function 29(256)
Point symmetry 25/2d(408)
Poisson distribution 25/2c(78, 95)
Polanret system 24(626)
Polar axis (off-axis German) mounting 29(326)
Polar optic mode
— and rule of mutual exclusion 25/2b(307)
— crystal without inversion 25/2b(308)
— crystal without inversion, splitting and activity 25/2b(308)
— splitting *LO-TO* 25/2b(307)

Polarisation of Zeeman components 28(301)
Polarisation, *p*- and *s*-type 25/1(226ff.)
Polarisation properties of diffraction fields 25/1(281)
Polarisation, quantum mechanical description 25/1(30ff.)
Polarisation state, azimuth 25/1(2)
Polarisation state, definition 25/1(1)
Polarisation state, ellipticity 25/1(2)
Polarised beams
— coherent addition 25/1(7–11)
— decomposition 25/1(19)
— incoherent addition 25/1(19ff.)
— phase difference 25/1(8–11)
— phase difference due to analysation 25/1(10–11)
Polarised light, analysis by photoelectric methods 25/1(49ff.)
Polarised light, nature of 25/1(31ff.)
Polarised light, production of 25/1(34ff.)
Polariser, definition 25/1(2)
Polarising microscope 25/1(169–172)
— for reflected light 25/1(171)
— shearing interference type 25/1(173ff.)
Polariton 25/2a(174)
Polariton, operators, definition 25/2b(338)
Polariton, scattering and Raman theory 25/2b(338)
Polaritons 25/2d(8, 92, 95, 123, 124, 468, 470)
Polarity of a bond 27(276, 279)
Polarity of crystals 25/2d(73)
Polarizability, *see also* Lattice polarizability, Raman scattering 25/2d(62, 63)
Polarizability 25/2a(24)
Polarizability, bond charge 25/2d(150)
Polarizability catastrophe 25/2a(138)
Polarizability, defects 25/2d(311, 339, 358–360, 363, 366, 371)
Polarizability, electronic 25/2d(28, 46, 74)
Polarizability ellipsoid (and tensor) 26(197–198, 203–204, 205)
Polarizability, expansion theory 25/2d(125)
Polarizability, inter-ionic 25/2d(150, 232, 237, 260, 262)
Polarizability, intra-ionic, quartic 25/2d(241, 243)
Polarizability, ionic 25/2d(60, 190)
Polarizability operator as second rank tensor field 25/2b(290)
Polarizability operator for Raman scattering 25/2b(288)
Polarizability operator, symmetry 25/2b(289)
Polarizability, oxygen 25/2d(60)
Polarizability, theory 25/2d(109, 115)
Polarizability variation 26(196–198, 200)
Polarization at two boundary surfaces 24(475)
Polarization behaviour 25/2a(226)
Polarization by gratings 29(443ff., 505, 575, 589)
Polarization by metal gratings 29(573)
Polarization degree of fluorescence 26(85–88)
Polarization effect in line broadening 27(73)
Polarization effects in grating boundary problems 29(571, 579)
Polarization effects, macroscopic electric field (Poulet effect) 25/2b(309)
Polarization energies 27(188, 190)
Polarization formula for hydrogen-like terms 27(125–129, 187)
Polarization in a magnetic field 25/2c(205, 211)
Polarization in infra-red absorption 25/2b(312)
Polarization in Raman scattering 25/2b(298, 301)
— two phonon 25/2b(304)
Polarization in Stark broadening 27(39)
Polarization methods 24(539)
— general 24(514)
— special 24(516)
Polarization microscope 29(398)
Polarization of active material 25/2c(175, 178)
Polarization of atomic orbitals 27(264, 271)
Polarization of IMO transitions 25/2a(289)
Polarization of lasing modes in a gas laser 25/2c(201)

Polarization of luminescence emission 26(18, 26, 85–89)
Polarization of radiation 25/2a(289)
— and nature of the involved electronic states 27(335, 336)
— in case of polyatomic molecules 27(348, 349)
Polarization of Raman scattered light 26(199–200)
Polarization of resonance radiation 27(412)
Polarization of the electronic charge 27(241)
Polarization, spontaneous 25/2a(136)
Polarization suppressed in blazed gratings 29(572, 581 ff.)
Polarization waves 25/2c(179)
Polarized modes, stability 25/2c(201)
Polaroid 25/1(34)
Polaron 25/2a(164, 286)
Polyatomic molecules, spectra 27(419–445)
Polycrystalline media 25/1(121 ff.)
Population analysis, Mulliken's method 27(290)
Population of molecular levels 27(446, 447)
Population pulsation, *see* Inversion
Porro prisms of first and second kind 29(217, 222)
Position group 28(266)
Positive and negative levels 27(322–324)
Positive crater of the carbon arc as radiation source 26(163)
Positive direction convention 25/2a(13)
Positive hole in filled lattice energy bands 26(6)
Positive hole migration, effect on fluorescence 26(73, 74)
Positive hole, production 26(71)
Positive rotational level 26(213, 216)
Positive sign convention 25/2a(15)
Positronium 28(39)
Potassium chlorate, iridescence 25/1(116 ff.)
Potassium halides 25/2d(347, 406)
Potassium, reflection spectra 25/1(117)
Potential barrier to internal rotation 26(194)
Potential curves 27(230, 231, 303, 412, 413)
— analytical methods of determination 27(304)
— asymptote 27(230)
— effective 27(305, 312, 347, 364, 377, 413)
— example NO 27(369)
— for luminescence centres, *see* Energy-configurational diagram
— importance in spectroscopy 27(387)
— in perturbed regions 27(369, 371)
— intersection 27(246–248, 257, 371)
— non intersection 27(246–248, 257, 265, 267, 362, 369, 371, 376)
— theory 28(162)
— with a double minimum 27(308, 432)
— with maximum 27(413)
Potential, effective 27(303, 304)
Potential energy of lattice
— and Born-Oppenheimer theory 25/2b(254)
— and general symmetry 25/2b(123)
— and translation symmetry 25/2b(118)
— classical 25/2b(115)
— positive semidefinite 25/2b(115)
Potential energy operator 27(269, 273)
Potential, lattice 25/2d(478)
— central 25/2d(24, 40, 171)
— cubic 25/2d(175)
— harmonic 25/2d(15, 34, 40, 477, 500, 505, 533)
— non-central 25/2d(25, 38, 40)
— quartic 25/2d(178)
Potential of perturbation 27(239, 244, 305, 309, 363, 375)
Potential, overlap 25/2a(93)
Potential, pseudo- 25/2a(106)
Potential, screened 25/2a(106)
Potential surfaces 27(300, 303, 378–380, 441)
Potential valley effect in line broadening 27(73, 74)
Potentially real representation of 𝔊, test 25/2b(176, 179)
Power of an optical surface 29(9, 49)
Power reflectance 25/2a(16)
Power transmittance 25/2a(17)
Poynting vector 25/1(59); 25/2a(7)
p-polarized wave 25/2a(15)

Preassociation 27(380, 381)
Precession, pure 27(317)
Predissociation 27(304, 305, 374–381); 28(165)
Predissociation, accidental 27(372–374, 380)
Predissociation and chemistry 27(381)
Predissociation and Franck-Condon principle 27(376, 377)
Predissociation and non-crossing of potential curves 27(380)
Predissociation and resonance 27(376)
Predissociation and symmetry 27(380)
Predissociation and vibronic interaction 27(379)
Predissociation by rotation 27(304, 305, 375–378)
Predissociation by tunnel effect 27(378, 380)
Predissociation, classification of the different cases 27(376–378)
Predissociation, critical value of J 27(377, 378)
Predissociation, determined by the weakening of emission 27(374, 379)
Predissociation, homogeneous and heterogeneous 27(375, 376)
Predissociation, induced 27(380, 381)
Predissociation, inverse 27(374, 380)
Predissociation, polyatomic molecules 27(378, 379, 388, 419, 436, 439, 441, 443)
Predissociation, quantum yield 27(375, 376)
Predissociation, selection rules 27(375, 376)
Pre-ionization 27(375)
Pre-ionization in Rydberg levels 27(417)
P-representation 25/2c(79)
Pressure 24(576)
Pressure effect on crystal spectra 28(210, 264)
Pressure effects 25/2d(147, 194, 195, 405, 439, 445, 450, 463)
Pressure shift 25/2a(188)
Preston rule 28(297, 298)
Primary aberration terms (fourth-power terms) 29(6, 14–18)
Primary approximation 29(103ff., 137ff., 189, 190)
Prime focus 29(259)
Prime focus correcting systems 29(268)
Principal angle of incidence 25/2a(18)
Principal azimuth angle 25/2a(18)
Principal electric axes 25/1(191)
Principal indices of refraction 25/2a(157)
Principal moments of inertia 26(215)
Principal quantum number 27(84, 85)
Principal ray 29(12, 22, 46, 47, 231)
Principle of maximum overlap 27(249)
Prism angles, test measurement 29(818)
Prism of Kösters 29(812ff., 823)
Prism, revolving 24(7)
Prismatic grooves 29(458)
Prisms 29(211ff.)
Product rule for isotopic molecules 26(191–192)
Product rule for rotational isomers 26(193–194, 237)
Profile imaging 29(755)
Profile of a measure bar 29(786)
Profile of grooves, *see* Grooves
Projection operator 25/2d(20)
Projection operator P^{l}_{mr}
— definition 25/2b(31)
— for translation group $\mathfrak{T}$ 25/2b(46)
Projectives 29(362)
Propagation constant 25/2a(240)
Propagation constant, complex 25/2a(4)
Propagation vector 24(463, 465)
Propagators, *see also* Green function 25/2d(5, 6, 474)
Pseudo-integral equation
— for magnetic surface current density 25/1(313)
— for scalar diffraction problem 25/1(301, 305)
Pseudo-potential 25/2a(106); 25/2d(46, 54, 68, 71–73)
Pseudo-real representation of $\mathfrak{G}$, test 25/2b(176, 179)
Pulse, 2π-pulse 25/2c(246–247)
Pulse, π-pulse 25/2c(245–246)
Pulse propagation
— in laser-active material 25/2c(236–246)
— simplified model 25/2c(243–245)
— steady state pulse 25/2c(241–243)
— transient build-up of the pulse 25/2c(239)

Pulse shape of mode-locked pulses 25/2c(222)
Pulse width of mode-locked pulses 25/2c(214, 222)
Pulses, ultrashort, optical 25/2c(213–224)
Pulsing, condition for self-pulsing 25/2c(239)
Pump power at threshold (rate equations) 25/2c(253, 254)
Pump process (*see also* Relaxation processes)
— quantum mechanical treatment 25/2c(45–50)
— semiclassical treatment 25/2c(175, 181, 209)
Pupil function 29(257)
Pupil imaging of tube lens 29(375)
Pupil perpendicular aberration 29(166ff.)
Pupil perpendicular, reduced 29(166ff.)
Pupil ray 29(11)
Pyramid defect in prisms 29(225ff.)
Pyramidal tetratomic molecules 26(232–233)
Pyrex 29(297)
Pyroelectric crystals 25/2d(108)

Q branch 26(217)
Q-spoiling 25/2c(267)
Quadratic field terms 25/1(246)
Quadratic Stark effect 27(34)
Quadrupolar deformability 25/2d(28, 32, 33)
Quadrupolar sum rule 25/2d(89)
Quadrupole energies 28(31, 32)
Quadrupole forces 25/2d(29)
Quadrupole moment 28(34, 40)
Quadrupole radiation 27(339, 340); 28(242)
Quadrupole transition 25/2a(193)
Quadrupole transitions in luminescence 26(84)
Quality of imaging in photography 29(209)
Quantization of the light field 25/2c(24–26)
Quantization, second, of the electron wave field 25/2c(27)
Quantum defect 27(87, 88, 185, 186, 417)
Quantum effects 25/2a(266)
Quantum efficiency for the luminescence 26(17, 18)
Quantum mechanical consistency 25/2c(44–45)
Quantum mechanical description of polarisation 25/1(30ff.)
Quantum numbers for molecules 27(223, 225, 226, 232, 252, 253, 302, 313–315, 424, 427)
Quantum numbers of levels determined by Zeeman measurements 28(326)
Quantum numbers, rotational 26(211, 213, 216)
Quantum theory of continuous electronic spectra 28(101)
Quantum theory of light scattering 26(200–205)
Quantum treatment of Stark broadening 27(35)
Quantum yield of fluorescence 27(382)
Quantum yield of perturbation 27(376)
Quantum yield of predissociation 27(375)
Quarter wavelength layer 24(485)
Quartz 25/2a(74); 29(297)
Quartz gauge 29(775)
Quartz, interference figures 25/1(133ff.)
Quartz, quadratic curves 25/1(135)
Quartz, vibrating 24(18)
Quartz wedge, effect on interference figures 25/1(183)
Quasi-cassegrainian systems 29(264)
Quasi-continuum of levels 28(95)
Quasi-harmonic approximation 25/2a(143, 147)
Quasi-harmonic frequency 25/2a(113, 121, 130)
Quasi-linearization (quantum mechanical) 25/2c(128)
Quasi-symmetrical rotator 27(318, 355, 428–431)
Quasi-triangular groove shape 29(570)
Quenching of fluorescence by magnetic field 27(412)
Quenching of lines in Stark broadening 27(39)
Quenching of luminescence 26(8, 18, 42, 58, 60, 78, 80)

Quenching of photoconductivity 26(93)
Q-value 25/2c(10)

R branch 26(214, 217)
Racah's formula 27(119)
Radiant emittance 26(131, 133, 134)
Radiant emittance of a black body 26(134, 141, 143)
Radiant energy, quantity of 26(131)
Radiant flux density 26(131)
Radiant flux (radiant power) 26(131)
Radiant intensity 26(131, 135, 154)
Radiation characteristics of carbon 26(161)
Radiation condition 25/1(220, 265, 338); 29(583)
— electromagnetic 25/1(248ff.)
— energy flow at infinity 25/1(249)
— scalar 25/1(236ff., 247ff.)
Radiation constants 26(139–141)
Radiation density 28(82)
Radiation from metals 26(129, 143–161)
Radiation from non-metals 26(161–170)
Radiation laws 26(133–137)
Radiation loss coefficient 25/1(529)
Radiation of a black body 26(129, 134, 141, 142, 143)
Radiation pattern
— connection with Fourier amplitudes 25/1(319ff., 331)
— for sphere 25/1(554ff.)
— of the far-field 25/1(316)
— Rayleigh limit 25/1(379ff.)
Radiation, polarization of 25/2a(289)
Radiation sources for temperature radiation 26(163, 168)
Radiation temperature 26(130, 160); 28(84)
Radiation theory 28(82)
Radiation-generating grating 29(440–443)
Radiationless processes in photoconducting phosphors 26(73–78)
Radiationless transitions, *see* Relaxation processes
Radiationless transitions in Stark broadening 27(39)
Radiative desactivation 27(449, 450)
Radiative transitions 27(330–360)
Radiative transitions and non-radiative transitions 27(374, 376)
Radicals, microwave resonance in 28(69, 70, 75)
Radicals produced by irradiation 28(71)
Radio astronomy 28(144, 157)
Radio frequency radiation 28(157)
Radioluminescence 26(2, 115–121)
Radiometric concepts 26(131)
Radio-telescopes 29(437)
Radius vector 29(79, 153)
Rainbow problem 25/1(573)
Raman effect
— classical interpretation 26(196–200)
— in liquid state 26(219)
— in solid state 26(219–222)
— of second order 26(222)
— quantum theory 26(200–205)
Raman scattering 25/2a(72, 81, 82, 124, 130); 25/2d(4, 5, 77, 108, 114, 143, 230, 231, 234, 409)
— active oscillator 25/2d(145, 339, 342)
— and the **A·A** term 25/2b(293)
— anomalies (Poulet), macroscopic field 25/2b(310)
— anti-Stokes processes 25/2d(79, 120, 144)
— anti-symmetric part 25/2d(123, 146)
— bond charge 25/2d(156)
— bond polarizability 25/2d(156, 253)
— by phonons (Hamiltonian) 25/2b(285)
— critical point analysis 25/2d(238)
— crystals, *see* Substances
— defect-induced 25/2d(266, 292, 341, 377, 384, 389, 393)
— diagrams 25/2d(153, 380–381)
— diamond crystal 25/2b(404)
— diamond (polarized) 25/2b(434)
— diamond structure and 2 phonon density of states 25/2b(400)
— dipole-dipole 25/2b(340)
— dynamic form factor 25/2d(116)
— external pressure 25/2d(147)
— field-induced 25/2d(147, 252)
— first-order 25/2d(144, 251)
— Fröhlich intraband, forbidden 25/2b(342)
— Germanium 25/2b(412)
— Green function theory 25/2d(116)

Raman scattering (cont.)
— higher-order 25/2d(144, 380)
— integrated scattered intensity 25/2d(343)
— ionic Raman effect 25/2d(108, 341, 381)
— model theory 25/2d(148, 358)
— multipole-dipole, forbidden 25/2b(341)
— ω^4-law 25/2d(120)
— perturbation theory 25/2d(111)
— perturbed crystals, *see* Raman scattering, defect-induced
— Placzek approximation 25/2b(287)
— polariton picture 25/2d(109, 113)
— polarizabilities 25/2d(60, 146)
— polarizability theory 25/2d(109, 115)
— polarization 26(199–200)
— polarizibility operator 25/2b(289)
— resonance 25/2b(339); 25/2d(124, 146, 230, 347, 381, 404)
— resonance and symmetry breaking 25/2b(340)
— rocksalt (polarized) 25/2b(431)
— scattering angle 25/2d(78)
— scattering efficiency 25/2d(80)
— scattering geometry 25/2d(79)
— scattering volume 25/2d(79)
— selection rules 25/2d(233)
— shell model 25/2d(153, 377)
— silicon 25/2b(407)
— Stokes, one phonon 25/2b(292)
— Stokes processes 25/2d(79, 120, 144)
— Stokes, two phonon 25/2b(292)
— Stokes, two phonon and derivative of density of states 25/2b(293)
— Stokes, two phonon cross section 25/2b(292)
— symmetry tensors 25/2d(145)
— tensor 25/2d(120, 122, 342, 552)
Raman scattering, cross section 25/2d(5)
— absolute 25/2d(80)
— differential 25/2d(81, 119, 341)
— power 25/2d(80)
— quantum 25/2d(80)
Raman scattering, many body formalism 25/2b(315)
— one phonon 25/2b(321)
— spectral density 25/2b(316)
— two phonon and density of states 25/2b(321)
Raman scattering, quantum theory 25/2d(109)
— Boch picture 25/2b(328)
— Boch picture, coupling, functions, electron radiation 25/2b(330)
— Boch picture, Hamiltonian 25/2b(330)
— Boch picture, tensor 25/2b(332)
— Exciton 25/2b(333)
— Exciton, tensor 25/2b(336)
— Polariton 25/2b(336)
Raman spectrum, activity 26(197, 206, 208)
Raman spectrum, rotational 26(214, 215)
Raman tensor
— and Clebsch-Gordan coefficients 25/2b(301)
— defined 25/2b(299)
— Loudon and Poulet conventions 25/2b(301)
— symmetry of 25/2b(332)
Raman-Laval theory 25/1(202ff.)
Ramaseshan halfshade analyser 25/1(43)
Ramsden disk 29(285)
Ramsden eyepiece 29(288)
Räntsch interferometer 29(784)
Rare earth ions, properties 28(63)
Rare earth halides 25/2d(271, 396, 436)
Rare gas spectra 28(187)
Rare-gas crystals 25/2d(331)
Rate equations 25/2c(249–271)
— approximate solution for small oscillations 25/2c(266–267)
— derivation of rate equations 25/2c(247–248)
— field equations 25/2c(249)
— in quantized form 25/2c(287)
— 4-level system 25/2c(266)
— 3-level system 25/2c(264)
— matter equations 25/2c(250)
— steady state solution 25/2c(250–251)
— threshold condition 25/2c(252)
Rate equations, completely homogeneous case 25/2c(251–255)
— 4-level system 25/2c(255)
— 3-level system 25/2c(252–253)
— pump power at threshold 25/2c(253)

Ray corepresentation of $\mathfrak{P}(\boldsymbol{k})$ 25/2b(197)
Ray data 29(79ff.)
Ray of light 29(2)
Ray propagation 25/1(56)
Ray representations of $\mathfrak{P}(\boldsymbol{k})$
— as multiplier 25/2b(69)
— associated 25/2b(70)
— definition 25/2b(69)
— gauge transformations 25/2b(70)
— or projective 25/2b(69)
Ray velocity surface 25/1(57)
Rayleigh criterion 29(32, 244, 344, 611ff., 764)
Rayleigh limit
— far field amplitude 25/1(379ff.)
— of diffraction 25/1(372)
— perturbation calculation 25/1(373ff.)
— quasi-static solution 25/1(373)
— static solution 25/1(373ff., 376ff., 428ff.)
Rayleigh modes 25/2d(473)
Rayleigh scattering 25/2d(90, 143, 342); 28(154)
— by a sphere 25/1(535, 552ff., 566ff.)
— of light 26(171, 172, 196, 202)
Rayleigh solution for circular disk 25/1(466ff.)
Rayleigh tolerance 29(479)
Rayleigh-Jeans formula 26(135)
Rayleigh-Jeans law 28(83)
Rayleigh's integral equations for diffraction 25/1(420ff., 426, 431ff.)
Rayleight's boundary problem of grating diffraction 29(570, 583–591)
Raytracing 29(9, 19)
RbCl 25/2d(32, 230, 237)
RbF 25/2d(69, 471)
Reading lens 29(353)
Real normal coordinates
— and corepresentations 25/2b(199)
— and crystal symmetry 25/2b(133)
— transformation 25/2b(134)
Real representation of $\mathscr{G}$ 25/2b(168, 169)
— and normal modes 25/2b(167)
— Frei test (modified) 25/2b(177)
— Herring-test 25/2b(176)
— test 25/2b(172)
Reciprocal height 29(112, 116)
Reciprocal lattice 25/2a(67)
— definition 25/2b(41)
— vectors $\boldsymbol{B}_H$ (rel vectors) 25/2b(41)
Reciprocal refractive index 29(112, 116)
Reciprocal vectors, definition of $\boldsymbol{b}_i$ 25/2b(40)
Reciprocity theorem 25/1(242, 400)
Recombination 28(86, 118)
Recombination coefficient 28(137)
Recombination coefficient for free carriers and empty centres 26(4)
Recombination emission 25/2a(197)
Recombination spectra, continuous 28(193)
Recombination spectra, molecular 28(172)
Reconstructive transition 25/2a(143)
Recording arrangement, direct, for Raman spectrum 26(173)
Rectangular aperture, diffraction pattern 29(452)
Rectangular groove profile 29(575, 581)
— unsuitability 29(581)
Recursion 24(469)
Recursion formulae 24(471)
Reduced matrix element
— and Wigner-Eckart theorem 25/2b(216)
— defined 25/2b(212)
— several 25/2b(214)
Reducible representation 28(213)
Reduction coefficient(s), *see also* Subgroup reduction coefficient 25/2b(107)
— *see* Star, reduction coefficient
— basis function for 25/2b(99)
— for products of representations 25/2b(33)
— for space groups 25/2b(87)
— for space groups, ordinary 25/2b(91)
— for space groups, symmetrized 25/2b(92)
— for stars 25/2b(94)
— full-group, defined 25/2b(91)
— little group method 25/2b(112)
— subgroup 25/2b(107)
Reduction group, definition 25/2b(97)
Reemission spectra of crystals 28(208)
Reference sphere 29(5, 58)

Reference wavefront 29(557ff.)
Reflectance, average 25/2a(50)
Reflectance charts 25/2a(40)
Reflectance, coefficient of the amplitude 25/2a(9)
Reflectance, constant circle relations 25/2a(37)
Reflectance extrapolation 25/2a(43)
Reflectance graphs 25/2a(35)
Reflectance of grating surface for x-rays 29(605)
Reflectance of the lamella, amplitude 25/2a(19)
Reflectance, power 25/2a(16)
Reflected wave 24(465)
Reflecting systems, design 29(39)
Reflection at boundaries of anisotropic media 25/1(110ff.)
Reflection, boundary surface without 24(473)
Reflection by an absorbing anisotropic crystal 25/1(117ff.)
Reflection coefficient 24(468)
Reflection coefficients of x-ray gratings 29(647ff.)
Reflection depth 24(636)
Reflection echelon 29(466ff.)
Reflection factor 24(468, 471); 26(131, 144, 153)
Reflection filter 24(512)
Reflection grating 29(433, 435, 446)
Reflection grating parameters 29(455, 456)
Reflection ideal 25/1(223)
Reflection, interband magneto-reflection 25/2a(311)
Reflection law 25/2a(10)
Reflection, magneto 25/2a(244)
Reflection, magneto-plasma 25/2a(281)
Reflection method with measurement of phase and amplitude, rigorous 24(540)
Reflection of light by lenses 29(209)
Reflection polarizer 24(484)
Reflection spectra 25/2a(204)
Reflection, total attenuated 25/2a(57)
Reflectivity, *see* Infrared absorption
Reflectivity 25/2a(196)
Reflectivity, infra-red 25/2a(134)
Reflectivity of gratings 29(435ff.)
Reflector corrector 29(269)
Reflex klystron 28(2)
Reflexion losses 25/2c(17)
Refracting surface 29(9)
Refraction at boundaries of anisotropic media 25/1(110ff.)
Refraction, effective indices of 25/2a(15)
Refraction formulae 29(82, 84ff., 88, 130, 155)
Refraction invariant 29(12)
Refraction law 24(467)
Refraction, principal indices of 25/2a(157)
Refraction, Snell's law of 25/2a(10)
Refractive, complex refractive index 25/2a(4)
Refractive index = refraction, complex 24(3, 463, 539); 25/2a(5, 92, 241); 25/2d(78, 175, 332); 29(72ff.)
— n, complex 24(463)
— determination 29(414)
— gradient of the 24(573)
— matrix 25/1(107ff.)
— measurement by immersion technique 25/1(175)
— measurement by simple rotation apparatus 25/1(178)
— *n*, real part of the 24(463)
— of absorbing crystals 25/1(93ff.)
— of optically active crystals 25/1(78ff.)
— pseudo 24(497)
— variation with temperature 25/1(189ff.)
Refractory power of an optical surface 29(9, 118ff.)
Regular doublets 27(165)
Relative term values of *d*-configurations 27(214, 217)
Relativistic CTP-invariance 25/2d(106)
Relativistic experiments 24(37)
Relativity of time 24(40)
Relaxation oscillations 25/2c(135–138, 266–267)
Relaxation processes in atoms
— quantum mechanical treatment 25/2c(46–49, 57–58)
— semiclassical treatment 25/2c(176, 180, 181, 196, 209)
Relaxation time τ 25/2a(280)
Rellich's theorem 25/1(249)
Renner coupling 27(308–310, 354)

Renner effect 27(308–310, 352)
— anharmonicity of ν_2 (triatomic molecules) 27(434)
— applications 27(310, 422–428, 433)
— higher order approximations 27(427)
— in π states 27(434)
— parameter ε 27(423, 427, 434)
Replica 25/2b(6)
Replicas of master gratings 29(599, 660, 727, 731–735)
Representation, ambiguous 28(220)
Representation, irreducible 25/2a(117, 128, 141); 28(213)
Representation of a group 26(182–183)
Representation of group as matrix group 25/2b(28)
Representation of Heisenberg 25/2a(115, 148)
Representation of $\mathfrak{G}(\boldsymbol{k})$
— based on complex normal coordinates 25/2b(155)
— based on eigenvectors 25/2b(148, 150)
— equivalence of $D^{(\boldsymbol{k})(e)}$ and $D^{(\boldsymbol{k})(j)}$ 25/2b(154)
Representation, spectral 25/2a(149)
Representation theory of finite groups 28(212)
Representations of group
— direct products 25/2b(32)
— direct products, reduction 25/2b(32)
— irreducible, definition 25/2b(29)
— unitarity 25/2b(30)
Repulsion of electronic states in resonance 27(368)
Repulsion of potential curves 27(248)
Repulsion of the nuclei 27(228, 230, 231)
Repulsive force constant 26(187)
Reservoir 25/2c(35)
Residual rays 28(209)
Resolution, definition of Lord Rayleigh 29(459ff.)
Resolving power 29(240)
— and degree of coherence 29(345)
— dependence on grating angle 29(615)
— increase by use of two gratings 29(616ff.)
— limited by energy available 29(497)
— of gratings 29(453, 459ff., 462)
— of gratings, conditions for high values 29(461)
— of gratings, modern successes 29(432)
— tests for gratings 29(611ff.)
Resonance, antiferromagnetic 25/2a(316); 28(68, 69)
Resonance, Azbel-Kaner 25/2a(269)
Resonance Azbel-Kaner cyclotron 25/2a(295)
Resonance between structures 27(226)
Resonance between two C–Br oscillators 26(177, 180)
Resonance between two C–H oscillators 26(180)
Resonance broadening 27(43ff.)
Resonance circle diagram 25/2a(27, 32)
Resonance condition for modes 25/2c(13)
Resonance, cyclotron 25/2a(234, 237, 239, 242, 251, 264); 28(66, 67, 68)
Resonance, diamagnetic 28(67)
Resonance, electron spin 28(65, 68, 77)
Resonance, ferromagnetic 28(68)
Resonance induced by irradiation 28(74)
Resonance, infra-red cyclotron 25/2a(267)
Resonance integral 27(240)
Resonance modes, *see also* Gap modes, Local modes 25/2d(271, 275, 288, 293, 320, 326, 337, 371, 406, 412, 416, 422, 424, 463)
— intensity 25/2d(326, 337, 417, 419)
— shift, *see also* Pressure 25/2d(424, 442)
— width 25/2d(326, 417, 431, 434, 435, 442)
Resonance, paramagnetic 28(37, 41)
Resonance series 27(345, 407, 408, 409)
Resonance transfer, theory of sensitisation 26(56, 82–85)
Resonance width theories 27(48, 49, 57)
Resonator, confocal 25/2c(19–21)
Resonator, Fabry-Perot 25/2c(3, 11–18)
Resonator, more general configuration 25/2c(22–24)

Resonator, optical 25/2c(9–24)
Resonator, stable and unstable regions 25/2c(24)
Response, time-dependent atomic, *see also* Inversion 25/2c(185)
Response, time-independent atomic, *see also* Inversion 25/2c(185)
Reststrahlen 28(209)
Reststrahlen bands 25/2d(159, 162, 396–405)
Reststrahlen oscillators, *see also* Dispersion oscillator 25/2d(7, 98, 135, 161, 335, 338, 342)
Reststrahlen peak 25/2a(33)
Reststrahlen peak, width of 25/2a(33)
Retarded Green function 25/2a(151)
Retrapping 26(78, 79)
Richartz half-shade analyser 25/1(42ff.)
Riemann space, multi-sheet 25/1(264, 266, 412)
Riemann surface, mathematical sheet 25/1(405)
Riemann surface of two sheets 25/1(404ff.)
Riemann surface, physical sheet 25/1(405)
Right circularly waves 25/2a(234)
Rigid ion model 25/2a(92)
Rigid rotator 26(211)
Rigid-ion model 25/2d(21, 24, 25, 52, 76)
Rigid-ion model, defects 25/2d(332)
Rigid-overlap approximation 25/2d(52)
Rigorous representation of image height aberration 29(166ff., 176, 187)
Rigorous solutions of boundary value problems 29(572, 583ff.)
Ring grating 24(562)
Ritchey-Chrétien System 29(39, 265)
Ritz formula 27(123–125, 127)
Ritz method 25/1(395ff.)
Ritz method to evaluate radiation patterns 25/1(321ff.)
Rock salt, vibration spectrum 26(222)
Rocket grating spectrometer 29(630)
Rocking frequencies, data 26(179)
Rocksalt, irreducible representations 25/2b(345)
— $\boldsymbol{\Gamma}$ 25/2b(347)
— $\star\boldsymbol{L}$ 25/2b(347)
— $\star\boldsymbol{X}$ 25/2b(351)
Rocksalt, NaCl and Raman, infra-red 25/2b(422)
Rocksalt, NaF 25/2b(426)
Rocksalt, phonon, critical points 25/2b(416)
Rocksalt, phonon, symmetry 25/2b(379)
Rocksalt, reduction of $\star\boldsymbol{L}^{(3-)} \otimes \star\boldsymbol{L}^{(3+)}$ 25/2b(359)
Rocksalt, reduction of $\star\boldsymbol{X}^{(4)} \otimes \star\boldsymbol{X}^{(5)}$ 25/2b(352)
Rocksalt, space group $Fm\,3\,m$, O_h^5 25/2b(21, 343)
Rocksalt structures 25/2d(22, 302, 397, 555)
Rocksalt, wave vector selection rules 25/2b(352)
Roof prism 29(212)
Ross correcting lens 29(268)
Rosseland weight function 28(156)
Rotatable microscope stage 29(407)
Rotating biplate 25/1(38)
Rotating mirror 29(757, 760)
Rotating wave approximation 25/2c(29–30, 176, 179)
Rotation, 3 dimensions 25/2b(16)
Rotation group 28(219)
Rotation spectra
— of asymmetric top molecules 26(215–216, 224)
— of linear molecules 26(210–212)
— of spherical top molecules 26(214, 224)
— of symmetric top molecules 26(213–214, 224)
Rotational absorption frequencies 28(14)
Rotational analysis
— example OH 27(395–404)
— perturbed, complex or superposed spectra 27(405, 406)
— polyatomic molecules 27(419, 420, 444, 445)
— successive stages 27(400)
Rotational branches 27(337)
— classification 27(400)
— intensity alternation 27(437, 439)
— main 27(399)
— number in different types of transition 27(338, 396, 398)

— *P*, *Q*, *R* 27(387, 399)
— perturbed 27(369)
— polyatomic molecules 27(420)
— satellites 27(337, 339, 399)
— schematic representation 27(341)
Rotational constants as functions of the geometrical parameters 27(429)
Rotational constants, example of determination 27(402, 404)
Rotational energy 27(299–312, 320)
Rotational entropy 26(227)
Rotational heat capacity 26(225–226)
Rotational isomerism 26(193, 219, 237)
Rotational levels 27(317, 318–320)
Rotational operators $\{\varphi|0\}$ 25/2b(14)
Rotational partition function 26(224–226)
Rotational quantum number, effect on line broadening 27(57, 60, 65)
Rotational quantum numbers 26(211, 213, 216)
Rotational Raman spectrum 26(214, 215)
Rotational resonance in molecular broadening 27(57–60)
Rotational spectra 24(30); 26(172)
— of diatomic hydrides 28(10)
— of diatomic or linear polyatomic molecules 28(10)
— pure 28(9, 10)
Rotational structure 27(319, 320, 337–339, 354–357, 395–406, 419–421)
Rotational structure of a vibration band 26(216)
Rotational symmetry and aberration classification 29(6)
Rotational transitions 28(14, 19, 21)
Rotations, composition 25/1(13)
Rotations, dyadic form, non-cartesian axes 25/2b(17)
Rotation-vibration lines, broadening 27(57)
Rotor, asymmetric 28(15)
Rovibrational interaction 27(223, 299, 300, 306)
Rovibronic interaction 27(223, 299, 300, 306)
Rovibronic levels 27(317, 319–320)
Rovibronic structure 27(319, 320, 337–339, 354–357, 395–406, 419–421)
Rowland circle 29(474, 477)
Rowland ghosts, *see also* Ghosts 29(549ff.)
Rowland method of ruling 29(429–431)
Rowland ruling engine 29(681)
Ruby, luminescence 26(30)
Ruling engines 29(429–431, 669–730)
Ruling engines, shock control 29(671–677)
Ruling engines, temperature control 29(678–681)
Ruling errors and wavefront aberrations 29(533ff., 541)
Ruling imperfections 29(610)
Ruling process of gratings 29(666)
Runge rule 28(297, 298)
Russell-Saunders coupling 27(94ff., 100ff.); 28(38, 42, 270, 303, 310)
Rydberg and non-Rydberg electronic states 27(417, 418)
Rydberg constant 27(85)
Rydberg series 27(123ff., 414–418)
Rydberg series, limit 27(417)
Rydberg transitions 27(343, 357, 358, 414–418, 453–455)

S branch 26(214, 217)
Sabattier effect 29(832)
Sagittal image height aberration 29(129, 130ff., 135ff., 138ff., 146ff., 180ff., 187, 190ff.)
Sagittal variation 29(127ff., 186)
Saha equation 28(97, 126)
Sampling method 25/2a(110)
Satellites 28(265); 29(540, 545, 560)
Saturated inversion, time-dependent 25/2c(140, 175, 180, 185)
Saturated inversion, time-independent 25/2c(131, 138, 140, 185)
Saturation effect in luminescence 26(111, 112)
Saturation of the spin 27(263, 265)
Saunders-Gross interference method of surface control 29(778)
Scalar boundary problem of grating diffraction 29(571, 600)
Scalar problems in grating diffraction 29(444, 446, 504)
Scalar product in lattice dynamics 25/2b(210)
Scalar product of an operator with eigenvectors 25/2b(213)

Scalar product, single dot and double dot, defined 25/2b(127, 128)
Scattered light from gratings 29(561–564, 650)
Scattering and diffraction 29(433)
Scattering, Brillouin 25/2a(74)
Scattering case, definition 25/1(252)
Scattering coefficient 28(155)
Scattering coefficient, definition 25/1(383)
Scattering coefficient of sphere 25/1(513, 535, 550, 555)
Scattering, coherent 25/2a(77)
Scattering, Compton 25/2a(76)
Scattering cross-section
— definition 25/1(383)
— of circular cylinder 25/1(526ff.)
— one-phonon 25/2a(129)
— optical theorem 25/1(385)
— plane screen 25/1(386ff.)
— scalar problem 25/1(383ff.)
— stationary representation 25/1(397ff.)
— vector problem 25/1(388ff.)
Scattering, incoherent 25/2a(78)
Scattering, inelastic neutron 25/2a(76, 145)
Scattering length 25/2a(77)
Scattering matrix, composite in cubic crystal 25/2b(305)
Scattering, neutron 25/2a(76, 124, 130)
Scattering of light, classical theory 26(196–200)
Scattering of light, quantum theory 26(200–205)
Scattering of x-rays 25/2a(74, 124, 130)
Scattering of x-rays, thermal diffuse 25/2a(75)
Scattering properties 25/2a(152)
Scattering, Raman 25/2a(72, 81, 82, 124, 130)
Scattering tensor, Raman, defined 25/2b(299)
Scattering tensors in cubic crystals 25/2b(302)
— and polarization 25/2b(303)
Scatter-plate interferometer 29(313)
Schliere 24(556)
Schlieren chamber 24(559)
Schlieren diaphragm 24(560)
Schlieren method 24(555, 556)
Schlieren method, geometric optic considerations of the 24(556)
Schmidt prism 29(218, 220ff.)
Schmidt telescope 29(18, 24, 39, 272ff.)
Schmidt-Pechan prism 29(215ff.)
Schrödinger equation for H_2^+ 27(222)
Schrödinger equation, formal solution 25/2c(298–299)
Schur lemma and corepresentations 25/2b(183)
Schur lemma, test for irreducibility 25/2b(29)
Schwarzschild iteration method 25/1(288)
— for interaction of diffracting objects 25/1(363ff.)
— for wide slit 25/1(437ff.)
Schwarzschild system 29(265)
Schwinger-Levine variational method, *see* Stationary representation
Scintillation counter crystals 26(19, 55, 115, 117)
Scintillation counters 26(115–117)
Screen, definition 25/1(225)
Screen, infinitely thin 25/1(220)
Screened potential 25/2a(106)
Screening 27(88, 89)
Screening defect (penetration) 27(89, 171, 173)
Screening doublets 27(156–160)
Screening, long-range forces 25/2d(215)
Screening numbers 28(271)
Screening parameters 27(168, 180, 181, 182, 183)
Screens, complementary 25/1(278)
Screws of ruling engines 29(670ff.)
Se, selenium 25/2d(214, 224, 226)
Second order Raman effect 26(222)
Secondary approximation 29(189, 191)
Secondary gauge 29(755)
Secondary spectrum 29(22, 255)
Second-order transition 25/2a(142)
Secular equation of molecular vibration 26(175)
Secular equation, solution 26(187–190)
Seidel approximation 29(7, 44, 51, 103, 139, 190, 232)
Seidel coefficients 29(236ff.)
Seidel difference formulae 29(13)
Seidel separation 29(107, 116)
Seidel sums 29(15ff., 51, 233ff.)

Seidel sums, extension to aspherical surfaces 29(17ff.)
Seismographic tests of ruling engine vibrations 29(676)
Selection rule(s) 25/2a(257, 258, 289, 290); 25/2b(278); 25/2d(18, 20, 266, 406, 408); 27(96, 330, 333–337); 28(103)
— and change of symmetry by excitation 27(350)
— and degeneracy 27(353, 354)
— and molecular symmetry 26(205–208)
— and nuclear symmetry 27(333)
— and symmetry operations 27(336, 350)
— electronic 27(349, 351, 352)
— for electric dipole radiations 27(335–337)
— for fields 25/2c(26)
— for infra-red absorption 25/2b(278)
— for magnetic dipole radiations 27(340)
— for quadrupole radiation 27(340)
— for Raman scattering 25/2b(290)
— for rotation spectra 26(212, 213, 216)
— for rotational and rovibronic transitions 27(354, 355)
— for symmetrical and asymmetrical rotators 27(355)
— in Back-Goudsmit effect 28(323)
— in case of perturbations 27(362–364, 368)
— in case of predissociation 27(375)
— in strong magnetic fields 28(313)
— in Zeeman transitions 28(305)
— of crystal terms 28(232, 240)
— of the infrared absorption 26(206)
— of the Raman effect, classical interpretation 26(196, 198)
— of the Raman effect, quantum theory 26(206, 208)
— of vibration-rotation spectra 26(217)
— rigorous and approximate 27(330, 339)
— violation 27(330, 335–342, 352, 364, 380)
Self broadening 27(43)
Selfconsistent field 27(244, 269, 270, 273, 281)
Self energy 25/2d(6, 7, 21, 109, 138–140, 163, 193, 380, 407, 413, 446, 486, 506, 507, 524)
— anharmonic (defects) 25/2d(327, 446)
— diagram 25/2a(118)
— diagrams 25/2d(502)
— effective (defects) 25/2d(410)
— harmonic 25/2d(336, 386, 404)
— one-phonon 25/2d(155)
— photon 25/2d(118)
— self-consistency 25/2d(503)
— shift 25/2b(321)
— vertex corrections 25/2d(516)
Self-luminous object 29(349)
Self-pulsing, condition 25/2c(239)
Semiclassical approach 25/2c(173–247)
Semiclassical approach, systematics 25/2c(183)
Semiclassical equations, gas laser 25/2c(194–197)
Semiclassical equations, solid-state laser 25/2c(51, 173–197)
Semiclassical treatment 25/2a(158)
Semiconducting crystals, spectral lines 28(292)
Semiconduction in phosphors 26(97, 98)
Semiconductor 25/2a(80, 100)
Semiconductor diodes in a magnetic field, emission from 25/2a(312)
Semiconductors, microwave resonance 28(64)
Semi-direct product group
— definition 25/2b(19)
— symmorphic space groups as 25/2b(74)
Sénarmont method 25/1(39ff.)
Sensitisation of phosphors 26(25, 81–85)
Sensitiser in luminescence 26(82)
Sensitivity limit 24(581)
Separation of variables 27(223–228)
Sequency of bands 27(385, 419)
Series formula, generalized 27(137, 141)
Series interactions 27(140–144, 153)
Series limit 28(87, 96, 104, 118)
Series limit, shift in plasma 28(96, 124, 129, 135)
Series of bands 27(345, 407, 408, 419)

Series of rovibronic levels 27(317, 318, 323)
Series, perturbed 27(136ff.)
Shading of bands or of a band sequence 27(385, 387, 430)
Shadow schlieren method 24(565, 588)
Shallow levels 25/2a(182)
Shearing interferometer 29(311)
Shearing polarization interferometer 29(410)
Shell model 25/2a(93, 97, 99, 132, 137); 25/2d(17, 26, 28, 30, 45, 57, 59, 74, 291, 299, 311, 345, 358, 364)
— breathing 25/2d(29, 30, 42, 291, 292, 345)
— defects 25/2d(332, 334, 358, 442)
— deformable 25/2d(33, 171, 365)
— dipole moments 25/2d(376)
— Lifshitz formalism 25/2d(363)
— non-linear 25/2d(149, 376)
— overlap 25/2d(36, 37, 50, 260)
— valence-overlap 25/2d(260)
Shenstone-Russell formula 27(137)
Shift direction 27(75, 76)
Shift function, *see also* Self energy 25/2d(418, 444)
Shift of high series members 27(72–76)
Shock isolation of ruling engines 29(671ff.)
Shock tube 28(134)
Shock wave emission continuum 28(193)
Shock wave luminescence 28(192)
Shock waves 28(119, 134, 157)
Shubnikov-de Haas 25/2a(294)
Si, a-(amorph) 25/2d(228)
Si, silicon 25/2a(82, 98, 113, 130, 266); 25/2d(45, 208, 211, 252ff.)
SiC, hexagonal 25/2a(279)
Sidebands 25/2c(106, 215–217); 25/2d(283, 286, 371, 418, 423, 434, 437, 441, 444)
Siegbahn's Nobel Institute ruling engine 29(727ff.)
σ-polarisation 28(305)
Sign convention 29(229)
Sign convention, positive 25/2a(15)
Signal, complex analytical 25/2c(73)
Signal velocity 24(3)
Signals, amplification 25/2c(124–125)
Signals, external signals applied to laser 25/2c(124–125)
Silica, luminescence 26(31)
Silicate phosphors 26(40–45)
Silicon carbide, electroluminescence 26(108)
Silver activated alkali halides 26(21)
Silver, emissivity 26(148)
Silver halides 25/2d(33, 198, 240, 436)
Silver halides, continuous absorption spectra 28(176)
Silver halides, luminescence 26(23)
Similarity law for diffraction patterns 29(543, 545)
Šimon charts 25/2a(40)
Sine condition 29(248)
Sine variables 29(86ff., 90, 102, 110, 124, 133)
Single bond, rotation about 26(193, 237)
Single crystals, electroluminescence 26(100)
Single crystals, excitation by high energy particles 26(115–121)
Single crystals of sulphides, preparation 26(38)
Single-limit series 27(129)
Single-mode operation, qualitative discussion 25/2c(116–120)
Single-mode operation, quantitative treatment
— above threshold (quantum mechanical), homogeneous line 25/2c(128–138)
— above threshold (quantum mechanical), inhomogeneous line 25/2c(144–146)
— above threshold (semiclassical) 25/2c(197–199)
— above threshold (semiclassical), exact semiclassical treatment 25/2c(201–203)
— above threshold (semiclassical), general features 25/2c(183–184)
— above threshold (semiclassical), homogeneous line 25/2c(185–186)
— above threshold (semiclassical), inhomogeneous line 25/2c(187–189, 190)
— below threshold (quantum mechanical) 25/2c(120–123)

Singlet-triplet transitions 27(339, 340, 360, 382, 451)
Singular axes 25/1(69)
Singular axes in absorbing crystals 25/1(90ff., 95ff., 106ff.)
Singular integral equations for diffraction by a slit 25/1(424)
Sinusoidal dielectric grating surface, boundary value problem 29(592–594)
Sinusoidal metallic grating surface, boundary value problem 29(586–591)
SISAM spectrometer of Connes 29(499, 502ff.)
Site group, definition 25/2b(19, 437)
Slit function 29(499ff.)
Slit of variable curvature 29(628)
Slit scanning function 29(499ff.)
Slit width, optimum 29(490)
Slit with curvature 29(469ff.)
Slits of grating infinitely narrow 29(520ff.)
Small particles, infrared absorption 25/2d(469)
Smith charts 25/2a(38)
Smith-Purcell experiment 29(434, 440–443)
Smooth objects
— diffraction theory 25/1(351ff.)
— resonance effects 25/1(351ff.)
— surface current 25/1(351ff., 362, 380)
— uniqueness of solution outside 25/1(356, 360ff.)
Smyth lens 29(196)
Sn, grey in 25/2d(44, 48, 208)
Snell's law 29(2ff.)
Snell's law of refraction 25/2a(10)
SnTe 25/2d(359)
SO_2, analysis of the spectrum 27(420–421, 428–432)
Sodium *D*-lines, *see* *D*-lines
Soft modes 25/2d(219, 359)
Soft x-rays, grating efficiency 29(603)
Soleil compensator 25/1(46ff.)
Solid radiators 29(387)
Solid state, Raman effect 26(219–222)
Solids, cohesion of ionic solids 25/2a(90)
Solids, inert-gas 25/2a(88)
Solid-state laser, semiclassical equations 25/2c(51, 173–197)
Solid-state laser (semiclassical treatment) 25/2c(173–194, 201)
Solvability of space groups 25/2b(85)
Sommerfeld's formula 27(156, 159, 161)
Sonnar 29(197)
Source conditions, effect on resolving power tests 29(614)
Source modulation 28(6)
Space group 28(266)
Space group $\mathfrak{G}$
— as extension 25/2b(19)
— conjugate subgroup of 25/2b(24)
— Definition 25/2b(7, 16)
— irreducible representation $D^{(*k)(m)}$ 25/2b(48)
— irreducible representation $D^{(*k)(m)}$, basis functions 25/2b(49)
— irreducible representation $D^{(*k)(m)}$ dimension 25/2b(48)
— non symmorphic 25/2b(22)
— order 25/2b(18)
— symmorphic 25/2b(21)
Space-time group, $\mathcal{G}$ defined 25/2b(160, 165)
Space-time point group $\mathfrak{P}(\boldsymbol{k})$
— defined 25/2b(196)
— ray corepresentation of 25/2b(199)
Spark discharge 28(119)
Spark discharge continuum 28(190)
Spatial dispersion 25/2a(194)
Spatial frequency variables 29(519)
Specific heat 25/2a(143, 144)
Spectra, channeled 25/2a(46)
Spectra of a symmetrical or quasi-symmetrical rotator 27(428–432)
Spectra of diatomic molecules 27(383–404, 455–460)
Spectra of linear molecules 27(421–428)
Spectra of periodic structures 24(591)
Spectra of polyatomic molecules 27(419–445)
Spectra, reflection 25/2a(204)
Spectral analogies 27(418, 419)
Spectral constants of hydrogen halides 28(12)
Spectral dispersion 29(487)
Spectral function 25/2a(125, 127)
Spectral image formation 29(527, 531ff.)
Spectral lines, curvature 29(451, 467–472)

Spectral representation(s) 25/2a(149); 25/2d(483)
Spectral-density function 25/2d(100, 484)
Spectrograph for the Raman measurement 26(173)
Spectrograph with reflection grating 29(455)
Spectrograph with transmission grating 29(455)
Spectroheliograms 29(629ff.)
Spectroheliograph 29(469)
Spectroheliograph of high resolution 29(626)
Spectrometer slit, *see* Entrance slit, Exit slit and Slit
Spectrometer, triple-axis crystal 25/2a(78)
Spectrometers 28(209)
Spectrometers for study of paramagnetic resonance 28(4, 6)
Spectrometers of high resolution 29(622ff.)
Spectrophotometer microscope 29(411ff.)
Spectrophotometry 27(446–450)
Spectrophotometry, application to the study of molecular structures 27(452–453)
Spectroscopic gratings 29(434–436, 444ff.)
Spectrum analyser 29(750ff.)
Spectrum phase, channeled 25/2a(53)
Speed of ruling 29(669)
Sphere, absorbing 25/1(558)
Sphere, acoustical scattering 25/1(533ff.)
Sphere, acoustically rigid 25/1(536)
Sphere, acoustically soft 25/1(536)
Sphere, asymptotic representation of scattering cross-section 25/1(544)
Sphere, back scattering 25/1(556ff.)
Sphere, diffraction problems 25/1(532)
Sphere, electric dipole 25/1(553)
Sphere, elektromagnetic scattering 25/1(544)
Sphere, integral equation for surface current 25/1(559)
Sphere, intensity of far field 25/1(569)
Sphere, magnetic dipole 25/1(553)
Sphere, Mie coefficients 25/1(548)
Sphere, Mie series 25/1(566ff.)
Sphere, polarisation of scattered radiation 25/1(550)
Sphere, radiation pattern 25/1(554ff.)
Sphere, Rayleigh scattering 25/1(535, 552ff., 566ff.)
Sphere, representation of total field 25/1(548ff.)
Sphere, scattering amplitude 25/1(552)
Sphere, scattering coefficient 25/1(535, 550, 556, 573)
Sphere, surface currents 25/1(563ff.)
Sphere, surface waves 25/1(537, 540ff., 570)
Sphere, Watson transformation 25/1(538ff., 567ff.)
Spherical aberration 29(16, 22, 43, 195, 234, 245ff., 265)
Spherical excess 25/1(8)
Spherical mirror 29(247)
Spherical part of polarizability 26(200)
Spherical surface 29(7, 50)
Spherochromatic aberration 29(256, 277, 285)
Spheroid, acoustic scattering 25/1(457)
Spheroidal wave functions 25/1(455ff.)
Spiking 25/2c(132, 134–138, 144, 266–267)
Spin $\frac{1}{2}$, equivalence with two-level atom 25/2c(31–33, 234–236)
Spin decoupling 27(315, 316)
Spin doublets 27(165–170)
Spin doubling 27(389, 396, 398, 403, 428)
Spin doubling, polyatomic molecules 27(425)
Spin, effective 28(53, 54)
Spin, fictious 25/2c(234–236)
Spin functions 28(229)
Spin intercombination 27(334, 335, 339, 340, 360)
Spin motion (classical treatment) 25/2c(228–229)
Spin motion (quantum mechanical treatment) 25/2c(229–231)
Spin multiplets 28(18)
Spin of electron 28(297, 320)
Spin of nuclei, effect on molecular rotation 26(211)
Spin of nucleus from Back Goudsmit effect 28(323f.)
Spin splitting 27(317)

Spin values, actual and effective 28(54)
Spin waves 25/2a(100)
Spin-echo 25/2c(231–232)
Spin-Hamiltonian 28(45ff.)
Spin-lattice relaxation 28(47)
Spin-orbit integral 27(107, 178–184)
Spin-orbit interaction 27(106–111, 339, 356, 396)
Spin-orbit parameters 27(110, 204, 205)
Spin-orbital coupling 27(339, 356, 396)
Spin-orbital coupling constant 27(253, 396, 399, 425, 427)
Spin-orbital operator 27(339, 356, 396)
Splitting factor 27(107)
Splitting of terms 27(102, 110, 111)
Splitting of the united atom 27(225, 227, 229)
s-polarized wave 25/2a(13)
Spontaneous emission 25/2c(88–90)
Spontaneous emission, coherence properties 25/2c(97–99)
Spontaneous linewidth 25/2c(97–99)
Spontaneous polarization 25/2a(136)
Spot diagram 29(21ff., 257)
SrF_2 25/2d(204ff.)
$SrF_2:H^-$ 25/2d(422)
SrO 25/2d(241)
$SrTiO_3$ 25/2d(60, 219ff., 245)
$SrTiO_3$, strontium titanate 25/2a(119, 136)
$Sr_xBa_{1-x}F_2$ 25/2d(396)
Stability conditions 25/2d(14)
Stability of amplitude 25/2c(134–138, 144, 238)
Stability of phase 25/2c(130)
Stability of polarized modes 25/2c(201)
Stability of solids 25/2d(10)
Staggering 27(426)
Standard approximations in the interruption theory 27(16)
Standard radiator 28(81)
Standard wavelength system 27(82)
Standing waves 24(22)
Standing waves in front of a mirror-like grating 29(579)
Star atmosphere 28(155)
Star, direct product of 25/2b(93)
Star, general 25/2b(61)
Star, reduction coefficients
— ordinary 25/2b(91)
— symmetrized 25/2b(92)
Star, special 25/2b(62)
Star spectra 28(151)
Stark broadening 27(29–43)
— by electrons 27(39)
— by ions 27(35, 41)
— quantum treatment 27(35)
Stark cells 28(5)
Stark effect 25/2d(451); 27(226, 229, 410); 28(24, 25, 26)
— broadening 28(124)
— linear 27(30, 34)
— modulation 28(5)
— of first order in symmetric-top molecules 28(25)
— of second order 28(26)
— quadratic 27(34, 38)
Stark energies of the asymmetric rotor 28(27)
Stars of principal sequence 28(152)
State, definition 27(100)
State I_a and I_b, according to Mulliken 27(268)
State Optical Institute ruling engine 29(712–716)
Stationary representation
— of diffraction fields 25/1(391ff.)
— of far field amplitude 25/1(399ff.)
— of far field amplitude of Hertz' vector 25/1(402ff.)
— of linear field quantities 25/1(400ff.)
— of scattering and transmission coefficient 25/1(397ff.)
— of vector diffraction fields 25/1(401ff.)
Stationary state, method of solution 25/2c(182–183)
Statistical broadening by dipole molecules 27(61, 62)
Statistical distribution of lines 27(36)
Statistical theory for molecular broadening 27(65–67)
Statistical theory of line broadening 27(2, 3, 7–20)
Statistical time average 27(10)
Statistical weight of a rotational level 26(211, 213)
Statistical weight of a term 27(100)
Statistical weight of rotational levels 27(338, 342, 347)

Stefan-Boltzmann constant 26(140)
Stefan-Boltzmann law 26(134); 28(83)
Step grating, Maxwell theory 29(576–578)
Stereo-microscopy 29(414)
Stereoscopic electron micrographs 29(656–659)
Sticking potential 26(113)
Stigmatism of spectral lines 29(470–472, 476)
Stigmatism of Wadsworth mounting 29(483)
Stimulated absorption 25/2c(88–93)
Stimulated emission, *see also* Emission 25/2c(4, 90–91, 120–125)
Stimulated emission, coherence properties 25/2c(97–99)
Stimulation of luminescence 26(78–81)
Stimulation of photoconductivity 26(93)
Stokes line 26(171, 202–203)
Stokes parameters 25/1(17ff., 23)
Stokes' relations 25/2a(15)
Stokes vector 25/1(17)
Stokes-MacCullagh method 25/1(36, 39ff.)
Strain, *see also* Lattice distortions
Strain, finite 25/2d(448)
Strain, homogeneous 25/2d(452, 453, 459, 460, 463)
Strain in mixed crystals 25/2d(405)
Strain, infinitesimal 25/2d(448)
Stray light 29(652)
Strehl definition 29(32ff., 241, 244, 258, 263)
Strength of the lattice absorption 25/2a(33)
Stress, *see* Pressure
Stress in optical glasses 29(294)
Stress optical coefficients 25/2d(247)
Stresses, unidirectional 25/1(209ff.)
Stretching force constant 26(187)
Stretching frequencies 26(178–180)
Stretching vibration 26(176, 177)
Strip grating 25/1(370ff.)
Strong's ruling engine 29(680–682)
Structure of glasses 28(294)
Structure of solutions 28(294)
Sub-bands of *K* structure 27(430, 431)
Subduces representation and compatibility 25/2b(86)
Subduces representation, definition 25/2b(49)
Subduces representation of $\mathfrak{G}$ onto $\mathfrak{T}$ 25/2b(52)
Subgroup reduction coefficient, compared to full group 25/2b(109, 111)
— defined 25/2b(107)
Substances
Sulfur 25/2d(214)
Sulphide phosphors, saturation 26(112)
Sum rule for isotropic molecules 26(191)
Sum rule for rotational isomers 26(193–194)
Sum rules 25/2d(50, 71, 75, 86, 86–88, 90, 101, 215, 321); 27(342, 448)
Sun continuum 28(122, 153)
Superachromat 29(23, 283, 359)
Superconvergence theorem 25/2d(89, 103)
Superposition of bands or band systems 27(384, 406, 407, 435)
Superposition of vibrations 28(260)
Super-radiance 25/2c(224–228)
Super-radiant emission, classical description 25/2c(231)
Super-radiant states 25/2c(228)
Super-Schmidt cameras 29(281)
Support systems for telescope mirrors 29(317)
Surface charge, magnetic 25/1(225)
Surface contribution 29(100ff., 104ff., 123, 129, 132, 134, 136, 138ff., 174, 177, 188ff.)
Surface current 25/1(220)
— circular cylinder 25/1(516ff.)
— determination by integral equation 25/1(299ff.)
— electric 25/1(225, 299)
— from variation principle 25/1(221, 401ff.)
— magnetic 25/1(225, 313, 329)
— method for solving diffraction problems 25/1(220ff., 298ff.)
Surface defects 29(299)
Surface magnetisation 25/1(226)
Surface measurement by interference methods 29(757)
Surface modes 25/2d(468, 471)
Surface reflectivity, effect upon efficiency 29(645)

Surface structure 26(133)
Surface temperature 26(133)
Surface waves 25/1(297ff., 495, 511ff., 540ff.)
Surface-resistance 24(494, 495)
Surfaces 25/2d(39, 289, 467)
Surfaces, ellipsoidal energy 25/2a(253)
Surfaces, Fermi 25/2a(79)
Surfaces, non-spherical energy 25/2a(277)
Surfaces of constant amplitude 25/2a(4)
Surfaces, of constant phase 25/2a(4)
Susceptibility, *see also* Infrared absorption 25/2d(65, 78, 101)
Susceptibility 25/2a(152)
Susceptibility, adiabatic 25/2d(104, 105)
Susceptibility, anharmonic 25/2d(137)
Susceptibility, electronic 25/2d(87, 88, 133, 143, 154, 341)
Susceptibility, isothermal 25/2d(104, 105)
Susceptibility, Kubo 25/2d(101)
Symbols denoting force constants 26(187)
Symbols denoting point groups 26(182)
Symbols denoting vibration types 26(185)
Symmetric objective 29(194)
Symmetric polynomials and symmetrized powers 25/2b(268)
Symmetric polynomials, elementary 25/2b(267)
Symmetric stretching vibration 26(177)
Symmetrical rotator 27(318, 355, 428–431)
Symmetries, common crystals 25/2d(8, 440, 551)
Symmetries, phonons 25/2d(18)
Symmetries, set of critical points 25/2d(19)
Symmetrized Kronecker product
— and harmonic lattice eigenfunction symmetry 25/2b(264)
— character of 25/2b(269)
— of hermite polynomials 25/2b(264)
Symmetry and transition propability 27(325, 326, 333–337, 348–354)
Symmetry breaking, external stresses 25/2b(455)
Symmetry character of a rotational level 26(211)
Symmetry coordinates 26(190)
Symmetry group of a crystal class 28(211)
Symmetry, importance for the classification of molecular levels 27(320–330)
Symmetry number 26(225)
Symmetry of crystals and index ellipsoid 25/1(66ff.)
Symmetry of crystals and optical activity 25/1(83–85)
Symmetry of electronic states
— diatomic molecules 27(322–324)
— linear molecules 27(324–327)
— non-linear molecules 27(327, 328)
Symmetry of molecular orbitals 27(229, 244, 245)
Symmetry of normal vibrations 26(184)
— standard symbols 26(185)
Symmetry of the vibronic and rovibronic levels 27(322, 323, 328, 329)
Symmetry of vibrational levels 27(321, 328, 329)
Symmetry operation(s) 26(180ff., 205); 27(322–329)
— definition, glide plane 25/2b(7)
— definition, screw axis 25/2b(7)
— $\{\varphi|t\}$ multiplication 25/2b(16)
— symmetry transformation, "imagined" 25/2b(6)
— symmetry transformation, "real" 25/2b(6)
Symmetry operator on functions 25/2b(26)
— group of 25/2b(27)
— idempotent 25/2b(31)
Symmetry properties 27(322–329)
Symmetry symbols 27(322–329)
Symmetry, translational $\{\varepsilon|t\}$ 25/2b(7)
Symmetry variation by vibrational excitation 27(325, 328, 349–354, 380)
Symmetry with respect to inversion at the origin 27(229, 245, 323)
Symmetry with respect to the exchange of nuclei 27(239, 323)
Synchrotron radiation 28(145)
Synthetic bands 27(429–431)

Synthetic spectra 27(429–431)
Systems in resonance 27(238)
Systems without axial symmetry 29(30)
Szigeti (effective) charge 25/2d(27, 67–69, 74, 188, 215)
Szivessy compensator 25/1(49)

T matrix 25/2d(315, 323)
T resonance 25/2d(316, 326)
Talbot's bands 25/2a(48)
Tangent approximation 25/1(506)
Tangential image height aberration 29(129ff., 135ff., 138ff., 180ff., 187, 190ff.)
Tangential variation 29(127ff., 186)
Tantalum, emissivity 26(156)
TE modes 25/2a(7, 8)
Te, tellurium 25/2a(279); 25/2d(214, 224)
Telecentric beam 29(760, 763)
Telecentric beam, perturbation 29(765)
Tele-objective 29(196, 199)
Television 29(390)
Teller-Redlich product rule 26(191–192)
Temperature, black 28(84, 142)
Temperature control of ruling engines 29(678–681)
Temperature dependence 25/2a(169)
— of effective mass 25/2a(275)
— of emissivity 26(145–152, 155, 156, 162, 169)
Temperature effect
— on crystal spectra 28(264)
— on length measurement 29(761)
— on optical activity 25/1(191)
— on the absorption spectra 27(407)
Temperature, effective 28(84, 152)
Temperature effects, dependence on crystal classes 25/1(189)
Temperature effects, measurement by interference method 25/1(190ff.)
Temperature effects on telescopes 29(320)
Temperature radiation, definition 26(129)
Temperature scale 26(130, 141)
Temperature shift 25/2a(188)
Tensor calculus and lattice dynamics 25/2b(210)
Tensor, conductivity 25/2a(240)
Tensor, effective mass 25/2a(161)
Tensor, mass 25/2a(260)
Tensor, permittivity 25/2a(240)
Term, definition 27(100)
Term series, perturbed 27(136ff.)
Term splitting 27(107, 110, 111)
Term structure in a single-limit series 27(129)
Term structure of *d*-electron configurations 27(213–218)
Term systems, isoelectronic 27(150–156)
Term values, absolute, definition 27(123)
Term values in *LS* coupling 27(100–104)
Term values of non-penetrating orbitals 27(125, 185, 186)
Tessar 29(196)
Test methods for telescopes 29(300ff.)
Tests of resolving power of gratings 29(611ff.)
Tetrahedral molecules, Raman spectra 26(234–235)
Tetratomic molecules, Raman spectra 26(231–233)
TE-wave 24(466, 473)
Thallium activated alkali halides 26(20, 21)
Thallium halides, continuous absorption spectra 28(176)
Thallous halides, luminescence 26(24)
Thermal conductivity, perturbed crystals 25/2d(391)
Thermal diffuse scattering of x-rays 25/2a(75)
Thermal equilibrium 27(447, 448, 458–462)
Thermal expansion 25/2a(113, 118, 123, 146); 25/2d(406, 427, 445, 450, 452, 466)
Thermal expansion of metal measures 29(785)
Thermal glow, *see* Thermoluminescence
Thermal neutrons 25/2a(77)
Thermodynamic functions derived from molecular spectra 26(222–228)
Thermodynamic Green's function 25/2a(115, 148)
Thermodynamic limit 25/2d(13)
Thermodynamic potentials 25/2d(491)

Thermodynamic properties 25/2a(143, 146)
Thermodynamic stability 25/2d(12, 14)
Thermodynamic theory of ferroelectricity 25/2a(143)
Thermoluminescence curves for halides 26(19, 21, 24)
Thermoluminescence curves for metastable states 26(62, 63)
Thermoluminescence curves for ZnS phosphors 26(33, 36)
Thermoluminescence, definition 26(2)
Thermoluminescence due to metastable states of luminescence centres 26(62–65)
Thermoluminescence in photoconducting phosphors 26(65–72, 89)
Thermoluminescence in the pure alkali halides 26(18, 19)
Thermoluminescence of alkaline earth sulphides 26(39–40)
Thermoluminescence of zinc silicate 26(42)
Θ matrices 25/1(28ff.)
Thick lamella 25/2a(46, 50)
Thick layers 24(527, 534)
Thickness, effective 25/2a(47)
Thickness gradient 24(573)
Thickness measurements of layers 24(539)
Thin films 25/2a(49)
Thin lamella 25/2a(49)
Thin layers, thickness of 24(513)
Thin lens approximation 29(17)
Thinnest layers 24(534)
Thomas-Fermi theory 25/2d(10, 11)
Thomson cross section 25/2d(91)
Thomson scattering 28(150, 154, 157)
Thomson's theorem of electrostatics 25/1(392)
Thomson-Whiddington law 26(113); 28(140)
Thoria, emissivity 26(165)
Three-dimensional problem of grating diffraction 29(449)
Three-level laser, quantum mechanical treatment 25/2c(125–137)
Three-mode operation, *see* Multimode operation
Threshold condition 25/2c(186, 188, 191, 252)
TH-wave 24(466, 473)
TH-wave of very oblique incidence 24(537)
Tilt errors 29(793, 808, 817)
Time of collision 27(9–11)
Time reversal
— operator *K*, applied to eigenvector 25/2b(161)
— operator *K*, defined 25/2b(160)
— test in diamond ⋆*X*, ⋆*W* 25/2b(372)
— test in rocksalt 25/2b(373)
— invariance 25/2d(476)
Tint 24(630)
$TiO_2:OH^-$ 25/2d(284)
TlBr 25/2d(51)
TM modes 25/2a(7, 8)
Tokyo ruling engine 29(722–724)
Tolansky method 24(544)
Tool half-shade analyser 25/1(41ff.)
Top molecules, *see* Rotation spectra
— asymmetric 28(15)
— symmetric 28(13)
— symmetric non-rigid 28(11, 14)
Töpler's schlieren image 24(618)
Töpler's schlieren method 24(557, 611)
Töpler's schlieren method as for cylindrical objects 24(618)
Topogon 29(198)
Torsional frequency 26(194)
Total angular momentum 27(90)
Total energy 27(319, 320)
Total reflection 24(513)
Total reflection at boundaries of anisotropic media 25/1(112ff.)
Total reflection, attenuated 25/2a(57)
Total representation as a direct product 25/2b(206)
Total representation, defined 25/2b(204)
Total representation, trace of 25/2b(206)
Transfer function 29(257)
Transfer of energy between different vibrational modes 27(420)
Transfer of energy between molecular systems 27(360)
Transfer of excitation energy 27(360, 374, 381, 382, 420)
Transform or transformation, Fourier-transform 25/2a(45, 83, 106)
Transformation, Ewald transform 25/2a(85)

Transformation operators for coordinates 25/2b(120)
Transformation operators for displacements 25/2b(119, 121)
Transformation operators of complex conjugation 25/2b(122)
Transition formulae 29(81, 84ff., 88, 155)
Transition probabilities 28(84)
Transition probabilities for an emission centre 26(62)
Transition probability, *see also* Intensity and Selection rules 27(330, 331, 345, 346, 388, 446–452)
— and perturbations 27(353, 361)
— H_2^+ 27(242)
— N_2 27(452)
— numerical data 27(464, 465)
— polyatomic molecules 27(348)
— spontaneous and induced 27(331)
Transition-polarizability 25/2d(90, 109, 115)
Transitions, allowed 25/2a(288)
Transitions between Hund's cases *a* and *b* 27(315, 316, 398)
Transitions between linear electronic states 27(421–428)
Transitions between states of a symmetrical or quasi-symmetrical rotator 27(428–432)
Transitions, classification according to atomic and molecular electronic configurations 27(414–417)
Transitions, competition between radiative and non-radiative transitions 27(374, 376)
Transitions, direct 25/2a(163)
Transitions, displacive 25/2a(137, 143)
Transitions, first-order 25/2a(142)
Transitions, forbidden 25/2a(288)
Transitions, indirect 25/2a(163, 190)
— exciton 25/2a(178, 192)
— interband 25/2a(165)
Transitions, interband 25/2a(159)
Transitions, intraband indirect 25/2a(168)
Transitions, magnetic dipole and electric quadrupol transitions 27(340)
Transitions of type donor-acceptor 27(358)
Transitions, polarization of IMO 25/2a(289)
Transitions, quadrupole 25/2a(193)
Transitions, reconstructive 25/2a(143)
Transitions, second-order 25/2a(142)
Transitions, standardization of the nomenclature 27(358)
Transitions with $\Delta n \neq 0$ 27(161–164)
Transitions with variation of molecular symmetry 27(432–435)
Transitions with variation of the multiplicity 27(360, 451)
Transitions, zero-phonon 25/2a(226)
Translation group $\mathfrak{T}$
— as direct product 25/2b(11)
— irreducible representation 25/2b(40, 42)
— normal subgroup of $\mathfrak{G}$ 25/2b(17, 23)
Translational entropy 26(227)
Translational heat capacity 26(225)
Translational partition function 26(225)
Transmissibility 24(468)
Transmissibility factor 24(471)
Transmission cross-section
— and complementary scattering cross-section 25/1(387)
— definition 25/1(383)
— scalar problem 25/1(387ff.)
— stationary representation 25/1(397ff.)
— vector problem 25/1(389ff.)
Transmission echelon 29(465, 466)
Transmission factor 26(131)
Transmission grating 29(433, 436, 446)
Transmission grating, parameters 29(455, 456)
Transmission of optical glasses 29(293)
Transmission-coefficient, definition 25/1(383)
Transmission-coefficient for narrow slit 25/1(436ff.)
Transmittance, amplitude transmittance coefficient 25/2a(9)
Transmittance, average 25/2a(50)
Transmittance of the lamella, amplitude 25/2a(20)
Transmittance, power 25/2a(17)
Transmittent illumination 29(377)
Transparency, self-induced 25/2c(237, 246–247)
Transparent plates, series of 25/1(12)

Transversal E-waves 24(465)
Transversal H-waves 24(466)
Transverse electric TE modes 25/2a(7, 8)
Transverse electric waves 25/2a(7)
Transverse excitons 25/2a(172, 193)
Transverse magnetic TM modes 25/2a(7, 8)
Transverse magnetic waves 25/2a(7)
Transverse magnification 29(8)
Transverse wave 25/2a(4)
Trapping 26(6, 8, 9, 36, 65–67, 75, 77)
Triangular groove profile, boundary value problem 29(597–599)
Triatomic molecules, Raman spectra 26(229–231)
Triple mirror 29(213, 782ff.)
Triple-axis crystal spectrometer 25/2a(78)
Tube lenses 29(373)
Tube of microscope 29(372)
Tungstate phosphors 26(47–49)
Tungstates, impurity activated 26(48)
Tungstates, temperature dependence of emission 26(47, 48)
Tungsten, color temperature 26(161)
Tungsten, emissivity 26(150, 151, 152, 156)
Tungsten filament lamp 28(81, 146)
Tungsten light source 29(387ff.)
Tungsten, luminance temperature 26(159)
Tunnel effect 27(304, 379, 380)
Tunneling 25/2d(285)
Tunneling, absorption 25/2d(282)
Tunneling, motion 25/2d(280)
Tunnelling of electrons 25/2a(79, 81, 130)
Twin plane reflection 25/1(117ff.)
Two-dimensional problem of diffraction 29(446)
Two-electron configurations 27(93–99)
Two-mirror system 29(230, 237, 261)
Two-mirror system, aplanatic 29(265)
Two-mode operation
— above threshold (quantum mechanical, semiclassical) 25/2c(141–144, 184–185, 189–190, 199–201)
— below threshold (quantum mechanical) 25/2c(139)
— near threshold (quantum mechanical) 25/2c(140–141)
Two-parameter Ritz formula 27(127)
Two-phonon bound states (in diamond)
— and Raman scattering 25/2b(310, 433)
— symmetry of 25/2b(311)
Twyman-Green interferometer 29(21)

U-centers, *see also* Substances 25/2d(264, 267, 269, 288, 289, 295ff., 347, 367ff., 389, 405, 415, 418, 422, 428, 432, 436ff., 443, 461)
Ultrashort optical pulses 25/2c(213–224)
Ultrasonic method for photoelastic measurements 25/1(216ff.)
Ultraviolet emitting phosphors 26(50)
Ultraviolet gratings 29(435, 473, 625, 651)
Ultraviolet microscopy 29(395)
Ultraviolet, optical materials 29(24)
Ultraviolet reflection of aluminum 29(601)
Ultraviolet reflection of platinum 29(602)
Uncertainty condition for waves 24(578)
Uncertainty relations 25/2c(83–88)
Uniaxial crystal 25/1(67ff.)
Uniaxial crystals, absorbing 25/1(89, 101)
Uniaxial stress 25/2a(193)
Unit of length 29(766)
Unitary group $SU_{(sl_j)}$ and lattice eigenfunction 25/2b(270)
United atom model 27(225–227, 436)
Universal broadening 27(39)
Unpolarised light, nature of 25/1(31ff.)
Unsaturated inversion 25/2c(121, 177)
UO_2, uranium dioxide 25/2a(99, 126)
Uranium activated alkali halides 26(23, 51)
Urey-Bradley field 26(186–187)
Usable magnification 29(347)

V, vanadium 25/2a(81)
Vacancies 25/2d(264)
Vacuum state of electrons 25/2c(27)
Vacuum state of field 25/2c(26)
Vacuum system for study of luminescence 26(127)

Vacuum ultraviolet grating spectrograph 29(625)
Vacuum wave length 24(463)
Valence electron 27(292–299, 437); 28(86, 104)
Valence force field 26(185, 186)
Valence forces 25/2d(44, 48, 49)
— bond-bending potential, Keating 25/2d(47)
— bond-stretching forces 25/2d(33)
Valence of bondage 27(243, 257, 265)
Valence of bondage, method of Heitler and London 27(260–266)
Valence state 27(273, 274)
Valency 25/2d(69)
van der Kolk method 25/1(176ff.)
Vanadate phosphors, emission characteristics 26(52)
van der Pol approximation, Fokker-Planck equation reduced in the 25/2c(158)
van der Pol equation 25/2c(144, 158)
van der Waals forces 28(162, 181, 205)
van der Waals interaction 25/2d(14, 22); 27(8, 9, 23)
Van Hove singularities 25/2a(209)
Van Vleck-Weisskopf line shape 28(24)
Variational approximation (H_2^+) 27(236–242)
Variational functions 27(235, 241)
Variational method 27(234, 235)
Variational parameters 27(235, 241)
Variational treatment of grating diffraction 29(571)
Vavilov treatment of polarization 26(86)
V-bands 28(291)
Vector, complex wave 25/2a(4)
Vector, eigen 25/2a(66, 119)
Vector model 28(302ff.)
Vector, Poynting 25/2a(7)
Vector, wave 25/2a(65, 240)
Vector wave equation 25/1(227ff.)
Vectorial boundary problem of grating diffraction 29(571, 600)
Vector-scalar 29(153ff.)
Vector-scalar, power series representation 29(155ff.)
Velocity of light
— astronomical methods, determination of the 24(5, 19)
— best value 24(31)
— constancy of the 24(37)
— direct methods, determination of the 24(4)
— electromagnetic methods, determination of the 24(21)
— in matter 24(32)
— in moving matter 24(34)
— in vacuum 24(463)
— in water 24(32)
— indirect methods, determination of the 24(19)
— interferometric measurement 29(675ff.)
— toothed wheel method, determination of the 24(6)
Verdet constant 25/1(199, 202)
Vertex part
— cubic 25/2d(508)
— diagrammatic 25/2d(512)
— functions 25/2d(510)
— higher-order 25/2d(508)
— integral equations 25/2d(509, 512)
Vertex perpendicular 29(91, 101, 102, 110)
— aberration 29(123, 136)
— reduced 29(120)
Vertex renormalization 25/2d(509)
Vertex-diffraction law 25/1(290)
Vertical illuminator of a polarising microscope 25/1(172)
Vibration, antisymmetric 27(321)
Vibration band, rotational fine structure 26(216)
Vibration, degenerate 27(321)
Vibration, normal 27(419)
Vibration of a molecule in classical mechanics 26(174)
Vibration of ruling engines 29(673)
Vibration, totally and non totally symmetrical 27(321)
Vibrational analysis, example LaO 27(389–395)
Vibrational analysis, polyatomic molecules 27(423–424, 427)
Vibrational constants, diatomic molecules 27(393, 394)
Vibrational energy 27(299–312, 320)
Vibrational entropy 26(228)
Vibrational heat capacity 26(225–226)
Vibrational levels 27(319, 320)

Vibrational levels, convergence 27(413)
Vibrational modes, attribution of the observed frequencies 27(419)
Vibrational modes, selective excitation 27(419)
Vibrational partition function 26(223–224)
Vibrational quantum number 26(216)
Vibrational spectra 26(172)
Vibrational structure 27(344–347, 384–394)
Vibrational symmetries 25/2d(552)
Vibration-rotation spectra 26(216–218)
Vibronic excitation 27(307)
Vibronic interaction 27(303, 306, 307, 309)
Vibronic levels 27(319, 320)
Vibronic processes 25/2d(8, 361, 367)
Vibronic structure 27(349, 350, 419–421)
Vickers imprint, interference test 29(795–797)
Video system 28(5)
Vignetting 29(208, 232, 262, 274, 279)
Vincent-Geisse and Lecomte, contruction 25/2a(56)
Violation of selection rules 27(330, 335–342, 352, 364, 380)
Virtual-crystal approximation 25/2d(382, 390, 399, 400)
Vogl interferometer 29(777)
Voigt configuration 25/2a(241, 243)
Voigt effect 25/2a(244, 280, 298)
Volume modes 25/2d(471)

Wadsworth mount of a concave grating 29(477ff., 483)
Wagging frequencies, data 26(179)
Walsh's method 27(292–299, 437)
Walsh's method, application
— to BO_2 and NH_2 27(428)
— to NCO 27(426)
Water molecule, spectrum 26(230)
Watson transformation 25/1(298, 507)
Wave aberration 29(243, 258)
Wave equation 24(463); 25/2c(174, 178–179)
Wave functions
— and nodal surfaces 27(225–230)
— and symmetry operators 27(322–329)
— monoelectronic 27(267–299)
— rovibronic 27(321)
Wave guide
— complementary problems 25/1(343ff.)
— coupled single-edge problems 25/1(344ff.)
— edge planes 25/1(337)
— elements, interaction 25/1(344ff.)
— open 25/1(337ff.)
— problems involving edges 25/1(337ff., 342ff., 348ff.)
— radiation 25/1(336ff.)
Wave length range for atomic spectra 27(81)
Wave propagation 25/1(56)
Wave propagation, analytical solution 25/1(65ff.)
Wave surface 25/1(57, 73ff.)
— for biaxial crystals 25/1(75)
— of optically active transparent crystals 25/1(84ff.)
Wave vector(s) 25/2a(65, 240)
— canonical, ***k*** of ****k*** 25/2b(59)
— class I, II, III, defined 25/2b(177)
— complex 25/2a(4)
— definition of ***k*** 25/2b(42)
— equivalent 25/2b(43)
— inequivalent 25/2b(54)
— reduction coefficients 25/2b(93)
— selection rules 25/2b(93)
— space group of ***k***, $\mathfrak{G}(\boldsymbol{k})$ 25/2b(53)
— star of, definition 25/2b(61)
Wave velocity 24(2)
Wave velocity surface 25/1(57, 73ff.)
Wavefront aberrations and grating blank imperfections 29(534ff., 541)
Wavefront aberrations and ruling errors 29(533ff.)
Wavefront calculation, programming 29(564ff.)
Wavefront deformation 29(7)
Wavefront interferograms 29(632ff., 635, 693ff.)
Wavefront, phase variations 29(638)
Wavefront reconstruction 29(736, 739, 743)
Wavelength dependence 24(501)
Wavelength dependence of emissivity 26(145–152)
Wavelength gauge of the meter 29(766ff.)

Wavelength on the velocity of light, influence of the 24(4)
Wavelength slide rules 29(770)
Wavelets 29(513)
Waves, anomalous 25/2a(194)
Waves, character of 25/2a(7)
Waves, homogeneous 25/2a(4, 6)
Waves, inhomogeneous 25/2a(4, 6)
Waves, intensity of plane 25/2a(7)
Waves, left circularly 25/2a(234)
Waves, longitudinal component 25/2a(5)
Waves, partial 25/2a(12)
Waves, *p*-polarized 25/2a(15)
Waves, right circularly 25/2a(234)
Waves, spin 25/2a(100)
Waves, *s*-polarized 25/2a(13)
Waves, transverse 25/2a(4)
Waves, transverse electric 25/2a(7)
Waves, transverse magnetic 25/2a(7)
Weak electronic transitions 27(358)
Weakly absorbing materials 25/2a(13, 51)
Wedge error of plane parallel plates 29(819)
Weisskopf interruption theory 27(9)
Weisskopf treatment of resonance broadening 27(48)
Welsbach mantle, emissivity 26(168)
Wide-angle objectives 29(199)
Wiener-Hopf method 25/1(222ff., 337ff.)
Wien's constant 26(141)
Wien's displacement law 26(136)
Wien's radiation law 26(135); 28(83)
Wigner distribution 25/2c(61–63, 80)
Wigner-Eckart formula, analogy for lattice dynamics 25/2b(212)
Wigner-Eckart formula and antiunitary elements 25/2b(216)
— case A wave vectors 25/2b(217)
— case B wave vectors 25/2b(218)
— case C wave vectors 25/2b(219)
Wire grating 29(443)
Wobbling 27(306)
Wollaston prism 26(174); 29(222, 410, 805)
Wood anomalies 29(574, 580, 583)
Wright-Väisälä System 29(278)
Wurtzite structures 25/2d(49, 224, 264, 271, 396ff.)

Xenon continuum in the ultraviolet 29(646)
Xenon high pressure arc 28(81, 101, 131, 208)
Xenon high-pressure lamp 29(389)
X-ray absorption continua 28(112)
X-ray gratings 29(429, 435, 473, 648)
X-ray gratings, efficiency 29(647)
X-ray gratings, reflection coefficients 29(647ff.)
X-ray scattering 25/2a(74, 124, 130)
X-ray-bremsstrahlung 28(87, 139)
X-rays of large wavelength, grating efficiency 29(603)
X-rays, thermal diffuse scattering 25/2a(75)

Young's principle 25/1(218, 267, 292)
Young's principle and Kirchhoff's theory 25/1(274ff.)
Young's principle, edge effect 25/1(267, 274)

Zeeman effect 25/2a(314, 315); 27(410–412)
— applications 28(298, 325f.)
— discovery 28(296)
— experimental methods 28(299)
— in atoms, classical theory 28(300)
— in crystals 28(209, 253, 258)
— intensities of components 28(317f.)
— of excitons 25/2a(297)
— of hyperfine structure 28(320f.)
— splitting of rotational lines 28(27, 43)
Zeeman energy 28(29)
Zeeman patterns, examples 28(306f., 332)
Zeeman patterns, unresolved 28(329)
Zehnder chamber 29(828)
Zenneck case 24(474)
Zero gap, *see* Origin of a band missing lines
Zero-phonon line 25/2a(187, 218)
Zero-phonon transitions 25/2a(226)
Zero-point energy 25/2a(90)
Zinc chalcogenides 25/2d(219, 259)

Zinc oxide, luminescence 26(28, 29)
Zinc silicate phosphors 26(40–43)
Zinc spectra, continuous 28(183)
Zinc sulphide phosphors 26(31–38)
— divalent impurities (Mn^{2+}, Fe^{2+}, Co^{2+}) 26(36, 37)
— impurity activated 26(34, 35)
— self-activated 26(34)
— structure theory 26(35)
ZnO 25/2d(224, 259, 261)
ZnS, zinc blende 25/2d(49, 55, 220, 259ff., 264, 271, 288, 307, 347, 396)
ZnS, zinc sulphide 25/2a(98)
ZnSe 25/2d(50)
ZnSe, zinc selenide 25/2a(278)
ZnS:Mn 25/2d(448)
$ZnS_{1-x}Se_x$ 25/2d(405)
ZnTe 25/2d(260)
Zoom lens 29(67, 200ff.)

Group 6

X-Rays and Corpuscular Rays

Abbe-Helmholtz sine condition 33(364)
Abbe's sine condition 30(329)
Aberration, chromatic, *see* Chromatic aberration
Aberration, chromatic 30(333)
Aberration coefficient 31(246)
Aberration coefficient for spherical sector analyzer 31(246)
Aberration function 30(326)
Aberration, longitudinal spherical 33(228)
Aberration, spherical 30(326, 328); 33(227, 399, 410, 414, 467, 477, 484, 611, 620, 630, 635)
Aberrations, curves of 33(226)
Aberrations, geometrical 33(218)
Aberrations in crossed systems 33(292)
Aberrations of the second order 33(315)
Absorbed photon energy in a medium 34(282)
Absorption coefficient 30(9, 19, 20, 87, 309, 339)
— for electrons in gases 34(4)
— for photons due to Compton effect 34(272, 273–278, 282, 283)
— for photons due to pair production 34(273–278, 282, 283)
— for photons due to photoelectric effect 34(273–278, 282, 283)
— for photons, total 34(273–278, 282, 283)
Absorption correction for x-rays 34(180)
Absorption cross section, average for photons 34(264)
Absorption curves for mono-energetic fast electrons 34(133, 134)
Absorption edge (or limit) 30(85, 86, 208–217, 222)
Absorption edge, wavelength and shape 30(208–211)
Absorption edges (tables) 30(214–217)
Absorption frequency, geometric mean 34(69)
Absorption instruments 31(145)
— double beam 31(145)
— double ion chamber 31(147)
— split beam 31(146)
Absorption of neutrons 32(567, 641)
Absorption of photons (x-rays) 32(153, 154, 182, 190, 218)
Absorption of radiation 31(129, 143)
— by excited atoms 31(178)
— by ions 31(178)
Absorption of slow electrons in gases 34(3, 4)
Absorption spectra of x-rays 30(85, 205–217)
Absorption spectroscopy 30(308)
Absorption spectrum of an inert gas 30(211)
Absorption transitions 30(221)
Abundances of ions, relative 33(550, 551)
Accelerators 32(577)
Acetaldehyde 31(417)
Acetylacetons 31(439)
Acetylenes 31(414, 434)
Acoustical branch of elastic waves 32(208)
Activation detector with photographic plate 32(586)
Activities, specific 33(678)
Additive shifts 31(433)
Adjustment of a bent crystal spectrograph 30(121)

Adjustment of an x-ray spectrograph 30(102ff.)
Adsorbate core line shapes 31(361)
Ag, PES of metal valence bands 31(459)
Ag, PES of outer valence shell 31(309)
Age determinations, geologic 33(585)
Age measurements of minerals, radioactive 33(569)
Air, photoelectron spectrum 31(233)
Algebraic space 32(224)
Alkali halide film 32(532)
Alkali halides 31(442, 444)
Alkali ions, collision cross section measurements in rare gases 34(33, 34)
Alkali ions, slow, beams of 34(31)
Alkali metals 31(127)
— PES or outer valence shell 31(309)
— photoionization cross sections 31(169ff.)
Al$K\alpha$ radiation 31(228)
Allenes 31(409)
Alloys 32(516)
Alloys, neutron diffraction 32(634)
Alpha particles, ranges 34(210–211)
Alpha particles, stopping cross section for 34(196–200, 204)
Ambiguity 32(128, 151)
Ammonium halides, neutron diffraction 32(605)
Amorphous aluminum oxide 32(548)
Amorphous solids, neutron diffraction 32(617)
Amorphous substances 32(524, 525)
Amorphous substances, evaporated films 32(539)
Amplifiers 33(560)
Amplitude object 33(348)
Analyzer
Analyzer, dispersive 31(244)
Analyzer lens 33(536)
Analyzer, undispersive 31(250)
Angle of maximum intensity of x-rays 30(354, 360, 363)
Angle of twist 31(438)
Angstrom unit 30(4, 92, 93, 239, 240)
Angular aperture 30(316)
Angular correlation of the three-photon annihilation radiation 34(143, 150)
Angular correlation of the two-photon annihilation radiation 34(161, 162)
Angular distribution 33(22)
— of Compton recoil electrons 34(254)
— of Compton scattered photons 34(253)
— of elastically scattered slow electrons 34(17–20, 27)
— of elastically scattered slow electrons, measuring methods 34(14–16)
— of elastically scattered slow ions 34(38–40)
— of inelastically scattered electrons 32(463)
— of scattered fast electrons 34(75, 76)
— of the electrons 33(494)
— of Thomson-scattered radiation 34(220, 221)
— of x-ray energy, theoretical 30(364)
— of x-rays 30(363)
— of x-rays from a thick target 30(348, 352)
— of x-rays from a thin target 30(353, 359, 360)
— of x-rays, relativistic correction 30(358, 359, 361)
— reduced (centre-of-mass system) 34(39, 40)
Angular distributions of Auger electrons
— coincidence experiments 31(554, 555)
— high-energy electron impact 31(548, 549)
— noncoincidence experiments 31(549ff.)
— photoionization 31(547, 548)
Angular distributions of photoelectrons (*see also* Asymmetry parameter β) 31(31, 101ff., 102, 104, 127, 128, 155, 199, 424)
— angular momentum transfer formulation 31(31ff.)
— approximation of vanishing final state interactions 31(39)
— Ar 31(201, 202)
— C 31(107)
— central potential approximation 31(40)
— Cl 31(105, 106)
— Cooper-Zare formulation 31(40)

— Cs 31(111, 205)
— from p-subshells 31(101 ff.)
— from s-subshells 31(106 ff.)
— interchannel interaction effects 31(103)
— Kr 31(201, 203)
— Ne 31(202)
— of adsorbates 31(455)
— relativistic effects 31(108 ff.)
— within a resonance 31(112)
— Xe 31(104, 110, 201, 203, 204, 207)
Angular magnification 33(184)
Angular momentum transfer 31(31)
— formulation for β 31(31 ff.)
— theory 31(426)
Anharmonicity 31(414)
Anisotropy 32(98)
Anisotropy coefficients 31(551, 552)
Annihilation of positronium 34(140–144)
Annihilation radiation, energy spectrum 34(143, 148, 150)
Annihilation rate of positrons 34(147)
Annihilation spectrum of ortho-positronium 34(150)
Anode cooling of an x-ray tube 30(30, 35, 38, 40)
Anode loading of an x-ray tube 30(14, 15, 35, 41, 57)
Anode of an x-ray tube, secondary 30(52)
Anode temperature 30(14, 15)
Anode, turning, of an x-ray tube 30(15, 35, 38–42, 52, 54, 57)
Anticathode, *see* Anode
Aperture, influence on the bundle 33(340)
Aperture optimum 33(488)
Aperture-diaphragms 33(411, 414)
Aplanatic systems 30(329)
Apparent structures 33(529)
Appearance-potential(s) 33(38, 572)
Approximate self-consistent molecular orbital theory
— CNDO 31(435)
— CNDO/S 31(421, 436)
— INDO 31(435)
— INDO/S 31(436)
— LNDO/S 31(436)
— MINDO 31(435)
— RSPT 31(421)
— SPINDO 31(436)
Ar, autoionizing structure in absorption spectrum 31(169)
Ar, β-parameter 31(102, 201, 202)
Ar, double photoionization cross section 31(119, 195)
Ar, ionization threshold 31(126)
Ar, KLM Auger spectrum 31(308)
Ar, partial photoionization cross section 31(53, 56, 62, 67, 71, 79, 81, 99)
Ar, 3s photoelectron spectrum 31(305, 306)
Ar, total photoionization cross section 31(164)
Arsabenzene 31(430)
AsH_3 31(407, 413, 415)
Assignment 33(655)
Astigmatic error 33(624)
Astigmatism 30(324, 326); 33(400, 409, 470)
Astigmatism, anisotropic 33(240)
Astigmatism, axial 33(259)
Astigmatism, isotropic 33(240)
Astigmatism of a general raybundle 33(234)
Aston's fundamental doublets 33(598)
Asymmetry parameter β (*see also* Angular distribution of photoelectrons) 31(40, 101 ff., 155, 200 ff., 209, 289, 311)
— for the rare gases 31(291)
— of the $n=2$ satellite in He 31(313)
— rotational state dependence 31(428)
— vibrational state dependence 31(427)
Asymptotic focal length 33(209)
Asymptotic image formation 33(191, 209)
Atom as scattering centre for x-rays 32(157 ff.)
Atomic absorption coefficient for x-rays 32(641)
Atomic binding energies 31(315)
Atomic constants, fundamental 30(366)
Atomic density, modified 32(235)
Atomic description 32(116)
Atomic form factor (*see* Scattering factor) 32(462)
Atomic groups, x-ray images 30(310)
Atomic ions, fraction of 33(62)
Atomic ions, production of 33(62)
Atomic levels, absolute values 30(229)

Atomic planes, artifical 30(143–145)
Atomic scattering factor for x-rays 32(105, 123, 161, 164, 641)
Atomic structure factor 34(291)
Atomic weight unit, physical 33(550)
Atom-volume 33(523)
Attachment of electrons at molecules 34(11)
Attenuation coefficient for photons
— due to Compton effect 34(270–281)
— total 34(272, 273–278, 282)
Au, PES of metal valence bands 31(459)
Au, PES of outer valence shell 31(309)
Auger kinetic energies for calibration 31(272)
Auger process 31(196)
Auger processes causing intensity anomalies 30(294)
Auger transition
— amplitudes 31(518, 538, 581ff.)
— component rate 31(519)
— definition 31(501)
— energy 31(501, 566ff.)
— group rate 31(519)
— intensity 31(523, 524)
— multiplet rate 31(519)
— of an atom 34(181, 186)
— term rate 31(519)
Autodetachment resonances 31(446)
Autoionization 31(132, 143, 159, 166, 399, 484ff.)
Autoionization, Fano theory of 31(63)
Autoionization state 31(428)
Average absorption cross section for photons 34(264)
Average collision cross section for photons 34(263, 275–281)
Average energy
— loss of fast electrons 34(61–71, 98, 99, 126)
— many-configuration *jj* 31(575)
— per Compton scattered photon 34(264)
— single-configuration *jj* 31(541)
Average ionization potential 34(62, 65, 69)
Average kinetic energy per Compton recoil electron 34(265, 282)
Average scattering cross section for photons 34(264)
Average transition energy 31(576)
Axial magnification 33(185)
Axial ratio of an ellipsoidal particle, determination 32(355)
Axis of symmetry of the system 33(16)
Axis, sixfold 32(135)
Axis, threefold 32(134)
Azimuthal intensity distribution 30(15–18, 355)
Azimuthal rotation 33(637)
Azomethane 31(441)

Ba 31(174, 175, 309)
Ba, photoelectron spectra 31(314, 315)
Ba, satellite photoelectron lines 31(314)
Background radiation 32(361)
Background scattering 32(365)
Backscattering of fast electrons 34(131, 132)
Backscattering of photons in Compton effect 34(239, 242)
Ball anode of an x-ray tube 30(40)
Ballast tank 33(653)
Balmer series 30(81, 82)
Band shape 30(261)
Band structure
— of Cu 31(456)
— of GaSe 31(453)
— three-dimensional 31(452)
— two-dimensional 31(450)
Barrel distortion 33(243)
Base spread 33(613)
Basic lines 33(278)
Basis vectors of a lattice 32(99, 107)
Be, double photoionization cross section 31(119)
Be, photoionization cross section 31(91, 116)
Beam producing systems, limits of efficiency 33(19)
Beam production 33(47)
Beam production and space charge 33(24)
Beams of carriers, potential in free 33(28)
Beer's law 31(130)
Beiduk-Konopinski field 33(621)
Bending device (for the crystal) 30(119, 120)
Bent crystal, quality tests 30(121)

Bent crystal spectrograph, *see* Crystal spectrograph
Benzene 31(359, 388, 389, 390, 411, 425, 428, 430)
Benzene, application of Fourier transforms 32(125)
Benzene, monosubstituted 31(430)
Benzene, shake up spectrum 31(359)
Benzene, valence photoelectron spectrum 31(387)
Benzene, vibrational energies 31(412)
Benzoyl chloride 31(395)
Berg and Barrett's x-ray microscope 32(432)
Beta- and gamma-ray spectroscopy, high precision 33(647)
Beta-particles, acceleration of 33(654)
Beta-ray spectrometers or spectrographs 33(609)
Beta-ray spectrum, continous 33(610)
Betatron as an x-ray source 30(3, 71–74)
Bethe-Bloch stopping power 34(62, 63, 98, 126)
Bethe-Heitler theory of radiation losses 34(63)
BF_3 31(348, 397)
B_2H_6 31(348)
Bhabha cross section for positron-electron scattering 34(59, 60, 63)
Binding energy(ies)
— for free atoms 31(316)
— for krypton 31(298)
— for xenon 31(298)
Binding energy
— adiabatic 31(254)
— atomic 31(293)
— correlation effects 31(565, 566)
— ΔSCF 31(285)
— for gases 31(272)
— for neon and H_2O 31(343)
— for solids 31(272)
— nuclear size effects 31(564, 565, 588)
— of electrons, effect on Compton scattering 34(284–294)
— of the *s*-electrons in the *K*, *L*, and *M* shells of all elements 34(287)
— perturbation analysis 31(560)
— quantum-electrodynamical effects 31(563, 564)
— relativistic effects 31(562, 563)
— vertical 31(284, 295)
Biochemical process 31(436)
Biological molecules 31(436)
Bipartition angles for Compton scattered photons 34(257, 261–263)
Biprisma method 32(508)
Blackening 33(429, 430, 431)
Bleakney and Hipple, exact focussing of 33(594)
Blunck-Leisegang correction 34(88–90, 96–100)
Body frame 31(476)
Bohm-Pines dispersion relation 34(76, 86)
Bohr radius 34(76, 116)
Boiling cooling 30(29, 30)
Boltzmann distribution 34(73)
Born approximation 34(102, 117)
Born approximation for inelastic collisions 34(168, 169, 170, 184)
Born-Haber cycle 31(354)
Born-Oppenheimer approximation 31(294)
Boron-tri-fluoride counter 32(584)
Boron-tri-methyl counter 32(584)
Bound electrons, energy transfer from fast electrons 34(88–90)
Bound electrons, scattering of photons 34(284–294)
Bound state intensity method 31(397)
Boundary conditions
— for radial wave functions 31(497, 583, 588)
— ingoing wave 31(475, 477, 486)
— outgoing wave 31(472, 477)
— standing wave 31(478)
Br_2^+ 31(412)
Bradley's correction for powder diffraction 32(591)
Bragg case 32(484)
Bragg condition 32(460, 494)
Bragg equation 32(101)
Bragg equation, modified 32(485, 494)
Bragg law 30(90)
Bragg law, deviations 30(138)
Bragg reflection 30(11, 54, 55, 68); 32(184)
Bragg scattering 32(402)
Bragg scattering, double 32(441)
Branching ratio 31(133, 152)
Bravais lattice 32(142, 152)
Breakdown field strength, electric, at high voltages 30(28)

Breit interaction 31(536, 538, 563)
Breit-Wigner formula 31(483, 487)
Breit-Wigner resonance formula 32(559)
Bremsstrahlung, *see* Continuous x-ray spectrum
Bremsstrahlung 34(60, 63, 100, 123, 184, 190–192)
Bremsstrahlung, directional effect 30(16, 17, 53)
$B\varrho$-factor 33(678)
$B\varrho$-value in gauss·cm 33(612)
$B\varrho$ vs keV tables 33(678)
Brillouin zones 30(252)
Brillouin's theorem 31(511)
BrO 31(442)
Broadening of emission edge 30(293)
Budini theory of density effect 34(67)
Butadiene 31(438)

C_2 31(447)
Calcite, anomalous dispersion 30(143)
Calculation of vibrational structure 31(418)
Calibration, absolute 33(648)
Calibration of photoelectron spectrum 31(268)
Canal-ray-discharge ion-sources 33(64)
Canal-ray-discharge tubes 33(595)
Capacitron 30(60, 61)
Capillary discharge for uv-radiation 31(237)
Carbons 32(387–389)
Carbonyl 31(434)
Cascade tubes 30(48, 49)
Cathode, powder 33(9)
Cathode rays 33(1)
Cathode rays, principles 30(1, 4)
Cathode sputtering 32(450, 528); 33(41)
Cathodes, life-time of 33(5, 6)
Cathodes, pulsed operation of 33(13)
Cathodes, self-heating oxide 33(9)
Cauchy's inequality 32(220)
Caustic condition 33(275)
Caustic tip 33(477)
Caustics 33(273)
Cd 31(152, 177)
Cd, photoionization cross section 31(178)
Ce, photoionization cross section of 4d-shell 31(51)
Cell, non-primitive 32(111)
Cell, primitive 32(107)
Cell, twofold primitive 32(111)
Cell volume 32(108)
Centering 33(400, 413, 477, 486)
Centering conditions 33(479)
Central field model
— nonrelativistic 31(496ff.)
— relativistic 31(533)
Central potential approximation 31(39)
— results for β 31(101)
— results for σ 31(40)
Centre of symmetry 32(132, 135)
Cerenkov effect and density effect 34(67, 71)
Cerenkov radiation 34(66, 67)
Cesium chloride, face centered 32(535)
Cestral-tube 30(44)
C_4F_6 31(437)
CH_2 31(442)
CH_3 31(441)
CH_4 31(305, 391, 393, 407, 426, 432)
C_2H_6 31(407)
C_3H_6 31(407)
C_4H_6 31(437)
Chance coincidences 33(669)
Channel electron multiplier 31(255)
Character 33(655)
Characteristic energy losses 32(463)
Characteristic function of a particle 32(327)
Characteristic function of Porod 32(327)
Characteristic length 32(355)
Characteristic radiation 30(6, 9–13)
Characteristic, static, of a Coolidge tube 30(34)
Characteristic, static, of a thermionic tube 30(24)
Characteristic, static, of an ion tube 30(22, 23)
Characteristic x-rays, excitation 34(178)
Characteristic x-rays, yields from various elements 34(179, 181)
Charge exchange 33(39)
Charge exchange cross section at low ion energy 34(48)
Charge exchange cross section for ions 34(29)
— measurements 34(41–45)

Charge exchange of doubly charged positive ions 34(49)
Charging of specimen 32(449)
CH_3Br 31(404, 405, 422)
$(CH_3)_2Cd$ 31(444)
$(C_2H_5)_2Cd$ 31(444)
CH_3Cl 31(404, 422)
Chemical bond 32(218)
Chemical reaction 31(438)
Chemical shift 31(221)
— between surface and bulk metal atoms 31(355)
— model of 31(339)
— of core electron lines 31(337)
— thermodynamic model 31(351)
Chemisorbed CO on tungsten 31(381)
$C_4H_2F_4$ 31(437)
CH_3Ga 31(444)
CH_3I^+ 31 (414, 422)
Ch_3I 31(404, 414)
CH_3In 31(444)
Chloroethenes 31(395)
CH_3OH 31(432)
$(CH_2OH_2)_2$ 31(433)
Chopper, helical 32(582)
$(C_2H_5)_4Pb$ 31(444)
Chromatic aberration 33(302, 400, 401, 414, 473, 476)
Chromatic aberration of electron lenses 33(248)
$(C_6H_5)_4Sn$ 31(444)
$C_5H_5X(X=N, P, As, Sb, Bi)$ 31(434)
Cinematography with normal x-ray tubes 30(58, 59, 60)
Cinematography with x-ray flashes 30(59, 69)
Circle of least confusion 33(238)
Circular axis, bundles of electrons and ions with 33(306)
Circularly polarized radiation, incident, in Compton scattering 34(249, 250, 295, 296)
Cl, β-parameter of $3p$-shell 31(105)
Cl, β-parameter of $3s$-shell 31(106)
Classical electron radius 34(220)
Clearing thickness 33(502)
ClO 31(442)
Close-coupling method 31(66, 91)
Cloud chamber 31(127, 128)
C_2N_2 31(420, 421)
CNDO, complete neglect of differential overlap method 31(435)
CNDO/S, screened CNDO method 31(421, 436)
CO 31(126, 368, 373, 396, 397, 426, 427, 432, 456, 457, 400)
CO_2 31(126, 374, 397, 401, 426, 427)
CO_2^+ 31(411, 422)
C_3O_2 31(412)
CO, absorbed on nickel 31(457)
CO, absorbed on tungsten 31(380)
CO_2, photoelectron spectrum 31(266, 268)
CO, shake-up spectrum 31(357)
Cobalt, neutron diffraction 32(633)
Coherent and Compton scattering, ratio 34(224, 291)
Coherent photon scattering 34(223, 224, 251, 258, 290–294)
— role in attenuation experiments 34(273–278)
Coherent (Rayleigh) and incoherent (Compton) collision cross section, total 34(293)
Coherent scattering cross section for photons 34(292)
Coherent scattering of neutrons 32(565)
Coherent scattering of photons 32(153)
Coil current supply 33(464)
Coils, closed in iron 33(404)
Coincidence measurements between recoil electron and scattered photon 34(225, 226)
Coincidence methods for electron-electron scattering 34(57, 58)
Coincidence methods in straggling experiments 34(97, 98)
Coincidence technique 33(668)
Coincidences, delayed 34(139, 146, 147)
Coincidences, genuine 33(669)
Coincident experiments 31(162)
Cold worked metals 32(423, 440–441)
Collector current 33(35)
Collector system 33(558)
Collimating pinhole 32(360, 369f.)
Collimation errors 32(348)
Collimator for a neutron beam 32(578)

Collision cross section
— differential, for plane polarized radiation 34(245–247)
— differential, for unpolarized radiation 34(250–254)
— for photons, average 34(263, 275, 281)
— for photons, correction for bound electrons 34(288–290)
— for photons, experimental verification 34(275–281)
— for photons, integral over finite angular intervals 34(255–257)
— for photons per average electron, total (coherent plus incoherent) 34(293)
— for slow electrons, measurements 34(6–11)
— for slow electrons, total 34(8)
— for slow ions, total 34(32, 33)
— for slow protons 34(34, 35, 36)
— of hydrogen for hydrogen ions 34(35, 36)
— of oriented electrons for circularly polarized radiation 34(295, 296)
— of rare gases for alcali ions 34(33, 34)
Collision of fast electrons with free electrons 34(55–60)
Collision time 34(61)
Collisions of fast electrons with the conduction electron plasma 34(71–86)
Colloidal particles, concentrated solutions 32(382f.)
Coma 30(327, 329)
Coma, isotropic and anisotropic 33(233)
Coma, rotation 33(234)
Composite molecule approach 31(433)
Compton absorption coefficient, experimental verification 34(283, 284)
Compton and coherent scattering, ratio 34(224, 291)
Compton attenuation 34(270–281)
Compton cross sections, classical 34(234)
Compton edge in scintillation spectroscopy 34(242, 267)
Compton effect 30(18, 19)
Compton effect, pair production, photoelectric effect, relative predominance at different energy and atomic number 34(218, 219)
Compton electrons 33(661)
Compton linear-absorption coefficient 34(272)
Compton linear-attenuation coefficient 34(270, 271)
Compton linear-scattering coefficient 34(272)
Compton lines 33(662)
Compton mass-absorption coefficient 34(272, 273–278, 283)
Compton mass-attenuation coefficient 34(271–281)
Compton mass-scattering coefficient 34(272, 273–278)
Compton recoil electrons, *see* Recoil electrons
Compton scattered photons, *see* Scattered photons
Compton scattering(s) 32(153, 161); 33(656)
— arrangement for two successive 34(231)
— by bound electrons 34(284–294)
— by magnetically oriented electrons 34(294–297)
— conservation of energy and momentum 34(225–229, 236)
— polarized radiation 34(231, 232, 244–250)
— relativistic quantum-mechanical model 34(244, 245)
— scattering plane 34(221, 236)
— simultaneous production of recoil electron and scattered radiation 34(225, 226)
— unpolarized radiation component 34(230, 233, 248)
Compton shift
— defect for bound electrons 34(286–288)
— discovery 34(223)
— in energy 34(237)
— in wavelength 34(237)
Compton shifted line, Doppler breadth 34(285)
Compton wavelength 34(236, 237)
Concave grating 30(324)

Condenser, electrical, for deflection 33(577)
Condenser lense 33(421, 476)
Conditions of operation, optimum 33(616)
Cone anode of an x-ray tub• 30(44, 63, 65, 68)
Configuration interaction 31(287, 401, 506, 520, 579, 604)
— form of the electric dipole transition matrix element 31(27, 28)
— form of the final-state wave function 31(60ff.)
— in the final continuum state 31(60ff., 288, 301, 310)
— in the final ionic state 31(288, 302, 304, 307, 308, 379, 400)
— in the final state 31(288, 302)
— in the initial state 31(288, 302, 308, 310, 400)
— interchannel interaction 31(61)
— intrachannel interaction 31(61, 62)
Connected domain 32(197)
Conservation laws in Compton effect 34(225–229, 236)
Constructional principles for x-ray tubes, general 30(36, 37)
Contact method 33(572)
Continuous x-ray spectrum (*see also* Bremsstrahlung and radiation) 30(6–9, 16, 17, 79, 337f., 358)
— at low voltages (1–2 kV) 30(343, 344)
— dependence on the form of the alternating voltage 30(342)
— energy distribution 30(7, 8, 343, 346)
— from ions 34(190–192)
— Kramers formula 30(361)
— maximum intensity 30(342, 344, 346)
— quantum theory 30(360, 362, 365)
— theoretical considerations 30(358)
— total intensity 30(7, 8, 363)
Continuum wave functions
— nonrelativistic 31(497, 498, 582)
— normalization 31(498, 583, 588, 591, 618)
— numerical procedures 31(589ff.)
— radial 31(497, 498, 536, 537)
— relativistic 31(536, 537, 587ff.)
Contrast 30(308, 312); 33(429, 432f., 468, 516, 522, 544)
Contrast as function of beam voltage and objective aperture 33(500)
Contrast by crystalline objects 33(502)
Contrast in extended objects 33(499)
Contrast near resolving limit 33(505)
Contrast with thin layers 33(501)
Control, process 33(574)
Convergence angle 33(519, 521)
Conversion factor 33(650)
Converter 33(657, 664)
Converter corrections 33(660)
Converter efficiency 33(660)
Converter, external 33(656)
Convolution integral 31(279)
Coolidge tube 30(32–34)
Cooling device 33(514)
Cooper Minimum 31(55)
Cooper-Zare formula for β 31(40)
— numerical results 31(101ff.)
Copper, analysis of its structure 32(103)
Core hole localization 31(346)
Core line multiplet structure 31(376)
— in molecules 31(376)
— in solids 31(376)
Core shift model including relaxation 31(345)
Correlation effects (*see also* Electron correlation) 31(123, 169)
— on binding energies 31(565, 566)
— on nonradiative transition amplitudes 31(604ff.)
— on transition energies 31(578, 581)
Correlation function for the electron density 32(394, 403)
Correspondence principle 30(360, 361)
COS 31(401, 426, 427)
Cosine-theorem of the electron-optical mapping, general 33(155)
Coster-Kronig transition
— amplitudes 31(518, 538, 581ff.)
— component rate 31(519)
— definition 31(501)
— energy 31(501, 566ff.)
— group rate 31(519, 520)
— intensity 31(523, 524)
— multiplet rate 31(519)
— term rate 31(519)
Coulomb approximation 31(89)
Coulomb excitation of the nucleus, effect on the total x-ray yield 34(168, 181)

Coulomb gauge 31(6, 490)
Coulomb matrix element 31(502, 505)
Coulomb phase shift 31(476, 497, 583)
Cr 31(309)
$Cr(CO)_6$ 31(397)
Cristalline objects, focussing of 33(529)
Critical wavelength 30(265)
Cross over 32(444)
Cross section(s)
— for ejection of delta rays 34(190)
— for inner shell ionization 34(168–177)
— for positron-electron scattering 34(59, 60, 63)
— for two- and three-photon annihilation 34(143)
Cross-over 33(19)
Crosstable 30(127)
Cryostat for neutron diffraction 32(595)
Crystal collimator 32(361 f.)
Crystal diffraction spectrometer 33(646, 647)
Crystal, disordered 32(399, 401)
Crystal, distorted 32(421–423)
Crystal faces 32(98, 140)
Crystal, finite 32(394–398)
Crystal gas 32(530)
Crystal gas, interference 32(525)
Crystal, ideally imperfect 32(405)
Crystal, imperfect 32(389, 620)
Crystal liquid 32(530)
Crystal monochromator(s) 30(54); 32(583)
Crystal, perfect 32(389)
Crystal plate, irradiation of a thin 32(470)
Crystal properties and intensities 30(235)
Crystal radius 30(110)
Crystal reflection 30(331)
Crystal reflection correction 30(340)
Crystal size and lattice constant 32(499)
Crystal spectrograph, cylindrical concave 30(110)
Crystal spectrograph with bent crystal 30(108, 111, 116, 117)
Crystal spectrometer 30(331)
Crystal structure factor 32(393)
Crystal surface, properties 30(119)
Crystallite growth in carbons 32(387)
CS 31(401, 426, 427)
Cs 31(127)
Cs, β-parameter of $6s$-shell 31(111, 205)
Cs, photoionization cross section 31(57, 170, 171, 173)
C1s shifts in fluoromethane 31(340)
Cu 31(309, 459)
Cubic system 32(115)
CuO, Cu_2O 31(363)
CuPd alloys 31(366, 464)
Current density per solid angle 33(23, 419 f.)
Current, dependence of x-rays 30(341)
Current difference, astigmatic 33(470)
Curvature of electron orbits 33(127)
Curvature of image field 33(240)
— meridional 33(241)
Curved axis, systems with 33(296)
Curved crystal collimator 32(367)
Cut-off voltage 33(18)
CX_4-compounds 31(404)
Cyclic group 32(141)
1,4-cyclohexadiene 31(430)
Cylinder condensor 33(577)
Cylindrical capacitor analyzer 31(155, 246)
Cylindrical mirror analyzer 31(155, 247, 248)

Damping 32(155)
Dark field 33(532)
Dark field focussing 33(508)
Dark field illumination
— axial 33(349)
— oblique 33(349)
de Broglie relation 32(445)
de Broglie wavelength 34(62, 75, 91, 102, 116)
Debye and Pirenne density function 32(326, 327, 357)
Debye length 34(72, 73, 75)
Debye-Scherrer line shape 32(418)
Debye-Scherrer method 32(103, 177)
Debye-Waller temperature factor 32(562, 571, 591)
Decay probability
— nonradiative 31(487, 488, 494)
— partial 31(487)
— radiative 31(492, 494)
— relativistic 31(536)

Decay rate of electron emission 31(283)
Deconvolution method 31(279)
Deflection aberration(s) 33(285); 33(289)
Deflection astigmatism 33(291)
Deflection coma 33(291)
Deflection distortion 33(291)
Deflection method, direct 33(610)
Deflection of electron beams 33(284)
Deflector field 33(286)
Deformation, static 32(212)
de Kronig structure, *see* Fine structure 30(217, 296)
Delayed coincidences 34(139, 146, 147)
Delta rays 34(187)
Delta rays, energy distribution 34(188, 189)
ΔE_{SCF} method 31(512, 558, 565, 575, 576)
Dempster's double focussing apparatus 33(582)
Density distribution, radial 32(326, 238, 357)
Density effect 34(63, 67–71, 100, 101)
Density function of Debye and Pirenne 32(326–328, 357)
Density functions 33(607)
Density of states 30(253)
— partial 30(261)
— structure 31(458)
Depolarization in Compton scattering, classical model 34(233)
Detachment of negative ions 34(49)
Detection methods for positronium 34(146, 148, 149)
Detection of x-rays 30(135)
Detector 33(653)
Detectors for neutrons 32(584)
Deuteration 31(417)
DF^+ 31(415, 424)
Diagram lines (tables) 30(181–197)
Diamond 30(274)
Diarsine 31(438)
Diazocyclo-hexadienone 31(439)
Dibromamine 31(440)
Dichloramine 31(440)
Dieder group 32(141)
Dielectric constant 34(75)
Dielectric frequency 34(66, 67)
Dielectric model 34(67)
Difference, astigmatic 33(238)
Differential collision cross section
— for electrons, definition 34(3)
— for plane polarized radiation 34(245–247)
— for unpolarized radiation 34(250–254)
Differential energy loss of charged particles 34(193, 194)
Differential scattering cross section 31(160)
— for photons, definition 34(220)
— for plane polarized radiation 34(247–249)
— for unpolarized radiation 34(257–259)
Diffraction 30(145 ff.)
Diffraction camera for electrons 32(444, 449)
Diffraction, elementary theory 32(459)
Diffraction error 33(399, 467)
Diffraction, geometric theory for thin crystal plates 32(475)
Diffraction groups 32(233)
Diffraction, highly resolving 33(526)
Diffraction image, defocused 33(528)
Diffraction image, primary 33(489)
Diffraction images 30(320)
Diffraction microscopy 33(348)
Diffraction of electron waves 32(550)
Diffraction of neutrons
— by a large perfect crystal 32(565)
— by a small perfect crystal 32(562)
— by an ideally imperfect crystal 32(567)
— by polycrystalline materials 32(569)
— structure factor 32(562)
Diffraction of x-rays 32(97)
Diffraction pattern 30(145–151, 153–156)
Diffraction-apparatus 33(524)
Diffuse spot 32(210, 212)
Diffusion cross section for slow electrons 34(8)
Diphosphine 31(438)
Dipole transitions in positronium 34(140)
Dirac electron 34(59, 103)
Dirac electron theory 34(249)
Dirac-Fock method
— general 31(543, 563, 571, 575)
— radial equation 31(587)

Dirac-Gordon cross section 34(235, 253)
Dirac-Slater calculation 31(406)
Direct light electron microscope 33(460)
Direct transitions in photoelectron spectra 31(448)
Direction focussing 33(550)
— curve 33(552)
— higher order 33(575)
— second order 33(588)
Discovery of positronium 34(146)
Discrete basis set theories 31(100)
Disintegration of coallayers 33(515)
Disk of confusion, radius 30(316)
Dislocation 32(426, 439)
Disorder by displacement 32(400–401, 407–409)
Disorder by point defects 32(415)
Disorder correlations 32(403)
Disorder, Fourier expansion 32(402)
Disorder, one-dimensional 32(406–412)
Disorder scattering 32(402, 403)
Disorder, substitutional 32(399–400, 412–415)
Dispersion 30(141); 33(614)
Dispersion, anomalous 30(142, 143)
Dispersion compensation 31(231)
Dispersion lines 33(597)
Dispersion terms 32(160, 161, 163, 164)
Dispersive analyzer 31(244)
Dissociation processes by electron rays 33(572)
Distant interferences 33(531)
Distortion 33(425, 435)
Distortion, anisotropic 33(243f.)
Distortion broadening of Debye-Scherrer lines 32(422)
Distortion, isotropic 33(243)
Distortion, rotational 33(244)
Distribution function 32(524)
Distribution of energy losses of fast electrons 34(87–90, 96–101)
Doggett and Spencer tabulation of Mott cross section 34(108–112)
Domain effects in neutron diffraction 32(629)
Doppler broadening 31(235)
Doppler broadening of HeI resonance radiation 31(235)
Dose, definition 30(74–76)
Double collector 33(559)
Double collector method 33(570, 583)
Double Compton effect 34(268–270)
Double condenser lense 33(424)
Double crystal spectrometer, parallel and antiparallel position 30(147–151)
Double electron excitation 31(159)
Double focussing 33(579)
— apparatus 33(581)
— apparatus, Bainbridge and Jordan's 33(583)
— condition 33(579, 782)
Double ion chamber 31(147)
Double photoionization 31(113, 193)
— Ar 31(119, 195)
— Be 31(119)
— He 31(113, 193)
— Ne 31(116, 194)
— Xe 31(199)
Double ray betatron 30(71)
Double reflection 32(182)
Double slit method 32(452)
Double spectrometers 30(104f.)
Double x-ray flash tube 30(69)
Doublet differences, *see* Spin doublets
Doublet measurements 33(583, 585)
Doublet method 33(579, 597)
Duane-Hunt law 30(4, 8)
Dynamical theory of electron interferences 32(467, 478)
Dynamical theory of x-ray scattering 32(566)

Edge, extrapolated high energy 33(658, 662, 665)
Edge spectrograph 30(102)
Efficiency 30(13–15, 342, 361, 365)
Efficiency dates, calculation of 33(56)
Eigenchannel method 31(97)
Einstein equation 30(4)
Elastic scattering of slow
— electrons 34(11, 13, 17–23)
— electrons, angular distribution 34(17–20)
— ions 34(28, 36–41)
— ions, angular distribution 34(38–40)
— ions, energy distribution 34(37)
Elastic wave 32(208, 212)
$E^{\frac{1}{2}}$-law 30(261, 278)
$E^{\frac{2}{3}}$-law 30(261, 273)
Electric moment 32(155)

Electric moment, oscillating 30(362, 363)
Electrical condensor 33(580)
Electric-dipole approximation 31(6, 289)
Electrolytically polished metals 32(525–527)
Electrometer, vibration-reed 33(560)
Electron affinities 31(180); 33(53)
Electron affinities of small molecules 31(445, 446)
Electron binding, effect on Compton scattering 34(285–294)
Electron binding energy 31(389)
Electron burst 30(61, 66)
Electron correlation 31(123, 169, 196)
Electron correlation theories for photoionization processes (*see also* Names of individual theoretical methods: close-coupling, configuration interaction, discrete basis set method, eigenchannel method, hyperspherical coordinate method, many-body perturbation theory, multiconfiguration Hartree-Fock theory, quantum defect theory, random phase approximation, *R*-matrix method) 31(59ff.)
Electron density 32(106, 116, 124, 187, 325, 394)
— continuous description 32(117)
— overlapping in lattice cells 32(116)
Electron detector 31(255)
Electron diffusion theory 34(124–128, 136, 137)
Electron emission 33(3)
— by ion impact 33(40, 41)
— of adsorption and oxide coated cathodes 33(7)
Electron energy analyzer 31(153, 243)
— cylindrical mirror analyzer 31(246)
— ellipsoidal mirror 31(251)
— parallel plate mirror 31(248)
— spherical capacitor 31(245)
— spherical grid retarding potential 31(250)
— toroidal condenser 31(250)
Electron flux in a medium 34(121–124)
Electron focussing in x-ray tubes, electric 30(2, 22, 24, 29, 32, 33, 35, 42, 48, 49, 55, 61, 65)
Electron focussing in x-ray tubes, magnetic 30(36, 43, 44, 49, 61, 72)
Electron impact ionization 31(399)
Electron interference by gases 32(520)
Electron interferometry 32(506)
Electron lenses, strong, with Newton's imaging equation 33(197)
Electron mirror 33(460)
Electron momentum 33(612)
Electron multiplier 33(559)
Electron neutron interaction 32(601)
Electron optical, axis 33(477)
Electron optics, limits of geometrical 33(317, 322)
Electron positron pairs 30(357)
Electron production
— by electric field strength 30(25, 26, 64)
— by impact ionization 30(21–23)
— by photoeffect 30(26, 27)
— in high vacuum 30(27–29, 64, 65)
— secondary 30(26, 27)
— thermionic 30(23–25, 33)
Electron radius 34(58, 60, 91)
Electron scattering
— at K atoms 34(18)
— in Cd vapor 34(18)
— in CO 34(20)
— in rare gases 34(19)
— in Zn vapor 34(18)
Electron sources 33(3, 418)
Electron sources, Gas discharge 33(14)
Electron spectrometer 31(215)
— iron-free double focusing 31(247)
— time-of-flight 31(252)
Electron spectroscopy, history of 31(218)
Electron-electron scattering 34(55–61)
Electrons, angular distribution of inelastic scattering 32(463)
Electrons, backscattered 33(652)
Electrons, fast, *see* Fast electrons
Electrons, inelastically scattered 32(503)
Electrons, magnetically oriented 34(249, 294–297)
Electrons, nearly free 30(250)
Electrons, slow, *see* Slow electrons
Electrons slow 32(505)
Electrons, tightly bound 30(251)
Electroplating 33(652)

Electrostatic lens in fine structure tubes 30(55)
Elements of neighboring Z, neutron diffraction 32(615)
Emission band, high energy
— edge 30(291)
— tail 30(295)
Emission band, low energy tail 30(289)
Emission control 33(17)
Emission intensity near long wavelength limit 30(261)
Emission lines (experimental data) 30(162–181)
Emission lines, shape and width 30(153)
Emission microscope 33(455)
Emission microscope, contrast in images of 33(459)
Emission spectroscopy 30(31, 51, 52)
Empirical rules of intensities of photoelectron bands 31(395)
Energy absorption (photon energy)
— experimental verification 34(283, 284)
— per gram (dose rate) 34(283)
— per unit volume 34(283)
Energy defect 34(45–48)
Energy defect, effect on charge exchange cross section 34(45, 46)
Energy diaphragm 33(581)
Energy dissipation of fast electrons
— around a point source 34(130, 131)
— in plane layers near a point source 34(131)
— in spherical shells around a point source 34(131)
— in thick layers 34(124–130)
Energy distribution
— half-value width of the 33(475f.)
— in the continuous x-ray spectrum 30(7, 8, 343, 346)
— of betatron and synchrotron radiation 30(357)
— of characteristic radiation 30(10, 11)
— of elastically scattered slow ions 34(37)
Energy function 33(678)
Energy gap 30(254, 283)
Energy level(s) 31(507)
— system, complete, graphical representation 30(84)
— of positronium 34(140)
— of the atoms 30(222–227)
— of the atoms, absolute values 30(229)
— of the atoms, (table) 30(224–227)
Energy loss 33(494, 499)
— characteristic 33(536, 539, 541)
— differential, of charged particles 34(193, 194)
— distribution for fast electrons 34(87–90, 96–101)
— of fast electrons in various materials 34(80–85)
— of fast electrons, mean 34(61–71, 98, 99, 126)
— of fast electrons, most probable 34(90, 93, 97, 98)
— of fast electrons to bound electrons 34(88–90)
— of slow electrons in gases 34(12, 16)
Energy of Compton
— recoil electrons 34(241–243, 265)
— recoil electrons scattered in any finite angular interval 34(260, 261)
— scattered photons 34(238, 239, 264)
— scattered photons in any finite angular interval 34(258, 260)
Energy resolution 33(614)
Energy spectra of electrons ejected from the K- and L-shell 34(188, 189)
Energy spectrum 33(61)
— of Compton recoil electrons 34(266–268)
— of Compton scattered photons 34(268)
— of the electrons 33(451f.)
Energy spread 33(577)
Energy term 31(503)
Energy-range relation for electrons 34(132–138)
Entrance slits 33(595)
Envelope 32(504)
Envelope of carbon 33(518)
Equation of motion 33(612)
Equilibrium system 31(438)
Equipment for small angle scattering experiments 32(359f.)
Error analysis 31(134)

ESCA diffraction 31(380, 383)
— for adsorption systems 31(385)
— single scattering model 31(385)
ESCA spectrometer for studies of solids and surfaces 31(264)
Ether 31(438)
Ethylene 31(401, 402, 417, 433)
Ethylene glycol 31(438)
Eu 31(309, 461)
Eu, photoelectron spectrum 31(337)
Evaporated films 32(532, 536)
Evaporation of active material 33(652)
Ewald and Sauermann's stigmatic double focussing mass spectrograph 33(592)
Ewald sphere 32(392, 460)
Excitation defect 32(476, 489)
Excitation energies of x-ray series 30(162, 163)
Excitation, indirect 32(464)
Excitation, magnetic 33(405)
Excitation of characteristic x-rays 34(178)
Excitation voltage of characteristic radiation 30(10)
Exciton levels 30(86)
Exit window, *see also* Lenard window 30(13, 51, 53, 54, 66, 68)
Extinction 32(103)
Extinction coefficient 32(194)
Extinction effects 32(593)
Extinction, integral 32(150)
Extinction law 30(9)
Extinction of x-rays 32(347)
Extinction, primary 32(196, 218, 404, 567)
Extinction, regular 32(148)
Extinction, secondary 32(196, 218, 404, 567)
Extinction, serial 32(150)
Extinction, zonal 32(150)
Extra translation 32(142, 143)

F_2^+ 31(415)
Factor group 32(140)
Fano line profiles 31(168, 485, 615)
Fast electrons
— absorption curves 34(133, 134)
— average energy loss 34(61–71, 98, 99, 126)
— distribution of energy losses 34(87–90, 96–101)
— maximum and minimum scattering angle 34(91, 116, 117)
— multiple nuclear scattering 34(115–121)
— radiation 30(358)
— radiation losses 34(60, 63, 100, 123)
— ranges 34(132–138)
— scattering at free electrons 34(55–60)
— scattering in the plasma 34(75–86)
— single nuclear scattering 34(101–116, 125)
— straggling distribution 34(87–101)
Feather analyzer 34(134)
Fermi energy 34(74–76)
Fermi level 30(229, 231, 232, 255)
Fermi theory of density effect 34(67, 101)
Fermi's Golden Rule 31(527)
Fermi-Thomas atom 34(73, 90, 102, 117)
Ferroelectric behavior of potassium dihydrogen phosphate 32(613)
Ferromagnetics, neutron diffraction 32(631, 637)
Feshbach projection operators 31(483, 556, 610ff.)
Field, electric-magnetic, in the moving trihedral 33(169)
Field electron emission 30(25, 26, 63, 64)
Field emitter sources 30(317, 319)
Field, homogeneous 33(611)
Field index 33(311)
Field strength, electric
— and work function 30(23–26)
— near points and edges 30(25, 26)
Field variation, adiabatic 33(129)
Field width 30(316, 327, 328)
Fields, crossed electric and magnetic 33(592)
Fields, electrostatic 33(611)
Fields, inhomogeneous, axially symmetric magnetic 33(633)
Fields, non-uniform 33(576)
Fields without rotational symmetry 33(257)
Figures of merit 33(640)
Filling device 33(653)
Films deposited from solutions 32(543)

Films, electrolytically deposited 32(542)
Films, fine-grained 30(314)
Films on amorphous substances 32(539)
Filter lense 33(543)
Filters for neutrons 32(584)
Filters, selective, for x-rays 32(360)
Final image luminescent-screen 33(428)
Final magnification 33(398, 425)
Final state density 31(490, 516, 517, 617)
Final state photon-electron interaction 31(492ff.)
Final state wavefunction 31(7ff.)
— incoming-wave boundary condition 31(8ff., 13, 20ff.)
— in-states 31(10)
— multichannel 31(15ff.)
— normalization 31(7, 12, 14, 17)
— out-states 31(10, 13, 20ff.)
— partial wave expansion 31(11, 14)
— single channel 31(8ff.)
Fine range diffraction 33(527)
Fine structure
— by formation of antiphase domains 32(514)
— constant 30(81)
— of diffraction spots by refraction (*see also* Interference double refraction) 32(486, 488)
— of *K*- and *L*-emission lines 34(86)
— of positronium 34(139, 152–158)
— of the x-ray absorption edge 34(87)
Fine structure of x-ray absorption spectra 30(217–222)
— dependence on temperature 30(220)
— methods of comparison 30(219)
Finite crystal 32(394–398)
Finsterwalder-condition 33(282)
Fluctuating dipoles 31(402)
Fluorescence 31(159)
Fluorescence of photons (x-rays) 32(153)
Fluorescence yield 34(181)
Flux of neutrons 32(576, 578, 580)
Focal distance 33(380)
Focal length(s) 33(302, 410, 413, 580)
— astigmatic difference of 33(401, 470, 473, 486)
— for x-rays 30(325)
— of the objective 33(425)
— osculating 33(194)
— real 33(197, 207)
Focal point abscissa 33(302)
Focal point, meridonal 33(237)
Focal point, sagittal 33(238)
Focal spot characteristic 30(317f.)
Fock operator 31(509, 525)
Focus after Goetze 30(35)
Focus, fine 30(36, 44, 55, 67)
Focus load 30(14, 15)
Focus, point focus 30(117)
Focussing, axial 33(576, 590)
Focussing by orthogonal systems 33(264)
Focussing condition 30(95)
Focussing defect 30(110, 111)
Focussing of electrons by lenses 32(445)
Focussing, point-to-point 30(123, 124)
Focussing, stigmatic 33(577, 592)
Focussing, twodirectional 33(301, 312)
Focussing twodirectional 33(576)
Folding (convolution) of functions (*see also* Fourier transform) 32(118)
"Folding" technique 33(647)
Forbidden zones 30(219, 251, 299)
Form factor (*see also* Scattering factor) 32(324)
Form factor for neutron scattering, magnetic 32(625, 626)
Form factor for the *K*- and *L*-shell 34(174)
Formation of positronium
— in gases 34(144, 151)
— in gases, effect of a static electric field 34(151)
— in liquids 34(158–162)
— in solids 34(158–162)
Formic acid 31(417)
Fourier coefficients 32(490)
Fourier coefficients of the potential energy 32(469)
Fourier integral 32(118, 201)
Fourier series 32(117, 119, 201, 205, 207, 218, 393, 394)
Fourier space 32(119)
Fourier transform 32(118, 123, 125)
Four-pole systems 33(268)
Four-spinor 31(532, 537)
Franck-Condon factors 31(418)
Franck-Condon principle 31(295)

Franck-Hertz experiment 34(12)
Franzini, proton-source of 33(117)
Fraunhofer approximation 32(468)
Fraunhofer plane 33(345)
Free electrons, collisions with 34(55–61)
Fresnel diffraction 30(315, 316, 320); 33(490)
— by an edge 32(551)
— fringes 33(334, 470, 473, 480, 486)
Fresnel integrals 33(333)
Friedel law 32(149)
Frozen-core approximation 31(501)
Fundamental constants 33(649)
Furan 31(430)
Furnace 32(447)

Ga 31(309)
Gamma radiation, multipole order of 33(655)
Gamma-Globulin 32(379)
Gamma-radiation 30(3, 18–20)
Gamma-ray energies 33(662)
Gamma-ray intensities 33(660, 663)
Gamma-ray polarimeter 34(247)
Gas analytic method, mass spectrometric 33(567)
Gas discharge, plasmas of 33(49)
Gas economy 33(55)
Gas inlet 33(42ff.)
Gas inlet systems, molecular flow in 33(561)
Gas pressure in ion tubes 30(22, 29)
Gas pressure in vacuum tubes 30(34, 48)
Gas pressure in x-ray flash tubes 30(64)
GaSe 31(451)
Gases, neutron diffraction 32(617)
Gases, separation of 33(45)
Gas-liquid chromatography 31(440)
Gas-phase reactions 31(439, 442)
Gathering power 33(614)
Gauge invariance of the Schrödinger equation 31(26)
Gauge transformation 31(531); 33(125)
Gauss' dioptric 33(174)
Gaussian radiation characteristic 33(375)
Gaussian scattering distribution 34(115–118, 120)
Gd 31(461)
Ge 31(309)
GeH_4 31(407)
Geiger counter 30(135, 137)
Geiger-Müller-counter 33(653)
Generalized function 31(608)
Generalized oscillator strength 31(161)
GeO 31(441)
Geometric interference theory 32(467)
— limits of validity 32(478)
Glancing angle 30(91); 32(101)
Glide reflection 32(132)
Goetze focus 30(17)
Goudsmit and Saunderson scattering theory 34(117)
Gradation 33(429, 430, 432)
Gradient method 31(421)
Grain 33(429)
Grain boundary 32(427)
Grain of the film 30(312)
Grain of the plate 33(501)
Grain size, neutron diffraction 32(620)
Graphite 30(274)
Graphite layer structure 32(411)
Graphite, neutron diffraction 32(616)
Grating formula 30(126)
Grating monochromator 31(237)
Grating spectrometer with concave grating 30(127)
Grating spectrometer with plane grating 30(124)
Gratings in x-ray spectroscopy 30(92, 345)
Grazing incidence 30(325, 327)
Grazing incidence, maximum angle 30(328)
Green function 33(355)
Green function method 31(294, 421)
Grinding of the crystal 30(119)
Grotrian-Siegbahn level diagrams 30(228)
Ground state potential model 31(343, 345, 348)
Group electronegatives 31(434)
Group property of the image formation 33(359)
Group shift analysis 31(350)
Group shift for carbon 31(351)

Group shift model 31(350)
Group theory and its applications on space transformations 32(137ff.)
Growth of films 32(539)
Guinier formula 32(329, 350)
Gyration radius 32(329, 350, 351, 354, 355)

H_2 31(422, 426, 429)
H_2^+ 31(429)
H_2, values of asymmetry parameter β 31(430)
Half width 30(151)
Half width of interference maxima 32(461)
Half-value layer 30(9, 20)
Hallucinogenic activity 31(435)
Halogene atoms, PES of outer valence shell 31(309)
Halogene diatomic molecule 31(403, 414)
Halpern and Hall theory of density effect 34(67)
Ham effect 31(422)
HAM/3, hydrogenic atoms in molecules method 31(436)
Hamilton function 33(139)
Hamiltonian, atomic 31(6)
Hamiltonian, interaction 31(6, 26)
Hamilton-Jacobi, differential equation of 33(139, 140)
Hamilton principal equation 33(138)
Hammett parameter 31(434)
Harker sections 32(151)
Harker-Kasper inequalities 32(614)
Hartree-Fock method 31(68)
— equations 31(284)
— multiconfiguration 31(80)
— radial equation 31(582)
— restricted 31(511)
— term averaged 31(511)
— unrestricted 31(509, 510, 560)
Hartree correction formula 33(647)
Hartree-Slater method 31(286)
Hayasi theory 30(297)
HBr^+ 31(415)
HCN^+ 31(411)
He^- 31(447)
He, autoionizing structure 31(169)
He, β-parameter 31(311)
He, double photoionization cross section 31(113, 193)
He, ionization threshold 31(126)
He, photoelectron spectrum 31(309ff.)
He, photoionization cross section 31(115, 164)
Heat motion 32(207)
Heat pipes 31(149)
Heat transfer of an electron 33(510)
Heating of specimen by electron beam 32(457)
Heavy atoms preponderant in the structure factor 32(152)
h/*e*-determination 30(366)
HeI excitation 31(388)
HeI resonance radiation 31(235)
HeII resonance radiation 31(237)
Heisenberg's uncertainty relation 33(339)
Helmholtz-Clausius theorem, generalized 33(363)
Helmholtz-Lagrange theorem 33(185)
Hemoglobin 32(377)
Henneberg's approximation for the *K*-shell ionization cross section 34(177)
Hermann-Mauguin notation 32(140)
Hermitian matrix 32(224, 225)
Hexafluorobutadiene 31(438)
HF 31(431, 415, 424)
Hg 31(152, 177)
Hg, photoelectron spectrum 31(186)
— of 4*f*-shell 31(266)
— of valence shells 31(187)
Hg, photoionization cross section 31(179)
High resolution analyzer 34(79, 81)
High resolution camera for electrons 32(452)
"High source" type 33(617)
"High source" vacuum equipment 33(465)
"High source" voltage equipment 33(463)
High vacuum breakdown 30(27–29, 64)
High vacuum x-ray flash tube 30(63–69)
High voltage sets 30(135)
Hill's differential equation 33(202)
H_2O 31(286, 370, 391, 393, 413, 415, 427, 431, 432, 441, 422, 424)
Hole configuration and wave functions 31(513ff.)

Hollow 33(530)
Hollow anode of an x-ray tube 30(37, 42, 43, 49)
Hollow cathode 30(44, 63, 65, 68)
Holographic grating 31(143)
Homogeneity, magnetic 33(408)
Homogenization by filters 30(8, 9)
Homometric structure 32(129)
Hopfield continuum 31(128, 140)
Hot band 31(412)
Hot cathodes, emission of 33(9)
H_2S 31(413, 415, 427, 422, 424, 417)
H_2Se 31(413, 415)
H_2Te 31(413, 415)
Hückel method 31(435)
Hückel method, extended 31(435)
Hydrazine 31(438)
Hydrides, neutron diffraction 32(607)
Hydrogen atom, contribution to structure factor 32(152)
Hydrogen atom, position determination by neutron diffraction 32(603–608)
Hydrogen atom, position of, Fourier synthesis 32(608)
Hydrogen cyamide 31(417)
Hydrogen ions, collision cross section measurements in hydrogen 34(35, 36)
Hydrogen isotopes 33(570)
Hydrogenic approximation 31(42)
Hydrogen-positronium molecule 34(140)
Hyperconjugation 31(435)
Hyperspherical coordinate method 31(115)

I_2^+ 31(412)
Ice, neutron diffraction 32(604)
ICl, photoelectron spectrum 31(266)
Ideal deflection 33(285)
Ignition electrode 30(64, 65, 68, 69)
Illumination, oblique 33(334)
Image aberrations of the third order 33(225)
Image formation, electron optical, wave mechanical 33(324, 329)
Image formation, wave mechanical, general problem 33(351)
Image of an axial point, wave mechanical 33(365)
Image, primary and secondary 33(346)
Image rotation 33(637)
Image surface, meridional 33(241)
Image surface, sagittal 33(242)
Image width, effective 33(616)
Imaging by a long magnetic coil 33(179)
Imaging, electron-optical, principal theorem for the 33(177)
Imaging equation 33(178)
Imaging of a circular aperture 33(335)
Imaging of a diffuse object 33(337)
Imaging of a rectangle 33(332)
Imaging of an object with periodic structure (grating) 33(342)
Immersion objective 33(456, 459)
Immersion objective, resolution of an 33(455)
Impact ionization 30(21–23, 27)
Impact parameter 34(55, 56, 61, 62, 72–75)
Impression method 33(450)
Incoherent scattering function 34(288–290)
Incoherent scattering of neutrons 32(564, 567)
Incoming wave boundary condition 31(8ff., 13, 20ff.)
Index of refraction 32(168, 184)
INDO, intermediate neglect of differential overlap method 31(435)
INDO/S, screened INDO model 31(436)
Inelastic mean free path, universal curve 31(275)
Inelastic scattering of slow
— electrons 34(11)
— ions 34(29)
Influencing function 33(349)
Initial density of states (IDOS) structure 31(459)
Inner shell ionization 34(86, 166ff.)
Inner-shell vacancy state
— energies 31(556ff.)
— notation 31(499)
In-states 31(10)
Insulator surface, polished, crystalline structure 32(531)
Insulators, absorption edge shift 30(299)
Insulators, polished 32(530)
Insymmetry, degree of 33(165)
Integrated reflection 32(564, 566, 568)

Intensities of characteristic lines 30(11, 12, 232, 234)
Intensities of satellites 30(237, 238)
Intensity, absolute value 32(231)
Intensity anomalies caused by Auger process 30(294)
Intensity decrease by lattice vibration 32(517)
Intensity distribution 33(652, 654)
Intensity fluctuation, periodic, behind a crystal edge 32(480)
Intensity function 32(393)
Intensity measurement 32(448, 463)
Intensity measurements of x-rays 34(179)
Intensity model for x-ray excited spectra 31(395)
Intensity of an electron beam 34(5)
Intensity of Auger transition 31(523, 524)
Intensity of continuous spectrum 30(7, 8, 343, 346)
Intensity of interference maxima 32(462)
Intensity of main band 31(393)
Intensity of multielectron transition 31(289)
Intensity of multiplets for open shell atoms 31(328)
Intensity of one-electron transition 31(289)
Intensity of photoelectron bands, empirical rules 31(395)
Intensity of photoelectron line 31(273)
Intensity of x-rays, dependence on tube voltage 30(341)
Intensity region 32(461, 486, 491)
Intensity, scattered 32(105, 151, 155, 157, 170)
Intensity space 32(119, 171, 198)
Intensity statistics 32(230)
Interaccelerator 33(414)
Interchannel interaction 31(61, 526, 527, 546, 605, 606)
— effects on β 31(103)
— effects on σ 31(81)
— limit of vanishing 31(36ff.)
Interference bands, cavities 32(507)
Interference by antiphase domains 32(513)
Interference by thin plates, dynamic considerations 32(478)
Interference fields 33(480)
Interference function 32(390, 393, 397)
Interference in a convergent beam 32(457)
Interference in double refraction (*see* Fine structure of diffraction spots) 32(454, 486, 488)
Interference maximum, half width 32(461)
Interference of electrons
— by amorphous substances 32(523)
— by gases 32(519)
— by intramolecular 32(521)
— by liquids 32(521)
— by mechanically deformed metal crystals 32(529)
Interference of fast electrons 32(504)
Interference of slow electrons 32(504)
Interferences 33(534)
Interhalogen molecules 31(414)
Intermediate coupling, nonrelativistic 31(506, 507, 520, 521)
Intermediate coupling, relativistic 31(540ff., 594)
Intermediate image 33(611)
Intermediate image focussing 33(668)
Intermediate image spectrometer 33(639)
Internal conversion 33(610, 655); 34(181, 187, 190)
— coefficient 33(655)
— peaks 33(670)
— probability 33(672)
Internal pair coefficient 33(656)
Internal potential determined by the modified Bragg equation 32(485)
Interparticle interference for iedentical particles 32(341)
Interparticle interference for polydisperse systems 32(345)
Interpolation function 33(678)
Interstitial atoms 32(415)
Intrachannel interaction 31(61)
Intramolecular field strength 31(444)
Intra-shell transition 30(289)
Invariant subgroup 32(140, 142)
Inversion 32(132)
Inversion barrier 31(415)

Ion beams, detection 34(32)
Ion beams, production 34(29–31)
Ion cathode 30(20, 21, 31)
Ion chambers 31(159)
Ion chambers, double 31(147)
Ion extraction 33(37, 47, 49)
Ion impact ionization 33(572)
Ion mass, effect on charge exchange cross section 34(47, 48)
Ion microscope 33(378)
Ion production 33(37)
Ion production, separation processes in 33(563)
Ion rays, energy distribution 33(548)
Ion retarding field 33(558)
Ion source, sample introduction to the 33(561)
Ion source, temperature at the electron impact 33(558)
Ion source, vacuum spark 33(549, 561, 583, 595)
Ion traps 33(12, 35)
Ion tube 30(2, 22, 29–32)
Ionic current, measurement of 33(59)
Ionic saturation current 33(49)
Ionization by electrons 33(37, 572)
Ionization by ions 33(38, 120)
Ionization chamber 30(135)
Ionization cross section
— energy dependence 34(175, 176, 184, 185)
— of the *K*-shell 34(172, 175, 177, 183, 184)
— of the *L*-shell 34(176, 186)
Ionization efficiency 31(132)
Ionization fraction 33(56)
Ionization of the inner shells of atoms 34(86, 166ff.)
Ionization potential, average, of an atom 34(62, 65–69)
Ionization potential of positronium 34(140)
Ionization range, extrapolated, of protons and α-particles 34(206, 207)
Ionization voltage 33(53)
Ionization x-ray spectrometer 30(337)
Ionization yield 31(148)
Ionization yield curves 33(572)
Ions, charge exchange 33(549)
Ions, dissociation of 33(549)
Ions, energy spread of the 33(553, 564)
Ions from solids 33(583)
Ions, negative 33(121)
Ions, slow, *see* Slow ions
Ion-sources 33(36)
— arc discharge, with magnetic fields 33(93)
— arc discharge, without magnetic fields 33(88)
— efficiency of 33(54f.)
— electron impact 33(100, 549, 557, 595)
— for mass separators 33(110)
— high frequency 33(71)
— pulsed 33(119)
— spark discharge 33(118)
— spectra, recording of 33(566)
— surface ionization 33(557)
— thermionic 33(51, 54f.)
Iron, neutron diffraction 32(633)
Iron tetracarbonylolefin complexes 31(395)
Iron, use of 33(624)
Irradiated region 33(419)
Irradiation microscope with electrostatic lenses 33(443)
Irradiation microscope with magnetic lenses 33(436)
Irradiation system 33(400, 420)
Isochromat 30(368)
Isophots 33(278)
Isotope dilution method 33(571)
Isotope replacement in neutron diffraction 32(617)
Isotope separation 33(570)
Isotope separation, electromagnetic 33(652)
Isotope separation processes in nature 33(570)
Isotopes, radiogenic 33(569)
Isotopic abundance measurements 33(567)
Isotopic abundances, relative 33(568, 579)
Isotopic composition of the elements, variations of 33(570)
Isotopic mass determinations 33(597)
Isotopic weights 33(579)

Jacobi's theorem 33(139)

Jahn-Teller interaction 31(389, 422)
— active mode 31(410)
— coupling parameter 31(407)
— distortions for point groups 31(408)
— effect, pseudo 31(409)
— theorem 31(406)
jj to LS transformation 31(509, 516)
jj-coupling 31(508, 517)

K 31(127, 128, 152, 171)
K, oxidation of films 31(382)
*K*α-satellites 30(197, 198)
Ketene 31(417)
Keto-enol tautomerism 31(439)
Kikuchi band 32(504)
Kikuchi lines 32(503)
Kirchhoff's formula 33(358)
Klein-Nishina cross sections
— for absorption, collision, and scattering, numerical values 34(265, 266)
— for polarized radiation 34(244–250)
— for unpolarized radiation 34(250–271)
Klein-Nishina formula, physical interpretation 34(248)
K-matrix 31(16ff., 479, 480)
— Hamiltonian eigenstates in terms of the 31(16)
— integral equation 31(16)
— on-the-energy-shell 31(17)
— relation to *S*-matrix 31(21)
Kohn variational method 31(582)
Koopmans' theorem 31(285, 510, 512, 560)
Koopmans' theorem, breakdown of 31(393)
Kossel lines 32(182, 503)
Kossel structure (*see also* Fine structure) 30(217)
Kr, autoionizing structure 31(143, 166)
Kr, β-parameter 31(201, 203)
Kr, ionization potential 31(126)
Kr, partial cross section 31(190)
Kr, photoionization cross section 31(164, 170)
K-radiation, fine structure of emission lines 34(86)
K-radiation, intensity measurements 34(178, 179)
K-radiation, thick target yields 34(182)
Kramers formula (intensity distribution of the continuous x-ray spectrum) 30(361)
K-series 30(10–12, 79, 80)
K-series, table of diagram lines 30(182–186)
K-shell
— corrections 34(65, 180)
— ionization 34(166, 169, 170, 178)
— ionization cross section 34(172, 175, 177, 183, 184)
— ionization cross section, dependence on *Z* 34(184)
K-spectra of light elements 30(271–274)
Kunsman anode 34(31)
Kunsmann sources 33(121)

L spectra of light elements 30(277–282, 286)
La, photoionization cross section of 4*d*-shell 31(53)
Lab frame 31(476)
Lagrange-Helmholtz-relation 31(252)
Lamb shift 31(491, 564, 574ff.)
Lamellar tube after Brasch and Lange 30(61)
Landau distribution 34(88–90, 93–101)
Landsberg model 30(290)
Large crystals as diffraction centres 32(193)
Latter tail correction 31(585)
Lattice 32(99, 107, 129)
Lattice, base centered 32(142)
Lattice, body-centered 32(144)
Lattice complex 32(140)
Lattice constant(s) 30(92)
— and crystal size 32(515)
— dependence on temperature 30(142, 143)
— determination with gauge substance 32(450)
— equal in film and substance 32(540)
— precision values 30(141)
Lattice, cubic 32(145, 203, 215)
Lattice defects 32(389f.)
Lattice defects by twinning 32(511)
Lattice, face-centered 32(144)
Lattice factor 32(468)

Lattice function 32(460)
Lattice, hexagonal 32(145)
Lattice, monoclinic 32(115, 142)
Lattice, orthorhombic 32(115, 143, 215)
Lattice plane 32(108)
Lattice, rhombohedric 32(145)
Lattice, tetragonal 32(144)
Lattice transformations 32(112)
Lattice, triclinic 32(142)
Laue case 32(484)
Laue conditions 32(391)
Laue group 32(148, 232, 148, 232)
Laue method 32(103, 180, 594)
Laue spikes 32(489, 491, 493, 512)
Laue theory 32(99)
Layer cathodes 33(7, 8)
Layer line, parallel to equatorial plane 32(102)
Lead equivalent 30(75, 76)
Leak detector 33(575)
Leak, viscous 33(562)
Least squares method 32(219)
Lenard window (*see also* Exit window) 30(1, 30, 48, 51, 61, 132, 134)
Length form for the transition matrix element 31(23ff.)
Lens equation 33(580)
Lens field, half-value width of a 33(405)
Lens, long 33(611)
Lens, short 33(611, 634)
Lens, spherical corrected 33(469)
Lens strength 33(405)
Lenses for electrons, electrostatic and magnetic 30(132f., 134)
Lenses for x-rays 30(322)
Level shift
Level shift, nonradiative 31(482, 556, 557)
Level shift, radiative 31(491)
Li 31(171)
Life time, *see* Mean life time
Ligand-field theory 31(444)
Light elements, combined with heavy, neutron diffraction 32(615)
Light sources 31(136)
— continuum 31(137, 140)
— dc glow 31(137)
— discrete 31(137)
— duoplasmatron 31(137)
— Hopfield continuum 31(140)
— spark 31(138)
— synchrotron 31(140)
Lilienfeld tube 30(32, 33)
Limegraph 33(628)
Lindemann glass 30(51, 53)
Line breadth, integral 32(180, 198, 201, 397, 421)
Line broadening by distortion 32(205)
Line broadening by particle size 32(205)
Line broadening by thermal motion 32(207, 212)
Line shape(s) 33(564)
— Fourier analysis 32(423)
— instrumental corrections 32(418)
— of Debye-Scherrer lines 32(418)
Line width 33(565)
Linear focus 30(3, 35, 39, 41, 42, 55, 64)
Lippich's theorem 33(179, 299)
Lippmann-Schwinger equation 31(479, 493)
Liquid ESCA 31(261)
Liquid function 32(522)
Liquid helium, neutron diffraction 32(618)
Liquid sample handling in photoelectron spectroscopy 31(260)
Liquids 32(524)
Liquids, neutron diffraction 32(617)
Lithium ion source 34(30)
LNDO/S, local neglect of differential overlap method/screened 31(436)
Local exchange potential, nonrelativistic 31(584ff., 598)
— relativistic 31(589)
Lone-pair orbital splitting 31(438)
Long range order 32(412)
Longitudinal wave 32(216)
Lorentz equation of motion 33(124)
Lorentz factor 32(176, 218); 34(62, 63)
L-radiation, fine structure of emission lines 34(86)
L-radiation, intensity measurements 34(178, 179)
L-radiation, thick target yields 34(185)
LS-coupling 31(403, 503ff., 517)
L-series 30(79, 80)
L-series, table of diagram lines 30(186–194)
L-shell corrections 34(65, 180)

L-shell ionization 34(166, 179)
— cross section 34(176, 186)
Luminescent screen 33(432)
Luminosity 31(157, 244); 33(615)
2,6-lutidine 31(430)

Magic angle 54° 44′ 31(156, 424)
Magnesium oxide, neutron diffraction 32(631)
Magnetic dipole moment of the electron 34(249, 257)
Magnetic electron lenses, optical characteristics of 33(216)
Magnetic field 33(564)
— effective boundary of the 33(564)
— in form of a power series 33(159)
Magnetic form factor for neutron scattering 32(625, 626)
Magnetic lens, calculation of a typical 33(203)
Magnetic lens, upper limit of chromatic aberration of 33(256)
Magnetic moment of the neutron 32(553)
Magnetic quenching of orthopositronium 34(154–156)
Magnetic reflection of neutrons 32(629)
Magnetic scattering amplitude of neutrons 32(624)
Magnetic scattering of neutrons 32(623)
— inelastic 32(636, 637f.)
Magnetized material, radiation transmission 34(297)
Magnetron ion-sources 33(117)
Magnification 33(178)
Magnification, determination of 33(434f.)
Magnification, maximum 33(424)
Magnification, useful 33(371)
Many-body perturbation theory 31(68ff., 294, 528, 529, 581, 605ff.)
— choice of zero order basis 31(69, 72–74)
— for double photoionization 31(116ff.)
— virtual double excitations in photoionization processes 31(74)
Many-configuration *jj*-average energy 31(575)
Many-electron transitions 31(400)
— wave function 31(505, 506, 508, 537, 538)
Mass attenuation coefficient(s)
— for photons in air 34(273)
— for photons in aluminum 34(275)
— for photons in copper 34(276)
— for photons in lead 34(278)
— for photons in sodium iodide 34(277)
— for photons in water 34(274)
— total 34(272, 273–278)
Mass dispersion 33(305, 314, 553, 584)
Mass, effective, of an electron 30(252)
Mass extinction coefficient 30(9, 20)
Mass filter, electric 33(601)
Mass of the neutron 32(553)
Mass scale 33(583ff., 594, 597)
Mass spectra, characteristic 33(573)
Mass spectrograph(s) 33(546)
— anastigmatic double focussing 33(590)
— Aston's 33(549)
— Aston's velocity focussing 33(577)
— Herzog's stigmatic double focussing 33(590)
— Mattauch and Herzog's double focussing 33(585)
— Ogata and Matsuda's double focussing 33(585)
— velocity focussing 33(577)
Mass spectrographic doublet measurements, errors in 33(596)
Mass spectrometer(s) 31(152); 33(546)
— applications of direction focussing 33(567)
— Dempster's direction-focussing 33(549)
— high frequency 33(599)
— Nier's 33(555)
— Nier's double focussing 33(588)
— radio frequency 33(599)
— time-of-flight 33(604)
— trochoidal time of flight 33(605)
— vacuum system 33(556)
Mass spectrometric chemical analysis 33(572)
Mass spectrometric examination of solids 33(563)
Mass spectrometric mass measurements, discrepancies in 33(598)

Mass spectrometry of molecules 33(572)
Mass spectroscopic device, direction focussing 33(549)
Mass spectrum 33(62)
Mass synchrometer 33(604)
Mass-absorption coefficient for photons
— due to Compton effect 34(272, 273–278, 283)
— due to pair production 34(273–278, 283)
— due to photoelectric effect 34(273–278, 283)
— total 34(273–278, 283)
Maxwell distribution of neutrons 32(554, 579)
McKinley and Feshbach cross section 34(103–107, 125)
MCl_4 31(395)
McLeod gauges 31(136)
Mean free path for loss of quantum 34(74, 83, 84)
Mean life time of
— orthopositronium 34(142)
— parapositronium 34(141, 142, 162)
— positrons, effect of temperature 34(160)
— positrons in liquids and solids 34(159–160)
Measurements, absolute 33(647)
Measurements, relative 33(647)
Membrane anode of an x-ray tube 30(41, 52)
Mercury streaming 31(136)
Mercury vapor stream 30(349)
Metal capillary cathodes 33(7, 8)
Metal films, evaporated 32(536)
Metal valence bands 31(459)
Metal vapour oven 31(259)
Metallic x-ray tube 30(3, 39)
Metallocenes 31(393)
Metals, neutron diffraction 32(634)
Metals, polished 32(525–528)
Methane 31(313, 367, 407, 427)
Methanol 31(417)
Methanol, photoelectron spectrum of liquid 31(261)
Methyl cyanide 31(417)
Methyl halides 31(405, 406)
Methyl radical 31(417)
MeV range of x-rays 30(357)
Mg$K\alpha$ radiation 31(228)
Mg$K\alpha$ radiation, monochromatized 31(259)
MgO, photoelectron spectrum 31(220)
Mica, anomalous dispersion 30(141, 144)
Mica, bent crystal 30(108)
Micro manometer 33(563)
Micro pipettes 33(562)
Micro-analysis, dichromatic 30(309)
Microbeam of x-rays 32(435)
Microchemical analysis 30(313)
Micrographs of atoms 30(310)
Microradiography 30(312f., 322)
Microtube 30(53)
Microwave discharge to produce transient species 31(259)
Mie theory 32(323)
Miller indices 32(99, 100, 391)
MINDO, modified INDO method 31(435)
Mirror for neutrons 32(584, 595)
Mirror plane 32(134, 232)
Mirror symmetric anisotropic medium 33(247)
Mirrors for x-rays 30(324)
— crossed aspherical 30(328)
— crossed spherical 30(324, 326, 327f., 328)
— metal coated 30(331)
— spherical, compound systems 30(328)
— surfaces 30(330, 331)
Mixed coupling 31(521)
Mixed set of twofold axes and screws 32(143)
Mn 31(309)
MnF_2 31(378, 379)
MO methods for characterization of spectra 31(431)
Modulated wave 32(189)
Molecular conformation 31(438)
Molecular subunit 31(432)
Molecular weight 32(355, 356)
Molecules, ionically bonded 31(442)
Molecules on metal surfaces 31(380)
Molière multiple scattering theory 34(93, 117–119, 126)
Möller cross section for electron-electron scattering 34(56–58, 63)
Möller formula 31(535, 536)
Molten metal 32(522)

Monochromatization by crystals 30(11, 54, 55)
Monochromatization by filters 30(11–13)
Monochromatized Al$K\alpha$ 31(230)
Monochromator for neutrons, mechanical 32(582)
Monochromator for neutrons, velocity resolution 32(583)
Monochromator with bent crystals 30(124)
Monochromators 31(143)
Monopole transition 31(293)
Monte Carlo calculation of straggling 34(93)
Morse potential 31(418)
Mosaic crystal 32(154, 169, 176, 195, 196)
Mosaic crystal, perfect 32(196)
Mosaic structure 30(147); 32(405)
Mosaic structure, ideal 32(405)
Moseley diagram, modified 30(201f., 229)
Moseley law 30(9–11)
Moseley law and diagram 30(201)
Mott cross section 34(108–112)
Mott theory of electron-electron scattering 34(56)
Mott theory of nuclear scattering 34(103–115)
M-series, table of diagram lines 30(195–196)
M-spectra of light elements 30(286)
Multichannel electron detection 31(255)
Multichannel plates 31(255)
Multichannel quantum defect theory 31(90ff.)
Multichannel *R*-matrix theory 31(98)
Multichannel scattering
Multichannel scattering, nonresonant 31(476–480)
Multichannel scattering, resonant 31(480–483)
Multichannel wavefunction 31(15ff.)
Multiconfiguration Hartree-Fock theory 31(80)
Multiconfiguration interaction method
— Dirac-Fock 31(543, 563, 575, 577, 578)
— Hartree-Fock 31(511, 574, 575, 578)
Multielectrode systems 33(15)
Multiple ionization 31(159, 193)
Multiple nuclear scattering 34(115–121)
Multiple scattering 32(323, 346–348, 389, 465, 511, 592)
Multiple scattering calculation 31(397, 426, 436)
Multiple scattering MS-$X\alpha$ approach 31(436)
Multiplet 31(524)
Multiplets 30(85)
Multiplicity factor 32(178)
Multiplier 30(137)
Multiplier for mass spectrometer 33(597)
Multipole transitions in x-ray spectra 30(82, 83)

N_2 31(126, 368, 369, 373, 374, 390, 391, 393, 396, 397, 401, 427, 432)
N_2^+ 31(400)
N, 1*s* spectrum of N_2 31(368)
N, 1*s* spectrum of NH_4NO_3 31(374)
N_2, shake-up spectrum 31(357)
N_2, vibrational energies 31(413)
Na 31(171, 172)
NaCl 31(384, 385)
"Natural equations" of principal orbits 33(298)
Natural excitation 31(524)
Natural line width 31(235)
— of HeI resonance radiation 31(235)
— of photoelectron line 31(281)
Ne, autoionizing structure 31(169)
Ne, β-parameter of 2*p*-shell 31(202)
Ne, binding energies 31(286)
Ne, double photoionization cross section 31(116, 194)
Ne, ionization threshold 31(126)
Ne, $p \rightarrow d$ transitions 31(56)
Ne, photoionization cross section 31(164, 184)
Ne, 1*s* photoelectron line 31(297)
Ne, 1*s* photoelectron spectrum 31(296ff.)
Ne, 1*s* satellite photoelectron spectrum 31(299)
Needle anode of an x-ray tube 30(44, 67)
Needle valves 33(595)
Negative ions 31(180, 444)
Negative ions, detachment 34(49)

Negative-energy states for the electron 34(244, 249, 257)
Net 32(99)
Net, reciprocal 32(125)
Neutron and x-ray intensities, comparison 32(554, 594)
Neutron magnetic moment 32(553)
Neutron mass 32(553)
Neutron proton scattering amplitude 32(598)
Neutron sources, natural 32(576)
Neutron spectrum 32(578)
Neutron wavelength 32(553)
Newton's imaging-fields 33(196, 198)
NH_3 31(126, 370, 407, 413, 415, 416, 432)
N_2H_2 31(442)
Nickel, neutron diffraction 32(633)
Nier's double focussing spectrometer 33(598)
NO 31(126, 368, 377, 378, 397, 426, 427, 414)
NO_2 31(126)
Non-diagram lines (*see also* Satellites) 30(85, 196–201)
Nondispersive analyzer 31(250)
Nonlocal potentials 31(24ff.)
Non-metals, neutron diffraction 32(635)
Nonradiative transition(s) 30(290)
— amplitudes, effect of bound orbital basis on 31(600, 601, 605, 606)
— amplitudes, effect of local potential and kinetic energy on 31(598, 599)
— energies, ab initio calculations 31(574ff.)
— energies, semiempirical treatments 31(566ff.)
Nonradiative transition rates
— closed-shell case 31(519, 520)
— nonrelativistic 31(519ff.)
— open-shell case 31(521ff.)
— relativistic 31(536ff.)
Normal congruence 33(235)
Normalization screening theory 31(46)
Normalization, wavefunction 31(7, 12, 14, 17)
Notations for energy levels 30(81)
NS 31(442)
N-series, table of diagram lines 30(197)
Nuclear effect, *see* Coulomb excitation
Nuclear potential 31(588)
Nuclear reactions 33(656)
Nuclear scattering of fast electrons 34(101–120)
Nuclear shell models 33(583)
Nuclear size, effect on scattering 34(91, 109, 112, 116)
Nuclear size effects on binding energies 31(564, 565)
Nuclear size effects on nonradiative transition energies 31(575)
Nucleation 32(541)
Nucleic acid 32(382)

O_2 31(126, 368, 377, 378, 396, 397, 404, 426)
O_2^+ 31(414, 415, 424)
O_3 31(126, 442, 447)
O_3^- 31(447)
Object 33(415)
Object and image field, curvature of 33(238)
Object diaphragm 33(415)
Object diaphragm, current density in the 33(419)
Object fouling 33(512, 540)
Object sluice 33(417)
Object stage 33(418)
Object temperature 33(512, 513)
Object warming 33(423, 510)
Objective lenses, electrostatic 33(411)
Objective lenses, magnetic 33(404, 422)
Objective pole shoes 33(404)
Obliquity of field 30(326, 328)
OH 31(442)
Oil film 32(522)
Omegatron 33(602)
Omegatron, ion paths in the 33(603)
Open-shell system 31(389)
Operating parameter 33(405)
Optical branch of elastic waves 32(208)
"Orange" spectrometer 33(628)
Orbital energy 31(284, 369)
— eigenvalues for Ar 31(298)
— shift 31(341, 342)
— shifts in second row elements 31(341)
— sum rule 31(435)
Orbits in the homogeneous electric-magnetic field, relativistic 33(141)
Orbits, paraxial 33(174)

Order x-ray determination 32(413)
Ore gap for positronium formation 34(144)
Orientation 32(139)
Orientation of crystals in metal films 32(537, 539)
Orientation of iron oxide layers 32(545)
Oriented electrons in Compton effect 34(249, 294–297)
Orthogonal system 33(258, 299)
Ortho-para transformation of positronium 34(145, 152, 160)
Orthopositronium, annihilation spectrum 34(150)
Orthopositronium, definition 34(139)
Orthopositronium, detection methods 34(149)
Orthopositronium, mean life time 34(142)
Orthopositronium, quenching 34(145, 152, 154–157)
Os 31(461)
Oscillations, axial and radial 33(311)
Oscillator description of the atomic scattering factor 32(163)
Oscillator strength 34(62, 68, 69)
— atomic 31(33ff.)
— generalized 31(35, 161)
— optical 31(133, 161)
— relation to photoionization cross section 31(34)
— relations of continuum and discrete spectra 31(89)
Osculating cardinal elements 33(194)
Out-states 31(10, 13, 20)
Ovalbumin 32(377)
Overall-luminosity *L* 33(615)
Overcorrection 33(227)
Overlap in energy 30(254, 281)
Overlap matrix element 31(512, 525)
Overlapping of the electron density 32(117)
Oxidation of potassium films 31(382)
Oxide coated cathodes 33(9, 10)
Oxide layer 32(528, 530, 543)
Oxide layers on alloys 32(549)
Oxide layers on molten metals 32(549)
Oxidized films, anodically 32(550)
Oxygen isotope ratios 33(561)
Oxygen isotopes 33(570)

P_2, vibrational energies 31(413, 447)
Pair correlation energy 31(288)
Pair formation 33(656)
Pair production 30(19, 20)
— mass-absorption coefficients 34(273–278)
— photoelectric effect, Compton effect, relative predominance at different energy and atomic number 34(218, 219)
Pair spectrometers 33(664, 666)
Paleo-temperatures of oceans 33(561)
Parabola spectrographs 33(546)
Parallax, stereoscopic 33(521)
Parallel plate electron analyzer 31(155, 248)
Parallel-perspective images 33(520)
Paramagentic scattering of neutrons 32(625)
Paramagentic scattering of neutrons, inelastic 32(638)
Paramagnetic molecules 31(376)
Parapositronium, definition 34(139)
Parapositronium, mean life time 34(141, 142, 162)
Parity 31(496, 504, 517, 537)
Parity change 30(83)
Partial photoionization cross sections 31(133, 152, 182)
— Ar 31(190, 195)
— Be 31(91, 116)
— Ce 31(51)
— Cs 31(57)
— He 31(192)
— Hg 31(188)
— Kr 31(190)
— Ne 31(184, 189, 194)
— Xe 31(188, 198, 199)
Partial wave method 34(22)
Particle mass distribution 32(343)
Particle shape determination 32(376f.)
Particle volume determination 32(355)
Path length distribution due to scattering 34(90–93)
Path, mean free 33(502)
Patterson function 32(126, 127, 151, 234, 326, 394, 403, 617)
Patterson map 32(128)
Patterson synthesis 32(219, 226)
Pauli matrices 31(532)
Pauli principle, random phase approximation violations of the 31(76)

Peak height 33(565)
Peak selector 33(574)
Pendellösung (Ewald) 32(479, 482, 484)
Penetration depth 32(485, 494, 495, 498)
Penetration factor 33(18)
Penning ion-sources 33(82)
Pentafluoropyridine 31(430)
Penumbra 30(315)
Penumbral blur 30(316, 320)
Perfluoro effect 31(432)
Permanent magnets 33(404, 408)
Perturbation method 33(219)
Perveance 33(25)
Petzval condition 33(248)
Petzval curvature 33(248)
Petzval theorem 33(247)
PH_3 31(407, 413, 415)
Phase contrast 33(502, 522)
Phase contrast method 33(349)
Phase conventions, Condon and Shortley 31(497)
Phase conventions, Fano and Racah 31(506)
Phase difference 32(104)
Phase factors 31(514, 515, 518, 521, 537)
Phase object 33(348)
Phase problem 32(107, 219)
Phase shift of the partial wave 34(23, 25)
Phase-sensitive detection 31(332)
Phonon broadening 31(374)
Phosphabenzene 31(430)
Phosphoryl chloride 31(395)
Photodetachment 31(180, 444, 446)
Photoelectric absorption 30(4, 87)
Photoelectric cross section 31(275)
— differential 31(289)
Photoelectric effect 31(219); 33(656)
Photoelectric effect, Compton effect, pair production, relative predominance at different energy and atomic number 34(218, 219)
Photoelectric mass-absorption coefficients 34(273–278)
Photoelectric yield, Al 31(154)
Photoelectron line, intensity 31(273)
Photoelectron line, natural width 31(281)
Photoelectron line, width 31(278)
Photoelectron spectrometer for gases and condensed matter 31(228)
Photoelectron spectroscopy 31(129, 154, 215)
— basic processes 31(215)
— experimental arrangement 31(215)
— gas cell arrangement 31(257)
— liquid sample handling 31(260)
— sample preparation 31(256)
— solid sample handling 31(262)
Photoelectron spectrum
— calibration of 31(268)
— deconvolution method 31(279)
— density of states structure 31(459)
— general features 31(265)
— of paraaminofluorobenzene 31(338)
— temperature dependence of 31(438)
— theoretical models 31(283)
Photoelectrons 33(657)
Photographic photometry 30(269)
Photographic plate(s) 30(135); 33(429, 433, 654)
Photoionization 31(123)
Photoionization cells 31(145)
Photoionization cross section 31(7, 29, 30, 132, 164)
— Al 31(174)
— Ar 31(73, 74, 80, 82, 98, 164)
— Ba 31(175)
— Cd 31(178)
— central potential approximation 31(39ff.)
— Cs 31(171, 173)
— delayed maximum at threshold 31(47ff.)
— He 31(164)
— Hg 31(179)
— high energy behaviour 31(42ff.)
— hydrogenic approximation 31(42ff.)
— K 31(171)
— Kr 31(164, 166, 170)
— Li 31(171)
— measurement of 31(396)
— minimum above threshold 31(47, 55ff.)
— Na 31(171, 172)
— Ne 31(164, 184)
— near threshold features 31(46ff.)
— Rb 31(171)
— vanishing final state interaction approximation 31(39ff.)
— Xe 31(164, 167)
— Zn 31(176)
Photoionization efficiency 31(132)

Photoionization from solids, selection rules 31(448)
Photoionization measurements 31(163)
Photomultiplier of Be-Cu type 30(267)
Photon source for photoelectron excitation 31(224)
Photon source, Mζ 31(234)
Photon source, uv 31(234)
Photon source, x-ray 31(227)
Photon-electron interaction 31(485, 490, 532)
Photons, interaction with matter *see under:* Absorption ..., Collision ..., Compton scattering, Rayleigh or coherent scattering, Scattered photons
π-band 30(276)
Pierce field 33(30)
Pierce system 33(15)
Pincushion distortion 33(244)
π-orbital splitting 31(438)
$\pi \rightarrow \pi$ transitions 31(359)
Pitch 32(131)
p$K\alpha$ activity 31(434)
Plasma energy 34(85, 86)
Plasma frequency 34(67, 71)
Plasma oscillation(s) 33(538); 34(76)
Plasma resonance, width of 34(85)
Plasmon excitation in photoelectron spectra 31(363)
Plasmon, extrinsic 31(364)
Plasmon, intrinsic 31(364)
Plastical bending of a crystal 32(428)
Plate anode of an x-ray tube 30(35, 40)
Plateholder 30(113, 127, 129, 130)
Plural scattering 34(117–120)
PN, vibrational energies 31(413)
Point defect 32(415)
Point defect, strain field 32(417)
Point group 32(140)
Point lattice 32(107)
Point source 30(316)
Poisoning 33(11)
Poisson distribution of energy losses 34(84, 86)
Poisson's equation 34(73, 74)
Polar diagram of x-ray emission 30(16, 17)
Polarization 31(144)
Polarization correlation of the three-photon radiation 34(143, 150)
Polarization effects, long-range 31(67)
Polarization factor 32(156, 218)
Polarization of Compton scattered radiation 34(229–233, 246)
Polarization of medium by the passing particle 34(67–71, 75, 76, 100)
Polarization of neutrons 32(624, 627, 628, 630)
Polarization of the Compton recoil electron 34(249)
Polarization of the incident radiation 34(245–250, 295, 296)
Polarization of x-rays 30(347f., 352f., 360, 362, 364)
— at the short wavelength limit 30(352)
— from thick target 30(347)
— theoretical dependence on emission angle 30(365)
Polarization rule for positronium 34(144)
Polarized radiation 31(238)
Pole shoes 33(408)
Polepieces, hyperbolic 33(640)
Polished rocksalt surface 32(531)
Polyacetylenes 31(414, 434)
Polyelectrons 34(139, 140)
Polyganization 32(427, 430)
Polymers 31(338, 339)
Positive ions, doubly charged 34(49)
Positron-electron collisions 34(59, 60, 63, 64)
Positronium annihilation 34(140–144, 150, 160, 161)
Positronium, detection methods 34(146, 148, 149)
Positronium, fine structure 34(139, 152–158)
Positronium, formation in gases 34(144, 151)
Positronium, formation in liquids and solids 34(158–162)
Positronium, polarization 34(143, 144, 150)
Positronium, singlet state (*see also* Parapositronium) 34(139, 153)
Positronium, stability 34(145, 151)
Positronium, triplet state (*see also* Orthopositronium) 34(139, 153)
Positrons 30(19)
Positrons, annihilation rate 34(147)
Positrons, mean life time 34(159–160)

Positrons, slowing down 34(158)
Potassium 31(127, 128, 152, 171)
Potassium bifluoride, neutron diffraction 32(610)
Potassium dihydrogen phosphate, neutron diffraction 32(609, 611)
Potassium films, oxidation of 31(382)
Potassium halides 31(376)
Potential as a closed expression 33(158)
Potential barriers 31(47ff.)
Potential, central atomic 31(39ff.)
Potential curve gradients 31(370)
Potential, Herman-Skillman 31(41)
Potential in form of a power series 33(156)
Potential, non-local 31(24ff.)
Potential, perturbation theory for screened Coulomb 31(45)
Potential, vector 31(6)
Powder anodes 33(121)
Powder diagram 32(501)
Powder diffraction
— apparatus 32(587)
— background 32(592)
— effect of the form of the sample 32(590)
— in transmission 32(589)
— thermal motion correction 32(592)
Precision measurements 33(624)
Predissociative state 31(415)
Preparation, technique of 33(516)
Preredardation electron lens 31(216, 252)
Preredardation of photoelectrons 31(254)
Principal plane 33(289)
Principal plane, real 33(197)
Prism and sectorfield spectroscopes 33(626)
Prism for x-rays 30(140)
Probability current density 33(318)
Probability density 33(318)
Production of x-rays
— by electrons 30(21–28)
— by ions 30(20–21)
— classical electromagnetic theory 30(359, 362)
— efficiency 30(342, 361, 365)
— thermal 30(21)
Profile of a line 32(179, 200, 202)
Projection, generalized 32(122)
Projection of the lattice structure 32(121, 232)
Projective lenses 33(424)
Projective magnification 33(425)
Propagation of x-rays in crystals 32(167)
Propagation vector 30(250)
Proportional counter 30(137)
Proportionality error 33(291)
Protein metabolism in the living body 33(571)
Protein solutions 32(372–380)
Proton bremsstrahlung 34(191)
Proton ranges 34(207, 208–210)
Proton source 34(31)
Protons, collision cross section for 34(34, 35, 36)
Protons, slow, beams of 34(30)
Protons, stopping cross section for 34(196–203)
Pseudoatom 32(127)
Pseudomorphy 32(540)
Pseudopotential 31(584)
Pseudopotential approaches 31(436)
Pullman's index 31(437)
Pulse height spectra of characteristic x-rays 34(180)
Purine 31(436)
Pyridazines 31(438)
Pyridine 31(393, 395, 430, 434)
Pyrrol 31(430)

Quality of an x-ray tube 30(14, 15)
Quality test of bent crystals 30(121)
Quantum defect theory 31(86ff.)
— multichannel 31(90ff.)
— single channel 31(86ff.)
Quantum-electrodynamical effects
— on binding energies 31(563ff.)
— on nonradiative transition energies 31(575, 576)
Quartz, anomalous dispersion 30(143)
Quasi-stationary states 30(298)
Quenching of orthopositronium 34(145, 152)
— in a magnetic field 34(154–156)

Radial eigenfunctions 34(24)
Radial Schrödinger equation 31(497)
— nonrelativistic 31(582, 583)
— relativistic 31(587)
Radiation by electrons 30(268)

Radiation by high energy electrons-(300 MeV) 30(274)
Radiation by relativistic electrons 30(266)
Radiation, characteristic 32(103)
Radiation characteristic of the object point 33(273)
Radiation length for fast electrons 34(91)
Radiation losses of electrons 34(60, 63, 100, 123)
Radiation protection 30(37, 38, 74–76)
Radiation transmission through magnetized material 34(297)
Radiationless transition (Auger transition) 34(181, 186)
Radiative correction to Compton effect 34(269)
Radiator 33(657)
Radicals 33(574)
Radius of gyration 32(329, 350–351, 354)
Radius of gyration for proteins 32(374)
Ramsauer effect, theory of 34(26–28)
Ramsauer-Townsend effect 34(2)
Random phase approximation 31(75ff., 294)
— configuration-interaction interpretations 31(77ff.)
— extensions 31(83ff.)
— interchannel interaction predictions 31(81ff.)
— Pauli principle violations 31(76, 77)
— with exchange 31(296, 303, 529, 530)
Random walk 32(228)
Range and stopping cross section 34(206)
Range-energy relation for electrons 34(132–138)
Ranges of α-particles 34(210–211)
Ranges of charged particles 34(206–212)
Ranges of protons 34(207, 208–210)
Rare earth elements 31(309, 461)
Raster microscope 33(447)
Raster microscope, resolution of the 33(447)
Ray constant 33(25)
Ray producing systems 33(15)
Rayleigh scattering 34(290–294)
Rayleigh scattering, role in attenuation experiments 34(273–278)
Rayleigh-Gans approximation 32(323, 324)
Rb 31(127, 171)
Re 31(461)
Reaction rate 31(440)
Reactions, charged particle 33(661, 664)
Reactors 32(578)
Reciprocal axes 32(113)
Reciprocal lattice 32(112, 119, 175f., 197, 392, 460)
Reciprocal net 32(125)
Reciprocal vector 32(114, 170f.)
Recoil electrons from Compton effect
— and scattered photons, angular relationship 34(239, 240)
— angular distribution 34(254)
— detection and first measurements 34(224, 225, 226)
— energy 34(241–243, 260, 261, 265)
— most probable direction 34(230)
— number scattered in any finite angular interval 34(257)
— polarization state 34(249)
Recording of doublets 33(589)
Recrystayllization in situ 32(427)
Reduced zone scheme 31(448)
Reference line(s) 30(123, 160, 161, 162); 33(573)
Reference lines using two x-ray tubes 30(101)
Reflecting power of the grating 30(264)
Reflection 32(130, 132)
Reflection abundance, explanation 32(470)
Reflection, asymmetrical 30(139)
Reflection coefficient 30(146, 340)
Reflection, forbidden 32(465)
Reflection on surfaces (*see also* Surfaces) 32(493, 495)
Reflexion images, contrasts of the 33(453)
Reflexion microscope 33(451)
Reflexion microscope, resolution of a 33(454)
Refraction effect 32(494, 497)
Refraction exponent 33(133, 137)
Refraction of neutrons 32(575)
Refraction of x-rays 30(138, 322)

Refractive index 30(138, 323)
Relativistic effects 31(393)
— on asymmetry parameter β 31(108ff.)
— on binding energies 31(562, 563)
— on nonradiative energies 31(571ff., 575)
— on nonradiative transition amplitudes 31(593ff.)
Relativistic units 33(612)
Relaxation 31(524, 560, 569, 602)
Relaxation energy 31(285, 342, 347, 370)
— division of 31(347)
— for boron 1s ionization 31(348)
— for s-subshells 31(287)
Renner active mode 31(410)
Renner effect 31(416, 422)
Reorganization energy 31(285)
Resolution 33(400, 424, 467, 480, 482, 485, 522, 544)
Resolution and uncertainty relation 33(370)
Resolution, angular 32(361)
Resolution, high 33(647)
Resolution limit 30(314, 322, 327); 33(369)
— experimental 33(486)
— limiting factor 33(488)
— of the human eye 33(371)
Resolution of spectrometer 32(581)
Resolution of spherical sector electron spectrometer 31(245)
Resolving distance, minimum 30(315, 327, 328, 333)
Resolving limit, theoretical 33(482)
Resolving power 33(553, 554, 557, 613)
Resolving power of crystal spectrographs 30(151, 155)
Resolving power of x-ray-microscopes 30(305, 310, 312, 314, 331)
Resolving time 33(673)
Resonance, autoionization profile of a 31(63ff., 73, 91ff., 99)
Resonance, β parameter within a 31(93, 107, 112)
Resonance lines 31(139)
Resonance loss 33(493)
Resonance quenching of ortho-positronium 34(156, 157)
Resonance scattering
— Auger electron emission 31(488)
— autoionization 31(486)
— general 31(480ff.)
— photon emission 31(489)
Resonance transfer of energy 34(61, 89)
Resonance transformer 30(48, 49)
Resonant cavity x-ray tube 30(61)
Reststrahlen 32(208)
Retardation 32(162)
Richardson effect 30(23, 24)
Richardson's constant 33(3, 7)
Ring focus 33(630)
R-matrix theory 31(94ff.)
— multichannel 31(98ff.)
— relation to eigenchannel method 31(97)
— single channel 31(95ff.)
Rocking curve 30(150, 151, 153–156, 332, 333); 32(364, 470, 496)
Rocksalt, analysis of its structure 32(103)
Roof angle 32(526)
Rotalix tube 30(39)
Rotating anode 31(228)
Rotating crystal method 30(94); 32(102, 174)
Rotating sector 32(520)
Rotation 32(130, 133)
Rotation center 33(479)
Rotation diagram 32(472)
Rotation, improper 32(133)
Rotational excitations 31(422)
Rotational symmetric field, mirror effect of the 33(181)
Rotational symmetry, perturbed 33(259)
Row of lattice points 32(99, 108)
Rowland circle 30(111, 115, 116, 117, 122, 127, 129, 130, 266); 31(230, 231)
RSPT, Rayleigh-Schrödinger Perturbation Theory method 31(421)
Ruling technique 30(127)
Russell-Saunders coupling 31(403, 503ff., 517)
Rutherford scattering 34(76, 101–111, 119)
Rutherford-Robinson mounting 33(617)
Rydberg constant 30(10)
Rydberg energy 34(65, 125)
Rydberg-Klein-Rees potential 31(419)

S_2^+ 31(415)
S matrix, nonresonant 31(474–479)
S, resonant 31(481, 482)
Sample preparation for photoelectron spectroscopy 31(256)
Samples, high molecular weight 33(574)
Satellite bands 30(289, 295, 296)
Satellite intensities 30(237, 238)
Satellite lines
— in adsorbate core photoelectron spectra 31(361)
— in benzene 31(360)
— in core electron spectra of solids 31(360)
— in the photoelectron spectrum 31(310)
— in transition metal core spectra 31(362)
Satellite structure of valence bands 31(400)
Satellite transitions in molecular spectra 31(357)
Satellites of x-ray lines 30(85, 196–201)
— graphical resolution 30(197, 199)
Saturation 33(406)
SbH_3 31(407, 413, 415)
SbH_3^+ 31(409)
Scanning microscope 30(321)
Scattered photons from Compton effect
— and recoil electrons, angular relationship 34(239, 240)
— angular distribution 34(253)
— bipartition angles 34(257, 261–263)
— energy 34(238, 239, 257, 258, 260, 264)
— fraction with more than any arbitrary energy 34(257)
— number scattered in any finite angular interval 34(255, 256)
— polarization state 34(229–233, 246)
— unpolarized component 34(230, 233, 248)
Scattering 33(499)
Scattering amplitude for neutrons 32(556, 639)
— bound nuclei 32(558)
— deuteron 32(601)
— magnetic 32(624)
— proton 32(598)
— sign 32(597)
— standard 32(597)
Scattering amplitude of an atom 32(469)
Scattering and energy loss in the object 33(492)
Scattering angle of fast electrons, maximum and minimum 34(91, 116, 117)
Scattering cross section, differential
— for plane polarized radiation 34(247–249)
— for unpolarized radiation 34(257–259)
Scattering cross section for neutrons 32(641)
— ortho- and parahydrogen 32(598)
Scattering cross section for photons
— average 34(264)
— coherent 34(292)
— integral between arbitrary angular limits 34(258–261)
Scattering cross section for slow electrons
— differential 34(23)
— total 34(21, 23)
Scattering cross section for slow ions 34(40)
Scattering, elastic 33(495, 506)
Scattering factor for electrons 32(156, 218)
Scattering function
— definition 32(330, 331)
— for cylinders 32(335–340)
— for ellipsoids of revolution 32(333–335, 355)
— for rectangular prisms 32(340)
— for spheres 32(331, 332)
Scattering, inelastic 33(496, 506)
Scattering length 32(464, 493)
Scattering of fast electrons
— at free electrons 34(55–60)
— at nuclei 34(101–120)
— in the plasma 34(75–86)
Scattering of neutrons
— by a bound nucleus 32(557)
— by a free neucleus 32(555)
— by a unit cell of a crystal 32(561)
— coherent 32(565)
— coherent elastic 32(571)
— coherent inelastic 32(573)
— incoherent 32(564, 567, 605)

— incoherent elastic 32(572)
— incoherent inelastic 32(572)
— inelastic 32(570, 622)
— magnetic 32(623)
— paramagnetic 32(625)
Scattering of photons, coherent, *see* Coherent photon scattering
Scattering of photons, incoherent, *see* Compton scattering
Scattering of positrons at electrons 34(59, 60)
Scattering of x-rays 30(88, 347); 32(97)
— by an atom, dispersion terms 32(160, 161, 163, 164)
— by an atom, Schrödinger equation and solution 32(157ff.)
— by identical particles 32(340–342, 353–358)
— due to disorder 32(402, 403)
— dynamical theory 32(566)
— from a polydisperse system 32(342–346, 358–359)
Scherrer constant 32(395)
Schneide method (Seemann) 30(101)
Schoenflies notation 32(140)
Schrödinger equation 33(318)
Schrödinger equation, paraxial 33(325)
Schumann plates 33(597)
Scintillation counter(s) 30(137); 32(585); 33(654)
Scintillation counter measurements of x-ray yields 34(179, 182)
Scintillation crystals, organic 33(670)
Scintillation detector, NaI(Tl) 33(669)
Scintillation spectrometer 33(654)
Screening constants 30(81)
Screening doublets 30(84, 203–205)
Screening effect 34(172–177)
Screening effect in plasma 34(73, 75, 76)
Screening numbers 34(173, 176)
Screw 32(131, 134)
Secondary electrons 33(558)
Secondary maxima 32(472, 477, 491)
Secondary maxima, distance in a convergent beam after dynamical theory 32(481)
Secondary radiation tube 30(132, 134)
Secondary structure 30(296)
Sector field
— cut-off 33(316)
— dispersion formulas of the magnetic 33(552)
— enlargement at the magnetic 33(553)
— general 33(301)
— image curve of the magnetic 33(551)
— magnetic 33(551)
— second order aberration of the magnetic 33(554)
Seidel image aberrations 33(219)
Selection rules for annihilation of positronium 34(140)
Selection rules in photoionization from solids 31(448)
Selective filter 30(12)
Selector-diaphragm 33(527)
Self-absorption of HeI resonance radiation 31(235, 236)
Self-consistent-field method 31(286, 510, 574ff.)
Semicircular focussing 33(610)
Semicircular spectrometers in coincidence 33(673)
Semicircular spectroscope 33(615)
Semi-classical harmonic oscillar description 31(418)
Sensitivity 33(429, 573)
Separation, statistical 33(667)
Separations in small crystals 32(516)
Serum Albumin 32(378)
SF_6 31(397)
Shade microscopy with x-rays 30(36, 43)
Shadow images 30(311)
Shadow microscope 33(445)
Shadow microscope, resolution of the 33(445)
Shadow projection microscopy 30(315f.)
Shake off 31(292)
Shake up 31(266, 292)
— spectrum benzene 31(359)
— spectrum of CO 31(359)
— spectrum of N_2 31(357)
Shape 33(516)
Shape function 32(172); 33(405)
Shape of a particle 32(322f., 377)
Shape of metal core electron line 31(365)
Shape parameter 33(405, 474)
Shape resonance 31(397, 428)

Shell 31(498, 499)
Shielding factor 33(480f.)
Shielding of nucleus 34(102)
Short range order 32(412, 523)
Short wavelength limit 30(338, 346, 347, 355, 360, 362, 363, 365)
— fine structure 30(368f.)
— polarization 30(352)
Short-lived species 31(440)
Siegbahn's mounting 33(619)
SiF_4 31(349)
σ-band 30(276)
Signal to background ratio in electron spectra 31(232)
SiH_4 31(407)
Similarity theorems 33(130)
Sine condition, deviation from the 33(364)
Single crystal 30(146, 147)
Single crystal diffraction method 32(592)
Single nuclear scattering 34(101–116, 125)
Single scattering model for ESCA diffraction 31(385)
Single-configuration *jj*-average energy 31(541)
Singlet state of positronium (*see also* Parapositronium) 34(139, 153)
Singularity index 31(365)
SiO 31(441)
Size distribution 32(344, 384)
Size of a particle 32(322, 344, 384, 422)
Size of a particle in solutions 32(355–357)
Slater determinant 31(284, 501)
Slater integral 31(503)
Slit collimation 32(362f.)
Slit corrections 32(348–353)
Slit height 32(352)
Slow electrons
— absorption in gases 34(3, 4)
— collision cross section 34(3, 6–11)
— definition 34(1)
— elastic scattering 34(11, 13, 17–23)
— scattering cross section 34(21–23)
Slow ions
— charge exchange 34(28, 29, 41–45, 48)
— collision cross section 34(32–36)
— definition 34(1)
— elastic scattering 34(28, 36–41)
— inelastic scattering 34(29)
— scattering cross section 34(40)
— stopping cross section for 34(214)
Slowing-down of positrons in liquids and solids 34(158)
Sm 31(309, 461)
Small angle scattering
— of neutrons 32(619)
— of x-rays 32(322f.)
Sn 31(309)
Snell's law of refraction 32(184)
SnH_4 31(407)
SO 31(442, 447)
Soft x-ray emission
— spectra 31(371)
— spectra of Co and N_2 31(373)
Solenoidal spectrometer 33(629)
Solid finely divided 32(384–386)
Solid sample handling in photoelectron spectroscopy 31(262)
Solid state x-ray spectra 30(230)
Soller slit system 32(588)
Solutions, concentrated 32(382)
Solutions, dilute 32(372–382)
Solutions, particle size in 32(355–357)
Sommerfeld model 30(250)
So^+ 31(415)
Space charge
— beam spreading in field free space 33(383)
— by carriers, compensation of 33(28)
— in a homogeneous magnetic field 33(387)
— of the electron beam 33(380)
— outside 33(379)
Space focussing 33(645)
Space group 32(139)
Space groups, 230 different 32(146)
Spatial distribution, *see* Angular distribution
Specimen holder 32(446)
Specimen region, small irradiated 32(455)
Spectra, resonance lines 31(139)
Spectral series in x-rays (*see also* *K*-, *L*-, etc. Series) 30(79)
Spectrograph 31(143)
Spectrograph (*see also* Spectrometer) 33(642, 644)
Spectrograph, image rotation 33(639)
Spectrographs for isotopic mass determinations 33(606)

Spectrometer (*see also* Spectrograph) 33(610, 620, 642, 644)
Spectrometer, double coincidence 33(669, 671)
Spectrometer, double focussing 33(611, 621)
Spectrometer, long lens 33(639, 676)
Spectrometer, placed "end to end", two short lens 33(676)
Spectrometer, proportional counter 33(654)
Spectrometer, short lens 33(636)
Spectrometer, single coincidence 33(669)
Spectrometer, third order focussing 33(610, 620)
Spectrometer with two crystals, *see* Double crystal spectrometer
Spectrometry, double channel coincidence 33(669)
Spectrometry, single channel coincidence 33(669)
Spectroscope, electrostatic field 33(644)
Spectroscope, flat 33(609)
Spectroscope, momentum selective 33(612)
Spectroscope, resolution of the 33(613)
Spectroscopy, beta- and gamma-ray 33(609)
Spencer-Fano theory of electron flux 34(121–124)
Sphere of reflection 32(171, 175)
Spherical analyzer (180°) 31(155, 245)
Spherical condensor 33(590)
Spherical harmonics 31(496, 497)
Spherical wave 32(155)
Spike of Laue 32(489, 491, 493, 512)
Spike-length 32(492)
Spin and magnetic dipole moment of the electron 34(249, 257)
Spin doublets 30(83, 84)
Spin doublets, relativistic 30(202, 203)
Spin flip scattering 32(565)
Spin orbital 31(498, 501)
SPINDO, spectroscopic-potentials-adjusted INDO method 31(436)
Spinel, neutron diffraction 32(615, 637)
Spinel-twins 32(509)
Spin-orbit coupling parameter 31(404)
Spin-orbit interaction 31(389, 444, 498, 499, 506, 562, 572)
Spin-orbit splitting 31(403)
Spiral orbit spectrometer 33(628)
Splitting of energy levels of positronium 34(153, 155)
Spreading of the beam 33(380)
Stability of positronium in gases 34(145, 151)
Stacking disorder, Fourier analysis 32(410)
Stacking faults 32(409, 425, 512)
Standard line 33(647, 649)
State, amorphous 32(525)
State, finest crystalline 32(525, 530)
State, solid-amorphous 32(525)
Stationary scattering wave function 31(472–475)
Statistical method 32(226)
Statistical weights 31(184, 185)
Statistical weights, Xe 31(185)
Stereomicroscopy 33(519)
Stereoscopic observation 30(312, 322)
Sternheimer theory of density effect 34(67–71, 101)
Stigmator 33(401, 409)
Stigmator, adjustment of the 33(472)
Stopping cross section
— and range 34(206)
— definition 34(194)
— energy dependence 34(195)
— for α-particles 34(196–200, 204)
— for heavy ions 34(205)
— for protons 34(196–203)
— for protons in metals 34(200, 202)
— for slow ions 34(214)
— molecular, of compounds 34(201, 203, 215)
Stopping electrons (*see also* Delta rays) 34(187, 188, 189)
Stopping number 34(195)
Stopping power
— non relativistic 34(62)
— of conduction electrons 34(71–74)
— of matter for electrons 34(61–64, 126)
— of matter for positrons 34(63)
— relativistic 34(63, 121–126, 128)
— temperature dependence 34(72)
Straggling, calculation by Monte Carlo method 34(93)

Straggling distribution for electrons 34(87–101)
Straggling distribution for positrons 34(90)
Strain field around a point defect 32(417)
Stray field(s) 33(289, 404, 407, 487)
— of magnets 33(564)
— of the sector field 33(577)
— screening 33(597)
Strength flow, closed expression for the 33(160)
Strength of a transition 31(524, 548)
Strong focussing 33(270, 311)
Strontium-method 33(585)
Structure, acentric 32(230)
Structure, centric 32(230)
Structure, centrosymmetric 32(219)
Structure factor 32(462, 469)
— for hollow spheres 32(326)
— for neutrons 32(562)
— for spheres with radially periodic density 32(326)
— for x-rays 32(104, 119, 124, 218)
— of a single particle 32(323–325)
— unitary 32(220, 234)
Structure, non-centrosymmetric 32(219)
Subshell 31(499)
Substituent effect on the vibrational structure 31(417)
Sub-structure in metals 32(427–435)
Sudden approximation 31(292)
Sulphide 31(438)
Sulphoxides 31(438)
Sulphur-containing molecules 31(342)
Sum rule for satellite intensities 31(293)
Super Coster-Kronig transition 31(288, 501)
— amplitudes 31(518, 538, 581 ff.)
— component rate 31(519)
— energy 31(501, 566 ff.)
— group rate 31(519, 520)
— in Xe 31(302)
— intensity 31(523, 524)
— multiplet rate 31(519)
— term rate 31(519)
Superconductors, neutron diffraction 32(637)
Superlattice 32(412, 512, 632)
Superlattice lines 32(412)
Superlattice reflections 32(513)
Supersonic jet technique 31(258)
Surface area of a particle 32(356)
Surface film 32(499)
— detection depending on roughness 32(499)
— thickness 32(500)
Surface, flat 32(494)
Surface layer, destroyed 32(526)
Surface, mechanically polished 32(528)
Surface mobility 32(540)
Surface, penetration depth into rough 32(498)
Surface reflection 32(493, 495)
Surface, rough 32(497)
s-wave 32(557)
Symmetry 32(98, 129)
Symmetry coordinates 31(419)
Symmetry elements 32(134, 231)
Symmetry species of vibronic levels in linear molecules 31(411)
Synchrotron radiation 31(136, 240)
— in excitation of electron spectra 31(240)
— universal curve 31(241)

T matrix, nonresonant 31(479, 480)
T matrix, resonant 31(482, 485, 486)
Tangent image formation 33(188)
Target current, total 30(316)
t-butyl 31(441)
Te_2^+ 31(404)
Teflon 33(556)
TeH_2 31(393)
Telefocus cathode 33(15, 20)
Teller-Redlich product rule 31(418)
Temperature, characteristic 32(217)
Temperature factor 32(210, 212, 215, 218, 219)
— artificial 32(234)
— for electron interferences 32(517)
— of neutron diffraction samples 32(595)
Temperature variation technique 31(412)
Temperature voltage 30(23, 24)
Term values 30(303)
Test-preparation 33(515)
Thallium halides 31(444, 445)
Theoretical models for photoelectron spectra 31(283)

Thermal motion (disorder) 32(207, 212, 400, 610)
Thermionic electron sources 30(317)
Thermodynamic equations 32(341)
Thermodynamic model 31(353)
Thermodynamic model of chemical shift 31(351, 354)
Thick target x-ray spectrum 30(339, 343)
— obtained from thin targets 30(350)
Thin crystal plates, diffraction 32(475)
Thin films, structure of 32(532)
Thin foils 30(350, 359)
Thin metal target 30(370)
Thin target, continuous x-ray spectrum 30(344, 348, 349f., 352, 361, 363)
Thin target results 30(351, 360)
Thin-section 33(516)
Thionanisole 31(438)
Thiophene 31(430)
Thomas-Fermi atom 34(73, 90, 102, 117)
Thomas-Reiche-Kuhn sum rule 31(34)
Thomson factor 32(156)
Thomson scattering 34(219–222)
— angular distribution 34(220, 221)
— deviations from 34(222)
— differential cross section 34(228)
Thorium deposit 33(653)
Three-beam problem 32(483, 484)
Three-photon annihilation of positronium 34(142)
— angular correlation 34(143, 150)
— polarization correlation 34(143, 150)
Three-step model 31(449)
Threshold law for multiple ionization 31(120)
Threshold maximum in σ 31(47ff.)
Threshold photoelectron spectroscopy 31(415)
Through-bond interaction 31(435)
Through-space interaction 31(435)
Time of exposure 30(15, 41, 57)
Time-of-flight electron spectrometer 31(252)
Time-of-flight spectrometer 32(580)
Time-of-flight technique 32(577)
Tolerance dose 30(75)
Toroidal condenser electron analyzer 31(250)
Toroidal condensor 33(591)
Total activity 33(615)
Total intensity of x-rays 30(346, 347)
— dependence on atomic number 30(342)
Total reflection 30(265, 324, 331); 32(194, 485)
Totalix tube 30(38)
Townsend method 34(9, 10)
Transformations of lattices 32(112)
Transient species 31(440)
Transition metal, photoionization cross section of 3*p*-shell 31(54)
Transition metals 30(285)
Transition method for calculation of SCF binding energies 31(286)
Transition operator method 31(346, 512, 575)
Transition potential model 31(348, 349)
Transitions, allowed 30(82, 83)
Transitions, forbidden 30(82, 83)
Translation 32(130, 139)
Transmission 33(614)
Transmission cross sections 32(621)
Transmission measurements 32(619)
Transmission of neutrons through two magnetized sample(s)
— double transmission 32(628)
— single transmission 32(627)
Transmission spectrograph 30(116)
Transverse wave 32(216)
Triangular field 33(635, 676)
Triode and tetrode systems 33(15)
Triple coincidence method 34(149)
Triplet state of positronium (*see also* Orthopositronium) 34(139, 153)
Trithiapentalene 31(370, 371)
Trochoidal mass spectrometer 33(594)
Tube spectrometer of Siegbahn and Larsson 30(99)
Tube voltage, dependence of x-rays 30(341)
Tubes for soft x-ray studies 30(262)
Tungsten anode 30(15, 318)
Turn anode 30(15, 35, 38–42, 52, 54, 57)
— after Bouwers 30(39)
— after Stintzing 30(40)
— after Thomson 30(35)
— after Ungelenk 30(35, 40)
— cooling 30(39, 40, 41, 42)

Twinning 32(508, 534)
Twinning in thin films 32(509)
Twins 32(532)
Twisted baffle system 33(633, 637)
Two wavelength microscopy 33(347, 489)
Two-beam problem 32(478, 485)
Two-center-problem, complex 33(146)
Two-dimensional band structures 31(450)
Two-dimensional systems 33(268)
Two-electron wave function 31(508)
Two-photon annihilation of positronium 34(140, 141, 143, 161, 162)

UF_6 31(393, 394)
UF_6^+ 31(403)
Undercorrection 33(227)
Unit of x-ray dose 30(74)
Units of wavelengths (Å and XU) 30(92, 93, 239, 240)
UV line source 31(236)
UV polarizer 31(239)
UV radiation source 31(234, 238, 240)

Va, photoionization cross section of 3*p*-shell 31(53)
Vacant lattice sites 32(415)
Vacuum spectographs for x-rays 30(97, 100, 107)
Valence bands
— alloy 31(461)
— in hydrocarbon 31(465)
— metal 31(459)
— of Au, Ag and Cu 31(460)
Valence photoelectron spectrum 31(387)
— of alkanes 31(466)
— of benzene 31(387)
Vapour pressure 33(42)
Variation of time 33(138)
Variational collapse 31(557ff.)
Vector set 32(219, 234)
Vector space 32(120, 126)
Velocity analyzer 33(535)
Velocity dispersion 33(302, 314, 552, 580)
Velocity filter 33(542)
Velocity focussing 33(549, 577)
Velocity focussing curve 33(578)
Velocity form for the transition matrix element 31(23ff.)
Velocity of propagation 32(208)
Vibration anode of an x-ray tube 30(42)
Vibrational coupling constants 31(421)
Vibrational energies of benzene 31(412)
Vibrational excitations 31(294, 338)
Vibrational excitations in core electron spectra 31(366)
Vibrational overlap integral 31(419)
Vibrational selection rules 31(409)
— calculation of 31(418)
Vibrational structure and orbital shape 31(412)
Vibronic interaction, type
— *a* 31(410, 428)
— *b* 31(410)
Villard-circuit 30(46, 47)
Virtual bound states in alloys 31(464)
Virus solutions 32(380–382)
Viruses, spherical 32(380, 381)
Voigt function 31(279)
Voltage, *see* Tube voltage
Volume of a cell 32(108)

W values for different gases 34(64, 129)
Wall charges 30(32, 33)
Wave function in field-free space 33(360)
Wave length of the neutron 32(553)
Wave mechanics 33(320)
Wave propagation in field-free space 33(330)
Wave surfaces 33(235)
Wave vector 32(166)
Wavelength limit (*see also* Short wavelength limit) 30(4, 5, 338, 346, 347, 355, 360, 362, 363, 365f.)
Wavelength measurements, absolute 30(125, 160)
Wavelength measurements, relative, by bent crystal method 30(122, 160)
Wavelength precision measurements 30(156–162)
— chemical effects 30(158–160)
Wavelengths of absorption edges (table) 30(303)
Wavelengths of emission edges (table) 30(301)
Wavelengths of emission lines (experimental data) 30(162–181)

Weisskopf-Wigner approximation 31(489)
Wentzel ansatz 31(516, 527, 528)
Wick theory of density effect 34(67)
Width (*see also* Half-width)
Width, empirical 30(291)
Width, nonradiative 31(494)
Width of an emission band 30(289, 302)
Width of bands (table) 30(302)
Width of field 30(316, 327, 328)
Width of photoelectron line 31(278)
Width of plasma resonance 34(85)
Width of the Compton shifted line 34(285, 286)
Width of x-ray lines 30(147, 232, 233)
Width, radiative 31(494)
Width, reduced 30(290)
Wien's velocity filter 33(585)
— with non-uniform magnetic field 33(593)
Window curve 33(613, 652)
W.K.B. method 34(102)
Work function 30(23–26); 33(3)
Work function correction 30(366, 367)

X unit (XU), in crystal scale 30(92, 93, 239)
Xe 31(126)
Xe, absorption coefficient 31(167)
Xe, β-parameter 31(94, 103, 109ff., 201ff.)
Xe, double ionization cross section 31(199)
Xe, partial cross section of $5s$-shell 31(83, 188)
Xe, photoionization cross section 31(164)
Xe, $4s$–$4p$ photoelectron spectrum 31(304)
Xe, super Coster-Kronig transitions 31(302)
XeF_2 31(444)
XeF_4 31(444)
X-ray absorption edge, fine structure 34(87)
X-ray flash interferences 30(67, 68)
X-ray flash time 30(62, 67)
X-ray flash tube 30(62–69)
X-ray monochromatization 31(230, 232)
X-ray monochromator 31(231)
X-ray scattering (*see also* Scattering) 32(152, 157ff.)
X-ray source 31(229, 230)
X-ray spark lines 30(56)
X-ray spectrometer 32(102)
X-rays, characteristic, *see* Characteristic x-rays
X-rays, continuous 34(190–192)
X-rays, discovery 30(1, 2)
X-rays, hard 30(18–20, 47–51)
X-rays of long wavelength 30(51, 55, 56)
X-rays, soft 30(51–55)
X-rays, ultra soft 30(55, 56)
X_α method 31(286)
X_α method, nonrelativistic 31(575)
X_α method, relativistic 31(575)

Yb 31(309)

Zeeman effect of positronium 34(153–158)
Zernike-Prins type equation 32(341)
Zn 31(152, 176)
Zone 32(116)
Zone axis 32(98)

Group 7

Atomic and Molecular Physics

Absorption bands 37/1(238)
Absorption cross-section 35(382)
Absorption of polyenes in the ultraviolet 37/2(216)
Absorption of waves within the plasma 37/1(348)
Absorption shift between even and odd-atom chains 37/2(216, 219)
Absorption spectra 35(341)
Abzählschema 36(68, 70, 81, 91)
Acceptor property of aza nitrogen 37/2(237)
Accidental degeneracy 37/2(13)
Accommodation coefficient 36(286)
Acetylene, infrared spectrum 37/1(248, 258, 278)
Acetylene, σ-bond 37/2(271)
Action variable 36(4)
Additivity of bonds 37/2(143)
Additivity of saturated bond properties 37/2(196)
Adiabatic approximation 37/2(1)
Adiabatic binding energy 37/2(128)
Adiabatic energy surface 37/2(2, 20)
Adiabatic transformation within rare-gas-like sequence 37/2(255)
Alcohols, ionization by electron impact 37/1(384–387)
Algebra of observables 37/1(178)
Aliphatic chains 37/2(183)
Alkali atoms 35(171); 36(28, 29)
Alkali atoms in atomic beams 37/1(6, 17, 29, 33, 67, 71, 73, 76, 97, 120)
Alkali borates in molecular beams 37/1(7)
Alkali cyanides in molecular beams 37/1(7)
Alkali halides in molecular beams 37/1(6, 8, 23, 30, 48, 124)
Alkali spectra 35(378)
Alkaline earth atoms 36(29)
Allene 37/2(276)
Allene infrared spectrum 37/1(299)
α-particles in helium, energy loss per ion pair 36(365)
Alternant molecules 37/2(193–194)
Alternant orbitals 37/2(150, 196–198, 200)
— in aromatic systems 37/2(230–233)
— supernumerary 37/2(202)
Alternating bonds 37/2(182)
Alternative resonance structures 37/2(206)
Aluminum in atomic beams 37/1(9, 29, 87, 100, 109)
Amino group, directing effect 37/2(237)
Amino-naphtalene 37/2(193)
Amino-quinoline 37/2(235)
Ammonia, infrared spectrum 37/1(256, 310)
Ammonia, ionization by electron impact 37/1(378)
Ammonia molecule, bond angles 37/2(248, 252, 259)
Ammonia molecule, LCAO approximation 37/2(69, 71)
Ammonium ion as perturbed neon system 37/2(261)
Ammonium ion, tetrahedral form 37/2(252–253)
Ammonium ion, twisting vibrations 37/2(262)
Angular correlation 37/2(36)
Angular distribution
— at large angles of inelastically scattered electrons 36(394)
— for electrons resulting from ionizing collisions 36(366)
— of Bremsstrahlung electrons 35(415)

Angular distribution (cont.)
— of electrons 36(345, 346, 347, 356, 357)
— of inelastically scattered electrons 36(358, 362, 363)
— of photons 35(417)
— of protons in polyhydrides 37/2(251–254)
— without retardation 35(394)
Angular frequency 35(334)
Angular momenta, addition 37/1(189–191)
Angular momentum 36(8, 75)
— commutation relations 37/1(185, 187, 224)
— distribution 36(180)
— in body-fixed coordinates 37/1(187)
— internal 37/1(227f.)
— internal, of nuclei 37/1(173)
— matrix elements 35(138, 434)
— matrix elements 37/1(186–189)
— of a molecule 37/1(174)
— of electrons 37/1(173)
— of molecular frame work 37/1(174)
— operator 35(138, 434)
— quantum number 36(26, 75)
— quantum theoretical description 37/1(184–191)
— spin contribution 37/1(173)
— total 37/1(173)
— vibrational 37/1(174, 214)
Anharmonic constants 37/1(174, 248, 250)
Anharmonic resonance of first order 37/1(282–288)
Anharmonic resonance of second order 37/1(290–291)
Anharmonic resonance of third and higher orders 37/1(291–292)
Anharmonic vibration 37/1(208)
Aniline 37/2(201, 238)
Annihilation operator 36(79)
Anomalous intensities in infrared bands 37/1(305–311)
Anomalous moment of the electron 35(179, 304)
Anomalous spin magnetic moment of the electron 37/1(81, 113)
Anthracene 37/2(207, 224, 225)
Anthracene, negative ion 37/2(237)
Anthracene, σ-electron box model 37/2(267)
Antibonding orbital 37/2(65, 196–198)
Antisymmetric operation 37/1(204, 206)
Antisymmetric wave function 37/2(40)
Antisymmetrizing operator 37/2(120, 161)
Appearance potential (or energy) 37/1(358, 365, 368)
Appearance potentials of ions
— from carbon tetrahalides 37/1(396–397)
— from cyanogen 37/1(389)
— from germanium tetrahalides 37/1(404)
— from silicon tetrahalides 37/1(404)
Approximation 36(18, 105)
Argon configuration 37/2(251)
Argon in atomic beams 37/1(14, 31, 118)
Aromatic negative ions 37/2(236)
Aromatic ring systems 37/2(181–183, 193, 198)
— box models 37/2(222–226)
Arsenic in atomic beams 37/1(101)
Artificial units 35(1)
A's, *see* Alternant molecules
AsH_3, infrared spectrum 37/1(256, 301, 304)
Asymmetric molecules, spectra 37/1(262–273)
Asymmetric rotator, *see* Rotator
Asymptotic hydrogen wave-function 35(108)
Asymptotic wave function 35(132, 242)
Atom in an external field 36(53)
Atom model 36(11, 13)
Atom with several electrons 36(80)
Atomic beam
— apparatus 37/1(53)
— method 36(311, 340)
— resonance method 35(310)
— spectroscopy 37/1(80ff.)
Atomic constants 35(2)
— history 35(6)
— in the period 1929–1949 35(17)
Atomic form factor 35(411); 36(193)
Atomic frequency standard 37/1(109–112)
Atomic *g*-values (tables) 37/1(115–120)
Atomic magnetism 37/1(112–123)
Atomic masses 37/1(145–148)
Atomic orbitals, coefficients in molecular orbitals 37/2(100–104)

Atomic population 37/2(84–86, 98)
Atomic populations of heteronuclear diatomic molecules 37/2(85)
Atomic radii 36(187)
Atomic spectra 36(184)
Atomic units 36(10); 37/2(6)
Atomic units (a.u.) 35(88)
Atomic wave functions, deformation in molecules 37/2(163, 165)
Atomic weights, chemical scale 35(14)
Atomic weights, physical scale 35(14)
Atoms in molecules, approximation 37/2(175)
Atoms in molecules, method of Moffitt 37/2(160)
Attachment of an electron 37/1(320)
Attractive potential energy curve 37/1(315, 316)
A-type bands 37/1(264, 265, 268–270)
Auger effect 35(211, 381)
Auger ejection of an electron 37/1(320)
Auger transitions 36(317)
Auger type process 37/1(319–320)
Aurora 36(400, 406)
Aurorae in high latitudes 36(408)
Auroral nitrogen spectrum 37/1(335–336)
Auroral spectrum 36(407)
Autoionization 36(300, 301, 312, 317, 349, 361)
Autoionization, inverse 36(301)
Auxiliary constants, 1955 adjustment 35(82)
Avogadro's number 35(14, 16)
Axial correlation 37/2(35)
Axially symmetric molecule 37/1(210–211)
— fundamental vibrations 37/1(257)
— spectra 37/1(252–257, 297)
Axially symmetric rotator, *see* Rotator
Axiom of maximum likelihood 35(53)
Aza nitrogen substitution
— in aromatic systems 37/2(235, 237)
— in odd-atom chain 37/2(219)
Azulene 37/2(193, 198, 226)

Back diffusion 36(318)
Balmer formula 35(94)
Balmer series 35(96)
Balmer series, radial integrals and transition probabilities 35(348)
Banana bonds 37/2(271)
Band head 37/1(174)
Band heads, progression 37/1(176)
Band spectrum 37/1(174)
Band structure of PH_3 (due to quantum number *K*) 37/1(253)
Band systems 37/1(176)
Bands of types *A*, *B* and *C* 37/1(264–267)
Barium in atomic beams 37/1(7, 30)
Barrier penetration factor 35(324)
Barrier to free rotation 37/2(269)
Barrier to twist 37/2(204, 205)
Base strength of aza nitrogens 37/2(235)
Basicity and excess charge density 37/2(236)
Beam deflection 37/1(35–53)
Beam detectors 37/1(25–35)
Beam geometry 37/1(18)
Beam intensity 37/1(19)
Beam spectrometer 37/1(2, 3)
Bending force constant 37/2(260)
Bent σ-bonds 37/2(270–272)
Benzaldehyde 37/2(237)
Benzene 37/2(182, 192, 193, 198)
Benzene, alternant carbons 37/2(150)
Benzene, density surface 37/2(223)
Benzene, LCAO approximation 37/2(70, 72)
Benzene, π-electron system 37/2(137)
Benzene, polar structures 37/2(137)
Benzene, resonance structure 37/2(129)
Benzene, σ-bonds 37/2(262)
Benzyl carbonium ion 37/2(277)
Benzyl cation 37/2(201, 207, 237)
Bethe-Heitler formula 35(422)
Bethe-Heitler formula, Coulomb corrections 35(428)
Bethe's approximation method 36(355, 357, 392)
Bethe's formula 36(364)
Bethe-Salpeter wave equation 35(282)
B-H bond 37/2(274, 275)
Binding distances in other molecules than hydrogen 37/2(66)
Binding energy 37/2(4, 97)
— adiabatic 37/2(128)
— of hydrogen molecule 37/2(21)
— values obtained by ICC 37/2(168)
Biphenyl 37/2(223, 224)
Biphenyl, negative ion 37/2(237)

Biphenylene 37/2(199)
Birge's method of evaluating data on atomic constants 35(18)
Bismuth in atomic beams 37/1(103)
Bjerrum's intensity rule in band spectra 37/1(175, 176)
Blatt-Jackson formula 36(266)
Body-fixed coordinates 37/1(173, 187, 192ff.)
Bohr energy relation 35(334)
Bohr magneton 35(142, 292)
Bohr radius 35(89); 37/2(5)
Boltzmann distribution 36(399)
Bond additivity 37/2(143)
Bond angle 37/2(175)
Bond angles, prediction by electrostatic method 37/2(258)
Bond densities in aromatic skeleton models 37/2(222)
Bond diagrams 37/2(131, 137)
Bond energies 37/2(142, 176)
Bond integral 37/2(191)
Bond lengths
— alternation 37/2(215, 216)
— and frequency, relationship 37/2(254–255)
— by Thomas-Fermi method 37/2(257)
— for polyhydrides, table 37/2(255)
— from benzene box model 37/2(267)
— of double bond 37/2(182)
— variations 37/2(192)
Bond orders 37/2(205, 207–209, 214)
Bonding electrons 37/2(65)
Bonding orbital 37/2(65)
Bonding power of a molecular orbital 37/2(81)
Born approximation 35(121, 131, 133, 385, 410); 36(232, 238, 246, 258, 270, 274, 278, 281, 282, 286, 293, 294, 295, 302, 306, 350, 351, 356, 361, 364); 37/1(321, 323, 324, 333–334, 340, 353–356)
Born approximation
— deviations 35(401, 427)
— first order 35(387)
— relativistic 35(398)
— zero order 35(387)
Born-Oppenheimer approximation 36(283, 369, 380); 37/1(340); 37/2(1)
Born's approximation
— and observed data for optically allowed excitations 36(367)
— application to complex atoms 36(360)
— first 36(239, 263, 264, 274, 294, 302)
— for the partial cross section 36(392)
— second 36(264, 294, 306, 393)
Boron hydrides (*see also* Diborane) 37/2(227, 273–275)
Boron in atomic beams 37/1(9, 33, 87, 100, 109)
Boron trichloride, ionization by electron impact 37/1(380–381)
Boron trifluoride, ionization by electron impact 37/1(378–379)
Bound states, number 36(250)
Boundary density 36(123, 140)
Boundary radius 36(123, 124, 126, 140, 148)
Bound-free transitions 35(381)
Bowen mechanism 36(406)
Box model(s) 37/2(184, 187, 192)
— and network, equivalence 37/2(192–193)
— for benzene σ-electrons 37/2(262–263)
— of aromatic systems 37/2(222–226)
Branch model of aromatic systems 37/2(221, 232)
Branch point 37/2(221)
Branch points in a network 37/2(188, 190, 192)
Branching relations 37/2(188, 209)
Breit equation 35(256, 257)
— reduced, in momentum space 35(264)
— reduced, in position space 35(267)
— without external field 35(278)
Breit operator 35(259)
Breit-Margenau corrections 35(308)
Breit-Rabi formula 37/1(84)
Bremsstrahlung, average energy loss due to 35(416, 426)
Bremsstrahlung, cross-section without retardation 35(413)
Bremsstrahlung electrons, angular distribution 35(415)
Bremsstrahlung, Elwert's expression 35(418)

Bremsstrahlung in the field of an electron 35(428)
Bremsstrahlung, polarization 35(414)
Bremsstrahlung, Sommerfeld differential cross-section 35(417)
Brillouin gap 37/2(216, 217, 220, 240)
Brillouin zone 37/2(183, 195, 215–217)
Brillouin's theorem 37/2(89)
Bromine in atomic beams 37/1(12, 30, 87, 101, 109, 118, 147)
B-type bands 37/1(264, 266, 271–272)
Butadiene 37/2(216)

Cadmium, ionization cross section 36(317)
Canonical tranformation, *see* Contact transformation
Capture of electrons from hydrogen atoms
— by positrons 36(284)
— by protons 36(283)
Carbon atoms as branch points 37/2(191–192, 199)
Carbon bonds, *see* C–C and C–H bonds
Carbon dioxide, infrared spectrum 37/1(259–260, 288–290, 309)
Carbon dioxide, LCAO self-consistent calculation 37/2(145)
Carbon monoxide, ionization by electron impact 37/1(365, 369–371)
Carbon tetrachloride, ionization by electron impact 37/1(359, 396–403)
Carbon tetrahalides, ionization by electron impact 37/1(396–403)
Carbon tetrahalides, scattering of electrons 37/1(322–323)
Carbon-halogen bonds 37/1(391, 396, 399)
Carbonyl group, directing effect 37/2(237)
Carotene 37/2(216)
Cascade transitions 35(354)
Cathode dark space 36(397)
Causality conditions 36(247)
Causality requirements 36(249)
Cavity resonator 37/1(76)
C–C bond in cyanogen 37/1(390)
C-C distance 37/2(267, 271)
C-C vibration frequency 37/2(216, 220)
Central field 36(21, 23)
Central field approximations 35(171)
Central potential 35(118)
Centrifugal distortion of a molecule 37/1(249, 264, 277, 288)
Centrifugal force 37/2(3)
Centrifugal potential 37/2(3)
Centrifugal stretching coefficients 37/1(174, 263–264, 279)
Cesium bromide in molecular beams 37/1(25, 148)
Cesium chloride in molecular beams 37/1(25, 47, 148, 151)
Cesium fluoride in molecular beams 37/1(25, 47, 148, 151)
Cesium frequency standard 37/1(110)
Cesium in atomic beams 37/1(7, 17, 29, 34, 79, 81, 102, 107, 109, 119)
Cesium iodide in molecular beams 37/1(47, 148)
C-H distance 37/2(267)
Character of representation 37/2(15–16)
Character tables of symmetry groups 37/2(169–171)
Charge defect 37/2(206, 232)
Charge distribution 35(231)
Charge excess 37/2(206, 218, 232)
Charge per unit length 37/2(190)
Charge transfer 36(287, 300)
Charge-to-mass ratio of the proton 35(33)
C–H-bond in prussic acid 37/1(390)
Chemical detection of a beam 37/1(34)
Chemical periodicity 37/2(174)
Chemical predictions, accuracy 37/2(174, 175)
Chemical structures 37/2(176)
Chemical symbolism and alternating bond lengths 37/2(215)
Chlorine in atomic beams 37/1(12, 15, 30, 87, 100, 107, 109, 147)
Chlorine molecules, electron scattering 37/1(330)
Chromium in atomic beams 37/1(9, 33, 118)
CI (correlation interaction) 37/2(96)
Circular box model 37/2(228, 229)
— of benzene σ-electrons 37/2(263, 265)
Cis and trans forms 37/2(204–205)
Class of conjugate operations 37/2(7)

Close coupling through exchange 36(380)
Closed shell of electrons 37/2(41, 42, 173, 174)
Closed-shell configuration 37/2(139)
Closure of a shell 36(30, 33, 56)
Collimating slit 37/1(18, 35)
Collimator 37/1(2, 3)
Collision broadening in beam spectroscopy 37/1(3, 16, 54)
Collision deactivation probability 36(404)
Collision strength 36(383, 389, 403, 404, 406)
Collision transition probability 36(404)
Collisions, effect on the 2S-state 35(372)
Combination principle of Rydberg and Ritz 36(2)
Commutation of observables 37/1(178, 181)
Commutation relations 35(133, 138); 37/1(179, 213)
Commutation relations of angular momenta 37/1(185, 187, 224)
Commutation rules 36(8)
Compressibility 36(208, 210)
Compressibility, dependence on pressure 36(211)
Compton effect 35(400)
Compton line, half-width 36(196)
Compton line, intensity distribution of the 36(195, 196)
Compton wavelength 35(151, 399)
Condensed-ring systems 37/2(203)
Conditional probability distribution 35(60)
Conditions, nearly adiabatic 36(299, 300)
Configuration 36(91); 37/2(212)
Configuration, definition 37/2(86)
Configuration interaction 36(101); 37/2(35, 86–90, 201, 217)
Configuration mixing 37/2(31)
Confluent hypergeometric function 35(98); 36(272)
Conjugate of a symmetry operation 37/2(7)
Conjugated chains 37/2(210)
Conservation of charge 36(381)
Conservation of particles 36(237, 290)
Consistency diagrams 35(19, 23)
Consistency measures and least squares adjusted values 35(79)
Constants of motion 37/2(178–179)
Contact transformation 37/1(183, 240–244, 280)
Continuity equation 36(178)
Continuous spectrum 35(107, 115, 126)
Continuous spectrum of hydrogen 37/1(344)
Convergence of Born expansions 36(242)
Convergence of series of partial cross sections 36(246)
Conversion constant $\Lambda=\lambda_g/\lambda_s$ 35(12)
Conversion factor voltage-wavelength 35(39)
Convex hydrocarbons 37/2(229)
Coordinates, body-fixed 37/1(173, 187, 192ff.)
Coordinates, cylindrical polar 37/1(210)
Coordinates, dimensionless 37/1(209)
Coordinates, normal 37/1(173, 193, 201–209)
Coordinates, space-fixed 37/1(173, 191)
Coordinates, spherical polar 37/1(210)
Coordinates, symmetry 37/1(203)
Copley-Bernstein arcs 37/2(271)
Copper in atomic beams 37/1(18, 34, 101, 118)
Coriolis coupling factor 37/1(174, 208–209, 229, 294, 299)
— sum rules 37/1(256–257, 261)
Coriolis interaction rule of Jahn 37/1(292–293)
Coriolis operator (and interaction) 37/1(209, 210–211, 241, 273, 276, 293, 306)
Coriolis resonance 37/1(294)
Coriolis type of intensity anomaly 37/1(308–309)
Corona, chemical composition 36(402)
Coronal lines 36(402)
Coronene 37/2(224, 225, 229, 230)
Coronium lines 36(400)
Correction of Fermi and Amaldi 36(137)
Correlation 36(119, 145)
Correlation between positions of electrons 37/2(30, 35, 36, 50)

Correlation coefficient(s) 35(65)
— graphically illustrated 35(61)
— and calculation of standard errors in a least square adjustment 35(58)
Correlation correction 36(145)
Correlation diagram 37/2(64–65, 67)
Correlation energy 36(119, 145, 146); 37/2(88, 94, 95)
Correlation interaction (CI) 37/2(96)
Correlation potential 36(176)
Correspondence of different theories, diagram 37/2(202)
Correspondence principle 36(3, 11, 298); 37/2(177)
Coulomb distortion phase 36(274)
Coulomb field 36(22, 23, 27, 271)
Coulomb field, modified 36(275)
Coulomb functions, irregular 35(110)
Coulomb integral 35(218); 37/2(19, 20, 51)
Coulomb integral, direct 35(219)
Coulomb interaction, instantaneous 35(257)
Coulomb scattering in large angles 35(408, 416)
Coulson bond orders 37/2(179, 207, 214)
Coulson-Rushbrook theorem 37/2(194)
Coulson's molecular orbital 37/2(29)
Coupling, close 36(287, 381)
Coupling constant 35(176)
Coupling, strong, in optically allowed transitions 36(392)
Covariance of a probability distribution 35(61, 62)
Covariant calculation 35(135, 177, 181)
Covariant two-body formalism 35(282)
Covering operations
— of an asymmetric rotator 37/1(218–219)
— of an XY_3 pyramidal molecule 37/1(205–206)
— of an XYZ_2 planar molecule 37/1(203–205)
Cracking patterns of molecules in mass spectra 37/1(373)
Creation operator 36(79)
Cross section
— differential 35(119); 36(233)
— for small angle inelastic collisions 36(365)
— for vibrational excitation or de-activation 36(286)
— near ionization potential 36(315)
— total 36(233)
Crossed fields 35(373)
Crossing point (of potential energy curves) 36(296, 298, 300, 306)
Crystal system, regular 36(105)
C-type bands 37/1(264, 267)
Cube for eight-electron state 37/2(249, 250)
Cubic lattice model of saturated chains 37/2(268)
Current conservation at a branch point 37/2(189)
Current potential 36(177)
Cyanine 37/2(220, 221)
Cyanine dyes 37/2(218)
Cyanogen, ionization by electron impact 37/1(389–391)
Cyclobutadiene 37/2(193, 198)
Cyclobutane 37/2(199, 267)
Cyclohexane 37/2(267)
Cyclo-octa-tetraene 37/2(198, 205, 228)
Cyclopropane, σ-bonds 37/2(271)
Cyclopropyl group 37/2(278)
Cyclotron frequency of charged particles 35(312)
Cyclotron frequency of the free electron 37/1(114)
Cyclotron frequency of the proton 35(33)
Cyclotron, inverse 35(35)

Darling-Dennison type of resonance 37/1(290–291)
Deactivation by electron impact 36(407)
Deactivation by superelastic impact 36(406)
Decay probability 35(337)
Defect of density 37/2(201)
Deflecting fields 37/1(41–53)
— electric 37/1(46–50)
— with four wires 37/1(45)
— with two wires 37/1(42, 43)
Deflecting iron magnet 37/1(44)
Deflecting properties of a molecular beam apparatus 37/1(44, 45)

Deformation of atomic wave functions in molecules 37/2(163, 165)
Degeneracy 37/1(180–182, 209)
Degeneracy, accidental 37/2(13)
Degenerate alternant orbitals 37/2(200)
Degenerate frequency 37/1(209, 211)
Degenerate perturbation methods 37/1(273)
d-electrons, their role in hydrides 37/2(242–243)
Delta function 35(266)
Density alternation 37/2(217)
Density alternation with period $2d$ 37/2(216–217, 220, 222, 238)
Density at a lattice point 37/2(195)
Density bond-order matrix 37/2(179)
Density disturbance by a proton in a rare-gas shell 37/2(247)
Density matrix 36(172, 193)
Density matrix of Dirac 37/2(177–179, 213, 239)
Density of electrical charge 36(6)
Density of electrical current 36(6)
Density of π-electrons along a polyene chain 37/2(213)
Density theorems 37/2(193)
Deslandres analysis of diatomic bands 37/1(176)
Detection of atoms by ionization potentials 37/1(28–30)
Detectors in molecular beam apparatus 37/1(25–35)
Deuterium (D_2) in molecular beams 37/1(10, 67, 124, 152, 154)
Deuterium in atomic beams 37/1(10, 13, 67, 95, 99, 100, 107, 117, 120, 122)
Dewar structure of benzene molecule 37/2(130)
Diamagnetic ring current 37/2(239)
Diamagnetism 35(304)
Diamagnetism of helium 35(313)
Diatomic homonuclear molecules, slow electron scattering 37/1(326, 328)
Diatomic molecules 37/1(124–126, 130, 132, 134, 136)
Diatomic molecules, resolved spectra 37/1(123, 138ff.)
Dibenzopentalene 37/2(228)
Diborane 37/2(260, 271, 273, 274)
Dielectric constant of helium 35(332)
Differential cross section for inelastic collisions 36(344)
Differential ionization cross section 36(351)
Diffraction at an edge 36(237, 278)
Diffraction effects in electron scattering 37/1(314)
Diffraction in molecular photoeffect 37/1(323)
Diffraction pattern 37/2(213, 215)
Diffusion, ambipolar 36(397)
Diffusion, coefficient of 36(288, 333)
Diffusion length 36(342)
Diffusion method for measuring excitation cross sections 36(332)
Diffusion of an electron swarm in a gas 37/1(345–346)
Diffusion of electrons in helium 36(342, 343)
Diphenyl-butadiene 37/2(203)
Dipolar molecule, excitation of rotation 37/1(355)
Dipolar odd-atom chain dye 37/2(220)
Dipolar structure in excited polyenes 37/2(216)
Dipole approximation 35(334f., 355)
Dipole moment(s)
— electric 35(366); 37/1(46–50)
— for hydrogen 35(350)
— magnetic 35(366); 37/1(41–46)
— nuclear 37/1(80, 83, 84ff., 103–108)
— of a neutral molecule 37/2(97)
— values for simple molecules 37/2(99)
Dipole radiation, magnetic 35(309, 356, 368)
Dipole transitions, magnetic, from $2S_{1/2}$ 35(371)
Dirac equation
— continuous spectrum 35(157)
— in momentum space 35(164)
— modified 35(136)
— quadratic 35(136, 141)
Dirac functions in the Coulomb field 35(155)
Dirac functions, irregular 35(156)
Dirac moment 35(293)
Dirac operators 35(133)
Dirac wave equation 35(133)
Dirac wavefunctions, small and large components 35(140)
Dirac-Breit equation 37/1(115)

Dirac's density matrix 37/2(177–179, 213, 239)
Direct product 37/2(17)
Direct product of groups 36(45)
Directed valence 37/2(173)
Directing effect in polyhydrides 37/2(245)
Directing effect of a substituent 37/2(236, 238)
Directing effect of substituents in aromatic molecules 37/2(237–239)
Direction cosine 37/1(173, 208)
Directional forces 37/2(174)
Discharge afterglow 36(330)
Discrete spectrum 35(113, 122)
Dispersion, anomalous 36(331)
Dispersion force between hydrogen atoms 37/2(37)
Dissociation cross section of hydrogen 37/1(341–342)
Dissociation energies
— of bonds 37/1(375, 390)
— of carbon tetrahalides 37/1(398)
— of carbon-halogen-bonds 37/1(391, 396, 399)
— of germanium halide radicals 37/1(404)
— of methyl halides 37/1(393)
— of silicon halide radicals 37/1(404)
Dissociation energy 37/1(318, 320); 37/2(4)
— of C–C-bond 37/1(390)
— of C–H-bond 37/1(390)
Dissociation of molecules by electron impact (*see also* Ionization) 37/1(365–372)
Distorted wave method 36(285, 296, 372, 379, 381, 387, 393, 396, 403)
Distribution function 36(112)
Donor property of amino nitrogen 37/2(237)
Doppler broadening 35(378)
Doppler broadening in beam spectroscopy 37/1(3, 16, 54, 75)
Double bond as bent single bonds 37/2(253)
Double bond between carbon atoms 37/2(182, 197)
Double bond, essential 37/2(203–205)
Double excitations 35(341)
Double ionization 36(319, 372)
Double quantum transitions in a beam 37/1(68, 69)
Double scattering 37/1(323)
Double scattering experiments 35(161)
Double σ-bond 37/2(271)
Doublet intervals 36(186)
Doublet splitting 36(35, 55, 57)
Doubly excited He-states 35(211)
Drift velocity 37/1(343, 346, 350)
Druyvesteyn distribution of electrons 37/1(346, 350)
DuMond and Cohen adjustment of 1953 (DC 53) 35(68–74)
Dyes, *see* Chain dyes

Effective charge 35(234)
Effective dipole moment 37/1(47)
Effective nuclear charge 37/2(22)
Effusion from a canal 37/1(5)
Effusion, ideal 37/1(4)
Eigenfunction 37/1(178, 181)
Eigenstate 37/1(178, 181)
Eigenvalue 37/1(178)
Einstein relation between probabilities of inverse processes 35(382)
Elastic scattering of electrons 36(295)
— by molecules 37/1(320–332)
Electric and magnetic fields crossed 35(373)
Electric charge, discrete nature 35(6)
Electric deflecting fields 37/1(46–50)
Electric dipole, *see* Dipole
Electric dipole moment 37/1(46–50, 141, 147)
Electric quadrupole
— interaction constants 37/1(151)
— interaction in molecular spectra 37/1(125, 129–132)
— interaction in molecular spectra, diatomic molecules 37/1(132)
— moment of the nucleus 37/1(80, 83, 84ff., 108–109, 125)
Electric resonance method 37/1(138–152)
Electric unit charge e in relationship to other atomic constants 35(9)
Electrodeless discharge for beam production 37/1(12)
Electron affinity 36(183, 185, 186); 37/1(30, 320)
Electron affinity of oxygen 37/1(365)

Electron bombardment for beam production 37/1(11)
Electron collisions with atoms 36(304)
Electron configuration, *see* Configuration
Electron density
— at the atom surface 36(123, 140, 145)
— distribution along a chain 37/2(185)
— in alternant molecules 37/2(195)
— three-dimensional 37/2(177, 178)
Electron diffraction by gaseous molecules 36(363)
Electron energies, electrostatic and magnetic methods of analysing them 36(344, 348)
Electron exchange 36(294, 368); 37/1(335)
Electron gas 36(110)
— density distribution of the 36(132, 142)
— equation of motion of the 36(178)
— flow equations 36(177)
— non static treatment of the 36(177)
— pressure of the 36(111, 113)
— vibrations 36(179)
— zero-point energy of the 36(111)
Electron impact spectrum 36(323)
Electron magnetic moment, *see* Magnetic moment
Electron pair bond 37/2(105, 141)
Electron pair, localized 37/2(183)
Electron pairs and density maxima 37/2(215)
Electron radius, classical 35(157, 399)
Electron scattering, elastic, by molecules 37/1(320–332)
Electron, spectroscopic mass 35(112)
Electron spin density 37/2(236)
Electron spin, g-value 35(38)
Electron temperature 36(329, 397)
Electronegativity 37/2(176, 242)
Electronic charge *e* 35(14, 16)
Electronic energy 37/2(2)
— for the hydrogen molecular ion 37/2(55)
— surface 37/2(2, 20)
Electronic state excitation 37/1(318)
Electronic transitions in band spectra 37/1(176)
Electronic wave functions, symmetry properties 37/2(9–17)
Electronics in beam spectroscopy 37/1(73–79)
Electron-ion collision 37/1(349)
Electrophilic substituent 37/2(236, 240)
Electrostatic approach 37/2(177, 179–180)
Electrostatic calculations with uniform spheres 37/2(258–259)
Electrostatic Coulomb and exchange interaction 36(115)
Elements of a group 37/1(205)
Ellipsoid of error 35(19)
Elliptic coordinates 35(232)
Elwert's expression for Bremsstrahlung 35(418)
Elwert's semi-empirical factor 35(427)
Emission spectra 35(341)
Empirical chemistry, bond rules 37/2(173, 175)
e/m-values, discrepancies 35(19)
End point in a network 37/2(188)
Energy balance method 36(332)
Energy depression due to correlation interaction 37/2(96)
Energy efficiency of atom production in glow discharge 37/1(343)
Energy formula in valence bond theory 37/2(123)
Energy level(s)
— determination by beam spectroscopy 37/1(54–79)
— diagrams 37/1(90)
— including hyperfine structure 37/1(141–145)
— of the asymmetric rotator 37/1(221–223)
Energy loss by hydrogen excitation 37/1(343)
Energy loss due to Bremsstrahlung 35(416, 426)
Energy of zero-point vibration 37/2(4)
Energy relations 36(135, 143, 201)
Energy shell, in momentum space 35(129, 130)
Equal particles 36(63, 80, 81, 84)
Equation of state 36(210, 220)
Equilibrium distance of nuclei in hydrogen molecule 37/2(21)
Equilibrium distances of nuclei in other molecules than hydrogen (*see also* Bond lengths) 37/2(66)

Equilibrium position of a molecular particle 37/1(192)
Equivalence of MO and resonance theory 37/2(201–209)
Equivalence of representations 37/2(11)
Equivalences among chemical bond theories 37/2(188–209)
Equivalent atomic orbitals 37/2(115)
Equivalent electrons 37/2(42)
Equivalent electrons, symmetry types of states 37/2(43)
Equivalent orbitals 37/2(253)
— in carbon dioxide 37/2(145)
— in methane 37/2(144)
— in water 37/2(144)
Error elements of DC 53 adjustment 35(74)
Error matrix 35(62, 65)
Error measures by external consistency 35(65)
Error measures by internal consistency 35(65)
Error measures of the output values of least squares adjustment 35(58)
Ethane, σ-bond 37/2(271, 274)
Ethyl alcohol, ionization by electron impact 37/1(385–387)
Ethyl halides, ionization by electron impact 37/1(394–395)
Ethylene 37/2(204, 253)
Ethylene, LCAO approximation 37/2(69, 71)
Ethylene, σ-bond 37/2(271)
Ethylene, σ-electron box model 37/2(267)
Eulerian angles 37/1(173, 184, 194, 199, 218)
Evaporation of solids for beam production 37/1(5–9)
Even 37/2(14)
Even operators 35(169)
Even ring system 37/2(226)
Even-atom benzenoid hydrocarbons 37/2(231)
Even-atom chain 37/2(216)
Exact resonance 36(287, 289)
Excess of density 37/2(201)
Exchange 36(304)
Exchange amplitudes 36(369)
Exchange correction 36(139)
Exchange correlation 37/2(90)
Exchange energy 36(115, 116, 117, 118, 143)
Exchange frequency 35(218)
Exchange integral 35(218, 221); 36(66); 37/2(19, 20, 51, 121)
Exchange interaction 35(285); 36(295)
Exchange operator 37/2(52)
Exchange polarization 37/2(76, 152–156)
Exchange potential 36(175, 176)
Excitation collisions 36(299)
Excitation conditions 35(376)
Excitation cross section 36(309, 352)
Excitation energy, average 35(185)
Excitation energy, mean 36(183)
Excitation function, optical 36(320)
Excitation in collisions of the second kind 36(321)
Excitation near energy threshold 36(327, 328, 329, 371)
Excitation of a positive ion 36(371)
Excitation of an electronic state 37/1(318)
Excitation of discrete states 36(292, 320, 352)
Excitation of heteronuclear gases by electron impact 37/1(352–353)
Excitation of hydrogen by electron impact 37/1(339–344, 350)
Excitation of metastable states 36(320, 335)
Excitation of nitrogen by electron impact 37/1(334–339, 351)
Excitation of oxygen by electron impact 37/1(352)
Excitation potential, mean 36(183)
Excitation transfer 36(287, 288, 323)
Excited singlet and triplet states 37/2(213)
Excited state of O^- 37/1(365–367)
Excited states of aromatic molecules 37/2(216)
Excited states of polyenes 37/2(212, 215–216)
Excited states of water ions 37/1(376)
Exclusion principle 36(56, 68, 84, 92, 110, 167)
Exclusion principle of occupied electron states 36(170)
Expansion in Legendre polynomials 35(247)

Expansion in powers of $1/Z$ 35(214, 239)
Expectation value 35(248, 250); 36(7)
Expectation values of r^ν 35(103)

Factorization of wave function 37/2(185, 187)
Faraday constant 35(16)
Fast electrons, scattering by molecules 37/1(321–322)
Faxén-Holtsmark formula 37/1(322, 325)
Fermi energy 36(112)
Fermi formula 37/1(104, 105)
Fermi hole 37/2(50, 258)
Fermi interaction 37/2(155)
Fermi level 37/2(196)
Fermi resonance 37/1(284, 286)
Fermi resonance and *l*-type doubling 37/1(288–289)
Fermi-Dennison resonance 37/1(284, 288)
Fermi-Dirac statistics 36(110)
Fermi-Thomas field 37/1(323, 324)
Feynman diagrams 35(176, 177)
Field aspect of matter 36(5)
Field of matter, proper states 36(14)
Filled orbitals 37/2(186)
Fine structure
— constant 35(170, 190); 36(9, 57); 37/1(122); 37/2(6)
— formula 35(169)
— intensities of H_α 35(360)
— of hydrogen 37/1(120–123)
— of x-ray absorption edges 37/1(323)
— separation in deuterium 35(36)
— splitting 35(170)
— splitting, numerical results 35(275)
— splitting of helium 35(269)
First order anharmonic resonance 37/1(282–288)
First order rotational resonance interaction 37/1(292–304)
Fluorbenzene 37/2(201)
Fluorescence yield 36(318)
Fluorine hydride 37/2(249, 255)
Fluorine in atomic beams 37/1(117)
Fock Hamiltonian 37/2(52)
Fock space 35(262)
Fock's method for helium 35(216, 226)
Fock's system of equations 36(108, 385, 389, 392)
Foldy Wouthuysen transformation 35(169)
Forbidden lines 35(332)
Forbidden transitions 35(364)
Force constant in a hydride 37/2(241, 242)
Force constants by Thomas-Fermi method 37/2(257)
Forced hybridization 37/2(63)
Formaldehyde, infrared spectrum 37/1(295)
Forward scattering amplitude 36(238, 249, 278)
Fourier transform 35(265)
Four-wire deflecting field 37/1(45)
Framework of a molecule 37/1(192)
Franck-Condon principle 35(224); 37/1(317–320, 340, 341, 363–365)
Free energy 36(160)
Free-electron
— calculations in polyene chains 37/2(211, 213, 215, 216, 219)
— method 37/2(184, 186–187)
— network 37/2(187, 191)
— network calculations in aromatic systems 37/2(221, 232)
— π-orbitals 37/2(211, 216)
Free-free absorptive transition 35(420)
Frequencies for polyhydrides, table 37/2(255)
Frequency and bond length, relationship 37/2(254–255)
Frequency measurements in beam spectroscopy 37/1(74–75)
Frequency of elastic collisions 36(344)
Frontier orbital 37/2(215)

g factor 36(57, 100)
Gallium in atomic beams 37/1(7, 15, 29, 81, 87, 101, 104, 107, 109, 118)
Gas kinetics in beam production 37/1(1, 3–5)
Gas recirculation system for beam production 37/1(12, 14)
Gas sources for beam production 37/1(10–13)
Gegenbauer function 35(125)
Geiger-Müller counter 36(364)
Genealogical construction of spin functions 37/2(132)

Generalized theory of least squares for non-Gaussian error distributions 35(56)
Generating function 35(106)
Germanium tetrahalides, ionization by electron impact 37/1(403–408)
Gibb's chemical potential 36(135)
Gold in atomic beams 37/1(9, 18, 33, 34, 71, 102, 119)
Gold L_{II} and L_{III} ionization of 36(319)
Graphical methods 35(19)
Graphite lattice 37/2(183, 192)
Grating constant 35(15)
Grating constant, effective 35(11)
Grating constant, true 35(11)
Grating measurements of x-ray wavelengths 35(11)
Grating space, effective 35(10)
Grating space, true 35(10)
Green's function 36(238)
Green's theorem 36(263, 280, 303)
Ground state 36(95)
Ground state configuration
— of carbon dioxide 37/2(145)
— of methane 37/2(144)
— of water 37/2(144)
Group C_{3v} 37/2(8)
Group $C_{\infty v}$ 37/2(9)
Group $D_{\infty h}$ 37/2(9)
Group T_d 37/2(8, 13)
Group theory 36(42, 43)
Group theory of covering operations 37/1(205)
Grüneisen constant 36(210)
Grüneisen relation 36(211)
g-values 37/1(42)
g-values, atomic 37/1(115–120)
g-values, nuclear 37/1(126–129)
Gyromagnetic ratio of the proton 35(32, 33)

Hainer, King and Cross, method of computation 37/1(221–222, 236, 240, 263, 306)
"Hairpin" arrangement 37/1(75)
Hall effect 37/1(346)
Hall's high energy expression for the photoeffect 35(401)
Halogens in aromatic rings 37/2(238)
Hamiltonian 37/1(177, 178)
Hamiltonian of a molecule
— classical 37/1(191–195)
— quantum mechanical 37/1(195–196)
— transformed under contact transformation 37/1(243–244, 280)
Hamiltonian, second order terms 37/1(240ff., 290, 294)
Hammett formula for different reactivities 37/2(234–235, 239)
Ham-Ruedenberg bond order 37/2(208)
Harmonic oscillator 37/1(174)
Harmonic oscillators in one, two, and three dimensions 37/1(210, 212ff.)
Harmonics, zonal 36(256)
Hartree approximation 37/2(28, 45–47)
Hartree densities in aromatic compounds 37/2(229–230)
Hartree ionization potential 35(230)
Hartree method, *see* Self consistent field
Hartree method 36(36, 107)
Hartree units, *see* Atomic units
Hartree wave function 35(230)
Hartree-Fock approximation 37/2(47–54)
h-determinations 35(50–52)
h/*e* determinations 35(49)
h/*e* values, discrepancy 35(19, 39, 40, 43)
Head of a band (*see also* Band head) 37/1(174)
Heat of formation of a molecule 37/2(142)
Heisenberg's method, first order for excited helium states 35(217)
Heisenberg's method for helium 35(215)
Heitler-London theory 37/2(17–22)
— extensions 37/2(22–26)
Helium, angular distribution of electrons scattered after exciting the $2\,^1P$ level of 36(366)
Helium atom(s) 37/2(27)
— metastable, diffusion in helium 36(288)
— mutual repulsion 37/2(77, 80, 139, 140)
— split-shell method 37/2(146)
Helium, deactivation of the $2\,^1S$ state 36(378)

Helium excitation 36(323, 361)
— of ^1S levels of 36(368)
— of the $4\,^1D$ and $5\,^1D$ levels of 36(367)
— of the double excited state 36(349, 361)
— of the $3\,^1P$ and $4\,^1P$ levels 36(365)
— of the $2\,^3P$ and $2\,^1P$-levels 36(306, 378)
— of the $2\,^1P$ level 36(378, 394)
— of the $2\,^3S$ and $2\,^1S$ levels of 36(296, 376)
— of the $2s$ state of He^+ 36(378)
Helium in atomic beams 37/1(2, 14, 16, 31, 65, 76, 96, 98, 99, 100, 106, 117, 120, 122)
Helium, inelastic collisions of electrons in 36(347)
Helium ion collisions in helium 36(287, 365)
Helium, ionisation of 36(365)
Helium, level scheme 35(210)
Helium levels, experimental energies 35(213)
Helium, metastable state of 36(336, 337, 340)
Helium, optical excitation 36(325)
Helium, optical transition probabilities for 36(326)
Helium, scattering of electrons 36(261, 265, 366)
Helium, total cross section of 36(327)
Helium triplet lines 37/1(335)
Helium wavefunctions, symmetry 35(204)
Helium-like ions 35(213)
Hellmann-Feynman theorem 37/2(177, 179–180, 234, 240, 245, 248, 257)
Hermite polynomial 37/1(212)
Hermitian matrix 37/1(182, 250)
Hermitian operator 37/1(178, 179)
Hertz diffusion method
— in hydrogen 37/1(342, 345, 350, 354)
— in nitrogen 37/1(345, 351)
Hetero-atoms 37/2(192, 193, 201)
Heteronuclear diatomic molecules 37/2(67, 83)
Heteronuclear gases, excitation by electron impact 37/1(352–353)
Hexatriene 37/2(182, 186, 190, 210, 213, 215, 216)
Hexatriene, density surface 37/2(223)
High pressure arc discharge 36(399)
Highly-excited S-states 35(227)
Hole theory 35(163)
Homonuclear molecule 37/1(127, 130)
Hulthén-Kohn variational procedure 37/1(331)
Hund's case (a) of molecular coupling 37/1(197, 198, 246)
Hund's case (b) of molecular coupling 37/1(197, 246)
Hund's rule 37/2(44, 212)
Hybridization 37/2(23, 63, 176, 177, 242)
Hybridization in lithium 37/2(106)
Hybridization in methane 37/2(109, 118)
Hybridization in water 37/2(107–109)
Hybridization of p^2d^2 type 37/2(117)
Hybridization of $2s$ and $2p$ states 37/2(63)
Hybridization of sp^3 type 37/2(115)
Hybridization of sp^2d type 37/2(117)
Hybridization of sp^3d^2 type 37/2(116)
Hydride molecules, binding energies 37/2(168)
Hydrides, diatomic 37/2(240–245)
Hydrocarbon chains, conjugated 37/2(181, 191)
Hydrocarbon chains, planarity 37/2(183)
Hydrocarbon chains, saturated 37/2(183)
Hydrocarbon spectra, general features 37/2(183)
Hydrogen atom(s) 36(23, 28)
— charge distribution 35(102)
— collisions with 36(261, 265, 278, 282, 294)
Hydrogen atomic
— differential cross section for excitation of lower states of 36(353)
— excitation functions for the $H\alpha$, $H\beta$ and $H\gamma$ lines of 36(329, 366)
— excitation of the $2p$ state of 36(306, 378, 394)
— excitation of the $2s$ state of 36(372, 381)
— excitation of the $2s$-$2p_{1/2}$ and $2s$-$2p_{3/2}$ transitions in 36(393)
— inelastic electron cross section for 36(282)
— ionization cross section 36(312)

— ionization from 2s and 2p states 36(358)
— metastable state of 36(340)
— proton collisions 36(287)
Hydrogen dissociation cross section 37/1(341–342)
Hydrogen, double ionization by electron impact 37/1(365)
Hydrogen excitation by electron impact 37/1(339–344, 350)
Hydrogen, excitation of the continuous spectrum 37/1(344)
Hydrogen (H) in atomic beams 37/1(2, 10, 11, 13, 26, 31, 34, 43, 67, 71, 80, 81, 95, 98, 100, 104, 106, 107, 113, 115, 117, 120, 122)
Hydrogen (H_2) in molecular beams 37/1(10, 26, 124, 152, 154)
Hydrogen (HD) in molecular beams 37/1(10, 33, 124, 152, 154)
Hydrogen ion, negative 35(240, 402); 36(269); 37/2(164)
Hydrogen, ionization by electron impact 37/1(363–365)
Hydrogen molecular ion 37/2(54–60)
Hydrogen, molecular orbitals 37/2(27)
Hydrogen molecule(s)
— approximation of James and Coolidge 37/2(37, 157, 245)
— electron scattering 37/1(329, 330–332)
— empirical constants 37/2(21)
— extensions of Heitler-London theory 37/2(22–26)
— Heitler-London theory 37/2(17–22)
— ionic structure contribution 37/2(163–164)
— its central role in theory 37/2(175)
— potential energy curve 37/2(20)
— Weinbaum's approximation 37/2(25)
Hydrogen, normalized eigenfunctions 35(101)
Hydrogen, radial eigenfunctions 35(101)
Hydrogen, rotational excitation 37/1(354–355)
Hydrogen selenide, ionisation by electron impact 37/1(376–377)
Hydrogen, specific primary ionization by electrons 36(364)
Hydrogen sulphide, ionization by electron impact 37/1(376–377)
Hydrogen, vibrational excitation 37/1(354–355)
Hydrogen wave-function, asymptotic expression 35(108)
Hydrogen-like approximation 35(241)
Hylleraas expansion 35(234)
Hyperbolic orbits, classical 36(271)
Hyperfine structure 35(193–200, 287–290); 36(186)
— anomalies 37/1(103, 106–108)
— Fermi formula 35(196)
— interaction constants 37/1(84–88)
— intervals, measurements 37/1(89–103)
— nuclear motion corrections 35(198)
— of electron spin resonance 37/2(236)
— of hydrogen 37/1(98, 99, 104–107)
— of the triplet S-state 35(289)
— operator 35(193)
— radiative corrections 35(197)
— theory 37/1(81–88)
Hypergeometric equation 37/1(216)
Hypergeometric function 36(272)

ICC (intraatomic correlation correction) 37/2(168)
Ice, tetrahedral proton arrangement 37/2(252)
Ideal aperture 37/1(8)
Ideal effusion 37/1(4)
Identical particle collisions 36(294)
Identical representation 37/2(13)
Identity operation 37/1(203–205); 37/2(7)
Impact parameter 36(236, 253)
Imperfect resonance 36(288)
In-and-out correlation 37/2(147)
Independent scattering model 37/1(322)
Indium in atomic beams 37/1(7, 15, 29, 81, 87, 101, 104, 107, 109, 118)
Induced emission 35(310)
Inelastic collisions between atoms 36(299)
Inelastic scattering, conservation theorem 36(276)
Inelastic scattering, cross sections 36(348)
Inertia, moment of inertia 37/1(173)
Inertia, moments of inertia of water vapor 37/1(273)
Inertia, principal axes 37/1(193)
Inertia, product of inertia 37/1(173)

Inertia, reciprocal of inertia 37/1(174)
Infrared band spectra 37/1(175, 251–273)
Infrared catastrophe 35(179)
Infrared spectra, anomalies 37/1(273–311)
Infrared spectra of axially symmetric molecules 37/1(257–260)
Inhomogeneity correction, kinetic 36(151)
Inhomogeneity correction of Weizsäcker 36(151)
Inner quantum number 35(139)
Inner shell ionization 36(308, 317, 318, 362, 365, 366)
Intensities, dynamical 35(353)
Intensities for hydrogen 35(353)
Intensities in parabolic coordinates 35(362)
Intensities in triplet spectra 35(361)
Intensities of electric quadrupole radiation 35(367)
Intensities of fine structure of H_α 35(360)
Intensities of Stark effect components 35(363)
Intensities, statistical 35(353)
Intensity anomalies in infrared bands 37/1(305–311)
Intensity distribution in a deflected beam 37/1(38–41)
Intensity formulas 35(359)
Intensity in infrared bands 37/1(237ff.)
Interaction 36(63)
Interaction constants 37/1(84–88)
Interaction energy 36(202)
Interaction energy, electrostatic 36(115)
Interaction of atoms and ions 36(165, 166)
Interaction of free electrons 36(114)
Interaction operator 37/2(19)
Interaction splitting 36(72)
Intercombination 36(328, 369)
Interference pattern of standing waves 37/2(215)
Intermediate field spectra of diatomic molecules 37/1(134–136)
Intermediate states 35(176); 36(306)
Internal angular momentum 37/1(173, 227ff.)
Intraatomic correlation correction (ICC) 37/2(168)
Intrabond non-orthogonality 37/2(126)
Iodine in atomic beams 37/1(87, 102, 109)
Iodine, ionization by electron impact 37/1(372)
Ion crystals 36(206)
Ionic character percentage 37/2(242, 277)
Ionic radii 36(187)
Ionic structure added to covalent structure 37/2(24–26, 31)
Ionization chamber for measuring appearance potentials 37/1(358)
Ionization coefficient 36(344)
Ionization cross section 36(327, 332, 356, 357, 398)
— apparent 36(309)
— experimental results for 36(315)
— near energy threshold 36(312, 315, 317)
— true 36(309)
Ionization degree 36(399)
Ionization energies, values for simple molecules 37/2(98)
Ionization energy 36(182); 37/2(97)
Ionization energy, vertical 37/1(319)
Ionization limit, classical 35(322)
Ionization, multiple 36(308)
Ionization of carbon monoxide by electron impact 37/1(365, 369–371)
Ionization of hydrogen by electron impact 37/1(363–365)
Ionization of iodine by electron impact 37/1(372)
Ionization of molecules by electron impact, mass spectrographic results 37/1(362)
Ionization of nitric oxide by electron impact 37/1(365, 369, 371)
Ionization of nitrogen by electron impact 37/1(365, 369, 371–372)
Ionization of oxygen by electron impact 37/1(365, 367–369)
Ionization of polyatomic molecules by electron impact 37/1(372–415)
Ionization of the beam by electron bombardment 37/1(32–33)
Ionization potential for helium-like ions 35(278)
Ionization potential, variational 35(236)

Ionization potentials for atom detection 37/1(28–30)
Ionizing collisions 36(307, 309, 372)
Iron deflecting magnet 37/1(44)
Iron: FeXIV and FeX 36(379, 402, 403)
Irreducibility of representations 37/2(12)
Irreducible determinant 37/1(207)
Isochromat 35(40)
Isoelectronic molecules other than hydrides 37/2(276–278)
Isoelectronic sequence of polyhydrides 37/2(255, 257)
Isoenergetic point 37/2(221)
Isomerization 37/2(205)
Isomerization of paraffins 37/2(275)
Isometric consistency chart 35(20, 23, 26)
Isomorphism of chemical bond theories 37/2(187–188)
Isopropyl group 37/2(278)
Isotope effect 37/2(217)
Isotope shift 35(255)
Isotropic oscillator in three dimensions 37/1(210, 212, 214–215, 233)
Isotropic oscillator in two dimensions 37/1(210, 212, 213, 273, 296)

$j=\frac{1}{2}$ states in an electric field 35(372)
Jahn's Coriolis interaction rule 37/1(292–293)
James-Coolidge approximation for the hydrogen molecule 37/2(37, 157, 245)
j-j coupling 35(283); 36(102)
j-j coupling, interaction energy 35(284)

K-absorption jump 35(393)
K-absorption limit 35(391)
Kekulé structure 37/2(197, 198)
— necessary for stability 37/2(203)
— of benzene molecule 37/2(130)
— sign 37/2(202, 205)
Kepler ellipse 36(23)
Ketene 37/2(276)
Kimball sphere 37/2(258–259)
Kinetic energy, local 35(320)
Klein-Gordon equation 35(136, 291)
Klein-Nishina formula 35(400)
Koopman's theorem 37/2(47)
Krypton in atomic beams 37/1(31)
Krypton, single ionization 36(316)
K-shell 35(174)
K-shell absorption coefficient 35(390)
Kuhn condition at branch points 37/2(188, 209)
Kuhn periodicity in long polyene chains 37/2(216, 220)

Lagendre polynomials 36(235, 240)
Lagrange determinant 37/1(202, 207)
Lagrange equations of motion 37/1(202)
Laguerre functions 35(99)
Laguerre functions, integral representation 35(107)
Lattice and box model, equivalence 37/2(192–193)
Lattice constants 36(207, 210)
Lattice energy 36(207, 208, 209, 210)
Lattice model 37/2(184, 187)
Laguerre polynomial 37/1(213, 214)
Lamb experiments 35(36, 190)
Lamb shift 37/1(72, 81, 98, 120–123)
— diagram 35(177)
— excitation energy 35(406)
— in helium 35(276)
— in lowest order 35(189)
— nonrelativistic 35(183)
— numerical values 35(193)
— plus Stark effect 35(326)
— splitting 35(371)
— with retardation 35(186)
Λ-type doubling in linear molecules 37/1(279–282)
Λ-type splitting for singlet states 37/1(280)
Lambert's law 37/1(238)
Landé factor 35(295, 307, 311)
Landé *g*-formula 36(100)
Landé interval rule 36(98)
Landé vector model 36(56)
Langer approximation 36(270)
Langmuir probe technique 36(331)
Lanthanum in atomic beams 37/1(102, 109)
Laplace operator 35(431)
Laporte's rule 35(341)
Larmor precession 35(292)
Latent heat of sublimation of carbon 37/1(383)
Latent heat of vaporization of carbon 37/1(389)

Lattice representation 37/2(187)
Lattice-LCAO parallels 37/2(190–192)
Lattice-network equivalence 37/2(188–190)
LCAO approximation (linear combination of atomic orbitals) 37/2(29, 60–73, 184)
LCAO coefficients 37/2(190)
LCAO method in hexatriene 37/2(186)
LCAO self-consistent method 37/2(73–76)
LCAO symmetry orbitals 37/2(61)
LCAO theory for aromatic systems 37/2(232)
LCAO-lattice parallels 37/2(190–192)
Least squares, adjusted
— constants 35(16)
— constants and consistency measures 1955 35(79)
— values 1953 (DC 53) 35(73)
Least squares adjustment
— discussion of data rejected in 1955 35(67)
— error measures of the output values 35(58)
— in four unknowns 35(63)
— linearization of primitive equations 35(54)
— measure of consistency 35(55)
— number of degrees of freedom 35(55)
— observationally correlated data 35(55)
— of the atomic constants 35(53)
— of the atomic constants, 1955 (DC 55) 35(66, 78–82)
— output values 1955 35(81)
— removal of correlations by diagonalization 35(56)
— residuals 35(55)
— section of input data, 1955 35(66)
Least squares condition for non-Gaussian distributions 35(54)
Least squares, significance 35(5)
Legendre polynomials 35(430–434)
Lenz-Jensen perturbation calculation 36(165)
Level scheme for large Z 35(282)
Lewis cube 37/2(194, 249)
Lewis cube for neon 37/2(194)
Lewis theory of electron pair bond 37/2(105)
Lifetimes in electric fields 35(375)
Limit of continuous x-ray spectrum 35(39)
Limit of x-ray spectrum, method of extrapolated tangent 35(41)
Limitations of chemical bond calculations 37/2(175)
Line shape, average over the entire velocity distribution of a beam 37/1(57–61)
Line shape, natural 37/1(55)
Line shape, theoretical 37/1(65)
Line spectra 36(2, 17, 28)
Line widths in beam spectroscopy 37/1(3, 16, 54, 121)
Linear combination of atomic orbitals, *see* LCAO approximation
Linear harmonic oscillator 37/1(210, 212, 296)
Linear molecules, spectra 37/1(257–260)
Linearized equations of observation, 1955 adjustment 35(79)
Lithium atom, exchange polarization 37/2(152, 153, 155)
Lithium bromide in molecular beams 37/1(25, 47, 132, 148, 151)
Lithium chloride in molecular beams 37/1(25, 47, 129, 132, 148, 151, 152)
Lithium fluoride in molecular beams 37/1(25, 47, 69, 144, 148, 149, 151, 152)
Lithium in atomic beams 37/1(29, 100, 107, 117)
Lithium iodide in molecular beams 37/1(6, 47, 132, 133, 148, 151)
Lithium (Li_2) in molecular beams 37/1(129, 134)
Lithium molecule 37/2(44)
— ionic structure contribution 37/2(163)
— valence bond method 37/2(106)
Local energy 35(246)
Local kinetic energy 35(320)
Localized electron pair bond 37/2(141, 183)
Localized orthogonal orbitals 37/2(138)
Lone pair 37/2(105)
Longitudinal correlation 37/2(35)
Longitudinal in-and-out effect 37/2(147)

Longuet-Higgins' and Brown's formula 37/2(262)
Longuet-Higgins' theorem 37/2(196–197)
Loop of constant potential 37/2(227)
Lorentz gauge condition 35(136)
Lorentz triplet 35(293, 295)
Lorentz triplet in the normal Zeeman-effect 35(340)
Low field spectra of diatomic molecules 37/1(136–138)
Löwdin's orthogonalization 37/2(139)
Lowest excited state of a polyene chain 37/2(213, 215–216)
Lozier method of measuring appearance potentials 37/1(358–360, 365, 409–412)
LS coupling 36(91, 96, 368, 383)
L-shell 35(174)
L-shell absorption coefficient 35(392)
l-type doubling
— and Fermi resonance 37/1(288–289)
— in acetylene bands 37/1(278)
— in axially symmetric molecules 37/1(273–277)
— in linear molecules 37/1(277–279)
l-type resonance 37/1(279, 286, 288)
l-type splitting for single states 37/1(280)
Luxembourg effect 37/1(346)
Lyman series 35(96)
— intensities in parabolic coordinates 35(363)
— mean oscillator strengths 35(349)
— radial integrals and transition probabilities 35(348)
Lyman-Birge-Hopfield bands of nitrogen 37/1(335)

Magic electron numbers 37/2(227, 230, 275)
Magnetic analysis of electron energies 36(312)
Magnetic and electric fields crossed 35(373)
Magnetic anomalous moment of the electron 35(38, 39, 143)
Magnetic deflecting fields 37/1(41–46)
Magnetic deflection analysis of ions 37/1(360–362)
Magnetic dipole, *see* Dipole
Magnetic dipole moment 37/1(41–46)
Magnetic dipole moment of the nucleus 37/1(80, 83, 84ff., 103–108)
Magnetic dipole-dipole interaction 37/1(126)
Magnetic moment
— of an atom 37/1(89, 93)
— of the electron 37/1(112–115)
— of the electron, Dirac term 35(38, 136)
— of the electron, Pauli term 35(136, 143)
— of the proton 35(34, 38)
— of the proton, diamagnetic correction 35(34)
— of the spin 36(56)
Magnetic octupole moment of the nucleus 37/1(80, 83, 84ff., 108–109)
Magnetic properties of atoms 36(103)
Magnetic quantum number 35(92)
— of molecule rotation 37/1(174)
Magnetic resonance method 37/1(126–138, 152–155); 37/2(239)
Magnetic susceptibility 35(313)
Majorana formula 37/1(57)
Manganese in atomic beams 37/1(100)
Maser 37/1(81, 111)
Mass, bare 35(184)
Mass correction, elementary 35(253)
Mass correction, exchange 35(253)
Mass correction, polarization 35(254)
Mass, observed 35(184)
Mass, reduced 35(252)
Mass renormalization 35(179)
Mass spectrograph 36(314)
Mass spectrograph for measuring appearance potentials 37/1(359, 360–363, 409)
Mass spectrometer 36(315)
Matrix elements
— for the asymetric rotator 37/1(217)
— for the linear harmonic oscillator 37/1(212)
— for the spherical rotator with internal angular momentum 37/1(233–236)
— for the three-dimensional isotropic oscillator 37/1(215)
— for the two-dimensional isotropic oscillator 37/1(213)
— of a dipole 35(338)
— of a gradient 35(338)
— of acceleration 35(338)

Matrix elements (cont.)
— of angular momentum 37/1(186–189)
Matrix formulation of quantum mechanics 36(4); 37/1(182)
Matrix representation, *see* Representation
Matter under high pressure 36(214, 222)
Maxima and minima of electron density 37/2(213, 214)
Maxwell distribution law 37/1(4, 20–23, 346–350)
Maxwell-Boltzmann distribution of rotational states 37/1(175)
Mean collision frequency 37/1(350)
Mean energy loss of electrons 37/1(346)
Mean energy of electrons 37/1(346, 350)
Mean excitation energy 36(363)
Mean ionization coefficient 36(402)
Mean lifetime 35(336)
Measure of consistency of a least squares adjustment 35(55)
Measure of consistency of the 1955 adjustment 35(79)
Mechanical models of aromatic systems 37/2(222ff.)
Mechanical models of double bonds 37/2(253)
Meinel system of ionized nitrogen bands 37/1(336, 338)
Mercury in atomic beams 37/1(35, 119)
Mercury in 6 ^3P state 36(332)
Mercury, ionization 36(312, 316, 317, 362)
Metal complexes 37/2(227)
Metal surface, potential and electron distribution 36(213)
Metallic atoms highly ionized 36(400)
Metallic bond 36(210)
Metals 36(208)
Metastable atoms, detection 37/1(30–32)
Metastable atoms, impact on a metal surface 36(339, 340)
Metastable ions formed by electron impact 37/1(373)
Metastable ions of SF_6 37/1(411, 415)
Metastable state of hydrogen and deuterium 35(36, 191, 354, 371)
Metastable state, quenching 36(341)
Metastable states of atoms in a beam 37/1(13–16, 98)
Methane bands showing Coriolis resonance 37/1(303–304)
Methane, electron scattering 37/1(332)
Methane, infrared spectrum 37/1(262)
Methane, ionization by electron impact 37/1(381–384)
Methane molecule 37/2(177, 251)
— as perturbed neon system 37/2(261)
— equivalent orbitals 37/2(144)
— LCAO approximation 37/2(69, 71)
— LCAO self-consistent method 37/2(78–80)
— Thomas-Fermi approximation 37/2(257)
— twisting vibrations 37/2(262)
— valence bond method 37/2(109, 118, 144)
Methane, self-consistent field 37/1(332)
Method of least squares (*see also* Least squares) 35(3, 53)
Methyl alcohol, ionization by electron impact 37/1(384–385)
Methyl cyanide, infrared spectrum 37/1(277)
Methyl cyanide, ionization by electron impact 37/1(387–389)
Methyl fluoride, perpendicular absorption band 37/1(255)
Methyl group 37/2(155)
Methyl group, hyperconjugation 37/2(238)
Methyl halides, ionization by electron impact 37/1(391–394)
Microphotometer 36(324)
Microwave break down 36(335, 342)
Microwave cavity 36(337)
Microwave conductivity 37/1(330)
Microwave method 36(331)
Microwave methods for measuring slow electron cross sections 37/1(330, 345, 351, 372)
Microwave region of band spectra 37/1(175)
Microwave resonance absorption method 35(310)
Millman effect 37/1(75)
Minimization conditions 37/2(191)
Minimum momentum transfer 35(423)
Mixed representation 35(164)

Mixing parameters 36(289)
MO, *see* Molecular orbitals
MO calculations, results 37/2(90–104)
Mobile electrons 37/2(183)
Modified bond order 37/2(208)
Moffitt's method of atoms in molecules 37/2(160)
Molecular beam apparatus, deflection properties 37/1(44, 45)
Molecular beam, composition 37/1(23–25)
Molecular beam spectroscopy 37/1(123ff.)
Molecular constants 37/1(145–148)
Molecular frequency standard 37/1(109–112)
Molecular ions, negative 37/1(320)
Molecular models
— spherical molecules of type YZ_4 37/1(261–262)
— XY_3 pyramidal 37/1(205–206, 256, 301)
— XY_4, tetrahedrally symmetric 37/1(231, 234)
— XYZ_2 planar 37/1(203–205, 293ff.)
Molecular orbital 37/2(27–30, 39ff., 183, 184)
Molecular orbital method
— coefficients of atomic orbitals 37/2(100–104)
— equivalence with resonance theory 37/2(201–209)
— free-electron approach 37/2(184, 186–187)
— results of calculations 37/2(90–104)
Molecular spectra, resolved 37/1(123, 138–155)
Molecular spectra, unresolved 37/1(123, 126–138)
Molecular spin-orbitals 37/2(39)
Molecules, heteropolar 36(202)
Molecules, homopolar 36(204)
Molecules, potential- and electron distribution 36(198)
Moment of inertia 36(206)
Moment of the atomic state 35(298, 303)
Moments of inertia 37/1(173)
Momentum distribution in the statistical atom 36(195)
Momentum, linear 36(8)
Momentum space 35(122, 248, 260)
— Dirac equation 35(164)
— reduced Breit equation 35(264)
— wave functions 35(125)
Mono-aza-aromatic amines, ionization constants 37/2(235)
Morse potential 37/2(4)
Morse's function for the potential energy 37/1(336)
MSO, *see* Molecular spin-orbitals
Multiple quantum transitions in a beam 37/1(61, 67–71)
Multiple scattering 37/1(321)
Multiplet 36(86, 89, 91, 93, 97, 186)
Multiplet, regular 35(272)
Multiplicity selection 37/2(154)
Multipole deflecting fields 37/1(51–53)
Multipole expansion 35(364)
Multipole radiation 35(356, 367, 368, 370, 371)

Naphtalene 37/2(182, 185, 190, 192, 198)
Naphtalene, bond orders 37/2(208, 209)
Naphtalene, carbon skeleton 37/2(221–222)
Naphtalene, density surface 37/2(223, 224)
Naphtalene, negative ion 37/2(237)
Naphtalene, orbitals 37/2(199)
Naphthyl-methylene carbanion 37/2(231)
Natural line width for a multiple quantum transition 37/1(70)
Nebulae, gaseous 36(396)
Nebular lines 35(370)
Negative bands of nitrogen 37/1(334–338)
Negative energy states 35(162, 259)
Negative hydrogen ion 37/2(164)
Negative ions
— excited states 37/1(320)
— formation from carbon tetrahalides 37/1(398)
— formation process 37/1(319)
— of aromatic molecules 37/2(236)
Negative molecular ions 37/1(320)
Negative oxygen atomic ion 37/1(366)
Neon configuration 37/2(247–248, 253)
Neon in atomic beams 37/1(14, 31, 100, 117)

Neon, ionization 36(362, 365)
Neon, Lewis cube 37/2(194)
Neon, metastable state 36(336, 340)
Neon, positive ion formation by electron impact 37/1(360)
Neptunium in atomic beams 37/1(103)
Network and box model, equivalence 37/2(192–193)
Network model 37/2(184, 187)
Network of σ-electrons 37/2(262–278)
Network-lattice equivalence 37/2(188–190)
Neutralization, mutual 36(306)
Nickel, K shell ionization 36(318, 365)
Nitric oxide, ionization by electron impact 37/1(365, 369, 371)
Nitrogen, collision strengths for NI 36(390, 407)
Nitrogen excitation by electron impact 37/1(334–339, 351)
Nitrogen, forbidden lines of NI 36(408)
Nitrogen in atomic beams 37/1(117)
Nitrogen, ionization by electron impact 37/1(365, 369, 371–372)
Nitrogen molecules, electron scattering 37/1(328–330)
Nitrogen (N_2) in molecular beams 37/1(11, 26)
Nitrogen, optical excitation 37/1(335–336)
Nitrogen substitution in chains 37/2(219)
Nodes 37/2(185, 186)
Non-adiabatic transition in a beam 37/1(41, 49)
Non-bonding electrons 37/2(237)
Non-bonding orbitals 37/2(65, 196–198)
Non-Coulombic field 37/1(107)
Non-crossing rule 37/2(58, 63)
Non-LCAO semi-localized orbital 37/2(34–35)
Non-observables 37/2(176, 181)
Non-orthogonality effect 37/2(123, 126)
Non-orthogonality integral 37/2(18)
Non-orthogonality matrix 37/2(139)
Non-polar solvent 37/2(221)
Normal coordinates 37/1(173, 193, 201–209)
Normal equations, 1955 adjustment 35(79)
Normal frequency 37/1(202)
Nuclear charge, effective 36(175)
Nuclear dipole moment, magnetic 37/1(80, 83, 84ff., 103–108)
Nuclear distance in hydrogen molecule 37/2(21)
Nuclear distance in other molecules than hydrogen 37/2(66)
Nuclear g-values 37/1(126–129)
— signs 37/1(128–129)
Nuclear magnetic moments observed by the molecular beam method 35(38)
Nuclear magnetic resonance test of π-electron densities 37/2(239)
Nuclear magneton 35(34, 195)
Nuclear moment 35(195)
Nuclear motions of amolecule 37/1(201)
Nuclear octupole moment, magnetic 37/1(80, 83, 84ff., 108–109)
Nuclear quadripole coupling 36(206)
Nuclear quadrupole moment, electric 37/1(80, 83, 84ff., 108–109, 125)
Nuclear radius 35(188)
Nuclear spin 35(287, 301)
Nuclear spin measurements 37/1(89, 96, 100–103)
Nuclear structure correction 35(188, 199)
Nuclear-molecular interaction 37/1(148–152)
Nucleons by nucleons, low energy scattering 36(267)
Nucleophilic substituent 37/2(236, 240)
Nucleus, motion of the 35(252)
Null-vibrations (= translations and rotations) 37/1(203)
Number of Kekulé structures, stabilizing effect 37/2(205, 207–209)
Number, symmetry quantum 36(25, 38)

Observable 37/1(177, 178ff.)
Observables in molecules 37/2(212)
Observational correlation (definition) 35(58)
Observational independence (definition) 35(54)
Occupation numbers 35(376)

Occupation rule of electronic states 36(168)
Octatetraene 37/2(225)
Octet theory of Lewis 37/2(251)
Octupole moment, nuclear 37/1(80, 83, 84ff., 108–109)
Octupole transitions, electric 35(370)
Odd 37/2(14)
Odd ring system 37/2(226, 227)
Odd-atom benzenoid hydrocarbons 37/2(231)
Odd-atom chain 37/2(216, 217–218)
Oil-drop experiment 35(6)
Omegatron 35(34)
One-electron jump 37/2(212)
One-electron potential 37/2(176, 185, 186)
One-electron potential along a chain 37/2(210)
One-electron wave function 37/2(184–185)
Opacity, stellar 35(403)
Operator 36(6, 7, 59)
Operator manipulation 35(346)
Optical absorption 36(331, 335)
Optical absorption method 37/1(335)
Optical band spectra 37/1(176)
Optical excitation, energy gap 37/2(183)
Optical excitation function 37/1(335)
Optical levels 36(184)
Optical transition, first 37/2(212)
Optically excited states in a beam 37/1(71–73, 96, 97)
— of atoms 37/1(13, 16)
Optimum amplitude of the oscillating field 37/1(60)
Orbital 37/2(176)
Orbital angular momentum
— mean square 36(180)
— operator 35(93, 138)
Orbital density 37/2(178)
Orthogonal functions 37/1(180)
Orthogonal orbitals 37/2(118, 126)
Orthogonality of matrix components 37/2(15)
Orthogonalization 37/2(138, 139)
Orthogonalized atomic orbitals 37/2(32)
Ortho-helium 35(205, 270)
Orthonormality conditions 37/2(190)
Ortho-state 35(206)
Oscillating fields in beam spectroscopy 37/1(54ff.)
Oscillator, *see* Harmonic oscillator, Linear oscillator etc.
Oscillator, harmonic 36(3)
Oscillator strength 35(336); 36(331)
— for hydrogen 35(351)
— asymptotic formula 35(351)
— mean, of the Lyman-series 35(349)
— of alkali-atoms 35(379)
— relativistic 35(380)
— summed 35(355, 394)
Output values, 1955 least-squares adjustment 35(81)
Ovanene 37/2(230)
Ovens for beam production 37/1(5–9)
— double chamber 37/1(8, 9)
— high temperatures 37/1(9)
— material 37/1(6)
Overlap charge density 35(226)
Overlap integral 37/2(18, 191)
Overlap population 37/2(81, 82, 98)
Overlap population of heteronuclear diatomic molecules 37/2(83)
Oxygen, allowed lines of OI 36(407)
Oxygen, Boltzmann distribution of OIII 36(406)
Oxygen, collision strength for OII and OIII 36(389, 390)
Oxygen, dissociation by electron impact 37/1(359–360)
Oxygen, excitation by electron impact 37/1(352)
Oxygen, excitation of allowed transitions 36(391)
Oxygen, excitation of OI 36(389, 407)
Oxygen, in atomic beams 37/1(70, 117, 120)
Oxygen, ionization by electron impact 37/1(365, 367–369)
Oxygen molecule(s) 37/2(44)
— electron scattering 37/1(330)
— exchange polarization 37/2(156)
— excited states 37/2(158–159)
— LCAO energy 37/2(147)
— valence states 37/2(112–114)
Oxygen (O_2) in molecular beams 37/1(10, 26, 69, 152, 155)
Oxygen, red and green forbidden lines 36(407, 408)
Oxygen, transitions between levels of OIII 36(390)

Pair annihilation, virtual 35(202)
Pair creation 35(400, 429)
Pair theory 35(163, 260)
Pairing theorem for an alternant molecule 37/2(193, 194, 199–201)
Parabolic coordinates 35(113)
Parabolic quantization, charge distribution 35(115)
Para-helium 35(205)
Parallel vibrations (and band) 37/1(239, 252ff., 258ff., 267, 297)
Parallel wire field 37/1(42)
Parameter of asymmetry (of a rotator) 37/1(221, 265–267)
Parameterization technique 36(370)
Para-state 35(206)
Parity 35(94, 314, 341); 36(72, 79, 92); 37/1(219)
Partial cross section 36(237, 273)
Partial *f* sum rule 35(342)
Partial transition regions of the Stark effect for weak fields 35(331)
Partial wave solutions 35(131)
Particle aspect of matter 36(3)
Particles, "bare" 35(261)
Paschen-Back effect
— complete 35(296, 306)
— general 35(297)
— partial 35(296)
Pauli approximation 35(142, 143, 264, 355)
Pauli exclusion principle 35(212)
Pauli principle for potential well model 37/2(212)
Pauli spinors 35(298)
Pauling bond order 37/2(205, 207–209, 214)
PbH_3, infrared spectrum 37/1(256)
P-branch 37/1(252, 261)
Pentalene, non-existence 37/2(198, 199)
Pentamethine cation 37/2(237)
Perimeter model 37/2(228)
Periodic lattice 37/2(185, 187)
Periodic potential, smoothing 37/2(183)
Periodic table of elements 36(31, 179)
Permanent electric dipole moments 37/1(47)
Perpendicular vibrations (and band) 37/1(239, 252ff., 260, 267, 276, 296)
Perturbation calculation 36(20, 164, 165, 191)
— in ring systems 37/2(234)
— of polyhydrides 37/2(245–247)
Perturbation energy 36(165)
Perturbation theory of quantum mechanics 37/1(183–184)
Perturbation theory, unsymmetric 35(209)
Perylene 37/2(223, 224, 229, 230)
PH_3 bands showing Coriolis resonance 37/1(301–302)
PH_3, sub-band structure (due to quantum number *K*) 37/1(253)
Phase equivalent 36(252)
Phase integral 36(4, 15)
Phase relation 35(376)
Phase shift 35(132)
Phase shifts for determination of the scattering potential 36(251)
Phase, stationary method of 36(257)
Phenanthrene 37/2(150, 151, 193, 207, 223, 224, 227)
Phenol 37/2(201)
Phosphorus in atomic beams 37/1(118)
Photodetachment of electrons form O^- 37/1(366)
Photoeffect, Hall's high-energy expression 35(401)
Photoelectric absorption coefficient 35(392)
Photoelectric cross-section, forward shift of the maximum 35(397)
Photoisomerization 37/2(205)
Photomultiplier 36(325, 337)
π electron 37/2(71, 72, 183, 184)
π orbitals 37/2(71, 129)
π-electron density
— check by nuclear magnetic resonance 37/2(239)
— in polyenes 37/2(213)
— in pyridine 37/2(237)
— plaster models 37/2(223)
π-electron network 37/2(190)
π-electron, nodal plane 37/2(186)
π-electrons in aromatic rings 37/2(129, 137)
π-electrons in hexatriene 37/2(211)
Pirani gauge 37/1(26–28)
π-transitions 37/1(94)
pKa value, *see* Base strength

Planarity of hydrocarbons 37/2(183)
Planck constant h (*see also* h and h/e determinations) 35(39); 36(1)
Plane rotator 37/2(227)
Plane wave plus incoming spherical wave 35(382, 409)
Planetary nebulae 36(403)
Plaskett's method 37/2(273)
Plasma conditions 36(397)
Plasma time constants 37/1(348)
Plaster models of π-electron densities 37/2(223)
Plutonium in atomic beams 37/1(103)
Point group 37/1(205)
Poisson bracket 37/1(179, 182, 185)
Polar solvent, reduction of Coulomb energy 37/2(221)
Polar structures of the benzene molecule 37/2(137)
Polarizability 36(103, 191)
Polarization 36(306)
Polarization effect on the elastic scattering of electrons 36(304)
Polarization energy 36(203, 208)
Polarization for excited helium states 35(223)
Polarization of an electron beam 35(161)
Polarization of atomic orbitals 37/2(24)
Polarization of Bremsstrahlung 35(414)
Polarization of impact radiation 36(349)
Polarization of photons 35(419)
Polyatomic molecule, ionization by electron impact 37/1(372–415)
Polyatomic molecule, Schrödinger equation 37/1(198–199)
Polyatomic molecules, LCAO approximation 37/2(68–73)
Polyene chain 37/2(182, 193, 203, 210, 212–213, 215, 216)
Polyhydrides 37/2(245–262)
Polymers 37/1(136)
Polymethine structures 37/2(218)
Polysubstitution 37/2(201)
Population analysis 37/2(80, 98)
Porphine 37/2(224, 225)
Positive bands of nitrogen 37/1(334–338)
Positive column 36(329, 336, 397)
Positive column, atom production 37/1(343)
Positive ion current analysis 36(314)
Positive oxygen atomic ion 37/1(367)
Positronium 35(200); 36(284); 37/1(73)
Positronium, fine structure 35(200)
Positronium ground state 35(203)
Positronium in Pauli approximation 35(201)
Positronium, lifetime 35(201)
Positronium, radiative corrections 35(203)
Post interaction 36(282)
Potassium, apparent ionization cross section 36(311)
Potassium bromide in molecular beams 37/1(47, 137, 144, 148, 151)
Potassium chloride in molecular beams 37/1(25, 47, 148, 150, 151)
Potassium, excitation of the 4 ^2P level of 36(332)
Potassium fluoride in molecular beams 37/1(25, 47, 144, 145, 148, 149, 150, 151, 152)
Potassium hydroxide in molecular beams 37/1(8, 128)
Potassium in atomic beams 37/1(8, 17, 21, 23, 29, 41, 61, 69, 70, 100, 103, 107, 109, 118, 147)
Potassium iodide in molecular beams 37/1(25, 47, 148, 151)
Potassium (K_2) in molecular beams 37/1(130)
Potassium-beryllium fluoride ($KF \cdot BeF_2$) in molecular beams 37/1(7)
Potential energy
— curve of the hydrogen molecule 37/2(20)
— curves for the hydrogen molecular ion 37/2(55)
— effective, for nuclear motion 37/1(315)
— function of a polyatomic molecule 37/1(195, 199, 200–201, 208, 240, 248)
— of a molecule 37/1(195, 199, 203, 204, 207)
Potential, modified 36(170, 176, 205, 209)
Potential, repulsive 36(168, 169)
Potential well, square well model 37/2(211, 213, 216)

Praseodyme in atomic beams 37/1(9, 29, 102, 109, 119)
Pressure density relation 36(210, 222)
— for iron 36(226)
Pressure-compressibility relation of the elements at high pressures 36(228)
Principal axes of inertia 37/1(193)
Principal quantum number 35(95)
— apparent 35(154)
Principal series 35(379)
Principal value 35(127)
Prior interaction 36(282)
Probability distribution, conditional 35(60)
Probability hill of two independent random variables 35(58)
Probable error 35(5)
Product of inertia 37/1(173)
Progression of band heads 37/1(176)
Projection operator 35(346)
Projection operators of Casimir 35(165)
Propagation law of independent errors 35(64)
Propagation of errors, generalized law 35(61, 65)
Propane 37/2(193, 194)
Propane, *K* shell and σ-electrons 37/2(270)
Proper functions 36(7, 14)
Proper phases 36(289, 303)
Proper values 36(7, 14)
Proton anomalous moment, structure 35(198)
Proton electron mass ratio 35(312)
Proton magnetic moment, *see* Magnetic moment of the proton
Proton magnetic resonance spectra in substituted benzenes 37/2(239)
Proton precession frequency 35(33)
Proton resonance experiments 35(312)
Proton spin *g*-value 35(38)
Prussic acid, infrared spectrum 37/1(291)
Purely electronic energy 37/2(2, 20)
Pyranene 37/2(230)
Pyrene 37/2(223–225, 229, 230)
Pyrene, negative ion 37/2(237)
Pyridine 37/2(237, 249)

Q-branch 37/1(252, 254, 256, 258, 261, 295, 297)
Quadrupole coupling 37/2(277)
Quadrupole intensities, electric 35(367)
Quadrupole moment
— electric 35(366)
— induced 36(192)
— nuclear 37/1(80, 83, 84ff., 108–109)
— of a molecule, effect upon rotational excitation 37/1(355–356)
Quantum electrodynamics 35(175)
Quantum number 36(14)
— *K* 37/1(216ff., 254ff., 258ff.)
— magnetic, of molecule rotation 37/1(174)
— of angular momentum, exclusive of spin 37/1(174, 214)
— of angular momentum, inclusive of spin 37/1(174)
— of electronic angular momentum 37/1(174)
— of spin angular momentum 37/1(174)
— of vibration 37/1(174)
Quenching of lines by strong fields 35(320, 321, 324)

Radial electron density in a hydride 37/2(241)
Radial filtering system of Lozier 37/1(358–360)
Radiation field, virtual 35(176)
Radiation width 35(370)
Radiative capture 35(408)
Radiative corrections 35(163, 175)
Radiative recombination 36(400, 401, 402)
Radicals, free, ionization potentials 37/1(374–375)
Radicals, unpaired electron distributions 37/2(200)
Radioactive atoms in beam spectroscopy 37/1(17–18)
Radioactive detection of a beam 37/1(33–34)
Ramsauer technique 37/1(332, 350)
Ramsauer-method 36(327)
Ramsauer-Townsend effect 36(269)
Rare earths 36(186)
Rare earths, hydrides 37/2(243)
Rare gas
— density perturbation by a proton 37/2(247–248)

— electron configuration 37/2(247–248, 253)
— system 37/2(227, 228)
— -like sequences 37/2(255)
Ratio of the electrical units 35(29)
R-branch 37/1(252, 254, 261)
Reaction technique for beam production 37/1(7)
Rearrangement collisions 36(281, 286)
Reciprocal of a symmetry operation 37/2(7)
Reciprocal of inertia 37/1(174)
Recombination 36(406)
Recombination, dissociative 36(301)
Recombination of hydrogen atoms 37/1(341)
Recombination processes 35(407)
Recombination radiation 36(330)
Rectangular molecule 37/2(185)
Red shift 35(319)
Reduced masses 37/1(106)
Reducible representation 37/2(12)
Reduction of a representation 37/2(16)
Reflection operation 37/1(203–205)
Refocussing of a beam 37/1(36, 37, 40, 48)
Regression line 35(60)
Rejection of certain DC 53 data 35(73)
Relativistic correction 36(161, 365)
Relativistic correction factors 37/1(87)
Relativistic effect in the total molecular energy 37/2(93)
Renormalization 35(178)
Representation 36(43)
Representation, irreducible 37/2(12)
Representation, its character 37/2(15–16)
Representation of observables 37/1(179–182)
Representation of the symmetry group 37/2(10)
Repulsion energy between helium atoms 37/2(77, 80)
Repulsive potential energy curve 37/1(315, 316)
Repulsive potential, non classical 36(168)
Residuals of a least squares adjustement 35(55)
Resolved molecular spectra
— electric resonance method 37/1(138–152)
— magnetic resonance method 37/1(152–155)
Resonance, anharmonic, of first order 37/1(282–288)
Resonance, anharmonic, of second order 37/1(290–291)
Resonance, anharmonic, of third and higher orders 37/1(291–292)
Resonance anomalies, definition 37/1(273)
Resonance denominator 37/1(273, 283, 290)
Resonance electron capture forming SF_6 37/1(373, 411)
Resonance frequency of particles in a beam 37/1(56)
Resonance frequency shift 37/2(239)
Resonance interaction, rotational
— of first order 37/1(292–304)
— of second order 37/1(292–304)
Resonance of covalent and ionic structures 37/2(24, 31)
Resonance of Darling-Dennison type 37/1(290–291)
Resonance radiation 36(321, 322, 323, 326, 327, 328)
Resonance structure 37/2(176, 182)
Resonance structures in hexatriene 37/2(216)
Resonance theory 37/2(184)
— equivalence with MO theory 37/2(201–209)
— of a single sign 37/2(184, 188, 203)
Resonance transfer of excitation 36(287, 288)
Resonating bond 37/2(173)
Rest-mass energy 35(133)
Retardation 35(335)
Retardation correction 35(268)
Retardation effects 35(396)
Retardation neglected 35(384)
Retarding potential difference method 37/1(368, 370, 372, 373, 411–412)
Retarding slit 37/1(360)
Ritz method 36(138)
Ritz variation method 35(207)
Rotation operation through the angle π 37/1(203–205)
Rotational constants 37/1(148)

Rotational energy 37/2(4, 5)
Rotational excitation theory 37/1(353–357)
Rotational frequencies of a molecule 37/1(175)
Rotational quantum number 37/2(3)
Rotational resonance interaction
— of first order 37/1(292–304)
— of second order 37/1(304–305)
Rotational symmetry 36(38)
Rotational-vibrational constants 37/1(148)
Rotator, asymmetric 37/1(215–216, 217–223, 226–227, 264)
Rotator, axially symmetric, with internal angular momentum 37/1(227–229)
Rotator, rigid 37/1(215–217)
Rotator, spherical, with internal angular momentum 37/1(229–236)
Rotator, symmetric 37/1(216–217, 223–226)
Rotator, symmetric, intensity relations 37/1(239)
Rubidium bromide in molecular beams 37/1(148)
Rubidium chloride in molecular beams 37/1(25, 148, 151)
Rubidium fluoride in molecular beams 37/1(25, 47, 69, 137, 143, 148, 149, 150, 151, 152)
Rubidium in atomic beams 37/1(7, 17, 29, 101, 107, 109, 118, 147)
Rubidium iodide in molecular beams 37/1(148)
Russell-Saunders approximation 35(305, 341, 357, 368)
Russell-Saunders coupling 35(305, 341, 357, 368); 36(91); 37/1(116)
Rutherford scattering 35(116); 36(273, 276, 359, 363)
Rutherford's atom model 36(1, 10)
Rydberg constant 35(98)
Rydberg constant for hydrogen 35(113)
Rydberg correction(s) 35(222); 36(185)
— Coulomb 35(220)
— exchange 35(221)
— for highly excited S-states 35(229)
Rydberg formula 36(17, 28)
Rydberg polarization correction 35(225)
Rydberg series 36(391)
Rydberg-Ritz formula 36(17)

Saha equation 36(399)
Satellite x-ray lines 36(319)
Saturated chains, cubic lattice model 37/2(268)
Saturated hydrocarbons, ionization potentials 37/1(373)
Saturated network 37/2(203)
Sauter formula, correction factors 35(398, 401)
SbH_3, infrared spectrum 37/1(301, 304)
Scale factor of Wang 37/2(22)
Scattering amplitude 35(119)
Scattering amplitudes, relativistic 35(160)
Scattering asymmetry factor 35(161)
Scattering by mean central fields 36(268)
Scattering, classical theory 36(236, 253)
Scattering cross-section of Thomson 35(398, 399)
Scattering, elastic 36(270, 306)
Scattering, elastic, of electrons by molecules 37/1(320–332)
Scattering, extreme relativistic 35(158)
Scattering intensity 36(192)
Scattering of fast electrons by molecules 37/1(321–322)
Scattering of light 35(400)
Scattering of medium energy electrons by molecules 37/1(322–324)
Scattering of slow electrons by molecules 37/1(324–332)
Scattering of x-rays and electrons 36(192)
Scattering (*S*) matrix 36(249, 302)
Schmidt procedure of orthogonalization 37/2(138)
Schrödinger's equation 35(90); 36(5, 6, 13, 25, 61); 37/1(179, 196, 198–199)
Schrödinger's variational principle 35(232)
Screened nuclear charge 37/2(22)
Screening 35(171, 383, 411, 420, 426)
Screening, atomic 35(420)
Screening, complete 35(426)
Screening constant, inner 35(383)
Screening factor, constant 35(214)

Screening, Sommerfeld's empirical 35(174)
Second order anharmonic resonance 37/1(290–291)
Second order corrections to molecular energy 37/1(245–250)
Second order Hamiltonian 37/1(240ff., 290, 294)
Secular determinant 37/2(186, 187, 197)
Secular determinant of Wang 37/1(217–221, 226, 236, 262)
Secular equation 36(20); 37/2(189, 191)
Selection of input data for least squares adjustment, 1955 35(66)
Selection rule(s) 35(338, 357); 36(38, 40, 42, 43, 68, 76, 90, 99)
— for *l*-type doubling 37/1(276, 277–278)
— for the asymmetric rotator 37/1(226–227, 264)
— for the axially symmetric rotator with internal angular momentum 37/1(228)
— for the magnetic quantum number 35(339)
— for the orbital quantum number 35(339)
— for the spherical rotator with internal angular momentum 37/1(230)
— for the symmetric rotator 37/1(223–226, 239, 252)
— in beam spectroscopy 37/1(94, 141–145)
— weakening by an electric field 35(332)
s-electrons 37/2(195, 196)
Self-consistent field 35(171, 215, 229, 231, 378); 36(172, 396)
— of the atom 37/1(322)
— with exchange 37/2(47)
— without exchange 37/2(46)
Self-consistent hydrogen molecule 37/1(331)
Self-consistent wave functions 37/2(28)
— of Hartree and Fock 36(385, 389, 392)
Self-energy diagram 35(177)
Self-energy, electrostatic 36(115)
Self-energy of the electron 35(184)
Self-energy, transverse 35(177)
Self-exchange 36(115)
Semi-classical treatment 36(296)
Semi-localized orbitals 37/2(34, 128, 141)
Separated atoms 37/2(56)
Separated oscillating fields in beam spectroscopy 37/1(64–67)
Separation of electronic and nuclear motions 37/2(1–7)
Separation of variables 37/1(199–200, 211)
Series formulae 36(17)
Series spectra 36(28)
Showers of the soft component 35(430)
Shutter methods 37/1(346)
Siegbahn-Avogadro number N_S' 35(15)
Siegbahn's scale of x-ray wavelengths 35(10)
σ transitions 37/1(94, 95)
σ-bond 37/2(196)
σ-bond configuration 37/2(250)
σ-bonds of a saturated chain 37/2(190)
σ-constant of Hammett 37/2(234–235, 239)
σ-electrons 37/2(129, 183)
σ-electrons in benzene 37/2(262)
σ-electrons in hexatriene 37/2(210)
σ-networks 37/2(262–278)
Silicon tetrahalides, ionization by electron impact 37/1(403–408)
Silver in atomic beams 37/1(9, 18, 33, 34, 101, 107, 118)
Silver, *K* and *L* shell ionization of 36(318, 319, 365)
Single configuration approximation 37/2(86)
Single-particle potentials 35(214)
Single-particle wave functions 35(171)
Single-sign resonance theory 37/2(184, 188, 203)
Singlet configuration in polyenes 37/2(212)
Singlet system 35(206); 36(66)
Singlet wave function 37/2(33)
Size of molecules 37/2(244–245)
Slater determinant 37/2(40, 48)
Slater exponents 37/2(261)
Slater orbitals in hydrides 37/2(241, 242, 244)
Slater's rule 37/2(77)
Slit, collimating 37/1(18, 35)

Slow electron scattering by molecules 37/1(324–332)
S-matrix 35(175); 36(249, 302)
S-matrix, relation to the bound states 36(249)
Sodium, apparent ionization cross section 36(311)
Sodium bromide in molecular beams 37/1(134, 138, 148, 151)
Sodium chloride in molecular beams 37/1(25, 47, 135, 138, 148, 151)
Sodium, excitation cross section 36(362)
Sodium, excitation of the *D* lines of 36(392)
Sodium fluoride in molecular beams 37/1(25, 136, 138, 151)
Sodium hydroxide in molecular beams 37/1(128)
Sodium in atomic beams 37/1(16, 17, 29, 71, 97, 98, 100, 109, 118)
Sodium iodide in molecular beams 37/1(25, 137, 148, 151)
Sodium (Na_2) in molecular beams 37/1(130, 133)
Sodium-aluminum chloride ($NaCl \cdot AlCl_3$) in molecular beams 37/1(7, 128)
Solar corona 36(396, 400)
Solar photosphere 36(400)
Solvent, effect on chain molecules 37/2(221)
Sommerfeld approximation 36(128)
Sommerfeld differential cross-section for Bremsstrahlung 35(417)
Sommerfeld fine structure constant 35(36)
Sound waves, scattering of 36(236)
Sources for atomic and molecular beams 37/1(3–18)
Sources for metastable and optically excited states 37/1(13–17)
Sources for radioactive atoms 37/1(17–18)
Space lattice of s^2p^6 configuration 37/2(249–250)
Space-charge detector 36(313)
Space-fixed coordinates 37/1(173, 191)
Spectra of aromatic ring systems 37/2(221)
Spectra of hydrocarbon chains, general properties 37/2(183)
Spectra of polyene chains 37/2(216, 219, 220)
Spectrograph with pulsed electron beam 36(313)
Spectroscopic absorption band 37/2(212)
Spherical Bessel functions 36(235)
Spherical harmonics 35(91, 430–435); 36(25)
Spherical harmonics addition theorem 35(435)
Spherical harmonics, normalized 35(432)
Spherical model of hydrides 37/2(256)
Spherical molecules, spectra 37/1(261–262)
Spherical rotator, *see also* Rotator 37/1(229–236)
Spherical wave, incoming 35(118)
Spherical wave, outgoing 35(116)
Spheroidal coordinates 37/2(29, 34, 54)
Spheroidal harmonics 37/1(327)
Spin 36(54)
Spin angular momentum 37/1(173)
Spin composition for split shells 37/2(148–150)
Spin coordinates 37/1(196)
Spin coupling in alternant orbitals 37/2(151)
Spin density 37/2(155, 157)
Spin direction 35(148)
Spin electron, Pauli theory 35(142, 143)
Spin flip 35(310)
Spin functions 37/2(132)
Spin magnetic moment of the electron 37/1(81, 113)
Spin matrices of Pauli 35(134)
Spin operators 36(59)
Spin operators of Dirac 35(134)
Spin orbit coupling 35(144, 268, 271)
Spin resonance, hyperfine structure 37/2(236)
Spin, total 35(273)
Spin wave functions 35(205)
Spinor components 36(59)
Spin-orbit interaction 36(91, 96, 98, 101, 162)
Spin-rotation interaction in molecular spectra 37/1(126, 129–132)
Spin-spin interaction 35(268, 272); 36(97)

Split notation (of Dirac matrices) 35(134, 166)
Split-shell method 37/2(146)
Split-shell method, composition of spins 37/2(148–150)
Square well model 37/2(211, 213, 216)
Standard ellipsoid 35(21)
Standard errors calculation and correlation ceofficients in a least squares adjustment 35(58)
Standing electron wave 37/2(185, 195)
Stark effect 35(225, 314–334); 36(37); 37/1(138–141)
— electric quantum number 35(316)
— for strong fields 35(320)
— linear 35(314f.)
— parabolic coordinates 35(315)
— partial transition regions for weak fields 35(331)
— plus Lamb shift 35(326)
— quadratic 35(318, 328)
— weakening of the selection rules 35(332)
States in the continuum, average energy 35(351)
States of observables 37/1(179–182)
Statistical atom, thermodynamics of 36(221)
Stereoisomeric carbon atoms 37/2(173)
Steric hindrance 37/2(173)
Steric hindrance for rotation about a bond 37/2(204–205)
Stern-Gerlach experiments 35(298)
Stilbene 37/2(198, 203, 204, 207)
Stobbe formula 35(390)
Stopping-power of atoms 36(196)
Stretching frequency 37/2(260)
Structures of opposite sign 37/2(184, 198)
Styrene 37/2(237–238)
Sub-band structure (due to quantum number *K*) 37/1(253, 255, 258 ff., 264)
Sublimation heat 36(210)
Subshell 36(32)
Substituents 37/2(201)
Substituents for methyl groups in a chain 37/2(269)
Substituents in aromatic systems 37/2(226)
Sulfur hexafluoride, LCAO approximation 37/2(69–71)
Sulfur ring S_8 37/2(193)
Sulphur hexafluoride, appearance of negative ions 37/1(368, 373)
Sulphur hexafluoride, ionization by electron impact 37/1(368, 373, 411–415)
Sum rules, examples 35(344)
Sum rules of quantum mechanics 35(341, 342, 358)
Sum rules, proof 35(345)
Superelastic collisions 36(371)
Supernumerary alternant orbitals 37/2(202)
Supernumerary orbitals 37/2(198)
Supraconductivity 37/2(217)
Surface energy 36(213)
Susceptibility, diamagnetic 36(189, 200)
Symmetric operation 37/1(204, 206)
Symmetric rotator, *see* Rotator
Symmetry, axial 36(90, 104)
Symmetry character(s) 36(25, 42, 44–47, 64, 75, 81, 84); 37/1(204, 205)
Symmetry coordinates 37/1(203)
Symmetry, cubic 36(51)
Symmetry, dihedral 36(105)
Symmetry group 37/2(7, 61)
Symmetry groups, character tables 37/2(169–171)
Symmetry, octohedral 36(51)
Symmetry of reflection 36(44, 47)
Symmetry operations of a molecule 37/2(7)
Symmetry orbitals 37/2(61)
Symmetry properties of electronic wave functions 37/2(9–17)
Symmetry, rotational 36(46, 47)
Symmetry species of an asymmetric rotator 37/1(219–220)
Symmetry, spherical 36(52, 90, 104)
Symmetry, tetrahedral 36(50)
Systems, non separable 36(52)
Systems of bands 37/1(176)

Tamm-Dancoff method 35(261, 262)
Tamm-Dancoff method, new 35(263)
Teller-Tisza selection rule 37/1(228, 296–297)
Temperature correction of the Thomas-Fermi model 36(156)

Temperature correction of the Thomas-Fermi-Dirac model 36(159)
Temperature of ovens for beam production 37/1(6, 8, 9)
Terminal atom, aza substitution 37/2(220)
Tetrahedral carbon atom 37/2(173)
Tetrahedral group T_d 37/2(8, 13)
Tetrahedral set 37/2(251)
Tetrahedrally symmetric molecule 37/1(211)
— electron scattering 37/1(332)
— vibration-rotation energies and transitions 37/1(233–234)
Tetra-methylene-ethylene 37/2(197)
Thallium chloride in molecular beams 37/1(7, 148, 151)
Thallium in atomic beams 37/1(7, 8, 16, 21, 23, 29, 34, 76, 87, 102, 107, 108, 109)
Thermal dissociation method for beam production 37/1(11)
Thermal isomerization 37/2(205)
Thermodynamic temperature scale, 1954 redefinition 35(72)
Thomas-Fermi approach 37/2(178)
— for methane 37/2(257)
— for PH_3 37/2(248)
Thomas-Fermi atom, energy of the 36(135)
Thomas-Fermi atom model 36(120, 121, 123, 362)
Thomas-Fermi equation 36(121, 122, 124, 125)
Thomas-Fermi equation, relativistic 36(161)
Thomas-Fermi-Dirac atom 36(139)
Thomas-Fermi-Dirac equation 36(140, 141)
Thomas-Reiche-Kuhn rule 35(342)
Thomson cross section 36(403)
Thomson scattering cross-section 35(398, 399)
Three-body recombination 36(401)
Three-centre bonding 37/2(272, 274)
Three-electron state, characters 37/2(43)
Three-parameter wave functions 35(241)
Threshold for of Xe^{++} 36(317)
Threshold for production of A^{++} 36(317)
Threshold law for single ionization 36(372)
Titanium tetrachloride, ionization by electron impact 37/1(408–411)
Topology of a molecule 37/2(209)
Total electron density 37/2(177, 199)
Total energy, apparent 35(384)
Total energy of a molecule 37/2(90)
Total energy of the atom 36(183)
Total inelastic cross section 36(364)
Total overlap population 37/2(82, 98)
Totally absorbing sphere 36(277)
Townsend ionization coefficient 36(335, 342)
Transformation of molecular vibrations to normal coordinates 37/1(193)
Transition fields in beam spectroscopy 37/1(73–79)
Transition frequencies between magnetic sublevels 37/1(91)
Transition metals 37/2(175)
Transition probabilities
— for hydrogen 35(352)
— for hydrogen in polar coordinates 35(348)
— optical 36(407)
Transition probability 35(336); 36(327)
— average 35(352)
— for quantum transitions in a beam 37/1(69)
Transition processes, analysis in a rotating coordinate system 37/1(62–63)
Transition processes in atomic or molecular beams 37/1(54–79)
Transition processes, method of separate oscillating fields 37/1(64–67)
Transition-density matrix of Löwdin 37/2(178)
Transitions between two stable levels 37/1(55–57)
Transitions, bound-free 35(381)
Transitions, density of 37/1(131)
Transitions, free-free 35(403, 420)
Transitions, free-free absorption 35(420)
Transitions involving optically excited states 37/1(71–73)
Transitions to unbound states (*see also* Ionization) 36(300)

Transitions within p^2, p^3 and p^4 configuration 36(382)
Transitions without change of principal quantum number 35(349)
Transverse correlation 37/2(35)
Transverse in-and-out effect 37/2(147)
Triple bond 37/2(253)
Triple σ-bond 37/2(271)
Triplet configuration in polyenes 37/2(212)
Triplet, inverted regular 35(272)
Triplet, regular 35(272)
Triplet system 35(206); 36(66)
Triplet wave function 37/2(33)
Tritium in atomic beams 37/1(10, 12, 100, 107)
Tropylium ion 37/2(277)
T-scale normalization 35(109)
Tungsten, L_{I}, L_{II} and L_{III} ionization 36(319)
Twist about a bond 37/2(204)
Twisting vibrations in MH_4 compounds 37/2(261)
Two-dimensional box model 37/2(228)
Two-fluid theory of water 37/2(252)
Two-quantum annihilation 35(200)
Two-quantum processes 35(371)
Two-wire deflecting field 37/1(42, 43)

Ultraionization 36(316)
Ultraviolet absorption of polyenes 37/2(216)
Ultraviolet band spectra 37/1(175)
Uncertainty principle 36(233)
United atoms 37/2(56)
— approximation of hydrides 37/2(241)
— Bingel's expansion of the approximation 37/2(262)
Units, artificial 35(1)
Units, atomic 35(88)
Unpaired electron distribution in radicals 37/2(200)
Unresolved spectra of molecules 37/1(123, 126–138)
Uranium hexafluoride, scattering of electrons 37/1(323)
Urey-Bradley field 37/2(260, 261)

Vacuum, bare 35(262)
Vacuum polarization 35(163)
Vacuum polarization diagram 35(177)
Vacuum, real 35(263)
Valence bond 37/2(105)
Valence cone 37/2(248, 250–251)
Valence cone in boron hydrides 37/2(273–274)
Valence states 37/2(110)
Valency 37/2(105)
van der Waals
— crystals 36(206)
— energy 36(203, 208)
— force 37/2(37)
Variance 35(62)
Variance analysis, DC 55 35(73)
Variation method 35(232)
Variation method for helium 35(216)
Variation perturbation method 35(208, 238, 333)
Variational equations for the mixing parameter 36(291)
Variational method(s) 36(19, 20, 36, 107, 258, 295, 303, 304, 381)
— of Coulson 37/2(29)
— of Hulthén 36(293, 374)
— of Schwinger 36(264, 266, 293, 305)
— for close coupling problems 36(289)
— for determining the phase shifts 36(258)
— for determining the total scattered amplitude 36(262)
Variational principle for classical scattering 36(267)
Variational solutions 36(295)
Variational wave functions 35(237)
VB method (*see also* Valence bond method) 37/2(105)
Vector model 36(56, 67, 68, 77, 81, 91)
Vegard-Kaplan band system of nitrogen 37/1(335)
Velocity distribution in the beam 37/1(20–23)
Velocity distribution in the beam, tables 37/1(57–61)
Velocity distribution of electrons 36(357)
Velocity of light 35(192)
Velocity of light measurements, modern methods of highest accuracy 35(29–32)
Vibrating string 37/2(211)
Vibration quantum number 37/1(174)
Vibration states of a molecule 37/1(201)

Vibrational angular momentum 37/1(174)
Vibrational energy 37/2(4, 5)
Vibrational excitation, theory 37/1(353–357)
Vibrational frequencies in polyhydrides 37/2(255, 259–261)
Vibrational frequencies of hydrogen molecule 37/2(21)
Vibrational zero-point energy 37/2(4)
Virial law 36(136, 144, 146, 156, 218, 219)
Virial theorem 35(104, 251); 37/2(23, 190, 254)
Viscosity of air 35(8)

Wang secular determinant 37/1(217–221, 226, 236, 262)
Wannier functions 37/2(32, 139)
Water ions, excited states 37/1(376)
Water molecule 37/2(177)
— bond angle 37/2(248, 249, 251, 259)
— equivalent orbitals 37/2(144)
— LCAO approximation 37/2(68, 69)
— valence bond method 37/2(107, 144)
Water, two-fluid theory 37/2(252)
Water vapour, infrared spectrum 37/1(268–273)
Water vapour, ionization by electron impact 37/1(375–376)
Wave absorption within the plasma 37/1(348)
Wave character of matter 36(2)
Wave equation 36(5)
Wave function in product form 35(270)
Wave interaction 37/1(346–350)
Wave number 35(97)
Weight matrix in a least squares adjustment 35(55)
Weinbaum's approximation for the hydrogen molecule 37/2(25)
Wentzel-Kramers-Brillouin method (WKB method) 35(104, 320, 323)
WKB method (Jeffreys' approximation) 36(15, 255, 270, 297, 299, 396)
Wood's discharge tube 36(329); 37/1(10)
Work function of a metal surface 37/1(29)

Xeneon in atomic beams 37/1(14, 119)
Xenon, single ionization 36(316)
X-ray absorption edge fine structure 37/1(323)
X-ray bond lengths 37/2(222)
X-ray emission 36(317)
X-ray levels 36(184)
X-ray spectra 35(380)
X-ray spectrum 36(30, 34)
XRCD method (x-ray crystal density) 35(16)
x-unit 35(10, 15)

Yttrium in atomic beams 37/1(101)

Zeeman effect 36(40, 99)
— anomalous 35(294); 36(40, 56)
— field dependence 35(298)
— intermediate field strengths 35(307)
— normal 36(40, 58)
— quadratic 35(303)
— quasi-normal 35(296)
— relativistic corrections 35(299, 307)
— total intensity of the Lorentz triplet 35(340)
Zero moment method 35(303)
Zero point pressure 36(120)
Zero rule 36(14)
Zero-energy orbitals, *see* Non-bonding orbitals
Zero-point energy of vibration 37/2(4)
Zinc, ionization cross section 36(317)

Group 8

Nuclear Physics

A coupling, *see* Axial vector interaction
Abramowitz asymptotic expansions of Coulomb functions 41/1(442–444)
Absolute neutron flux determinations 45(479–484, 497–499, 513)
Absolutely coherent scattering 38/2(410)
Absorbed dose of neutrons 38/2(89)
Absorbed dose of radiation 38/2(674)
Absorbing boundary 38/2(775, 779)
Absorption coefficient
— inside a nucleus 40(460, 462)
— mutual, of two resonances 38/2(390)
— of nucleons in nuclear matter 41/1(258)
— of radiation energy 38/2(673)
Absorption constant in the optical model 41/1(245)
Absorption cross section(s)
— at high energies 40(490f.)
— at high energies (table) 40(493)
— for thermal neutrons 44(424–426)
— for thermal neutrons, conventional definition 38/2(41)
— for thermal neutrons, experimental values 38/2(594–596, 619, 629)
— macroscopic 38/2(213)
Absorption curve of epi-cadmium neutrons
— in cadmium 38/2(358)
— general properties 38/2(386–392)
Absorption curves of thermal neutrons 38/2(653)
Absorption density 38/2(496)
Absorption line widths 38/1(50)
Absorption measurement of resonance integrals 38/2(369–375)
Absorption of a nuclear magnetic substance 38/1(174)
Absorption of neutrons, *see* Capture
Absorption of neutrons, kinematical frame of theory 38/2(27ff.)
Absorption of π-mesons by deuterium 42(542)
Absorption of radiation 42(390)
Absorption, photoelectric, of x-rays 38/2(667–668)
Absorption potential 39(369, 381, 385, 446)
Absorption process 39(446)
Absorption resonance integral 38/2(53)
Absorption signal 38/1(188)
Absorption spectra of actinide elements 42(215, 216)
Absorption spectra, pure rotational 38/1(38)
Absorption voltage 38/1(186, 187)
Abundance, relative 39(260)
Abundance total 39(259)
Accelerating tube of a Van de Graaff generator 44(82–89)
Acceleration by induction 44(193)
Accelerators for negative ions 44(103)
Acceptance area 44(274–275)
Actinide concept, chemical evidence 42(207)
Actinide concept, crystallographic evidence 42(209)
Actinide concept, physical evidence 42(215)
Actinide elements
— absorption spectra 42(215, 216)
— crystallographic data 42(209, 211, 212, 264)
— fluorescence spectra 42(215, 216)
— metallic state 42(211)

Actinide elements (cont.)
— oxidation states 42(207, 209, 210, 229, 241)
— solvents for extraction 42(235, 251, 259, 268, 274)
— spectroscopic data 42(221)
Actinium 42(212)
Actinium neutron source 38/2(122)
Activation analysis 44(424)
Activation cross section for thermal neutrons 44(422–424)
Activation equation 44(403)
Activation measurement of resonance absorption integral 38/2(368–369)
Activation of detector 38/2(41)
Activation of detector, difference from counting rate 38/2(354)
Activities on both sides of detector, arithmetic average 38/2(357, 640)
Activity, deuteron induced 40(132)
Activity, measured, of a detector 38/2(354)
Addition of angular momenta 41/1(296)
Addition theorem of spherical harmonics 41/1(302)
Adiabatic damping 44(195–196, 203)
Adiabatic damping of betatron oscillations 44(307, 315)
Adiabatic demagnetization 41/2(70)
Adiabatic expansion producing supersaturation 45(266–270)
Adiabatic functions 41/1(379, 380)
Adiabatic precession, *see* Precession
Adiabatic switching off 39(136)
Adjustment, orthogonal 38/1(109)
Admittance 44(264, 270, 304, 307)
Advantages of heavy particle studies 41/1(368)
Afterpulses in mulipliers 45(94)
AG, *see* Alternating gradient
Age correction for thermal neutrons 38/2(493)
Age determinations, archaeological 45(210, 211)
Age equation 38/2(270)
— as a special case of the Boltzmann equation 38/2(287–289)
— solution for plane source 38/2(272–273, 274–275)
— solution for point source 38/2(273–274, 275–276)
Age increase from indium resonance to thermal energies 38/2(497–499)
Age measurements in graphite-iron moderator 38/2(335)
Age measurements in heavy water 38/2(329)
Age measurements in water-metal mixtures 38/2(315, 318, 326)
Age of indium resonance neutrons 38/2(311, 321)
Age of neutrons 38/2(255, 305)
— definition 38/2(269)
— generalized 38/2(500)
— methods of measurement 38/2(306)
Age of thermal neutrons in graphite 38/2(499)
Age of thermal neutrons in water 38/2(498–499)
Age theory 38/2(268ff., 287)
— as a limiting case 38/2(475, 490)
— elementary improvements 38/2(276–278)
— range of applicability 38/2(289–291)
Age to absorption of neutrons 38/2(496)
AGS (alternating gradient synchrotron, Brookhaven) 44(326–335)
Air, dielectric strength 44(65)
Air, energy absorption coefficient of γ-rays 38/2(674)
Air in Van de Graaff generators 44(76, 97)
Air-core magnet 44(235–236)
Albedo 38/2(535, 545–552)
Albedo defect 38/2(548)
Albedo, experimental 38/2(549, 596)
Albedo in Wick's approximation 38/2(574–575)
Albedo of radiation 38/2(789, 791)
Albedo of thermal neutrons 38/2(69)
Alignment, nuclear 42(558, 568)
Allowed infavoured transition 41/2(15, 41)
Almost isotropic scattering, diffusion of thermal neutrons 38/2(524)
Alpha emission channel 41/1(361)
(α, n) process 38/2(3, 13)
(α, n)-reactions as neutron sources 38/2(107–132)
Alpha nucleus 39(460, 461)

Alpha particle
— capture 40(47)
— counter 45(52, 77)
— energies 42(124–131)
— model 39(366, 460–464); 40(11, 185, 187)
— scattering: determination of nuclear size 39(216, 217)
— sources, absolute strength determination 45(44)
Alpha radioactivity, decay energies 40(342)
Alpha radioactivity, hindrance (or departure) factor 40(345)
Alpha radioactivity, lifetimes 40(343)
α-α interaction from scattering 39(464)
α-decay 38/1(5); 41/1(355ff.)
α-decay and ℛ-matrix 41/1(356)
α-decay as a Coulomb excitation process 41/1(357)
α-decay determinations of nuclear size 39(218)
α-decay, systematics of decay energies 38/1(31–32)
α-decay energy systematics 42(288, 289, 297)
α-decay, favored 42(122, 167, 177, 182, 199, 200)
α-decay from aligned nuclei 42(571)
α-decay in odd nucleon types of transuranium elements 42(237)
α-decay, non-central interactions 42(159, 162, 184, 202)
α-decay rate of transuranium elements 42(295)
α-decay rates, numerical formulas 42(155)
α-decay, selection rules 42(183)
α-decay, theory of Gamov 41/1(1, 32)
α-decay, time-dependent treatment 42(149)
α-disintegration energies 42(111)
α-distribution of photoprotons or photoneutrons at low energies 42(326)
α-distribution of radiation from aligned nuclei 42(568, 575)
α-emitters among rare earths 42(120)
α-energies, systematics 42(117)
α-energy surface 42(112)
α-γ cascades in radioactive nuclei 42(413–417)
α-γ-angular correlation 42(186, 188, 189, 192, 201)
α-groups of long range 42(123)
α-half lives 42(124–131)
α-potential energy, intranuclear 42(172)
α-spectra 42(120, 123)
Alternating gradient accelerators, theory 44(300–319)
Alternating gradient focussing principle 44(115, 215, 301)
Aluminium, (α, n) reaction 38/2(129)
Aluminium cross section for high energy neutrons 45(491)
Aluminium, directional distribution of photon flux 38/2(703)
Aluminium, energy absorption coefficient of γ-rays 38/2(674)
Aluminium, minimum of γ-ray attenuation 38/2(675)
Aluminium, radiation albedo 38/2(791, 794)
Aluminium, radiation attenuation coefficients 38/2(667)
Aluminium, spectral distribution of radiation flux 38/2(700)
Aluminium-heavy water arrangement 38/2(330)
Aluminium-water arrangement 38/2(318)
Alvarez-type accelerator 44(366ff., 386)
A-matrix 40(412)
Americium 42(254–264)
Americium, chemical properties 42(258)
Americium, decay schemes 42(255)
Americium, isotopes 42(256)
Amplification of ionization chamber signals 45(31)
Amplitude balance 38/1(188)
Amplitude discriminator 45(184, 189, 246, 247)
Analogue states 40(141)
Angular correlation(s) 38/1(399, 400); 41/1(11, 522–524, 531)
— anisotropy 42(13, 14, 27)
— between β- and γ-rays 41/2(51–52)
— between β-particles and circularly polarized γ-quanta 41/2(71–74)
— between electron and neutrino 41/2(24–25)
— coefficient for neutron decay 41/2(76, 81, 86, 91)

Angular correlation(s) (cont.)
— coefficients, numerical results 41/1(532)
— for barium 139 decay 41/2(49)
— for several successive radiations 42(399)
— for three successive radiations 42(397)
— of γ-rays from oriented Co^{60} 42(581)
— of successive radiations 40(54)
— parameter 41/2(25)
Angular deflection by small-angle scattering 45(372)
Angular distribution
— experiments 40(54, 59)
— in a γ-γ-cascade 41/1(521–524, 531)
— in heavy particle reactions, calculations 41/1(395)
— in heavy particle reactions, experiments 41/1(397–398)
— in neutron scattering 45(438, 440)
— law in Wick's approximation 38/2(574–575)
— law of Fermi for thermal neutrons 38/2(561, 574, 654)
— of β-particles from polarized nuclei 41/2(68–71)
— of capture radiation 40(60)
— of elastically scattered neutrons 41/1(249–250)
— of electrons scattered 40(509)
— of gamma rays 41/1(437, 515, 517)
— of gamma rays, coupling to atomic electrons 41/1(556–558)
— of inelastically scattered neutrons by a single crystal 38/2(455)
— of neutrons from B(d, n) 38/2(205)
— of neutrons from Be(α, n) 38/2(115, 117)
— of neutrons from D(d, n) 38/2(190, 192–193)
— of neutrons from Li(d, n) 38/2(200)
— of neutrons from Li(p, n) 38/2(181)
— of neutrons from T(d, n) 38/2(196–197)
— of neutrons from T(p, n) 38/2(176–177)
— of neutrons scattered 40(488f.)
— of neutrons scattered by beryllium 38/2(75)
— of neutrons scattered by carbon 38/2(77)
— of neutrons scattered by deuterium 38/2(72–73)
— of neutrons scattered by heavy elements 38/2(81)
— of neutrons scattered by oxygen 38/2(78)
— of photoneutrons from beryllium 38/2(144, 145)
— of photoneutrons from bismuth 38/2(157)
— of photoneutrons from deuterium 38/2(138, 141)
— of products 40(217)
— of protons from D(d, p) 38/2(190, 193)
— of radiation 38/2(702–703, 712, 722, 749, 804, 809)
— of radiation, moments 38/2(749, 763)
— of reaction products 41/1(293–304)
— of reflected x-rays 38/2(793–794)
— of scattered neutrons, molecular interference effects 38/2(427–431)
— of target x-rays 44(192)
— of thermal neutrons emerging from a free surface 38/2(539)
Angular frequencies of betatron oscillations 44(194–195)
Angular momenta, addition using Racah coefficients 41/1(296)
Angular momenta, coupling 39(390, 391)
Angular moments of neutron current 38/2(511, 519)
Angular momentum 42(160, 177, 181, 183)
— considerations in stripping 38/2(186–187)
— expansion 38/2(28, 30)
— of collective oscillation 39(420)
Angular number flux, *see* Flux density
Angular resolution of tracks 45(343)
Angular shape of nuclei 39(229, 237)
Angular shape of nuclei near closed shells 39(237)
Angular spread of neutrons, effect on self-absorption curve 38/2(390–391)
Anharmonicity of the potential function 38/1(41)
Anisotropic scattering, diffusion of thermal neutrons 38/2(510–511, 538)

Anisotropy in angular correlation 42(13, 14, 27)
Anisotropy of γ-rays 42(583, 586)
Anisotropy of scattering, effect on validity of slowing-down theory 38/2(289, 295)
Annihilation in flight 41/2(5, 61)
Annihilation radiation 38/2(665, 752, 755, 793)
Annihilation radiation from stopped positrons 41/2(4, 33, 60–64)
Anomalies of isotope shifts 38/1(96)
Anomaly, magneto-mechanical 38/1(274)
Anthracene crystal scintillators 45(144, 145, 148, 149, 152)
Anthracene response 45(148, 149)
Anticoincidence 45(81–82, 224, 231–234)
Anticoincidence arrangement of proportional counters 45(189, 213)
Anticoincidence stage 45(232)
Anti-damping due to radiation 44(186)
Antimony-beryllium source of neutrons 38/2(307, 321, 335, 337)
Anti-neutrino 41/2(89)
Antiparticles 41/2(4)
Antiproton, detection in emulsions 45(348)
Antiradioactive states 41/1(44)
Antishielding 38/1(413, 415)
Antishielding factor 38/1(414, 416, 417, 430)
Antishielding in ionic crystals 38/1(416)
Aperture at injection 44(234)
Application of Green's functions to reactions 41/1(80–83)
Arc sheath (plasma boundary) 44(147, 148)
Archaeological age determinations 45(210, 211)
Area analysis of a neutron resonance line 40(378, 380, 381)
Area analysis of resonances 44(417)
Argon 37
— electron capture 41/2(21–22)
— generation by neutrino absorption 41/2(33)
— internal bremsstrahlung 41/2(104)
— L/K capture ratio 41/2(97)
— recoil investigations 41/2(23, 29)
Argon 35, recoil investigations 41/2(25, 28)
Argonne heavy water pile, internal measurements 38/2(362–364)
Assembly of different moderators 38/2(279)
Asymmetry coefficient, forward 39(6, 113)
Asymmetry constant 38/1(386–389)
Asymmetry of γ-ray emission from polarized nuclei 41/2(70–71)
Asymmetry of slow neutron scattering due to electron interaction 38/2(9–10)
Asymmetry parameter 38/1(386, 387); 39(434); 41/1(475)
Asymptotic behavior of the Coulomb functions 41/1(409, 440–449)
Asymptotic distribution of monoenergetic neutrons 38/2(521)
Asymptotic solution 38/2(528 ff.)
Atomic beams 38/1(77–83)
Atomic beams, magnetic resonance method 38/1(77)
Atomic effects 38/1(114)
Atomic mass calculations 42(119, 152)
Atomic shielding and Coulomb excitation 41/1(559)
Atomic spectra, experimental methods 38/1(75)
Atomic spectra, optical instruments 38/1(75)
Atomic spectra, sources 38/1(75)
Attachment of electrons 45(25, 177)
Attenuation 38/1(131)
Attenuation coefficient 40(462, 491, 521)
— of radiation 38/2(672, 756)
— of radiation, mimimum 38/2(675, 707, 751, 755)
Attenuation factor for correlations 42(522–525)
Attenuation length of neutrons 38/2(297, 309)
Attenuation length of thermal neutron waves 38/2(579–580)
Attenuation of broad-beam radiation 38/2(675)
Attenuation of narrow-beam radiation 38/2(665 ff.)
Attractive Coulomb interaction 41/1(409)

Auger effect following electron capture 41/2(95)
Auger electrons 41/2(21, 24); 42(23); 45(9, 203, 217)
Ausstrahlungsbedingung 38/2(683)
Automatic triggering of cloud chamber expansions 45(277)
Avalanche 45(2, 53, 54, 58)
Average electrical gradient in accelerator tube 44(368, 370)
Average energy decrement per collision 38/2(218)
Average level distance 38/2(367)
Average logarithmic energy decrement per collision 38/2(218, 227)
— corrected for thermal motion 38/2(400)
Average neutron cross section 44(418, 430)
Average square distance travelled by neutrons 38/2(259, 278)
Average velocity of thermal neutrons 38/2(40)
Average width of many levels 38/2(367)
AVF cyclotron, *see* Azimuthally varying field cyclotron
Axial focussing in a betatron 44(195, 202)
Axial oscillations, *see* Vertical oscillations
Axial vector interaction 41/2(9, 88–93)
Azimuthal harmonics in constant gradient synchrotron 44(241)
Azimuthal trimming 44(243)
Azimuthally varying field (AVF) 44(112, 115–117)

Back scattering of electrons 41/2(40)
Background counting rate in proportional counters 45(208–214)
Background counts 45(54)
Background due to γ-ray interaction 45(209)
Background eradication 45(395–398)
Background grains 45(357)
Background matrix (resonance-free part of scattering matrix) 41/1(45)
Background term in the one-body treatment 41/1(48)
Background, ultimate 45(211)
Backscattering coefficient 38/2(792)
Backscattering of beta-rays 45(202)
Backscattering of radiation 38/2(742, 760, 789, 804, 809, 811)
Balanced chamber method 45(43)
Band width 38/1(198, 200, 230)
Bare reactor 44(391, 392)
Barium 139, anisotropy 41/2(49)
Barrier around square well example 41/1(139–148)
Barrier penetration 41/1(1, 53, 62, 142, 145)
Bartlett interaction 39(425)
Beads on central wire 45(84)
Beam control in synchrotrons 44(289–293)
Beam deflection in the cyclotron 44(150–153)
Beam deflection in the synchrocyclotron 44(153–154)
Beam movement around axis of Van de Graaff tube 44(85)
Beam of polarized neutrons 41/2(82); 44(439–440)
Beam survival (*see* Gas scattering) 44(271–273, 282–283)
Belt, *see* Charging belt
Belt arrangement in a generator 44(74–76)
Belt, field at its surface 44(75)
Belt, mechanical forces on it 44(76)
Belt replacement 44(76)
Belt separation 44(76)
Belt stability 44(76)
Berkeley Bevatron, *see* Bevatron
Berkeley grid system 40(479)
Berkelium 42(270–274)
Berkelium, chemical properties 42(273)
Berkelium, decay schemes 42(270, 272)
Berkelium isotopes 42(271)
Beryllium 10 decay 41/2(44)
Beryllium 8 disintegration during lithium 8 recoil 41/2(29–30)
Beryllium 7, *K*-capture recoils 41/2(20)
Beryllium 7, *L*-capture 41/2(97)
Beryllium, albedo 38/2(596)
Beryllium, (α, n) reaction 38/2(113–123)
Beryllium decay curve of thermal neutrons 38/2(626)
Beryllium, diffusion parameters of thermal neutrons, measurements 38/2(582, 584, 605–607, 624)
Beryllium, (d, n) reaction 38/2(201–203)

Beryllium, elastic scattering of neutrons 38/2(75–76)
Beryllium moderator, resonance neutron distributions 38/2(330–331)
Beryllium, nuclear photoeffect 38/2(132, 141–145)
Beryllium oxide
— albedo 38/2(596)
— diffusion parameters of thermal neutrons, measurements 38/2(584, 607, 627–628)
— filter producing cold neutrons 38/2(436)
— polycrystal, neutron cross section 38/2(439)
Beryllium photoneutron sources, technical details 38/2(161)
Beryllium, (*p*, *n*) reaction 38/2(171, 183)
Beryllium polycrystal, neutron cross section 38/2(438, 445–446, 476)
Beryllium, reactions induced by neutrons 38/2(75–76)
Beryllium, threshold for photodisintegration 38/2(141–143)
Beryllium used in (*α*, *n*) neutron sources 38/2(108–109)
Beryllium, zero-energy scattering cross section 38/2(65)
Beryllium-antimony source of neutrons 38/2(307, 321, 335, 337)
Bessel function(s)
— notation used 41/1(338, 383)
Beta activities, induced in indium by slow neutrons 38/2(56)
Beta activities, induced in rhodium by slow neutrons 38/2(52)
Beta activity, induced in gold by slow neutrons 38/2(58)
Beta decay 38/1(4–5); 40(34)
— of the neutron 38/2(8)
— selection rules 40(35)
— systematics of decay energies 38/1(20, 32–33)
Beta particle counter (*see also* Windows) 45(76, 77)
Beta stability 42(119)
Beta stability, maximum line 38/1(19)
Beta stable isotopes of transuranium elements 42(293)
Beta stable nuclides 38/1(17–20)
β vibrations 42(195)
β-decay (*see also* *ft* values) 39(270ff.)
β-decay, angular correlation 42(505)
β-decay, electron spin polarization 42(571, 573)
β-decay energy systematics 42(291)
β-decay matrix elements 39(408)
β-*γ*-correlations 42(506, 508, 511)
Beta-ray backscattering 45(202)
Beta-ray scattering in a counter 45(195)
Beta-ray, spectrometer transmission 42(20)
Beta-rays from aligned nuclei 42(571)
β-transitions 39(296)
Betatron 40(468, 472, 511)
Betatron acceleration in synchrotrons 44(189)
Betatron bremsstrahlung producing photoneutrons 38/2(168)
Betatron, different types 44(205–206)
Betatron, historical remarks 44(193)
Betatron orbits, linearized theory 44(196–202)
Betatron oscillations 44(194)
— in FFAG synchrotrons 44(336)
— in the synchrotron 44(185–188, 222, 227, 300–307, 446)
Bethe-Weizsäcker formula 38/1(26–30)
Bevatron, description 44(219, 230–233)
Bevatron, field index 44(238–239)
Bevatron, gas scattering 44(262–263)
Bevatron, injector and pre-accelerator 44(266, 268, 294, 382)
Bevatron, magnet 44(237, 243, 246)
Bevatron, phase oscillations 44(279)
Bevatron, power supply 44(251, 253)
Bevatron, radiofrequency apparatus 44(287–289)
Bevatron, radiofrequency parameters 44(279)
Bevatron, survival of injected beam 44(273, 282)
Bevatron, target problems 44(294, 299)
Bevatron, vacuum chamber 44(259)
Bhabha scattering of polarized positrons by electrons 41/2(53, 59–60)
Biased sampling of photon histories 38/2(783, 786)
Binding effect, chemical 39(43, 46)
Binding effect on the neutron-proton scattering cross section 38/2(417)

Binding energies 39(258, 295, 370, 371, 373)
— from the α-particle model 39(461–464)
— of atomic electrons 42(24)
— of nuclei, semiempirical formula 39(415, 416)
— of nucleons 42(113)
— of transuranium elements 42(293)
Binding energy
— definition 38/1(2, 3)
— effect of magic numbers 38/1(35–36)
— formula, semi-empirical 38/1(24–26)
— of a nucleus, higher order corrections 39(440–442)
— of a nucleus in the Fermi model 39(426–429)
— of H^3 and He^3 39(153)
— of the deuteron 38/2(5); 39(2, 32)
— per nucleon 38/1(14)
— surfaces 38/1(16)
— surfaces, isobaric section 38/1(15, 20–21)
— tables for all nuclides 38/1(7–13)
Binding forces in molecules, effect on neutron scattering cross section 38/2(401 ff., 416 ff.)
Binding of atoms in a crystal, effect on Doppler broadening 38/2(349)
Biological effectiveness 44(441)
Birmingham synchrotron
— beam after injection 44(272–273, 282)
— Cee 44(286–287)
— description 44(219, 230–233)
— field index 44(238)
— gas scattering 44(263)
— injector and pre-accelerator 44(265, 268, 281, 294)
— magnet 44(237, 242, 245)
— phase oscillations 44(279)
— power supply 44(251–252)
— radiofrequency apparatus 44(283–287)
— radiofrequency parameters 44(279, 287)
— target problems 44(294)
— vacuum chamber 44(257)
Bismuth fission as high-energy neutron detector 45(445, 504–506, 512)
Bismuth loaded scintillant 45(506, 516)
Bismuth, photodisintegration 38/2(157)
Black body for thermal neutrons 38/2(543–545)
Black detector 38/2(654)
Black sphere 41/1(304)
Black sphere as a nuclear model 38/2(81, 86)
Black-box treatment of the nucleus 41/1(5–6, 8, 24, 53, 73)
Black-box treatment, principal limitations 41/1(88)
B_L-approximation 38/2(286)
Blob density 45(356, 379)
Blob density, definition 45(419)
Blob length measurement 45(423–425)
Blob structure of track, its influence on curvature measurements 45(373)
Blobs in emulsion tracks 45(352)
Bloch's differential equations 38/1(130)
Bloch's differential equations, quasistationary solutions 38/1(178)
Blocking oscillator 45(229, 246)
Body system 39(482, 519, 527)
Bohr hypothesis of compound nucleus 40(12, 203, 204)
Bohr magneton 38/1(260)
Bohr theory 38/1(41)
Bohr-Mottelson strong coupling model 42(135, 166)
Bohr's model of resonance reactions 38/2(46, 84–85)
Bohr's model of resonance reactions, explanation of giant resonances 38/2(152–153, 155)
Bohr-Weisskopf factor 38/1(74)
Boltzmann equation
— approximate solution by method of moments 38/2(297–299)
— approximate solution by spherical harmonics method 38/2(284–287)
— for neutrons 38/2(211–216)
— for neutrons below 1 eV energy 38/2(489–491)
— for thermal neutrons 38/2(505, 507–511, 517–520, 522)
— reduction to age equation 38/2(287–289)
— time dependent, for thermal neutrons 38/2(575)
Boltzmann factor 38/1(125)
Bombarding particles, polarized 42(592)
Bond, covalent 38/1(424, 426)

Bond, covalent, radii 38/1(425)
Bond hybridization 38/1(421, 423)
Bond, intermolecular 38/1(428, 429)
Bond radii, single, double, and triple 38/1(425)
Bond structure 38/1(423)
Bonding electrons 38/1(422)
Boric acid dissolved in water, neutron spectrum 38/2(471–472)
Born approximation for slow neutron scattering 38/2(402, 418)
Boron, absorption cross section for thermal neutrons 38/2(594–596, 619, 629)
Boron, absorption of slow neutrons 38/2(60–62)
Boron, (α, *n*) reactions 38/2(123–126)
Boron as a detector of thermal neutrons 38/2(633)
Boron as a neutron detector 45(442–444, 463)
Boron as a standard for absorption cross sections 38/2(60)
Boron assumption 38/2(62)
Boron, (*d*, *n*) reactions 38/2(203–206)
Boron filter for epicadmium neutrons 38/2(351)
Boron impregnation of emulsions 45(347, 395)
Boron lined proportional counter 44(415)
Boron, (*p*, *n*) reaction 38/2(171, 183)
Boron poisoning method 38/2(590–592)
Boron, reactions by slow neutrons 38/2(25–26, 42)
Boron, scattering cross section 38/2(61, 62)
Boron trifluoride counters 45(460)
Bose-Einstein statistics 38/1(91, 94)
Boundary, absorbing 38/2(775, 779)
Boundary condition(s)
— at nuclear surface 38/2(34)
— depending on energy 41/1(163–167, 186)
— determining phase 41/1(110–111, 112, 134, 180, 220)
— for thermal neutron diffusion 38/2(541–542, 548)
— in age theory 38/2(271)
— in P_L approximation 38/2(566–567)
— in the Weisskopf approach 41/1(195)
— of Teichmann and Wigner 41/1(156, 161–163, 167, 180)
— of the photon transport equation 38/2(683)
— of Wigner and Eisenbud 41/1(112, 147, 150, 151, 161, 278)
Boundary of the nucleus 41/1(109, 257)
Boundary problems of radiation solved by Monte Carlo methods 38/2(789–801)
Bragg reflection of neutrons 44(435)
Bragg-Gray equation of dose 45(37)
Branching ratio 41/2(14)
Branching ratio of reaction channels 38/2(47–48)
Branching ratios in rotational transitions 39(493)
Breit-Wigner dispersion formula 39(446)
Breit-Wigner formula 40(20, 123, 403, 405, 406); 41/2(87); 44(415)
Breit-Wigner formula of slow neutron capture 41/1(37)
Breit-Wigner resonance formula 38/2(45, 49)
Bremsstrahlung, *see* Internal, External bremsstrahlung
Bremsstrahlung 40(472f., 476, 509, 524)
Bremsstrahlung, cross section formulas 38/2(167)
Bremsstrahlung in polarization measurements 41/2(65)
Bremsstrahlung, inner 45(219–220)
Bremsstrahlung spectrum of synchrotron x-ray beam 44(190)
Bremsstrahlung used for photoneutron production 38/2(145, 147–149, 166–170)
Bridge methods for the nuclear magnetic high-frequency spectroscopy 38/1(185–204)
Brillouin function 38/1(126)
Broad-beam attenuation 38/2(675)
Broch's correction factor to perturbation theory 38/1(102)
Brookhaven alternating gradient synchrotron (AGS) 44(326–335)
— parameters 44(309, 317, 326, 333)
Brookhaven cosmotron, *see* Cosmotron
Brookhaven electron analog 44(318, 320–325)

Brookhaven graphite pile, internal measurements 38/2(362–363)
Brookhaven linear accelerator for protons 44(343, 372)
Brueckner theory 39(374, 380, 445); 40(461); 41/1(14–24, 151)
Bubble chamber
— applications 45(335–341)
— as high-energy neutron detector 45(511)
— as moderate-energy neutron detector 45(466)
— development in recent years 45(314)
— principle 45(314)
— sensitive time 45(322–325, 327, 328)
Bubble count 45(333)
Bubble density 45(317)
Bubble growth rapidity 45(316)
Bubble nucleation by ions, mechanism 45(328–330)
Bubble track curvature 45(333)
Buckshot hypothesis 41/1(406)
Build-up factor of secondary x-rays 38/2(676, 678, 721–722, 724, 740, 752–753, 756–758, 760, 772, 795)
Build-up resonance, aperiodic and periodic 38/1(170)
Bunching in linear accelerators 44(354–357, 379)
Butane, scattering of slow neutrons 38/2(417, 425)
B-values of rotational spectra, equilibrium values 38/1(45)

C group of neutrons 38/2(43, 63)
Cadmium, absorption of thermal neutrons 38/2(43–44, 54)
Cadmium, absorption of water moderated neutrons 38/2(358–359)
Cadmium as a detector of thermal neutrons 38/2(634)
Cadmium cut-off 44(409)
Cadmium, cut-off energy 38/2(56, 342)
Cadmium difference method 38/2(351, 634–635)
Cadmium, energy dependence of slow-neutron total cross section 38/2(54–55)
Cadmium ratio of a detector 38/2(360, 364)
Cadmium, resonance parameters 38/2(54)
Cadmium tungstate as a scintillant 45(143)
Cadmium-covered detector foil 38/2(311, 317)
Calcium tungstate as a scintillant 45(143)
Calibration of a synchrotron 44(190–192)
Calibration of counters by radioactive sources 45(184)
Californium 42(274–278)
Californium, chemical properties 42(278)
Californium, decay schemes 42(275, 277)
Californium isotopes 42(276)
Californium, spontaneous fission 38/2(210)
Calorimetric mean energy measurements 41/2(17–19)
Canberra Synchrotron, magnet 44(235–236)
Capacitor input of a rectifier 44(2–20)
Capacitor shell for voltage control 44(93)
Capillary arc ion source 44(89)
Capture cross section
— effect of Doppler broadening 38/2(347–348)
— effective, per pile atom 38/2(376)
— for neutrons (*see* Neutron capture) 40(303f., 307)
— for protons 40(260)
— of boron for thermal neutrons 38/2(594–596, 619, 629)
— of protons for thermal neutrons 38/2(594–596)
— $1/v$ law 38/2(21, 24, 41, 42, 51)
— varying rapidly with energy, effect on slowing-down process 38/2(244–246)
— varying slowly with energy, effect on slowing-down process 38/2(240–242)
Capture γ-radiation 42(429)
Capture γ-ray intensity in dependence of slowing-down time 38/2(341)
Capture γ-ray studies at a reactor 44(428–429)
Capture of neutrons 45(447–449)
— by beryllium 38/2(65, 76)
— by carbon 38/2(65, 77)

— by deuterons 38/2(65, 73, 75)
— by oxygen 38/2(65, 79)
— by protons 38/2(68–70)
— Fermi formula 38/2(70)
Capture of thermal neutrons, measurements in steady state assemblies 38/2(592–596)
Capture probability for nucleons 40(130, 133)
Capture processes, direct 40(68)
Capture, radiative 38/2(18, 41ff., 48, 82)
Capture-to-fission ratio 44(421)
Carbon 12, excited levels 38/2(116, 120–122)
Carbon, (α, *n*) reaction 38/2(127–128)
Carbon as a high-energy neutron detector 45(450, 507, 508, 513)
Carbon cross section for high energy neutrons 45(491)
Carbon, (*d*, *n*) reactions 38/2(206–208)
Carbon, elastic scattering of neutrons 38/2(77)
Carbon, inelastic scattering of neutrons 38/2(77, 85)
Carbon ions inducing reactions 41/1(405–406)
Carbon, radiation attenuation coefficients 38/2(664)
Carbon, reactions induced by neutrons 38/2(76)
Carbon, scattering mean free path 38/2(293)
Carbon, slowing-down length, calculated 38/2(267)
Carbon tetrachloride in Van de Graaff generators 44(67)
Carbon, thermal absorption cross section 38/2(77)
Carbon, zero-energy scattering cross section 38/2(65)
Carbon-tetra-fluoride, scattering of slow neutrons 38/2(420–421, 428, 430)
Carnegie tech. synchrocyclotron 44(134)
Cascade decay of Ξ^-- and Σ^0-particles 42(531)
Cascade detection in emulsions 45(348)
Cascade excitation 41/1(549)
Cascade generator 40(24)
— as a band pass 44(57, 58)
— as pre-accelerator 44(378–379, 382)
— comparison with measurements 44(40)
— computational principles 44(39)
— different types 44(38, 46, 47)
— load characteristic 44(42)
— optical step number 44(30, 33)
— ripple 44(27)
— stray capacities 44(26, 34, 36–38)
— substitutional rectifier 44(30)
— symmetrical type 44(43–45)
— under load 44(25, 30–34, 38)
— voltage drop due to stray capacities 44(26, 34, 37)
— voltage drop under load 44(27, 33, 42)
— with open circuit 44(24, 25, 35)
Cascade radiation 42(485)
Cascade shower observation in a cloud chamber 45(311)
Cascade shower with a scintillation counter 45(164–169)
Cascade theory of nuclear reactions 40(457, 464f., 504, 518, 522)
Casimir operator 39(291)
Casimir function 38/1(385)
Causality considerations 41/1(365)
Cavities, effect upon thermal neutron distribution 38/2(544)
Cavity design for linear accelerator 44(347)
Cavity lengths in electron accelerator 44(351)
Cavity lengths in ion accelerator 44(370–371)
Cavity parameters in linear accelerators 44(350–353, 368–371)
Cee (accelerating electrode) 44(257, 286)
Cell length on a curved track 45(375)
Cement of high strength 44(82, 84)
Center-of-mass frame of reference 38/2(18, 91ff.)
Central field, correspondence principle 41/1(94)
Central force in the two-particle interaction 39(403)
Central forces 39(23, 24, 27, 30, 31, 32, 96, 290, 332, 403, 405)
Central potential, effective 39(74, 87)

Centre of mass motion, correction 39(329)
Centrifugal barrier 41/1(53, 55)
Centrifugal barrier, effect on particle width 38/2(50)
Centrifugal distortion 38/1(41)
Centrifugal force on particles in a linear accelerator 44(359)
Centrifugal reduction factor for α-emission 42(149)
Ceramic tube 44(73, 82)
Ceramic walls of a vacuum chamber 44(257)
Čerenkov counter design 45(86, 158–164, 171–173)
Čerenkov counter for high-energy recoil protons 45(495, 498)
Čerenkov detector 40(475f., 506)
Čerenkov directional counters 45(159–160)
Čerenkov focusing counters 45(161–164)
Čerenkov gas counters for high energy particles 45(164)
Čerenkov radiation in emulsions 45(358)
Čerenkov radiation, properties 45(154–157)
Čerenkov threshold counters 45(157–159)
Cerium isotopes 139 and 141, decay schemes 42(586, 587)
CERN injector to proton synchrotron 44(383)
CERN-PS (proton synchrotron), parameters 44(317, 326, 333)
Cesium fluoride as a scintillant 45(139, 140)
Cesium iodide as a scintillant 45(138, 139)
c.f.p. (coefficients of fractional parentage) 39(332, 398, 399)
Channel 40(13)
Channel, definition 41/1(39–41)
Channel index 42(588)
Channel of decay 40(403f., 407f., 410)
Channel radius 40(410); 41/1(167, 183, 234)
Channel spin 40(21, 56); 42(421, 446, 588)
Channel wave function 41/1(160, 219)
Channels, closed 41/1(160, 168–171)
Channels of reaction 38/2(18, 47–48)
Channels, open 41/1(168, 172)
Characteristic of a counter 45(54, 55, 68)
Characteristic of a rectifier, non-linear 44(2, 3)
Characteristic of a rectifier with capacitor input, exponential law 44(17–19)
Characteristic of a rectifier with capacitor input, power law 44(10–17)
Characteristic x-ray emission 41/2(7)
Characteristic x-ray following electron capture 41/2(95, 97)
Charge conjugation 41/2(104)
Charge coordinates 39(16)
Charge density of a nucleus 41/1(255)
Charge determination by Čerenkov counters 45(160–161)
Charge determination for an emulsion track, charge ratio 45(377)
Charge distribution 39(195, 196, 416, 431)
Charge distribution of recoil nuclei 41/2(24)
Charge distribution within the nucleus 40(451, 493, 503, 505, 526f.)
Charge independence in nuclear forces 42(349)
Charge independence of nuclear forces 40(486); 41/1(11, 13)
Charge independent interaction 39(17, 88, 91, 117, 124, 125, 126, 127)
Charge multiplet 40(5)
Charge of β-particles 41/2(2)
Charge of daughter nucleus, experimental determination 41/2(8)
Charge radius 39(454, 455)
Charge space 39(16)
Charge, specific, of the electron 38/1(271)
Charge, specific, of the proton 38/1(271)
Charge spin (*see also* Isotopic spin) 39(16)
Charge symmetry 41/1(13, 236, 237, 253)
Charge symmetry of nuclear forces at high energies 39(170)
Charge Z of the nucleus, effective values 38/1(407, 411)
Charged meson theory 39(17, 18, 24)

Charge-to-mass ratio for β-particles 41/2(3)
Charging belt 44(74–79)
Charging current in a generator 44(74)
Charging of a belt 44(76–78)
Chemical binding
— effect on neutron scattering cross section 38/2(410ff., 416ff.)
— effect on the neutron-proton cross section 38/2(417)
— effect on total neutron cross section 38/2(420ff.)
— of beryllium 7, influence on half-life 41/2(97)
Chemical bond 38/2(212)
Chemical composition of an emulsion 45(381–384)
Chemical factor 38/2(403)
Chemical influence on internal conversion 41/2(6)
Chemical shift 38/1(291, 293, 296)
Chemical valence bond 39(422)
Chew-Goldberger momentum distribution of nucleons 40(452, 498, 500)
Chirality (*see also* Helicity) 41/2(100)
Chlorine 37, absorption of a neutrino 41/2(33)
Chlorine 34, β-decay and polarization 41/2(68–69)
Chlorine 36, decay and Fermi plot 41/2(45–46)
Chlorine 35, reactions by slow neutrons 38/2(26)
Choke input of a rectifier 44(20–24)
Chopper 38/2(89)
Chopper for injection 44(271)
Choppers, *see* Fast and Slow chopper

Chronotron 45(248)
Circular corrugated waveguide, *see* Waveguide, Disk-loaded
Circular polarization of γ-rays 41/2(53, 64, 71–74)
— angular correlation with β-particles 41/2(71–74)
— detection by Compton scattering 41/2(73)
Circular resonance 38/1(186)
Circularly polarized γ-rays 42(573, 575, 581)
Classical approximation to Coulomb excitation 41/1(525–526)
Classical dynamics and correspondence principle 41/1(92, 100)
Classical limit of radial matrix elements 41/1(537)
Classical phases 41/1(98)
Classical turning point 41/1(445)
Clean bubble chamber 45(319)
Cleaning expansion of a cloud chamber 45(275)
Clearing field applied to a cloud chamber 45(277, 289)
Clebsch-Gordan coefficients 39(389, 390); 41/1(115); 42(160, 370, 543)
Clebsch-Gordan series 41/1(302, 519)
Closed channels 41/1(160, 168–171)
Closed shell 39(220, 221)
Closed shell interaction 39(405)
Cloud chamber
— as high-energy neutron detector 45(510–511, 512)
— as moderate-energy neutron detector 45(465–466)
— expanding radially 45(281)
— falling freely 45(281)
— tracks, *see* Tracks
— triggered by counters 45(82)
Cloud limit of supersaturation 45(265, 268)
Cloudy crystal ball, *see* Optical model
Cloudy crystal ball model of the nucleus 38/2(367–368)
C-magnet 44(238)
C.m.s. = center-of-mass system 38/2(18, 91ff.)
Co-axial cylinder electrodes 44(71–72, 97)
Cobalt 60
— alignment 42(575)
— β-γ-polarization correlation 41/2(74)
— decay 41/2(44)
— polarization of electrons 41/2(56)
Cobalt 56, anisotropy near energy maximum 41/2(52)
Cobalt 56, γ-ray anisotropy 42(583)
Cobalt 57, decay scheme 42(585)
Cobalt 58, decay scheme 42(582)
Cockroft-Walton generator, *see* Cascade generator
Coefficients of angular correlation 41/1(522, 524, 531–532)

Coherent radiation from synchrotron accelerated electrons 44(180)
Coherent scattering 41/1(242, 298)
— amplitudes 44(434–436)
— amplitudes, average 44(438)
— cross section 38/2(411–413)
— cross section of neutrons in crystals 38/2(451)
— length 38/2(68, 74)
— of neutrons 38/2(6); 44(430, 432, 434)
— of x-rays 38/2(669–670)
Coincidence arrangement of proportional counters 45(189)
Coincidence beta spectrometry 45(255)
Coincidence circuit 42(29); 45(80–82)
— diode 45(233, 240–242)
— fast 45(236–249)
— multiple 45(231)
— of Bell, Graham and Petch 45(242–244)
— of De Benedetti and Richings 45(241)
— of Garwin 45(238)
— of moderate resolution 45(225–236)
— parallel type 45(222, 225)
— series type 45(222, 239)
Coincidence, delayed 42(28)
Coincidence measurements 42(24)
Coincidence selector, bridge type 45(245)
Coincidence selector, diode 45(230, 231)
Coincidence technique 45(52, 80–83)
Coincidence technique in neutron detection 45(464)
Coincidences, photographic recording 45(222)
Cold neutrons 38/2(436, 476); 44(430–431)
Collective kinetic energy 39(419)
Collective model 39(350, 361, 480); 40(10); 42(6, 298, 339, 463, 493)
Collective model of a rotating deformed nucleus 38/1(430)
Collective oscillations, classical 39(418, 419)
Collective oscillations, quantized 39(420)
Collective rotation 38/1(435, 436); 39(494)
Collimated point source 38/2(215)
Collision cross section 41/1(123–124)
Collision density
— for small lethargy 38/2(233)
— in hydrogen 38/2(231, 234)
— integral equation 38/2(229)
— of radiation 38/2(778, 782)
— total 38/2(215, 216)
Collision function 38/2(213, 257–259)
Collision matrix 41/1(123, 186)
Collisions with large impact parameter 41/1(99)
Collisive momentum 40(458, 462, 491, 510, 532)
Color centers, *see F*-centers
Column gradient establishment 44(73–74)
Column of a Van de Graaff generator (region of uniform field) 44(72)
Comparative half-life 41/2(11)
Compensating magnets in alternating gradient accelerators 44(319)
Compensation chamber 45(42)
Competitive modes of disintegration 41/1(53)
Complex diffusion length 38/2(630)
Complex eigenvalues 41/1(1, 8, 29, 184)
Complex eigenvalues in many-body α-decay theory 41/1(358)
Complex energy treatment of nuclear reactions 41/1(28–34, 42–47)
Complex potential 39(446, 449ff.); 41/1(22, 240, 241, 245, 255, 257, 265)
— model of nuclei 38/2(47)
— of neutron-nucleus interaction 38/2(367–368)
— polarization of scattered particles 41/1(486–495)
Complex relaxation distance for thermal neutrons 38/2(579)
Complexity of angular correlation 42(395f., 400)
Composite barrier 38/2(779, 796)
Compound elastic cross section 40(208)
Compound elastic scattering cross section 41/1(242)
Compound nuclei, table 41/1(210–216)
Compound nucleus 38/2(46ff.); 39(446, 448); 40(11, 13, 203f., 403f.); 41/1(4, 6–7, 18, 56); 42(422, 453, 454)
Compound nucleus cross section 41/1(242)

Compound system 39(448)
Compound-elastic scattering 39(448)
Compressibility, nuclear 39(203, 232)
Compton effect 38/2(668–669)
Compton effect inside a counter 45(185, 203)
Compton law of wavelength shift 38/2(668–669)
Compton scattering 41/2(73)
Compton scattering by protons 40(510)
Compton-scattered γ-rays 42(450)
Concrete, passage of radiation 38/2(676, 760, 797)
Condensation nuclei 45(263, 266)
Condition of no self-protection of a detector 38/2(354)
Conduction-induction charging 44(78)
Configuration mixing 39(310, 357, 358, 381); 40(165); 42(184)
Configuration mixing in electron levels 38/1(118)
Configurational interaction 40(179)
Confluent hypergeometric functions 41/1(409)
Conical source of x-rays 38/2(683, 762)
Conservation laws applied to nuclear reactions 38/2(90–101)
Conservation of parity 41/1(477, 495)
Conservation theorems of nuclear reactions 40(22)
Conserved vector current 41/2(51)
Constant background approximation 41/1(138)
Constant gradient proton synchrotron 44(220–300)
Contaminants in accelerating tube 44(88)
Continuity equation of neutron current 38/2(269)
Continuous cloud chamber 45(287–289)
Continuous energy spectrum in β-decay 41/2(5, 9)
Continuum model, *see* Optical model
Continuum theory of nuclear reactions 40(14, 58, 193); 41/1(11)
Contour diagram showing yield of radio-nuclides 40(504)
Control of cyclotron operation 44(158–160)
Control rod 44(91)
Conventional Fermi plot 41/2(13, 45, 46)
Convergence of angular momentum expansions 38/2(29, 30)
Conversion coefficient 42(2, 5, 10, 19, 476)
Conversion electron-beta ray coincidence 45(206)
Conversion electrons, tensor parameters 42(475)
Conversion, external 45(205)
Conversion, internal 42(2, 5, 8, 12, 24, 25, 477); 45(201, 205, 206, 219)
Conversion line, photoelectric 42(21)
Conversion of time to pulse-height 45(248, 249)
Converter plate method (neutron converter) 44(407, 428)
Cooling by diffusion 38/2(608–609)
Cooling of magnet coils in the cyclotron 44(136)
Cooling of RF system in the cyclotron 44(145)
Coordinates describing a hyperbolic path 41/1(511–512)
Copper 64, β-decay and polarization 41/2(68–69)
Copper, cross section for photoneutron production 38/2(149)
Core isomerism 42(16, 17)
Coriolis force in nuclear rotation 39(497, 537)
Cornell synchrotron 44(172ff., 181)
Corona current along the column in Van de Graaff generators 44(73)
Corona current belt charging 44(78)
Corona triode for voltage control 44(93)
Corrected age 38/2(278)
Corrected average logarithmic energy loss 38/2(400)
Correction factor for absorption of epicadmium neutrons in cadmium filter 38/2(358–359)
Correction factors for forbidden transitions 41/2(10)
Correction for first collision 38/2(277, 291)
Correction for hardening of thermal neutrons in the detector 38/2(636, 657)

Correction for obliqueness of neutron beam in the detector 38/2(636, 638–640, 654)
Correction for self-protection of the detector 38/2(636)
Corrector coil 44(211)
Correlated sampling 38/2(787, 788, 794)
Correlation, *see* Angular correlation
Correlation between the position of the nucleons 42(338)
Correlation effects 39(422, 443–445)
Correlation function 38/1(142)
Correlation in position of nuclear particles 41/1(20)
Correlation, perturbed 42(517, 522, 523, 524)
Correlation, theoretical, for several overlapping levels of the compound nucleus 42(454)
Correlation time 38/1(134); 42(516, 525)
Correlation, true 42(526)
Correspondence picture of radiation theory 41/1(66–70)
Correspondence principle
— and Coulomb excitation 41/1(496, 497–508)
— and Hamilton-Jacobi partial differential equation 41/1(92, 104–107)
— and WKB method 41/1(91, 93, 96, 100, 104)
— central fields 41/1(94)
— Coulomb field 41/1(95–96, 100–101)
— general relations 41/1(90)
— one-dimensional problems 41/1(93)
Cosmic ray applications of ionization chambers 45(43, 44)
Cosmic ray neutrons 38/2(209–210)
Cosmotron, betatron core 44(277)
Cosmotron, description 44(219, 230–233)
Cosmotron, field index 44(238, 240)
Cosmotron, gas scattering 44(262)
Cosmotron, injector and pre-accelerator 44(265–266, 268, 294)
Cosmotron, magnet 44(237, 242–243, 246, 250)
Cosmotron, median surface 44(241)
Cosmotron, orbits of injected protons 44(270–271)
Cosmotron, phase oscillations 44(279)
Cosmotron, power supply 44(251–253)
Cosmotron, radiofrequency apparatus 44(287–288)
Cosmotron, radiofrequency parameters 44(279)
Cosmotron, shape of proton bunch 44(280)
Cosmotron, survival time 44(272)
Cosmotron, target problems 44(294, 296–298)
Cosmotron, vacuum chamber 44(257–258)
Cosyns stage 45(411)
Coulomb barrier 40(294); 41/1(53, 55, 255)
— determinations of nuclear size 39(217, 218)
— effect on particle width 38/2(50, 82)
— reduction by heavy particle passage 41/1(368)
Coulomb correction for electron emission 41/2(9, 11)
Coulomb correction for *K*-capture 41/2(12)
Coulomb correction for positron emission 41/2(12)
Coulomb effect on polarization 41/1(493)
Coulomb effects in proton-deuteron scattering 39(175)
Coulomb energy 39(370)
— difference 40(153)
— of nuclei 41/1(12–13)
Coulomb excitation 38/1(109, 433–436); 39(224, 495, 498); 40(48, 88, 108, 164, 328); 41/1(205); 42(33, 501)
— and stripping 41/1(310–312, 319)
— by a muon 40(532)
— by heavy ion bombardment 41/1(400)
— cross section 38/1(434, 435, 436); 42(34)
— discovery 41/1(497)
— effects of higher order 41/1(548–558)
— of both colliding particles 41/1(560)
— parameters 42(503)
— pure 41/1(496, 504, 514)
— reduced transition probabilities (table) 40(332–334)
— virtual, in transfer reactions 41/1(369, 391–400)

Coulomb field, correspondence principle 41/1(95, 96, 100, 101)
Coulomb field included in the Wigner-Eisenbud method 41/1(124)
Coulomb field induced in the Wigner-Eisenbud method 41/1(127–128)
Coulomb forces 39(289)
Coulomb functions, *see* Regular and irregular Coulomb function 42(147, 156)
— derivatives 41/1(413)
— expansions for small η 41/1(434–439)
— integral representations 41/1(414–416, 427–429)
— list of tables 41/1(460–465)
— near the classical turning point 41/1(445–449)
— numerical treatment 41/1(454)
— recurrence relations 41/1(416–418)
— WKB approximation 41/1(440–442, 450–453)
Coulomb multiple scattering 40(486, 531); 45(345, 352, 366–367, 371–379)
Coulomb phase shift 39(58, 59); 40(15, 21)
Coulomb radius 39(430, 471)
Coulomb scattering 40(484f., 531)
— and correspondence principle 41/1(95, 100)
— wave 41/1(101–103)
Counter at high temperature 45(215)
Counter characteristic 45(54, 55, 68)
Counter construction 45(73–76, 84, 85)
Counter efficiency 45(65–67, 255)
Counter, electric field 45(56)
Counter envelopes 45(72–74)
Counter, high field region 45(58)
Counter impurities 45(68–69, 75)
Counter, photoelectron production 45(58, 64)
Counter, photon production 45(58, 62, 64)
Counter, recovery time 45(54, 55)
Counter, reignition 45(55, 60)
Counter, resolving time 45(55)
Counter, self-quenching 45(62, 63, 65, 67–68, 71, 84)
Counter, sensitive volume 45(55)
Counter, space charge 45(59, 60)
Counter telescope 45(82, 223, 224)
Counter telescope in neutron detection 45(464)
Counter threshold 45(55)
Counter window, cut-off by its finite thickness 41/2(39)
Counter with end window 45(77)
Counter with external cathode 45(72)
Counter with low background 45(182)
Counting losses 45(252)
Counting rate, difference from activation of detector 38/2(354)
Counting rate of proportional counter 45(189)
Counts 45(54)
Counts, spurious 45(55, 60)
Coupled cavities 44(351)
Coupler connecting crystal to photocathode 45(95, 96)
Coupling anisotropy in heavy particle reactions 41/1(385)
Coupling between oscillatory modes 44(222, 224, 276, 310–312)
Coupling constants
— determination by neutron decay 41/2(81–86)
— determination from forbidden transitions 41/2(49)
— for A and V coupling 41/2(91, 95)
— found from neutron decay 41/2(91)
— general determination procedure 41/2(88–95)
— (ps-ps and ps-pv) 39(15, 27)
Coupling, intermolecular 38/1(324)
Coupling, intramolecular 38/1(324)
Coupling, normal 39(475)
Coupling not depending upon derivatives 41/2(52)
Coupling of angular momenta 39(390)
Coupling schemes in nuclei 39(474, 484)
Coupling schemes in resonance reactions 42(420)
Coupling, strong 39(314)
Coupling system of cyclotron Dees 44(142–143)
Coupling types 41/2(9, 88)
Coupling, weak 39(314)
Covalent bond 38/1(424, 426)
Covalent bond, radii 38/1(425)
Crane's neutrino experiment 41/2(33)
Cranking model 38/1(109)
Crenellated poles 44(245)

Critical angle for neutron reflection 44(436)
Critical energy of shower formation 45(165, 167)
Critical size of a reactor 44(391–392)
Cross section(s)
— at energy zero 39(43, 46, 51)
— averaged 39(426)
— compound elastic (incoherent elastic scattering) 40(208, 422)
— definitions 40(207, 208)
— definitions of several types 41/1(241–242)
— depending on polarization 41/2(53)
— differential (angular distribution) 40(54)
— effective 38/2(397)
— elastic scattering 40(207)
— for compound nucleus formation 40(17)
— for Coulomb excitation 38/1(434, 435)
— for direct reactions 40(222)
— for giant resonances 38/1(440)
— for inelastic scattering 41/1(529)
— for nuclear reactions 40(13)
— for photoneutron production, dependence on mass number 38/2(151)
— for pion reactions 40(535)
— for protons in iron 40(222)
— for protons in tin 40(223)
— for specific reactions 40(214)
— for thermal neutron capture, experimental values 38/2(594–596, 619, 629)
— formula 40(20)
— formula of Fernbach, Serber and Taylor 40(464, 492)
— formula of Klein and Nishina 38/2(672)
— inelastic (scattering and reactions) 40(207)
— integrated 42(314)
— macroscopic 38/2(213, 397)
— measurement by activation methods 42(332)
— measurement of (γ, n) and (γ, p) by neutron or proton detection 42(334)
— near reaction threshold 41/1(274–293)
— nonelastic (scattering only) 40(207)
— of B(α, n) reactions 38/2(124)
— of beryllium photodisintegration 38/2(143, 144)
— of coherent neutron scattering in crystals 38/2(451)
— of D(d, n) reaction 38/2(188, 192)
— of D(d, p) reaction 38/2(188)
— of deuteron photodisintegration 38/2(136–138, 140)
— of deuteron photodisintegration, experimental values 38/2(139, 140)
— of diffusion 38/2(257)
— of (γ, n) reaction 42(332)
— of incoherent neutron scattering in crystals 38/2(450)
— of inverse β-decay 41/2(32)
— of Li(p, n)-reaction 38/2(180)
— of lithium reactions under deuteron bombardment 38/2(199–200)
— of neutron transfer 41/1(389, 399)
— of (p, n) reactions as a function of Z 38/2(174)
— of T(d, n) reaction 38/2(195–197)
— of T(p, n) reaction 38/2(176–177)
— shape elastic (coherent elastic scattering) 40(21, 208, 422)
— total, at high energies 40(481, 491)
— total, of Coulomb excitation 41/1(497, 510–514, 527, 530)
Cross-over transition 41/1(549)
Crystal, coherence effects 38/2(414)
Crystal field gradients 38/1(325)
Crystal growth (sodium iodide) 45(103)
Crystal lattice imperfections, sensitivity to radiation 44(405)
Crystal monochromator 38/2(89)
Crystal monochromator for neutrons 44(411–413, 430–431, 434)
Crystal mounting (sodium iodide) 45(132, 133)
Crystal size, effective 38/2(415)
Crystal structure 38/1(312–326)
Crystallographic data of actinide elements 42(209, 211, 212f., 264)
Cumulative probability 38/2(776)
Curie law 38/1(126)
Curium 42(264–270)
Curium, chemical properties 42(267)
Curium, decay schemes 42(267)
Curium isotopes 42(266)
Current albedo 38/2(545)

Current probe 38/2(637, 642)
Current regulators for cyclotron magnet 44(137–138)
Current shimming 38/1(235)
Curvature of a track by a magnetic field 45(303)
Curvature of a track without field 45(306–308)
Cusp phenomenon 41/1(176, 278)
Cut-off by absorption in counter window 41/2(39)
Cut-off energy of cadmium 38/2(56, 342)
Cut-off of potential 39(29, 30)
Cut-off wavelength 38/2(435)
Cut-off wavelength for coherent neutron scattering 44(430, 432)
Cyclotron 38/1(268–270); 40(24)
Cyclotron control systems 44(158–160)
Cyclotron frequency of electrons 38/1(260–270)
Cyclotron theory 44(107–132)
Cyclotrons, list of characteristics 44(160–161, 162–165)
Cylindrical boundary, radiation behavior 38/2(798)
Cylindrical ionization chamber 45(36)

(*d*, *n*) reaction in deuterium 38/2(187–193)
(*d*, *n*) reaction in tritium 38/2(193–198)
(*d*, *n*) reactions, angular distribution 40(275)
(*d*, *n*) reactions as neutron sources 38/2(185–209)
(*d*, *p*) reactions, angular distribution 40(275)
(*d*, *p*) reactions, dependence on energy 40(278)
(*d*, *p*) reactions, dependence on spin 40(278)
(*d*, *p*) reactions, purity of states 40(280)
(*d*, *p*) reactions, theories of Butler, Bhathia et al. 40(276)
(*d*, *p*) reactions with Ca and Sr 40(279)
(*d*, *p* γ) reactions 40(280)
Damping, adiabatic, of betatron oscillations 44(307, 315)
Damping matrix in radiation problems 41/1(64)
Damping of betatron oscillations 44(195–196, 201, 227, 307, 315)
Damping of oscillations in a linear accelerator 44(356, 372)
Damping of phase oscillations 44(223, 229)
Damping of synchrotron oscillations 44(182–184, 446)
Damping parameter 39(22, 27)
Damping theory 39(139)
Danger coefficient method
— for determination of effective neutron temperature 38/2(482, 488–489)
— of absorption measurement 38/2(373–374)
— of neutron absorption measurement 44(424)
Daughter nucleus 41/2(2)
D-D reaction 42(600)
Dead time in coincidence arrangements 45(252)
Dead time of a counter (*see also* Recovery time) 45(54)
Dead time of an expansion cloud chamber 45(280)
Dead volume of counter 45(179)
Debye model of a crystal 38/2(414, 441, 458)
Debye-Scherrer rings 38/2(434)
Debye-Waller factor 38/2(415, 429); 44(433, 435, 438)
Decay channel 40(403f., 407f., 410)
Decay curves of thermal neutrons 38/2(623, 626)
Decay energy cycles 42(286/287, 288)
Decay formula for polarized neutrons 41/2(75)
Decay rate, formula 41/2(10)
Decay schemes of transuranium elements 42(228, 238, 255, 267, 270, 272, 275, 277, 279, 283)
Decay time of anthracene 45(144)
Decay time of sodium iodide 45(99)
Dee system 44(142–143)
Deep penetration of neutrons 38/2(295)
Deep penetration of x-rays, obliquity effects 38/2(713–714)
Dees of a cyclotron 44(105, 142)
Deflection of x-ray flux 38/2(714, 764)
Deflector for the cyclotron beam 44(151)

Defocussing in linear accelerators 44(344, 374)
Defocussing phase effects in the cyclotron 44(125–126)
Deformability of nuclei 39(237)
Deformation parameters 39(219, 518)
Deformation potential 39(503, 513, 522)
Deformation potential and two-body forces 39(511, 512)
Deformed nuclei
— examples of decay schemes 40(349)
— levels in spheroidal well 40(347)
— lifetime of low energy $E1$ γ-rays 40(355)
— odd states in even-even nuclei 40(353)
— selection rules 40(346)
Degradation integral 38/2(213)
Degradation of radiation energy 38/2(699–706)
— and slowing down of neutrons 38/2(700)
Degrees of freedom in unified model 39(541)
Delay demodulator 45(248, 249)
Delay line 45(245)
Delayed coincidence(s) 42(28); 45(224, 231–234, 257)
Delayed curve 45(257)
Delayed neutrons, effect on reactor period 44(393)
Delta rays 45(4, 7)
Delta rays in emulsions 45(352, 354, 435)
Demagnetization, adiabatic 38/1(129)
Density albedo 38/2(545)
Density anisotropy, calculated from potential 39(504)
Density, calculation from potential 39(453, 454)
Density, central, of the nucleus 39(193)
Density discontinuity of radiation at an interface 38/2(801)
Density effect in stopping of particles 45(353)
Density effect of Fermi 45(261)
Density from individual particle states 39(235)
Density matrix 42(375, 385, 436, 520, 557)
Density, nuclear 39(178, 416, 431)
Density, nuclear, below the surface 39(433, 459)
Density of an emulsion 45(360, 383)
Density of neutrons, relative perturbation by probe 38/2(643, 653)
Density of nuclei, theory 39(231, 233)
Density probe 38/2(637, 638–641)
Density scaling method 38/2(802)
Density, statistical, of excited levels 39(437)
Density, statistical, of single particle levels 39(435)
Depolarization of electrons in the source 41/2(54)
Depolarizing effect 41/1(557)
Depression of neutron flux in the vicinity of a detector 38/2(356)
Derivatives of the Coulomb functions 41/1(413)
Detailed balance principle 38/2(396)
Detection efficiency as a function of neutron beam angle 38/2(639)
Detection efficiency of proportional counters for x-rays 45(198, 202–203)
Detector of finite thickness, neutron self-absorption 38/2(390)
Detector of photons indicating their direction 38/2(807)
Detectors following the $1/v$ law 38/2(635)
Detectors for epicadmium neutrons 38/2(352–359)
Detectors of thermal neutrons 38/2(633–637)
Deterioration of targets 38/2(182)
Deuterium, (α, n) reaction 38/2(126)
Deuterium, capture of slow neutrons 38/2(65, 73, 75)
Deuterium cross section for high-energy neutrons 45(491)
Deuterium cross section for neutrons below 20 MeV 45(439, 440)
Deuterium, (d, n) reaction 38/2(187–193)
Deuterium, distintegration by neutrons 38/2(71–72, 74)
Deuterium, elastic scattering of neutrons 38/2(71–74)
Deuterium, inelastic scattering of neutrons 38/2(71–72, 74)
Deuterium loaded emulsions 45(347, 395)

Deuterium, (p, n) reaction 38/2(171, 174)
Deuterium, total cross section for neutrons 38/2(71–72)
Deuterium, zero energy scattering cross section 38/2(65)
Deuteron, binding energy 38/2(5, 67, 69, 134–135, 185); 39(1, 32, 91)
Deuteron break-up in Coulomb field of a nucleus 38/2(186, 191)
Deuteron capture 40(47)
Deuteron disintegration 40(63, 97)
Deuteron, effective range 39(86, 91)
Deuteron, electric disintegration, Dancoff's theory 41/1(310–312)
Deuteron, electric disintegration, Landau-Lifshitz theory 41/1(313–318)
Deuteron, electric quadrupol moment 39(67, 91)
Deuteron, excited states 39(33, 35)
Deuteron formation by pick-up 40(452, 455, 499f., 518)
Deuteron, ground state 39(32, 36, 66)
Deuteron, magnetic moment 39(32, 69, 74)
— exchange current correction 39(71, 72)
— non-additivity correction 39(71, 72, 73)
— relativistic correction 39(71, 72, 73)
Deuteron, photodisintegration 38/2(4, 132, 133–141)
Deuteron photodisintegration at high energies 40(511f.)
Deuteron radius 39(35, 40, 75)
Deuteron reactions, energy balance 40(267)
Deuteron reactions in nuclei of mass 3 41/1(233–239, 291)
Deuteron size 38/2(67)
Deuteron spin 39(2, 32)
Deuteron stripping 40(487, 492, 499f.)
— reaction 41/1(161)
— reactions at high energies 41/1(304–312)
Deuteron, threshold for photodisintegration 38/2(134–135)
Deuteron, virtual singlet state 38/2(68)
Deuteron, wave function 39(33, 66, 74, 76, 91); 41/1(15, 18, 307)
Developers 45(403–406)
Development of emulsions 45(401–406)
Diamagnetic corrections of neutral atoms 38/1(290)
Diamagnetic field strength at the nucleus 38/1(288)
Diamagnetic shielding constant 38/1(289)
Diamagnetic shielding of protons 38/1(253)
Diamagnetism of electron shells 38/1(281–293)
Diatomic molecules, *see* Molecules
Dielectric strenght of air 44(65)
Dielectric strenght of nitrogen 44(102)
Difference method 42(314)
— applied to boron filter 38/2(351)
— applied to resonance filters 38/2(351)
Difference of neutron and proton distribution 39(435, 454, 455)
Differential albedo of radiation 38/2(791)
Differential cross section for inelastic scattering 41/1(550, 553)
Differential equation for the Coulomb field 41/1(409)
— numerical integration 41/1(455)
Differential ionization chamber 41/2(6–7)
Differential yield 38/2(103)
Diffraction maxima 40(93)
Diffraction maxima and polarization 41/1(490)
Diffraction of a neutron wave by a nucleus 40(481, 488f.)
Diffraction of neutrons 44(432–435)
Diffraction pattern of neutrons from a crystal 38/2(416)
Diffraction pattern of neutrons from molecular gases 38/2(427–431)
Diffraction pattern of scattered neutrons 38/2(81)
Diffraction polarizing neutrons 44(440)
Diffraction stripping 40(503)
Diffraction theory 39(165ff.)
Diffuseness of nuclear boundary 39(236)
Diffuseness parameter of nuclear surface 41/1(257)
Diffusion 38/1(134, 136)
Diffusion approximation 38/2(539–545)

Diffusion approximation, results 38/2(549)
Diffusion cloud chamber 45(284, 287–289, 334)
Diffusion coefficient in age theory 38/2(268)
Diffusion coefficient of thermal neutrons 38/2(516, 580)
Diffusion coefficient of thermal neutrons, temperature dependence 38/2(613, 624–625)
Diffusion cross section 38/2(257)
Diffusion equation of thermal neutrons 38/2(491, 493–494)
Diffusion kernel, time dependent 38/2(578)
Diffusion kernels 38/2(553–554)
Diffusion length 44(426)
— complex 38/2(630)
— measurements in steady state assemblies 38/2(596–608)
— of thermal neutrons 38/2(69, 493, 514, 518, 580)
Diffusion mean free path 38/2(257)
Diffusion of ions 45(17)
Diffusion of ions in a cloud chamber track 45(305, 308)
Diffusion of radiation 38/2(730–733)
Diffusion of thermal neutrons
— elementary theory 38/2(511–517, 552–557)
— elementary theory, as P_1 approximation 38/2(514–517)
— elementary theory, boundary problems 38/2(552–557)
— one-dimensional model 38/2(557–562)
— one-group picture 38/2(504–507)
— with energy loss 38/2(527)
Diffusion theory, second order 38/2(562–575)
Diffusion time 38/2(577)
Diluted emulsions 45(392)
Di-neutron 40(146)
Diode, germanium and silicon junctions 44(18)
Diode, selenium rectifier 44(18, 19)
Diode, semiconductor 44(17)
Dipole Coulomb excitation 41/1(527, 538–540)
Dipole decay, magnetic 38/1(432)
Dipole emission 41/1(67, 70–73)
Dipole moment
— electric 38/1(380); 39(68)
— electric, of a molecule 38/1(424)
— magnetic 39(252, 277, 294, 311, 407)
— magnetic, absolute measurement 38/1(243–244, 249)
— magnetic, absolute measurement, Cotton scales 38/1(249)
— magnetic, absolute measurement for the proton 38/1(248–259, 268)
— magnetic, relative measurement 38/1(245, 259–270, 275–336)
Dipole oscillation of a nucleus 39(419)
Dipole relaxation 38/1(150)
Dipole resonance, *see* Giant resonance
Dipole transitions, electric 42(316)
Dipole transitions, magnetic 42(318)
Dipole-dipole coupling, direct 38/1(138)
Dirac equation for nucleons 38/2(7)
Dirac-particle 41/2(35, 36, 100)
Dirac's frequency shift 41/1(71)
Direct interaction 40(105, 148, 205, 220ff., 224)
Direct nuclear interaction theories 41/1(9–10)
Direct photoeffect 38/2(156)
Direct transitions 40(13, 71, 80)
Directional Čerenkov counters 45(159–160)
Directional coincidence arrangements 45(82)
Directional correlation, *see* Angular correlation
Directional correlations of decay particles 41/2(19)
Directional detector of photons 38/2(807)
Directional distribution, *see* Angular distribution
Dirty bubble chamber 45(300–321)
Disadvantage factor 38/2(384)
Discharge in accelerating tube 44(86–89)
Discharge propagation along the counter wire 45(64, 65)
Discharging of a belt 44(76–78)
Discontinuity of radiation density at an interface 38/2(811)
Discovery of artificial radioactivity 38/2(13–14)
Discovery of (*n*, α) reactions 38/2(12–13)

Discovery of the neutron 38/2(1, 2–3, 12–13)
Discriminator 45(184, 248)
Disinfection of gelatine 45(409)
Disintegration of deuterons by neutrons 38/2(71–72, 74)
Disk spacing in a corrugated waveguide 44(351–352)
Disk-loaded linear accelerator 44(348)
Dispersion of a nuclear magnetic substance 38/1(174)
Dispersion signal 38/1(188)
Dispersion sum formula 41/1(60)
Dispersion theories of nuclear reactions 41/1(4, 5)
Dispersion theory of nuclear reactions 40(14)
Dispersion voltage 38/1(186, 187)
Displaced equilibrium orbit 44(307–309)
Dissipation of radiation energy 38/2(673)
— near a boundary 38/2(798, 800)
— near an interface 38/2(800)
Distance of levels, average 38/2(367)
Distant collisions 41/1(99)
Distortion 39(505, 529, 530)
Distortion of emulsions by processing 45(401ff.)
Distortion of nuclei by passage of heavy particles 41/1(368)
Distortion of the correlation function 42(404)
Distribution of neutrons, synthesis from its moments 38/2(335)
Distribution of pulse sizes 45(190)
Distribution of radiation flux, construction from moments 38/2(739–749)
Distribution of thermal neutrons in beryllium as a function of time 38/2(626)
Distribution of thermal neutrons in water as a function of time 38/2(617)
Disturbances in the synchrotron 44(224–226, 278)
Diverging wave type of Green's functions 41/1(78, 80, 84)
Donut (vacuum chamber) 44(174, 211–217)
Doppler broadening 38/1(50); 42(37)
— no effect on resonance integral 38/2(365)
— of neutron resonance 38/2(51, 346–349)
— of neutron resonance lines 40(376f., 379, 381)
— of resonance lines 44(417)
Doppler effect 38/2(105)
— for a monatomic moderator gas 38/2(397–401)
— in a Debye crystal 38/2(348–349)
Doppler shift 42(37)
Doppler shift of γ-ray 41/2(87)
Doppler width of a resonance line 38/2(238–239, 347, 348)
Dose meters 45(38–41)
Dose of radiation 38/2(674)
Dose units 38/2(89–90)
Dosimetry 38/2(89)
Double β-decay 41/2(2, 99–102)
Double Compton scattering 38/2(688)
Double hypergeometric series of Appell 41/1(534)
Double K-capture 41/2(100)
Double quadrupole excitation 41/1(549)
Double reflection of neutrons 44(440)
Double scattering
Double scattering (*see also* Polarization) 40(494); 41/2(57, 58)
— experiment 41/1(473); 42(472)
— of spin $^1/_2$ particles at spinless nuclei 41/1(475–481)
Double-resonance experiments 38/1(333)
Doubling of tracks 45(277)
Doughnut pressure 44(178)
Doughnut (vacuum chamber) 44(174, 211–217)
Drift tube 44(369, 373)
Drift velocity
— of electrons 45(18, 56, 57)
— of electrons, diagrams 45(19–24)
— of electrons in the counter gas 45(187, 188)
— of ions 45(18)
Droplet count 45(295, 299)
Droplet formation 45(266)
Droplet growth, classical theory 45(264)
Droplet growth in various gases and liquids (table) 45(269)

Drying emulsions 45(406–408)
D-state probability in the deuteron 39(70, 73, 74, 91)
Dulono synchrophasotron, *see* Synchrophasotron
Dynode structure multipliers 45(88)
Dyson transformation 39(19)
Dysprosium as a detector of thermal neutrons 38/2(634)

Earnshaw's theorem 44(344, 374)
Eckart-Wigner theorem 41/1(510, 514)
Eddy currents in betatron magnet 44(208, 210)
Eddy currents in synchrotron magnet 44(235, 239, 242–243, 329)
Effective capture cross section due to Doppler effect 38/2(347–348)
Effective capture cross section of an assembly 38/2(376)
Effective charge, loss along track 45(371)
Effective cross section 38/2(397)
Effective crystal size 38/2(415)
Effective energy of neutrons absorbed 38/2(496)
Effective neutron temperature 38/2(471, 477–489)
— experimental determination 38/2(480–489)
— inside a reactor 38/2(488–489)
Effective range 39(38, 51, 54, 59–61, 83, 85, 91)
— of proton-neutron system 38/2(67–68)
— theory 38/2(67)
— theory of deuteron photodisintegration 38/2(136)
Effective resonance integral 44(409)
— of an assembly 38/2(376–385)
Efficiency matrix 42(436)
Efficiency of counters 45(65–67)
Efficiency of injection 44(273)
Efficiency of Monte Carlo calculations 38/2(781, 789)
Efficiency of sodium iodide scintillators 45(104–131)
Efficiency of the radiation detector 42(404)
Efficiency tensor 42(384, 386, 387, 436, 454, 475, 510, 520)
Ehrenfest's theorem 41/1(501)
Eigenfunctions of the age equation 38/2(276)
Einstein model of a crystal 38/2(414, 457–458, 502)
Einsteinium 42(278–282)
Einsteinium, chemical properties 42(281)
Einsteinium, decay scheme 42(279)
Einsteinium isotopes 42(280)
Elastic scattering 41/1(11)
Elastic scattering (*see also* Cross section) 40(14, 20f., 35)
— by moving nuclei, momentum balance 38/2(98–100)
— by nuclei at rest, momentum balance 38/2(98–100)
— cross section 41/1(241–242)
— cross section, vanishing 41/1(300)
— of heavy particles 41/1(400–402)
— of neutrons 45(437)
— of neutrons by beryllium 38/2(75–76)
— of neutrons by carbon 38/2(77)
— of neutrons by deuterium 38/2(71–74)
— of neutrons by heavy elements 38/2(81)
— of neutrons by hydrogen 38/2(65–68)
— of neutrons by oxygen 38/2(78)
— of slow neutrons by polycrystalline materials 38/2(431–440)
Electric break-up of a deuteron passing a nucleus 38/2(186, 191)
Electric dipole excitation 41/1(527, 538)
Electric dipole transition 38/2(136, 143, 153, 156)
Electric disintegration of the deuteron 41/1(310–318)
Electric field effects in transfer reactions 41/1(369, 391–400)
Electric field gradients perturbing the angular correlation 42(515)
Electric field in a counter 45(56)
Electric multipol transition 39(100, 102, 103, 104, 110)
Electric quadrupole excitation 41/1(513, 514–516, 522, 527, 542)
Electric quadrupole transition 38/2(141, 153); 39(490, 546)
Electric vertical focussing in the cyclotron 44(117–119)

Electrodes in Van de Graaff generators 44(71–74, 83, 97)
Electrodes of flat disk shape 44(85)
Electrodisintegration of nuclei 42(356)
Electromagnetic moments 39(252)
Electromagnetic transition moment 41/1(496)
Electromagnetic transitions for H^3 and He^3 39(176)
Electron absorption inside the detector 38/2(641, 655)
Electron affinity 45(177)
Electron analog at Brookhaven 44(318, 320–325)
Electron attachment 45(25)
Electron beam intensity from a synchrotron 44(447–449)
Electron bombardment as a neutron source 38/2(169–170)
Electron capture, (*see also* *K*-capture) 41/2(2, 21, 95–99)
— internal bremsstrahlung 41/2(103–104)
Electron configuration assignment using isotope shift data 38/1(117–118)
Electron density in an emulsion 45(386)
Electron drift velocity 45(18, 19–24, 56, 57)
— in the counter gas 45(187, 188)
Electron linear accelerators 44(347–366)
Electron loading of an accelerating tube 44(86–88)
Electron resonance 38/1(333)
Electron scattering 38/1(104, 437); 39(416)
— at nuclei, elastic 40(505f.)
— at nuclei, inelastic 40(455, 505, 508f.)
— by nuclei 39(186, 228ff.)
— by nuclei, Born approximation 39(187)
— by nuclei, cross sections 39(195)
— by nuclei, high energies 39(191)
— by nuclei, inelastic 39(197)
— by nuclei, low energies 39(188)
— by protons 40(508)
— determinations of nuclear size 39(192–195)
— from aspherical charge 39(228)
Electron shells 42(9)
Electron source, monoenergic 45(185)
Electron spectra produced by radiation 38/2(678)
Electron spin system, saturation 38/1(335)
Electron synchrotron 40(472)
Electron synchrotron, invention 44(170, 449)
Electronbremsstrahlung 42(312)
Electronegativity of atoms 38/1(425, 426)
Electronic computers for magnetic design 44(249)
Electronic configurations of the heaviest elements 42(205, 206, 207)
Electronic scanner for counting track blobs 45(418, 419)
Electron-neutrino correlation for the neutron decay 41/2(76, 81, 86, 91)
Electron-neutron interaction 38/2(9–12)
Electron-nucleon interaction 38/1(106)
Electrons, contribution to the molecular moment of inertia 38/1(42, 43)
Electrons, distortion of closed shells 38/1(41)
Electrons liberated from the cathode by ion impact 45(9, 10)
Electrons of high energy, photoneutron production 38/2(166–170)
Electrostatic analyser 40(28)
Electrostatic beam analyzer 44(92–93)
Electrostatic β-ray spectrometer 41/2(37)
Electrostatic generator 40(24, 375)
Electrostatic inflector 44(265–267)
Elementary charge, identity with β-particle charge 41/2(2)
Elementary diffusion theory of thermal neutrons 38/2(511–517, 552–557)
Elementary physical constants, determination 38/1(243, 270, 275)
Ellipsoidal vibrations 39(518, 533)
Emission solution 41/1(69)
Emission, spontaneous 38/1(133)
Emulsion blocks (stacks) 45(349, 393)
Emulsion sensitivity 45(342, 343, 389)
Emulsion techniques 40(479f.)
Emulsions as detectors for high-energy neutrons 45(508–510)
Emulsions as detectors for moderate-energy neutrons 45(466–468)

Emulsions, chemical composition 45(381–384)
Emulsions, contact with metals 45(389)
Emulsions, densities 45(360, 383)
Emulsions, diluted 45(392)
Emulsions in high-energy research 45(348–351)
Emulsions in low energy research 45(344)
Emulsions not commercially available 45(394)
Emulsions, principal commercial 45(389)
Emulsions, storage 45(394)
End arrangement of counters 45(73)
End effect in a counter tube 45(178)
Endo-energetic reactions 38/2(96)
Endoergic reactions as neutron detectors 45(449–451)
Energies of photoneutrons from various sources 38/2(160)
Energy absorption coefficient of radiation 38/2(673)
Energy albedo of radiation 38/2(791)
Energy balance of *K*-capture 41/2(98)
Energy build-up factor 38/2(752)
Energy change kernel of crystals 38/2(474)
Energy cycles 42(113, 119)
Energy decrement per collision, average (logarithmic) 38/2(218, 227)
— corrected for thermal motion 38/2(400)
Energy degradation of radiation 38/2(699–706)
Energy dependence
— of neutron flux 38/2(344)
— of neutron width 38/2(50)
— of polarization 41/2(56, 59)
— of radiation width 38/2(50)
Energy dependent boundary conditions 41/1(163–167, 186)
Energy dependent parameters 41/1(58)
Energy dissipation of x-rays 38/2(673)
Energy dissipation of x-rays near a boundary 38/2(798, 800)
Energy dissipation of x-rays near an interface 38/2(800)
Energy distribution after *n* collisions in hydrogen 38/2(220–222)
Energy distribution after *n* collisions with nuclei heavier than hydrogen 38/2(222–226)
Energy for the formation of one ion pair 45(11–15)
Energy gain, losses by imperfections 44(361–363)
Energy gain of an electron in a linear accelerator 44(353, 361–363)
Energy gain per turn 44(221, 227)
Energy levels in closed shell nuclei
— double closed shell nuclei 40(356ff.)
— Harvey's experiment 40(361)
— single closed shell nuclei 40(356ff.)
— special cases: Ca; Zr; Pb; Bi; Po; Tl 40(357–368)
Energy levels in nuclei with $N<90$, even-even nuclei 40(368f.)
Energy levels in nuclei with $N<90$, odd mass number nuclei 40(368)
Energy loss
— density correction 45(354)
— in emulsions, electrons 45(355, 370–371)
— in emulsions, protons 45(355, 369)
— of charged particles 45(261–263)
— of electrons, theory 45(353)
— of heavy particles, theory 45(352)
— per collision 38/2(217, 614)
— per ion pair 45(193, 194)
— rate, semi-empirical data 45(369)
Energy matrix 39(332)
Energy of neutrons from Li(*p*, *n*) 38/2(179)
Energy of neutrons from T(*d*, *n*) 38/2(194)
Energy of radiation at minimum attenuation coefficient 38/2(675, 707, 751, 755)
Energy of recoil nuclei 41/2(20)
Energy reflection coefficient 38/2(792)
Energy release 40(23)
Energy release of a nuclear reaction (*see also* Reaction energies) 38/2(92)
Energy resolution in neutron resonance measurements 40(376, 379)
Energy resolution of an electrostatic analyzer 44(93)
Energy resolution of scintillation spectrometers 45(131, 133–137)
Energy shift of bound states 39(199)

Energy spacing and second sum rule 41/1(194, 196, 200)
Energy spacings of nuclear states 42(93)
Energy spectrum
— of bismuth photoneutrons 38/2(157)
— of delta rays 45(4)
— of neutrons from B(*d*, *n*) 38/2(204–205)
— of neutrons from Be(α, *n*) 38/2(120–121)
— of neutrons from D+D reactions at higher deuteron energies 38/2(191)
— of neutrons from Li(*d*, *n*) 38/2(200–201)
— of neutrons from Poα+B 38/2(125)
— of recoil protons from Be(*d*, *n*) 38/2(202)
— of slowed-down neutrons, infinite medium 38/2(228, 394–397)
— of uranium fission neutrons 38/2(310)
Energy, total static 38/1(126)
Energy width of stripping cross section 41/1(309, 310)
Enriched reactor 44(395)
Entrance channel 38/2(18)
Epi-cadmium neutrons 38/2(342)
— absorption in cadmium 38/2(358–359)
— detector 38/2(356)
Equations of orbital stability 44(108–109)
Equilibrium *B*-values in molecular rotation 38/1(45)
Equilibrium distribution 38/1(129, 334)
Equilibrium of primary and secondary radiation 38/2(676ff.)
Equilibrium orbit, fixed 44(195, 202)
Equilibrium orbit in the synchrotron 44(220, 268–269, 301)
Equilibrium orbit, instantaneous 44(194, 201)
Equilibrium temperature 38/2(612, 628)
Equipartition 38/1(129, 334)
Equivalent quanta (x-ray intensity) 44(191)
Equivalent radius 38/1(101)
Equivalent thickness of target 38/2(103)
Equivalent two-body problem for three nucleons 39(172)
Eradication of background 45(395–398)
Escape peak 45(204)
Euler angles 39(483); 41/1(302, 303); 42(374)
Europium 152, *K*-capture 41/2(86–88, 92)
Evaporation of nucleons 40(397, 464, 466f., 503, 515, 523)
Evaporation of protons 41/1(18)
Evaporation theory of neutron emission 38/2(84)
Even-even nuclei 39(293)
— first exited states 40(320)
— second exited states 40(323)
Even-odd staggering (of isotope shifts) 38/1(96, 105, 115)
Ewald sphere 38/2(433)
Exchange broadening 38/1(306)
Exchange current correction 39(98, 114, 115, 116, 130)
Exchange currents and magnetic moments of H^3 and He^3 39(177)
Exchange forces between nucleons 41/1(252)
Exchange integral 40(6)
Exchange processes 38/1(144)
Exchange processes, chemical 38/1(298, 300, 303)
Excitation curve 38/2(101)
— of B(α, *n*) reactions 38/2(109, 125)
— of B(*d*, *n*) reactions 38/2(204)
— of Be(α, *n*) reaction 38/2(109, 114–115)
— of C(*d*, *n*) reactions 38/2(206)
— of F(α, *n*) reaction 38/2(109)
— of Li(α, *n*) reactions 38/2(127)
— of Mg(α, *n*) reactions 38/2(129)
— of (*p*, *n*) reactions in several elements 38/2(183)
— of Rb(*p*, *n*) reaction 38/2(173)
— of several (α, *n*) reactions 38/2(109, 111)
— of several (α, 2*n*) reactions 38/2(111)
Excitation energies of compound nuclei, table 41/1(210–216)
Excitation energies of first excited states 39(260)

Excitation of vibrational states 40(225)
Excitation potential 45(5)
Excitation spectra of nuclei 39(477–479)
Excited state energies 42(124–131)
Exciting coils of cyclotron magnet 44(135–136)
Exo-energetic reactions 38/2(96)
Exo-thermic reactions as neutron detectors 45(442–445)
Expansion, adiabatic, to produce supersaturation 45(266–270)
Expansion cloud chamber 45(270–284, 334)
Expansion in orders of scattering, approach to solve the photon transport equation 38/2(686ff., 706–707, 711, 733, 745, 767ff., 771)
Expansion ratio 45(268)
Experimental albedo 38/2(549, 596)
Experimental methods for determining (α, n) excitation curves 38/2(112)
Exponential theory of scattering mean free path 38/2(266)
Exposure dose of radiation 38/2(674)
Exterior region of the nucleus 41/1(108)
External bremsstrahlung 41/2(102)
External quenching of a counter 45(54, 61)
External scattering, *see* Potential scattering
External slow-neutron experiments 38/2(344)
Extraction of beam from a betatron 44(206–208)
Extra-nuclear fields 42(416, 513)
Extrapolated end point measurements 38/2(585–589)
Extrapolated end point of thermal neutron distribution 38/2(536, 581)
Extrapolation chamber 45(38)
Extrapolation length of thermal neutron distribution 38/2(536, 547)
Extrapolation length of thermal neutron distribution for black bodies 38/2(544)
Extreme geometry ionization chamber 45(35–37)
Eyepieces for scanning microscopes 45(412)
Factorization in the one-level formula 41/1(7)
Factorization of $\mathcal{R}$-matrix numerator 41/1(119, 129, 155–161, 177)
Fading of the latent image 45(387–388)
Falling cloud chamber 45(281)
Faraday number 38/1(272)
Fast chopper 44(413–414, 431)
Fast effect 44(391)
Fast neutron(s) 44(399–400)
— capture cross section 44(407, 428)
— doses in water 38/2(301–302, 320)
— spectroscopy with ionization chambers 45(49)
Favored α-decay 42(122, 167, 177, 182, 199, 200)
F-centers (color centers) 38/1(149)
Fermi angular distribution law of thermal neutrons 38/2(392, 561, 574, 654)
Fermi approximation 39(45)
Fermi coupling 41/2(9, 25, 49, 58, 68, 89, 92, 105)
Fermi gas model of the nucleus 39(421); 41/1(16, 19, 195, 255)
Fermi gas of nucleons 40(456, 463)
Fermi integral 41/2(10–11)
Fermi integral for K-capture 41/2(98)
Fermi interaction 42(507)
Fermi intercept 41/1(328)
Fermi momentum 39(426)
Fermi momentum sphere 40(466, 519)
Fermi plateau in ionization loss curve 45(353, 356)
Fermi plot 41/2(13)
— as a means to determine neutrino mass 41/2(36–37)
— examples 41/2(38, 39, 44)
— of the neutron decay 41/2(78)
— of tritium 41/2(37, 42)
Fermi pseudopotential 38/2(402, 407, 418)
Fermi resonance 38/1(45)
Fermi-Dirac distribution 38/1(150)
Fermi-Dirac statistics 38/1(91, 94); 38/2(6)
Fermi's one-dimensional model of thermal neutron diffusion 38/2(570)
Fermi-Segrè factor 38/1(66, 72)
Fermi-Segrè formula 38/1(111)
Fermium 42(283–284)

Fermium, chemical properties 42(284)
Fermium, decay scheme 42(283)
Fermium isotopes 42(284)
Feshbach-Porter-Weisskopf model 41/1(240–257)
FFAG: fixed-field alternating-gradient
Fick's law 38/2(268, 490–491, 512, 514, 519)
— derived from the transport equation of thermal neutrons 38/2(513, 516, 519, 520–527)
— generalizations 38/2(521, 526–527)
Field biased betatron 44(206)
Field deviation according to the error function 38/1(159)
Field distortion by charging belt 44(74–75)
Field effects of isotope shifts
— finite charge distribution 38/1(100)
— for spherical incompressible nuclei 38/1(100)
— non-Coulombic forces 38/1(106)
— nuclear charge distribution 38/1(103)
— nuclear polarization 38/1(105)
— nuclear volume effect 38/1(100)
Field emission 44(86)
Field gradients, effective, at the nucleus 38/1(406, 418)
Field gradients in a molecule 38/1(418)
Field imperfections in a betatron 44(203)
Field imperfections in a synchrotron 44(224–226, 307–312)
Field index 44(198, 221, 238–240, 300)
Field inhomogeneity 38/1(159, 161)
Field, quasistatic 38/1(146)
Field stabilization, direct 38/1(236–240)
Field stabilization, indirect 38/1(240–243)
Field strength measurement, relative errors 38/1(244, 248–255)
Field tube 45(179)
Fierz interference 41/2(19, 25, 50, 96); 42(507)
— for the neutron decay 41/2(76)
— smallness due to polarization 41/2(50–51)
— vanishes 41/2(90)
Fierz plot 41/2(50)
Filter for slow neutrons 44(430, 436)
Filters for epicadmium neutrons 38/2(350–352)
Fine structure 38/1(407); 42(343)
— of alpha rays 41/1(357)
— of nuclear magnetic spectra 38/1(296, 298)
— optical 39(202)
Finite nuclear size, influence on decay probability 41/2(13)
Finite range corrections to deuteron photoeffect formulae 38/2(137–138)
Finite thickness of detector, neutron self-absorption 38/2(390)
First forbidden transition 41/2(10, 15, 45)
First sum rule 41/1(192)
First-order theory of Coulomb excitation, quantum mechanical 41/1(527–533)
Fission 39(231); 40(504, 517)
Fission as a detector of thermal neutrons 38/2(633)
Fission chamber 44(420)
Fission channel 42(464)
Fission cross section data 45(446)
Fission detection of high-energy neutrons 45(492, 504–506)
Fission detection of moderate-energy neutrons 45(445–447)
Fission energy distribution 44(400)
Fission fragments from oriented nuclei 42(571)
Fission induced by pions 40(535)
Fission interference 44(419)
Fission investigations with emulsions 45(347, 390)
Fission neutron irradiation 44(408)
Fission neutrons inside reactor lattice 44(403)
Fission neutrons, penetration in water 38/2(301)
Fission of nuclei 38/2(17)
Fission of transuranium elements, multiplicity of neutrons 42(303, 305)
Fission of transuranium elements, number of released neutrons 42(301, 304)
Fission of uranium, energy spectrum of neutrons 38/2(310)
Fission pulses from ionization chambers 45(49)

Fission reactions, neutron induced 40(257)
Fission source in graphite, distribution of resonance neutrons 38/2(333)
Fission source in heavy water, distribution of resonance neutrons 38/2(329)
Fission source in water, distribution of resonance neutrons 38/2(317–318, 323–324)
Fission, spontaneous 38/2(210–211)
Fissionable nuclides 44(419–422)
Fixed-field alternating-gradient (FFAG) accelerators 44(301, 336–337)
Fixed-field alternating-gradient (FFAG) betatron 44(198, 201, 214–217)
Fixing emulsions 45(406–408)
Fizeau wheel 38/2(39)
Flash-over along insulator surface 44(80)
Floquet's theorem 44(305)
Flow counter 45(78, 182, 183)
Fluctuations in energy loss 45(263, 354, 358)
Fluctuations in gas gain 45(191)
Fluctuations in photon flux 38/2(683)
Fluctuations, statistical, in ionization 45(15–17)
Fluorescence radiation 38/2(664, 811)
Fluorescence spectra of actinide elements 42(215, 216)
Fluorescence x-ray yield 45(204, 205)
Fluorescence x-rays as a means of counter calibration 45(185)
Fluorescence yield 45(8)
Fluorescence yield from *K*-capture 45(219)
Fluorescent response of sodium iodide 45(99–104)
Fluorescent yield 42(23)
Fluorine 17, Fermi plot 41/2(50)
Fluorine, (α, *n*) reaction 38/2(109, 128)
Fluorine, (*d*, *n*) reaction 38/2(209)
Flux biased betatron 44(206)
Flux density of neutrons 38/2(215, 216)
Flux depression in the neutron detector 38/2(636, 642–652)
Flux distribution
— of neutrons 44(400, 402, 410)
— of resonance neutrons 38/2(323–324, 330–333, 337)
— of thermal neutrons 38/2(322)
— of thermal neutrons around a point source 38/2(530–531)
— of thermal neutrons in a gas moderator 38/2(464–467)
Flux moments 38/2(255, 297)
Flux of fast neutrons in water 38/2(301–302)
Flux of neutrons, energy dependence 38/2(344)
Flux plots of betatron magnets 44(209)
Focusing Čerenkov counters 45(161–164)
Focussing coupling in phase and radius 44(344, 374)
Focussing in the accelerating tube 44(83, 90)
Focussing of linear accelerator beam 44(359, 373–374)
Focussing of orbits in a betatron 44(194–195)
Focussing of orbits in a cyclotron, electric vertical 44(117–119)
Focussing of orbits in a cyclotron, magnetic vertical 44(109)
Foil activation by neutrons 45(468)
Foil effect upon neutron density 45(469)
Forbidden transitions 41/2(10, 15, 44–50)
Force on belts 44(76)
Force radius of nucleus (*see also* Radius, Size etc.) 39(212ff., 236)
Forced oscillations in the synchrotron 44(224)
Forces, non-Coulombic 38/1(106)
Fork in the cloud chamber 38/2(12)
Form factor, nuclear 40(458, 485)
Form factor to betatron oscillations 44(305, 306, 322, 327)
Four-component theory of the neutrino 41/2(35, 105)
Fourier transform inversion 38/2(765–768)
Fourier transform of a resonance curve 38/1(320)

Fourier transform of photon transport equation 38/2(690ff., 724)
Fourth forbidden transition 41/2(46)
Fowler-Perkins parameter of gap length distribution 45(419–422, 427)
FPW, *see* Feshbach Porter and Weisskopf
Fractional parentage coefficients 39(332, 398, 399)
Frame of reference, rotating 38/1(152)
Free atom cross section 38/2(64)
Free surface, boundary condition for neutron diffusion 38/2(541)
Freon in Van de Graaff generators 44(76, 96, 102)
Frequency, accuracy of measurement 38/1(230, 232, 236, 276)
Frequency factor 42(144, 173)
Frequency spectrum of lattice vibrations 38/2(414)
Frisch grid in an ionization chamber 45(34, 35, 455)
ft-value 41/2(8, 11, 38, 40)
ft-value from K-capture 41/2(98–99)
ft-value of allowed transitions 41/2(42)
ft-value of mirror nuclei 41/2(92, 94)
ft-value of the neutron decay 41/2(76, 86, 91)
ft-value of unfavoured transitions 41/2(43)
ft-values, allowed transitions in odd-nuclei 39(273)
ft-values, first forbidden transitions in odd nuclei 39(274)
ft-values, second forbidden transitions 39(276)
ft-values, superallowed transitions in odd nuclei 39(272)
ft-values, their classification 41/2(15)
Full polarization 41/2(53, 104)
Function fitting 38/2(735, 746ff., 758, 763, 766)
Fundamental translation vectors of a lattice 38/2(431)
Gadolinium as a detector of thermal neutrons 38/2(634)
Gallium 66, β-decay and polarization 41/2(68–69)
(γ, α) reactions 40(293)
Gamma emission in radiative capture 41/1(35–37)
Gamma emission, self-consistent model 41/1(26–27)
(γ, γ') reactions 40(294)
(γ, n) and (γ, p) reactions, energy and angular distribution 40(287)
(γ, $2n$) and (γ, pn) reactions 40(291)
(γ, n) reactions as neutron sources 38/2(132–170)
Gamma ray
— counter 45(79)
— detection by sodium iodide scintillators 45(100, 107–131)
— detection in emulsions 45(347)
— emission following neutron capture 38/2(24–25)
— spectra of radioactive elements used in photoneutron sources 38/2(158–159)
— yield from Be(α, n) 38/2(114)
Gamma rays (*see also* Capture, Neutron capture, etc.) 40(32, 97)
(γ, t) reactions 40(293)
γ-cascades, experimental methods 42(401)
γ-decay 39(223)
γ-γ angular correlation 41/1(522–524, 531)
γ-γ cascade 41/1(517, 521–524, 531)
γ-γ-correlation 42(410–413)
γ-γ-directional correlation 42(406)
γ-matrix 40(411)
γ-radiation from aligned nuclei 42 (568)
γ-radiation from polarized Co^{60} 42(581)
γ-radiation, polarized 42(441, 442, 573, 575, 581)
γ-ray after Coulomb excitation 41/1(517)
γ-ray angular distribution 41/1(437, 515, 517)
γ-ray anisotropies for Co^{56} 42(583)
γ-ray polarimeter 45(256)
γ-rays accompanying (p, α)-reactions as neutron sources 38/2(162–163)
γ-rays, Compton-scattered 42(450)
γ-rays from radioactive neutron sources 38/2(107–108, 114)
γ-rays, giant resonances 38/1(438)
γ-rays of high energy, photoneutron production 38/2(162–166)
γ-rays, selection rules 42(593)

γ-rays strength, absolute, measurement of 42(19)
γ-rays used in nuclear photoeffect observations 38/2(132–133, 145)
γ-ray-sources 42(311)
γ-transitions 38/1(109, 432, 433)
— matrix elements 39(408)
— of mixed multipolarity 42(406)
— probability 38/1(109, 432, 433)
γ-vibrations 42(195)
γ-vibrations of spheroidal nuclei 39(544–546)
Gamma-width, *see* Radiation width
Gamow formula 40(16, 70, 74, 117, 118)
Gamow-Teller coupling 41/2(9, 25, 49, 58, 68, 89, 92, 105)
Gamow-Teller interaction 42(507)
Gap length 45(419)
— distribution 45(419–422, 426)
— measurement 45(423–425)
Gaps, effect upon thermal neutron distribution 38/2(544)
Gas amplification (multiplication) 45(2, 53, 54)
Gas Čerenkov counters for high energy particles 45(164)
Gas fillings of Van de Graaff generators 44(67, 76, 78, 96–97, 102)
Gas gain, dependence on counter voltage 45(187)
Gas multiplication factor 45(176)
Gas purification 45(30, 453)
Gas, radioactive, in a counter 45(194–199)
Gas scattering in synchrotron chamber 44(229–230, 261–263)
Gas supply to ion source 44(90)
Gas target of tritium 38/2(176–177)
Gases as scintillants 45(86)
Gases with quenching properties (*see also* Quenching) 45(75)
Gate circuit 45(213, 234–236)
Gauges 44(150)
Gauss integration formula 38/2(569, 570)
Gauss shape 39(356)
Geiger counter, first discovery 45(52)
Geiger-Müller counter (*see also* Counters) 45(52, 176, 177)
Geiger-Nuttall relation 41/1(355)
Gel preparation 45(393)
Generalization of Fick's law
— for monoenergetic neutrons 38/2(521)
— for polyenergetic neutrons 38/2(526–527)
Generalized neutron age 38/2(500)
Generator noise voltage 38/1(189, 193)
Geometrical amplitude factor 38/2(435)
Geometrical buckling 38/2(276, 611, 623)
Geometry corrections in neutron distribution measurements 38/2(307–308)
g-factor of the free electron 38/1(273)
g-factor of the free proton 38/1(273)
Ghoshal's experiment 40(206)
Giant resonance(s) 39(470); 40(12, 50, 68, 150f., 282, 284f., 423, 510, 517, 521f.); 41/1(230, 265–274)
— dependence on mass number 38/2(150)
— for γ-rays 38/1(438)
— in copper 38/2(149)
— parameters 42(334)
— theory 38/2(152–154); 40(285, 423); 42(337)
Glass as insulator material 44(80, 85)
Glass envelope counters 45(72, 74)
G.M. counter, *see* Geiger-Müller counter or Counter
Goertzel-Greuling procedure of slowing-down calculation 38/2(243–244)
Gold 198, β-γ polarization correlation 41/2(74)
Gold 198, general data and polarization 41/2(66–67)
Gold, activation by slow neutrons 38/2(58)
Gold as a detector of slow neutrons 45(448)
Gold as a detector of thermal neutrons 38/2(634)
Gold as a standard for absorption cross section 38/2(60)
Gold detector covered with cadmium 38/2(311)
Gold, energy-dependence of slow-neutron total cross section 38/2(59)
Gold, resonance parameters 38/2(59)
Gold, thermal absorption 38/2(60, 634)

Goldhaber-Sunyar experiment 41/2(86–88, 92)
Good geometry 40(481, 484)
Gradient errors 44(309–310)
Gradient index, *see* Field index
Graetz connection of a capacitor input rectifier 44(3)
Grain density
— definition 45(419)
— dependence on residual range 45(379)
— dependence on temperature 45(388, 389)
— relativistic rise 45(355, 357)
Grain sizes in emulsion photographs 45(387, 420)
Grains of silver along an emulsion 45(345, 352, 354, 387)
Granularity of nuclear charge 39(198)
Graphite, age of thermal neutrons 38/2(499)
Graphite, albedo 38/2(552, 596)
Graphite, diffusion parameters of thermal neutrons, measurements 38/2(582, 585, 607–608, 624, 628)
Graphite, measurement of the extrapolated end point 38/2(589)
Graphite moderated research reactors 44(394–395)
Graphite moderator 38/2(331–338)
Graphite, thermalization time 38/2(628)
Graphite, total cross section for slow neutrons 38/2(444)
Graphite, transport mean free path of thermal neutrons 38/2(633)
Green's function
— and $\mathcal{R}$-matrix 41/1(158–159)
— construction 41/1(75–80)
— in many dimensions 41/1(83–85)
— of elementary diffusion theory 38/2(553)
— of radiation distribution 38/2(692ff.)
— of radiation distribution, moments 38/2(725ff.)
Greinacher connection of a capacitor input rectifier 44(3, 9)
Gridded chamber 45(34, 35, 45, 455)
Gridded chamber, fast, for nuclear reaction analysis 45(45–47)
Gridded ion chamber 45(174)
Grid-focussing 44(374, 375–376, 383)
Ground states of H^3 and He^3, empirical data 39(149)
Group *C* of slow neutrons 38/2(43, 63)
Group picture 38/2(279–283)
G-Stack-Experiment 45(349)
Guard electrodes in counters 45(179)
Guard tube 45(179)
Guide rods for belts 44(76)
Gyromagnetic ratio 38/1(123, 127, 195)
— absolute measurement for protons 38/1(248, 252, 253, 254, 257)
— microphonics 38/1(249)
— relative measurement 38/1(275–336)

H^3, binding energy 39(153)
H^3, electromagnetic transitions 39(176)
H^3, ground state, empirical data 39(149)
H^3, magnetic moment 39(150, 177)
H^3, role of tensor force in binding 39(155)
Half width, relative, of the inhomogeneity of a field 38/1(161)
Half-density radius of a nucleus 38/1(104)
Half-integral resonance 44(225, 238, 310)
Half-life of the neutron 38/2(8)
Half-lives in double β-decay 41/2(101)
Half-lives, isomeric, measurement 42(28)
Half-lives, partial 42(18)
Half-space, thermal neutrons emerging from it 38/2(536–539, 541)
Half-value layer 38/2(666)
Half-value resonance breadth 41/1(36)
Half-value response in resonance 41/1(27)
Halide crystals in emulsions 45(343, 386–387)
Halo of nucleons at the nuclear surface 41/1(368)
Halogen filled counters 45(71)
Hamiltonian for the Coulomb excitation problem 41/1(508, 528)
Hamiltonian of collective vibration 39(519)
Hamiltonian theory, application to orbits 44(311, 313–314)

Hamilton-Jacobi partial differential equation 41/1(497–499, 503)
— and correspondence principle 41/1(92, 104–107)
Hard core cut-off, *see* Cut-off
Hard core potential 41/1(149–151, 201, 353)
Hard core (repulsive potential) 39(30, 31, 43, 53, 56, 63, 76, 88, 90, 92, 96, 117, 118, 122, 124, 127, 130, 156)
Hard sphere scattering 40(36)
Hardening of thermal neutrons in the detector 38/2(636, 657)
Hard-sphere scattering 41/1(55, 131)
Harmonic order to the radiofrequency 44(229, 316)
Harmonic oscillator as molecular model 38/2(404, 419, 430)
Hartree central field approximation 38/1(98, 112)
Hartree-Fock potential 39(456)
Hartree-Fock self consistent field method 39(376)
Harwell electron accelerator, block diagram 44(365)
Harwell electron accelerator, waveguide 44(353, 365)
Hauser and Feshbach theory 40(217, 247)
He^3, binding energy 39(153)
He^3, electromagnetic transitions 39(176)
He^3, ground state, empirical data 39(149)
He^3, magnetic moment 39(150, 177)
He^3, role of tensor force in binding 39(155)
Health protection 45(37–41)
Heating by diffusion 38/2(608–609)
Heavy hydrogen, *see* Deuterium
Heavy ice target 38/2(189, 191)
Heavy ion accelerators 44(384–387)
Heavy particle
— collision 41/1(345)
— collisions, semiclassical approximation 41/1(370–373)
— Coulomb excitation 41/1(548)
— reactions 41/1(367–406)
— stripping 41/1(343–344)
— studies, advantages 41/1(368)
Heavy water
— albedo 38/2(596)
— diffusion coefficient of thermal neutrons 38/2(631)
— diffusion parameters of thermal neutrons, measurements 38/2(582, 584, 592)
— flux distribution of neutrons below 1 eV energy 38/2(468–469)
— measurement of the extrapolated end point 38/2(587–588)
— moderated normal uranium reactors 44(394–397)
— moderator 38/2(327–330)
— relaxation length measurements 38/2(591)
— scattering of slow neutrons 38/2(427)
— slowing down length, calculated 38/2(267)
— transport mean free path of thermal neutrons 38/2(631)
— -water mixture 38/2(329)
Heisenberg interaction 39(425)
Helicity of neutrino 41/2(86–88, 92)
Helium 3 as a neutron detector 45(442–444, 471)
Helium 3 counter 45(181, 185)
Helium 3, reactions by slow neutrons 38/2(26)
Helium 6, recoil investigations 41/2(25, 28, 29)
Helium (α, n) reaction 38/2(126)
Helium bubble chamber 45(326–327, 334)
Helium cross section for neutrons below 20 MeV 45(439, 440)
Helium ions in Van de Graaff generator 44(90, 91)
Helium, negative ions accelerated 44(103)
Helmholtz coils 38/1(235)
Heterogeneous moderator 38/2(345, 380)
Heterogeneous reactor 44(394)
Heterogeneous source of neutrons 38/2(87)
High energy detecting device 40(473f., 483)
High energy machines 40(468)
High energy neutron beams from several cyclotrons 45(490)

High field region in a counter 45(58)
High frequency
— field strength, optimal 38/1(175)
— spectroscopy, nuclear magnetic, bridge methods 38/1(185–204)
— spectroscopy, nuclear magnetic, magnetic field 38/1(231–243)
— spectroscopy, nuclear magnetic, oscillators 38/1(204–216)
— spectroscopy, nuclear magnetic, pulse circuiting 38/1(216–230)
High potential electrode 44(72)
High pressure cloud chamber 45(282)
High pressure operation of ionization chambers 45(49, 50)
High temperature studies at reactors 44(406)
High-energy electrons for photoneutron production 38/2(166–170)
High-energy γ-rays for photoneutron production 38/2(162–166)
High-energy nucleon scattering 41/1(257–262)
High-energy reactions 41/1(14ff., 304–312, 492–493)
High-energy scattering, sign of polarization 41/1(492, 493)
Higher order Coulomb excitation effects 41/1(548–558)
High-momentum components in nuclear wavefunctions 41/1(17, 24)
Hill's equation 44(302, 305)
Hindrance factor 42(142, 175, 181, 190)
Hindrance factor, calculations 42(176)
History of the cyclotron and synchrocyclotron 44(167)
H-magnet 44(239)
Holmium 166, β-decay and polarization 41/2(66–67)
Homogeneity of a neutron beam 38/2(106)
Homogeneous ionization chamber 45(452)
Homogeneous moderator 38/2(345)
Homogeneous reactor 44(394)
Homogenous moderator 38/2(376)
Homogenous source of neutrons 38/2(87, 104–107)
Hot laboratory 44(405, 444)
Houston-Lamb shift 38/1(105)
Hulthén potential 39(31)
Hulthén wave function 39(33, 34, 35, 140)
Humidity, *see* Moisture
Humidity effects in open-air Van de Graaff generators 44(66)
Hund-operator 39(329)
Hutchison-Scarrott kicksorter 45(192)
Huth-Kühn push-pull circuit 38/1(220)
Hybridization 38/1(420, 423, 425, 426)
Hydrocarbons, scattering of slow neutrons 38/2(417, 425–427)
Hydrodynamical model 42(339)
Hydrogen bubble chamber 45(321–326, 334)
Hydrogen, capture cross section for thermal neutrons 38/2(68–70, 594–596)
Hydrogen cross section for high energy neutrons 45(491, 495)
Hydrogen cross section for neutrons below 20 MeV 45(439)
Hydrogen moderator, energy distribution after *n* collisions 38/2(220–222)
Hydrogen molecules at low temperature, scattering of slow neutrons 38/2(422)
Hydrogen molecules, scattering of slow neutrons 38/2(417–418, 421, 425–426, 428)
Hydrogen, negative ions accelerated 44(103)
Hydrogen, scattering isotropy 38/2(66)
Hydrogen, slowing-down spectrum of neutrons 38/2(231–239)
Hydrogen, slow-neutron absorption cross section 38/2(65)
Hydrogen, theory of neutron scattering 38/2(66–67)
Hydrogen, total neutron cross section 38/2(65, 66)
Hydrogen, zero-energy scattering cross section 38/2(65)
Hydrogeneous moderators, slow neutron distribution results 38/2(311–327)
Hydrogeneous substances, discovery of their effect on neutrons 38/2(18–22)
Hylleraas wave functions 38/1(99)
Hyperbolic orbits 41/1(372, 378, 511)
Hyperfine coupling 42(519)
Hyperfine effects of second order, magnetic 38/1(405)

Hyperfine energy, magnetic 38/1(407)
Hyperfine interactions in molecules 38/1(418)
Hyperfine splitting 38/1(40)
Hyperfine structure 38/1(59, 63, 67, 68, 72, 96, 111, 279)
— due to two or more nuclei in the same molecule 38/1(389, 391)
— magnetic 38/1(59)
— of hydrogen 38/1(272)
— of nuclear magnetic spectra 38/1(296)
Hyperfragments 40(456, 472, 536)
Hyperon detection in emulsions 45(348)
Hyperon mass estimations 45(379)
Hyperons 40(456, 524f., 536); 42(530, 536)
Hyperons, ranges in emulsions 45(362–363, 368)
Hysteresis of a counter 45(55)

Identity of electrons and β-particles 41/2(2–8)
Ignitron 44(251, 253–254)
Illumination of scanning microscopes 45(412, 418)
Imaginary part of complex potential 41/1(22, 240, 241, 245, 255, 257, 265)
Imaginary part of potential 39(446, 451, 452)
Impact parameter 41/1(98, 100)
Imperfections in a linear accelerator 44(361–364, 377, 379–381)
Importance function 38/2(784, 787)
Importance sampling 38/2(783–785, 793)
Impulse approximation 39(165); 40(457, 461f., 533); 41/1(9–10, 260)
Impulsive raction 40(454)
Impurities of counters 45(68–69, 75)
Incoherent approximation to inelastic cross section 38/2(441, 443, 446, 473)
Incoherent scattering cross section 38/2(411–413)
— of neutrons in crystals 38/2(450)
Incoming Coulomb wave 41/1(346)
Independent (individual) particle model 39(236, 245, 257, 317)
Independent neutron source 38/2(214)
Independent particle model 42(14)
Independent particle shell model 40(7, 8, 131)
Indium 114, Fermi plot 41/2(38)
Indium activation by neutrons 44(409–410)
Indium, activation by slow neutrons 38/2(56)
Indium as a detector of slow neutrons 45(448)
Indium as a detector of thermal neutrons 38/2(634)
Indium detector covered with cadmium 38/2(311, 317)
Indium detector foil, relative perturbation of activation 38/2(651–652)
Indium difference method 44(411)
Indium, early work on selective absorption 38/2(43)
Indium, energy dependence of slow neutron total cross section 38/2(57)
Indium filter for epicadmium neutrons 38/2(351)
Indium resonance neutrons, distribution measurements 38/2(313ff., 323ff.)
Indium, resonance parameters 38/2(58)
Individual (independent) particle model 39(236, 245, 257, 317)
Individual particle model, application to light nuclei 39(335)
Individual particle shell model 40(7, 8, 131)
Induction acceleration 44(193)
Induction electrode 44(279)
Inelastic collisions 39(168)
Inelastic processes 40(14)
Inelastic scattering 40(30); 41/1(529)
— cross section 41/1(251)
— cross section formula 38/2(83)
— differential cross section 41/1(550, 553)
— effect on neutron age 38/2(326–327)
— of neutrons 38/2(15, 18, 64); 40(487f.); 45(451)
— of neutrons by an aluminium single crystal 38/2(455)
— of neutrons by bound protons 38/2(404)
— of slow neutrons by polycrystalline materials 38/2(440–456, 476)
— table of cross sections in several elements 38/2(85)

Inertial parameters 39(526–532)
Inflector 44(267, 268)
Inflector, missing it during later revolutions 44(267–274)
Inhomogeneity of a magnetic field 38/1(232)
Initial self-absorption coefficient 38/2(388)
Initiating channel 38/2(18, 47)
Injection coordinates 44(269)
Injection into AG proton synchrotrons 44(332–333)
Injection of electrons into a linear accelerator 44(355–357)
Injection of ions into a linear accelerator 44(378–379)
Injection of protons into synchrotrons 44(263–274, 333)
Injector for a synchrotron 44(176)
Injector of a betatron, design 44(213–214)
Injector of a betatron, hit by electrons 44(200–201)
Inner bremsstrahlung 45(219–220)
Inorganic scintillants 45(86, 97–143)
Instability of cyclotron orbits due to first harmonic 44(109–110)
Insulators in Van de Graaff generators 44(80, 84)
Integral effects in the epi-cadmium region 38/2(343)
Integral equation for diffusion of thermal neutrons with isotropic scattering 38/2(507–511, 517–520)
Integral equation of slowing-down spectrum, stationary case 38/2(228–230, 394–397)
Integral of pure resonance absorption 38/2(365)
Integral representations of the Coulomb functions 41/1(414–416, 427–429)
Integral resonance 44(224, 238, 308)
Integral resonance in the AVF cyclotron 44(115–117)
Integral spectrum of neutrons
— below 1 eV in boric acid solutions 38/2(471–472)
— below 1 eV in heavy water 38/2(468–469)
— below 1 eV in water 38/2(468–470)
Integral transform of photon transport equation 38/2(690ff., 724)
Integral yield 38/2(103, 118, 124)
Integration, numerical, of radiation moment equations 38/2(737ff.)
Intensity distribution in nuclear magnetic spectra 38/1(303)
Intensity function 38/1(142)
Intensity of the electron beam from a synchrotron 44(447–449)
Interaction assemblies of nuclei in crystals 38/1(312)
Interaction between nucleons 40(2)
Interaction between the nucleus and the electrons
— electrostatic 38/1(60)
— magnetic 38/1(60–63)
— spin-orbital 38/1(64–67)
Interaction, direct 40(105, 148)
Interaction energy of two dipoles 38/1(139)
Interaction forces, finite range 39(307)
Interaction Hamiltonian for β-decay 41/2(9, 88–93)
Interaction, intramolecular 38/1(144)
Interaction, non-static 39(23)
Interaction of neutrons with electrons 38/2(9–12)
Interaction radius 40(16)
Interaction representation 39(135)
Interface, boundary condition for neutron diffusion 38/2(542)
Interface of two media, radiation propagation 38/2(800ff.)
Interference 40(408)
Interference between two overlapping levels 42(459)
Interference minimum 39(7, 8, 59)
Interference, nuclear magnetic 38/1(155)
Interference of Coulomb and nuclear scattering 40(490)
Interference of scattered neutron waves 38/2(394)
Interior region of the nucleus 41/1(108, 185, 187)
Intermediate coupling 40(9, 131, 152, 156, 161, 174, 175)
— departures 39(355)
— model 39(335, 338, 354)
— parameter 40(177)

Intermediate model of Lane, Thomas and Wigner 40(12, 423)
Intermediate type nuclear model 41/1(268)
Intermolecular bond 38/1(428, 429)
Intermolecular coupling (*see also* Coupling) 38/1(324)
Internal bremsstrahlung 41/2(95, 98, 102–104)
Internal coincidence 45(205)
Internal conversion 40(47f.); 42(2, 5, 8, 477); 45(201, 205, 206, 219)
— coefficient 42(2, 5, 10, 19, 476)
— definition 41/2(2)
— electron lines 41/2(6)
— ratios 42(12, 24)
— reduced matrix elements for electric and magnetic multipole radiation 42(477)
— spectra 42(25)
Internal pair creation 41/2(2)
Internal radiators in a counter 45(79)
Internal scattering, *see* Resonance scattering
Internal slow-neutron experiments 38/2(344, 352, 356, 636)
Interval factor for atoms with more than one electron 38/1(66)
Interval factor for hydrogenlike atoms 38/1(63–65, 72)
Interval rule 38/1(62–68)
Interval theory of scattering mean free path 38/2(266–267)
Intramolecular coupling, *see* Coupling
Intrinsic crystal width (in scintillation spectrometer) 45(136)
Invariance principles and polarization 41/1(474, 484, 495)
Invariance to time reversal 41/1(41–42)
Invariants in betatron oscillations 44(307, 312, 315)
Inverse β-decay as a means of neutrino detection 41/2(32, 33)
Inverse mass tensor in scattering by bound nuclei 38/2(404–406, 467)
Inverse processes of nuclear reaction 40(21, 116)
Inversion of Fourier transform 38/2(765–768)
Iodine as a detector of slow neutrons 45(448–449)
Iodine detector covered with cadmium 38/2(311)
Ion diffusion 45(17)
Ion exchange method 42(214, 237, 252, 268, 273, 278, 279)
Ion gauge for synchrotron chamber 44(261)
Ion limit of supersaturation 45(265, 268)
Ion linear accelerators 44(366–388)
Ion reactions (*see also* Reaction of heavy ions) 40(148)
Ion recombination 45(26, 27)
Ion source(s) 40(25)
— for cyclotrons 44(146–149)
— for pulsed currents 44(90)
— for Van de Graaff generators 44(89–91)
Ionic character 38/1(423, 424, 425, 428)
Ionic crystals, calculation of ΔV 38/1(430)
Ionization along a cloud chamber track, its determination 45(295)
Ionization along an emulsion track 45(352ff.)
Ionization by radioactive recoil 45(13)
Ionization chamber 40(103, 105)
— application in cosmic ray investigation 45(43, 44)
— applications 45(37–50)
— average current operation 45(33)
— fast pulse operation 45(33, 34)
— for neutron detection 45(451–457)
— operation at high pressure 45(49, 50)
— operation principles 45(2)
— saturation 45(28–30)
— signal generation 45(27)
— slow pulse operation 45(33)
— types 45(34–37, 41–43, 46)
Ionization efficiency 45(7)
Ionization fluctuations 45(15–17)
Ionization induced by α-decay 42(152)
Ionization process 45(3, 4)
Ionization, specific 45(66)
Ionization theory, results 45(352–354)
Ionization, total 45(10–15, 261)
Ionizing event, primary (or initial) 45(3, 53, 56)
Ionizing event, secondary 45(7)
Ions, drift velocity 45(18)
Ions, mobility 45(18, 56)

Ions, paramagnetic 38/1(137, 149)
I-R (ionization-range) method of mass determination 45(378)
Iron, annihilation radiation 38/2(752)
Iron, build-up factors 38/2(756, 772)
Iron, energy absorption coefficient of γ-rays 38/2(674)
Iron, minimum of γ-ray attenuation 38/2(675)
Iron, radiation attenuation coefficients 38/2(667)
Iron, radiation energy flux 38/2(767)
Iron, reflection of radiation 38/2(792–793)
Iron, spectral distribution of radiation flux 38/2(700)
Iron, total reaction cross section 38/2(86)
Irregular Coulomb function G_L
— asymptotic behavior 41/1(409, 440–449)
— Bessel expansion for low energies 41/1(421–425)
— expansions for large η 41/1(425–434)
— power series 41/1(411–412)
Irregular function 41/1(226)
Irrotational flow 39(495, 531)
I-S (ionization-scattering) method of mass determination 45(379)
Isobaric corrections 40(152, 160, 161, 169)
Isobaric spin, *see* Isotopic spin
Isobars, relative binding energy positions 38/1(15, 20–21)
Isomer formation as a neutron detector 45(451)
Isomer formation by neutrons 38/2(15)
Isomeric state(s) 39(282); 40(181)
Isomeric states, activation by slow and fast neutrons 40(313, 314)
Isomeric transitions 39(296)
Isomerism, definition 42(1)
Isomerism in nuclei, discovery 42(1)
Isomers, data 42(39)
Isomers, empirical classification 42(5)
Isotope shift 38/1(38, 59, 380)
— anomalies 38/1(96)
— even-odd staggering 38/1(96, 105, 115)
— field effects 38/1(59, 96, 100–106)
— mass effects 38/1(59, 96, 97–100)
— nuclear deformation effects 38/1(106–109)
— of x-rays 39(201)
— optical 39(202, 224)
Isotope spin 39(209, 319)
Isotope spin, reduced 39(289, 326)
Isotopic incoherence of scattering 38/2(412)
Isotopic spin 40(3, 42, 71, 73f., 82, 107, 111, 124, 151, 173, 175)
— rules 40(120, 189)
— selection rule 41/2(74, 93)
— selection rules 42(349, 350)
Isotropic distribution of scattered neutrons 38/2(32)
Isotropic neutron source 38/2(214)
Isotropic scattering, diffusion of thermal neutrons 38/2(507–511, 517–520, 523)
Isotropy coefficient 39(6, 114, 129)
Isotropy factor 39(14)
Iteration approach to the solution of the photon transport equation 38/2(686ff., 706–707, 711, 733, 745, 767ff.)

j-j and *L-S* coupling, transformation between them 39(327)
j-j coupling 39(3, 25, 474); 40(9)
jj-doublet 40(194, 195, 196)
JWKB, *see* WKB

K mesons 40(525, 536)
Kapur-Peierls black-box treatment of the nucleus 41/1(5–6, 8, 24)
K-capture 41/2(2, 95, 99)
K-capture, Coulomb correction 41/2(12)
K-capture detection 45(174, 185, 217–219)
K-capture, energy balance 41/2(98)
K-capture, Fermi integral 41/2(98)
K-capture fluorescence yield 45(219)
K-capture recoils 41/2(20)
K-conversion 42(475)
Kernel of energy change for neutrons in crystals 38/2(474)
Kernel, symmetric, of the Wigner-Wilkins integral equation 38/2(396, 459)
Kernels for diffusion problems 38/2(553–554)

Kernels in slowing-down theory 38/2(253–254, 278, 283)
Kicksorter 45(184, 189, 246, 247)
Kicksorter types 45(192)
Kinetic energy of a configuration 39(334)
K-isomerism 42(16)
Klein-Nishina formula 38/2(668–669, 672); 42(450)
K/*L*-ratio curves, empirical 42(5)
K/*L*-ratio in argon 37 electron capture 41/2(21)
Klystron 44(351, 353)
K-matrix of Heitler 41/1(173)
K-meson mass estimations 45(378–379)
Knight resonance shift 38/1(330)
Knock-on electrons 45(309–310)
Kovar seal of counters 45(73)
Kraushaar-Goldhaber rule 40(369)
Krypton 85, recoil investigations 41/2(23, 24)
K-selection rules 39(492); 42(138, 197)
K-shell conversion 42(5)
K-shell vacancy 41/2(97)
Kurie plot, *see* Fermi plot

Laboratory frame of reference 38/2(18, 91 ff.)
Lacunarity 45(420, 422)
Laguerre polynomial expansion 38/2(740 ff.)
Lamb shift 38/1(105); 39(201)
Lambert law of angular neutron distribution 38/2(391, 560, 574, 654)
Landau distribution of energy losses 45(263)
Langevin formula 38/1(126)
Lanthanide elements 42(206, 210, 215, 216)
Laplace transformation 38/2(704, 707, 715)
Larmor frequency 38/1(123, 128)
Larmor frequency of protons 38/1(260–270)
Last-flight correction 38/2(255, 278)
Lattice of a substance 38/1(129)
Lattice of reactor 44(394)
Lattice vibrations 38/2(414)
Lauritsen electroscope with boron lining 45(39)

L-capture detection 45(185, 217–219)
Lead, annihilation radiation 38/2(752, 755)
Lead, build-up factors 38/2(753, 756, 772, 797)
Lead, diffraction of neutrons 38/2(416)
Lead, energy absorption coefficient of γ-rays 38/2(674)
Lead glass counter 45(171, 172)
Lead, minimum of γ-ray attenuation 38/2(675)
Lead, radiation attenuation coefficients 38/2(664, 667)
Lead, radiation energy flux 38/2(751, 767)
Lead reflection of radiation 38/2(792–793)
Lead, spectral distribution of radiation flux 38/2(700, 754–755)
Lead, total reaction cross section 38/2(86)
Leakage currents 45(179)
Left-handedness of electrons 41/2(90, 104)
Legendre coefficients 38/2(741 ff.)
Legendre polynomial expansion approximating solution of Boltzmann equation 38/2(259)
Legendre series, *see* Spherical harmonics expansion
Lepton conservation 41/2(100)
Lethal dose (LD) 44(443)
Lethargy 38/2(217, 219)
Lethargy as a function of neutron age 38/2(225)
Level broadening by emission of light 41/1(64)
Level densities
— Bethe's formula 40(297)
— dependence on neutron and proton numbers 40(299)
— dependence on spin 40(318)
— effect of closed shells 40(305)
— experimental 40(18, 106, 191, 194, 298, 436–443)
— from reaction data 40(18, 106, 191, 194, 300, 302, 436–443)
— Lang and Le Couteur's formula 40(301)
— Newton's theory 40(300)
— theoretical 40(296)
Level density 39(262)

Level density (*see also* Level spacing) 41/1(208)
Level diagram of Zr^{90} 39(309)
Level schemes from α-particle model 39(462–464)
Level shift 41/1(62, 135, 154, 164, 225, 227, 234)
Level spacing, *see* Spacing of levels
— and second sum rule 41/1(194, 196, 200)
— in compound nuclei, table 41/1(210–216)
— in the optical model 41/1(243, 249, 251)
Level systematics of α-emitting transuranium elements 42(297, 299)
Level width 38/1(131); 42(343, 346)
— broadening, instrumental 38/1(231, 236)
— neutron partial width 40(378, 428)
— partial (table) 40(382–395)
— radiation partial width 40(432f.)
— reduced for neutron emission 42(483)
— reduced partial width 40(409, 428)
— in compound nuclei, table 41/1(210–216)
Levels of the compound nucleus 41/1(57)
Levinger-Bethe sum rule 38/2(154–155)
Levinger's theory 42(352)
Lifetime of a self-quenching counter 45(67–68, 71, 84)
Lifetime of transitions 42(2, 5)
Lifetime, physical change of 42(17, 18)
Lifetime, reduced 39(284)
Lifetimes, comparative 42(89)
Lifetimes in double β-decay 41/2(101)
Lifetimes of excited nuclear states 40(41, 51)
Lifetimes of radiative transitions 40(54)
Light pipe in a Čerenkov counter 45(95)
Light sources for cloud chamber photography 45(294)
Lightest nuclei (up to helium) 39(422)
Limit of stability in a synchrotron 44(223, 274, 278)
Limitation of electron energy imposed by radiation loss in a synchrotron 44(181)
Linac, *see* Linear accelerator
Line spectrum in β-decay 41/2(6)
Line spectrum, nuclear magnetic 38/1(295)
Line width, *see* Resonance and Radiation width
Line width 38/2(46, 48, 50)
Linear accelerator(s) 40(468, 472)
— as pre-accelerators 44(266, 382)
— construction elements 44(360)
— for electrons 40(26); 44(347–366)
— for ions 44(366–388)
— for ions, list 44(343)
— historical remarks 44(342)
— theory 44(343–347)
Linear theory of betatron orbits 44(196–202)
$Li^7(p, n)$ Be^7 reaction 42(602)
Lippmann-Schwinger equation 39(135)
Liquid drop model 39(362, 414)
Liquid hydrocarbon mirror for neutrons 38/2(68)
Liquid mirror reflection of neutrons 39(3, 49)
Liquid scintillation counter types 45(169, 170)
Lithium 8, recoil measurements 41/2(29–30)
Lithium, absorption of slow neutrons 38/2(61)
Lithium, (α, *n*) reactions 38/2(126–127)
Lithium as a neutron detector 45(442, 443, 463, 467)
Lithium carbonate solution in heavy water, thermal neutron distribution 38/2(604–605)
Lithium, (*d*, *n*) reactions 38/2(198–201)
Lithium iodide as a scintillant 45(140–143)
Lithium loaded emulsions 45(347, 395, 467)
Lithium, (*p*, *n*) reaction 38/2(88, 171, 174, 178–182)
Lithium, reactions by slow neutrons 38/2(25–26)
Lithium targets 38/2(181–182)
Livingston-Jones ion source 44(147, 149)
L/*K* capture ratio 41/2(96–98)
Load characteristic of a cascade generator 44(42)

Load characteristic of a rectifier with choke input 44(23)
Loaded guide, theory 44(348–350)
Lock-in amplifier 38/1(199)
Logarithmic derivative, *see* Boundary conditions
Long counter for neutron flux measurement 45(481)
Long tube effect on voltage limit 44(87)
Longitudinal motion in the ion accelerator 44(371–372)
Longitudinal polarization 41/2(53)
Longitudinal variables of the radiation field 41/1(65, 71)
Loop gain of magnet regulator 44(138)
Lorentz invariance 41/2(52)
Los Alamos large Van de Graaff generator 44(101–103)
Loss of memory in angular correlation 42(14)
Losses in a valve 44(16)
Losses in energy-gain by imperfections 44(361–363)
Low energy states, systematics 40(319)
Low intensity sources 45(200, 207)
Low pressure cloud chamber 45(282)
Low temperature studies at reactors 44(406)
Low temperature target for slow neutrons 38/2(422, 446)
Low voltage counters 45(69)
L-S and *j-j*-coupling, transformation between them 39(327)
L-S-coupling 39(117, 119); 40(6, 9)
L-S-coupling scheme 42(419, 423, 429)
L-shell conversion 42(6)
L-shell vacancy 41/2(97)
l.s. = laboratory system 38/2(18, 91 ff.)
Lucite as insulator material 44(80)
Lucite coupler 45(95, 96)
Lutetium 177, *β*-decay and polarization 41/2(66–67)

Macroscopic absorption cross section of thermal neutrons, measurement 38/2(592–596)
Macroscopic cross section 38/2(213, 397)
Macroscopic transport cross section 38/2(257)
Magic numbers 38/1(16); 39(249, 466, 468); 40(8); 41/1(207); 41/2(15); 42(4)
— anomalies 39(258)
— effect on binding energy 38/1(35–36)
Magnesium, (*α*, *n*) reactions 38/2(129)
Magnesium, total cross section for slow neutrons 38/2(444)
Magnet construction in cyclotrons 44(132–134)
Magnet edge, rapid field fall-off 44(110–112)
Magnet parameters 44(234–235)
Magnetic analysis 40(141, 474, 477)
Magnetic beam analyzer 44(93)
Magnetic curvature
— of a bubble or cloud chamber track 45(303, 333)
— of an emulsion track 45(352, 366–367, 380)
— of an emulsion track, compared to multiple scattering curvature 45(380)
— of an emulsion track, sandwich method 45(380)
Magnetic dipole decay 38/1(432)
Magnetic dipole excitation 41/1(539)
Magnetic dipole transition 38/2(136, 143, 153); 39(491)
Magnetic effect in scattering 41/2(53)
Magnetic field(s) (*see also* Field) 38/1(231–243)
— applied on a counter 45(196–198)
— coils in a linear accelerator 44(359, 373)
— inhomogeneity (*see also* Field) 38/1(159)
— measurement 44(249)
— methods of modulation 38/1(172)
— strength drift as a function of temperature 38/1(238)
Magnetic guide field 44(170)
Magnetic length of a quadrant 44(249)
Magnetic moment(s), *see* Nuclear magnetic moment
— absolute measurement 38/1(243–244)
— anomalous 38/1(106)
— deviations from Schmidt values 41/2(94)
— matrix elements 39(407)

— molecular 38/1(43, 44)
— of H^3 and He^3 39(150, 177)
— of neutrino 41/2(31)
— of nuclei 39(475, 487)
— of stable nuclei of odd mass number in the ground state (table) 39(263ff.)
— of the deuteron 39(2, 32, 69)
— of the neutron 38/2(7, 8)
— of the proton 38/2(7)
— quantum electrodynamical corrections 38/1(272, 274)
— relative measurement 38/1(245)
Magnetic multipole transitions 39(100, 102, 103, 104, 110)
Magnetic ripple filters 44(255–256)
Magnetic scattering of thermal neutrons 42(594)
Magnetic spectrometer 40(27, 125, 141)
Magnetic susceptibility, experimental studies 42(218)
Magnetic susceptibility, theoretical consideration 42(217)
Magnetic vertical focussing in the cyclotron 44(109)
Magnetomechanical anomaly 38/1(274)
Magnetron 44(352)
Magnets for betatrons, design 44(206, 208–211)
Magnets for electron synchrotrons, design 44(172–175)
Magnets for proton synchrotrons, design 44(249–250)
Majorana interaction 39(425, 438)
Majorana particle 41/2(35, 36, 100)
Malter effect 44(86)
Manganese 52, polarized 42(587)
Manganese 55, polarized, neutron cross section 42(606)
Manganese 54, variation of anisotropy 42(586)
Manganese as a detector of slow neutrons 45(448–449)
Manganese as a detector of thermal neutrons 38/2(634–635, 637)
Manganese bath method 45(477)
Many-body features of alpha decay 41/1(355, 358)
Many-body forces 40(453)
Many-body problem with non-central operators 39(400)
Many-channel formalism 41/1(52–64, 111–116)
Many-level resonance formulas 41/1(5)
Mass attenuation coefficient 38/2(666)
Mass determination
— by cloud chamber tracks 45(297, 309, 310)
— by microwave methods 38/1(3)
— for an emulsion track, mass ratio 45(377, 378)
— from ionization change for long tracks 45(380)
Mass difference, odd-even 38/1(49)
Mass difference of neutron and proton 38/2(4–5)
Mass difference ratios 38/1(46, 54)
Mass, effective 39(234)
Mass, effective, of bound nucleons 39(440)
Mass effects of isotope shifts
— exchange 38/1(97); 38/1(99)
— normal 38/1(97)
— normal in hydrogen 38/1(98)
— polarization 38/1(99)
— relativistic 38/1(99)
— specific 38/1(59, 97, 98)
— specific in excited states of helium 38/1(100)
— specific in ground state of helium 38/1(99)
Mass excess, definition 38/1(3)
Mass excess, tables for all nuclides 38/1(7–13)
Mass expansion kernel 38/2(474–475)
Mass formula, semiempirical 39(370)
Mass of the neutrino 41/2(36–38)
Mass of the neutron 38/2(4, 8)
Mass ratio 38/1(41, 52)
Mass ratio proton/electron 38/1(272)
Mass region, intermediate 38/1(100)
Mass spectrometers 38/1(3)
Mass spectrometric methods for neutron absorption measurements 44(426)
Mass surface 42(114)
Mass tensor, inverse, in scattering by bound nuclei 38/2(405–406, 467)
Mass thickness of a layer 38/2(802)
Mass, variation with velocity 41/2(4)
Masses of nuclei, semiempirical formula 39(231, 233)

Masses of transuranium elements 42(293)
Materials for vacuum chambers 44(212–213)
Mathieu equation 44(225)
Matrix elements
— contribution to β-decays 41/2(10)
— determination 41/2(93–94)
— for the free neutron 41/2(93)
— reduced 42(382, 383, 386, 455, 477)
Matrix theory of alternating gradient accelerator 44(302–303, 305, 311)
Matter radius of nucleus (*see also* Radius, Size etc.) 39(186ff., 454, 455)
Maximum detectable momentum 45(307)
Maximum energy, definition 41/2(10, 38)
Maximum energy determined from Fermi plot 41/2(14)
Maximum energy of neutron β-rays 38/2(8)
Maximum non-conservation of parity 41/2(90)
Maximum permissible dose (MPD) 38/2(89)
Maximum possible absorption 38/2(30–31)
Maxwell distribution of neutrons, establishment 38/2(396, 397)
Maxwell distribution of velocities 44(401–402, 410, 423, 427–430)
Maxwellian distribution of velocities 38/2(40, 347)
Maxwellian neutron beam, absorption curve 38/2(656–659)
Mayer-Jensen shell model 42(135)
Maze proportional counter 45(183, 184)
Mean energy of β-decay, calorimetric determination 41/2(17–19)
Mean excitation potential 45(5)
— table of values 45(6)
Mean free path 39(369, 421); 45(57, 69)
— of a nucleon inside a nucleus 40(220, 456, 463, 482, 521)
— of neutrons in paraffin 38/2(63)
— of neutrons in the moderator 38/2(213)
— of photons 38/2(666)
— of pions inside nuclei 40(534)
Mean lacunarity 45(419–422)
Mean life for capture of neutrons 38/2(580, 592, 623, 626)
Mean obliquity of x-ray flux 38/2(714, 764)
Mean orbit of semi-classical approximation 41/1(544)
Measured activity of a detector 38/2(354)
Mechanical analogue for synchrotron oscillations 44(171)
Mechanism of bubble nucleation by ions 45(328–330)
Median surface deformations 44(241, 242)
Mendelevium 42(285, 286)
Mercury-cooled plutonium reactor 44(396)
Mesic atoms 40(455, 526f.)
Mesic atoms, isotope shift 38/1(103)
Mesic x-rays, detection in proportional counter 45(221)
Meson capture 41/1(17, 19)
Meson detection in bubble chambers 45(336–341)
Meson detection in emulsions 45(348, 378, 379)
Meson field explanation of magnetic moment anomaly 38/2(7)
Meson pair terms 39(20, 21)
Meson production, *see* Pion production
Meson reactions 40(524f.); 42(538)
Meson source 40(471)
Meson theory, neutral 39(17, 18)
Meson theory, symmetrical 39(17, 18, 24, 90, 91)
Mesonic atoms, binding energies 39(206ff.)
Mesonic atoms, doublet splitting 39(206ff.)
Mesonic effects 42(321)
Mesonic levels, quadrupole effects 39(227, 228)
Mesonic quadrupole effects 38/1(440)
Mesons, heavy 42(530, 536)
Mesons, ranges in emulsions 45(362, 368)
Metal-heavy water mixture, slow neutron distribution 38/2(330)
Metallic state of actinide elements 42(211)

Metals, nuclear magnetic analysis 38/1(329–333)
Metal-to-glass seal 45(73)
Metal-water mixture, slow neutron distribution 38/2(314–318, 326)
Metastable atoms or molecules 45(9, 60, 70)
Methane scattering of slow neutrons 38/2(405–406, 408–409, 423, 431)
Method of discrete ordinates 38/2(569)
Method of moments (*see also* Moment method) 38/2(297–299)
Microprojection of emulsion tracks 45(417)
Microscopes for nuclear track investigation 45(411–417)
Microwave gas absorption cell 38/1(50)
Microwave mass measurements 38/1(3)
Migration area of slowed-down neutrons 44(392)
Migration area of thermal neutrons 38/2(311, 319, 493, 498)
Migration energy 38/2(493, 495)
Miller indices 38/2(432)
Milne's problem 38/2(536–538, 541, 560, 569, 572)
Minimum of attenuation coefficient of radiation 38/2(675, 707, 751, 755)
Minnesota linear accelerator 44(383)
Mirror balancing techniques 44(437–439)
Mirror nuclei 38/2(172); 39(211, 417); 40(7)
Mirror nuclei, ft-values 41/2(92, 94)
Mirrors for neutrons 44(435–439)
Mixed Fermi and Gamow-Teller transitions 41/2(25, 58)
Mixed radiation 42(9, 11, 12)
Mixing 39(505, 529, 530)
Mixing parameter 39(7, 9, 80, 81, 82, 85)
— α-particle model 39(366, 460–464)
— by Rainwater 39(477, 498)
— collective 39(350, 361, 480)
— Fermi gas model 39(421)
— independent (individual) particles 39(236, 245, 257, 317, 335, 412, 421)
— intermediate coupling 39(335, 338, 354)
— liquid drop 39(362, 414)
— optical 39(372, 445–450)
— rotational model 39(480, 537)
— shell model 39(230, 236, 242, 368, 464–476)
— spheroidal core 39(477, 498–501, 514)
— strong coupling 39(446)
— unified 39(349, 362, 476–543)
Mobility of electrons 45(56)
Mobility of ions 45(18, 56)
Model magnet 44(138, 249)
Model, one-dimensional, of thermal neutron diffusion 38/2(557–562, 570)
Model wavefunction of Brueckner 39(445)
Models of nuclei 42(3, 6, 14, 135, 150, 166, 173, 185, 290, 298, 339, 343, 463, 493)
— brief survey 39(412–414)
— with strong interaction 39(412)
Moderating ratio 38/2(227)
Moderation of neutrons 40(397)
Moderation time, *see* Slowing-down time
Moderator 38/2(22); 40(398); 44(391)
Moderator model of monatomic heavy gas 38/2(462)
— compared to heavy polycrystal 38/2(476)
Moderator model of monatomic hydrogen gas 38/2(458–465)
Moderator surface, thermal neutrons emerging 38/2(536–539, 541)
Moderators, different, assembled 38/2(279)
Moderators, properties of some 38/2(228, 244)
Modified synchrotrons 44(338)
Modulation coil 38/1(231)
Modulation of magnetic fields 38/1(172)
Moisture of an emulsion 45(360, 384–386)
— effect on fading 45(388)
Molecular beam 38/1(50)
— electric resonance 38/1(50)
— method 38/2(37–39)
— microwave emission spectrometer 38/1(50)
— spectroscopy 38/1(279)
Molecular binding, effect on neutron scattering 38/2(401ff., 416ff.)

Molecular crystal, calculation of ΔV 38/1(430)
Molecular hyperfine structure 38/1(418)
Molecular spectra 38/1(83–94)
Molecular spectra, nuclear magnetic (*see also* Nuclear magnetic spectra) 38/1(302)
Molecular structure effect at very low neutron energy 38/2(62–63)
Molecule, linear triatomic type 38/1(48)
Molecule, magnetic moment 38/1(43, 44)
Molecule, polyatomic 38/1(45)
Molecule, polyatomic linear 38/1(45)
Molecules, diatomic 38/1(38, 40, 52)
— homonuclear, with non-zero nuclear spin 38/1(89–94)
— homonuclear, with zero nuclear spin 38/1(87–89)
Møller scattering of polarized electrons by electrons 41/2(53, 59–60)
Moment equations of radiation 38/2(724)
— numerical solution 38/2(737ff.)
Moment fit of neutron spectra 38/2(478)
Moment, magnetic 38/1(125)
Moment, magnetic, static 38/1(126)
Moment method as an approach to solve the photon transport equation 38/2(724, 737ff.)
Moment of inertia 39(494–497, 527); 40(341)
— effective 38/1(431)
— of a molecule, electronic contribution 38/1(42, 43)
— of a nucleus 38/1(109)
Moment, second 39(384, 387)
Moments, infinitesimal 38/1(133)
Moments, method of 38/2(297–299)
Moments of Green's function of radiation distribution 38/2(725ff.)
Moments of indium resonance neutrons in graphite from a fission source 38/2(333–334)
Moments of indium resonance neutrons in water from a point fission source 38/2(317, 324)
Moments of neutron flux 38/2(255, 297)
Moments of the angular distribution 38/2(749, 763)
Moments of the space distribution of radiation 38/2(724ff., 735ff.)
Momentum balance in elastic scattering 38/2(98–101)
Momentum compaction factor 44(175, 220, 227, 301, 335, 336)
Momentum conservation in nuclear reactions 38/2(95)
Momentum determination by cloud chamber tracks 45(302–308)
Momentum distribution
— in a nucleus 41/1(16)
— of nucleons 40(451f., 461, 466, 497f., 533)
— of nucleons after Chew and Goldberger 40(452, 498, 500, 533)
— of nucleons after Fermi 40(452, 463, 519, 531, 533)
— of nucleons, Gaussian 40(452, 489, 519f., 533)
Momentum exchange in pion absorption 41/1(17–18)
Momentum space approach to the optical model 41/1(261)
Momentum space representation 39(139)
Momentum transfer in dense nuclei 39(441, 442)
Monatomic heavy gas moderator model 38/2(462)
Monatomic hydrogen gas moderator model 38/2(458–465)
Monitoring proton synchrotron operation 44(293)
Monitors for slow and fast neutrons 45(40, 41)
Monochromator 42(314)
Monochromator for slow neutrons 40(375)
Monoenergetic electrons from a synchrotron 44(191)
Monoenergetic neutrons at high energies 45(487, 488)
Monopole (compressive) oscillation of a nucleus 39(419)
Monopole, electric 42(9)
Monopole transition 40(509)
Monte Carlo calculations, efficiency 38/2(781, 789)
— estimates and errors 38/2(779ff.)

Monte Carlo method 40(464f., 494, 499, 503)
— for neutrons 38/2(295, 327, 459–460, 463, 479)
— of calculating photon histories 38/2(685, 769ff.)
Monte Carlo solution of radiation boundary problems 38/2(789–801)
Most probable energy of thermal neutrons 38/2(40)
Most probable velocity of thermal neutrons 38/2(40)
Mott scattering of polarized electrons by nuclei 41/2(53–58)
Mott-Gordon Coulomb wave 41/1(102)
Mounting of emulsion stacks before development 45(399–401)
Mounting of emulsion stacks before exposure 45(398–399)
Multi-channel discriminator 45(184)
Multigroup picture, *see* Group picture
Multiparticle wave functions 39(288)
Multiphonon processes with neutrons 38/2(441, 451)
Multiplate cloud chamber 45(271–273, 282)
Multiplate scintillator counter 45(170)
Multiple collisions 39(448, 450)
Multiple collisions within a nucleus 40(459)
Multiple Coulomb scattering 45(345, 352, 358, 366–367, 371–379)
— curvature, compared to magnetic track curvature 45(380)
— results of theory 45(372)
Multiple delay discriminator 45(234)
Multiple scattering 39(445, 450); 40(453)
Multiple scattering in a cloud chamber track 45(305, 306)
Multiplication process in a proportional counter 45(175)
Multiplier 45(87–94)
Multiply charged particles in emulsions 45(370–371)
Multipolarity 40(43, 46, 47, 49)
Multipole assignments for Co^{56} 42(583)
Multipole moment(s) 38/1(379); 39(67ff., 113)
Multipole operator 41/1(35)
Multipole order 42(7)
Multipole transitions 42(316, 318, 319)
Multipole transitions (*see also* Transition probabilities) 40(43, 46)
Multipole transitions: selection rules and matrix elements 39(100, 102, 103, 104, 110)
Multiprong stars 40(503, 522)
Multi-wire counter tube 45(217, 218)
Muon 40(525)
Muon capture 41/2(17, 21)
Muon capture within a nucleus 40(528f.)
Muon decay 40(525, 529); 41/2(17)
Muon pair production 40(532)
Muon scattering 41/1(263)
Muons as nuclear probe particles 39(204)
Muons scattered by non-spherical nuclei 38/1(438)
Mutual absorption of two resonances 38/2(390)
Mykroy insulators (mica bonded with glass) 44(85)
μ-mesic atoms, isotope shifts 38/1(103)
μ-meson, *see* Muon

(n, α)-process, discovery 38/2(12–13)
(n, γ)-process 38/2(15)
(n, $2n$) reactions 40(255)
(n, n') reactions 40(245)
(n, $2n$) reactions, cross sections 40(217)
(n, $2n$) reactions, multiplication measurements 40(256)
(n, np) reactions 40(254)
(n, n')-process 38/2(15–16, 64)
(n, $2n$)-process 38/2(15–16, 17)
(n, $2n$)-process, as a secondary neutron source 38/2(64)
(n, p) and (n, α) reactions 40(251)
(n, p)-process 38/2(15)
Narrow-beam attenuation 38/2(665ff.)
Natural width of neutron resonance 38/2(346)
Needle valve 44(90)
Needles for charging belts 44(77–78)
Negative ion accelerators 44(103)
Negative ions as discharging carriers 44(86)
Negative ions in a counter discharge 45(177)

Negative point source of thermal neutrons 38/2(644)
Negative source of neutrons 38/2(532, 597)
Neher-Harper quenching circuit 45(61)
Neodymium 147, decay scheme 42(587)
Neon 19, recoil investigations 41/2(25, 26–28)
Neon 23, recoil investigations 41/2(25, 28)
Neptunium 42(224–237)
Neptunium, chemical properties 42(229)
Neptunium, decay schemes 42(228)
Neptunium in nature 42(227)
Neptunium isotopes 42(226)
Neutral theory 39(17, 18)
Neutralization of ions in self-quenching counters 45(62)
Neutrino 38/2(8); 45(207)
Neutrino absorption by a proton 41/2(32, 33)
Neutrino, detection by inverse β-decay 41/2(32)
Neutrino, direct detection experiments 41/2(31–35)
Neutrino, emission in mesonic processes 41/2(16–17)
Neutrino γ-ray polarization correlation in *K*-capture 41/2(86–87)
Neutrino, helicity 41/2(86–88, 92)
Neutrino, magnetic interaction 41/2(31–32)
Neutrino momentum, determination by recoil experiments 41/2(19)
Neutrino, rest mass 41/2(36–38)
Neutron age, *see also* Age 44(392)
Neutron beam at high energies, contaminants 45(489)
Neutron beam polarization 45(490, 515–516)
Neutron beams from a reactor 44(398)
Neutron, β-decay 41/2(75–86)
Neutron binding energies (table) 40(268–270); 41/1(213–216)
Neutron booster 38/2(170)
Neutron capture 40(47, 303–312, 397)
— as a neutron detector 45(447–449)
— by deuterons 39(176)
Neutron capture cross section 39(262); 42(300)
— theoretical 40(303f., 307)
Neutron capture gamma rays 40(308f., 311f., 397)
— cascade 40(311)
— multiplicity 40(312)
— spectrum 40(312)
— transition probabilities 40(308, 309)
Neutron converter 44(406–407)
Neutron cross section(s)
— nonelastic 40(242, 243)
— nonelastic, energy dependence 40(244)
— theoretical, optical model 40(228)
— theoretical, strong coupling 40(227)
— total 40(226)
— total, at high energies 45(491)
Neutron decay, best values of constants 41/2(86)
Neutron decay, determination of coupling constants 41/2(91)
Neutron detection 40(30)
— by bismuth fission 45(445, 504–506, 512)
— by induced radioactivity, high-energy neutrons 45(506–508, 512)
— by induced radioactivity, slow neutrons 45(447–449, 468)
— in emulsions 45(347, 395, 466–468, 508–510)
— in the bubble chamber 45(466, 511)
— in the cloud chamber 45(465–466, 510–511, 512)
— technique above 700 MeV 45(493)
— technique below 20 MeV 45(437–485)
— technique between 20 and 100 MeV 45(491)
— technique between 100 and 700 MeV 45(492–493)
Neutron, determination of half-life 41/2(77, 80)
Neutron diffusion 44(392, 426–427)
Neutron dosage measurements 45(483–485, 516)
Neutron double scattering experiments 41/1(483–484)
Neutron efficiency of thick scintillants 45(501)
Neutron emission, delayed 39(259)
Neutron, Fermi plot 41/2(78)

Neutron filter 45(489)
Neutron fission cross section 42(300, 302)
Neutron flux, *see* Flux
— absolute determination at high energies 45(497–499, 513)
— absolute determination at thermal energies 45(479–480)
— absolute determination for several MeV energy 45(480–483, 484)
— of a reactor 44(398, 399)
Neutron life cycle in a reactor 44(391)
Neutron mechanical monochromator 44(432)
Neutron mirror(s) 38/2(6, 10, 68); 44(435–439)
Neutron monitor 45(40, 41, 469–470)
Neutron polarization 44(439–440)
Neutron polarization in reactions 41/1(484)
Neutron probe 38/2(344, 637)
Neutron production by spontaneous fission 38/2(210–211)
Neutron production threshold 40(29)
Neutron proton competition 40(215)
Neutron proton scattering cross section at high energies 40(463)
Neutron, recoil experiments 41/2(80–81)
Neutron resonance energies, table 41/1(213–216)
Neutron resonance strength 40(231, 424, 427f.)
Neutron scattering 45(437–442, 451)
— at low energies by nuclei 42(598)
— by bound protons 41/1(314, 327, 330)
— determination of nuclear size 39(213, 214)
Neutron, shape of β-spectrum 41/2(81)
Neutron source, *see also* Source 40(24)
— high energy 40(470)
— strength 45(475–479)
Neutron spectra at high energies 45(488–489, 506, 513–515)
Neutron spectra at moderate energies 45(471–475)
Neutron spectrometers 40(374f.)
— at high energies 45(514)
— at moderate energies 45(474–475)
— using He^3 45(181)
Neutron spectroscopy 44(411–415)
Neutron tail correction to thermal cross sections 44(423)
Neutron temperature 44(427)
Neutron total cross sections, result 41/1(246–248)
Neutron transfer cross sections 41/1(389, 399)
Neutron width in the optical model 41/1(243, 249, 251)
Neutron width of a resonance level 38/2(50, 83)
Neutron width, reduced 40(428)
Neutron yield(s)
— from Be(α, n) reaction 38/2(114–115, 118–119)
— from Be(γ, n) reaction 38/2(143–145)
— from copper photodisintegration 38/2(149)
— from D(γ, n) reaction 38/2(135, 139)
— from different sources, tables 38/2(110)
— from Poα sources 38/2(108, 118–119, 124)
— from Raα+B source 38/2(125)
— from Raα+Be source 38/2(119)
— of rhodium photo-disintegration 38/2(147)
Neutron-deuteron scattering
— at low energies 39(170)
— empirical data 39(158)
— perturbation theory 39(159)
Neutron-electron interaction 44(438–439)
Neutron-proton capture cross section 42(322)
Neutron-proton capture, exchange terms 42(324)
Neutron-proton capture, thermal 39(5, 115)
Neutron-proton cross section measured by mirror balancing technique 44(438)
Neutron-proton scattering 38/2(6, 65–68)
— at high energies 39(9, 116)
— at low energies 39(3, 4, 83)
— double 39(11, 121, 124)
Neutrons from stripping reactions 41/1(309)

Neutrons in cosmic rays 38/2(209–210)
Neutrons, polarized 42(606, 607)
Neutrons, thermal, magnetic scattering 42(594)
Nilsson potential 39(508)
Nitrogen 13, β-decay and polarization 41/2(68–69)
Nitrogen, (α, n) reactions 38/2(128)
Nitrogen as a neutron detector 45(442, 445)
Nitrogen, (d, n) reactions 38/2(208)
Nitrogen, dielectric strength 44(102)
Nitrogen ions inducing reactions 41/1(404–405)
Nitrogen molecules, scattering of slow neutrons 38/2(424)
Nitrogen, reactions by slow neutrons 38/2(26)
3n-j-symbols 39(391)
Noise 38/1(189, 193)
Noise and resolving power 45(32, 33)
Noise, magnetic 38/1(133)
Noise spectrum, intensity 38/1(135, 200)
Noise spectrum, thermal 38/1(133, 200)
Noise, thermal 38/1(133, 141, 200)
Non-central forces (*see also* Tensor force) 39(23, 24)
Non-central forces in nuclear photoeffect 42(320)
Non-central operators 39(400)
Non-linear effects in alternating gradient theory 44(312)
Non-retarded dipole emission 41/1(67, 72)
Nordheim number 39(298)
Nuclear boundary 41/1(109, 257)
Nuclear charge distribution 38/1(103); 39(195, 196, 416, 431)
Nuclear charge radius 40(505)
Nuclear compressibility 38/1(110, 114)
Nuclear deformation 38/1(106, 113, 115, 432, 437)
Nuclear deformation by scattered muons 38/1(438)
Nuclear density, *see* Density
Nuclear dipole resonance, pure 38/1(322)
Nuclear distance in molecules 38/1(425)
Nuclear emulsions 40(26, 51)
Nuclear energy levels 38/1(109)
Nuclear forces 41/1(11)
— charge symmetry at high energies 39(170)
— from high energy experiments 40(453f., 513f.)
— saturation 41/1(20–22)
Nuclear induction 38/1(111)
Nuclear induction (excitation and observation with crossed coils) 38/1(193, 195, 196)
Nuclear interaction 39(330)
Nuclear level density (*see also* Level density) 40(18, 106, 191, 194)
Nuclear magnetic moment(s), *see* Magnetic
— determination 38/1(59, 70–75, 82–83)
— measurement 42(528)
— report, explanation of content 38/1(336, 337)
— report, table of all isotopes 38/1(338–376)
— sign 38/1(70)
Nuclear magnetic spectra
— of molecules 38/1(302)
— structure for gases 38/1(293–311)
— structure for liquids 38/1(293–311)
— structure for polycrystals 38/1(315)
— structure for solids 38/1(311–336)
Nuclear magnetism, basic equations 38/1(128)
Nuclear magneton 38/1(265); 38/2(7)
Nuclear mass effects 38/1(97)
Nuclear models, *see* Models
Nuclear orientation, *see* Oriented nuclei
Nuclear Overhauser effect 38/1(335)
Nuclear parameters 40(482f., 486, 534)
Nuclear phase shift 39(58, 59)
Nuclear photoeffect
— as a secondary neutron source 38/2(64)
— as primary neutron source 38/2(97–98)
— giant resonances 38/2(149ff.)
— in beryllium 38/2(132, 141–145)
— in deuterium 38/2(132, 133–141)
— in medium and heavy elements 38/2(145–158)
— releasing more than one neutron 38/2(152)
— sum rules 38/2(154–155)

Nuclear polarization 38/1(105, 115)
Nuclear potential, *see* Potential
Nuclear quadrupole coupling
— in crystals 38/1(324)
— in solids 38/1(392, 428, 429)
Nuclear quadrupole moments, *see* Quadrupole moment
— electric 38/1(323, 325, 377, 378, 380)
— evaluation from their coupling in molecules 38/1(417)
— report on experimentally determinated values 38/1(443–446)
— systematic behaviour 38/1(449–451)
Nuclear quadrupole resonance, pure 38/1(322, 329)
Nuclear radii 38/1(23–24)
Nuclear radius (*see also* Radius) 38/2(27); 40(15, 131, 143, 152, 451, 482f., 505, 521, 526f.); 41/1(11, 34, 82, 206–207, 209, 228, 343, 362, 400); 42(15, 139, 143, 144, 145, 153); 44(419)
Nuclear reaction(s)
— energies 38/1(4)
— inside the ionization chamber 45(47–49)
— outside the ionization chamber 45(45–47)
— statistical theory 40(17)
Nuclear reactor 40(25)
Nuclear resonance 38/1(186)
— absorption 38/1(195)
— fluorescence 42(37)
— frequency, dependence on the macroscopic susceptibility 38/1(281–286)
— frequency, dependence on the microscopic screening of the electronic shells 38/1(286)
Nuclear rotational energy levels 38/1(431)
Nuclear shape, *see* Shape of nucleus
Nuclear shape, effects on scattering 38/1(436)
Nuclear shell model 40(7, 8, 131, 415)
— and β-decay 41/2(15–16)
Nuclear sign voltage 38/1(187)
Nuclear signal 38/1(188)
Nuclear size, *see* Size of nucleus
Nuclear size, finite, effect on magnetic moments 38/1(72)
Nuclear spectrometer 38/1(279)
Nuclear spectrometer sensitivity 38/1(279)
Nuclear spin(s) 38/1(124); 42(8, 300, 301)
— determination 38/1(59, 67–70, 80–82)
— measurement with high frequency spectroscopic methods 38/1(245)
— of even-even nuclei 38/1(89)
— of homonuclear diatomic molecules 38/1(87–94)
— system, excited, temperature 38/1(177)
Nuclear statistics 38/1(91, 94)
Nuclear surface 41/1(205, 257)
Nuclear surface phenomena 38/1(110)
Nuclear temperature 38/2(84); 40(216)
Nuclear transition quadrupole matrix 41/1(529)
Nuclear volume effects 38/1(72, 100)
Nucleation of boiling on boundary surfaces 45(319)
Nucleation of bubbles 45(314, 316)
— mechanism 45(328–330)
Nucleon binding energy 38/1(33–35)
Nucleon scattering 39(371)
Nucleon transfer by virtual Coulomb excitation 41/1(369, 391–400)
Nucleon tunneling in transfer reactions 41/1(369, 370, 379–390)
Nucleon width, reduced 39(409)
Nucleon-nucleon interaction, direct 42(491)
Nucleus, deformed 38/1(432, 437)
Nucleus, spherical 38/1(438)
Nucleus, spheroidal 38/1(107)
Number albedo of photons 38/2(789)
Number flux, *see* Flux
Number of betatron wavelengths per revolution 44(227, 303, 306)
Number of collisions until capture of neutron 38/2(546, 575, 592, 596)
Numerical integration of the Coulomb-field differential equation 41/1(455)
Numerical solution of radiation moment equations 38/2(737ff.)

Oak Ridge cyclotron, radiofrequency system 44(143–145)
Objectives for scanning microscopes 45(412)

Obliquity of neutron beam in the detector 38/2(636, 638–640, 654)
Obliquity of x-ray flux 38/2(714, 764)
— effect on deep penetration 38/2(713–714)
Obliquity of x-ray source emission 38/2(679, 713, 723, 746, 760)
Occupation numbers of energy levels 38/1(125, 126)
Octuply charged oxygen ions accelerated 41/1(367–368)
Octupole excitation 39(523–525)
Octupole moment, magnetic 39(255, 281)
Octupole moments 38/1(138)
Odd-odd nuclei 39(293, 298)
Odd-odd nuclei, electromagnetic moments 39(298)
Omegatron 38/1(265)
One level formula of Breit and Wigner 40(20, 45, 123, 403ff.)
One-body model of alphy decay 41/1(355)
One-body resonance, illustration 41/1(36)
One-channel specialization 41/1(47–50)
One-dimensional model of thermal neutron diffusion 38/2(557–562, 570)
One-group picture 38/2(282)
One-level approximation 41/1(136–138)
One-level formula for two channels, applications 41/1(231–239)
One-level resonance formula of Breit and Wigner 38/2(45, 49)
One-level terms in the Wigner-Eisenbud method 41/1(131)
One-level with background approximation 41/1(45, 61, 63, 163–167, 223)
One-phonon processes with neutrons 38/2(447, 449)
One-velocity group treatment of thermal neutrons 38/2(504–507)
— results of transport theory 38/2(527–539)
Opacity, nuclear 39(447); 40(462)
Opaque nucleus, stripping reaction 41/1(309)
Open channels 41/1(168, 172)
Operating cycle of a linear accelerator 44(360, 381–382)
Operation of a cyclotron 44(159–160)
Optical distance 38/2(527)
Optical model
— and $\mathscr{R}$-matrix 41/1(243, 265, 267)
— and polarization 41/1(261, 263)
— for nucleon reactions 40(12, 86, 210, 228, 415, 459f., 492)
— for pion reactions 40(534f.)
— of nuclear reactions 41/1(23, 240–265, 400)
— of nucleus 39(372, 445–459)
— parameters 40(460, 482f., 486)
— with diffuse surface, parameters 41/1(256)
Orbit parameters for heavy particle collisions 41/1(372–373)
Orbital stability in cyclotrons 44(107–121)
Orbital stability in the AVF cyclotron 44(112–115)
Orbital states, group theoretical classification 39(396)
Organic crystal scintillators, use and properties 45(145)
Organic liquid bubble chamber 45(319–321)
Organic liquid scintillators, use and properties 45(146, 149–152)
Organic plastic scintillators, preparation 45(153, 154)
— use and properties 45(147, 149–152)
Organic scintillants 45(86, 144–154)
Orientation of initial particles 42(592)
Orientation parameters 42(564)
Orientational degree of freedom, magnetic 38/1(133)
Oriented nuclei
— alignment 42(558, 568)
— detection 42(567)
— polarization 42(568, 573, 588)
— production by low temperature methods 42(559, 562)
— production by microwave methods 42(563)
— production by optical methods 42(563)
Orthohydrogen 39(48, 49)
Orthohydrogen, neutron scattering 38/2(422)
Orthomolecules 38/1(323)
Orthostate 38/1(145)
Oscillation damping in a linear accelerator 44(356, 372)

Oscillation frequencies of cyclotron orbits 44(109)
Oscillator for cyclotron radiofrequency voltage 44(140–143)
Oscillator potential, anisotropic 39(501–506)
Oscillator potential for nucleons 39(468–470)
Oscillator strength 40(285)
Oscillator well, infinite harmonic 39(246)
Oscillator with cathode coupling 38/1(209)
Oscillatory motion in a linear accelerator 44(346–347, 372)
Oszillators for the nuclear magnetic high-frequency spectroscopy 38/1(204–216)
Outgoing channel 38/2(18, 47)
Outgoing Coulomb wave 41/1(346)
Outgoing wave type of Green's functions 41/1(78, 80, 84)
Overcompression ratio 45(283)
Overhauser effect 38/1(333)
Overlap integral 38/1(421)
Overlap of channels 41/1(160)
Overlapp model 42(173, 185)
Overloading by big signals 45(188)
Overvoltage of a counter 45(55)
Oxidation states of lanthanide and actinide elements 42(207, 209, 210, 229, 241)
Oxygen, (α, n) reaction 38/2(128)
Oxygen, capture of neutrons 38/2(65, 79)
Oxygen, (d, n) reactions 38/2(208–209)
Oxygen molecules, scattering of slow neutrons 38/2(424)
Oxygen, negative ions accelerated 44(103)
Oxygen quenching in organic phosphors 45(150)
Oxygen, reactions induced by neutrons 38/2(78–79)
Oxygen, zero-energy scattering cross section 38/2(65)
Ozone formation in Van de Graaff generators 44(97)

(p, α) reactions accompanied by γ-rays as neutron sources 38/2(162–163)
p electrons, number of unbalanced 38/1(423, 424, 427)
(p, γ) reactions as photo-neutron sources 38/2(162–166)
(p, n) reactions 40(261)
— as neutron sources 38/2(170–184)
— energy distribution 40(263)
— excitation of isomeric states 40(262)
(p, p') reactions 40(264)
(p, $2p$) reactions 40(267)
(p, pn) and (p, $2n$) reactions 40(265)
Packing fraction curve 38/1(6)
Paddles 38/1(194)
Pair damping 39(22, 95)
Pair internal conversion 40(47)
Pair production 38/2(670)
Pair spectrometer 40(33, 98, 125)
Pair-creation coefficient 42(496)
Pair-creation, internal 42(496)
Pairing energy 39(268, 436)
Palladium tube 44(90)
P_1-approximation and elementary diffusion theory 38/2(514–517)
P_1-approximation in Wick's method 38/2(570)
— time dependent problems 38/2(575–580)
Paraffin, measurement of the extrapolated end point 38/2(586)
Paraffin, slow neutron distribution results 38/2(311–327)
Parahydrogen 39(48, 49)
Parahydrogen, neutron scattering 38/2(6, 422)
Parallel beam of neutrons, self-absorption curve 38/2(389)
Parallel plate ionization chamber 45(34)
Paramagnetic catalyzers 38/1(137)
Paramagnetic ions 38/1(149)
Paramagnetism of the electronic shell 38/1(281–293)
Parameters describing a hyperbolic path 41/1(511–512)
Paramolecules 38/1(323)
Parastate 38/1(145)
Parent nucleus 41/2(2)
Parent state (of a decay) 40(44)
Parity 39(25); 40(4); 42(4, 7)
Parity conservation 39(489); 41/1(477, 495)

Parity non-conservation 41/2(9, 52, 54, 89, 104)
Parity non-conservation maximum 41/2(90)
Parity, non-normal 39(354)
Parity, normal 39(337); 40(152)
Parity of collective oscillations 39(420)
Partial decay rate 40(403)
Partial resonance width 41/1(63–64, 147, 164, 198)
Partial scattering cross section 38/2(29)
Partial widths of a compound state 38/2(48, 83)
Partial widths of levels 41/1(62–63, 164)
Particle accelerators 40(24)
Particle detection by sodium iodide scintillators 45(102)
Particle model, independent 42(343)
Particle parameters 42(397)
Particle, primary ionizing 45(3)
Particle transfer reactions 41/1(310, 369)
Particles at different velocities, table of properties 45(366–367)
Paschen-Back effect of the hyperfine structure 38/1(69)
Pathlength of photons 38/2(706ff.)
Pauli term 39(71, 97, 98, 130)
Pedestal of a coincident signal pulse 45(235)
Peeler 44(153, 207–208)
p-electron contribution 38/1(112)
Pellicle stacks 45(349, 393–394)
Pellicle stacks, mounting prior to exposure 45(398–399)
Penetrability 40(15, 16)
Penetrability of barrier 41/1(1, 53, 62, 142, 145)
Penetrability of Coulomb barrier 41/1(1)
Penetration factor 41/1(53, 86, 88, 219, 223); 42(144)
Penetration factor for a gamma-ray 41/1(72)
Penetration in heavy particle reactions 41/1(384)
Penetration of potential barriers 40(15, 16)
Percentage of expansion 45(268)
Perfect microcrystal 38/2(437)
Persistence of velocity 38/2(256)
Perturbation of neutron density by the probe 38/2(643, 653)
Perturbation of neutron probe activation 38/2(643, 651–652)
Perturbed correlation
— by applied magnetic field 42(524)
— interaction with free atom 42(523)
— liquid sources 42(524)
— micro-crystalline solid 42(522)
Phase balance 38/1(188)
Phase comparison method of beam control 44(292–293)
Phase equation of the cyclotron 44(124–127)
Phase grouping in the cyclotron 44(121–123)
Phase history of an ion in the cyclotron 44(123–124)
Phase motion in a linear accelerator 44(346)
Phase oscillation in a synchrotron 44(182–185, 222–224, 226, 228, 276, 301, 315–319)
Phase oscillation in the synchrocyclotron 44(128–130)
Phase oscillation parameters 44(279)
Phase plots 44(275–277, 291)
Phase rotation 38/1(190)
Phase shift 38/2(31); 39(4, 5, 36, 37, 78, 79, 80, 81, 82)
— analysis 39(36, 57, 77)
— evaluation with WKB approximation 41/1(259)
— in betatron oscillations 44(303, 306)
— in channel 41/1(47, 49, 55)
Phase spreading by radiation 44(185)
Phase stability in a synchrotron 44(170–171, 223, 274, 278, 315)
Phase stability in alternating-gradient accelerators 44(315)
Phase stability in the synchrocyclotron 44(130–132)
Phase transition 44(319, 324, 337)
Phase transition probability 38/1(146)
Phase velocity for ion acceleration 44(366)
Phase velocity in a waveguide depressed by loading 44(348–350)
Phase-amplitude method of Milne 41/1(439)

Phillips Ionization Gauge (PIG) as an ion source 44(90)
Phonons, energy exchange between neutrons and crystal lattice 44(432)
Phosphorus 32+33, β-spectra 41/2(43, 66–67)
Phosphorus 32, internal bremsstrahlung 41/2(103)
Photoabsorption inside a counter 45(185, 198, 203)
Photocathode material 45(90, 91, 94)
Photodifference method 38/2(149–150)
Photodisintegration of a fast deuteron in a Coulomb field 40(502)
Photodisintegration of beryllium 38/2(132, 141–145); 42(355)
Photodisintegration of the deuteron 38/2(4, 132, 133–141); 42(451)
— angular distribution of neutrons 38/2(138, 141)
— at high energies 39(13, 128); 40(511 f., 519 f.); 42(329)
— at intermediate energies 42(327)
— at low energies 39(5, 108–114); 42(322)
— at moderate energies 40(49)
— cross section formulae 38/2(136–138, 140)
— cross section measurements 38/2(139, 140)
— neutron yield 38/2(139)
— quadrupole effect 38/2(141)
— ratio of magnetic to electric effect 38/2(138–139)
— theory 42(315)
Photodisintegration, threshold measurement 42(331)
Photodisintegration, total cross section measurements at low energies 42(325)
Photoeffect, nuclear, at high energies 40(461, 515 f.)
Photoelectric absorption of x-rays 38/2(667–668)
Photoelectric cross section 39(6)
Photoelectric effect in the deuteron 38/2(136)
Photoelectron production in a counter 45(58, 64)
Photofission 40(517); 42(359, 463)
Photofraction 38/2(798, 806)
Photographic emulsion 40(478 f., 503, 524)
Photography of bubble chambers 45(331)
Photography of cloud chamber tracks 45(289–295)
Photoionization 45(7, 8)
Photomagnetic cross section 39(6)
Photomagnetic effect in the deuteron 38/2(70, 136)
Photomeson production, (*see also* Pion production) 42(540)
Photomicrography of emulsion tracks 45(417)
Photomultiplier 38/2(805); 40(475, 477); 45(236, 243)
Photomultiplier characteristics 45(87, 90, 91)
Photomultiplier types 45(87, 90, 91)
Photon absorption process 42(310)
Photon detection in a counter 45(66)
Photon histories, sampling, *see* Monte Carlo method
Photoneutron energies from various sources 38/2(160)
Photoneutron production by high-energy
— electrons 38/2(166–170)
— γ-rays 38/2(162–166)
Photoneutron sources 38/2(158–162); 44(408)
Photoneutron yields from various sources 38/2(160)
Photoneutrons accompanying passage of γ-rays through matter 38/2(665)
Photoneutrons from beryllium 38/2(143–145)
Photoneutrons from bismuth 38/2(157)
Photoneutrons from copper 38/2(149)
Photoneutrons from deuterium 38/2(133)
Photoneutrons from rhodium 38/2(147)
Photons from the avalanche in a counter 45(58, 62, 64)
Photons, separation of gas and wall absorption 45(198)
Photonuclear cross section, integrated 40(285)
Photonuclear interaction, direct interaction 40(289)
Photonuclear reactions 42(431)

Photonuclear reactions, experimental methods 40(281)
Photonucleons, direct, angular distribution 42(341)
Photoprocesses at high energies 42(352, 353)
Photoprocesses, direct 42(341)
Photoproduction of deuterons, tritons, and α-particles 40(518f.)
Photoproduction of neutrons 38/2(97–98, 162–170)
Photoproduction of nucleons 40(515f.)
Pick-up electrode 44(279–280, 323)
Pick-up of a proton by a neutron, theory of Thomas 41/1(322–326)
Pick-up reaction(s) 39(218); 40(281, 452, 455, 499f., 518); 41/1(9, 15, 161, 320–350); 42(487)
Pile, *see* Reactor
Pile neutrons 44(398)
Pile oscillator 44(425)
Pile oscillator method of absorption measurement 38/2(374)
π-meson, *see* Pion
π-meson scattering by nucleons 42(539, 542)
Pion decay 41/2(16, 21)
Pion mass, dependence of deuteron photoeffect upon it 38/2(139–140)
Pion production 40(513, 532f.)
Pion reactions with nuclei 40(534f.)
Pion scattering and nuclear potential 41/1(263)
Pion-nucleon scattering 41/1(264–265)
Pions as probe particles 39(218, 219)
Placzek procedure of slowing-down calculation 38/2(240–243)
Plane collimated source of x-rays 38/2(682, 758ff.)
Plane isotropic source of x-rays 38/2(683, 758ff., 764–765)
Plane oblique source of x-rays 38/2(682, 758ff.)
Plane perpendicular source of x-rays 38/2(683, 735, 742ff., 750ff., 764–765, 767)
Plane slowing-down kernel 38/2(254)
Plane source of thermal neutrons, flux distribution 38/2(530–533, 550)
Plane transport kernel 38/2(510)
Plane waves within the nucleus 39(422)
P_L-approximation 38/2(286)
P_L-approximation, boundary conditions 38/2(566–567)
Plasma of the ion source 44(147)
Plateau flatness 45(68–69)
Plateau of counter characteristic 45(55, 68, 183)
Plateau slope (normalized and relative) 45(55)
Plural scattering 41/2(54, 55, 57)
Plutonium 42(237–254)
Plutonium, chemical properties 42(240)
Plutonium, decay schemes 42(238)
Plutonium in nature 42(237)
Plutonium isotopes 42(239)
Plutonium isotopes of high mass, production by neutron irradiation 42(301, 304)
Plutonium neutron sources 38/2(108, 118–119, 132)
Plutonium reactor 44(396)
Plutonium, spontaneous fission 38/2(210)
Pneumatic tubes for short-lived activities 44(424)
Pocket ionization chamber 45(38, 39)
Point collimated source of x-rays 38/2(681, 761ff.)
Point diffusion kernel 38/2(553)
Point isotropic source of x-rays 38/2(683, 756ff.)
Point slowing-down kernel 38/2(254)
Point source of thermal neutrons, flux distribution 38/2(530–531)
Point transport kernel 38/2(510)
Point-source kernel 38/2(254, 278)
Poisoned pile 38/2(335)
Polarimeter for gamma-rays 45(256)
Polarizability of a nucleus 38/1(402)
Polarization 39(10, 12, 31, 120–122)
Polarization and optical model 41/1(261, 263)
Polarization, channel-spin representation 42(446)
Polarization correlation 40(55, 58, 191)
Polarization correlation measurements 42(435)
Polarization detection 42(450, 542)
Polarization, differential 42(590)
Polarization effects in Coulomb scattering 42(595)
Polarization efficiency 42(440)

Polarization experiments 42(322)
Polarization for γ-radiation 42(441, 442)
Polarization in double scattering 41/1(474)
Polarization in scattering 42(471)
Polarization in stripping reactions 41/1(349–350)
Polarization in transmutation reactions 42(600)
Polarization measurement 40(90)
Polarization measurements, experimental methods 41/2(53)
Polarization, nuclear 42(558, 573, 588)
Polarization of a beam of neutrons 40(104)
Polarization of β-particles, determination of coupling constants 41/2(89–90)
Polarization of neutron beams 45(490, 515–516)
Polarization of neutrons 44(439–440)
— from D(d, n) reaction 38/2(193)
— in nuclear reactions 41/1(484)
Polarization of nuclei 42(438, 486)
Polarization of nucleon scattering 40(41, 454, 460, 490, 494f.)
Polarization of nucleus by a mesonic level 39(210)
Polarization of photoneutrons from beryllium 38/2(145)
Polarization of protons scattered
— from carbon 41/1(492, 493)
— from helium 41/1(466, 478, 488)
Polarization of radiation from aligned nuclei 42(568, 573)
Polarization of radium E electrons 41/2(48)
Polarization of spin-$^1/_2$ particles 42(439, 443)
Polarization of spin-s particles 42(443, 457)
Polarization of x-rays 38/2(671–673, 684, 696, 788)
Polarization originated by Coulomb excitation 41/1(497)
Polarization potential 41/1(550)
Polarization produced by a complex potential 41/1(486–495)
Polarization, properties 42(591)
Polarization sign in high-energy scattering 41/1(492, 493)
Polarization, transversal 38/1(130)
Polarization vector 42(440, 444, 445, 457, 588, 590)
Polarization vector and statistical matrix 41/1(472–473)
Polarized bombarding particles 42(592)
Polarized neutrons 42(606, 607)
Polarized neutrons, correlation experiments 41/2(76, 82–86)
Polarized nuclei, angular distribution of β-particles 41/2(68–71)
Polarized particles, detection 42(594)
Polarized particles, interaction with polarized nuclei 42(606)
Polarized particles, production 42(589, 594)
Polarized radiation 42(434, 436, 442)
Polarized target nuclei 42(592)
Pole face windings 44(241–244, 331)
Pole faces in the cyclotron 44(132–134)
Poleless C-magnet 44(238, 240)
Polonium α-emitters 42(174)
Polonium neutron sources 38/2(108–109, 117ff., 124, 131)
Polonium-beryllium source, distribution of resonance neutrons in moderators 38/2(313, 315, 326, 331)
Polycristalline materials, elastic scattering of slow neutrons 38/2(431–440)
Polycristalline materials, inelastic scattering of slow neutrons 38/2(440–456, 476)
Polycrystalline moderator 38/2(473–476)
Polycrystalline spectra, nuclear magnetic 38/1(322)
Polymerization under discharge conditions 45(68)
Polynomial expansion 38/2(739ff.)
Polynomial method 38/2(298)
Polyvinyl alcohol as emulsion matrix 45(394)
Portable dose meters 45(39)
Position correlation of nuclear particles 41/1(20)
Positive ions in a counter discharge 45(177)
Positron emission, Coulomb correction 41/2(12)
Positronium formation by polarized positrons 41/2(63)

Potassium iodide as a scintillant 45(143)
Potential acting on a nucleon 41/1(255)
Potential, anharmonicity 38/1(41)
Potential barrier resonance 41/1(150)
Potential, calculation from density 39(455)
Potential, complex 40(459)
Potential cut-off 39(29, 30)
Potential, electrostatic, at the nucleus 38/1(289)
Potential, imaginary part 41/1(22, 240, 241, 245, 255, 257, 265)
Potential, nuclear 39(30, 422, 446)
— for a proton 39(431)
— parameter values 39(431)
— velocity dependence 39(429, 439, 442, 456–459)
— well depth 39(429)
Potential of fourth order 39(22, 27, 96)
Potential of second order 39(21, 27, 94)
Potential, realistic, of nucleus 39(470, 506–512)
Potential scattering 38/2(34–35, 48–50, 366); 41/1(55); 42(464, 465)
— cross section 44(418–419)
— length 40(423)
Potential, tapered, of Woods and Saxon 40(232)
Potential, velocity dependent 41/1(22)
Potential well depth 40(483f.)
Potential well for hyperons 40(536)
Potential well, tapered 40(459, 497, 507)
Potentials, singular 39(31, 56)
Powder diffraction of neutrons 44(432, 433–435)
Power series for the irregular Coulomb function G_L 41/1(411–412)
Power series for the regular Coulomb function F_L 41/1(411)
Power supply to proton synchrotrons 44(251–254, 332)
Power supply to Van de Graaff generator 44(91)
Praseodymium 144, β-decay and polarization 41/2(68–69)
Pre-acceleration in linear accelerators 44(354–357)
Pre-acceleration of synchrotron protons 44(263)
Pre-accelerator 44(264, 354)
Precession, forced 38/1(151, 165)
Precession, free 38/1(151)
Precession in a constant magnetic field 38/1(151)
Precession in a modulated magnetic field 38/1(172)
Precession measurement for free precession 38/1(227)
Precession, non-adiabatic 38/1(182)
Precession, nuclear 42(14)
Predissociation in self-quenching counters 45(63, 65, 71)
Preparation of radioactive neutron sources 38/2(130–132)
Pressure defined cloud chamber 45(273)
Pressure effect in electron synchrotrons 44(176–178)
Pressure effect on accelerating tube performance 44(87–88)
Pressure, gas counter at high 45(215)
Pressure tank of a Van de Graaff generator 44(95–96)
Primary ionization process 45(3)
Principle of detailed balance 38/2(396)
Prismatic moderator, solution of age equation for point source 38/2(275–276)
Probability of a neutron energy after *n* collisions 38/2(221–224)
Probability of Coulomb excitation 41/1(510)
Probability of non-capture in slowing-down 38/2(234)
Probable error of a counting experiment 45(252)
Probe for neutrons 38/2(344)
Probe for thermal neutrons 38/2(637)
— flux depression 38/2(636, 642–652)
Production, associated, of deuteron and π-meson 42(542)
Profile method of track scanning 45(435)
Projection factor 39(220, 488)
Prometheum 147, Fermi plot 41/2(45)
Prompt curve, *see* Resolution curve
Prongs of stars 40(455, 466, 503)
Propagation of discharge along the counter wire 45(64, 65)

Propagation velocity of thermal neutron waves 38/2(579–580)
Proportional counter(s) 40(34, 474); 45(52, 55, 174)
— coincidence and anticoincidence arrangement 45(189)
— construction 45(180–186)
— for neutron detection 45(457–460)
— for neutrons 44(415)
— resolving power 45(189–192)
Proportional region 45(55)
Protactinium 42(209, 212)
Proton, *see* Hydrogen
Proton capture 40(45, 260); 42(427)
Proton cross sections, nonelastic 40(260)
Proton, polarized, production 42(605)
Proton scattering, determination of nuclear size 39(215)
Proton scattering of neutrons, binding effect 38/2(417)
Proton scattering, optical model results 41/1(256)
Proton-carbon scattering 41/1(466, 478, 488, 492, 493)
Proton-deuteron scattering, Coulomb effects 39(175)
Proton-deuteron scattering, empirical data 39(158)
Proton-helium-scattering 41/1(466, 478, 488)
Proton-proton scattering
— angular distribution at high energies 39(124)
— at high energies 39(7, 116, 118)
— at low energies 39(97)
— cross section at high energies 40(463)
— double 39(11, 121, 122, 124)
Protons, capture cross section for thermal neutrons 38/2(594–596)
Protons, cyclotron frequency of electrons and protons 38/1(259)
Protons, diamagnetic shielding 38/1(253)
Protons, Larmor frequency 38/1(259)
Protons of high energy, ranges in emulsions 45(363, 368)
Protons of moderate energy, ranges in emulsions 45(361–362, 368)
Proton-tritium scattering 41/1(288, 289)
Pseudopotential 39(45, 48)
Pseudopotential of Brueckner 39(445)
Pseudopotential, saturating 39(428, 429)
Pseudo-quadrupole effects 38/1(402)
Pseudo-quadrupole interaction 38/1(406)
Pseudo-random numbers 38/2(776)
Pseudoscalar interaction 41/2(9, 88)
Pseudoscalar interaction, smallness 41/2(47)
Pseudoscalar meson theory 39(14)
Pulley used as inductor 44(77, 79)
Pulse amplitude analyser (kicksorter) 45(184, 189)
Pulse circuiting 38/1(216–230)
Pulse equipment 38/1(217–231)
Pulse forming stage 45(228, 245)
Pulse generator, fast 45(250)
Pulse groups, generator 38/1(220)
Pulse height selector 45(228, 229)
Pulse height spectrum 45(45)
Pulse lengthening diode 45(242)
Pulse running, coherent 38/1(218, 224)
Pulse running, incoherent 38/1(218)
Pulse shape 45(188)
Pulse shape from ionization chambers 45(31)
Pulse size distribution 45(189, 190)
Pulse transmitter 38/1(220)
Pulsed nature of cyclotron beam 44(123)
Pulsed proton synchrotrons 44(325–335)
Pulsed source of neutrons
— in a finite medium 38/2(619–629)
— in an infinite medium 38/2(614–619)
— slowing down process 38/2(246–253, 303–306)
Pure Coulomb excitation 41/1(496, 504, 514)
Pure resonance absorption integral 38/2(365)
Pure scattering without absorption 38/2(31–33)
Purification of gases 45(30, 453)
Push-pull-oscillator 38/1(222)

Q values for determination of decay energies 41/2(14, 16)
Q-matrix of Teichmann and Wigner 41/1(55, 173)

Quadrature calculation of the irregular Coulomb function G_L 41/1(457–458)
Quadrupolar deformation 38/1(438)
Quadrupole coupling constant 38/1(384, 406, 423, 429)
— sign 38/1(397)
Quadrupole coupling
— in angular correlation 38/1(399)
— in molecular crystals 38/1(428)
— in poly-electron spectra 38/1(412)
— in solids 38/1(429)
— tensor 38/1(425)
— terms 42(165)
Quadrupole distortion 39(270, 281)
Quadrupole effect in deuteron photodisintegration 38/2(141)
Quadrupole effects, false 38/1(402)
Quadrupole energy 38/1(380)
Quadrupole excitation 41/1(513, 514–516, 522, 527)
Quadrupole excitation, yield function 41/1(532)
Quadrupole focussing 44(375–377, 383)
Quadrupole hyperfine energy for a symmetric top 38/1(384)
Quadrupole hyperfine interaction, μ-mesonic 38/1(440)
Quadrupole hyperfine structure
— in asymmetric rotors 38/1(385)
— in atoms 38/1(413)
— in electronic paramagnetic resonance 38/1(396, 398)
— relativistic correction factor 38/1(408)
Quadrupole interaction 42(178)
Quadrupole interaction, nuclear, in an atomic or molecular system 38/1(378, 381, 382)
Quadrupole matrix of nuclear transition 41/1(529)
Quadrupole moment(s) 39(219, 222, 475, 488)
Quadrupole moment(s) (*see also* Nuclear quadrupole moment) 38/1(39, 108, 113, 138, 377, 378, 428, 431, 435, 436)
— electric 39(254, 295, 298, 313, 317, 380); 42(529)
— electric, of the deuteron 39(2, 67, 68)
— induced 38/1(413)
— intrinsic 38/1(108, 431, 437, 439); 39(488); 42(137, 169, 192)
— matrix elements 38/1(432); 39(408)
— of an excited state 41/1(539, 551–552)
— of stable nuclei of odd mass numbers in the ground state (table) 39(263ff.)
— report on experimentally determined values 38/1(443–446)
— systematic behaviour 38/1(449–451)
Quadrupole radiation, electric 38/1(432, 433)
Quadrupole relaxation 38/1(150)
Quadrupole splitting of nuclear magnetic resonance levels 38/1(324)
Quadrupole states in a crystal 38/1(394)
Quadrupole transitions, electric 42(319)
Quadrupole vibration 39(523, 524)
Quantameter 44(192)
Quantum mechanical and semi-classical treatment of Coulomb excitation, relation between them 41/1(498–499, 502, 503, 505, 540–548)
Quantum number, magnetic 38/1(124)
Quantum numbers, asymptotic 42(197)
Quantum theoretical corrections in heavy particle collisions 41/1(375–378)
Quantum theory of neutron collisions 38/2(27–35)
Quantum theory of radiation 41/1(64–66)
Quartz counter 45(215)
Quasideuteron 40(452, 460, 516, 518f., 522)
Quasideuteron model of the nucleus 41/1(18–19)
Quasielastic scattering 40(452, 460)
Quasi-equilibrium of neutron generations in slowing-down 38/2(277, 290, 296)
Quasi-equilibrium of primary and secondary radiation 38/2(676, 700, 708, 710, 720, 738, 765, 767, 768)
Quasistationary state 40(403)
Quenching by halogens 45(71)
Quenching by organic vapors 45(53, 62, 75)
Quenching circuits 45(61)
Quenching of counter discharge 45(54, 55)

Quenching of pulses 45(176)
Q-values (decay energies) 42(119, 134)

Racah coefficients 41/1(294, 303, 497, 510, 514); 42(372, 381, 544)
Racah function 39(391)
Racah sum 41/1(294, 333)
Rad (radiation dose unit) 38/2(89); 45(38)
Radial betatron oscillations 44(195, 199–200)
— in a synchrotron 44(186–187)
Radial buckling 38/2(590)
Radial defocussing by phase focussing 44(344, 374)
Radial dependence of betatron magnetic field 44(198)
Radial dependence of nuclear density 39(231)
Radial focussing in a betatron 44(195)
Radial matrix elements in Coulomb excitation 41/1(533–538)
Radial missing of inflector 44(267)
Radial moments 39(189)
Radial motion in a linear accelerator 44(347, 358–359, 373–378)
Radial oscillation of cyclotron orbits 44(109, 119–121)
Radial oscillations in a synchrotron 44(171, 184)
— anti-damping 44(187)
Radial phase oscillation 44(223)
Radially expanding cloud chamber 45(281)
Radiation absorption 42(390)
Radiation by electrons undergoing acceleration in a synchrotron 44(178–182, 449)
Radiation corrections of quantum electrodynamics for magnetic moments 38/1(274)
Radiation damage 44(404–406)
Radiation equilibrium 38/2(275, 676)
Radiation exposure 44(441–444)
Radiation exposure limits 44(442)
Radiation field 40(42)
Radiation following decay as a means of determining recoil momentum 41/2(29–30)
Radiation from aligned nuclei 42(568, 573, 575, 590)
Radiation length for various materials 45(165, 167)
Radiation loss in betatron 44(199)
Radiation parameters 42(381, 382, 391, 392, 394, 436, 437, 440, 475)
— for γ-radiation 42(392, 475)
— for heavy particles 42(391)
— for spin-s particles 42(437, 439, 443, 457)
Radiation, polarized 42(434, 436, 442)
Radiation protection at a cyclotron 44(157–158)
Radiation protection at reactors 44(444–446)
Radiation survey 45(37–41)
Radiation theory 41/1(64–73)
Radiation theory, classical correspondence 41/1(66–70)
Radiation transport equation, *see* Transport equation of radiation
Radiation width 40(44, 314–317, 382–395, 432f.); 41/1(67)
— constancy thereof 40(315)
— dependence on atomic weight 40(432f.)
— experimental 40(314, 382–395)
— for dipole emission 41/1(70–73)
— independence of spin 40(316)
— numerical estimates 40(317)
— of a resonance level 38/2(50)
Radiation-induced anti-damping 44(186)
Radiation-induced betatron oscillations in a synchrotron 44(185–188, 446)
Radiation-induced phase oscillations in a synchrotron 44(185)
Radiative capture (*see also under* Capture and Absorption) 38/2(18, 41ff., 48, 82); 41/2(103)
Radiative capture of neutrons by protons 38/2(68–70)
Radiative corrections 39(198)
Radiative corrections of μ-mesonic levels
Radiative decay 41/2(102)
Radiative equilibrium 38/2(295, 676)
Radiative lifetime 42(14, 15, 36)
Radiative process, direct 40(47)
Radiative transition, its order 40(43)
Radiative transition probability 42(15)

Radiative transition, selection rules 40(42)
Radiative width 40(44, 314–317, 382–395, 432f.)
Radiative width, average 38/2(367)
Radiator-counter for detection of high-energy neutrons 45(499–502, 512, 514)
Radii of nuclei 38/1(23–24, 72, 104)
Radioactive (α, *n*) sources of neutrons 38/2(107, 130–132)
Radioactive gas in a counter 45(194–199)
Radioactive source of neutrons 38/2(88, 107, 130–132)
Radioactive sources, calibration 45(255)
Radioactive sources, routine measurements 45(42, 43)
Radioactive states 41/1(29)
Radioactivity induced by high-energy neutrons 45(506–508, 512)
Radioactivity induced by slow neutrons 45(447–449, 468)
Radioactivity of the neutron 38/2(8)
Radioactivity produced by α-particles 38/2(13–14)
Radioactivity produced by neutrons 38/2(14)
Radiocarbon analysis 45(211)
Radiofrequency ion source 44(90)
Radiofrequency oscillator, master oscillator power amplifier (MOPA) 44(141–142)
Radiofrequency oscillator, push-pull system 44(141)
Radiofrequency oscillator, self-excited with grounded grid 44(140)
Radiofrequency system of a cyclotron 44(139–146)
Radiofrequency system of a synchrocyclotron 44(145–146)
Radiofrequency system of a synchrotron 44(188–190, 274–293)
Radium E decay
— calorimetric investigation 41/2(17–18)
— electron polarization 41/2(48)
— Fermi plot 41/2(46, 47)
Radium general data and polarization 41/2(68–69)
Radium, internal bremsstrahlung 41/2(103)
Radium neutron sources 38/2(108, 118–119, 122, 125, 130–131)
Radium-beryllium source, distribution of resonance neutrons in moderators 38/2(313–314, 331, 334)
Radius, equivalent 39(181)
Radius of nuclear potential 39(453)
Radius of nucleus, (*see also* Size of nucleus) 38/2(27); 39(416, 417); 42(15, 139, 143, 144, 145, 153)
— determined by neutron scattering 39(213, 214)
— determined from mass defects 39(212)
Radon neutron sources 38/2(122)
Rainwater's nuclear model 39(477, 498)
Ramsauer effect in noble gases 45(24)
Ramsauer effect, nuclear 39(186)
Random coincidences 45(253)
Random numbers 38/2(775)
Random sampling techniques 38/2(773, 775, 777)
Range correction 38/2(67, 70)
Range, effective 39(37, 38, 40, 51, 54, 59–61, 83, 85, 91)
Range, effective, mixed 39(39, 51)
Range effects in counters 45(195)
Range, intrinsic 39(41, 52)
Range measurement 40(26)
Range measurement in a cloud chamber 45(308–309)
Range of particles in target of finite thickness 38/2(102–103)
Range reduction by a magnetic field 45(196)
Range-energy data, tables 45(366–368)
Range-energy index 45(365–367, 369)
Range-energy relation in emulsion tracks 45(358–371)
— for protons of high energies 45(363ff.)
— for protons of moderate energies 45(361–362)
— for slow electrons 45(370–371)
Range-energy relations in air 45(456)
Rare earth elements 42(206, 210, 215, 216)
Rate of energy loss 45(369)

Rayleigh scattering of x-rays (*see also* Giant resonances) 38/2(669–670); 40(510)
RBE (=relative biological effectiveness) 44(441–443)
RBE (relative biological effectiveness) 45(38)
Reactance matrix Q 41/1(55, 173)
Reaction amplitude 41/1(59–64)
Reaction channel 38/2(18, 47–48)
Reaction channels 40(13)
Reaction cross section(s) 38/2(86); 41/1(241)
— for charged particles 40(213)
— for neutrons 40(212)
— kinematical frame of theory 38/2(29–31, 33)
— total incoherent 40(21, 208)
Reaction energies, high voltage scale for comparison 40(30)
Reaction energies of (p, n) reactions, table 38/2(171)
Reaction matrix 42(420)
Reaction products, detection and analysis 40(26)
Reaction scheme 39(447, 448)
Reaction threshold 41/1(274, 278)
Reactions, *see* (n, p)-, (p, α)-, (γ, n)-reactions etc.
Reactions induced by carbon ions 41/1(405–406)
Reactions induced by nitrogen ions 41/1(404–405)
Reactions of heavy ions 40(148)
Reactivity of a reactor 44(392–393)
Reactor controlling ionization chambers 45(41)
Reactor neutrinos 41/2(35)
Reactor period 44(393)
Reactor types 44(393–399)
Reciprocal lattice 38/2(432)
Reciprocal velocity law
— in boron 38/2(60, 62)
— in gold 38/2(60)
— in lithium 38/2(60)
— ($1/v$-law) 38/2(21, 24, 41, 42, 51, 60–62)
Recoil detection of neutrons below 20 MeV 45(438, 471)
Recoil effect in Coulomb excitation 41/1(559)
Recoil energy spectra 41/2(19)
Recoil experiments for neutron decay 41/2(80–81)
Recoil experiments on neutrino emission 41/2(19)
Recoil ionization 45(13)
Recoil method 42(36)
Recoil nucleus 41/2(2, 19)
Recoil proton telescopes 45(494, 498, 512)
Recoil protons from high-energy neutrons 45(492, 493–502)
Recoil protons from neutrons below 20 MeV 45(438–441, 471)
Recoil spectrometer 41/2(23)
Recombination 45(25–27, 50, 58)
Recombination of electrons in the track 45(177)
Recombination of ions and electrons 45(26)
Recombination of positive and negative ions 45(27)
Recording, magnetic 38/1(203)
Recording, photographic 38/1(203)
Recovery time of a counter 45(55)
Rectification, phase sensitive 38/1(197)
Rectifier, *see* Valve
Rectifier tube, gas filled 44(3)
Rectifier with capacitor input 44(2–20)
— characteristics 44(10, 17)
— connections 44(3, 9)
— loading 44(4, 9ff.)
— one-way type 44(3, 9)
— peak current 44(2)
— two-way type 44(3, 19)
— voltage drop 44(2, 9)
Rectifier with choke input 44(20–24)
Recurrence relations between Coulomb functions 41/1(416–418)
Reduced energy width 41/1(269)
Reduced level width (*see also* Width, reduced) 40(409, 428)
Reduced mass effect on scattering cross section 38/2(393, 402)
Reduced $\mathscr{R}$-matrix and resonance formulas 41/1(176–187)
Reduced $\mathscr{R}$-matrix, relation to scattering matrix 41/1(168–176)
Reduced neutron width of a resonance level 38/2(50)
Reduced width
— average 38/2(367)
— for a single particle 41/1(270)

Reduced width (cont.)
— in alpha-decay 41/1(363)
— in heavy-particle reactions 41/1(385)
— modified definition of Thomas 41/1(364)
— of a level 41/1(63, 198, 218)
Reflected radiation, spectral and angular distribution 38/2(792–794)
Reflecting coats on sodium iodide crystals 45(132)
Reflecting microscope of Burch 45(415–416)
Reflection coefficient of radiation 38/2(792)
Reflection of neutrons from mirrors 44(435–439)
Reflection of radiation, *see* Backscattering
Reflection polarizing neutrons 44(440)
Reflectivity for neutrons 44(437)
Reflectivity of different materials 45(96, 97)
Reflector surrounding a reactor 44(392, 428)
Refraction index for neutrons 44(435)
Refraction, index of 40(460, 483, 492)
Refractive index 39(449–451)
Refractive (real) potential 39(369, 372, 447, 449–451)
Regenerative peeler 44(153)
Regular Coulomb function F_L
— asymptotic behavior 41/1(409, 440–449)
— Bessel expansion for low energies 41/1(418–421)
— power series 41/1(411)
Regular function 41/1(226)
Reignition in counters 45(55, 60)
Reines-Cowan neutrino experiment 41/2(33–35)
Reinitiation of counter discharge 45(60)
Rejection technique 38/2(776)
Rejuvenation of a counter 45(68)
Relative perturbation of neutron
— density by probe 38/2(643, 653)
— probe activation 38/2(643, 651–652)
Relative sign of A and V coupling constants 41/2(91, 94–95)
Relative sign of Fermi and Gamow-Teller coupling constants 41/2(49)
Relatively coherent scattering 38/2(411)
Relativistic current correction 39(98, 114, 115, 116, 127, 128, 130)
Relaxation constant 42(516)
Relaxation in gases 38/1(148)
Relaxation in liquids 38/1(148)
Relaxation in metals 38/1(150)
Relaxation in solids 38/1(148)
Relaxation length of neutrons in paraffin 38/2(313)
Relaxation length of photons 38/2(666)
Relaxation length of thermal neutrons 38/2(556, 590–592)
Relaxation theory, nuclear magnetic 38/1(138)
Relaxation time 38/1(131)
— longitudinal 38/1(129)
— of thermal neutrons in various materials 38/2(627)
— transversal 38/1(130)
Rem (dose unit) = Roentgen equivalent man 38/2(89–90)
Remanent field 44(240, 329)
Reorientation effect 41/1(551, 556–557)
Rep (dose unit) = Roentgen equivalent physical 38/2(89)
Representation coefficients of a coordinate rotation 41/1(126)
Reproduction factor of fission cycles 44(391)
Repulsive Coulomb interaction 41/1(408–409)
Repulsive force 39(438)
Repulsive potential, *see* Hard core
Residual interactions 39(512–516, 530, 531)
Residual range 45(378, 379)
Resistance voltmeter 44(92)
Resnatron 44(382, 383)
Resolution curve 45(249–251, 257)
Resolution of energy with scintillation spectrometer 45(131, 133–137)
Resolution of neutron resonance lines 40(376, 379, 397)
Resolution of neutron velocities 40(374f.)
Resolution of time-of-flight measurements at high energies 45(503)
Resolving power (*see also* Frequency, accuracy of measurement) 38/1(230, 232, 236, 276)

— of proportional counters 45(189–192)
— of proportional counters, energy dependence 45(192)
Resolving time 45(55, 227, 237, 249)
Resonance absorption
— discovery 38/2(45)
— integral, definition 38/2(360, 364)
— integral, measurement by absorption 38/2(369–375)
— integral, measurement by activation 38/2(368–369)
— integrals, tables of values 38/2(370–371, 374)
Resonance and barrier penetration 41/1(143, 145)
Resonance, coupling 44(310)
Resonance curve, Fourier transform 38/1(320)
Resonance denominators in complex energy treatment 41/1(30, 42–47)
Resonance detectors of neutrons 44(410–411)
Resonance diagram of constant gradient synchrotron 44(228)
Resonance diagram of electron analog 44(323)
Resonance, elementary definition 41/1(25–26)
Resonance energies 40(30)
Resonance escape probability 38/2(235, 239)
— for dilute absorber solution in water 38/2(238)
— of a neutron 44(391)
Resonance filters 38/2(350–352)
Resonance fluorescence, nuclear 42(37)
Resonance flux 44(400–401)
Resonance formula, extension to many levels 38/2(46)
Resonance formula of Breit and Wigner 38/2(45, 49)
Resonance formulas for nuclear reaction cross sections 41/1(3)
Resonance frequency measurements by pulse technique 38/1(226)
Resonance frequency, nuclear magnetic, splitting 38/1(293)
Resonance in fissionable nuclides 44(419–422)
Resonance integral 38/2(364–384); 44(409, 425)
Resonance integral, effective
— in an heterogeneous system 38/2(380–385)
— in an homogeneous system 38/2(376–380)
Resonance integral, independence of Doppler effect 38/2(365)
Resonance, isolated 42(427)
Resonance level, nuclear magnetic 38/1(278)
Resonance level, single 42(418, 465)
Resonance levels for neutron capture 39(446)
Resonance neutrons 44(399, 400–401, 408–422)
Resonance neutrons, old definition 38/2(343)
Resonance parameters 41/1(132–134)
— determination 44(415–418)
— of nuclei (table) 40(382–395)
— table 41/1(210–216)
Resonance passage, adiabatic 38/1(172)
Resonance passage, rapid adiabatic 38/1(177)
Resonance processes with neutrons 40(373)
Resonance reactions 42(418)
— absorption followed by successive emissions 42(424)
— involving γ-radiation 42(425)
— with nucleonic particles 42(431)
Resonance scattering 38/2(35, 51); 41/1(25–28, 38); 41/2(86); 42(37, 465, 466)
Resonance scattering integral 38/2(366, 375–376)
Resonance theory of nuclear reactions 40(18)
Resonance width, partial 41/1(63–64, 147, 164, 198)
Resonance with phase angle 90° 41/1(145, 147, 225)
Resonances in the synchrotron 44(224, 225, 228, 238, 308, 310, 312–313, 324)
Resonances, non-linear 44(313)
Resonances of radial oscillations in the AVF cyclotron 44(115–117)
Resonant absorption, cross section 42(38)
Resonant level of compound nucleus 38/2(49)

Resonant scattering 40(36, 51)
Resonating group structure 41/1(20)
Response of anthracene 45(148, 149)
Rest mass of the neutrino 41/2(36–38)
Restricted rate of loss in emulsions 45(354–355)
Retardation factor 39(14, 129)
Retardation time for thermal neutrons 38/2(576, 578)
Reversal dip 40(400)
$\mathcal{R}$-function 41/1(109–110, 176, 202)
$\mathcal{R}$-function, statistical 41/1(203)
Rhenium 186, β-decay and polarization 41/2(66–67)
Rhodium, activation by slow neutrons 38/2(52)
Rhodium as a detector of thermal neutrons 38/2(634–635)
Rhodium detector covered with cadmium 38/2(311)
Rhodium, early work on selective absorption 38/2(43)
Rhodium, energy dependence of slow-neutron total cross section 38/2(53)
Rhodium, photoneutron yield 38/2(147)
Rhodium, resonance parameters 38/2(52–53)
Riccati equation for Coulomb functions 41/1(456–457)
Right-handedness of positrons 41/2(90, 104)
Rigid molecule, scattering of slow neutrons 38/2(408, 429)
Rigorous solution of the transport equation of thermal neutrons 38/2(517–520)
Ripple of a cascade generator 44(27)
Ripple of synchrotron magnetic field 44(254–256, 278)
Rise-time of pulse amplifier 38/1(226)
R-matrix 40(411, 424); 42(457)
$\mathcal{R}$-matrix 41/1(3, 5, 24, 53, 81, 107)
$\mathcal{R}$-matrix and alpha decay 41/1(356)
$\mathcal{R}$-matrix and Green's function 41/1(158–159)
$\mathcal{R}$-matrix degenerate to a scalar (*see also* $\mathcal{R}$-function) 41/1(109–110, 176, 202)
$\mathcal{R}$-matrix, factorization of numerator 41/1(119, 129, 155–161, 177)
$\mathcal{R}$-matrix in the many-channel case 41/1(111–116)
$\mathcal{R}$-matrix, reduced, relation to scattering matrix 41/1(168–176)
$\mathcal{R}$-matrix, relation to scattering matrix 41/1(119–123, 168–176)
$\mathcal{R}$-matrix, statistical 41/1(202–205)
$\mathcal{R}$-matrix theory and optical model 41/1(243, 265, 267)
$\mathcal{R}$-matrix theory and string model 41/1(88)
Roentgen equivalent(s) 38/2(89–90)
— man (Rem) 44(443)
— physical (rep) 44(443)
— Rep and rem 45(38)
Roentgen, unit of dosage 44(442)
Root-mean-square velocity of thermal neutrons 38/2(40)
Rosenfeld interaction 39(337)
Rossi circuit 45(237)
Rossi stage 45(225)
Rossi stage, cathode coupled 45(230)
Rotary condenser 44(146)
Rotation, collective 38/1(435, 436)
Rotation, hindered 38/1(429)
Rotation homogeneisation 38/1(234)
Rotation matrices 42(374)
Rotation matrix 41/1(301–304, 519)
Rotation of coordinate system, representation coefficients 41/1(126)
Rotational absorption spectra, pure 38/1(38)
Rotational bands 42(137, 196, 298)
Rotational constant B_0 38/1(45, 47)
Rotational diffusion 38/1(134)
Rotational energy 42(136, 191)
Rotational fine structure 40(10)
Rotational levels 40(335f.)
Rotational model 39(480)
Rotational model, self-consistency 39(537)
Rotational spectra 39(485)
Rotational spectrum of a nucleus 38/1(109)
Rotational states 42(7)
— energies in even-even nuclei 40(336)
— energies in odd-mass nuclei 40(337)
— of nuclei 39(222, 229, 481–516)
— of nuclei, matrix elements 39(481–516)
— theoretical transition probabilities 40(338)

Rotation-particle coupling 39(533–535)
Rotation-vibration constant 38/1(45)
Rotation-vibration interaction 39(486, 532)
Rotor asymmetric 38/1(46)
Rotor, asymmetric, quadrupole hyperfine structure 38/1(385)
Roughness of a charging belt 44(74)
Rubber diaphragm chamber 45(281)
Rubidium 87, a highly forbidden transition 41/2(46)
Rubidium, excitation curve of (*p*, *n*) reaction 38/2(173)
r-unit 45(38)
Russell-Saunders coupling 39(319); 40(6, 9); 42(419, 423, 429)
— and *j*-*j*-coupling, transformation between 39(327)
Rutherford scattering 40(484f.); 41/1(103); 41/2(59); 42(466)

Saclay, synchrotron à protons 44(219, 238, 244–245)
Saddle point method(s) 38/2(706, 709–710, 765–766)
— in calculating Coulomb functions 41/1(453–454)
Safe limit of exposure to radiation 44(442)
Samarium 153, β-decay and polarization 41/2(66–67)
Samarium 149, polarized, neutron cross section 42(607)
Sample inside counter 45(78, 79)
Sampling, correlated 38/2(787, 788, 794)
Sampling techniques 38/2(773)
Sandwich method of magnetic bending 45(380)
Sandwiching of emulsions 45(395, 416)
Satellite pulses in multipliers 45(94)
Saturated activity 44(403); 45(449)
Saturation 38/1(173)
Saturation degree of a vapour 45(264)
Saturation effect 40(400)
Saturation factor 38/1(175)
Saturation in ionization chambers 45(28–30, 452)
Saturation of nuclear binding 39(422)
Saturation of nuclear forces 40(454); 41/1(20–22)
Saturation requirements 39(438)
Saxon-Woods potential 41/1(255–256)
Scalar interaction 41/2(9, 88–91)
Scandium 46, β-γ-polarization correlation 41/2(74)
Scandium 46, general data and electron polarization 41/2(66–67)
Scandium, neutron yield of (*p*, *n*) reaction 38/2(184)
Scanning, differential and static 38/1(197)
Scanning equipment 45(411–419)
Scattering at a potential 40(21)
Scattering at high energies 40(458)
Scattering by bound nuclei
— effect of binding forces 38/2(401ff., 416ff.)
— reduced mass effect 38/2(393, 402)
Scattering by small angles, individual process 45(372)
Scattering constant or coefficient 45(375)
Scattering cross section(s) *see* Cross section
— and statistical matrix 41/1(472)
— definition 38/2(28); 41/1(241–242, 251)
— differential 38/1(437)
— formula of Klein and Nishina 38/2(672)
— in the Weisskopf approach 41/1(195)
— inelastic 41/1(529)
— macroscopic 38/2(213)
— of high-energy neutrons by protons, dependence on angle and energy 45(495)
Scattering, elastic 40(14, 20, 21, 35); 41/1(11, 241–242, 300, 400–402)
— of alpha-particles 40(239)
— of charged particles 42(466, 468)
— of deuterons 40(239)
— of neutrons 40(232, 425ff.)
— of photons 42(356)
— of protons 40(235)
— of uncharged particles 42(464–466)
Scattering, inelastic 40(30)
— of charged particles 42(500)
— of electrons by nuclei 42(501)
— of neutrons, angular dependence 40(250)
— of neutrons, energy dependence 40(248, 249)
— of neutrons, theoretical 40(247)
— of protons 42(495)

Scattering length 38/2(394); 39(3, 37, 38, 39, 47, 51, 60, 61, 85, 91); 40(423); 41/1(328)
— coherent 39(3, 47, 48, 49, 51)
— geometrical meaning 38/2(32)
— imaginary part 38/2(33)
— of beryllium for neutrons 38/2(75–76)
— of deuteron-neutron system 38/2(73–74)
— of proton-neutron system 38/2(67–68)
Scattering matrix 39(78, 79, 82, 135); 40(412f.); 41/1(8, 42, 123); 42(588)
— relation to $\mathscr{R}$-matrix 41/1(119–123, 168–176)
— relation to reactance matrix 41/1(173, 175)
Scattering mean free path
— exponential theory 38/2(266)
— in carbon 38/2(293)
— in water 38/2(293)
— interval theory 38/2(266–267)
Scattering measurement, method of constant
— cell 45(373, 375)
— sagitta 45(373, 376–377)
Scattering of α-particles 39(464); 42(469)
Scattering of β-rays in a counter 45(195)
Scattering of electrons in source 41/2(40)
Scattering of epithermal neutrons by bound protons 41/1(314, 327, 330)
Scattering of high-energy nucleons 41/1(257–262)
Scattering of light by bubbles 45(331)
Scattering of light by droplets in a cloud chamber 45(293, 331)
Scattering of muons 40(531)
Scattering of neutrons
— by parahydrogen 38/2(6)
— by protons 38/2(6)
— coherence effects 38/2(6, 409ff.)
— experimental 42(467)
— kinematical frame of theory 38/2(27ff.)
Scattering of nucleons 39(246, 371)
Scattering of nucleons by nuclei of spin zero 42(596)
Scattering of polarized electrons by electrons 41/2(53, 59–60)
Scattering of polarized electrons by nuclei 41/2(53–58)
Scattering of polarized positrons by electrons 41/2(53, 59–60)
Scattering of protons by carbon 41/1(466, 478, 488, 492, 493)
Scattering of protons by helium 41/1(466, 478, 488)
Scattering of protons by tritium 41/1(288–289)
Scattering of unpolarized nucleons producing polarization 41/1(474)
Scattering, quasi-elastic 40(497f.)
Scattering resonance 44(411)
— elementary theory 41/1(25–28)
— integral 38/2(366, 375–376)
Scattering surface 38/2(447, 448, 450)
Scattering, thermal 40(102)
Scattering without absorption 38/2(31–33)
Scattering-order expansion of the solution of the photon transport equation 38/2(686ff., 706–707, 711, 733, 745, 767ff., 771)
Scattering-properties of some moderators 38/2(228)
Schmidt diagrams (Schmidt lines) 39(277, 295, 316, 407)
Schmidt model 38/1(116)
Schmidt moment 39(311)
Schmidt values of magnetic moments 41/2(94)
Schmitt trigger circuit 45(230)
Scintillants, ionorganic 45(86, 97–143)
Scintillants, mean life of light emission 45(223)
Scintillants, organic 45(86, 144–154)
Scintillation counter 40(474f., 481, 510); 44(415, 420)
— design 45(86, 133, 169, 170)
— for high energy β- and γ-rays 45(164)
Scintillation counting 45(223)
Scintillation detector for fast coincidences 45(236–238)
Scintillation response of sodium iodide 45(99–104)
Scintillation spectrometer 40(27, 32, 77, 125)

Scintillator with bismuth loading 45(506, 516)
Scintillators for neutron detection 45(460–464, 469)
Scintillators, neutron efficiency 45(501)
Scott diagram of angular correlation coefficient 41/2(92)
Screening 38/1(420)
Screening correction 42(141, 151, 156)
Screening effect in β-decay 41/2(13)
Screening effect in Mott scattering 41/2(55)
Screening effects 38/1(112)
SCT, *see* Semi-classical theory
Search coil 44(249)
Second forbidden transition 41/2(10, 15, 45)
Second moment
— measurement in water 38/2(313)
— measurements in water-metal mixtures 38/2(314)
— of neutron distribution in hydrogeneous moderators 38/2(259–264)
— of neutron distribution in non-hydrogen moderators 38/2(264–268)
Second sum rule 41/1(193–194, 194–201)
Secondary electron spectrum 38/2(678)
Secondary electrons from radiation 38/2(665, 753)
Secondary electrons in β-ray spectrometer 41/2(40)
Secondary emission tube 45(246)
Secondary ions 45(5)
Secondary neutron sources in moderators 38/2(64)
Secondary x-rays, build-up factor 38/2(676, 678, 721–722, 724, 740, 752–753, 756–758, 760, 772, 795)
Secularisation by thermal noise 38/1(141)
Selection rule(s)
— for α-decay 42(183)
— for isotopic spin 41/2(74, 93)
— in pick-up theory 41/1(334)
— of β-decay 41/2(9, 10, 40, 89)
Selective capture of slow neutrons, early work 38/2(41 ff.)
Selengut-Goertzel procedure 38/2(291, 293)
Selenium rectifier 44(18, 19)
Self-indication 38/2(387), 40(397, 402)
Self-absorption 38/2(387)
— curve 38/2(387–392)
— curve, initial slope 38/2(388)
— of β-rays in the detector 38/2(355)
Self-consistent field of the nucleus 41/1(20–21, 23)
Self-consistent model of γ-emission 41/1(26–27)
Self-mirror nuclei of spin zero 41/1(494)
Self-protection correction 38/2(636)
Self-quenching counters 45(62, 63, 65, 67–68, 71, 84)
— life time 45(67–68, 71, 84)
Self-shielding proportional tube 45(210)
Self-sustaining discharge 45(53, 58)
Semi-classical and quantum mechanical treatment of Coulomb excitation, relation between them 41/1(498–499, 502, 503, 505, 540–548)
Semi-classical approximation to heavy particle collisions 41/1(370–373)
Seniority number 39(289, 326)
Sensitive time of a bubble chamber 45(322–325, 327, 328)
Sensitive time of a cloud chamber 45(270)
Sensitive volume of a counter 45(55)
Sensitivity of emulsions 45(342, 343, 389)
Separation of belts 44(76)
Septum between last two cyclotron orbits 44(151, 152)
Serber force 39(117, 337)
Serber interaction 39(428, 439)
Serber mixture of nuclear forces 40(454, 456 f.)
Serber's approach to the optical model 41/1(257–259)
Serber's semiclassical picture of a nuclear collision 40(456 f., 461 f., 482)
Serber's theory of high-energy stripping 41/1(304–312)
Serber's theory of stripping, extension to low energies 41/1(319)
Shape correction of synchrotron magnet 44(242)
Shape dependent parameter 39(40, 52, 53, 60, 63, 85, 87)
Shape elastic cross section 41/1(241)
Shape elastic scattering 39(448)

Shape factor of continuous spectrum 41/2(40)
Shape independent approximation 39(40, 51)
Shape independent theory of scattering 38/2(67)
Shape of continuous spectrum, formula 41/2(9)
Shape vibrations, *see* γ-vibrations
Sharp resonance 41/1(143)
Shell model(s) 40(7, 8, 131, 415); 41/1(151, 260)
— and stripping theory 41/1(349)
— applied to high-energy scattering 41/1(466, 486)
— comparison of atoms and nuclei 39(467)
— history 39(242)
— of nuclei 38/2(47); 42(3, 290)
— of nucleus 39(230, 236, 464–476)
— theoretical basis 39(368)
— with a single particle 39(468–472)
— with many particles 39(472)
Shell transport kernels 38/2(510)
Shellac bonded textolite 44(80–82)
Shield rings in Van de Graaff generators 44(73, 84–85)
Shielding 38/1(413)
Shielding by a Frisch grid 45(34, 35)
Shielding effect in Coulomb excitation 41/1(559)
Shielding effects due to closed shells of electrons 38/1(412)
Shielding factor 38/1(414, 436)
Shielding for radiation protection 44(444–446)
Shielding of a cyclotron 44(156–158)
Shielding of counters 45(210, 212)
Shimming procedures 44(138–139)
Shims 38/1(233); 44(134)
Short collision time approximation 38/2(418)
Short range limit 39(291)
Shower 40(476, 478, 522)
Shower detection in emulsions 45(348)
Shower development within a scintillation counter 45(164–169)
Shrinkage of emulsions during processing 45(401 ff.)
Side band echo 38/1(165)
Siegert's theorem 42(315)
Σ^0- and Ξ^--particles, cascade decay 42(531)
Sigma pile 38/2(600)
Sign of charge determination for emulsion tracks 45(380)
Sign of charges of β-particles 41/2(2)
Sign of high-energy scattering polarization 41/1(492, 493)
Sign of tensor force 39(27)
Signal at the wire 45(188)
Signal forms, absorption signal 38/1(188)
Signal forms, dispersion signal 38/1(188)
Signal generation in an ionization chamber 45(27)
Signal-to-noise ratio 38/1(189, 195, 197, 226, 230, 281)
— of ionization chambers 45(31–33)
Silver as a detector of thermal neutrons 38/2(634–635)
Silver detector covered with cadmium 38/2(311)
Silver, early work on selective absorption 38/2(42–44)
Silver, neutrons released by α-rays 38/2(111)
Simulator cavity 44(380)
Single crystal spectra, nuclear magnetic 38/1(322, 325)
Single level formula 40(20, 45, 123, 403 ff.)
Single particle
— levels 40(141)
— model, *see* Individual particle model
— shell model 40(7, 8)
— states 40(166, 171)
— transitions 42(14)
Singlet state of proton-neutron system 38/2(67)
Singlet states (even and odd) 39(27, 30, 31)
Singly scattered x-rays 38/2(738, 755, 761, 764–765)
Singularities of the radiation transport equation 38/2(707, 713 ff.)
Singularities of the x-ray flux 38/2(687, 700, 702, 738, 740, 745–746, 749)
Size of nuclei, determination
— by α-particles 39(216–218)
— by bound electrons 39(199)

— by neutron scattering 39(213, 214)
— by proton scattering 39(215)
— from electron scattering 39(192–195)
Size of nuclei, early studies 39(178)
Size of very light nuclei 39(199)
Size resonances 40(229)
Skin effect 38/1(32)
Slater determinant 39(467, 475)
Slater integrals 39(291)
Slow chopper 44(431)
Slow electrons in emulsions 45(370–371)
Slow expansions of a cloud chamber 45(275)
Slow neutron detection 45(447–449, 463)
Slow neutron threshold method 38/2(113, 127)
Slow neutrons 41/1(191)
Slowing down
— area 38/2(270)
— by heavy nuclei 38/2(243)
— density 38/2(216, 462); 44(400)
— density in hydrogen 38/2(231, 234)
— density in hydrogen-free absorbing media 38/2(239–246)
— dependence on neutron age 38/2(246–253)
— efficiency of some moderators 38/2(228)
— for weak resonance capture 38/2(245)
— for widely spaced narrow resonances 38/2(245)
— kernels 38/2(253–254, 278, 283)
— length 38/2(255, 282)
— length in water 38/2(264)
— length, theoretical values in carbon and heavy water 38/2(267)
— of neutrons and degradation of radiation energy 38/2(700)
— of neutrons, discovery 38/2(18–22, 35ff.)
— of neutrons, stationary case 38/2(228, 394–397)
— power of a moderator 38/2(22, 227)
— process 40(397, 398)
— time 38/2(338–342)
— time spectrometer 38/2(339–341)
Slow-neutron filter 44(430)
S-matrix 40(412); 42(385)
Sodium 22
— β-γ-polarization correlation 41/2(74)
— decay 41/2(44)
— decay scheme 41/2(96)
— general data 41/2(68)
— polarization of positrons 41/2(64, 69)
Sodium 24, radiation from recoil nucleus 41/2(30)
Sodium, (α, n) reaction 38/2(129)
Sodium iodide 45(97–131)
— absolute efficiency 45(104–105)
— crystal 38/2(805–806)
— decay times 45(99)
— detection efficiency diagrams 45(110–129)
— fluorescent response 45(99–104)
— fluorescent response for heavy particles 45(102)
— fluorescent response for photons 45(100)
— γ-ray absorption 45(109)
— influence of thallium added 45(99, 102)
— methods to grow crystals 45(103)
— peak-to-total ratio 45(127–131)
— photopeak efficiency 45(109, 127–131)
— radiation length 45(167)
— scintillation spectra 45(98)
— stopping power for particles 45(105–107)
— total γ-ray detection efficiency 45(107, 108, 127–131)
Soft beta-spectra 45(214)
Solenoid of high accuracy 38/1(254)
Solenoidal field counter 45(197)
Solenoidal focussing 44(359, 374)
Solid angle, finite, correction of the observation 42(426)
Solid radioactive source in a counter 45(199)
Solids, nuclear quadrupole coupling 38/1(392, 428, 429)
Solvents for extraction of actinide element ions 42(235, 251, 259, 268, 274)
Sommerfeld's fine structure constant 38/1(274)
Source, effective, for thermal neutrons 38/2(508, 517)

Source of fission neutrons, distribution of slow neutrons
— in beryllium 38/2(330)
— in graphite 38/2(333)
— in heavy water 38/2(329)
— in hydrogeneous moderators 38/2(317–318)
Source of fission neutrons, energy spectrum 38/2(310)
Source of from beryllium, technical details 38/2(161)
Source of heterogeneous neutrons 38/2(87)
Source of homogeneous neutrons 38/2(87, 104–107)
Source of neutrons
— from antimony-beryllium 38/2(307, 321, 335, 337)
— harmonically modulated 38/2(629–633)
— independent 38/2(214)
— isotropic 38/2(214)
— using α radioactivity 38/2(88, 107–132)
— using α radioactivity, preparation 38/2(130–132)
Source of photoneutrons 38/2(158–162)
Source sign 38/2(532, 597, 644)
Source strength
— absolute determination 45(44)
— of neutrons 38/2(214)
— of neutrons in diffusion approximation 38/2(545)
— routine measurements 45(42, 43)
Source support effects 45(200–202)
Source technique, preparation and mounting 45(199–202)
Sources for γ-cascades 42(403)
Sources of x-rays
— obliquity effects 38/2(679, 713, 723, 746, 760)
— symmetry types 38/2(742ff., 748, 761)
— with special geometry 38/2(674–675, 681ff., 688ff., 690, 693, 726ff., 735, 749ff., 762, 764ff., 767)
Space charge in a counter 45(59, 60)
Space distribution of radiation
— found by Fourier inversion 38/2(765–768)
— moments 38/2(724ff., 735ff.)
Space reflection invariance 41/1(484)
Space rotation invariance 41/1(484)
Space-charge effects in a betatron 44(203–205)
Spacing of levels, *see* Level spacing, Level density
— and second sum rule 41/1(194, 196, 200)
— in compound nuclei, table 41/1(210–216)
Spallation 38/2(17); 40(455, 468, 504, 523)
Spark-over along charging belt 44(67)
Sparks from open-air generators 44(66, 96)
Spatial distribution of slow neutrons around point source 38/2(313ff.)
Spatial moment (*see also* Moment) 38/2(308)
Spatial reconstruction of cloud chamber tracks 45(301)
Specific ionization 45(66)
Specific ionization, studies by proportional counters 45(220–221)
Spectra, calculation 39(334)
Spectra, nuclear magnetic, *see* Nuclear magnetic spectra
Spectra of sodium iodide scintillators 45(98)
Spectra of states 39(291)
Spectral analysis 38/1(117)
Spectral analysis of x-rays, experimental techniques 38/2(804)
Spectral distribution of x rays, integral over space and direction 38/2(699)
Spectrometer for fast neutrons 40(141)
Spectrometer for heavy particles 45(44, 45)
Spectrometer for high-energy neutrons 45(514)
Spectrometer for moderate-energy neutrons 45(474–475)
Spectrometer for neutrons of intermediate energy above 1 keV, time of flight method 40(375)
Spectrometer for particle energies 45(131–138)
Spectrometer for slow neutrons
— crystal spectrometer 40(375, 378)
— fast chopper 40(374)
— time of flight method 40(374)

Spectrometer measurements of effective neutron temperature 38/2(482–488)
Spectrometers for β-rays 41/2(13, 37, 39)
Spectrometers for neutrons 44(411–415)
Spectroscopic data of the actinide elements 42(221)
Spectroscopy of slow neutrons 38/2(88–89)
Spectrum of internal bremsstrahlung 41/2(98)
Spectrum of radiation from synchrotron accelerated electrons 44(180)
Spherical boundary, radiation behavior 38/2(797–798)
Spherical core plus single nucleon 39(499)
Spherical harmonics
— addition theorem 41/1(302)
— expansion approximating solution of Boltzmann equation 38/2(259, 284–287, 490–491)
— expansion of photon flux 38/2(691, 702, 724)
— expansion of wave function 38/2(27–28)
— method 38/2(562–566, 568)
— notation 41/1(112–113)
Spherical moderator, solution of age equation for central point source 38/2(273–274)
Spherical shell model 41/1(73–75, 85)
Sphericity, departure from 39(314)
Spheroidal box potential 39(506, 507)
Spheroidal core model 39(477, 498–501, 514)
Spheroidal core plus several nucleons 39(500)
Spin angular function 41/1(126, 297)
Spin echo 38/1(155)
Spin eigenstates for three nucleons 39(145)
Spin flip 40(406)
Spin flip scattering 40(39)
Spin Hamiltonian 42(599)
Spin incoherence of scattering 38/2(412)
Spin, isotopic 39(289, 319); 40(3, 42, 71, 73, 74, 82, 107, 111, 124, 151, 173, 175)
Spin matrices 42(437, 439)
Spin multiplet 40(5)
Spin, nuclear 42(8, 300, 301)
Spin of β-particles 41/2(4)
Spin of electron 38/1(274)
Spin of the Neutron 38/2(6, 8)
Spin orbit coupling 39(23, 250, 325, 404, 471, 472); 40(4, 118, 154, 155); 42(595)
Spin orbit potential 39(369, 378)
Spin quantum number, total isotopic 40(4)
Spin *s*-particles 42(437, 443, 457)
Spin, statistical weight 40(403)
Spin temperature 38/1(174, 177)
Spin tensor 42(436, 457)
Spin tensor of rank 1 42(444)
Spin term in binding energies 38/1(15, 21–23)
Spin twisting 41/2(53, 55, 57)
Spin-lattice relaxation time 38/1(129, 147)
Spin-lattice, transition probability 38/1(143, 144)
Spinor 41/1(468)
Spin-orbital interaction, magnetic 38/1(292)
Spin-orbital interaction of the electron with the nucleus 38/1(64–65)
Spin-orbit-interaction 41/1(260–263, 466, 474, 486)
Spin-orientation exchange by pairs 38/1(149)
Spins of shells 42(4)
Spins of stable nuclei of odd mass number in the ground state 39(263 ff.)
Spins of unstable nuclei of odd mass number in the ground state 39(270)
Spin-spin exchange interaction 38/1(147)
Spin-spin interaction 38/1(293, 296)
Spin-spin interaction, coupled by electrons 38/1(147)
Spin-spin relaxation time 38/1(130, 146)
Spiral or shimming homogeneisation 38/1(235)
Splitting technique 38/2(787, 796)
Spontaneous fission 42(301, 305)
Spontaneous fission producing neutrons 38/2(210–211)

Spurious pulses from a counter 45(55, 60)
Spurious states 39(329)
Square distance, average, traveled by neutrons 38/2(259, 278)
Square well
— infinite 39(248)
— potential 40(459f.)
— potential, radius, effective 40(460)
— with barrier example 41/1(139–148)
S-R (scattering-range) method of mass determination 45(378)
Stability limits for the ion accelerator 44(372, 376)
Stability of betatron oscillations in alternating gradient accelerator 44(301–304, 315)
Stability of orbits in the AVF cyclotron 44(112–115)
Stability of orbits in the cyclotron 44(107–121)
Stability requirement 44(200, 222)
Stabilization of a magnetic field 38/1(236–243)
Stacks of emulsions 45(348, 349, 393–394)
Stage micrometer 45(417)
Stages for scanning microscopes 45(411–417, 423)
Standard absorption cross sections 38/2(60)
Standard indium foils 44(410)
Standard problems of the one-velocity transport theory 38/2(535–539)
Standing wave method of acceleration 44(352)
Stanford electron accelerator
— buncher 44(357)
— klystrons 44(351, 353)
— layout 44(364, 366)
— waveguide subsections 44(364)
Star formation 38/2(17)
Star formation by pions 40(534)
Star production as a high-energy neutron detector 45(492, 493, 509)
Stark modulation 38/1(55)
Stars induced by photons 40(521f.)
States, classification 39(290)
States of inequivalent particles 39(323)
States of normal parity 39(337)
States of nuclei in the region of Pb^{208} 39(302)
States of several particles in different shells 39(296)
States of several particles in the same shell 39(287)
Static approximation for nucleons 39(18, 20, 23)
Static approximation to scattering of neutrons by molecules or crystals 38/2(429, 451)
Static interaction 39(7, 18, 23)
Static quadrupole moment of an excited state 41/1(551–552)
Statistical error in random sampling 38/2(780)
Statistical errors in coincidence counting 45(251–252)
Statistical fluctuations in energy loss 45(263, 354, 358)
Statistical fluctuations in scintillants 45(237)
Statistical $\mathcal{R}$-function 41/1(203)
Statistical ionization fluctuations 45(15–17)
Statistical matrix 41/1(468)
Statistical $\mathcal{R}$-matrix 41/1(202–205)
Statistical mixture of substates 41/1(468)
Statistical model for α-emission 42(170)
Statistical nuclear model 42(462)
Statistical tensor 42(384, 386, 454, 520)
Statistical theory
— of neutron emission 38/2(84)
— of nuclear levels 41/1(11)
— of nuclear reactions 40(17, 58, 105, 204, 212f.)
— of nuclear reactions, experimental conformation 40(218)
Statistics obeyed by neutrons 38/2(6, 8)
Steel for magnets 44(236, 248, 331)
Steep trajectories in emulsions 45(346)
Steepest descent, *see* Saddle point method
Steepest descent approximation, *see* Saddle point
Stilbene crystal scintillators 45(144, 145, 152)
Stokes' parameter(s) 38/2(671, 697); 42(442, 443)
Stokes-Einstein relations 38/1(134)
Stopbands 44(310, 312, 323)
Stopping formula for electrons 45(6)

Stopping formula for heavy particles 45(5)
Stopping power (*see also* Energy loss, Range-energy relation) 45(5)
Stopping power, curves 45(7)
Stopping power of sodium iodide for particles 45(105–107)
Straight cut-off, *see* Cut-off
Straight sections 44(226–229)
Straight-ahead approximation 38/2(706, 711, 713)
Stratified medium, passage of radiation 38/2(800ff.)
Stray inductivities of a transformer, effect on a rectifier 44(23)
Strength function 41/1(202–203, 218, 228, 253, 254, 267)
— measurement 41/1(228–230)
— of a resonance 40(231, 424, 427f.)
— of resonances 44(418)
Strictly incoherent scattering 38/2(411)
String model 41/1(54, 68–69, 72, 88, 155, 219, 283, 286)
String model and $\mathcal{R}$-matrix theory 41/1(88)
Stripped emulsion stacks 45(348, 381, 393–394)
Stripping as Coulomb excitation 41/1(310–312)
Stripping cross section 41/1(306, 309)
Stripping experiment 42(487, 604)
Stripping in B(*d*, *n*) reactions 38/2(205)
Stripping in Be(*d*, *n*) reaction 38/2(202)
Stripping in C(*d*, *n*) reactions 38/2(207)
Stripping in D(*d*, *n*) reaction 38/2(193)
Stripping in T(*d*, *n*) reaction 38/2(197)
Stripping mechanism 42(462)
Stripping of deuterons, neutron spectrum 45(488, 506)
Stripping of heavy particles 41/1(343)
Stripping process 38/2(186)
— energy dependence 40(271)
— Serber's theory 40(272)
Stripping reactions 41/1(8–9, 304–319, 320–350); 42(480, 482, 483)
— by deuterons 40(129, 487, 492, 499f., 502f.)
— by high-energy deuterons 41/1(304–312)
— extension of Serber theory to low energies 41/1(318–319)
Stripping theory and shell model 41/1(349)
Strong binding of atoms, Doppler effect 38/2(349)
Strong coupling 39(314)
Strong coupling model 42(135, 166)
— of N. Bohr 39(446)
Strong focussing 44(215, 301)
Strong interaction, Brueckner theory 40(461)
Strontium 90, β-decay and polarization 41/2(66–67)
Structural analysis, crystallographic 38/1(321)
Structure amplitude 38/2(415, 434)
Structure factor 38/2(435)
Structure factor for neutron diffraction 44(433)
Structure of liquids 38/1(144)
Structure of nuclear magnetic spectra
— of gases 38/1(293–311)
— of liquids 38/1(293–311)
— of solids 38/1(311–336)
Structure of single crystal spectra 38/1(325)
Structure problems, chemical 38/1(311)
Subdivision of accelerating tube 44(82)
Subresonances of third order 44(315)
Subshell 42(118, 198)
Sulfur 35, Fermi plot 41/2(39)
Sum rules 41/1(187–194, 270); 42(316, 337)
— applicability limitations 41/1(205)
— for nuclear photoeffect 38/2(154–155)
— of Thomas and Reiche 41/1(394)
Summation effects in counters 45(205)
Superallowed transition 41/2(15, 41)
Supercritical state of a reactor 44(392–393)
Superheated liquid 45(315)
Supermultiplet 40(5, 155, 157, 161, 163, 175, 187, 193, 185)
Supermultiplet of Wigner 39(322)
Super-regenerative methods 38/1(210)
— incoherent 38/1(211, 212)
— precession coherent 38/1(211, 212)
— vibration coherent 38/1(211, 213)
Supersaturation by adiabatic expansion 45(266–270)
Supersaturation by temperature gradient 45(284–287)

Supersaturation of a vapour 45(264)
Support of a solid source in a counter 45(200)
Support structures of Van de Graaff generators 44(79–82, 97)
Surface, behaviour of thermal neutrons 38/2(536–552)
Surface energy of nucleus 39(233, 370, 415, 432–434)
Surface harmonics 39(388)
Surface interactions 42(493)
Surface of a nucleus 38/2(33–34)
Surface of nucleus, deformed 39(418)
Surface oscillations of nuclei 39(419, 520)
Surface probability of the α-particle 42(143, 153, 154, 158, 178, 180, 181, 185)
Surface slope of density 39(193, 233)
Surface thickness of a nucleus 38/1(104)
Surface well model 42(150, 171)
Survival time after injection 44(271)
Suryan stabilization 38/1(239)
Susceptibility, complex 38/1(173)
Susceptibility, dynamic 38/1(173)
Susceptibility, macroscopic, dependence on the nuclear resonance frequency 38/1(281–286)
Susceptibility, nuclear paramagnetic, static 38/1(126)
Sweep field, *see* Clearing field
Swelling of emulsions during processing 45(401ff.)
Swimming pool reactor 44(396)
Swirl angle in AVF cyclotron 44(112, 115)
Symmetric top 38/1(38); 39(481)
Symmetrical cascade generator 44(43–45)
Symmetry energy, nuclear 39(233, 370, 371, 434–436)
Symmetry of nuclear wave functions 39(482–485)
Symmetry of x-ray flux 38/2(694, 739)
Symmetry of x-ray flux, applications 38/2(742ff., 748, 761)
Symmetry operators for three nucleons 39(145)
Symmetry properties of homonuclear diatomic molecules 38/1(83–89)
Symmetry properties of Σ states 38/1(87)
Symmetry relations of vector addition coefficients 41/1(294–295, 303)
Symmetry term of the binding energy 38/1(24)
Symmetry types of x-ray sources 38/2(742ff., 748, 761)
Symon's smooth approximation (FFAG) 44(216)
Symplectic group 39(326)
Symplectic matrix 44(311)
Synchrocyclotron 40(468f., 530)
Synchrocyclotron, list of characteristics 44(161, 166)
Synchrocyclotron phase effects 44(127–132)
Synchrocyclotron, theory 44(107, 132)
Synchronization, forced 38/1(130)
Synchronous orbit 44(171, 183)
Synchrophasotron, description 44(219, 230–233)
Synchrophasotron, field index 44(238–240)
Synchrophasotron, injector and preaccelerator 44(267, 268, 294)
Synchrophasotron, magnet 44(237, 244, 247)
Synchrophasotron, phase oscillations 44(279)
Synchrophasotron, power supply 44(251, 253–254)
Synchrophasotron, radiofrequency apparatus 44(287)
Synchrophasotron, radiofrequency parameters 44(279)
Synchrophasotron, survival of injected beam 44(273, 282)
Synchrotron 40(468f., 472, 511f.)
Synchrotron oscillations, damping 44(182–184)
Synchrotron oscillations, mechanical analogue 44(171)
Synchrotron, vacuum chamber 44(260)
Synthesis of neutron distribution from its moments 38/2(335)
Synthetic kernels 38/2(283)
Systematics of α-decay energies 38/1(31–32)
Systematics of β-decay energies 38/1(20, 32–33)
Szilard-Chalmers method 45(449)

Target deterioration under ion bombardment 38/2(182)
Target, methods to strike it 44(294–299)
Target nuclei, polarized 42(592)
Target of tritium 38/2(176–178)
Target thickness 38/2(102–103)
Targets in cyclotrons 44(154–156)
Targets of heavy ice 38/2(189, 191)
Targets of lithium 38/2(181–182)
τ^+-meson, decay 42(533, 535)
Technetium, isomer 41/2(6)
Teflon 45(39)
Teichmann-Wigner boundary conditions 41/1(156, 161–163, 167, 180)
Telegraph equation for thermal neutrons 38/2(576)
Telescope, *see* Counter telescope
Telescope for particle detection 40(474, 481, 488, 515)
Telescopes for recoil protons 45(494, 498, 512)
Telescoping electrodes 44(83)
Tellurium 130, possibility of double β-decay 41/2(101)
Temperature, *see* Effective neutron temperature
Temperature coefficients of the thermal neutron diffusion coefficient 38/2(613, 624–625)
Temperature, counter at elevated 45(215)
Temperature dependence
— of diffusion length 38/2(603–604)
— of effective resonance integral 38/2(378, 385)
— of grain density 45(388, 389)
— of graphite cross section for slow neutrons 38/2(444)
— of magnesium cross section for slow neutrons 38/2(444)
— of multipliers 45(93)
— of organic scintillators 45(152)
Temperature diffuse scattering 38/2(412, 413)
Temperature effect on linear accelerators 44(362)
Temperature, effective, of neutrons 38/2(463, 471, 477–489)
Temperature effects in expansion cloud chambers 45(280–281)
Temperature of an excited nuclear spin system 38/1(177)
Temperature of an excited nucleus 38/2(84)
Temperature of neutrons, measurements 38/2(602)
Temperature of neutrons, time dependence 38/2(504)
Tensor force 39(23, 24, 26, 27, 30, 31, 64ff., 77, 83, 87, 96, 117, 332, 337, 355, 360, 405, 438); 40(177)
Tensor force effect of second order 39(379)
Tensor force in the binding of H^3 and He^3 39(155)
Tensor force model, singular 39(117, 119, 122, 123, 124, 127)
Tensor interaction 41/2(9, 88–91)
Tensor operator 39(23, 394)
Test body 38/1(190, 196, 231)
Textolite tubes 44(80–82)
Thallium 204, β-decay and polarization 41/2(66–67)
Thallium chloride Čerenkov counter 45(171)
Thallium influence on sodium iodide scintillators 45(99, 102)
Thermal column of a reactor 44(398, 402, 427)
Thermal cross section 44(402, 427)
Thermal effect, discovery 38/2(35ff.)
Thermal equilibrium of neutrons and moderator 38/2(396)
Thermal excitation of a nucleus 40(466f.)
Thermal flux 44(400, 402, 422)
Thermal group of neutrons 38/2(492)
Thermal inelastic scattering 44(431)
Thermal neutron(s) 44(399, 401–402, 432–440)
— absorption cross section 38/2(65)
— beam 44(427–430)
— definition 38/2(504)
— detectors 38/2(633–637)
— distribution at a surface 38/2(536–545)
— elementary diffusion theory 38/2(511–517, 552–557)
— flux 45(479–480)
— flux distribution around a point source 38/2(530–531)

Thermal neutron(s) (cont.)
— one-velocity group method 38/2(504–507)
— plane source 38/2(530–533)
— results of transport theory 38/2(527–539)
— space distribution in hydrogeneous moderators 38/2(312, 319–322)
— transport equation 38/2(505); 38/2(508)
— waves 38/2(576, 579–580, 631)
Thermal utilization in a reactor cycle 44(391)
Thermalization time 38/2(339, 502–504)
Thermalization time in graphite 38/2(628)
Thick detector 38/2(640, 653)
Thin target 38/2(103)
Third forbidden transition 41/2(15)
Thomas coupling 39(378)
Thomas focussing in the AVF cyclotron 44(114)
Thomas term 41/1(487)
Thomson scattering 40(510f.)
Thorium 42(209, 212)
Thorium pile, effective resonance integral 38/2(381, 384)
Thorium, spontaneous fission 38/2(210)
Three-body forces 39(156, 352)
Three-particle cluster correction 41/1(24)
Three-quarter wave system 44(145)
Threshold behaviour data 41/1(291–292)
Threshold Čerenkov counter 45(157–159)
Threshold channel 41/1(281, 282, 287)
Threshold detector 38/2(86, 157)
Threshold energy for photoneutrons
— dependence on mass number 38/2(150)
— from beryllium 38/2(141–143)
— from deuterium 38/2(134–135)
— table for various elements 38/2(148, 152)
Threshold energy of an endoenergetic reaction 38/2(96)
Threshold for emission of monoenergetic neutrons 38/2(97)
Threshold measurements 42(331)
Threshold of a reaction 41/1(274, 278)
Threshold of Li(*p*, *n*) reaction 38/2(171, 178)
Threshold potential of a counter 45(55, 184)
Threshold reaction 44(403)
Threshold values in high-energy neutron detection 45(492)
Threshold values in moderate-energy neutron detection 45(450)
Thresholds of (*p*, *n*) reactions, table 38/2(171)
Thulium 170, β-decay and polarization 41/2(66–67)
Time channel analyser 42(32)
Time constant (*see also* Resolving time) 45(188)
Time dependence of the thermal neutron distribution
— in beryllium 38/2(626)
— in water 38/2(617)
Time dependent diffusion kernel 38/2(578)
Time dependent problems in P_1 approximation 38/2(575–580)
Time dependent slowing-down problem 38/2(246–253, 303–306)
Time development operator 42(518, 520)
Time distribution of synchrotron x-ray pulse 44(190)
Time for slowing down 38/2(338–342)
Time for thermalization of neutrons 38/2(339, 502–504, 628)
Time intervals, coincidence measurement 45(257–259)
Time lag in a counter (*see also under* Delay) 45(84)
Time lapse from indium resonance to thermal region 38/2(499–504)
Time of diffusion 38/2(577)
Time of injection 44(271)
Time resolution of a coincidence circuit 42(30)
Time reversal 41/1(41–42, 117–118)
— and polarization 41/1(474, 484, 495)
— operator 42(393)
Time-dependent Hamiltonian 41/1(508)
Time-integral and time-constant formulations of radiation transport theory 38/2(680, 699)
Time-of-flight distribution of argon 37 recoils 41/2(22)

Time-of-flight measurements for neutrons 40(374f.)
Time-of-flight method 38/2(37–39); 40(473, 475)
Time-of-flight neutron detectors
— around 1 MeV 45(474)
— at high energies 45(502–503, 512, 514)
Time-of-flight neutron spectrometer 38/2(423); 44(413–415)
Time-reversal invariance 41/2(47)
Timing circuits for cloud chamber expansions 45(278, 279)
Tin, minimum of γ-ray attenuation 38/2(675)
Tin, radiation attenuation coefficients 38/2(667)
Tin, reflection of radiation 38/2(792–793)
Tissues, effect of radiation 44(441–442)
Titanium hydride bonded porcelain 44(82)
Tolerance rate 44(443)
Tolerances in linear accelerator construction 44(361–364, 379–381)
Tolerances in synchrotron construction 44(247–249, 276)
Top, accidentally symmetric 38/1(388)
Top of oblate shape 38/1(388)
Top of prolate shape 38/1(388)
Top, symmetric 38/1(39)
Top, symmetric, quadrupole hyperfine structure 38/1(384)
Torsional vibration 38/1(429)
Total activation of a detector 38/2(353)
Total angular momentum quantum number 40(4)
Total collision density 38/2(215, 216)
Total cross section(s) 38/2(49)
— in the chemical bond region 38/2(444–446)
— measurement 40(39)
— of Coulomb excitation 41/1(497, 510–514, 527, 530)
— of nuclear reactions 40(14)
Total macroscopic cross section 38/2(213)
Total reflection of neutrons 38/2(6, 10, 68)
Total scattering cross section 38/2(28–29)
Total width of resonance level 38/2(48)
Townsend coefficient, first 45(58)
Trace techniques 41/1(470–472)
Track formation in emulsions, theory 45(425–429)
Track noise (in a cloud chamber) 45(308)
Track photometry 45(435)
Track properties in different chambers, comparison 45(334)
Tracks in a cloud chamber
— curvature 45(303)
— ionization determination 45(295)
— photography 45(289–295)
— spatial reconstruction 45(301)
Tracks in emulsions, characteristic properties 45(345–346)
Transfer by tunneling 41/1(369, 370, 379–390)
Transfer of a nucleon by virtual Coulomb excitation 41/1(369, 391–400)
Transfer reactions 41/1(310, 369)
Transformation from laboratory to centre-of-mass frame 38/2(91–95)
Transformations, intrinsic, in solids
— investigation by nuclear resonance spectra 38/1(332–336)
— processes of diffusion 38/1(332)
— processes of rotation 38/1(332)
— reversal of orientation 38/1(332)
Transformer, design of input transformer 44(49–54)
Transformer, effect of stray inductance 44(20, 23)
Transformer, losses 44(19)
Transformer, resistivity 44(2, 19)
Transformer, small d.c. magnetization 44(20)
Transient phenomena, periodic and aperiodic 38/1(170)
Transit time of electrons in the counter 45(84)
Transition energy of phase angle shift 44(301, 317–319, 324)
Transition half-life 42(17)
Transition matrix 39(138)
Transition period through chemical bond region 38/2(500)

Transition probabilities 39(252, 255, 317)
— delayed coincidences 40(325)
— $E2$ transitions 40(327, 331, 339)
— measurement 40(324)
— $M1$-transitions 40(326, 340)
— reduced 42(178, 180, 181, 185, 187, 188)
— resonance fluorescence 40(325)
Transition probability 42(35)
— intermolecular 38/1(145)
— intramolecular 38/1(145)
— reduced 39(489)
— systematic behavior of 42(90)
Transitions between rotational states 39(490–493)
Transitions, electromagnetic 42(8)
Transitions, favoured 39(276)
Transitions, isomere 39(296)
Transitions of non-unique shape 41/2(45–46)
Transitions of unique shape 41/2(11, 44)
Transitions, unfavoured 39(276)
Transitron circuiting 38/1(206–208)
Translational diffusion 38/1(134)
Transmission coefficient for thermal neutrons of an one-dimensional slab 38/2(559)
Transmission measurements for slow neutrons 44(432–433)
Transmission of light in Čerenkov counters 45(95)
Transmission polarizing neutrons 44(439)
Transmission problems of radiation solved by Monte Carlo methods 38/2(793–796)
Transparency of nuclear matter at higher energies 41/1(257)
Transparency of nuclei for high-energy neutrons 38/2(12)
Transparent nucleus, stripping reaction 41/1(309)
Transport cross section 38/2(257); 44(426)
Transport equation of radiation 38/2(680–682, 707)
— numerical solution 38/2(737ff.)
— singularities 38/2(707, 713ff.)
— solution by Fourier transform 38/2(690ff., 724, 765–768)
— solution by iteration 38/2(686ff., 706–707, 711, 733, 745, 767ff., 770ff.)
— solution by method of moments 38/2(724, 736ff.)
— solution by orders-of-scattering expansion 38/2(686ff., 706–707, 711, 733, 745, 767ff., 771)
— solution by spherical harmonics expansion 38/2(691, 702, 724)
— transformations 38/2(686ff., 690ff., 704, 724, 762–763)
Transport equation without energy loss for thermal neutrons 38/2(505, 508)
Transport kernel 38/2(278)
Transport kernels for thermal neutrons 38/2(510)
Transport mean free path 38/2(256, 305)
Transport mean free path measurements, steady state assemblies 38/2(585–592)
Transuranic element formation by heavy ion bombardment 41/1(368, 406)
Transuranium elements, *see* Actinide elements
Transverse motion in a linear accelerator 44(357–360, 373–378)
Transverse polarization 41/2(53); 42(435)
Traveling wave 44(343–344, 356)
Traveling wave helix 44(386)
Triangle condition 42(371, 373, 455)
Triggered oscilloscope method 42(33)
Triggering 38/1(219)
Triggering of cloud chamber expansions 45(277)
Triple correlation 42(458)
Triple scattering (*see also* Polarization) 40(497)
Triple scattering experiments 41/1(482–483)
Triplet state of proton-neutron system 38/2(67)
Triplet states (even and odd) 39(27, 30, 31)
Tritium beta rays 45(174, 195, 207, 215)
Tritium, (*d*, *n*) reaction 38/2(193–198)
Tritium decay, calorimetric investigation 41/2(19)

Tritium, determination of neutrino mass 41/2(37)
Tritium estimation 40(144)
Tritium, (*p*, *n*) reaction 38/2(175–177)
Tritium target 38/2(176–178)
True absorption coefficient of radiation 38/2(673)
Tube aperture effect on performance 44(88)
Tube count, *see* Count
Tungsten, minimum of γ-ray attenuation 38/2(675)
Tunneling of the nucleon in a transfer reaction 41/1(369, 370, 379–390)
Twists of the magnetic field 44(224)
Two-body interaction, residual 38/1(116)
Two-body problem, equivalent 40(459)
Two-body resonance treated by Green's function 41/1(87)
Two-body spin-orbit forces 39(332, 379, 404)
Two-body vector interaction 39(346)
Two-body wavefunction, relativistic 39(140)
Two-channel specialization 41/1(50–52)
Two-component theory of the neutrino 41/2(35, 105)
Two-nucleon model of the nucleus 41/1(19)
Two-particle interaction 39(405, 423, 425, 438)
— central force 39(403)
— Coulomb force 39(405)
— matrix element 39(402)
— radial integral 39(403)
— spin-orbit force 39(404)
Two-particle wavefunction 39(424)
Two-quantum annihilation of positronium 41/2(63–64)

Undersaturation of a vapour 45(264)
Unfavoured factor 41/2(42)
Unified model and shell model, relations 39(538–543)
Unified model of nucleus 39(349, 362, 476–543)
Unified nuclear model (of Bohr and Mottelson) 38/1(106, 115)
Unique shape transition 41/2(11, 44)
Unit cell of ion linear accelerator 44(367, 370)
Unitary symmetric matrix 41/1(299)
Univibrator 45(229, 233)
Uranium 42(212)
Uranium 235 as a slow neutron detector 45(446)
Uranium fission, energy spectrum of neutrons 38/2(310)
Uranium impregnation of emulsions 45(347, 395)
Uranium, minimum of γ-ray attenuation 38/2(675)
Uranium pile, effective resonance integral 38/2(381, 383–384)
Uranium, spontaneous fission 38/2(210)
Uranium, standard absorption cross section 38/2(60)
Uranium under electron bombardment as a neutron source 38/2(170)

V coupling, *see* Vector interaction
Vacancy after electron capture 41/2(95, 97)
Vacuum chamber of a betatron 44(211–217)
Vacuum chamber of a synchrotron 44(174, 256–261, 332)
Vacuum polarization effect in μ-mesonic levels 39(209)
Vacuum system of a cyclotron 44(149–150)
Valence electrons, non-bonding 38/1(422)
Valence shell 38/1(422)
Valve characteristic 44(2, 3, 10–17, 17–19)
Valve current, peak value 44(2, 5)
Valve current, r.m.s. value 44(2, 8)
Valve resistance 44(2, 13–17)
Van de Graaff generators
— as synchrotron injectors 44(176)
— design 44(68–69, 70, 83, 96, 98–104)
— early open-air type 44(66–67)
— horizontal 44(80)
— injector for the cosmotron 44(87–88)
— vertical 44(79–80)
— with high pressure gas 44(67–71)
Vanadium, neutron yield of (*p*, *n*) reaction 38/2(184)
Van de Graaff-*p*-*n*-sources 40(375)

Van der Waals force between α-particles 39(461)
Van der Waals interaction 38/1(428)
Van Vleck equation for the second or fourth moment 38/1(320–321)
Vapour bubbles in a superheated liquid 45(314)
Vapour condensation on ions 45(263–266)
Variance of gas gain 45(191)
Variational principles of Hulthén and Schwinger 39(131ff.)
Vector addition coefficients 41/1(115, 497, 514)
— relation to representation coefficients of rotational group 41/1(302)
— symmetry relations 41/1(294, 303)
Vector current conservation 41/2(51)
Vector interaction 41/2(9, 88–93)
Vector-coupling coefficients 39(389)
Vector-polarization 42(440, 444, 445, 457)
Vector-polarization of spin-s particles 42(457)
Velocity averages of thermal neutrons 38/2(40)
Velocity dependence of β-particle mass 41/2(4)
Velocity dependence of potential 41/1(22)
Velocity selector 38/2(37–39)
Vertical betatron oscillations 44(188, 194, 202–203, 446)
Vertical missing of inflector 44(274)
Vertical oscillation of cyclotron orbits 44(109)
Vessels for processing of emulsions 45(410)
Vibrating molecule, scattering of slow neutrons 38/2(408, 430)
Vibrational excitation, collective 39(516–525)
Vibrational function 39(485)
Victoreen R-meter 45(37)
Virtual Coulomb excitation in heavy particle collisions 41/1(369, 391–400)
Virtual excitation by electric field in transfer reactions 41/1(369, 391–400)
Virtual levels 40(11)
Virtual ^1S state 39(56)
Virtual state formation by Coulomb excitation 41/1(205, 497)
Virtual state of the deuteron 38/2(68)
Virtual states 41/1(35)
Voltage control in Van de Graaff generator 44(92–95)
Voltage drop in a cascade generator 44(26, 27, 33, 34, 37, 42)
Voltage drop in a rectifier with capacitor input 44(2, 9)
Voltage drop in a rectifier with choke input 44(22)
Volume advantage factor 38/2(376)
Volume defined cloud chamber 45(270–273)
Volume energy 39(370, 373)
von Weizsäcker's hypothesis 42(2)

Wall effect elimination in counters 45(196)
Washing emulsions 45(406–408)
Water, age of thermal neutrons 38/2(498–499)
Water, albedo 38/2(552, 596)
Water, annihilation radiation 38/2(752)
Water boiler reactor 44(396)
Water, build-up factors 38/2(676, 753, 757, 760, 795)
Water, decay curve of thermal neutrons 38/2(623)
Water, diffusion length measurement 38/2(582–583, 598–599, 624)
Water, diffusion length, temperature dependence 38/2(603–604, 626)
Water, diffusion parameters of thermal neutrons, measurements 38/2(582–583, 589, 596, 598, 624)
Water, directional distribution of radiation energy flux 38/2(746, 759)
Water, dissipation of radiation energy 38/2(762, 798–799)
Water, distribution of thermal neutrons as a function of time 38/2(617)
Water, energy absorption coefficients of γ-rays 38/2(674)
Water, flux distribution of neutrons below 1 eV energy 38/2(468–470)
Water, mean life of thermal neutrons 38/2(615–619, 624)
Water, measurement of the extrapolated end point 38/2(589)

Water, minimum of γ-ray attenuation 38/2(675)
Water moderator, slowing-down length 38/2(264)
Water, penetration of fast neutrons 38/2(301–302)
Water, polarization effect on radiation flux 38/2(757)
Water, radiation attenuation coefficients 38/2(667)
Water, radiation energy flux 38/2(751, 759)
— moments 38/2(738)
Water, reflection of radiation 38/2(792–793)
Water, scattering mean free path 38/2(293)
Water, scattering of slow neutrons 38/2(418, 426)
Water, slow neutron distribution results 38/2(311–327)
Water, spectral distribution of radiation flux 38/2(700, 702)
Water, temperature dependence of diffusion coficient 38/2(624–625)
Water-heavy water mixture 38/2(329)
Watson's approach to the optical model 41/1(257, 259)
Wave functions of nuclear top motion 39(481–485, 543)
Wave packet approximation to heavy particle collisions 41/1(370–372)
Wavefunction formalism 42(378)
Waveguide 44(345, 348–350)
Waveguide, disk-loaded 44(348–350)
Waveguide gas absorption cell 38/1(49)
Waveguide imperfections 44(361–364, 379–381)
Wavelength of thermal neutron waves 38/2(579–580)
Wavelength shift by Compton effect 38/2(668–669)
Waves of thermal neutrons 38/2(576, 579–580, 631)
Weak binding of atoms, Doppler effect 38/2(349)
Weak coupling 39(314)
Weak coupling of channels to resonating element 41/1(63, 73)
Weak interaction 40(458, 525, 529)
Weber function 41/1(422, 424)
Weight function 38/2(734ff., 739ff., 746ff., 754, 764–765)
Weisskopf approach to the second sum rule 41/1(194–201)
Weisskopf formula 42(15)
Weisskopf-Wigner treatment of radiation 41/1(69, 89, 271)
Well depth parameter 39(41, 52)
Whittaker differential equation 41/1(409)
Whittaker functions 41/1(332, 409–410)
Whittaker functions, integral representations 41/1(413)
Whole-body radiation 44(443)
Wick eigenvalue equation 38/2(716)
Wick's second order diffusion theory 38/2(562, 569–572)
Width formulas 41/1(272–273)
Width of neutron resonance lines 38/2(46, 48, 50)
Width of the resonance level 42(422)
Width, radiative, *see* Radiative (or radiation) width
Width, reduced 40(19, 20, 22f., 114, 119, 129f., 133, 139, 141, 409, 428)
Width, reduced, in α-decay 42(149, 151, 153, 154, 158, 181)
Widths of absorption lines 38/1(50)
Widths of levels in compound nuclei table 41/1(210–216)
Wiggle 38/1(184)
Wigner approach, short description 41/1(107–111)
Wigner approximation for nuclear interaction 40(4)
Wigner coefficients 39(389, 390)
Wigner interaction 39(425)
Wigner vector addition coefficients 41/1(115, 294, 302, 497, 514)
Wigner-Eckart-theorem 42(382, 383)
Wigner-Eisenbud boundary conditions 41/1(112, 147, 150, 151, 161, 278)
Wigner-Eisenbud formalism 40(410f.)
Wigner-Eisenbud method 41/1(119–120, 124, 127, 139)
Wigner-Eisenbud method and Coulomb field 41/1(124, 127)
Wigner-j-symbols 42(373, 545)
Wigner's 3-j symbol 41/1(303)
Wigner's 6-j symbol 41/1(303)
Wigner's $\mathscr{R}$-matrix 41/1(3, 5, 24, 53, 81, 107)

Wigner-Wilkins integral equation 38/2(396)
— reduction to differential equation in special cases 38/2(458–459)
Wilkins' equation 38/2(461)
Wilkinson kicksorter 45(192)
Wilkinson's model 42(340)
Wilson chamber, *see* Cloud chamber
Wilson vacuum seal 44(150)
Windows in counters 45(76–78)
"Wine-bottle" potential 39(250)
Wing regions of a resonance line 38/2(51)
Wire diameter (in a counter) 45(69, 74)
Wire material 45(74)
Wire of counter, uniformity 45(185)
Wisconsin high-pressure Van de Graaff generators 44(68–70, 83, 96)
WKB approach to heavy particle collisions 41/1(374)
WKB approximation in Coulomb excitation 41/1(502, 504ff., 541)
WKB approximation of Coulomb functions 41/1(440–442, 450–453)
WKB method and correspondence principle 41/1(91, 93, 96, 100, 104)
WKB-approximation 42(148)
Wolfenstein-Ashkin relations 41/1(484–485)
Wu's experiment with polarized nuclei 41/2(70)

Xenon 131, recoil investigations 41/2(23, 24)
Xenon 130, relative abundance 41/2(101)
Xenon bubble chamber 45(327, 334)
Xenon, scattering of slow neutrons 38/2(9)
Ξ^-- and Σ^0-particles, cascade decay 42(531)
X-ray counter 45(80)
X-ray fine structure 39(199)
— and isotope shift 38/1(105)
X-ray pulse of synchrotron, time distribution 44(190)
X-ray sources for photoneutron production 38/2(134–135, 145–146)
X-ray spectrum of synchrotron 44(190)
X-rays following electron capture 41/2(95, 97)
X-rays from muon-atoms 40(507, 526f.)
X-rays from μ-mesonic atoms 39(204, 227)
X-rays, isotope shift 39(201)
X-rays scattered by nuclei 40(510f.)
X-ray-sources

Yield function of quadrupole excitation 41/1(532)
Yield of neutrons from
— C(*d*, *n*) 38/2(207)
— heavy ice targets under deuteron bombardment 38/2(189)
— Li(*p*, *n*) 38/2(179–180)
— Sc(*p*, *n*) 38/2(184)
— T(*d*, *n*) 38/2(195–197)
— V(*p*, *n*) 38/2(184)
Yield of photoneutrons from
— beryllium 38/2(143–145)
— copper 38/2(149)
— deuterium 38/2(135, 139)
— rhodium 38/2(147)
— various sources 38/2(160)
Yoke of cyclotron magnet 44(136–137)
Young tableau 39(321)
Yttrium 90, β-decay and polarization 41/2(66–67)
Yttrium 91, β-decay and polarization 41/2(66–67)
Yttrium 91, Fermi plot 41/2(44)
Yukawa potential 39(31, 356)

Zeeman effect 38/1(395)
Zeeman energy levels 38/1(322)
Zeeman levels in a paramagnetic material 38/1(396, 398)
Zero cut-off, *see* Cut-off
Zero energy cross section, *see* Cross section
Zero energy scattering cross section 38/2(64)
Zero range 39(306)
Zero-degree neutron telescope 45(495–496, 512)
Zero-point vibrations 38/1(45)
Zinc sulfide as a scintillant 45(143)
Zinn type ion source 44(90)
Zirconium x-rays from tritium β-particle capture 41/2(4)
Zirconium-tritium target 38/2(177–178)

Group 9

Cosmic Rays

Absorption by pair production 46/2(276)
Absorption mean free path (absorption length), *see also* Attenuation mean free path 46/2(78, 212, 386, 387, 394, 404)
— for star production 46/2(394, 404)
— of neutrons 46/2(437, 444, 446, 449, 450, 463)
— of protons 46/2(463)
Absorption of high energy photons in galactic and intergalactic space 46/2(276)
Absorption of primary particles 46/1(149–150)
Abundance, relative
— odd-even charge 46/2(230, 247)
— of Be and B 46/2(231)
— of nuclei with $Z \geqq 30$ 46/2(229)
— of various charge components 46/2(228–237)
Abundances, isotopic, in meteorites 46/2(616–617, 621, 630)
Abundances of nuclei in stars 46/1(1, 5, 74)
Abundances of nuclei in the primary radiation 46/1(5, 74, 257)
Acceleration by magnetic fields 46/1(24–28)
Acceleration of cosmic ray particles 46/1(9, 24)
Acceleration processes 46/2(239, 252, 254–256, 339)
— critical energy 46/1(43)
— fast 46/2(254–256)
— slow 46/2(254–256)
Acceleration, statistical, *see also* Fermi acceleration 46/2(252, 255, 359)
Acceptance one 46/2(326)
Accretion of irradiated extraterrestrial material 46/2(552, 586, 587)
Achondrites 46/2(615)
Adiabatic change of magnetic field 46/1(21–22)
Adiabatic compression of a gas 46/1(27)
Adiabatic deceleration 46/2(366)
Adiabatic deceleration, invariants 46/2(315, 355, 359, 360–364, 366, 367)
Adiabatic invariance of motion 46/1(22, 23)
Aerosols 46/2(587, 593, 603)
Age determination, potassium-argon method 46/2(623)
Age determination, rubidium-strontium method 46/2(623)
Age determination, uranium-lead method 46/2(623)
Age theory 46/1(42, 46)
Air shower(s), *see* Extensive air showers
— cores, localisation 46/1(235–238)
— essential processes 46/1(216–218)
— high-energy nuclear interactions 46/1(211–212)
Albedo 46/1(49); 46/2(195, 555)
Albedo cones 46/1(126–128)
Albedo of atmosphere 46/1(138)
Albedo, reentrant 46/2(197)
Alfvén velocity 46/2(343)
Alfvén waves 46/1(29)
Allowed cone 46/1(88 ff., 116)
Allowed regions in the meridian plane 46/1(93)
α Herculis 46/1(57)
α particles, *see also* Helium nuclei 46/2(343)
α-particles, vertical intensity 46/1(150, 151)

Altitude dependence of
— neutron intensities 46/1(154)
— neutrons in the atmosphere 46/2(427, 435, 440, 448)
— nuclear disintegrations in the atmosphere 46/2(563–566)
— protons in the atmosphere 46/2(414, 418, 420)
— star formation 46/1(145)
— stars in the atmosphere 46/2(393–397)
— the hard component in the atmosphere 46/2(420)
— the ionization 46/1(135)
— the vertical intensity 46/1(131, 132)
Analytic continuation, method 46/2(29, 90)
Andromeda Nebula 46/1(75)
Angular distribution of cosmic rays in space 46/1(41)
Angular distribution of penetrating shower secondaries 46/1(188–191)
Angular spread of electron shower particles, *see also* Angular structure functions 46/2(22)
Angular structure functions of electrons and photons 46/2(43–44, 47–48, 100)
Anisotropies 46/2(340, 344, 346, 352, 357, 358)
Anisotropy of cosmic rays 46/1(4, 38–39)
Annihilation, matter-antimatter 46/2(266, 278, 284)
Anti-matter 46/1(81)
Approximation A in electron shower theory 46/2(11, 15, 23–27)
Approximation B in electron shower theory 46/2(11, 15, 28–33)
Ariel I results 46/2(337, 338, 363, 364, 365)
Aroos 46/2(616, 617, 618, 620, 624, 628)
Astronomical setting for cosmic rays 46/1(50–82)
Asymmetry in high energy interactions 46/2(130, 132)
Asymptotic cone of acceptance 46/2(326, 327)
Asymptotic directions 46/2(325, 326, 327, 333, 341, 345)
Asymptotic latitude 46/2(327)
Asymptotic longitude 46/2(328, 333)
Asymptotic orbits in the earth's magnetic field 46/1(99, 106ff., 113ff.)
Atmosphere, model 46/2(492)
Atmosphere, standard 46/2(376, 377, 491)
Atmosphere, top 46/2(373, 377, 483)
Atmospheric effects 46/2(326, 327)
Atmospheric effects threshold 46/2(326)
Atmospheric influence on air showers 46/1(230–235, 263–265)
Atmospheric latitude effect 46/1(291)
Attenuation length 46/2(78, 152, 212, 386, 387, 394, 404, 430, 431, 484)
Attenuation length, experimental determination 46/1(166–168)
Attenuation mean free path of penetrating shower primaries 46/1(163, 164, 167)
Aurora 46/1(82)
Auroral zone 46/2(320)
Azimuth dependence of the directional intensity 46/1(133)

Band width 46/1(37)
Barometer coefficient, *see* Pressure coefficient
Baryons, excited, *see also* Isobar model 46/2(164–165, 559)
Baryons, persisting 46/2(390)
Bernstein solution in electron shower theory 46/2(5, 34–35)
Beta decay of fission fragments 46/2(316)
Betatron acceleration 46/1(24, 25)
Betatron deceleration 46/2(359–360, 361–362)
Bethe formula for the stopping power 46/2(270)
Bethe-Heitler formula, corrections 46/2(4–10)
Biosphere 46/2(553, 588, 595, 596, 599, 601, 602, 607)
Bismuth fission chamber 46/2(461–462)
Black tracks in stars 46/1(139, 142)
Bombardment age 46/2(618, 627)
Born approximation, corrections for deviation 46/2(9–10)
Boson, intermediate 46/2(176)
Bremsstrahlung 46/1(13); 46/2(266, 271, 274, 284, 290, 292, 294)
Bremsstrahlung, electron-electron 46/2(274)

Bremsstrahlung electrons 46/1(192)
Bremsstrahlung in the atmosphere 46/1(137)
Bremsstrahlung in the intergalactic medium 46/2(284)
Bremsstrahlung, inner 46/2(266, 292)
Bremsstrahlung loss rate 46/2(266, 271, 290)
Bremsstrahlung models for nucleon-nucleon collisions 46/1(255, 256)
Bremsstrahlung production by non-thermal electrons 46/2(292)
Brownian motion 46/1(46)
Bruderheim 46/2(618, 622, 624, 629)
Budini-Molière method, modification 46/1(306–307)
Budini-Molière treatment of the hard component 46/1(286–288, 295, 299)
Bush's computing machine 46/1(110–112)

C^{14} dating 46/2(574–575, 607, 613)
C^{14} decay rate 46/2(501, 503, 510)
C^{14} production 46/2(498–517, 554, 562, 566, 571–573, 613, 617)
— absolute normalisation 46/2(498–503, 504–505)
— by solar protons 46/2(510–517, 580)
— during the solar cycle 46/2(509, 571–573)
Californium decay 46/1(68)
Calorimetric method, *see also* Ionisation calorimeter 46/2(69, 130, 131, 384)
Carbo 46/2(620, 623, 624, 627, 628)
Cascade showers, *see also* Electron showers, Nucleonic cascade 46/2(125, 126, 139, 144, 198)
Cascade showers 46/1(137)
Cascade theory 46/1(13, 219ff.)
Cassiopeia A 46/1(63, 69)
Castagnoli method 46/2(130, 132)
Catastrophic energy loss, *see* Sudden energy loss
Cauchy-Riemann equation 46/2(88)
C^{13}/C^{12} ratio 46/2(231)
C^{14}/C^{12} ratio 46/2(575)
Characteristic exponents of principal periodic orbits 46/1(103, 104, 109)
Charge composition of the primary radiation 46/2(195–210, 213–219, 223–224, 233–237)
Charge distribution in nuclei 46/1(174–175)
Charged to neutral shower secondaries, ratio of 46/1(191–192, 194, 205, 208)
Chemical evolution of stars 46/1(73)
Chondrites 46/2(615, 619, 623, 625, 630)
Chromosphere 46/1(50)
Circulation time 46/2(596)
Clark County 46/2(620, 624)
Cloud chamber 46/2(118, 120, 131, 136–137)
Cloud chamber evidence of explosive showers 46/1(158)
Cloud chamber evidence of mesons in penetrating showers 46/1(161)
Cloud scattering 46/1(48)
Clusters of galaxies 46/1(75)
Cocconi-Köster-Perkins model (CKP model) 46/2(387, 389)
Coefficient of mixing 46/2(596)
Coefficient to exchange 46/2(589)
Colliding galaxies 46/1(79, 271)
Collision lifetime 46/1(39)
Collision mean free path, *see also* Interaction mean free path 46/1(149); 46/2(151, 386, 387)
Collision number 46/1(44)
Collision of nucleon with a nucleus 46/1(171)
Collision volume 46/1(181, 182, 199)
Collisions of protons with nuclei 46/1(14)
Collisions with photons 46/1(15–16)
Composite-collision model 46/1(187, 200, 207)
Compton effect in the atmosphere 46/1(137)
Compton process (Compton scattering) 46/2(3, 16, 266, 272, 290)
Compton scattering 46/1(17)
Compton-Getting effect 46/1(270)
Compton-synchrotron radiation 46/2(291)
Conductivity, electrical, of the interplanetary plasma 46/2(344)
Continuity equation in energy 46/1(44, 47)
Continuity equation in space 46/1(41)
Convection 46/2(357–363, 364–365)
Cooling time 46/2(293, 294)
Core of air showers 46/1(235–238)

Core selectors 46/1(237)
Corona 46/1(50, 52); 46/2(300, 344, 350)
Cosmic dust 46/2(552, 587, 623)
Cosmic ray age 46/2(622–625)
Cosmic rays produced by the sun 46/1(51–53)
Cosmological theories, tests 46/2(284–285)
Coulomb multiple scattering 46/1(12–13)
Coulomb scattering in air showers 46/1(221–223)
Crab Nebula 46/1(49, 63–69); 46/2(289, 290, 291, 294–299, 304–306)
Crab Nebula, lunar occulation 46/2(295, 306)
Critical energy 46/2(10–11)
— for acceleration processes 46/1(43)
— of cascade theory 46/1(13, 219)
Critical frequency of emission 46/1(36)
Cross section for nuclear interactions 46/1(144)
Cross section for nucleon-nucleon collisions at highest energies 46/1(175, 176, 178)
Cross sections of protons at nuclei 46/1(14)
Cut-off energy, magnetic 46/1(122–127, 304, 311)
Cut-off energy of spectrum 46/1(6–7, 10)
Cyclotron frequency 46/1(18)
Cygnus A 46/2(302)
Cygnus A, radio source 46/1(79)

Daily intensity variation 46/2(326–327, 333, 350, 351, 354, 365)
Daudin effect 46/1(271)
Debye distance 46/1(12, 13)
Decay of π^0-mesons 46/2(266)
Deep Springs 46/2(620, 629)
Demography of neutrons 46/2(496, 507–508, 520)
Density of cosmic rays in space 46/1(10)
Density of matter in nuclei 46/1(174)
Density of muons in air showers 46/1(247)
Density of *N*-particles in air showers 46/1(243, 245)
Density spectrum of air showers 46/1(227–230)
Detectors for muons 46/1(246)
Detectors for *N*-particles 46/1(242)
Deuterium formation 46/2(266)
Deuterons, primary abundance 46/2(204)
Difference equations in electron shower theory 46/2(96, 100–108)
Diffusion coefficient 46/2(244, 347–348)
Diffusion equation of cosmic rays in space 46/1(41–45)
Diffusion equations for electron showers 46/2(11–18)
Diffusion equations for muons 46/1(281)
Diffusion extrapolation 46/2(209)
Diffusion mechanism of propagation 46/2(244–250)
Diffusion of a cascade through a nucleus 46/1(197)
Diffusion, regular 46/2(246–250)
Diffusion theory of cosmic rays 46/1(38–50)
Dipolar region on the sun 46/1(50)
Dipole approximation of the geomagnetic field 46/2(311–314)
Dipole field, magnetic 46/1(22)
Dipole field, magnetic, particle orbits 46/1(3, 90, 91)
Dipole model, centered 46/2(311, 313, 314, 537)
Dipole model, eccentric 46/2(311, 313, 538)
Dipole moment, solar 46/2(349)
Directional intensity 46/1(130, 133)
Discrete sources of high energy photons 46/2(285–302)
Doppler effect 46/1(15)
Doubly asymptotic orbits 46/1(114)
Drift velocity 46/1(20)
Dust grain hypothesis 46/1(257)
Dwarfs of solar type 46/1(59)

Earth, magnetic field, *see also* Geomagnetic field 46/1(3, 22); 46/2(311–313, 325, 378, 537–538)
Earth magnetic field, effect on muon component 46/1(292)
Ehole 46/2(618, 622, 623, 629)
Elastic scattering in the atmosphere 46/1(136)

Electro-magnetic cascade theory 46/1(219ff.)
Electron pairs from π^0 decay 46/1(192)
Electron production in the intergalactic medium 46/2(273–275)
Electron production spectrum 46/2(270)
Electron scattering experiments 46/1(174–175)
Electron shower theory
— Approximation A 46/2(11, 15, 23–27)
— Approximation B 46/2(11, 15, 28–33)
— diffusion equations 46/2(11–18)
— elementary solutions 46/2(19)
— one-dimensional 46/2(2, 14–15)
— three-dimensional 46/2(2, 15, 38–49)
Electron showers, energy measurements 46/2(69–74, 112–113)
Electrons, atmospheric secondaries 46/2(74–75, 220)
Electrons galactic primary 46/2(219–223, 267, 283)
Electrons, galactic secondary 46/2(220–223, 274–275)
Electrons in penetrating showers 46/1(165, 191)
Electrons in the atmosphere 46/1(154, 155)
Element abundances in cosmic rays 46/1(5, 74, 257)
Eleven-year cycle 46/2(185, 332, 374, 577–579, 614, 630)
Eleven-year solar cycle 46/1(3, 53)
Emission-line stars 46/1(58)
Emulsion chamber 46/2(70, 72–73, 113, 118, 125, 126, 138)
Emulsion spectrograph 46/1(274, 281)
Emulsion stack 46/2(119, 121, 132)
Energy corrections due to excentric dipole field 46/1(127)
Energy cut-off, geomagnetic 46/1(122–127, 304, 311)
Energy density of cosmic rays 46/1(8); 46/2(240)
Energy dependence of lateral distribution in air showers 46/1(223–224)
Energy distribution of the secondaries in air showers 46/1(217)
Energy drift 46/1(46)
Energy limit for the penumbra 46/1(120)
Energy loss
— by ionization 46/1(11–12)
— by nuclear collisions 46/1(42)
— due to expansion 46/2(274)
— of an air shower by ionization 46/1(235)
— of electrons in photon flux 46/1(16–17)
— of mesons 46/1(282)
— rate of electrons in the galaxy 46/2(270, 273–275)
— rate of muons 46/2(148)
Energy of particles producing pions 46/1(183–187)
Energy spectrum
— of cosmic rays 46/1(6–8, 40)
— of electrons and photons in an air shower 46/1(220, 221)
— of Fermi theory 46/1(182)
— of Heisenberg theory 46/1(182)
— of muons 46/2(147, 149)
— of neutrinos 46/2(174–175)
— of neutrons in the atmosphere 46/2(417, 447)
— of *N*-particles in air showers 46/1(244)
— of primaries 46/2(143, 153, 195–227, 378–380)
— of protons in the atmosphere 46/2(200–204, 382–391, 417)
— of shower electrons and photons 46/2(22, 24, 33, 34, 142–143, 149)
— of the muon component in air showers 46/1(250, 251)
— of the primary radiation 46/1(260–263)
— peak in the differential 46/2(238)
— values of the exponent 46/2(384)
Energy transfer to charged secondaries 46/2(130, 134)
Energy transfer to γ-rays 46/2(130, 134)
Energy unit of Störmer 46/1(91)
e-neutrinos 46/2(174–176, 307)
Equation of continuity, *see* Continuity equation
Equator, cosmic ray 46/2(314, 538)
Equatorial ring current 46/2(320, 321, 322)
Equipartition 46/1(51, 77)

Equivalent band width of radiated energy 46/1(37)
Erosion hypothesis 46/2(623, 627)
Escape of particles from the source region 46/2(254)
Evaporation neutrons, source spectrum 46/2(427, 430)
Evolution of stars 46/1(73)
Excentric dipole, energy corrections 46/1(127)
Excess of positive muons, *see also* Positive excess 46/1(294)
Expanding cloud model of pion production 46/1(183)
Experimental methods to determine attenuation lengths 46/1(166–168)
Explorer X results 46/2(343)
Explosive showers 46/1(157, 158)
Exponent of pion production spectrum 46/1(289–290)
Extensive air showers 46/2(75–84, 139, 144, 150, 224–227, 281)
— lateral structure function 46/2(77–82)
— size spectrum 46/2(224–225)
Extensive shower 46/1(159)
External galaxies 46/1(75–81)
Extragalactic discrete sources of high energy photons 46/2(301–302)
Extragalactic origin of cosmic rays 46/1(269, 270, 271)

Fall-out 46/2(589–595, 606)
Families of electron showers 46/2(76, 126, 140, 142–147, 165)
Faraday rotation of polarization plane 46/1(37)
Fermi acceleration process, *see also* Statistical acceleration 46/2(359)
Fermi deceleration process 46/2(361–362, 366)
Fermi effect 46/1(11)
Fermi relation of energy loss 46/1(40, 43, 47)
Fermi collision theory 46/1(181–182, 199, 212, 254)
— comparison with experiment 46/1(189, 190)
Fermi thermodynamical model of multiple production 46/2(155, 168)
Fireballs 46/2(157–168, 169, 560)
Flare particles 46/2(325)
Flare stars 46/1(55)
Fluctuations in energy 46/1(45)
Flügge-Yuan density maximum of slow neutrons 46/2(374, 426, 427, 438, 439, 483)
Flux of latitude sensitive primaries 46/2(378)
Flux of primaries, integral 46/2(380)
Fokker-Planck approximation 46/2(12, 17, 91)
Fokker-Planck equation 46/1(46)
Forbidden regions in the meridian plane 46/1(93)
Forbush effect (Forbush decrease) 46/2(310, 320, 323, 330, 331, 335, 338–340, 341, 346, 347, 350–352, 354, 357, 362, 368, 576)
Force-free field 46/1(49, 66)
Fourier transformation 46/2(39, 85)
Four-momentum transfer 46/2(158, 160, 162, 163, 169)
F-plot of angular distribution 46/1(186, 188, 209)
Fractionation process 46/2(599)
Fragmentation 46/1(138, 146–148)
Fragmentation of heavier nuclei 46/2(209, 247–249)
Fragmentation parameters 46/2(210–212)
Free path, *see* Mean free path
Funnel model of nucleon-nucleus interaction 46/1(196, 200–202, 207)

Galactic center 46/2(299)
Galactic gas density 46/2(270)
Galactic magnetic fields 46/2(240)
Galactic medium 46/1(11)
Galactic nucleus 46/1(70)
Galactic origin of cosmic rays 46/1(269, 270, 271)
Galactic rotation 46/1(8, 70)
Galaxies in collision 46/1(79)
Galaxy 46/1(69–73)
Galaxy rotation, energy density 46/1(8)
γ-ray spectrum 46/2(140, 142–143)
γ-rays, *see* Photons, Photon component
Gaunt approximation to the bremsstrahlung cross section 46/2(292)
Gauss coefficient 46/2(313)
Geomagnetic cavity 46/2(320–321, 324, 343)

Geomagnetic effects 46/2(311, 330)
Geomagnetic energy cut-off 46/1(122–127, 304, 311)
Geomagnetic equator 46/2(313)
Geomagnetic field, *see also* Earth magnetic field
Geomagnetic field, influence on air showers 46/1(224)
Geometrical cross section for shower
— production 46/1(168, 175)
— secondaries 46/1(179)
Geosphere 46/2(553, 587–607)
Global inventory of radioisotopes 46/2(602–607)
Global map of nuclear disintegrations 46/2(564–566)
Global production rate of radioisotopes 46/2(572–573, 581–583)
Globular clusters 46/1(69)
Gould belt, *see* Local system
Grant 46/2(620, 623, 624, 625)
Gray tracks in stars 46/1(139)
Green function method 46/2(35, 97)
Greisen approximation of the lateral structure function 46/2(57–59, 150)
Growth curves of primary components 46/2(247)

Hankel transformation 46/2(46, 85, 87)
Hard cloud 46/1(48)
Hard component 46/1(272)
Hard radiation emitted through radioactivity 46/2(288)
Hardening of spectrum with increasing zenith angle 46/1(303)
Harleton 46/2(618, 619, 622, 629)
Hasegawa theory of multiple-fireball emission 46/2(165–168)
Heavy nuclei in the primary radiation 46/1(258, 259)
Heisenberg's explosion theory 46/1(182, 212)
Helical motion in a magnetic field 46/1(18)
Helium nuclei
— energy spectrum 46/2(205–207)
— isotopic composition 46/2(208)
— primary 46/2(204–208)
— rigidity spectrum 46/2(207)
He/M ratio 46/2(212)
He/S ratio 46/2(232, 234–235)
High energy stars 46/1(145)
High-energy cut-off 46/1(7, 10)
H/M ratio 46/2(212, 218–219, 227, 232)
hot universe cosmological model 46/2(284)
H-quantum 46/2(165–168)
Hubble constant 46/1(11)
Hubble radius 46/2(273, 277, 283)
Hydromagnetic waves 46/1(28–30)
Hydromagnetics 46/1(2)
Hydrosphere 46/2(553, 595–599, 602, 606–607)
Hyperon/nucleon ratio 46/2(391)
Hyperons 46/1(195)
Hysteresis 46/2(185)

IGY 46/2(334, 338, 339)
Image dipole 46/2(324)
Imp I results 46/2(341–342)
Impact zones 46/2(325, 344–346, 356, 368)
Incoming beam, element composition 46/1(6)
Inelastic nucleon-nucleon interaction 46/1(164)
Inelasticity in nucleon interactions 46/2(132, 135, 151, 159, 162, 167, 385, 387, 388)
Inelasticity of high-energy collisions 46/1(208–211)
Inhomogeneities, magnetic 46/2(342)
Injection of particles becoming cosmic rays 46/1(9)
Inner periodic orbits in the earth's magnetic field 46/1(99)
Instrumental bias 46/1(281)
Integral representation method 46/2(29, 90)
Integral spectrum 46/1(7)
Intensity, directional 46/1(130, 133)
Intensity of air showers as a function of air pressure 46/1(230–232, 263)
Intensity, unidirectional 46/2(181)
Intensity, vertical, alteration with altitude, latitude and longitude 46/1(131, 132)
Interaction mean free path, *see also* Collision mean free path 46/2(123, 211)
— of penetrating shower primaries 46/1(163, 169)
— of penetrating shower primaries in hydrogen 46/1(176–178)

Interaction mean free path (cont.)
— of penetrating shower secondaries 46/1(178–180)
Intergalactic medium, electron production and energy losses 46/2(273–275)
Internuclear cascade 46/1(181, 197)
Interplanetary fields 46/2(310, 318, 324, 325, 330, 341, 342, 345, 347, 353, 357, 362, 368)
Interplanetary medium as x-ray source 46/2(300)
Interplanetary plasma 46/2(310, 313, 325, 330)
Interstellar dust, *see also* Cosmic dust 46/2(552, 587)
Interstellar hydrogen, density 46/2(243, 270)
Ionisation calorimeter 46/2(69, 384)
Ionisation chamber 46/2(188, 195, 318, 333, 340, 384)
Ionisation loss processes 46/2(3, 13, 270)
Ionisation regression curve 46/2(186–187)
Ionization generating cosmic rays 46/1(11)
Ionization of the atmosphere 46/1(135, 136, 154)
Ionized state of galactic medium 46/1(11)
Ionizing and not ionizing *N*-particles, ratio 46/1(244)
Iron meteorites 46/2(615, 618, 619–621, 623, 625, 627, 629, 630)
Isobar model 46/2(389)
Isotope production in the atmosphere 46/2(553, 561–585)
Isotope production, time variations 46/2(573–580)
Isotropy of cosmic radiation 46/1(37, 41)
Isotropy of secondaries 46/1(182, 190)
Isotropy of the primary radiation 46/1(266)

Jaws to trap, *see* Trapping mechanism
Jet(s) 46/1(185, 191, 197, 206, 210); 46/2(118, 129)
Jet energies, degree of inelasticity 46/1(209)

Kalos and Blatt method 46/2(104–105)
Kepler's 1604 supernova 46/2(272)
Kepler's nova 46/1(63)
K-figures 46/2(333, 343)
K-fluorescence yield 46/2(292)
Klein-Nishina 46/2(272)
K-meson decay 46/1(291)
K-meson effects on zenith angle results 46/1(307)
K-meson in penetrating showers 46/1(196)
K-meson production 46/1(290, 291, 309)
K-mesons, effect on particles at large zenith angles 46/1(309–311)
K-mesons, effect on sea-level spectrum 46/1(291, 301)
K/π ratio 46/2(121, 150, 157, 169)

λ-meson hypothesis 46/1(162)
Landau approximation, theories in 46/2(12, 14, 18, 44)
Landau approximation, theories without 46/2(12, 15, 18, 44–49, 61, 64)
Landau effect 46/2(6–8, 18, 44)
Landau-Belenkii-Amai model 46/1(254)
Landau's hydrodynamical model of multiple production 46/2(156)
Landau's semiclassical collision theory 46/1(182, 212, 254)
Laplace transformation 46/2(85, 86, 99, 101–104)
Large zenith angles, theoretical predictions 46/1(307)
Larmor frequency 46/2(272)
Larmor radius 46/1(18, 22, 48); 46/2(271, 352, 355–356, 361)
Lateral distribution of air showers 46/1(221–225, 238–240, 245)
Lateral spread of shower particles 46/2(17, 22)
Lateral structure function
— at extremely high energies 46/2(75–84)
— Greisen approximation 46/2(57–59)
— of electrons and photons 46/2(43–44, 48–49, 62, 63, 73)
— of the energy flow of shower particles 46/2(83–84)
Latitude dependence
— of neutron intensities 46/1(154)
— of star formation 46/1(145)

— of the ionization 46/1(154)
— of the vertical intensity 46/1(131, 132)
Latitude effect 46/1(3); 46/2(196, 199–200, 311, 313, 375, 427, 473, 562)
— in inclined directions 46/1(311)
— of fast neutrons 46/2(442, 451)
— of neutrons during sunspot minimum and maximum 46/2(438, 442, 451)
— of neutrons, total 46/2(444, 452, 477, 520)
— of slow neutrons 46/2(436)
— of the ionisation in the atmosphere during a solar cycle 46/2(476)
— of the primary cosmic radiation 46/2(473)
— on sea level spectrum 46/1(291)
Leakage lifetime 46/1(39, 42)
Leakage out of the galactic halo 46/2(273)
Lifetime of cosmic ray particles 46/2(245)
Light curves of novae 46/1(60)
Limiting energy for acceleration in a magnetic field 46/1(31)
Liouville's theorem 46/2(184, 312, 354, 360)
Lithosphere 46/2(553, 585–587, 601)
L-nuclei, differences in the spectra 46/2(236)
L-nuclei, differential spectrum 46/2(214)
L-nuclei, integral intensities 46/2(213)
L-nuclei solar modulation of the spectrum 46/2(215)
Local group of galaxies 46/1(75, 76)
Local shower 46/1(159)
Local system within galaxy 46/1(59)
Localisation of air shower cores 46/1(235–238)
log tan ϑ plots 46/2(130, 157, 158, 160, 164, 165)
Longitude dependence of the vertical intensity 46/1(131, 132)
Longitude effect 46/1(122, 128)
Longitudinal adiabatic invariant 46/1(28)
Longitudinal development of air showers 46/1(219–221)
Lorentz force 46/1(20)
Loss of cosmic rays 46/1(9)
Low-energy cut-off, *see also* Cut-off energy 46/1(6)
Lower atmosphere, behavior of air showers 46/1(230–232, 235)
L/S ratio 46/2(212, 213, 227, 232, 235, 248)

Magellanic clouds 46/1(75)
Magnetic bremsstrahlung 46/1(31, 44)
Magnetic centre of the earth 46/1(122)
Magnetic cut-off energy 46/1(122–127, 304, 311)
Magnetic diffusion 46/1(48)
Magnetic dipole field of the earth 46/1(3, 90, 91)
Magnetic field of the sun 46/1(54–55)
Magnetic field, particle paths 46/1(17ff.)
Magnetic fields in galaxy 46/1(72–73)
Magnetic kink 46/2(352, 356, 362)
Magnetic mirror 46/1(20, 26)
Magnetic moment 46/1(20)
Magnetic scattering centres 46/2(350, 363)
Magnetic spectrometer 46/2(412)
Magnetic storms 46/2(317, 318–320, 321, 323, 330, 364)
Magnetic tongue 46/2(350, 351–352, 354, 356)
Magnetic variable stars 46/1(59)
Magnetosphere 46/2(323, 324)
Main cone 46/1(88, 105, 117); 46/2(312)
Main phase 46/2(323, 324)
Mariner II results 46/2(341–342, 344)
Mars I results 46/2(340)
Mass ejection by stars 46/1(58–58)
Mass spectrometer 46/2(619)
Mass spectrum of the primary radiation 46/1(259)
Maximum detectable momentum 46/1(273, 274)
Maxwell distribution 46/1(182)
McIlwain parameter 46/2(315)
m.d.m., *see* Maximum detectable momentum
Mean free path, *see also* Interaction, Transport mean free path 46/1(39, 41)
— for nuclear collisions 46/1(149)
— in magnetic diffusion 46/1(48)
— of a nulceon in nuclear matter 46/1(172, 200, 204)

Mean free path (cont.)
— of air shower primaries 46/1(258)
— of nucleons in hydrogen 46/1(176–178)
— of penetrating shower primaries 46/1(160, 163, 164, 167, 169)
Mean lifetime 46/1(39)
Mean-square angle in multiple scattering 46/1(13)
Mellin transformation 46/2(23, 85, 86)
Meson telescopes 46/2(325, 329, 330, 332)
Meson-nucleon cascade in air showers 46/1(241 ff.)
Mesons, *see* *K*-mesons, Muons, Pions, Slow mesons
Messier 31, Population 46/1(75)
Messier 87, radio emission 46/1(77)
Meteorites 46/2(552, 573–574, 587, 613–631)
Meteorites, radiation ages 46/2(553, 618, 622–625)
Meteorites, radionuclide production rate 46/2(625–630)
Meteoroids 46/2(552)
Meteorological effects 46/2(330)
Meteorological effects on muon component 46/1(312–315)
Minimum grain density 46/2(119, 372)
Mirror, magnetic 46/1(20, 26)
Mirror point(s) 46/1(22); 46/2(315)
Missed interaction 46/1(170)
Model experiments on geomagnetic effects 46/2(325)
Modulation, aperiodic 46/2(431)
Modulation aperiodic, periodic 46/2(431)
Modulation function 46/2(337–338, 362)
Modulation mechanism 46/2(330–368)
Modulation of nucleons in the atmosphere 46/2(374, 375)
Modulation of the helium componente 46/2(191–192)
Modulation of the *L*-nuclei 46/2(215)
Modulation of the proton component 46/2(189–190, 328)
Modulation of the *S*-nuclei 46/2(192)
Modulation, solar 46/2(186, 189–195, 374, 375)
Modulation, solar, of neutrons in the atmosphere 46/2(424–429, 471, 478–508, 567)
Molecular exchange, removal of the radioactivity 46/2(594–595, 598)
Molière scattering theory 46/2(44, 68, 93–96)
Molière unit 46/2(58)
Momentum spectrum
— of deuterons 46/2(400, 401)
— of muons at sea level 46/1(274–280)
— of protons 46/2(400, 463, 555, 556)
— of secondary particles 46/2(400, 401)
Mott formula of the scattering probability 46/2(92)
Moving magnetic mirror 46/1(26)
Multiple process 46/1(181)
Multiple production, Fermi's thermodynamical model 46/2(155, 268–270, 385)
Multiple production, Landau's hydrodynamical model 46/2(156)
Multiple production of secondaries 46/2(115–118, 153–169)
Multiple scattering 46/1(12–13); 46/2(91–96, 119)
Multiple-fireball emission 46/2(165–168)
Multiplicity, *see* Pion multiplicity
Multiplicity and average energy 46/1(177)
Multiplicity, dependence on primary energy 46/1(182)
Multiplicity distribution 46/1(206–207)
Multiplicity, energy dependence 46/1(189)
Multiplicity of intranuclear cascade 46/1(198–199, 207)
Multiplicity of penetrating shower secondaries 46/1(188–191)
Multiplicity of secondary production 46/2(135, 151, 167, 169, 269, 385)
Multiply charged components 46/1(151)
Mu-mesons, *see* Muons
Muon component of air showers above ground 46/1(246–248)
Muon component of air showers underground 46/1(248–252)
Muon decay 46/1(282, 288)
Muon momentum spectrum 46/1(274–279)
— at high energies 46/1(279–281, 289)
— at large zenith angles 46/1(307)

Muon production spectrum 46/1(284–285)
Muon survival probability 46/1(283, 288, 306)
Muons in inclined directions 46/1(301)
Muons in the atmosphere 46/1(153)
Muons, low energy spectrum 46/1(293)
Muons, meteorological effects 46/1(312–315)
Muons (μ-mesons)
— energy loss rate 46/2(148)
— energy spectrum 46/2(147, 149)
— intensity-depth relation 46/2(148)
— interactions 46/2(135, 160, 585)
— positive excess 46/2(391)
— range fluctuations 46/2(108–112)
Muons, positive excess, *see also* Positive excess 46/1(294)
Muons, residual range spectrum 46/1(283)
Muons, vertical momentum spectrum 46/1(281–291)
μ-component, *see* Muon component
μ-neutrinos 46/2(119, 174–176)

Narrow muon showers 46/1(251, 252)
N-component 46/1(166, 208, 216, 217, 233, 234, 239, 242–246, 255)
N-component of air showers 46/1(216, 217, 233, 234, 239, 242–246, 255)
Near-horizontal particles 46/1(310)
Neon hodoscope 46/2(80, 151)
Neutral pion 46/1(143, 191, 192)
Neutral to charged shower secondaries, ratio of 46/1(191–192, 194, 205, 208)
Neutrino astronomy 46/2(266)
Neutrino experiments, cosmic ray 46/2(173–177)
Neutrino pair emission 46/2(303)
Neutrino production spectrum 46/2(304)
Neutrino sources 46/2(302–307)
Neutrino telescope 46/2(305)
Neutrinos, energy spectrum 46/2(174, 175)
Neutrinos, solar, radioisotope production 46/2(586)
Neutron albedo, *see also* Neutron leakage flux 46/2(527)
Neutron capture 46/2(498–517)
Neutron density, Flügge-Yuan maximum 46/2(374, 426, 427, 438, 439, 483)
Neutron flux
— equilibrium spectrum 46/2(465)
— in the atmosphere 46/2(374, 434–453, 464–535)
— Simpson maximum 46/2(374, 450, 453)
— S_n-method calculation 46/2(497)
— unexplained fluctuations 46/2(443)
Neutron leakage
— angular distribution 46/2(521–527, 533)
— aperiodically emitted 46/2(532–535)
— during the solar cycle 46/2(520)
— flux 46/2(427, 428, 452, 453, 508, 518–535)
— flux, experimental values 46/2(527–532)
— latitude effect 46/2(524, 531, 533)
— spectrum 46/2(519)
Neutron monitor 46/2(185, 313, 315, 325, 329, 332–336, 338–340, 345, 462–463)
Neutron production
— by solar protons 46/2(510–517)
— by stars 46/1(144)
— in the atmosphere 46/2(428, 472, 494)
— rate 46/2(472, 474–478)
Neutron reaction cross section in air 46/2(499)
Neutron source distribution 46/2(489)
Neutron source, normalisation 46/2(431, 496)
Neutron source spectrum 46/2(428, 429, 472)
Neutron stars 46/2(287, 288–290)
Neutron transport equation 46/2(479–489, 495–498)
Neutrons, age theory 46/2(425, 482)
Neutrons, angular distribution 46/2(463–464)
Neutrons, decay density in the interplanetary space 46/2(526)
Neutrons, demography 46/2(498, 507–508, 520)
Neutrons, diffusion theory 46/2(425, 480–489, 493–495)
Neutrons, energy spectrum 46/2(434, 447, 448, 464–472, 567)

Neutrons, fast 46/2(432, 441–453, 463–464, 521–531)
Neutrons in the atmosphere 46/1(153, 154)
— absolute number 46/2(502)
— absorption mean free path 46/2(444, 446, 449, 450)
— density 46/2(439, 440, 485)
— latitude dependence 46/2(438, 451)
— modulation 46/2(424–429, 508)
— origin 46/2(424–429)
— source spectrum 46/2(432)
Neutrons, primary solar 46/2(426, 583–584)
Neutrons, slow 46/2(434–439, 452, 466–468, 483, 524–526, 527–529)
Neutrons, slowing down length and slowing down time 46/2(481)
Neutrons, transition effect in water 46/2(441)
Neutrons, velocity distribution 46/2(425)
Nishimura-Kamata function 46/2(56–63, 150)
Niu's theory of two-fireball emission 46/2(161–165)
Non-dipole terms of the geomagnetic field 46/2(313, 325)
Novae 46/1(57, 59–61)
Nuclear cascade (nucleon cascade) 46/2(151, 372, 383–391, 428, 559, 564)
Nuclear interactions in light elements 46/1(170–172)
Nuclear interactions, neutral-charged ratio 46/2(121, 122, 457)
Nuclear reactions in the atmosphere 46/1(137, 144)
Nuclear transformations, cosmic ray induced 46/2(551–557, 571, 614–615, 625–630)
Nuclear-active (nucleoactive) component 46/2(136, 138, 379, 385)
Nucleo-active component 46/1(166, 208, 216, 217, 233, 234, 239, 242–246, 255)
Nucleon collision with a nucleus 46/1(171)
Nucleon pairs 46/1(195)
Nucleonic cascade 46/1(164)
Nucleon-nucleon cross section at highest energies 46/1(175–176, 178)
Nucleon-nucleus collisions at very high energies 46/1(196–202)
Nucleon-pion cascade 46/1(286)
Nucleons, elastic scattering 46/2(171–173)
Nucleons in the atmosphere, geomagnetic effects 46/2(374, 379, 422, 431, 508)
Nucleons in the atmosphere, inelastic scattering, altitude dependence 46/2(427)
Nucleons, modulation in the atmosphere 46/2(372–377, 385, 422, 431, 508, 557–559)
Number spectrum of air showers 46/1(229, 230, 240, 241)

Odd-even abundance ratio 46/2(230, 247)
Odessa 46/2(619, 620, 625)
Ohmic dissipation time of interplanetary plasma 46/2(344)
Olbers' paradox 46/1(81)
Omnidirectional intensity 46/1(134)
One-dimensional approximation 46/1(282)
One-fireball emission 46/2(158–161)
Opaque nucleus 46/1(171)
Operator calculus 46/2(97)
Optical emission of electrons 46/1(36)
Outer periodic orbits in the earth's magnetic field 46/1(99)
Overabundance of *L*-nuclei 46/2(230)

P Cygni 46/1(57)
Pair creation, *see also* Pair production 46/2(3, 6–9, 16, 276)
— effect of atomic electrons 46/2(8–9)
— in photon-photon collisions 46/2(276)
— Landau effect 46/2(6–8)
— Migdal's treatment 46/2(7)
— of nucleons 46/2(558)
Pair production in the atmosphere 46/1(137)
Pallasites 46/2(616, 619, 620)
Paths of particles in magnetic fields 46/1(17ff.)
Penetrating shower(s)
— definition 46/1(203)
— discovery 46/1(157)
— primaries 46/1(160)

— research, results 46/1(212)
— secondaries, interaction mean free path 46/1(178–180)
Penumbra 46/2(312, 315, 318)
Penumbra orbits 46/1(88, 120, 121, 123, 124, 125)
Periodic orbits in the earth's magnetic field 46/1(97, 99, 103)
Persisting baryons 46/2(290)
Pfotzer maximum 46/1(134, 137); 46/2(117, 373)
Photo-emission of nucleons 46/1(16)
Photon astronomy, high energy 46/2(267)
Photon collisions 46/1(15–16)
Photon producing processes 46/2(266, 267–285)
Photon production spectrum 46/2(277)
Photon spectrum, high energy 46/2(279)
Photons, high energy, in the atmosphere 46/2(74–75)
Photons, high energy primary 46/2(265)
Photons in air showers 46/1(216, 218, 221)
Photons in the atmosphere 46/1(154, 155)
Photopion production 46/2(278)
Photosphere 46/1(50)
Piccioni's experiment to distinguish pions and muons 46/1(161–162)
π-meson, *see* Pion
π^0-meson, *see* Pion, neutral
Pinkau's method for determination of shower energies 46/2(112)
Pion distribution 46/2(388)
Pion multiplicity distribution 46/1(289, 294, 297, 298, 301)
Pion multiplicity in the isobar model 46/2(390, 559)
Pion, neutral 46/1(191, 192)
Pion production 46/2(121, 153–168, 268–270, 385, 557–561)
— determination of primary energy 46/1(183–187)
— in nucleon-nucleon collisions 46/1(180–183)
— spectrum 46/1(282, 286–287, 299–301)
— spectrum, exponent 46/1(289–290)
Pioneer V results 46/2(340, 341–342, 356)
Pionisation 46/2(385)
Pion/neutron ratio 46/2(463)
Pion/proton ratio 46/2(463)
Pions in air showers 46/1(217, 218, 253)
Pions in penetrating showers 46/1(161–162)
Pions in the atmosphere 46/1(151)
Pions, neutral, in showers 46/1(169)
Pions, neutral, production spectrum 46/2(198, 269, 297)
Pitch angle of helix 46/1(18, 20)
Planets as x-ray sources 46/2(300)
Plasma frequencies 46/2(274)
Plasma interaction 46/1(35)
Plasma, interplanetary 46/2(310, 313, 349)
Plasma oscillation(s) 46/1(12); 46/2(274)
Plasma, solar 46/2(313, 321, 323, 332, 341–342, 349)
Plateau grain density 46/2(119, 392)
Plural process 46/1(181, 196)
Plural production of secondaries 46/2(116)
Plural scattering effects 46/2(92)
Point sources of cosmic rays 46/1(5)
Polar cap 46/2(320)
Polar ice cap, radioisotopes 46/2(576, 587, 601)
Polarization of light from Crab Nebula 46/1(66)
Polarization of light from stars 46/1(72)
Polarized radiation 46/1(37–38)
Population I 46/1(58, 59, 70)
Population II 46/1(70)
Population types 46/1(71)
Positive excess
— in cosmic rays 46/1(294)
— in inclined directions 46/1(312)
— theory 46/1(295–296)
— variation with momentum 46/1(294–295)
Positron annihilation cross section 46/2(278)
Positron/electron ratio (P/e ratio) 46/2(222, 238, 242)
Power radiated in a magnetic field 46/1(33–34)

Power-law spectrum of energies 46/1(6–8, 40, 47, 83)
Precipitation 46/2(587, 591–594, 606)
Pre-existing field 46/1(180, 181)
Pressure coefficient of muon component 46/1(314)
Pressure coefficient of penetrating showers 46/1(160)
Primaries of penetrating showers 46/1(160)
— attenuation 46/1(163, 164, 167)
Primary component 46/2(311, 337, 377–383)
Primary energy in pion production 46/1(183–187)
Primary nuclei, total flux 46/2(374)
Primary of an air shower 46/1(215, 216, 217)
Primary radiation, changes in 46/1(265–271)
Primary radiation, nature of 46/1(256–260)
Primary radiation of air showers 46/1(256ff.)
Primary spectrum, *see* Energy spectrum Rigidity spectrum 46/2(143, 153, 195–227, 337, 380–383, 387)
Primary spectrum of cosmic rays 46/1(286, 287, 289)
Principle periodic orbits in the earth's magnetic field 46/1(99, 103)
Production spectrum, inhomogeneous 46/2(389)
Production spectrum of neutrons and protons (Rossi modell) 46/2(416)
Project Argus 46/2(316)
Propagation of cosmic rays 46/2(239–251)
Proton intensity in the atmosphere 46/1(152)
Proton showers 46/1(253, 260)
Proton spectrum, sunspot minimum 46/2(203)
Proton/helium ratio (P/He ratio) 46/2(197, 232, 234–235)
Protons and nuclei, inelastic cross-sections 46/1(14)
Protons, angular distribution 46/2(422–424)
Protons, differential energy spectrum 46/2(200–204, 413, 414)
Protons, geomagnetic energy cut-off 46/1(125, 126)
Protons in the atmosphere, energy spectrum 46/2(382–391, 410–422)
Protons in the atmosphere, experimental methods 46/2(200–204, 383, 411–413, 463)
Protons, percentage in the total penetrating component 46/2(419)
Protons, solar 46/2(315, 317, 318, 320, 325, 365, 512, 556)
Protons, total flux 46/2(374)
Protons, vertical intensity 46/2(418–420)

Queen orbit 46/1(114)
Quiescent cosmic radiation 46/2(377)

Radiation damping 46/1(181)
Radiation length 46/1(13, 44, 219); 46/2(10–11)
Radiation loss of energy 46/1(32–36, 44)
Radiation of a relativistic particle in a magnetic field 46/1(32–36)
Radiation of particles in a magnetic field 46/1(31)
Radiation process 46/2(3, 6–9)
— effect of atomic electrons 46/2(8–9)
— Landau effect 46/2(6–8)
— Migdal's treatment 46/2(7)
Radio brightness temperature 46/2(282)
Radio emission 46/2(241–242, 282, 292)
— meter waves 46/1(72)
— of electrons 46/1(36)
— of the sun 46/1(50)
Radio flux 46/1(36)
Radio outbursts of type IV 46/1(52)
Radio polarization 46/1(38)
Radio sources 46/1(63, 79)
Radioactivity, cosmic ray induced 46/2(553, 561–562, 602–607, 613–614, 620–622, 630)
Radius of the galactic halo 46/2(277)
Radius of universe 46/1(11)
Range fluctuations of high energy muons 46/2(108–112)
Red shift of distant galaxies 46/1(76)
Reentrant albedo 46/2(197)
Reentrant orbits 46/1(120, 121)

Reflecting boundary, magnetic 46/1(49)
Registration stations, cosmic ray 46/2(535–537)
Regression curve of atmospheric ionization 46/2(186–187)
Relativistic particle in a magnetic field 46/1(32)
Residence time 46/2(598)
Residual range spectrum of muons 46/1(283)
Response functions, differential 46/2(337, 339)
Rigidity, cut off 46/2(183)
Rigidity, magnetic 46/2(182, 311, 312, 318, 555–556)
Rigidity spectrum 46/2(183, 374, 556–557)
Rigidity, threshold 46/2(311, 313, 318, 321–325)
Ring current, equatorial 46/2(320–322, 324)
Riometer 46/2(317)
Rossi curve 46/1(159)
Rotation of galaxy, energy density 46/1(8)

Saddle point method 46/2(31, 88)
Scattering, method of relative 46/2(120, 125)
Scattering, multiple 46/2(119)
Scattering, plural 46/2(119)
Scattering, single 46/2(119)
Scattering, spurious 46/2(120)
Scattering technique 46/1(274)
Sco XR-1 46/2(285, 301, 306)
Screening cross section
— complete 46/2(4, 13)
— incomplete 46/2(5)
Screening effect 46/2(4–5, 13)
Sea level spectrum of cosmic rays, *see* Spectrum at sea level
Secondaries in air showers, produced by the first collision 46/1(217)
Secular equilibrium 46/2(616)
Sedimentation of radioisotopes 46/2(575, 587, 601, 602, 607)
Sgr XR-1 46/2(286)
Shadow cone 46/1(88, 117–119)
Shell stars 46/1(58)
Shock front 46/1(29)
Shock wave 46/2(352–355, 357)
Shower, *see* Air shower, Penetrating shower, Cascade shower, Electron shower
Shower age 46/2(21)
Shower axis 46/2(60–69)
Shower diagrams 46/2(398, 399)
Shower frequency at great altitudes 46/1(165)
Shower functions 46/2(23–33)
Shower maximum, depth of the 46/2(19, 26)
Shower particles, total number 46/2(18, 27, 66)
Shower secondaries 46/2(121–122, 268, 385, 557–559)
Shower secondaries, neutral-charge ratio 46/2(121–122)
Sidereal variations of the radiation 46/1(268)
Sikhote-Alin 46/2(620, 624, 625)
Similarity relation 46/2(65)
Simple albedo cones 46/1(126–128)
Simple shadow cones 46/1(118, 119)
Simple shadow orbits 46/1(105)
Simpson maximum of neutron flux 46/2(374, 450, 483)
Simpson pile, *see* Neutron monitor
Single scattering, contribution to the lateral spread 46/2(67, 92)
Slope of energy spectrum, *see also* Exponent 46/1(8, 40)
Slow mesons, latitude effect 46/1(292)
S-nuclei 46/2(216–218)
Soft cloud 46/1(48)
Soft component 46/1(155)
Solar cycle of eleven years 46/1(3, 53)
Solar energetic particles, *see also* Protons 46/2(344–347)
Solar flare(s) 46/1(3, 25, 51); 46/2(315, 330, 341, 346, 353, 355, 366, 368, 590)
Solar modulation, *see also* 11-years cycle, 27-days variation 46/2(374, 378)
Solar neutrinos 46/2(586)
Solar particle bursts, variation in radioisotope production 46/2(579–580)
Solar plasma 46/2(313, 321, 323, 341–342, 349, 359, 368)
Solar system, main properties 46/1(50)
Solar system x-ray sources 46/2(299–300)

Solar type stars 46/1(55–56, 59)
Solar wind 46/2(320–321, 323–324, 342, 344, 349, 355, 359, 365)
Solar-flare produced cosmic rays 46/1(51)
Source and sink functions for isotope production 46/2(553–555, 562–575, 587, 589, 591, 594–595, 597)
Source regions 46/2(245, 257–260)
Sources and sinks of high energy electrons 46/2(275)
Sources spectrum of evaporation neutrons 46/2(427, 430)
South African anomaly 46/2(337)
Spallation reactions 46/2(557, 559, 566, 574, 614, 616, 621)
Spark chamber 46/2(120)
Spatial density of cosmic rays 46/1(10)
Spatial diffusion 46/1(41)
Spectral distribution of radiation in a magnetic field 46/1(34)
Spectrograph, rotatable 46/1(302)
Spectrographs for cosmic rays 46/1(274–277)
Spectrum at sea level 46/1(273–280)
— effect of *K*-mesons 46/1(291, 309)
— latitude effect 46/1(291)
— of slow muons 46/1(293)
Spectrum of muons at large zenith angles 46/1(307–311)
Spectrum of muons at small zenith angles 46/1(301–307)
Spiral arm 46/1(59)
Spiral galaxies 46/1(76)
S-star 46/1(195, 196)
Stability of periodic orbits 46/1(103)
Standard atmosphere 46/2(376–377, 491)
Star production in the atmosphere 46/2(394, 396)
Star production mean free path 46/2(454)
Star production reaction rates 46/2(454, 564, 582)
Star size distribution 46/2(397, 398, 402, 404, 568, 571)
Starlight, energy density 46/1(8)
Stars 46/1(138, 139–146)
Stars, analysis of 15300 observations 46/1(140)
Stars in nuclear emulsions 46/2(391–410, 453–458, 565)
— absolute frequency 46/2(396–397)
— altitude dependence 46/2(393)
— percentage of deuteron, triton, and He tracks 46/2(401)
Stars, latitude and altitude dependence 46/1(145)
Stars of solar type 46/1(55–56, 59)
Steady state cosmological theory 46/2(284)
Stellar coronae, hard radiation from 46/2(300–301)
Stirring process of cosmic rays 46/1(9)
Stone meteorite 46/2(615, 616, 620–621)
Storage of cosmic rays 46/1(9)
Störmer cone 46/1(88)
Störmer parameter 46/1(92)
Störmer theory 46/2(312, 313, 318, 321)
Störmer unit of length 46/1(91); 46/2(311)
Störmer formula 46/1(92, 93)
Störmer unit of energy 46/1(91)
Strange particles in penetrating showers 46/1(193–196)
Stratosphere 46/2(396, 555, 589–590, 599, 601, 603–605, 614, 616, 621)
Stripping of electrons 46/1(11)
Structure functions
— at large distance from the shower axis 46/2(67–69)
— in electron shower theory 46/2(2, 39, 43–44, 47–49, 49–69)
— mean square deviations 46/2(53–56)
— near the shower axis 46/2(60–69)
— normalized 46/2(56)
Structure of the nucleus, neutral to charged secondaries ratio in penetrating showers 46/1(205)
Structure of the nucleus, penetrating shower production cross section 46/1(202–204)
Sudden change of motion in a magnetic field 46/1(24)
Sudden commencement 46/2(318, 332, 346, 353)
Sudden energy loss 46/1(41, 47)
Sun as x-ray source 46/2(299–300)
Sun, magnetic field 46/1(54–55)
Sun, main properties 46/1(50)

Sunspot cycle, *see also* 11-years cycle 46/2(185, 329, 330, 333, 336–340)
Sunspots 46/1(25)
Supergalaxy 46/1(81)
Supergiants 46/1(57)
Supernovae 46/1(61–69); 46/2(257–258, 287–288, 615)
Supernovae remnants 46/2(285)
Survival probability of muons 46/1(283, 306)
Synchrotron radiation 46/1(31, 34, 44); 46/2(240, 266, 271, 290)

T Tauri stars 46/1(56)
Takagi model of nucleon interactions 46/2(161)
Tamm and Belenky, track length solutions of 46/2(37–38)
Tau XR-1 46/2(285)
Temperature coefficient of muon component 46/1(313–314)
Theory of zenith angle variation of muons 46/1(304)
Thermal neutrons, nuclear disintegrations induced by 46/2(566, 572–573, 585, 614)
Thermodynamical model of pion production 46/1(182, 183)
Thickness of matter traversed by primary cosmic rays 46/2(246, 584)
Thin tracks in stars 46/1(139, 142)
Thompson limit 46/2(272)
Thomson scattering cross-section 46/1(16–17)
Threshold rigidity, *see also* Cut of rigidity 46/2(538)
Tidal disruption of stars 46/1(58)
Time constancy of cosmic rays 46/1(3, 37)
Time distribution of air showers 46/1(225–227)
Time variation in cosmic rays, solar dependence 46/1(2–4)
Time variation of air showers 46/1(263–271)
Tracer studies 46/2(554, 597, 599)
Track length, differential 46/2(18, 25, 38)
Track length, total (integral) 46/2(16, 18–19, 25, 35–38)
Tracks in nuclear emulsions 46/2(392, 403)
Trajectory of the third kind in the earth's magnetic field 46/1(116)
Transition effect(s) 46/1(157, 158, 167)
— atmospheric 46/1(164)
— of neutrons in water 46/2(441)
— of stars 46/2(405–410, 564)
Transparency 46/2(313)
Transparency of light nuclei 46/1(171, 202–203)
Transport mean free path 46/1(41)
Transverse momentum 46/2(76, 113, 125, 127–129, 145, 157, 166, 169–170, 172, 386)
Trapped radiation 46/2(315, 316, 321)
Trapping mechanism between magnetic mirrors 46/1(27, 30)
Trays of counters for penetrating showers 46/1(159, 163)
Treysa 46/2(620, 623, 624, 627, 628, 629)
Tritium, solar 46/2(584)
Troposphere 46/2(554, 591–595, 599, 601, 605–606)
Tunnel model of nucleon-nucleus interaction 46/1(196–199, 207)
Two-fireball model 46/2(161–165)
Tycho's nova 46/1(63)

Underground depth-intensity curve 46/1(289, 290)
Underground showers 46/1(157)
Universal cosmic ray spectrum 46/2(267)

van Allen radiation 46/1(82)
Variation, daily 46/2(326–327, 333, 350, 351, 354, 365)
Variation, 27-day 46/2(330)
Variation, long-term 46/2(310, 332, 357, 363, 366)
Variation, rigidity dependence 46/2(332, 336–340, 362)
Variation, short-term 46/2(341)
Variational coefficient 46/2(326)
Velocity of Alfvén waves 46/1(29)
Vertical intensity, alteration with altitude, latitude and longitude 46/1(131, 138)
Vertical intensity of α particles 46/1(150, 151)

Vertical momentum spectrum 46/1(281–291, 312)
Virgo clusters 46/1(77)

Weizsäcker-Williams method 46/1(161, 181)
Williamstown 46/2(620, 621, 624)
Wolf-Rayet stars 46/1(58)

X-particles 46/2(122, 124, 127)
X-ray emission by electron-ion radiative recombination 46/2(293)
X-ray emission by inelastic collisions followed by radiative de-excitation 46/2(293)
X-ray films for energy measurements on electron showers 46/2(69, 139–140)
X-rays, characteristic 46/2(266, 292)
X-rays from external galaxies 46/2(281–282)
X-rays production in discrete sources 46/2(290–294)
X-rays sources 46/2(286–302)

Yield functions 46/2(187, 311, 328, 329, 331, 337)
Yield of isotopes in spallation reactions 46/2(566–573)

Zatzepin-Greisen's theory 46/1(232, 233)
Zenith angle
— dependence of positive excess 46/1(312)
— dependence of the directional intensity 46/1(133)
— large, muon spectrum 46/1(307–311)
— variation of muon spectrum 46/1(301–311)

Group 10

Geophysics

A priori knowledge 49/2(209)
Absolute calibration in the solar spectrum 49/6(312, 323)
Absolute ionization 49/1(164)
Absolute magnetic theodolite 49/3(309)
Absorption 49/4(478); 49/5(214)
Absorption at normal incidence 49/2(88ff.)
Absorption, atmospheric 49/7(269 Figs. 33a, b, 273 Figs. 38a, b)
Absorption, carbon-dioxyde 48(228)
Absorption, Čerenkov- 49/4(444, 458); 49/5(238)
Absorption coefficient 49/6(16)
Absorption, collisionless 49/4(447)
Absorption, collisionless Čerenkov 49/4(447)
Absorption cross section 49/6(16)
— 10–100 mm 49/6(21)
— 175–100 nm 49/6(20)
— above 175 nm 49/6(17)
— atomic nitrogen 49/6(25)
— atomic oxygen 49/6(24)
— below 60 nm 49/6(23)
— metals 49/6(19)
— N_2 49/6(20, 23)
— NO 49/6(19, 21)
— O_2 49/6(17, 18, 20, 22, 179, 180)
— ozone 49/6(19)
— water 49/6(21, 22)
— x-rays in air 49/6(25)
Absorption decrement 49/2(264ff.)
— at oblique incidence 49/2(128)
— Chapman-Elias layer 49/2(99)
— parabolic layer 49/2(98)
Absorption in clouds 48(213)
Absorption in line center 49/6(332, 333)
Absorption in solar spectrum 49/6(316)
Absorption in the infrared 48(171)
Absorption index 49/2(25, 35, 89)
— diagrams 49/2(80ff.)
— for whistler propagation 49/2(93)
Absorption, influence of pressure and temperature 48(230)
Absorption measurements 49/2(262–266)
Absorption, multifrequency measurement 49/2(321)
Absorption of cosmic radio waves 49/2(380)
Absorption of electromagnetic waves 49/2(24)
Absorption of EUV solar radiation 49/7(459)
Absorption of high-frequency transverse waves 49/4(458)
Absorption of radio waves in the ionosphere 49/1(71ff.)
Absorption of sound 48(460)
Absorption of the atmosphere 48(155, 160)
Absorption, ozone 48(229, 373)
Absorption parameters 49/2(266)
Absorption vector 49/2(122ff.)
Absorption, water vapor 48(224)
Absorption, winter anomaly 49/2(330)
Abundance (chemical) 49/6(354, 371–373)
Acceleration 49/4(102)
Acceleration, local of particles 49/4(217)
Acceleration, neutral sheet 49/4(221)
Acceleration, resonant 49/4(168)
Acceleration, synchronous 49/4(168)
Acceleration term 49/6(253)
Accelerometer 49/7(247 Figs. 13–16, 250, 322, 323 Figs. 82, 83)
Accelerometer, electrostatic 49/7(250)

Accommodation, partial 49/5(244)
Accuracy of height readings 49/2(210, 215)
Acoustic cut-off frequency 49/5(168, 186)
Acoustic domain 49/5(189)
Acoustic duct 49/5(200)
Acoustic, gravito-, wave modes 49/5(188)
Acoustic power density vector 49/2(111)
Acoustic wave(s) 49/2(10, 18, 61 ff., 349); 49/5(240 ff.)
— spectrum 49/7(256)
Acoustical internal gravity wave 49/5(168)
Acoustical structure 49/5(166)
Actinium 47(300)
Actinometer 48(250, 251)
Actinometer, shielded 48(250)
Action 49/5(212, 215, 216)
Active regions (sun) 49/6(344, 345, 348)
Activity, geomagnetic 49/4(93)
Adaptation 48(267)
Addition law of ARM 49/1(295)
Addition law of TRM 49/1(274)
Additional solar-daily variations on quiet days 49/3(18)
Adiabatic atmosphere 49/7(228)
Adiabatic compression 49/3(156)
Adiabatic, contraction 49/7(472, 478 Fig. 175)
Adiabatic, equilibrium 49/7(226)
Adiabatic heating 49/6(296)
Adiabatic invariants of particle motion 49/1(87, 135)
Adiabatic law 49/5(180)
Adiabatic, temperature gradient 49/7(227)
Adriatic Sea, tides 48(913)
Aerial mollusks 48(266)
Aerials, *see* Antennae
Aerodynamic braking 49/7(233–235)
Aeronomical models 49/6(83, 95)
Aeronomy 49/6(2, 83)
AEROS satellites 49/7(3, 17, 30, 35, 44, 242, 333, 345 Fig. 99a, 368 Fig. 113, 373 Fig. 116, 409 Figs. 135, 137, 411 Fig. 138, 412 Fig. 139, 421 Fig. 143, 529)
Aerosol 48(255, 565)
Aerosol, natural 48(488)
Aerosol near the ground 48(169)
Aerosol particles 48(547)
Aerosol, size distribution 48(170, 190, 201)
Aerosol, stratification 48(260)
Aerosol, vertical distribution 48(179, 181, 200)
Afterglow 49/6(60, 68, 121)
Afterglow, flowing 49/6(61)
Age determination, decay of uranium and thorium to helium and lead 47(319)
Age determination, radioactive 47(288)
Age of earth 47(294, 333)
— assumptions made in calculating 47(334)
— early physical arguments 47(291)
— from rate of evolution 47(291)
— from sedimentation and weathering 47(291)
— Lord Kelvin 47(291)
Age of minerals 47(288)
Age of the crust 47(335)
Age of the mantle 47(335)
Age of the universe 47(294)
Air, collision frequencies 49/2(12, 14)
Air density 48(439, 441)
Air, dry 48(2)
Air glow 48(199, 216)
Air light 48(262)
Air light, Koschmieder's theory of 48(262)
Air mass
— absolute 48(157)
— distribution 47(17)
— relative 48(157)
— synoptic 48(178)
Air, moist 48(4)
Air pressure 48(439)
Air pressure variations 48(928)
Air, unsaturated 48(5)
Airborne magnetometer 47(500)
Air-earth-current 48(543)
Airglow 49/6(43, 121, 189, 278)
Airy functions 49/2(181 ff.)
Airy's channel theory 48(893)
Aitken-nuclei 48(160, 164, 489)
Aitken-nuclei counter 48(489, 566)
Albedo 48(196, 207, 214, 299)
Albedo, limiting value 48(211)
Albedo neutrons 49/4(65)
Albedo, planet earth 48(216)

Albedo, planetary 48(155)
Albedo, snow 48(215)
Albedo, soil surfaces 48(214)
Albedo, water surfaces 48(216)
Alfvén oscillations 49/4(482)
Alfvén shock 49/4(35)
Alfvén speed 49/4(202)
Alfvén velocity 49/5(221)
Alfvén wave(s) 49/4(477, 529–531); 49/5(235)
— modified 49/5(236)
— ELF 49/5(250, 251)
— VLF, modified 49/5(250)
Alfvén's electric field model of aurora 49/1(136)
Algonkian 47(292)
Allard's law 48(276)
Alpha activity 47(310)
α-particle in the solarwind 49/4(86, 92)
α-particle spectrum 49/1(226)
α-particles 49/4(10, 152)
Alps 47(239, 259, 268)
Alternating field demagnetization 49/1(275)
Ambipolar (condition field) 49/7(208)
Ambipolar diffusion 49/2(518); 49/6(99, 145, 152, 193)
Amphidromic point 48(967)
Amphidromy 48(907)
Amplification, ion-cyclotron 49/5(305)
Amplitude modulation 48(766)
Amplitude of echo 49/2(215)
Amplitude record 49/2(245, 249)
Amplitudes of signals 49/2(209)
Andean (mountain) 48(972)
Anderson, faulting theory 47(264)
Angle of arrival 49/2(216)
Angle of incidence 49/2(445)
Angle of shear 47(76)
Ångström's formula 48(237)
Ångström-scale 48(251)
Angular character, multi-lobe 49/5(288)
Angular momentum 49/5(213)
— absolute 48(140)
— absolute, geostrophic flux 48(149)
— absolute, in a cyclone 48(150)
— absolute, mean meridional flux 48(145, 146)
— quantum number 49/7(65)
— wave 49/5(216)
Angular width 49/5(290)
Anhysteretic remanent magnetization ARM 49/1(272, 294–297)
Anisotropic processes 49/5(205)
Anisotropy coefficient 48(158, 160)
Anisotropy in velocity space 49/2(9, 10, 14)
Anisotropy of magnetic susceptibility 49/1(268 ff.)
Anisotropy of plasma 49/2(2, 63)
Annual means of cosmic-ray intensity 49/1(186, 194, 230)
Annual variation of activity 49/3(255)
Anomalies, isostatic 47(233, 578)
Anomalies of magnetic field intensity 49/3(414)
Anomalous Doppler effect 49/5(228 ff.)
Anomaly, gravitational 47(537)
Anomaly, summer/winter 49/6(287)
Antarctic bottom water 48(641, 644, 667)
Antarctic circumpolar current 48(663)
Antarctic circumpolar water 48(646)
Antarctic intermediate water 48(644, 648, 667)
Antenna, impedance in plasma 49/2(478)
Antennae 49/2(205, 484)
Anticyclones, *see* Cyclones
Anti-cyclones 48(112)
Anti-ferrimagnetic minerals 49/1(248, 258)
Anti-ferromagnetism 47(472)
Anti-hermitean tensor 49/2(7)
ap index for the activity of magnetic storms 49/1(55, 66)
Apogee 48(746)
Appalachians 47(268)
Apparent size, sun and moon 48(738)
Apparent velocity 49/2(428)
Appleton-Hartree dispersion formula 49/2(489)
Appleton-Lassen dispersion formula 49/2(30, 40, 59, 72, 89, 116, 143, 489)
— approximation for weak absorption 49/2(83)
Approximation, Boussinesq 49/5(186)
Approximation, linear 49/4(439)
Approximation, neutral 49/5(252, 253)
Approximation, quasi-gasdynamic 49/4(420, 429)
Approximation, static 49/5(162)

Approximations near zeros of refractive index 49/2(86)
Aquarids 48(435)
Arago-point 48(300)
Arc, reversed 47(259)
Arch of enhanced ionization 49/2(515ff.)
Archeozoic 47(292)
Arctic Water 48(645)
Argence approximation 49/2(257ff.)
Argon-40 47(297, 336); 49/6(177)
Argument number 48(750)
Argus 49/4(138)
Arietids 48(435, 437, 438)
ARM, *see also* Anhysteretic remanent magnetization 49/1(272, 294–297)
Artificial aurora 49/1(120ff.)
Artificial bodies, flow around 49/5(227)
Artificial star 48(280)
Artificially stimulated emissions (ASE) 49/5(332)
Ash-eruptions, volcanic 48(201)
Askania-Gravimeter 48(805)
Aspiration method 48(552)
Associative detachment 49/6(97)
Asthenosphere 47(246)
Astronomisches Jahrbuch 47(1)
Asymmetry effect 48(559)
Atenuation decrement 49/4(490, 492, 518, 526, 529, 533)
Atenuation decrement 49/4(460)
Atlantic Ocean, water masses 48(644)
Atlantic tides 48(918)
Atmosphere, angular momentum 47(20)
Atmosphere, emission of the 48(155)
Atmosphere, isothermal 48(28); 49/5(186)
Atmosphere, Martian neutral 49/5(374)
Atmosphere of Venus, neutral 49/5(374)
Atmosphere, optical standard 48(265)
Atmosphere, polytropic 48(28)
Atmosphere, supply of heat in the 48(945)
Atmosphere, unified 49/5(173)
Atmosphere, upper, structure of the Venusian 49/5(376)
Atmosphere, vertical motions in the 48(97)
Atmospheres, autobarotropic 48(953)
Atmospheric circulation 47(17)
Atmospheric currents 49/7(445 Fig. 155, 451 Fig. 158)
Atmospheric electric current circuit 48(601)
Atmospheric electricity 48(540)
Atmospheric Explorer satellites 49/7(255 Fig. 22, 529)
Atmospheric gravity waves 49/2(345, 350)
Atmospheric heating, *see also* Heating 49/7(8, 36, 40–44)
Atmospheric modelling 49/7(32, 34)
Atmospheric parameters 49/4(401)
Atmospheric tides 48(928)
— diurnal wave 48(971)
— eight-hourly wave 48(969)
— gravitational excitation 48(933, 959)
— lunar waves 48(960)
— oropgraphic influences 48(972)
— resonance curves 48(959)
— resonance theory 48(930, 965, 967, 969)
— seasonal variations 48(961, 967)
— semidiurnal progressing wave 48(965)
— semidiurnal standing wave 48(967)
— six-hourly wave 48(970)
— solar waves 48(962)
— thermal excitation 48(935)
Atmospherics 48(576, 587)
Atom interchange reactions 49/6(45)
Atom-bomb-explosion 47(82)
Atomic bombs 48(976)
Atomic bombs, sound 48(476)
Atomic ion(s) 48(550); 49/6(118)
Atomic nitrogen, absorption cross section 49/6(25)
Atomic nitrogen, excited states 49/6(123)
Atomic nitrogen ion, reactions 49/6(51, 92, 118, 123)
Atomic oxygen 49/7(330)
— excited states 49/6(121)
— ion, reactions 49/6(49)
— profile 49/6(72)
— removal 49/6(183)
Attachment 49/2(268); 49/6(53, 107, 126)
Attachment coefficient 48(450, 568, 570)
Attachment in *F*1 layer 49/2(285)
Attachment in *F*2 layer 49/2(289)

Attachment of small ions 48(567)
Attachment, three-body 49/6(51, 53, 103, 107)
Attennuation 49/5(228)
Attennuation, spatial 49/5(307)
Attraction of type bodies 47(228)
Audibility, threshold 48(455)
Auger-electrons 47(307)
Aurora 49/3(124, 152, 158); 49/4(99, 111, 215, 216); 49/6(43, 121); 49/7(380 Fig. 121c, 497)
Aurora and main phase of a magnetic storm 49/1(62–64)
Aurora and polar magnetic substorms 49/1(56–62)
Aurora, artificial 49/4(139)
Aurora, artificial production 49/1(120ff.)
Aurora australis 49/1(2)
Aurora borealis 49/1(2)
Aurora, electric fields 49/7(506, 507 Figs. 199, 200, 517 Fig. 213)
Aurora of 11 February, 1958 49/1(63–65)
Aurora of 23 September 1957 49/1(82ff.)
Aurora, theories 49/1(124–140)
Aurora, theories of 49/4(217)
Aurora tropicalis 49/1(2, 36ff., 120)
Aurorae 49/2(529ff.); 49/4(207)
Aurorae, energetics of 49/4(209)
Auroral Activity Index 49/3(245)
Auroral cap 49/1(2, 10ff.)
Auroral distribution in the polar cap 49/1(10ff.)
Auroral distribution, time variation 49/1(13–15)
Auroral E-layer (Ea) 49/7(502)
Auroral electrojet(s) 49/1(56–62, 88, 90); 49/3(97); 49/4(101, 105, 107)
Auroral electrons 49/7(498)
Auroral Es 49/2(337)
Auroral morphology 49/1(1)
Auroral oval 49/4(60, 99–102, 111); 49/7(499–521)
Auroral oval belt 49/3(79)
Auroral precipitation 49/4(101, 103)
Auroral processes 49/4(222)
Auroral radar echo(es) 49/1(75–81); 49/3(123)
Auroral radar technique 49/1(70)
Auroral regions 49/1(2)
Auroral spectrum 49/1(67, 100–124)
— analysis 49/1(106–113)
— helium lines 49/1(112ff.)
— hydrogen lines 49/1(110–112)
— inference on temperature 49/1(121–124)
— nitrogen lines and bands 49/1(108–110, 115)
— oxygen lines and bands 49/1(106–108, 115)
— production 49/1(113–117)
— special types 49/1(117–121)
— typical examples 49/1(102–105, 114)
Auroral type absorption of radio waves 49/1(72)
Auroral zone(s) 49/1(2, 6ff.); 49/2(367); 49/4(99, 100, 105–109, 209, 211); 49/7(497–523)
— electron production rates 49/7(522)
— irregularities 49/2(401)
— theoretical 49/1(3–10)
Autobarotropy 48(2, 12)
Auto-correlation 49/2(426, 430)
Auto-correlation function 49/2(334)
Auto-correlogram 48(692)
Automatic observatory 49/3(325)
Automatic tuning 49/2(200)
Average Poynting vector 49/2(110)
Averaged motion 48(35)
Averaged transport collision frequency 49/2(10, 11ff., 14, 15, 20, 22, 97)
Axford and Hines model of the aurora 49/4(219)
Axial matrix tensors 49/7(214)
Axial tensors (Appendix A) 49/7(211ff.)
Axial-hypothesis (Cortie) 49/3(257)

B, *L* coordinates 49/4(126)
Babinet point 48(300)
Babinet's theorem 48(340)
Background noise 48(466)
Backscatter 49/2(231, 337, 360ff., 405, 442)
Ba-ion clouds 49/4(101)
Balance equation 49/2(251ff., 377, 379, 518); 49/6(352)
Balance equation for *E* region 49/2(282)
Balance equation for *F*2 layer 49/2(288ff.)

Balancing magnet 49/3(285, 316)
Ball lightning 48(544)
Balmer emissions 49/4(215)
Balmer lines 49/1(110–113)
Balmer lines, velocity profiles 49/1(110, 113)
Baloon measurement 49/4(82)
Baltic 48(911)
Band model, Elsasser's 48(173, 223, 245)
Band model, Goody's 48(224)
Band spectra 48(220, 222)
Band-pass-filter 47(155)
Bands of aurora 49/1(16)
— dimensions 49/1(18–23)
— orientation 49/1(25ff.)
Bandwidth 49/2(206, 211, 219)
Barium cloud (artificial) 49/4(82); 49/6(214)
Baroclinic current, waves in a 48(85)
Baroclinicity vector 48(11)
Barometric effect 49/1(166, 216)
Barometric formula (law) 49/6(1, 251, 265)
Barometric variation, semi-diurnal 47(7)
Barometrical differential equation 49/7(224)
Barometrical formula 49/7(224)
Baropause 49/6(237, 239–241)
Barotropic current, waves in a 48(82)
Barotropic flow, vorticity equation for 48(83)
Barotropic model 48(101)
Barotropy 48(11)
Bartels' index 49/3(208, 212)
Bartels' index *u* 49/1(223, 230)
Bartels' *Kp* (planetary magnetic character figure) 49/7(330)
Basalt 47(475)
Base line value 49/3(279, 298)
Basic rocks 47(313)
Basis vectors of polarized coordinates 49/2(31)
Bates-Walker temperature profile 49/7(327)
Bathymetric charts 48(615)
Bathythermograph 48(611)
Bay disturbance 49/4(91)
Beacon satellites 49/2(473)
Beam experiments 49/6(65)
Beam instability 49/5(226)
Beat method for differential Doppler effect 49/2(454, 459)
Beat spectra 49/2(220)
Beer's law 48(156, 172, 291, 376, 394)
Belt on other planets 49/4(222)
Belt, outer 49/4(152, 173)
Belt, outer, electrons 49/4(165)
Bemporad's tables 48(157)
Bennet mass spectrometer 49/6(74)
Bennet spectrometer 49/7(237)
Bent ionospheric model 49/7(351–353)
Beryllium-7 47(301)
Bessel functions 48(953)
Beta plane 49/5(184)
Betatron effect 49/4(76)
Biblical accounts 47(289)
Biconical dipole 49/2(478)
Bifilar-Gravimeter 48(802)
Billow clouds 48(82)
Bimetalactinometer 48(250)
Binary mixture 49/6(135, 138)
Bishop's ring 48(201)
Bismuth 47(299)
Bit-number 49/2(212)
Bjerknes, circulation theorems 48(10, 12, 22)
BKG model 49/4(468)
Black body 48(250)
Black body temperature 49/6(184)
Black bulb thermometer 48(250)
Black Sea 48(912)
Blackout 49/2(450); 49/3(82)
Blanketing *Es* 49/2(250, 265)
Blanketing frequency 49/2(225)
Blasts 48(466, 468)
Blocking high 48(129)
Blue sun 48(168)
BMZ 47(499); 49/3(315)
Bobrov's portable magnetograph 49/3(291)
Bodies, boundary of 49/5(243)
Body at rest, small 49/5(276)
Body, cylindrical 49/5(271)
Body, point 49/5(220)
Body potential 49/5(245)
Body, shape 49/5(254)
Body, slowly moving 49/5(282)
Body, small 49/5(220, 274, 277)
Body's velocity 49/5(219)
Bohr radius 49/6(143); 49/7(69)
Bolometer 48(183)

Boltzmann equation 49/2(7ff.); 49/6(130, 135, 237); 49/7(229, 526)
Bomb, fusion 49/4(137)
Booker quartic 49/2(128ff., 141, 151, 444)
Booker quartic, approximate factorization 49/2(132)
Booker's complex *Q* plane 49/2(44–46, 57)
Bore hole logging methods 47(312)
Bore, internal 48(880)
Born approximation 49/6(352)
Bottom boundary condition 49/5(206)
Bottom friction 48(728)
"Bottom" of the magnetosphere 49/4(9)
Bottom samples, geothermal gradients 48(614)
Bouguer-anomaly 47(223, 254)
Bouguer's and Lampert's law 48(156)
Bouguer's reduction 47(579)
Bouguer's refraction law 49/2(123)
Bounce period 49/4(119)
Boundary 49/4(26)
Boundary condition(s) 49/6(236, 256); 49/7(481, 485)
— bottom 49/5(206)
— upper 49/5(201)
Boundary layer friction 48(850)
Boundary, magnetospheric 49/4(206)
Boundary of bodies 49/5(243)
Boundary of plasma 49/2(51)
Boundary sea-air 48(625)
Boundary surface of solar plasma 49/1(129ff.)
Boussinesq approximation 49/5(186)
Bow shock 49/3(138, 178); 49/4(41, 202)
Bowen-ratio 48(630)
Bowie effect 47(577)
Bracking of satellites 49/7(233–236)
Branch cut of the Riemann surface of the polarization parameter 49/2(47ff., 53, 75ff.)
Branch, high-frequency 49/5(233)
Branch, low-frequency 49/5(234)
Branch point of the Riemann surface of the polarization parameter 49/2(47)
Branching ratio 47(298, 305, 306)
Breakdown of invariants of particle motion 49/1(135)
Breaker types 48(729)
Breakers, spilling, plunging, surging, asymmetry 48(730, 732)
Breaking drop effect 48(560)
Breaking strength 47(78)
Break-up of an aurora 49/1(42, 46, 47, 56ff.)
Breakwater 48(729)
Breathing, diurnal, of atmosphere 49/6(295)
Bremsabsorption 49/4(467)
Bremsstrahlung in the ionosphere 49/1(82ff., 90)
Bremsstrahlung photons 49/4(210)
Brewster point 48(300)
Bridgman relation 49/7(182)
Brightness coefficient for aurora, international 49/4(209)
Broad-band 49/5(335)
Broadening, thermal 49/6(335)
Brunt's formula 48(237)
Brunt-Vaisälä frequency 49/5(168, 186)
Buckingham potential 49/7(79, 80 Fig. 3, 124, 143, 146ff. Figs. 26a–d)
Buckling 47(274)
Building sites 47(178)
Bulk velocity 49/6(240, 242, 250, 343); 49/7(261)
Bulk-modulus, incompressibility 47(79)
Bulk-motion 49/4(492)
Busch's lemniscate 48(304)

CACTUS accelerometer 49/7(250 Figs. 17, 18, 323 Fig. 83)
Calcium-40 47(336)
Calcium plage index 49/7(16)
Calculated current systems 49/5(174)
Calibration curve 49/6(362)
Calibration of solar flux measurements 49/6(29, 312, 323)
Calibration of spectra, absolute 49/6(312, 323)
Caloi's Pa, Sa 47(134)
Cambridge pendulum 47(210)
Cap range 47(259)
Capacitance of ion sheath 49/2(485)
Capacitor in plasma 49/2(480)
Capillarity 48(703)
Carbon-14 47(296, 301, 308)
Carbon dioxide 48(171, 242, 246)
Carbon dioxide atmospheres 49/7(456)
Carbon dioxide, isotopes 48(171)
Carnallite 47(336, 349)

Carnegie (ship) 48(580)
Cartesian geometry 49/5(182)
Cartesian geometry, rotating 49/5(183)
Case histories 47(153)
Casimir's generalisation 49/7(183)
Catastrophies 47(289)
Caustic 49/5(293ff.)
CCIR mapping system 49/7(390)
Center of gravity, coordinates 49/6(133)
Center-of-mass velocity 49/2(8, 10)
Center-to-limb variation (Sun) 49/6(329)
Central water 48(643, 645, 648)
Central wave 48(680)
Čerenkov absorption 49/4(512, 546); 49/5(238)
Čerenkov attenuation 49/4(484)
Čerenkov damping 49/5(228)
Čerenkov dissipation 49/4(542)
Čerenkov effect 49/4(537, 550–552)
Čerenkov electron absorption 49/4(481)
Čerenkov excitation 49/5(228)
Čerenkov oscillations 49/4(497)
Čerenkov radiation 49/5(228ff.); 49/7(463)
Čerenkov-Vavilov effect 49/5(228)
Cetids 48(435)
c.g.s system 49/3(1, 2); 49/4(1, 2)
Chain formula for matrizants 49/2(160ff.)
Chain lightning 48(544)
Chandler's period 47(609); 48(790, 813, 818)
Chandrasekhar, principle of invariance 48(296)
Changes of state, dry-adiabatic 48(4)
Changes of state in the atmosphere 48(945)
Changes of state, moist-adiabatic 48(6)
Chang's extended effective range expansion 49/7(98)
Chapman and Enskog method 49/4(431)
Chapman distribution 49/2(579, 583, 586)
Chapman function 49/7(268)
Chapman-Elias layer 49/2(97ff., 471)
— absorption decrement 49/2(99)
— equivalent height 49/2(118)
— model 49/5(121)
Chapman-Enskog method 49/7(63)
Chapman-Ferraro theory 49/4(217)
Chappuis band 48(172, 199, 375, 385); 49/6(19)
Character figure(s) 49/3(27, 28, 207)
Characteristic frequencies 49/2(222ff.)
Characteristic height of $F2$ layer 49/2(307–318)
Characteristic height parameters 49/2(226ff.)
Characteristic polarization ellipses 49/2(463)
Characteristic polarizations 49/2(28ff., 34, 47ff., 221)
Characteristic value 49/3(207)
Charge exchange 49/3(165)
Charge exchange (cross section) 49/7(114, 119)
Charge separation model 49/1(138)
Charge transfer (exchange) 49/6(4, 43, 98, 114, 117, 118, 228); 49/7(434, 442, 446, 497 Fig. 187)
Charged particles, collisions 49/7(105–121, 162, 163, 190, 466, 527)
Charged-particle trap 49/5(353ff., 358)
Charge-exchange 49/5(379)
Chemical abundance 49/6(354)
Chemical composition 49/7(236–244, 308–314 Figs. 70–75, 340, 343, 344 Figs. 97, 98)
Chemical composition (solar atmosphere) 49/6(375)
Chemical release 49/6(160, 161, 163)
Chemical remanent magnetization CRM 49/1(272, 284–286)
Chemistry, ions 49/6(7)
Chemistry, neutral 49/6(7)
Chemistry-oriented models 49/7(433–443)
Chirp radar 49/2(220)
Chiu ionospheric model 49/7(386 Fig. 123c)
Chlorinity, definition 48(617)
Chorus 49/5(309ff., 332ff.)
Chorus spectrum 49/5(325)
Chromosphere 49/6(351, 365)
Chrystal's method 48(852)
CIRA 1961 49/7(315)
CIRA 1965 49/7(316)
CIRA 1972 49/7(316–326)
Circular polarization 49/2(34, 43, 47)

Circular vortex 48(64)
— meridional circulation 48(139)
— stability 48(64)
Circulation, absolute 48(11)
Circulation, general 48(137)
Circulation, meridional 48(666)
Circulation, stratospheric 49/5(117, 135)
Circulations 48(248)
Clairaut's differential equation 47(603)
Clairaut's theorem 47(552, 554, 570)
Classical geomagnetic measuring technique 49/3(323)
Classical MUF, *see also* MUF 49/2(227)
Classification of emissions 49/5(302)
Cleft 49/7(408)
Clemmow-Mullaly-Allis (CMA) diagram 49/2(70, 105)
Climate, photochemical 48(422)
Climate values, synthetic 48(937)
Closed orbit 49/5(277)
Closest approach 49/7(72)
Closest approach point 49/2(447)
Cloud, albedo 48(211, 213)
Cloud, barium 49/6(214)
Cloud, cylindrical plasma- 49/6(208, 210, 211)
Cloud droplets 48(492)
Cloud drops 48(210)
Cloud (due to release) 49/6(162, 167, 204, 206, 208)
Cloud, elliptical plasma- 49/6(212)
Cloud of droplets 48(351)
Cloud, spherical plasma- 49/6(206)
Clouds 48(359, 479); 49/5(162)
Clouds as black body radiators 48(236)
Clouds, cumuliform 48(484)
Clouds, layer- 48(482)
Clouds, nacreous 49/5(162, 164)
Clouds, noctilucent 49/5(162, 163, 165)
Clouds, orographic 48(482, 520)
Clouds, physical classification 48(481)
Clouds, radiation 48(240)
Clouds, water contents 48(480, 493)
Cloud-seeding 48(529, 534)
Cluster ions 49/6(16); 49/7(378–383 Figs. 119–122, 437, 438)
Clusters 48(543, 550)
CO_2, dissociation 49/5(375)
Coagulation 48(160, 551)
Coalescence efficiency 48(497)
Coalescence of droplets 48(495, 498, 519)
Coated surface 49/6(310)
Co-cummulative energy spectrum 48(697, 715, 723)
Coefficient curve, lift 49/5(168)
Coefficient, displacement 49/4(160)
Coercive force 47(475); 49/3(318)
Coercive force, dependence on grain size 49/1(266)
Coherence 48(344)
Coherent detection 49/2(217)
Coherent scatter sounding 49/7(307)
Coil magnetometer 47(499)
Coincidence 49/6(350, 359)
Co-latitude, geocentric 47(549)
Co-latitude, reduced 47(544)
Cold dome 48(129)
Cold front 48(136)
Cold plasma approximation 49/4(421)
Collection efficiency 48(497)
Collectors, radioactive 48(578)
Collimation angle 49/3(299, 300)
Collision damping 48(221)
Collision efficiency 48(497)
Collision, electron-ion 49/4(544)
Collision energy 49/7(64)
Collision frequency 49/2(1, 4ff., 10ff., 21, 247, 267ff.); 49/4(17, 19, 20, 61, 401); 49/6(40, 47, 230); 49/7(265–267 Figs. 30–32)
— average transport 49/4(434)
— averaged transfer 49/6(42, 250)
— averaged transport 49/4(5)
— effective 49/4(436)
— gas kinetic 49/6(44, 151)
— ion 49/5(170)
— mean transport 49/4(5)
— most probable 49/4(5)
— transfer 49/6(41)
— transport 49/4(5); 49/5(169)
Collision integral 49/2(8, 10); 49/4(419, 425, 435); 49/5(225)
Collision, ionic 49/4(550–554)
Collision, ion-ion 49/4(539)
Collision length 49/7(72)
Collision parameters 49/6(40)
Collision processes 49/6(39)
Collision to gyro frequency ratio 49/7(448)
Collisional absorption 49/4(467)
Collisional dissociation 49/6(103)

Collisional heating 49/2(393)
Collisional modulation 49/2(16)
Collision-dominated conditions 49/6(237)
Collisionless conditions 49/6(237, 264)
Collisions 49/4(30, 66, 435)
Collisions, elastic 49/4(437, 457, 459, 462)
Collisions, elastic and inelastic 49/2(5, 16)
Collisions of suprathermal electrons 49/7(527–530)
Collisons, inelastic 49/6(43)
Color temperature 49/6(309)
Color value 48(201)
Colorimetry 47(311)
Colors of objects, apparent 48(278)
Columnar resistance 48(572, 575)
Combination laws 49/7(122)
Combined multiple ion frequency 49/5(236)
Combined temperature 49/7(106, 129, 134)
Combustion products 48(492)
Cometary orbits 48(427)
Commencement, SC, sudden 49/4(184)
Committee on the Measurement of Geological Time 47(294)
Common lead correction 47(321)
Compensating masses 47(622)
Compensation, correction for 47(234)
Compensation, depth of 47(238)
Compensation, isostatic 47(254)
Compensation pyrheliometer 48(183, 251)
Complex height plane 49/2(112, 119)
Complex refractive index 49/5(224)
Complex vectors 49/2(5, 6, 23, 24, 31)
Component, main 49/5(373, 378)
Components, meridional 49/5(136)
Composition 49/6(288, 298, 300)
Composition, chemical (solar atmosphere) 49/6(375)
Composition measurement 49/5(371)
Composition profile (neutrals) 49/7(264 Fig. 28, 273 Fig. 38)
Composition, seasonal changes 49/6(288)
Composition, wind-induced changes 49/6(301)
Compressibility 47(79, 170, 173, 179)
Compressibility of plasma 49/2(10, 18, 61)
Compressibility-pressure hypothesis 47(115, 375)
Compression 47(77)
Compression of the geomagnetic field 49/3(169)
Compressional waves 47(153)
Compton scattering 49/6(26)
Compton-Bennett ionization chamber 49/1(160–164)
Compton-Getting effect 49/1(178)
Computerized models 49/7(327–340)
Condensation 48(248)
Condensation anomaly 47(578)
Condensation nuclei 48(160, 189, 487, 489, 491, 550, 565, 573)
Condensation nuclei, coagulation 48(491)
Condensation on droplets 48(496)
Condensation region 49/5(261)
Condition, steady state 49/4(176)
Conditions, quiet and disturbed 49/4(182)
Conduction current 48(569, 590)
Conduction of heat in the soil 48(937)
Conductivities, Pedersen and Hall 49/5(170)
Conductivity 49/4(414)
Conductivity and tidal wind 49/2(575, 589)
Conductivity, contribution of ions 49/2(4)
Conductivity eigenvalues 49/2(4, 6); 49/7(185, 190 Fig. 39, 200)
Conductivity, electric, of soil 47(410)
Conductivity, electric, of the core 47(521)
Conductivity, electric, of the mantle 47(518)
Conductivity, finite 49/4(28)
Conductivity matrix 49/2(148)
Conductivity maximum 49/2(6)
Conductivity of air 48(543, 555, 557)
Conductivity relation 49/2(3)
Conductivity tensor (electrical) 49/2(2, 3, 5ff., 18, 21, 28, 30, 61ff., 270ff., 288, 377); 49/7(176)
Conductivity, thermal 49/4(547)
Cone of confidence 47(480)
Confinement 49/4(534)

Confluent hypergeometric function (series) 49/6(174, 175)
Conformal mapping 49/2(47, 57)
Conjugate diameters 49/2(6, 43)
Conjugate points of the magnetic surface 49/1(9, 52, 95)
Conjugate zone 49/2(495)
Connaissance des Temps 47(1)
Constant of gravitation 47(545)
Contact discontinuity 49/4(35)
Continental drift 47(493)
Continental extension of an aurora 49/1(46–52)
Continental margins 47(253)
Continental nuclei 47(357)
Continents 47(19, 246)
Continents, growth 47(286)
Continuity, equation 48(8, 25); 49/2(7); 49/5(180); 49/6(84, 97, 99, 135, 162, 187, 195, 202, 238, 259, 261); 49/7(232, 449, 470, 479)
Continuum emission 49/6(309)
Contraction 49/6(295)
Contraction hypothesis 47(274, 282)
Contrast, apparent 48(264)
Contrast, influence of air 48(262)
Contrast, inherent 48(264)
Contrast rendition by telescopes 48(275)
Contravariant components 49/2(32)
Control magnet 49/3(281)
Convection 48(170, 248, 641); 49/4(77, 79)
Convection currents 47(279, 282)
Convection electric field 49/7(516)
Convection in the mantle 47(495)
Convection, magnetodynamic 49/4(77)
Convection, octahedral 47(284)
Convection of the magnetosphere 49/3(146, 166)
Convective current 48(569, 590, 597)
Convective heating 49/6(296)
Convective model of the magnetosphere 49/1(136ff.)
Convective motion 49/4(60, 79, 88, 102, 111)
Conversion ratio 49/2(164)
Cooling 49/6(83)
Cooling of electrons 49/7(532)
Cooling rate 49/7(468, 469 Fig. 171, 472)
— at night 48(238)
— of the atmosphere 48(239)
Coordinates, solar magnetospheric 49/4(40)
Core 47(19, 261, 370)
Core coil 49/3(344)
Core composition 47(378)
Core density 47(373)
Core gravity 47(373)
Core hydrodynamics 47(528)
Core movements 47(402)
Core pressure 47(373)
Core temperature 47(388)
Core, westward drift 47(526)
Coreless coil 49/3(344)
Core-method 48(644)
Coriolis force 48(9, 649, 859); 49/6(259, 277); 49/7(171–173, 232)
Coriolis parameter 48(15); 49/5(184)
Cornu's Photopolarimeter 48(310)
Corona 49/1(16, 20); 49/6(351, 365, 373)
Coronal discharges 48(583, 590, 598)
Coronal electron density distribution 49/6(356)
Coronal emission 49/6(343)
Coronal holes 49/6(343)
Coronal lines 49/6(355, 356)
Coronal temperature 49/6(3455)
Coronograph 49/6(345)
Corotation 49/4(79, 83–85)
Corpuscular flux from the sun 49/3(77)
Corpuscular heating 49/7(331, 335 Fig. 90, 490)
Corpuscular hypothesis 49/3(200)
Corpusenlar bombardment 49/7(497, 502)
Corrected geomagnetic coordinates 49/3(41)
Correction, geological 47(240)
Correlation, *see* Auto-correlation
Correlation analysis 49/2(422–433)
— apparent velocity 49/2(428)
— drift velocity 49/2(427)
— fading velocity 49/2(426)
— graphic time table 49/2(427)
— spreading velocity 49/2(428)
Correlation ellipsoid 49/2(424)
Correlation function 49/7(63)
Correlation function, definition 49/2(423)
Correlation, normalization 49/2(424, 426)
Correlation, spatial and temporal behaviour 49/2(425)

Cosine law in *D* region 49/2(322)
Cosmic noise absorption 49/2(367, 384); 49/4(215)
Cosmic radiation 48(545)
Cosmic radio noise 49/1(241)
Cosmic radio waves, absorption 49/2(380ff.)
Cosmic radio noise absorption 49/1(89)
Cosmic ray decrease 49/3(133)
Cosmic rays 49/6(96)
— diurnal variations 49/1(171–180, 217)
— effect of muon decay 49/1(169, 216)
— effect of solar flares 49/1(204–216, 217)
— galactic intensity 49/1(234)
— high-altitude observations 49/1(224–227)
— latitude curve 49/1(196, 202, 224)
— results since the Geophysical Year 49/1(218–244)
— results up to the Geophysical Year 49/1(216ff.)
— seasonal variations 49/1(170ff., 217, 230)
— tides in 48(932, 933)
— variation with sunspot cycle 49/1(193–198, 217)
— world-wide variations 49/1(159, 175, 180–204)
Coulomb collisions 49/6(43); 49/7(105–121, 162, 163, 190, 466, 527)
Coulomb drag 49/7(235)
Coulomb forces between ions 49/2(356)
Coulomb interaction 49/2(12ff.); 49/7(73, 76, 162)
Coulomb logarithm 49/4(536); 49/7(466)
Coulomb potential 49/7(90, 157)
Coulomb potential, screened 49/7(88, 91, 127, 155–158 Figs. 31a, b, 162)
Coulomb scattering decay 49/4(147)
Counter current 48(663)
Counter telescope 49/1(164ff., 169)
Counter-telescope observation of 24-hour variation 49/1(174ff.)
Coupled wave equations 49/2(170ff.)
Coupling angle 49/2(70)
Coupling angle of incidence 49/2(125ff.)
Coupling, between ionized and neutral particle fluxes 49/7(183–185)
Coupling collision frequency 49/2(73, 197)
Coupling cone 49/2(71)
Coupling, electromagnetic, of mantle and core 47(19)
Coupling frequency 49/2(54)
Coupling level 49/2(131)
Coupling matrix 49/2(150, 154ff., 171ff.)
Coupling, neutral and ionized atmospheres 49/6(4)
Coupling of characteristic polarizations 49/2(126ff.)
Courant-Friedrichs-Lewy-criterion 48(99)
Covariant components 49/2(32)
Covington index 49/2(270); 49/6(29, 320, 368, 369); 49/3(251)
Covington index (10.7 cm solar radio noise) 49/7(16, 22–27, 29–31, 285, 295, 299, 319, 324, 327, 329, 387)
CP approximation (circular polarization) 49/2(56)
CRAND 49/4(127, 132, 134, 136)
Crest angle 48(685, 730)
Crest length 48(685, 723)
Criterion of trapping 49/4(184)
Critical angle 48(461)
Critical frequency 49/2(193ff., 222, 223, 267)
Critical frequency for *E* region 49/2(278)
Critical latitudes 48(947)
Critical level, for escape 49/6(216)
Critical wind speed 48(628)
CRM, *see also* Chemical remanent magnetization 49/1(272, 284–286)
Cross sections of ionization and excitation 49/1(115)
Crossed beam experiments 49/6(65)
Cross-field diffusion 49/4(161)
Cross-field plasma motion 49/7(505 Fig. 197)
Crossing beam array 49/3(374)
Cross-*L* drift 49/4(58)
Crossover frequency 49/2(55); 49/5(236, 237ff.)
Cross-section, effective, differential 49/5(286)
Cross-tail potential 49/7(493)
Crustal rocks 47(331)
Crystal optics 49/2(22ff., 30, 48ff., 63)

Crystallization remanent magnetization 49/1(284)
Cumulus, seeding with silver iodide 48(531)
Curie point 47(470); 48(1004)
Curie temperature 49/1(252ff., 256, 260, 262)
Current and wind 48(655)
Current, asymmetric ring 49/4(101)
Current, computation 48(651)
Current density 49/2(2, 4, 10, 22, 27)
Current, dynamo-electric 49/4(82)
Current, earth 49/4(215)
Current, equatorial 48(663)
Current, field-aligned 49/4(104–109)
Current, maintenance 48(659)
Current meter 48(613)
Current, ring 49/4(70, 92, 103, 185, 218)
Current system(s)
— calculated 49/5(174)
— dynamo 49/5(173)
— equivalent 49/4(104)
— Sq 49/4(112)
— three-dimensional 49/4(104)
Current tensor 49/4(446)
Currents, dynamo 49/5(171)
Currents in polar magnetic substorms 49/1(60–62)
Currents, permanent 48(613, 659)
Curvature drift 49/4(49, 62, 111)
Curved magnetic field 49/4(556)
Cusp 49/4(71)
Cusp of ray path 49/2(125)
Cut-off at low particle energies 49/1(197ff.)
Cut-off cyclone 48(125)
Cut-off frequency 49/5(237, 301, 316, 321)
Cut-off frequency, acoustic 49/5(168, 186)
Cut-off radius 49/2(13)
Cut-off wave 49/5(235)
Cycles, intermittent 47(281)
Cyclone conservation 48(151)
Cyclone formation
— adiabatic 48(121)
— by condensation 48(136)
— by cutting-off 48(128)
— by shearing instability 48(81)
Cyclone in a basic flow 48(116)
Cyclone, temperature field 48(115)
Cyclone, tropopause 48(116)
Cyclones 48(112)
Cyclones, migrating, vertical motion in 48(119)
Cyclones, tropical 48(129)
Cyclotron absorption 49/4(467)
Cyclotron instability 49/4(5023)
Cyclotron resonance 49/4(464)
Cyclotron wave 49/5(235)
"Cyclotron-oscillations" 49/4(487)
Cylindrical body 49/5(271)

D echo 49/2(236, 253)
D region 49/2(222, 318–332)
— cosine law 49/2(322)
— cosmic noise absorption (method A2) 49/2(331)
— deviative absorption 49/2(322, 327)
— diurnal variation 49/2(322, 327)
— drift 49/2(413)
— electron density profiles 49/2(331, 383ff., 391)
— equatorial anomaly 49/2(325)
— fluctuations of absorption 49/2(331)
— ionograms at low frequencies 49/2(318)
— irregularities 49/2(400)
— lunar tides 49/2(565ff., 591)
— partial reflection 49/2(319)
— residual absorption 49/2(326)
— seasonal variation 49/2(325, 327)
— solar activity effect 49/2(324)
— sun rise phenomena 49/2(330)
— winter anomaly 49/2(330)
Daily international character figure 49/3(237)
Daily mean storm-time effect 49/3(60)
Daily means of cosmic-ray intensity 49/1(188ff., 192, 228ff.)
Daily means of neutron component 49/1(192)
Daily planetary character figure 49/3(227)
Daily variation of auroral distribution 49/1(13ff., 35)
Daily variation of radio wave auroral absorption 49/1(73)
DAIRCHEM model 49/7(440, 441 Fig. 152)
Damping 49/3(295)
Damping, Čerenkov 49/5(228)
Damping coefficient, spatial 49/5(239)
Damping decrement 49/5(224)

Damping decrement, temporal 49/5(238)
Damping factor, spatial 49/5(224)
Damping, gyroresonance 49/5(228)
Damping, Landau 49/4(204, 452, 456, 465); 49/5(228, 238)
Damping, Larmor 49/5(228)
Damping, spatial 49/5(224)
Damping, temporal 49/5(224)
Darbyshire spectrum 48(714)
Darwin's development 48(771)
Data flux of geomagnetic observatories 49/3(382)
Data reduction 49/2(212)
Dayglow, emission rate 49/6(122)
Daylight, colour of 48(212)
Daytime echoes from auroras 49/1(79)
Daytime-meteor showers 48(435)
Dazzle source 48(269)
$\mathbf{D}_{CF}$ 49/3(77)
De Boer-Bird expansion 49/7(69, 91, 115)
Deactivation, collisional (=quenching) 49/6(54)
Debye length 49/2(13, 482, 487); 49/4(447, 450); 49/6(80)
Debye radius λ_D 49/4(429); 49/5(218, 220, 238)
Debye screening 49/4(30, 548)
Debye-Hückel length 49/7(105–107 Fig. 18, 155, 162)
Debye-Hückel, screening 49/7(156 Figs. 31a, b, 157)
Decameter radiation 49/4(228)
Decay constant 47(295)
Decay of the disturbance 49/3(88)
Decay, quiet time 49/4(182)
Decay series 47(296)
Decca 47(622)
Decibel (dB) 48(352); 49/2(264)
Decile 49/2(269)
Decision act 49/2(209, 212)
Declination 47(499)
Declination (magnetic) 49/6(290, 291)
Declination tides 48(753)
Decrement, damping 49/5(224)
Deep sea trenches 47(253)
Deep Water 48(644)
— circulation 48(666)
— waves 48(671, 676, 704, 724)
Defant's method 48(853)
Definite integrals 49/7(215ff.)
Deflection 49/3(303)
Deflection angle 49/7(70)
Deflection, capped 47(259)
Deflection, fractured 47(259)
Deflection of the vertical 48(769, 787)
Deformation 47(75, 170)
Deformation of earth 48(785)
Deformograph 48(816)
Delta antenna 49/2(205)
Demagnetisation curves 49/3(319)
Demagnetization 47(475, 480)
Demagnetization in alternating field 49/1(275)
Demagnetizing factor 49/3(320)
Densities, neutral particle 49/5(372)
Density $\mathscr{L}$, Lagrangian 49/5(212, 215)
Density currents 48(616)
Density determination 49/7(247–251, 264, 268–272 Figs. 33–36)
Density, energy 49/5(343ff., 344)
Density field 48(11)
Density in situ 48(618)
Density, maps (incl. partial densities) 49/7(330 Fig. 86c, 339 Fig. 93c)
Density, profile 49/7(249 Fig. 15, 318 Fig. 77b)
Density profiles, electron 49/5(365, 368ff., 369)
Density, scale height 49/7(324, 326)
Density, wave momentum 49/5(214)
Density within the earth 47(114, 371, 376)
Deposition, *see* Detrite
Depth, greatest 48(615)
Depth, relative 48(725, 732)
Description of auroral displays 49/1(32–37)
Descriptive letter 49/2(228)
Detachment, associative 49/6(97)
Detail contrast 48(274)
Detectability range 49/3(433)
Determination of scale values 49/3(288)
Determination of the horizontal force 49/3(315)
— by the method of Gauss-Lamont 49/3(301)
Determination of the inclination 49/3(300)
Determination of the magnetic declination 49/3(298)

Determination of the vertical force 49/3(315)
Detrital remanent magnetization (DRM) 49/1(272, 286–290)
Deuterium 49/5(377)
Development of an aurora 49/1(32–37)
Deviations from equilibrium 49/2(9)
Deviative absorption 49/2(92, 225, 265ff.)
Deviative absorption in *D* region 49/2(322, 327)
Diagnostic diagram 49/5(187, 188)
Diagram, diagnostic 49/5(187, 188)
Diamagnetic minerals 49/1(248)
Diastrophism 47(260)
Dielectric eigenvalues 49/2(55, 59, 72)
Dielectric eigenvalues, effect of ions 49/2(41)
Dielectric matrix 49/2(148)
Dielectric permeability 49/5(228ff.)
Dielectric susceptibility 49/2(1, 22); 49/5(227)
Dielectric tensor 49/2(23ff., 29, 34, 149); 49/4(465, 466)
Dielectricity tensor 49/4(467, 468)
Differential absorption 49/2(319, 477)
Differential cross section 49/7(64, 92, 210)
Differential Doppler effect, *see also* Doppler effect 49/2(451–463)
Differential Doppler method 49/2(319)
Differential transport 49/5(175)
Diffraction 48(728)
Diffraction by droplets 48(257)
Diffraction by water drops 48(187)
Diffraction effect 48(256)
Diffraction filter 49/6(317)
Diffraction rings 48(187)
Diffuse boundaries 48(266, 269)
Diffuse elastic reflection 49/5(244)
Diffuse radiation, flux of 48(218)
Diffusing medium 48(336)
Diffusion 49/2(343); 49/6(5, 84, 99, 129, 145, 153, 154, 161, 170, 171, 204, 206, 208, 212, 301); 49/7(225, 301–305)
Diffusion, ambipolar 49/6(99, 145, 149, 193)
Diffusion, ambipolar theory 48(451)
Diffusion coefficient 48(451); 49/7(301, 303)
— ambipolar 49/6(149)
— eddy- 49/5(375, 376, 378)
— effective 49/6(167)
— molecular 49/6(84, 132, 134, 138, 139, 163, 165, 176, 188)
— temperature dependence 49/6(139)
— thermal 49/6(132)
— thermal ambipolar 49/6(149)
Diffusion, eddy 49/6(144, 155, 178, 185, 186, 260)
Diffusion, electron/ion controlled 49/6(212)
Diffusion equation 49/6(161)
Diffusion factor, thermal ambipolar 49/6(150)
Diffusion functions 48(332)
Diffusion in balance equation 49/2(252, 290)
Diffusion, molecular 49/5(126)
Diffusion of radiations, electromagnetic theory 48(330)
Diffusion of radiations in the atmosphere 48(329)
Diffusion operator 49/2(517)
Diffusion, plasma, effect on neutral wind 49/6(301, 302)
Diffusion, radial 49/4(76, 77)
Diffusion regime 49/7(225)
Diffusion tensor 49/4(427); 49/7(175)
Diffusion, thermal 49/6(84, 217)
Diffusion thermo effect (Peltier effect) 49/7(179, 191)
Diffusion velocity 49/4(168); 49/6(85, 99, 131)
Diffusive equilibrium 49/6(152, 154, 169, 170, 219)
Diffusive separation 49/6(158, 169, 174, 244)
Digamma function 49/7(216)
Digital magnetic character 49/3(235)
Digitalisation of data sequences 49/3(389)
Digitalisation of registrations curves 49/3(382)
Dilatation 47(76)
Diminishing factor 48(787, 791, 796, 829)
Dingle integrals 49/7(196–198, 200)
Dip angle of magnetic field and lunar tides 49/2(556)
Dip equator 49/6(288); 49/7(449, 482)
Dip equator belt 49/2(316)
Dipole 49/3(3)

Dipole antennae of different shapes 49/2(478ff., 484)
Dipole density 49/2(22, 28)
Dipole field 49/4(120)
Dipole, field of a 49/4(3)
Dipole field, reversal 47(532)
Dipole moment 49/7(98)
Direct determination
— of $\mathscr{H}$ 49/3(301, 305, 307, 308, 309, 314)
— of $\mathscr{I}$ 49/3(300)
— of $\mathscr{Z}$ 49/3(308, 309, 315)
— of geomagnetic elements 49/3(297)
— of scale values 49/3(289)
Disagreement between theoretical and observational results 49/5(160)
Disappearance range meter 48(286)
Discharge-flow method 49/6(58)
Discontinuity, rotational 49/4(26)
Discontinuity surfaces 49/4(34)
Dispersion 48(674, 719)
Dispersion equation(s) 49/4(32, 472); 49/5(196, 224, 227)
Dispersion equation, general 49/4(471)
Dispersion formula(e)
— near zeros of refractive index 49/2(87)
— of Appleton and Hartree 49/2(489)
— of Appleton and Lassen 49/2(30, 40, 59, 72, 89, 116, 143, 489)
— of Försterling 49/2(37ff., 68, 72, 89, 143)
— of Sellmeier 49/2(41, 93, 446, 454)
— of Sen and Wyller 49/2(39ff., 59, 62, 72, 89, 143)
Dispersion, frequency 49/5(227)
Dispersion of sound 48(460)
Dispersion range of statistics 49/2(251, 269)
Dispersion relation 49/2(29ff.); 49/4(405, 496, 499, 525, 528, 531)
Dispersion, spatial 49/4(447); 49/5(227, 238)
Dissipation law 48(571)
Dissipation of energy 48(711, 715, 728)
Dissipation, turbulent 49/5(206)
Dissipative processes 49/5(205)
Dissociation 49/6(103, 177)
Dissociation of CO_2 49/5(375)
Dissociation threshold 49/6(26)
Dissociative charge transfer 49/6(119)
Dissociative photoionization 49/6(24)
Dissociative recombination 49/2(285, 373); 49/6(4, 46, 67, 97, 117, 121, 194); 49/7(433, 434, 446)
Distance conversion factor of $F2$ layer 49/2(308)
Distribution 49/3(303)
Distribution coefficients 49/3(303, 304)
Distribution function(s) 49/5(225, 226); 49/7(229)
— gas kinetic 49/6(265)
— Maxwellian 49/5(226ff.)
— for trapped particles 49/1(88, 98)
Distribution of aurora in the polar cap 49/1(10ff.)
Distribution of charged particles around the earth 49/1(81)
Distribution of electron density, *see* Electron density distribution
Disturbance, daily variation 49/3(55, 56, 78)
Disturbance effect on neutrals 49/7(313 Figs. 74, 75, 332–335 Figs. 90, 91, 346, 347 Figs. 100, 101)
Disturbance field 49/3(26)
Disturbance field, magnetic 49/2(271)
Disturbance force vector 49/3(62)
Disturbance local-time variation 49/3(56)
Disturbance longitudinal inequality 49/3(56)
Disturbance, magnetic 49/4(59, 215)
Disturbances by magnetic storms 49/2(366–377)
Disturbances by solar flares 49/2(361–366)
Disturbances of the ionosphere 49/2(241)
Disturbing magnetic field D, subdivision and symbols 49/1(55)
Disturbing potential 47(589)
Diurnal heat wave 49/5(150, 154)
Diurnal meridional winds 49/5(157)
Diurnal temperature variations 49/5(152–154, 161)
Diurnal tide 49/7(338)
Diurnal variation(s) 49/2(251, 289); 49/5(119); 49/6(201, 286)
— of activity 49/3(270)
— of cosmic-ray intensity 49/1(171–180, 217)
— of cosmic-ray intensity and magnetic activity 49/1(178)

— of *D* region 49/2(322, 327)
— of *F*2 layer 49/2(293, 309)
Diurnal wind fields 49/5(158)
Diurnal zonal winds 49/5(159)
$\mathbf{D_m}$ 49/3(60, 62, 63)
Dobson spectral photometer 48(378, 407)
Domain, acoustic 49/5(189)
Doppler broadening 48(221); 49/6(80)
Doppler curve 49/2(449)
Doppler effect 49/2(356, 435, 437, 444, 447ff.); 49/5(228, 366)
— anomalous 49/5(228ff.)
— application of rockets 49/2(457–462)
— beat method 49/2(454, 459)
— differential 49/2(451–463)
— normal 49/5(228)
— of radar echoes 49/1(79)
Doppler measurements 49/7(265–267, 280 Fig. 46)
Doppler of oxygen lines in aurora 49/1(122)
Doppler shift 49/5(228); 49/6(278, 350); 49/7(173, 256)
Doppler shift of $H\alpha$ line in aurora 49/1(110)
Doppler temperature 49/1(122)
Doppler-shift 49/4(495)
Double bifilar gravimeter 48(802)
Double core probe 49/3(348)
Double refraction of gyrotropic plasma 49/2(28)
$\mathbf{D_p}$ 49/3(77, 97)
DP 2 current system 49/4(80, 88, 96)
$\mathbf{D_p}$ substorm 49/3(100)
$\mathbf{D_R}$ 49/3(59, 77)
Drag coefficient 49/7(234)
Drapery 49/1(16)
D-region 49/6(15, 37, 53, 62, 69, 77, 96, 100, 104, 125); 49/7(358–364, 378–383 Figs. 119–122, 434–440 Figs. 146–152)
D-region, electron density profiles 49/7(359–364 Figs. 110, 111)
D-region, ion composition 49/6(100); 49/7(378–383 Figs. 119–122)
D-region, photoionization 49/6(38)
Drift 49/4(21, 74)
Drift across field lines 49/4(119)
Drift current 48(717)
Drift, *D* region 49/2(413)
Drift, dominant component 49/2(438)
Drift, *E* region 49/2(413)
Drift, electric field 49/4(62, 63)
Drift, electrodynamic 49/4(97)
Drift, *Es* clouds 49/2(442)
Drift, *F* region 49/2(420)
Drift, full correlation methods 49/2(409, 433)
Drift, galactic noise 49/2(433)
Drift, large scale irregularities 49/2(438)
Drift, lifetime of fades 49/2(407)
Drift, lunar 12 h component 49/2(416)
Drift mechanism 49/4(163)
Drift, meteor echoes 49/2(435)
Drift motion 49/4(18, 49, 50, 61, 67, 71, 72, 78, 81, 84, 90, 101–103, 111, 112)
Drift motion of a charged particle(s) 49/1(84, 86); 49/2(547, 580)
Drift observation, fading method 49/2(405–422)
Drift period 49/4(120)
Drift, prevailing 49/2(418)
Drift process 49/4(154)
Drift, pulse radar method 49/2(435)
Drift, radar echoes 49/2(438)
Drift, random component 49/2(413)
Drift, satellite methods 49/2(435)
Drift, similar fade method 49/2(407, 410)
Drift, solar 12 h component 49/2(415, 417)
Drift, solar 24 h component 49/2(419)
Drift tube 49/6(63)
Drift velocity 49/2(427); 49/4(18, 61, 67)
Drift, vertical displacement 49/2(432)
Drift, zonal system 49/2(419)
Drifting force 47(279)
Drifts in sporadic *E* layer 49/2(337)
Drizzle 48(483, 519)
DRM, *see* Detrital remanent magnetization
Drop charge 48(597)
Droplet-growth, observations 48(492)
Droplet-growth, theory 48(495)
Droplets, accretion 48(497)
Droplets, concentration 48(518)
Droplets, salty 48(369)
Droplets spectra 48(493)
Dropout 49/2(450)
$\mathbf{D_S}$ **(SC)** 49/3(48, 74, 99)

Duct 49/5(197)
Duct, acoustic 49/5(200)
Duct inside the ionosphere 49/4(521)
Ducted modes 49/5(200)
Dust 48(160)
Dust particles 48(164)
$\mathbf{D}_W$ 49/3(77)
Dye magnetometer 49/3(314)
Dykes 47(265, 484, 487)
Dynamic metre 48(650)
Dynamic scale value 49/3(294)
Dynamic thermal control 49/5(133)
Dynamics of auroral displays 49/1(37)
Dynamo current systems 49/5(173)
Dynamo currents 49/5(171)
Dynamo region 49/7(444, 450)
Dynamo region, high latitude 49/5(173)
Dynamo theory 48(931); 49/3(24)
— of diurnal variation 49/2(289)
— of magnetic tidal effects 49/2(570ff., 576, 587)
Dynamo-electric effect 49/4(112)

E region 49/2(222, 278–285)
— critical frequency 49/2(278)
— drift 49/2(413)
— height of maximum electron density 49/2(278)
— ionization origin 49/2(285)
— irregularities 49/2(401)
— latitude variation 49/2(282ff.)
— lunar tides 49/2(280, 558, 561, 568, 579, 591)
— noon depression 49/2(281)
— solar activity effect 49/2(283)
— thickness 49/2(278)
— virtual height 49/2(278)
Earth, bodily tides 47(21, 114)
Earth, central core 47(89, 114)
Earth, core 47(19, 261, 370); 48(821)
— liquid 48(791)
— rigidity of 48(821)
Earth, crustal layers 47(104)
Earth current 47(406)
Earth, density variation 47(112)
Earth, dimensions 47(633)
Earth, division into regions 47(103, 370)
Earth, elastic yielding 47(19)
Earth ellipsoid, three-axial 47(599, 634)
Earth, ellipticity 47(93, 553, 560, 592, 635)
Earth, fluid zone 47(89)
Earth, incompressibility 47(104)
Earth, inner core 47(89, 104, 110, 376, 382, 388)
Earth, inner core solid 47(110, 116)
Earth, lower mantle 47(104)
Earth, mantle 47(89, 252, 261, 370)
Earth, mass 47(555, 636)
Earth, mean density 47(561, 604)
Earth, models *A* and *B* 47(114, 115)
Earth, moment of inertia, change 47(10)
Earth, moments of inertia 47(636)
Earth, origin 47(380, 399)
Earth, outer core 47(104)
Earth, radius 47(561)
Earth, rigidity 47(104)
Earth, rotation 47(1)
— irregularities 47(17)
— seasonal variations 47(15, 17)
Earth storm 49/3(12)
Earth, surface 47(561)
Earth, temperature 47(395)
Earth, thermal history 47(389)
Earth tides, seasonal changes 47(21)
Earth, upper mantle 47(104)
Earth, volume 47(561)
Earth-inductor 49/3(300)
Earthquake, asymmetry near focus 47(83)
Earthquake, dimensions of strained region 47(82)
Earthquake, dislocation theory 47(83)
Earthquake, energy 47(81)
Earthquake, epicentre 47(82)
Earthquake, focal region 47(82)
Earthquake, focus 47(75, 82)
Earthquake, initial conditions 47(81, 88)
Earthquake, kinematic effect 47(77)
Earthquake, magnitude 47(81)
Earthquake, release of energy 47(83)
Earthquake, tectonic 47(81)
Earthquake, volcanic 47(81)
Earthquake-belt, circum-Pacific 47(85)
Earthquakes, aftershocks 47(84)
Earthquakes, anticentre 47(89)
Earthquakes, causes 47(81)
Earthquakes, deep focus 47(84, 89)

Earthquakes, first ground movement 47(83)
Earthquakes, focal depth 47(84)
Earthquakes, foreshocks 47(84)
Earthquakes, major 47(84)
Earthquakes, mediterranean belt 47(85)
Earthquakes, *P*-waves 47(85)
Earthquakes, *S*-waves 47(85)
Earth's centre, density 47(116, 373)
Earth's crust 47(75, 81, 246, 261, 370, 460, 534)
— composition 47(257)
— continental 47(122, 128)
— definitions 47(246, 257)
— heterogeneity 47(88)
— oceanic 47(124, 128, 251)
— oscillations 47(19)
— shortening 47(258, 267)
— silicic 47(246)
Earth's current, vertical 48(594)
Earth's elasticity, characteristic numbers 48(788)
Earth's figure 47(535)
Earth's figure, magnitudes 47(563)
Earth's interior
— composition 47(117)
— compressibility 47(114)
— density 47(603)
— elastic constants 47(374)
— electromotive forces 47(532)
— gravity 47(114, 603)
— level surfaces 47(634)
— pressure 47(114)
— rigidity 47(114)
— temperature 47(117)
Earth's magnetic field 49/2(1, 2, 10, 32)
— angle with wave normal 49/2(33, 49)
— causing anisotropy of plasma 49/2(2)
— longitudinal and transverse components 49/2(40, 49)
Earth's magnetic tides 48(931, 932)
Earth's mantle, inhomogeneous 47(113)
Earth's mass 47(555, 636)
Earth's oblateness 47(93, 553, 560, 635)
Earth's relief development in spherical harmonics by Pres 47(582, 625)
Earth's rotation, deceleration by Moon 48(924)
Earth's rotation, secular retardation 47(3)
Earth's shadow 48(217)
Earth's tide(s) 48(783)
— accuracy 48(782)
— and length of day 48(844)
— direct 48(784)
— diurnal 48(797)
— geological influences 48(841)
— indirect effects 48(784)
— influence of earth's cosmic motion 48(844)
— in the interior 48(822)
— meteorological influences 48(835)
— ocean influence 48(825, 833)
— secondary effects 48(784)
East Chinese Border Seas 48(912)
East Indian Archipelago 48(912)
Eastward drift motion 49/1(43ff., 48, 49)
East-West effect 49/4(135)
East-West motion of auroras 49/1(38ff., 78ff.)
$E \times B$ plasma drift 49/6(303)
Ebb, *see* Tides
Echo amplitude 49/2(215)
Echo from *D* region 49/2(236, 253)
Echo polarization 49/2(216)
Echo sounding 49/2(199)
Echo traces above and below gyrofrequency 49/2(197)
Echo traces, reflection and polarization designations 49/2(197)
Echoes of high-frequency signals from auroras 49/1(75–81)
Echoes of topside sounding 49/2(85)
Echoes of vertical sounding 49/2(82, 84)
Eclipse 47(622); 49/6(344, 347)
Eclipses 48(747)
Ecliptical coordinates 48(748, 750)
Eddies 48(613, 621); 49/6(166, 168)
Eddy conductivity 48(44)
Eddy diffusion 49/6(85, 130, 144, 154, 155, 178, 185, 186, 260); 49/7(302–304 Fig. 65)
— coefficient 49/5(375, 376, 378); 49/6(86, 145, 192, 260)
— transport coefficient 49/5(126)
Eddy diffusivity 48(44)
Eddy energy 48(35)
Eddy flux of angular momentum 48(77)
Eddy flux of internal energy 48(44)
Eddy flux of sensible heat 48(44)

Eddy transport 49/5(123, 127, 134); 49/6(166)
Eddy viscosity 48(39); 49/5(127)
Eddy viscosity coefficient 48(718)
Eddy-transported heat, vertical 49/5(128)
Effect, Doppler 49/5(228, 366)
Effect of focusing 49/5(266, 267)
Effect, orographic 49/5(144)
Effect, resonance 49/5(321)
Effective dielectric tensor 49/2(23, 30, 34); 49/4(510)
Effective differential cross-section 49/5(286)
Effective displacement vector for plasma 49/2(23, 27, 37)
Effective mass 49/5(236)
Effective ray 48(347)
Effective temperature 49/6(27)
Effects, viscous 49/5(128)
Eiconal 49/4(509)
Eiconal equation 49/4(516, 540, 543, 544, 547, 555)
Eigen values 48(948, 953)
Eigenbase 49/7(212)
Eigenperiod, influence of friction 48(850)
Eigenvalue representation of a tensor 49/2(3, 6, 21, 31)
Eigenvalues (general) 49/7(211)
Eigenvalues of dielectric tensor 49/2(24, 34)
Eigenvectors 49/7(211)
Eigenvectors of waves 49/2(151ff.)
Eikonal 49/2(121)
Eikonal approximation 49/2(24ff., 57–62)
Eikonal equation 49/2(133, 137); 49/5(195, 196)
Einstein coefficients 49/6(39, 352)
Einstein relation 49/7(61)
Ejecta 48(982)
Ejecta spiral 48(39, 653)
E.J.F., *see* Estimated junction frequency
Elastic after-effect 49/3(312)
Elastic collisions 49/7(466, 527)
Elastic parameters 47(78)
Elastic parameters, adiabatic or isothermal 47(87)
Elastic vibrations, damping 47(180)
Elastic waves, theory 47(365)
Elasticity, aelotropy 47(80, 86)
Elasticity, dependent on pressure and porosity 47(175)
Elasticity, Hysteresis 47(170)
Elasticity, imperfect 47(86)
Elasticity, isotropy 47(78)
Elasticity of rocks 47(169)
Elasticity theory 47(75)
E-layer and aurora 49/1(70)
E-layer, ionization by meteors 48(453)
E-layer (region) 49/7(364)
Electric conductivity 49/3(6)
Electric current density 49/2(273)
Electric current structure 49/5(173)
Electric current-systems for $\mathbf{D}_{st}$- and $\mathbf{S}_D$-fields 49/3(68)
Electric field(s) 49/4(50, 60); 49/6(99, 146, 151, 212–214, 236, 262, 288, 304)
— effects 49/7(444, 446 Fig. 156c, 448, 450, 457 Fig. 165)
— influence 49/2(274ff.)
— influence, transverse component 49/2(49)
— model of aurora 49/1(136)
— over the magnetic equator 49/2(578)
Electric mode 47(525)
Electric potential fields 49/5(173)
Electrical charge of the earth 48(541)
Electrical conductivity 47(405); 49/7(185–191)
— of air 48(572)
— structure 49/5(170)
Electrical field(s) 48(542, 576)
— during precipitation 48(583)
— during thunderstorms 48(583)
— in the free atmosphere 48(589)
Electric-field drift 49/4(75, 84)
Electrification effects 48(584)
Electro jet, equatorial 49/6(120)
Electrode effect 48(570, 588, 594, 596)
Electrodynamic force 49/7(448)
Electrodynamic lift 49/7(457 Fig. 165)
Electrofiltration 47(433)
Electrojet 49/2(270); 49/3(245); 49/4(215)
Electrojets in auroras 49/1(56–62, 88, 90)
Electromagnetic braking 49/7(235)
Electromagnetic drift 49/6(279)
Electron acoustic wave 49/2(61ff., 70)
Electron attachment 49/6(53, 126)

Electron beam 49/4(496)
Electron belt, decay of the 49/4(144)
Electron content 49/7(408 Fig. 134)
Electron content of a layer 49/2(268, 316, 377, 446, 452ff., 467, 473)
Electron content, total 49/4(97); 49/5(81, 358)
Electron density
— coronal 49/6(356)
— distribution 49/2(252–262, 267)
— distribution according to Chapman's theory 49/2(579, 583, 586)
— distribution, distortion by electron drift 49/2(583)
— distribution in *F*2 layer 49/2(558)
— diurnal variation 49/6(201, 286)
— in the magnetosphere 49/2(500ff.)
— inhomogeneities 49/6(213)
— layer shape 49/6(202)
— maximum 49/5(369, 370)
— peak 49/6(200, 203, 285, 287)
— profile(s) 49/2(119, 359); 49/6(12, 194, 195, 198, 199, 201, 202, 226, 244); 49/7(38, 347–367 Figs. 102–112, 395 Fig. 127, 407 Fig. 133, 409 Fig. 136, 520 Fig. 216); 49/5(365, 368ff., 369)
— profiles by Doppler rocket method 49/2(461, 462)
— profiles by field-strength measurement 49/2(476)
— topside profile 49/2(509, 512–514, 520ff.)
Electron drift velocity 49/1(60)
Electron energy flow 49/7(461 Fig. 167, 462)
Electron energy spectra 49/4(140)
Electron energy spectrum 49/1(90–93, 95); 49/7(369 Fig. 114)
Electron flux 49/4(64, 65)
Electron flux, auroral 49/4(214)
Electron fluxes 49/4(164)
Electron fluxes in tail, islands of 49/4(195)
Electron focusing 49/5(269)
Electron gas, loss rate 49/7(469 Fig. 171)
Electron gyrofrequency 49/5(220, 241)
Electron heat flux 49/4(432)
Electron impact (collisions) 49/6(352)
Electron islands 49/4(52)
Electron oscillations, high frequency Langmuir 49/4(450)
Electron plasma 49/4(450); 49/7(185–192)
Electron plasma pulsation 49/5(221)
Electron population, inner zone 49/4(135)
Electron precipitation 49/7(462)
Electron sound wave 49/2(18, 483, 509)
Electron spectrum, auroral 49/4(212)
Electron temperature 49/2(5, 8, 15ff., 485, 523ff.); 49/7(517 Fig. 213)
— profile 49/6(13); 49/7(36–374 Figs. 112, 113, 116, 38, 412 Fig. 139, 413, 477 Fig. 174)
— vs. electron density 49/7(370–373 Fig. 115, 413, 476)
— world maps 49/7(411 Fig. 138, 414–417 Figs. 140a–d)
Electron whistler wave 49/5(236)
Electron-acoustic wave 49/5(239, 329)
Electronic collisions 49/4(550–553)
Electronic cyclotron wave 49/4(489)
Electronic polarization 49/2(49, 52, 61ff.)
Electronic wave 49/5(236)
Electronically excited states 49/6(89)
Electron-ion collisions 49/2(13, 15)
Electrons 48(543, 574); 49/4(164)
Electrons, auroral 49/1(95); 49/4(216)
Electrons, collisions with neutrals 49/7(99–105, 162)
Electrons, flux of precipitating 49/4(172)
Electrons in radiation belts 49/1(82)
Electrons, Langmuir oscillations 49/5(221)
Electrons of high energies in the outer belt 49/1(96)
Electrons of low energies and aurora formation 49/1(100)
Electrons of medium energies and aurora formation 49/1(97–99)
Electrons, trapped 49/4(216)
Electrostatic accelerometer 49/7(250)
Electrostatic potential 49/5(243)
Electrostatic wave 49/5(240)
ELF Alfvén waves 49/5(250, 252)
ELF waves 49/5(250, 251)
ELF whistlers 49/5(302)
Elias-Chapman layer 49/5(121); 49/7(348, 350)
Elias-Chapman profile 49/6(35, 201, 202)

Elias-Chapman theory 49/2(251)
Elleswortbite 47(349)
Ellipse of polarization 49/2(27, 42, 44, 48)
Ellipsoid, formula 47(568)
Ellipsoid, international 47(564, 633)
Ellipsoidal coordinates 47(544, 626)
Elliptical polarization 49/2(35)
Ellipticity 47(93, 553, 560, 635); 48(290); 49/2(43)
Ellipticity, change with height 47(560)
Ellipticity corrections 47(96)
Elsasser, radiation chart 48(235)
Emanation (Radium, Thorium) 48(546)
Emanometer 48(549)
Emersio 47(89)
Emission 49/5(339)
Emission from the wake 49/5(299)
Emission, Hiss 49/5(322)
Emission measure 49/6(351, 353–356)
Emission, narrow-band 49/5(338)
Emission rate 49/7(282 Fig. 47)
Emissions, artificially stimulated (ASE) 49/5(332)
Emissions, Classification 49/5(302)
Emissions, saucer-shaped 49/5(322, 324)
Emissions, very low frequency 49/4(215)
Emissions, whistler-triggered, (WTE) 49/5(331)
Emulsion observations of solar cosmic rays 49/1(242)
Energetic electrons 49/3(154); 49/7(462, 463 Fig. 168)
Energization 49/4(72, 75)
Energy 49/4(70, 88, 102, 103)
Energy balance in wave optics 49/2(166)
Energy balance of plasma 49/2(5, 10, 11, 15ff.)
Energy densities in tail and neutral sheet 49/4(197)
Energy densities, ratio of 49/4(24, 25)
Energy density 49/5(343ff., 344)
Energy density, wave 49/5(210)
Energy, eddy 48(35)
Energy equation(s) 48(15, 23); 49/6(259)
Energy exchange, water–air 48(635)
Energy exchange with space 48(155)
Energy, flow in electron gas 49/7(461 Fig. 167, 462)
Energy flux 49/2(11); 49/5(343)
— balance 49/7(58)
— vector 49/7(59)
— velocity 49/5(248)
Energy, internal, of dry air 48(2)
Energy, internal, of moist air 48(5)
Energy, kinetic, dissipation 48(16, 20, 36, 41)
Energy level diagram, N_2 49/6(21, 48)
Energy level diagram, O_2 49/6(18, 48)
Energy, meridional flux 48(153)
Energy-momentum tensor 49/5(212)
Energy of ring current 49/1(132)
Energy spectra of electron fluxes 49/7(530 Fig. 222)
Energy spectra of trapped elecrons 49/4(165)
Energy spectra, steady state 49/4(164)
Energy spectrum 48(691, 693, 695, 700, 714, 720, 722); 49/4(161)
— of electrons 49/1(90–93, 95)
— of protons 49/1(95, 226, 237)
Energy transfer equation 49/7(233)
Energy transfer, short to long waves 48(715)
Energy transfer, wind to waves 48(710, 713, 715, 717, 728)
Energy transformation(s) 48(18)
— between mean and eddy motion 48(36)
— eddy-motion 48(36, 43)
— for zonally averaged flow 48(151)
— in baroclinic waves 48(87)
Energy, wave 48(52); 49/5(209)
Engineering seismics 47(168)
English Channel 48(911)
Enhancement of x-rays 49/2(366)
Enskog-Chapman gas-kinetic theory 49/6(264)
Enthalpy, meridional flux 48(153)
Enthalpy of moist air 48(5)
Enthalpy of saturated air 48(6)
Enthalpy of water-vapor 48(4)
Entropy 48(3, 146)
Entropy, meridional eddy flux 48(146)
Eötvös-Effekt 47(212)
Ephemerides 47(1)
Ephemeris-time 47(2, 22)
Epstein layer 49/2(187ff., 192, 194ff.)
Epstein step 49/2(187ff.)
Epstein step function 49/7(324–326, 348, 351)

Epstein transition function 49/7(352)
Epstein's index fraction 49/2(187ff.)
Equalizing layer 48(544, 580, 586)
Equation, continuity 49/5(180)
Equation, dispersion 49/5(196, 224, 227)
Equation, eikonal 49/5(195, 196)
Equation, Euler-Lagrange 49/5(212, 215)
Equation, inhomogeneous 49/5(204)
Equation, kinetic 49/5(225, 253)
Equation, Navier-Stokes 49/5(127)
Equation of continuity 49/2(7)
Equation(s) of motion 48(8, 22, 25, 31, 35); 49/4(17, 19); 49/5(180); 49/6(204, 260, 262); 49/7(232, 448, 470)
— for electrons 49/2(272)
— hydro-dynamic 48(649)
Equation of time 48(748)
Equation, Poisson 49/5(225, 227)
Equation, Sellmeier 49/5(355)
Equation, Whitaker's 49/5(193)
Equatorial anomaly 49/2(515ff.)
Equatorial anomaly of *D* region 49/2(325)
Equatorial electrojet 49/3(17, 113); 49/6(120)
Equatorial *Es, see also Es* 49/2(235)
Equatorial ring current 49/3(77, 100)
Equatorial water 48(647, 648)
Equilibrium, *see* termal equilibrium 49/2(8); 49/4(26, 28)
Equilibrium at boundary 49/4(38)
Equilibrium luminance 48(265)
Equilibrium tides 48(737, 780, 933)
Equinoctial events 49/5(149)
Equinoctial-hypothesis 49/3(257)
Equi-potential surface, curvature 47(217)
Equivalent depth 48(948)
Equivalent height, *see also* Virtual height 49/2(113, 196)
— for a Chapman-Elias layer 49/2(118)
— for a parabolic layer 49/2(119)
— for an Epstein layer 49/2(195)
— in polar regions 49/2(198)
Equivalent ionospheric current 49/3(19)
Equivalent monotonous profile 49/2(261)
Equivalent overhead current-system 49/3(18)
Equivalent overhead electric current 49/3(14, 47)
— -system 49/3(76, 114)
Equivalent path 49/2(112, 117, 142)
Equivalent planetary amplitude 49/3(226)
Equivalent thickness determined by Faraday effect 49/2(470ff.)
E-region 49/6(35, 97, 110)
E-region, ion chemistry 49/6(110, 111)
E-region (layer) 49/7(440–442 Fig. 153a)
E-region, photoionization 49/6(35)
Erosion 49/5(245)
Error function 49/5(259)
Ertel's theorem 48(13)
Erythem 48(423)
Es 49/2(224)
Es clouds 49/2(442)
Es layer, lunar tides 49/2(560–565, 586–589, 595)
Es layer, lunar tides near magnetic equator 49/2(564ff., 591)
Escape 49/6(215–217, 225, 265)
Escape, critical level 49/6(216)
Escape parameters 49/6(217)
Escape problem 49/7(231)
Escape time 49/4(152)
Escape velocity 49/6(3)
ESRO-4 model 49/7(331, 334, 335 Fig. 89, 340–346)
ESRO-4, satellite 49/7(331–333)
Establishment of a port 48(881)
Estimated junction frequency 49/2(218, 227, 228)
Eulerian equations 48(7)
Euler-Lagrange equation 49/5(212, 215)
Euler's equation 49/7(232)
Euler's equations for deformable earth 47(607, 615)
Euler's period 47(609); 48(790)
EUV emissions, quiet intensity 49/7(19)
EUV emissions, two component model 49/7(18, 20, 21)
EUV (extreme ultra violet) 49/6(309, 365)
EUV flux (total) 49/7(11)
EUV indices 49/6(365, 369, 371)
EUV range, radiated power (total flux) 49/6(361, 363, 365)
EUV spectral range 49/7(2, 8)
EUV-indices 49/7(33–35)

EUV-spectrograph 49/6(312, 340)
EUV-spectrometer (spectroscope) 49/6(311, 360)
Evanescent eigenwaves 49/2(153ff.)
Evanescent waves 49/2(24, 60, 131)
Evaporation 48(248, 628, 631, 633, 659); 49/5(245)
Evaporation flux 49/5(245)
Evaporation rate 49/5(245)
Evaporative theory 49/4(13)
Evaporative theory of polarwind 49/4(81)
Evection 48(747, 749, 752)
Evectional tides 48(753)
Evening bulge 49/4(58)
Events, equinoctial 49/5(149)
ex Echo 49/2(504)
Exchange 48(547, 937, 963, 974)
Exchange coefficient 48(39, 399, 406)
Excitation 49/5(228, 320); 49/6(352, 355, 374)
Excitation as a source of auroral spectra 49/1(115–117)
Excitation by electron impact 49/6(352)
Excitation by photoelectrons 49/6(39)
Excitation (by suprathermal electrons) 49/7(528)
Excitation, Čerenkov 49/5(228)
Excitation, gyroresonance 49/5(228)
Excitation, ion-cyclotron 49/5(305)
Excitation of waves 49/5(295)
Excited species 49/6(53, 57, 120, 121, 123, 124); 49/7(491–493)
Excited states 49/6(24, 53, 57, 65, 89)
Exhalation 48(546)
Exner-function 48(3)
Exobase 49/6(237, 264)
Exosphere 49/6(3, 9, 86, 220, 237, 264); 49/7(226, 231)
Exosphere, lunar effects 49/2(595)
Exospheric temperature 49/6(220, 252); 49/7(42, 43, 262–264 Figs. 27a–c, 284–286 Figs. 50, 51, 293–295 Figs. 58a, b, 316–321 Figs. 76–80, 329–332 Figs. 86, 87, 337–342 Figs. 94–96, 345–347)
Expansion 49/6(295)
Exploration for ores 47(244)
Exploration for petroleum 47(242)
Explorer 48(382)
Explorer I & III 49/1(81)
Explorer VI 49/1(233)
Explorer XII 49/1(81)
Explosion earthquakes 48(993, 997)
Explosion, high altitude 49/4(143)
Explosions 48(464)
Explosive warming 49/5(118, 131, 133ff.)
Exponential potential 49/7(78, 127)
Extension 47(76)
Extension of recording range 49/3(290)
Extensometer 48(816)
External particle fluxes, velocity 49/5(219)
External sources 49/5(225)
Extinction 48(156, 254)
Extinction, anomalous 48(164, 170, 259)
Extinction by spherical particles 48(255)
Extinction coefficient 48(156, 158, 258, 259)
Extinction coefficient in clouds 48(209)
Extinction coefficient (sea) 48(622)
Extraordinary wave 49/5(232, 236)
Extreme low frequency branch 49/5(234)
Extreme Ultra Violet (EUV) 49/6(309, 365)
Eye and ear method 49/3(338)
Eye, properties 48(267)
Eye, threshold 48(267)
ez trace 49/2(239)

*F*1 layer 49/7(349, 364)
— critical frequency 49/2(285)
— effect of solar cycle 49/2(286)
— height of its center 49/2(287)
— lunar tides 49/2(558–560)
— recombination 49/2(285)
— virtual height 49/2(287)
*F*1½ layer 49/2(346)
*F*2 layer 49/6(193, 285, 287); 49/7(299 Fig. 61)
— attachment coefficient 49/2(289)
— characteristic heights 49/2(307–318)
— diffusion 49/2(290)
— distance conversion factor 49/2(308)
— disturbances 49/4(97)
— diurnal heating 49/2(289)
— diurnal variation 49/2(293, 309); 49/6(286)
— electron content 49/2(316)

— electron density profile 49/2(290–292, 316)
— evening maximum 49/2(294)
— geomagnetic control 49/2(298)
— height in parabolic approximation 49/2(308)
— ionization balance 49/2(288 ff.)
— ionization maps 49/2(302, 310)
— longitude effect 49/2(298)
— lunar tides 49/2(296, 315, 549–558, 580, 584)
— magnetic field influence 49/2(316)
— maximum electron density 49/2(292)
— minimum virtual height 49/2(307)
— real height of maximum density 49/2(314)
— seasonal variation 49/2(302, 313); 49/6(287)
— solar activity effect 49/2(303, 316)
— solar cycle effect 49/2(298, 305, 313)
— thickness 49/2(315)
— tidal effect 49/2(289)
— trough at magnetic equator 49/2(517)
— world-wide index 49/2(306)
F_2 model 49/5(373)
F region 49/2(222, 285–318)
— drift 49/2(420)
— equatorial phenomena 49/2(514)
— irregularities 49/2(401)
Fabry-Perot interferometer 49/7(273 Figs. 39a–c)
Factorization of Booker's quartic 49/2(132)
Fading method 49/2(405–422)
Fading patterns 49/2(431)
Fading velocity 49/2(426)
Failure 47(263, 273)
Falling body 47(205)
Falling tone emission 49/5(332)
Fallstreaks 48(518, 520)
Far zone 49/5(253, 269, 274)
Faraday effect 49/2(56–59, 85, 463–475, 477)
— determination of equivalent thickness 49/2(470–472)
— indetermination 49/2(468 ff., 472)
Faraday fading frequency 49/2(472 ff.)
Faraday rotation 49/2(464)
Fast electron-acoustic wave 49/5(240)
— ion-acoustic wave 49/5(241, 251 ff., 296, 316–318, 320, 329)
Fast magneto-acoustic wave 49/5(236)
Fast particles 49/4(60, 63, 72)
Fault, dynamics 47(263)
Fault, normal 47(258, 264)
Fault, reversed 47(258, 264)
Fault, transcurrent 47(258, 264)
Faulting 47(83)
Fe XVI (33.5 nm solar emission) 49/7(53–55)
Features, small scale 49/5(124)
Feldspars 47(336, 349)
Fermat's principle 47(83, 90); 49/2(144 ff.)
Fermi acceleration 49/3(155)
Fermi process 49/4(75, 76)
Ferrimagnetic minerals 49/1(248, 250 ff., 254, 258, 266 ff.)
Ferrimagnetic oxides and sulphides 49/1(250)
Ferrimagnetism 47(470)
Ferrimagnetism of *N*-type 49/1(280 ff.)
Ferrite 47(470)
Ferromagnetism 47(470)
Fetch 48(709, 714, 720, 729)
Fick's diffusion law 49/7(61, 175)
Field changes 48(587)
Field components 49/3(277)
Field compression 49/3(179)
Field, distorted 49/4(189)
Field, electric 49/4(82–84, 95, 106–109, 112)
Field equations 49/4(396, 397, 405)
Field, geomagnetic 49/5(175)
Field line, merging 49/4(25)
Field line shell 49/4(67–70)
Field lines, Frozen-in magnetic 49/4(21)
Field mill 48(577)
Field variance 49/4(204)
"Field-aligned" 49/4(544)
Field-aligned currents 49/7(476)
Fields, electric 49/4(103)
Fieldstrength determination 49/2(475)
Fieldstrength difference measuring device 49/3(351)
Fieldstrength meter 49/3(350)
Filter 48(252)
Filter, interference 48(176, 251, 253)
Filter method 48(588)
Filtering approximations 48(92)
Fine structure 49/5(334, 339); 49/7(467, 468, 474)

Finite orbit 49/5(277ff.)
Fireballs 48(429)
First order term 47(591)
Fission reaction 49/4(137)
Flame-photometer 47(341)
Flare, *see also* Solar flare 49/7(6)
Flare (solar eruption) 49/6(340, 348)
Flares, *see* Solar flares
Flash spectrum (at eclipse) 49/6(331)
Flexural waves 47(139, 184)
Flocculi (Sun) 49/6(347)
Flood, *see* Tides
Flow 47(273)
Flow around artificial bodies 49/5(227)
Flow density 49/2(9)
Flow drift tube 49/6(64)
Flow, linear shear 49/5(192)
Flow, zero-order 49/5(191)
Fluid-dynamical model 49/4(547)
Fluorescence 49/6(38)
Fluorimetry 47(311)
Flute-instability 49/4(557)
Flux, energy 49/5(343)
Flux, evaporation 49/5(245)
Flux, maximum trapped 49/4(176)
Flux of electrons 49/1(91, 99)
Flux of low energy protons 49/4(186)
Flux, thermal 49/5(126)
Fluxgate magnetometer (Förster-Probe) 47(500); 49/3(348, 395, 399)
Fluxmeter 49/3(297)
Flux-preservation 49/4(25)
Flying ionospheric observatory 49/7(500–502)
Focusing, effect 49/5(266, 267)
Focusing, electron 49/5(269)
Focusing, ion 49/5(269)
Focusing, particle 49/5(261)
Focusing, wave 49/5(293)
Focussing 49/5(284)
Fog 48(163, 259, 362)
Fog, diffusion by 48(266)
Fokker-Planck equation 49/4(145, 160, 161, 427)
Folding 47(259)
Folding, concentric 47(269)
Folding, dynamics 47(267)
Folding scale model experiments 47(271)
Forbidden lines, coronal 49/6(355, 356, 374)
Forbidden lines of nitrogen 49/1(108ff.)
Forbidden lines of oxygen 49/1(106–108)
Forbush decreases 49/1(227–234)
Force density, ratio 49/4(20)
Forces of non-electric origin 49/2(274)
Forcing, orographic 49/5(206)
Forecast of temperature minimum 48(238)
Forecasting, numerical 48(91)
Formulation, Macroscopic 49/4(18)
Försterling equations 49/2(173)
Försterling's dispersion formula 49/2(37ff., 68, 72, 89, 143)
Fossiliferous rocks 47(290)
"Fountain" (at the magnetic equator) 49/7(450, 452 Figs. 159, 160, 457 Fig. 165)
Four ion model 49/7(441, 446)
Fourier transform(s) 49/2(5); 49/4(4, 398, 406)
Fourier-Bessel integral 49/6(173)
Fourier's law 49/7(61, 176)
F2-peak 49/6(285–287)
Frame, absolute 48(8)
Fraunhofer absorption lines 49/6(371)
Free air anomaly 47(222, 254, 578)
Free air reduction 47(222, 578)
Free precession 49/3(363)
Free-bound transitions 49/1(111)
Freezing (droplets) 48(501)
Freezing nuclei 48(499)
F-region 49/6(35, 98, 116, 193); 49/7(442, 443 Figs. 153b, 154)
F-region, ion chemistry 49/6(116)
F-region, photoionization 49/6(35)
Frequency, acoustic cut-off 49/5(168, 186)
Frequency, (Brunt-) Vaisälä 49/5(186)
Frequency, Brunt-Vaisälä 49/5(186)
Frequency characteristic method 49/3(332)
Frequency, cross-over 49/5(236, 237ff.)
Frequency, cut-off 49/5(237, 301, 316, 321)
Frequency delivering magnetometer 49/3(324)
Frequency dispersion 49/5(227)
Frequency, intrinsic 49/5(192)
Frequency markers 49/2(214)
Frequency measurement 49/3(365)
Frequency scale of ionograms 49/2(214)

Frequency spectrum 49/4(460, 490, 533)
Frequency variation method 49/2(219)
Fresnel equation of wave normals 49/2(32)
Fresnel's formulae 49/2(177, 192)
Friction 48(652); 49/2(274)
Friction electricity 48(564)
Friction force 49/4(430–432)
Friction, internal 48(711, 715, 850)
Friction, lateral 48(657)
Friction layer 48(38, 41)
Friction on sea bottom 48(902)
Frictional effect 49/4(427)
Frictional force 49/6(249, 250, 254)
Frontal cyclones 48(132)
Frontal formation 48(124)
Frontal wave 48(134)
Frontal zone 48(122)
Fronts 48(10)
Frost Gravimeter 48(803)
Frozen-in field lines 49/4(54)
F_1-type model 49/5(375)
Full wave (propagation solutions) 49/7(348)
Function, distribution 49/5(225, 226)
Function, Whitaker 49/5(193)
Functions, Hough 49/5(183)

Gain-control, automatic 47(155)
Gal 47(203, 537)
Galactic radio emission 49/1(241)
Galactic rotations 49/1(178, 234)
Galena 47(326, 350)
Galitzin's seismograph 48(800)
Gamma function 49/6(174)
Gamma (γ) 49/3(277)
γ-Whistlers 49/5(325, 328)
Garnet 49/1(249ff.)
Gas kinetic theory 49/6(131, 132, 264)
Gas-dynamic beam instability 49/4(497, 503, 506)
Gas-dynamic equation 49/4(420)
Gasdynamics of one fluid 49/4(466)
Gasmagnetic waves 49/3(129, 141)
Gauss Magnetometer 47(499)
Gauss-Lamont method 49/3(301)
GBK-integral 49/4(436)
Geiger counter telescopes 49/1(164ff., 169)
Geiger-Mueller-counter 47(310)
Geminids 48(435, 437)
Generalization of harmonics 49/2(399)
Generalized emission measure 49/6(353–356)
Generalized Ohm's law 49/2(273)
Generation of ring current 49/1(132ff.)
Geochemical basis for the distribution of uranium and thorium 47(314)
Geochronology 47(292)
Geochronology, absolute 47(351)
Geodetic function 48(751)
Geographical coordinates 49/2(277)
Geoid 47(535)
Geoid determinations from gravity anomalies 47(593)
Geoid, theory of determination 47(541)
Geoid, warping of 47(539, 571)
Geoidmeter 47(622)
Geological time, sub-divisions 47(290)
Geology and physics 47(288)
Geomagnetic activity 49/3(27, 137)
Geomagnetic activity, effect on cosmic rays 49/1(184–193)
Geomagnetic bay 49/3(104)
Geomagnetic coordinates 49/2(277); 49/3(39)
Geomagnetic disturbance 49/7(41, 44)
Geomagnetic field 49/3(277); 49/5(175)
— dynamo-theory 47(401)
— semidiurnal lunar variations 49/2(572, 592)
Geomagnetic latitudes or longitudes, *see* Latitudes or longitudes
Geomagnetic micropulsations 49/1(67ff.)
Geomagnetic pulsations 49/3(115ff., 119, 129)
Geomagnetic storm 49/4(10, 65, 70)
Geomagnetic storms 49/3(131)
Geomagnetic tail 49/3(140, 143, 146)
Geomagnetic tides 48(931, 932)
Geometric approximations to index surfaces 49/2(87)
Geometrical optics 49/2(22ff.); 49/5(195, 211)
Geometrical optics approximation 49/4(508)
Geometry, Cartesian 49/5(182)
Geophone 47(154, 188)
Geophone, calibration curve 47(155)
Geopotential 47(536); 48(9, 650)
Geostrophic approximation 48(31)
Geostrophic departure 48(32)
Geostrophic wind 48(29, 651)

Geosynchrotron 49/4(168, 191)
Giacobinids 48(435, 453)
Ginzburg-Gurevič functions 49/7(204–207)
Glaciation 48(486)
Global radiation 48(205, 209, 251)
Glow 49/1(16, 111 ff., 120)
$\mathscr{H}_{gm}$ 49/3(58)
Gradient drift 49/4(62, 63, 75)
Gradient wind 48(33, 714, 718)
Gradiometer magnetic 49/3(405)
Grad's method 49/4(431)
Grains of large size 49/1(267 ff.)
Grains of one single domain of magnetization 49/1(269)
Granitic rocks 47(313)
Grating 49/6(310)
Gravimeter 47(213); 48(797, 799)
Gravimeter, drift 47(214)
Gravimeter, scale-value 47(214)
Gravimetric factor 48(787, 797, 809)
Gravitational field, second derivatives 47(230)
Gravitational oscillations 49/5(162)
Gravitational-compressibility waves 48(51)
Gravito-acoustic wave, modes 49/5(188)
Gravity, absolute 47(621)
Gravity, absolute determination 47(203)
Gravity, absorption of 48(845)
Gravity anomaly, apparent and true 47(540)
Gravity change 48(787)
Gravity field, analysis 47(235)
Gravity, formulae 47(563, 635)
Gravity gradient, normal 47(636)
Gravity, horizontal gradient 47(217)
Gravity, international formula 47(221, 564, 636)
Gravity, local anomaly 47(219)
Gravity, mean 47(561, 634)
Gravity measurements at sea 47(210)
Gravity mode 49/5(189)
Gravity, normal 47(536, 546, 548, 554)
Gravity, potential of 47(536)
Gravity pseudomodes 49/5(203)
Gravity, reduction 47(577)
Gravity reduction, indirect effect 47(577, 622)
Gravity surveys, local 47(242)
Gravity, vector of total variations 48(811, 840)
Gravity wave(s) 49/5(167); 49/6(153)
— acoustical internal 49/5(168)
— internal 49/5(168, 177, 188, 200)
— in the atmosphere 49/2(345, 350)
Gravity-anomalies, isolation 47(229)
Gravity-anomalies, negative 47(236, 241, 255)
Gravity-network 47(219–, 221)
Gravity-observations, reductions 47(221)
Grazing incidence 49/6(310, 311, 340, 345)
Green house effect 48(248)
Grenville 47(337)
Ground attenuation 49/2(264)
Ground inversion 48(237)
Ground roll 47(138)
Group delax 49/2(195)
Group index 49/2(111, 113 ff., 252–254)
Group index, whistler propagation 49/2(117, 498)
Group path, *see* Ray path
Group surface 49/2(109)
Group travel time 49/2(111, 135, 142, 145, 167 ff., 194 ff.)
Group velocity 48(678, 714, 718, 720, 722); 49/2(108 ff., 136); 49/4(410); 49/5(188, 189, 198, 247)
Group velocity, intrinsic 49/5(211)
Growth increment 49/5(224)
Grüneisen's constant 47(385)
Guided waves 47(119)
Guiding center 49/4(118)
Guiding center approximation 49/4(116, 118)
Gulf of Mexico 48(912)
Gulf of Saint Lawrence 48(913)
Gulf stream 48(652, 664, 666)
G-waves 47(120, 125)
Gyration 49/4(118)
Gyration of a charged particle 49/1(84 ff., 87)
Gyro frequency, *see also* Cyclotron frequency 49/7(58, 167)
Gyro-absorption 49/2(221, 395)
Gyro-controlled tripel probe 49/3(357)
Gyrofrequencies 49/4(18)
Gyrofrequency 49/4(31)
Gyrofrequency, electron 49/5(220)
Gyrofrequency, ion 49/5(170, 220)

Gyrofrequency, proton 49/5(316)
Gyromagnetic ratio 49/3(360)
Gyropulsation 49/2(2)
Gyroresonance 49/2(6, 10, 18–20, 30, 62, 67, 92, 394, 479, 511, 513); 49/5(336)
Gyroresonance damping 49/5(228)
Gyroresonance, electron 49/5(241)
Gyroresonance excitation 49/5(228)
Gyroresonance, ion 49/5(247)
Gyroresonance of higher order 49/2(511)
Gyroresonances, multiple proton 49/5(316)
Gyrotropic optics 49/2(23, 28, 30ff., 57)

H^+ ions 49/7(425, 428 Figs. 145b, e)
Haarmann, undation theory 47(280)
Hadley regime 48(150)
Haematite 47(471); 49/1(251, 258)
Hahn-Mason-Smith parameter 49/7(75 Fig. 1)
Hail 48(486, 524, 527)
Hail stones 48(528)
Half life 47(298)
Half width 48(221, 226, 231)
Hall conductivity 49/3(7)
Hall current(s) 49/3(7); 49/4(218); 49/7(504 Fig. 196)
Hall drift 49/6(214)
Hamilton equations of the group path 49/2(137ff.)
Hamilton's variational principle 49/5(211, 212)
Harmonic analysis 48(896)
Harmonic constants 48(781)
Harmonic dial 48(763, 934, 939)
— for 27-day period 49/1(200)
— for 10-year period 49/1(222ff.)
— of Bartels 49/1(171ff.)
Harmonic procedure 48(899)
Hartley band(s) 48(172, 373); 49/6(19, 32, 87, 122)
Haruna rock 47(476)
Haselgrove's equations 49/2(144)
Haurwitz's formula 48(78, 106)
Hawaii 47(238)
Hayford-zones 47(224)
Haze 48(359)
Haze atmosphere, homogeneous 48(166)
Haze extinction 48(160, 163, 181)
— dependence on wave length 48(161, 164)
— observations 48(166)
Haze particles 48(164, 261)
Haze particles, composition 48(162)
Haze scattering 48(174, 188)
Haze, vertical distribution 48(200)
He^+ ions 49/7(426, 429 Figs. 145c, f)
Heat balance (budget) 49/7(469–478, 479–483 Fig. 177, 525–532)
Heat budget of the oceans 48(630)
Heat conduction 49/7(470, 472, 474)
Heat conduction equation 49/6(259)
Heat conductivity 49/7(470)
— matrix 49/7(176)
— tensor 49/7(177, 185–191)
Heat contact 49/7(261)
Heat deposit 49/7(459)
Heat engine, atmospheric 48(156)
Heat flow 47(257, 395)
Heat flux 48(629); 49/2(10, 11, 15)
— balance 49/7(167–174)
— transport 49/5(127)
Heat, latent (evaporation) 48(4)
Heat, primeval 47(262)
Heat radiation 48(217)
Heat, vertical eddy-transported 49/5(128)
Heat wave 49/5(151)
Heat wave, diurnal 49/5(150, 154)
Heating 49/4(215); 49/6(12, 196)
Heating by collisions 49/2(393)
Heating by electric currents 49/2(393)
Heating by precipitating electrons 49/7(508–514 Figs. 208–210)
Heating in *F*2 layer 49/2(289)
Heating of electron gas 49/6(12)
Heating of plasma 49/2(5ff.)
Heating of the stratopause 49/5(121)
Heating profile 49/7(464 Fig. 169)
Heating rate 49/7(472)
Heat-production, radiogenic 47(257)
Height above sea-level 47(602)
Height determination 49/2(226ff.)
Height markers 49/2(214)
Height of an auroral band 49/1(19–23)
Height of *F*1 layer 49/2(287)
Height of reflection, real 49/2(252–256)
Height parameters 49/2(226ff.)

Height readings, their accuracy 49/2(210)
Height-integrated electric conductivity 49/3(8)
Heights, dynamic 47(602)
Heights levelling 47(602)
Heights orthometric 47(602)
Heights, practical 47(603)
Heligoland blast 48(474)
Heligoland-explosion 47(247)
Heliogram 49/6(336, 340, 342)
Helium-3 47(308); 49/5(378)
Helium bulge 49/6(224); 49/7(322 Fig. 81)
Helium escape 49/6(215, 225)
Helium ion profile 49/6(225, 233)
Helium ion reactions 49/6(50, 119, 232)
Helium method 47(293)
Helium profile 49/6(217, 220–223)
Helium, seasonal variation 49/6(224)
Helix antenna 49/2(206, 216)
Helmert's estimate 47(576)
Helmholtz coil 47(499)
Helmholtz, principle of reciprocity 48(296)
Helmholtz-Gaugain coil 49/3(289)
Hemispheric similarities 49/5(147)
Hemispheric systems, interaction between 49/5(174)
Hemispheric tidal system 49/5(151)
Hemo-ilmenite 49/1(258–262)
Hermite polynomials 49/2(9)
Hermitean tensor and its conjugate 49/2(7, 24)
Herpolhody-cone 47(608)
Herzberg band system 49/6(55, 123)
Herzberg continuum 49/6(17, 31, 181)
Heterosphere 49/6(10)
HF Langmuir wave 49/5(238 ff., 337, 340)
HF waves 49/5(248)
High energy approximation 49/7(116)
High latitude dynamo region 49/5(173)
High latitude ionograms 49/2(237)
High latitudes 49/6(292)
High moment of inertia approximation 49/7(98)
High water 48(881)
High-altitude observations of cosmic rays 49/1(224–227)
High-altitude red arc 49/1(21, 117)
Higher hybrid frequency 49/5(231, 337)
High-frequency approximation 49/2(20)
High-frequency approximation of refractive index 49/2(41)
High-frequency branch 49/5(233)
High-latitude dynamical model 49/7(521–524)
High-pass 49/5(335)
High-sensitivity gradiometer 49/3(437)
Hindcast method 48(713)
Hinteregger-type spectrometer 49/6(364)
Hiran 47(622)
Hiss 49/5(309 ff., 332 ff.)
Hiss emission 49/5(322)
Hiss, VLF 49/5(322)
Historical survey of auroral theories 49/1(124–129)
Höiland's frequency formula 48(57)
Homogeneous arc 49/1(16)
Homopause 49/7(225)
Homosphere 49/6(11)
Hooke's law 47(78, 170)
Hooks 49/5(331)
Horizontal double pendulum 48(794)
Horizontal Force ($\mathscr{H}$) 49/3(277)
Horizontal intensity, *see* Magnetic horizontal intensity
Horizontal pendulum 47(42); 48(793)
Horizontal scale 49/5(125)
Hot avalanche 48(985)
"Hot" coronal radiations 49/6(371)
Hot plasma, *see also* Plasma 49/2(62); 49/5(228); 49/7(157)
Hough extension mode 49/7(292)
Hough functions 48(950, 952, 971); 49/5(183); 49/7(288)
H-Theorem 49/4(436)
Huggins band(s) 48(172, 374); 49/6(19)
Humidity effect on quartz suspensions 49/3(313)
Huronian 47(292)
Hurricanes 48(129)
Hybrid frequency 49/5(236)
— higher 49/5(231, 337)
— lower 49/5(230, 231, 234, 236, 301, 320 ff., 321 ff.)
Hybrid resonances 49/2(67)
Hydrated hydronium 49/6(48)
Hydrated ions, recombination coefficients 49/6(49)
Hydration 49/6(103, 105, 108)

Hydrodynamic approximation, quasi- 49/5(228)
Hydrodynamic equations 49/6(135, 237)
Hydrodynamic theory 49/4(12, 43)
Hydrodynamic whistlers 49/5(302, 303)
Hydrodynamical approximation 49/6(130)
Hydrogen, atomic ion, profile 49/6(225, 233, 234, 242)
Hydrogen, atomic ion, reactions 49/6(232)
Hydrogen atoms, neutral 49/4(186)
Hydrogen, Balmer lines 49/1(110–113)
Hydrogen, escape 49/6(215)
Hydrogen glow 49/1(111ff.)
Hydrogen, peroxide 49/6(33)
Hydrogen, profile 49/6(217, 219–221)
Hydrostatic equation 48(22, 25, 650); 49/7(223)
Hydrostatic equilibrium 49/6(1, 137)
Hydrostatic model 49/6(9)
Hydrostatic pressure 49/6(137)
Hydrostaticity 49/5(185)
Hydroxyl (radicals) 49/6(45, 89, 189)
Hygrometry 48(176)
Hyperfinestructure 49/3(370)
Hypergeometric differential equation 49/2(186ff.)
Hypergeometric series 49/7(192)
Hysteresis curve of high permeability alloy 49/3(347)
Hysteresis cycle 49/1(270, 296)

Ice blink 48(212)
Ice crystals 48(511)
— growth in supercooled water clouds 48(517)
— habit-effect of temperature and supersaturation 48(516)
— in the laboratory 48(515)
Ice nuclei 48(504, 508)
Ice nuclei, artificial 48(509)
Ice particles 48(507)
Ice-pyrheliometer 48(251)
Identification of solar emissions 49/6(350, 359)
Igneous rock 47(290)
Igneous rocks, contining titano-magnetites 49/1(257)
Illumination 48(207, 212, 216)
Illumination, extra-terrestrial 48(186)
Ilmenite 47(471); 49/1(258–260)
Images, solar 49/6(335, 340, 343)
Impact parameter 49/7(64, 70, 114–117)
Impact zones of cosmic-ray particles
— observations 49/1(212–216)
— theory 49/1(209–212)
Impedance 48(857)
Impedance of an antenna in plasma 49/2(478)
Impedance probes 49/2(484–486)
Impetus 47(89)
Incidence 49/1(2, 6, 10, 37)
Inclination ($\mathscr{I}$) 49/3(277)
Inclination, dip 47(499)
Inclination error 49/1(286)
Inclination, magnetic (=dip) 49/6(289)
Incoherence 48(294)
Incoherent scatter 49/2(321, 356, 358)
Incoherent scatter sounding 49/6(79, 278, 282); 49/7(256–267)
Incoherent scatter, spectrum 49/6(82)
Incompressibility 49/5(185)
Increment 49/4(408)
Increment, growth 49/5(224)
Index, absorption 49/4(491, 493)
Index fossils 47(352)
Index function 49/2(181, 183)
— of Epstein 49/2(187ff.)
— of Rawer 49/2(186, 188)
Index of magnetic activity 49/1(55)
Index, refractive 49/4(32, 454, 491, 493)
Index surface 49/2(35ff., 59ff., 70, 104, 106)
Index surface, geometric approximations 49/2(87)
Index vector 49/2(25, 59); 49/4(409)
India, gravity anomalies and plumb line deflections 47(596)
Indian Ocean, water masses 48(647)
Indices, absorption 49/4(454)
Induced dipole 49/6(41)
Induction factor 49/3(302, 304, 312)
Induction-coil with amplifier 49/3(342)
Induction-coil with galvanometer 49/3(342)
Inelastic collisions 49/2(5, 11, 16); 49/7(527–530)
Inelastic reflection 49/5(244)
Inequality, daily 48(882)
Inequality, elliptic 48(752)
Inequality, parallactic 48(882)

Inertia, moment of 47(555)
Inertia, spatial 49/5(227)
Inertia, temporal 49/5(227)
Inertial waves 48(876)
Inferred parameters 49/7(261)
Infinite orbit 49/5(277)
Infinite particles 49/5(281)
Infinite trajectories 49/5(278)
Inflation of the geomagnetic field 49/3(182)
Inflection point 49/7(357 Fig. 109)
Influence effect 48(558)
Information content 49/2(212)
Information theory 49/2(209ff., 221, 251)
Infra-gravity waves 48(672)
Infrared emissions 49/6(125); 49/7(468, 472, 474)
Infrasonics 48(455)
Inhomogeneous equation 49/5(204)
Inhomogenity parameter 49/4(509)
Inhomogenity scales 49/4(513)
Initial nuclei 48(565)
Initial phase 49/4(184)
Initial phase of a magnetic storm 49/1(55)
Initial phase of the magnetic storm 49/3(131, 160)
Initial waves 48(703, 705)
Injun I 49/1(81)
Inner auroral zone 49/1(11)
Inner radiation belt 49/1(81)
Instabilities, aperiodical 49/4(542)
Instabilities, gasdynamic 49/4(549)
Instability, *see* Stability 49/4(408, 411, 503, 555)
Instability, beam 49/5(226)
Instability, condition for 49/4(176)
Instability, conditional 48(51)
Instability, gasdynamic 49/4(498)
Instability, kinetic 49/4(498)
Instability of auroral rays 49/1(24ff.)
Instability of water surface 48(704)
Instability producing aurora 49/4(217)
Instable system 49/4(492)
Integral, collision 49/5(225)
Integral equation for the matrizant 49/2(176)
Integrallogarithmus 48(219)
Integrated control 49/3(328)
Integrated electric conductivity of the ionosphere 49/3(7)
Intensity, total 48(289)
Interaction between hemispheric systems 49/5(174)
Interaction between the neutral and electrical components 49/5(171)
Interaction of magnets 49/3(303)
Interaction of particles 49/5(228)
Interaction potential 49/7(65, 71, 111–113 Figs. 20, 21, 210)
Interchange of magnetic lines of force 49/3(146)
Intercomparison of instruments 49/3(321)
Intercomparison of models 49/7(340–347)
Interface conditions 48(10)
Interface modes 49/5(199, 200)
Interference filter 48(176)
Interference gravimeter 48(804)
Interference methods 49/2(217ff.)
Intermediate ions 48(543, 551)
Intermediate molecules, state 49/7(122)
Intermediate rocks 47(313)
Intermediate zone 49/5(220, 252)
Internal gravity wave(s) 49/5(168, 177, 188, 200); 49/6(153)
Internal gravity wave, acoustical 49/5(168)
International quiet days 49/3(56)
International Reference Atmosphere (COSPAR'S CIRA) 49/7(315–326)
International Reference Ionosphere 49/7(351–377 Figs. 105, 109, 112, 118, 413)
Interplanetary magnetic field 49/7(503)
Intramolecular force 49/6(138)
Intrinsic frequency 49/5(192)
Intrinsic group velocity 49/5(211)
Intrinsic magnetic field 49/5(380)
Invariant, first 49/4(154)
Invariant, integral 49/4(124)
Invariant latitude 49/2(521); 49/7(332)
Invariant, longitudinal 49/4(123)
Invariant, second 49/4(155)
Invariant, third 49/4(154)
Invariant violation mechanism 49/4(156)
Invariants, adiabatic 49/4(68, 69, 123)
Invariants of particle motion in the radiation belt 49/1(87, 135)
Inverse power potential 49/7(72, 90, 125, 133)

Inverse thermo-remanent magnetization (ITRM) 49/1(272, 282ff.)
Inversion 48(243)
Inverted **SC, SC*** ($\mathscr{D}$) 49/3(43)
Io echo 49/2(504)
Ion beams (in polar caps) 49/7(511 Fig. 204)
Ion chemistry 49/6(7, 100, 110, 111, 116)
Ion clouds, artificial 49/4(81, 82, 101)
Ion composition 49/6(13, 15, 71)
— D-region 49/6(100)
— E-region 49/6(112)
— F-region 49/6(112)
— of outer ionosphere 49/2(525)
— precipitation effects 49/7(515 Fig. 211)
— profile(s) 49/6(14, 112); 49/7(264 Figs. 28, 29, 374–383 Figs. 117–122, 419–430 Figs. 142–145)
Ion density profile 49/2(521); 49/7(441 Fig. 152, 481 Fig. 176, 484 Fig. 178, 486 Figs. 179c, d, 516, 519 Fig. 215)
Ion drag 49/7(506)
Ion drift velocity 49/1(60)
Ion effect on dielectric eigenvalues 49/2(41)
Ion gyrofrequency 49/2(55)
Ion mass spectrometer(s) 49/6(72); 49/7(240)
Ion poles of refractive index 49/2(68)
Ion sheath 49/2(482ff., 485)
Ion sound waves 49/4(522, 533)
Ion source (in neutral mass spectrometers) 49/7(241)
Ion temperature(s) 49/4(58); 49/7(258, 366 Fig. 112, 374, 419, 517 Fig. 213)
Ion temperature profile 49/6(13); 49/7(38)
Ion velocity 49/6(250)
Ion-acoustic wave 49/5(240ff.)
IONCAP (ionospheric model) 49/7(349, 350 Fig. 103)
Ion-collision frequency 49/5(170)
Ion-counting 48(552, 554)
Ion-cyclotron amplification 49/5(305)
Ion-cyclotron excitation 49/5(305)
Ion-cyclotron focusing 49/5(269)
Ion-cyclotron frequency, combined multiple 49/5(236)
Ion-cyclotron gyrofrequency 49/5(170, 220)
Ion-cyclotron gyroresonance 49/5(247)
Ion-cyclotron plasma pulsation 49/5(221)
Ion-cyclotron resonance wave 49/5(305)
Ion-cyclotron waves 49/5(221)
Ion-cyclotron whistler 49/5(306)
Ion-cyclotron-whistler wave 49/5(235)
Ion-electron collisions 49/6(351)
Ionic polarization 49/2(49, 52, 63)
Ionic reactions 49/7(435, 436)
Ionic spectrum 48(553)
Ionic wave 49/5(235)
Ion-ion recombination 49/6(70)
Ionisation, production of 49/7(36, 39)
Ionization 48(544, 549)
Ionization as a source of auroral spectra 49/1(113–115)
Ionization balance of the *E* region 49/2(282)
Ionization chamber(s) 47(312); 48(549); 49/1(160–164, 166–169, 219); 49/6(349); 49/7(269–272 Figs. 34–36)
Ionization chambers, observation of diurnal variation 49/1(175–177)
Ionization degree 49/6(228, 352)
Ionization equilibrium 48(570)
Ionization patches 49/2(519)
Ionization potential 49/6(23, 26, 353)
Ionization sources 49/7(437)
Ionization temperature 49/6(353, 355, 372)
Ionization threshold 49/6(23, 26, 27, 29)
Ionization thresholds, N_2 49/6(24)
Ionization, total 48(549)
Ionizators 48(544)
Ionized column 49/1(75)
Ion-molecule reactions 49/6(59)
Ion-neutral rections 49/6(49)
Ionogram 49/2(64, 193, 196ff., 212)
Ionogram, panoramic 49/2(218)
Ionogram, recording 49/2(212ff.)
Ionogram, typical examples 49/2(228–250)
Ionograms of *D* region at low frequencies 49/2(318)
Ionograms on low frequencies 49/2(239)
Ionosonde 49/2(199)
Ionosphere 48(575); 49/6(3, 7, 11)
Ionosphere, electric currents 49/1(59)
Ionosphere of Mars 49/5(360, 363, 380)
Ionosphere of Venus 49/5(365)

Ionosphere sound 48(474)
Ionosphere, top of 49/6(233)
Ionosphere, upper 49/5(376)
Ionosphere winds 48(475)
Ionospheric cross modulation, *see* Luxembourg effect
Ionospheric disturbance 49/2(241, 345ff., 362)
Ionospheric layers, classification 49/2(222)
Ionospheric regions 49/6(11)
Ionospheric storm 49/2(367, 519)
— average disturbance effect 49/2(372)
— decrease of electron density 49/2(371)
— explanation of $F2$ effects 49/2(373–377)
— heating 49/2(373)
— increase of characteristic height 49/2(372)
— solar component 49/2(372)
Ionospheric tides 48(932, 933)
Ionospherics 49/2(502)
Ions 48(160, 164, 543)
Ions, collision with neutrals 49/7(107–121, 163–165)
Ions, contribution to conductivity 49/2(4)
Ions, density profiles 49/6(230)
Iovian ionosphere 49/7(456 Fig. 164)
Irish channel 48(912)
IRM, *see* Isothermal remanent magnetization
Irreducible tensor 49/2(18)
Irregularities 49/4(544)
Irregularities in ionospheric layers 49/2(355)
Isentropic analysis 48(27)
Island arcs 47(236, 255, 598)
Islands 47(259)
Isobaric surface 48(619, 651)
Isochasms and isoaurores 49/1(2ff.)
Isoclinics 47(506)
Isodense lines 49/6(340)
Iso-dynamics 47(506)
Isoelectronic sequences 49/6(358, 376)
Isogonics 47(506)
Isolated rays 49/1(16)
Isoporic focus 47(510)
Isopors 47(506)
Isostasy 47(232, 253)
Isothermal atmosphere 49/5(186)
Isothermal remanent magnetization (IRM) 49/1(268, 272)
Isotope dilution method 47(311)
Isotopic analysis 47(293, 350)
Isotropic velocity distribution 49/2(9, 18)
ITRM, *see* Inverse thermo-remanent magnetization
Iz echo 49/2(503, 504)
Iz trace 49/2(236)

Jacchia model 49/7(340–347)
Jacobi's Ellipsoid 47(600)
Jacobsite 49/1(262ff.)
Jeffreys stresses 48(146)
Jet, stratospheric tidal 49/5(155)
Jet-Stream 48(122, 414, 665)
Jodrell Bank 48(434, 437)
Joule heating 49/4(403); 49/7(462, 506, 508 Fig. 202, 515 Fig. 211b, 518)
Joule heating of plasma 49/2(5ff., 16, 62, 393)
Jump conditions 49/4(206)
Jump-correlation 47(160)
Junge, frequency distribution of particle sizes 48(164, 170, 259)
Jupiter 49/4(228)
Jupiter radio waves 49/4(224)

Kappa-number 48(896)
Kármán's high pressure cylinder 47(173)
Katmai 48(168, 288, 323)
Keewatin 47(292)
Kelvin's waves 48(859)
Keweenawan 47(292)
Keweenawan ratio 47(490)
Key wavelength 49/7(17, 53–55)
Kihara distance 49/7(81)
Kihara potential 49/7(83 Fig. 5, 151)
Kinematic models 49/7(432–458)
Kinetic equation 49/4(514); 49/5(225, 253)
Kinetic equation, linearized 49/4(536)
Kinetic theory 49/7(229)
Kinetic theory of gases, fundamental equations 49/2(7ff., 21ff.)
Kirchhoff's law 48(218, 225)
Knee 49/4(56)
Knee of the cosmic-ray latitude curve 49/1(196ff., 198, 224ff.)

Knee (=plasmapause) 49/5(330, 341, 371)
Knudsen number 49/7(234)
Kopfwellen 47(193)
Kp (magnetic character figure) 49/7(330)
Kp-index for the activity of magnetic storms 49/1(55, 62)
Krakatoa eruption 48(288, 323, 476, 975)
Kronecker symbol 49/5(227)
Kuroshio 48(652, 664)
K-Variations 49/3(210)

L lines of McIlwain 49/1(9ff.)
La Cour magnetograph 49/3(294)
Lacoste-Romberg Gravimeter 48(803)
Lagrange's wave velocity 48(847)
Lagrangian density $\mathscr{L}$ 49/5(212, 215)
Laguerre polynomials 49/7(131, 132, 194, 195, 197)
Lamb modes 49/5(199, 200)
Lamb waves 49/5(189, 200)
Lambert's constant 48(788)
Lambert's law of reflexion 48(299)
Lamé 47(78)
Landau damping 49/4(204, 465); 49/5(228, 238)
Landau formula 49/4(428)
Landau length 49/2(13); 49/7(106)
Land-sea-distribution 48(938)
Langevin cross section 49/6(42)
Langevin's theory of paramagnetism 49/1(249ff.)
Langmuir frequency 49/4(414)
Langmuir ion oscillation, low frequency 49/5(240)
Langmuir oscillation 49/4(451–453, 457, 497)
Langmuir oscillations of electrons 49/5(221)
Langmuir probe 49/2(486, 524)
Langmuir wave, HF 49/5(238ff., 337, 340)
Langmuir waves 49/4(452)
Langmuir-Tonks wave 49/5(240)
Laplace's speed of sound 48(456)
Laplace's tidal equation 48(947); 49/5(156, 183)
Laplace's tidal theory 48(887)
Laporte's rule 49/6(358)
Lapse rate 48(485)
Lapse rate, dry-adiabatic 48(29, 50)
Lapse rate, moist-adiabatic 48(50, 485)
Large body 49/5(220)
Large body potential 49/5(280)
Large ions 48(543, 550, 553, 565, 573)
Large-scale circulation 49/7(488–495 Figs. 180–185)
Larmor damping 49/5(228)
Larmor drift 49/4(524)
— frequency 49/4(534)
— of charged particles 49/4(513)
— velocity 49/4(515)
Larmor frequency 49/3(360)
Larmor radius 49/5(218, 220, 231)
Lateral displacements 48(788)
Lateral mixing 48(150)
Lateral oscillations 48(859)
Latitude, change with height 47(560)
Latitude changes 47(606)
Latitude effect (cosmic radiation) 48(581)
Latitude effect of cosmic-ray intensity 49/1(196, 202, 224)
Latitude effect of neutron intensity 49/1(165, 202ff.)
Latitude, generalized magnetic 49/4(126)
Latitude, geocentric 47(93, 549)
Latitude, geographic 47(93)
Latitude in ecliptic 48(750)
Latitude, invariant 49/4(126)
Latitude, mean 47(562, 634)
Latitude, reduced 47(544)
Latitude service, international 47(17, 606)
Latitude variation of *E* region 49/2(282ff.)
Latitudes of particle impact 49/1(210)
Latitudinal distribution of lunar tides 49/2(549ff.)
Latitudinal variation of neutron intensity 49/1(165, 202ff.)
Lattice constants 49/1(253, 255, 262)
Laurentian 47(292)
Lava, crystallization 48(986)
Lava flows 48(617)
Lava, gas content 48(1005)
Lava, kinetic energy 48(984)
Lava, temperature 48(988)
Lava, viscosity 48(986)
Law, adiabatic 49/5(180)
Law, Maxwell-Boltzmann 49/5(227)

Layer, boundary 49/4(109)
Layer clouds, seeding 48(533)
Layer thickness 49/7(349)
Layered model 49/5(194)
Layers of the ionosphere 49/2(220ff.)
L-B coordinates 49/3(41)
L-diffusion of alpha particles 49/4(163)
Leaching 47(323)
Lead-210 47(325)
Lead abundance, anomalous 47(329, 332)
Lead abundance, common 47(319)
Lead abundance, primeval 47(330)
Lead dating methods 47(328)
Lead iodide 48(510)
Lead, isotopic analysis 47(320)
Lead, meteoric 47(327)
Lead, primeval 47(326)
Lead, radiogenic 47(319)
Lead-ratio 47(324)
Leaky-Bucket model 49/4(129, 174)
Leaky-Bucket model of the aurora 49/1(81, 94)
Ledge 49/5(379)
Ledges 49/2(245)
Lee waves 49/5(208)
Legendre functions (polynomials and associated polynomials) 49/7(65, 328, 389, 390, 411)
Legendre polynomials 49/2(12, 14)
Length of an auroral band 49/1(18)
Length of day, fluctuations 47(1, 15, 17, 527)
Length of the geomagnetic tail 49/3(151)
Lennard-Jones potential 49/7(84 Fig. 6, 109, 151 Fig. 28)
Leonids 48(435, 437)
Lepidolites 47(344, 350)
Levelling and gravity 47(601)
Lewisian 47(335)
LF waves 49/5(249)
Lg-waves 47(120, 250)
LiF (Lithium fluorid) 49/6(309)
Life times of trapped electrons 49/4(183)
Lifetime (chemical) 49/6(97, 153)
Lifetime, radiative 49/6(39, 54, 89, 122)
Lift coefficient curve 49/5(168)
Light, penetration in water 48(614, 622)
Lightning 48(586, 599)
Limb brightening (Sun) 49/6(344, 347)
Limiting mobility 48(552)
Limiting ray (sound) 48(467)
Line broadening 48(175)
Line deformation 49/6(335)
Line excitation 49/6(351)
Line form 48(221)
Line formation (solar atmosphere) 49/6(351)
Line intensity 48(221)
Line reversal (optical) 49/6(334)
Line spectrum 49/6(310)
Line structure 48(173)
Linear polarization 49/2(34, 43, 47)
Linear shear flow 49/5(192)
Line-preservation 49/4(25)
Linewidth 49/6(331); 49/7(272)
Linkage 47(259)
Liouville's theorem 49/4(156, 157)
Liquid core 48(791)
Lithosphere 47(246)
LMT (=local mean time) 49/2(277)
Loads, uncompensated 47(235)
Local activity indices 49/3(217)
Local extinction 49/2(475)
Local spectrum 49/4(543)
Logarithmic-periodic antennae 49/2(205)
Lognormal noise distribution 49/2(210)
Lomonosov ridge 48(647)
Longitude effect 49/2(270)
Longitude effect (ionosphere) 49/7(405 Fig. 131)
Longitude effect of *F*2 layer 49/2(298)
Longitude in ecliptic 48(748)
Longitudes invariant of oscillatory motion 49/1(87, 135)
Longitudes, mean 48(746)
Longitudes of particle impact 49/1(211)
Longitudinal component of electric field 49/2(27, 30, 32, 61)
Longitudinal component of the earth's magnetic field 49/2(40, 49)
Longitudinal plasma oscillation 49/5(233)
Longitudinal, quasi- 49/5(230)
Longitudinal resonance oscillation 49/5(238)
Longitudinal waves, *see also* Acoustic waves 49/2(10)
Longshore current 48(732)

Long-term variations (of solar activity) 49/7(9, 27)
Lorentz force 49/4(396); 49/6(116, 261, 303)
Lorentz force density 49/2(2)
Lorentz gas 49/2(270); 49/4(475)
Lorentz gas, kinetic theory 49/2(17ff.)
Lorentz method 49/7(63, 192)
Lorentz-transformation 49/4(494)
Loss 49/4(67, 71)
Loss by diffusion He, Ar 47(340)
Loss of electrons 49/6(97, 285)
Loss rate, specific 49/6(43)
Loudness 48(455)
Love, dispersion 47(120, 250, 252)
Love waves 47(119, 145)
Low energy particle cut-off 49/1(197ff.)
Low frequency
— branch 49/5(234)
— branch, extreme 49/5(234)
— Langmuir ion oscillation 49/5(240)
— range, very 49/5(235)
Low latitude (ionosphere) 49/7(406 Fig. 132)
Low water 48(881)
Lower hybrid frequency 49/5(230, 231, 234, 236, 298, 301, 320ff., 321ff.)
Lower ionosphere 49/7(358–364, 378–383)
Low-frequency drifting-oscillations 49/4(534)
Low-frequency longitudinal waves 49/4(523)
LP approximation (linear polarization) 49/2(56)
L-shell splitting 49/4(201)
Lunar age and *Es* layer 49/2(564ff.)
Lunar age and *F*2 layer 49/2(555)
Lunar age and twilight sky radiance 49/2(593)
Lunar current system 49/2(573, 578)
Lunar day 48(747); 49/2(548)
Lunar distance, mean 48(736)
Lunar drift motions in the *F*2 layer 49/2(584–586)
Lunar echos 49/2(470)
Lunar eclipse 48(747)
Lunar eclipses, green shadow border 48(386)
Lunar stratification 49/2(552)
Lunar tidal wind 49/2(547, 567ff., 591)
Lunar tidal wind, computation 49/2(575–578)
Lunar tides
— and earth's magnetic field 49/2(556, 570ff.)
— and lunar age 49/2(555)
— and surface winds 49/2(568–570)
— in *E* region 49/2(280)
— in *F*2 layer 49/2(296, 315, 549–558)
— of the ionosphere 49/2(547ff.)
— seasonal variations 49/2(557, 593)
— semi-diurnal variations 49/2(550ff.)
— theory 49/2(579–589)
Lunar time, apparent 48(743)
Lunar time, mean 48(747, 750)
Lunar-daily variation 49/3(20)
Lupolen-H 48(252)
Lutetium 47(299)
Luxembourg effect, double pulse method 49/2(396)
Luxembourg effect, frequency variation 49/2(395)
Luxembourg effect, modulation of collision frequency 49/2(393, 395)
L-value 49/4(56, 69, 70)
Lyman continuum 49/6(29, 316, 370)
Lyman-alpha emission 49/7(3, 7, 20, 21–23)
Lyman-alpha of hydrogen 49/6(20, 26, 29, 31–33, 37, 72, 96, 100, 112, 121, 125, 189, 219, 310, 312, 313, 317, 340, 344, 348)
Lyman-alpha, profile 49/6(332, 333)
Lyman-beta, solar emission 49/7(53–55)
Lyman-beta of hydrogen 49/6(29, 37, 112)
Lyman-beta, profile 49/6(333)
Lyman-Birge-Hopfield bands 49/6(20)
Lyrids 48(435)

Mach cone 49/5(261)
Macrometeorology 48(457)
Maghemite 47(471)
Magma 48(1007)
Magma, phase-transition 48(1008)
Magnetic activity and diurnal variation of cosmic rays 49/1(178)
Magnetic activity, waves of 27-day period 49/1(198)
Magnetic cleaning 47(481)

Magnetic declination 49/6(290)
— effect 49/7(405)
— effect on plasma drift 49/6(291)
Magnetic deflection mass spectrometer 49/7(238)
Magnetic dip (=inclination) 49/6(152, 288, 289)
Magnetic dipole field 49/2(277, 518)
Magnetic disturbances 49/7(296, 313 Figs. 74, 75, 332, 335 Figs. 90, 91, 346, 347 Figs. 100, 101, 511–518)
Magnetic equator 49/7(449, 482)
— electric fields 49/2(578)
— "fountain" 49/7(450, 452 Figs. 159, 160, 457 Fig. 165)
— lunar tides of *Es* layer 49/2(564ff., 591)
Magnetic field 49/4(11)
— core theory 47(524)
— correlation with lunar effects 49/2(556, 570ff.)
— dipole 47(507)
— dynamo theory 47(530)
— effect upon *F*2 layer 49/2(316)
— effects of terrestrial 49/6(152, 240)
— elements 47(498)
— energy 47(518)
— exterior part 47(508)
— ferromagnetic theory 47(522)
— fundamental theory 47(524)
— galactic 47(509)
— gyroscopic effects 47(522)
— intrinsic 49/5(380)
— local anomalies 47(500)
— maps 47(502)
— multipoles 47(516)
— needed to hold a magnetosphere 49/4(223)
— non-potential part 47(508)
— produced by ring current 49/1(131)
— rotation as cause 47(522)
— spherical harmonic analysis 47(506)
— strong 49/4(500)
— westward drift 47(511)
Magnetic fields in spherical shells 47(525)
Magnetic flux 49/4(124)
Magnetic horizontal intensity 49/1(181, 183, 186, 190, 195)
— auroral effects 49/1(56–60, 64, 66)
Magnetic inclination (=dip) 49/2(270); 49/6(289)
Magnetic influence on recording 49/2(254, 256, 258)
Magnetic latitude 49/2(277)
Magnetic materials 49/3(284, 318)
Magnetic measurements 47(499)
Magnetic meridian 49/3(277)
Magnetic micropulsations 49/2(388)
Magnetic minerals 47(470)
Magnetic mode 47(525); 49/2(487)
Magnetic moment 49/3(279, 303, 321); 49/4(3, 118, 123)
— of gyratory motion 49/1(87, 135)
— of ions 49/1(249)
— of the ring current 49/1(75)
— of trapped particles 49/2(526)
Magnetic observatories 49/3(276)
Magnetic parameters of rock formation 49/3(439)
Magnetic rigidity 49/1(86, 210ff.)
Magnetic sector field mass spectrometer 49/6(73)
Magnetic shell parameter 49/3(42)
Magnetic storm(s) 49/1(53ff.); 49/3(33, 42)
— and cosmic-ray intensity 49/1(180–183)
— disturbances in the ionosphere 49/2(366–377)
— effects of 49/6(293, 300)
— of 23 September 1957 49/1(82ff.)
— variation 49/3(56)
— with sudden commencements (ssc) 49/1(53ff., 65, 94)
Magnetic substorms 49/1(54, 56–62, 65ff.)
Magnetic surface 49/1(87)
Magnetic surface of the earth 49/1(8)
Magnetic susceptibility, *see* Susceptibility
Magnetic tape recording 47(157)
Magnetic tension 49/4(49)
Magnetic time 49/2(370)
Magnetic (true) equator (=dip equator) 49/6(288)
Magnetic viscosity 49/1(272, 290)
Magnetically quiet and disturbed days, cosmic-ray intensity 49/1(184–186, 220–222)
Magnetism, induced 48(1001)
Magnetism, thermo-remanent 48(1001)
Magnetite 47(470)
Magnetite, coercive force 49/1(267)

Magnetites 49/1(250)
Magnetization, isothermal remanent 47(473)
Magnetization, measurement 47(480)
Magnetization of sedimentary rocks 47(468, 478)
Magnetization of single-domain grains 49/1(269)
Magnetization of varved clays 47(479)
Magnetization, partial thermo-remanent 47(473)
Magnetization, remanent 47(473)
Magnetization, reverse thermo-remanent 47(476)
Magnetization, self-reversal 47(492)
Magnetization, thermo-remanent 47(473, 475)
Magneto gas-dynamic equation 49/3(183ff.)
Magneto-acoustic waves 49/5(251)
Magneto-"Bremsstrahlung" 49/4(464)
Magnetocrystalline anisotropy 49/1(269)
Magnetodynamic 49/4(5)
Magnetofluid dynamics 49/4(416)
Magneto-gas-dynamic 49/4(5)
Magneto-gas-dynamic system with two components 49/4(435)
Magnetograph 49/3(277, 289, 291)
Magnetometer, airborne 49/3(408)
Magnetometer, astatic 47(480)
Magnetometer, direct-reading 49/3(425)
Magnetometer, optical absorption 49/3(428)
Magnetometer, Overhauser 49/3(427)
Magnetometer, proton precession 49/3(423)
Magnetometer results of space probes 49/1(233)
Magnetometer, rubidium 49/3(400)
Magnetometer, spin-precession 49/3(427)
Magnetometer, theodolite 49/3(304)
Magnetometer, three-component airborne 49/3(384, 441)
Magnetometer, transit 49/3(300)
Magnetometers with photoeffect 49/3(327)
Magnetopause 49/2(530ff.); 49/3(139, 173); 49/4(8, 9, 25–27, 37–39, 41, 42, 53, 54, 86, 109–111, 187, 190); 49/5(219)
Magnetopause boundary 49/4(191)
Magnetopause, drag 49/3(149)
Magnetosheath 49/3(138); 49/4(40–42, 45, 59, 202, 206); 49/5(219)
Magnetosheath electrons 49/4(51)
Magnetosheath plasma 49/4(109)
Magnetosphere 49/1(128); 49/3(8, 10ff., 54, 138); 49/5(219); 49/7(498)
Magnetosphere, "bottom" of the 49/4(78, 82)
Magnetosphere by the solar wind, distortion of the 49/4(199)
Magnetosphere, closed 49/2(530)
Magnetosphere, convective model 49/1(136ff.)
Magnetosphere, electron density models 49/2(500)
Magnetosphere, flow around 49/4(190)
Magnetosphere, lunar effects 49/2(595)
Magnetosphere, open 49/2(531); 49/4(49, 53, 66, 192)
Magnetosphere, open, model 49/4(193, 218)
Magnetosphere, propagation of whistlers 49/2(491)
Magnetosphere, pseudo- 49/5(376)
Magnetosphere shape 49/4(188)
Magnetosphere, spaghetti structure 49/2(498)
Magnetospheric convection 49/7(493)
Magnetospheric reflected whistlers 49/5(325, 327)
Magnetospheric tail 49/3(12, 32); 49/7(497)
Magneto-tellurics 47(440, 457)
Main component 49/5(373, 378)
Main phase 49/3(58); 49/4(91–93, 101, 185)
— inflation 49/3(164, 168)
— of a magnetic storm 49/1(55, 62, 100)
— of the magnetic storm 49/3(131, 161, 162, 202)
— relaxation 49/3(165)
Maintenance, nighttime 49/6(287)
Mantle 47(89, 252, 261, 370)
Mantle, composition 47(377)
Mantle, density 47(372)
Mantle, gravity 47(372)
Mantle, pressure 47(372)
Mantle Rayleigh-waves 47(130)

Maps of $F2$ layer ionization 49/2(302, 310)
Marginal seas 48(615)
Margules' formula 48(31)
Mars 48(424); 49/4(223)
Martens polarimeter 48(310)
Martian ionosphere 49/5(360, 363, 380); 49/7(454–458 Figs. 162a, b, 163b)
Martian ionosphere, model 49/5(371)
Martian neutral atmosphere 49/5(374)
Martin-Grahan filter 49/3(447, 448)
Masked layers 47(164)
Mason-Schamp-Kihara potential 49/7(79, 109, 109 Fig. 19, 148ff. Figs. 27a–d)
Mass, effective 49/5(236)
Mass fluxes 49/7(167)
Mass spectrometer(s) 49/6(63, 66, 72–75, 78); 49/7(237–241); 47(319, 324, 332)
— after Bennet 49/7(237)
— calibration 49/7(241)
— magnetic deflection 49/7(238)
— quadrupole 49/7(239 Figs. 6, 7)
— sources 49/6(79)
Mass, topographic 47(577)
Mass transport by currents 48(651)
Material field equation 49/4(397)
Material time derivative 49/2(273)
Matrix method 49/2(255, 260)
Matrizant 49/2(157ff.)
Matrizant, chain formula 49/2(160ff.)
Matrizant, integral equation 49/2(176)
Maximum heating 49/2(7)
Maximum of electron density 49/5(369, 370)
Maximum usable frequency (M.U.F.) 49/2(219, 227)
Maxwell distribution 49/2(8ff., 14, 480); 49/4(422); 49/7(194, 230 Figs. 2a, b)
Maxwell equations for plasma 49/2(22ff.)
Maxwell interaction potential 49/2(12, 15, 20ff.)
Maxwell potential 49/7(91, 94, 96, 108, 128, 135)
Maxwell transport equation 49/2(9); 49/6(263)
Maxwell-Boltzmann law 49/5(227)
Maxwell distribution function 49/5(226ff.)
Maxwellian view 48(280)
Maxwell's equations 49/3(2)
McIlwain parameter 49/2(521, 523)
Mead model 49/4(161, 195)
Mean collision frequency 49/2(1, 4ff., 10ff., 21, 247, 267ff.); 49/7(58)
Mean free path 49/7(58)
— independent of velocity 49/2(12)
— method 49/2(5)
Mean peculiar velocity 49/2(10, 18)
Mean thermal velocity (speed) 49/7(58, 167)
Mean transport collision frequency 49/2(14, 97)
Measurement, composition 49/5(371)
Measurement of components with additional fields 49/3(376)
Median value, definition 49/2(250ff.)
Mediterranean 48(912)
Mediterranean seas 48(615)
Mediterranean water 48(644, 667)
Medium, interplanetary 49/4(205)
Meinel bands 49/6(46)
Meinel bands of N_2^+ 49/1(110)
Melting band 48(521)
Melting point 47(381, 383, 388)
Merged beam 49/6(70)
Merging (of field lines) 49/4(55, 56, 93, 102)
Merian's formula 48(850)
Meridional components 49/5(136)
Meridional (NS) winds 49/5(144, 148); 49/6(258, 283, 289, 305)
— profiles 49/5(138, 139)
— diurnal 49/5(157)
Mesopause 49/6(8, 89); 49/7(292)
Mesosphere 49/6(2, 8, 84)
Metal ions 49/6(13, 113, 120, 235)
Metal oxide ions 49/6(114)
Metal vapour, absorption cross sections 49/6(19)
Metal vapour (atoms) 49/6(98, 114)
Metallic ions 49/7(375)
Metamorphic rocks 49/1(266)
Metastable electronic states 49/6(70)
Metastable ions 49/6(24)
Metastable oxygen reactions 49/6(55)
Metastable species 49/7(492, 524)
Meteor, deceleration 48(440)
Meteor mass, total 48(453)
Meteor photography 48(440)
Meteor radar 49/7(292)
Meteor showers, age 48(437)

Meteor trail(s) 49/2(401); 49/6(204)
— diffusion 48(451)
— radar reflection 49/1(75)
— recombination 48(449)
Meteoric, ions of origin 49/6(113)
Meteorite 48(428)
Meteorites, radioactivity 47(317)
Meteorological noise 48(92)
Meteorological pocket network (MRN) 49/5(117, 119, 120)
Meteors 48(427, 933)
Meteors, ablation 48(431)
Meteors, binding energy of atoms 48(428)
Meteors, compression cap 48(428, 429)
Meteors, daily and seasonal variation 48(434)
Meteors, evaporation 48(427)
Meteors, evaporation equations 48(428)
Meteors, Fresnel zone patterns 48(431)
Meteors, heat transfer coefficient 48(429)
Meteors, heights 48(437)
Meteors, ionization 48(429)
Meteors, latent heat of evaporation 48(428)
Meteors, line density of electrons 48(429, 432, 450, 454)
Meteors, luminosity 48(429)
Meteors, mass 48(427)
Meteors, mass distribution 48(436)
Meteors, observing technique 48(430)
Meteors, parabolic velocity limit 48(452)
Meteors, radio echo technique 48(431)
Meteors, shape-factor 48(428)
Meteors, sporadic 48(427)
Meteors, total influx 48(435, 452)
Method A1 49/2(264, 325)
Method A2 49/2(331)
Method A3 49/2(264, 328)
Method of Chapman and Enskog 49/4(538)
Metonic cycle 48(747)
M-factor (M(3000)F2) 49/7(365)
Mica 47(336, 349)
Microbar 48(455)
Microbarograph 48(465)
Micro-meteorites 48(430)
Micro-meteorology 48(457)
Microphone gauge 49/7(247)
Micropulsations 49/4(34, 60, 215)
Micropulsations of the magnetic field 49/1(67ff.)
Micro-pulsations, pearl-type 49/5(302)
Microscopic formulation 49/4(17)
Microseismic 47(92, 140)
Microwave mass-spectrometer afterglow 49/6(68)
Microwaves 48(189, 222)
Mie 48(330)
Mie-Debye-theory 48(293)
Mie's theory of scattering 48(161, 188, 255)
Milligal 47(203, 537); 48(774)
Minauroral belt 49/1(2, 11ff., 80, 118)
Mine disasters 48(843)
Minimum frequency 49/2(225)
Minimum height values 49/2(226)
Minimum virtual height 49/2(226ff.)
Minimum-pendulum 47(207)
Minor constituents 49/7(312–314, 440 Fig. 151)
— diffusion 49/6(171, 215)
— profile 49/6(159, 215, 225)
Minor ions, profiles of 49/6(225)
Minor species (constituents) 49/6(86, 159, 171, 215, 235)
Mintrop-waves 47(193)
Mirror image effect 48(597)
Mirror point for trapped particles 49/2(526)
Mirror points 49/1(85)
Mixed nuclei 48(488, 491, 565)
Mixing 49/6(91, 152, 153, 170, 176, 218, 221)
Mixing length 48(38)
Mixing, time constant 49/6(91)
Mobility 48(543, 554); 49/6(64)
Mobility eigenvalues 49/2(18–20); 49/7(199 Fig. 40)
Mobility (higher orders) 49/7(180)
Mobility tensor 49/2(18, 30, 274); 49/7(60)
Mode, gravity 49/5(189)
Model, aerodynamical 49/4(203)
Model atmosphere 49/6(157, 176, 251, 252)
Model *C* ionization chamber 49/1(160)
Model, Chapman-Elias layer 49/5(121)
Model, closed 49/4(193)

Model, F_1-type 49/5(375)
Model, layered 49/5(194)
Model of the Martian ionosphere 49/5(371)
Model of type
— E 49/5(373)
— F_2 49/5(373)
— Venusian ionosphere 49/5(371)
Model, splash catcher 49/4(174)
Model temperature structure 49/5(155)
Modes 49/5(196)
Modes, ducted 49/5(200)
Modes, gravitoacoustic wave 49/5(188)
Modes, gravity pseudo 49/5(203)
Modes, interface 49/5(199, 200)
Modes, Lamb 49/5(199, 200)
Modes, normal 49/5(197, 198)
Modes of oscillation 49/5(202)
Modified Alfvén VLF waves 49/5(250)
Modified Alfvén wave 49/5(236)
Modified dip 49/7(389, 392–394 Figs. 125, 126)
Modulation 48(748, 764)
Modulation effect 49/2(16)
Modulation pulsation 49/2(16)
Modulation transfer 49/2(394)
Mohorovičič-discontinuity 47(120, 246, 251, 253)
Mohr's diagram 47(263)
Molecular diffusion 49/5(126); 49/6(130, 136, 153, 154, 176); 49/7(304, 305)
Molecular diffusion coefficient 49/6(86, 165)
Molecular ion(s) 48(550); 49/6(116)
Molecular mass 49/7(225)
Molecular mass, average 49/6(137)
Molecular oxygen, excited states 49/6(121)
Molecular scattering 48(158)
Molecular velocity 49/7(229)
Möller, radiation chart 48(234)
Moment method 49/7(63, 167–192)
Momentum, angular, *see* Angular momentum
Momentum balance 49/7(58, 59, 168)
Momentum balance of electrons 49/2(10)
Momentum density, wave 49/5(214)
Momentum, energy-, tensor 49/5(212)
Momentum integral of trapped particles 49/2(526)
Momentum transfer 49/6(41, 135)
Momentum transfer collision frequency 49/6(250)
— coefficients 49/6(251)
— ion-neutral, averaged 49/6(42)
Momentum transfer cross section 49/7(66, 99–105 Figs. 11–17, 115, 121)
Momentum transport collision frequency 49/7(158–160 Figs. 32–38)
Momentum, wave, angular 49/5(216)
Monazite 47(349)
Monoclinic pyroxene 49/1(249 ff.)
Monodromy group 49/2(178)
Monoenergetic transfer collision frequency 49/7(127, 196)
Monsoonal character 49/5(117, 135, 140)
Month, anomalistic, draconitic, synodic, tropical 48(747)
Monthly means of cosmic-ray intensity 49/1(195, 219)
Monthly median value 49/2(269)
Moon 49/4(198)
Moon, birth 47(282)
Moon, fictitious 48(757)
Moon light 48(217)
Moon light, ashcolored 48(216)
Moon, orbital motion 48(748)
Moon's longitude, secular acceleration 48(924)
Moon's motion, fluctuations 47(8)
Moon's motion, secular acceleration 47(3)
Moon's motion, theory 47(8)
Moon's orbit 47(621)
Morse potential 49/7(77, 118, 140 ff. Figs. 25 a–h)
Most probable thermal speed 49/2(12)
Most probable transport collision frequency 49/2(12–14, 16, 20, 97)
Motion, equation 49/5(180); 49/6(204, 249, 260, 262)
Motion, laminar 48(623)
Motion of electrons 49/2(272)
Motion of plasma 49/2(276, 288, 342, 373)
Motion of radar echoes 49/1(78)
Motion of structures 49/7(280)
Motion, turbulent 48(623)
Motion, vertical 49/6(248, 261, 288)
Motions of auroras 49/1(37–45)
Mountain chains, submarine 47(253)

Mountain ranges 47(259)
Mountain roots 47(239, 254)
Mountain waves 49/5(206)
Mouth correction 48(856)
Moving coils 49/3(339)
M-Region 49/3(32, 254, 266)
MRN (meteorological rocket network) 49/5(117, 119, 120)
MSIS atmospheric model 49/7(36, 333, 338–347 Figs. 92–100)
MSIS, improved model 49/7(340)
M.U.F., *see* Maximum usable frequency
Mügge's weather types 48(121)
Multicomponent plasma 49/5(236)
Multicomponent system 49/6(132)
Multigrid diffraction filter 49/6(317, 364)
Multilayer structure 49/2(259)
Multi-lobe angular character 49/5(288)
Multiparameter interaction potentials 49/7(71, 139–162)
Multiple pendulum apparatus 47(209)
Multiple proton gyroresonances 49/5(316)
Multiple reflexions 47(163)
Multiple-reflection gravimeter 48(804)
Multiples in ionograms 49/2(230)
Multiplets 49/6(357)
Multiplett structure 49/6(350)
Multiplicity of auroral arcs 49/1(27)
Multiplier, degradation 49/6(364)
Multipole expansion of magnetic field 49/4(121)
Multipole expansion of velocity distribution 49/2(17ff.)
Muon-decay effect on cosmic-ray intensity 49/1(169)

N_2^+ first negative bands 49/1(67, 100, 110, 113, 121)
N_2^+, Meinel bands 49/1(110)
N_2 positive bands 49/1(109ff., 117)
Nacreous clouds 49/5(162, 164)
Narrow-band emission 49/5(338)
Natural frequencies of solid earth 48(784)
Natural oscillations of the atmosphere 48(476, 930, 965)
Nautical Almanac 47(1, 8, 9)
Navier-Stokes equation 49/5(127); 49/6(249, 256, 257)
Navier-Stokes stress tensor 48(8)
Neap tide 48(739, 882)
Near zone 49/5(253, 266)
Néels theory 47(474)
Negative ions 49/6(14, 126); 49/7(379, 383 Fig. 122, 436, 439)
Negative ions, reactions 49/6(53, 62, 107, 109)
Negative sudden commencement 49/3(160)
Negative to positive ion ratio (λ) 49/7(439)
Neodym 47(299)
Neper (nep) 49/2(264)
Nettleton-Profile 47(600)
Neumann-spectrum 48(696, 698, 700, 702)
Neutral and electrical components interaction between 49/5(171)
Neutral approximation 49/5(252, 253)
Neutral atmosphere, Martian 49/5(374)
Neutral atmosphere of Venus 49/5(374)
Neutral composition 49/6(71, 78, 84)
Neutral composition, wind effects 49/6(298, 300)
Neutral constituents 49/6(10)
Neutral gas 49/4(7, 8)
Neutral gas, continuity equation 49/6(259)
— energy equation 49/6(259)
— equation of motion 49/6(260)
— transport equations 49/6(259, 263)
Neutral line discharge model 49/1(139ff.)
Neutral lines 48(303); 49/4(54)
Neutral particle densities 49/5(372)
Neutral point 49/3(150); 49/4(25, 39, 54, 55, 66, 77, 84, 103, 109, 110)
Neutral reactions 49/6(56)
Neutral sheet 49/3(11, 154); 49/4(37, 42, 47–50, 53)
Neutral temperature 49/6(259)
Neutral temperature, wind effects 49/6(294, 300)
Neutral winds 49/6(282, 286, 288, 301, 303)
Neutrals, collision between 49/7(122–124 Table 10, 166)
Neutron decay, cosmic ray albedo 49/4(127)
Neutron decay injection 49/4(132)

Neutron intensity, diurnal variation 49/1(177)
Neutron intensity, effect of solar flares 49/1(209)
Neutron intensity, latitude curve 49/1(165, 202ff.)
Neutron meters, observation of diurnal variation 49/1(175–177)
Neutron monitor 49/1(165, 169, 219, 231, 234ff.)
Neutron point discharge model 49/1(139)
Neutron variation with latitude 49/1(165, 202ff.)
Neutrons, polar cap 49/4(132)
Newcomb's period 47(610)
Nichols Diagram 49/3(332)
Night clouds, luminous 48(422)
Night *E* 49/2(368)
Night *Es* 49/2(337)
Nitrate 48(618)
Nitric oxides 49/6(92)
— absorption cross sections 49/6(19)
— asorption cross sections 49/6(21)
— ion, formation 49/6(52)
— ion, hydration 49/6(103, 105)
— ion reactions 49/6(51, 52)
— photodissociation 49/6(34)
— photoionization 49/6(96)
Nitrogen, atomic ion
— profile 49/6(233)
— reactions 49/6(51, 232)
— production profile 49/6(119)
Nitrogen, collision frequencies 49/2(12)
Nitrogen, energy level diagrams 49/6(21, 48)
Nitrogen, excited states 49/6(93, 123)
Nitrogen, forbidden atomic lines 49/1(108ff., 117)
Nitrogen, molecular
— absorption cross sections 49/6(20, 23)
— absorption of 49/6(316)
— bands 49/1(67, 100, 109ff.)
— ion, reactions 49/6(51)
— ionization threshold 49/6(24)
Nitrogen, permitted atomic lines 49/1(109)
Niveau-Ellipsoid 47(555)
NO^+ formation 49/1(119)
Noctilucent clouds 49/5(162, 163, 165)
Node, ascending 48(747, 749)
Node factor 48(780)
Nodes 48(850)
Noise 49/2(210)
No-*K*-variations 49/3(210)
Non-conservative waves 49/4(527)
Non-cyclic change 49/3(63)
Non-deviative approximation 49/2(91, 265)
Non-divergence, level of 48(86, 102)
Non-divergent waves 49/5(190)
Non-fully developed state 48(714)
Non-harmonic procedure 48(899)
Non-ideal plasma 49/7(157)
Non-isothermal sound velocity 49/5(240ff.)
Non-linear phenomena 49/2(393–400); 49/4(439)
Non-linear processes 49/5(204)
Non-linear resonance effects 49/2(18)
Non-Maxwellian electron population 49/7(369 Fig. 114)
Non-relativistic case 49/4(495)
Non-selective absorption 49/2(0, 91, 100)
Nonsymmetric inflation 49/3(193)
Nonthermal sound velocity 49/5(221)
Noon depression of *E*-region 49/2(281)
Normal Doppler effect 49/5(228)
Normal gravity 47(568, 636)
Normal gravity, vertical gradient 47(558)
Normal incidence 49/2(169ff.)
Normal mode, stationary 49/4(408)
Normal modes 49/5(197, 198)
Normalization energy 49/7(107, 119)
Normalization length 49/7(106, 119)
Normalized transfer cross section 49/7(82 Fig. 4, 83 Fig. 5, 85 Fig. 6, 86 Fig. 7, 87 Fig. 8, 89 Fig. 9, 94 Fig. 10b)
North Polar Basin 48(913)
North Sea 48(911)
Northern light 49/1(2)
North-south motion of auroras 49/1(37)
Norwegian Sea 48(646)
Nose frequency 49/2(117)
Nose whistler 49/2(499ff.)
N-type ferrimagnetism 49/1(280ff.)
Nuclear bombs, high altitude 49/4(137)
Nuclear detonations 49/2(347)

Nuclear resonance in Earth's magnetic field 49/3(361)
Nucleation 48(1009)
Nucleation, spontaneous 48(499)
Nuclei counter 48(567)
Nuclei, critical radii 48(488)
Nuclei, giant 48(489)
Nuclei, hygroscopic 48(488, 490)
Nuclei, large 48(489)
Nuclei, man-made 48(551)
Nuclei, maritime 48(551)
Nuclei, size 48(161)
Nuée ardente 48(985)
Nullbahnen 49/1(209)
Number density 49/2(2, 4)
Number distribution of Zeeman-terms 49/3(369)
Numerical mapping 49/7(388–403 Figs. 125–129)
Nutation 47(620)

O^+ ions 49/7(424, 427 Figs. 145a, d)
Oblique echoes 49/2(221, 319)
Oblique incidence, absorption decrement 49/2(128)
Occlusion 48(136)
Occultation 47(622)
Occultation method, radio- 49/5(353ff., 354, 359, 360)
Ocean currents, heat transport 48(636)
Ocean tides 48(914)
Ocean tides, effect on earth currents 47(435)
Ocean wave(s)
— analyzer 48(691)
— measurement 48(682)
— refraction 48(726)
— spectrum 48(691, 713, 720, 722, 727)
Oceanic circulation 47(18)
Oceanography 48(608)
$O(^1D)$, excited atom 49/6(54, 124)
Odd nitrogen 49/6(93, 94)
— production 49/6(94)
— profiles 49/6(94)
— reactions 49/6(93)
— species 49/7(523 Fig. 220)
OGO satellites 49/7(275, 327, 329 Fig. 87, 333)
OGO-6 temperature model 49/7(329–331 Fig. 86)
Ohm's law 49/7(61, 175)
Ohm's law, generalization 49/2(273)
OH-radical 49/6(33)
Olivine 49/1(249ff.)
O'Malley's modified range expansion 49/7(92 Fig. 10, 97, 98, 128)
Omega integrals 49/7(131)
One dimensional aeronomic theory 49/7(315)
One fluid gasdynamic model 49/4(469)
One fluid gasdynamics 49/4(482)
Onsager relations 49/7(181, 182, 191, 196)
Onsager-Casimir relations 49/7(62)
Opacity of sporadic *E* layer 49/2(339)
Open orbit 49/5(277)
Optical depth 48(156); 49/6(16, 26, 344)
Optical pumping 49/3(368)
Optical scale value 49/3(279)
Optical slant range 48(265)
Optical thickness 48(295, 352)
Optics, geometrical 49/5(195, 211)
Optics in sea water 48(621)
Orbit, closed 49/5(277)
Orbit, finite 49/5(277ff.)
Orbit, open 49/5(277)
Orbital path 48(674, 676, 681, 731)
Orbits, infinite 49/5(277)
Ordinary polarization 49/2(50)
Ordinary wave 49/5(232)
O-reflection (ordinary r.) 49/2(63)
Orientation of variometer magnets 49/3(288, 289)
Orientation probe 49/3(355)
Orionids 48(435)
Orogenesis 47(273, 331, 358)
Orogenetic shell 47(262)
Orographic effect 49/5(144)
Orographic forcing 49/5(206)
Orthogonal series expansion of velocity distribution 49/2(9, 11, 17ff.)
Orthopyroxene 49/1(249)
$O(^1S)$, excited atom 49/6(54)
Oscillation, longitudinal resonance 49/5(238)
Oscillation, modes 49/5(202)
Oscillation of a charged particle 49/1(84ff., 87)
Oscillation of the servo-system 49/3(332)
Oscillation spectra 49/4(514)
Oscillations 49/3(301); 49/4(508)
Oscillations, drift 49/4(524, 539, 540, 548)

Oscillations, electronic Langmuir 49/4(521)
Oscillations (Gauss-Lamont) 49/3(301)
Oscillations, gravitational 49/5(162)
Oscillations, longitudinal 49/4(502)
Oscillations, longitudinal (potential) 49/4(524)
Oscillations, long-wave drift 49/4(542, 546)
Oscillations, long-wave kinetic 49/4(550–553)
Oscillations of coupled systems 48(857)
Oscillations of lake levels 48(852)
Oscillations, slowly increasing 49/4(541)
Oscillator strength 49/6(39, 354)
Outer ionosphere 49/2(520–531); 49/5(219, 236)
Outer radiation belt 49/1(81 ff., 96)
Overhauser effect 49/3(379)
Overhauser magnetometer 49/3(427)
Over-seeding 48(536)
Oxidation of titanomagnetites 49/1(254–257)
Oxides with ferrimagnetic properties 49/1(250)
Oxygen 48(171, 618)
Oxygen, atomic
— ion, profile 49/6(242)
— ions, reactions 49/6(49)
— loss 49/6(181)
— metastable reactions 49/6(55, 121)
Oxygen atoms, profile 49/6(185, 188, 191, 192, 244)
Oxygen atoms, removal 49/6(183)
Oxygen, collision frequencies 49/2(12)
Oxygen contents 48(649, 668)
Oxygen, dissociation 49/7(440)
Oxygen, energy level diagrams 49/6(18, 48)
Oxygen, excited states 49/6(121, 123)
Oxygen, forbidden atomic lines 49/1(106–108, 115, 118, 120, 121)
Oxygen ions: O^+ profiles 49/7(453 Fig. 161)
Oxygen, molecular
— absorption cross sections 49/6(17, 18, 20, 22)
— absorption of 49/6(316)
— bands 49/1(108, 115)
— ionization threshold 49/6(23)
— ions, reactions 49/6(50)
— metastable reactions 49/6(55)
— profile 49/6(185, 188, 190, 192)
Oxygen, permitted atomic lines 49/1(108, 117)
Oxygen, photodissociation 49/6(32, 177)
Oxygen, thermal diffusion 49/6(143)
Oxygen-hydrogen chemistry 49/6(86, 88)
Ozone 48(158, 172, 217, 237, 242, 246, 370); 49/6(2, 19, 31, 32, 45, 46, 86, 124, 182)
Ozone, absorption cross section 49/6(17–20, 22, 32, 179, 180)
Ozone, advection 48(399)
Ozone, amount 48(372)
Ozone and pressure distribution 48(408)
Ozone and vertical turbulence 48(405)
Ozone, annihilation 48(405)
Ozone, bioclimatic significance 48(422)
Ozone, blue colour 48(375)
Ozone clouds 48(408)
Ozone commission 48(371)
Ozone concentration 48(372)
Ozone content, absolute 48(372)
Ozone dissipation, thermal 48(395, 405)
Ozone distribution, vertical
— causes 48(399)
— measurement 48(380)
— variations 48(396)
Ozone, formation of 49/6(182)
Ozone formation, reaction constants 48(391)
Ozone, infra-red bands 48(376)
Ozone layer, optical upper limit 48(422)
Ozone layer, temperature 48(382, 416)
Ozone, lunar eclipses 48(384)
Ozone maximum 48(381, 394)
Ozone measurement, chemical 48(388)
Ozone measurement, optical 48(377)
Ozone, near the ground 48(372, 387)
— variations 48(403)
Ozone, photo-chemical theory 48(389)
Ozone, photodissociation 49/6(31)
Ozone, polar night 48(395, 415)
Ozone profile 49/5(122)
Ozone radiosonde 48(379, 384)
Ozone, removal 49/6(183)
Ozone value 48(372, 377)
Ozone variations, interdiurnal 48(407)

Ozone variations, meridional 48(412)
Ozone variations, seasonal 48(412)
Ozone-oxygen-window 48(393)
Ozonosphere 49/5(120)
Ozonosphere sound 48(473)

Pacific Ocean, water masses 48(647)
Packet, wave 49/5(214, 215)
Palaeomagnetism 49/1(273)
Paleomagnetism, interpretation 47(492)
Paleomagnetism, mesozoic 47(486)
Paleomagnetism, metamorphic rocks 47(490)
Paleomagnetism, paleozoic 47(487)
Paleomagnetism, precambrium 47(489)
Paleomagnetism, quaternary lava flows 47(482)
Paleomagnetism, teriary 47(483)
Paleomagnetism, varved clays 47(481)
Panoramic ionogram 49/2(218)
Parabolic cylinder functions 49/2(184ff.)
Parabolic layer 49/2(97ff., 119ff.)
— absorption decrement 49/2(98)
— equivalent height 49/2(119)
— model 49/7(348)
— profile 49/2(258, 308)
Parallaxe 48(744)
Parallel propagation 49/2(34, 58, 62, 198)
Paramagnetic electron resonance 49/3(379)
Paramagnetic minerals 49/1(249)
Parameter 49/5(184)
Parameter space 49/2(60, 64)
Parent gas, collisions in 49/7(114–121, 163–166)
Parity 49/7(111)
Partial absorption cross section 49/6(17)
Partial accommodation 49/5(244)
Partial heat fluxes 49/7(167)
Partial pressure tensors 49/7(167)
Partial reflection 49/2(192, 194ff.)
— from *D* region 49/2(319)
— on a thin layer 49/2(247)
Partial thermo-remanent magnetization 49/1(273)
Partial tide(s), *see also* Tides 48(780, 885)
— phase 48(896)
Particle acceleration 49/3(155); 49/4(66, 71)
Particle bombardment 49/7(497–502)
Particle containment 49/4(215)
Particle, energies 49/4(60)
Particle focusing 49/5(261)
Particle loss 49/4(65, 66)
Particle method 48(49, 65, 70)
Particle, motion of a charged 49/4(116)
Particle reflection 49/5(243, 244)
Particle sources 49/4(65)
Particle trajectories 49/4(221)
Particles collision 49/4(425)
Particles, drift, trapped 49/4(189)
Particles, heavier 49/4(136)
Particles, infinite 49/5(281)
Particles, interaction 49/5(228)
Particles, low-energy 49/4(70, 184)
Particles, motion of individual 49/4(412)
Particles, motion of, near a magnetic null 49/4(219)
Particles, precipitated 49/4(211)
Particles, production of new 49/5(243, 245)
Particles, size-distribution 48(164, 201)
Particles, slow 49/4(72)
Particles, trapped 49/4(60, 63, 71, 218)
Particles, vertical distribution 48(200)
Patches 49/1(27ff., 67)
Path, phase 49/5(354, 355, 357)
pc, pi 49/3(116)
PCA (=polar cap absorption) 49/2(388–393)
Peak altitude (*hmF*2) 49/7(365)
Peak electron density (world maps) 49/7(392–403 Figs. 125–129)
Peak (*F*2-, ionosphere) 49/6(200, 203, 285, 287)
Peak (of an ionospheric layer) 49/7(348, 364–367, 385)
Peak (plasma frequency/electron density) 49/7(348, 365 Fig. 112, 384, 388 Fig. 124, 404–406 Figs. 130–132)
Peak power 49/2(199)
Pearl-type micro-pulsations 49/5(302)
Peculiar velocity 49/2(8, 10)
Pedersen and Hall conductivities 49/5(170)
Pedersen conductivity 49/3(7, 25)

Pedersen current 49/3(7); 49/7(504 Fig. 196)
Pegmatite 47(352, 355)
Peltier effect (tensor) 49/7(179, 191)
Pendulum clocks 48(797)
Pendulum, disturbed 47(208)
Pendulum, fictitious 47(210)
Pendulum, invariable 47(206)
Pendulum observations, corrections 47(212)
Pendulum, reverse 47(46)
Pendulum, reversible 47(203)
Pendulum stations, distribution 47(590)
Penetration coefficient 49/2(179)
Penetration decrement of Epstein layer 49/2(194)
Penetration method 47(187)
Penetration of low-frequency transverse waves 49/4(461)
Penetration of solar radiations 49/6(26, 27)
Penetration ratio 49/2(94ff., 128, 164, 182)
Pennstate model 49/7(485)
Percolation 48(728)
Performance of ionization chambers 49/1(161ff.)
Perigee 48(748); 49/7(234, 236)
Perihelion 48(747)
Period length measuring device 49/3(337)
Period-amplitude diagram of geomagnetic variations 49/3(341)
Periodic motion, nearly 49/4(67)
Periodicities, 27-day 49/4(180)
Permanent magnets 49/3(284, 318)
Permeability 49/2(1)
Permeability, dielectric 49/5(228ff.)
Permeability, tensor of complex dielectric 49/4(399)
— of dielectric 49/4(494)
— plasma dielectric 49/5(225)
Permittivity 49/4(414); 49/5(228)
Permittivity, dielectric 49/4(444)
Permittivity, longitudinal (dielectric) 49/4(449, 470)
Permittivity, tensor of 49/4(463)
Permittivity, transverse 49/4(449)
Perpendicular propagation 49/2(34, 58)
Perseids 48(435, 437, 438)
Persian Gulf 48(912)
Perturbation, solar, of moon's orbit 48(749)
Perturbation theory 49/6(271)
Phantom horizon 47(162)
Phase angle 49/2(43)
Phase borders of water, electric effects 48(560)
Phase difference 49/3(295)
Phase function 48(292)
Phase integral 49/2(95, 183)
Phase matrix 48(295)
Phase measurement 49/2(457)
Phase methods of observation 49/2(443–475)
Phase path 49/5(354, 355, 357)
Phase shift 49/2(42)
Phase space 49/2(7); 49/6(265)
Phase trajectories 49/2(125, 134)
Phase velocity 49/2(28, 100); 49/4(410); 49/5(188, 189, 224, 231)
Phase velocity surface (polar diagram) 49/2(492)
Phase-coherent detection 49/2(219)
Phases of a magnetic storm 49/1(55)
Phosphate 48(618)
Phosphors 49/6(309)
Photo electrons 49/7(35, 459–465 Figs. 167, 168, 170, 525–532)
Photo-chemical equilibrium 48(372, 392)
Photochemical processes 49/6(11)
Photodetachment 49/6(53, 97, 108)
Photodissociation 49/6(2, 87, 93, 108)
Photodissociation, NO 49/6(34)
Photodissociation, O_2 49/6(31, 32)
Photodissociation, ozone 49/6(31)
Photodissociation, rates 49/6(31, 88, 93, 110, 187, 188)
Photodissociation, water vapour 49/6(33)
Photoelectric detection 49/6(311, 317, 328, 355, 360)
Photo-electric recording device 49/3(295)
Photographic recording 49/3(277, 289, 291)
Photographic recording of Spectra 49/6(312, 316, 348)
Photoionization 49/6(34, 36, 118, 125)
Photoionization, cross sections 49/6(36)
Photoionization, H 49/6(118)

Photoionization, NO 49/6(96)
Photoionization rates 49/6(93)
— D-region 49/6(38)
— E- and F-regions 49/6(35)
Photoionization yields 49/6(17, 125)
Photolysis 49/6(56)
Photometric decrement 49/2(263)
Photometry 49/6(343, 367)
Photosphere 49/6(373)
Photosynthesis 48(618)
Physical constants 49/7(220)
Physical geodesy, boundary problem 47(576)
Physical geodesy, differential equation 47(571)
Physical quantity 49/4(3)
Piezo-remanent magnetization, *see* Pressure-remanent magnetization 49/1(298ff.)
Piezotropy 48(1)
Pilansberg dykes 47(487)
Piling-up of water by wind 48(627, 654)
Pinhole camera 49/6(335)
Pioneer I & III 49/1(81)
Pioneer V 49/1(233)
Piscun's photometric screens 48(287)
Piston corer 48(609, 616)
Pitch angle 49/4(118)
— diffusion 49/4(175)
— of trapped particle path 49/2(526ff.)
— scattering 49/4(67, 77)
Pitchblende 47(323, 350)
Placers 47(315)
Planck's law 49/6(309)
Plane, beta 49/5(184)
Plane waves, inhomogeneous 49/4(409)
Planes of constant amplitude 49/2(122)
Planetary index *Kp* 49/3(209, 222ff.)
Planetary index *Kp*-value 49/3(209, 222, 253)
Planetary ionospheres 49/7(451, 454–458 Figs. 162–164, 166)
Planetary scale of auroras 49/1(52)
Plasma 49/4(395); 49/5(169)
Plasma, collisionless 49/4(420, 424, 440, 441, 461, 462, 537)
Plasma, completely ionized 49/4(424, 428, 429, 457, 459, 462, 482, 523, 526, 527, 532)
Plasma, continuity equation 49/6(261)
Plasma density 49/7(258)
Plasma dielectric permeability tensor 49/5(225)
Plasma drift, effect of magnetic declination 49/6(291)
Plasma drift, effect of magnetic inclination 49/6(289)
Plasma, electromagnetic drift 49/6(279)
Plasma, equation of motion 49/6(262)
Plasma flow around solid bodies 49/5(217, 252)
Plasma, flow of 49/4(25)
Plasma flux 49/4(51)
Plasma frequency 49/7(157)
Plasma, gyrotropic (magneto-active) 49/4(462, 466)
Plasma, hot 49/5(228)
Plasma, ionospheric 49/4(522, 537)
Plasma line 49/7(260 Figs. 26, 27)
Plasma model, fluid-dynamic 49/4(416)
Plasma model, quasi-hydrodynamic 49/4(416)
Plasma, momentum equation 49/6(261)
Plasma motion(s) 49/2(276, 288, 342, 373, 517ff.); 49/6(5)
— diffusive 49/6(172)
— effect on winds 49/6(301)
— induced by neutral wind 49/6(5, 247, 279, 284, 288, 289, 292)
Plasma, multicomponent 49/5(236)
Plasma, non-isothermal 49/4(485, 542)
Plasma, non-isothermal hot 49/4(422)
Plasma oscillation, longitudinal 49/5(233)
Plasma parameters 49/2(3)
Plasma, polarization field 49/6(262)
Plasma potential 49/5(273)
Plasma pulsation 49/2(2)
— circular frequency 49/7(157)
— electron 49/5(221)
— ion 49/5(221)
Plasma, quiescent 49/5(225)
Plasma refraction 49/5(355)
Plasma resonance frequencies 49/2(509ff.)
Plasma resonance in meteor trail 48(433)
Plasma resonance, limiting angles 49/2(511)
Plasma sheet 49/4(43, 51, 52)

Plasma temperature(s) 49/6(151); 49/7(369–374 Figs. 115, 116, 410–419 Figs. 137–141)
— profiles 49/5(364)
Plasma, thermal 49/4(8, 60, 61)
Plasma, wave trapping in 49/5(235)
Plasma, waves in warm 49/4(481)
Plasma, weakly ionized 49/4(435, 523, 526, 527, 532)
— gyrotropic 49/4(468)
Plasma, weakly magnetized 49/4(500)
Plasmapause 49/4(56, 104); 49/5(219, 224, 330, 341, 376, 381)
Plasmasphere 49/4(56, 61); 49/5(353)
Plasmasphere, oxygen ions 49/7(512 Fig. 205)
Plasmasphere, Plasmasheet, Plasmapause 49/7(512 Fig. 206)
Plasticity 47(178, 279)
Platform gyro-stabilized 49/3(386)
Plumb line(s) 47(537)
— absolute deviation 47(579)
— curvature 47(560)
— deflection 47(539)
Poeverlein construction of ray paths 49/2(124)
Poincaré sphere 49/2(44, 46ff., 58)
Poincaré's waves 48(860)
Point body 49/5(220)
Point, conjugate 49/4(119)
Point, magnetic neutral 49/4(192)
Point, mirror 49/4(119, 145)
Point, singular 49/5(228)
Point-discharges 48(569, 583, 590, 598, 602)
Pointing 49/6(312)
Point-source 48(269)
Point-sources, thresholds for 48(270)
Poisson equation 49/5(225, 227)
Poisson's ratio 47(79, 170, 173)
Poisson's relation 47(79)
Polar black out 49/1(70, 71); 49/2(325, 367ff., 386)
Polar cap(s) 49/7(497–521)
— absorption 49/6(77)
— absorption of radio waves 49/1(74)
— absorption (PCA) 49/2(388–393); 49/7(380 Fig. 121c)
— electron density distribution 49/7(508 Fig. 201)
— electron temperature 49/7(514 Fig. 210)
— ion beams 49/7(511 Fig. 204)
— neutral winds 49/7(510 Fig. 203)
— O^+ ions 49/7(520 Fig. 217, 522, 524 Fig. 221)
— plasma convection 49/7(515, 519)
— plasma drift 49/7(510 Fig. 203)
Polar cusps 49/4(70, 81, 109)
Polar cusps (of magnetosphere) 49/4(53, 55)
Polar darkening (sun) 49/6(344)
Polar disturbance field ($\mathbf{D}_p$) 49/3(77)
Polar elementary storms 49/3(99)
Polar glow 49/1(16, 111ff., 120)
Polar magnetic substorm 49/1(54)
Polar motion 47(15, 613)
Polar region, equivalent height 49/2(198)
Polar regions, winter 49/5(134)
Polar Sea 48(646)
Polar substorm 49/3(134)
Polar wandering 47(287, 493, 495)
Polar wind 49/4(14, 81); 49/6(225, 236, 243)
Polar zone ionograms 49/2(238)
Polarisabilities 49/6(42)
Polarisation current 49/3(365)
Polarisation state 49/3(368)
Polarizability 49/7(95, 96)
Polarization 49/7(120)
Polarization and Faraday effect 49/2(463)
Polarization changes, *see also* Faraday effect 49/2(477)
Polarization, degree 48(290)
Polarization, direct experimental determination 49/2(267)
Polarization ellipse 49/2(6, 27, 42ff., 52)
Polarization, elliptical 48(289)
Polarization field 49/6(262)
Polarization force 49/6(42)
Polarization, noticed by insects 48(325)
Polarization of echo 49/2(216)
Polarization parameter 49/2(38, 47ff., 52)
Polarization, partial elliptic 48(289)
Polarization, plane of 48(290)
Polarization, positive and negative 48(300)
Polarization, *see also* Dipole density 49/2(22)
Polarization, spontaneous 47(430)
Polarized coordinates 49/2(31ff., 36)

Polarography 47(311)
Pole movement 48(813)
Pole of inertia, instantaneous 47(611)
Pole shifts, periodic 47(606)
Poles of refractive index 49/2(61, 66, 479)
Polhody 47(608)
Poloidal vector fields 47(526)
Polychroism 48(315)
Polyhedron 47(534)
Population of electrons 49/4(51, 52)
Portable magnetograph 49/3(290)
Portable magnetograph, type
— Askania 49/3(290)
— Bobrov 49/3(291)
Positive ion density profile 49/2(521)
Positive ions 49/6(14, 15)
Post-disturbance period 49/7(509)
Potassium-40 47(297, 305, 307, 336)
Potassium-40, heat-production 47(305)
Potassium, contents of rocks 47(317)
Potassium, measuring methods 47(312)
Potassium method 47(335)
Potassium-argon-ages 47(338)
Potential body 49/5(245)
Potential, electrostatic 49/5(243)
Potential field, electric 49/5(173)
Potential fields, interpretation 47(225)
Potential gradient 48(542, 576)
Potential gradient, daily variation 48(581)
Potential, large body 49/5(280)
Potential, plasma 49/5(273)
Potential scattering 49/2(12)
Potential, small body 49/5(280)
Potential temperature as vertical coordinate 48(26)
Potential vorticity and vortex structure 48(113)
Potsdam g 47(204)
Power density 49/2(5ff.)
Poynting vector 49/2(110, 152)
Prandtl number 49/7(472)
Precambrium 47(289, 291)
Precambrium, age 47(335)
Precambrium, nomenclature and correlation 47(292)
Precession 47(613)
Precession, luni-solar 47(619)
Precipitating electrons 49/2(528, 529); 49/7(499, 508)
Precipitating particles 49/6(96, 199)
Precipitation 48(517, 583); 49/4(101, 174, 175)
Precipitation, electron 49/4(171, 173, 209)
Precipitation forecast 48(181)
Precipitation, formation 48(517)
Precipitation of electrons from x-ray bursts 49/1(90–95)
Precipitation, particle 49/4(215)
Precission, luni-solar 48(739)
Prediction of solar emission line intensities 49/6(350, 358)
Prediction of solar EUV irradiance 49/7(28–32)
Predissociation 49/6(31, 34, 190)
Preionization 49/6(22)
Preliminary reverse impulse 49/3(43, 47)
Preselection (optical) 49/6(317)
Pressure and pressure tensor 49/2(10)
Pressure as vertical coordinate 48(24)
Pressure, balance, hydromagnetic 49/4(187)
Pressure balance, neutral sheet 49/4(195)
Pressure broadening 48(175)
Pressure field 48(650)
Pressure gauges 49/7(244–247 Figs. 11, 12)
Pressure gradient 49/2(10, 61); 49/6(248, 251, 255)
Pressure, load on earth 48(837)
Pressure remanent magnetization (PRM) 49/1(272, 297–299)
Pressure, solar wind 49/5(378)
Pressure, total kinetic 49/4(513)
Prevailing wind 49/6(275)
Prey's reduction 47(579)
Primary alpha-particles 49/1(226)
Primary electrons 49/1(113)
Primary spectrum of cosmic rays 49/1(197ff., 226, 234)
Principal axes of a tensor 49/2(31ff.)
Principal axes of polarization, *see also* Principal polarizations 49/2(42)
Principal directions of wave propagation in gyrotropic optics 49/2(34)
Principal group indices 49/2(114ff.)
Principal polarizations 49/2(34ff., 42ff.)
Principal refractive indices 49/2(34ff.)
Principal resonances 49/2(66)
Principal stresses 47(77)

Prism binoculars, transmittance 48(275)
PRM, *see* Pressure remanent magnetization
Probe sensitivity 49/3(352)
Processes, anisotropic 49/5(205)
Processes, dissipative 49/5(205)
Processes, nonliner 49/5(204)
Production of auroral spectrum 49/1(113–117)
Production of new particles 49/5(243, 245)
Production rates, of ions 49/6(37)
Production term 49/2(251)
Production/loss balance 49/7(496 Figs. 186, 187)
Profile, of electron and ion temperature 49/6(13)
Profile of electron density 49/6(12)
Profile of ion composition 49/6(14)
Profile of recorded layer 49/2(256, 261)
Profile, ozone 49/5(122)
Profile parameters 49/2(262)
Profile, vertical temperature 49/5(130, 131)
Profiles, meridional wind 49/5(138, 139)
Profiles of refractivity 49/5(356, 361)
Profiles, plasma temperature 49/5(364)
Profiles, zonal wind 49/5(126, 136, 137)
Projectors 49/7(211)
Prominence 49/6(342)
Propagating eigenwaves 49/2(153ff.)
Propagation attenuation 49/2(223)
Propagation, parallel 49/4(473)
Propagation parallel or perpendicular to earth's magnetic field 49/2(34ff.)
Propagation, perpendicular 49/4(473)
Propagation vector 49/2(24, 27)
Proportional control 49/3(328)
Prospection, electric 47(421, 448)
Proterozoic era 47(291)
Proton decay 49/4(131)
Proton density 49/4(11, 14)
Proton distribution up to 6 earth radii 49/1(130)
Proton energy spectrum 49/1(95, 226, 237); 49/4(134)
Proton flux 49/4(64, 65)
Proton gyrofrequency 49/5(316)
Proton magnetometer 49/3(364)
Proton whistler 49/5(306)
Proton-beam spectrometer 49/6(66)
Protonosphere 49/6(199)
Protons 49/4(152); 49/6(119, 242)
Protons, characteristic energy 49/4(153)
Protons, energetic 49/4(64)
Protons from the sun 49/2(392)
— space probe results 49/1(233)
— spectrum during solar flare 49/1(237)
— summary of origin 49/1(243ff.)
Protons in radiation belts 49/1(82)
Protons in the aurora 49/1(95)
Protons, precipitation of 49/4(175)
Pseudo break-up of an aurora 49/1(42, 45, 89)
Pseudomagnetosphere 49/5(376)
Pulsation, geomagnetic 49/4(104)
Pulsation recorder, type Grenet 49/3(296)
Pulse compression radar 49/2(220)
Pulse delay 49/2(475)
Pulse receiver 49/2(206ff.)
Pulse repetition 49/2(199, 210)
Pulse transmitter 49/2(204ff.)
Pumping mechanism 49/4(158)
Purplish-red lower border 49/1(22ff., 117)
Pütter circle 49/2(406)
Pyranometer 48(206, 250, 251)
Pyranometer, effective 48(252)
Pyranometer, spherical 48(251)
Pyrgeometer 48(236, 252)
Pyrheliometer 48(183, 251)
Pyrheliometric scale 48(253, 1956)
Pyroxene 49/1(249)
Pyrrhotite 47(470)
Pyrrhotites 49/1(264ff.)

QHM 47(499)
QHM-magnetometer 49/3(309)
Q-Index 49/3(101, 104, 243)
Quadrantids 48(435, 437)
Quadrature 48(739, 882)
Quadrupole mass spectrometer 49/6(75); 49/7(239 Figs. 6, 7)
Quadrupole moment 49/7(98)
Qualifying letter 49/2(228)
Quantization 49/4(524, 525)
Quantizations error 49/3(381)
Quantization rule 49/4(519, 526)
Quantum corrections 49/7(68, 91, 107)
Quantum efficiency 48(390)

Quarks 49/6(376, 377)
Quarter thickness of a parabolic layer 49/2(97)
Quartile 49/2(269)
Quartz horizontal magnetometer 47(499)
Quartz-crystal-clocks 47(2, 15)
Quase-static approximation 48(20)
Quasi-geostrophic approximation 48(93)
Quasi-hydrodynamic approximation 49/5(228)
Quasi-longitudinal 49/5(230)
Quasineutral 49/4(395)
Quasi-periodic structure 49/5(259)
Quasi-quiescent 49/5(220)
Quasi-solenoidal approximation 48(100)
Quasi-three-dimensional wind calculations 49/6(270)
Quasitrapping 49/4(70)
Quenching 49/7(472)
Quenching (collisional de-excitation) 49/6(53–55, 87, 123, 124)
Quick-run magnetograms 49/3(44)
Quick-run magnetograph 49/3(291)
Quick-run recorder, type La Cour 49/3(394)
Quiescent plasma 49/5(225)
Quiet arcs, lower limit of latitude 49/1(62ff.)

Radar 48(521, 528)
Radar echoes and aurora 49/1(75–77)
Radiant intensity 48(156)
Radiated power, EUV spectrum 49/6(361)
Radiation, absorbed 48(936)
Radiation, atmospheric 48(236)
Radiation balance 48(247, 250)
Radiation balance meters 48(236, 252)
Radiation balance of the earth 48(155)
Radiation belt(s) 49/2(341, 520, 525); 49/4(9, 10, 60, 63, 70, 103)
— artificial 49/4(137)
— inner region 49/4(64, 67)
— of the earth 49/1(81ff.)
— outer region 49/4(65, 67)
Radiation, Čerenkov 49/5(228ff.)
Radiation charts 48(234)
Radiation conductivity 48(243)
Radiation, effective 48(236)
Radiation equilibrium 48(243)
Radiation fluxes 48(204, 210, 241)
Radiation, harmonics of the diurnal variation 48(935)
Radiation in clouds 48(208)
Radiation in the free atmosphere 48(239)
Radiation instruments 48(250)
Radiation, Jupiter's non-thermal 49/4(226)
Radiation, measuring methods 48(250)
Radiation, neutral 48(289)
Radiation, sea 48(631)
Radiation, short wave 48(180)
Radiation surplus 48(636)
Radiation temperature of the earth 48(155)
Radiative processes 49/6(182)
Radiative transfer 49/6(16)
Radiative transition 49/6(351)
Radio carbon 47(308, 346)
Radio frequency studies of the aurora 49/1(68–81, 89)
Radio star scintillations 49/1(81); 49/2(402, 433)
Radio wave propagation 49/4(519)
Radioactive decay, theory 47(295)
Radioactive equilibrium 47(324)
Radioactive heating 47(19)
Radioactive isotopes, natural 47(298)
Radioactivity 47(288, 389); 48(546)
Radioactivity, methods 47(310)
Radioactivity of rocks 47(308)
Radioactivity of single batholites 47(313)
Radio-frequency mass spectrometer 49/6(74)
Radionoise temperature 49/6(373)
Radio-occultation method 49/5(353ff., 354, 359, 360)
Radiosonde 48(241)
Radio-waves, scattering from ionized meteor trail 48(432)
Radio-waves, short 48(354)
Radium in uranium 47(297)
Radius λ_D, Debye 49/5(218, 220, 238)
Radius, Larmor 49/5(218, 220, 231)
Radius, typical 49/5(275)
Radon 47(310, 323)
Rain 48(484, 519)
Rain current 48(597)
Rain, diffusion 48(359)

Rain shower 48(486, 522, 524, 528)
Rain-bands 48(172)
Rainbow 48(187, 346)
Rainbow, white 48(350)
Rainbows, multiple 48(347)
Raindrop multiplication, chain reaction 48(524)
Ramsauer effect 49/7(93)
Ramsey's hypothesis 47(378)
Random changes 49/2(429ff.)
Random phase 48(700, 703)
Rarefaction region 49/5(262)
Rate coefficients 49/6(47, 55)
Rate, evaporation 49/5(245)
Ratio, vertical-horizontal 49/5(125)
Rationalization problem 49/3(2)
Rawer layer 49/2(188)
Rawer step 49/2(188)
Rawers' index function 49/2(186, 188)
Ray direction 49/2(101, 103ff.)
Ray index 49/2(101)
Ray paths 49/2(106, 124, 135)
Ray tracing 49/5(195)
Ray velocity 49/2(100, 102)
Rayed arc 49/1(16, 23ff.)
Rayleigh waves 47(120, 125, 145, 250)
Rayleigh waves, M_1 and M_2 branches 47(126)
Rayleigh-extinction 48(177)
Rayleigh-friction 48(64)
Rayleigh-scattering 48(158, 186, 254)
Rays, geometry and nature 49/1(23–25)
Reaction schemes for ions 49/7(435, 436)
Real height of $F2$ layer density maximum 49/2(314)
Real reflection height 49/2(252–256)
Rearrangement of ionization 49/2(341)
Receiver tuning 49/2(200)
Reciprocal tensor 49/2(4, 29); 49/7(213)
Recombinance, *see* Recombination coefficient
Recombination 49/2(268, 285, 343); 49/6(44, 46, 49, 69)
Recombination coefficient 48(454, 564, 570); 49/2(251, 380); 49/6(49)
Recombination, dissociative 49/6(4, 46)
Recombination, ion-ion 49/6(70)
Recombination of ions 48(564)
Recombination of nitrogen 49/1(115)
Recombination, radiative 49/6(46, 181)
Recombination, relaxation time 49/2(364ff.)
Recommended scale values 49/3(290)
Recorder 49/3(277)
Recording balloons 48(371, 383)
Recording ionization chamber observations 49/1(162ff.)
Recording of characteristics 49/2(218, 241ff., 267)
Recording of pulsations 49/3(345)
Recording of rapid geomagnetic pulsation 49/3(296)
Recording of rapid geomagnetic variations 49/3(294)
Recording of rapid variations 49/3(294)
Recording of time derivatives 49/3(297)
Recording of time derivatives of magnetic elements 49/3(297)
Recovery phase 49/4(185)
Recovery time 48(600)
Recurring storm 49/3(133)
Red Sea 48(912)
Red Sea Water 48(647)
Red veil 49/1(27, 29–32, 118)
Reduced mass 49/7(64)
Reduction factor (potential) 48(579)
Reduction, isostatic 47(578, 591)
Reduction, topographic 47(600)
Reference ellipsoid 47(538)
Reference spectra (flux) 49/7(17, 45–55)
Reflection 49/5(301)
Reflection by droplets 48(187, 257)
Reflection coefficient 49/2(179)
Reflection coefficient of Epstein layers 49/2(192)
Reflection conditions 49/2(64ff., 193ff., 197, 221)
Reflection decrement of Epstein layer 49/2(194)
Reflection, diffuse, elastic, diffuse 49/5(244)
Reflection factor 48(462)
Reflection, inelastic 49/5(244)
Reflection matrix 48(296)
Reflection of sound 48(461)
Reflection, particle 49/5(243, 244)
Reflection ratio 49/2(94ff., 128, 164, 182)
Reflection, specular 49/5(244)
Reflexion method 47(187)
Reflexion shooting 47(153, 247)

Reflexion-seismics, amplifier 47(155)
Reflexion-seismics, dip-shooting 47(160)
Reflexion-seismics, field operations 47(154)
Reflexion-seismics, frequency 47(154)
Reflexion-seismics, recording apparatus 47(154)
Refraction, astronomical 48(157)
Refraction by droplets 48(187, 257)
Refraction diagram 48(727)
Refraction law
— of Bouguer 49/2(123)
— of Snellius 49/2(123, 125)
— of Sommerfeld and Runge 49/2(121, 136)
— of Witham 49/2(122, 136)
Refraction, plasma 49/5(355)
Refraction shooting 47(153, 163, 247)
Refraction-seismics, fan-shooting 47(166)
Refraction-seismics, field procedures 47(165)
Refractive index 49/2(25, 38, 50, 59ff.)
— complex 49/4(472); 49/5(224)
— diagrams 49/2(80ff.)
— high-frequency approach 49/2(41)
— ion poles 49/2(68)
— poles 49/2(66, 479)
— surface (polar diagram) 49/2(489ff.)
— zeros 49/2(65, 86)
Refractive indices, principal 49/2(34ff.)
Refractivity of air 48(158)
Refractivity, profiles 49/5(356, 361)
Region, transitional 49/5(224)
Regions: *D*, *E*, *F*1, *F*2 49/6(12)
Regions *E*, *F*, and *D* 49/2(222)
Regions of the Earth's interior 47(104)
Registration of fluctuating local gradients 49/3(335)
Regular magnet 49/3(303)
Relativ sun spot number 49/3(251)
Relative velocity 49/2(11, 12)
Relaxation temperature 49/1(276)
Relaxation time 49/1(271)
Relaxation time of recombination 49/2(364ff.)
Release experiments 49/6(160, 161, 163, 166)
Remanent magnetic induction 49/3(318)
Remanent magnetization of minerals and rocks, different types 49/1(272ff.)
Remanent magnetometer 49/3(440)
Replenishment, continual 49/4(182)
Repulsive (exponential) potential 49/7(87, 153–155 Fig. 30)
Residual absorption 49/2(326)
Residual geomagnetic agitation 49/3(38)
Residual torsion 49/3(299)
Resistance coefficient 48(627)
Resistivity, apparent 47(426)
Resistivity eigenvalues 49/2(4, 19ff.)
Resistivity of soil 47(410)
Resistivity tensor 49/2(3, 18, 21, 29); 49/7(185)
Resonance 49/2(10); 49/5(297)
Resonance absorption 49/6(123)
Resonance acceleration 49/3(156)
Resonance approximation 49/7(202)
Resonance at plasma frequency 49/2(509ff.)
Resonance charge transfer 49/6(43)
Resonance condition 49/4(151); 49/5(297)
Resonance effect 49/5(321)
Resonance frequency 49/4(479, 481); 49/6(334)
Resonance limit 49/7(205 Table 11)
Resonance lines 49/6(357, 371, 375)
Resonance oscillation, longitudinal 49/5(238)
Resonance rectification probe 49/2(486)
Resonance scattering 49/6(39, 72)
Resonance transition 49/6(16)
Resonant charge exchange 49/7(114)
Resonant magnetodynamic oscillation 49/4(104)
Retarding potential analyzer 49/7(241 Figs. 8–10)
Return spectrum 49/7(256–261 Figs. 24–26)
Reversal, of spectral line 49/6(334)
Reverse thermo-remanent magnetization (RTRM) 49/1(278ff.)
Reversible pendulum 47(203)
Revolution without rotation 48(742)
Reynolds number 49/6(166)
Reynold's number, magnetic 49/4(29, 30)
Reynolds stresses 48(35)

Rhenium 47(299)
Ricco's law 48(268)
Richardson number 49/5(192, 193); 49/7(472)
Richardson's equation 48(24)
Richardson's forecasting 48(91)
Richardson's number 48(45)
Riemann sphere 49/2(44, 46ff., 58)
Riemann surface of polarization parameter 49/2(47ff., 50ff., 71, 73ff.)
Rift valleys 47(265)
Rigid sphere 49/2(12)
Rigid sphere model 49/6(141); 49/7(73, 76)
Rigidity 47(79, 170)
Rigidity, rocks 47(87)
R-Index 49/3(244)
Ring current 49/1(64, 74ff., 131–133); 49/3(59)
— effect on cosmic rays 49/1(184, 231)
— its generation 49/1(132ff.)
— its introduction by Störmer 49/1(126)
— its magnetic field 49/1(131)
— values after Kertz 49/3(240)
— values after Sugiura 49/3(242)
— values after Vestine 49/3(240)
Riometer 49/1(72ff., 241); 49/2(380ff.)
Rip-current 48(732)
Ripples 48(703); 49/2(405)
Rise-time of **SC**'s 49/3(46, 55)
Rising tone 49/5(332)
Roble's semi-empirical aeronomic theory 49/7(511)
Rock bursts 47(247)
Rock magnetism 47(470)
Rock pressure 47(200)
Rock, radioactivity 47(396)
Rock-dating 47(347)
Rocket 47(622)
Rocket application to Doppler effect measurement 49/2(457–462)
Rocket measurements 48(184)
Rocket panel air densities 48(452)
Rockets 49/1(81, 83, 130)
Rock-forming minerals of paramagnetic properties 49/1(249)
Rockoon (balloon-borne rocket) 49/1(81); 49/6(348)
Rocky Mountains 47(268)
Rod, vibrations of 47(170, 181)
Rossby regime 48(150)
Rossby-waves 48(77, 84)
Rossby-waves in a baroclinic current 48(85)
Rotary motion 49/4(78–80, 83–86, 111)
Rotating Cartesian geometry 49/5(183)
Rotating sphere 49/5(183)
Rotation 49/4(84)
Rotation spectrum 48(225)
Rotational discontinuity 49/4(35)
Rotational excitation 49/7(467)
Rotational temperature 49/1(123ff.)
Roughness length 48(626)
RTRM, *see* Reverse thermo-remanent magnetization
Rubidium-87 47(295, 304, 307)
Rubidium, contents of rocks 47(317)
Rubidium, measuring methods 47(312)
Rubidium method 47(343)
Rubidium-strontium ages 47(345)
Rubidium-strontium method 47(303)
Rudzki, inversion method 47(579)

Salinity at the sea surface 48(638)
Salinity, definition 48(617)
Salt nuclei 48(490, 492)
Samarium 47(299)
Samarskite 47(323, 349)
San Marco satellites 49/7(125, 248 Figs. 13–16)
Sand transport 48(732)
Sandström's theorem 48(19)
Saros period 48(747)
Satellite ALOUETTE 49/7(352)
Satellite, artificial 47(622)
Satellite drag 49/7(233–236)
Satellite ECHO 49/7(345 Fig. 99b)
Satellite ISS-b 49/7(396–403 Figs. 128, 129, 414–417 Figs. 140a–d, 420 Fig. 142, 422–429 Figs. 144, 145)
Satellite-borne rubidium magnetometer 49/3(440)
Satellites 49/1(81, 96–100, 218, 232ff., 242)
Satellites: AEROS 49/7(3, 17, 30, 35, 44)
Satellites: Atm. Explorer 49/7(18, 19, 30, 31, 35, 44)
Satellites: OGO 49/7(42)
Satellites: OSO 49/7(9, 15)
Satellites: SOLRAD 49/7(14, 32)
Sato potential (approximation) 49/7(87, 117)

Saturation core probe 49/3(348)
Saturation magnetization 47(472)
Saucer-shaped emission 49/5(322, 324)
Savart's polariscope 48(309)
Sawtooth generator 49/2(213, 220)
SC 49/4(186)
SC (= sudden commencement) 49/2(388)
Scale height 48(83, 157, 428, 439, 943); 49/2(268); 49/6(11, 155, 157, 193, 198, 252); 49/7(224, 228, 297, 298)
Scale height, gradient 49/6(193, 198)
Scale, horizontal 49/5(125)
Scale value 49/3(290)
Scale value determination 49/3(288)
Scanning of resonance line 49/3(372)
Scattered light 48(159); 49/6(310)
Scattering 48(186); 49/5(225, 286, 377)
Scattering amplitudes 49/7(64)
Scattering area ratio 48(256)
Scattering by droplets 48(189)
Scattering by haze 48(159)
Scattering by water vapor 48(159)
Scattering coefficient 48(291)
Scattering cross section 48(162, 209); 49/6(345)
Scattering crystal 49/6(326)
Scattering function 48(190, 201, 256, 260); 49/5(287, 288)
Scattering in infra-red 48(160)
Scattering, instruments for measuring 48(284)
Scattering layer 48(614)
Scattering length 49/7(97)
Scattering matrix 49/2(163ff.)
— for Airy functions 49/2(182)
— for hypergeometric functions 49/2(191)
— for parabolic cylinder functions 49/2(185)
— transformation formula 49/2(166)
Scattering, multiple 48(186, 193, 195, 206, 294)
Scattering of radio waves 49/2(350–361)
Scattering potentials 49/2(12)
Scattering, primary 48(192)
Scattering probability 49/4(426)
Scattering spatial lobe 49/5(289)
Scattering theory 49/2(13)
Schematic magnet 49/3(303)
Schlieren-Methode 47(193)
Schmidt vertical magnetometer 49/3(315)
Scholz counter 48(566)
Schrödinger equation 49/4(616)
Schuman emulsion 49/6(311)
Schumann-Runge band system 49/6(17, 31, 33, 34, 87, 124, 191)
Schumann-Runge continuum 48(373); 49/6(20, 26, 31, 33, 89, 121, 178)
Schuster-Smith coil magnetometer 49/3(307)
Schwarzian derivative 49/2(186)
S.C.I., *see* Stratospheric circulation index
Scintillation counter 47(310); 49/6(348)
Scintillation twinkling 48(266)
Scintillations 49/2(402, 404, 433, 449, 475)
Scintillations of radio waves from stars 49/1(81)
SCNA, *see* Sudden cosmic noise absorption
Screening 49/4(455)
Screw radius of trapped particle path 49/2(526)
S_D 49/3(56, 64, 65, 70)
Sea bottom slope 48(724, 729, 732)
Sea, fully arisen 48(690, 695, 709, 714, 716)
Sea ice 48(625)
Sea level 47(535)
Sea level, changes 47(18)
Sea, short-crested 48(700, 702, 713, 720, 724)
SEA, *see* Sudden enhancement of atmospherics
Sea water
— absorption of radiation 48(621)
— colour 48(623)
— compressibility 48(620)
— density 48(618, 639)
— diffusion 48(621, 629)
— electrical conductivity 48(621)
— elements in 48(617)
— freezing point depression 48(620)
— gases in 48(618)
— latent heat of vaporization 48(620)
— potential temperature 48(620, 641)
— surface tension 48(621)
— temperature 48(609)
— thermal conductivity 48(620)
— viscosity 48(621, 623)

Search-light beam, modulated 48(285)
Search-light, N.R.L. method 48(285)
Season number 48(772)
Seasonal lunar variations of the magnetic field 49/2(593)
Seasonal variation of auroral distribution 49/1(14ff.)
Seasonal variations 49/6(287, 288, 300)
— of cosmic-ray intensity 49/1(170ff., 230)
— of *D* region 49/2(325, 327)
— of *F*2 layer 49/2(302, 313)
— of sporadic *E* layer 49/2(334ff.)
Seaway, mathematical representation 48(701, 703)
Second 47(1, 22)
Second adiabatic invariant 49/3(42)
Second heat source 49/7(316)
Second order term, zonal 47(592)
Sector structure 49/4(14, 93)
Sector structure of the interplanetary magnetic field 49/3(136, 266, 269)
Sectors, stable, unstable 48(68)
Secular change 47(405, 509)
— impulses 47(514)
— in the Pacific 47(514)
— induction theory 47(529)
— radial dipoles 47(516)
— spectrum 47(513)
— westward drift 47(512)
Sediment carpet 48(616)
Sedimentary deposits, magnetism 49/1(286–290)
Sedimentary rock 47(290)
Seiches 48(850, 852)
Seismic phases 47(88)
Seismic ray, parameter 47(91, 95)
Seismic rays, radius of curvature 47(98)
Seismic waves
— abnormal velocities 47(99)
— amplitude theory 47(105)
— amplitudes 47(83)
— angle of emergence 47(91, 109)
— core waves 47(89)
— damping 47(86)
— diffraction 47(83, 110)
— energy 47(81, 106, 109)
— longitudinal 47(88)
— notation 47(89, 97)
— plane 47(87, 106)
— polarisation 47(88)
— reflection 47(90, 107)
— refraction 47(90, 107)
— relations between travel times and velocities 47(102)
— scattering 47(87)
— total reflection 47(109)
— transverse 47(88)
— transmission, fundamental equation 47(80)
— velocities 47(97, 103)
Seismogram 47(75, 92)
Seismogram, pulses 47(89)
Seismograph 47(75, 92)
Seismograph, astatisation 47(31, 40)
Seismograph, damping 47(31)
Seismograph, friction 47(37)
Seismograph, magnification 47(36)
Seismograph, resonance 47(36)
Seismological observatories, direction-cosines 47(96)
Seismological Summary, International 47(75)
Seismology and tides 48(820)
Seismology, history 47(117)
Seismometry 47(24)
Selective absorption 49/2(96)
Self consistent field 49/4(418)
Self orientating follower system 49/3(323)
Self oscillating single suspension 49/3(374)
Self-coupling 49/2(150)
Self-interaction 49/2(397)
Sellmeier dispersion formula 49/2(41, 93, 446, 454); 49/4(516)
Sellmeier equation 49/5(355)
Sellmeier formula 49/4(447)
Semiannual variation 49/5(147)
Semi-diurnal lunar variations 49/2(548)
— of cosmic noise attenuation 49/2(590)
— of the *D* region 49/2(566, 591)
— of the *E* region 49/2(561, 579)
— of the *Es* layer 49/2(560, 563, 586ff., 595)
— of the *F*1 layer 49/2(560)
— of the *F*2 layer 49/2(550ff.)
— of the magnetic field 49/2(572, 592)
— seasonal dependence 49/2(574)
Semidiurnal tide 49/7(338)
Semitransparent layer 49/2(225)
Sensation (eye) 48(267)

Sensible heat, meridional flux 48(153)
Sensitivity 49/2(211)
Sensitivity of aeromagnetic gradiometers 49/3(455)
Sen-Wyller dispersion formula 49/2(39ff., 59, 62, 72, 89, 143)
Separation, diffusive 49/6(158, 169, 174, 244)
Separation of variables 49/5(181)
Sequential *Es* 49/2(344)
Servo-system amplifier 49/3(331)
Shadow zone 47(110)
Shadowing 49/5(284)
Shallow water waves 48(671, 704, 713, 728, 732)
Shape anisotropy 49/1(269)
Shape of the body 49/5(254)
Shaver-bands 48(374)
Shear flow, linear 49/5(192)
Shear fracture 47(264)
Shearing instability 48(79, 89)
Shear-waves 47(153)
Sheet, current 49/4(194)
Sheet, neutral 49/4(85, 111, 194, 220)
Shell parameter 49/4(124)
Shida's number 48(788)
Shields, exposed 47(240)
Shift, Doppler 49/5(228)
Shifting of the auroral zone 49/3(89)
Shimmer 48(266)
Shkarofsky functions 49/7(207)
Shock 49/3(140)
Shock, collisionless 49/4(203, 204)
Shock front 49/2(531); 49/4(10, 34, 36, 37, 91); 49/5(381)
— hydrodynamic theory of 49/4(43–46)
— stationary 49/4(41–43)
Shock wave 48(456); 49/4(191, 206); 49/5(261, 271)
"SHORAN" 49/3(406)
Shoran 47(622)
Short cylindrical dipole 49/2(479)
Short electron whistler 49/5(307, 330)
Short ion whistler 49/5(307)
Short wave fadeout (SWF) 49/2(362)
Short-hop whistlers 49/5(325)
Shower meteors 48(427)
Shrinkage of photographic paper 49/3(289)
Shrinkage, tetrahedral 47(283)
Shuman-Runge continum 49/7(3, 4, 40)
Siberian Meteor 48(476, 976)
SID, *see* Sudden ionospheric disturbance)
Sidereal day 48(747)
Sidereal diurnal variations of cosmic-ray intensity 49/1(178–180, 217)
Sidereal time 48(744)
Signal quality 49/2(211)
Signal suppression ratio 49/2(210)
Signal to noise ratio 48(466); 49/2(211); 49/3(362)
Silicate 48(618)
Silver iodide 48(509)
Silver iodide generators 48(510)
Silverdisc-actinometer 48(183, 251)
Similarities, hemispheric 49/5(147)
Simpson, wave pattern 48(583)
Sine-galvanometer 49/3(305)
Singular point 49/5(228)
Sittkus ionization chamber 49/1(164)
Skeleton function (Booker) 49/7(351)
Skin effect 47(416, 464)
— abnormal 49/4(461)
— anomalous 49/4(488)
— normal 49/4(461)
Skip-phenomenon (sound) 48(471)
Skirt 49/4(71)
Sky, brightness of the 48(193, 197)
Sky, colour 48(159, 193, 198)
Sky light, circumsolar 48(200)
Sky light polarization 48(196, 288)
— anomalies 48(323)
— dispersion 48(316)
— in twilight 48(323)
— photoelectric measurement 48(311)
Sky radiation, diffuse 48(186, 203, 295)
Slab thickness 49/2(453, 470); 49/7(299–301 Figs. 62, 63)
Slate 47(336)
Sliding means 49/2(269)
Slitless spectrograph 49/6(340, 342)
Slow electron-acoustic wave 49/5(240)
Slow ion-acoustic wave 49/5(241, 251ff., 318, 319)
Slow magneto-sound wave 49/4(486)
Slowly moving body 49/5(282)
Small body 49/5(220, 274, 277)
— at rest 49/5(276)
— potential 49/5(280)
Small ions 48(543, 550, 552, 573)
Small ions, average life 48(572)
Small scale features 49/5(124)

Small scale structure 49/5(119, 123)
Smbols of magnetic storm fields 49/1(54ff.)
Smith transformation 49/2(218)
Smithsonian Institution 48(166, 183, 247)
Smithsonian revised scale 48(183)
Smithsonian scale 48(251, 1913)
Smoke 48(168)
Smoke column 48(984)
Snellius' refraction law 49/2(123, 125)
Snell's law 47(90, 164)
Snell's law, vector form 48(459)
Snow 48(484)
Snow cover 48(209)
Snow crystals 48(511)
Snow flakes 48(483, 511, 515, 521)
Snow, granular 48(483)
Sodium clouds, artificial 49/2(401)
Sofar channel 48(1001)
Sofar-waves 47(120, 135)
Soft hail 48(527)
Soft radiation 49/1(82)
Soft x-rays 49/6(30, 348, 349)
Soil particles 48(492)
Solar activity 48(183); 49/3(135); 49/4(180)
— and x-ray emission 49/6(30, 348)
— cycle (Sunspot-) 49/6(351, 365, 367, 368)
— effect upon *D* region 49/2(324)
— effect upon *E* region 49/2(283)
— effect upon *F*2 layer 49/2(303, 316)
Solar atmosphere 49/6(365, 371)
Solar constant 48(182); 49/7(1, 2)
Solar constant, variations 48(185, 247)
Solar continum radiation 49/7(20)
Solar corona 49/6(351)
Solar corpuscular radiation 49/3(169, 204)
Solar cycle 49/6(351, 365, 368); 49/7(1, 5, 9, 296)
— changes 49/4(134)
— effect upon *F*1 layer 49/2(286)
— effect upon *F*2 layer 49/2(298, 313)
Solar day 48(747)
Solar eclipse 48(747); 49/2(377–380); 49/6(344); 49/7(438 Fig. 149d)
Solar eclipse, absorption 49/2(380)
Solar eclipse, ionization decrease 49/2(378)
Solar EUV activity indices 49/7(11, 33, 34, 35)
Solar EUV-spectrum 49/6(314, 321, 322, 324–327, 366, 367)
Solar flare(s) 49/1(74); 49/4(132); 49/7(438 Figs. 149a, b)
— effects 49/2(361–366, 385)
— effect on nucleonic component 49/1(208ff., 233)
— large increases of cosmic ray intensity 49/1(204–209)
— of February 23, 1956 49/1(234–238)
— small ones 49/1(216, 238)
— type IV 49/2(388, 392)
Solar flux
— below 10 nm 49/6(30, 325, 348)
— intensities 49/6(27, 184, 355, 363, 370)
— 5–120 nm 49/6(30, 364)
— 10–130 nm 49/6(29, 322, 325, 366, 367)
— 100–300 nm 49/6(28, 184, 367)
Solar irradiance 49/7(1)
Solar limb 49/6(328, 346)
Solar limb spectrum 49/6(330, 360)
Solar maximum 49/7(5, 14)
Solar minimum 49/7(5, 8, 14)
Solar noise 49/2(270)
Solar noise storm 49/2(385)
Solar plasma 49/1(129–131)
Solar plasma flow, types 49/1(52ff.)
Solar plasma stream 49/3(54)
Solar proton(s)
— event 49/2(392)
— of low energies 49/1(74)
— space probe results 49/1(233)
— summary 49/1(243ff.)
Solar radiation 48(156, 295)
— extraterrestrial 48(182)
— interaction with atmosphere 49/6(16, 72)
— interaction with atmospheric gases 49/7(431)
— pressure 49/7(235)
Solar radio noise 49/6(347)
Solar relationships with auroras 49/1(52ff.)
Solar rotation 49/6(351, 367, 368); 49/7(6)
Solar temperature 48(185)
Solar tidal wind 49/2(567)
Solar time, apparent 48(743)
Solar time, mean 48(747)

Solar wind 49/1(54); 49/2(530); 49/3(11, 33, 54, 135ff., 138, 140, 204); 49/4(10, 86, 95, 159, 187, 208, 215); 49/5(222, 381); 49/6(343)
— composition 49/4(10)
— distortion of geomagnetic field by 49/4(180)
— pressure 49/5(378)
— sectors 49/4(14, 93)
— supersonic 49/4(43, 49)
— velocity 49/4(95)
Solar XUV-spectrum 49/6(4, 309)
Solar zenith angle 49/6(34)
Solar-daily variations on quiet days 49/3(16)
Solar-flare effect (s.f.e.) 49/3(20, 26, 211, 215)
Solenoids 48(11, 652)
Solid bodies, plasma flow around 49/5(217, 252)
Solid-source mass spectrometer 47(342)
Sommerfeld-Runge refraction law 49/2(121, 136)
Sonagram 49/2(117, 496ff.)
Sonine polynomials 49/7(131, 194, 197)
Sonogram 49/5(303, 306)
Soret effect 49/7(178)
Sound diffraction 48(463)
Sound focus 48(471)
Sound, group velocity 48(461)
Sound intensity 48(455)
Sound, phase velocity 48(461)
Sound propagation, anamalous 48(416)
Sound propagation in air 48(455)
Sound propagation, normal and abnormal or anomalous 48(471)
Sound ray 48(459)
Sound receivers 48(463)
Sound refraction 48(458)
Sound refraction in the troposphere 48(466)
Sound sources 48(463)
Sound speed 49/5(166, 167, 201)
Sound transmission channel 48(463, 1001)
Sound velocity
— in the sea 48(609, 621)
— non-isothermal 49/5(240ff.)
— nonthermal 49/5(221)
— sound speed 48(459)
Sound, wave forms 48(456)
Sound wave propagation 49/7(252 Fig. 19)
Soundings 48(609)
Sources 49/5(202)
Sources, external 49/5(226)
Southern light 49/1(2)
SPA, *see* Sudden phase anomaly
Space charge in thunderstorm 48(585)
Space charges 48(544, 587)
Space charges, artificial 48(581, 588, 593)
Space probes 49/1(218, 233ff.)
Space research and ionosphere 49/2(442)
SPAND 49/4(132)
Spatial attenuation 49/5(307)
Spatial damping 49/5(224)
— coefficient 49/5(239)
— factor 49/5(224)
Spatial decrease
Spatial dispersion 49/2(62); 49/5(227, 238)
Spatial inertia 49/5(227)
Spatial lobe, scattering 49/5(289)
Special functions 49/7(216ff. Appendix B)
Specific heat 47(386)
Spectral density 49/5(343)
Spectral line 48(221)
Spectral, measurements, summary of 49/4(213)
Spectrograph 49/6(309)
Spectroheliogram 49/6(341, 346)
Spectrum of aurora, *see also* Auroral spectrum 49/1(67, 100–124)
Spectrum of probe output voltage 49/3(349)
Specular reflection 49/5(244)
Speed of sound 49/5(166, 167, 201)
Sphere 47(539)
Sphere, rotating 49/5(183)
Spherical harmonics 47(542, 623, 626, 630); 48(744, 941, 948)
— addition theorem 47(631)
— expansion of lunar effects 49/2(575, 577)
— second kind 47(545, 628, 630)
Spheroid 47(538, 550, 554, 635)
Spheroid, Bruns' 47(555)
Spikes 49/2(512); 49/5(336)
Spinel 47(470)
Spinel structure 49/1(254, 257)

Spiral pattern of geomagnetic disturbance 49/3(80)
Spiral waves 49/4(478, 485)
Splintering (crystals) 48(521)
Split-spread 47(159)
Splitting factor 49/3(370)
Splitting, *L*-shell 49/4(70, 77)
Spontaneous magnetization 49/1(252, 259)
Sporadic E 49/6(98, 113)
Sporadic *E* layer 49/1(70)
— auto-correlation function 49/2(334)
— blanketing frequency 49/2(332)
— classification 49/2(338)
— drifts 49/2(337)
— equatorial *Es* 49/2(338)
— lunar control 49/2(335)
— minimum virtual height 49/2(332)
— opacity 49/2(339)
— reflection coefficient 49/2(339)
— seasonal variation 49/2(334ff.)
— *see also Es* layer 49/2(332–344)
— sequential *Es* 49/2(344)
— theories 49/2(341–344)
— top frequency 49/2(332ff.)
— transparency 49/2(339)
— meteors 48(454)
S_q^p enhancement 49/3(86)
Spread echoes 49/2(350)
Spread *F* 49/2(352ff., 402, 404, 520)
Spreading velocity 49/2(428)
Spring tide 48(739, 882)
Sputnik III 49/1(81)
S_q current systems 49/3(17)
S_q-Value 49/3(211)
ssc magnetic storms (=with sudden commencements) 49/1(53ff., 65, 94)
ssc, ssc*, SC, SC* 49/3(42, 45, 46, 54)
Stability 48(640); 49/4(490, 496)
Stability, baroclinic 48(87, 110)
Stability criteria 49/4(491)
Stability criterion 48(73)
Stability, frontal zone 48(71)
Stability, linear flow 48(72)
Stability of ionospheric layers 49/2(222)
Stability, static 48(47)
Stability, static, inertial, dynamic 48(66, 71)
Stability tensor 48(67)
Stabilization 49/4(500, 542)
Stabilizing effect of the magnetic field 49/4(508)
Standard flux 49/7(9, 16)
Standard Meridian 48(753)
Standard MUF, *see also* MUF 49/2(227)
Standard sphere 47(93)
Standard symbols for ionograms 49/2(228ff.)
Standard time 48(753)
Standard types of *Es* traces, *see also Es* 49/2(229)
Stand-off distance 49/4(45)
Starfish 49/4(138, 144)
State of sea 48(681, 684, 701)
Static approximation 49/5(162)
Statistical fluctuations of cosmic rays 49/1(229)
Steady state, equation 49/6(218, 359)
Step-sounding 49/2(205)
Stiles-Crawford effect 48(281)
Still water 48(868)
Still water level 48(681, 730)
Stimulus 48(267)
Stokes' differential equation 49/2(180ff.)
Stokes' dissipation function 48(16)
Stokes' formula 47(573, 579)
Stokes' function 47(575)
Stokes' parameters 49/2(45)
Stokes' polarization parameters 48(289)
Stoneley-waves 47(125)
Stored energy in the radiation belt 49/1(94)
Stored energy in the ring current 49/1(132)
Storm, geomagnetic 49/4(86, 90, 96, 101)
Storm magnetograph 49/3(290)
Storm time 49/2(370); 49/3(56)
Storm time variation 49/3(55, 56, 60, 62)
Storm variations 49/4(162)
Störmer's theory 49/7(500)
Stormines 49/3(243)
Storms, magnetic 49/4(96, 97, 101)
Strain, deviation tensor 47(77)
Strain distribution 48(991)
Strain measurements 48(815)
Strain, principal axes 47(76)
Strain, rotation tensor 47(77)
Strain seismometer 48(816)

Strain tensor 47(76)
Strain-energy 47(79)
Strain-energy function 47(80)
Stratification 49/2(344, 346)
Stratified medium 49/2(88, 120, 122, 141, 145ff.)
Stratonull surface 49/5(135)
Stratopause 49/6(8)
Stratopause, heating 49/5(121)
Stratosphere 48(416); 49/6(8)
Stratosphere, radiation balance 48(418)
Stratosphere temperature 48(244)
Stratospheric circulation 49/5(117, 135)
Stratospheric circulation index (S.C.I.) 49/5(139–149)
Stratospheric tidal jet 49/5(155)
Stream-convective instability 49/4(557)
Streamlines 49/4(80)
Streamlines in the equitorial plane 49/4(80)
Strength 47(180)
Stress 47(77, 170)
Stress, deviator 47(78)
Stress release 48(843)
Stress, shear component 47(78)
Stress tensor 47(77); 49/2(10)
Stress-strain-relations 47(78)
String-gravimeter 47(212)
Strong absorption 49/2(19, 30ff., 111)
Strong coupling 49/2(173)
Strontium-87 47(295, 344)
Structure, acoustical 49/5(166)
Structure, electric current 49/5(173)
Structure, model temperature 49/5(155)
Structure of the Venusian upper atmosphere 49/5(376)
Structure, quasi-periodic 49/5(259)
Structure, small scale 49/5(119, 123)
Structure, wave 49/5(164)
Structures, two-dimensional 47(229)
Sub-Arctic Water 48(645, 649)
Subauroral zone 49/1(2, 11ff.)
Sublimation nuclei 48(499)
Subprotonospheric whistlers 49/5(325, 327)
Substantial time derivative 49/2(10, 273)
Substituted magnetites 49/1(263)
Substorm, auroral 49/4(100, 101)
Substorm, magnetic 49/4(91, 92, 100–103, 109)
Substorm, magnetospheric 49/4(58, 101)
Substorm, polar magnetic 49/4(92, 101, 109)
Sub-surface current 48(659)
Subvisual red arc 49/1(118–120)
Sudden commencement of magnetic storms 49/3(42, 127, 131, 160)
Sudden commencement (SC) 49/2(388)
Sudden cosmic noise absorption (SCNA) 49/2(363)
Sudden enhancement of atmospherics (SEA) 49/2(363)
Sudden impulse (si) 49/3(42, 46, 53, 54, 212)
Sudden ionospheric disturbance (SID) 49/2(241, 362)
Sudden phase anomaly (SPA) 49/2(363)
Sudden storm commencement 49/3(212)
Sulphides with ferrimagnetic properties 49/1(250)
Sulphite mineralization 47(326)
Summer-winter anomaly 49/6(287)
Sun, orbital motion 48(748)
Sun rise, effect upon *D* region 49/2(330)
Sun-follower, photoelectric 48(312)
Sunlit aurora 49/1(23, 117)
Sunspot cycle 49/1(15, 53, 193–198, 217, 219)
Sunspot effect on particle spectra 49/1(226ff.)
Sunspot minimum of 1954 49/1(225)
Sunspot number (R_z) 49/2(269); 49/6(349); 49/7(8, 16, 20, 21, 28, 29)
Sunspot numbers, annual means 49/1(187)
Sunspot numbers, monthly means 49/1(219)
Sunspots 48(183); 49/2(554)
Super Schmidt Meteor Camera 48(431, 442)
Supercooling (droplets) 48(499)
Superparamagnetism 49/1(270ff.)
Superposition 48(764)
Supersaturation 48(480, 487, 498)
Supersonic 49/4(11, 13)
Supersonic flow 49/5(253)
Supersonic motion, zone 49/5(220)
Supersonic velocity 49/5(220)

Suprathermal electrons 49/7(36, 37, 369 Fig. 114, 525–530)
Surf 47(151); 48(672, 723, 729)
Surf beats 48(672, 732)
Surf, interior 48(880)
Surf zone 48(729, 731)
Surface current 48(659, 661)
Surface features 47(258)
Surface of constant amplitude 49/2(88, 122)
Surface of no motion 48(651, 661, 662)
Surface, stratonull 49/5(135)
Surface temperature, daily variation 48(633)
Surface temperature (sea) 48(638)
Surface waves 47(119, 153, 250); 48(671)
— energy of progressive 48(677, 695, 710, 720, 722, 732)
— generation by wind 48(703, 728)
Surface wind 48(42)
Surface wind, lunar tides 49/2(568–570)
Surfaces of aurora 49/1(27–32)
Surfaces of discontinuity 47(89)
Surge 49/1(41)
Susceptibilities of oxides of iron 49/1(266)
Susceptibility, anisotropy 49/1(268 ff.)
Susceptibility, dependence on grain size 49/1(266, 268)
Susceptibility, dielectric 49/5(227)
Susceptibility, longitudinal dielectric 49/4(446)
Susceptibility, magnetic 47(496)
Susceptibility tensor 49/2(22)
Susceptibility, transverse dielectric 49/4(446)
Suspended matter 48(623)
Sutherland potential 49/7(85, 86 Fig. 7, 108–110, 152 Fig. 29)
Swell 48(672, 688, 717, 728)
Swell, angular spread 48(717, 720)
Swell decay distance 48(717)
Swell, filter 48(722)
Swell, forecast 48(723)
Swell, travel time 48(720)
SWF, *see* Short wave fadeout
Sylvite 47(336, 349)
Symbol, Kronecker 49/5(227)
Symbols for ionograms 49/2(228 ff.)
Synchronization 49/2(213)
Synchrotron radiation 49/4(141, 226)
Synge's principle 49/2(144)
Synthetic ionograms 49/2(257)
System, non-rationalized 49/4(2)
System of units, rationalized 49/4(1, 2)
Système international 49/3(1); 49/4(1)
Syzygies 48(739)

Tail 49/4(40, 42, 46, 66, 74, 86–88, 90, 93, 102, 111); 49/5(369)
Tail field 49/4(221)
Tail, geomagnetic 49/4(193, 221)
Tail, length of the geomagnetic 49/4(197)
Tail, magnetospheric 49/4(219)
Tail of the magnetosphere 49/7(497, 500)
Tail, width of the 49/4(195)
Tangential discontinuity 49/4(35, 51)
Taurids 48(435)
Taylor's theorem 48(975)
Taylor's waves 48(861)
TCRM, *see* Thermo-chemical remanent magnetization
TΔT method 47(162)
Teddington g 47(204)
Telephotometer 48(279, 363)
Telephotometry, base line 48(284)
Temperature 49/2(268); 49/4(11); 49/5(372)
Temperature coefficient of variometer 49/3(279, 283)
Temperature dependence of diffusion 49/6(139)
Temperature dependence of rate coefficients 49/6(110)
Temperature effect (Gauss-Lamont) 49/3(302, 304)
Temperature effect on variometers 49/3(279)
Temperature, effective 49/4(436)
Temperature, effects of winds upon 49/6(294, 295, 300)
Temperature, electron/ion ratio 49/6(82, 227)
Temperature from optical measurements 49/7(272–275)
Temperature from velocity of sound 49/7(252 Fig. 19, 253 Figs. 20a, b)
Temperature gradient 49/2(11)
Temperature gradient, adiabatic 47(381, 385)

Temperature in the high atmosphere 49/1(121–124)
Temperature maps 49/7(275 Figs. 40, 41, 278 Fig. 43, 321 Fig. 80, 330 Figs. 86a, b, 339 Figs. 93a, b)
Temperature measurements 49/7(252–256, 275–281)
Temperature measurements (results) 49/7(275–281 Figs. 40–46)
Temperature, neutral 49/6(259, 296, 297)
Temperature of electrons 49/2(5, 8)
Temperature of ions 49/6(227)
Temperature of the exosphere 49/6(220, 252)
Temperature, potential 48(3, 25)
Temperature, profile(s) 49/6(9, 252, 267); 49/7(253 Fig. 20a, 255 Fig. 22, 486 Figs. 179a, b)
— neutral 49/7(226 Fig. 1, 253 Fig. 20a, 255 Fig. 22, 283, 284, 316, 318 Fig. 77a)
— plasma 49/5(364)
— stratosphere 49/7(325 Fig. 85)
— vertical 49/5(130, 131)
Temperature ratio of electrons and ions 49/2(358ff., 524)
Temperature, reduced average 49/4(448)
Temperature salinity diagram 48(642)
Temperature, solar corona 49/6(345)
Temperature structure, model 49/5(155)
Temperature variation, harmonics of the diurnal 48(937)
Temperature variation, radiation term 48(964, 966, 968, 974)
Temperature variations, diurnal 49/5(152–154, 161)
Temperature, virtual 48(5)
Temperature wave, lunar 48(962)
Temperature wave, planetary 48(936, 940)
Temperature wave with universal time 48(941)
Temperature-pressure hypothesis 47(386)
Temporal damping 49/5(224)
Temporal damping decrement 49/5(238)
Temporal inertia 49/5(227)
Tendency equation 48(23)
Tendency equation for barotropic flow 48(83)
Ten-point method 49/2(256ff., 260)
Tension fracture 47(265)
Tension, magnetic 49/4(90, 111)
Tensor 49/4(4)
Tensor, antihermitean 49/2(7)
Tensor, eigenvalue representation 49/2(3, 6, 21, 31)
Tensor, energy-momentum 49/5 (212)
Tensor, hermitean 49/2(7, 24)
Tensor of conductivity, *see also* Conductivity tensor 49/2(3)
Tensor of dielectricity 49/2(23ff., 29, 34, 149)
Tensor of resistivity, *see* Resistivity tensor
Tensor of susceptibility 49/2(22)
Tensor, principal axes 49/2(31ff.)
Tensor, reciprocal 49/2(4, 29)
Term, great empirical 47(8)
Ternary mixture 49/6(134, 148, 185)
Ternary systems, magnetic properties 49/1(251–262)
Terrain reduction 47(223)
Terrella 49/1(125)
Terrestrial atmosphere 49/4(402)
Terrestrial magnetic field, *see* Earth's magnetic field
Theodolite magnetometer 49/3(304)
Theoretical and observational results, disagreement between 49/5(160)
Theory, magnetodynamic 49/4(28, 30)
Theory, "magnetodynamic" or "hydromagnetic" 49/4(17)
Thermal broadening 49/6(335)
Thermal conduction coefficient 49/6(141)
Thermal conductivity 47(386); 49/2(5)
Thermal control, dynamic 49/5(133)
Thermal deformation 48(835)
Thermal diffusion 49/6(85, 132, 140, 142, 143, 217)
Thermal diffusion (tensor) 49/7(177–179, 191)
Thermal diffusivity 47(391)
Thermal effects 47(262)
Thermal equilibrium 49/2(8, 358); 49/7(459)
— deviations 49/2(9); 49/6(150)
Thermal expansion, coefficient 47(386)

Thermal flux 49/5(126)
Thermal stratification in the atmosphere 48(944)
Thermal structure 49/5(129)
Thermal tides 49/5(150, 161)
Thermal transport 49/5(126)
Thermal velocity of electrons 49/5(220)
Thermal velocity of ions 49/5(220)
Thermal wind equation 48(32)
Thermal wind relation 49/5(156)
Thermo-chemical remanent magnetization (TCRM) 49/1(273, 284–286)
Thermocline 48(639)
Thermodynamic equilibrium 48(218)
Thermoelectric currents 47(401)
Thermoelectric heat flow 49/7(475)
Thermographs (sea) 48(609)
Thermometer, reversing 48(610)
Thermopause 49/6(264, 266); 49/7(42, 43, 226)
Thermo-remanent magnetization (TRM) 49/1(272, 273–283)
Thermosound 48(612)
Thermosphere 49/2(523); 49/6(2, 86, 87, 280, 297, 300); 49/7(226)
Thermosphere, seasonal variation 49/6(300)
Thermospheric heat budget 49/7(459–478)
Thermospheric model 49/6(270)
Thermospheric winds 49/6(280, 297)
Thick layer, critical frequency 49/2(224)
Thickness of F2 layer 49/2(315)
Thickness parameter 49/2(278)
Thin gas approximation 49/4(418)
Thin layers, *see* Sporadic layers
Tholeiite dykes 47(484)
Thomson formula 49/2(357)
Thorite 47(325)
Thorium-232 47(298, 302)
Thorium, contents of igneous rocks 47(312)
Thorium, measuring methods 47(310)
Thoron 47(310)
Three-body attachment 49/6(51, 96, 103, 107)
Three-body process (collision) 49/6(44, 51, 87, 180, 182)
Three-component gas 49/4(437)
Three-hourly equivalent planetary amplitude 49/3(63)
Thuillier's atmospheric model 49/7(331–333 Figs. 87, 88, 335, 340–348)
Thunderstorm 48(544, 583, 585, 602, 604)
Thunderstorm project 48(584, 586)
Tidal accelerations, harmonic analysis 48(768)
Tidal components 48(775)
Tidal computing machines 48(901)
Tidal constant 48(736, 896)
Tidal current 48(613, 901, 904)
Tidal currents, effect on waves 48(719)
Tidal effect of the galaxy 48(777)
Tidal effects of planets 48(734, 777)
Tidal elastic factors 48(817)
Tidal equation, Laplace's 49/5(156, 183)
Tidal field on rotating earth 48(745, 778)
Tidal flood 48(770, 881)
Tidal forces 48(734, 883)
— greatest possible 48(776)
— harmonic analysis 48(746, 780)
— meaning 48(742)
— nomograms 48(770)
— primary 48(734)
— secondary 48(734, 771, 785, 811)
Tidal friction 47(3); 48(823, 910, 923)
Tidal friction, energy dissipation 47(14)
Tidal gravimeter 48(806)
Tidal jet, stratospheric 49/5(155)
Tidal modes 49/7(288)
Tidal movements, *see also* Lunar tide 49/2(277)
Tidal potential, harmonic analysis 48(746, 884)
Tidal potential, lunar 48(735)
Tidal potential, solar 48(737)
Tidal system, hemispheric 49/5(151)
Tidal theory 49/5(161, 205)
Tidal waves 48(672, 676, 846)
— forced 48(886)
— in channels 48(846)
— quasistatic theory 48(943, 971)
Tidal wind(s) 48(962, 967); 49/2(547, 567 ff., 575 ff.)
Tide, age 48(898)
Tide, argument 48(752)
Tide, co-oscillational 48(906)
Tide effects of the moon 49/2(547 ff.)
Tide for zenith and nadir 48(740)

Tide, form quotient 48(899)
Tide, independent 48(906)
Tide, Main 48(753)
Tide producing force 48(776)
Tide tables 48(899)
Tides 48(881); 49/6(282); 49/7(287 Figs. 52, 53, 290 Fig. 55, 310 Figs. 72, 73, 338)
Tides, classes 48(753)
Tides, diurnal 48(746, 759)
Tides, dynamical theory 48(886)
Tides, elliptic 48(753)
Tides, groups 48(753)
Tides in border seas 48(906)
Tides in closed basins 48(890)
Tides in lakes 48(791)
Tides in pipes 48(792)
Tides in volcanoes 48(842)
Tides in wells 48(842)
Tides, latitude variations 48(813)
Tides of long periods 48(746, 753)
Tides of the atmosphere 48(928)
Tides of the earth's crust 48(919)
Tides of the sea, energy dissipation 48(925)
Tides of the solid earth 48(775, 972)
Tides, prediction 48(899)
Tides, semi-diurnal 48(746, 761)
Tides, superposition 48(769)
Tides, symbols 48(751, 773)
Tides, table 48(754)
Tides, ter-diurnal 48(746, 763)
Tides, thermal 49/5(150, 161)
Tilt meter 48(839)
Tilt of the ground 48(993)
Time, astronomical 47(22)
Time constants 49/6(91, 187, 301)
Time marks 49/3(289)
Time, unit 47(1, 22)
Time variations of cosmic rays, earliest observations 49/1(159)
Time-averaged Poynting vector 49/2(152)
Time-constant (atmospheric electric field) 48(592, 595)
Time-scale, geological 47(352, 361)
Time-scale, paleontological 47(352)
Titano-heamatite 49/1(258, 261)
Titano-maghemite(s) 47(472); 49/1(255–257)
Titano-magnetites 49/1(251–258)
TMA (trimethyl aluminium) 49/6(160)
Top frequency 49/2(225)
Topography, dynamic 48(651, 662)
Topography, influence upon atmospheric motions 48(59, 96, 106)
Topside ionogram 49/2(249, 504–508)
Topside (of the ionosphere) 49/7(353)
Topside profile of electron density 49/2(509, 512–514, 520ff.)
Topside sounding 49/2(82ff., 250, 503–520)
Topside sounding, spikes 49/2(512)
Topside sounding, technical data of sounders 49/2(503)
Toroidal vector fields 47(526)
Torsion balance 47(215)
Torsion constant 49/3(279, 281)
Torsional waves 47(184)
Total electron content 49/5(81, 358)
Total flux (EUV) 49/6(363)
Total reflection 48(461)
Total reflection and index function 49/2(183)
Total thermo-remanent magnetization 49/1(273)
T-phases 47(120, 135)
Trace 49/2(221)
Trace velocity 49/5(188, 189)
Trace-free tensor 49/2(18)
Trajectories 48(726)
Trajectories, infinite 49/5(278)
Transducer 47(154)
Transfer collision
— frequency 49/6(41); 49/7(67, 124–130, 162, 201, 204, 217)
— interval 49/7(202)
— rate 49/7(67)
— tensor 49/7(192–194)
Transfer cross section 49/6(42); 49/7(66, 68, 70, 74, 76, 78 Fig. 2, 80 Fig. 3, 94 Fig. 10b, 120, 124–131, 134);
Transfer, radiative 48(294, 365)
Transformation of longitudinal waves 49/5(301)
Transit magnetometer 49/3(300)
Transition height 49/6(224, 229)
Transition layer 49/6(365)
Transition level 49/7(374)
Transition ratio 49/2(164)
Transitional region 49/5(224)
Transitorial layers 49/2(222, 344ff.)

Transits of Mercury 47(10)
Transmission display 49/2(225)
Transmission function, Ladenburg and Reiche 48(222)
Transmission functions 48(220)
Transmission matrix 48(296); 49/2(163ff.)
Transmission matrix, transformation formula 49/2(166)
Transmissivity 48(156, 177, 352)
Transmissometer 48(282)
Transparency 48(622)
Transparency of sporadic *E* layer 49/2(339)
Transparency range 49/4(517, 518, 520, 528)
Transport 49/6(98, 110, 166, 218, 259, 295)
Transport coefficient, eddy diffusion 49/5(126)
Transport coefficients 49/7(132, 158–166 Figs. 32–38, 183–211)
Transport collision frequency 49/2(10, 11ff., 19, 97); 49/5(169); 49/7(130–132, 166 Figs. 22–31, 217)
Transport, differential 49/5(175)
Transport, eddy 49/5(123, 127, 134)
Transport eigenvalues 49/7(195, 204–208)
Transport enthalpy 49/2(11)
Transport equation(s) 49/2(9); 49/6(259, 263); 49/7(60, 174–183)
Transport functions 49/7(197)
Transport heat flux 49/5(127)
Transport (horizontal) 49/7(301)
Transport integrals 49/7(196, 200)
Transport phenomena 49/4(435)
Transport, thermal 49/5(126)
Transport, vertical 49/6(298)
Transport-oriented models 49/7(444–458)
Trans-tidal waves 48(672)
Transverse component of electric field 49/2(27, 30, 49, 61)
Transverse components of the earth's magnetic field 49/2(40)
Transverse relaxation time 49/3(361)
Transverse waves 49/2(10, 18, 26); 49/4(457)
Transverse whistlers 49/5(325, 327)
Trapped particles 49/2(525); 49/3(158, 165)
Trapped particles in the magnetic dipole field 49/1(84ff., 99, 128)
Trapped waves 49/5(340)
Trapping, diurnal variation of the upper limit of 49/4(199)
Trapping, high latitude limit of 49/4(198)
Trapping, outer radius limit of 49/4(200)
Traps, charged-particle 49/5(353ff., 358)
Travel times 47(91); 48(467)
Travelling disturbances 49/2(345, 347, 405)
Travel-time curve 47(163)
Travel-time tables, Jeffreys-Bullen 47(95, 366)
Travel-time tables, Zöppritz-Turner 47(93)
Trichel pulses 48(599)
Tritium 47(296, 301, 308, 346)
TRM, *see* Thermo-remanent magnetization
Trochoid 48(674, 681)
Tropical aurora 49/1(2, 36ff., 120)
Tropical subvisual red arc 49/1(120)
Troposphere 49/6(8)
Trough 49/4(56); 49/7(408 Fig. 135, 503)
Troughs 49/4(59)
T-S-diagram 48(642)
Tsubokawa magnetometer 49/3(308)
Tsunami 48(672)
Tuning 49/2(200)
Turbidity, coefficient, Ångström's 48(181)
Turbidity, daily variation 48(179)
Turbidity factor 48(176)
— for Kurzstrahlung 48(179)
— new 48(179)
— virtual variation 48(180)
Turbidity quotient 48(181)
Turbidity, variation with height 48(178)
Turbopause 49/6(85, 118, 160, 161, 244); 49/7(301, 471)
Turbulence 48(623, 711, 715, 718, 723, 728, 732); 49/2(401); 49/6(152, 154, 161, 163, 166, 260); 49/7(471)
Turbulence, atmospheric 48(33)
Turbulence in the upper atmosphere 48(442)
Turbulence inversion 48(44)
Turbulence, mechanical 48(37)

Turbulence, thermal 48(37)
Turbulence, turbulent velocity 49/6(168, 260)
Turbulence, vertical and horizontal 48(624, 639)
Turbulent dissipation 49/5(206)
Turbulent friction 48(908)
Turning points 49/4(517, 520, 522)
Twilight 48(209, 216)
Twilight sky radiance depending upon lunar age 49/2(593)
Two-body process 49/6(43)
Two-channel receiver for Doppler beat method 49/2(454)
Two-parameter model 48(107)
Two-stream approach 49/7(464, 465 Fig. 170, 486 Fig. 179, 492, 527)
Type *A* aurora 49/1(21, 117)
Type *B* aurora 49/1(22, 117)
Type of oscillation 49/4(460, 490, 492)
Typhoons 48(129)
Typical radius 49/5(275)

Ultrasonics 48(455)
Ultraviolett excess 49/7(16)
Ulvöspinel 47(471)
u-measure of geomagnetic activity 49/1(223, 230)
Umkehreffekt 48(371, 380)
Unified atmosphere 49/5(173)
Unifilar variometer (theory) 49/3(279)
Uniformitarians 47(290)
Unit(s) 49/2(2, 22, 535ff.); 49/3(1); 49/4(1)
Units, conversion of 49/7(220)
Units of measurement 49/3(277)
Universal time (UT) 47(2); 49/2(277)
Universal-torsion-magnetometer 49/3(313)
Updraught 48(525)
Upper atmosphere nomenclature 49/7(226)
Upper boundary condition 49/5(201)
Upper face of ionosphere, *see also* Topside 49/2(222, 249)
Upper hybrid resonance 49/2(67, 479, 485, 511ff.)
Upper ionosphere 49/5(376)
Upwelling 48(654)
Uraninite 47(319, 349)
Uranium-235 47(298, 302)
Uranium-238 47(297, 302)
Uranium, contents of igneous rocks 47(312)
Uranium, contents of sedimentary rocks 47(314)
Uranium, measuring methods 47(310)
Uranium-lead method 47(324)
Ursids 48(435)
UT, *see* Universal Time
UV-correction 48(184)
UV-dosimeter 48(208)

Vaisälä frequency, (Brunt-) 49/5(186)
Valley (in electron density profile) 49/7(349, 354–357 Figs. 106–109)
Valley regions in the ionosphere 49/2(222, 259ff.)
Value 49/3(206)
Van Allen belt, *see also* Radiation belt 49/2(520, 525)
Van Allen radiation 49/4(215)
van der Waals force 49/6(143)
Vapour pressure over sea water 48(620)
Vapour trail, cloud 49/6(161, 164, 280)
Variability class index 49/7(17, 45–52)
Variability index 49/7(17, 45–52)
Variability of daily means 49/1(186, 189–192)
Variability of $\mathbf{S_q}$ 49/3(18)
Variables, separation 49/5(181)
Variation 48(749, 752)
Variation, annual 48(750)
Variation, diurnal 49/5(119)
Variation, semiannual 49/5(147)
Variation, solar cycle 49/4(182)
Variational principle, Hamilton's 49/5(211, 212)
Variations, diurnal temperature 49/5(152–154, 161)
Variations, local time 49/4(178)
$\mathscr{X}$-Variometer 49/3(285)
$\mathscr{Y}$-Variometer 49/3(285)
$\mathscr{Z}$-Variometer 49/3(285)
Variometer eigenfrequency 49/3(333)
Variometer for $\mathscr{D}$ 49/3(277, 280)
Variometer for $\mathscr{H}$ 49/3(282)
Variometer for $\mathscr{X}$ 49/3(285)
Variometer for $\mathscr{Y}$ 49/3(285)
Variometer for $\mathscr{Z}$ 49/3(285)
Variometer for $\mathscr{X}$ or $\mathscr{Y}$ 49/3(284)
Variometer with direct view registration 49/3(336)

Variometers 49/3(277)
Vavilov-Čerenkov effect 49/5(228)
Vector magnetometer probe 49/3(356)
Vegard-Kaplan bands of nitrogen 49/1(110)
Veil 49/1(27, 29–32, 118)
Velocity 49/4(10)
Velocity, Alfvén 49/4(21, 24, 32, 36, 43, 55, 91, 466, 476); 49/5(221)
Velocity average 49/2(8)
Velocity body's 49/5(219)
Velocity distribution 49/2(7ff., 9)
Velocity distribution function 49/6(240)
Velocity, group 49/5(188, 189, 198)
Velocity, guiding center 49/4(18)
Velocity, intrinsic group 49/5(211)
Velocity, non-isothermal sound 49/5(240ff.)
Velocity of electrons, thermal 49/5(220)
Velocity of energy flux 49/5(248)
Velocity of external particle fluxes 49/5(219)
Velocity of ions, thermal 49/5(220)
Velocity of random changes 49/2(430)
Velocity of sound 49/4(423)
Velocity, phase 49/5(188, 189, 224)
Velocity profiles of Balmer lines 49/1(110, 113)
Velocity space 49/2(9, 14)
Velocity, supersonic 49/5(220)
Velocity, trace 49/5(188, 189)
Velocity, vertical 49/6(262, 279, 289, 291, 295–297)
Velocity-depth curves 47(366)
Venus 49/4(223)
Venus, ionosphere 49/5(365)
Venus, neutral atmosphere 49/5(374)
Venusian ionosphere 49/7(454–458 Figs. 162c, 163a, 166)
Venusian ionosphere, model 49/5(371)
Venusian upper atmosphere, structure 49/5(376)
Vertical eddy-transported heat 49/5(128)
Vertical force ($\mathscr{Z}$) 49/3(277)
Vertical motion 49/2(377)
Vertical motion due to surface friction 48(42)
Vertical rhombic antenna 49/2(205, 217)
Vertical temperature profile 49/5(130, 131)
Vertical wave equation 49/5(182, 192)
Vertical-horizontal ratio 49/5(125)
Very low frequency range 49/5(235)
Vibrational energy transfer 49/6(125)
Vibrational excitation 49/6(89, 123, 126); 49/7(442, 443, 467)
Vibrational temperature 49/1(123)
Vibration-rotation spectrum 48(226)
Vibrations 47(181)
Video frequency amplifier 49/2(207)
Virga 48(518, 520)
Virtual height, *see also* Equivalent height 49/2(113, 199, 219, 226ff.)
— of *E* region 49/2(278)
— of *F*1 layer 49/2(287)
— of *F*2 layer 49/2(307)
— of sporadic *E* layer 49/2(332)
Virtual temperature 48(457)
Viscosity 47(87, 180); 49/4(547); 49/6(169, 255, 273)
Viscosity (anisotropic) 49/7(209)
Viscosity coefficients 49/4(433)
Viscosity, eddy 49/5(127)
Viscosity tensor 49/4(430)
Viscous effects 49/5(128)
Viscous fluid 47(279)
Viscous remanent magnetization (VRM) 49/1(272, 290–294)
Visibility from aircraft 48(276)
Visibility meter, Gold's 48(287)
Visibility meter, Wigand's 48(286)
Visibility meters 48(279, 285)
Vision along the horizontal 48(262)
Vision, foveal 48(268)
Vision, non-horizontal 48(263)
Vision, parafoveal 48(267)
Vision through telescopes 48(270)
Visual range 48(166, 168)
— estimates of meteorological observers 48(271)
— in clouds 48(209)
— in fog 48(274)
— in natural light 48(271)
— instruments 48(279)
— nomograms 48(273)
— of colored objects 48(274)
— of extensive objects 48(273)
— of light sources 48(276)
— of objects in a searchlight beam 48(277)
Vlasow equation 49/2(8)
VLF hiss 49/5(322)

VLF waves 49/5(250)
Voigt effect 49/2(58, 463, 465)
Volcanic dust 48(168)
Volcanic earthquakes, magnitude-frequency relation 48(999)
Volcanic eruption, prediction 48(993)
Volcanic eruptions 48(323, 982)
Volcanic tremors 48(993, 997)
Volcanoes, change of geomagnetic field 48(1002)
Volcanoes, crustal deformation 48(992)
Volcanoes, depression of ground 48(990)
Volcanoes, detonations 48(1000)
Volcanoes, electric phenomenae 48(1004)
Volcanoes, energetics 48(1006)
Volcanoes, gravimetry 48(1004)
Volcanoes, seismic prospection 48(1004)
Volcanoes, upheaval 48(991)
Volcanology 48(982)
Volume, specific 48(619)
Vorticity equation 48(12, 23, 656)
Vorticity equation in quasi-geostrophic approximation 48(94)
Vorticity geostrophic 48(32, 93)
Vorticity, potential 48(13)
Vorticity, potential, conservation 48(14, 27)
Voss' polariscope 48(309)
VRM 49/1(272, 290–294)

Wake 49/5(254); 49/7(246)
Wake, emission from the 49/5(299)
Wake of a satellite in plasma 49/2(483, 512)
Wanted signals 49/2(209)
Warm front 48(135)
Warm plasma 49/2(61)
Warm plasma, conditions of body inserted 49/2(480ff.)
Warming, explosive 49/5(118, 131, 133ff.)
Washington g 47(204)
Water bottle, reversing 48(612)
Water budget of the oceans 48(635)
Water cluster ions 49/6(14, 52, 66, 77, 101, 106)
— formation 49/6(102, 105)
— profile 49/6(106)
Water content of minerals, effect on magnetism 49/1(250)
Water droplets, absorption in 48(257)
Water masses, classification 48(642)
Water, reactions 49/6(66)
Water sky 48(212)
Water vapour 48(172, 181)
— absorption cross sections 49/6(21, 22)
— enthalpy 48(4)
— flux 48(628)
— photodissociation 49/6(33, 87)
— profile 49/6(90)
Wave, acoustic 49/5(240ff.)
Wave age 48(711)
Wave, Alfvén 49/4(32); 49/5(235)
Wave amplitude, apparent 48(693)
Wave amplitude, maximum 48(694)
Wave angular momentum 49/5(216)
Wave channels 47(119)
Wave components, harmonic 48(688, 713, 717, 724, 728)
Wave crests, curvature 48(848)
Wave cut-off 49/5(235)
Wave, cyclotron 49/5(235)
Wave direction 48(685, 700, 703, 721, 728)
Wave, electron whistler 49/5(236)
Wave, electron-acoustic 49/5(239, 329)
Wave, electronic 49/5(236)
Wave, electrostatic 49/5(240)
Wave, energy 48(52, 695, 716); 49/5(209)
Wave energy density 49/5(210)
Wave, equation, vertical 49/5(182, 192)
Wave, extraordinary 49/5(232, 236)
Wave, fast
— electron-acoustic 49/5(240)
— ion-acoustic 49/5(241, 251ff., 296, 316–318, 329)
— magneto-acoustic 49/5(236)
— magnetodynamic 49/4(32)
Wave focusing 49/5(293)
Wave form 48(685, 731)
Wave generation
— area 48(696, 712, 717, 722)
— by sheltering effect 48(705)
— "gust" theory 48(707)
— minimum wind speed 48(704, 706)
Wave group 48(678, 679)
Wave group propagation 49/2(100ff.)

Wave height 48(673)
— apparent 48(688, 693, 716, 721)
— characteristic 48(689, 701, 710, 717, 720, 729)
— maximum 48(687, 689)
Wave, HF Langmuir 49/5(238ff., 337, 340)
Wave, internal gravity 49/5(168, 177, 188, 200)
Wave, ion-acoustic 49/4(32); 49/5(240ff.)
Wave, ion-cyclotron resonance 49/5(305)
Wave, ionic 49/5(235)
Wave, ion-whistler 49/5(235)
Wave, Langmuir-Tonks 49/5(240)
Wave length, apparent 48(688, 690, 699)
Wave length, limiting 48(703)
Wave length, maximum 48(687)
Wave, LF 49/5(249)
Wave, magnetodynamic 49/4(31, 32, 91)
Wave, modified Alfvén 49/5(236)
Wave momentum density 49/5(214)
Wave motion, irrotational 48(673, 732)
Wave normal 49/2(26)
— angle with earth's magnetic field 49/2(33, 49)
— equation in gyrotropic optics 49/2(32ff.)
— equation of Fresnel 49/2(32)
Wave, ordinary 49/5(232)
Wave packet 49/5(214, 215)
Wave period, apparent 48(688, 690, 698, 719, 721, 727)
Wave period, characteristic 48(689, 701, 710, 717, 729)
Wave period of maximum energy 48(696, 699, 714, 722)
Wave profile 48(674, 685, 732)
Wave recording 48(691, 702)
Wave, shock 49/5(261, 271)
Wave, slow electron-acoustic 49/5(240)
Wave, slow ion-acoustic 49/5(241, 251ff., 318, 319)
Wave, solitary 48(730, 732)
Wave spectrum 48(691, 719, 728)
Wave statistics 48(690)
Wave steepness 48(673, 685, 705, 711, 718, 725, 729)
Wave structure 49/5(164)
Wave train, conservative 48(681, 688, 710, 724)
Wave trapping in plasma 49/5(235)
Wave vector 49/2(26); 49/4(460, 509)
Wave-fixed coordinates 49/2(37, 42)
Wave-guides in the upper mantle 47(134)
Wavelet 47(88)
Waves, action on coastal engineering 48(732)
Waves, air resistance 48(718, 723)
Waves, apparent 48(688)
Waves, backwash 48(729, 731)
Waves, bodily 47(75)
Waves, capillary 48(672, 704, 706)
Waves, cellular 48(871)
Waves, circular 48(677)
Waves dispersion 47(120)
Waves, electron acoustic 49/4(477)
Waves, ELF 49/5(250, 251)
Waves, ELF-Alfvén 49/5(250, 251)
Waves, energy flow 48(678, 710, 723, 725, 728)
Waves, excitation 49/5(295)
Waves, external 48(55, 866)
Waves, fast magneto-sound 49/4(483, 529, 532)
Waves, free 48(886)
Waves, frequency 48(677, 695, 701)
Waves, frontal 48(89)
Waves, gravity 48(672, 706, 942); 49/5(167)
Waves, group velocity 47(119)
Waves, HF 49/5(248)
Waves hydromagnetic 49/4(31)
Waves, in anisotropical bodies 47(194)
Waves in coastal waters 48(724)
Waves in isothermal atmosphere 48(54)
Waves, internal 48(55, 613, 658, 865, 867, 870, 873, 876)
Waves, ion cyclotron 49/4(177); 49/5(221)
Waves Lamb 49/5(189, 200)
Waves, lee 48(59); 49/5(208)
Waves, long 48(846)
Waves, longitudinal 47(170); 49/4(521, 525)
Waves, long-period 48(672)
Waves, magneto-acoustic 49/5(251)
Waves, modified Alfvén VLF 49/5(250)
Waves, mountain 48(59); 49/5(206)

Waves, non-divergent 49/5(190)
Waves of 27-day period 49/1(198–204, 217)
Waves of 10-year period 49/1(222–224)
Waves on an interface 48(81, 865)
Waves, phase velocity 47(119)
Waves, progressive 48(859)
Waves, reflexion 48(732)
Waves, resistance coefficient 48(705, 715, 718)
Waves, rotational 48(680)
Waves, seismic 47(75)
Waves, slow magneto-sound 49/4(483)
Waves, standing 48(850)
Waves, total energy 48(695, 697, 716)
Waves, transformation of longitudinal 49/5(301)
Waves, transitional 48(724, 727, 732)
Waves, transverse 47(170)
Waves, trapped 49/5(340)
Waves, true 48(688)
Waves, ultra gravity 48(672)
Waves, VLF 49/5(250)
Weak absorption 49/2(18ff., 21, 29ff., 39, 52ff., 60, 90ff.)
Weak absorption, dispersion formula 49/2(73)
Weathered layer 47(160, 165)
Weber's differential equation 49/2(183ff.)
Wegener, drift-theory 47(280)
Well-shooting 47(154, 162)
Westward drift of geomagnetic field 47(19)
Westward traveling folds 49/1(40)
Whispering gallery 47(134)
Whistler 49/4(57, 81, 151)
Whistler growth rate 49/4(176)
Whistler, ion-cyclotron 49/5(306)
Whistler mode 49/2(83, 487–503); 49/4(176, 478, 530)
— dispersion 49/2(501)
— ducting 49/2(490, 498)
— guiding action of magnetic field 49/2(490, 498)
— nose frequency 49/2(499)
— propagation through magnetosphere 49/2(491)
— radiation, whistler waves 49/4(139, 485)
— wave, interaction of a 49/4(150)
Whistler propagation 49/2(84ff., 107ff.)
Whistler propagation, group index 49/2(117)
Whistler, proton 49/5(306)
Whistler, short ion 49/5(307)
Whistler signal, travel time 49/2(498)
Whistler statistics 49/2(501)
Whistler triggered emissions (WTE) 49/5(331)
Whistler wave, electron 49/5(236)
Whistlers, ELF 49/5(302)
Whistlers hydromagnetic 49/5(302, 303)
Whistlers, magnetospheric reflected 49/5(325, 327)
Whistlers, short-hop 49/5(325)
Whistlers, subprotonspheric 49/5(325, 327)
Whistlers, transverse 49/5(325, 327)
Whistlers, *v* 49/5(325, 328)
Whitaker function 49/5(193)
Whitaker's equation 49/5(193)
White noise 49/2(210)
Width of an auroral band 49/1(19)
Wiener's calculation 48(257)
Wilson-plate 48(577, 590)
Wilson's theory 48(542)
Winches 48(608)
Wind currents 48(653)
Wind direction 48(700, 702, 721)
Wind, effects 49/6(99, 212, 213, 247, 284, 288)
— on composition 49/6(298)
— on temperature 49/6(294, 295)
Wind field 49/7(493–495 Figs. 184–185)
Wind fields, diurnal 49/5(158)
Wind gradient in the upper atmosphere 48(444)
Wind, meridional (NS) 49/6(258, 283, 289, 305)
Wind pressure (on waves) 48(705, 707)
Wind, prevailing 49/6(275)
Wind profile(s) 49/6(276); 49/7(291 Figs. 56, 57, 447 Fig. 157)
— zonal 49/5(126, 136, 137)
— meridional 49/5(138, 139)
Wind relation, thermal 49/5(156)
Wind sea 48(672, 687, 695, 708, 710, 713, 720, 722, 729)
Wind shear 48(459); 49/2(342ff., 435); 49/5(127); 49/6(116)

Wind shield wiper effect 49/4(147)
Wind speed 48(696, 712, 714)
Wind stress, normal 48(710)
Wind stress, tangential 48(626, 656, 710, 713, 715, 718)
Wind system, thermospheric 49/6(280)
Wind velocity 49/6(268–270, 272–274, 278,280, 282)
Wind velocity, Fourier analysis 49/6(271)
Wind waves 48(672)
Wind, zonal (EW) 49/6(257)
Wind-induced plasma motion 49/7(444–446 Fig. 156b)
Window (optical) 49/6(309, 348)
Winds 49/7(279 Figs. 44, 45, 281 Fig. 46, 283 Fig. 49, 287–289 Figs. 52–54, 292 Fig. 57, 306–308 Figs. 67–69)
Winds, affected by plasma motion 49/6(301)
Winds, diurnal meridional 49/5(157)
Winds, diurnal zonal 49/5(159)
Winds in strato-, meso-, and lower thermosphere 49/7(253 Fig. 20b)
Winds in the upper atmosphere 48(442, 447)
Winds, meridional 49/5(144, 148)
Winds, semi-diurnal variation 48(447)
Winter anomaly 49/7(380 Fig. 121b)
Winter anomaly of *D* region 49/2(330)
Winter polar regions 49/5(134)
Winter storm period 49/5(117, 131, 133, 137, 140, 147)
Wire ropes 48(609)
Witham's refraction law 49/2(122, 136)
WKB approximation 49/2(160, 176)
WKB-solution 49/7(68, 114)
Wolfram 47(299)
Worden-Gravimeter 47(214); 48(804)
World thunderstorm activity 48(542)
World-wide index of *F*2 layer 49/2(306)
Wronski matrix 49/2(157ff., 165)

X-ray bursts in the auroral zone 49/1(88–90)
X-ray bursts in the subauroral zone 49/1(88)
X-rays 49/2(366, 390); 49/4(210, 211, 215); 49/6(37)
X-rays, absorption cross section, air 49/6(25)
X-rays, auroral-zone 49/4(212)
X-rays, effect of solar activity 49/6(30)
X-rays from aurora 49/1(83)
X-rays (soft) 49/7(14)
X-rays spectroheliogram 49/6(341)
x-reflection (extraordinary r.) 49/2(63)
XUV 49/6(335, 343)

Yamamoto, radiation chart 48(235)
Year, anomalistic, tropical 48(747)
Year, division of 48(772)
Year, siderical 47(22)
Year, tropical 47(22)
Yellow substance 48(623)
Yielding of the solid earth 48(783)
Young's modulus 47(170, 185)

Z magnetometric balance 47(499)
Zeeman effect 49/3(358)
Zenith distance of the sun 49/2(251)
Zero spring 48(801)
Zero-order flow 49/5(191)
Zeros of refractive index 49/2(50, 60, 62, 64ff., 86)
Zeta function (generalized 49/7(216)
Zircon 47(315)
Zonal (East-West) winds 49/6(257)
Zonal flow, mean, kinetic energy 48(151)
Zonal index 48(144)
Zonal wind profile 49/5(126, 136, 137)
Zonal winds, diurnal 49/5(159)
Zone, intermediate 49/5(220, 252)
Zone of silence 48(463)
Zone of supersonic motion 49/5(220)
Zürich Sunspot number 49/6(368)

Group 11

Astrophysics

A stars
— classification 50(8, 26, 31, 46–51)
— model of the photosphere 50(319, 320)
— peculiar spectra 50(94)
— white dwarfs 50(172, 173)
Aberration in aerial systems 54(77)
Ablation of meteoroids 52(519, 521, 527, 529, 532)
Absolute gradients 50(276, 325, 380)
Absolute magnitude(s) 50(24, 30, 34, 42, 60, 378)
— definition 53(2)
— of classical Cepheids 53(462)
— of external galaxies 53(314, 315, 316–317, 437)
— of galactic Novae 53(466)
— of globular clusters 53(167–168)
— of nearby galaxies 53(468–474)
— of planetary nebulae 50(141, 142)
— of RR Lyrae stars 53(453, 470)
Absolute photographic magnitude
— of novae at maximum 51(752–753, 763)
— of supernovae 51(763, 778–779)
Absorption and coherent scattering combined 52(44–45)
Absorption as a function of gas pressure 50(311)
Absorption by hydrogenic ions in planetary nebulae 50(399)
Absorption coefficient
— continuous 51(332)
— continuous, for helium and ionized helium 50(297)
— continuous, for negative hydrogen ions 50(296, 328, 329)
— continuous, for neutral hydrogen 50(295)
— for lines 50(340–346)
— of stellar matter 50(299)
— of the photosphere, continuous 52(23–26, 42, 47)
— of the photosphere for lines 52(47, 50, 51)
Absorption, continuous, in stellar atmospheres 50(281, 294–300, 309, 310)
Absorption continuum of molecules 50(123–126)
Absorption cross section of the ionosphere 52(347)
Absorption in the Earth's atmosphere 50(275)
Absorption, interstellar 51(86–89); 53(449, 460, 464, 468–469, 471, 473–474, 479)
Absorption lines 50(276, 278, 340–364)
— central intensities 50(346, 358)
— overlap 50(364)
— profiles 50(354–358)
— role of scattering 50(349, 351, 355, 357)
— total intensity 50(359–364)
Absorption of energy by the Earth's atmosphere 52(368–373)
Absorption of energy in the Venus atmosphere 52(389, 390)
Absorption of light from galactic clusters 53(137–138)
Absorption of light from globular clusters 53(168)
Absorption of light from spirals 53(335)
Absorption of radio radiation by interstellar neutral hydrogen 53(227, 272)
Absorption of radio radiation by the corona 52(294–295)

Absorption of the Earth's atmosphere in the infrared 52(372)
Absorption of the nebular shells 50(408)
Absorption spectra of novae 51(756, 757, 758, 759)
Absorption spectra of white dwarfs 50(168)
Absorption spectrum of the Earth's atmosphere 52(36, 37)
Absorption, true 50(347, 351, 355)
Abundance anomalies in metallic-line stars 50(97)
Abundance anomalies in stars 51(268–272, 346–350, 707–709)
Abundance deficiencies of light nuclei in the Sun 52(342)
Abundance determination(s)
— of elements in the Sun 52(61, 64–65)
— by means of model atmospheres 51(337–339, 341)
— from curves of growth 51(333–337, 341)
— observational data 51(325–328)
Abundance effects in the spectra 51(89)
Abundance lines 51(321–322)
Abundance of nitrogen in late type stars 50(131)
Abundance ratio(s)
— C:O in late type stars 50(130)
— Li^6: Li^7 in the Sun 50(120)
— in white dwarfs 50(169, 182)
— of homologous elements 51(305–307)
Abundance rules 51(297–298)
Abundance, total, of atoms of one kind 50(293)
Abundance variation in variable stars, phase delay towards pulsation period 51(485–487)
Abundances and magic numbers 51(322)
Abundances, chemical, determination 50(369)
Abundances during evolutionary stages 51(250–275)
Abundances in early type stars 51(343, 344)
Abundances in globular clusters 51(113–114, 214–216)
Abundances in isotopes 51(345–346)
Abundances in novae 51(762)
Abundances in peculiar *A* stars 51(707–709)
Abundances in peculiar stars 50(97, 100, 105)
Abundances in planetary nebulae 51(343, 344)
Abundances in stellar atmospheres 50(371)
Abundances in the solar atmosphere 50(373)
Abundances in the Sun 51(343, 344)
Abundances in white dwarfs 51(268, 271, 347)
Abundances, molecular, in late type stars 50(128, 130)
Abundances, molecular, theoretical considerations 50(133, 143)
Abundances of elements (table) 51(316)
Abundances of nuclides (table) 51(317–320)
Abundances, relative, of elements in late type stars 50(132)
Abundances, relative, of isotopes in stars 50(120, 132)
Abundances, relative to solar ones 51(341)
Acceleration mechanism for particles emitted from the Sun 52(223)
Acceleration of the tail particles of comets by solar corpuscles 52(507)
Acceleration parameter of Hubble 53(446, 480, 481, 483)
Accretion of matter by stars 51(238–241)
Accretion of matter by the corona 52(282)
Accuracy of photometric measurement 54(143)
Achondrite 52(562)
Achromatic coronagraph 54(6)
Acoustic waves in the chromosphere and corona 52(149)
Activation of filaments 52(210)
Activation of quiescent prominences 52(237–239)
Active corona 52(271–283)
Active dark flocculi 52(210, 244)
Active prominences 52(225, 237–239)
Active sunspots 52(354)
Activity centres 52(183–188)

Activity centres, spectrum 52(298–300)
Activity centres, thermal radio radiation 52(296–301)
Activity regions of the corona 52(189, 273–279)
Adams-Williamson equation 52(433, 434)
Adiabatic conditions, departure from 51(487–490)
Adiabatic exponent 51(12)
Adiabatic gradient 51(12–14)
Adiabatic model 51(635)
Adiabatic pulsations in a spiral arm 53(95)
Adiabatic radial oscillations 51(458–466)
Adiabatic temperature gradient
— of the Earth's atmosphere 52(373, 374)
— of the Jupiter atmosphere 52(409)
Adjacency effects in photographic plates 54(133, 134)
Aerial characteristics 53(210)
Aerial parameters 54(61–64)
Aerial theorems 54(66–67)
Aerials for radio waves 54(60–87)
Aerials, main types 54(68–75)
Aerials, mounting and driving 54(82–87, 88–89)
Aerolite (=stony meteorite) 52(559, 560)
Age determinations of meteorites 52(564)
Age of craters 54(195)
Age of globular clusters 53(190)
Age of stars 53(98)
Age of stars in galactic clusters 53(145–146)
Age of Universe 53(486, 487)
Age of white dwarfs 50(185)
Age zero main sequence 51(95, 96, 98)
Ages of galactic clusters and associations 51(177, 201–204, 223)
Ages of globular clusters 51(177, 213–214, 216, 224)
Aging effects in chemical composition 51(264–267)
Air, constituents 52(367)
Albedo of Mercury 54(209)
Albedo of Moon and Mecury 52(404)
Albedo of some asteroïds 54(212)
Alfvén waves in prominences 52(248)
Algol-type eclipsing binaries 50(251, 252, 257, 258, 263, 272)
Alkali metals, first use in photoelectric cells 54(242)
Aller-Pierce model 52(59)
AlO bands in variable stars 50(119)
α Cygni stars 50(388, 389)
α Virginis, spectra 50(247, 248)
α-process for element synthesis 51(254, 271)
Altazimuth mounting of aerials 54(82)
Alternating current methods in photometry 54(276)
Aluminium, abundance 51(304, 316, 317, 343)
Ammonia in the atmospheres of the major planets 52(407, 410, 411)
Amplification of a photomultiplier current 54(255–258, 261–262)
Amplifiers for photoconductive cells 54(267–269)
Amplitude correlation between velocity and light curves of variable stars 51(588–589)
Amplitude limitation by coupling 51(547–548, 588)
Amplitude-asymmetry correlation in variable stars 51(592–593)
Amplitude-period diagram for RR Lyrae stars and cepheids 51(370, 371)
Amplitude-period relation in explosive variables 51(420, 576)
Amplitude-period relation of recurrent novae and U Geminorum stars 51(764)
Andromeda Nebula, *see also* Mesier 31 53(379)
Andromeda Nebula, radio frequency radiation 53(250–252)
Andromeda source, apparent angular diameter 53(226)
Ångström's pyrheliometer 54(2–3)
Angular diameter, apparent, of radio sources 53(224–227)
Angular diameter of galactic clusters 53(137)
Angular distribution of brightness across a radio source 53(213, 214)
Angular distribution of radio sources 53(219–220)

Angular extent of a discrete radio source 54(118–119)
Angular momentum of contracting protostars 51(152–154)
Angular size measurements of radio galaxies 53(273, 274)
Angular size of clusters of galaxies 53(408–409)
Angular spectrum 54(66)
Angular spectrum of an interferometer 54(100)
Angular speed of galactic rotation 53(24, 53)
Anharmonic oscillations 51(659)
Anisotropy in the turbulent field 52(116)
Anisotropy of convective motion 52(96)
Anomalous chemical abundances in stars 51(268–272, 346–350, 707–709)
Antimony, abundance 51(311, 316, 318, 343)
Antimony-cesium cathode 54(243, 244, 247)
AO Cassiopeiae, spectra 50(265)
AO Cassiopeiae, velocity curve 50(268)
Aparaboloid aberration 54(82)
Aperiodic expansion or contraction 51(458)
Aperture distribution 54(65)
Aperture efficiency 54(74, 120)
Apex, solar 53(3)
Apparent angular diameter of radio sources 53(224–227)
Apparent axial ratio of spirals 53(307, 308)
Apparent brightness of eclipsing binaries 50(228–232)
Apparent coordinates of a galaxy 53(425)
Apparent diameters of clusters of galaxies 53(431)
Apparent diameters of galaxies 53(313)
Apparent disk temperature 52(286); 54(91)
Apparent distance modulus 53(449, 468–474)
Apparent distance of a galaxy 53(423, 426)
Apparent distribution of galactic clusters 53(132–138)
Apparent distribution of globular clusters 53(166–172)
Apparent flattening of galaxies 53(303–308, 322, 328)
Apparent magnitude(s) 51(81, 82)
— of galaxies 53(411, 435–439, 448, 469–475, 479)
— of globular clusters 53(167–168)
— of planetary nebulae 50(140, 141)
Apparent radius of clusters of galaxies 53(440)
Apsidal motion 51(41–42)
Arches of the corona 52(272, 273, 281)
Area velocity in a system of rotational symmetry 53(25–27)
Argon, abundance 51(304, 316, 317, 343)
Arizona meteor expedition 52(534)
Array theorem 54(66–67)
Arsac array 54(117)
Arsenic, abundance 51(308, 316, 318, 343)
Ascending filaments 52(240)
Aspect effects in *A* type spectra 51(697)
Associations, definition and properties 51(197–201, 203)
Associations of stars 53(69–73, 165–166)
Asteroidal origin of meteors 52(519, 529, 530, 554, 556, 559)
Asteroïds 52(431, 432)
Asteroïds, brightness variations 54(211, 212)
Asteroïds, diameter 54(211)
Asteroïds, number 54(211)
Asteroïds, photometry 54(211)
Asteroïds, radio echoes 52(462)
Asteroïds, surface properties 54(211–215)
Astrometric results for eclipsing binaries 50(220)
Astrometric results for spectroscopie binaries 50(220)
Astrometric study of photographically unresolved binaries 50(210–218)
Astronomical positions, definition 53(1)
Asymmetrical drift of stellar motion in the Galaxy 53(6, 22–23, 29–31)
Asymmetry in light and velocity variations of variable stars 51(542, 544, 549, 553, 554, 576, 592–593)
Asymmetry of spirals 53(330, 333–335)
Atlas of cometary spectra 52(477)
Atmosphere of comets 52(478–483)
Atmosphere of Jupiter 52(406–410)

Atmosphere of Mars 52(393–404); 54(227–228, 235)
Atmosphere of Mercury 52(404–406); 54(209)
Atmosphere of Neptune 52(406, 410–412)
Atmosphere of Pluto 52(412, 414)
Atmosphere of Saturn 52(406, 410)
Atmosphere of the Earth 52(363, 366–383)
Atmosphere of the Jupiter satellites 54(219, 220)
Atmosphere of the Moon 52(404–406)
Atmosphere of the Satellites of major planetes 52(412, 414)
Atmosphere of the sun, *see* Chromosphere, Corona, Photosphere
Atmosphere of the Venus 52(383–393)
Atmosphere of Uranus 52(406, 410–412)
Atmospheres of eclipsing binaries 50(234, 238)
Atmospheres of stars, general, *see* Stellar atmospheres
Atmospheres of the Sun, *see* Solar atmosphere
Atmospheres of variable stars 51(593–601)
Atmospheres of white dwarfs 50(167)
Atmospheric extinction 54(149–154)
Atmospheric layers of variable stars
— bounded atmosphere 51(492–496)
— influence on pulsation 51(492–499)
— isothermal layer 51(497–499)
— unbounded atmosphere 51(496–499)
Atmospheric models 54(151)
Atmospheric refraction 50(204, 205)
Atmospherics 52(351, 352, 462)
Atomic clock 53(485)
Atomic levels, population 52(109–112)
Attenuation device 54(164)
Aureols around the big craters 54(195–197)
Auriga source, apparent angular diameter 53(226, 232)
Aurora 52(357, 358)
Autocollimating spectroscope 54(21, 38)
Auxiliary solar instruments 54(29–33)
Axial ratio of spirals 53(307, 308)
Axial-symmetric perturbations 51(528, 529–530)
B Cassiopeiae (Nova Tychonis) as radio source 53(231)
B stars, classification 50(8, 26, 31, 43–46)
B stars, model of the photosphere 50(320, 321)
B stars, white dwarfs 50(170, 171)
Ba II, chemical abundance 50(105)
Ba II stars 50(104–106)
Baade's criterion 51(595–598)
Babcock's magnetograph 54(32)
Background fog 54(139)
Background light of night sky 54(250, 254, 276)
Background radiation of the solar radioemission 52(286)
Bailey's types of RR Lyrae stars 51(367)
Baily beads 52(106)
Balloons 54(37)
Balmer continuum in planetary nebulae 50(153)
Balmer continuum in the low chromosphere 52(119)
Balmer decrement in planetary nebulae 50(409–412)
Balmer discontinuity 50(78, 276, 323, 390); 52(16)
Balmer discontinuity in white dwarfs 50(173, 176, 177)
Balmer line(s)
— in stellar spectra 50(366)
— in the low chromosphere 52(120, 121)
— in the solar spectrum 50(365, 368)
— of white dwarfs 50(180, 182)
— profiles of flares 52(202)
Baluns 54(68)
Band emission of comets 52(484, 487)
Band shapes 54(50)
Band spectra of sunspots 52(160)
Band widths of bursts 52(304)
Bands, *see* Molecular bands
Bands of polyatomic molecules 50(118)
Bandwidth 54(44)
Bandwidth effect in an interferometer 54(104–105)
Barbier model of the solar photosphere 50(338)
Barium, abundance 51(311, 316, 318, 343)

Barium stars, abundance anomalies 51(271–272, 349)
Barotropic relations 51(451)
Barred spiral structure in the central region of the Galaxy 53(43, 90–92)
Barred spirals 53(277, 279, 283, 287, 289, 290, 294–295)
Basic solar motion 53(16)
Be stars 50(388, 389)
Beam efficiency 54(74, 120)
Beam ratio 54(74, 120)
Beat phenomenon in β Cephei stars 51(579–582)
Beat phenomenon in RR Lyrae stars 51(582, 583)
Beating effects in receivers 54(47)
Bemporad's model of atmosphere 54(151)
Bergedorfer Spektraldurchmusterung 50(12)
Beryllium, abundance 51(303, 316, 317, 343)
Bessel-Bredikhin theory of comets 52(496, 497)
β Aurigae, velocity curves 50(245)
β Canis Maioris stars, *see* β Cephei stars
β Cephei stars
— beat phenomena 51(579–582)
— cepheids, RR Lyrae stars, comparison 51(400)
— characteristic data 51(356, 365)
— double periodicity 51(398, 399)
— evolutionary sequence 51(279–280)
— light curves 51(398–400)
— line spectrum 51(400–402)
— list 51(399)
— non-radial oscillations 51(580–581, 589, 596)
— period-luminosity relation 51(402)
— periods 51(579–582)
— phase correlation between light and velocity curves 51(591)
— rotation 51(399, 400, 579–582)
— variable line broadening 51(401, 579–581)
— velocity curves 51(398–400)
Betatron effect in stellar atmospheres 50(132)
Biermann-Schlüter mechanism of protostar formation 51(145, 146)
Bifurcation points 51(613)
Binaries, *see* Eclipsing binaries, spectroscopic binaries, visual binaries
Binaries, close 50(237, 257, 263, 272)
Binaries, common envelope 50(258, 263, 271)
Binaries, evolution 50(271, 272)
Binaries, frequency 50(223)
Binaries, orbital motion 50(194–202)
Binaries, origin 50(272, 273)
Binaries, planetary companions 50(218, 219)
Binaries, prominences 50(270, 271)
Binaries, synchronization of rotation and revolution 50(253)
Binaries, visual and photographic measurements 50(189, 190)
Binary 61 Cygni 50(219)
Binary 99 Herculis, orbital motion 50(212–213)
Binary hypothesis of variable stars 51(359–361, 363–364, 432, 573)
Binary Ross 614 50(217, 218)
Binary stars in galactic clusters 53(141)
Binary systems, formation 51(153)
Binary systems of early spectral types 50(265, 266, 272)
Binary systems of late spectral types 50(265, 266)
Bipolar groups of sunspots 52(167)
Bipolar magnetic regions
— of the rôle in the development of a centre of activity 52(184–185)
— of the Sun 52(181, 185, 230, 248)
Birch's homgeneity test 52(434)
Bismuth, abundance 51(316, 320, 343)
Black-body radiation 54(52, 90)
Blanketing effect 50(319); 51(265)
Blazhko effect 51(384–388)
Blends 50(364)
Block diagram of radio wave reception 54(45)
Blue dwarf region in the H-R diagram 51(107)
Blue supergiants, variable 51(280–281)
b_n coefficients of the hydrogen levels 50(411)
Böhm's model 52(59)
Böhm-Vitense model of the solar photosphere 50(338)
Bolides 52(520, 527, 556–559)
Bolometer 54(241)

Bolometric correction 50(325)
Bolometric magnitude 51(84)
Boltzmann equation 50(288)
Boltzmann excitation temperature 52(47, 74)
Bondi variables 51(48)
Bondi-Gold theory of the Universe 53(487)
Border effect 54(134)
Boron, abundance 51(303, 316, 317, 343)
Bottlinger diagram 53(8)
Bottlinger-Lohmann method for mass determination of galaxies 53(357)
Boundary conditions at the surface 51(58, 71, 179, 180)
Boundary conditions for the differential equations of stellar structure 51(14)
Boundary conditions in cosmology 53(493–494, 506)
Bound-free opacity 51(20–22)
Brewer effect 52(372, 376, 379, 389, 398, 399)
Bridges between neighbouring galaxies 53(380–384, 389)
Bright galaxies, radio emission (table) 53(253)
Bright regions on the Mars 54(224–229)
Bright ring of sunspots 52(152, 158)
Brightest galaxies 53(393–394)
Brightest galaxy of a cluster 53(479, 481)
Brightest stars 53(467–470, 472)
Brightness, absolute, of novae 51(752–753, 763)
Brightness, absolute of supernovae 51(763, 778–779)
Brightness, apparent, of eclipsing binaries 50(228–232)
Brightness distribution across the Andromeda Nebula 53(250)
Brightness distribution of the quiet Sun 52(292, 293)
Brightness maxima of the corona 52(272)
Brightness of a radio source 53(209)
Brightness of comets 52(468, 469)
Brightness of granules 52(82, 85)
Brightness of Jupiter, relation to sunspot number 52(408)
Brightness temperature 54(87, 90, 122)
— definition 53(102)
— of radio sources 53(209, 210)
— of radio sources at metre wave lengths 53(123, 124)
— of the Sun 52(286, 289)
Brightness variations of asteroïds 54(211, 212)
Brightness variations of novae 51(752–753, 755, 756, 761)
Broadside arrays of dipoles 54(69–70)
Bromine, abundance 51(306, 308, 316, 318, 343)
Brooks' comet 1889 V 52(474)
Brown and Twiss system 54(111–113)
Brusts in the solar radioemission 52(222, 286, 296, 301, 302–308, 311–322)
Butterfly diagram of sunspots 52(326)

C_2 bands in carbon stars 50(116)
C regions of the corona 52(189)
C stars, *see* Carbon stars
CA, *see* Centre of activity
Ca II emission lines in cepheid spectra 51(397)
Ca^+ spectroheliograms 52(129–131)
CaCl bands in variable stars 50(119)
Cadmium, abundance 51(310, 316, 318, 343)
CaH bands in *M*-type dwarfs 50(115)
CaI resonance line in *M*-type dwarfs 50(115)
Calcium abundance 51(305, 316, 317, 343)
Calcium chromosphere of a binary component 50(266)
Calibration of receiver 54(45, 52)
Callisto, atmosphere 54(219)
Callisto, diameter 54(215, 216)
Camouflaged eclipsing binaries 50(235–239)
Canals on the Mars 54(224)
CaO bands in variable stars 50(119)
Capri, Fraunhofer-Institut 54(38)
Carbon, abundance 51(303, 316, 317, 343)
— in the atmosphere of R Coronae Borealis 51(423)
— isotope effects 51(345)

Carbon arc standard 52(14, 16)
Carbon dioxide and monoxide in the atmosphere of Venus 52(383)
Carbon dioxide, equilibrium pressure 52(365)
Carbon dioxide in the Martian atmosphere 52(394)
Carbon group, abundance anomalies 51(347, 348)
Carbon stars 50(101–103)
— abundance anomalies 51(269, 271, 272)
— classification 50(61)
— hydrogen-poor 50(103, 118, 172)
— molecular bands 50(66, 102, 116–118)
— spectra 50(66)
Carbon-nitrogen cycle 51(28, 166, 250–253)
— cross sections 51(32–34)
— energy production functions 51(35, 36)
Carbon-poor stars 50(107)
Cartes Synoptiques 52(186–187)
Cascaded image tube 54(286–287)
Cassegrainian reflecting telescopes 54(10)
Cassiopeia A source 53(209, 236)
— apparent angular diameter 53(225)
— distance 53(227)
— polarisation of radiation 53(229)
— spectrum 53(227–229)
Catalogues of comets 52(473)
Cathodes, quantum efficiency 54(247, 249)
Cathodes, response 54(247, 249)
$C^{12}C^{13}$ bands in R and N stars 50(121)
$C^{13}C^{13}$ bands in R and N stars 50(121)
Celestial coordinates of a radio source 53(213)
Centaurus A source 53(267)
Centimetre bursts of the solar radio-emission 52(301)
Central condensation in a star 51(467, 468)
Central intensities of strong Fraunhofer lines 52(75–77)
Central intensities of strong lines 50(358)
Central nucleus of a galaxy 53(287)
Central reversals in H_α spectra of flares 52(202)
Central reversals in the Fraunhofer lines 52(76)
Central stars of planetary nebulae 50(139, 141, 151, 152, 153, 159, 160, 426–431)
Centre of activity, definition 52(183)
Centre of activity, development of a typical 52(184–188)
Centre of activity, thermal radio radiation 52(296–301)
Centre of attraction in a protuberance 52(237)
Centre-limb studies of Fraunhofer lines 52(72–73)
Centre-limb variation 52(143, 147)
— in the local Doppler shift 52(71)
— of activity centres 52(298–300)
— of the granules 52(83,, 88)
Cepheids, amplitude correlation between light and velocity curves 51(588, 589)
Cepheids, amplitude-period diagram 51(371)
Cepheids as evolutionary stage 51(281–282)
Cepheids, characteristic data 51(122, 355, 365)
Cepheids, classical 53(457–458, 462, 466, 468, 471–473, 481)
— absolute magnitude 53(462)
— amplitude of light variation 53(465–466, 473)
— distribution 53(87)
— globular clusters 53(184)
— in galactic clusters 53(142, 143)
— in M 31 53(462)
— period luminosity curve 53(458–466, 474)
Cepheids, classification according to the form of light curves 51(370)
Cepheids, color temperatures and radiation temperatures 51(389)
Cepheids, comparison of theoretical and observed periods 51(577)
Cepheids, continuous spectrum 51(388–391)
Cepheids, correlation between asymmetry and amplitude of light and velocity curves 51(592–593)

Cepheids, double or multiple periodicity 51(384–388, 400)
Cepheids in globular clusters 51(120)
Cepheids, light curves 51(369, 370)
Cepheids, line spectrum 51(391–397)
Cepheids of type II, *see also* W Virginis and RV Tauri stars 51(120, 366, 569)
Cepheids of Type II in globular clusters 53(184, 450, 455–457)
Cepheids of Type II, period luminosity diagram 53(456)
Cepheids, period-luminosity relation 51(376–379)
Cepheids, phase correlation between light and velocity curves 51(374–376, 589)
Cepheids, place in the H-R diagram 51(120–122)
Cepheids, RR Lyrae stars, β Cephei stars, comparison 51(400)
Cepheids, secular changes in period 51(381–384)
Cepheids, spectral type and its variation with phase 51(379, 380, 391–392)
Cepheids, spectral type-period relation 51(379–381)
Cepheids, velocity curves 51(375)
Cepheids with large outer convective zone 51(499–506, 685)
Ceres, diameter 54(211)
Ceres, polarisation of light 54(214, 215)
Cesium, abundance 51(311, 316, 318, 343)
Cesium oxide-on-silver cathode 54(243)
CH bands in *G* and *K* stars 50(114)
CH stars 50(91, 101, 102)
CH stars, luminosity 50(102)
Chandrasekhar approximations 50(303)
Chandrasekhar approximations, numerical data 50(306)
Chandrasekhar mean absorption 50(310)
Chandrasekhar's analysis of the velocity distribution 53(33–36)
Chandrasekhar's solution 52(52)
Chandrasekhar's theory of completely degenerate configurations 51(739–740)
Charge integration of phototube current 54(263–267)
Chemical abundances, *see* Abundances
Chemical composition
— and age of stars 51(36, 264–267)
— and population type 51(128, 129, 214, 288, 346)
— differences between stars 51(346–350)
— discontinuities 51(65–66, 175, 183)
— of a star 50(277, 278, 279, 300, 327, 369, 371, 373)
— of early type stars 50(371)
— of globular cluster stars 51(214–216)
— of late type stars 50(61)
— of meteorites 51(300–302)
— of normal stars 51(324–345)
— of novae 51(762)
— of peculiar stars 51(346–350)
— of sun and planets 51(298–300)
— of the Earth's atmosphere 52(366–368)
— of the Martian atmosphere 52(394–395)
— of the nebular shells 50(408)
— of the solar atmosphere 50(372)
— of the Sun 52(40, 41)
— of the Venus atmosphere 52(383–388)
Chemical evolution of stars 51(249–275)
Chemical inhomogeneity between core and envelope 51(65–66, 175, 183)
Chlorine, abundance 51(304, 316, 317, 343)
Chlorine-bromine abundance ratio 51(306)
Chondritic stones 52(562)
Chord construction for restoration 54(127–128)
Christiansen's interferometer 54(113–115)
Chromatic aberration 54(146)
Chromatic factors in image formation 54(146–149)
Chromium, abundance 51(305, 316, 317, 343)
Chromosphere, *see also* Low, Medium, Upper Chromosphere 52(106–150)
Chromosphere at the limb 52(106–124)

Chromosphere, emission lines 52(106, 107, 115, 144–148)
Chromosphere, models 52(139, 140, 142)
Chromosphere of spectroscopic binaries 50(387)
Chromosphere of stars 50(386)
Chromosphere of the bright binary component 50(266)
Chromosphere of the Sun 50(385)
Chromosphere on the disk 52(124–135)
Chromosphere, turbulence 52(112–116)
Chromosphere, variation during the solar cycle 52(330)
Chromospheric faculae, size and life time 52(177–179)
Chromospheric faculae, spectra and temperatures 52(179–181)
Chromospheric heights of emission of Fraunhofer lines 52(124–126)
Chromospheric mass motion 52(112–116)
Chromospheric mottling 52(126, 127, 129, 133–135)
Chromospheric spicules 52(108, 109, 112, 113, 114, 123)
Chromospheric stability regions 52(151)
Chromospheric surface brightness 52(106, 108)
Circulation, internal, of the Sun 52(342)
Circulation of planetary atmospheres 52(372)
— Venus 52(392)
Clairaut's equation 52(422)
Class I and class II of radio sources 53(222, 237)
Classical Cepheids, *see* Cepheids
Classical cepheids, *see* Cepheids
Classification, *see* Special entries under *A*-stars, *B*-stars etc.
Classification, accuracy 50(20)
Classification at small dispersion 50(71)
Classification by photometry 50(83, 84)
Classification, choice of parameters 50(379)
Classification, comparison of different systems 50(12, 15, 35–40, 59)
Classification from the continuous spectrum 50(74)
Classification in the near infrared 50(72)
Classification of Barbier and Chalonge 50(77, 81, 379)
Classification of Bergedorf, *see* Bergedorfer Spektraldurchmusterung
Classification of Elvius 50(33)
Classification of external galaxies 53(275–286)
Classification of filaments 52(225)
Classification of flares 52(192)
Classification of galactic clusters 53(135–136)
Classification of galaxies 51(237)
Classification of globular clusters 53(171)
Classification of Harvard 50(2, 7–10)
Classification of Morgan and Keenan 50(43, 59, 379)
Classification of planets 52(431)
Classification of prominences 52(225)
Classification of spot groups 52(166–167)
Classification of the McCormick Observatory 50(14)
Classification of Westerlund 50(29)
Classification parameter 50(4)
Classification, photo-electric 50(31, 34)
Classification, Swedish 50(25)
Classification, three-dimensional 50(81, 380)
Close binaries 50(237, 257, 263, 272)
Close binaries, emission lines 50(264, 265)
Cloud bands on Jupiter 52(409)
Cloud formation in the Galaxy, effect on velocity dispersion 53(77–80)
Clouds of Mars 52(396–398)
Clouds of Venus 52(385)
Clouds on Mars 54(229)
Cluster dynamics 53(80–84)
Cluster variables, *see* RR Lyrae stars
Clusters, *see* Galactic clusters, Globular clusters
Clusters of galaxies 53(390–414, 417–444)
— angular size 53(408)
— catalogues and surveys 53(390–396)
— center 53(419)
— characteristics 53(423)
— counts 53(408)
— distances 53(411–414)

— distribution in breadth 53(409–410)
— distribution of apparent diameters 53(431)
— far 53(402)
— higher order clustering 53(410, 416, 433, 443–444)
— hydrogen line emission 53(265)
— interlocking 53(433–435)
— internal structure 53(397–406, 428, 429, 431, 436)
— kinematics and dynamics 53(406–408)
— masses 53(407)
— near 53(402)
— observational data (with table) 53(477–480)
— radio emission 53(260–265)
— statistics 53(423, 439–443)
— systematic selection by the observer 53(481)
— table 53(395, 478)
— types of individual galaxies 53(398)
— velocity dispersion 53(384)
— velocity of recession 53(425, 430)
Clusters of stars, *see* Galactic clusters, Globular clusters
Clusters of stars, stability 53(84–86)
CN bands in carbon stars 50(66)
CN bands in comets 52(487)
CN bands in *G* and *K* stars 50(113)
$C^{13}N^{14}$ bands in *R* and *N* stars 50(121)
Coarse mottling 52(126, 127, 129, 133–135, 174, 178)
Coaxial cables 54(75–76)
Cobalt, abundance 51(305, 316, 317, 343)
Coelostat 54(12, 13, 14)
Cohenite 52(560)
Coherence in an interferometer 54(102–104)
Coherence in sky light 54(93)
Coherent scattering in the photosphere 52(43, 75)
— combined with absorption 52(44–45)
Cold bodies, mass-radius relation 52(420–431)
Collection of meteoritic dust 52(532–533)
Collective behaviour of electrons in white dwarfs 51(728)
Colliding galaxies as radio sources 53(225, 233, 265–266, 270, 272, 385–389)
Collision damping 50(343, 344, 375)
Collision excitation of forbidden lines in nebular spectra 50(403–407)
Collision excitation parameter for metastable levels 50(407)
Collisional broadening of lines 50(343, 344)
Collisional broadening of the Fraunhofer lines 52(46, 49, 65)
Collisional ionization in the corona 52(257)
Color coefficient 54(147)
Color equations 54(146–147)
Color excess 51(87)
Color index 50(30, 60)
— of a star 54(159, 174, 242)
— of atmosphere 54(153)
— of the asteroïds 54(213)
— of Venus 52(386)
— of white dwarfs 50(176, 178)
— period relation for variable stars 51(186)
Color indices 51(81, 83, 87, 88)
Color, intrinsic 50(30)
Color magnitude diagram for *DA* stars 50(166)
Color of the Jupiter atmosphere 52(408)
Color of the Venus atmosphere 52(388)
Color photometry 54(158–163)
Color temperature 50(276, 278, 280)
Color temperatures of cepheids 51(389)
Colorado High Altitude Observatory 54(8, 9)
Colored glass filters 54(17)
Colorimetry of the lunar surface 54(182)
Color-magnitude diagram 51(79)
— of galactic clusters 51(89–107, 201–213); 53(147)
— of globular clusters 51(107–119, 213–216); 53(177–179, 452, 454)
— of high velocity stars 51(130)
— of white dwarfs 51(723)
Colour distribution in elliptical galaxies 53(323)
Colour distribution in lenticular galaxies 53(324)

Colour distribution in spirals 53(330–333)
Colour index, definition 53(2)
Colour-luminosity asymmetry of spirals 53(330, 333–335)
Colours of external galaxies 53(317–319)
Coma Berenices cluster of stars 53(132, 153–154)
Coma Berenicis cluster, H-R diagram 51(102)
Coma cluster of galaxies 53(400, 401, 478)
Coma of a comet 52(465)
Coma of a meteor 52(526, 537)
Combination spectrum of stars 50(243)
Comet Arend-Roland 52(503)
Comet Brooks 1889 52(474)
Comet Cunningham (1940c) 52(478, 479)
Comet Daniel 52(469)
Comet Delavan 52(471)
Comet Donati 52(469, 499)
Comet Encke 52(476, 546–547, 554)
Comet groups 52(474)
Comet Halley 52(466, 473)
Comet heads 52(465, 497, 498)
Comet Morehouse 52(467, 484, 502)
Comet Mrkos 52(467, 481, 482, 483, 513, 514, 515)
Comet September 1882 II 52(474, 499)
Comet tails 52(467, 468, 500–501, 506, 507)
Cometary nuclei 52(465, 491–495)
Cometary origin of meteors 52(519, 529–530, 554, 559)
Comets, brightness 52(468, 469)
Comets, collision with the Moon 54(202)
Comets, cosmogony 52(509–518)
Comets, discoveries 52(465)
Comets, orbits 52(496–476)
Comets, periods 52(470, 471, 511, 516)
Comets, shapes 52(465–469, 496–509)
Comets, spectra 52(477–495)
Common novae and supernovae, distinction 51(766–767)
Communication between two points on the Earth via the Moon 52(454, 459)
Compact clusters of galaxies 53(395)
Complete degeneracy in white dwarfs 51(739–742)
Completely-mixed models 51(173, 174, 177, 217–218)
Composition of stars, *see* Chemical composition
Compressibility effects on stability 51(629–636)
Compressible sphere, oscillations 51(514–516, 527, 541)
Concave gratings 54(20–21)
Concentration class of globular clusters 53(171)
Condensation, central, in a star 51(467, 468)
Condensation, coronal 52(189, 209)
Condensation mechanism of the prominences 52(245)
Condensation of protogalaxies and protostars 51(138–141, 660)
Conductivity in the solar convection zone 52(102, 103)
Conservation of energy 51(445–452)
Conservation of mass 51(434–435)
Conservation of momentum 51(435–445)
Conservative mechanical system 51(608, 611)
Constituents of the air 52(367)
Constitution of cometary nuclei 52(491–495)
Constitutive equations of stellar material 51(5, 15–38)
Contact surface lobes of binaries 50(253, 255, 258, 263, 268)
Continuity equation 50(284)
Continuous absorption
— coefficient of the photosphere 52(23–26, 42, 47)
— in molecules 50(123–126)
— in stellar atmospheres 50(281, 294–300, 309, 310)
— of helium 50(297)
— of neutral and ionized hydrogen 50(295, 296, 328, 329)
Continuous corpuscular radiation from the Sun 52(362)
Continuous emission at the Sun's limb 52(117, 119)
Continuous emission coefficient of the photosphere 52(45)

Continuous emission spectrum of T Tauri 51(426)
Continuous mass loss from stars 51(245–247, 275)
Continuous opacity 50(124, 125)
Continuous radio emission 53(102–107, 116, 239)
Continuous radio emission of the Magellanic Clouds 53(245–248)
Continuous spectra of planetary nebulae 50(152)
Continuous spectrum
— of flares 52(206–208)
— of novae 51(761)
— of quiescent prominences 52(235–237)
— of sunspots 52(157)
— of the corona 52(250–256)
— of the photosphere 52(1 ff.)
— of variable stars 51(388–391, 594–598)
— reception 54(47–49)
Continuous stellar spectra 50(152, 274, 322, 389)
Continuum source function 52(47, 51)
Contracting stars, dynamical stability 51(660)
Contracting universe 53(425)
Contraction of globular clusters 53(191–192)
Contraction of stars, *see* Gravitational contraction
Contrast of spectroscopic emulsions 54(130, 132)
Contribution curve of the line depth 52(59–61)
Convection flux 51(11)
Convection in stellar atmospheres 50(329, 332)
Convection in sunspots, inhibited by a magnetic field 52(172, 173)
Convection in the atmospheres of late type stars 50(126)
Convection in the solar atmosphere 52(98–105)
Convection, model photosphere with 52(104–105)
Convection zone of the Sun 52(98–102)
— upper limit 52(102–103)
Convective core models 51(57, 61–62, 66, 67, 68, 175, 178)
Convective energy transport 51(6, 7, 10–12)
Convective instability in the presence of a magnetic field 51(718)
Convective mixing in main sequence stars 51(172)
Convective motion, anisotropic 52(96)
Convective stability 51(665, 667, 670, 672, 680)
Convective zone, outer
— in giants 51(500)
— in globular cluster stars 51(71)
— in red dwarfs 51(62–64)
— in white dwarfs 51(742–743)
— models 51(180)
— of hydrogen and (or) helium in variable stars 51(499–501, 505, 506, 585, 590, 660, 685)
Convolution, definition 54(66)
Convolution, pentagram notation 54(49, 67)
Convolution theorem 54(67, 94)
Cool stars of abnormal composition 51(347–349)
Cooling as energy source of white dwarfs 51(750)
Cooling down of the Moon 54(200, 203, 204)
Copper, abundance 51(307, 316, 317, 343)
Core variables 51(49)
Cores of planets, instability 52(423)
Corona 50(386); 52(248–322)
Corona, activity regions 52(189, 273–279, 301)
Corona Borealis Cluster of galaxies 53(477–478)
Corona, components 52(249)
Corona, density variation 52(265–266)
Corona, emission lines 52(259–261)
Corona, heating 52(150–151)
Corona, ionization 52(257–259)
Corona, model 52(271)
"Corona" of radio emission around the Galaxy 53(126–128, 251, 256, 257)
Corona polarimeter 54(8)
Corona, polarization 52(255)
Corona, reflection of radio waves 52(460)
Corona, rôle in the development of a centre of activity 52(184–185)

Corona, temperatures 52(269–271)
Corona, thermal radio radiation 52(283–286)
Corona, variation during the solar cycle 52(331, 332)
Corona, x-ray spectrum 52(256, 261–264, 278)
Coronagraph 54(5–6)
Coronal arches 52(181, 272, 273, 281)
Coronal clouds 52(242, 244)
Coronal condensations 52(189, 209, 243, 257, 273, 276–279)
— permanent 52(278)
— sporadic 52(279)
Coronal ellipticity during the solar cycle 52(333)
Coronal rays 52(251, 272, 273, 280–283)
Coronal streamers 52(280–283)
Coronograph 52(249)
Corpuscular radiation from the Sun 52(222–224, 354–362)
Corrections to observed limb darkening curves 52(6–12)
Correlation coefficient in radio source measurements 53(214)
Cosecans law of interstellar absorption 53(449, 469, 471–473)
Cosmic noise absorption 52(350)
Cosmic radiation from the Sun 52(215–221, 223, 322)
Cosmic radio source 51(783)
Cosmic ray pyranometer 54(4)
Cosmic rays originating from supernovae 51(784)
Cosmogony of comets 52(509–518)
Cosmological aspects of radio astronomy 53(272–274)
Cosmological principle 53(418, 425)
Cosmological problems in physics 53(489–490, 521–522)
Cosmology, empirical test 53(509–515)
Cosmology of Hoyle 53(525–530)
Cosmology of Jordan 53(522–525)
Cosmology of Milne 53(530–537)
Counts of clusters of galaxies 53(408–409)
Counts of galaxies 53(420–422)
Coupling between different modes of radial pulsation 51(547, 578, 579, 583)
Coupling frequencies in dwarf cepheids 51(578)
Coupling frequencies in RR Lyrae stars 51(583)
Coupling, limiting of the amplitude by 51(547–548, 588)
Cowling model 51(57–61, 166)
Crab Nebula
— as radio source 52(294); 53(231)
— as remnant of a supernova 51(767, 783)
— magnetic field 51(713)
— Supernova of 1054, magnetic field 53(236)
Crater aureols 54(195–197)
Crater formation by meteors 54(189, 192)
Craters, distribution on the lunar surface (map) 54(190–191)
Craters, new and old 54(193)
Craters on Mercury 54(208, 210)
Craters, photometric curves 54(181)
Craters, topography 54(193–194)
Creation of matter in the Universe 53(487)
Crimean Astrophysical Observatory 54(15, 27, 29, 38)
Critical mass for condensation
— of a galaxy 51(138)
— of protostars 51(145)
Critical mass for vibrational instability 51(483)
Critical mass of white dwarfs 50(184)
Critical velocity of a star 50(384)
Crochets 52(350, 351)
Cross interferometer 53(223, 224)
Cross section of a classification volume 53(288)
Cross sections for hydrogen-burning reactions 51(32–34)
Cross-over effect in the spectra of magnetic 51(702)
Curve(s) of growth 52(61–64)
— abundance determination from 51(333–337, 341)
— for cepheids 51(394)
— for the Sun 50(362)
— information on microturbulence 52(95, 97)
— theoretical 50(360, 361, 375)
Cut-off theorem for an aerial 54(97)
Cut-off wavelength 54(253)
Cyanogen discrepancy 50(99)

Cyclonic and anticyclonic motions around sunspots 52(163)
Cygnus A source 53(212, 233, 234, 265–266, 387)
— apparent angular diameter 53(224)
— distance 53(227, 484, 485)
— polarisation of radiation 53(229)
— spectrum 53(228, 229)
Czerny spectroscope 54(22–24, 38)

D layer of the ionosphere 52(347, 350–354)
D lines, central intensities 52(75)
D of sodium in the solar spectrum 52(53, 54)
DA stars 50(172)
DA stars, opacity 50(182)
DAe stars 50(173)
d'Alembert's principle, generalized 51(641)
Damping and Doppler broadening, combined 52(50)
Damping broadening in the solar spectrum 52(46, 48, 49)
Damping coefficient 51(477, 585, 586)
Damping, collisional 50(343, 344, 375)
Damping constants 50(341, 361, 377); 51(331, 332)
Damping, radiational 50(343)
Damping times (due to molecular viscosity) for radial oscillations 51(507)
Damping times for non-radial oscillations 51(526)
Dark emission 54(250, 269)
Dark hemisphere of Venus 52(391)
Dark intergalactic matter 53(399)
Dark matter in spirals 53(345)
Dark regions on the Martian surface 54(229, 237)
Dark surges 52(244)
Dating of clusters 51(177, 201–204, 213–214, 216, 223, 224)
DB stars 50(170, 171)
DC stars 50(172)
de Jager model of the solar photosphere 50(339)
Decay time(s)
— of a magnetic field 51(714)
— of flares 52(222)
— of solar magnetic field 52(341)
— of spot groups 52(167)
Decline rate of novae (*m*/day) 51(752–753)
Degeneracy in white dwarfs 51(739–742)
Degeneracy, non-relativistic 51(726)
Degeneracy, relativistic 51(726)
Degenerate electron gas 51(724–727)
Degenerate models 51(56, 70–73, 176, 179–181, 739–742)
Degenerate models, isothermal 51(56)
δ Aquilae, spectra 50(246)
δ Aquilae, velocity curves 50(245)
δ Orionis, orbital elements 50(248)
δ Scuti variables 51(284)
Dense matter (white dwarfs) 51(731–732)
— state equation 51(724–728)
— thermal conductivity 51(729–731)
Density and masses of globules 51(149)
Density distribution in globular clusters 53(174–175, 189)
Density fluctuations in the photosphere 52(86–92)
Density of a meteoroid 52(531, 542, 544, 554)
Density of interstallar hydrogen 53(113, 115)
Density of interstallar matter 53(49, 80)
Density of jovian planets 52(445, 446, 447)
Density of meteoritic material in space 52(531)
Density of planets at zero pressure 52(439, 440)
Density of stars above and below the galactic plane 53(45–52)
Density of terrestrial planets 52(437)
Density of the Earth's atmosphere 52(368, 373)
Density of the Earth's interior 52(432, 434)
Density of the Jupiter satellites 54(219)
Density of the low chromosphere 52(118–122, 140)
Density of the medium chromosphere 52(122–124, 140)
Density variations 51(471)
Density variations in white dwarfs 51(737–738)
Dependence method of Schlesinger 50(207, 208)

Depolarisation of light from the Moon 54(183–184)
Detectivity of a photoconductive cell 54(248)
Detector, first *see* Mixer
Detector law 54(48)
Detector, second 54(47, 48, 51)
Detonating fire balls, *see* Bolides
Deuterium in flares 52(208)
DF stars 50(174)
DG stars 50(175)
Diameter of asteroïds 54(211)
Diameter of Mars 54(220–221)
Diameter of Mercury 54(205)
Diameter of the Moon 54(205, 206)
Diameter of the satellites of Jupiter 54(215–216)
Diameters of clusters of galaxies 53(431)
Diameters of galaxies 53(311–315)
Diameters of globular clusters 53(169)
Dicke's method of steady noise suppression 54(57–59)
Differential equations of stellar structure 51(5–15)
Differential galactic rotation 53(24)
Diffraction of light from spirals 53(336)
Diffraction of radio waves at meteoroids 52(536, 538)
Diffuse clouds of dust 51(141)
Diffuse enhanced spectrum of novae 51(757, 758, 761)
Diffuse nebulosities in globular clusters 53(185)
Diffusion in the Earth's atmosphere 52(376)
Dimensions of nearby galaxies 53(314)
Dimensions of quiescent prominences 52(227–228)
Dimming term for galaxies 53(423, 424)
Diode as noise generator 54(54–55)
Diode detector 54(51)
Dipole aerials 54(68)
Dipole arrays 54(69)
Direct isophotes of lenticular galaxies 53(324, 325)
Direct isophotes of spirals 53(329)
Direct-current amplifier following multiplier tube 54(261–262)
Direction of rotation of spirals 53(343–346)
Directional distribution of meteors 52(552–553)
Directivity achievement factor 54(77)
Directivity factor 54(68)
Directivity of an aerial 54(62, 67)
Disappearance of filaments 52(210)
Discontinuities in chemical composition 51(65–66, 175, 183)
Discontinuities in density 51(471)
Discontinuities, propagation 51(556–561, 568)
Discoveries of comets 52(465)
Discrete interval theorem 54(43, 97–98, 115)
Discrete mass ranges in star formation 51(173)
Discrete radio sources 53(208–238); 54(117–120)
— catalogue 53(218)
— origin and nature 53(236–238)
— table of surveys 53(216–217)
Disk distribution of the radio radiation 52(289–293)
Disk temperature, apparent 52(286)
Dispersion of stellar velocities 53(5–6, 29–31, 73–86)
Dispersion orbit 53(63–66, 88–90)
Displacements of Fraunhofer lines 52(91–92)
Disruption of galactic clusters 53(160–163)
Dissipation, negative, in the atmospheres of variable stars 51(506, 585)
Dissipation, viscous 51(475)
Dissipative forces in the atmosphere of variable stars 51(495)
Dissociation constants 50(291)
Dissociation equation 50(290)
Dissociation of molecules in stars 50(109, 133, 290)
Dissociation of nitrogen and oxygen in the Earth's atmosphere 52(370)
Dissolution of a local distribution in the Galaxy 53(63)
Dissolution of galactic clusters 51(211–212)
Distance determination
— of galactic clusters 53(138–140)
— of galaxies from the luminosity 53(446, 470, 475, 483, 484–485)

— of galaxies from the red shift 53(474ff., 483–484)
— of globular clusters 53(172–174)
— of the Pleiades 53(151)
Distance in cosmology, types of 53(446, 470, 475)
Distance, local 53(447)
Distance modulus 51(90); 53(448, 449, 468–474)
Distance modulus, apparent 53(449, 468–474)
Distance scale of the universe 53(474, 481, 484)
Distances of clusters of galaxies 53(411–414)
Distances of eclipsing binaries 50(204, 242)
Distances of galactic clusters 51(98)
Distances of galaxies 53(423, 426, 468–474)
Distances of globular clusters 53(457)
Distances of radio sources 53(227)
Distant companions of stars 50(223)
Distant galaxies, stellar evolution 51(235–236)
Distortion of the velocity curves of binaries 50(251, 252, 256, 262, 263)
Distribution index
— for galaxies in clusters 53(404)
— of ionized hydrogen 53(116–119)
— of mass in the Galactic System 53(52–57)
— of neutral hydrogen 53(107–115)
— of radio radiation at metre wavelengths 53(121–126)
— of radio sources in galactic latitude 53(219, 220)
Distribution of Cepheids 53(87)
Distribution of clusters of galaxies 53(409–410)
Distribution of dark matter in spirals 53(345)
Distribution of galactic clusters 53(132–138)
Distribution of galaxies 53(416–444)
Distribution of globular clusters 53(166–172)
Distribution of novae 51(755)
Distribution of semi-major axes of comets 52(512)
Disturbed corona 52(271–283)
Ditches, radial, in the lunar seas 54(198–199, 201, 203)
DK stars 50(175)
DN Orionis group of eclipsing binaries 50(236, 239)
Doppler and damping broadening combined 52(50)
Doppler broadening in stellar spectra 50(341, 342, 375)
Doppler broadening in stormbursts 52(317)
Doppler broadening of the Fraunhofer lines 52(46, 48, 65)
Doppler effect in planetary nebulae 50(157)
Doppler effect in radio echoes 52(451, 455–456, 459)
Doppler shift, local 52(67, 71–72)
Doppler width of metallic Fraunhofer lines 52(48)
Double cluster in Perseus 53(154–155)
Double galaxies 53(285, 374, 377, 380–382)
Double or multiple periodicity of variable stars 51(384–388, 398–400, 408)
Double star catalogues 50(192)
Double star theories of variable stars 51(359–361, 363–364, 432, 573)
Double stars, optical 50(187)
Double stars, physical, *see* Binaries
Drag coefficient of a meteoroid 52(521, 542)
Drag equation for meteoroids 52(521, 542)
Driving of aerials 54(86–87)
Duration of a meteor stream 52(547)
Dust clouds on the Venus 52(387)
Dust concentration leading to protostar formation 51(141–144, 149)
Dust formation in a galaxy 51(141)
Dust, meteoritic 52(519, 530–534, 559)
Dust on the Moon 54(181, 182, 184, 188, 189)
Dust-tail of a comet 52(467)
Dwarf cepheids 51(125, 126)
Dwarf cepheids, periods and modulation 51(577–579)
Dwarf galaxies 53(280)
Dwarf novae, *see* U Geminovum stars
Dwarfs and giants, distinction 50(32)
Dwarfs of type *M* 50(63)
Dwarfs, white, *see* White dwarfs

Dynamic spectra 52(309)
Dynamical ellipticity of a planet 52(427)
Dynamical friction, effect on velocity dispersion 53(76–77)
Dynamical instability 51(466, 509, 519, 541, 605, 616–620, 684)
Dynamical stability 51(645–647, 659–668)
— and secular stability relation 51(619, 620)
— local 51(647)
— of white dwarfs 51(746)
— sufficient condition 51(618)
— towards non-radial oscillations 51(664–668)
— towards radial oscillations 51(645–647, 660–664)
Dynamical viscosity coefficient 51(436)
Dynamics of multiple galaxies 53(384–385)
Dynamics of star clusters 53(80–84)
Dynamo current 52(351)
Dynamo problem, generalized, in the theory of magnetic stars 51(714–716)

E region of the ionosphere 52(346, 349)
Eagle mounting of a grating 54(20, 38)
Early spectral types of galaxies 53(338–339)
Early type stars, abundance of elements 51(343, 344)
Early type stars, evolution 51(67–69)
Earth, age 52(380)
Earth, core 52(432, 440)
Earth, density and pressure 52(432, 434)
Earth, models of the interior 52(432–435)
Earth, primitive atmosphere 52(380)
Earth's atmosphere 52(363, 366–383)
— chemical composition 52(366–368)
— density, pressure, temperature 52(368, 369, 374)
— escape of hydrogen and helium 52(375–379)
— origin 52(379–382)
East-West asymmetry of spot groups 52(167–170)
Ebert spectrograph with order-sorter 54(25)
Ebert spectroscope 54(24)
Echelle spectroscope 54(26, 38)
Eclipse observations of the disk distribution 52(290, 296)
Eclipses of the Sun 54(33–35)
Eclipsing binaries
— apparent brightness 50(228–232)
— astrometric results 50(220)
— atmospheres 50(234, 238)
— camouflaged 50(235–239)
— definition 50(225)
— distance 50(240, 242)
— distribution on the Galaxy 50(239, 240, 241)
— emission lines 50(235, 237, 239, 264)
— gaseous ring of a component 50(266)
— history 50(225)
— in other galaxies 50(240)
— light curves 50(237–238)
— luminosity-spectrum relation 50(235)
— magnitudes of the components 50(228–232)
— masses of the components 50(228–232, 236–237)
— period of revolution 50(228–232, 236–237)
— primary minimum 50(228–232)
— radii of the components 50(228–232, 236–237)
— relative luminosity of a component 50(228–232)
— secondary minimum 50(228–232)
— spectral type of the components 50(228–232, 236–237)
— spectroscopic study 50(251)
— velocity curves 50(252, 256, 262)
Eddies in the solar nebula, condensation into planets 51(192)
Eddington-Barbier relations 52(20, 21, 31)
Eddington's standard model 51(45, 50)
Edgewise systems 53(284, 287, 298–300)
Effective aerial temperature 54(92, 120, 122)
Effective area of an aerial 53(210); 54(63–64)
Effective gravity 51(597, 600)
Effective level of origin of photospheric lines 52(61)
Effective radius of an elliptical galaxy 53(320)

Effective surface of a subsystem 53(31, 99)
Effective temperature 50(4, 22, 278, 321, 367, 374, 379, 383); 51(82–84)
Effective temperature of sunspots 52(157)
Efficiency of an aerial 54(63)
Einstein *A*-values 51(328)
Einstein tower of Potsdam 54(15)
Einstein's gravitational theory 53(499, 501)
Ejection of matter by stars 51(174, 177, 184, 189, 190, 217–218, 245–249, 275, 565, 666, 667)
— explosive (e.g. novae) 51(247–249, 275, 758, 761, 763)
— steady 51(245–247, 275)
Electric conductivity of the lunar surface 54(187, 188)
Electrical conductivity of the solar convection zone 52(103)
Electromagnetic forces in spiral arms 53(93–96)
Electrometer of Hoffmann 54(243)
Electrometer of Lindemann 54(242, 243)
Electron capture in white dwarfs 51(734, 736, 737)
Electron corona 52(250)
Electron degeneracy in stellar matter 51(17, 724–727)
Electron densities in planetary nebulae 50(152, 154, 412)
Electron densities in the corona 52(265, 266, 271)
Electron pressure in metallic-line stars 50(97)
Electron pressure in stellar atmospheres 50(22, 278, 279, 292, 339)
Electron temperature in the photosphere 52(22, 23, 32)
Electron temperatures in planetary nebulae 50(152, 154, 412)
Electronography 54(284)
Electron-scattering opacity 51(183)
Element abundances in the Sun, early type stars and planetary nebulae 51(343, 344)
Element abundances (table) 51(316)
Element enrichment in the Galaxy 51(274–275)
Element loss from planetary atmospheres 52(363, 364, 375–379, 392, 393, 400–403, 405–406, 414)
Element separation in white dwarfs 51(743–746)
Element synthesis in stars 51(250–263, 268–275, 320–321)
Elementary oxygen in the Earth's atmosphere 52(379)
Elements of the orbits of comets 52(469, 470)
Ellerman bombs 52(206–207)
Ellipsoidal frequency function in a limited region around the Sun 53(31–32)
Ellipsoidal hypothesis of stellar motion 53(5, 21)
Ellipsoidal luminosity law 53(322)
Elliptical galaxies 53(276, 283, 287, 290)
— apparent and true flattening 53(303–305, 322)
— energy distribution 53(341)
— flattening curve 53(322)
— luminosity and colour distribution 53(319–323)
— relation to globular clusters 53(192–193)
— stellar evolution 51(232–234)
Ellipticity of a planet 52(427)
Ellipticity of globular clusters 53(170)
Ellipticity of the corona 52(253, 254)
Emden equation 52(422)
Emission bands 50(120, 122)
Emission bands in supernova spectra 51(781)
Emission coefficient of the photosphere 52(43)
Emission from hydrogenic ions in planetary nebulae 50(399)
Emission heights of Fraunhofer lines 52(124–126)
Emission lines
— from external galaxies 53(318, 339–341)
— in close binaries 50(264, 265)
— in early-type binary systems 50(265, 266)
— in eclipsing binaries 50(235, 237, 239)
— in flare stars 51(430)
— in late-type binary systems 50(265, 266)

Emission lines (cont.)
— in long-period variables 51(409–414)
— in R Coronae Borealis stars 51(423)
— in red semiregular variables 51(418)
— in RR Lyrae spectrum 51(394, 395)
— in RV Tauri stars and yellow semiregular variables 51(416)
— in stars with extended atmospheres 50(388, 389, 391, 392)
— in the W Virginis spectrum 51(395)
— of the chromosphere 52(106, 107, 115, 144–148)
— of the corona 52(256, 259–261)
Emission of the nebular shells 50(408)
Emission spectra of novae 51(757, 759)
Emission spectra of RW Aurigae and T Tauri stars 51(424, 425)
Emissivity 50(282, 350)
Empirical model of the chromosphere 52(139–143)
Empirical solar model of Aller and Pierce 52(58)
Encke's comet 52(476, 546–547, 554)
Encounters between individual stars 53(73–76)
Energy dissipation in the chromosphere and corona 52(150)
Energy distribution
— for ellipticals 53(341)
— for spirals 53(342)
— in stellar spectra 50(322)
— in the solar spectrum 52(1, 13–18)
Energy emitted in nova and supernova outbursts 51(763, 784)
Energy generation rate
— of helium burning processes 51(253)
— of hydrogen burning processes 51(250, 251)
Energy levels of forbidden lines in nebular spectra 50(150)
Energy loss by the Earth's atmosphere 52(368)
Energy method for studies of stability 51(638–645, 647, 657–659)
Energy per unit mass in a system of rotational symmetry 53(25–27)
Energy producing shell of a star 51(176, 179)
Energy production
— constant 51(60)
— functions for carbon cycle and proton-proton chain at different temperatures 51(35, 36)
— in stars 51(25–38)
— in stars and element synthesis 51(250–263, 268–275, 320–321)
— in stars, influence on vibrational stability 51(481–485)
— in stars, phase delay towards pulsation period 51(485–487)
— in supernovae and novae 51(764–765, 784–785)
— in white dwarfs 51(732–738, 748–750)
Energy source of the Sun 51(165)
Energy sources, history of the problem 51(161–163)
Energy transfer within the Earth's atmosphere 52(373–375)
Energy transport by convection 51(6, 7, 10–12)
Energy transport by radiation 51(6, 7–10)
Energy transport, effect on vibrational stability 51(481–485)
Enhanced radio intensity of the Sun 52(286, 308, 315, 316, 320)
Entropy variations and stability 51(643, 644)
Envelope, common, of binary components 50(258, 263, 271)
Envelope, convective, of a star 51(62–64)
Envelope of a galaxy 53(287)
Envelope, radiative, of a star 51(57, 59, 60, 175, 179)
Epicyclic obits in the galactic plane 53(57–61)
e-process (equilibrium) for element synthesis 51(254–255)
Equation of continuity in the theory of stellar deformations 51(434–435)
Equation of mass motion in the gravitational field of the Galaxy 53(24, 25)
Equation of motion in the theory of stellar deformations 51(434–445)
Equation of state for dense matter (white dwarfs) 51(724–728)
Equation of state, polytropic 51(13, 16–17, 45)
Equation of transfer 52(42–43, 46, 52)

Equation of transfer, numerical integration 52(54–58)
Equatorial mounting of aerials 54(82, 88–89)
Equatorial reflector 54(9–11)
Equatorial refractor 54(5)
Equatorial telescopes 54(5–10)
Equatorial tropopause of Mars 52(399)
Equatorial tropopause of the Earth 52(374)
Equilibrium configurations, linear series 51(613–616, 624)
Equilibrium density, distribution 51(42)
Equilibrium, hydrostatic 51(39)
Equilibrium, hydrostatic, in stellar atmospheres 50(317, 332)
Equilibrium of ionization 51(449)
Equilibrium pressure of carbon dioxide 52(365)
Equilibrium, radiative 51(39)
Equilibrium, radiative, in stellar atmospheres 50(313, 317, 329)
Equilibrium states, stability 51(608)
Equilibrium, thermodynamical, in stellar atmospheres 50(135, 286, 287, 317, 332)
Equipartion of energy between stars in clusters 53(84)
Equipartition between magnetic and turbulent energy 51(715)
Equivalent width(s) 52(61, 62); 54(49)
— abundance determination from 51(325, 333–337)
— calculation by the weighting function method 52(56, 58)
— of faint lines 52(62)
— of Fraunhofer lines 52(35)
— of lines 50(30, 276)
Erosion on the Moon 54(192)
Errors in aerials 54(77–82)
Errors in counts of galaxies 53(421, 429)
Eruption hypothesis of the solar cycle 52(325)
Eruptive prominences 52(225, 239–241)
Escape current 52(377)
Escape layer, definition 52(375)
Escape of atmosphere from the Satellites of Jupiter 52(414)
Escape of gases
— from the Earth's atmosphere 52(364, 375–379)
— from the Martian atmosphere 52(400–403)
— from the Moon and Mercury 52(405–406)
— from the Venus atmosphere 52(392–393)
Escape of stars from galactic clusters 53(162)
Euclidean region, local, round the Galaxy 53(447, 473–474)
Euler equation 51(645)
Eulerian notations in the theory of stellar deformations 51(433ff.)
Eulerian perturbations 51(453)
Europe, diameter 54(215, 216)
Evanescent waves 54(65)
Evaporation of matter from the corona 52(282)
Evershed effect 52(162–164)
Evolution along and off the main sequence 51(172–184)
Evolution and secular stability 51(650–657)
Evolution, chemical, of stars 51(249–275)
Evolution of binary stars 50(271, 272)
Evolution of early-type stars 51(67–69)
Evolution of globular clusters 53(190)
Evolution of, historical sketch of theories 51(160–165)
Evolution of massive stars 51(183, 184)
Evolution of quiescent prominences 52(229)
Evolution of solar-neighborhood field stars 51(212)
Evolution of stars in
— associations 51(197–201)
— external galaxies 51(230–236)
— galactic clusters 51(201–210, 221–225); 53(145–146)
— globular clusters 51(70–73, 178–182, 185–187, 213–216, 221–225)
Evolution of the Earth's atmosphere 52(379–382)
Evolution of the Martian atmosphere 52(403–404)
Evolution of the Sun 51(191–195)
Evolution of the Venus atmosphere 52(393)
Evolutionary aspects in stellar dynamics 53(96–99)

Evolutionary models for population II stars 51(178)
Evolutionary scheme for close binaries 51(189)
Evolutionary scheme for novae and related variables 51(188–190)
Evolutionary scheme for white dwarfs 51(190, 191, 750–751)
Evolutionary sequence of galaxies 51(236–238)
Excitation, collisional 50(403–407)
Excitation of cometary spectra 52(485–488)
Excitation of nebular lines 50(147, 403)
Excitation temperature (Boltzmann) 52(47, 74)
Excitation temperature of the chromosphere 52(141)
Excitation temperature of the solar chromosphere 50(112)
Expanding associations 53(70)
Expanding *O* and *B* associations 51(146)
Expansion age of *O* associations 51(199)
Expansion of planetary nebulae 50(156)
Expansion of the universe 53(425–426, 428, 429, 430–432, 440)
Exploding planets 52(423)
Explosive ejection of matter by stars 51(247–249, 275, 758, 761, 763)
Explosive variables 51(249, 419–422, 574–577, 763)
Explosive variables, evolutionary stage 51(188)
Extended atmosphere, *see also* Shells 50(278, 384–397)
Extended atmosphere of early spectral types 50(387)
Extended radio sources 54(121–125)
External galaxies, *see* Elliptical galaxies, Lenticular galaxies, Spirals
— absolute magnitude 53(314, 315, 316–317, 437, 468–474)
— apparent magnitude 53(411, 435–439, 448, 469–475, 479)
— classification 53(275–286)
— containing globular clusters 53(193–194)
— distances 53(423, 426, 468–474)
— luminosities 53(315–317, 319–321, 324, 326–330, 360, 405, 428, 435)
— masses 53(348–360, 407)
— morphology 53(287–310)
— radio frequency radiation 53(239–274, 385–389)
— redshift of spectral lines 53(318, 384, 411–414, 423, 431, 435–439, 440, 446, 474–484, 486, 509)
— sequences 53(279, 283, 294)
— spectra 53(338–341, 362–366, 412)
— stellar evolution in 51(230–238)
— table 53(310)
External gravitational field causing instability 51(480, 509, 607, 680–684)
Extinction 50(281, 282, 350, 352)
Extinction equation 54(149–152)
Extinction in the atmosphere 54(149–154)
Extrafocal errors 54(81)
Extreme halo population, aging effect 51(264–265)

F corona 52(249, 250–256)
F region of the ionosphere 52(346, 349, 350, 453, 454)
F stars classification 50(8, 26, 31, 51)
F stars, peculiar spectra 50(97)
F stars, white dwarfs 50(174)
Fabry-Pérot étalon 54(31)
Faculae, *see also* Chromospheric faculae, photospheric faculae 52(173–190, 287)
Faculae, H_α profiles 52(200)
Faculae, rôle in the development of a centre of activity 52(184–185)
Facular granules 52(175)
Facular region, model 52(176, 177)
Facular spectra 52(179–181)
Feedback in direct-current amplifiers 54(261)
Feeder, *see* Transmission line
Feedhorn at reflector focus 54(73)
Fermi gas 52(419, 435)
Ferraro's law of isorotation 51(721)
Ferrite in waveguide 54(59, 60)
Fibrils, solar 52(179)
Field equations of Einstein's graviational theory 53(501)
Field galaxies 53(420, 422–423, 480, 482)
Field galaxies, apparent magnitude 53(435–439)

Field, transfer of photometric scale to another one 54(168)
Filament connections between neighboring galaxies 53(380–384, 389)
Filaments 52(224–248)
Filaments, activation by flares 52(210)
Filaments, classification 52(225)
Filaments, disappearance 52(210)
Filaments, H_α profiles 52(200)
Filaments, origin 52(229, 230)
Filaments, poleward motions 52(181, 231)
Filaments, rôle in the development of a centre of activity 52(184–185)
Filaments, upward motion 52(240)
Filter photometry for several colors 54(279–280)
Filtergrams 52(134, 140)
Filters for U, B, V photometry 54(161)
Filters of light 54(16–19)
Final stage of a supernova 51(783)
Fine mottling 52(126, 129, 133, 178)
Finite amplitude by coupling 51(547–548, 588)
Finite oscillations, energy method 51(657–659)
Fireballs 52(519, 520, 527, 556–557)
Fission as source of vibrational instability 51(485)
Fission theory of double star formation 51(629)
Fixed-aperture photometer 54(139–141)
Flare effect in geomagnetism 52(350, 351)
Flare indicator 52(352)
Flare spectra 52(199–210)
Flare stars 51(426, 429, 430)
Flare storm 52(356)
Flare surges 52(211–214, 222)
Flare-filaments 52(210, 212, 214)
Flare-like faculae 52(210, 211, 240)
Flare-puff 52(215)
Flares 52(191–224, 278)
Flares and particle emission 52(354–359)
Flares, classification 52(192)
Flares, continuous spectra 52(206–208)
Flares, effects in the Earth's atmosphere 52(350–354)
Flares, H_α profiles 52(200, 202–205)
Flares in connection with radiophenomena 52(308–322)
Flares, influence on filaments and prominences 52(210)
Flares, model 52(205)
Flares, origin 52(221–222)
Flares, relation to faculae and spots 52(194)
Flares, rôle in the development of a centre of activity 52(184–185)
Flash spectrum of the Sun 52(106, 107, 115, 123)
Flash stars 51(429, 430)
Flattening of galaxies 53(303–308, 322, 328)
Flattening of stars 50(249, 382)
Flip-flop 54(265)
Fluctuation of the brightness of comets 52(469)
Fluctuations in the light curves of long-period variables 51(408)
Fluctuations of radiation flux 54(249, 282)
Fluorescence effects in Fraunhofer lines 52(75)
Fluorescence spectra of long-period variables 51(414)
Fluorine, abundance 51(303, 316, 317, 343)
Flux at the surface of the Sun 52(22)
Flux density 54(87, 91, 119)
— of a radio source 53(209, 213)
— of a radio source, limiting value 53(215)
— of the Magellanic Clouds 53(246–248)
— of the Sun 52(286)
Flux of radiation energy 50(280, 281, 313, 357)
Fog-image interaction 54(140)
Folds in the lunar seas 54(204)
Forbidden lines in nebular spectra 50(150, 403–407, 416)
Forbidden lines in nova spectra 51(757, 761, 762)
Forbush type decrease 52(216)
Formation of stars 51(135–157)
Formation of terrestrial planets 52(364)
Fornax A source 53(269)
Fourier transform 54(64–66)

Fractionation of elements by escape from planetary atmospheres 52(363, 364)
Fragmentation process leading to star formation 51(138, 139)
Fraunhofer corona 52(530, 531)
Fraunhofer diffraction region 54(61, 74)
Fraunhofer lines
— calculation of line profiles 52(51, 56, 58–59)
— central intensities 52(75–77)
— central reversals 52(76)
— identification 52(40–42)
— observational 52(36–42)
— widening and displacements in the inhomogeneous model 52(89–92)
Fraunhofer spectrum, use of faint lines for analyse 52(55)
Free neutrons for element synthesis 51(255–260)
Free-free opacity 51(22)
Free-free transitions in an ionized gas 53(102–104, 117, 234–235, 239)
Frequency change receiver 54(46)
Frequency of binaries 50(223)
Frequency of galaxy types 53(277–278, 285–286)
Frequency of supernovae 51(772–775)
Fresnel diffraction at meteoroids 52(536, 538)
Fresnel diffraction region 54(61)
Fringe visibility 54(118)
Frost point of Mars 52(395)
Frozen lines of force 51(535, 537, 677)
Fundamental mode of radial oscillations for the standard model 51(470, 473)
Funnel effect in the H-R diagram of galactic clusters 51(105, 225)
Funnel-type prominences 52(225, 242, 243)
f-values for atomic lines 51(328, 330, 331)

G stars classification 50(9, 31–34, 51–58)
G stars, molecular bands 50(113)
G stars, peculiar spectra 50(97)
G stars, white dwarfs 50(175)
Gain calibration 54(45)
Gain factor, *see* Directivity factor
Gain in an aerial 54(63)
Galactic centre 53(22, 52)
Galactic centre, radio emission 53(119, 121–122)
Galactic clusters 53(129, 132–166)
— and globular clusters, comparison 51(129)
— apparent distribution 53(132–138)
— catalogue 53(194–203)
— classification 53(135–136)
— composite H-R diagram 51(104–105, 202, 225)
— dating 51(177, 201–204, 223)
— disruption 53(160–163)
— dissolution 51(211–212)
— distance determination 53(138–140)
— distances 51(98)
— evolution and age 53(145–146)
— H-R diagrams 51(89–107, 201–213)
— luminosity functions 51(220–224)
— of intermediate age 51(209)
— old 51(209–210)
— spectra and colours 53(144, 146)
— stellar content 53(140–144)
— stellar evolution 51(201–210, 221–225)
— very young 51(205–207); 53(165)
— young 51(207–209)
Galactic corona 53(126–128)
Galactic distribution of planetary nebulae 50(143, 144)
Galactic Novae 53(466–467)
Galactic Novae (table) 51(752–753)
Galactic orbits of globular clusters 53(186–187)
Galactic plane, distribution of stars above and below 53(45–52)
Galactic radio-frequency radiation 53(100–128, 256–259)
— spectrum 53(117)
Galactic rotation 53(6–11, 22, 24, 53)
— from globular clusters 53(186–187)
— from radio observations 53(108, 115–116)
Galactic System
— brightest stars 53(467)
— evolution 53(96–99)
— mass distribution 53(52–57, 116)
— models 53(54, 55)
— radio-frequency radiation 53(100–128, 256–259)
— total mass 53(56)

Galaxies, *see* Eliptical galaxies, Lenticular galaxies, Spirals
Galaxies, absolute magnitude 53(314, 315, 316–317, 437, 468–474)
Galaxies, apparent and true flattening 53(303–305, 322)
Galaxies, apparent distance 53(423, 426)
Galaxies, apparent magnitudes 53(411, 435–439, 448, 469–475, 479)
Galaxies, brightest, table of 53(393–394)
Galaxies, catalogues and surveys 53(390–396)
Galaxies, classification 51(237); 53(275–286)
Galaxies, clustering, *see also* Clusters of galaxies 53(390–414, 417–444)
Galaxies, counts 53(420–422)
Galaxies, distances 53(468–474)
Galaxies, distant, spectra 53(412)
Galaxies, distribution 53(416–444)
Galaxies, distribution in clusters 53(400–405)
Galaxies, interconnected 53(378–384)
Galaxies, limiting apparent magnitude 53(424, 429, 431)
Galaxies, luminosities 53(315–317, 319–321, 324, 326–330, 360)
Galaxies, luminosity function 53(405, 428, 429, 431, 435, 436, 437)
Galaxies, masses 53(348–360, 407)
Galaxies, morphology 53(287–310)
Galaxies, red shift of spectral lines 53(318, 384, 411–414, 423, 431, 435–439, 440, 446, 474–484, 486, 509)
Galaxies, segregation within the clusters 53(399–400)
Galaxies, spectra 53(338–341, 362–366, 412)
Galaxies, stellar evolution 51(230–238)
Galaxies, types in clusters 53(398)
Galaxies, velocity dispersion within the clusters 53(407)
Galaxy collisions 53(225, 233, 265–266, 270, 272, 385–389)
Galaxy, enrichment of elements 51(274–275)
Galaxy pairs 53(285, 374, 377, 380–382)
Gallium, abundance 51(307, 316, 317, 343)
γ-process (rapid) for element synthesis 51(38, 258–260, 272–274)
Gamov factor 51(30, 733)
Gamov's theory of stellar evolution 51(163–165)
Ganymed, diameter 54(215, 216)
Gas discharge as noise generator 54(55, 59, 60)
Gas pressure in stellar atmospheres 50(292)
Gaseous ring of a binary component 50(266, 271)
Gaseous streams in binary systems 50(262, 263)
Gaseous tail of a comet 52(467)
Gaunt factor 51(22, 23)
Gaussian distribution of atomic velocities 52(48)
Gegenschein 52(462, 531)
Gemini A source, apparent angular diameter 53(226, 232)
Geminid meteor stream 52(525, 533, 545–547, 554)
— luminous efficiency 52(554)
Geminid meteoroid densities 52(554)
General magnetic field of the Sun 52(340–344)
Generalized specific heats 51(456)
Genetic relation between bright blue stars and *M*-type supergiants 51(207)
Genetic relation between T Tauri stars and dense nebulae 51(197)
Geomagnetic disturbances, relation to sun-spots 52(190)
Geomagnetic storm 52(354, 355)
Geometric methods for distance determination of galactic clusters 53(138)
Germanium, abundance 51(308, 316, 318, 343)
Giants and dwarfs, distinction 50(32)
Giants in globular clusters 51(112, 177, 180–182)
Giants, luminosity functions 51(224–225)
Giants, mass-luminosity relation 51(170)
Giants, models with outer convective zone 51(500)
Giants of type *M* 50(62)
Giants, variable 51(280–283)
Giants, vibrational stability 51(484)

Giants with anomalous chemical abundances 51(268, 271–272)
Globular cluster stars
— chemical composition 51(113–114, 214–216)
— helium burning processes 51(180, 181, 182)
Globular clusters 53(129, 166–194)
— absolute magnitudes 53(455–457)
— and galactic clusters, comparison 51(129)
— apparent distribution 53(166–172)
— brightest stars 53(457)
— catalogue 53(204–206)
— cepheids 51(120)
— Cepheids of Type II 53(184, 450, 455–457)
— classification 53(171)
— color-magnitude diagrams 51(107–119, 213–216)
— color-magnitude diagrams, zero point of absolute magnitude 51(109, 118)
— colour-magnitude diagrams 53(177–179, 452, 454)
— correction for interstellar absorption 53(173)
— correlation of characteristics 51(113)
— dating 51(177, 213–214, 216, 224)
— difference from solar neighborhood stars 51(213)
— distance determinations 53(172–174, 457)
— evolution and age 53(190–192)
— giant branch 51(112, 177, 180–182)
— horizontal branch 51(181, 185–187, 214)
— in external galaxies 53(193–194, 462, 468, 469, 470, 473)
— in M 31 53(462, 470)
— in M 87 53(473)
— in the Galaxy 53(451–457)
— in the Magellanic Clouds 53(468, 469)
— integrated spectra 51(114)
— luminosity functions 51(221–224)
— main sequence 51(114, 117–118)
— masses and densities 53(188–189)
— number 53(166–167)
— proper motions 53(185–188)
— relation to elliptical galaxies 53(192–193)
— RR Lyrae stars 51(110); 53(183, 450–457, 469)
— stellar content 53(174–185)
— stellar evolution 51(70–73, 178–182, 185–187, 213–216, 221–225)
— subgiant branch 51(180)
— ultraviolet color indices 51(116, 117)
Globules 51(145, 149, 150)
g-modes 51(521, 537, 573, 664, 665, 666)
Gödel cosmos 53(490, 496–497, 516)
Gold, abundance 51(313, 316, 319, 343)
Gradients, absolute 50(276, 325, 380)
Granularity of spectroscopic plates 54(130, 133)
Granulation, solar 52(13, 19, 30, 31, 67, 80–86)
Granules, brightness 52(82, 85)
Granules, centre-limb variation 52(83, 88)
Granules, diameter 52(82, 85, 98, 99)
Granules, life time 52(83)
Granules, radial velocities 52(83–86)
Granules, umbral and penumbral 52(151, 153)
Grating mountings 54(20–21, 26, 35, 38)
Gravitational clustering leading to star formation 51(136)
Gravitational contraction as a source of stellar energy 51(25–26, 98, 144)
Gravitational contraction of protostars 51(157–160)
Gravitational instabiity of a gas cloud under external pressure 51(146, 147)
Gravitational interaction with a companion, effect on stability 51(480, 509, 607, 680–684)
Gravitational lenses 53(384–385)
Gravitational pressure in spiral arms 53(94)
Gravitational red shift 52(78)
Gravitational stability 51(608)
Gravitational theory, foundations of Newtonian 53(491–494)
Gravitational theory of Einstein 53(499, 501)
Gravitational time-scale 51(607)
Gravity at the surface 50(3, 4, 374, 379, 383)
Gravity darkening 50(384)
Gravity, effective 51(597, 600)
Gray atmosphere 50(285, 300)
Grid current 54(243)

Groups of sunspots, *see* Spot groups
Guillotine factor 51(24)
Gurney-Mott two-stage photochemical mechanism 54(136, 137)
Gyro frequency 52(286, 314)

h and χ Persei, H-R diagram 51(97, 207, 208)
H II regions in planetary nebulae, radii 50(421)
H lines of CaII in the solar spectrum 52(76)
Hafnium, abundance 51(312, 316, 319, 343)
Hale Solar Laboratory, Pasadena 54(15, 32)
Half-Filter 54(164, 165)
Halley's comet 52(466, 473)
Halm's catenary 54(135)
H_α of surges 52(212–214)
H_α photoheliograph 54(18–19)
H_α spectra of flares 52(199–205)
H_α spectroheliograms 52(131–135)
Hamilton's principle 51(438)
Hard component of cosmic radiation 52(215)
Hard self-excited oscillations 51(550, 587, 607)
Harkins' rule 51(297, 323)
Harmonic analysis of spectrum variables 51(705–707)
Harvard program of meteor photography 52(539, 541)
Haze on Mercury 52(404)
He^3 accumulation leading to vibrational instability 51(575)
Head of a comet 52(465, 497, 498)
Heat balance in sunspots 52(165)
Heating of the chromosphere 52(149–151)
Heating of the corona 52(149–151, 281)
Heights of meteors 52(535)
HeIII-HII stratification in planetary nebulae 50(421)
Heiligenschein 52(531)
Helical aerials 54(70)
Heliostat 54(11)
Helium, abundance 51(302, 303, 316, 317, 343)
Helium abundance in white dwarfs 51(736)
Helium burning processes
— and element synthesis 51(253–254, 268–271)
— energy generation rate 51(253)
— in globular cluster stars 51(180, 181, 182)
Helium escape from the Earth's atmosphere 52(375–379)
Helium ionization in the non-adiabatic region of variable stars 51(505, 506, 585, 660)
Helium levels, population 52(109–112)
Helium lines of the medium chromosphere 52(123, 129)
Helium, physical properties under high pressure 52(441–445)
Helium reactions in white dwarfs 51(735)
Helium, solar models with 52(28)
Helium, solid 52(441, 443, 444)
Helmholtz-Kelvin time scale 51(26)
Henry Draper 50(7, 59)
Herbig-Haro objects 51(150, 151, 425)
Hercules A source 53(269)
Hercules cluster of galaxies 53(478, 484, 485)
Herring-bone structures of the solar radio-radiation 52(305)
Hertzsprung gap 51(106)
Hertzsprung relationship 51(369, 370)
Hertzsprung-Russell diagram
— definition and terminology of sequences and regions 51(75, 196)
— distribution of variable stars 51(121, 125, 570–572)
— for heavy stars 51(184)
— for solar-neighborhood field stars 51(212–213)
— for the Pleiades 53(151)
— forbidden regions 51(106)
— of galactic clusters 51(89–107, 201–213)
— of globular clusters 51(107–119, 178, 213–216)
— of 96 Hyades stars 51(96)
— of nearby stars ($\pi > 0\farcs050$) 51(93)
Heterogeneous incompressible sphere, oscillations 51(509, 516)
Hierarchy of clustering 53(416, 443–444)
High energy electrons in magnetic fields 53(102, 104–106, 126–127, 239, 260)

High modes of non-radial oscillations (*p*, *g*) 51(521, 537, 573, 664, 665, 666)
High photosphere 52(31)
High velocity stars 50(90, 97–100); 53(6, 23, 97)
— abundance anomalies 51(346)
— chemical abundances 50(100)
— color magnitude diagram 51(130)
— population type 51(130–131)
— spectrum 50(98)
Higher order clustering 53(416, 433, 443–444)
H II regions as radio sources 53(232)
H II regions in the Galaxy 53(86, 116–119, 258, 467, 470)
Hoffmann duant electrometer 54(243)
Holtsmark theory 52(49, 50)
Homodyne 54(267–268)
Homogeneity-postulate, *see* World postulate
Homogeneous compressible sphere, oscillations 51(514–516, 527, 541)
Homogeneous incompressible sphere, oscillations 51(509, 535)
Homogeneous magnitude scale 54(167–168)
Homogeneous models 51(173, 174, 177)
Homologous stars 51(45–47)
Homology invariant functions 51(55, 56, 57, 58)
Horizontal branch in globular cluster diagrams 51(181, 185–187, 214)
Horizontal branch variables 51(283)
Horizontal telescopes 54(13–14)
Horns for electromagnetic waves 54(70–71)
Hot regions in the corona 52(263)
Hot subdwarfs, evolutionary stage 51(187–188)
Hot subdwarfs, variable 51(284)
Hoyle's cosmology 53(525–530)
Hoyle-Schwarzschild models 51(178–182)
H-R diagram, *see* Hertzsprung-Russell diagram
Hubble acceleration parameter 53(446, 480, 481, 483)
Hubble constant 53(436, 446, 479–480, 482, 486, 514)
— definition 53(480)
— in customary units 53(480)
— values 53(482–483)
Hubble parameter, *see* Hubble constant
Hugoniot relation 51(560)
Hugoniot-Rankine conditions 51(557)
Hyades 53(157–159)
Hyades, color indices and magnitude of each star (table) 51(92, 94)
Hyades, color-magnitude diagram 51(95, 101)
Hyades subdwarfs 51(106, 107)
Hydra cluster of galaxies 53(478, 481, 484, 485)
Hydra A source 53(269)
Hydra A source, irregularities 53(230)
Hydrogen, 21 cm line 53(10, 52, 86, 102, 106, 107, 109)
Hydrogen, abundance 51(302, 303, 316, 317, 343)
Hydrogen abundance in Jupiter and Saturn 52(447)
Hydrogen burning processes
— and element synthesis 51(250–253, 268–271)
— cross sections 51(32–34)
— energy generation rate 51(250, 251)
Hydrogen, 21 cm line 52(144)
Hydrogen, continuous absorption 50(295, 296, 328, 329)
Hydrogen deficient stars 51(347, 348)
Hydrogen distribution in the Galaxy 53(107–115, 467, 470)
Hydrogen escape from the Earth's atmosphere 52(364, 375–379, 380)
Hydrogen escape from the Venus atmosphere 52(392)
Hydrogen in galactic clusters 53(144)
Hydrogen ionization in the non-adiabatic region of variable stars 51(499–501, 585, 590, 660)
Hydrogen ionization in the solar atmosphere 52(29)
Hydrogen levels b_n coefficients 50(411)
Hydrogen levels, population 52(109–112)
Hydrogen line profiles, photospheric 52(31, 86, 87)
Hydrogen lines, *see* Balmer lines
— from clusters of galaxies 53(265)
— from the Magellanic Clouds 53(240–245)
— in stellar spectra 50(366–368)

— in the solar spectrum 50(342, 365)
— in the upper chromosphere 52(131–135)
Hydrogen nebula 50(416)
Hydrogen, negative ions 50(296, 328, 329)
Hydrogen, physical properties under high pressure 52(441–445)
Hydrogen, solid 52(441, 443, 444)
Hydrogen spectrum of planetary nebulae 50(409–412)
Hydrogen-helium ratio in stellar atmospheres 51(344–345)
Hydrogen-poor stars 50(103)
Hydromagnetic waves in a star 51(535, 719–720)
Hydromagnetic waves in prominences 52(248)
Hydrostatic equilibrium 50(332); 51(6, 39)
Hydrostatic equilibrium, application to cold bodies 52(420, 429)
Hyperbolic meteor orbits 52(550)
Hypersensitizing of photographic emulsions 54(135)

Ice on the Mars 54(224, 228, 235)
Icy conglomerate model of cometary nuclei 52(494)
Identification of clusters of galaxies 53(441)
Identification of Fraunhofer lines 52(40–42)
Identification of radio sources 53(231–238)
Image converters 54(7, 283–288)
Image frequency 54(47)
Image rejection 54(47)
Image scales 54(138)
Image-forming factors 54(144–146)
Impact regions for cosmic rays 52(217, 219)
Inclination of the axis of spot groups 52(169)
Incompressible sphere, oscillations 51(509, 516, 535)
Indicial equation 51(460)
Indium, abundance 51(310, 316, 318, 343)
Inertial system, local 53(491, 500)
Influx of meteors in the atmosphere 52(548)
Infrared lines from Mars 54(225, 230–231)
Infrared lines from the Sun 54(7)
Infrared radiation of the Sun 52(17)
Inhibition of convection in a magnetic field 52(172, 173)
Inhomogeneities in chemical composition 51(65, 66, 175, 183)
Inhomogeneities in the photosphere 52(66–71, 86–92)
Inhomogeneous models 51(65–66, 70, 175–178, 183–184, 218–220)
— of the photosphere 52(31, 78, 86, 87–92, 105)
— of the photosphere, comparison of various models 52(92)
Inhomogeneous world models 53(498, 517)
Initial luminosity function for main sequence stars 51(218, 219)
Initial mass function for main sequence stars 51(217, 218, 219)
Instability of planets 52(423)
Integrated brightness contours of the Magellanic Clouds 53(241, 242)
Integrated colours of globular clusters 53(168)
Integrated magnitudes of galactic clusters 53(136–137)
Integrated spectra of globular clusters 53(179–181)
Integrator 54(51)
Intense radio source in Cassiopeia 53(209)
Intensities of the Fraunhofer lines 52(75–77)
Intensity measurements of the solar continuum 52(16, 17)
Intensity profile of sunspots 52(157)
Intensity variations in peculiar *A* stars 51(703, 704)
Intensity-chromatic factors of star image formation 54(144)
Interconnected galaxies 53(378–384)
Interfaces between regions of different chemical composition 51(15, 65–66)
Interference filters 54(17)
Interferometer for radio waves 54(43, 98–117, 122)
Interferometer observation of discrete radio sources 54(117)

Interferometer observations of the Sun 52(290, 291, 297)
Interferometer of Christiansen 54(113–115)
Interferometer of Payne-Scott and Little 54(106–108)
Interferometer of Ryle 54(108)
Interferometer with grating spectrograph 54(31)
Interferometers for radio observations 53(213, 223, 224)
Intergalactic extinction 53(423, 424, 432, 475–476)
Intergalactic globular clusters 53(173)
Intergalactic magnetic fields 53(265)
Intergalactic matter 53(398–399, 432)
Intergalactic space, population 53(389)
Intergalactic star clusters, stellar evolution 51(234–235)
Interior of Jupiter 52(446)
Interior of Mars 52(439)
Interior of Mercury 52(438)
Interior of Neptune 52(448)
Interior of Saturn 52(446)
Interior of the Earth 52(434)
Interior of the Moon 54(199, 200)
Interior of Uranus 52(448)
Interior of Venus 52(439)
Interlocking of clusters of galaxies 53(433–435)
Intermediate frequency amplifier 54(46)
Intermediate zone as seat of vibrational instability 51(585)
Intermediate zone between convective core and radiative envelope 51(68–69, 183, 503)
Internal circulation of the Sun 52(342)
Internal motion of sunspots in spot groups 52(170)
Internal structure of clusters of galaxies 53(397–406, 428, 429, 431, 436, 440)
Internal structure of white dwarfs 51(739–750)
International system of photometric standard 54(157)
Interstellar absorption 50(83); 51(86–89); 53(449, 460, 464, 468–469, 471, 473–474, 479)
Interstellar hydrogen
— 21 cm line 53(102, 106, 107, 109)
— absorption of radio radiation 53(227)
— density 53(113, 115)
— distribution 53(107–115)
Interstellar matter, density 53(49, 80)
Interstellar origin of comets 52(510)
Interstellar origin of meteors 52(519, 538, 554)
Interstellar polarization 53(94)
Interstellar reddening 51(86–89, 391)
Invisible distributions for an aerial 54(95, 124)
Io, diameter 54(215, 216)
Io, yellow color 54(218)
Iodine, abundance 51(311, 316, 318, 343)
Ion transitions, forbidden, in nebular spectra 50(146–150, 403–407)
Ionization energy of the isoelectronic sequences 52(258)
Ionization equilibrium 51(449)
Ionization formula by Woolley and Allen 52(145)
Ionization in planetary nebulae 50(146)
Ionization in the corona 52(257–259)
Ionization of helium in the non-adiabatic region of variable stars 51(505, 506, 585, 660)
Ionization of hydrogen in the non-adiabatic region of variable stars 51(499–501, 585, 590, 660)
Ionization of hydrogen in the solar atmosphere 52(29)
Ionization of the air by meteoroids 52(522, 526)
Ionization potentials of atoms in stellar atmospheres 50(22, 290)
Ionization spectra of novae 51(756, 757, 761)
Ionization temperature of the chromosphere 52(141)
Ionized hydrogen clouds in the Galaxy 53(86, 116–119, 258, 467, 470)
Ionized hydrogen regions as radio sources 53(232)
Ionosphere, absorption cross section 52(347)
Ionosphere during solar eclipses 52(348–350)
Ionosphere, F_2 region 52(453, 454)
Ionospheric disturbances 52(350–362); 53(230)

Ionospheric fade outs 52(351)
Ionospheric layers of the Earth 52(344–348, 453, 454)
Iridium, abundance 51(313, 316, 319, 343)
Iris, polarisation of light 54(214)
Iris-diaphragm photometer 54(141–143)
Iron, abundance 51(305, 316, 317, 343)
Iron core of the Earth 52(432, 440)
Iron oxides on Mars 54(226)
Iron peak in abundance distribution 51(323)
Irregular galaxies 53(277, 280, 283, 284, 287, 296–297, 302)
Irregular variables 51(282, 283, 355, 403, 417–419); 53(472–474)
Isoelectronic sequences, ionization energy 52(258)
Isophotal surfaces in eliptical galaxies 53(323)
Isophotes of lenticular galaxies 53(324, 325)
Isophotes of spirals 53(329)
Isorotation 52(342)
Isorotation, Ferraro's law 51(721)
Isothermal core, limiting mass 51(66, 67, 176)
Isothermal core models 51(66, 67, 70–73)
Isothermal layer in the atmosphere of variable stars 51(497–499)
Isothermal models 51(55–57)
Isothermal shock 51(559, 561, 568, 590)
Isotope abundances 51(345–346)
Isotopes, relative abundance in stars 50(120, 132)
Isotopic effects in molecular spectra 50(120)
Isotropic expansion of the substratum 53(497, 503–504, 506–509)
Iteration method (non-gray atmosphere) 50(314)

Jacobi ellipsoids, stability 51(622, 623, 624, 627, 681)
Jacobi equation 51(646)
Jeans' criterion for gravitational instability 51(135)
Jeffrey's inequality 52(430)
Jet 52(272)
Jordan's cosmology 53(522–525)
Jovian planets 52(431, 440–448)
Jovian planets, models 52(445–448)
Jovian planets, pressure and density 52(446, 447)
Juno, diameter 54(211)
Jupiter atmosphere 52(406–410)
Jupiter, effect on meteors 52(529, 530, 548)
Jupiter, interior 52(445–448)
Jupiter, satellites 54(215–220)

K corona 52(249, 250–256, 265)
K criterion of Whipple 52(555–556)
K line of Ca^+ in the solar spectrum 52(53)
K lines of Ca II in the solar spectrum 52(76)
K, molecular bands 50(113)
K, peculiar spectra 50(97)
K stars 50(9, 31–34, 58)
K, white dwarfs 50(175)
Kanzelhöhe (Graz) 54(38)
$\varkappa$ Crucis cluster 53(157)
Kapteyn's typical stellar system 53(22)
K-correction 53(475–476)
Kelvin contraction 53(98)
Kelvin-Helmholtz contraction 51(158)
Kepler problem for binaries 50(195)
Kepler's Nova as radio source 53(231)
Kepler's Nova of 1604 51(768)
Kiloparsec, definition 53(448)
Kirchhoff law 50(287); 52(44, 284)
Kitt Peak Observatory 54(36, 37)
Knot prominences 52(225, 242, 243)
Kostinsky effect 54(133)
Kourganoff approximation 50(307)
Kramers' opacity 51(24, 165)
Kron's law of reciprocity failures 54(134)
Krypton, abundance 51(307, 308, 316, 318, 343)
K-term of Campbell 53(3, 5)
Kukarkin-Parenago relation 51(420, 576)

L corona 52(249, 256–259, 273)
Ladenburg *f*-value 51(328, 330)
Lagrangian density 51(438)
Lagrangian perturbations 51(453)
Lagrangian point 50(263)
Lagrangian representation for deformation of stars 51(432ff.)

Lake Angelus Observatory 54(38)
Lambda term 53(494, 501, 526)
Landé g-factor 52(154)
Lane-Emden equation 51(51, 52)
Langley's method of determining the solar constant 52(2)
LaO bands in *S* stars 50(116)
Laplace law 52(422)
Large Magellanic Cloud
— brightest stars 53(470)
— distance and absolute magnitude 53(469–470)
— mass 53(243)
— period-luminosity curve 53(458, 461)
Late type stars, *see also* *M*, *C* and *S* stars 50(61)
— classification 50(129)
— continuous opacity 50(123–126)
Latent image 54(137)
Late-type giants with magnetic field 51(712)
Late-type stars of abnormal composition 51(347–349)
Latitude drift of sunspots 52(339)
Latitude effect of cosmic radiation 52(216, 220)
Lead, abundance 51(314–315, 316, 320, 343)
Least detectable signal 54(56)
Legendre condition 51(647)
Lejeune-Dirichlet theorem 51(608, 610)
Lennard-Jones potential 52(441)
Lens of a galaxy 53(287)
Lenticular galaxies 53(283, 287, 290, 292, 299)
— apparent and true flattening 53(305–306)
— luminosity and colour distribution 53(324–326)
Lenz' law 52(221)
Leonid meteor shower 52(534)
Level effect in velocity curves 51(396, 397)
Libration of the Moon 52(454, 455)
Life on Mars 52(403–404); 54(231, 236, 237, 239)
Life on Venus 52(393)
Life time of a cluster 53(83)
Life time of chromospheric faculae 52(178, 179)
Life time of facular granules 52(175)
Life time of granules 52(83)
Life time of the magnetic fields of spot 52(171)
Life time of white dwarfs 51(749, 750)
Life times of elements in white dwarfs 51(735)
Light and velocity curves of variable stars, amplitude correlation 51(588–589)
Light and velocity curves of variable stars, phase correlation 51(374–376, 399, 400, 505, 589–592)
Light curves of
— β Cephei stars 51(398–400)
— binaries 50(237–238)
— classical cepheids 51(369, 370)
— flares 52(198, 199)
— long-period variables 51(405–410)
— magnetic variables 51(428)
— meteors 52(523–524)
— novae 51(755, 756)
— peculiar *A* stars 51(696)
— R Coronae Borealis 51(422)
— RR Lyrae stars 51(367, 368)
— RR Tauri 51(425)
— RV Tauri stars and yellow semiregular variables 51(415–417)
— SS Cygni 51(420)
— supernovae 51(776–778)
— T Orionis 51(425)
— variable stars, asymmetry in magnitude and shape 51(371–374)
— variable stars, double or multiple periodicity 51(384–388, 399, 400)
— variable stars, progressive phase shift for different colors 51(594)
— variable stars, relation between shape and period 51(366–370)
— Z Camelopardalis 51(422)
Light filters 54(16–19)
Light pressure in comet tails 52(506)
Light, velocity of 53(480)
Limb darkening 50(280, 302, 326)
— correction to observed curves 52(6–12)
— effect on velocity curve 51(598, 599)
— observations 52(1, 5–6, 12–13, 18–22)
Limb flares 52(193)
Limb observations of the chromosphere 52(108)
Limb of the Sun 52(117–118)
Limit circle for pulsation 51(549, 550, 658)

Limitation of the amplitude by coupling 51(547–548, 588)
Limiting apparent magnitude of galaxies 53(424, 429, 431)
Limiting magnitude of meteors 52(528)
Limiting mass of the isothermal core 51(66, 67, 176)
Lindemann electrometer 54(242, 243)
Line absorption coefficients in the photosphere 52(47, 50, 51)
Line broadening
— by collision 50(343, 344)
— by rotation 50(252, 381, 384)
— Doppler effect 50(342, 375)
— in flare spectra 52(203, 204)
— mechanism in the solar atmosphere, *see also* Collision, Doppler, Pressure, Statistical broadening, Radiation damping 52(48–51)
— Stark effect 50(344, 346)
— theories 51(331, 332)
— variable, in β Cephei spectra 51(401, 579–581)
Line depth in white dwarf spectra 50(179)
Line doubling in β Cephei spectra 51(401, 580–581)
Line doubling in RR Lyrae spectra 51(394, 395)
Line doubling in the W Virginis spectrum 51(395)
Line emission coefficient of the photosphere 52(44)
Line emission in radio astronomy 53(102, 107–109, 239)
Line intensification in sunspot spectra 52(159, 160)
Line intensities
— anormal distribution in shell spectra 50(395, 396)
— central 50(346, 358)
— curve of growth 50(360–362, 375)
— in planetary nebulae 50(147, 151, 412)
— total 50(42, 359)
— weak lines 50(354)
Line profiles 50(277, 354–358, 361, 375)
— according to the inhomogeneous model of the photosphere 52(86–89)
— calculation by the weighting function method 52(51, 56, 58–59)
— for spectroscopic binaries 50(252, 253, 254, 262)
— from the Magellanic Clouds 53(241)
— in the spectra of peculiar *A* stars 51(697–702)
— information on turbulence 52(95, 97)
— of facular spectra 52(180)
— of spicules 52(114)
Line spectra, *see* Absorption lines, Emission lines
Line spectra, effect of pulsations 51(598–601)
Line spectra of cepheids and RR Lyrae stars 51(391–397)
Line spectrum
— of quiescent prominences 52(232–235)
— of sunspots 52(159)
— of the chromosphere 52(106, 107, 115, 144–148)
— of the corona 52(256, 259–261)
— of the photosphere 52(31, 33 ff.)
Line splitting in planetary nebulae 50(156, 158)
Line strengths, empirical values 51(330, 331)
Line strengths, theoretical calculations 51(328, 330)
Line width, equivalent 50(30, 276)
Line width, natural 50(340)
Line widths in the spectra of *A* stars 51(698)
Linear detector 54(48)
Linear series of equilibrium configurations 51(613–616, 624)
Linear Stark effect 52(49)
Liouville's theorem 53(27–28)
Liquid helium, lowering of noise 54(53)
Liquid rotating masses, stability 51(611, 620–629)
Lithium, abundance 51(303, 316, 317, 343)
Lithium in *M*, *S*, and *C* stars 50(132)
Local distance 53(447)
Local Doppler shift 52(67, 71–72)
Local Doppler shift, asymmetries 52(71–72)
Local Group 53(397)
— dimensions 53(314)
— distances and absolute magnitudes 53(472)
— stellar evolution 51(230–235)

Local inertial system 53(491, 500)
Local instability in external layers of stars 51(605, 664, 669, 677, 684, 685)
Local oscillator 54(46)
Local solar motion 53(16)
Local standard of rest in the Galaxy 53(11, 15)
Local Supergalaxy 53(397)
Local Supergalaxy, radio emission 53(263–265)
Local thermodynamic equilibrium 52(44, 108)
— deviations from 52(73–75)
Logarithmic spiral 53(289)
Lohmann-Bottlinger method for mass determination of galaxies 53(357)
Long period comets 52(471, 511–516)
Long scale fluctuations of the solar activity 52(328–329)
Long-focus photographic astrometry 50(190, 204)
Long-period fading in echoes from the Moon 52(457, 458)
Long-period variables 51(123–125, 282, 283, 355, 356)
— emission lines 51(409–414)
— light curves 51(405–410)
— molecular bands 51(410–414)
— periods 51(402–405, 408, 584)
— population type 51(405)
— spectral type 51(402–405, 410–414)
— velocity curves 51(409–410)
Loop prominences 52(225, 241, 242)
Loss of elements from planetary atmospheres 52(363, 364, 375–379, 392–393, 400–403, 405–406, 414)
Low chromospheres, density 52(118–122)
Low chromospheres, mass motion 52(96)
Low chromospheres, spectroheliograms 52(126–129)
Lumen (unit) 54(244)
Luminosities of bright galaxies 53(360)
Luminosities of external galaxies 53(315–317)
Luminosities of novae 51(752, 753, 754, 755)
Luminosities of solar-neighborhood stars 51(165–172)
Luminosity 50(24, 42, 43, 46, 51, 58)
Luminosity classes 50(36, 40, 378)
Luminosity classes, definition 53(3)
Luminosity criteria 51(84–86)
Luminosity determinations 51(4)
Luminosity distance 53(446, 470, 475, 483, 484–485)
Luminosity distribution in elliptical galaxies 53(319–321)
Luminosity distribution in lenticular galaxies 53(324)
Luminosity distribution in spirals 53(326–330)
Luminosity equation of a star in equilibrium 51(6)
Luminosity function(s)
— definition 51(216)
— for *K* giants 51(224–225)
— of cluster galaxies 53(405, 428, 429, 431, 435, 436, 437); 51(221–224)
— of galactic clusters 51(220–224)
— of globular clusters 53(175–177)
— of solar neighborhood field stars 51(216–220)
— of supernovae 51(776)
Luminosity gradient 51(14)
Luminosity law for elliptical galaxies 53(320–322)
Luminosity maximum of protostars 51(159, 160)
Luminosity of peculiar stars 50(95, 96, 102)
Luminosity of spectroscopic binaries 50(256)
Luminosity of white dwarfs 50(165)
Luminosity profiles of spirals 53(327, 328, 331)
Luminosity, relative, on an eclipsing component 50(228–232)
Luminosity-colour asymmetry of spirals 53(330, 333–335)
Luminosity-mass-temperature relation for white dwarfs 51(743)
Luminosity-spectrum relation for eclipsing binaries 50(235)
Luminous efficiency of a meteor 52(522, 542, 554)
Luminous flux from a star 54(244)
Lunar surface
— composition 54(182, 184, 185, 188)
— photometry 54(180–182)
— properties 54(180–205)
— ruggedness 54(181, 184, 187, 188, 190)

Ly_α line of the chromosphere 52(146, 147)
Lyman discontinuity 50(324)
Lyot's coronagraph 54(5–6)

M 3 evolutionary track 51(222, 223)
M 67, evolutionary track 51(222, 223)
M 3, H-R diagram 51(108, 111, 115, 116, 129)
M 11, H-R diagram 51(103)
M 13, H-R diagram 51(111)
M 25, H-R diagram 51(100)
M 41, H-R diagram 51(103)
M 67, H-R diagram 51(104, 129, 209)
M 92, H-R diagram 51(111, 115, 129)
M 31, stellar evolution 51(230–231)
M 33, stellar evolution 51(231)
M regions of the Sun 52(190, 223)
M regions of the Sun and particle emission 52(359–362)
M stars, classification 50(9, 31–34, 58, 61, 62, 63, 72)
M stars, molecular bands 50(114, 115)
M stars of high dispersion 50(115)
M storms 52(359–362)
Mach number 51(561)
Maclaurin spheroids, stability 51(622, 623, 624, 627)
Macroturbulence 50(376)
Macroturbulence from line profiles 52(95, 97)
Macroturbulence in the atmosphere of cepheids 51(394)
Macroturbulence in the photosphere 52(93, 98)
Macroturbulent velocities 52(71)
Magellanic Clouds
— comparison of radio and optical data 53(248–250)
— distances and absolute magnitudes 53(468–470)
— period-luminosity curve 53(458, 461, 463)
— radio frequency radiation 53(240–250, 259)
— rotational curves 53(244)
— stellar evolution 51(231–232)
— total masses 53(243)
Magellanic irregulars 53(283, 284, 287, 299)
Magic numbers and nucelar abundances, relation 51(322)
Magnesium, abundance 51(304, 316, 317, 343)
Magnetic axes of the Sun 52(341)
Magnetic braking in protostars 51(153)
Magnetic classification of spot groups 52(166–167)
Magnetic effects in the spectra of peculiar *A* stars 51(699–700)
Magnetic energy and turbulent energy, equipartition 51(715)
Magnetic energy of sunspots 52(343)
Magnetic evaporation of ions and electrons in sunspots 52(173)
Magnetic field
— decay time 51(714)
— effect on non-radial oscillations 51(532–538)
— effect on stability 51(480, 509, 524, 607, 666, 675–680)
— influence on protostar formation 51(154–157)
— inhibiting convection in sunspots 52(172, 173)
— of faculae 52(181–182)
— of Sun, general 52(340–344)
— of sunspots 52(152–157, 164, 171)
— of the Crab Nebula 53(236)
— of the solar convection elements 52(103–104)
— and stellar evolution 51(284–286)
Magnetic flux of sunspots 52(167)
Magnetic foreshortening in sunspots 52(154)
Magnetic intensification of lines 51(699, 700)
Magnetic mapping of the Sun 52(181)
Magnetic polarity, definition 52(340)
Magnetic regions of the Sun 52(181, 185, 190, 248)
Magnetic solar fields, observations 54(32–33)
Magnetic stabilization of laminar flow 51(722)
Magnetic stars, *see* Peculiar *A* stars, Spectrum variables
— abundance anomalies 51(349–350, 707–709)
— observations 51(690ff.)
— theory 51(714ff.)
Magnetic storms 52(190, 222, 223, 354, 355)

Magnetic variables
— periodic 51(426–429)
— periodic, light curves 51(428)
— periodic, non-radial oscillations 51(537, 573)
— periodic, period 51(535, 536, 573)
— periodic, table of data 51(695)
— periodic, velocity curves 51(428)
Magnetic viscosity 51(722)
Magnetoheliograph 52(154)
Magnetohydrodynamic shock waves in the corona 52(306)
Magnetohydrodynamic wave in magnetic variables 51(535)
Magnetohydrodynamic waves in spiral arms 53(93)
Magnetohydrodynamical steady states of stars 51(720–722)
Magnetohydrostatic equilibrium of stars 51(716–720)
— stability 51(717–718)
Magneto-ionic effect in the ionosphere 52(457)
Magnitude, absolute 50(24, 30, 34, 42, 60, 378)
— of classical Cepheids 53(462)
— of external galaxies 53(314, 315, 316–317, 437)
— of galactic Novae 53(466)
— of nearby galaxies 53(468–474)
— of planetary nebulae 50(141, 142)
— of RR Lyrae stars 53(453, 470)
Magnitude, apparent 51(81, 82)
— of galaxies 53(411, 435–439, 448, 469–475, 479)
— of planetary nebulae 50(140, 141)
Magnitude, average, of a star cluster 53(450)
Magnitude, bolometric 51(84)
Magnitude compensation for visual binaries 50(190, 191)
Magnitude, mean, of variable stars 53(449)
Magnitude, median, of a star cluster 53(450)
Magnitude, observed apparent 53(449, 460, 480)
Magnitude of globular clusters 53(167–168)
Magnitude scale, establishment by photography 54(163ff.)
Magnitude scale, zero-point 54(244)
Magnitudes of the components of eclipsing binaries 50(228–232)
Main beam 54(74)
Main sequence for age zero 51(95, 96, 98)
Main sequence stars
— in galactic clusters 53(140)
— mass-luminosity relation 51(167–170)
Major planets, atmosphere 52(406–414)
Maksutov telescope 54(10)
Manganese, abundance 51(305, 316, 317, 343)
Manned balloons 54(37)
Mars, effect on meteors 52(529, 548)
Mars, maps 54(222–223)
Mars, model of interior 52(439)
Mars, radius 52(399)
Mars, satellites 54(221)
Mars, surface properties 54(220–237)
Martian atmosphere 52(393–404)
— chemical composition 52(394–395)
— escape of gases 52(400–403)
— origin 52(403–404)
— polarization 52(394, 397)
— structure 52(398–400)
— temperatures 52(395–396)
Maser 54(53)
Mass accretion by stars 51(238–241)
Mass balance of the Sun 52(282)
Mass, critical, for vibrational instability 51(483)
Mass density above and below the galactic plane 53(45–52)
Mass determination of stars 51(4)
Mass distribution function in star formation 51(148)
Mass distribution in the Galactic System 53(52–57, 116)
Mass equation for meteoroids 52(521–522)
Mass equation of a star in equilibrium 51(5)
Mass exchange in binary systems 51(242–245)
Mass function for a binary component 50(216, 218, 250)
Mass gradient 51(6)
Mass loss from stars 51(174, 177, 184, 189, 190, 217–218, 245–249, 275, 565, 666, 667)

— continuous 51(245–247, 275)
— explosive (novae) 51(247–249, 275, 758, 761, 763)
Mass loss, models with 51(174, 177, 178, 184)
Mass luminosity ratio of galaxies 53(360–362)
Mass motion in the chromosphere 52(112–116)
Mass motion in the gravitational field of the Galaxy 53(24ff.)
Mass of the planets (table) 52(431)
Mass of the Venus atmosphere 52(391)
Mass ratio of binary components 50(212–213, 251, 258, 260)
Mass transfer from one binary component to the other 50(272)
Masses of binary components 50(221, 228–232, 236–237, 249, 268)
Masses of clusters of galaxies 53(407)
Masses of globular clusters 53(188–189)
Masses of globules 51(149)
Masses of individual galaxies 53(348–360, 407)
Masses of novae 51(762)
Masses of solar-neighborhood stars 51(165–172)
Masses of the Magellanic Clouds 53(243)
Masses of variable stars 51(357)
Masses of white dwarfs 50(183)
Massive stars, evolution off the main sequence 51(183, 184)
Massive stars, stability 51(607)
Mass-luminosity relation 51(4, 47, 58, 60)
— departures from 50(257–259)
— for giants 51(171)
— for main sequence stars 51(167–170)
— for spectroscopic binaries 50(249, 257)
— for subdwarfs 51(171–172)
— for subgiants 51(170)
— for visual binaries 50(220, 221)
— theoretical 51(46)
Mass-luminosity-temperature relation for white dwarfs 51(743)
Mass-radius diagram for terrestrial planets 52(438)
Mass-radius relation for white dwarfs 51(740, 741)
Mass-radius relation of cold bodies 52(420–431)
Mass-to-light ratio in elliptical galaxies 51(233, 234, 235)
Matching velocities 51(701)
Maximum brightness of novae 51(752, 753, 757, 763)
Maximum corona 52(252, 271–283)
Maximum radius of a cold body 52(420)
Maxwellian velocity distribution 53(76)
Mayall and Wyse method for mass determination of galaxies 53(350–354)
McMath-Hulbert Observatory 54(15, 16, 23, 29, 33)
Mean absolute magnitudes of external galaxies 53(316–317)
Mean colour indices of external galaxies 53(317–318)
Mean density of the Earth 52(432)
Mean density of the planets (tables) 52(431, 437)
Mean molecular weight of stellar matter 51(15–16)
Mean optical depth 52(97)
Mean polarisation of external galaxies 53(337)
Mean spectral types of external galaxies 53(338)
Mean surface brightness of external galaxies 53(315)
Mechanical properties of external galaxies 53(343–366)
Medium chromosphere, density, spectrum temperature 52(122–124)
Medium chromosphere, spectroheliograms 52(129–131)
Medium compact clusters of galaxies 53(395)
Megaparsec, definition 53(448)
Melon seed theory 52(223)
Mercury, abundance 51(314, 316, 319, 343)
Mercury, albedo 52(404)
Mercury and Moon, comparison 54(210)
Mercury, atmosphere 52(404–406)
Mercury, interior 52(438)

Mercury, properties of the surface 54(205–210)
Meridian mounting of aerials 54(82)
Meridional currents of the Sun 52(342, 343)
Messier 11, colour-magnitude diagram 53(155)
Messier 67, colour-magnitude diagram 53(156)
Messier 31, distance and absolute magnitude 53(470–472)
Messier 33, distance and absolute magnitude 53(472)
Messier 81, distance and absolute magnitude 53(473)
Messier 87, distance and absolute magnitude 53(473)
Messier 31, radio emission 53(259)
Messier 31, stellar content 53(462, 470)
Metal content and age of stars 53(98)
Metal line spectroheliograms 52(126–129)
Metal lines in the low chromosphere 52(121, 126–129)
Metallic ammonium in Jupiter and Saturn 52(448)
Metallic phase of solid
— helium 52(443)
— hydrogen 52(441, 443, 444)
Metallic transitions of solids 52(435)
Metallic-line stars 50(89, 94–97); 51(711)
— abundance deficiencies 50(96, 100)
— classification 50(95)
— color measurements 50(95)
— electron pressure 50(97)
— luminosity 50(95, 96)
— turbulence 50(97)
Metastability in stars 51(606)
Meteor orbis 52(527–530)
Meteor showers 52(544–548)
Meteor spectra 52(524–527, 537)
Meteor streams 52(544)
Meteor streams, list 52(545–546)
Meteoric dust 52(519, 530–534, 559)
Meteoric processes theory 52(520–524)
Meteorite falls 52(557, 559)
Meteorites 52(431, 559–564)
Meteorites, abundances of lead and its isotopes 51(315)
Meteorites, age 52(564)
Meteorites, chemical composition 51(300–302)
Meteorites, definitions of different types 52(520)
Meteorites, explosion on the lunar surface 54(182, 189–190, 192, 195)
Meteorological conditions 54(36)
Meteors, crater formation 54(189, 192)
Meteors falling on the lunar surface 54(190, 195)
Meteors falling on the Martian surface 54(227)
Meteors, formation of lunar seas 54(198–200)
Meteors, origin 52(519, 529–530, 538, 554–559)
Meteors, radiant points 52(534–537, 542, 551–552)
Meteors, velocities (velocity) 52(527, 532, 557, 558); 54(193)
Meudon astrophysial observatory 54(13–14)
Michelson interferometer for radio waves 54(43)
Micrometeorites 52(520, 527, 529, 530–533, 559)
Micrometric diameters of galaxies 53(311–313)
Microphotometric diameters of galaxies 53(313–315)
Microturbulence 50(376)
Microturbulence from curve of growth 52(95, 97)
Microturbulence from line profiles 52(95, 97)
Microturbulence in sunspots 52(160)
Microturbulence in the atmosphere of cepheids 51(394)
Microturbulence in the photosphere 52(93–98)
Microturbulent velocities 52(71, 73, 90)
Mie-Lennard-Jones potential 52(441)
Migration of spot zones 52(327)
Milky Way, *see* Galactic System
Mills cross 52(291); 53(223, 224); 54(108–111)
Milne-Eddington model 50(352, 356, 358); 52(51, 62)
Milne's cosmology 53(530–537)
Minimum corona 52(252, 267, 269)
Minimum principle of potential energy 51(608, 610)

Minimum temperature of Mars 52(399)
Minor planets 52(431, 432)
Mira Ceti stars 51(354, 365, 403, 406)
— emission lines 51(411)
— light curves 51(405, 406)
— velocity curves 51(409)
Mixed models, completely 51(173, 174, 177, 217–218)
Mixer 54(46, 47)
Mixing in main sequence stars 51(172, 173)
Mixing length 51(10, 440)
Mixture zone in white dwarfs 51(744)
MKK spectral classification 51(81)
Model atmosphere methods for abundance determination 51(337–339, 341)
Model atmospheres according to the energy distribution in the continuous and line spectrum of variable stars 51(597, 598, 600)
Model chromospheres 52(139, 140, 142)
Model of a flare 52(205)
Model of a typical large sunspot 52(164)
Model of cometary nuclei 52(494)
Model of jovian planets (interior) 52(445–448)
Model of photospheric faculae 52(176, 177)
Model of prominences 52(235)
Model of terrestrial planets, (interior) 52(432, 433, 436–440)
Model of the corona 52(271)
Model of the Earth's atmosphere 52(369)
Model of the Earth's interior 52(432–435)
Model of the Venus atmosphere 52(390–392)
Model photospheres 52(26–31, 47, 59, 72, 73, 78, 86, 87–92, 104–105)
Model stars 51(49–74)
Model universe 53(446)
Models, adiabatic 51(635)
Models, degenerate 51(56, 70–73, 739–742)
Models, isothermal 51(55–57)
Models, non-degenerate 51(56, 180, 181)
Models of stellar atmospheres 50(316–339)
Models of the Galactic system 53(54, 55)
Models, partially degenerate 51(176, 179–181)
Models with complete mixing 51(173, 174, 177, 217–218)
Models with convective core 51(57, 61–62, 66, 67, 68, 175, 178)
Models with homogeneous chemical composition 51(173, 174, 177)
Models with inhomogeneities in chemical composition 51(65–66, 70, 175–178, 183–184, 218–220)
Models with intermediate zone between convective core and radiative envelope 51(68–69, 183)
Models with isothermal core 51(66, 67, 70–73)
Models with mass loss 51(174, 177, 178, 184)
Models with outer convective zone 51(180, 500)
Modulation in light and velocity curves
— of *β* Cephei stars 51(398–400)
— of cepheids and RR Lyrae stars 51(384–388, 400)
Mögel-Dellinger effect 52(350)
Molecular bands
— emission bands 50(120, 122)
— from polyatomic molecules 50(118)
— in *C* stars 50(116–118)
— in *G* and *K* stars 50(113, 114)
— in *M* and *S* stars 50(114, 115, 116)
— in peculiar stars 50(98, 100, 102, 104, 106)
— in *R* and *N* stars 50(121)
— in the solar spectrum 50(110–113); 52(41, 42)
— in the spectrum of long-periodic variables 51(410–414)
— in variable stars 50(119)
— intensity difference in population types I and II 50(136)
— isotopic effects 50(120)
— rotation and vibration temperature 50(110, 127)
Molecular continua 50(126)
Molecular weight, mean, of stellar matter 51(15–16)
Molybdenum, abundance 51(309, 316, 318, 343)

Monochromatic flux 51(7)
Monochromatic interferometer theory 54(99–102)
Monochromatic limb darkening 52(1)
Monochromatic photoheliograph 54(18–19)
Monochromatic receiver operation 54(45–47)
Monte Mario (Roma) 54(38)
Moon, albedo 52(404)
Moon and Mercury, comparison 54(210)
Moon, atmosphere 52(404–406)
Moon, general map 54(190)
Moon, interior 54(199, 200)
Moon, photometry of the surface 54(180–182)
Moon, radio echoes 52(452–460)
Moon, structure as recognised in the telescope 54(191–205)
Moon, surface, *see* Lunar surface
Moon, temperature 54(184–186)
Moon, tidal origin 52(424)
Morgan-Keenan classification 50(43, 59, 379)
Morgan's natural groups 50(73)
Morphology of external galaxies 53(287–310)
Moscow, Sternberg Astronomical Institute 54(38)
Motion of infinitely many randomly distributed particles 53(417)
Motions in sunspots 52(162–165)
Mottling, chromospheric 52(126, 127, 129, 133–135)
Mount Wilson Observatory 54(13, 14, 21–22, 32)
Mounting of a grating 54(20–21)
Mounting of aerials 54(82–86, 88–89)
"Moustaches" in flare spectra 52(202, 207, 222, 223)
Moving clusters 53(157–160)
Moving prominences 52(237–248)
Multicolor photometry 51(87–89); 54(174–175, 279)
Multi-element interferometers 54(113–117)
Multiple clustering of galaxies 53(443–444)
Multiple galaxies 53(373–389)
Multiple galaxies, dynamics 53(384–385)
Multiple periodicity of variable stars 51(384–388, 398–400, 578)
Multiple systems, formation 51(153)
Multiple systems (triple stars etc.) 50(223)
Multiplication of secondary emission 54(244)
Multiplier phototubes in photometry 54(244, 258–261)
Multipolar groups of sunspots 52(167)
Multipolar magnetic regions of the sun 52(181)
Multi-slit spectrograph 54(28)

N stars, classification 50(6, 7, 8, 10)
N stars, molecular bands 50(121)
Natural velocity ellipsoid 53(62)
Naur-Osterbrock criterion for convective core 51(61, 62)
Nearby galaxies, dimensions 53(314)
Nearby galaxies, distances and absolute magnitudes 53(468–474)
Nearby stars, color-magnitude diagram 51(93)
Nearby stars, table of spectral types, color indices and parallaxes 51(90–92)
Nebula NGC 3587 50(141)
Nebula NGC 7293 50(140)
Nebula NGC 7027, spectrum 50(145)
Nebula, planetary, *see* Planetary Nebula
Nebular stage of the nova spectrum 51(757, 758, 761)
Nebular variables 51(424)
Negative dissipation as cause of vibrational instability 51(585)
Negative dissipation in external atmospheric layers of variable stars 51(506)
Neon, abundance 51(303, 316, 317, 343)
Neon-sodium reaction 51(38, 250)
Neptune, atmosphere 52(406, 410–412)
Neptune, composition of interior 52(448)
Network 52(126, 129)
Network of faculae 52(174, 178)
Neumann bands on meteorites 52(560–561)
Neutral filters 54(164)
Neutral hydrogen in galactic clusters 53(144)
— 21 cm line 53(10)

Neutron capture processes for element synthesis 51(255–260)
Neutron core, theory of supernova outbursts 51(784, 785)
Neutron detector 54(4)
Newtonian gravitational theory 53(491–494)
NGC 2264 cluster in Monoceros 53(165)
NGC 4321, distance and absolute magnitude 53(473)
NGC 752, H-R diagram 51(102, 209)
NGC 2264, H-R diagram 51(99, 205)
NGC 2362, H-R diagram 51(97)
NGC 6087, H-R diagram 51(100)
NGC 6530, H-R diagram 51(205)
NGC 7789, H-R diagram 51(209)
Nickel, abundance 51(305, 316, 317, 343)
Night sky brightness 54(250, 254, 276)
Niobium, abundance 51(309, 316, 318, 343)
Nitrogen, abundance 51(303, 316, 317, 343)
Nitrogen, dissociation in the Earth's atmosphere 52(370)
Nitrogen flaring in novae 51(759, 761)
Nitrogen in the Martian atmosphere 52(395, 402)
Nitrogen in the Venus atmosphere 52(383, 390)
Noise factor 53(211); 54(52)
Noise in multiplier phototubes 54(260)
Noise of receiver 54(44, 51–53)
Noise sources 54(53–55)
Noise storms 52(286, 302, 308, 315–320, 354)
Noise temperature 54(52)
Noise-free receiver, monochromatic operation 54(45–47)
Non-adiabatic radial oscillations 51(475–478)
Non-adiabatic region in variable stars 51(487–490)
— influence on stability 51(499–503)
Non-circular motions in the Galaxy 53(16–19, 89)
Non-coherent scattering in the photosphere 52(45–47, 53–54, 75)
— observational tests 52(53–54)
Non-conservative system, stability 51(618)
Non-degenerate models 51(56, 180, 181)
Non-degenerate models, isothermal 51(56)
Non-homogeneous models 51(70, 175–178, 183–184, 218–220)
Non-linear oscillations, energy method 51(657–659)
Non-linear radial oscillations 51(538–554, 592)
— non-adiabatic case 51(546–547)
Non-radial oscillations 51(509–538)
— adiabatic 51(511–514)
— dynamical stability towards 51(664–668)
— high modes (*g*, *p*) 51(521, 537, 573, 664, 665, 666)
— in *β* Cephei-Sternen 51(580–581, 589, 596)
— periods 51(523, 581)
Non-relativistic degeneracy 51(726)
Non-spherical symmetry of the Sun 52(148)
Non-thermal radio emission 53(116, 117, 125, 126–127, 234, 235–237)
— from the Galaxy 53(258)
— from the Sun 52(301–308)
Normal galaxies, radio frequency radiation 53(239, 255–260)
Normal spirals 53(276, 277, 279)
North polar sequence 54(147, 154, 157)
North-South asymmetry of spot groups 52(166)
Nova explosions 51(263, 275, 555, 686)
Nova Ophiuchi (Kepler's Nova) as radio source 53(231)
Nova Tychonis as radio source 53(231)
Nova Tychonis of 1572 51(768)
Nova WZ Sge 50(174)
Novae 53(466–469, 471)
Novae, absolute photographic magnitude at maximum 51(752–753, 763)
Novae and supernovae, distinction 51(766–767)
Novae, binary nature of 50(272)
Novae, brightness variations 51(752–753, 755, 756, 761)
Novae, decline rate (*m*/day) 51(752–753)
Novae, distribution and population type 51(755)
Novae, energy emitted in outbursts 51(763)

Novae, evolutionary stage 51(188)
Novae, galactic, characteristic data (table) 51(752–753)
Novae in nearby galaxies 51(754)
Novae in the Galaxy 53(466, 467)
Novae, luminosities 51(752, 753, 754, 755)
Novae, magnetic fields 51(286)
Novae, mass loss 51(248, 758, 761, 763)
Novae, masses and chemical composition 51(762)
Novae, radial velocities 51(757, 761)
Novae, relation to other stars 51(762–764)
Novae, spectrum variations 51(755–762)
Novae, speed classes 51(752–753, 761)
Novae, theory of outbursts 51(764–765)
Novae with expanding nebular shells 51(754)
Novae with extremely rapid light variations 51(429)
Nova-like variables, composite spectra (symbiotic) 51(419)
Novulae, *see* U Geminorum stars
Nuclear abundances and shell structure of nuclei, relation 51(322)
Nuclear abundances (table) 51(317–320)
Nuclear chain reactions in supernovae 51(779)
Nuclear equilibrium 51(662)
Nuclear reactions
— in carbon-nitrogen cycle 51(28, 166, 250–253)
— in production of heavier elements 51(37–38, 255–263)
— in proton-proton chain 51(28, 165–166, 250–253)
— in stars 51(25, 27–38, 250–263)
— in the atmospheres of peculiar *A* stars 51(709–711)
Nucleus of a comet 52(465, 491–495)
Nucleus of a galaxy 53(287)
Nucleus-electron interaction in white dwarfs 51(727–728)
Number of globular clusters 53(166–167)
Number-flux-density relation of radio sources 53(220–224)
O II doublet in planetary nebulae 50(154, 416)
O stars, classification 50(8, 41, 43)
O stars, model of the photosphäre 50(320, 321)
O-associations, definition 51(197)
O-associations, expansion 51(198, 199)
O-associations, star formation 51(200, 201)
Objective photometry 54(241)
Oblate spheroid approximation for mass determination of galaxies 53(354–357)
Oblateness of Mars 52(400); 54(221)
Oblique rotator model for a spectrum variable 51(428, 694)
Obscuration of spicules 52(136)
Observer effect in counts of galaxies 53(424, 427)
Occurrence rate of meteors 52(535)
Oceans on Mars 52(404)
Oceans on Venus 52(387)
Of stars 50(159, 389)
Offset guiding of faint stars 54(274–275)
Onset of cosmic radiation 52(218, 219)
Oort approximation 53(14, 15)
Oort's analysis of the velocity distribution 53(32–33)
Oort's constant 51(98)
Oort's constant of differential galactic rotation 53(24)
Oort-Spitzer mechanism of protostar formation 51(145, 146)
Opacity, absorption edges 51(21)
Opacity, continuous, in the infra-red 50(125)
Opacity, continuous, in the photographic and visual region 50(124)
Opacity contribution of bound-free processes 51(20–22)
Opacity, contribution of free-free processes 51(22)
Opacity, contribution of scattering by free electrons 51(23)
Opacity in dense matter 51(731–732)
Opacity in stellar atmospheres *see also* Continuous absorption 50(123–126, 182, 309)
Opacity of the lunar surface for wavelengths of about 1.4 mm 54(185–186, 188)

Opacity, Rosseland mean 50(309); 51(9, 20, 23–25)
Opacity, selective 50(124)
Open clusters of galaxies 53(399)
Open star clusters, *see* Galactic clusters
Optical arrangements for photoelectric photometry 54(272–276)
Optical depth 50(283, 313, 321, 329, 337); 52(42)
Optical depth, mean 52(97)
Optical double star 50(187)
Optical properties of external galaxies 53(311–343)
Optical thickness of planetary nebulae 50(417, 422, 425, 429, 430)
Optical thickness of the chromosphere 52(75)
Optimum image point 54(144, 145, 169)
Optimum light intensity for photographic photometry 54(134)
Orbit, apparent, of a visual binary 50(187, 197, 198, 209)
Orbit, photocentric, of an unresolved binary 50(210, 214, 217)
Orbit, true, of a visual binary 50(196, 197)
Orbital elements of spectroscopic binaries 50(243)
Orbital elements of visual binaries 50(196–202)
Orbits of comets 52(469–476)
Orbits of meteors 52(527–530)
Ordinary instability, *see* Dynamical instability
Ordinary novae and supernovae, distinction 51(766–767)
Ordinary spirals 53(283, 287, 293–294, 298)
Orientation, effect on the appearance of spirals 53(287, 300–301)
Origin of binary stars 50(272, 273)
Origin of comets 52(510)
Origin of elements in stars 51(37–38, 250–263, 320–321)
Origin of flares 52(221–222)
Origin of globular clusters 53(190–191)
Origin of life on the Earth 52(382–383)
Origin of meteors 52(519, 529–530, 538, 554–559)
Origin of planetary nebulae 51(247)
Origin of quiescent prominences 52(228–230, 244–248)
Origin of radio emission 53(102, 236–238)
Origin of the Earth's atmosphere 52(379–382)
Origin of the energy of novae 51(764–765)
Origin of the lunar craters 54(192–194)
Origin of the lunar seas 54(198–202)
Origin of the Martian atmosphere 52(403–404)
Origin of the solar system 51(192)
Origin of the supernovae 51(784–785)
Origin of the Venus atmosphere 52(393)
Origin of the white dwarfs 51(748–750)
Original orbits of comets 52(472)
Orion arm of the Galaxy 53(112)
Orion Nebula Cluster 53(164)
— H-R diagram 51(204)
Orion spectrum of novae 51(757, 758, 761)
Orion type variables 51(424)
Orography of the Moon 54(192)
Orthicon (image converter) 54(287–288)
Oscillations, hydromagnetic, of a star 51(535, 719–720)
Oscillations, non-radial 51(509–538, 580–581, 589, 596, 664–668)
Oscillations of a compressible sphere 51(514–516, 527, 541)
Oscillations of an incompressible sphere 51(509, 516, 535)
Oscillations, radial, adiabatic 51(458–466)
Oscillations, radial, non-adiabatic 51(475–478)
Oscillations, radial, non-linear 51(538–554, 592, 657–659)
Oscillator strengths 51(328, 330, 331)
Osmium, abundance 51(313, 316, 319, 343)
Outbursts in the solar radioemission 52(286, 302–303, 308, 311–315, 356)
Outbursts of novae, theory 51(555, 686, 764–765)
Outbursts of supernovae, theory 51(555, 686, 784–785)
Overlap of spectral lines 50(364)
Overstability, *see* Vibrational instability

Oxford solar reflector 54(10)
Oxidation in planetary atmospheres 52(364, 365)
Oxidizing atmosphere 52(381)
Oxygen, abundance 51(303, 316, 317, 343)
Oxygen, dissociation in the Earth's atmosphere 52(370, 371)
Oxygen in the Martian atmosphere 52(395, 404)
Oxygen in the Venus atmosphere 52(384, 385)
Ozone region in the Earth's atmosphere, absorption of energy 52(371, 372)

P Cygni stars 50(388, 389)
Pagel's model 52(32, 33)
Pairs of galaxies 53(285, 374, 377, 380–382)
Palladium, abundance 51(309, 316, 318, 343)
Pallas, diameter 54(211)
Parabolic cylinder 54(111)
Paraboloidal reflector 54(71, 82)
Parallaxes, definition 53(2)
Parallaxes, spectroscopic 50(24)
Parametric amplifier 54(53)
Parsec, definition 53(448)
Partial instability in the central layer of the Galaxy 53(88)
Partially degenerate models 51(176, 179–181)
Particle emission from the Sun 52(222–224, 354–362)
— effect on planets 52(371, 390)
Partition functions 52(57)
Paschen discontinuity 50(324)
Pawsey stub 54(69)
Payne-Scott and Little's interferometer 54(106–108)
Pear-shaped configurations 51(624, 636)
Pecker method 51(337–339, 341)
Peculiar *A* stars
— abundance anomalies 51(707–709)
— color-magnitude diagram 51(695, 696)
— light curves 51(696)
— nuclear reactions 51(709–711)
— spectrum 51(694–702)
— velocity variations 51(702–704)
Peculiar galaxies, radio-frequency radiation 53(233, 234, 265–274)
Peculiar interval 54(120, 121)
Peculiar lenticulars 53(287, 302)
Peculiar nebulosities as radio sources 53(232)
Peculiar stars
— Ba II stars 50(104–106)
— carbon stars, *see also* Carbon stars 50(101–103)
— carbon-poor 50(107)
— CH stars 50(101–102)
— classification 50(88, 95, 380)
— high-velocity stars, *see also* HV stars 50(90, 97–100)
— hydrogen-poor 50(103)
— luminosity 50(95, 96, 102)
— metallic-line stars, *see also* ML stars 50(89, 94–97)
— molecular bands 50(98, 100, 102, 104, 106)
— semi-carbon and strong-CN stars 50(106)
Pentagram notation of convolution 54(49, 67)
Penumbra of sunspots 52(151)
Perek's method for mass determination of galaxies 53(354–357)
Perfect cosmological principle (PCP) 53(526)
Periastron, advance of 50(248)
Periastron, longitude of 50(244, 261)
Period-amplitude relation of recurrent novae and U Geminorum stars 51(764)
Periodic magnetic variables 51(426–429, 535–537, 573, 695)
Periodicity, multiple of variable stars 51(384–388, 398–400)
Period-line width relation for spectrum variables 51(698)
Period-luminosity curve 53(456–457, 459–461, 463–466, 474, 481)
— adopted 53(461, 465)
— correction to zero point 53(458, 461, 463–465)
— for classical Cepheids 53(458–466, 474)
Period-luminosity diagram for the Magellanic clouds 53(459)
Period-luminosity diagram for Type II Cepheids 53(456)

Period-luminosity relation 51(109, 124, 356, 376–379)
— for β-Cephei stars 51(402)
— for classical cepheids 51(577)
— for RV Tauri stars and yellow semiregular variables 51(415)
Periods of cepheids, secular changes 51(381–384)
Periods of comets 52(470, 471, 511, 516)
Periods of for radial oscillations 51(473)
Periods of pulsation for non-radial oscillations 51(523, 581)
Periods of rotation of spirals 53(346–348)
Periods of variable stars, comparison between theoretical and observed values 51(574–585)
Period-spectral type relation
— for cepheids 51(379–381)
— for red semiregular variables 51(418)
Period-spectrum relation for RV Tauri stars and yellow semi-regular variables 51(415)
Permanent coronal condensations 52(189, 300)
Permanent multiple galaxies 53(374, 375–384)
Perseid meteor stream 52(526, 545–547)
Perseus A source 53(270)
Perseus A source, apparent angular diameter 53(226)
Perseus arm of the Galaxy 53(112)
Perseus cluster of galaxies 53(386, 478, 484, 485)
Perseus double cluster 53(154–155)
Perseus moving cluster 53(159–160)
Perturbation method for study of stability 51(478–481, 531, 536, 610, 611, 636, 647, 657)
Perturbations by rotation and apsidal motion 51(38–42)
Pfund spectroscope 54(24, 38)
Phase anomaly between ground wave and sky wave 52(351, 353)
Phase correlation between velocity and light curves of variable stars 51(374–376, 399, 400, 505, 589–592)
Phase diagrams for pulsations 51(549)
Phase errors in aerials 54(77–78)
Phase sensitive detector 54(58)
Phase shift between light curves for different colors 51(594)
Phase transformation pressure 52(443)
Phase transitions of solids 52(435–436)
Phase-sensitive synchronous rectifier (homodyne) 54(267–268)
Phase-shift broadening of the Fraunhofer lines 52(47, 49)
Phase-switching interferometer 53(214)
Φ-shaped barred spiral 53(290)
Phosphorus, abundance 51(304, 316, 317, 343)
Photocenter of an unresolved astrometric binary 50(210, 211, 217)
Photochemical mechanism 54(136)
Photoconductive cells 54(248–249)
Photodissociation and photoionization in comets 52(488–489)
Photoelectric cells, use in solar observations 54(29–30, 32)
Photoelectric detectors 54(241)
Photoelectric dinstinction of white dwarfs 50(165)
Photoelectric scanning of spectra 54(280–282)
Photoelectric techniques 51(80)
Photographic atlas of the Moon 54(192)
Photographic dimensions of external glalaxies 53(311–315)
Photographic effects 54(133)
Photographic emulsions, properties 54(130–133)
Photographic meteor techniques 52(538–542)
Photographic recording 54(29)
Photoheliograph 54(10, 18)
Photo-ionization in the corona 52(257)
Photometers 54(138–143, 240)
Photometric classification 50(83, 84)
Photometric measurement, accuracy 54(143)
Photometric measurement, principal steps 54(139)
Photometric methods for distance determination of galactic clusters 53(139)
Photometric sieve 54(175, 177)
Photometry of asteroïds 54(211)
Photometry of seasonal changes on Mars 54(234, 236)

Photometry of the dark regions on the Martian surface 54(230)
Photometry of the lunar surface 54(180–182)
Photometry of the satellites of Jupiter 54(218)
Photometry of the surface of Mercury 54(208–209)
Photomultiplier, *see* Multiplier
Photo-recombination in the corona 52(257)
Photosphere, continuous absorption coefficient 52(23–26, 42, 47)
Photosphere, density 52(86–92)
Photosphere, incoherent scattering 52(45–47, 53–54, 75)
Photosphere, line broadening mechanism 52(48–51)
Photosphere of a star 50(278, 319, 320, 334)
Photosphere, temperatures 52(22, 23, 32, 47, 86–92)
Photosphere, turbulence 52(93–98, 102, 103, 104)
Photospheres, models 50(319–339)
Photospheric faculae, development and relation to sunspots 52(173–175)
Photospheric faculae, model 52(176, 177)
Photospheric inhomogeneities 52(86–92)
Photospheric models 52(26–31, 47, 59, 72, 73, 78, 86, 87–92, 104–105)
Photovisual system 54(158, 160)
Photovisual technique 50(190, 204)
Physical photometers 54(138, 240–241)
Physics of comets 52(477–495)
Piezotropic relations 51(451)
Pip 52(316)
Planck's law 52(47)
Planck's radiation formula 54(90, 245)
Plane-layer model of atmosphere 54(151)
Planetary atmospheres, *see also* Earth, Jupiter, Mars, Venus etc. 52(363 ff.)
Planetary atmospheres, constituents (table) 52(413)
Planetary companions as binary components 50(218, 219)
Planetary constants 52(365)
Planetary nebula in M 15 53(184)
Planetary nebulae 53(467–468, 471)
— absolute magnitudes 50(141, 142)
— abundance of elements 51(343, 344)
— apparent magnitudes 50(140, 141)
— central stars 50(139, 141, 151, 152, 159, 160)
— central stars, temperatures 50(151, 153, 159, 160, 426–431)
— continuous spectra 50(152)
— Doppler effect 50(157)
— electron densities 50(152, 154, 412)
— electron temperatures 50(152, 154, 412)
— evolutionary stage 51(187)
— expansion 50(156)
— forbidden lines 50(150, 403–407, 416)
— galactic distribution 50(143, 144)
— hydrogen spectrum 50(153, 409–412)
— line intensities 50(147, 151, 412)
— line spectra 50(144–152, 156, 158, 403–407)
— line spectra, excitation 50(147, 403–407)
— line splitting 50(156, 158)
— nebular shells, absorption and emission 50(399, 408 ff.)
— nebular shells, chemical composition 50(408)
— optical thickness 50(417, 422, 425, 429, 430)
— origin 51(247)
— photometric detection 54(175–176)
— size distribution 50(142)
— stratification 50(416, 425)
— two-photon emission 50(155, 156)
Planetary origin of comets 52(510)
Planetary system, formation conditions 50(271)
Planets, chemical composition 51(298, 300)
Planets, classification 52(431)
Planets, mass-radius relation 52(420–431, 438)
Planets, pressure-density relation 52(434, 440, 444, 447)
Planets, radio echoes 52(449, 460)
Planisphere of the satellites of Jupiter 54(217)
Plaskett's model 52(30)
Plaskett's star, spectra 50(254, 258)
Plasma, emission of thermal radio-radiation 52(283–286)

Plasma frequency 52(284)
Plasma oscillations as a cause for radio emission 53(235)
Plasma oscillations in the corona 52(305)
Plasticity of the lunar rocks 54(195)
Plate effect in counts of galaxies 53(424, 427)
Plateau's experiment 51(624)
Platinum, abundance 51(313, 316, 319, 343)
Platinum filter 54(164)
Pleiades 53(149–153)
Pleiades, H-R diagram 51(96, 97, 208)
Pluto, atmosphere 52(412, 414)
p-modes 51(521, 537, 664, 665, 666)
Poincaré's coefficients of stability 51(627)
Point mass approximation for mass determination of galaxies 53(348–349)
Poisson equation 53(25)
Poisson equation, existence problems 53(39–41)
Polar caps of Mars 54(224, 228, 235)
Polar faculae 52(177)
Polar plumes of the corona 52(272)
Polar prominences 52(225, 230–231)
Polar rays of the corona 52(177, 268)
Polarisation of light from
— Mercury 54(209)
— spirals 53(336–338)
— the asteroïds 54(214, 215)
— the bright regions of Mars 54(226, 228)
— the dark regions on Mars 54(236–237)
— the Moon 54(182, 183, 197)
Polarisation of starlight, photometer 54(277)
Polarisation of the radiation from radio sources 53(229)
Polarised light from Crab Nebula 53(231, 236)
Polarising monochromatic filters 54(17)
Polarity law of sunspots 52(166)
Polarity of spot groups 52(328)
Polarization of outbursts 52(311–315)
Polarization of stormbursts 52(318, 319)
Polarization of the clouds of Venus 52(385, 386)
Polarization of the corona 52(255)
Polarization of the Martian atmosphere 52(394, 397)
Polarization of the radiation of comets 52(489–491)
Poleward motion of filaments 52(181)
Poleward motion of of planetary atmospheres 52(372, 392)
Poleward motion of quiescent prominences 52(230–231)
Poloidal magnetic field of the Sun 52(181, 340)
Polyatomic molecules in stellar atmospheres 50(118, 291)
Polymorphic transitions of solids 52(436)
Polytropes, classes of 51(50–55)
Polytropic equation of state 51(13, 16–17, 45)
Polytropic index 51(12, 16, 42)
Polytropic index of protostars 51(158, 159)
Population II variables, emission lines and line doubling 51(396, 397)
Population of atomic levels 52(109–112)
Population of intergalactic space 53(389)
Population type(s)
— and chemical composition 51(128, 129, 214, 228, 346); 53(55, 96)
— concept and subdivision 51(128–131, 225–226, 365)
— of novae 51(755)
— of variable stars 51(365, 366, 405, 418)
— physical differences 53(97)
Populations, *see* Stellar populations
Position, relative, of binary components 50(188, 189, 196)
Positional accuracy 50(192, 205, 206)
Post-exposure effects 54(167)
Postnova stage 51(755)
Potassium-rubidium abundance ratio 51(306)
Potsdamer Spektraldurchmusterung 50(13)
Poynting vector 52(306)
Poynting-Robertson effect 52(530, 531, 548)
p-process for element synthesis 51(261)
Praesepe cluster 53(153)

Praesepe, H-R diagram 51(102)
Prandtl's mixing length 52(98)
Preamplifier 54(45–46)
Precision, *see* Signal-to-noise ratio
Precision limits in radio wave detection 54(49–53)
Pre-exposure effects 54(167)
Preferential motion of stars 53(5)
Premaximum spectrum of novae 51(755, 756, 757, 761)
Prenova stage 51(755, 762, 764)
Pressure at the surface of Mars 52(393–394)
Pressure broadening of the Fraunhofer lines 52(46, 49)
Pressure equality between corona and prominences 52(244)
Pressure equation of a star in equilibrium 51(5)
Pressure gradient 51(6)
Pressure in spiral arms 53(94)
Pressure in the atmospheres of Uranus and Neptune 52(411)
Pressure in the Earth's atmosphere 52(368, 369, 373)
Pressure in the Earth's interior 52(434)
Pressure on the lunar surface 52(406)
Pressure-density relation of planetary matter 52(434, 440, 444, 447)
Pre-stellar nuclei 51(147)
Primordial turbulent gas 51(135)
Principal solution 54(123, 125)
Principal spectrum of novae 51(757, 758, 761)
Probability density function for clusters of galaxies 53(419, 437, 438)
Profiles *see* Line profiles
Profiles of facular lines 52(180)
Profiles of Fraunhofer lines 52(42–47, 56, 58–59, 75–77)
Profiles of limb radiation 52(117, 118)
Profiles of the 21 cm line 53(109, 110)
Progressive waves in the atmosphere of variable stars 51(497, 554–570, 588, 590, 592, 593, 600, 686, 687)
Prominences 52(224–248)
Prominences, classification 52(225)
Prominences during the solar cycle 52(329, 330)
Prominences, H_α profiles 52(200)
Prominences, heights 52(227)
Prominences in binaries 50(270, 271)
Prominences, model 52(235)
Prominences, origin 52(228–230, 244–248)
Prominences, rôle in the development of a centre of activity 52(184–185)
Prominences, spectra 52(226, 232–237)
Prominences, temperatures 52(232, 233)
Proper motion
— definition 53(1)
— of globular clusters 53(185–186)
— of stars in galactic clusters 53(147–148)
— of stars in the galactic plane 53(17)
— of white dwarfs 50(165, 166)
— perturbed 50(189, 214)
Properties of planets, life enabling 52(382)
Properties of the terrestrial atmosphere 52(369)
Protogalaxy 51(135)
Proton-deuterium reactions in white dwarfs 51(736)
Proton-proton chain 51(28, 165–166, 250–253)
— cross sections 51(32–34)
— energy production functions 51(35, 36)
— in white dwarfs 51(734, 736)
Protostar formation
— in the absence of stars and dust 51(137–141)
— in the presence of a magnetic field 51(154–157)
— influence of turbulence 51(148)
— loss of angular momentum 51(152–154)
— mass distribution 51(148)
— out of prestellar nuclei 51(147–148)
— through concentration of dust 51(141–145)
— through influence of other stars 51(145–147)
Protostars, gravitational contraction 51(157–160)
— for different masses 51(160)
Protostars, opacity 51(138, 157, 158)
Protostars, polytropic index 51(158, 159)
Pseudo-radial oscillations 51(528)
Pulkovo 54(6, 7)
Pulsating universe 53(425)

Pulsation caused by vibrational instability 51(585–587)
Pulsation characteristics for different stellar models 51(471–474, 502)
Pulsation periods for non-radial oscillations 51(523, 581)
Pulsation periods for radial oscillations 51(473)
Pulsation theory of variable stars 51(431 ff.)
Pulsational line profiles 51(598, 599)
Pulse counting following multiplier tube 54(264)
Puppis A source 53(232)
Pure absorption in the photosphere 52(44, 75)
Pure hydrogen stars, evolution 51(182)
Purkinje coefficient 54(147)
Purkinje effect 54(146, 240)
P,V-system 54(158, 160)
Pyranometer 54(4)
Pyrheliometer 52(2); 54(1–4, 241)

Quadrant electrometer of Lindemann 54(242, 243)
Quadratic phase error 54(78–79)
Quadratic Stark effect 52(49)
Qualitative morphology of galaxies 53(287–310)
Quantum efficiency of photographic plates 54(250)
Quasi-uniform distribution of particles 53(417–418)
Quiescent prominences 52(225, 226–237, 245)
Quiet corona 52(250–271, 295, 296)

R Coronae Borealis stars 51(283, 422–424)
R stars, classification 50(6, 7, 8, 10)
R stars, molecular bands 50(121)
Radar echoes on the Moon 54(187)
Radar equation 52(449–450)
Radar magnitude of meteors 52(522)
Radau-Darwin approximation 52(428, 430)
Radau's transformation 52(427)
Radial distribution of galaxies in clusters 53(400–402)
Radial oscillations
— adiabatic 51(458–466)
— dynamical stability towards 51(645–647, 660–664)
— non-adiabatic 51(475–478)
— non-linear 51(538–554, 592, 657–659)
— properties for different stellar models 51(471–474, 502)
— vibrational stability towards 51(647–650, 684–685)
Radial velocities
— definition 53(1)
— of globular clusters 53(186)
— of granules 52(83–86)
— of individual stars in galactic clusters 53(148)
— of individual stars in globular clusters 53(187–188)
— of novae 51(757, 761)
— of stars in the galactic plane 53(16, 17)
Radial velocity curves, *see* Velocity curves
Radiant distribution of sporadic meteors 52(534–535, 537, 557–552)
Radiant distribution of stream meteors 52(537)
Radiant motions 52(547)
Radiant point of meteors 52(534, 536–537, 542)
— concentration near apex, Sun and antihelion point 52(552)
Radiants of meteor streams, table 52(545–546)
Radiation damping in flare spectra 52(203, 204)
Radiation damping, natural 50(343)
Radiation damping of the Fraunhofer lines 52(46, 48)
Radiation effect on interplanetary dust particles 52(530)
Radiation equilibrium 50(313, 317, 329)
Radiation field 51(442)
Radiation flux 50(280, 281, 313, 357)
Radiation from the entire sky 54(3)
Radiation interchange in the photosphere 52(52)
Radiation law of Planck 54(90, 245)
Radiation pattern of an aerial 54(61, 62, 64–68)
Radiation pressure 51(44–45)
Radiation pressure in stellar atmospheres 50(318)

Radiation temperature of the disk centre 52(143)
Radiation temperature of the photosphere 52(47)
Radiation temperatures of cepheids 51(389)
Radiation thermocouple 54(241)
Radiation, total, from the Sun 54(1–4)
Radiative capacity, *see* Opacity
Radiative energy transport 51(6, 7–10)
Radiative envelope of a star 51(57, 59, 60, 65, 175, 179)
Radiative equilibrium 51(39)
Radiative equilibrium in the atmospheres of white dwarfs 51(742–743)
Radiative flux 51(10)
Radiative gradient 51(7–10)
Radiative gradient, stability 51(10, 13)
Radiative transfer in the solar atmosphere 52(42–43)
Radiative viscosity coefficient 51(445)
Radii of stars, determination 51(4)
Radii of the components of eclipsing binaries 50(228–232, 236–237)
Radii of the components of spectroscopic binaries 50(251)
Radii of white dwarfs 50(176, 178, 183)
Radio echoes 52(449 ff.)
Radio echoes, conditions for detection for different astronomical bodies (table) 52(461)
Radio emission
— at metre wavelengths 53(121–126)
— dependence on galaxy type 53(255–256)
— from the Sun, non-thermal 52(301–308)
— from the Sun, thermal 52(283–286, 296–301)
— non-thermal 53(104–106, 116, 117, 125, 126–127, 234, 235–237, 258)
— spatial extent 53(271)
— thermal 53(102–104, 117, 234–235, 258)
Radio frequency radiation from
— clusters of galaxies 53(260–265)
— external galaxies 53(239–274)
— ionized hydrogen regions 53(232)
— normal galaxies 53(233, 255–260)
— peculiar galaxies 53(233, 234, 265–274)
— peculiar nebulosities 53(232)
— supernovae 53(231, 260)
— the Andromeda Nebula 53(250–252, 259)
— the Galactic System 53(100–128, 256–259)
— the Local Supergalaxy 53(263–265)
— the Magellanic Clouds 53(240–250, 259)
Radio galaxies 53(239)
— characteristics 53(270–272)
— cosmological aspects 53(272–274)
— identifications 53(265–270)
— spectra 53(271)
Radio magnitude scale 53(247)
Radio magnitudes of the Magellanic Clouds 53(246–248)
Radio meteor techniques 52(536–538)
Radio meteors 52(520, 527, 528)
Radio observations, information on galactic rotation 53(115–116)
Radio phenomena, solar, connected with flares 52(308–322)
Radio receiver 54(44)
Radio source(s)
— Cassiopeia A 53(209, 236)
— Centaurus A 53(267)
— Cygnus A 53(212, 233, 234, 265)
— Fornax A 53(269)
— Gemini A 53(226)
— Hercules A 53(269)
— Hydra A 53(269)
— in Andromeda 53(226)
— in Auriga 53(226)
— Perseus A 53(270)
— Puppis A 53(232)
— Sagittarius A 53(119, 233)
— Taurus A 53(225, 228, 229)
— Virgo A 53(268)
— apparent angular diameter 53(224–227)
— brightness 53(209, 213, 214)
— cosmic 51(783)
— catalogue 53(218)
— counts 53(273)
— distances 53(227)
— flux density 53(209, 213, 215)
— identification 53(231–238)
— luninosity distances 53(485)
— mechanism of radiation 53(234–237)
— of class I and II 53(222, 237)
— origin 53(102, 236–238)
— polarisation of radiation 53(229)

— scintillation 53(212, 230)
— spectra 53(227–229)
— table of surveys 216–217
Radio spectra of discrete sources 53(227–229)
Radio spectra of the Magellanic Clouds 53(246–248)
Radio spectrum of the chromosphere 52(143–144)
Radio spectrum of the corona 52(283–286)
Radio spectrum of the quiet Sun 52(286–289)
Radio stars 53(237)
Radio surveys 53(100, 101, 216–217)
Radio telescopes for solar observations 54(4–5)
Radio waves, rotation of polarization plane in the ionosphere 52(457)
Radioactive isotopes in meteorites 52(563)
Radioactive substances on Jupiter 52(409)
Radioactivity in the interior of the asteroïds 54(213)
Radioactivity of the lunar interior 54(195)
Radioemission from the Moon 54(186, 188)
Radiomaps of the Sun 52(250)
Radiometer 54(241)
Radiometric light curves of long-period variables 51(408)
Radiotelescope 53(211)
Radiotelescope, sensitivity 53(214–215)
Radius mounting of a grating 54(20)
Radius of clusters of galaxies 53(440)
Radius of Mars 52(399)
Radius of terrestrial planets 52(437)
Radius of the Earth, determination from echoes from the Moon 52(459)
Radius of the planets (table) 52(431)
Radius-mass relation for white dwarfs 51(740, 741)
Radius-mass relation of cold bodies 52(420–431)
Rainfall due to meteoric dust 52(519, 533, 548)
Random errors in aerials 54(79–81)
Range-cycle relation, *see* Amplitude-period relation
Rare earth elements, abundances 51(311–312, 316, 319, 343)
Rare gases, abundance ratios 51(306–307, 343)
Ray structure of comet tails 52(468)
Rayleigh distribution 54(47, 48)
Rayleigh number 51(672, 679)
Rayleigh-Jeans approximation 54(91)
Rayleigh-Jeans law 52(284)
Rayleigh-Ritz method 51(469–470)
Rayleigh's criterion for convective stability 51(667)
Rayleigh's principle 51(462, 531, 534)
Rays of the corona 52(251, 272, 273, 280–283)
Reabsorption in stellar shells 50(392)
Receiver noise 54(44, 51–53)
Receiver noise in radio echoes 52(450)
Receiver of radio waves 54(44)
Receiver parameters 54(44–45)
Reception filters, list of types 54(50)
Reception pattern of an aerial 54(61)
Reciprocity failure in exposure of photographic plates 54(134, 135)
— at high intensities 54(137)
— at low intensities 54(136, 137)
Reciprocity theorems for aerials 54(61–62)
Recombination coefficient 52(347)
Recombination lines in spectra of nebulae 51(344)
Recurrence of magnetic storms 52(190, 222, 359, 360, 361)
Recurrence tendency of flares 52(197)
Recurrence time pattern of magnetic storms 52(360)
Recurrent novae, properties 51(763)
Red dwarf model 51(62–64)
Red giant models 51(69–70)
Red giants in galactic clusters 51(105)
Red semiregular variables, *see* Semiregular variables
Red spot of Jupiter 52(409)
Reddening, interstellar 51(86–89)
Red-infrared photometric system 54(163)
Redistribution of absorbed frequencies 52(46)
Redshift in the spectra of external galaxies 53(318, 384, 411–414, 423, 431, 435–439, 440, 446, 474, 477–478, 486, 509)

Redshift of the Fraunhofer lines 52(77–78, 91)
Redshift, relativistic 52(91)
Reduced corona activity 52(191)
Reduced prominence activity 52(191)
Reduction curves 54(166)
Reduction process with a half-filter 54(165–167)
Re-emission mechanism in the solar atmosphere 52(43–47)
Reference sources of noise 54(59, 60)
Reference stars 50(5, 206–208)
Reflecting power of the asteroïds 54(212–213, 215)
Reflecting power of the Jupiter satellites 54(219)
Reflecting power of the Martian surface 54(225)
Reflection diffraction gratings 54(19)
Reflection effect in binaries 50(264)
Reflection of radio waves on the Moon 54(187)
Reflection process in echoes from the Moon 52(456)
Reflectors for radio waves 54(71–75)
Refraction, atmospheric 50(204, 205)
Refractive index 50(326)
Refrigeration of phototubes 54(269–272)
Regraduation of clocks in Milne's cosmology 53(533)
Regular logarithmic spiral 53(289)
Regular velocity ellipsoid 53(62)
Relative sensitivity of photographic emulsions 54(132)
Relativistic degeneracy 51(726)
Relativistic electrons radiating in a magnetic field 53(102, 104–106, 126–127, 239, 260)
Relativistic red-shift 52(91)
Relaxation oscillations 51(552)
Relaxation time of a galactic cluster 53(161–162)
Relaxation time of a stellar system 53(74–76)
Relaxation times for the non-adiabatic temperature field 51(490, 491, 575)
Relief of the Martian surface 54(228–229)
Remnants of supernovae 51(767, 783)
Remnants of supernovae as radio sources 53(231, 236, 237)
Repulsive forces in comets 52(496, 505–509)
Residual central intensities of lines 50(346)
Resistive noise 54(53–54)
Resolved astrometric binaries 50(199, 203)
— analysis of path 50(208–210)
Resolving power of photographic emulsions 54(132)
Resolving power of solar spectroscopes 54(19)
Resonance between atmosphere and interior of variable stars 51(494)
Response functions in color photometry 54(153, 155, 156)
Response functions of various cathodes 54(246)
Restoration 54(125–128)
Reversals, central, in the Fraunhofer lines 52(76)
Reversing layer in a stellar atmosphere 50(278)
Revised classification of external galaxies 53(281–286)
Rhenium, abundance 51(313, 316, 319, 343)
Rhodium, abundance 51(309, 316, 318, 343)
Rigidity, magnetic, of particle orbits 52(216, 219)
Rills, radial, in the lunar seas 54(189–199, 201, 203)
Ring current around the Earth 52(357)
Ring formation in a binary system 50(266, 271)
Ring Nebula in Lyra 50(139, 140)
Ringed type of a Galaxy 53(283, 284, 292)
Roche's model 51(516, 631–634)
Roche's model, generalized 51(563)
Rock faults on the lunar surface 54(204)
Rocket observation of the solar spectrum 52(118)
Rockets 54(37, 40)
Rockets, effect of meteoritic dust 52(531)
Rosseland mean opacity 50(309); 51(9, 20, 23–25, 731)
Rosseland variables 51(49)
Rotating disc in waveguide 54(59)

Rotating masses, stability for the compressible case 51(629ff., 636ff.)
Rotating masses, stability for the incompressible case 51(619–629)
Rotating planets, figure 52(424–429)
Rotating sub-systems in the Galaxy 53(23, 29, 44–45, 96, 98)
Rotating-lobe interferometer 53(214)
Rotation and mixing 51(38–40)
Rotation, effect on non-radial oscillations 51(509, 524, 526–532)
Rotation, effect on stability 51(480, 607, 616–636, 666, 669–674)
Rotation, galactic 53(6–11, 22, 24, 53)
Rotation, galactic, from globular clusters 53(186–187)
Rotation hypothesis (asymmetric body) for variable stars 51(358, 432, 573)
Rotation of β-Cephei stars 51(400, 579–582)
Rotation of giants 51(277–278)
Rotation of main sequence stars 51(276–277)
Rotation of Mercury 54(205)
Rotation of spirals, direction 53(343–346)
Rotation of spirals, periods and velocities 53(346–348)
Rotation of stars 50(251, 252, 380, 383)
Rotation of stars in galactic clusters 53(149)
Rotation of the asteroïds 54(211, 212)
Rotation of the Earth 52(455)
Rotation of the plane of polarisation in a radio source 53(229)
Rotation of the satellites of Jupiter 54(218)
Rotation of the substratum 53(496–497, 503–504)
Rotation of the Sun 52(337–340)
Rotation period of Venus 52(387, 389)
Rotation temperature 50(110, 127)
Rotation velocity, upper limit 51(621)
Rotational broadening of lines 50(252, 381, 384)
Rotational effects in the spectra of peculiar *A* stars 51(697)
Rotational flattening of binary components 50(249)
Rotational methods for mass determination of galaxies 53(348–359)
Rotational mixing in main sequence stars 51(173)
Rotational motion of the Magellanic Clouds 53(244)
Rotational symmetry, characteristic diagram 53(26)
Rotational symmetry, quasistationary system 53(28–29)
Rotational symmetry, special types of potential functions 53(36)
Rotational verlocity of the Sun 52(338)
Rotation-vibration bands of H_2 in the atmospheres of Uranus and Neptune 52(411)
Rowland circle 54(20)
Rowland intensity scale 51(325, 333)
Rowlands's tables of solar spectrum wavelengths 52(34, 37, 79)
RR Lyrae
— amplitude-period diagram 51(370, 371)
— as zero point in H-R diagrams of globular clusters 51(109, 118, 126, 127)
— β Cephei stars, classical cepheids comparison 51(400)
— characteristic data 51(355, 365)
— classification according to the shape of light curves 51(367)
— correlation between asymmetry and amplitude of variations 51(592–593)
— double or multiple periodicity 51(384–388, 400)
— emission lines and line doubling 51(394, 395)
— light curves 51(367, 368)
— magnetic field 51(711)
— period-luminosity relation 51(376–379)
— phase correlation between light and velocity curves 51(589)
— region in globular cluster H-R diagrams 51(110, 181, 283)
— secular changes in period 51(381–384)
— spectral type and its variation with phase 51(392–393)
— stars, absolute magnitude 53(453, 470)
— stars, amplitude correlation between light and velocity curves 51(588)

RR Lyrae (cont.)
— stars in globular clusters 53(183, 450–457, 469)
— stars, in M 31 53(462, 470)
— stars in the Magellanic Clouds 53(468, 469)
— stars, statistical parallax 53(453)
— theoretical and observed periods 51(582–583)
— velocity curves 51(376)
RR Pictoris, spectrum 51(760)
RT Andromedae or RZ Ophiuchi group of eclipsing binaries 50(236, 237, 239)
Rubidium, abundance 51(306, 308, 316, 318, 343)
Rudkjöbing's model for white dwarfs 51(741)
Ruggedness of the lunar surface 54(181, 184, 187, 188, 190)
Ruggedness of the Martian surface 54(227)
Ruggedness of the surface of asteroïds 54(213–214)
Ruggedness of the surface of the Jupiter satellites 54(218–219)
Rumford spectroheliograph 54(13)
Russell and Hertzsprung's evolutionary scheme 51(160–161)
Ruthenium, abundance 51(309, 316, 318, 343)
RV Tauri
— emission lines 51(416)
— light curves 51(415–417)
— period-luminosity and period-spectral type relation 51(415)
— stars, characteristic data 51(123, 365)
— theoretical and observed periods 51(583–584)
— velocity curves 51(416–417)
RW Aurigae stars 51(424–426)
Ryle's interferometer 54(108)
Ryle-Vonberg system 54(59–60)
RZ Scuti, spectra 50(255)
RZ Scuti, velocity curves 50(256)

S Andromedae of 1885 51(769)
S stars 50(93, 105, 106)
— abundance anomalies 51(271–272, 348, 349)
— classification 50(10, 61, 69, 72)
— molecular bands 50(115)
Sacramento Peak Observatory 54(8, 9, 28)
Sagittarius A source 53(119, 233)
Sagittarius arm of the Galaxy 53(112)
Saha equation 50(278, 289)
Saha equation for planetary nebulae 50(417)
Saha ionization temperature 52(47)
Saha theory 50(22)
Salpeter's luminosity function 51(218–220)
Sandage-Schwarzschild models 51(176, 177)
Satellites 54(40)
Satellites, artificial, effect of meteoritic dust 52(531)
Satellites of Jupiter 54(215–220)
Satellites of Mars 54(221)
Satellites of the major planets 52(412, 414)
Saturation function 50(363); 51(339); 52(57)
Saturn, atmosphere 52(406, 410)
Saturn, model of interior 52(445–448)
Saturn, satellite Titan 54(215)
SBO objects 53(278, 279)
Scale-of-two circuit 54(265)
Scandium, abundance 51(305, 316, 317, 343)
Scanning device in solar observations 54(30)
Scanning of spectra by photoelectric methods 54(280–282)
Scatter in radiants of meteors 52(547)
Scattered light, effect on limb darkening 52(9–11)
Scattering coefficient 50(299, 350)
Scattering, coherent 50(349, 351, 355, 357)
Scattering, coherent, in the photosphere 52(43, 75)
Scattering, incoherent 50(349)
Scattering, non-coherent, in the photosphere 52(45–47, 53–54, 75)
Scattering of radio-radiation by the corona 52(294–295)
Scattering, role of, in line spectra 50(349–357)
Schönberg-Chandrasekhar limit 51(66, 67, 176)
Schreibersite 52(560)
Schuppmann telescope 54(10)

Schuster-Schwarzschild model 50(352, 356, 358); 52(51, 62)
Schwarzschild exponent 54(134)
Schwarzschild variables 51(48)
Schwarzschild's criterion for convective stability 51(665, 666, 670)
Schwarzschild's method for mass determination of galaxies 53(357–358)
Schwarzschild's progressive wave theory 51(554–570)
Scintillation noise of bright stars 54(254)
Scintillation of a radio source 53(212, 230)
Scorpio-Centaurus cluster 53(70, 72, 160)
Sea interferometer 54(105)
Seas of Mercury 54(210)
Seas on the Moon, distribution 54(190)
Seas on the Moon, origin 54(198–202)
Seasons on Mars 54(228, 233, 234, 236, 237)
Secondary electron emission 54(244)
Secular changes of the orbits of comets 52(476)
Secular parallaxes 53(5)
Secular stability 51(607, 616–620, 627, 645, 650–657, 684, 746)
— and dynamical stability, relation 51(619, 620)
— and evolution 51(650–657)
Segregation of Galaxy types within clusters 53(399–400)
Seismic vibrations on the Moon 54(202, 203)
Selection factor in counts of Galaxies 53(424, 428)
Selenium, abundance 51(305, 308, 316, 318, 343)
Selenium cell 54(241, 248)
Self-absorption in chromospheric lines 52(115, 116, 121)
Self-excited oscillations 51(550–554)
— hard, as cause of pulsation 51(550, 587, 607)
Semi-amplitude of the velocity curve 50(245, 259)
Semi-carbon stars 50(106)
Semi-major axes of comets 52(512)
Semiregular variables 51(123, 282, 283, 355, 365, 403, 417–419, 584)
— red, emission lines 51(418)
— red, periods 51(584)
— red, period-spectral type relation 51(418)
— red, population type 51(418)
— yellow, emission lines 51(416)
— yellow, light curves 51(415–417)
— yellow, period-luminosity relation 51(415)
— yellow, velocity curves 51(416–417)
Sensitivity of a radio telescope 53(214–215)
Sensitivity of photographic emulsions 54(131, 132)
September comet 1882 II 52(474, 499)
Sequences of spirals 53(279, 283, 294)
Sequences of the H-R diagram 51(75)
Shape factor of a meteorid 52(521, 542)
Shape of spiral arms 53(308–310)
Shapes of comets 52(465–469, 496–509)
Shedding of matter in the equatorial plane 51(635, 636, 669)
Shell source, *see* Energy-producing shell
Shell spectra, anormal distribution of line intensities 50(396)
Shell spectra, coincidence of lines 50(395)
Shell spectra, profiles of emission lines 50(391, 392)
Shell stars 50(388, 389)
Shell stars, Balmer discontinuity 50(390)
Shell structure of nuclei and nuclear abundances, relation 51(322)
Shells, stellar
— continuous spectrum 50(389)
— emission lines 50(388, 389, 391, 392)
— physical conditions 50(396)
— reabsorption 50(392)
— theory of cycles 50(393–394)
Shock waves
— in novae 51(765)
— in prominences 52(248)
— in the atmospheres of variable stars 51(554–570, 686, 687)
— in the chromosphere 52(150)
— in the corona 52(150, 306)
— in the meteor after the impact 54(193)

Short period cepheids in clusters 51(583–584)
Short period comets 52(470, 516–518)
Short wave radio fade-out 52(350, 351)
Shot noise 54(53, 260, 263)
Side reception 54(73, 75)
Side-radiation 54(81, 82)
Siderite (=iron meteorite) 52(559, 564)
Siderolite (=stony iron meteorite) 52(559, 560)
Siderostat 54(11, 12)
Signal, least detectable 54(56)
Signal-to-noise ratio 54(249, 251)
— in radio echoes 52(450)
Silver, abundance 51(309, 316, 318, 343)
Silver disk pyrheliometer 54(3)
Similarity theorem 54(66)
Simple clustering of galaxies 53(417–443)
Simple clustering of galaxies, fundamental formula 53(426–430, 433–435)
Sirius, measurement of its diameter 54(43)
Six-color photometry 51(390, 594); 54(279)
Six-colour photometry of bright galaxies 53(341)
Size distribution of planetary nebulae 50(142)
Slit width and height, effect on limb darkening 52(11)
Slowly varying component of the solar radioemission 52(286, 296–301)
Small cores of planets, instability 52(423)
Small Magellanic Cloud
— distance and absolute magnitude 53(468–469)
— mass 53(243)
— period-luminosity curve 53(458, 461, 463)
Smoothing filter 54(45, 49)
Snow telescope 54(13, 14)
SO objects 53(278)
Sodium, abundance 51(304, 316, 317, 343)
Soft component of cosmic radiation 52(215)
Solar activity, long scale fluctuations 52(328–329)
Solar activity, measurements at various observing stations 52(333–337)
Solar apex 53(3)
Solar chromosphere 50(385)
Solar chromosphere, temperatures 50(112, 386)
Solar constant, determination 52(2–4)
Solar constant, variability 52(4)
Solar cycle 52(322–337)
— 11 year cycle 52(322–326, 331)
— 22 year cycle 52(328, 340)
Solar eclipse observation 54(33–35)
Solar energy at top of atmosphere 54(245)
Solar evolution 51(191–195)
Solar magnetism 52(340–344)
Solar models, *see also* Models 51(193, 194)
Solar motion 53(3–5)
Solar motion, local 53(11)
Solar neighborhood stars, luminosity functions 51(216–220)
— masses and luminosities 51(165–172)
Solar photosphere, models 50(319, 320, 334–339)
Solar radiation total 54(1–4)
Solar red shift 52(78)
Solar rotation 52(337–340)
Solar spectroscopes, resolving power 54(19)
Solar spectrum
— centre limb variation 50(112, 113)
— excitation temperature of the lower chromosphere 50(112)
— hydrogen lines 50(342, 365)
— molecular bands 50(110–113)
— spectrum of sun spots 50(112)
Solar system, origin 51(192)
Solar tower telescopes 54(14–16)
Solar units, definitions 51(1)
Solid hydrogen and helium 52(441, 443, 444)
Solid particles, clouds of 51(412, 423)
Solid particles in the corona 52(266)
Soot particles, condensation in the atmosphere of R Coronae Borealis 51(423)
Sound waves in the chromosphere and corona 52(149)
Source function 50(283, 315)

Source function for the solar atmosphere 52(5, 9, 13, 22, 42, 74, 75)
Space motion, *see* Proper motion
Space-charge separation in the corona 52(305)
Spaghetti process 52(103, 171)
Spallation in meteorites due to cosmic rays 52(563–564)
Spatial distribution of stars 51(226–229)
Spatial frequency 54(94)
Spectra, *see* Stellar spectra, Solar spectrum
Spectra, integrated, of globular clusters 53(179–181)
Spectra of cepheids, continuous 51(388–391)
Spectra of cepheids, line spectra 51(391–397)
Spectra of chromospheric faculae 52(179–181)
Spectra of comets 52(477–495)
Spectra of discrete radio sources 53(227–229)
Spectra of external galaxies 53(338–341, 362–366, 412)
Spectra of flares 52(199–210)
Spectra of globular clusters, individual 53(181)
Spectra of Herbig-Haro objects 51(150)
Spectra of meteors 52(524–527, 537)
Spectra of noise storms 52(316–318)
Spectra of nova-like variables 51(419)
Spectra of outbursts 52(311–315)
Spectra of quiescent prominences 52(232–237)
Spectra of radio galaxies 53(271)
Spectra of sunspots 52(157, 159, 160)
Spectra of supernovae 51(779–782)
Spectra of surges 52(212–214)
Spectra of the corona 52(259–264, 278, 283–286)
Spectra type of
— β-Cephei stars 51(400)
— classical cepheids 51(379, 380, 391–392)
— long-periodic variables 51(402–405, 410–414)
— RR Lyrae stars 51(392–393)
Spectral classification, *see* Classification
Spectral energy distribution of radiation from the Sun 52(13–18)
Spectral energy of bright galaxies 53(342)
Spectral sensitivity function 54(43, 96, 100, 114)
Spectral sensitivity of photographic plates 54(130–131, 144)
Spectral sensitivity theorem 54(43, 96)
Spectral type, general, *see also* Classification, *A* stars, *B* stars etc. 50(1–8)
Spectral type-period relation
— for cepheids 51(379–381)
— for RV Tauri stars and yellow semi-regular variables 51(415)
Spectral types, correlation to spatial and velocity distribution of stars 51(227)
Spectral types, definition 53(2)
Spectral windows in the Earth's atmosphere 52(15, 36, 37)
Spectro-enregistreur (Deslandres) 54(27)
Spectroheliograms 52(124–135)
Spectroheliograph 54(13, 26–27, 38)
Spectrohelioscope 54(26, 38)
Spectrophotometry of the bright regions of Mars 54(224–225)
Spectrophotometry of the dark regions on Mars 54(230–231)
Spectroscopes for solar observations 54(19–26)
Spectroscopic binaries
— astrometric results 50(220)
— axial rotation 50(251, 252)
— chromospheres 50(387)
— contact surface lobes 50(253, 255, 258, 263, 268)
— distortion of the velocity curve 50(251, 252, 256, 262, 263)
— double lined 50(243, 248, 255, 260)
— gaseous ring of a component 50(266)
— line profiles 50(252, 253, 254, 262)
— luminosity 50(256)
— mass ratio of the components 50(251, 258, 260)
— masses of the components 50(249, 268)
— mass-luminosity relation 50(249, 257, 259)

Spectroscopic binaries (cont.)
— motion of gaseous streams 50(262, 263)
— orbital elements 50(243)
— radii of the components 50(251)
— reflection effect 50(264)
— semi-amplitude of the velocity curve 50(245, 259)
— single lined 50(243, 251, 255, 257, 259, 260)
— threshold curve of the secondary component 50(256)
— velocity curves 50(244, 245, 250, 268)
Spectroscopic emulsions, contrast 54(130, 132)
Spectroscopic methods for distance determination of galactic clusters 53(139)
Spectrum of an activity centre 52(298–300)
Spectrum of Comet Cunninghams (1940c) 52(478, 479)
Spectrum of Comet Mrkos (1957d) 52(481)
Spectrum of galactic radio-frequency radiation 53(117, 257)
Spectrum of Neptune and Uranus 52(412)
Spectrum of the 27 days component 52(300)
Spectrum of the low chromosphere 52(118–122, 126)
Spectrum of the primary cosmic radiation 52(220)
Spectrum of the radio source Cassiopeia A 53(227–229)
Spectrum variables 51(426–429, 695ff.)
— abundance anomalies 51(349–350)
— harmonic analysis 51(705–707)
— oblique rotator model 51(428, 694)
— period-line width relation 51(698)
— table of data 51(695)
Spectrum variations of novae 51(755–762)
Speed classes of novae 51(752–753, 761)
Spicules, chromospheric 52(108, 109, 112, 113, 114, 123, 225)
Spicules, diameter and number 52(135–139)
Spillover 54(71, 75)
Spiral arm objects 51(229)
Spiral arms, rotation 53(343)
Spiral arms, shape 53(308–310)
Spiral sequences 53(279, 283, 294)
Spiral structure in the Galaxy 53(86–87, 90–96)
Spiral structure of hydrogen regions 53(111, 112)
Spirals 53(275, 276, 277, 279)
Spirals, absorption, diffraction and polarisation 53(333–338)
Spirals, apparent and true flattening 53(306–308, 328)
Spirals, energy distribution 53(342)
Spirals, flattening curve 53(322)
Spirals, luminosity and colour distribution 53(326–333)
Spirals, rotation 53(343–348)
Spirals, tilt 53(344)
Splitting of nebular lines 50(156, 158)
Sporadic coronal condensations 52(189)
Sporadic meteors 52(534–535, 544, 548–554)
Spot areas, measurements 52(322)
Spot groups, decay times 52(167)
Spot groups, magnetic classification 52(166–167)
Spot prominences 52(225, 241–244, 245)
Spots on the Mars 54(236)
Spots on the surface of Mercury 54(205–207)
Sprays 52(223)
s-process (slow) for element synthesis 51(38, 256–258, 271–272)
Sproul refractor 50(191, 204, 212)
Square-law detector 54(48, 49)
Squint angle 54(78)
SS Cygni, light curve 51(420)
S-shaped spiral 53(283, 289)
Stability, generalized definition, *see also* Convective, Dynamical, Secular, Vibrational stability 51(609)
Stability of a magnetohydrostatic equilibrium 51(718)
Stability of a spiral arm 53(94–96)
Stability of radio receiver 54(57)
Stability of rotating masses, compressible case 51(629ff., 636ff.)
— incompressible case 51(619–629)
Stability of star clusters 53(84–86)
Stability of white dwarfs 51(657, 663, 746–748)

Stability regions of the chromosphere 52(151)
Standard carbon arc 52(14, 16)
Standard classification of galaxies 53(276–278)
Standard composition of stars 51(343)
Standard field 54(168)
Standard frame 50(206)
Standard main sequence 53(144)
Standard matter of Vitense 50(300)
Standard models of stellar atmospheres 50(317)
Standard solar apex 53(5)
Standard stars 50(5, 206–208)
Standard system in photometry 54(155–157)
Standing oscillations in the atmosphere of variable stars 51(496, 497, 593)
Star capture 50(273)
Star chains 51(151)
Star clusters, *see* Galactic clusters, Globular clusters
Star counts in globular clusters 53(174)
Star formation 51(135–157); 53(99)
Star formation in *O*- and *T*-associations 51(200, 201)
Star image, photographic 54(143–144)
Star streaming in a local region 53(37–39)
Star streams, theory of 53(5, 21)
Stark broadening in the solar spectrum 52(49, 50)
Stark effect in flare spectra 52(203, 204)
Stark effect in stellar atmospheres 50(344, 346)
Star-like nuclei in Herbig-Haro objects 51(150, 151)
Stars, age 53(98)
Stars with trigonometric parallaxes greater than 0″.050 (table) 51(90–93)
State equation for dense matter (white dwarfs) 51(724–728)
State equation of solid H_2 and He 52(444)
State equation, polytropic 51(13, 16–17, 45)
Static universe 53(425, 429, 430)
Stationary state of a stellar system, deviations from 53(41–43)
Stationary stochastic process(es) (representing the distribution of galaxies) 53(419, 433)
Statistical broadening of the Fraunhofer lines 52(46, 49)
Statistical fluctuations of radiation flux 54(249, 282)
Statistical parallax 53(448, 453, 460, 462–463, 466–467)
Statistical relationships (light and velocity curves) 51(374–376)
Statistically uniform distribution of particles 53(418)
Statistics of clusters of galaxies 53(423, 439–443)
Steady ejection of matter by stars 51(245–247, 275)
Steady-state oscillations in a star 51(540–542)
Stebbins-Whitford effect 53(475, 480)
Stellar associations 53(69–73, 165–166)
Stellar atmospheres
— absorption coefficient for lines 50(340–346)
— absorption coefficient in the continuum 50(295, 296, 297, 299, 328, 329)
— absorption lines 50(276, 278, 340–364)
— chemical abundances 50(293, 369, 371, 373)
— chromosphere 50(385, 386, 387)
— continuous absorption 50(281, 294–300, 309, 310)
— convection 50(329, 332)
— damping 50(343, 344, 375)
— electronic pressure 50(278, 279, 292, 339)
— emission lines 50(278, 388, 389, 391, 392)
— extended, *see also* Shell stars, *Be* stars 50(278, 384–397)
— gas pressure 50(292, 311)
— photosphere 50(316–339)
— reversing layer 50(278)
— thermodynamical equilibrium 50(286, 287, 317, 332)
— transfer problems 50(300ff.)
— turbulence 50(375–378)
Stellar energy production 51(25–38, 250–263)
Stellar evolution, *see also* Evolution 53(96–99)

Stellar gradients 50(74, 84)
Stellar magnetism, effect on evolution 51(284–286)
Stellar material, constitutive equations 51(5, 15–38)
Stellar material, equations of state 51(16–20)
Stellar models, *see* Models
Stellar populations 50(1, 114, 136, 159); 53(55, 96, 97)
— concept and subdivision 51(128–131, 225–226)
— type I, early types 50(159)
— type II 50(81, 88, 91, 97, 159)
Stellar radii, *see* Radii
Stellar rotation 51(276–278)
Stellar spectra
— classification, *see* Classification
— continuous 50(152, 275, 322, 389)
— energy distribution 50(322)
— hydrogen lines 50(366–368)
— line broadening 50(252, 341–346, 375, 381, 384)
— line spectra, *see* Absorption lines, Emission lines
— shell spectra 50(388–396)
Stellar structure, differential equations 51(5–15)
Stellar variability and stage of evolution 51(278–284)
Stone spectroheliograph 54(28)
Stoner-Anderson relation 51(726)
Stormbursts of the solar radio-radiation 52(302–303, 308, 315–320)
Störmer orbits 52(217)
Stratification in planetary nebulae 50(416–425)
Stratification of the solar atmosphere 52(55, 62)
Stray light 54(5, 6)
Streamers of the corona 52(280–283)
Strip integration 54(93, 123)
Strömgren's method of model atmospheres 52(27)
Strong-CN stars 50(106)
Strong-line stars 50(88, 99)
Strontium, abundance 51(308, 316, 318, 343)
Structural index for clusters of galaxies 53(401)
Structure, internal, of white dwarfs 51(739–750)
Structure of clusters of galaxies 53(397–406, 428, 429, 431, 436, 440)
Structure of CN bands in comets 52(487)
Structure of lunar surface 54(182, 184, 185, 188)
Structure of the Martian atmosphere 52(398–400)
Structure of the Martian surface 54(225–226)
Structure of the Venus atmosphere 52(388–390)
Sturm-Liouville problem 51(459, 461)
Subdivision of spiral sequences 53(283, 286)
Subdwarfs, abundance anomalies 51(268, 346)
Subdwarfs as globular cluster main sequence 51(114, 117)
Subdwarfs, hot evolutionary stage 51(187–188)
Subdwarfs in galactic clusters 51(106, 107)
Subdwarfs, mass-luminosity relation 51(171–172)
Subgiant components of binary systems 50(258)
Subgiants in globular clusters 51(180)
Subgiants, mass-luminosity relation 51(170)
Subharmonic resonance 51(545, 548, 593)
Substratum, expansion 53(497, 503–504, 506–509)
Substratum, rotation 53(496–497, 503–504)
Sub-systems in the Galaxy 53(23, 29, 44–45, 96, 98)
Sudden commencement storm 52(354)
Sudden cosmic noise absorption 52(350)
Sudden disappearances of prominences 52(237, 239)
Sudden enhancements of atmospherics 52(351, 352)
Sudden fieldstrength anomalies 52(351, 353)
Sudden phase anomalies 52(351, 353)
Sulphur, abundance 51(304, 316, 317, 343)
Sulphur-selenium abundance ratio 51(305)

Sum rule for isobaric nuclear abundances 51(321)
Sun, *see also* Solar
Sun, abundance of elements 51(343, 344)
Sun as photometer standard 54(244)
Sun, chemical composition 51(298–300)
Sun, energy source 51(165)
Sun, model atmospheres 51(339, 340)
Sun, radio echoes 52(449, 460)
Sun's limb, *see also* Limb
Sun's limb, mass motions at the 96
Sunspot groups, *see* Spot groups
Sunspot number 52(322–326, 331)
Sunspot prominences 52(225, 241–244, 245)
Sunspot theory of variable stars 51(359)
Sunspots, band spectra 52(160)
Sunspots, butterfly diagram 52(326)
Sunspots, continuous radiation 52(157)
Sunspots, diameter 52(151, 154)
Sunspots, effective temperature 52(157)
Sunspots, intensity profile 52(157)
Sunspots, internal motion in spot groups 52(170)
Sunspots, latitude drift 52(339)
Sunspots, line spectra 52(159)
Sunspots, magnetic field 52(152–157, 164, 343)
Sunspots, model of the umbra 52(159, 161)
Sunspots, periodicity of number 52(325)
Sunspots, rôle in the development of a centre of activity 52(184–185)
Sunspots, theory of 52(171–173)
Sunward tails of a comet 52(467, 502–505)
Supercluster 53(410, 444)
Supergiants of type *M* 50(62)
Supergiants, variable 51(125, 280–283)
Superheterodyne receiver 54(46)
Supernova of 1054 51(767)
Supernovae, absolute brightness 51(763, 778–779)
Supernovae and common novae, distinction 51(766–767)
Supernovae, energy emitted in outbursts 51(763, 784)
Supernovae, evolutionary stage 51(188)
Supernovae, explosions 51(258–260, 272–275, 555, 686)
Supernovae, frequency 51(772–775)
Supernovae, history of discoveries 51(766–772)
Supernovae, initial and final stage 51(782–784)
Supernovae, light curves 51(776–778)
Supernovae, magnetic fields 51(285)
Supernovae, mass loss 51(249)
Supernovae, radio emission 53(260)
Supernovae, spectra 51(779–782)
Supernovae, table of data 51(774–775)
Supernovae, theory of outbursts 51(784–785)
Supernovae, type I and II 51(763, 776, 781, 782)
Super-Schmidt Meteor Camera 52(539–541, 544)
Support of prominences 52(247)
Suppression of steady noise component 54(57–59)
Surface brightness of external galaxies 53(315)
Surface brightness of the chromosphere 52(106, 108)
Surface condition 51(58, 71, 179, 180)
Surface density of planet, upper limit 52(430)
Surface frost point of Mars 52(395)
Surface layers of white dwarfs 51(742–746)
Surface of Mars, properties 54(220–237)
Surface of Mercury, properties 54(205–210)
Surface of the asteroïds 54(211–215)
Surface of the Jupiter satellites, properties 54(218–220)
Surface of the Moon, *see* Lunar surface
Surface pressure of Mars 52(393–394)
Surface structure of meteorites 52(560–561)
Surface temperature 50(348, 349)
— of Mars 52(395–396, 397)
— of Venus 52(391)
Surges 52(211–214, 223, 225, 242, 244)
Surges, rôle in the development of a centre of activity 52(184–185)

Swan bands of the C_2 molecule 50(66)
SX Cassiopeiae, spectra 50(265)
Synchronization of rotation and revolution in binary systems 50(253)
Synchrotron radiation 52(321)
Synchrotron radiation as a cause for radio emission 53(102, 104–106, 126–127, 235–237, 260)
Syndynames in comet tails 52(500, 501)
Synoptic maps 52(186, 187)
Synthetic brightest galaxy of a cluster 53(479, 481, 482)

T Tauri, genetic relation to dense nebulae 51(197)
T Tauri stars 51(149, 197, 198, 278, 424–426)
Tails of comets 52(467–468, 500–501, 506, 507)
Tangential motions of radio sources 52(320, 321)
Tantalum, abundance 51(312, 316, 319, 343)
T-associations, definition 51(197)
T-associations, expansion 51(199)
T-associations, star formation 51(200, 201)
Taurid meteor stream 52(546–547, 554)
Taurus A radio source 52(294)
— apparent angular diameter 53(225)
— polarisation of radiation 53(229)
— spectrum 53(228)
Taylor number 51(673)
Technetium in *M*, *S*, and *C* stars 50(116, 117, 132)
Technetium lines in *S* stars 51(349, 414)
Tectites 54(181)
Tektite (=glassy meteorite) 52(559)
Telescope area and photometric precision 54(252)
Telescopic observations
— of the lunar surface 54(191–204)
— of the Martian surface 54(222–223, 228–229, 232–234)
— of the surface of Mercury 54(207–208)
— of the surface of the Jupiter satellites 54(216–218)
Television tubes 54(287–288)
Telluric bands 50(110)
Tellurium, abundance 51(311, 316, 318, 343)
Temperature distribution in various inhomogeneous models of the photosphere 52(92)
Temperature distribution, solar, versus depth 52(22, 23, 32)
Temperature, effective 50(4, 22, 278, 321, 367, 374, 379, 383)
Temperature for formation of terrestrial planets 52(364)
Temperature gradient for convective transport of energy 51(6, 7, 10–12)
Temperature gradient for radiative transport of energy 51(6, 7–10)
Temperature gradient of the Earth's atmosphere 52(373, 374)
Temperature in the interior of the asteroïds 54(213)
Temperature of central stars of planetary nebulae 50(426–431)
Temperature of interstellar clouds 53(107)
Temperature of the high atmosphere of Venus 52(392)
Temperature of the lunar surface 54(184–186)
Temperature of the solar chromosphere 50(112, 386)
Temperature of the surface of Mercury 54(210)
Temperature of white dwarfs 50(176, 178)
Temperature, rotational 50(110, 127)
Temperature, surface 50(348, 349)
Temperature variations in white dwarfs 51(737–738)
Temperature, vibrational 50(68, 110, 127)
Temperature-altitude curve for Mars 52(400)
Temperature-altitude curve for Venus 52(391)
Temperature-luminosity-mass relation for white dwarfs 51(743)
Temperatures in quiescent prominences 52(232, 233)
Temperatures in Saturn 52(410)
Temperatures in sunspots 52(157, 172, 173)

Temperatures in the Earth's atmosphere 52(368, 369, 373)
Temperatures in the photosphere 52(22, 23, 32, 47, 86–92)
Temperatures in the transition region chromosphere-corona 52(146)
Temperatures in the Venus atmosphere 52(388–390, 391)
Temperatures of chromospheric faculae 52(179–181)
Temperatures of Jupiter 52(407)
Temperatures of Mars 52(395–396, 397, 399)
Temperatures of Moon and Mercury 52(405)
Temperatures of the corona 52(269–271)
Temperatures of the low chromosphere 52(118–122, 140)
Temperatures of the medium chromosphere 52(122–124, 140)
Temperatures of the minimum corona 52(269–271)
Terminal velocity of a meteor 52(527, 532)
Terrestrial atmosphere, *see* Earth's atmosphere
Terrestrial planets, atmospheres 52(363, 366–383)
Terrestrial planets, interior 52(431, 432–440)
Terrestrial planets, masses and diameters 52(437)
Test particles in world-models 53(518)
Thallium, abundance 51(314, 316, 320, 343)
Theoretical models of the Sun 52(59)
Thermal conductivity in dense matter (white dwarfs) 51(729–731)
Thermal conductivity of the lunar material 54(185, 188)
Thermal conductivity of the solar convection zone 52(102)
Thermal detectors 54(241, 253)
Thermal energy available 51(636, 640, 643)
Thermal noise, theory of Nyquist 54(52)
Thermal radiation from Mercury 54(209–210)
Thermal radiation from the lunar surface 54(184–186)
Thermal radio emission from
— an ionized gas 53(102–104, 117, 234–235)
— the Galaxy 53(258)
— centres of activity 52(296–301)
— the corona 52(283–286)
Thermal tensions in the lunar rocks 54(188–189, 203)
Thermodynamic equilibrium 52(32, 33)
— departure from 52(59, 72, 73)
— local 50(286, 287, 332)
Thermodynamical properties of dense matter (white dwarfs) 51(729)
Thermonuclear explosions 51(185, 258–260, 274)
Thermonuclear reactions 51(29–32)
— in white dwarfs 51(732–738)
— in white dwarfs, response to temperature and density variations 51(737–738)
Thiele-Innes method 50(201)
Thin disk approximation for mass determination of galaxies 53(350–354)
Thin-film interference filters 54(17)
Thorium, abundance 51(316, 320)
Three-color photometry 54(159)
Three-dimensional classification of external galaxies 53(282)
Three-stream model of the photosphere 52(72, 73, 88)
Threshold curve of the secondary component of spectroscopic binaries 50(256)
Tidal origin of the Moon 52(424)
Tilt of spirals 53(287, 308)
Time measurement in the Universe 53(485–488)
Time-displacement curves for visual binaries 50(198, 199)
Tin, abundance 51(310, 316, 318, 343)
TiO bands in *M* stars 50(114, 115)
Titan, satellite of Saturn 54(215)
Titanium, abundance 51(305, 316, 317, 343)
Titanium-zirconium abundance ratio 51(305)
Titius-Bode sequence 52(432)
Tolerances of aerials 54(76–82)
Top heights of flares 52(192)
Topographic map of Mercury 54(205–207)

Topography of the Martian surface 54(221–223)
Topology of world models 53(515–516)
Tornado prominences 52(225)
Torsional oscillations in magnetic variables 51(535)
Torsional oscillations of a star 51(720, 721)
Total mass of the Galactic System 53(56)
Total masses of the Magellanic Clouds 53(243)
Total solar radiation 54(1–4, 245)
Tower telescopes for solar observations 54(14–16, 29, 38–39)
Train of a meteor 52(526, 536, 539, 543, 544)
Train of a meteor, coasting velocity 52(543–544)
Transfer equation 50(282, 283, 350, 352); 52(42–43, 46, 52, 54–58)
Transfer of photometric scale 54(168, 172)
Transfer problems in a gray atmosphere 50(300ff.)
Transition probabilities of forbidden lines 50(406, 416)
Transition region chromosphere-corona 52(146–148)
Transition spectrum of novae 51(758, 759)
Transmission line 54(75)
Transmission of the Earth's atmosphere 52(345)
Travel times of particles from the Sun 52(362)
Traving model 50(321)
Tri-alkali cathode 54(247)
Triple stars, frequency 50(223)
Troilite 52(560)
Tropopause height for the Earth 52(374)
Tropopause of Venus 52(389, 390)
True flattening of galaxies 53(303–308)
Tungsten, abundance 51(313, 316, 319, 343)
Turbidity effect 54(134)
Turbulence 50(97, 375–378)
Turbulence and curves of growth 50(375)
Turbulence and line profile 50(375)
Turbulence and protostar formation 51(148)
Turbulence element 52(93)
Turbulence in the chromosphere 52(112–116)
Turbulence in the photosphere 52(93–98, 102, 103, 104)
Turbulence, solar models with 52(26, 104–105)
Turbulence theory of spiral structure 53(92–93)
Turbulent and magnetic energy, equipartition 51(715)
Turbulent velocity 52(48, 65, 69, 70, 71, 141)
Turbulent velocity of novae 51(758)
Turbulent viscosity 51(441, 506–508)
Turbulent viscosity and damping of oscillations 51(585)
Two-aerial interferometer 53(213)
Two-channel pulse-counting photometer 54(278)
Two-element interferometers 54(98–113)
Two-photon emission in planetary nebulae 50(155, 156)
Two-stream model of the photosphere 52(31, 78, 89)
Two-stream theory of stellar motion 53(5, 21)
Tycho's Star of 1572 51(768)
Type I and II of Supernovae 51(763, 776, 781, 782)
Type I bursts of the solar radio-radiation 52(302–303, 315–320)
Type II bursts of the solar radio-radiation 52(222, 302–303, 311–315)
Type II Cepheids, *see* Cepheids of Type II
Type III burst of the solar radio-radiation 52(222, 302–308, 311)
Type IV phenomenon of the solar radioradiation 52(302–303, 320–322)
Type V phenomenon of the solar radioradiation 52(302–303, 320–322)
Typical *S*-shaped barred spiral 53(289)

U, B, V-System 54(159–161, 280)
U Cephei, velocity curves 50(250, 262)
U Geminorum stars 51(419–421, 763, 764)
— periods 51(575–577)

U type bursts of the solar radio-radiation 52(302–303, 304)
Ultraviolet excess in globular cluster stars 51(117)
Ultraviolet radiation of the sun 52(14, 15, 144–148, 264, 344, 350)
Ultraviolet spectrum of flares 52(208)
Umbra of sunspots 52(151)
Umbra of sunspots model 52(159, 161)
Underground continuum in planetary nebulae 50(155)
Unequal interferometer 54(105–106)
Uniform model universe 53(446)
Unipolar groups of sunspots 52(167)
Unipolar magnetic regions of the Sun 52(181, 190, 248)
Universal redshift 53(318, 384, 411–414, 423, 431, 435–439, 440, 446, 474, 477–478, 486, 509)
Universal time 53(502)
Unresolved astrometric binaries 50(199, 203, 210, 222)
— analysis of path 50(210–218)
Unstable layer in the photosphere 52(30)
Unstable modes 51(515, 516)
Upper chromosphere, spectroheliograms 52(131–135)
Upward motion of erruptive prominences 52(240)
Uranium, abundance 51(316, 320)
Uranostat 54(11, 13)
Uranus, atmosphere 52(406, 410–412)
Uranus, composition of interior 52(448)
Ursa Major cloud 53(392, 397)
Ursa Major cluster 53(159)
UV Ceti variables 51(429–430, 574)
UX Monocerotis or SX Cassiopeiae group of eclipsing binaries 50(236, 237, 239)
UX, spectroscopic study 50(269, 270)

V 444 Cygni or DQ Herculis group of eclipsing binaries 50(237, 239)
Variable giants and supergiants 51(125, 280–283)
Variable line broadening in β-Cephei spectra 51(401, 579–581)
Variable radio sources 53(230)
Variable stars
— classification 51(364, 365)
— distribution in the HR diagram 51(570–572)
— evolutionary stage 51(278–284)
— gravitationally contracting 51(278–279)
— history of discoveries 51(354–357)
— history of theories 51(357–364)
— in galactic clusters 53(142)
— in globular clusters 53(181–184, 450–457, 469)
— in the horizontal branch 51(283)
— light and velocity curves, *see also* Velocity curves 51(366ff.)
— mass 51(357)
— molecular bands 50(119)
— of population I 53(457–466)
— of population II 53(450–457)
— on and above the main sequence 51(279–280)
— period-luminosity relation, *see also* Period-luminosity relation 51(356)
— population type 51(365, 366)
— relation between class and stellar dynamics 51(365)
— tables of data 51(365, 574)
— theoretical and observed periods 51(574–585)
— with double or multiple periodicity 51(385, 400)
— with extremely rapid light variations 51(429–431)
Variables, eclipsing, *see* Eclipsing binaries
Vegetation on Mars 54(231, 236, 237, 239)
Veil theories of variable stars 51(359, 587)
Veil theory for long-period variables 51(423)
Velocities of meteors 52(527, 532, 557–558)
Velocities of particles emitted from the Sun 52(222–224)
Velocities of rotation of spirals 53(346–348)
Velocity and light curves of variable stars, amplitude correlation 51(588–589)

Velocity and light curves of variable stars, phase correlation 51(374–376, 399, 400, 589–592)
Velocity curves of
— β Cephei stars 51(398–400)
— binaries 50(244, 245, 250, 252, 256, 262)
— binaries, rotational distortion 50(251, 252, 256)
— classical cepheids 51(375)
— long-period variables 51(400–410)
— magnetic variables 51(428)
— RV Tauri stars and yellow semiregular variables 51(416–417)
— variable stars, asymmetry 51(376)
— variable stars, double or multiple periodicity 51(398–400)
Velocity dispersion of
— clusters of galaxies 53(384)
— galaxies 53(358–359, 366)
— galaxies in clusters 53(407)
— stars 51(228); 53(5–6, 29–31, 73–86)
Velocity distribution, asymmetrical in the Galaxy 53(6, 22–23, 29–31)
Velocity distribution in the neighbourhood of the Sun 53(32–36)
Velocity distribution of stars 51(226–229)
Velocity ellipsoid 53(31)
— and dispersion orbits 53(66–69)
— theories 53(61–63)
Velocity of escape 53(45, 52)
Velocity of light, determination from echoes from the Moon 52(459)
Velocity of recession of clusters of galaxies 53(425, 430)
Velocity space, equidensity curves 53(42)
Velocity variations in peculiar *A* stars 51(702–704)
Venus atmosphere
— chemical composition 52(383–388)
— escape of gas 52(392–393)
— evolution 52(393)
— models 52(390–392)
— temperature and structure 52(388–390)
Venus, model of interior 52(439)
Venus, radio echoes 52(462)
Vertex deviation 53(67)
Vesta, albedo 54(212)
Vesta, diameter 54(211)
Vesta, polarisation of light 54(214)
Vibration temperature of stellar molecules 50(68, 110, 127)
Vibrational instability 51(605, 673, 674, 678–680)
Vibrational instability as cause of pulsation 51(585–587)
Vibrational stability
— of stars 51(475–508, 684–685)
— of stars, effects of energy generation in the star 51(481–485)
— of white dwarfs 51(747–748)
— towards radial oscillations 51(647–650, 684–685)
Vibration-rotation bands 50(126)
Violet layer of Mars 52(396–398)
Virginid meteor stream 52(545–546, 554–555)
Virgo A source 53(268)
Virgo A source, apparent angular diameter 53(226)
Virgo Cluster of galaxies 53(392, 397, 473–474, 478, 481, 484, 485)
Virial theorem 51(137, 465, 530); 53(80, 160–161)
Viscosity of the solar convection zone 52(102)
Viscous damping of non-radial oscillations 51(525–526)
Viscous dissipation 51(475)
Visual binaries
— catalogues 50(192)
— discovery 50(189)
— Kepler problem 50(195)
— magnitude compensation 50(190, 191)
— mass function 50(216, 218)
— masses 50(221)
— mass-luminosity relation 50(220, 221)
— mass-ratio of the components 50(212–213)
— orbital elements 50(196, 198, 200–202)
— photocenter 50(210, 211, 217)
— photographic measurements 50(189, 190)
— relative position of the components 50(188, 189, 196)
— resolved 50(199, 203)
— resolved, orbital analysis 50(208–210)
— spectra of the components 50(193)
— time-displacement curves 50(198, 199)

— unresolved astrometric 50(199, 203, 210, 222)
— unresolved astrometric, orbital analysis 50(210–218)
Visual magnitude scale 54(244)
Visual meteor studies 52(534)
Visual photometry 54(138)
VO bands in *M* stars 50(114)
Vogt-Russell theorem 51(42–43, 74, 75, 129)
Volcanic ashes on the Moon 54(182, 183, 219)
Volcanoes on Mars 52(397)
Volcanos on the Moon 54(202, 203)
Von Zeipel's theorem 51(39, 526, 534)
Vortices, solar 52(179)

W Orionis or XZ Sagittarii group of eclipsing binaries 50(237–239)
W Ursae Maioris eclipsing binaries 50(257, 261, 262, 271)
W Virginis stars 51(282, 366)
— emission lines and line doubling 51(395)
— theoretical and observed periods 51(583–584)
Wadsworth mounting of a grating 54(21, 26, 35)
Water content of the Earth's atmosphere 52(372)
Water in the Martian atmosphere 52(395, 404)
Water vapor in the Venus atmosphere 52(384, 393)
Water-flow pyrheliometer 54(2, 4)
Waveguide as feeder 54(75–76)
Wavelength dependence of extinction 54(152)
Wavelength dependence of the absorption coefficient 52(65, 66)
Wavelength displacements of the Fraunhofer lines 52(77–78)
Wavelengths of the solar spectrum 52(37–39)
Weak signals 54(56–57)
Weak stormbursts 52(317)
Weak-line stars 50(90)
Weaver's model 53(15)
Weierstrass condition 51(647, 657)
Weighting functions for the stratification of the solar atmosphere 52(55, 56, 59)
Whipple's equation 51(142)
Whirlpool Nebula 53(375, 378)
Whistlers 52(462)
White corona 52(249)
White dwarf(s) 52(419)
— absorption lines 50(168, 179)
— abundance anomalies 51(268, 271, 347)
— abundance ratios 50(169, 182)
— age 50(185)
— as a final evolution stage 51(190, 191, 750–751)
— atmospheres 50(167)
— Balmer discontinuity 50(173, 176, 177)
— Balmer lines 50(180, 182)
— color index 50(176, 178)
— color magnitude diagram 51(723)
— energy production 51(732–738, 748–750)
— Grw + 70° 8247, spectrum 50(171)
— in galactic clusters 51(209); 53(141)
— internal structure 51(739–750)
— luminosity 50(165)
— masses 50(183, 184)
— mass-radius relation 51(740, 741)
— models 51(73–74)
— opacity 50(182)
— photoelectric distinction 50(165)
— proper motion 50(165, 166)
— R 640, spectrum 50(174)
— radii 50(176, 178, 183)
— spectral types 50(170–175)
— stability 51(657, 663, 746–748)
— table of observational data 50(162, 163)
— temperatures 50(176, 178)
— W 219, spectrum 50(174)
Widening of Fraunhofer lines 52(89–90)
Widmannstätten figures on meteorites 52(560–561)
Wiechert models 52(432)
Wien's approximation 54(245)
Wigner-Seitz method 52(442)
"Windows" in the Earth's atmosphere 52(15, 36, 37)
Wire grating as attenuation device 54(164)
Wolf number (sunspot number) 52(322–326, 331)

Wolf-Rayet stars 50(388, 389)
— abundance anomalies 51(269, 346, 347)
— evolutionary stage 51(187)
World models 53(495–499, 506–509)
— inhomogeneous 53(498, 517)
— topology of 53(515–516)
World-postulate 53(495, 504–506, 526, 531)
Wyse and Mayall method for mass determination of galaxies 53(350–354)

Xenon, abundance 51(311, 316, 318, 343)
x-process for element synthesis 51(261–263)
X-ray radiation of the Sun 52(344, 347, 350, 353)
X-ray spectrum of flares 52(208, 209)
X-ray spectrum of the corona 52(256, 261–264, 278)

Yagi aerials 54(69)
Yellow semiregular variables, *see* Semiregular variables
Yerkes Observatory 54(10)
YO bands in *M* stars 50(115)
υ Sagittarii or β Lyrae group of eclipsing binaries 50(237, 239)
Yttrium, abundance 51(308, 316, 318, 343)

Z Camelopardalis stars 51(421–422)
Zanstra theory of planetary nebulae 50(426)
Zeeman effect in stellar spectra 51(690–692)
Zeeman splitting of Fraunhofer lines 52(153)
Zenith extinction coefficient 54(153)
Zero point of classical cepheids 51(127–128)
Zero point of period luminosity curve 53(458, 461, 463–465, 474)
Zero point of RR Lyrae stars 51(109, 118, 126–127)
Zero-point of scale of magnitudes 54(244)
Zinc, abundance 51(307, 316, 317, 343)
Zirconium, abundance 51(309, 316, 318, 343)
Zirconium-hafnium abundance ratio 51(305)
Zodiacal cloud 52(530)
Zodiacal light 52(253, 254, 255, 431, 520, 530–531)
Zonally stratified atmospheres in *A* stars 51(697)
ZrF bands in variable stars 50(119)
ZrO bands in *S* and *M* stars 50(116)
Zwiers method 50(200)

Index of Contributors

The group number is given in boldface, the volume number in brackets and the final number gives the first page of the article.

Åberg, T., Howat, G. (1982): Theory of the Auger Effect. **6** (31), 469

Akasofu, S.-I., Chapman, S., Meinel A.B. (1966): The Aurora. **10** (49/1), 1

Alburger, D.E. (1957): Nuclear Isomerism. **8** (42), 1

Aller, L.H. (1958): The Abundances of the Elements in the Sun and Stars. **11** (51), 324

Allis, W.P. (1956): Motions of Ions and Electrons. **4** (21), 383

Alpert, D. (1958): Production and Measurement of Ultrahigh Vacuum. **3** (12), 609

Al'pert, J.L. (1976): Wave-Like Phenomena in the Near-Earth Plasma and Interactions with Man-Made Bodies. **10** (49/5), 217

Amaldi, E. (1959): The Production and Slowing Down of Neutrons. **8** (38/2), 1

Anderegg, J.W. → Beeman, W.W. et al. (1957)

Antman, S.S. (1972): The Theory of Rods. **3** (6a/2), 641

Arnold J.R. → Honda M. (1967)

Arp, H.C. (1958): The Hertzsprung-Russell Diagram. **11** (51), 75

Auster V. → Schmidt, H. (1971)

Bagley, A. (1967): Frequency and Time Measurements. **4** (23), 289

Bahner, K. (1967): Teleskope. **5** (29), 227

Baldinger, E. (1959): Kaskadengeneratoren. **8** (44), 1

Balsley, J.R. (1971): Aeromagnetic Surveying with the Fluxgate Magnetometer. **10** (49/3), 395

Barbier, D. (1958): Théorie générale des atmosphères stellaires. **11** (50), 274

Bardeen, J. (1956): Theory of Superconductivity. **3** (15), 274

Barker, L.M. → Nunziato, J.W. et al. (1974)

Barone, A. (1962): Generation, Detection and Measurement of Ultrasound. **3** (11/2), 74

Barschall, H.H. (1958): Detection of Neutrons. **8** (45), 437

Bartels. J. (1957): Gezeitenkräfte. **10** (48), 734

Baule, H., Müller E. (1956): Messung elastischer Eigenschaften von Gesteinen. **10** (47), 169

Beeman, W.W., Kaesberg, P., Anderegg, J.W., Webb, M.B. (1957): Size of Particles and Lattice Defects. **6** (32), 321

Bell, E.E. (1967): Optical Constants and their Measurement. **5** (25/2a), 1

Bell, J.F. (1973): The Experimental Foundations of Solid Mechanics. **3** (6a/1), 1

Benedetti, S. de, Findley, R.W. (1958): The Coincidence Method. **8** (45), 222

Berger, M.J. → Fano, U. et al. (1959)

Bergmann, P.G. (1962): The Special Theory of Relativity. **2** (4), 109

Bergmann, P.G. (1962): The General Theory of Relativity. **2** (4), 203

Bergstrand, E. (1956): Determination of the Velocity of Light. **5** (24), 1

Berker, R. (1963): Intégration des équations du mouvement d'un fluide visqueux incompressible. **3** (8/2), 1

Berry, D.S. → Sneddon, I.N. (1958)
Bethe, H.A., Salpeter, E.E. (1957): Quantum Mechanics of One- and Two-Electron Systems. **7** (35), 88
Bilz, H., Strauch D., Wehner, R.K. (1984): Vibrational Infrared and Raman Spectra of Non-Metals. **5** (25/2d), 1
Birkhoff, R.D. (1958): The Passage of Fast Electrons Through Matter. **6** (34), 53
Birman, J.L. (1974): Theory of Crystal Space Groups and Infra-Red and Raman Lattice Processes of Insulating Crystals. **5** (25/2b), 1
Bishop, G.R., Wilson, R. (1957): The Nuclear Photoeffect. **8** (42), 309
Blackman, M. (1955): The Specific Heat of Solids. **3** (7/1), 325
Blin-Stoyle, R.J., Grace, M.A. (1957): Oriented Nuclei. **8** (42), 555
Borgnis, F.E., Papas, C.H. (1958): Electromagnetic Waveguides and Resonators. **4** (16), 285
Bosch, J.C. van den (1957): The Zeeman-Effect. **5** (28), 296
Bouman, J. (1957): Theoretical Principles of Structural Research by X-rays. **6** (32), 97
Bracewell, R.N. (1962): Radio Astronomy Techniques. **11** (54), 42
Breene jr., R.G. (1964): Line Width. **5** (27), 1
Breit, G. (1959): Theory of Resonance Reactions and Allied Topics. **8** (41/1), 1
Breit, G., Gluckstern, R.L. (1959): Coulomb Excitation. **8** (41/1), 496
Breit, G., McIntosh, J.S. (1959): Polarization of Nucleons Scattered by Nuclei. **8** (41/1), 466
Breit, G. → McAllister jr., H.H. (1959)
Bremmer, H. (1958): Propagation of Electromagnetic Waves. **4** (16), 423
Breus, T.K. → Gringauz, K.I. (1976)
Bricard, J. (1957): Diffusion des radiations par les gouttes d'eau en suspension dans l'atmosphère. **10** (48), 329
Brown jr., W.F. (1956): Dielectrics. **4** (17), 1
Brown, R.H. (1959): Discrete Sources of Cosmic Radio Waves. **11** (53), 208
Brown, S.C. (1956): Breakdown in Gases: Alternating and High-Frequency Fields. **4** (22), 531
Bückner, H. (1955): Moderne Rechenmaschinen. **1** (2), 471
Buff, F.P. (1960): The Theory of Capillarity. **3** (10), 281
Bullen, K.E. (1956): Seismic Wave Transmission. **10** (47), 75
Burbidge, E.M., Burbidge, G. (1958): Stellar Evolution. **11** (51), 134
Burbidge, G. → Burbidge, E.M. (1958)
Burcham, W.E. (1957): Nuclear Reactions, Levels, and Spectra of Light Nuclei. **8** (40), 1

Cabannes, H. (1960): Théorie des ondes de choc. **3** (9), 162
Cagniard, L. (1956): Electricité tellurique. **10** (47), 407
Carlson, D.E. (1972): Linear Thermoelasticity. **3** (6a/2), 297
Carr jr., W.J. (1966): Secondary Effects in Ferromagnetism. **4** (18/2), 274
Chapman, S. → Akasofu, S.-I. et al. (1966)
Chen, P.J. (1973): Growth and Decay of Waves in Solids. **3** (6a/3), 303
Christensen, C.J. → Kofoed-Hansen, O.M. (1962)
Claussen, H.C. (1967): Mikroskope. **5** (29), 343
Cocconi, G. (1961): Extensive Air Showers. **9** (46/1), 215
Cochran, W., Cowley, R.A. (1967): Phonons in Perfect Crystals. **5** (25/2a), 59
Cohen, B.L. (1959): Cyclotrons and Synchrocyclotrons. **8** (44), 105
Cohen, E.R., DuMond, J.W.M. (1957): The Fundamental Constants of Atomic Physics. **7** (35), 1
Collatz, L. (1955): Numerische und graphische Methoden. **1** (2), 349
Collings, S.C. (1956): Helium Liquefiers and Carriers. **3** (14), 112
Corrsin, S. (1963): Turbulence: Experimental Methods. **3** (8/2), 524
Coulomb, J. (1956): Séismométrie. **10** (47), 24

Coulomb, J. (1956): L'agitation microséismique. **10** (47), 140
Courant, E.D. → Green, G.K. (1959)
Cowley, R.A. → Cochran, W. (1967)
Cox, E.F. (1957): Sound Propagation in Air. **10** (48), 455
Craggs, J.D., Massey, H.S.W. (1959): The Collisions of Electrons with Molecules. **7** (37/1), 314
Curran, S.C. (1958): The Proportional Counter as Detector and Spectrometer. **8** (45), 174
Curtis, D.A. → Verstelle, J.C. (1968)

Darmois, E. (1957): Électrochimie. **4** (20), 392
Daunt, J.G. (1956): The Production of Low Temperatures Down to Hydrogen Temperature. **3** (14), 1
Defant, A. (1957): Flutwellen und Gezeiten des Wassers. **10** (48), 846
Dehlinger, U. (1958): Umwandlungen und Ausscheidungen im kristallinen Zustand. **3** (7/2), 211
Deutsch, A.J. (1958): Magnetic Fields of Stars. **11** (51), 689
Devons, S., Goldfarb, L.J.B. (1957): Angular Correlations. **8** (42), 362
Dollfus, A. (1962): La nature de la surface des planètes et de la Lune. **11** (54), 180
Döring, W. (1966): Mikromagnetismus. **4** (18/2), 341
DuMond, J.W.M. → Cohen, E.R. (1957)

Edlén, B. (1964): Atomic Spectra. **5** (27), 80
Edmondson, F.K. (1959): Kinematical Basis of Galactic Dynamics. **11** (53), 1
Elbaum, Ch. → Truell, R. (1962)
Eliassen, A., Kleinschmidt, E. (1957): Dynamic Meteorology. **10** (48), 1
Eller, G. von → Guinier, A. (1957)
Elliott, J.P., Lane, A.M. (1957): The Nuclear Shell-Model. **8** (39), 241
Engel, A. von (1956): Ionization in Gases by Electrons in Electric Fields. **4** (21), 504
Ericksen, J.L. (1960): Appendix. Tensor Fields. **2** (3/1), 794

Evans, R.D. (1958): Compton Effect. **6** (34), 218
Ewald, H. (1956): Massenspektroskopische Apparate. **6** (33), 546
Ewing, W.M., Press, F. (1956): Surface Waves and Guided Waves. **10** (47), 119
Ewing, W.M., Press, F. (1956): Seismic Prospecting **10** (47), 153
Ewing, W.M., Press F. (1956): Structure of the Earth's Crust. **10** (47), 246

Falk, G. (1955): Algebra. **1** (2), 1
Falk, G., Jung, H. (1959): Axiomatik der Thermodynamik. **2** (3/2), 119
Fano, U., Spencer, L.V., Berger, M.J. (1959): Penetration and Diffusion of X-rays. **8** (38/2), 660
Fan, H.Y. (1967): Photon-Electron Interaction, Crystals Without Fields. **5** (25/2a), 157
Fehrenbach, Ch.F. (1958): Les classifications spectrales des étoiles normales. **11** (50), 1
Ferraro, V.C.A. → Parker, E.N. (1971)
Fichera, G. (1972): Existence Theorems in Elasticity. **3** (6a/2), 347
Fichera, G. (1972): Boundary Value Problems of Elasticity with Unilateral Constraints. **3** (6a/2), 391
Fick, E., Joos, G. (1957): Kristallspektren. **5** (28), 205
Findley, R.W. → Benedetti, S. de (1958)
Finkelnburg, W., Maecker, H. (1956): Elektrische Bögen und thermisches Plasma. **4** (22), 254
Finkelnburg, W., Peters, T. (1957): Kontinuierliche Spektren. **5** (28), 79
Fisher, G.M.C. → Leitman, M.J. (1973)
Forbush, S.E. (1966): Time-Variations of Cosmic Rays. **10** (49/1), 159
Forsbergh jr., P.W. (1956): Piezoelectricity Electrostriction and Ferroelectricity. **4** (17), 264
Fournet, G. (1957): Etude de la structure des fluides et des substances amorphes au moyen de la diffusion des rayons X. **6** (32), 238
Fowler, G.N., Wolfendale, A.W.

(1961): The Hard Component of μ-Mesons in the Atmosphere. **9** (46/1), 272
Fowler, R.G. (1956): Radiation from Low Pressure Discharges. **4** (22), 209
Francis, G. (1956): The Glow Discharge at Low Pressure. **4** (22), 53
Françon, M. (1956): Interférences, diffraction et polarisation. **5** (24), 171
Franke, G. (1967): Das photographische Objektiv. **5** (29), 192
Franz, W. (1956): Dielektrischer Durchschlag. **4** (17), 155
Freudenthal, A.M. (1958): Fatigue. **3** (6), 591
Freudenthal, A.M., Geiringer, H. (1958): The Mathematical Theories of the Inelastic Continuum. **3** (6), 229
Fujimoto, Y., Hayakawa, S. (1967): Cosmic Rays and High-Energy Physics. **9** (46/2), 115
Fukushima N. → Nagata, T. (1971)
Fulbright, H.W. (1958): Ionization Chambers in Nuclear Physics. **8** (45), 1

Gaposchkin, S. (1958): The Eclipsing Binaries. **11** (50), 225
Garland, G.D. (1956): Gravity and Isostasy. **10** (47), 202
Garlick, G.F.J. (1956): Photoconductivity. **4** (19), 316
Garlick, G.F.J. (1958): Luminescence. **5** (26), 1
Geiringer, H. (1973): Ideal Plasticity. **3** (6a/3), 403
Geiringer, H. → Freudenthal, A.M. (1958)
Gendrin, R. (1971): Phénomènes T.B.F. d'origine magnétosphérique. **10** (49/3), 461
Gerholm, T.R. (1956): Beta-ray Spectroscopes. **6** (33), 609
Gerritsen, A.N. (1956): Metallic Conductivity, Experimental Part. **4** (19), 137
Geschwind, S. (1958): Determination of Atomic Masses by Microwave Methods. **8** (38/1), 38
Gilbarg, D. (1960): Jets and Cavities. **3** (9), 311
Ginzburg, V.L., Ruhadze, A.A (1972): Waves and Resonances in Magnetoactive Plasma. **10** (49/4), 395
Givaudon, P. → Maréchal, A. et al. (1967)
Glaser, D.A. (1958): The Bubble Chamber. **8** (45), 314
Glaser, W. (1956): Elektronen- und Ionenoptik. **6** (33), 123
Gluckstern, R.L. → Breit, G. (1959)
Goldberg, L., Pierce, A.K. (1959): The Photosphere of the Sun. **11** (52), 1
Goldfarb, L.J.B. → Devons, S. (1957)
Gombás, P. (1956): Statistische Behandlung des Atoms. **7** (36), 109
Good, R.H., Müller, E.W. (1956): Field Emission. **4** (21), 176
Gordy, W. (1957): Microwave Spectroscopy. **5** (28), 1
Gould, R.J., G.R. Burbidge (1967): High Energy Photons and Neutrinos from Cosmic Sources. **9** (46/2), 265
Grace, M.A. → Blin-Stoyle, R.J. (1957)
Grad, H. (1958): Principles of the Kinetic Theory of Gases. **3** (12), 205
Greenstein, J.L. (1958): The Spectra of the White Dwarfs. **11** (50), 161
Green, G.K., Courant, E.D. (1959): The Proton Synchrotron. **8** (44), 218
Green, H.S. (1960): The Structure of Liquids. **3** (10), 1
Gringauz, K.I., Breus, T.K. (1976): Some Characteristic Features of the Ionospheres of Near-Earth Planets. **10** (49/5), 351
Guggenheim, E.A. (1959): Thermodynamics, Classical and Statistical. **2** (3/2), 1
Guinier, A., Eller, G. von (1957): Les méthodes expérimentales des déterminations de structures cristallines par rayons X. **6** (32), 1
Gurtin, M.E. (1972): The Linear Theory of Elasticity. **3** (6a/2), 1

Haken, H. (1970): Laser Theory. **5** (25/2c), 1
Handel, J. van den (1956): Low Temperature Magnetism. **3** (15), 1
Hartshorn, L., Saxton, J.A. (1958): The Dispersion and Absorption of Electromagnetic Waves. **4** (16), 640

Harvey, A.F. (1967): Measurement Methods and Instruments for Microwave Frequencies. **4** (23), 84
Hawkins, G.S. → Whipple, F.L. (1959)
Hayakawa, S. → Fujimoto, Y. (1967)
Heckmann, O.H.L., Schücking, E. (1959): Newtonsche und Einsteinsche Kosmologie. **11** (53), 489
Heckmann, O.H.L., Schücking, E. (1959): Andere kosmologische Theorien. **11** (53), 520
Herb, R.G. (1959): Van de Graaff Generators. **8** (44), 64
Hess, W.N. (1972): The Earth's Radiation Belt. **10** (49/4), 115
Hill, D.L. (1957): Matter and Charge Distribution within Atomic Nuclei. **8** (39), 178
Hogg, H.S. (1959): Star Clusters. **11** (53), 129
Honda M., Arnold J.R. (1967): Effects of Cosmic Rays on Meteorites. **9** (46/2), 613
Hönl, H., Maue, A.W., Westpfahl, K. (1961): Theorie der Beugung. **5** (25/1), 218
Hood, P.J. (1971): Geophysical Applications of High Resolution Magnetometers. **10** (49/3), 422
Howarth, L. (1959): Laminar Boundary Layers. **3** (8/1), 264
Howat, G. → Åberg, T. (1982)
Huang, S.-s. → Struve, O. (1958)
Hughes, D.J. (1959): Reactor Techniques. **8** (44), 390
Hughes, V.W. → Kusch, P. (1959)
Hugues, E.D. → Maréchal, A. et al. (1967)
Hulthén, L., Sugawara, M. (1957): The Two-Nucleon Problem. **8** (39), 1
Hund, F. (1956): Quantenmechanik der Atome. **7** (36), 1
Hunsinger, W. (1967): Temperaturmessung. **4** (23), 373
Hyde, E.K., Seaborg, G.T. (1957): The Transuranium Elements. **8** (42), 205

Ingard, K.U. → Morse, P.M. (1961)
Ingram, J.E. (1968): Electron Spin Resonance. **4** (18/1), 94
Irwin, G.R. (1958): Fracture. **3** (6), 551

Jacobs, J.A. (1956): The Earth's Interior. **10** (47), 364
Jaeckel, R. (1958): Allgemeine Vakuumphysik. **3** (12), 515
Jager, C. de (1959): Structure and Dynamics of the Solar Atmosphere. **11** (52), 80
Jagodzinski, H. (1955): Kristallographie. **3** (7/1), 1
Jessop, H.T. (1958): Photoelasticity. **3** (6), 127
Jones, F.L. (1956): Ionization Growth and Breakdown. **4** (22), 1
Jones, H. (1956): Theory of Electrical and Thermal Conductivity in Metals. **4** (19), 227
Jones, W.L. (1976): Linear Internal Gravity Waves in the Atmosphere. **10** (49/5), 177
Joos, G. → Fick, E. (1957)
Jung, H. → Falk, G. (1959)
Jung, K. (1956): Figur der Erde. **10** (47), 534

Kaesberg, P. → Beeman, W.W. et al. (1957)
Källén, G.A.O. (1958): Quantenelektrodynamik. **2** (5/1), 169
Kamke, D. (1956): Elektronen- und Ionenquellen. **6** (33), 1
Kamp, P. van de (1958): Visual Binaries. **11** (50), 187
Karlsson, L. → Siegbahn, H. (1982)
Karpenko, B.V. → Vonsovsky, S.V. (1968)
Kayama, K. → Kotani, M. et al. (1961)
Keenan, Ph.C. (1958): Stars with Peculiar Spectra. **11** (50), 93
Keesom, P.H., Pearlman, N. (1956): Low Temperature Heat Capacity of Solids. **3** (14), 282
Keffer, F. (1966): Spin Waves. **4** (18/2), 1
Kelly, F.M. (1958): Determination of Nuclear Spins and Magnetic Moments by Spectroscopic Methods. **8** (38/1), 59
Kerr, F.J. (1959): Radio Echoes from Sun, Moon and Planets. **11** (52), 449
Kerst, D.W. (1959): The Betatron. **8** (44), 193
Kertz, W. (1957): Atmosphärische Gezeiten. **10** (48), 928
King, R.W.P. (1958): Quasi-Stationary

and Nonstationary Currents in Electric Circuits. **4** (16), 165
Kinsey, B.B. (1957): Nuclear Reactions, Levels, and Spectra of Heavy Nuclei. **8** (40), 202
Kirkpatrick, P., Pattee jr., H.H. (1957): X-ray Microscopy. **6** (30), 305
Kleinschmidt, E. → Eliassen, A. (1957)
Klemens, P.G. (1956): Thermal Conductivity of Solids at Low Temperatures. **3** (14), 198
Klerk, D. de (1956): Adiabatic Demagnetization. **3** (15), 38
Kneller, E. (1966): Theorie der Magnetisierungskurve kleiner Kristalle. **4** (18/2), 438
Kneser, H.O. (1961): Schallabsorption und -dispersion in Gasen. **3** (11/1), 129
Knops, R.J., Wilkes, E.W. (1973): Theory of Elastic Stability. **3** (6a/3), 125
Kofoed-Hansen, O.M., Christensen, C.J. (1962): Experiments on β-decay. **8** (41/2), 1
Kollath, R. (1956): Sekundärelektronen-Emission fester Körper bei Bestrahlung mit Elektronen. **4** (21), 232
Kollath, R. (1958): Durchgang langsamer Elektronen und Ionen durch Gase. **6** (34), 1
Kondo, S. → Ono, S. (1960)
Korff, S.A. (1958): Geiger Counters. **8** (45), 52
Kotani, M., Ohno, K., Kayama, K. (1961): Quantum Mechanics of Electronic Structure of Simple Molecules. **7** (37/2), 1
Kusch, P., Hughes, V.W. (1959): Atomic and Molecular Beam Spectroscopy. **7** (37/1), 1

Laitone, E.V. → Wehausen, J.V. (1960)
Lal, D., Peters, B. (1967): Cosmic Ray Produced Radioactivity on the Earth. **9** (46/2), 551
Lane, A.M. → Elliott, J.P. (1957)
Laukien, G. (1958): Kernmagnetische Hochfrequenz-Spektroskopie. **8** (38/1), 120
Laursen, V., Olsen, J. (1971): Classical Methods of Geomagnetic Observations. **10** (49/3), 276
Lecomte, J. (1958): Spectroscopie dans l'infrarouge. **5** (26), 244
Ledoux P. (1958): Stellar Stability. **11** (51), 605
Ledoux, P., Walraven, T. (1958): Variable Stars. **11** (51), 353
Leibfried, G. (1955): Gittertheorie der mechanischen und thermischen Eigenschaften der Kristalle. **3** (7/1), 104
Leisegang, S. (1956): Elektronenmikroskope. **6** (33), 396
Leitman, M.J., Fisher, G.M.C. (1973): The Linear Theory of Viscoelasticity. **3** (6a/3), 1
Lense, J. (1956): Grundbegriffe der klassischen Analysis, gewöhnliche Differentialgleichungen, Funktionentheorie. **1** (1), 1
Lense, J. (1956): Partielle Differentialgleichungen. **1** (1), 90
Lense, J. (1956): Elliptische Funktionen und Integrale. **1** (1), 120
Leonard, R.W. (1962): Generation and Measurement of Sound in Gases. **3** (11/2), 1
Lewis, H.W. → Merzbacher, E. (1958)
Lidiard, A.B. (1957): Ionic Conductivity. **4** (20), 246
Lindblad, B. (1959): Galactic Dynamics. **11** (53), 21
Lin, C.-C., Reid, W.H. (1963): Turbulent Flow, Theoretical Aspects. **3** (8/2), 438
Little, P.F. (1956): Secondary Effects. **4** (21), 574
Loeb, L.B. (1956): Formation of Negative Ions. **4** (21), 445
Loeb, L.B. (1956): The Recombination of Ions. **4** (21), 471
Loeb, L.B. (1956): Electrical Breakdown of Gases with Steady or Direct Current Impulse Potentials. **4** (22), 445
Lohrmann, E. → Schopper, E. et al. (1967)
Lovell, A.C.B. (1957): Geophysical Aspects of Meteors. **10** (48), 427
Ludlam, F.H., Mason, B.J. (1957): The Physics of Clouds. **10** (48), 479

MacCunn Farquhar, R. → Wilson, J.T. et al. (1956)
MacDonald, D.K.C. (1956): Electrical Conductivity of Metals and Alloys at Low Temperatures. **3** (14), 137
Madelung, O. (1957): Halbleiter. **4** (20), 1
Maecker, H. → Finkelnburg, W. (1956)
Marcus, W.C. de (1959): Planetary Interiors. **11** (52), 419
Maréchal A. (1956): Optique géométrique générale. **5** (24), 44
Maréchal, A., Hugues, E.D., Givaudon, P. (1967): Méthode de calcul des systèmes optiques. **5** (29), 43
Marx, H. (1967): Theorie der geometrisch-optischen Bildfehler. **5** (29), 68
Mason, B.J. → Ludlam, F.H. (1957)
Mason, W.P. (1961): Dispersion and Absorption of Sound in High Polymers. **3** (11/1), 361
Massey, H.S.W. (1956): Theory of Atomic Collisions. **7** (36), 232
Massey, H.S.W. (1956): Excitation and Ionization of Atoms by Electron Impact. **7** (36), 307
Massey, H.S.W. → Craggs, J.D. (1959)
Mathieu, J.P. (1957): Activité optique naturelle. **5** (28), 333
Matsushita, S. (1967): Lunar Tides in the Ionosphere. **10** (49/2), 547
Mattis, D.C. → Methfessel, S. (1968)
Mauck, G. → Schopper, E. et al. (1967)
Maue, A.W. → Hönl, H. et al. (1961)
Mayer, J.E. (1958): Theory of Real Gases. **3** (12), 73
McAllister jr., H.H., Breit, G. (1959): Coulomb Wave Functions. **8** (41/1), 408
McIntosh, J.S. → Breit, G. (1959)
McMath, R.R., Mohler, O.C. (1962): Solar Instruments. **11** (54), 1
McVittie, G. (1959): Distance and Time in Cosmology: The Observational Data. **11** (53), 445
Meinel A.B. → Akasofu, S.-I. et al. (1966)
Meixner, J. (1956): Spezielle Funktionen der mathematischen Physik. **1** (1), 147
Meixner, J., Reik, H.G. (1959): Thermodynamik der irreversiblen Prozesse. **2** (3/2), 413
Mendelssohn, K. (1956): Liquid Helium. **3** (15), 370
Merzbacher, E., Lewis, H.W. (1958): X-ray Production by Heavy Charged Particles. **6** (34), 166
Methfessel, S., Mattis, D.C. (1968): Magnetic Semiconductors. **4** (18/1), 389
Meyer, R.E. (1960): Theory of Characteristics of Inviscid Gas Dynamics. **3** (9), 225
Middleton, W.E.K. (1957): Vision Through the Atmosphere. **10** (48), 254
Mills, B.Y. (1959): Radio Frequency Radiation from External Galaxies. **11** (53), 239
Mizushima, S.-I. (1958): Raman Effect. **5** (26), 171
Mohler, O.C. → McMath, R.R. (1962)
Möller, F. (1957): Strahlung in der unteren Atmosphäre. **10** (48), 155
Morrison, P. (1961): The Origin of Cosmic Rays. **9** (46/1), 1
Morse, P.M., Ingard, K.U. (1961): Linear Acoustic Theory. **3** (11/1), 1
Moszkowski, S.A. (1957): Models of Nuclear Structure. **8** (39), 411
Mott, W.E., Sutton, R.B. (1958): Scintillation and Čerenkov Counters. **8** (45), 86
Mühleisen, R. (1957): Atmosphärische Elektrizität. **10** (48), 541
Müller, E.W. → Good, R.H. (1956)
Müller E. → Baule, H. (1956)
Münster, A. (1959): Prinzipien der statistischen Mechanik. **2** (3/2), 176
Münster, A. (1962): Statistische Thermodynamik kondensierter Phasen. **3** (13), 1

Nagata, T. (1966): Magnetic Properties of Rocks and Minerals. **10** (49/1), 248
Nagata, T., Fukushima N. (1971): Morphology of Magnetic Disturbance. **10** (49/3), 5
Nagata, T. → Sakuma, S. (1957)

Naghdi, P.M. (1972): The Theory of Shells and Plates. **3** (6a/2), 425
Naumann, H. (1967): Spiegelnde Prismen. **5** (29), 211
Neyman, J., Scott, E.L. (1959): Large Scale Organization of the Distribution of Galaxies. **11** (53), 416
Nielsen, H.H. (1959): The Vibration-rotation Energies of Molecules and their Spectra in the Infra-red. **7** (37/1), 173
Nikol'skij, G.M (1982): Extreme Ultraviolet Observational Data on the Solar Spectrum. **10** (49/6), 309
Nishimura, J. (1967): Theory of Cascade Showers. **9** (46/2), 1
Noll, W. → Truesdell, C. (1965)
Noltingk, B.E. (1962): The Effects of Intense Ultrasonics in Liquids. **3** (11/2), 258
Nottingham, W.B. (1956): Thermionic Emission. **4** (21), 1
Nunziato, J.W., Walsh, E.K., Schuler, K.W., Barker, L.M. (1974): Wave Propagation in Nonlinear Viscoelastic Solids. **3** (6a/4), 1

Ohno, K. → Kotani, M. et al. (1961)
Olsen, J. → Laursen, V. (1971)
Ono, S., Kondo, S. (1960): Molecular Theory of Surface Tension in Liquids. **3** (10), 134
Oort, J.H. (1959): Radio-frequency Studies of Galactic Structure. **11** (53), 100
Oswatitsch, K. (1959): Physikalische Grundlagen der Strömungslehre. **3** (8/1), 1

Paetzold, H.-K., Regener, E. (1957): Ozon in der Erdatmosphäre. **10** (48), 370
Papas, C.H. → Borgnis, F.E. (1958)
Parker, E.N., Ferraro, V.C.A. (1971): Theoretical Aspects of the World-wide Magnetic Storm Phenomenon. **10** (49/3), 131
Pattee jr., H.H. → Kirkpatrick, P. (1957)
Pauli, W. (1958): Die allgemeinen Prinzipien der Wellenmechanik. **2** (5/1), 1
Payne-Gaposchkin, C. (1958): The Novae. **11** (51), 752
Pearlman, N. → Keesom, P.H. (1956)
Perlman, I., Rasmussen, J.O. (1957): Alpha Radioactivity. **8** (42), 109
Peters, B. → Lal, D. (1967)
Peters, T. → Finkelnburg, W. (1957)
Phillips, M. (1962): Classical Electrodynamics. **2** (4), 1
Pierce, A.K. → Goldberg, L. (1959)
Platt, J.R. (1961): The Chemical Bond and the Distribution of Electrons in Molecules. **7** (37/2), 173
Poeverlein, H. (1972): The Earth's Magnetosphere. **10** (49/4), 7
Press, F. → Ewing, W.M. (1956)
Press, F. → Ewing, W.M. (1956)
Press F. → Ewing, W.M. (1956)

Quenby, J.J. (1967): The Time Variations of the Cosmic Ray Intensity. **9** (46/2), 310

Raether, H. (1957): Elektroneninterferenzen. **6** (32), 443
Rainwater, J. (1957): Resonance Processes by Neutrons. **8** (40), 373
Ramachandran, G.N., Ramaseshan, S. (1961): Crystal Optics. **5** (25/1), 1
Ramakrishnan, A. (1959): Probability and Stochastic Processes. **2** (3/2), 524
Ramaseshan, S. → Ramachandran, G.N. (1961)
Rasmussen, J.O. → Perlman, I. (1957)
Rawer, K. (1976): Introductory Remarks. **10** (49/5), 1
Rawer, K. (1982): Introduction: A Summary Description of Aeronomy. **10** (49/6), 1
Rawer, K. (1984): Modelling of Neutral and Ionized Atmospheres. **10** (49/7), 223
Rawer, K., Suchy, K. (1967): Radio-Observations of the Ionosphere. **10** (49/2), 1
Ray, E.C. (1961): Experimental Results of Flights in the Stratosphere. **9** (46/1), 130
Regener, E. → Paetzold, H.-K. (1957)
Reid, W.H. → Lin, C.-C. (1963)
Reik, H.G. → Meixner, J. (1959)

Reiner, M. (1958): Rheology. **3** (6), 434
Ringo, G.R. (1957): Neutron Diffraction and Interference. **6** (32), 552
Roll, H.U. (1957): Oberflächen-Wellen des Meeres. **10** (48), 671
Rosen, B. (1964): Spectroscopie électronique moléculaire. **5** (27), 221
Rowlinson, J.S. (1958): The Properties of Real Gases. **3** (12), 1
Ruhadze, A.A → Ginzburg, V.L. (1972)
Runcorn, S.K. (1956): Magnetization of Rocks. **10** (47), 470
Runcorn, S.K. (1956): The Magnetism of the Earth's Body. **10** (47), 498
Russell, R.D. → Wilson, J.T. et al. (1956)
Rutgers, G.A.W. (1958): Temperature Radiation of Solids. **5** (26), 129

Sakuma, S., Nagata, T. (1957): Physical Volcanology. **10** (48), 982
Salpeter, E.E. → Bethe, H.A. (1957)
Samson, J.A.R. (1982): Atomic Photoionization. **6** (31), 123
Sandström, A.E. (1957): Experimental Methods of X-ray Spectroscopy: Ordinary Wavelengths. **6** (30), 78
Saxton, J.A. → Hartshorn, L. (1958)
Schaaffs, W. (1957): Erzeugung von Röntgenstrahlen. **6** (30), 1
Schaaf, S.A. (1963): Mechanics of Rarefied Gases. **3** (8/2), 591
Schatzman, E. (1958): Théorie des naines blanches. **11** (51), 723
Scheidegger, A.E. (1956): Forces in the Earth's Crust. **10** (47), 258
Scheidegger, A.E. (1963): Hydrodynamics in Porous Media. **3** (8/2), 625
Schiffer, M. (1960): Analytical Theory of Subsonic and Supersonic Flows. **3** (9), 1
Schlichting, H. (1959): Entstehung der Turbulenz. **3** (8/1), 351
Schlögl, F. (1956): Randwertprobleme. **1** (1), 218
Schmidtke, G. (1984): Modelling of the Solar Extreme Ultraviolet Irradiance for Aeronomic Applications. **10** (49/7), 1
Schmidt, H., Auster V. (1971): Neuere Meßmethoden der Geomagnetik. **10** (49/3), 323
Schonland, B.F.J. (1956): The Lightning Discharge. **4** (22), 576
Schopper, E., Lohrmann, E., Mauck, G. (1967): Nukleonen in der Atmosphäre. **9** (46/2), 372
Schreiner, F. (1967): Temperaturregelung. **4** (23), 457
Schuler, K.W. → Nunziato, J.W. et al. (1974)
Schulze, R. (1967): Optische Meßmethoden. **5** (29), 755
Scott, E.L. → Neyman, J. (1959)
Seaborg, G.T. → Hyde, E.K. (1957)
Seeger, A. (1955): Theorie der Gitterfehlstellen. **3** (7/1), 383
Seeger, A. (1958): Kristallplastizität. **3** (7/2), 1
Sekera. Z. (1957): Polarization of Skylight. **10** (48), 288
Selzer, E. (1972): Variations rapides du champ magnétique terrestre. **10** (49/4), 231
Serin, B. (1956): Superconductivity. Experimental Part. **3** (15), 210
Serrin, J. (1959): Mathematical Principles of Classical Fluid Mechanics. **3** (8/1), 125
Serson, P.H., Whitham, K. (1971): Three-Component Airborne Magnetometers. **10** (49/3), 384
Sette, D. (1961): Dispersion and Absorption of Sound Waves in Liquids and Mixtures of Liquids. **3** (11/1), 275
Shapiro, M.M. (1958): Nuclear Emulsions. **8** (45), 342
Siebert, M. (1971): Maßzahlen der erdmagnetischen Aktivität. **10** (49/3), 206
Siegbahn, H., Karlsson, L. (1982): Photoelectron Spectroscopy. **6** (31), 215
Siegel, R.T. (1958): High Energy Neutron Detectors. **8** (45), 487
Simons, L. (1958): Positronium. **6** (34), 139
Sitte, K. (1961): Penetrating Showers. **9** (46/1), 157
Slater, J.C. (1956): The Electronic Structure of Solids. **4** (19), 1
Smith, L. (1959): Linear Accelerators. **8** (44), 341

Smith, S.D. (1967): Magneto-Optics in Crystals. **5** (25/2a), 234
Sneddon, I.N. (1955): Functional Analysis. **1** (2), 198
Sneddon, I.N., Berry, D.S. (1958): The Classical Theory of Elasticity. **3** (6), 1
Spencer-Jones, H. (1956): The Rotation of the Earth. **10** (47), 1
Spencer, L.V. → Fano, U. et al. (1959)
Starace, A.F. (1982): Theory of Atomic Photoionization. **6** (31), 1
Staverman, A.J (1962): Thermodynamics of Polymers. **3** (13), 399
Stephenson, S.T. (1957): The Continuous X-ray Spectrum. **6** (30), 337
Stevels, J.M. (1957): The Electrical Properties of Glass. **4** (20), 350
Stevels, J.M. (1962): The Structure and the Physical Properties of Glass. **3** (13), 510
Strauch D. → Bilz, H. et al. (1984)
Stroke, G.W. (1967): Diffraction Gratings. **5** (29), 426
Struve, O., Huang, S.-s. (1958): Spectroscopic Binaries. **11** (50), 243
Stubbe, P. (1982): Interaction of Neutral and Plasma Motions in the Ionosphere. **10** (49/6), 247
Suchy, K. (1984): Transport Coefficients and Collision Frequencies for Aeronomic Plasmas. **10** (49/7), 57
Suchy, K. → Rawer, K. (1967)
Suess, H.E., Urey, H.C. (1958): Die Häufigkeit der Elemente in den Planeten und Meteoriten. **11** (51), 296
Sugawara, M. → Hulthén, L. (1957)
Sutton, R.B. → Mott, W.E. (1958)
Sverdrup, H.U. (1957): Oceanography. **10** (48), 608
Swings, P. (1958): Les bandes moléculaires dans les spectres stellaires. **11** (50), 109
Synge, J.L. (1960): Classical Dynamics. **2** (3/1), 1

Tamm, K. (1961): Schallabsorption und -dispersion in wäßrigen Elektrolytlösungen. **3** (11/1), 202
Thomas, L. (1982): The Neutral and Ion Chemistry of the Upper Atmosphere. **10** (49/6), 7
Thurston, R.N. (1974): Waves in Solids. **3** (6a/4), 109
Tietz, H. (1955): Geometrie. **1** (2), 117
Timman, R. (1960): Linearized Theory of Unsteady Flow of a Compressible Fluid. **3** (9), 283
Ting, T.W. (1973): Topics in the Mathematical Theory of Plasticity. **3** (6a/3), 535
Tol, J.F. van → Verhagen, C.J.D.M. (1967)
Tomaschek, R. (1957): Tides of the Solid Earth. **10** (48), 775
Tomboulian, D.H. (1957): The Experimental Methods of Soft X-ray Spectroscopy and the Valence Band Spectra of the Light Elements. **6** (30), 246
Toupin, R.A. → Truesdell, C. (1960)
Townes, C.H. (1958): Determination of Nuclear Quadrupole Moments. **8** (38/1), 377
Truell, R., Elbaum, Ch. (1962): High Frequency Ultrasonic Stress Waves in Solids. **3** (11/2), 153
Truesdell, C., Noll, W. (1965): The Non-Linear Field Theories of Mechanics. **2** (3/3), 1
Truesdell, C., Toupin, R.A. (1960): The Classical Field Theories. **2** (3/1), 226

Uenohara, M. (1967): Low Noise Amplification. **4** (23), 1
Urey, H.C. (1959): The Atmospheres of the Planets. **11** (52), 363
Urey, H.C. → Suess, H.E. (1958)

Vallarta, M.S. (1961): Theory of the Geomagnetic Effects of Cosmic Radiation. **9** (46/1), 88
Vassy, A.T., Vassy, E. (1976): La luminescence nocturne. **10** (49/5), 5
Vassy, E. → Vassy, A.T. (1976)
Vaucouleurs, G. de (1959): Classification and Morphology of External Galaxies. **11** (53), 275
Vaucouleurs, G. de (1959): General Physical Properties of External Galaxies. **11** (53), 311
Verde, M. (1957): The Three-body Problem in Nuclear Physics. **8** (39), 144

Verhagen, C.J.D.M., Tol, J.F. van (1967): Recording of Measuring Data. **4** (23), 217
Verstelle, J.C., Curtis, D.A. (1968): Paramagnetic Relaxation. **4** (18/1), 1
Vonsovsky, S.V., Karpenko, B.V. (1968): Austauschwechselwirkung in Isolatoren. **4** (18/1), 265

Waldmann, L. (1958): Transporterscheinungen in Gasen von mittlerem Druck. **3** (12), 295
Walraven, T. → Ledoux, P. (1958)
Walsh, E.K. → Nunziato, J.W. et al. (1974)
Wapstra, A.H. (1958): Atomic Masses of Nuclides. **8** (38/1), 1
Wattenberg, A. (1957): Nuclear Reactions at High Energies. **8** (40), 450
Weaver, H. (1962): Photographic Photometry. **11** (54), 130
Webber, W.R. (1967): The Spectrum and Charge Composition of the Primary Cosmic Radiation. **9** (46/2), 181
Webb, M.B. → Beeman, W.W. et al. (1957)
Webb, W.L. (1976): Dynamic Structure of the Stratosphere and Mesosphere. **10** (49/5), 117
Wehausen, J.V., Laitone, E.V. (1960): Surface Waves. **3** (9), 446
Wehner, R.K. → Bilz, H. et al. (1984)
Weissinger, J. (1963): Theorie des Tragflügels bei stationärer Bewegung in reibungslosen, inkompressiblen Medien. **3** (8/2), 385
Weissler, G.L. (1956): Photoionization in Gases and Photoelectric Emission from Solids. **4** (21), 304
Weiss, H. (1967): Messung von Magnetfeldern. **4** (23), 198
Welford, W.T. (1967): Optical Calculations and Optical Instruments, an Introduction. **5** (29), 1
Wendt, G. (1958): Statische Felder und stationäre Ströme. **4** (16), 1
Wertz, J.E. (1968): Structural Information from Paramagnetic Resonance. **4** (18/1), 145
Westpfahl, K. → Hönl, H. et al. (1961)
Whaling, W. (1958): The Energy Loss of Charged Particles in Matter. **6** (34), 193
Whipple, F.L., Hawkins, G.S. (1959): Meteors. **11** (52), 518
Whitford, A.E. (1962): Photoelectric Techniques. **11** (54), 240
Whitham, K. → Serson, P.H. (1971)
Wilets, L. (1958): Isotope Shifts. **8** (38/1), 96
Wilkes, E.W. → Knops, R.J. (1973)
Wilson, J.T., Russell, R.D., MacCunn-Farquhar, R. (1956): Radioactivity and Age of Minerals. **10** (47), 288
Wilson, R.R. (1959): Electron Synchrotrons. **8** (44), 170
Wilson, R. → Bishop, G.R. (1957)
Wolfendale, A.W. → Fowler, G.N. (1961)
Wolter, H. (1956): Optik dünner Schichten. **5** (24), 461
Wolter, H. (1956): Schlieren-, Phasenkontrast- und Lichtschnittverfahren. **5** (24), 555
Wrubel, M.H. (1958): Stellar Interiors. **11** (51), 1
Wurm, K. (1958): Die Spektren der planetarischen Nebel. **11** (50), 139
Wurm, K. (1958): Theorie der planetarischen Nebel. **11** (50), 399
Wurm, K. (1959): Die Kometen. **11** (52), 465

Yonezawa, T. (1982): Diffusion in the High Atmosphere. **10** (49/6), 129
York, C.M. (1958): Cloud Chambers. **8** (45), 260

Zwicky, F. (1958): Supernovae. **11** (51), 766
Zwicky, F. (1959): Multiple Galaxies. **11** (53), 373
Zwicky, F. (1959): Clusters of Galaxies. **11** (53), 390

Index of Contributions

The Index of Contributions gives a complete listing of the contents of all volumes of *Handbuch der Physik/Encyclopedia of Physics*. Please note that the page numbers given refer to the numbers in the original volumes, *not* to the general Index.

Group 1 Mathematical Methods

Vol. 1	Grundbegriffe der klassischen Analysis, gewöhnliche Differentialgleichungen, Funktionentheorie. By J. Lense	1
	Partielle Differentialgleichungen. By J. Lense	90
	Elliptische Funktionen und Integrale. By J. Lense	120
	Spezielle Funktionen der mathematischen Physik. By J. Meixner	147
	Randwertprobleme. By F. Schlögl	218
Vol. 2	Algebra. By G. Falk	1
	Geometrie. By H. Tietz	117
	Functional Analysis. By I.N. Sneddon	198
	Numerische und graphische Methoden. By L. Collatz	349
	Moderne Rechenmaschinen. By H. Bückner	471

Group 2 Principles of Theoretical Physics

Vol. 3/1	Classical Dynamics. By J.L. Synge	1
	The Classical Field Theories. By C. Truesdell, R.A. Toupin	226
	Appendix. Tensor Fields. By J.L. Ericksen	794
Vol. 3/2	Thermodynamics, Classical and Statistical. By E.A. Guggenheim	1
	Axiomatik der Thermodynamik. By G. Falk, H. Jung	119
	Prinzipien der statistischen Mechanik. By A. Münster	176
	Thermodynamik der irreversiblen Prozesse. By J. Meixner, H.G. Reik	413
	Probability and Stochastic Processes. By A. Ramakrishnan	524
Vol. 3/3	The Non-Linear Field Theories of Mechanics. By C. Truesdell, W. Noll	1
Vol. 4	Classical Electrodynamics. By M. Phillips	1
	The Special Theory of Relativity. By P.G. Bergmann	109
	The General Theory of Relativity. By P.G. Bergmann	203
Vol. 5/1	Die allgemeinen Prinzipien der Wellenmechanik. By W. Pauli	1
	Quantenelektrodynamik. By G. Källén	169

Group 3 Mechanical and Thermal Behaviour of Matter

Vol. 6	The Classical Theory of Elasticity. By I.N. Sneddon, D.S. Berry	1
	Photoelasticity. By H.T. Jessop	127
	The Mathematical Theories of the Inelastic Continuum. By A.M. Freudenthal, H. Geiringer	229
	Rheology. By M. Reiner	434
	Fracture. By G.R. Irwin	551
	Fatigue. By A.M. Freudenthal	591
Vol. 6a/1	The Experimental Foundations of Solid Mechanics. By J.F. Bell	1
Vol. 6a/2	The Linear Theory of Elasticity. By M.E. Gurtin	1
	Linear Thermoelasticity. By D.E. Carlson	297
	Existence Theorems in Elasticity. By G. Fichera	347
	Boundary Value Problems of Elasticity with Unilateral Constraints. By G. Fichera	391
	The Theory of Shells and Plates. By P.M. Naghdi	425
	The Theory of Rods. By S.S. Antman	641
Vol. 6a/3	The Linear Theory of Viscoelasticity. By M.J. Leitman, G.M.C. Fisher	1
	Theory of Elastic Stability. By R.J. Knops, E.W. Wilkes	125
	Growth and Decay of Waves in Solids. By P.J. Chen	303
	Ideal Plasticity. By H. Geiringer	403
	Topics in the Mathematical Theory of Plasticity. By T.W. Ting	535
Vol. 6a/4	Wave Propagation in Nonlinear Viscoelastic Solids. By J.W. Nunziato, E.K. Walsh, K.W. Schuler, L.M. Barker	1
	Waves in Solids. By R.N. Thurston	109
Vol. 7/1	Kristallographie. By H. Jagodzinski	1
	Gittertheorie der mechanischen und thermischen Eigenschaften der Kristalle. By G. Leibfried	104
	The Specific Heat of Solids. By M. Blackman	325
	Theorie der Gitterfehlstellen. By A. Seeger	383
Vol. 7/2	Kristallplastizität. By A. Seeger	1
	Umwandlungen und Ausscheidungen im kristallinen Zustand. By U. Dehlinger	211
Vol. 8/1	Physikalische Grundlagen der Strömungslehre. By K. Oswatitsch	1
	Mathematical Principles of Classical Fluid Mechanics. By J. Serrin	125
	Laminar Boundary Layers. By L. Howarth	264
	Entstehung der Turbulenz. By H. Schlichting	351

Vol. 8/2	Intégration des équations du mouvement d'un fluide visqueux incompressible. By R. Berker	1
	Theorie des Tragflügels bei stationärer Bewegung in reibungslosen, inkompressiblen Medien. By J. Weissinger	385
	Turbulent Flow, Theoretical Aspects. By C.-C. Lin, W.H. Reid	438
	Turbulence: Experimental Methods. By S. Corrsin	524
	Mechanics of Rarefied Gases. By S.A. Schaaf	591
	Hydrodynamics in Porous Media. By A.E. Scheidegger	625
Vol. 9	Analytical Theory of Subsonic and Supersonic Flows. By M. Schiffer	1
	Théorie des ondes de choc. By H. Cabannes	162
	Theory of Characteristics of Inviscid Gas Dynamics. By R.E. Meyer	225
	Linearized Theory of Unsteady Flow of a Compressible Fluid. By R. Timman	283
	Jets and Cavities. By D. Gilbarg	311
	Surface Waves. By J.V. Wehausen, E.V. Laitone	446
Vol. 10	The Structure of Liquids. By H.S. Green	1
	Molecular Theory of Surface Tension in Liquids. By S. Ono, S. Kondo	134
	The Theory of Capillarity. By F.P. Buff	281
Vol. 11/1	Linear Acoustic Theory. By P.M. Morse, K.U. Ingard	1
	Schallabsorption und -dispersion in Gasen. By H.O. Kneser	129
	Schallabsorption und -dispersion in wäßrigen Elektrolytlösungen. By K. Tamm	202
	Dispersion and Absorption of Sound Waves in Liquids and Mixtures of Liquids. By D. Sette	275
	Dispersion and Absorption of Sound in High Polymers. By W.P. Mason	361
Vol. 11/2	Generation and Measurement of Sound in Gases. By R.W. Leonard	1
	Generation, Detection and Measurement of Ultrasound. By A. Barone	74
	High Frequency Ultrasonic Stress Waves in Solids. By R. Truell, Ch. Elbaum	153
	The Effects of Intense Ultrasonics in Liquids. By B.E. Noltingk	258
Vol. 12	The Properties of Real Gases. By J.S. Rowlinson	1
	Theory of Real Gases. By J.E. Mayer	73
	Principles of the Kinetic Theory of Gases. By H. Grad	205
	Transporterscheinungen in Gasen von mittlerem Druck. By L. Waldmann	295
	Allgemeine Vakuumphysik. By R. Jaeckel	515
	Production and Measurement of Ultrahigh Vacuum. By D. Alpert	609

Vol. 13 Statistische Thermodynamik kondensierter Phasen. By A. Münster 1
Thermodynamics of Polymers. By A.J Staverman 399
The Structure and the Physical Properties of Glass. By J.M. Stevels 510

Vol. 14 The Production of Low Temperatures Down to Hydrogen Temperature. By J.G. Daunt 1
Helium Liquefiers and Carriers. By S.C. Collings 112
Electrical Conductivity of Metals and Alloys at Low Temperatures. By D.K.C. MacDonald 137
Thermal Conductivity of Solids at Low Temperatures. By P.G. Klemens 198
Low Temperature Heat Capacity of Solids. By P.H. Keesom, N. Pearlman 282

Vol. 15 Low Temperature Magnetism. By J. van den Handel 1
Adiabatic Demagnetization. By D. de Klerk 38
Superconductivity. Experimental Part. By B. Serin 210
Theory of Superconductivity. By J. Bardeen 274
Liquid Helium. By K. Mendelssohn 370

Group 4 Electric and Magnetic Behaviour of Matter

Vol. 16 Statische Felder und stationäre Ströme. By G. Wendt 1
Quasi-Stationary and Nonstationary Currents in Electric Circuits. By R.W.P. King 165
Electromagnetic Waveguides and Resonators. By F.E. Borgnis, C.H. Papas 285
Propagation of Electromagnetic Waves. By H. Bremmer 423
The Dispersion and Absorption of Electromagnetic Waves. By L. Hartshorn, J.A. Saxton 640

Vol. 17 Dielectrics. By W.F. Brown jr. 1
Dielektrischer Durchschlag. By W. Franz 155
Piezoelectricity Electrostriction and Ferroelectricity. By P.W. Forsbergh jr. 264

Vol. 18/1 Paramagnetic Relaxation. By J.C. Verstelle, D.A. Curtis 1
Electron Spin Resonance. By J.E. Ingram 94
Structural Information from Paramagnetic Resonance. By J.E. Wertz 145
Austauschwechselwirkung in Isolatoren. By S.V. Vonsovsky, B.V. Karpenko 265
Magnetic Semiconductors. By S. Methfessel, D.C. Mattis 389

Vol. 18/2 Spin Waves. By F. Keffer 1
Secondary Effects in Ferromagnetism. By W.J. Carr jr. 274
Mikromagnetismus. By W. Döring 341
Theorie der Magnetisierungskurve kleiner Kristalle. By E. Kneller 438

Vol. 19 The Electronic Structure of Solids. By J.C. Slater 1
Metallic Conductivity, Experimental Part. By A.N. Gerritsen 137
Theory of Electrical and Thermal Conductivity in Metals. By H. Jones 227
Photoconductivity. By G.F.J. Garlick 316

Vol. 20 Halbleiter. By O. Madelung 1
Ionic Conductivity. By A.B. Lidiard 246
The Electrical Properties of Glass. By J.M. Stevels 350
Électrochimie. By E. Darmois 392

Vol. 21 Thermionic Emission. By W.B. Nottingham 1
Field Emission. By R.H. Good, E.W. Müller 176
Sekundärelektronen-Emission fester Körper bei Bestrahlung mit Elektronen. By R. Kollath 232
Photoionization in Gases and Photoelectric Emission from Solids. By G.L. Weissler 304
Motions of Ions and Electrons. By W.P. Allis 383
Formation of Negative Ions. By L.B. Loeb 445
The Recombination of Ions. By L.B. Loeb 471
Ionization in Gases by Electrons in Electric Fields. By A. von Engel 504
Secondary Effects. By P.F. Little 574

Vol. 22 Ionization Growth and Breakdown. By F.L. Jones 1
The Glow Discharge at Low Pressure. By G. Francis 53
Radiation from Low Pressure Discharges. By R.G. Fowler 209
Elektrische Bögen und thermisches Plasma. By W. Finkelnburg, H. Maecker 254
Electrical Breakdown of Gases with Steady or Direct Current Impulse Potentials. By L.B. Loeb 445
Breakdown in Gases: Alternating and High-Frequency Fields. By S.C. Brown 531
The Lightning Discharge. By B.F.J. Schonland 576

Vol. 23 Low Noise Amplification. By M. Uenohara 1
Measurement Methods and Instruments for Microwave Frequencies. By A.F. Harvey 84
Messung von Magnetfeldern. By H. Weiss 198
Recording of Measuring Data. By C.J.D.M. Verhagen, J.F. van Tol 217
Frequency and Time Measurements. By A. Bagley 289
Temperaturmessung. By W. Hunsinger 373
Temperaturregelung. By F. Schreiner 457

Group 5 **Optics**

Vol. 24
Determination of the Velocity of Light. By E. Bergstrand . . 1
Optique géométrique générale. By A. Maréchal 44
Interférences, diffraction et polarisation. By M. Françon . . 171
Optik dünner Schichten. By H. Wolter 461
Schlieren-, Phasenkontrast- und Lichtschnittverfahren. By H. Wolter 555

Vol. 25/1
Crystal Optics. By G.N. Ramachandran, S. Ramaseshan . . 1
Theorie der Beugung. By H. Hönl, A.W. Maue, K. Westpfahl 218

Vol. 25/2a
Optical Constants and their Measurement. By E.E. Bell . . . 1
Phonons in Perfect Crystals. By W. Cochran, R.A. Cowley 59
Photon-Electron Interaction, Crystals Without Fields. By H.Y. Fan 157
Magneto-Optics in Crystals. By S.D. Smith 234

Vol. 25/2b
Theory of Crystal Space Groups and Infra-Red and Raman Lattice Processes of Insulating Crystals. By J.L. Birman . . . 1

Vol. 25/2c
Laser Theory. By H. Haken 1

Vol. 25/2d
Vibrational Infrared and Raman Spectra of Non-Metals. By H. Bilz, D. Strauch, R.K. Wehner 1

Vol. 26
Luminescence. By G.F.J. Garlick 1
Temperature Radiation of Solids. By G.A.W. Rutgers . . . 129
Raman Effect. By S.-I. Mizushima 171
Spectroscopie dans l'infrarouge. By J. Lecomte 244

Vol. 27
Line Width. By R.G. Breene jr. 1
Atomic Spectra. By B. Edlén 80
Spectroscopie électronique moléculaire. By B. Rosen 221

Vol. 28
Microwave Spectroscopy. By W. Gordy 1
Kontinuierliche Spektren. By W. Finkelnburg, T. Peters . . . 79
Kristallspektren. By E. Fick, G. Joos 205
The Zeeman-Effect. By J.C. van den Bosch 296
Activité optique naturelle. By J.P. Mathieu 333

Vol. 29
Optical Calculations and Optical Instruments, an Introduction. By W.T. Welford 1
Méthode de calcul des systèmes optiques. By A. Maréchal, E.D. Hugues, P. Givaudon 43
Theorie der geometrisch-optischen Bildfehler. By H. Marx . . 68
Das photographische Objektiv. By G. Franke 192
Spiegelnde Prismen. By H. Naumann 211
Teleskope. By K. Bahner 227
Mikroskope. By H.C. Claussen 343
Diffraction Gratings. By G.W. Stroke 426
Optische Meßmethoden. By R. Schulze 755

Group 6 X-Rays and Corpuscular Rays

Vol. 30	Erzeugung von Röntgenstrahlen. By W. Schaaffs	1
	Experimental Methods of X-ray Spectroscopy: Ordinary Wavelengths. By A.E. Sandström	78
	The Experimental Methods of Soft X-ray Spectroscopy and the Valence Band Spectra of the Light Elements. By D.H. Tomboulian	246
	X-ray Microscopy. By P. Kirkpatrick, H.H. Pattee jr.	305
	The Continuous X-ray Spectrum. By S.T. Stephenson . . .	337
Vol. 31	Theory of Atomic Photoionization. By A.F. Starace	1
	Atomic Photoionization. By J.A.R. Samson	123
	Photoelectron Spectroscopy. By H. Siegbahn, L. Karlsson . .	215
	Theory of the Auger Effect. By T. Åberg, G. Howat	469
Vol. 32	Les méthodes expérimentales des déterminations de structures cristallines par rayons X. By A. Guinier, G. von Eller . . .	1
	Theoretical Principles of Structural Research by X-rays. By J. Bouman	97
	Etude de la structure des fluides et des substances amorphes au moyen de la diffusion des rayons X. By G. Fournet . . .	238
	Size of Particles and Lattice Defects. By W.W. Beeman, P. Kaesberg, J.W. Anderegg, M.B. Webb	321
	Elektroneninterferenzen. By H. Raether	443
	Neutron Diffraction and Interference. By G.R. Ringo . . .	552
Vol. 33	Elektronen- und Ionenquellen. By D. Kamke	1
	Elektronen- und Ionenoptik. By W. Glaser	123
	Elektronenmikroskope. By S. Leisegang	396
	Massenspektroskopische Apparate. By H. Ewald	546
	Beta-ray Spectroscopes. By T.R. Gerholm	609
Vol. 34	Durchgang langsamer Elektronen und Ionen durch Gase. By R. Kollath	1
	The Passage of Fast Electrons Through Matter. By R.D. Birkhoff	53
	Positronium. By L. Simons	139
	X-ray Production by Heavy Charged Particles. By E. Merzbacher, H.W. Lewis	166
	The Energy Loss of Charged Particles in Matter. By W. Whaling	193
	Compton Effect. By R.D. Evans	218

Group 7 Atomic and Molecular Physics

Vol. 35 The Fundamental Constants of Atomic Physics. By E.R. Cohen, J.W.M. DuMond 1
Quantum Mechanics of One- and Two-Electron Systems. By H.A. Bethe, E.E. Salpeter 88

Vol. 36 Quantenmechanik der Atome. By F. Hund 1
Statistische Behandlung des Atoms. By P. Gombás 109
Theory of Atomic Collisions. By H.S.W. Massey 232
Excitation and Ionization of Atoms by Electron Impact. By H.S.W. Massey 307

Vol. 37/1 Atomic and Molecular Beam Spectroscopy. By P. Kusch, V.W. Hughes 1
The Vibration-rotation Energies of Molecules and their Spectra in the Infra-red. By H.H. Nielsen 173
The Collisions of Electrons with Molecules. By J.D. Craggs, H.S.W. Massey 314

Vol. 37/2 Quantum Mechanics of Electronic Structure of Simple Molecules. By M. Kotani, K. Ohno, K. Kayama 1
The Chemical Bond and the Distribution of Electrons in Molecules. By J.R. Platt 173

Group 8 Nuclear Physics

Vol. 38/1 Atomic Masses of Nuclides. By A.H. Wapstra 1
Determination of Atomic Masses by Microwave Methods. By S. Geschwind 38
Determination of Nuclear Spins and Magnetic Moments by Spectroscopic Methods. By F.M. Kelly 59
Isotope Shifts. By L. Wilets 96
Kernmagnetische Hochfrequenz-Spektroskopie. By G. Laukien 120
Determination of Nuclear Quadrupole Moments. By C.H. Townes 377

Vol. 38/2 The Production and Slowing Down of Neutrons. By E. Amaldi 1
Penetration and Diffusion of X-rays. By U. Fano, L.V. Spencer, M.J. Berger 660

Vol. 39 The Two-Nucleon Problem. By L. Hulthén, M. Sugawara . . 1
The Three-body Problem in Nuclear Physics. By M. Verde 144
Matter and Charge Distribution within Atomic Nuclei. By D.L. Hill 178
The Nuclear Shell-Model. By J.P. Elliott, A.M. Lane 241
Models of Nuclear Structure. By S.A. Moszkowski 411

Vol. 40	Nuclear Reactions, Levels, and Spectra of Light Nuclei. By W.E. Burcham	1
	Nuclear Reactions, Levels, and Spectra of Heavy Nuclei. By B.B. Kinsey	202
	Resonance Processes by Neutrons. By J. Rainwater	373
	Nuclear Reactions at High Energies. By A. Wattenberg	450
Vol. 41/1	Theory of Resonance Reactions and Allied Topics. By G. Breit	1
	Coulomb Wave Functions. By H.H. McAllister jr., G. Breit	408
	Polarization of Nucleons Scattered by Nuclei. By G. Breit, J.S. McIntosh	466
	Coulomb Excitation. By G. Breit, R.L. Gluckstern	496
Vol. 41/2	Experiments on β-decay. By O.M. Kofoed-Hansen, C.J. Christensen	1
Vol. 42	Nuclear Isomerism. By D.E. Alburger	1
	Alpha Radioactivity. By I. Perlman, J.O. Rasmussen	109
	The Transuranium Elements. By E.K. Hyde, G.T. Seaborg	205
	The Nuclear Photoeffect. By G.R. Bishop, R. Wilson	309
	Angular Correlations. By S. Devons, L.J.B. Goldfarb	362
	Oriented Nuclei. By R.J. Blin-Stoyle, M.A. Grace	555
Vol. 44	Kaskadengeneratoren. By E. Baldinger	1
	Van de Graaff Generators. By R.G. Herb	64
	Cyclotrons and Synchrocyclotrons. By B.L. Cohen	105
	Electron Synchrotrons. By R.R. Wilson	170
	The Betatron. By D.W. Kerst	193
	The Proton Synchrotron. By G.K. Green, E.D. Courant	218
	Linear Accelerators. By L. Smith	341
	Reactor Techniques. By D.J. Hughes	390
Vol. 45	Ionization Chambers in Nuclear Physics. By H.W. Fulbright	1
	Geiger Counters. By S.A. Korff	52
	Scintillation and Čerenkov Counters. By W.E. Mott, R.B. Sutton	86
	The Proportional Counter as Detector and Spectrometer. By S.C. Curran	174
	The Coincidence Method. By S. de Benedetti, R.W. Findley	222
	Cloud Chambers. By C.M. York	260
	The Bubble Chamber. By D.A. Glaser	314
	Nuclear Emulsions. By M.M. Shapiro	342
	Detection of Neutrons. By H.H. Barschall	437
	High Energy Neutron Detectors. By R.T. Siegel	487

Group 9 **Cosmic Rays**

Vol. 46/1 The Origin of Cosmic Rays. By P. Morrison 1
Theory of the Geomagnetic Effects of Cosmic Radiation. By M.S. Vallarta 88
Experimental Results of Flights in the Stratosphere. By E.C. Ray 130
Penetrating Showers. By K. Sitte 157
Extensive Air Showers. By G. Cocconi 215
The Hard Component of μ-Mesons in the Atmosphere. By G.N. Fowler, A.W. Wolfendale 272

Vol. 46/2 Theory of Cascade Showers. By J. Nishimura 1
Cosmic Rays and High-Energy Physics. By Y. Fujimoto, S. Hayakawa 115
The Spectrum and Charge Composition of the Primary Cosmic Radiation. By W.R. Webber 181
High Energy Photons and Neutrinos from Cosmic Sources. By R.J. G.R. Burbidge Gould 265
The Time Variations of the Cosmic Ray Intensity. By J.J. Quenby 310
Nukleonen in der Atmosphäre. By E. Schopper, E. Lohrmann, G. Mauck 372
Cosmic Ray Produced Radioactivity on the Earth. By D. Lal, B. Peters 551
Effects of Cosmic Rays on Meteorites. By M. Honda, J.R. Arnold 613

Group 10 **Geophysics**

Vol. 47 The Rotation of the Earth. By H. Spencer-Jones 1
Séismométrie. By J. Coulomb 24
Seismic Wave Transmission. By K.E. Bullen 75
Surface Waves and Guided Waves. By W.M. Ewing, F. Press 119
L'agitation microséismique. By J. Coulomb 140
Seismic Prospecting By W.M. Ewing, F. Press 153
Messung elastischer Eigenschaften von Gesteinen. By H. Baule, E. Müller 169
Gravity and Isostasy. By G.D. Garland 202
Structure of the Earth's Crust. By W.M. Ewing, F. Press . . 246
Forces in the Earth's Crust. By A.E. Scheidegger 258
Radioactivity and Age of Minerals. By J.T. Wilson, R.D. Russell, R. MacCunn Farquhar 288
The Earth's Interior. By J.A. Jacobs 364
Electricité tellurique. By L. Cagniard 407
Magnetization of Rocks. By S.K. Runcorn 470
The Magnetism of the Earth's Body. By S.K. Runcorn . . . 498
Figur der Erde. By K. Jung 534

Vol. 48 Dynamic Meteorology. By A. Eliassen, E. Kleinschmidt . . . 1
Strahlung in der unteren Atmosphäre. By F. Möller 155
Vision Through the Atmosphere. By W.E.K. Middleton . . . 254
Polarization of Skylight. By Z. Sekera. 288
Diffusion des radiations par les gouttes d'eau en suspension dans l'atmosphère. By J. Bricard 329
Ozon in der Erdatmosphäre. By H.-K. Paetzold, E. Regener 370
Geophysical Aspects of Meteors. By A.C.B. Lovell 427
Sound Propagation in Air. By E.F. Cox 455
The Physics of Clouds. By F.H. Ludlam, B.J. Mason 479
Atmosphärische Elektrizität. By R. Mühleisen 541
Oceanography. By H.U. Sverdrup 608
Oberflächen-Wellen des Meeres. By H.U. Roll 671
Gezeitenkräfte. By J. Bartels. 734
Tides of the Solid Earth. By R. Tomaschek 775
Flutwellen und Gezeiten des Wassers. By A. Defant 846
Atmosphärische Gezeiten. By W. Kertz 928
Physical Volcanology. By S. Sakuma, T. Nagata 982

Vol. 49/1 The Aurora. By S.-I. Akasofu, S. Chapman, A.B. Meinel . . 1
Time-Variations of Cosmic Rays. By S.E. Forbush 159
Magnetic Properties of Rocks and Minerals. By T. Nagata 248

Vol. 49/2 Radio-Observations of the Ionosphere. By K. Rawer, K. Suchy 1
Lunar Tides in the Ionosphere. By S. Matsushita 547

Vol. 49/3 Morphology of Magnetic Disturbance. By T. Nagata, N. Fukushima 5
Theoretical Aspects of the Worldwide Magnetic Storm Phenomenon. By E.N. Parker, V.C.A. Ferraro 131
Maßzahlen der erdmagnetischen Aktivität. By M. Siebert . . 206
Classical Methods of Geomagnetic Observations. By V. Laursen, J. Olsen 276
Neuere Meßmethoden der Geomagnetik. By H. Schmidt, V. Auster 323
Three-Component Airborne Magnetometers. By P.H. Serson, K. Whitham 384
Aeromagnetic Surveying with the Fluxgate Magnetometer. By J.R. Balsley 395
Geophysical Applications of High Resolution Magnetometers. By P.J. Hood 422
Phénomènes T.B.F. d'origine magnétosphérique. By R. Gendrin 461

Vol. 49/4	The Earth's Magnetosphere. By H. Poeverlein	7
	The Earth's Radiation Belt. By W.N. Hess	115
	Variations rapides du champ magnétique terrestre. By E. Selzer	231
	Waves and Resonances in Magneto-active Plasma. By V.L. Ginzburg, A.A Ruhadze	395
Vol. 49/5	Introductory Remarks. By K. Rawer	1
	La luminescence nocturne. By A.T. Vassy, E. Vassy	5
	Dynamic Structure of the Stratosphere and Mesosphere. By W.L. Webb	117
	Linear Internal Gravity Waves in the Atmosphere. By W.L. Jones	177
	Wave-Like Phenomena in the Near-Earth Plasma and Interactions with Man-Made Bodies. By J.L. Al'pert	217
	Some Characteristic Features of the Ionospheres of Near-Earth Planets. By K.I. Gringauz, T.K. Breus	351
Vol. 49/6	Introduction: A Summary Description of Aeronomy. By K. Rawer	1
	The Neutral and Ion Chemistry of the Upper Atmosphere. By L. Thomas	7
	Diffusion in the High Atmosphere. By T. Yonezawa	129
	Interaction of Neutral and Plasma Motions in the Ionosphere. By P. Stubbe	247
	Extreme Ultraviolet Observational Data on the Solar Spectrum. By G.M Nikol'skij	309
Vol. 49/7	Modelling of the Solar Extreme Ultraviolet Irradiance for Aeronomic Applications. By G. Schmidtke	1
	Transport Coefficients and Collision Frequencies for Aeronomic Plasmas. By K. Suchy	57
	Modelling of Neutral and Ionized Atmospheres. By K. Rawer	223

Group 11 **Astrophysics**

Vol. 50	Les classifications spectrales des étoiles normales. By Ch.F. Fehrenbach	1
	Stars with Peculiar Spectra. By Ph.C. Keenan	93
	Les bandes moléculaires dans les spectres stellaires. By P. Swings	109
	Die Spektren der planetarischen Nebel. By K. Wurm	139
	The Spectra of the White Dwarfs. By J.L. Greenstein	161
	Visual Binaries. By P. van de Kamp	187
	The Eclipsing Binaries. By S. Gaposchkin	225
	Spectroscopic Binaries. By O. Struve, S.-s. Huang	243
	Théorie générale des atmosphères stellaires. By D. Barbier	274
	Theorie der planetarischen Nebel. By K. Wurm	399

Vol. 51	Stellar Interiors. By M.H. Wrubel	1
	The Hertzsprung-Russell Diagram. By H.C. Arp	75
	Stellar Evolution. By E.M. Burbidge, G. Burbidge	134
	Die Häufigkeit der Elemente in den Planeten und Meteoriten. By H.E. Suess, H.C. Urey	296
	The Abundances of the Elements in the Sun and Stars. By L.H. Aller	324
	Variable Stars. By P. Ledoux, T. Walraven	353
	Stellar Stability. By P. Ledoux	605
	Magnetic Fields of Stars. By A.J. Deutsch	689
	Théorie des naines blanches. By E. Schatzman	723
	The Novae. By C. Payne-Gaposchkin	752
	Supernovae. By F. Zwicky	766
Vol. 52	The Photosphere of the Sun. By L. Goldberg, A.K. Pierce	1
	Structure and Dynamics of the Solar Atmosphere. By C. de Jager	80
	The Atmospheres of the Planets. By H.C. Urey	363
	Planetary Interiors. By W.C. de Marcus	419
	Radio Echoes from Sun, Moon and Planets. By F.J. Kerr	449
	Die Kometen. By K. Wurm	465
	Meteors. By F.L. Whipple, G.S. Hawkins	518
Vol. 53	Kinematical Basis of Galactic Dynamics. By F.K. Edmondson	1
	Galactic Dynamics. By B. Lindblad	21
	Radio-frequency Studies of Galactic Structure. By J.H. Oort	100
	Star Clusters. By H.S. Hogg	129
	Discrete Sources of Cosmic Radio Waves. By R.H. Brown	208
	Radio Frequency Radiation from External Galaxies. By B.Y. Mills	239
	Classification and Morphology of External Galaxies. By G. de Vaucouleurs	275
	General Physical Properties of External Galaxies. By G. de Vaucouleurs	311
	Multiple Galaxies. By F. Zwicky	373
	Clusters of Galaxies. By F. Zwicky	390
	Large Scale Organization of the Distribution of Galaxies. By J. Neyman, E.L. Scott	416
	Distance and Time in Cosmology: The Observational Data. By G. McVittie	445
	Newtonsche und Einsteinsche Kosmologie. By O.H.L. Schücking, E. Heckmann	489
	Andere kosmologische Theorien. By O.H.L. Schücking, E. Heckmann	520

Vol. 54 Solar Instruments. By R.R. McMath, O.C. Mohler 1
Radio Astronomy Techniques. By R.N. Bracewell 42
Photographic Photometry. By H. Weaver 130
La nature de la surface des planètes et de la Lune.
By A. Dollfus 180
Photoelectric Techniques. By A.E. Whitford 240

MIX
Papier aus verantwortungsvollen Quellen
Paper from responsible sources
FSC® C105338

If you have any concerns about our products,
you can contact us on
ProductSafety@springernature.com

In case Publisher is established outside the EU,
the EU authorized representative is:
Springer Nature Customer Service Center GmbH
Europaplatz 3, 69115 Heidelberg, Germany

Printed by Libri Plureos GmbH
in Hamburg, Germany